13th European Workshop on Advanced Control and Diagnosis (ACD 2016)

Journal of Physics: Conference Series Volume 783

Lille, France
17 - 18 November 2016

ISBN: 978-1-5108-3560-3
ISSN: 1742-6588

Printed from e-media with permission by:

Curran Associates, Inc.
57 Morehouse Lane
Red Hook, NY 12571

Some format issues inherent in the e-media version may also appear in this print version.

Printed by Curran Associates, Inc. (2017)

For permission requests, please contact the Institute of Physics
at the address below.

Institute of Physics
Dirac House, Temple Back
Bristol BS1 6BE UK

Phone: 44 1 17 929 7481
Fax: 44 1 17 920 0979

techtracking@iop.org

Additional copies of this publication are available from:

Curran Associates, Inc.
57 Morehouse Lane
Red Hook, NY 12571 USA
Phone: 845-758-0400
Fax: 845-758-2633
Email: curran@proceedings.com
Web: www.proceedings.com

Table of contents

Volume 783

13th European Workshop on Advanced Control and Diagnosis (ACD 2016)
17–18 November 2016, Lille, France

Accepted papers received: 2 December 2016
Published online: 19 January 2017

Preface

Papers

IOP Conf. Series: Journal of Physics: Conf. Series 783 (2017) 011001 doi:10.1088/1742-6596/783/1/011001

13th European Workshop on Advanced Control and Diagnosis

The 13th European Workshop on Advanced Control and Diagnosis (ACD 2016), www.acd2016.eu took place at Hautes Etrudes d'Ingénieur (HEI), Lille, France, on November 17-18, 2016.

The annual European Workshop on Advanced Control and Diagnosis has been organized since 2003 by Control Engineering departments of several European universities in Germany, France, the UK, Poland, Italy, Hungary, Denmark and Czech Republic, to bring together senior and junior academics and engineers from diverse fields of automatic control, fault detection, and signal processing. The workshop provides an opportunity for researchers and developers to present their recent theoretical developments, practical applications, or even open problems. It also offers a great opportunity for industrial partners to express their needs and priorities and to review the current activities in the fields.

A total of 100 papers have been submitted for ACD 2016. Based on the peer reviews 71 papers were accepted for oral presentation, 4 papers withdrawn, 3 papers not presented. The accepted papers covered areas of Fault Diagnosis, Estimation and Observation, Fault Tolerant Control, Predictive Control , Robust Control, Control Theory, fractional order systems, Modeling and Data °Processing, Software tools and code generation. In addition, four excellent plenary lectures were delivered by experts on their domain: Ir. Serge Legonidec from the company Airbus Safran Launchers, France : An overview of connections between scientific automatic topics and their applications in the propulsive systems, Dr. Ir. Jan-Willem Van Wingerden, Technical University of Delft , Nederland : Wind Energy Control Research, Prof. Sarah Spurgeon, University College London, UK: On

discontinuous Observers: From Basic Properties to a Robust Fault Detection and Condition Monitoring Tool, Dr. Joseph-Julien Yame, University of Lorraine, Nancy, France : Data Driven Fault Tolerant Control: a behavioral approach.
The ACD 2016 has been co-sponsored technically by the International Federation of Automatic Control (IFAC) .

On behalf of the ACD 2016 organizing committee, we would like to thank all those who prepared and submitted papers, participated in the peer review process, supported, and attended the workshop.

Abdel AITOUCHE	Josef KORBICZ	Vincent COCQUEMPOT
General Chair	IPC Chair	IPC Co-Chair

IOP Conf. Series: Journal of Physics: Conf. Series **783** (2017) 011001 doi:10.1088/1742-6596/783/1/011001

Workshop Organisation

General Chair

Abdel Aitouche

International Program Committee Chair

Józef Korbiçcz (Poland)

International Program Committee Co-Chair

Vincent Cocquempot (France)

Steering Committee

Józef Korbiçcz (Poland|
Hans Henrik Niemann (Denmark)
Horst Schulte (Germany)
Silvio Simani (Italy)
Ondrej Straka (Czech Republic)
Didier Theilliol (France)

International Program Committee

Pedro Albertos (Spain)
Mireille Bayart (France)
Michał Bartyś (Poland)
Boulaid Boulkroune (Belgium)
Alessandro Caavola (Italy)
Jean Philippe Cassar (France)
Paolo Castaldi (Italy)
Olivier Colot (France)
Maan El Badaoui (France)
Ahmed El Hajjaji (France)
Peter Fogh Odgard (Denmark)
David Henry (France)
Ghaleb Hoblos (France))
Jan M. Kościelny, (Poland)
Dusan Krokavek (Slovak Republic)
Balázs Kulcsár, (Sweden)
Didier Maquin (France)
Driss Mehdi (France)
Rochdi Merzouki (France)
Hans Henrik Niemann (Denmark)
Mustapha Ouladsine (France)
Belkacem Ould Bouamama (France)
Ronald J. Patton (United Kingdom)
Wilfrid Perruquetti (France)
Thomas Parisini (United Kingdom)
Dumitru Popescu (Romania)
Niels Kjølstad Poulsen (Denmark)
Vicenç Puig (Spain)
Ole Ravn (Denmark)
Jean Pierre Richard (France)
Dominique Sauter (France)
Horst Schulte (Germany)
Silvio Simani (Italy)
Ondřej Straka (Czech Republic)
Fernando Tadeo (Spain)
Achraf Jabeur Telmoudi (Tunisia)
Didier Theilliol (France)
Haoping Wang (China)
Marcin Witczak (Poland)
Elena Zattoni (Italy)
Janan Zaytoon (France)
Ali Zemouche (France

Local Arrangement Chairs

Dhaker Abbes (France)
Patrick Debay (France)
Christine Yvoz (france)
Judy Espin-Cunat (France)
Michel Pollart (France)

Hosting Institution

Workshop Web Page
http://www.acd2016.eu/

Sponsors

Technical Co-sponsor

Plenary Lectures

Serge Le Gonidec: Airbus Safran Launchers, France

Title: An overview of connections between scientific automatic topics and their applications in the propulsive systems

Abstract: In the space propulsion applications the automatic topics have a significant contribution. Indeed, from development to post flight analysis, through the overall life cycle, the automatic tools allow or contribute (or could contribute in the future) to: Reduce the development and production cost, improve the trust accuracy and the reliability, make the engine control management more robust, improve the capability of flight mission agility, facilitate the post flight analysis, ...Of course all the ACD technologies are not currently in use or should not be applied even in future space context ! So a small "travel" in the space propulsion world from past studies through current applications to potential needs should illustrate the primary importance of the ACD technologies.

Biography: **Serge Le Gonidec** was born in France, in 1964. He started its carrier in 1985 at the "Société Européenne de Propulsion" with a BTS CIRA (French acronym for Industrial Automatic Control). In the support team "control, measurement and command system" on Ariane rocket engine test benches, he participated to the post-V18 return to flight. Early in 1990, the Ariane 5 program gave him the opportunity to work on the test bench control engineering (flow control, tank pressurization, turbine speed, etc.). These new A5 test benches allowed introducing the digital control and its generalization for the engine or component tests under a high required safety. At that time, he started the CNAM School and received the engineer certificate in 1997, with a specialization in "Industrial Control". The topic of its memory was "System Identification in transient phases: Genetic Algorithm development and application". In 1994, he rejoins the team of System department. Its activities are oriented modeling development and stability analysis for system developments or for the support for understanding the bad behaviors. After 1997, he contributed to the engine design (e.g. Vinci) or evolution (e.g. Vulcain2) in introducing the Multi-variable Predicted Control. In the same time, in relation with ESA or CNES, he worked on each space engine control project. In Space Engine Division at Snecma, its activities are mainly oriented on the R&T roadmap dedicated to automatic application and the support on the engine development. These R&T activities are focused on the preparation of the integration of the engine automatic control (system, engineering methodology, equipment, electrification, ...) and on the health monitoring system for the future launchers (reusable launchers, control and monitoring interaction / IVHM / AFTC, automatic support to the flight analysis, ...). He participated to the deployment of the first HIL test bench dedicated to the beginning of electric space engine demonstration. During 5 years he managed the R&T project dedicated to monitoring for space engine application (7 countries) in the frame of ESA activities. In parallel, the Snecma diversification activities brought him to work on the Fuel Cell control. In these related thematic, he deployed the methodology for system stability analysis for the application of the anomaly investigation or for support to the design of the multi-physic systems. Since 2008, he is referent expert « Expert Senior » on the « space systems control & monitoring » in the Safran group. In the frame of the new company « Airbus Safran Launchers » entity, he is a member of the Audit team. Currently, its preoccupations are concentrated on the electric and the control functions deployment on the space propulsion systems and the associated engineering system development methodology. Up to now, he is the author of about 40 edited or in proceed filing patents..

Dr.ir Jan-Willem van Wingerden: TU Delft / DCSC, Room 34 C-2-320 Mekelweg 2, 2628 CD Delft, Nederland

Title: Wind Energy Control Research

Abstract: In this presentation I will give an overview of my research in the area of control of wind turbines and farms. After a brief history on how developments in control theory and technologies contributed to the development of large-scale variable-speed wind turbines, novel control technologies and the corresponding research challenges will be presented. This will include: Lidar control, control of floating wind turbines, control of wind farms and "smart" rotor control.

Biography: **Dr.ir. Jan-Willem van Wingerden**, was born on December 9, 1980 in Ridderkerk, The Netherlands. In 1999 he started his study Mechanical Engineering at the Delft University of Technology, where he graduated cum laude in December 2004 at the Control Engineering group, nowadays part of the Delft Center for Systems and Control (DCSC). His graduation project was carried out at Philips Applied Technologies in Eindhoven. In January 2005, he started his Ph.D. project entitled: `Smart' dynamic rotor control for large offshore wind turbines within the Delft Center for Systems and Control and the Delft University Wind Energy Research Institute. In 2008, he received his Ph.D. degree Cum Laude from the Delft Center of Systems and Control, Delft University of Technology. In 2010 he was a visiting scientist at the National Renewable Energy Lab. (USA). Currently he is an associate professor at the Delft University of Technology. His main research interests include system identification, robust control, and control and identification wind energy systems..

Prof. Sarah Spurgeon : Department of Electronic and Electrical Engineering, University College London, UK.

Title: On discontinous Observers: From Basic Properties to a Robust Fault Detection and Condition Monitoring Tool.

Abstract: Sliding mode observers have generated a ground swell of interest in recent years. These observers have unique properties. Their ability to force the error between the measured plant output and the output of the observer to be identically zero produces a set of state estimates that are precisely commensurate with the actual output of the plant. It is also the case that analysis of the applied observer injection signal, the so-called equivalent injection signal, contains useful information about the mismatch between the model used to define the observer and the actual plant. This lecture presents an overview of both linear and non-linear sliding mode observer paradigms. The use of the equivalent injection signal in problems relating to fault detection and condition monitoring is demonstrated. A number of applications specific results are also described.

Biography Prof. Sarah Spurgeon OBE, FREng, FInstMC, FIET, FIMA is Professor of Control Engineering and Head of the Department of Electronic and Electrical Engineering at University College London and President of the Institute of Measurement and Control in the UK. Sarah Spurgeon's research interests are in the area of systems modelling and analysis, robust control and estimation in which areas she has published over 270 refereed research papers. She was awarded the Honeywell International Medal for 'distinguished contribution as a control and measurement technologist to developing the theory of control' in 2010 and an IEEE Millenium Medal in 2000. She is currently a member of the Council of the International Federation of Automatic Control (IFAC) and a member of the General Assembly of the European Control Association. Within the UK she is an independent member of the Defence Scientific Advisory Council (DSAC) which provides independent advice to the Secretary of State for Defence on science, technology, engineering, analysis and mathematics matters and is also a Board Member of EngineeringUK.

IOP Conf. Series: Journal of Physics: Conf. Series **783** (2017) 011001 doi:10.1088/1742-6596/783/1/011001

Dr. Ir. Joseph Yamé : CRAN, University of Lorraine, Nancy, France.
Title: Data Driven Fault Tolerant Control: a behavioral approach

Abstract: In this talk we will review our recent work on data-driven fault-tolerant control (FTC) from a behavioral system-theory perspective. We will go through the basics of the mathematical tools of Willems' behavioral system theory and formulate within this framework a control problem as an interconnection of two dynamical systems. Thanks to this fundamental viewpoint on controls and the key concept of behavior, we will address the fault-tolerant control problem when no much a priori information about a plant is available in real-time. Here, we deal simply with the trajectories that are generated by the plant. These trajectories determine the so-called behavior of the plant in various faulty/healthy modes and based on the desired controlled behavior, only those trajectories are selected that obeys certain laws. We will show that the vantage point of the behavioral approach is that the fault-tolerant control problem now becomes completely a real-time model-free or data-driven scheme. We will discuss in this framework two methodologies under the broad classification of FTC systems, namely projection-based approach and online redesign approach. We will also discuss the practical issue of transients management during controller reconfiguration in order to guarantee real-time smooth interconnection between the controller and the unknown faulty plant.

Biography: **Dr. Ir. Joseph Yamé** received the degree of Doctor in Applied Sciences with the Highest Distinction from the Université Libre de Bruxelles (ULB), Brussels, Belgium in 2001. He previously graduated from Ecole Polytechnique of the ULB in Electrical and Mechanical Engineering with the title of "Ingénieur Civil" and received also the Post-Graduate Engineering degree in Automatic Control. From 2000 to 2005, he was a Research Associate in the Control Engineering and Systems Analysis Department of the ULB, Brussels. In September 2005, he joined the University of Lorraine, Nancy, France, as an Associate Professor in control engineering and computer science. Over the several past years his research activities have focused on different topics in systems theory and advanced control engineering with special interests in dual-adaptive control of stochastic systems, mathematical control theory with an emphasis on sampled-data control and infinite-dimensional discrete-time systems, and the analytical aspects of fuzzy control. During these last years, his research interest has been mainly concentrated on fault tolerant control and control for energy-efficient buildings. He has authored and co-authored over 80 publications, including book contributions, journal and conference papers, and technical reports. He is a member of the Institute of Electrical and Electronics Engineers (IEEE), the American Mathematical Society (AMS), and a senior member of the American Institute of Aeronautics and Astronautics (AIAA).

Peer review statement

All papers published in this volume of *Journal of Physics: Conference Series* have been peer reviewed through processes administered by the proceedings Editors. Reviews were conducted by expert referees to the professional and scientific standards expected of a proceedings journal published by IOP Publishing.

Robust unknown input observer design for state estimation and fault detection using linear parameter varying model

Shanzhi Li[1,3], Haoping Wang[1,†], Abdel Aitouche[2], Yang Tian[1], Nicolai Christov[3]

[1] Automation School, Nanjing University of Technology & Science, Nanjing, 210094, China
[2] CRIStAL UMR CNRS 9189, HEI-Lille,Lille, 59014, France
[3] CRIStAL UMR CNRS 9189, University of Lille 1, Lille, 59650, France

E-mail: hp.wang@njust.edu.cn

Abstract. This paper proposes a robust unknown input observer for state estimation and fault detection using linear parameter varying model. Since the disturbance and actuator fault is mixed together in the physical system, it is difficult to isolate the fault from the disturbance. Using the state transforation, the estimation of the original state becomes to associate with the transform state. By solving the linear matrix inequalities (LMIs)and linear matrix equalities (LMEs), the parameters of the UIO can be obtained. The convergence of the UIO is also analysed by the Layapunov theory. Finally, a wind turbine system with disturbance and actuator fault is tested for the proposed method. From the simulations, it demonstrates the effectiveness and performances of the proposed method.

1. Introduction

Since some states of the plant cannot be measured directly, observers are applied for estimating the states, faults or disturbances using the plant input and the output signals. Observer not only improves performance, but also enhances reliability in the control system.

As for the linear or nonlinear systems , there are many different types of observers to be developed and designed, such as Luenberger observer [1], high gain observer [2], extend state observer [3] etc. For these observers, it is necessary to obtain all input signals. However, in the physical system, some of the input signals cannot be measured and obtained directly. Since these input signals are restricted or unknown in some control system, an unknown input observer (UIO) is put forward [4]. In the UIO, the uncertainty or disturbance can be treated as an unknown input for and be estimated. In the closed-loop system, using these state estimation , the controller can regulate the real-time states in a desired trajectory.

Over the years, many different types of the UIO and its evolutionary schemes have been developed for the nonlinear systems. It shows very well performance in dealing with the corresponding problems including noise, uncertainty and disturbance etc [5, 6, 7] . In the [8], using a linear matrix inequality (LMI)approach, a full order nonlinear UIO for a class of Lipschitz nonlinear systems with unknown inputs. Also, the reduce order UIO is developed for one-sided Lipschitz nonlinear systems in [9].

As a model-based observer, the performance of UIO is challenged by the modelling errors, parameter variations or other uncertain factors, since the nominal model of a physical system has to be involved for designing UIO. It is required that the UIO is robust for the measuring noise, uncertain factors and disturbance. Considering the factors, robust UIO is designed for unknown inputs Takagi-Sugeno models [10]and linear parameter varying (LPV) system [11]. In the UIO, using a state transformation, the original states of the real system are transferred into canonical forms and the estimation states can be calculated from the canonical forms . By solving the LMIs and linear matrix equalities(LMEs), the solution of the UIO can be obtained. However, since there are many conditions and hypothesis, it is not easy to find a feasible solution for LMIs and LMEs. Hence, reference [10]and [12] etc give the sufficient and necessary conditions for the existence of UIO.

In the industrial systems, the components challenged by a variety of extreme environments and have to subject to the faults. If no action is taken after the faults occur, it will cause the system performance degradation, even unable to work. In order to overcome the sensors or other components weakness, many approaches to fault detection and diagnosis (FDD) and fault tolerant control (FTC) are developed and designed[13, 14, 15, 16]. In the fault tolerant control, observer or estimation algorithm plays a key role in the fault diagnosis and isolation methods. In the active fault tolerant control system, faults are required to be estimated and compensated. Using the UIO for the fault diagnosis and isolation , many research works and significant results about the fault with sensor noise, disturbance or uncertainty can be found. In these fault tolerant control based-UIO, they mainly deal with output sensor fault , actuator fault , both sensor and actuator fault . For the linear parameter varying (LPV) system, there is a little research about the fault tolerant control based-UIO with actuator fault and disturbance. Despite some researches related to this issue, the task for compromising the LMIs' solution with fault estimation algorithms is not easily, especially the impact of the disturbance. When the actuator fault occurs, the input disturbance and fault will mix together such that it is difficult to isolate the fault. This motivates our present research.

In this paper, we will design a UIO for state and fault estimation algorithm with LPV model. Firstly, since the UIO is a based -model observer, the polytopic LPV model is utilized for modeling. Secondly, using the state transforation and state error equation, the value of the UIO's parameters become the solution of the LMIs and LMEs. By solving the LMIs, we can obtain the other values in turns. Considering the disturbance and actuator fault, the convergence of the UIO is also analysed using the Layapunov and H_∞ theory. At last, a wind turbine system with disturbance and actuator fault is utilized to illustrate the performance. From the results, it indicates that the proposed UIO and fault estimation algorithm are of effectiveness.

This paper is organized as follows. In Section 2, the problem formulation and preliminaries are presented. Design and analysis the UIO is presented in section 3. The simulation results are presented in Section 4. Finally, some conclusions are given in Section 5.

2. Problem formulation and preliminaries

Considering a nonlinear system, the input ,the output and the unknown input vectors are $u \in \Re^m, y \in \Re^p, d \in \Re^l$ respectively. Assume that it can be expressed in the LPV form as following

$$\begin{aligned} \dot{x}(t) &= A(\alpha(t))x(t) + B(\alpha(t)u(t) + D(\alpha(t))d(t) + Ff(t) \\ y(t) &= Cx(t) \end{aligned} \tag{1}$$

where $x \in \Re^n$ is the state vector, $A(\alpha) \in \Re^{n\times n}, B \in \Re^{n\times m}, C \in \Re^{p\times n}, D(\alpha) \in \Re^{n\times l}$ are the matrices of α. α is a time varying parameter and is bounded. $d(t)$ and $f(t)$ are the disturbance and the fault vector respectively. Assumes that the matrices $A(\alpha)$, $B(\alpha)$, $D(\alpha)$ can be written as polytopic matrices and they are given

$$\begin{aligned} A(\alpha) &= \sum_{i=1}^{r} \rho_i A_i \\ B(\alpha) &= \sum_{i=1}^{r} \rho_i B_i \\ D(\alpha) &= \sum_{i=1}^{r} \rho_i D_i \end{aligned} \tag{2}$$

where ρ_i are weights of the LPV subsystem, supposed it that satisfied

$$\sum_{i=1}^{r} \rho_i = 1, \quad 0 \leq \rho_i \leq 1 \tag{3}$$

The polytopic representation of the system (1) becomes

$$\begin{aligned} \dot{x}(t) &= \sum_{i=1}^{r} \rho_i(A_i x(t) + B_i u(t) + D_i d(t)) + Ff(t) \\ y(t) &= Cx(t) \end{aligned} \tag{4}$$

where A_i, B_i, D_i are the time invariant matrices. Assume that the matrices D_i is of full column rank.

3. Design and analysis the unknown input observer for LPV system

In this section, we will design the unknown input observer for LPV system. The convergence of the unknown input observer will be analysed.

3.1. Design the unknown input observer

For system (4), the unknown input observer can be written as

$$\begin{aligned} \dot{z} &= \sum_{i=1}^{r} \rho_i(N_i z(t) + G_i u(t) + L_i y(t)) + T\hat{f} \\ \hat{x} &= z(t) - Ey(t) \\ \dot{\hat{f}} &= \Gamma S e_y \end{aligned} \tag{5}$$

where $e_y = Cx - C\hat{x} = Ce$. $\hat{x}$ and $\hat{f}$ are the estimated state vector and estimated fault vector. N_i, G_i, L_i, E are the unknown matrices with appropriate dimension. And they satisfy the following conditions:

$$\begin{aligned} N_i &= MA_i - K_i C \\ G_i &= MB_i \\ L_i &= K_i(I_1 + CE) - MA_i E \\ MD_i &= 0 \\ MF &= T \end{aligned} \tag{6}$$

Define the state error vector $e(t) = x(t) - \hat{x}(t)$, we obtain that

$$e(t) = (EC + I_2)x(t) - z(t) \triangleq Mx - z \tag{7}$$

where $M = EC + I_2$. I_1 and I_2 are unit matrices with apposite dimensions.

If the conditions (6) are satisfied , then the derivative of error becomes

$$\begin{aligned}\dot{e}(t) &= M\dot{x} - \dot{z} \\ &= M(\sum_{i=1}^{r}\rho_i(A_ix(t) + B_iu(t) + D_id(t) + Ff(t)) - (\sum_{i=1}^{r}\rho_i(N_iz(t) + G_iu(t) + L_iy(t)) + T\hat{f}(t)) \\ &= \sum_{i=1}^{r}\rho_i[N_ix + (MA_i - N_iM - L_iC)x + (MB_i - G_i)u + MD_id + (MF - T)f + T\tilde{f}]\end{aligned} \tag{8}$$

Define a matrix $K_i = L_i + N_iE$, we can obtain

$$\begin{aligned}&MA_i - N_iM - L_iC \\ &= MA_i - N_iM - (K_i - N_iE)C \\ &= MA_i - (MA_i - K_iC)M - K_iC + (MA_i - K_iC)EC \\ &= MA_i(I_2 - M + EC) + K_iC(M - I_2 - EC) \\ &= (MA_i - K_iC)(I_2 - M + EC) \\ &= 0\end{aligned} \tag{9}$$

Submitting equation 9 into equation 8, the error dynamic equation can be simplified as

$$\dot{e} = \sum_{i=1}^{r}\rho_i(N_ie + T\tilde{f}) \tag{10}$$

3.2. Stability and convergence analysis for observer

Lemma 1 If there exists a symmetric positive definite matrices P, Γ, such that the following LMI is satisfied for $i = 1, ...r$

$$\begin{aligned}N_i^TP + PN_i < 0 \\ T^TP - SC = 0\end{aligned} \tag{11}$$

and the estimated algorithm of fault is selected as

$$\dot{\hat{f}} = \Gamma S e_y \tag{12}$$

then the state error and the fault error converge to zero.

Proof : The candidate Lyapunov function is selected as

$$V = e^TPe + \tilde{f}^T\Gamma^{-1}\tilde{f} \tag{13}$$

then the derivative of Lyapunov function becomes

$$\begin{aligned}\dot{V} &= \dot{e}^TPe + e^TP\dot{e} + \dot{\tilde{f}}^T\Gamma^{-1}\tilde{f} + \tilde{f}^T\Gamma^{-1}\dot{\tilde{f}} \\ &= (\sum_{i=1}^{r}\rho_i(N_ie + T\tilde{f}))^TPe + e^TP(\sum_{i=1}^{r}\rho_i(N_ie + T\tilde{f})) + 2\tilde{f}^T\Gamma^{-1}\dot{\tilde{f}} \\ &= \sum_{i=1}^{r}\rho_i(e^T(N_i^TP + PN_i)e + \tilde{f}^TT^TPe + e^TPT\tilde{f} + 2\tilde{f}^T\Gamma^{-1}\dot{\tilde{f}} \\ &= \sum_{i=1}^{r}\rho_i(e^T(N_i^TP + PN_i)e + 2\tilde{f}^T(T^TPe + \Gamma^{-1}\dot{\tilde{f}})\end{aligned} \tag{14}$$

If the fault is assumed as a step[17], we find

$$\dot{\tilde{f}} = \dot{f} - \dot{\hat{f}} = -\dot{\hat{f}} = -\Gamma S e_y \tag{15}$$

Replace equation (15) to equation (14), the derivative of Lyapunov function becomes

$$\begin{aligned}\dot{V} &= \sum_{i=1}^{r} \rho_i (e^T(N_i^T P + PN_i)e + 2\tilde{f}^T(T^T Pe + \Gamma^{-1}\dot{\tilde{f}}) \\ &= \sum_{i=1}^{r} \rho_i (e^T(N_i^T P + PN_i)e + 2\tilde{f}^T(T^T Pe - \Gamma^{-1}\Gamma S e_y) \\ &= \sum_{i=1}^{r} \rho_i (e^T(N_i^T P + PN_i)e + 2\tilde{f}^T(T^T P - SC)e \\ &= \sum_{i=1}^{r} \rho_i (e^T(N_i^T P + PN_i)e\end{aligned} \tag{16}$$

According to the lemma 1, if the conditions are satisfied, the derivative of Lyapunov function will be negative. It ensures that the state errors and the fault errors will be convergent. Now the stability of the LPV system is to find a suitable matrix by solving the LMIs.

Until now, the estimation algorithm for the fault is obtained from equation (12), the estimated value of the fault is

$$\hat{f} = \int \Gamma S e_y dt = \int \Gamma S C e dt \tag{17}$$

After obtaining the fault signal, it can be compensated in the closed-loop system.

3.3. Steps for calculating the solution of UIO

In this subsection, we will introduce the methods for computing the parameters of the UIO system. From the foregoing condition for UIO (Equation (6)), it is rewritten as follows

$$\begin{aligned} M &= EC + I_2 \\ MD_i &= 0 \end{aligned} \tag{18}$$

The development of equation (18) gives

$$\begin{bmatrix} M & E \end{bmatrix} \begin{bmatrix} I_2 & D_1 & \dots & D_r \\ -C & 0 & \dots & 0 \end{bmatrix} = \begin{bmatrix} I_2 & 0 & \dots & 0 \end{bmatrix} \tag{19}$$

Defined matrix $H = \begin{bmatrix} M & E \end{bmatrix}$, $Y = \begin{bmatrix} I_2 & D_1 & \dots & D_r \\ -C & 0 & \dots & 0 \end{bmatrix}$, and $Q = \begin{bmatrix} I_2 & 0 & \dots & 0 \end{bmatrix}$. Y, Q are known matrices and H is a unknown matrix and requires to be solved. The condition of solvability of equation (19) is that Y is of full column rank [18], then it satisfies

$$rank(Y) = rank(I_2) + rank(D_1) + \cdots + rank(D_r) \tag{20}$$

Notice that the condition ensures the solvability of equation (20). The solution is given

$$H = QY^+ \tag{21}$$

where Y^+ is a pseudo inverse of Y. If Y is of full column rank, it is

$$Y^+ = (Y^T Y)^{-1} Y^T \tag{22}$$

The matrix Y^+ can be partitioned into two parts: H_1, H_2 with apposite dimension. The equation (21) can be rewritten as

$$H = \begin{bmatrix} M & E \end{bmatrix} = Q \begin{bmatrix} H_1 & H_2 \end{bmatrix} \tag{23}$$

Thus, the solution of M and E are given as follows

$$\begin{aligned} M &= QH_1 \\ E &= QH_2 \end{aligned} \tag{24}$$

Substituting equation (24) to the condition (6), we can obtain that

$$\begin{aligned} N_i &= QH_1A_i - K_iC \\ L_i &= K_i(I_1 + CE) - QH_1A_iQH_2 \\ T &= QH_1F \end{aligned} \tag{25}$$

Theorem 1 If there exists a positive symmetric definite matrix P and matrices $\bar{K}_i$, for $i = 1, \ldots, r$. such that

$$\begin{aligned} & P = P^T > 0 \\ & (QH_1A_i)^TP + PQH_1A_i - C^T\bar{K}_i^T - \bar{K}_iC < 0 \end{aligned} \tag{26}$$

Then the UIO (Equation (5)) exists and the estimation of error converges to zero.

Proof: Replacing the matrices $\bar{K}_i = PK_i$ (In equation (26)), according to the **Lemma 1**, it can be easily proved.

Notice that by solving the LMI (26), the solution of matrices P, $\bar{K}_i$ can be obtained. Then the remaining matrices are calculated in turns. The main steps for solving the matrices can be summarized as following:

(i) Compute H from equation (21), and obtain the matrices M, E;
(ii) Solve the LMI from inequalities (26) and obtain the matrices P, Then calculate $K_i = P^{-1}\bar{K}_i$ and $S = T^TPC^+$;
(iii) Calculate the matrices G_i from equation (6);
(iv) Calculate the matrices N_i, L_i ,T from equation (25).

4. Simulation example

In this section, a wind turbine system is used for designing the UIO. Wind turbine system as an energy conversion system, usually contains the aerodynamic subsystem, drive-train subsystem, pitch mechanism subsystem and generator system. In reference [19] , a four-order mathematical model in the LPV form is given by

$$\begin{aligned} \dot{x} &= A(v)x + Bu + D(v)v \\ y &= Cx \end{aligned} \tag{27}$$

Where $A(v) = \begin{bmatrix} 0 & 1 & -1/n_g & 0 \\ -K_s/J_t & -(B_s + B_r(v))/J_t & B_s/(n_gJ_t) & k_{rb}(v)/J_t \\ K_s/(n_gJ_g) & B_s/(n_gJ_g) & -(B_s/n_g^2 + B_g)/J_g & 0 \\ 0 & 0 & 0 & -1/\tau_1 \end{bmatrix}$, $B = \begin{bmatrix} 0 & 0 & B_g/J_g & 0 \\ 0 & 0 & 0 & 1/\tau_1 \end{bmatrix}^T$, $C = \begin{bmatrix} 0 & 0 & 1 & 0 \\ 0 & 0 & 0 & 1 \end{bmatrix}$, $D(v) = \begin{bmatrix} 0 & k_{rv}(v) & 0 & 0 \end{bmatrix}^T$, $B_r(v)$, $k_{rb}(v)$ and $k_{rv}(v)$ are the linearization parameters of the aerodynamic torque. The other parameters and their names are listed in the table 1, where their values are taken from the reference [20]. In this paper, we will calculate the solutions of the proposal method for a 5MW variable pitch wind turbine benchmark [21].

Notice that during normal operation, the wind speed varies within a certain range, i.e. $v \in [v_{\min}, v_{\max}]$. In our LPV system, there is only one varying parameter, i.e. wind speed

Table 1. Parameters of the $5MW$ wind turbine system

Parameter Description	Value	Parameter Description	Value
Rated Power	$P_N = 5.2755MW$	Stiffness coefficient	$K_s = 867637KNm/rad$
Rotor Radius	$R = 63m$	Damping coefficient	$B_s = 6120KNm/(rad/s)$
Gear Box Ratio	$n_g = 97$	Rating speed	$\omega_{gN} = 1173.7rpm$
rotor inertia	$J_t = 4.38E + 07kg.m^2$	Damping of generator	$B_g = 337.698Ns/m$
generator inertia	$J_g = 543.2kg.m^2$	Time constant	$\tau_1 = 0.1$

v. Thus, in our wind turbine system, the polytopic representation of LPV model has two subsystems. According to the range of wind speed, the weight of subsystem can be calculated as following

$$\begin{aligned} \rho_1(v) &= \frac{v_{\max} - v}{v_{\max} - v_{\min}}, \\ \rho_2(v) &= \frac{v - v_{\min}}{v_{\max} - v_{\min}} \end{aligned} \tag{28}$$

In a wind turbine system, the pitch subsystem starts to work only if the wind speed is up to the rating value. And when the wind speed is over the cut-off wind speed, the pitch subsystem will stop working. In this paper, the variation range of wind speed is chosen in $(6, 25)$, during the pitch subsystem's working. According to the detail parameters of wind turbine system, the system matrix can be written as a polytopic representation and given as follows

$$\begin{aligned} A_1 &= \begin{bmatrix} 0 & 1 & -0.1031 & 0 \\ -22.39 & -0.2196 & 1.652 \times 10^{-3} & -0.00627 \\ 16744.0 & 119.84 & -1.8677 & 0 \\ 0 & 0 & 0 & -10.0 \end{bmatrix}, D_1 = \begin{bmatrix} 0 \\ 0.0249 \\ 0 \\ 0 \end{bmatrix}, \\ A_2 &= \begin{bmatrix} 0 & 1 & -0.1031 & 0 \\ -22.39 & -0.7017 & 1.652 \times 10^{-3} & -0.02876 \\ 16744.0 & 119.84 & -1.8677 & 0 \\ 0 & 0 & 0 & -10.0 \end{bmatrix}, D_2 = \begin{bmatrix} 0 \\ 0.0390 \\ 0 \\ 0 \end{bmatrix} \end{aligned} \tag{29}$$

Appling theorem 1 and following the steps, we can obtain the parameters of UIO. The LMIs can be solved by using Yalmip toolbox[22]. The solutions of matrixes are

$$N_1 = \begin{bmatrix} 0 & 1.0 & -8379.5 & 0 \\ -22.344 & 0.0029615 & -5832.9 & 0 \\ 8372.1 & 59.922 & -2.1612 & 0 \\ 0 & 0 & 0 & -0.67499 \end{bmatrix}, L_1 = \begin{bmatrix} 12569.0 & 0 \\ 8749.4 & 0.0041041 \\ 2.3079 & -0.00001 \\ 0 & -3.9875 \end{bmatrix},$$

$$N_2 = \begin{bmatrix} 0 & 1.0 & -8379.5 & 0 \\ -22.344 & -0.7004 & -5832.9 & 0 \\ 8372.1 & 59.922 & -2.1612 & -0.00006 \\ 0 & 0 & 0.00006 & -0.67499 \end{bmatrix}, L_2 = \begin{bmatrix} 12569.0 & 0 \\ 8749.4 & -0.028716 \\ 2.3079 & 0.00008 \\ -0.00008 & -3.9875 \end{bmatrix},$$

$$G_1 = G_2 = T = \begin{bmatrix} 0 & 0 \\ 0 & 0 \\ 0.31608 & 0 \\ 0 & 5.0 \end{bmatrix}, S = \begin{bmatrix} 0.90314 & 0 \\ 0 & 14.287 \end{bmatrix}. \Gamma = \begin{bmatrix} 1.0 & 0 \\ 0 & 0.002 \end{bmatrix}.$$

For the UIO, the wind speed is the main disturbance which is considered as an unknown input vector $d(t)$ (Shown in figure (1)). The initial conditions are considered as the $x(0) = [0, 1, 100, 0]$ and $\hat{x}(0) = [0, 0, 0, 0]$. In this paper, a fault for the pitch actuator system is considered as following

$$f_1(t) = \begin{cases} -0.25u_2(t), 50 \le t, \\ 0, otherwise. \end{cases}, f_2(t) = \begin{cases} 0.3u_1(t), 40 \le t, \\ 0, otherwise. \end{cases} \tag{30}$$

IOP Conf. Series: Journal of Physics: Conf. Series **783** (2017) 012001 doi:10.1088/1742-6596/783/1/012001

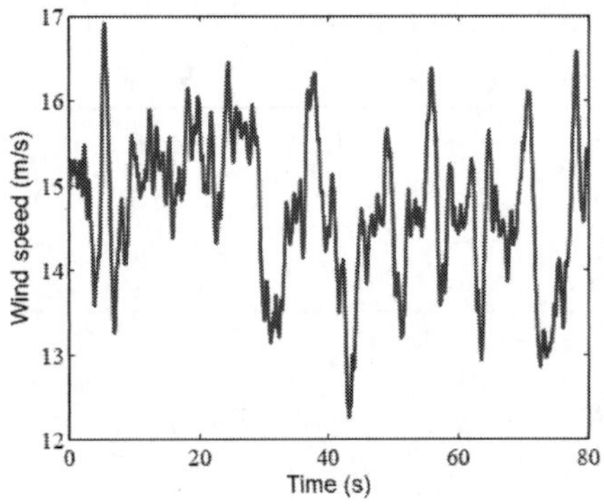

Figure 1. Wind speed

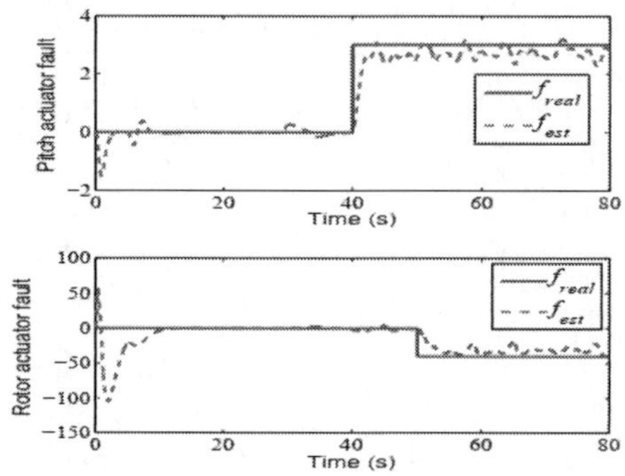

Figure 2. Simulation and estimation values of the actuator faults

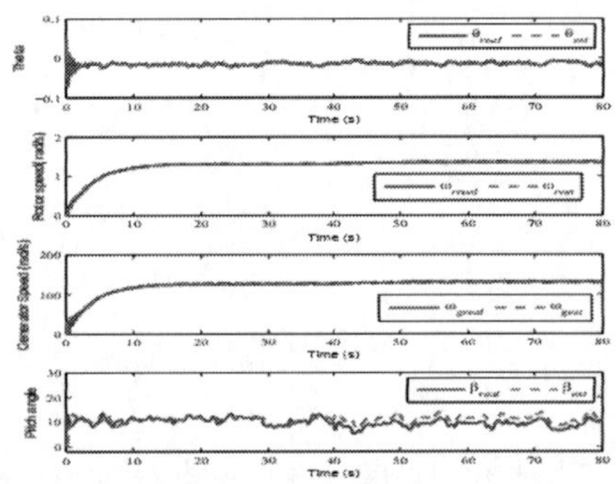

Figure 3. Simulation and estimation values of the states

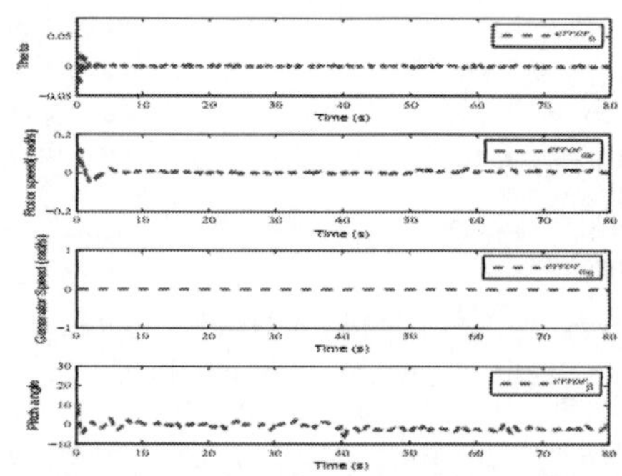

Figure 4. State errors between simulation and estimation

The simulation results are illustrated as following. Figure (2) shows the reference and the estimated values of faults . Figure (3) displays the simulated and the estimated values of the system states. From Figure (2), we can see that the estimated value tends to the simulated value rapidly and asymptotically in the absence of fault. The corresponding errors are illustrated in Figure (4). From these results, we can see that the estimation value of the states tend to the simulation value. And the estimation faults change slightly within the scope of the simulation value after the faults occur. It illustrates that the proposed UIO is effective.

5. Conclusions

In this paper, an robust UIO is designed for actuator fault diagnosis based on LPV model systems. Considering the disturbance, actuator fault in the wind turbine system, we design and analyze the convergence of the observer using Layapunov and H_∞ theory. By solving the LMIs and LMEs, the parameters of theUIO can be found. Computing the residual, the fault signal is generated easily. In addition, a wind turbine system with pitch actuator fault is tested to verify the UIO's performance. The simulation results show that the proposed method is efficient.

Acknowledgments
This work is partially supported by the National Natural Science Foundation of China (61304077), by International Science & Technology Cooperation Program of China (2015DFA01710), by the Natural Science Foundation of Jiangsu Province (BK20130765), by the Chinese Ministry of Education Project of Humanities and Social Sciences (13YJCZH171), by the Chinese Ministry of Education Project of Humanities and Social Sciences (13YJCZH171), by the 11th Jiangsu Province Six talent peaks of high level talents project (2014_ZBZZ_005), and by the Jiangsu Province Project Blue: young academic leaders project. This wrok is also supported by Campus France through the Program Partenariats Hubert Curien, Cai Yuanpei.

References

[1] Szabat K, Tran-Van T, Kami M, x, ski. A Modified Fuzzy Luenberger Observer for a Two-Mass Drive System. IEEE Transactions on Industrial Informatics. 2015, 11, 531-9.
[2] Prasov AA, Khalil HK. A Nonlinear High-Gain Observer for Systems With Measurement Noise in a Feedback Control Framework. IEEE Transactions on Automatic Control. 2013, 58, 569-80.
[3] Yao J, Jiao Z, Ma D. Extended-State-Observer-Based Output Feedback Nonlinear Robust Control of Hydraulic Systems With Backstepping. IEEE Transactions on Industrial Electronics. 2014, 61, 6285-93.
[4] Guan Y, Saif M. A novel approach to the design of unknown input observers. IEEE Transactions on Automatic Control. 1991, 36, 632-5.
[5] Hui S, Zak SH. Stress estimation using unknown input observer. 2013 American Control Conference. IEEE2013. p. 259-64.
[6] Mondal S, Chakraborty G, Bhattacharyy K. LMI approach to robust unknown input observer design for continuous systems with noise and uncertainties. International Journal of Control, Automation and Systems. 2010, 8, 210-9.
[7] Xiong Y, Saif M. Unknown disturbance inputs estimation based on a state functional observer design. Automatica. 2003, 39, 1389-98.
[8] Chen W, Saif M. Unknown input observer design for a class of nonlinear systems: an LMI approach. IEEE American Control Conference. 2006, 834-838.
[9] Zhang W, Su H, Zhu F, et al. Unknown input observer design for one-sided Lipschitz nonlinear systems. Nonlinear Dynamics, 2015, 79, 1469-1479.
[10] Chadli M, Karimi H R. Robust observer design for unknown inputs Takagi-Sugeno models. IEEE Transactions on Fuzzy Systems, 2013, 21,158-164.
[11] Kulcsar B, Bokor J, Shinar J. Unknown input reconstruction for LPV systems. International Journal of Robust and Nonlinear Control. 2010, 20, 579-95.
[12] Hammouri H, Tmar Z. Unknown input observer for state affine systems: A necessary and sufficient condition. Automatica. 2010, 46, 271-8.
[13] Chen J, Patton RJ, Zhang H-Y. Design of unknown input observers and robust fault detection filters. International Journal of Control. 1996, 63, 85-105.
[14] Chen W, Saif M. Fault detection and isolation based on novel unknown input observer design. 2006 American Control Conference. IEEE2006. p. 6 pp.
[15] Marx B, Koenig D, Ragot J. Design of observers for Takagi-Sugeno descriptor systems with unknown inputs and application to fault diagnosis. IET Control Theory and Applications. 2007, 1, 1487-95.
[16] Wang D, Lum KY. Adaptive unknown input observer approach for aircraft actuator fault detection and isolation. International Journal of Adaptive Control and Signal Processing. 2007, 21, 31-48.
[17] Hassanabadi AH, Shafiee M, Puig V. UIO design for singular delayed LPV systems with application to actuator fault detection and isolation. International Journal of Systems Science 2016; 47(1): 107-121
[18] Hamdi H, Rodrigues M, Mechmeche C, Theilliol D, Braiek NB. Fault detection and isolation in linear parameter-varying descriptor systems via proportional integral observer. International Journal of Adaptive Control and Signal Processing 2012; 26(3): 224-240.
[19] Bianchi FD, De Battista H, Mantz RJ. Wind turbine control systems: principles, modelling and gain scheduling design. Springer Science & Business Media, 2007.
[20] Shirazi FA, Grigoriadis KM, Viassolo D. Wind turbine integrated structural and LPV control design for improved closed-loop performance. International Journal of Control 2012; 85(8): 1178-1196.
[21] Jonkman JM, Butterfield S, Musial W, Scott G. Definition of a 5-MW reference wind turbine for offshore system development. National Renewable Energy Laboratory Golden, CO, USA, 2009.
[22] Löfberg J. YALMIP: A toolbox for modeling and optimization in MATLAB, 2004 IEEE International Symposium on Computer Aided Control Systems Design. IEEE, 2004;284-289.

IOP Conf. Series: Journal of Physics: Conf. Series **783** (2017) 012002 doi:10.1088/1742-6596/783/1/012002

H_-/H_∞ fault detection observer design based on generalized output for polytopic LPV system

Meng Zhou[1,2], Mickael Rodrigues[2], Yi Shen[1], Didier Theilliol[3]

[1] School of Astronautics, Harbin Institute of Technology, Harbin 150001, P. R. China

[2] Univ Lyon, Université Claude Bernard Lyon 1, CNRS, LAGEP UMR 5007, 43 boulevard du 11 novembre 1918, F-69100, Villeurbanne, France

[3] Centre de Recherche en Automatique de Nancy (CRAN), Université de Lorraine, CNRS UMR 7039, F-54506 Vandoeuvre-les-Nancy, France

E-mail: `didier.theilliol@univ-lorraine.fr`

Abstract. This paper proposes an H_-/H_∞ fault detection observer design method by using generalized output for a class of polytopic linear parameter-varying (LPV) system. First, with the aid of the relative degree of output, a new output vector is generated by gathering the original and its time derivative. The actuator fault is introduced into the measurement equation of the new system. An H_-/H_∞ observer is designed for the new LPV polytopic system to guarantee the robustness against disturbances and to improve the fault sensitivity, simultaneously. The existence conditions of the H_-/H_∞ observer are given and solved by a set of linear matrix inequalities (LMIs). Finally simulation results are given to illustrate the effectiveness of the proposed method.

1. Introduction

Most nonlinear control systems can be approximated into a linear parameter varying (LPV) systems, besides, it is convenient to extend the methods and theories of linear systems into LPV system models. Control theories for LPV systems have received considerable attention [1]. In recent decades, with the increasing demand of safety and reliability for modern complex control system, fault detection for LPV systems has become more and more important and fruitful literatures on this topic can be found, see, e.g. [2-4] and the references therein.

A robust fault detection observer should guarantee the generated residual significantly robust to unknown disturbances and sensitive to faults, simultaneously. In order to achieve a proper trade-off between the sensitivity to faults and robustness to disturbances, mixed H_-/H_∞ observer method has received many researchers' attention, in which the H_∞ norm is regarded as the criterion to evaluate the influence of the disturbances and the H_- index is used to measure the sensitivity of the faults. Mixed H_-/H_∞ fault detection observer is firstly proposed in [5], in which the H_- norm is defined as the smallest nonzero singular value of the transfer function matrix at the particular frequency $\omega = 0$. In [6, 7], the H_- index is defined as the minimum singular value of the transfer function matrix. Besides, the H_- index definition is also suitable for a finite frequency domain situation. Based on this H_- index definition, many literatures focus on H_-/H_∞ observer design for LTI system can be found in [8], [9], [10] and so on. Recently, based on the H_- index definition in full-frequency domain and finite-frequency domain, H_-/H_∞

fault detection observe design method has been extended into nonliear systems [3], [11], [12], [13], [14], [17], just to name a few.

It should be mentioned that mixed H_-/H_∞ observer design method is not feasible for strictly proper systems in full-frequency domain, as the H_- index over $[0,\ \infty)$ is always zeros[18]. Recently, in [19], a new actuator fault diagnosis approach is proposed with the relative degree of the output respect to fault. It is proved that with the aid of the relative degree of output respect to fault, the actuator fault vector can be considered into the output equation. Hence mixed H_-/H_∞ observer is feasible for the new system. However, in [19], the authors only address the fault diagnosis approach considering the relative degree of output into linear system. Besides, the H_- index do not be taken into consideration, which motives the work of this paper.

Based on the idea of [19], the main contribution of this paper lies in the following aspects. First, an H_-/H_∞ fault detection observer method is designed to a class of polytopic LPV system for actuator fault detection. Thanks to the relative degree of output respect to fault, it is feasible to consider H_- index for system with only actuator faults in full-frequency domain. Second, the H_- index is considered in observer design to improve the fault sensitivity. Besides, both the system unknown disturbance and measurement noise are considered in this paper.

The rest of this paper is organised as follows. The preliminaries are given in Section II. The problem statements is described in Section III. In Section IV, a new system is first constructed with generalized output, then a mixed H_-/H_∞ observer is designed for the new polytopic LPV system. At last, the residual evaluation and decision making logic are proposed. Simulation results are given in Section V to demonstrate the proposed approach. Finally, conclusions are given in Section VI.

Notation 1. Throughout the paper, $P > 0$ and $Q < 0$ represent that P is positive-definite and Q is negative-definite, respectively. The star symbol $(\star)$ in a symmetric matrix denotes the transposed block in the symmetric position. $\overline{\sigma}(X)$ denotes the maximum singular value of X and $\underline{\sigma}(X)$ is the minimum singular value of X. If $\mathcal{L}_2$ is a set of square integrable functions, then the $\mathcal{L}_2$ norm of $x(t) \in \mathcal{L}_2$ is defined as $\|x(t)\|_2 = (\int_0^\infty x^T(t)x(t)dt)^{1/2}$.

2. Preliminaries

The following definitions and assumption are given and used throughout the paper.

Definition 1. [13] For a stable linear parameter-varying system described as

$$\begin{cases} \dot{x}(t) = A(\rho(t))x(t) + B(\rho(t))u(t) \\ y(t) = C(\rho(t))x(t) + D(\rho(t))u(t) \end{cases} \tag{1}$$

H_∞ norm for LPV system (1) is defined as

$$\|G(\rho(t))\|_\infty = \sup_{\forall \rho,\ \omega} \overline{\sigma}[G(j\omega)] \tag{2}$$

and H_- index is defined as

$$\|G(\rho(t))\|_- = \inf_{\forall \rho,\ \omega} \underline{\sigma}[G(j\omega)] \tag{3}$$

where $G(\rho(t))$ represents the transfer function from input $u(t)$ to output $y(t)$.

Definition 2. (Relative Degree [21]). Considering a linear control system as

$$\begin{cases} \dot{x}(t) = Ax(t) + Ef(t) \\ y(t) = Cx(t) \end{cases} \tag{4}$$

where $x \in \mathbb{R}^{n_x}$, $f \in \mathbb{R}^{n_f}$ and $y \in \mathbb{R}$ are the state vector, actuator fault signal and measurement output, respectively. The relative degree of the output $y(t)$ respect to fault $f(t)$ is λ_f satisfying

$$\begin{cases} CA^{i-1}E = 0, \ \forall i = 1, \cdots (\lambda_f - 1) \\ CA^{\lambda_f - 1}E \neq 0 \end{cases} \tag{5}$$

The relative degree of output $y(t)$ respect to fault $f(t)$ is λ_f means that the λ_fth time-derivative of output $y^{\lambda_f}(t)$ depends on fault explicitly, while all lower order time-derivatives of output do not depend on the fault explicitly, such as

$$y^{\lambda_f}(t) = CA^{\lambda_f}x(t) + \underbrace{CA^{(\lambda_f-1)}E}_{\neq 0} f(t) \tag{6}$$

Assumption 1. In this paper, it is assumed that the relative degree of output $y(t)$ respect to actuator fault $f(t)$ is 1, such that $CE(\rho(t)) \neq 0$.

3. Problem formulation

Considering the following LPV systems subject to actuator faults and unknown disturbances as

$$\begin{cases} \dot{x}(t) = A(\rho(t))x(t) + B(\rho(t))u(t) + E(\rho(t))f(t) + D_x(\rho(t))w(t) \\ y(t) = Cx(t) + D_y v(t) \end{cases} \tag{7}$$

where $x(t) \in \mathbb{R}^{n_x}$, $u(t) \in \mathbb{R}^{n_u}$, $y(t) \in \mathbb{R}^{n_y}$ are the state vector, input vector and measurement output vector, respectively. $f(t) \in \mathbb{R}^{n_f}$ is the actuator fault which needs to be detected. $w(t) \in \mathbb{R}^{n_w}$ is the system unknown disturbance and $v(t) \in \mathbb{R}^{n_v}$ is the measurement noise which are assumed bounded. $A(\rho(t)) \in \mathbb{R}^{n_x \times n_x}$, $B(\rho(t)) \in \mathbb{R}^{n_x \times n_u}$, $E(\rho(t)) \in \mathbb{R}^{n_x \times n_f}$, $D_x(\rho(t)) \in \mathbb{R}^{n_x \times n_w}$, $C \in \mathbb{R}^{n_y \times n_x}$, $D_y \in \mathbb{R}^{n_y \times n_v}$ are system matrices. $\rho(t) = [\rho_1(t), \cdots, \rho_s(t)]$ is the scheduling vector assumed measurable online, s is the number of the scheduling vector. It is assumed that the scheduling vectors are measurable. This paper mainly focuses on the actuator fault, i.e. $B(\rho(t)) = E(\rho(t))$, so each of the output $y_i(t)$, $i = (1, 2, \cdots, n_y)$ has the same relative degree respect to the input $u(t)$ and fault $f(t)$. Without loss of generality, it is assumed the system $(A(\rho(t)), B(\rho(t)))$ is controllable, $(A(\rho(t)), C)$ is observable, and the rank condition is satisfied as $\text{rank}(CB(\rho(t))) = \text{rank}(B(\rho(t)))$.

In this paper, the LPV system (7) is assumed that the time-varying parameter vector $\rho(t)$ varies within a polytope Θ of vertices $\{\rho_1, \cdots, \rho_s\}$ with $\rho_i \in [\underline{\rho}_i, \overline{\rho}_i]$, $i = (1, 2, \cdots, s)$. The system matrices range in a polytope of matrices defined as the convex hull, each polytope vertex corresponds to a particular value of scheduling variable [23]. Then the polytopic LPV system (7) is expressed as

$$\begin{cases} \dot{x}(t) = \sum_{i=1}^{N} h_i(\rho(t))(A_i x(t) + B_i u(t) + B_i f(t) + D_{xi} w(t)) \\ y(t) = Cx(t) + D_y v(t) \end{cases} \tag{8}$$

where $A(\rho(t)) = \sum_{i=1}^{N} h_i(\rho(t))A_i$, $B(\rho(t)) = \sum_{i=1}^{N} h_i(\rho(t))B_i$, $D_x(\rho(t)) = \sum_{i=1}^{N} h_i(\rho(t))D_{xi}$. N is the number of weighting functions. A_i, B_i, D_{xi} are time-invariant matrices of the ith model. The polytopic system is scheduled through the weighting functions that lie in a convex set and satisfy the following properties

$$\sum_{i=1}^{N} h_i(\rho(t)) = 1, \quad 0 \leq h_i(\rho(t)) \leq 1, \ i = (1, 2, \cdots, N) \tag{9}$$

Classically, a mixed H_-/H_∞ observer is generated as

$$\begin{cases} \dot{\hat{x}}(t) = \sum_{i=1}^{N} h_i(\rho(t))[A_i\hat{x}(t) + B_iu(t) + L_i(y(t) - \hat{y}(t))] \\ \hat{y}(t) = C\hat{x}(t) \\ r(t) = M[y(t) - \hat{y}(t)] \end{cases} \tag{10}$$

where $\hat{x}(t) \in \mathbb{R}^{n_x}$ is the state estimation of observer (10), $L_i \in \mathbb{R}^{n_x \times n_y}$, $i = (1, 2, \cdots, N)$ and $M \in \mathbb{R}^{n_y \times n_y}$ are the gain matrices of residual generator based on an observer (10) that have to be designed.

Defining the state estimation error vector as

$$e(t) = x(t) - \hat{x}(t) \tag{11}$$

Subtracting (8) from (10), the estimation error dynamic system is obtained as

$$\dot{e}(t) = \sum_{i=1}^{N} h_i(\rho(t))[(A_i - L_iC)e(t) + B_if(t) + (\bar{D}_{xi} - L_i\bar{D}_y)d(t)] \tag{12}$$

$$r(t) = MCe(t) + M\bar{D}_yd(t) \tag{13}$$

where

$$d(t) = \begin{bmatrix} w(t) \\ v(t) \end{bmatrix}, \bar{D}_{xi} = \begin{bmatrix} D_{xi} & 0 \end{bmatrix}, \bar{D}_y = \begin{bmatrix} 0 & D_y \end{bmatrix} \tag{14}$$

According to [5], observer (10) is called an H_-/H_∞ observer if the following conditions hold

(i) The state estimation error system (12) is asymptotically stable;

(ii) The residual $r(t)$ (13) is robust to the unknown disturbance $d(t)$, i.e.

$$\|r(t)\|_2 < \gamma\|d(t)\|_2 \tag{15}$$

(iii) The residual $r(t)$ (13) is sensitive to the actuator fault $f(t)$, i.e.

$$\|r(t)\|_2 > \beta\|f(t)\|_2 \tag{16}$$

Obviously, the goal (i) and (ii) can be satisfied easily based on the H_∞ theory. While considering the fault sensitivity condition (16), the observer matrices should satisfy the following LMI

$$\begin{bmatrix} C^TM^TMC - P(A_i - L_iC) - (A_i - L_iC)^TP & -PB_i \\ \star & -\beta^2I \end{bmatrix} > 0 \tag{17}$$

where P is a symmetrical positive definite matrix.

Unfortunately, due to term $-\beta^2I < 0$, there is no feasible solution for LMI (17). As a result, the H_-/H_∞ observer cannot be obtained for system in (8) in full frequency domain. It is known that if a system consists a sensor fault, such that $y(t) = Cx(t) + Df(t) + D_yv(t)$, then the generated residual signal will be influenced by the sensor fault, the H_- condition will become

$$\begin{bmatrix} C^TM^TMC - P(A_i - L_iC) - (A_i - L_iC)^TP & C^TM^TMD - PB_i \\ \star & D^TM^TMD - \beta^2I \end{bmatrix} > 0 \tag{18}$$

There may be a feasible solution for (18) with the matrix D. Therefore, the mixed H_-/H_∞ observer is theoretically valid for a system with faults in the measurement equations. Actually, the actuator fault also influence the outputs. If the actuator fault can be transformed into the measurement output equation, the H_-/H_∞ observer is theoretically feasible. Fortunately, relative degree of output is a useful method to bring the actuator fault to the measurement output. In this paper, we mainly focus on H_-/H_∞ fault detection observer design based on the relative degree of output.

4. Main Results

4.1. H_-/H_∞ observer design based on generalized output

According to [19], for multiple output systems, each output $y_i(t)$, $i = (1, \cdots, n_y)$ may have different relative degree respect to fault $f(t)$. The first step is to generate a new output component $\tilde{y}_i(t)$ for each output $y_i(t)$. Then gather all the new component $\tilde{y}_i(t)$ as a whole to become the new output $\tilde{y}(t)$. However in this paper, under assumption 1, the relative degree of the each output respect to fault is all assumed to be 1, such that the output $y(t)$ can be treat as a whole and the time derivative $\dot{y}(t)$ is expressed as

$$\dot{y}(t) = \sum_{i=1}^{N} h_i(\rho(t))[CA_i x(t) + CB_i(u(t) + f(t)) + CD_{xi}w(t)] + D_y\dot{v}(t) \tag{19}$$

A new generalized output $\tilde{y}(t)$ is generated by gathering the original output $y(t)$ and the time derivative $\dot{y}(t)$

$$\tilde{y}(t) = \sum_{i=1}^{N} h_i(\rho(t))(\tilde{C}_i x(t) + R_i f(t) + \bar{D}_{yi}d(t)) \tag{20}$$

where

$$\tilde{y}(t) = \begin{bmatrix} y(t) \\ \dot{y}(t) - \sum_{i=1}^{N} h_i(\rho(t))CB_i u(t) \end{bmatrix}, \tilde{C}_i = \begin{bmatrix} C \\ CA_i \end{bmatrix}, R_i = \begin{bmatrix} 0 \\ CB_i \end{bmatrix},$$
$$\bar{D}_{yi} = \begin{bmatrix} 0 & D_y & 0 \\ CD_{xi} & 0 & D_y \end{bmatrix} \text{ and } d(t) = \begin{bmatrix} w(t) \\ v(t) \\ \dot{v}(t) \end{bmatrix} \tag{21}$$

By letting,

$$\bar{D}_{xi} = \begin{bmatrix} D_{xi} & 0 & 0 \end{bmatrix} \tag{22}$$

a new LPV system can be expressed as

$$\begin{cases} \dot{x}(t) = \sum_{i=1}^{N} h_i(\rho(t))[A_i x(t) + B_i(u(t) + f(t)) + \bar{D}_{xi}d(t)] \\ \tilde{y}(t) = \sum_{i=1}^{N} h_i(\rho(t))[\tilde{C}_i x(t) + R_i f(t) + \bar{D}_{yi}d(t)] \end{cases} \tag{23}$$

For system (23), an H_-/H_∞ observer with LPV structure is designed as

$$\begin{cases} \dot{\hat{x}}(t) = \sum_{i=1}^{N} h_i(\rho(t))[A_i\hat{x}(t) + B_i u(t) + L_i(\tilde{y}(t) - \hat{\tilde{y}}(t))] \\ \hat{\tilde{y}}(t) = \sum_{i=1}^{N} h_i(\rho(t))[\tilde{C}_i\hat{x}(t)] \\ r(t) = M[\tilde{y}(t) - \hat{\tilde{y}}(t)] \end{cases} \tag{24}$$

Define the state estimation error vector as

$$e(t) = x(t) - \hat{x}(t) \tag{25}$$

Subtracting (23) from (24), the estimation error dynamic system is obtained as

$$\dot{e}(t) = \sum_{i=1}^{N}\sum_{j=1}^{N} h_i(\rho(t))h_j(\rho(t))[(A_i - L_i\tilde{C}_j)e(t) + (B_i - L_iR_j)f(t) + (\bar{D}_{xi} - L_i\bar{D}_{yj})d(t)] \quad (26)$$

$$r(t) = \sum_{i=1}^{N} h_i(\rho(t))(M\tilde{C}_ie(t) + M\bar{D}_{yi}d(t) + MR_if(t)) \quad (27)$$

The main objective of this paper is to design the matrices L_i, $i = (1, 2, \cdots, N)$ and M of the H_-/H_∞ observer (24) to guarantee the residual $r(t)$ robust to disturbance $d(t)$ and sensitive to fault $f(t)$, simultaneously, such that

(i) The estimation error system (26) is asymptotically stable;

(ii) The residual $r(t)$ (27) is robust to the disturbance $d(t)$, i.e.

$$\|r(t)\|_2 < \gamma\|d(t)\|_2 \quad (28)$$

(iii) The residual $r(t)$ (27) is sensitive to the actuator fault $f(t)$, i.e.

$$\|r(t)\|_2 > \beta\|f(t)\|_2 \quad (29)$$

The following Theorem is proposed to synthesize the matrices L_i and M of observer in (24).

Theorem 1. For given positive scalars γ and β and θ_1, the observer (24) is called an H_-/H_∞ observer, if there exist a symmetrical positive definite matrix $P \in \mathbb{R}^{n_x \times n_x}$, matrices $W_i \in \mathbb{R}^{n_x \times n_y}$ and symmetrical non-negative matrix $U \in \mathbb{R}^{n_y \times n_y}$ to satisfy the following LMIs hold

$$\begin{aligned} &\Psi_{ii} < 0, i = 1, \cdots N \\ &\Psi_{ij} + \Psi_{ji} < 0, 1 \le i < j \le N \end{aligned} \quad (30)$$

$$\begin{aligned} &\Pi_{ii} < 0, i = 1, \cdots N \\ &\Pi_{ij} + \Pi_{ji} < 0, 1 \le i < j \le N \end{aligned} \quad (31)$$

. where

$$\Psi_{ij} = \begin{bmatrix} \psi_{11} & \theta_1 P\bar{D}_{xi} - \theta_1 W_i\bar{D}_{yj} + \tilde{C}_i^T U\bar{D}_{yj} \\ \star & \bar{D}_{yi}^T U\bar{D}_{yj} - \gamma^2 I \end{bmatrix} \quad (32)$$

$$\psi_{11} = \theta_1 PA_i - \theta_1 W_i\tilde{C}_j + \theta_1 A_i^T P - \theta_1\tilde{C}_j^T W_i^T + \tilde{C}_i^T U\tilde{C}_j \quad (33)$$

$$\Pi_{ij} = \begin{bmatrix} \pi_{11} & PB_i - W_iR_j - \tilde{C}_i^T UR_j \\ \star & -R_i^T UR_j + \beta^2 I \end{bmatrix} \quad (34)$$

$$\pi_{11} = PA_i - W_i\tilde{C}_j + A_i^T P - \tilde{C}_j^T W_i^T - \tilde{C}_i^T U\tilde{C}_j \quad (35)$$

$$W_i = PL_i, U = M^T M \quad (36)$$

Proof: (i) *Disturbance attenuation condition*

The robustness of the residual signal $r(t)$ (27) against to disturbance $d(t)$ is first considered. Making $f(t) = 0$ in (26) and (27), then it becomes

$$\begin{cases} \dot{e}(t) = \sum_{i=1}^{N}\sum_{j=1}^{N} h_i(\rho(t))h_j(\rho(t))[(A_i - L_i\tilde{C}_j)e(t) + (\bar{D}_{xi} - L_i\bar{D}_{yj})d(t) \\ r(t) = \sum_{i=1}^{N} h_i(\rho(t))[M\tilde{C}_ie(t) + M\bar{D}_{yi}d(t)] \end{cases} \quad (37)$$

Choosing a Lyapunov function as

$$V_1(t) = e(t)^T P_1 e(t) > 0 \tag{38}$$

where $P_1 \in \mathbb{R}^{n_x \times n_x}$ is a symmetrical positive definite matrix.

Then the time derivative of $V_1(t)$ is

$$\begin{aligned}\dot{V}_1(t) = &\sum_{i=1}^{N}\sum_{j=1}^{N} h_i(\rho(t))h_j(\rho(t))\{e^T(t)[P_1(A_i - L_i\tilde{C}_j) + (A_i - L_i\tilde{C}_j)^T P_1]e(t) \\ &+ 2e^T(t)[P_1(\bar{D}_{xi} - L_i\bar{D}_{yj})]d(t)\}\end{aligned} \tag{39}$$

To satisfy the H_∞ index $\|r(t)\|_2 < \gamma\|d(t)\|_2$, a criterion is defined as

$$J_\infty = \int_0^\infty [r^T(t)r(t) - \gamma^2 d^T(t)d(t)]dt \tag{40}$$

Then

$$J_\infty = \int_0^\infty [r^T(t)r(t) - \gamma^2 d^T(t)d(t) + \dot{V}_1(t)]dt - V_1(t) \tag{41}$$

Due to the fact that $V_1(t) > 0$, if

$$r^T(t)r(t) - \gamma^2 d^T(t)d(t) + \dot{V}_1(t) < 0 \tag{42}$$

then it is satisfied $J_\infty < 0$. Substitute (27) into (42), it becomes

$$\begin{aligned}&\sum_{i=1}^{N}\sum_{j=1}^{N} h_i(\rho(t))h_j(\rho(t))\{[e^T(t)(P_1(A_i - L_i\tilde{C}_j) + (A_i - L_i\tilde{C}_j)^T P_1)e(t)] \\ &+ 2e^T(t)[P_1(\bar{D}_{xi} - L_i\bar{D}_{yj})]\}d(t) + \sum_{i=1}^{N}\sum_{j=1}^{N} h_i(\rho(t))h_j(\rho(t))[e^T(t)\tilde{C}_i^T M^T M\tilde{C}_j e(t) \\ &+ 2e^T(t)\tilde{C}_i^T M^T M\bar{D}_{yj}d(t) + d^T(t)\bar{D}_{yi}^T M^T M\bar{D}_{yj}d(t)] - \gamma^2 d^T(t)d(t) \\ &= \sum_{i=1}^{N}\sum_{j=1}^{N} h_i(\rho(t))h_j(\rho(t))\xi^T(t)\Omega_{ij}\xi(t) < 0\end{aligned} \tag{43}$$

where

$$\xi(t) = \begin{bmatrix} e(t) \\ d(t) \end{bmatrix}, \quad \Omega_{ij} = \begin{bmatrix} \omega_{11} & P_1(\bar{D}_{xi} - L_i\bar{D}_{yj}) + \tilde{C}_i^T M^T M\bar{D}_{yj} \\ \star & \bar{D}_{yi}^T M^T M\bar{D}_{yj} - \gamma^2 I \end{bmatrix} \tag{44}$$

$$\omega_{11} = P_1(A_i - L_i\tilde{C}_j) + (A_i - L_i\tilde{C}_j)^T P_1 + \tilde{C}_i^T M^T M\tilde{C}_j \tag{45}$$

To make $J_\infty < 0$, it is required

$$\Omega_{ij} < 0, \quad i,j = (1,2,\cdots,N) \tag{46}$$

Note that (43) can be re-written as

$$\sum_{i=1}^{N} h_i(\rho(t))^2\xi^T(t)\Omega_{ii}\xi(t) + \sum_{i=1}^{N}\sum_{i<j}^{N} h_i(\rho(t))h_j(\rho(t))\xi^T(t)(\Omega_{ij} + \Omega_{ji})\xi(t) < 0 \tag{47}$$

Then we have the following conditions

$$\begin{cases} \Omega_{ii} < 0, & i = 1, \cdots N \\ \Omega_{ij} + \Omega_{ji} < 0, & 1 \le i < j \le N \end{cases} \tag{48}$$

According to [22], the conditions in (48) is more relaxed than that in (46).

(ii) *Fault sensitivity condition*

Next the fault sensitivity of the residual $r(t)$ in (27) is considered, letting $d(t) = 0$ in (26) and (27), then

$$\begin{cases} \dot{e}(t) = \sum_{i=1}^{N}\sum_{j=1}^{N} h_i(\rho(t))h_j(\rho(t))[(A_i - L_i\tilde{C}_j)e(t) + (B_i - L_iR_j)f(t) \\ r(t) = \sum_{i=1}^{N} h_i(\rho(t))(M\tilde{C}_ie(t) + MR_if(t)) \end{cases} \tag{49}$$

The proof for fault sensitivity condition is similar to disturbance attenuation condition. A Lyapunov function is chosen as

$$V_2(t) = e^T(t)P_2e(t) > 0 \tag{50}$$

To maintain the H_- index $\|r(t)\|_2 > \beta\|f(t)\|_2$, defining a criterion as

$$J_- = \int_0^{\infty} [r^T(t)r(t) - \beta^2 f^T(t)f(t)]dt \tag{51}$$

which is equivalent to

$$J_- = \int_0^{\infty} [r^T(t)r(t) - \beta^2 f^T(t)f(t) - \dot{V}_2(t)]dt + V_2(t) \tag{52}$$

As $V_2(t) > 0$, if

$$r^T(t)r(t) - \beta^2 f^T(t)f(t) - \dot{V}_2(t) > 0 \tag{53}$$

Then the H_- index $J_- > 0$ is satisfied. Combining (27), then (53) can be written such that

$$\sum_{i=1}^{N}\sum_{j=1}^{N} h_i(\rho(t))h_j(\rho(t))\eta^T(t)\Upsilon_{ij}\eta(t) > 0 \tag{54}$$

where

$$\eta(t) = \begin{bmatrix} e(t) \\ f(t) \end{bmatrix}, \quad \Upsilon_{ij} = \begin{bmatrix} \upsilon_{11} & -P_2(E_i - L_iR_j) + \tilde{C}_i^T M^T MR_j \\ \star & -R_i^T M^T MR_j + \beta^2 I \end{bmatrix}$$
$$\upsilon_{11} = -P_2(A_i - L_i\tilde{C}_j) - (A_i - L_i\tilde{C}_j)^T P_2 + \tilde{C}_i^T M^T M\tilde{C}_j \tag{55}$$

For similar reasons explained in [22], the inequality (54) can be replaced by

$$\sum_{i=1}^{N} h_i(\rho(t))^2\eta^T(t)\Upsilon_{ii}\eta(t) + \sum_{i=1}^{N}\sum_{i<j}^{N} h_i(\rho(t))h_j(\rho(t))\eta^T(t)(\Upsilon_{ij} + \Upsilon_{ji})\eta(t) > 0 \tag{56}$$

As a result,

$$\begin{cases} \Upsilon_{ii} > 0, & i = 1, \cdots N \\ \Upsilon_{ij} + \Upsilon_{ji} > 0, & 1 \le i < j \le N \end{cases} \tag{57}$$

Noted that there is coupling between (48) and (57), in order to facilitate the observer solving process, letting

$$P_1 = \theta_1 P, \quad P_2 = P, \quad W_i = PL_i, \quad U = M^T M \tag{58}$$

Then (48) and (57) is equivalent to (30) and (31), respectively. And the observer gain matrices L_i and M can be obtained by $L_i = P^{-1}W_i$, $M = \sqrt{U}$.

4.2. Residual evaluation and decision making

After the H_-/H_∞ observer is designed, a residual evaluation function and fault decision making scheme are given. Due to the presence of exogenous disturbances, the residual signals are different from zero even under fault-free operation. Since the proposed H_-/H_∞ observer guarantees that $\|r(t)\|_2 < \gamma\|d(t)\|_2$, the $\mathcal{L}_2$ norm of the residual signal $\|r(t)\|_2$ can be used as the residual evaluation function. However, since it is usually unrealistic to calculate the evaluation function over the whole time [20], in this paper, the Root Mean Square (RMS) function is chosen as the evaluation function

$$J = \|r(t)\|_{\text{RMS}} = \sqrt{\frac{1}{\Delta t}\int_{t-\Delta t}^{t} r^T(\tau)r(\tau)d\tau} \tag{59}$$

where J measures the average energy of the residual over a time interval $(t-\Delta t, t)$. Δt is a large enough time window which is usually determined depends on the applications [11].

Once the evaluation function is selected, a threshold J_{th} is needed. The threshold should represent the maximum influence of the unknown disturbances $d(t)$ respect to the residual $r(t)$ under a fault-free situation, therefore the threshold can be set as

$$J_{th} = \sup_{d(t), f(t)=0} \|r(t)\|_{\text{RMS}} \tag{60}$$

Then comparing the residual evaluation function J with the predefined threshold J_{th}, the decision logic is formulated as

$$\begin{cases} J > J_{th} \Rightarrow \text{alarm} \\ J \leq J_{th} \Rightarrow \text{no alarm} \end{cases} \tag{61}$$

5. Simulation Results

Numerical simulations are proposed to demonstrate the effectiveness of the proposed H_-/H_∞ observer design method. Considering a Vertical Takeoff and Landing (VTOL) aircraft model in the vertical plane described as an LPV form from [23]

$$\begin{cases} \dot{x}(t) = A(\rho(t))x(t) + B(\rho(t))(u(t) + f(t)) + D_x w(t) \\ y(t) = Cx(t) + D_y v(t) \end{cases} \tag{62}$$

where the state vectors $x(t) = [V_h \; V_v \; q \; \theta]^T$ are horizontal velocity, vertical velocity, pitch rate and pitch angle, respectively, and the control input $u(t) = [u_c \; u_l]^T$ are collective pitch control and longitudinal cyclic pitch control, respectively. $w(t)$ represents the model uncertainties and unknown disturbances, $v(t)$ denotes the measurement noise. The system matrices with varying parameters is expressed as

$$A(\rho(t)) = \begin{bmatrix} -9.9477 & -0.7476 & 0.2632 & 5.0337 \\ 52.1659 & 2.7452 & 5.5532 & -24.4221 \\ 26.0922 & 2.6361+\rho_1(t) & -4.1975 & -19.2774+\rho_2(t) \\ 0 & 0 & 1 & 0 \end{bmatrix}, B(\rho(t)) = \begin{bmatrix} 0.4422 & 0.1761 \\ 3.5446+\rho_2(t) & -7.5922 \\ -5.5200 & 4.4900 \\ 0 & 0 \end{bmatrix}$$

The output and the unknown input distribution matrices are

$$C = \begin{bmatrix} 1 & 0 & 0 & 0 \\ 0 & 1 & 0 & 0 \\ 0 & 0 & 1 & 0 \\ 0 & 1 & 1 & 1 \end{bmatrix}, D_x = \begin{bmatrix} 0.001 \\ 0 \\ 0 \\ 0 \end{bmatrix}, D_y = \begin{bmatrix} 0.001 & 0 & 0 & 0 \\ 0 & 0 & 0.001 & 0 \\ 0 & 0 & 0 & 0.001 \\ 0 & 0.001 & 0 & 0 \end{bmatrix}$$

Obviously the relative degree of each output respect to actuator faults is 1, based on (21), the new output and its corresponding matrices can be obtained. In this paper, it is assumed that $\rho(t) = [\rho_1(t) \ \ \rho_2(t)]^T$ are the scheduling vectors with $\rho_1(t) \in [-0.5, \ 0.5]$ and $\rho_2(t) \in [-2, \ 2]$. The scheduling variables $\rho_1(t)$ and $\rho_2(t)$ are shown as figure 1 and figure 2. Then four local models are derived and the weighting functions $h_i(\rho(t))$, $i = (1, \cdots, N)$, $N = 4$ are assumed as

$$\begin{aligned} h_1(\rho(t)) &= \frac{\overline{\rho}_1 - \rho_1(t)}{\overline{\rho}_1 - \underline{\rho}_1} \cdot \frac{\overline{\rho}_2 - \rho_2(t)}{\overline{\rho}_2 - \underline{\rho}_2}, \quad h_2(\rho(t)) = \frac{\overline{\rho}_1 - \rho_1(t)}{\overline{\rho}_1 - \underline{\rho}_1} \cdot \frac{\rho_2(t) - \underline{\rho}_2}{\overline{\rho}_2 - \underline{\rho}_2} \\ h_3(\rho(t)) &= \frac{\rho_1(t) - \underline{\rho}_1}{\overline{\rho}_1 - \underline{\rho}_1} \cdot \frac{\overline{\rho}_2 - \rho_2(t)}{\overline{\rho}_2 - \underline{\rho}_2}, \quad h_4(\rho(t)) = \frac{\rho_1(t) - \underline{\rho}_1}{\overline{\rho}_1 - \underline{\rho}_1} \cdot \frac{\rho_2(t) - \underline{\rho}_2}{\overline{\rho}_2 - \underline{\rho}_2} \end{aligned} \tag{63}$$

The local weighting function $h_i(\rho(t))$ of each local model is depicted in figure 3. According to the separation principle of the controller and the observer design, the controller has been synthesized with a state feedback structure so as to maintain the closed-loop system stable.

Generally speaking, the disturbance reduction performance γ and fault sensitivity performance β are need to be made a trade-off. The principle of determining these parameters in this paper is described as follows. First, the robustness performance index γ is fixed based on experience, then finding the largest value of the fault sensitive index β by tuning the parameter θ_1 to satisfy the LMIs of Theorem 1. In this paper, the robustness index is chosen as $\gamma = 0.08$, and the largest values of the sensitivity performance is obtained as $\beta = 2.83$ with the parameter $\theta_1 = 30$ by solving the LMIs in Theorem 1, and the gain matrices of the proposed H_-/H_∞ observer for the new system are obtained as

$$L_1 = \begin{bmatrix} 0.8715 & 0.7565 & 0.9320 & 0.7549 & -0.1939 & 0.2457 & -0.9359 & 0.2418 \\ 0.0641 & 2.0349 & 0.1363 & 1.8852 & -0.0315 & 0.6062 & -0.3151 & 0.4274 \\ -0.0516 & -0.4609 & -0.1530 & 0.1742 & -0.1402 & -0.0592 & 0.3670 & 0.1549 \\ -0.0278 & 0.0194 & -0.0130 & 0.0390 & 0.0275 & -0.0080 & 0.0166 & -0.0045 \end{bmatrix}$$

$$L_2 = \begin{bmatrix} 0.7548 & 1.1492 & 1.0942 & 1.1406 & 0.5589 & -0.5021 & -0.0639 & 0.5281 \\ -0.0484 & 2.1406 & 0.1995 & 1.9285 & 0.4204 & 0.3959 & -0.0583 & 0.5371 \\ -0.0166 & -0.4570 & -0.2079 & 0.2097 & -0.1869 & 0.0686 & 0.2528 & 0.1135 \\ -0.0069 & 0.0039 & 0.0019 & 0.0462 & -0.0413 & 0.0187 & -0.0336 & -0.0130 \end{bmatrix}$$

$$L_3 = \begin{bmatrix} 0.9134 & 0.7841 & 0.9556 & 0.7921 & -0.5989 & 0.2851 & -1.0435 & 0.1972 \\ 0.1258 & 2.0508 & 0.2880 & 1.9941 & -0.1817 & 0.5794 & -0.4080 & 0.4595 \\ -0.0677 & -0.4371 & -0.1755 & 0.1941 & -0.0765 & -0.0581 & 0.3888 & 0.1552 \\ -0.0282 & 0.0196 & -0.0110 & 0.0428 & 0.0489 & -0.0107 & 0.0204 & -0.0013 \end{bmatrix}$$

$$L_4 = \begin{bmatrix} 1.0496 & 0.7056 & 0.7324 & 0.9308 & -1.9622 & -0.1131 & -0.5318 & 0.0875 \\ 0.1454 & 2.0137 & 0.2207 & 2.0076 & -0.5766 & 0.4963 & -0.3034 & 0.4136 \\ -0.0861 & -0.4153 & -0.1269 & 0.1993 & 0.0462 & -0.0027 & 0.3099 & 0.1804 \\ -0.0295 & 0.0249 & 0.0010 & 0.0405 & 0.1112 & 0.0021 & 0.0023 & 0.0054 \end{bmatrix}$$

$$M = \begin{bmatrix} 71.2831 & 7.1480 & 5.7026 & -6.5965 & 3.8840 & -2.2201 & -1.6248 & 1.8577 \\ 7.1480 & 61.0252 & -5.0354 & -5.9652 & 0.7846 & 2.2000 & 1.0432 & -1.7922 \\ 5.7026 & -5.0354 & 37.6864 & 3.3263 & 0.7685 & -4.8937 & -3.8885 & 4.6469 \\ -6.5965 & -5.9652 & 3.3263 & 59.1665 & -0.2383 & -1.8249 & -0.0181 & 1.4833 \\ 3.8840 & 0.7846 & 0.7685 & -0.2383 & 27.9935 & -2.0691 & 3.6357 & 2.3351 \\ -2.2201 & 2.2000 & -4.8937 & -1.8249 & -2.0691 & 10.7606 & 8.0019 & -10.3884 \\ -1.6248 & 1.0432 & -3.8885 & -0.0181 & 3.6357 & 8.0019 & 12.2915 & -7.8525 \\ 1.8577 & -1.7922 & 4.6469 & 1.4833 & 2.3351 & -10.3884 & -7.8525 & 10.6633 \end{bmatrix}$$

In the simulations, initial states of the systems is chosen as $[0 \ 0.01 \ 0 \ 0]^T$, the unknown inputs are set as random signals bounded by $[-1, \ 1]$. The time window of the the residual evaluation function is chosen as $\Delta t = 10s$ and according to (60), the threshold is calculated as $J_{th} = 0.15$. In

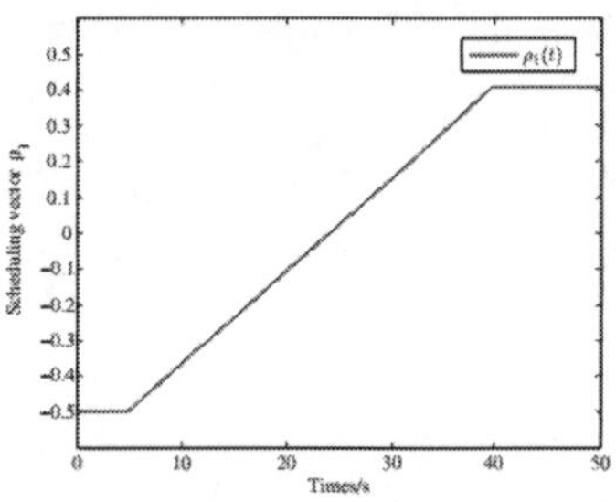

Figure 1. $\rho_1(t)$

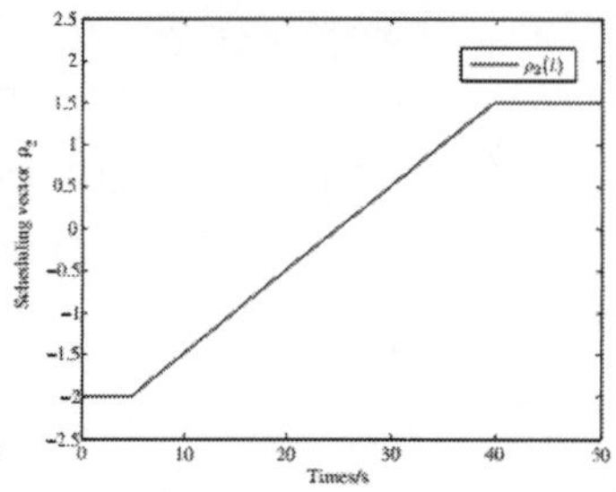

Figure 2. $\rho_2(t)$

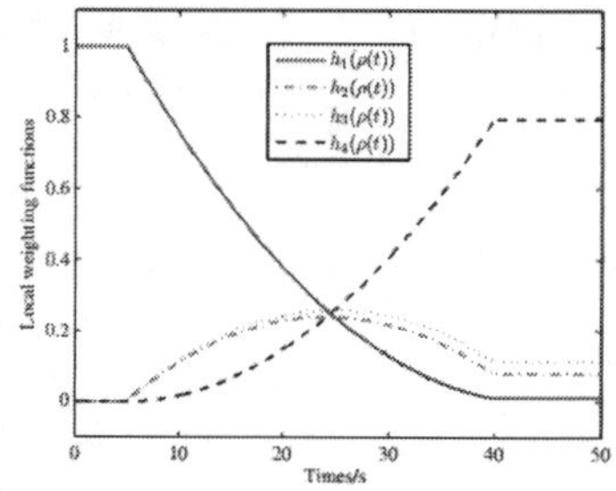

Figure 3. $h_i(\rho)$

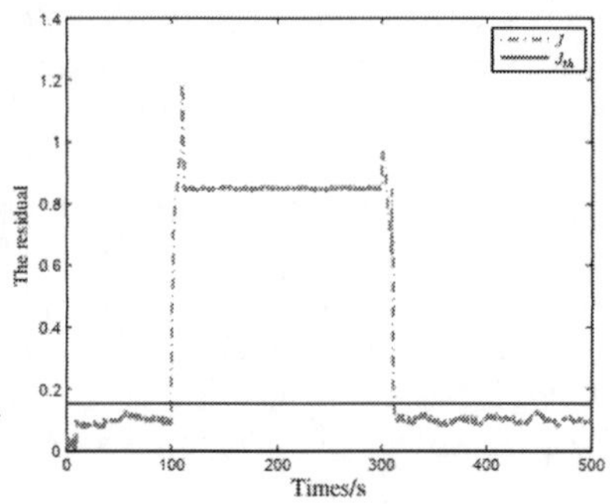

Figure 4. Residual of fault (64)

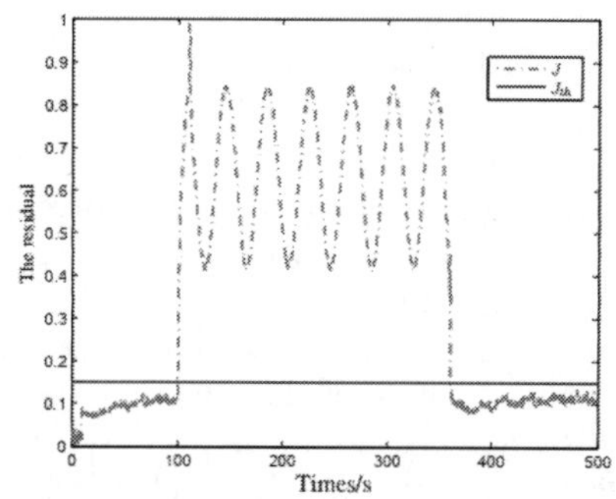

Figure 5. Residual of fault (65)

order to demonstrate the effectiveness of the proposed method, two fault scenarios are assumed to be detected, i.e. abrupt fault and time-varying fault. First, an abrupt fault is formulated as

$$f(t) = \begin{cases} [0 \quad 0]^T & 0s < t \leq 100s \\ [0.1 \quad 0]^T & 100s < t \leq 300s \\ [0 \quad 0]^T & 300s < t \leq 500s \end{cases} \tag{64}$$

In the second scenario, it is assumed that a time-varying fault occurs with the formulation as

$$f(t) = \begin{cases} [0 \quad 0]^T & 0s < t \leq 100s \\ [0 \quad 0.05 + 0.02\sin(0.1\pi(t-50))]^T & 100s < t \leq 350s \\ [0 \quad 0]^T & 350s < t \leq 500s \end{cases} \tag{65}$$

The simulation results are depicted as in figures 4–5. Figure 4 shows the residual generated by the proposed method with abrupt fault (64). From the picture it can be seen that the residual exceeds the predefined threshold between $100s < t \leq 300s$, the fault (64) can be detected by the proposed H_-/H_∞ observer successfully. Figure 5 depicts the residual respect to the time-varying fault (65), which shows that the time-varying fault (65) is also detected successfully.

6. Conclusions

In this paper, an H_-/H_∞ fault detection observer is investigated for a class of polytopic LPV systems with the aid of relative degree of output. This paper proposes a new method to consider the H_- index for systems only consider actuator faults in full-frequency domain, which may has some conservatism compared with finite-frequency domain method. In the future, we will combine the proposed method with finite-frequency domain method. Besides, note that the proposed observer in this paper is depended on numerical derivation of the outputs, which is difficult to be available in practice, and it may enlarge the effect of the noise. As a result, in our future work, we will consider H_-/H_∞ fault detection observer design methods without using the numerical derivation.

Acknowledgements

This work was partially supported by National Natural Science Foundation of China (Grant No. 61273162, 61403104) and the Fundamental Research Funds for the Central Universities under Grant HIT.KLOF.2015.076.

References

[1] O. Sename, P. Gaspar and J. Bokor 2013 *Robust control and linear parameter varying approaches* (Springer Heidelberg, New York, London)
[2] Z. H. Wang, M. Rodrigues, D. Theilliol and Y. Shen 2015 Actuator fault estimation observer design for discrete-time linear parameter-varying descriptor systems **29** 242–258
[3] A. Farhat, D. Koenig and O. Sename 2015 Fault detection for LPV systems : Loop shaping H_- approach *1st IFAC Workshp on LPV systems* (Grenoble, France)
[4] M. Rodrigues, H. Hamdi, N. B. Braiek and D. Theilliol 2014 Observer-based fault tolerant control design for a class of LPV descriptor systems *Journal of the Franklin Institute* **351** 3104–3125
[5] M. Hou and R. J. Patton 1996 An LMI approach to H_-/H_∞ fault detection observers, *Proceedings of the UKACC International Conference on Control* (Exeter UK) p 305–310
[6] J. Liu, J. L. Wang and G. H. Yang 2003 An LMI approach to worst case analysis for fault detection observers. *In proceedings of the American Control Conference*, USA, 2985–2990.
[7] J. Liu, J. L. Wang and G. H. Yang 2005 An LMI approach to minimum sensitivity analysis with application to fault detection *Automatica* **41** 1995–2004
[8] J. L. Wang, G. H. Yang and J. Liu 2007 An LMI approach to H_- index and mixed H_-/H_∞ fault detection observer design *Automatica* **43** 1656–1665
[9] S. C. Jee, H. J. Lee and Y. H. Joo 2012 H_-/H_∞ sensor fault detection and isolation in linear time-invariant systems *International Journal of Control, Automation and Systems* **10** 841–848.
[10] K. S. Lee and T. G. Park 2015 Robust fault detection observer design under fault sensitivity constraints *Journal of Franklin Institute* **352** 1791–1810
[11] X. Wei and M. Verhaegen 2011 LMI solutions to the mixed H_-/H_∞ fault detection observer design for linear parameter-varying systems *International Journal of Adaptive Control and Signal Processing* **25** 114–136
[12] S. Grenaille, D. Henry and A. Zolghadri 2008 A method for designing fault diagnosis filters for LPV polytopic systems, *Journal of Control Science and Engineering*, 1–11
[13] D. Henry 2012 Structured fault detection filters for LPV systems modelled in a LFR manner *International Journal of Adaptive Control and Signal Processing*, **26** 190–270
[14] M. Chadli, A. Abdo and S. X. Ding 2013 H_-/H_∞ fault detection filter design for discrete-time Takagi-Sugeno fuzzy system *Automatica* **49** 1996–2005
[15] S. C. Jee, H. J. Lee and D. W. Kim 2014 H_-/H_∞ sensor fault detection and isolation of uncertain time-delay systems *Journal of Electrical Engineering and Technology* **9** 313–323.
[16] A. Farhat and D. Koenig 2014 H_-/H_∞ fault detection observer for switched systems *53rd IEEE Annual Conference on Decision and Control* (Los Angeles, United States)
[17] X. J. Li and G. H. Yang 2014 Fault detection in finite frequency domain for Takagi-Sugeno fuzzy systems with sensor faults, *IEEE Transactions on Cybernetics*, **44** 1446–1458
[18] H. Wang and G. H. Yang 2008 A finite frequency domain approach to fault detection observer design for linear continuous-time systems *Asian Journal of Control* **10** 1–10
[19] D. Ichalal, B. Marx, D. Maquin and J. Ragot 2016 Actuator fault diagnosis: H_∞ framework with relative degree notion, *International Conference on Intelligent Control and Automation Science*, (Reims, Framce) p 327-332
[20] S. X. Ding 2013 *Model-based fault diagnosis techniques* (Springer Heidelberg, New York, London)
[21] A. Isidori 1995 *Nonlinear control systems* (Communications and Control Engineering, Springer)
[22] K. Tanaka, H. O. Wang 2001 *Fuzzy control systems design and analysis* (John Wiley & Son)
[23] Q Jia, W Chen, Y Zhang and X Chen 2015 Fault Reconstruction and Accommodation in Linear Parameter-Varying Systems via Learning Unknown-Input Observers *Journal of Dynamic Systems, Measurement and Control* **137** 061008-061008-9

A process fault estimation strategy for non-linear dynamic systems

Marcin Pazera, Józef Korbicz

Institute of Control and Computation Engineering, University of Zielona Góra, Podgórna 50, 65-246 Zielona Góra, Poland

E-mail: {M.Pazera,J.Korbicz}@issi.uz.zgora.pl

Abstract. The paper deals with the problem of simultaneous state and process fault estimation for non-linear dynamic systems. Instead of estimating the fault directly, its product with state and the state itself are estimated. To derive the fault from the product, a simple algebraic approach is proposed. The estimation strategy is based on the quadratic boundedness approach. The final part of the paper presents an illustrative example concerning a laboratory multi-tank system. The real data experiments clearly exhibit the performance of the proposed approach.

1. Introduction

In the last decades, the problem of Fault Diagnosis (FD) [3, 17, 20, 11, 14, 27, 26] takes important part in the correct functioning of the industrial systems. It also plays a significant role in conditions monitoring of the plants. The FD issue can be divided into three tasks, i.e., detection, isolation and identification [11, 5]. Industrial systems consist of process dynamics, sensors and actuators. Thus, it leads to fault classes [3, 8, 28] such as process faults (or component faults), sensor faults and actuator faults. The fault estimation is also basis for the implementation of so-called Fault-Tolerant Control (FTC) [8, 19, 21, 16, 25, 29, 23, 22] which can be divided into passive and active one. The active FTC needs the fault estimation to calculate current control signal, and hence, to compensate the influence of occurred fault. In contrast to active one, the passive FTC is designed to be robust for predefined faults. The faults are treated as disturbances. Among Fault Detection and Isolation (FDI) methodologies available in the literature, strategies such as minimum variance input and state estimator [9], sliding mode high-gain observers [24, 2], Kalman filter [13], adaptive estimation [32] and also $\mathcal{H}_\infty$ approach [28, 18, 12, 4, 30, 15] draws considerable attention in the field of control engineering. However, there are no many FDI solutions dedicated to the problem of process fault estimation, especially for non-linear systems. Most of them are concentrated to the problem of sensor or actuator faults. The paper aims at providing a novel process fault estimation strategy for non-linear dynamic systems. The proposed strategy is based on Quadratic Boundedness (QB) [1, 6, 7] approach allowing the convergence analysis.

The paper is organized as follows: section 2 presents the problem formulation, subsequently in section 3 methodology proceeding due to design the observer is shown. Final part of the paper shows an illustrative example (section 4) wherein the algorithm has been implemented to the multi-tank system, and section 5 concludes the paper.

2. Preliminaries

Let us consider a non-linear discrete-time system:

$$\boldsymbol{x}_{k+1} = \boldsymbol{A}\boldsymbol{x}_k + \sum_{i=1}^{n_f} \boldsymbol{A}_{f,i}\boldsymbol{f}_{i,k}\boldsymbol{x}_k + \boldsymbol{B}\boldsymbol{u}_k + \boldsymbol{g}(\boldsymbol{x}_k) + \boldsymbol{W}_1\boldsymbol{w}_k, \tag{1}$$

$$\boldsymbol{y}_k = \boldsymbol{C}\boldsymbol{x}_k + \boldsymbol{W}_2\boldsymbol{w}_k, \tag{2}$$

where $\boldsymbol{x}_k \in \mathbb{X} \subset \mathbb{R}^n$, $\boldsymbol{y}_k \in \mathbb{R}^m$, $\boldsymbol{u}_k \in \mathbb{R}^r$, $\boldsymbol{f}_k \in \mathbb{R}^{n_f}$ are the state, output, control input and process fault vectors, respectively and $\boldsymbol{g}(\boldsymbol{x}_k)$ is the non-linear function which describes the behaviour of the system with respect to state. Note that $\boldsymbol{A}_{f,i}$ denote the distribution matrix of ith process fault $\boldsymbol{f}_i$, i.e., it describes the way in which the fault influences the system matrix. Moreover, $\boldsymbol{W}_1$ and $\boldsymbol{W}_2$ denote the noise distribution matrices and $\boldsymbol{w}_k$ expresses the exogenous disturbance vector. Furthermore, $\boldsymbol{w}_k$ can be split as follows $\boldsymbol{w}_k = \left[\boldsymbol{w}_{1,k}^T, \boldsymbol{w}_{2,k}^T\right]^T$ where $\boldsymbol{w}_{1,k}$ and $\boldsymbol{w}_{2,k}$ represent the process and measurement uncertainties, respectively.

The problem is to design an observer which makes it possible to simultaneous estimate the state and process fault. Throughout the paper, the following substitution is applied:

$$\boldsymbol{z}_{i,k} = \boldsymbol{f}_i\boldsymbol{x}_k, \tag{3}$$

which leads to the new representation of the system:

$$\boldsymbol{x}_{k+1} = \boldsymbol{A}\boldsymbol{x}_k + \sum_{i=1}^{n_f} \boldsymbol{A}_{f,i}\boldsymbol{z}_{i,k} + \boldsymbol{B}\boldsymbol{u}_k + \boldsymbol{g}(\boldsymbol{x}_k) + \boldsymbol{W}_1\boldsymbol{w}_k, \tag{4}$$

$$\boldsymbol{y}_k = \boldsymbol{C}\boldsymbol{x}_k + \boldsymbol{W}_2\boldsymbol{w}_k. \tag{5}$$

Thus, the problem is to design an estimator that will be able to estimate $\boldsymbol{z}_{i,k}$ and $\boldsymbol{x}_k$ simultaneously. While the idea to estimate the fault is based on recovering it from $\boldsymbol{z}_{i,k}$ and $\boldsymbol{x}_k$. For the aim of further deliberations, let us consider the following assumptions:

Assumption 1:

$$\begin{aligned} \boldsymbol{\varepsilon}_{i,k} &= \boldsymbol{z}_{i,k+1} - \boldsymbol{z}_{i,k}, \\ \boldsymbol{\varepsilon}_{i,k} \in \mathcal{E}_{\varepsilon,i}, \quad \mathcal{E}_{\varepsilon,i} &= \{\boldsymbol{\varepsilon} : \boldsymbol{\varepsilon}^T\boldsymbol{Q}_{\varepsilon,i}\boldsymbol{\varepsilon} \le 1\}, \boldsymbol{Q}_{\varepsilon,i} \succ 0. \end{aligned} \tag{6}$$

Assumption 2:

$$\boldsymbol{w}_k \in \mathcal{E}_w, \quad \mathcal{E}_w = \{\boldsymbol{w} : \boldsymbol{w}^T\boldsymbol{Q}_w\boldsymbol{w} \le 1\}, \boldsymbol{Q}_w \succ 0. \tag{7}$$

These assumptions are required for the subsequent fault estimation algorithm. First of them (*Assumption 1*) concerns the system behaviour in case of fault, which means that all real faults and states are bounded. Consequently, all $\boldsymbol{z}_{i,k}$ are bounded as well. The second one, i.e., *Assumption 2*, similarly as *Assumption 1*, states that the external disturbances are unknown, but bounded.

To handle the aforementioned problem of simultaneous estimation of the state $\boldsymbol{x}_k$ and $\boldsymbol{z}_{i,k}$, it is proposed to use the Process Fault Estimator (PFE) of the following structure:

$$\hat{\boldsymbol{x}}_{k+1} = \boldsymbol{A}\hat{\boldsymbol{x}}_k + \sum_{i=1}^{n_f} \boldsymbol{A}_{f,i}\hat{\boldsymbol{z}}_{i,k} + \boldsymbol{B}\boldsymbol{u}_k + \boldsymbol{g}(\hat{\boldsymbol{x}}_k) + \boldsymbol{K}(\boldsymbol{y}_k - \boldsymbol{C}\hat{\boldsymbol{x}}_k), \tag{8}$$

$$\hat{\boldsymbol{z}}_{i,k+1} = \hat{\boldsymbol{z}}_{i,k} + \boldsymbol{L}_i(\boldsymbol{y}_k - \boldsymbol{C}\hat{\boldsymbol{x}}_k), \qquad i = 1 \ldots s. \tag{9}$$

Note that ith fault estimate obeys

$$\hat{\boldsymbol{z}}_{j,k} = \hat{\boldsymbol{f}}^{j}_{i,k}\hat{\boldsymbol{x}}_{j,k}, \quad i = 1, \ldots, s, \quad j = 1, \ldots, n. \tag{10}$$

Indeed, there is no single estimate $\hat{\boldsymbol{f}}_{i,k}$ satisfying n equations of the above form (10). Thus, rather than having one single estimate, a set of n estimates will be obtained $\hat{\boldsymbol{f}}^{j}_{i,k}$, $j = 1, \ldots, n$. Thus, the final step is to calculate the Mean Value (MV) of these estimates

$$\hat{\boldsymbol{f}}_{i,k} = \frac{1}{n}\sum_{j=1}^{n}\hat{\boldsymbol{f}}^{j}_{i,k}. \tag{11}$$

A complete scheme of the proposed estimation strategy is depicted in figure 1.

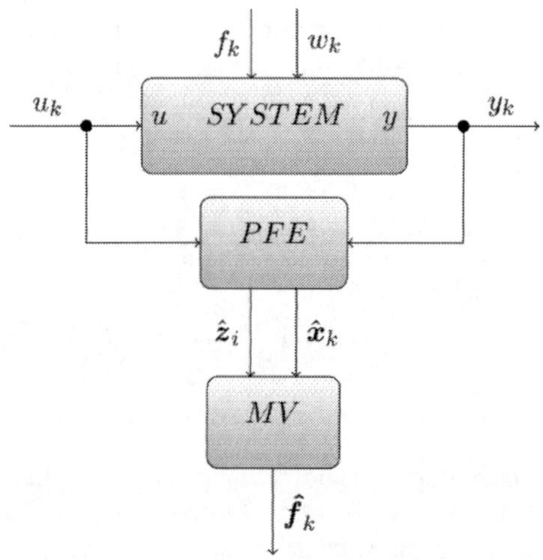

Figure 1. Diagram of proposed approach

3. Observer design

The main goal of this section is to present the design strategy of the process fault estimator. The estimator should be robust to the process and measurement uncertainties. From (4), (5) and (8), the state estimation error can be described by

$$\begin{aligned}\boldsymbol{e}_{k+1} &= \boldsymbol{x}_{k+1} - \hat{\boldsymbol{x}}_{k+1} \\ &= [\boldsymbol{A} - \boldsymbol{K}\boldsymbol{C}]\,\boldsymbol{e}_k + \sum_{i=1}^{n_f}\boldsymbol{A}_f\boldsymbol{e}_{z,i,k} + \boldsymbol{g}(\boldsymbol{x}_k) - \boldsymbol{g}(\hat{\boldsymbol{x}}_k) + [\boldsymbol{W}_1 - \boldsymbol{K}\boldsymbol{W}_2]\,\boldsymbol{w}_k,\end{aligned} \tag{12}$$

where $\boldsymbol{e}_{z,i,k} = \boldsymbol{z}_{i,k} - \hat{\boldsymbol{z}}_{i,k}$.
Subsequently, following the Differential Mean Value Theorem (DMVT) [31], it can be shown that

$$\boldsymbol{g}(\boldsymbol{x}_k) - \boldsymbol{g}(\hat{\boldsymbol{x}}_k) = \boldsymbol{M}_x(\boldsymbol{x}_k - \hat{\boldsymbol{x}}_k), \tag{13}$$

with

$$\boldsymbol{M}_x = \begin{bmatrix} \frac{\partial g_1}{\partial x}(\boldsymbol{c_1}) \\ \vdots \\ \frac{\partial g_n}{\partial x}(\boldsymbol{c_n}) \end{bmatrix}, \tag{14}$$

where $\boldsymbol{c}_1, \ldots, \boldsymbol{c}_n \in \mathrm{Co}(\boldsymbol{x_k}, \hat{\boldsymbol{x}}_{\boldsymbol{k}})$, $\boldsymbol{c}_i \neq \boldsymbol{x}_k$, $\boldsymbol{c}_i \neq \hat{\boldsymbol{x}}_k$, $i = 1, \ldots, n$. Following that all states are bounded in real system, $\boldsymbol{x}_k \in \mathbb{X}$ let

$$\bar{x}_{i,j} \geq \frac{\partial \boldsymbol{g}_i(\boldsymbol{x})}{\partial x_j} \geq \underline{x}_{i,j}, \quad i = 1, \ldots, n, \quad j = 1, \ldots, n, \tag{15}$$

it is clear that there exist $\boldsymbol{M}_x \in \mathbb{M}$ such that

$$\mathbb{M}_x = \left\{ \boldsymbol{M} \in \mathbb{R}^{n \times n} | \bar{x}_{i,j} \geq m_{x,i,j} \geq \underline{x}_{i,j}, \, i, j = 1, \ldots, n, \right\}. \tag{16}$$

Thus, the equation (12) can be rewritten as

$$\begin{aligned} \boldsymbol{e}_{k+1} &= \boldsymbol{x}_{k+1} - \hat{\boldsymbol{x}}_{k+1} \\ &= [\boldsymbol{A} - \boldsymbol{KC} + \boldsymbol{M}_x] \, \boldsymbol{e}_k + \sum_{i=1}^{n_f} \boldsymbol{A}_f \boldsymbol{e}_{z,i,k} + [\boldsymbol{W}_1 - \boldsymbol{K}\boldsymbol{W}_2] \, \boldsymbol{w}_k. \end{aligned} \tag{17}$$

Moreover, the dynamics of the estimation error of the product of the state and the fault can be expressed by

$$\begin{aligned} \boldsymbol{e}_{z,i,k+1} &= \boldsymbol{z}_{i,k+1} + \boldsymbol{z}_{i,k} - \boldsymbol{z}_{i,k} - \hat{\boldsymbol{z}}_{i,k+1} \\ &= \boldsymbol{\varepsilon}_{i,k} + \boldsymbol{z}_{i,k} - \hat{\boldsymbol{z}}_{i,k+1} \\ &= \boldsymbol{\varepsilon}_{i,k} + \boldsymbol{z}_{i,k} - \hat{\boldsymbol{z}}_{i,k} - \boldsymbol{L}_i \boldsymbol{C} \boldsymbol{x}_k - \boldsymbol{L}_i \boldsymbol{W}_2 \boldsymbol{w}_k + \boldsymbol{L}_i \boldsymbol{C} \hat{\boldsymbol{x}}_k \\ &= \boldsymbol{\varepsilon}_{i,k} + \boldsymbol{e}_{z,i,k} - \boldsymbol{L}_i \boldsymbol{C} \boldsymbol{e}_k - \boldsymbol{L}_i \boldsymbol{W}_2 \boldsymbol{w}_k, \end{aligned} \tag{18}$$

where $\boldsymbol{\varepsilon}_{i,k} = \boldsymbol{z}_{i,k+1} - \boldsymbol{z}_{i,k}$.
By constructing the following extended vectors:

$$\bar{\boldsymbol{e}}_{k+1} = \left[\boldsymbol{e}_{k+1}^T, \boldsymbol{e}_{z,1,k+1}^T, \ldots, \boldsymbol{e}_{z,n_f,k+1}^T \right]^T, \tag{19}$$

$$\boldsymbol{v}_k = \left[\boldsymbol{w}_k^T, \boldsymbol{\varepsilon}_{1,k}^T, \ldots, \boldsymbol{\varepsilon}_{n_f,k}^T \right]^T, \tag{20}$$

the state and fault estimation error dynamics can be formulated as

$$\bar{\boldsymbol{e}}_{k+1} = \begin{bmatrix} \boldsymbol{A} - \boldsymbol{KC} + \boldsymbol{M}_x & \boldsymbol{A}_{f,1} & \boldsymbol{A}_{f,2} & \ldots & \boldsymbol{A}_{f,n_f} \\ -\boldsymbol{L}_1 \boldsymbol{C} & \boldsymbol{I} & 0 & \ldots & 0 \\ -\boldsymbol{L}_2 \boldsymbol{C} & 0 & \boldsymbol{I} & \ldots & 0 \\ \vdots & \vdots & \vdots & \ddots & \vdots \\ \boldsymbol{L}_{n_f} \boldsymbol{C} & 0 & 0 & 0 & \boldsymbol{I} \end{bmatrix} \bar{\boldsymbol{e}}_k + \begin{bmatrix} \boldsymbol{W}_1 & \boldsymbol{0} & \boldsymbol{0} & \ldots & \boldsymbol{0} \\ -\boldsymbol{L}_1 \boldsymbol{W}_2 & \boldsymbol{I} & \boldsymbol{0} & \ldots & \boldsymbol{0} \\ -\boldsymbol{L}_2 \boldsymbol{W}_2 & \boldsymbol{0} & \boldsymbol{I} & \ldots & \boldsymbol{0} \\ \vdots & \vdots & \vdots & \ddots & \vdots \\ -\boldsymbol{L}_{n_f} \boldsymbol{W}_2 & \boldsymbol{0} & \boldsymbol{0} & \ldots & \boldsymbol{I} \end{bmatrix} \boldsymbol{v}_k. \tag{21}$$

Thus, equation (21) can be rewritten in an equivalent compact form

$$\bar{\boldsymbol{e}}_{k+1} = \boldsymbol{X}_x \bar{\boldsymbol{e}}_k + \boldsymbol{Z} \boldsymbol{v}_k, \tag{22}$$

with

$$\boldsymbol{X}_x = \bar{\boldsymbol{A}}_x - \bar{\boldsymbol{K}}\bar{\boldsymbol{C}} = \begin{bmatrix} \boldsymbol{A}+\boldsymbol{M}_x & \boldsymbol{A}_{f,1} & \boldsymbol{A}_{f,2} & \dots & \boldsymbol{A}_{f,n_f} \\ \boldsymbol{0} & \boldsymbol{I} & \boldsymbol{0} & \dots & \boldsymbol{0} \\ \boldsymbol{0} & \boldsymbol{0} & \boldsymbol{I} & \dots & \boldsymbol{0} \\ \vdots & \vdots & \vdots & \ddots & \vdots \\ \boldsymbol{0} & \boldsymbol{0} & \boldsymbol{0} & \boldsymbol{0} & \boldsymbol{I} \end{bmatrix} - \begin{bmatrix} \boldsymbol{K} \\ \boldsymbol{L}_1 \\ \boldsymbol{L}_2 \\ \vdots \\ \boldsymbol{L}_{n_f} \end{bmatrix} \begin{bmatrix} \boldsymbol{C} & \boldsymbol{0} & \boldsymbol{0} & \dots & \boldsymbol{0} \end{bmatrix}, \quad (23)$$

$$\boldsymbol{Z} = \bar{\boldsymbol{W}} - \bar{\boldsymbol{K}}\bar{\boldsymbol{V}} = \begin{bmatrix} \boldsymbol{W}_1 & \boldsymbol{0} & \boldsymbol{0} & \dots & \boldsymbol{0} \\ \boldsymbol{0} & \boldsymbol{I} & \boldsymbol{0} & \dots & \boldsymbol{0} \\ \boldsymbol{0} & \boldsymbol{0} & \boldsymbol{I} & \dots & \boldsymbol{0} \\ \vdots & \vdots & \vdots & \ddots & \vdots \\ \boldsymbol{0} & \boldsymbol{0} & \boldsymbol{0} & \dots & \boldsymbol{I} \end{bmatrix} - \begin{bmatrix} \boldsymbol{K} \\ \boldsymbol{L}_1 \\ \boldsymbol{L}_2 \\ \vdots \\ \boldsymbol{L}_{n_f} \end{bmatrix} \begin{bmatrix} \boldsymbol{W}_2 & \boldsymbol{0} & \boldsymbol{0} & \dots & \boldsymbol{0} \end{bmatrix}. \quad (24)$$

Therefore, the ellipsoidal set bounding $\boldsymbol{v}_k$ can be described by:

$$\mathbb{E}_v = \{\boldsymbol{v} : \boldsymbol{v}^T\boldsymbol{Q}_v\boldsymbol{v} \leq 1\}, \quad (25)$$

with

$$\boldsymbol{Q}_v = \frac{1}{2+n_f}\text{diag}\left(\boldsymbol{Q}_w, \boldsymbol{Q}_{\epsilon,1}, \dots, \boldsymbol{Q}_{\epsilon,s}, \boldsymbol{Q}_w\right). \quad (26)$$

For the purpose of further deliberations, the so-called Quadratic Boundedness (QB) [1, 6, 7] approach can be use. The Lyapunov function can be formulated as follows

$$\boldsymbol{V}_k = \bar{\boldsymbol{e}}_k^T\boldsymbol{P}\bar{\boldsymbol{e}}_k, \quad (27)$$

with $\boldsymbol{P} \succ 0$.

To introduce the QB approach, the following definitions are recalled [1]:

Definition 1: System (22) is strictly quadratically bounded for all allowable $\boldsymbol{v}_k \in \mathbb{E}_v$, if $\bar{\boldsymbol{e}}_k^T\boldsymbol{P}\bar{\boldsymbol{e}}_k > 1$ implies $\bar{\boldsymbol{e}}_{k+1}^T\boldsymbol{P}\bar{\boldsymbol{e}}_{k+1} < \bar{\boldsymbol{e}}_k^T\boldsymbol{P}\bar{\boldsymbol{e}}_k$ for any $\boldsymbol{v}_k \in \mathbb{E}_v$.

It should be highlighted that the strict quadratic boundedness of (22) ensures that $\boldsymbol{V}_{k+1} < \boldsymbol{V}_k$ for any $\boldsymbol{v}_k \in \mathbb{E}_v$ when $\boldsymbol{V}_k > 1$.

Definition 2: A set $\mathbb{E}$ is a positively invariant set for (22) for all $\boldsymbol{v}_k \in \mathbb{E}_v$ if $\bar{\boldsymbol{e}}_k \in \mathbb{E}$ implies $\bar{\boldsymbol{e}}_{k+1} \in \mathbb{E}$ for any $\boldsymbol{v}_k \in \mathbb{E}_v$

The results in [1] and the above definitions determine the following lemma for (22):

Lemma 1: The following statements are equivalent:

(i) The system (22) is strictly quadratically bounded for all $\boldsymbol{v}_k \in \mathbb{E}_v$ and $\boldsymbol{M}_x \in \mathbb{M}$.

(ii) The ellipsoid

$$\mathbb{E} = \{\bar{\boldsymbol{e}}_k : \bar{\boldsymbol{e}}_k^T\boldsymbol{P}\bar{\boldsymbol{e}}_k \leq 1\}, \quad (28)$$

is an invariant set for (22) for any $\boldsymbol{v}_k \in \mathbb{E}_v$.

(iii) There exists a scalar $\alpha \in (0,1)$ such that:

$$\begin{bmatrix} \boldsymbol{X}_x^T\boldsymbol{P}\boldsymbol{X}_x - \boldsymbol{P} + \alpha\boldsymbol{P} & \boldsymbol{X}_x^T\boldsymbol{P}\boldsymbol{Z} \\ \boldsymbol{Z}^T\boldsymbol{P}\boldsymbol{X}_x & \boldsymbol{Z}^T\boldsymbol{P}\boldsymbol{Z} - \alpha\boldsymbol{Q}_v \end{bmatrix} \preceq 0. \quad (29)$$

Note that the above lemma was initially developed for linear systems [1], and hence, it is straightforwardly extended to the class of systems being considered. As a consequence, the following theorem is proposed:

Theorem 1: The system (22) is strictly quadratically bounded for all $\boldsymbol{v}_k \in \mathbb{E}_v$ and $\boldsymbol{M}_x \in \mathbb{M}$ if there exist matrices $\boldsymbol{P} \succ 0$, $\boldsymbol{U}$ and a scalar $\alpha \in (0,1)$, such that the following inequality is satisfied:

$$\begin{bmatrix} -\boldsymbol{P}+\alpha\boldsymbol{P} & \boldsymbol{0} & \bar{\boldsymbol{A}}_x^T\boldsymbol{P}-\bar{\boldsymbol{C}}^T\boldsymbol{U}^T \\ \boldsymbol{0} & -\alpha\boldsymbol{Q}_v & \bar{\boldsymbol{W}}^T\boldsymbol{P}-\bar{\boldsymbol{V}}^T\boldsymbol{U}^T \\ \boldsymbol{P}\bar{\boldsymbol{A}}_x-\boldsymbol{U}\bar{\boldsymbol{C}} & \boldsymbol{P}\bar{\boldsymbol{W}}-\boldsymbol{U}\bar{\boldsymbol{V}} & -\boldsymbol{P} \end{bmatrix} \preceq 0. \tag{30}$$

Proof: Inequality (29) can be rewritten to the following form:

$$\begin{bmatrix} \boldsymbol{X}_x^T \\ \boldsymbol{Z}^T \end{bmatrix} \boldsymbol{P} \begin{bmatrix} \boldsymbol{X}_x & \boldsymbol{Z} \end{bmatrix} + \begin{bmatrix} -\boldsymbol{P}+\alpha\boldsymbol{P} & \boldsymbol{0} \\ \boldsymbol{0} & -\alpha\boldsymbol{Q}_v \end{bmatrix} \preceq 0. \tag{31}$$

Then, using the Schur complement and multiplying left and right side by diag$(\boldsymbol{I},\boldsymbol{I},\boldsymbol{P})$ give

$$\begin{bmatrix} -\boldsymbol{P}+\alpha\boldsymbol{P} & \boldsymbol{0} & \boldsymbol{X}_x^T\boldsymbol{P} \\ \boldsymbol{0} & -\alpha\boldsymbol{Q}_v & \boldsymbol{Z}^T\boldsymbol{P} \\ \boldsymbol{P}\boldsymbol{X}_x & \boldsymbol{P}\boldsymbol{Z} & -\boldsymbol{P} \end{bmatrix} \preceq 0. \tag{32}$$

Introducing

$$\begin{aligned} \boldsymbol{P}\boldsymbol{X}_x &= \boldsymbol{P}\left(\bar{\boldsymbol{A}}_x-\bar{\boldsymbol{K}}\bar{\boldsymbol{C}}\right) \\ &= \boldsymbol{P}\bar{\boldsymbol{A}}_x-\boldsymbol{P}\bar{\boldsymbol{K}}\bar{\boldsymbol{C}} = \boldsymbol{P}\bar{\boldsymbol{A}}_x-\boldsymbol{U}\bar{\boldsymbol{C}}, \end{aligned} \tag{33}$$

$$\boldsymbol{P}\boldsymbol{Z} = \boldsymbol{P}\bar{\boldsymbol{W}}-\boldsymbol{P}\bar{\boldsymbol{K}}-\bar{\boldsymbol{V}} = \boldsymbol{P}\bar{\boldsymbol{W}}-\boldsymbol{U}\boldsymbol{V}, \tag{34}$$

and substituting (33) and (34) into (32) completes the proof. ■

Finally, the design procedure brings down to solve (30) and then calculate

$$\bar{\boldsymbol{K}} = \begin{bmatrix} \boldsymbol{K} \\ \boldsymbol{L}_1 \\ \boldsymbol{L}_2 \\ \vdots \\ \boldsymbol{L}_{n_f} \end{bmatrix} = \boldsymbol{P}^{-1}\boldsymbol{U}. \tag{35}$$

Solving (30) is equivalent to solve (for $i = 1,\ldots,N$)

$$\begin{bmatrix} -\boldsymbol{P}+\alpha\boldsymbol{P} & \boldsymbol{0} & \bar{\boldsymbol{A}}_{x,i}^T\boldsymbol{P}-\bar{\boldsymbol{C}}^T\boldsymbol{U}^T \\ \boldsymbol{0} & -\alpha\boldsymbol{Q}_v & \bar{\boldsymbol{W}}^T\boldsymbol{P}-\bar{\boldsymbol{V}}^T\boldsymbol{U}^T \\ \boldsymbol{P}\bar{\boldsymbol{A}}_{x,i}-\boldsymbol{U}\bar{\boldsymbol{C}} & \boldsymbol{P}\bar{\boldsymbol{W}}-\boldsymbol{U}\bar{\boldsymbol{V}} & -\boldsymbol{P} \end{bmatrix} \preceq 0, \tag{36}$$

and then determine

$$\bar{\boldsymbol{K}} = \begin{bmatrix} \boldsymbol{K} \\ \boldsymbol{L}_1 \\ \boldsymbol{L}_2 \\ \vdots \\ \boldsymbol{L}_{n_f} \end{bmatrix} = \boldsymbol{P}^{-1}\boldsymbol{U}. \tag{37}$$

A complete design procedure of the Process fault estimator can be presented as follows:

(i) Select

$$\boldsymbol{Q}_v = \frac{1}{2+n_f}\text{diag}\left(\boldsymbol{Q}_w, \boldsymbol{Q}_{\epsilon,1}, \ldots, \boldsymbol{Q}_{\epsilon,n_f}, \boldsymbol{Q}_w\right),$$

(ii) Obtain $\boldsymbol{U}$, $\boldsymbol{P}$ by solving (for $i = 1, \ldots, N$)

$$\begin{bmatrix} -\boldsymbol{P}+\alpha \boldsymbol{P} & \boldsymbol{0} & \bar{\boldsymbol{A}}_{x,i}^T \boldsymbol{P} - \bar{\boldsymbol{C}}^T \boldsymbol{U}^T \\ \boldsymbol{0} & -\alpha \boldsymbol{Q}_v & \bar{\boldsymbol{W}}^T \boldsymbol{P} - \bar{\boldsymbol{V}}^T \boldsymbol{U}^T \\ \boldsymbol{P}\bar{\boldsymbol{A}}_{x,i} - \boldsymbol{U}\bar{\boldsymbol{C}} & \boldsymbol{P}\bar{\boldsymbol{W}} - \boldsymbol{U}\bar{\boldsymbol{V}} & -\boldsymbol{P} \end{bmatrix} \preceq 0,$$

$\boldsymbol{P} \succ 0, \quad 0 < \alpha < 1$

(iii) Calculate

$$\bar{\boldsymbol{K}} = \begin{bmatrix} \boldsymbol{K} \\ \boldsymbol{L}_1 \\ \boldsymbol{L}_2 \\ \vdots \\ \boldsymbol{L}_{n_f} \end{bmatrix}.$$

4. Illustrative example

To verify proposed approach for the process fault estimation task, it was implemented for the multi-tank system which is depicted in figure 2. Such a system is designed for simulating the real industrial multi-tank system in the laboratory conditions. It can be regularly used to practically examine both, linear and non-linear control, identification and diagnostics methods. The considered system consists of three separate tanks which are placed each above other and equipped with drain valves and level sensors based on a hydraulic pressure measurement. Those tanks are differently shaped to reflect the nonlinearities of the system. The lower bottom tank is a water reservoir for the system. The multi-tank system is fitted with a DC water pump which is used to fill the upper tank. The water outflows from the tanks due to gravity. The multi-tank system exchanges data with the level sensors, it also communicates with valves and a pump with a PC-based digital controller through the dedicated I/O board and the power interface. Real time software is controlled by the I/O board with MATLAB/SIMULINK environment. For more information the reader is referred to [10]. The non-linear model of the multi-tank system can be presented as:

$$\frac{\mathrm{d}H_1}{\mathrm{d}t} = \frac{1}{\beta_1(H_1)} q - \frac{1}{\beta_1(H_1)} C_1 {H_1}^{\alpha_1}, \tag{38}$$

$$\frac{\mathrm{d}H_2}{\mathrm{d}t} = \frac{1}{\beta_2(H_2)} C_1 {H_1}^{\alpha_1} - \frac{1}{\beta_2(H_2)} C_2 {H_2}^{\alpha_2}, \tag{39}$$

$$\frac{\mathrm{d}H_3}{\mathrm{d}t} = \frac{1}{\beta_3(H_3)} C_2 {H_2}^{\alpha_2} - \frac{1}{\beta_3(H_3)} C_3 {H_3}^{\alpha_3}, \tag{40}$$

where H_i, $\beta_i(H_i)$, α_i and C_i is the liquid level in ith tank, the surface area of the ith tank with the liquid level equal to H_i, flow coefficient of the ith tank and outflow resistance of the ith tank, respectively. The parameters were obtained by identification with using real data as follows: $C_1 = 1.0057 \cdot 10^{-4}$, $C_2 = 1.1963 \cdot 10^{-4}$, $C_3 = 9.8008 \cdot 10^{-5}$, $\alpha_1 = \alpha_2 = \alpha_3 = 0.5$. The above mentioned equations which are describing the system were discretized using Euler method with sampling time $T_s = 0.01[s]$. Furthermore, the output equation is characterised by

$$\boldsymbol{C} = \begin{bmatrix} 1 & 0 & 0 \\ 0 & 0 & 1 \end{bmatrix}, \tag{41}$$

which means that liquid level in the second tank is unavailable during the experiment.

Figure 2. Multi-tank system

According to the methodology of the PFE design described in section 3, by solving the inequality (36) the following matrices were obtained:

$$\boldsymbol{K} = \begin{bmatrix} 1.0031 & -0.0000 \\ -0.0720 & 0.0000 \\ -0.0000 & 0.0679 \end{bmatrix}, \tag{42}$$

$$\boldsymbol{L} = \begin{bmatrix} 0.002 & -0.0000 \\ -0.0000 & 0.0003 \\ 0.0000 & 0.9739 \end{bmatrix}, \tag{43}$$

for $\alpha = 0.4$. The parameter α is chosen for the greatest value of the trace of matrix $\boldsymbol{P}$ as the best value. In figure 3 is presented the evolution of the trace.

Let us assume that the initial state for the system and observer are $\boldsymbol{x}_0 = [0.001, 0.001, 0.001]$, $\hat{\boldsymbol{x}}_0 = [0.01, 0.01, 0.01]$, respectively while the input is $\boldsymbol{u}_k = 5 \cdot 10^{-5}$.

Moreover, let us consider the following fault scenario

$$\boldsymbol{f}_k = \begin{cases} -3 \cdot 10^{-4}, & 5000 \leq k \leq 10000, \\ 0, & \text{otherwise}, \end{cases} \tag{44}$$

with the fault distribution matrix

$$\boldsymbol{A}_f = \begin{bmatrix} 0 & 0 & 0 \\ 0 & 0 & 0 \\ 0 & 0 & 1 \end{bmatrix}, \tag{45}$$

which can be perceived as a temporary leakage in the third tank.

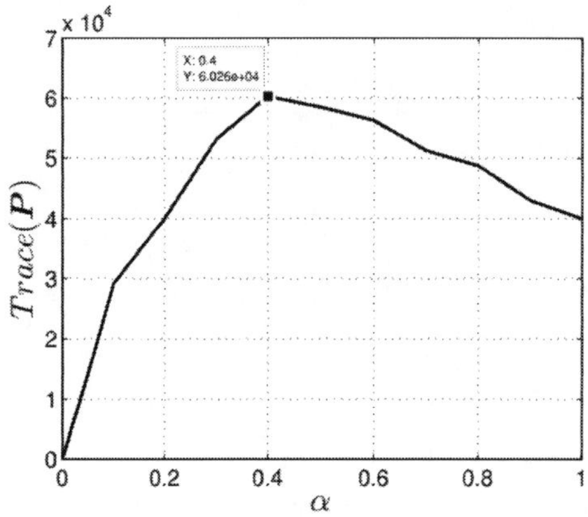

Figure 3. Evolution of the trace of $\boldsymbol{P}$

All experimental results were performed exclusively with the real system. Figures 4–6 present the state for the first, second and third tank, respectively and their estimates. It can be seen that all states were estimated very well. The estimates are following the real states with high accuracy even in case of leakage from the third tank. The unmeasurable liquid level (second

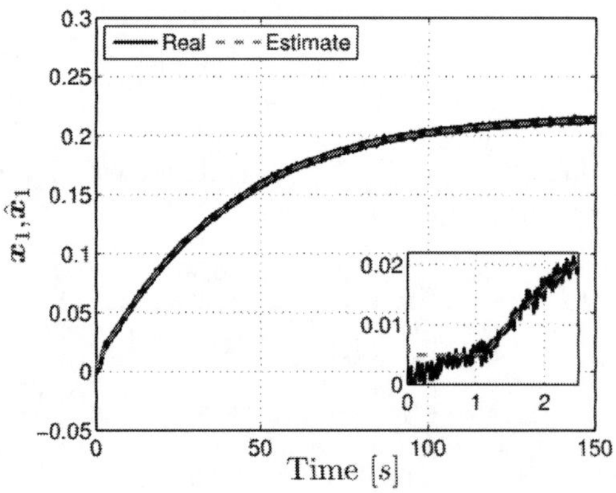

Figure 4. State $\boldsymbol{x}_1$ and its estimate $\hat{\boldsymbol{x}}_1$ – first tank

tank) was also estimated quite well. Taking into account that the level varies from 0 to 0.35m, the state estimation error (figure 5 on the right) which varies from 0.5 to 1.5cm should be viewed as highly satisfactory. Figure 7 depicts the product of the real state $\boldsymbol{x}_3$ and the fault as well as its estimate. Figure 8 show the real fault and its estimate (left figure) and the fault estimation error (right figure). The obtained results clearly exhibit that the proposed approach can be successfully applied to the non-linear systems.

5. Conclusions

The main objectives of this paper was to propose the strategy to handle with the problem of simultaneous state and process fault estimation for a class of non-linear systems. At first, description of the systems which includes the process fault and adequate observer structure is

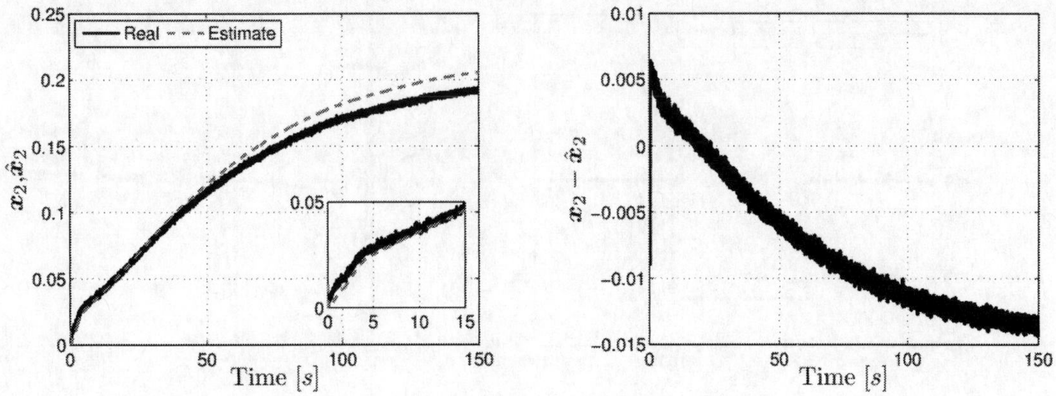

Figure 5. State x_2 and its estimate $\hat{x}_2$ (left figure) and state estimation error (right figure) – second tank

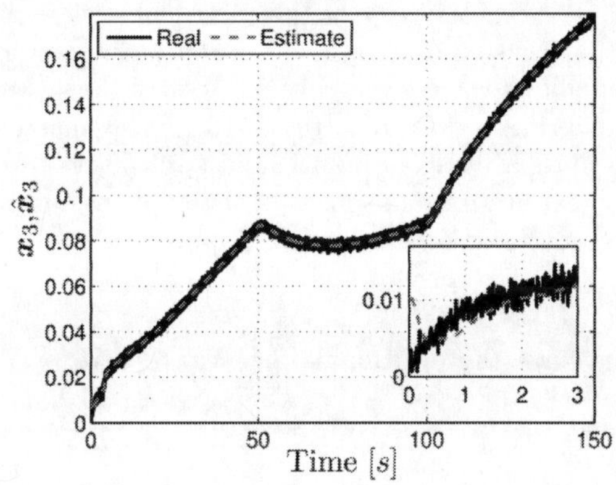

Figure 6. State x_3 and its estimate $\hat{x}_3$ – third tank

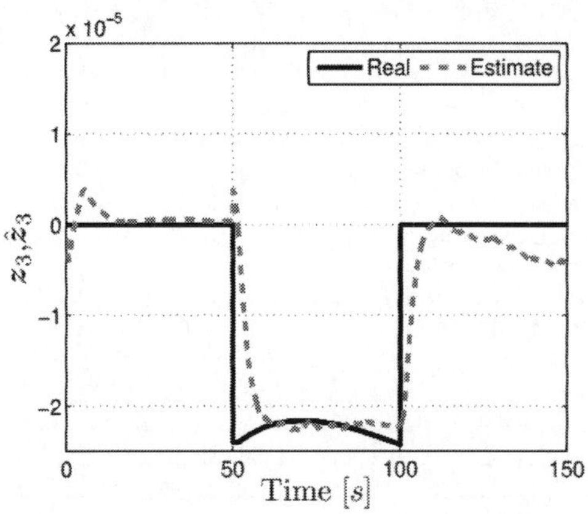

Figure 7. Product of the state and fault z_3 and its estimate $\hat{z}_3$

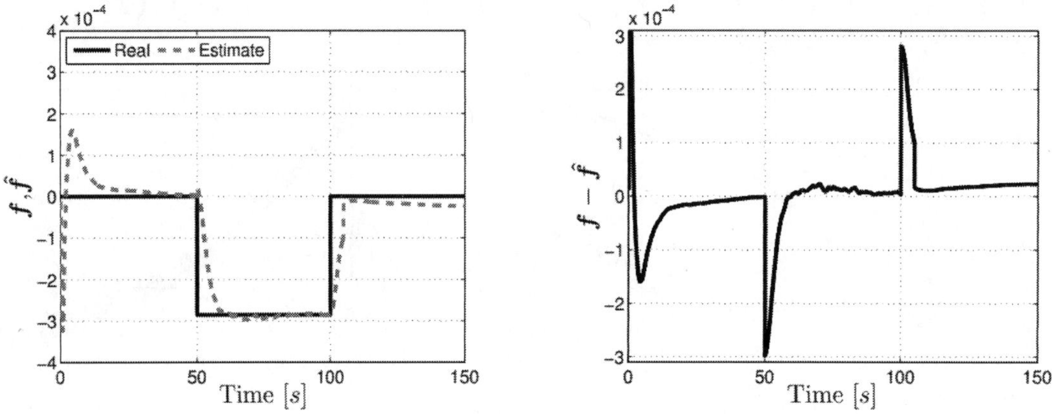

Figure 8. Process fault f and its estimate $\hat{f}$ (left figure) and fault estimation error (right figure)

presented. Subsequently, the design strategy of the robust observer, which takes into account uncertainties influences onto the system is shown. The proposed approach is based on Quadratic Boundedness strategy which allows the convergence analysis of the system. The design strategy is quite simple and at last it brings down to solve a set of linear matrix inequalities. The final part of the paper shows an illustrative example with an application to multi-tank system. The results were obtained with real system, exclusively. They show huge quality of the received estimates and confirm the correctness of the proposed approach.

Acknowledgments

The work was supported by the National Science Centre of Poland under grant: 2013/11/B/ST7/01110.

References

[1] A. Alessandri, M. Baglietto, and G. Battistelli. Design of state estimators for uncertain linear systems using quadratic boundedness. *Automatica*, 42(3):497–502, 2006.

[2] A. Brahim, S. Dhahri, F. Hmida, and A. Sellami. An h_∞ sliding mode observer for Takagi–Sugeno nonlinear systems with simultaneous actuator and sensor faults. *International Journal of Applied Mathematics and Computer Science*, 25(3):547–559, 2015.

[3] J. Chen and R.J. Patton. *Robust Model-Based Fault Diagnosis for Dynamic Systems.* Kluwer Academic Publishers, 1999.

[4] Lejun Chen, Ron Patton, and Philippe Goupil. Robust fault estimation using an lpv reference model: Addsafe benchmark case study. *Control Engineering Practice*, 49:194–203, 2016.

[5] Wei Chen, Abdul Khan, Muhammmad Abid, and Steven Ding. Integrated design of observer based fault detection for a class of uncertain nonlinear systems. *International Journal of Applied Mathematics and Computer Science*, 21(3):423–430, 2011.

[6] B Ding. Constrained robust model predictive control via parameter-dependent dynamic output feedback. *Automatica*, 46(9):1517–1523, 2010.

[7] B. Ding. Dynamic output feedback predictive control for nonlinear systems represented by a Takagi–Sugeno model. *Fuzzy Systems, IEEE Transactions on*, 19(5):831–843, 2011.

[8] Guillaume JJ Ducard. *Fault-tolerant flight control and guidance systems: Practical methods for small unmanned aerial vehicles.* Springer Science & Business Media, 2009.

[9] Steven Gillijns and Bart De Moor. Unbiased minimum-variance input and state estimation for linear discrete-time systems. *Automatica*, 43(1):111–116, 2007.

[10] INTECO. *Multitank System - User's manual.* www.inteco.com.pl, 2013.

[11] Rolf Isermann. *Fault-diagnosis applications: model-based condition monitoring: actuators, drives, machinery, plants, sensors, and fault-tolerant systems.* Springer Science & Business Media, 2011.

[12] Qingxian Jia, Huayi Li, Yingchun Zhang, and Xueqin Chen. Robust observer-based sensor fault reconstruction for discrete-time systems via a descriptor system approach. *International Journal of Control, Automation and Systems*, 13(2):274–283, 2015.

[13] JY Keller and Mohamed Darouach. Two-stage kalman estimator with unknown exogenous inputs. *Automatica*, 35(2):339–342, 1999.

[14] J Korbicz, JM Koscielny, Z Kowalczuk, and W Cholewa. Fault diagnosis. models, artificial intelligence, applications, 2004.

[15] Marcel Luzar, Marcin Witczak, Mariusz Buciakowski, and Vicenç Puig. Robust $\mathcal{H}_\infty$ actuator fault diagnosis and fault-tolerant control for a multi-tank system. In *2013 Conference on Control and Fault-Tolerant Systems (SysTol)*, pages 758–763. IEEE, 2013.

[16] Mufeed Mahmoud, Jin Jiang, and Youmin Zhang. *Active fault tolerant control systems: stochastic analysis and synthesis*, volume 287. Springer Science & Business Media, 2003.

[17] M. Mrugalski. *Advanced Neural Network-based Computational Schemes for robust Fault Diagnosis.* Springer-Verlag, 2014.

[18] Euripedes G Nobrega, Musa O Abdalla, and Karolos M Grigoriadis. Robust fault estimation of uncertain systems using an lmi-based approach. *International Journal of Robust and Nonlinear Control*, 18(18):1657–1680, 2008.

[19] Hassan Noura, Didier Theilliol, Jean-Christophe Ponsart, and Abbas Chamseddine. *Fault-tolerant control systems: Design and practical applications.* Springer Science & Business Media, 2009.

[20] Ron J Patton, Paul M Frank, and Robert N Clark. *Issues of fault diagnosis for dynamic systems.* Springer Science & Business Media, 2013.

[21] Vicenç Puig. Fault diagnosis and fault tolerant control using set-membership approaches: Application to real case studies. *International Journal of Applied Mathematics and Computer Science*, 20(4):619–635, 2010.

[22] Damiano Rotondo, Fatiha Nejjari, and Vicenç Puig. Robust quasi-lpv model reference ftc of a quadrotor uav subject to actuator faults. *International Journal of Applied Mathematics and Computer Science*, 25(1):7–22, 2015.

[23] Lothar Seybold, Marcin Witczak, Paweł Majdzik, and Ralf Stetter. Towards robust predictive fault–tolerant control for a battery assembly system. *International Journal of Applied Mathematics and Computer Science*, 25(4):849–862, 2015.

[24] Kalyana C Veluvolu, MY Kim, and Dongik Lee. Nonlinear sliding mode high-gain observers for fault estimation. *International Journal of Systems Science*, 42(7):1065–1074, 2011.

[25] M. Witczak. *Fault Diagnosis and Fault-Tolerant Control Strategies for Non-Linear Systems: Analytical and Soft Computing approaches.* Springer International Publishing, Heidelberg, Germany, 2014.

[26] M. Witczak, M. Buciakowski, V. Puig, D. Rotondo, and F. Nejjari. An lmi approach to robust fault estimation for a class of nonlinear systems. *International Journal of Robust and Nonlinear Control*, 2015.

[27] Marcin Witczak. *Modelling and estimation strategies for fault diagnosis of non-linear systems: from analytical to soft computing approaches*, volume 354. Springer Science & Business Media, 2007.

[28] Marcin Witczak. Fault diagnosis and fault-tolerant control strategies for non-linear systems. *Lecture Notes in Electrical Engineering*, 266, 2014.

[29] Marcin Witczak, Mariusz Buciakowski, and Christophe Aubrun. Predictive actuator fault-tolerant control under ellipsoidal bounding. *International Journal of Adaptive Control and Signal Processing*, 30(2):375–392, 2016.

[30] Marcin Witczak, Mariusz Buciakowski, and Marcin Mrugalski. An $\mathcal{H}\infty$ approach to fault estimation of non-linear systems: Application to one-link manipulator. In *Methods and Models in Automation and Robotics (MMAR), 2014 19th International Conference On*, pages 456–461. IEEE, 2014.

[31] A. Zemouche and M. Boutayeb. Observer design for Lipschitz non-linear systems: the discrete time case. *IEEE Trans. Circuits and Systems - II:Express Briefs*, 53(8):777–781, 2006.

[32] Xiaodong Zhang, Marios M Polycarpou, and Thomas Parisini. Fault diagnosis of a class of nonlinear uncertain systems with lipschitz nonlinearities using adaptive estimation. *Automatica*, 46(2):290–299, 2010.

IOP Conf. Series: Journal of Physics: Conf. Series **783** (2017) 012004 doi:10.1088/1742-6596/783/1/012004

Fault Detection and Isolation using Viability Theory and Interval Observers

Majid Ghaniee Zarch[1], **Vicenç Puig**[2] **and Javad Poshtan**[1]

[1]Electrical Engineering Department, Iran University of Science and Technology (IUST), Narmak, 16846-13114 Tehran, Iran, [2]Institut de Robtica i Informtica Industrial (CSIC-UPC). Carrer Llorens Artigas, 4-6, 08028 Barcelona

E-mail: majidghaniee@iust.ac.ir, vicenc.puig@upc.edu, jposhtan@iust.ac.ir

Abstract. This paper proposes the use of interval observers and viability theory in fault detection and isolation (FDI). Viability theory develops mathematical and algorithmic methods for investigating the adaptation to viability constraints of evolutions governed by complex systems under uncertainty. These methods can be used for checking the consistency between observed and predicted behavior by using simple sets that approximate the exact set of possible behavior (in the parameter or state space). In this paper, fault detection is based on checking for an inconsistency between the measured and predicted behaviors using viability theory concepts and sets. Finally, an example is provided in order to show the usefulness of the proposed approach.

1. Introduction

Conventional feedback control systems are vulnerable to malfunctions in sensors, actuators or other system components. So, diagnosing which kind of faults are developing is an important task to prevent physical damage and performance degradation. Fault detection and isolation (FDI) could also lead to more reliable and efficient systems. Set theory has been started to be used in fault detection and fault tolerant context since more that one decade [12]. There are two main approaches in using set theory for fault detection: the set-membership and the set-invariance approach. One of the most used techniques in the set-membership approach is based on interval observers. Interval observer-based fault detection (FD) consists in generating adaptive intervals for system outputs by considering the bounds of uncertainties, propagating their effect through the mathematical models of the system and testing the consistency between the predicted output intervals and the corresponding measurements of outputs [5], [8]. Another set-theoretic fault detection and isolation (FDI) approach is to consider invariant sets. A system can switch among several modes (a healthy one and at least a faulty one). For each mode, an invariant set for the residual can be obtained [11]. Once the system operates in steady state, it is possible to confine the residual to one of these invariant sets and, as long as all the invariant sets are disjoint, FDI can be performed. Most importantly, in the case that the invariant sets intersect, FD can still be done whenever the residual exits its healthy invariant set [15], [11]. An extensive comparison between these two approaches has been done in [17].

Viability theory develops mathematical and algorithmic methods for investigating the adaptation to viability constraints of evolutions governed by complex systems under uncertainty [2]. Viability is a theory that until now mostly used in safety verification in control systems [7]. It provides some concepts that is actually more general than what is used in set theory for fault detection. Viability kernel is an acceptable tool for safety verification. However, the problem with this theory is how to compute the involved sets. Nowadays, there have been proposed some algorithms that can approximate this kernel effectively. The algorithms are surveyed in this paper. This theory is used in different areas of study like economics, biology and etc. Finding the concepts that can be used in fault detection is major contribution of this paper. Also, we compare these concepts with the concepts in set theory that are used for fault detection.

The main contribution of this paper is to propose the combined use of interval observers and viability theory in fault detection and isolation (FDI). Fault detection is based on checking for an inconsistency between the measured and predicted behaviors using viability theory concepts and sets. Finally, an example is provided in order to show the usefulness of the proposed approach.

This paper is organized as follows. In Section 2, some definitions and preliminary concepts are provided in context of viability theory. Also, some algorithms in finding some important sets as the viability kernel are investigated. How viability theory can be used in fault detection and isolation is a task that is carried out in Section 3. In Section 4, the viability based FDI approach is integrated with the interval observer approach. An illustrative example is provided in Section 5 in order to show the concepts. Finally, in Section 6 concluding remarks are drawn.

2. Preliminary Concepts

In this section, some concepts of viability theory are recalled that can be used in fault detection and isolation [2]. Assume a simple nonlinear autonomous system in differential inclusion form:

$$\dot{x}(t) \in f(x(t)) \tag{1}$$

It is assumed that the above system is defined in a proper open set $\mathrm{O} \subseteq \mathbb{R}^n$ and that there exist a globally defined solution for every initial condition $x(0) \in \mathrm{O}$. The evolutionary system:

$$S : X \to \mathbb{C}(0, +\infty; X)$$

maps any initial state $x \in X$ to the set $S(x)$ of evolutions $x(.)$ starting from $x(0)$ and governed by differential inclusion (1).

2.1. Concepts definition

Definition 1 (Invariance Kernel) Let $K \subset X$ be an environment and $C \subset K$ be a target. The subset $Inv_S(K,C)$ of initial states $x(0) \in K$ such that all evolutions $x(t) \in S(x)$ starting at $x(0)$ are viable in K for all $t \geqslant 0$ or viable in K until they reach C in finite time is called the invariance kernel of K with target C under S.

The above definition is similar to the invariance set definition in set theory that has been used extensively for fault detection recently, see for example [11], [15]. In viability theory, there are some more general concepts that can be used in fault detection and isolation.

Definition 2 (Viability Kernel). The viability kernel of K under the evolutionary system S is the set $Viab_S(k)$ of initial states $x(0) \in K$ from which starts at least one evolution $x(t) \in S(x)$ viable in K for all times $t \geqslant 0$:

$$Viab_S(K) := \left\{ \begin{array}{l} x(0) \in K | \exists x(.) \in S(x) \\ such\ that \quad \forall t \geqslant 0,\ x(t) \in K \end{array} \right\} \tag{2}$$

Definition 3 (Capture Basin). The capture basin of C (viable in K) under the evolutionary system S is the set $Capt_S(K,C)$ of initial states $x(0) \in K$ from which starts at least one evolution $x(t) \in S(x)$ viable in K on $[0,T)$ until the finite time T when the evolution reaches the target at $x(T) \in C$.

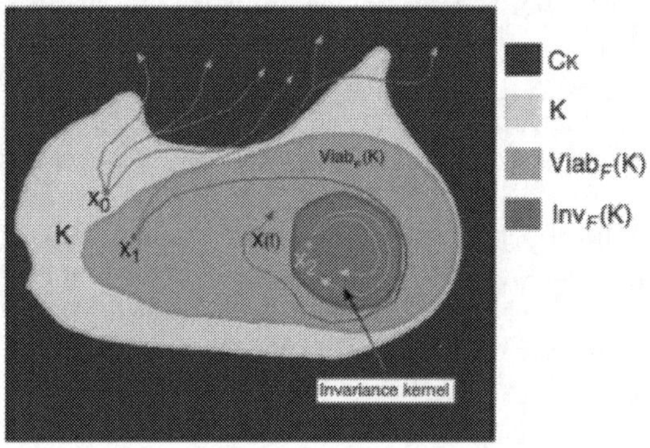

Figure 1. A sample of invariance kernel [2]

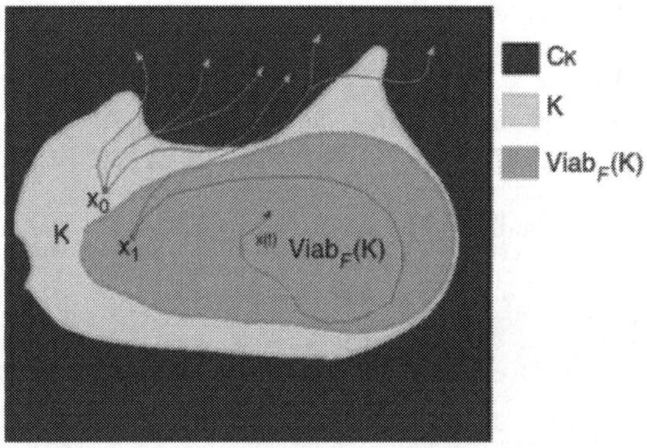

Figure 2. A sample of viability kernel [2]

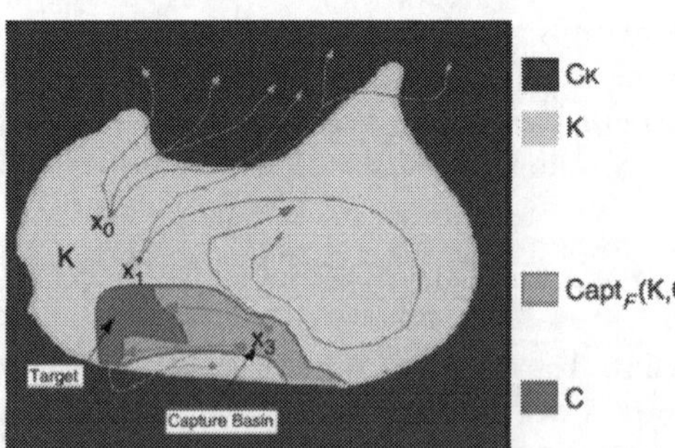

Figure 3. A sample of capture basin [2]

2.2. *Algorithms*

The viability kernel has been approximated using two main group of methods: Eulerian methods and Lagrangian methods. First Eulerian algorithm was proposed by Saint Pierre [13]. In this one, the so-called *viability algorithm*, the continuous dynamical system of the form

$$\dot{x}(t) = f(x(t), u(t)) \tag{3}$$

is considered. This algorithm provides an approximation of the viability kernel when f is Lipschitz. The idea of the algorithm is first to discretize the dynamical system in time. Choosing a time step dt and using the Euler approximation of system (3):

$$\begin{aligned} &x(t+dt) = x(t) + f(x(t), u(t))\, dt = G(x(t), u(t)) \\ &u(t) \in U(x(t)) \end{aligned} \tag{4}$$

then using discretization in the state space and defining a grid of points K_h of resolution h covering K such that:

$$\forall x \in K, \exists x_i \in K_h \quad such\ that \quad \|x - x_i\| \leqslant h$$

Then, Saint-Pierre considers the sequence

$$K_h^{n+1} = \left\{ \begin{array}{l} x_k \in K_h^n, \exists u \in U(x_k) \\ such\ that \quad G_h(x_k, u) \in K_h^n \end{array} \right\}$$

with $K_h^0 = K_h$. It can be shown that

$$Viab_{G_h}(K_h) = \bigcap_{n=0}^{+\infty} K_h^n$$

And there exits one integer p such that

$$K_h^p = Viab_{G_h}(K_h)$$

This sequence gives an iterative algorithm to approximate viability kernels by using a discrete approximation. In summary, the algorithm can be described as:

$$\begin{aligned} &K_h^0 = K \\ &K_h^{n+1} = \{x \in K_h^n | K_h^n \cap F(x) \neq \emptyset\} \end{aligned} \tag{5}$$

where K is initial guess and $F(x) = \{f(x,u) | u \in U\}$. Moreover, Saint-Pierre shows that when f is μ-Lipschitz, the discrete viability kernel tends to the viability kernel of the initial system when the resolution of the grid h tends to 0.

Level set approach is also an Eulerian algorithm that is proposed in [9]. Eulerian methods requires gridding state space and time, hence, their run time and memory complexity grow exponentially with the state dimension. In practice, this approach is infeasible for systems with more than three or fourt states.

Lagrangian methods also have been applied to the computation of viability kernels, for example in [3], but the implementation has relied on polyhedral set representations that also do not scale well with the number of states. Lagrangian methods have been applied successfully to the computation of reachable sets [6]. In contrast to Eulerian methods, Lagrangian methods use representations that follow the vector field's flow. Since Lagrangian methods do not depend on gridding the state space, it is computationally feasible to analyse high-dimensional systems. In [7], an algorithm is proposed to approximate viability kernel using maximal reach sets. It is shown that (5) can be reformulated in terms of backward reach set over one discrete time step:

$$\begin{aligned} &K_h^0 = K \\ &\text{Reach}_\tau(K_k) = \left\{x_0 \in \mathbb{R}^d | \exists u_0 : x(\tau) \in K_k\right\} \\ &K_h^{n+1} = K_h^0 \cap \text{Reach}_1(K_n) \end{aligned} \tag{6}$$

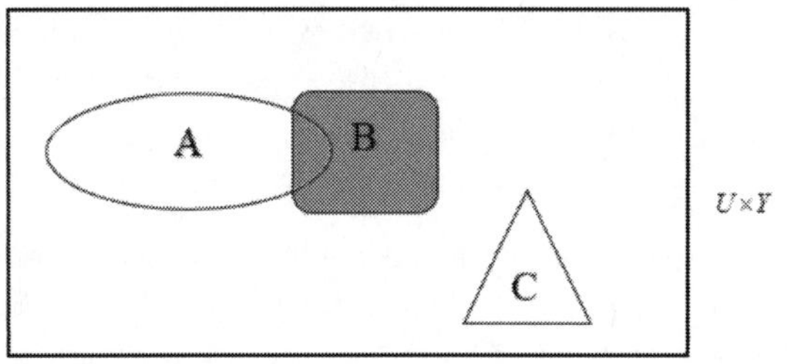

Figure 4. System behavior with and without fault

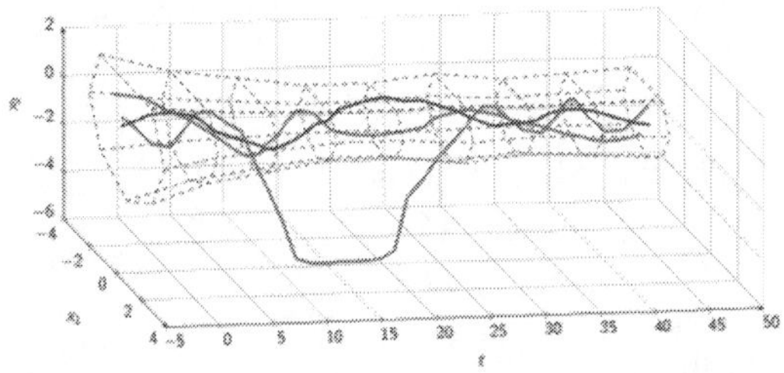

Figure 5. Basic idea of fault detection using sets [16]

3. Fault Detection and Isolation using Viability Theory

3.1. Review of FDI using sets

A fault in a dynamical system is a deviation of the system structure or the system parameters from the nominal situation. The principle of model-based fault detection is to test whether the measured system inputs and outputs are consistent with the system behavior described by a fault-less model. If the measurements are inconsistent with the model of the healthy system, the existence of a fault is proved.

For a dynamical system, consider that the output $y(t)$ is the reaction of the plant to the input $u(t)$. The pair (u, y) is called input/output (I/O) pair. (U, Y) is the set of all possible I/O pairs. Faults leads to deviations of the dynamical input/output (I/O) properties of the plant from the nominal ones and hence, change the performance of the closed-loop system which further results in a degradation or even the loss of the system function.

A fault changes the system behavior as illustrated in Fig. 4. If the system works in white set (A), it is working in healthy mode. The system behavior can be moved by a fault toward the grey set (B). If a common input u is applied to the healthy and faulty systems, then both systems answer with different outputs $A = (u, y_A)$ and $B = (u, y_B)$. This change in the system behavior makes the detection of the fault possible, unless the faulty I/O pair lies in the intersection of A and B.

For isolation of more than one fault, in each faulty situations, a proper set must be determined. For example, in Fig. 4, there are two faulty cases, f_1 and f_2. If the I/O pair lies in the set b, our suggestion is that the system is working in faulty mode f_1. But, if the I/O pair lies in set C, it is more probable that the system works in the faulty mode f_2. If the I/O pair is in the intersection and outside of these predefined sets, nothing can be said about the system status.

The basic idea of using sets in fault detection is to find a predefined set that can assure safety if the system works on that set (see Fig. 5).

In the framework of the analytical redundancy concept, the process model that is driven by the same process input will run in parallel to the process. It is reasonable to expect that, in the fault-free case, both process and its model shows similar behavior. Comparing outputs leads to a signal that can be used

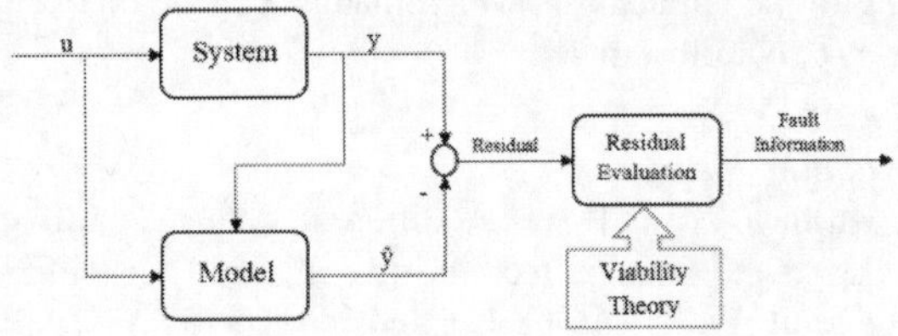

Figure 6. A general fault detection scheme

for fault detection. The difference between the measured process variables and their estimates is called residual. This residual signal carries information about faults that have occurred in the system.

Residual signal information is generally affected by model uncertainties and unknown disturbances. Moreover, fault isolation and identification techniques needs additional residual analysis. So, the main problem with the application of model based fault diagnosis techniques can be expressed as extracting the needed and useful information about the faults of interest from residual signal. This step is called residual evaluation and can be performed using set-based approaches [12].

3.2. Fault detection using viability theory

Using the viability theory concepts introduced in Section 2, the FDI problem presented in previous section using sets can be addressed using viability theory concepts. These concepts can be used for evaluating the residual as it is shown in Fig. 6 and illustrated in the following subsections.

3.2.1. Steady State As with all the methods based on the set invariance approach, we can use invariance kernel for fault detection. Hence, residual signal must lie in the invariance kernel of the system residual under healthy functioning, i.e.,

$$r(k) \in Inv_{S_r}(K,C) \tag{7}$$

in which C can be any arbitrary target around equilibrium point. If the residual exits its healthy invariant set, it indicates that a fault has occurred. Otherwise, it is assumed that the system is still healthy.

3.2.2. Transient In transient state, because changes in system states are somewhat unpredictable and also, we do not know initial state of system, we cannot use invariance kernel in fault detection appropriately. In this situation, the question to be answered is: Is there any possible action that bring our system near steady state? Translating this question to the concepts in viability theory, we can use viability kernel for fault detection

$$r(k) \in Viab_{S_r}(K,C) \tag{8}$$

That is while the system is in the viability kernel means that the system can find a way to be safe. Actually in this manner, we can not say anything about convergence of the system toward steady-state. Note that in the constructing viability kernel, system constraints are considered. So, being in viability kernel also means the system do not violate constraints. Satisfying constraints and have the opportunity to come back to steady state can ensure us that our system is in healthy state.

In most practical cases, we have time constraint in the transient mode, which means system must come back near steady state in finite time. In this situation, we can use capture basin for fault detection.

$$r(k) \in Capt_{S_r}(K,C) \tag{9}$$

It means that the system has a limited time T to go to steady state. So, being in capture basin means that the system can find a way to come back to target in limited time instants. The point is that a target can be chosen arbitrarily according to the application, which means we can choose invariance kernel as target or have two different targets (or equilibriums).

3.3. Fault isolation using viability theory

Consider that our system can have $i = 0, 1, 2, ..., N$ different states: The first one is healthy mode and others are different faulty modes to be detected. In this situation, considering that we can construct a kernel for every types of fault, faults can be isolated if kernels are separable. Therefore, residual equation is written for every fault as follows:

$$r_i(k) = y_i(k) - \hat{y}(k) \tag{10}$$

So, the condition for faults to be isolable is that those kernels are separable:

$$\begin{aligned} &Invs_{S_{r0}}(K,C) \cap Invs_{S_{r1}}(K,C) \cap ... = \phi \\ &Viabs_{S_{r0}}(K,C) \cap Viabs_{S_{r1}}(K,C) \cap ... = \phi \\ &Capts_{S_{r0}}(K,C) \cap Capts_{S_{r1}}(K,C) \cap ... = \phi \end{aligned} \tag{11}$$

Considering this condition, if

$$r(k) \in Invs_{S_{r1}}(K,C) \tag{12}$$

means that the system have works in mode 1 in steady state, that is actually a faulty situation.

4. Integration with Interval Observers

4.1. Interval observers

The application of the viability theory to FDI requires the generation of the residual signal. Residuals can be generated in many ways as discussed in [4]. A particular well established way of residual generation is based on the use of observers. Viability theory is well suited for dealing with FDI in non-linear systems since most of the concepts have been developed in this context (see Section 2). Designing observers for nonlinear systems is a difficult problem. A possible approach to deal with the observer design for for nonlinear systems by approximating them as linear parameter varying (LPV) system [14] and apply LMI based designs. LPV is a class of nonlinear systems whose state space matrices depend on a set of time-varying parameters that are not known in advance, but it can be measured or estimated upon operation of the system [1]. The system to be monitored can be described by a nonlinear uncertain dynamic model in a state-space form as follows:

$$\begin{aligned} x(k+1) &= A(\rho)x(k) + B(\rho)F_i u(k) + \omega(k) \\ y(k) &= C(\rho)x(k) + \eta(k) \end{aligned} \tag{13}$$

where $x(k)$ is state, $y(k)$ is plant measured output, $u(k)$ is control input, $\omega(k)$ is bounded perturbation, $\eta(k)$ is measurement noise and F_i is actuator fault matrix: if $F_i = 1$, the system is working in healthy mode, otherwise there is fault in the actuator. The parameter ρ is an a priori unknown time-varying parameter whose measurement is available.

For design purposes, the LPV system (13) can be expressed in polytopic form

$$[A(\rho), B(\rho), C(\rho)] = \sum_{j=1}^{N} \zeta_j(\rho)[A_j, B_j, C_j]$$

for certain constant matrices A_j, B_j, C_j and continuous functions ζ_j such that $\zeta_j(\rho) \geqslant 0$ and $\sum_{j=1}^{N} \zeta_j(\rho) = 1$ for all ρ. Assume that the pairs A_j, B_j are stabilisable and the pairs A_j, C_j are detectable for $j = 1, ..., N$.

An LPV observer for this system is defined as:

$$\begin{aligned}\hat{x}(k+1) &= A(\rho)\hat{x}(k)+B(\rho)u(k)+\\ &\quad L(\rho)(C(\rho)x(k)-\hat{y}(k))+\breve{\omega}(k) \\ \hat{y}(k) &= C(\rho)\hat{x}(k)+\breve{\eta}(k)\end{aligned} \tag{14}$$

where $u(k)\in U$ is the measured system input vector, $\hat{x}(k)\in\hat{X}$ is the estimated system state vector, $\hat{y}(k)$ is the estimated system output vector, the uncertain variables $\breve{\omega}(k)$ and $\breve{\eta}(k)$ are used to describe the effect of variable $\omega(k)$ and $\eta(k)$ on the plant (13), respectively. The uncertain variables $\breve{\omega}(k)$ and $\breve{\eta}(k)$ are different from $\omega(k)$, but are defined to have the same bounds, respectively (i.e. $\breve{\omega}(k)\in W$ and $\breve{\eta}(k)\in V$).

According to [10], taking into account uncertainty bounds when obtaining the observer estimation, intervals that bound the estimated state and output can be generated. This type of observer is known as an interval observer and it is well accepted approach in robust FDI.

The observer gains L_j have to be designed in order to stabilize the observer given by (14). According to [10], observer gain can be determined through the following LMIs:

$$\begin{bmatrix} -rX_j & aX_j+X_j^TA_j(\rho)-W_j^TC_j(\rho) \\ \left(a+A_j(\rho)^T\right)X_j-C_j(\rho)^TW_j & -rX_j \end{bmatrix}<0 \tag{15}$$

where $L_j=\left(W_jX_j^{-1}\right)^T$.

In order to be able to detect a fault, a residual is generated:

$$r(t)=y(t)-\hat{y}(t)=C(\rho)\tilde{x}(k)+\eta(k)-\breve{\eta}(k) \tag{16}$$

where the state estimation error $\tilde{x}(k)$ of the observer is defined as

$$\tilde{x}(k)=x(k)-\hat{x}(k)$$

Furthermore, by using (13) and (14), the dynamics of $\tilde{x}(k)$ can be derived as

$$\begin{aligned}\tilde{x}(k+1) &= (A(\rho)-L(\rho)C(\rho))\tilde{x}(k)+B(F_i-I)u(k)\\ &\quad +\omega(k)-\breve{\omega}(k)\end{aligned} \tag{17}$$

Consider the evolutionary system $S_r: X\to\mathbb{C}(0,+\infty;X)$ maps any initial state $x\in X$ to the set $S_r(x)$ of evolutions $x(.)$ starting from $x(0)$ and governed by 16.

4.2. Interval observers and set invariance

Recently, a lot of effort have been done in order to use interval observers and set invariance in fault detection and isolation, see for example [8], [14]. In [17], a relationship between these two methods is constructed. Interval observers have the advantage of considering noise and uncertainty in on-line phase of fault detection, which makes this method more robust, but with more computational demand. Set invariance is more conservative than interval observer, but it has guaranteed fault detection capability. Interval observers and set invariance are used for fault detection in steady state. Here, we propose using viability theory in order to make it possible to detect faults in transient states combining interval observers for generating residuals and viability theory concepts for evaluating them in FDI as described in Section 4. The main motivation for this integration is because viability offers a general framework for dealing with nonlinear systems. Another thing is that in viability theory, sets can be defined without considering any specific shape (as e.g., ellipsoids or zonotopes), which makes it less conservative and more general implementation can be achieved.

5. Illustrative Example

In this section, the fault detection scheme developed in previous sections is applied to a two-tank nonlinear SISO system described by [14]

$$\begin{aligned} \dot{h}_1(t) &= -\left({}^{s}/_{S}\right)\sqrt{2g}\sqrt{h_1(t)} + \left({}^{\kappa}/_{S}\right)u(t) + \omega_1(t) \\ \dot{h}_2(t) &= \left({}^{s}/_{S}\right)\sqrt{2g}\left[\sqrt{h_1(t)} - \sqrt{h_2(t)}\right] + \omega_2(t) \end{aligned} \tag{18}$$

where $u(t)$ is the voltage applied to the pump, $h_1(t)$, $h_2(t)$ are system states, $\omega_1(t)$, $\omega_2(t)$ are bounded state perturbation and the parameters are as follows: $S = 15.5179cm^2$ is the cross section area of the tanks; $s = 0.1781cm^2$ the cross section of the tanks outflow orifice; $\kappa = 3.3cm^3/Vs$ the gain of the pump; and $g = 981cm/s^2$ the gravitational constant. After Euler discretisation with sampling period $\tau = 1\ s$, the whole system is formulated in its quasi-LPV form through parameter non-linear embbeding approach

$$\begin{aligned} x(k+1) &= A(\rho(k))x(k) + BF_i u(k) + E\omega(k) \\ y(k) &= Cx(k) + \eta(k) \end{aligned} \tag{19}$$

where actuator fault is modelled by F_i and

$$\begin{aligned} x(k) &= \begin{bmatrix} h_1(k) & h_2(k) \end{bmatrix}^T \\ A(\rho(k)) &= I + \tau \begin{bmatrix} -\rho_1(k) & 0 \\ \rho_1(k) & -\rho_2(k) \end{bmatrix} \\ B &= \tau \begin{bmatrix} \kappa/S & 0 \end{bmatrix}^T \\ C &= \begin{bmatrix} 0 & 1 \end{bmatrix} \\ E &= \tau \begin{bmatrix} 1 & 1 \end{bmatrix}^T \\ \eta(k) &= \begin{bmatrix} \eta_1(k) & \eta_2(k) \end{bmatrix}^T \end{aligned}$$

$\eta(k)$ is measurement noise matrix. Disturbances and noises are considered bounded by means of zonotopes with center in 0.01 and segment of 0.01. The varying parameters can be shown as

$$\rho_i(k) = \frac{s}{S}\sqrt{\frac{2g}{h_i(k)}}, \quad i = 1,2$$

The system states and parameters are bounded as

$$\begin{aligned} h_i^{\min} &= 0.5 \to \rho_i^{\min} = 0.0928 \\ h_i^{\max} &= 30 \to \rho_i^{\max} = 0.7189 \end{aligned}$$

Hence, an LPV model with convex polytopic description can be obtained from (19) by taking $A_1 = A\left(\underline{\rho}, \underline{\rho}\right), A_2 = A\left(\underline{\rho}, \bar{\rho}\right), A_3 = A\left(\bar{\rho}, \underline{\rho}\right), A_4 = A(\bar{\rho}, \bar{\rho})$ and $A(\rho(k)) = \sum_{i=1}^{4} \alpha_i A_i$. These weights can be computed using following equations, which is valid for the case of polytopes with four vertices:

$$\begin{aligned} \rho &= \frac{h_1^{\max} - h_1(k)}{h_1^{\max} - h_1^{\min}} \\ \lambda &= \frac{h_2(k) - h_2^{\min}}{h_2^{\max} - h_2^{\min}} \\ \alpha_1 &= \rho\lambda, \quad \alpha_2 = (1-\rho)\lambda \\ \alpha_3 &= \rho(1-\lambda), \quad \alpha_4 = (1-\rho)(1-\lambda) \end{aligned}$$

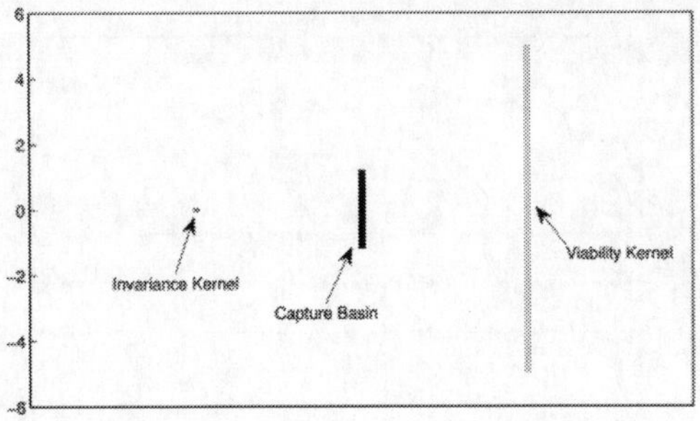

Figure 7. Kernels for residual signal in healthy mode

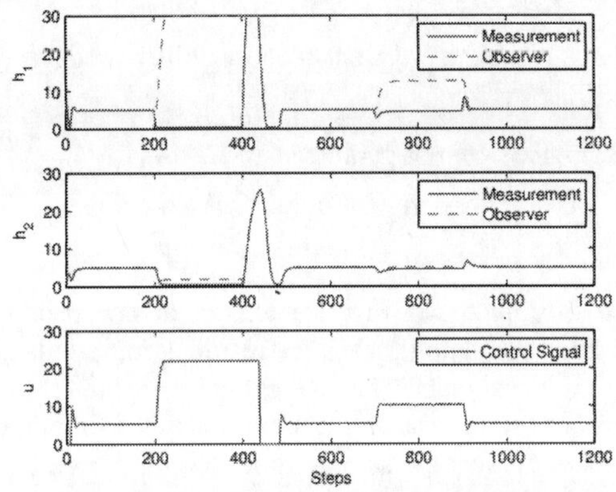

Figure 8. Simulation results

Considering (14) as observer and using (15), observer gains are derived as:

$$L_1 = L_2 = L_3 = L_4 = \begin{bmatrix} 0.8288 \\ 1.4975 \end{bmatrix}$$

Initial condition for the system is $\begin{bmatrix} 10 & 0 \end{bmatrix}^T$ and for the observer is $\begin{bmatrix} 8 & 3 \end{bmatrix}^T$ .Residual signal can be defined as

$$r(k) = y(t) - \hat{y}(t) = C\tilde{x}(k)$$
$$\tilde{x}(k+1) = (A(\rho(k)) - LC)\tilde{x}(k)$$

Invariance kernel, viability kernel and capture basin for residual signal in healthy mode are shown in Fig. 7. This figure is only for comparing the size of the kernels for residual signal with one graded axes. As it can be see, residual in steady state must lie in $[-0.05, 0.05]$, otherwise fault detection scheme raises a fault alarm. For transient, if no time constraint is considered, then viability kernel must be used for fault detection. In this case, if we consider that $\tilde{x}(k)$ starts in a box with zero origin and length 5, then residual must lie in $[-5, 5]$. In this application, it is reasonable, because there is always a way to bring the system to steady state as long as system constraints satisfied. If we consider a time limitation of five sample times and start from invariance kernel as initial set, as we consider in capture basin derivation, interval $[-1.2, 1.2]$ can be used for fault detection. It means that if residual becomes more than 1.2, there is no possibility to come back to steady state in less than five sample times. In Fig. 10 simulation results are

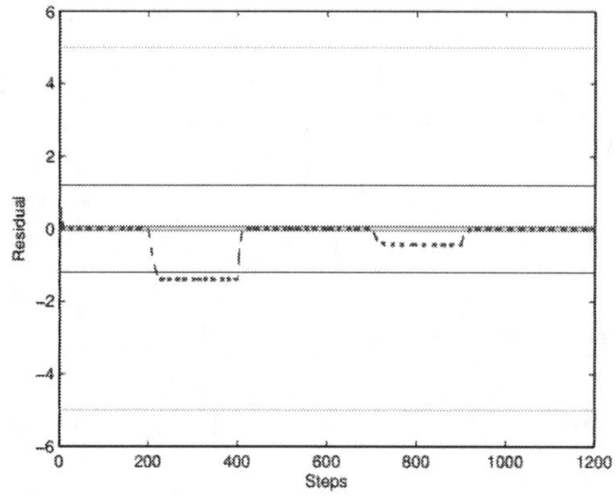

Figure 9. Invariance kernel, capture basin and viability kernel for residual signal

presented including two faults in the actuator parameterised as in (13)

$$F_1 = 0, \qquad 200 \leqslant k \leqslant 400$$
$$F_2 = 0.5, \qquad 700 \leqslant k \leqslant 900$$

The residual signal from (16) is depicted in Fig. 9. It is clear that residual in steady state is inside the invariance kernel in healthy mode. Fault makes the residual going outside invariance kernel, which means a fault occur. The first fault also makes the residual going outside the capture basin, which means there is no possibility to come back to invariance kernel in less that five sample times. In both situations, when fault disappears, it takes more than ten sample times to come back to invariance kernel. But, for second fault, we can not say that there is a fault in transient mode, as long as it lies in capture basin. During the whole simulation time, residual lies in viability kernel, which means system is safe and have the opportunity to come back to invariance kernel.

Invariance kernel for the applied faults can be detemined to be in $[-1.2, -1.45]$ for the first fault and to be $[-0.3, -0.5]$ for the second fault. Because, invariance kernel in healthy and faulty modes are separable according to Section 3.3, so the faults can be isolated. In Fig. 10, it can be seen that when fault occurs, residual signal goes from healthy invariance kernel to one of faulty invariance kernels.

6. Conclusions

In this paper, the application of viability theory to fault detection and isolation has been developed regarding its application to nonlinear systems. Concepts of invariance kernel, viability kernel and capture basin are introduced and adapted to be used in fault detection and isolation. It is shown that fault detection and isolation in steady state and transient mode can be done using those proposed concepts. Moreover, the integration of the proposed fault detection and isolation approach with interval observers has been presented. Finally, a simulation has been done in order to show effectiveness of the proposed approach.

The main advantage of this approach is that fault detection can be carried out in transient mode. Using interval observer, noise can be considered in residual generation phase. Therefore, the proposed method is more robust against noises. The difficulty with fault detection in transient mode is residual evaluation that can be done by viability theory concepts. The main drawback of this method is the difficulty with computation of these kernels.

References

[1] Ali Abdullah and Mohamed Zribi. Model reference control of lpv systems. *Journal of the Franklin Institute*, 346(9):854–871, 2009.

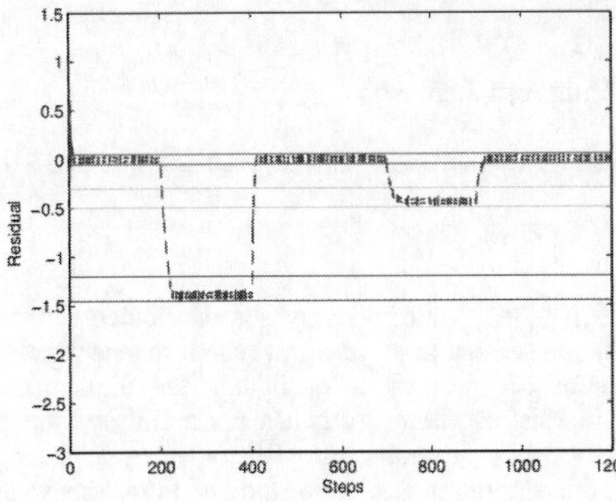

Figure 10. Invariance kernels for different faults

[2] Jean-Pierre Aubin, Alexandre M Bayen, and Patrick Saint-Pierre. *Viability theory: new directions*. Springer Science & Business Media, 2011.

[3] Franco Blanchini and Stefano Miani. *Set-theoretic methods in control*. Springer, 2008.

[4] Mogens Blanke, Michel Kinnaert, Jan Lunze, and Marcel Staroswiecki. *Diagnosis and fault-tolerant control*. Springer, 2006.

[5] Pedro Guerra, Vicenc Puig, and Marcin Witczak. Robust fault detection with unknown-input interval observers using zonotopes. In *Proceedings of the 17th IFAC World Congress, Seoul, Korea*, pages 5557–5562, 2008.

[6] Colas Le Guernic and Antoine Girard. Reachability analysis of linear systems using support functions. *Nonlinear Analysis: Hybrid Systems*, 4(2):250–262, 2010.

[7] John N Maidens, Shahab Kaynama, Ian M Mitchell, Meeko MK Oishi, and Guy A Dumont. Lagrangian methods for approximating the viability kernel in high-dimensional systems. *Automatica*, 49(7):2017–2029, 2013.

[8] Frédéric Mazenc and Olivier Bernard. Interval observers for linear time-invariant systems with disturbances. *Automatica*, 47(1):140–147, 2011.

[9] Ian M Mitchell, Alexandre M Bayen, and Claire J Tomlin. A time-dependent hamilton-jacobi formulation of reachable sets for continuous dynamic games. *Automatic Control, IEEE Transactions on*, 50(7):947–957, 2005.

[10] Saúl Montes de Oca, Vicenç Puig, and Joaquim Blesa. Robust fault detection based on adaptive threshold generation using interval lpv observers. *International Journal of Adaptive Control and Signal Processing*, 26(3):258–283, 2012.

[11] Sorin Olaru, Jose Adrian De Doná, MM Seron, and Florin Stoican. Positive invariant sets for fault tolerant multisensor control schemes. *International Journal of Control*, 83(12):2622–2640, 2010.

[12] V. Puig. Fault diagnosis and fault tolerant control using set-membership approaches: Application to real caes studies. *International Journal of Applied Mathematics and Computer Science*, 20:619635, 2010.

[13] Patrick Saint-Pierre. Approximation of the viability kernel. *Applied Mathematics and Optimization*, 29(2):187–209, 1994.

[14] María M Seron and José A De Doná. Robust fault estimation and compensation for lpv systems under actuator and sensor faults. *Automatica*, 52:294–301, 2015.

[15] María M Seron, Xiang W Zhuo, José A De Doná, and John J Martínez. Multisensor switching control strategy with fault tolerance guarantees. *Automatica*, 44(1):88–97, 2008.

[16] Florin Stoican. *Fault tolerant control based on set-theoretic methods*. PhD thesis, Supélec, 2011.

[17] Feng Xu, Florin Stoican, Vicenc Puig, Carlos Ocampo-Martinez, and Sorin Olaru. On the relationship between interval observers and invariant sets in fault detection. In *Control and Fault-Tolerant Systems (SysTol), 2013 Conference on*, pages 49–54. IEEE, 2013.

A methodology for distributed fault diagnosis

V. Gupta, V. Puig and J. Blesa

[1] Advanced Control System Research Group, Universitat Politècnica de Catalunya (UPC), Barcelona, Spain

Abstract. In this paper, a methodology for distributed fault diagnosis is proposed. The algorithm places the sensors in a system in such a manner that the partition of a system into various subsystems becomes easier facilitating the implementation of a distributed fault diagnosis system. This algorithm also reduces or minimized the number of sensors to be used or install thus reducing overall cost. Binary integer linear programming is used for optimization in this algorithm. Real case study of Barcelona water network has been used to demonstrate and validate the proposed algorithm.

1. Introduction

The model based diagnosis is all about diagnosing faults in the system by comparing the set of system observations with a model. A suitable approach basically depends on the system model and the faults to diagnose since there is no general framework for such diagnosis systems. In any time during their working life the processes can suffer from faults or malfunctions. A diagnosis system works in parallel with the process. Observations provided by the process measurements and the control or operating signals are the diagnosis system inputs. The diagnosis system tries to infer which behaviour described by the model is actual in the process from these observations.

Serious damage on systems or a loss of process performance can be prevented by diagnosing faults correctly. This makes fault diagnosis [1] an important research area but it is rare to find novel techniques in industrial processes. The main reasons are first that the achieved diagnosis performance is not the expected one and the second is, design techniques are cumbersome and difficult to implement. The present work tries to solve or simplify these problems by choosing the location of sensors in the process. Generally performance of diagnosis systems strongly depend on the on-line information acquired from the process. Therefore, appropriate sensor location will lead to better implementation facilities and diagnosis performance.

Sensors are the components of the system by means of which process observations are obtained. Therefore, the number of observations increases and consequently so does the analytical redundancy degree of the model by installing extra sensors in the process. There are processes where installing extra sensors is not indicated, (e.g. the economical impact on electronic devices, insufficient space or extra weight in aircrafts). However, the presence of a non-diagnosed fault can cause serious problems in many industrial processes; to improve fault diagnosis the installation of extra sensors is fully recommended. The analysis by which it is determined that which sensors are needed to achieve certain specifications is called *sensor placement* analysis. There are some research areas where the sensor placement problem has been widely studied (e.g. observability in the control field) but this is not the case in diagnosis field, where there are only few isolated works present in the literature. However, this topic is gaining an increasing interest from last few years.

Till now, just few works had been dedicated to this problem [6] [7] [8] [9]. In [10] a part of the work proposed in this paper is already done which specifically focuses on analyzing which sensors should be installed in a process, to facilitate its design and at the same time improve the diagnosis capabilities of a model based diagnosis system. In [5] an alternate approach is shown to reduce number of links or communication between various nodes of a system. In the present work a novel algorithm is proposed to place the sensors in a system in such a manner that the partition of a system into various subsystems becomes easier, thus facilitating the implementation of a distributed fault diagnosis system.

The structure of the paper is the following: Section 2 describes the entire problem description. Section 3 describes the proposed methodology. Section 4 describes the implementation of the solution. Finally, in Section 5, conclusions are presented.

2. Problem Description

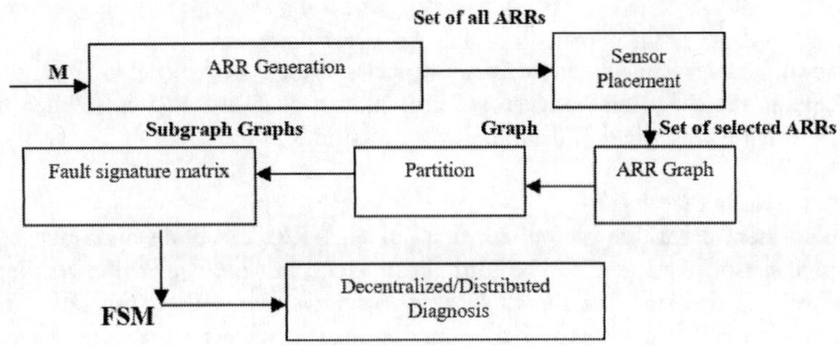

Figure 1. Various blocks of decentralized/distributed fault diagnosis algorithm using ARRs

Figure 1 shows a block diagram of a decentralized/distributed fault diagnosis algorithm described in [2] [3]. The proposed algorithm in this paper is trying to find an optimal sensor placement solution such that when the set of selected ARRs are generated they have minimum coupling so that when applying the partitioning algorithm various subsystems are easily decoupled and partitioning of a system into various subsystem becomes easy. In the off-line phase, this fault diagnosis approach starts from obtaining the analytical redundancy relations (ARRs) using the system model and the set of available sensors. These ARRs are converted into a graph. This graph is further divided into various subgraphs using a partition algorithm. Each subgraph corresponds to a subsystem. From various subgraphs, different local fault signature matrices for various subsystems are obtained. Finally, in the on-line phase, using various local fault signature matrices, a set of diagnoser agents are created that allow the global diagnosis in a large scale system. For each subsystem, a local fault signature matrix is then obtained as next step. Using each local fault signature matrix, a local diagnoser is implemented by means of an agent. This agent is responsible for local diagnoses in a subsystem and communicating with other agents of others subsystems to perform global diagnosis. In order to demonstrate the applicability of the proposed approach, a case study based on the Barcelona drinking water network (DWN) was used. However, the main problem of this approach is that the set of sensors used for the generation of ARRs cannot be placed in such a manner that the ARRs generated are decoupled and thus ARRs produced by this approach are highly coupled and it is difficult to partition such highly coupled ARRs into various subsystems, to allow the distributed/decentralized implementation. Therefore, sensor placement in fault diagnosis is generally the first block or first step in a decentralized/distributed fault diagnosis algorithm so that the ARRs generated are decoupled and partition of ARRs into various subsystems becomes easier.

IOP Conf. Series: Journal of Physics: Conf. Series **783** (2017) 012005 doi:10.1088/1742-6596/783/1/012005

3. Proposed Algorithm

The proposed methodology for distributed fault diagnosis algorithm starts from discrete-time space state model. This model is the starting point for obtaining a set of analytical redundancy relations (ARRs) described below in section 3.1. The proposed methodology has 6 Blocks.

3.1 Block 1: Formation of set of all ARRs

ARRs are obtained from a model $M(z, x)$ by using *ranking algorithm* [13]. Let $z=\{z_1, z_2, z_3, z_4......z_m\}$ be the set of the constraints which represent the system model and let $x=\{x_1, x_2, x_3, x_4......x_n\}$ be the set of the variables which contains three subsets: let $k=y\cup u$ be the set of known variables: u is the subset of input variables, y is the subset of the output variables and k is the subset of the unknown (non-measured) variables. The structure of the model is described by the binary relation:

$$M: z\times x \rightarrow \{0, 1\}$$

where: $(z_i, x_j) \rightarrow M(z_i, x_j)=1$ if z_i applies to x_j and $M(z_i, x_j)=0$, otherwise.
The unknown variables are replaced by known variables of the system model to obtain set of all ARRs or parity matrix P. From the set of all ARRs, a set of minimum coupled ARRs is selected by sensor placement algorithm which is described in detail below in section 3.2.

3.2 Block 1: Sensor Placement

Sensor placement algorithm is applied to set of all ARRs to obtain minimum coupled set of ARRs. The optimal sensor placement can be formulated as a *Binary Integer Programming* (BIP) problem and in literature it is described as *Mixed Integer Programming.* BIP optimization restricts the value of optimization variable to {0, 1}.The optimal sensor placement problem when solved, the sensor installation is represented by a binary vector, q, and each sensor $s_i \in S$ belongs to a binary element q_i of q, with $q_i \in \{0, 1\}$. When q_i equals 1, means the sensor must be installed but if q_i equals 0, the sensor does not have to be installed.

The new work or constraints added in already existing sensor placement algorithm [4] [10] are the constraints to minimize coupling between ARRs which are described in equations (1) and (2) below.

3.2.1 Minimizing the coupling between ARRs

In order to facilitate the distributed implementation of the fault diagnosis systems, the sensors should be placed such that the coupling between ARRs is minimized. This is achieved by adding additional constraints that minimizes the number of common links between ARRs. First, a constraint that reduces the number of row links coupling is written in compact form as

$$\begin{pmatrix} W_{i\times j} & (0)_{i\times p} & -I_{i\times i} & (0)_{i\times u} \end{pmatrix} \quad (r)_{i\times 1} \leq 0_{i\times 1} \tag{1}$$

An analog constraint could be added to minimize the row links coupling as follows

$$\begin{pmatrix} \left(W_{i\times j}\right)^T & (0)_{j\times y} & -I_{j\times j} \end{pmatrix} \quad (c)_{j\times 1} \leq 0_{j\times 1} \tag{2}$$

Rest of the constraints of already existing sensor placement algorithm [4] [10] in the literature are defined below.

3.2.2 Constraint Formulation for BIP

The proposed algorithm deals with fault detectability and isolability for process, sensors faults and also minimization of number of links in a system to ease partition of the system into various subsystems. To obtain this, it is assumed that the complete set Ω_S of all the analytical redundancy relations (ARRs) is available, when all the candidate sensors S are installed in the system.
Let n be the number of ARRs available in $\Omega_S = \{w_1, w_2,\ldots,w_n\}$, and let k be the number of candidate sensors to be installed, i.e. $k = |S|$. A binary matrix $W = [w_{ij}]$ of size $n\times k$ be the set of ARRs (the row set) and sensors (the column set) is formed. Matrix W is built from the set of ARRs sets as

$$w_{ij} = \begin{cases} 1 & if \quad e_{sj} \in w_i \\ 0 & otherwise \end{cases} \tag{2}$$

All $w_i \in \Omega_S$ and all $s_j \in S$ and $w_{ij} = 1$ means that ARR_i depends on the sensor measurement s_j. V is binary matrix which relates the process faults with the ARR sets. l is the number of process faults that have to be detected and isolated (i.e. $l = |F_{DP}|$). The binary matrix $V = [v_{ij}]$ of size $n \times l$ relates process faults (the column set) and the set of ARRs (the row set),

$$v_{ij} = \begin{cases} 1 & if \quad e_{ff_P} \in w_i \\ 0 & otherwise \end{cases} \tag{3}$$

The matrix V is known as the fault signature matrix (FSM). FSM includes both process and sensor faults and is represented as $(V, \; W)$.The first l columns represent process faults and the remaining k columns represent sensor faults.

3.2.3 AAR set selector
The set of all ARRs Ω_S cannot be used to diagnose when a sensor is selected for installation. P is the *ARR set selector* vector, it is n elements binary vector (one for each ARR set) that indicates whether the ARR set is valid, for a candidate sensor set S ⊆S, i.e.

$\rho_i = 1 \longleftrightarrow$ the ARR set w_i is valid

$\rho_i = 0 \longleftrightarrow$ the ARR set w_i is not valid

(4)

$w_i \in \Omega_S$ be an ARR set and $s_j \in S$ be any of the sensors that depends on this MSO set w_i (i.e., $e_{sj} \in w_i$). The ARR set is not valid till any of its corresponding sensors is not installed (i.e. $q_j = 0$). Otherwise, the ARR set is valid. ρ_i is computed as

$$\rho_i = \sum_{j=1}^{k} [w_{ij} q_j + (1 - w_{ij})] \tag{5}$$

If ρ_i=1 the ARR set is valid and the ARR set is not valid if ρ_i=0. Now, λ_i a binary variable is introduced in the inequality. λ_i is a *dummy* variable in the optimization problem and it is zero when λi is a non-valid ARR set. So,

$$\begin{pmatrix} -W & kI_n \end{pmatrix} \begin{pmatrix} q \\ \lambda \end{pmatrix} \leq \beta \tag{6}$$

Where $\lambda = (\lambda 1, \ldots, \lambda_n)^T$ is the set of linear ARR set selectors and $\beta = (\beta_1 \ldots\ldots \beta_n)^T$ is a vector of independent coefficients.

3.2.4 Process and sensor fault detectability
The constraint for process fault detectability written in compact form using the linear MSO set selector as

$$\begin{pmatrix} 0_{l \times k} & - & V^T \end{pmatrix} \begin{pmatrix} q \\ \lambda \end{pmatrix} \leq \; -1_{l \times 1} \tag{7}$$

The constraint for sensor fault detectability written in compact form using the linear ARR set selector as

$$\begin{pmatrix} I_k - & W^T \end{pmatrix}\begin{pmatrix} q \\ \lambda \end{pmatrix} \leq \quad 0_{k\times 1} \tag{8}$$

3.2.5 Isolability between process and sensor faults

Let matrix $V_{ff}= [v_{ff\,im}]$ be an $n\times C^l_2$ ($C^n_r = \frac{n!}{(n-r)!r!}$) matrix such that (9)

$v_{ff\,im} = |t_{ij1} - t_{ij2}|\ \forall\, j_1, j_2 \in \{1, \ldots, l\} : j_1 < j_2$

Where m indexes, in alphabetic order, the set of pairs (j_1, j_2)

The constraint for Isolability between Process Faults written in compact form as

$$\begin{pmatrix} 0_{C^l_2} - & V^T_{ff} \end{pmatrix}\begin{pmatrix} q \\ \lambda \end{pmatrix} \leq \quad -1_{C^l_2\times\ 1} \tag{10}$$

The constraint for isolability between process and sensor faults written in compact form as

$$\begin{pmatrix} G_2 - & V^T_{fs} \end{pmatrix}\begin{pmatrix} q \\ \lambda \end{pmatrix} \leq \quad 0_{l.k\times 1} \tag{11}$$

Where G_2 is an $(l\cdot k) \times k$ matrix, given as

$$G_2 = \left(I_k\ I_k \ldots\ I_k\right)^T \tag{12}$$

and $V_{fs}= [V_{fsip}]$ is an $n\times(l.k)$ matrix such that

$$V_{fsip} = \left(v_{ij1} - \quad w_{ij2}\right)^2 \forall j_1 \in (1.....l), \forall j_2\,(1....k) \tag{13}$$

Let the matrix $V_{ss} = [V_{ssit}]$ of size $n\times C^k_2$ be

$$V_{ssit}= \left(W_{ij1} - \quad W_{ij2}\right)^2 \forall j_1, j_2 \in (1,....k) : j_1 < j_2 \tag{14}$$

The constraint for isolability between sensor faults written in compact form as

$$\left(G_3 - \quad V_{ss}\right)^T\begin{pmatrix} q \\ \lambda \end{pmatrix} \leq \ 1_{C^k_2\times 1} \tag{15}$$

the matrix G_3 is an $C^k_2 \times k$ matrix and built as

$$G_3 = \begin{pmatrix} 1 & & & & \\ \vdots & & I_{k-1} & & \\ 1 & & & & \\ 0 & 1 & & & \\ \vdots & \vdots & & I_{k-2} & \\ 0 & 1 & & & \\ & & \vdots & & \\ 0 & \cdots & 0 & 1 & 1 \end{pmatrix} \tag{16}$$

3.2.6 Formulation of final optimization problem

The optimal sensor placement for fault detection, isolation and minimization of links for better partition is formulated as

IOP Conf. Series: Journal of Physics: Conf. Series **783** (2017) 012005 doi:10.1088/1742-6596/783/1/012005

$$\min_{q,\lambda} (c^T \quad 0_{1\times n}) \begin{pmatrix} q \\ \lambda \end{pmatrix}$$

Subject to:

$$\begin{pmatrix} -W & & kI_n & \\ 0_{l\times k} & & -V^T & \\ I_k & & -W^T & \\ 0_{C_2^l\times k} & & -V_{ff}^T & \\ G_2 & & -V_{fs}^T & \\ G_3 & & -V_{ss}^T & \\ W & 0 & -I & 0 \\ W^T & 0 & -I & \end{pmatrix} \begin{pmatrix} q \\ \lambda \\ r \\ c \end{pmatrix} \leq \begin{pmatrix} \beta \\ -1_{l\times 1} \\ 0_{k\times 1} \\ -1_{C_2^l\times 1} \\ 0_{l.k\times 1} \\ 1_{C_2^k\times 1} \\ 0_{i\times 1} \\ 0_{j\times 1} \end{pmatrix} \tag{17}$$

Where q, λ have binary values while rows and columns have integer values

After applying sensor placement algorithm, a set of minimum coupled set of ARRs are obtained. The ARR graph is obtained from the set of minimum coupled ARRs. Method of ARR graph generation is described in section 3.3 described below

3.3 Block 3: ARR graph formation

Graph is generally defined as an abstract representation of a group of objects from a collection, where few pairs of objects are joint by links. The elements which are interconnected are typically called vertices while the connected links are called edges. If any1or -1 present in rows of parity matrix P makes that particular row a vertex of the graph and all the 1 or -1 present at same location of two rows is connected edge between the vertexes. From ARR matrix P, the vertex and edges of graph are obtained and finally graph $G(V, E)$, where V denotes the set of vertices, E is the set of edges. The graph $G(V, E)$ can be represented in form of incidence matrix denoted as I_M, which is defined such that

$$I_{Mij} = \begin{cases} -1 & if \quad the \quad edge \quad e_j \quad leaves \quad vertex \quad v_i \\ 1 & if \quad the \quad edge \quad e_j \quad enters \quad vertex \quad v_i \\ 0 & otherwise \end{cases} \tag{18}$$

This matrix has dimensions $n_v \times n_e$, where n_v corresponds with the total number of vertices and n_e denotes the total number of edge. The graph obtained from the set of minimum coupled ARRs is further divided into various subgraphs or subsystems using partition algorithm defined below in section 3.4.

3.4 Block 4: Partition of ARR graph

The first step to implement partition algorithm [16] is to find the strongly connected vertices. A strongly connected vertex is the one which has maximum number of edges. This vertex will be the basis for forming the first subsystem being the core of the first subsystem. Second subsystem is formed by second strongly connected vertex. The important condition is that no two subgraphs can have same vertex but same edge can be shared. Together all the subgraphs must contain all the vertices of a system, that is, no vertex must be left. Every vertex must be part of any one subsystem and the subsystem should be least connected. The maximum weight ω for each vertex is equal to number of edges each vertex have. The heaviest vertex is the vertex which has maximum number of edges, the heaviest vertex forms the first subgraph and the centre of the first subgraph $G_1$is defined. Those vertices which are connected to this heaviest vertex are included in G_1 or in the first subgraph. The set of non-selected vertices are defined as $V_r = \{v_j \in V: v_j \notin V_1\}$. The above procedure is repeated for all vertices $v_j \in V_r$(now j = $\{1, 2, \ldots, n_v\}$) until V_r is empty. The subgraph of higher connectivity is highlighted by the above method. The subgraphs which have only one vertex are merged to the closest subgraph and thus a set of subgraphs $G_i(V_i, E_i)$, for $i=1, 2 \ldots k$, is obtained, algorithm is defined below

1:$I_M \leftarrow$ System topology

% Start up

2:$G(V, E) \leftarrow I_M$

3: **for** $j = 1$ to ϕ **do**

4: Compute ω_j

5: **end for**
% Partitioning
6: $V_r \leftarrow V_i = 1$
7: **repeat**
8: Find $v \in V_r$ with maximum ω
9: $V_i \leftarrow v$ and all its neighbour vertices
10: $V_r = V - \{ \bigcup_{h=1}^{i} V_h \}$
11: $i = i + 1$
12: **until** $V_r = \emptyset$

After obtaining various subsystems, fault signature matrices or local diagnosers are obtained by the method defined below in section 3.5.

3.5 Block 5: Fault signature matrices of local diagnosers

The subsystem or sets of minimum coupled ARRs are represented as

$$R = \{r_i | r_i = \Psi_i(y_k, u_k, \theta_k), i = 1, \ldots, n_r\} \quad (19)$$

ψ_i is the mathematical expression for ARR sets and n_r is the ARR sets number obtained. Fault diagnosis is done by identifying the set of consistent ARR sets

$$R_0 = \{r_i | r_i = \Psi_i(y_k, u_k, \theta_k), = 0, i = 1, \ldots, n_r\} \quad (20)$$

and inconsistent ARR sets

$$R_1 = \{r_i | r_i = \Psi_i(y_k, u_k, \theta_k), \neq 0, i = 1, \ldots, n_r\} \quad (21)$$

when some inconsistency in (19) at time instant k is detected, the process of fault isolation starts by obtaining the observed fault signature, where each single fault signal indicator is defined as follows:

$$\varphi_i(k) = \begin{cases} 0 & if \quad r_i(k) \in R_0, \\ 1 & if \quad r_i(k) \in R_1 \end{cases} \quad (22)$$

Fault isolation is the binary relation between the considered fault hypothesis set $\{f_1(k), f_2(k) \ldots \ldots f_{nf}(k)\}$ and the fault signal indicators $\varphi_i(k)$, stored in the *Fault Signature Matrix F*. Till Block 5 the methodology is in offline mode. Block 6 or distributed fault diagnosis works in online mode and the method is described below in section 3.6.

3.6 Block 6: Distributed fault diagnosis

A distributed system compose of a set of agents $\mathcal{A} = \{A_1, \ldots, A_n\}$and each agent includes a diagnostic system. A local diagnosis in such distributed systems is a diagnosis that is determined by violated residuals in one agent, while a global diagnosis is determined by all violated residuals in all agents. The violated residuals are generated from responded diagnostic tests. A set of local diagnoses in agent $\mathcal{A}_i$ is denoted by $\mathcal{D}^{A_i}$ and a set of global diagnoses is denoted by $\mathcal{D}$. The set of global diagnoses can be computed from the sets of violated residuals in all agents. The global diagnoses can also be computed by extracting the diagnoses in the set resulting from a merge of all sets of local diagnoses. This idea of computing the global diagnoses from the sets of local diagnoses will be used in the algorithm which computes the set of diagnoses in each agent. The diagnosis in one agent should only include the variables that the agent uses. To be able to decide which variable that an agent uses, the variables are divided into internal variables $\mathcal{U} \subseteq \mathcal{C}$ and shared variables $\mathcal{V} \subseteq \mathcal{C}$ (where $\mathcal{C}$ is set of entire variables including both internal as well as shared variable). A internal variable is only used by one agent, while a shared variable is used by two or more agents. The set of internal variables is therefore further divided into different sets belonging to different agents, where the set $\mathcal{U}^{A_i} \subseteq \mathcal{U}$ is used by agent $\mathcal{A}_i$.

The outline of the algorithm is that each agent first computes its set of local diagnoses, and then transmits, to all other agents, the local diagnoses that include variables used by other agents. After this, each agent merges the received sets with its own set of local diagnoses which results in the set of diagnoses.

The transmitting algorithm is described in detail in algorithm 1, and the receiving and merging algorithm is described in algorithm 2. Finally, the main algorithm is described in algorithm 3.

Algorithm 1– Transmit $(\mathcal{D}^{Ai})$. Transmit subsets of local diagnoses.

Input: A set of local diagnoses $\mathcal{D}^{Ai}$
Output: Set of tuples TX^{Ai} representing the diagnoses in $\mathcal{D}^{Ai}$ that should be transmitted.
1: $\mathcal{D}^{TX} := \{\mathcal{D} \in \mathcal{D}^{Ai} : \mathcal{D} \cap \mathcal{V} \neq \emptyset\}$
2: $\mathcal{TX}^{Ai} := \emptyset$
3: **for each** $\mathcal{D}_i \in \mathcal{D}^{TX}$ **do**
4: $\mathcal{TX}^{Ai} := \mathcal{TX}^{Ai} \cup \{\mathcal{D}_i\}$
5:**end for**

It is seen that a local diagnosis in one agent is of interest for the other agents if it includes variables used by some other agent. The local diagnoses including such variables should be transmitted to the other agents. Before the diagnoses are transmitted, the internal variables can be removed since these are not used by any other agent.

Algorithm 1 performs the steps described above. Row 1decides which local diagnoses that include variables used by other agents, resulting in the set $\mathcal{D}^{TX}$. Rows 2–7 construct a tuple for each local diagnosis $\mathcal{D} = \mathcal{U} \cup \mathcal{V}$ in the set $\mathcal{D}^{TX}$, where the set $\mathcal{U}$ is the internal variable, $\mathcal{V}$ is the shared variable included in the local diagnosis. Each tuple in the set $\mathcal{TX}^{Ai}$ includes a set consisting of $\mathcal{V}$ which is taken directly from the local diagnosis which depends on any of their moved local variables in $\mathcal{U}$.

Algorithm 2 – Receive ($\{\mathcal{TX}^{A1}, \ldots, \mathcal{TX}^{An}\}$). Compute the diagnoses in agent $\mathcal{A}_i$.

Input: For each agent $\mathcal{A}_j$ except $\mathcal{A}_i$, a received set $\mathcal{TX}^{Aj}$ resulting from the evaluation of Transmit($\cdot$).
The set of local diagnoses $\mathcal{D}^{Ai}$
Output: The set of diagnoses D_c^{Ai}.

1: **for each** $\mathcal{A}_j$ except $\mathcal{A}_i$ **do**
2: $\mathcal{RX}^{Aj} := \emptyset$
3: **for each** $\mathcal{D}_i \in \mathcal{TX}^{Aj}$ **do**
4: $\mathcal{RX}^{Aj} := \mathcal{RX}^{Aj} \cup \{D_i\}$
5: **end for**
6: **end for**
7: $\mathcal{RX}^{Ai} := \{\mathcal{D}. \mathcal{D} \in \mathcal{D}^{Ai}\}$
8: D_c^{Ai} := CrossProductTuple($\{\mathcal{RX}^{A1}, \ldots, \mathcal{RX}^{An}\}$)

Algorithm 2 performs the steps described above. Rows 1–7 transform each set of transmitted tuples, such as $\mathcal{TA}^{Ai}$ transmitted by agent $\mathcal{A}j$, into sets of received tuples, such as $\mathcal{RA}^{Aj}$. The received tuple $\{(\mathcal{V})\}$ is thereafter constructed and stored in the set of tuples $\mathcal{RA}^{Aj}$. Row 8 transforms the local diagnoses in agent A_i to the same format as the received tuples. In rows 8 in the algorithm, the sets $\mathcal{RA}^{A1}, \ldots, \mathcal{RA}^{An}$ are merged. The function CrossProductTuple for a set $\mathcal{M}$ of sets of tuples performs a normal cross product with respect to the sets included in the tuples.

Algorithm 3 – *Main*. Computation of the set of diagnoses in each agent.
Input: The set of violated residuals Π^{Ai} in each agent $\mathcal{A}_i$.
Output: The set diagnoses D_c^{Ai} in each agent $\mathcal{A}_i$.
1: **for each** agent $\mathcal{A}_i$ **do** $\mathcal{D}^{Ai}$:=Diagnoses (Π^{Ai}) in $\mathcal{A}_i$
2: **for each** Agent A_i **do**
3: compute $\mathcal{TA}^{Ai}$:= Transmit($\mathcal{D}^{Ai}$) in agent $\mathcal{A}_i$
4: broadcast $\mathcal{TA}^{Ai}$ on the network
5: **end for**
6: **for each** Agent $\mathcal{A}_i$ **do**
7: receive $\mathcal{TA}^{Aj}$ in $\mathcal{A}_i$ from all agents $\mathcal{A}_j$ except $\mathcal{A}_i$
8: compute D_c^{Ai} :=Receive $(\{\mathcal{TA}^{A1}, \ldots, \mathcal{TA}^{An}\} \setminus \mathcal{TA}^{Ai})$ in $\mathcal{A}_i$
9: **end for**

Algorithm 3 used to compute the set diagnoses in each agent. For each agent, the algorithm computes the set of minimal local diagnoses $\mathcal{D}^{Ai}$ in row 1 using a minimal hitting set algorithm. The algorithm then evaluate rows 2–5 which include calls to Algorithm 1, where each call has a set of local diagnoses as input and gives a set of transmitted diagnoses $\mathcal{TA}^{Ai}$ as output. Finally, rows 6–9 are evaluated which includes calls to Algorithm 2, where each call has the set of transmitted sets $\mathcal{TA}^{A1}, \ldots, \mathcal{TA}^{An}$ except $\mathcal{TA}^{Ai}$ as input. After that Algorithm 3 has been evaluated, each agent has a unique set of minimal condensed diagnoses D_c^{Ai}

4. Implementation of Algorithm

The proposed algorithm is implemented on real Barcelona water network shown in Figure 2 below. The Barcelona DWN, managed by Aguas de Barcelona, S.A. (AGBAR), supplies drinking water to Barcelona city and also to the metropolitan area. The sources of water are the Ter and Llobregat rivers, which are regulated at their head by some dams with total capacity of 600 cubic hectometres. Currently, there are four drinking water treatment plants (WTP): the Abrera and Sant Joan Despı′ plants, which extract water from the Llobregat river, the Cardedeu plant, which extracts water from Ter river, and the Beso`s plant, which treats the underground flows from the aquifer of the Beso`s river. Several underground sources (wells) are also there that can provide water through pumping stations. Different water sources currently provide a flow of around 7 m^3/s. The water flow from each source is limited and with different water prices depending on legal extraction canons and water treatments. The Barcelona DWN consists of two layers. The upper layer links the water treatment plants with the reservoirs distributed all over the city and named as transport network. The lower layer is sectorised in sub networks and named distribution network. Each sub network generally links a reservoir with each consumer. Focus of this paper is on the transport network. Each sub network of the distribution network is modelled as a demand sector. The demand of each sector can be predicted by using a time-series model and is characterised by a demand pattern. The control system of the transport network is also consists of two layers. The upper layer establishes the set-points of the regulatory controllers at the lower layer and also is in charge of the global control of the network. The supervisory layer controller is of MPC type and regulatory controllers are of proportional–integral–derivative (PID) type. Regulatory controllers hide the network non-linear behaviour to the supervisory controller and thus allow the MPC supervisory controller to use a control oriented linear model.

This modelling methodology has been applied to the Barcelona DWN aggregate network in Figure 2. From this figure, it can be seen that the network comprises 17 tanks (state variables), 61 actuators (26 pumping stations and 35 valves), 11 nodes and 25 main sectors of water demand (model disturbances). The model has been simulated and compared against real behaviour assessing its validity. The detailed information about physical parameters and other system values are reported in [15].

The following equation or set of all ARRs describes the aggregated Barcelona water network used for real case study shown in figure 2.

Tank 1: $x_1(k+1) = x_1(k) + \Delta_t(u_{19}(k) - d_8(k))$ ***C1***

Tank 2: $x_2(k+1) = x_2(k) + \Delta_t(u_{43}(k) + u_{15}(k) - d_{16}(k))$ ***C2***

Tank 3: $x_3(k+1) = x_3(k) + \Delta_t(u_{42}(k) + u_{35}(k) + u_{52}(k) + u_9(k) + u_{13}(k) + u_{23}(k) - u_6(k) - u_{15}(k) - u_{16}(k) - u_{47}(k) - d_{12}(k))$ ***C3***

Tank 4: $x_4(k+1) = x_4(k) + \Delta_t(u_{32}(k) + u_{21}(k) + u_{10}(k) - u_{24}(k))$ ***C4***

Tank 5: $x_5(k+1) = x_5(k) + \Delta_t(u_2(k) + u_{20}(k) - d_{11}(k))$ ***C5***

Tank 6: $x_6(k+1) = x_6(k) + \Delta_t(u_{16}(k) - u_{53}(k) - d_{20}(k))$ ***C6***

Tank 7: $x_7(k+1) = x_7(k) + \Delta_t(u_{47}(k) + u_5(k) + u_{24}(k) + u_8(k) - u_{13}(k) - u_{14}(k) - u_{59}(k) - u_{61}(k) - u_{57}(k) - d_4(k))$ ***C7***

Tank 8: $x_8(k+1) = x_8(k) + \Delta_t(u_1(k) + u_7(k) + u_{12}(k) - u_{50}(k) - d_9(k))$ ***C8***

Tank 9: $x_9(k+1) = x_9(k) + \Delta_t(u_{39}(k) + u_4(k) + u_{50}(k) + u_{17}(k) - u_{12}(k) - u_7(k) - u_{46}(k) - d_{25}(k))$ ***C9***

Tank 10: $x_{10}(k+1) = x_{10}(k) + \Delta_t(u_{40}(k) + u_{56}(k) + u_{51}(k) - u_{18}(k) - d_{21}(k))$ ***C10***

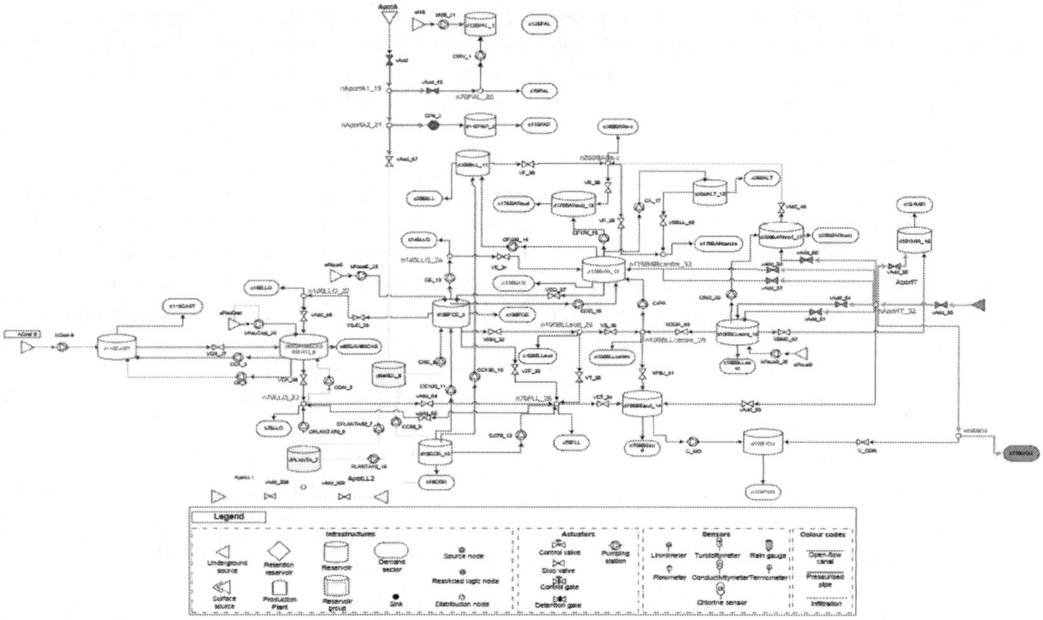

Figure 2. Barcelona water network

Deposit 11: $x_{11}(k+1) = x_{11}(k) + \Delta_t(u_{28}(k) + u_{29}(k) - u_{26}(k) - u_{21}(k) - u_{22}(k))$ **C11**

Deposit 12: $x_{12}(k+1) = x_{12}(k) + \Delta_t(u_{26}(k) - u_{11}(k) - u_{9}(k) - u_{8}(k) - u_{10}(k)) - d_{7}(k))$ **C12**

Deposit 13: $x_{13}(k+1) = x_{13}(k) + \Delta_t(u_{6}(k) - u_{45}(k) - d_{17}(k))$ **C13**

Deposit 14: $x_{14}(k+1) = x_{14}(k) + \Delta_t(u_{41}(k) + u_{44}(k) - d_{6}(k))$ **C14**

Deposit 15: $x_{15}(k+1) = x_{15}(k) + \Delta_t(u_{37}(k) + u_{34}(k) + u_{3}(k) - u_{44}(k) - u_{25}(k) - u_{48}(k) - d_{2}(k))$ **C15**

Deposit 16: $x_{16}(k+1) = x_{16}(k) + \Delta_t(u_{18}(k) + u_{49}(k) - d_{10}(k))$ **C16**

Deposit 17: $x_{17}(k+1) = x_{17}(k) + \Delta_t(u_{54}(k) + u_{25}(k) + u_{36}(k) - d_{18}(k))$ **C17**

Node 1: $u_{27}(k) = u_{31}(k) + u_{19}(k) + u_{32}(k)$ **C18**

Node 2: $u_{31}(k) = u_{20}(k) + d_{24}(k)$ **C19**

Node 3: $u_{59}(k) = u_{39}(k) + d_{5}(k)$ **C20**

Node 4: $u_{33}(k) + u_{46}(k) + u_{22}(k) = u_{38}(k) + u_{17}(k) + d_{23}(k)$ **C21**

Node 5: $u_{14}(k) = u_{52}(k) + d_{14}(k)$ **C22**

Node 6: $u_{57}(k) = u_{58}(k) + u_{60}(k) + d_{3}(k)$ **C23**

Node7: $u_{60}(k) + u_{61}(k) + u_{38}(k) + u_{11}(k) = u_{51}(k) + u_{33}(k) + d_{22}(k)$ **C24**

Node 8: $u_{53}(k) = u_{54}(k) + u_{43}(k) + u_{55}(k) + d_{19}(k)$ **C25**

Node 9: $u_{45}(k) + u_{55}(k) = d_{15}(k)$ **C26**

Node 10: $u_{48}(k) + u_{58}(k) = u_{23}(k) + u_{56}(k) + d_{1}(k)$ **C27**

Node 11: $u_{30}(k) = u_{41}(k) + u_{36}(k) + u_{35}(k) + u_{42}(k) + u_{34}(k) + u_{37}(k) + u_{40}(k) + u_{49}(k) + d_{13}(k)$ **C28**

The proposed distributed fault diagnosis algorithm starts from discrete-time space state model

$$x(k+1) = A\,x(k) + B_u\,u(k) + B_p d(k)$$
$$y(k) = C\,x(k) \qquad (23)$$

where $A \in R^{nxn}$, $B_u \in R^{nxm}$, $C \in R^{rxn}$ are the state space matrices and $B_p \in$ is R^{nxp} the disturbance known, $x \in R^{n}$ is the state vector corresponding to the volume of deposits, $u \in R^{m}$ is the vector of input variables, $d \in R^{p}$ corresponds the vector of known disturbances, in this case are the water demands, $y \in R^{r}$ is the vector of outputs.

IOP Conf. Series: Journal of Physics: Conf. Series **783** (2017) 012005 doi:10.1088/1742-6596/783/1/012005

After obtaining model from the sets of equations or set of all ARRs, sensor placement algorithm is applied on the given model. W matrix represents sensor faults while V matrix represents process faults. W and V matrix is obtained using equations C1 upto C28 described above. W matrix has constraints (*C1, C2...etc*) or ARRs in rows and variables in column (x_1, u_1, d_1 ...etc), total W matrix has 28 rows and 98 columns. If a variable present in an equation or row, that particular column has 1 otherwise the column contains zero, similarly V matrix has 28 rows and 4 columns. V matrix has constraints (*C1, C2...etc*) or ARRs in rows and process or main supply variables in columns (u_{27}, u_{70} (u_{28}, u_{29}), u_{30}).
Sensor placement algorithm is applied on the W and V matrix (Barcelona water network) described above, the algorithm chosen 4 ARRs C4, C11, C18, C28 as shown below in Table1 from the set of all ARRs. The chosen system or set of ARRs has minimum coupling. The system obtained (shown in Table 1) is first converted into graph shown in Figure 3 and this graph is partitioned into different subsystem or subgraph using partition algorithm. The system is divided into 3 subsystems. The first subsystem consists of ARRs C4, C11, the second system consists of ARR C18 and the third subsystem consists of ARR C28. The details of the 3 subsystems are shown in Fig 2, Fig 3 and Table 2.

Table 1. Set of ARRs (Barcelona water network)

	X 4	X 11	U 10	U 19	U 21	U 22	U 24	U 26	U 27	U 30	U 31	U 32	U 34	U 35	U 36	U 37	U 40	U 41	U 42	U 49	U 70	d 13
C4	1	0	1	0	1	0	1	0	0	0	0	1	0	0	0	0	0	0	0	0	0	0
C11	0	1	0	0	1	1	0	1	0	0	0	0	0	0	0	0	0	0	0	0	1	0
C18	0	0	0	1	0	0	0	0	1	0	1	1	0	0	0	0	0	0	0	0	0	0
C28	0	0	0	0	0	0	0	0	0	1	0	0	1	1	1	1	1	1	1	1	0	1

From the 3 subsystems, 3 fault signature matrices are created shown in Table 3, 4, 5. These fault signature matrices are represented by 3 agents A_1, A_2, A_3. Agents A_1, A_2, A_3 communicate with each other to detect and isolate a fault or faults. A scenario is explained below

Till now all steps performed are offline, the fault detection is done online. A scenario is taken that there are 3 subsystems (*3, 4, 5*) represented by *3* agents A_3, A_4, A_5. Suppose in the subsystem *3*, *MSO 10* and variable *51* is faulty or violated and similarly in subsystem 4, *MSO* 24 and variable *51* is faulty or violated and in subsystem 5, no *MSO* is faulty or violated. So how the 3 agents will communicate with each other to diagnose and isolate the fault using concept of distributed fault diagnosis (local diagnosis with minimum global diagnosis) following traditional

FDI approach, this entire process is explained below in point wise manner. Everything happening together and parallel shown in Fig 4.4 to 4.8

1. First all Agents A_3, A_4 and A_5 do local diagnosis ($\mathcal{D}^{Ai}$) according to algorithm *1*. After doing local diagnosis agent A_3 finds out that *MSO 10* is faulty and the single fault can occur in shared variable *51* or internal variable *56*. Similarly A_4 finds out that *MSO 24* is faulty and single fault can occur in shared variables *33 38* or *51* and A_5 finds out that no *MSO* is faulty after local diagnoses. A_3 sends a communication message ($\mathcal{TA}^{Ai}$) containing shared variable number

51 as data to agent A_4 and A_5. Similarly at the same time A_4 will send a message containing shared variable numbers *33 38 51* to A_3 and A_5 and A_5 will send a message containing *00* indicating no fault to A_3 and A_4. Entire process is explained with help of arrow diagram below.

1. for Agent A_3 **do**
Checks residual value of each $\mathcal{MSO}$ of
Agent A_3 whether 0 or not according to (3.11)
end for

2. A_3 $(\mathcal{TA}^{A1} = 51,56)$ $A_4, A_5 \rightarrow$

3. for Agent A_4 **do**
Checks residual value of each $\mathcal{MSO}$ of
Agent A_4 whether 0 or not according to (3.11)
end for

4. A_4 $(\mathcal{TA}^{A2} = 33\ 38\ 51)$ $A_3, A_5 \rightarrow$

5. for Agent A_5 **do**
Checks residual value of each $\mathcal{MSO}$ of
Agent A_5 whether 0 or not according to (3.11)
end for

6. A_5 $(\mathcal{TA}^{A3} = 00)$ $A_3, A_4 \rightarrow$

2. In reply of message from A_3, A_4 and A_5 will first perform local diagnoses $(\mathcal{D}^{Ai})$ according to algorithm *2*. A_3 will receive message $(\mathcal{RA}^{Aj})$ containing data *00* from A_5, indicating that there is no error or fault and will receive message containing data *(33, 38, 51)* from A_4. A_4 will receive message containing data *00* from A_5 and will receive message containing data *51* from A_3.

3. After each agent receives the information, on the basis of this information the agents A_3, A_4 and A_5 will detect and isolate the variable in which the single fault occurred(D_c^{Ai}) according to algorithm *3*. In this scenario the agent A_3 detect that single fault has occur in shared variable *51* and not in internal variable *56* and agent A_4 detect that single fault has occur in shared variable *51* and not in shared variables *33 38*.

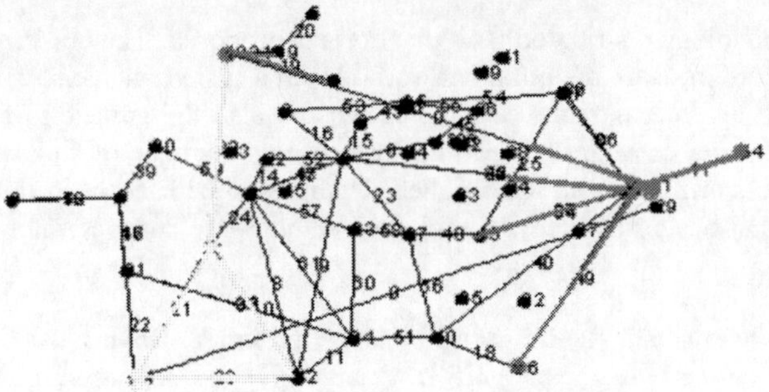

Figure 3. Showing three subsystems obtained

Table 2. Three subsystems obtained

Subsystem (ARRs)	Color
C4, C11	Yellow
C18	Red
C28	Blue

Table 3. Local fault signature matrix of subsystem 1diagnosed by agent A_1

	X 4	X 11	U 10	U 19	U 21	U 22	U 24	U 26	U 27	U 30	U 31	U 32	U 34	U 35	U 36	U 37	U 40	U 41	U 42	U 49	U 70	d 13
C4	1	0	1	0	1	0	1	0	0	0	0	1	0	0	0	0	0	0	0	0	0	0
C11	0	1	0	0	1	1	0	1	0	0	0	0	0	0	0	0	0	0	0	0	1	0

Table 4. Local fault signature matrix of subsystem 2 diagnosed by agent A_2

	X 4	X 11	U 10	U 19	U 21	U 22	U 24	U 26	U 27	U 30	U 31	U 32	U 34	U 35	U 36	U 37	U 40	U 41	U 42	U 49	U 70	d 13
C18	0	0	0	1	0	0	0	0	1	0	1	1	0	0	0	0	0	0	0	0	0	0

Table 5. Local fault signature matrix of subsystem 2 diagnosed by agent A_3

	X 4	X 11	U 10	U 19	U 21	U 22	U 24	U 26	U 27	U 30	U 31	U 32	U 34	U 35	U 36	U 37	U 40	U 41	U 42	U 49	U 70	d 13
C28	0	0	0	0	0	0	0	0	0	1	0	0	1	1	1	1	1	1	1	1	0	1

5. Conclusion

In this paper, a methodology distributed fault diagnosis is proposed. The objective of this methodology is to place the sensor to measure the unknown variables in a system in such a manner that detection of sensor faults as well as process faults become easier and also the partition of a system into various subsystems become more easy and simplified by reducing number of links within a system. This algorithm also reduces or minimized the number of sensors to be used or install thus reducing overall cost. The algorithm has been satisfactorily implemented on Barcelona water network.

References

[1] V. Vekatasubramanian, R. Rengaswamy, K. Yin, and S. Kavuri, 2003. A review of process fault detection and diagnosis part I: Quantitative model-based methods. *Computer and Chemical Engineering*, 27:293–311.

[2] Vikas Gupta, Vicenç Puig, 2016. "Decentralized Fault Diagnosis using Analytical Redundancy Relations: Application to a Water Distribution Network," in proceedings of 15th *European Control Conference ECC*, Aalborg, Denmark.

[3] Vikas Gupta, Vicenç Puig, 2016. "Distributed Fault Diagnosis using Analytical Redundancy Relations: Application to a Water Distribution Network," in proceedings of 3rd *International Conference on Control and Fault-Tolerant Systems*, Barcelona, Spain.

[4] Vikas Gupta, Vicenç Puig, 2016. "Sensor placement in Distributed Network," in proceedings of 3rd *International Conference on Control and Fault-Tolerant Systems*, Barcelona, Spain.

[5] H. Khorasgani, G. Biswas and D. Jung, 2015. "Minimal Structurally Overdetermined Sets Selection for Distributed Fault Detection," Proceedings of the *26th International Workshop on Principles of Diagnosis*, 75-82.

[6] R. Raghuraj, M. Bhushan, and R. Rengaswamy, 1999. "Locating sensors in complex chemical plants based on fault diagnostic observability criteria," *AIChE J.*,45(2):310–322.

[7] M. Basseville, A. Benveniste, G. Moustakides, and A. Rougee, 1987. "Optimal sensor location for detecting changes in dynamical behavior," *IEEE Transactions on Automatic Control*, 32(12):1067 – 1075.

[8] M. Bagajewicz, 2000. "Design and Upgrade of Process Plant Instrumentation," *Technomic Publishers, Lancaster, PA.*

[9] L. Travé-Massuyès, T. Escobet, and X. Olive, 2006. "Diagnosability analysis based on component supported analytical redundancy relations," *IEEE Transactions on Systems, Man, and Cybernetics-Part A*, 36(6):1146–1160.

[10] A. Rosich, 2011. "Sensor Placement for Fault Diagnosis based on Structural Models: Application to a Fuel Cell Stack System," PhD Thesis, *UPC.*

[11] A. Rosich, R. Sarrate, and F.Nejjari, 2009. "Optimal sensor placement for FDI using binary integer programming," *International Workshop on Principles of Diagnosis.*

[12] H. Khorasgani, D. E. Jung, G. Biswas, E. Frisk, and M.Krysander, 2014. "Off-line robust residual selection using sensitivity analysis," *International Workshop on Principles of Diagnosis (DX-14)*, Graz, Austria.

[13] M.Blanke, M. Kinnaert, J. Lunze, M.Staroswiecki, 2016. "Diagnosis and Fault-Tolerant Control," *Springer*, 3rd Edition.

[14] Laurence A.Wolsey, 1998. "Integer programming," volume 42, *Wiley, New York.*

[15] Fambrini, V, Ocampo-Martinez, C, 2009. 'Modelling and decentralized model predictive control of drinking water networks'. *Technical report IRI-TR-04-09, Institut de Robo`tica i Informa`tica Industrial (CSIC-UPC).*

[16] J. Biteus, EC. Ocampo-Martinez, S. Bovo, and V. Puig, 2011. "Partitioning approach oriented to the decentralised predictive control of large-scale systems," *Journal of Process Control*, 21(5):775–786.

[17] V. Puig, C. Ocampo-Martinez, 2015. "Decentralised fault diagnosis of large-scale systems: Application to water transport networks," *26th International Workshop on Principles of Diagnosis.*

Sensor fault detection and isolation over wireless sensor network based on hardware redundancy

Jingjing Hao and Michel Kinnaert

Department of Control Engineering and System Analysis, Universit libre de Bruxelles (ULB), CP 165/55, 50 Av. F.D. Roosevelt, Belgium

E-mail: Jingjing.Hao@ulb.ac.be, Michel.Kinnaert@ulb.ac.be

Abstract. In order to diagnose sensor faults with small magnitude in wireless sensor networks, distinguishability measures are defined to indicate the performance for fault detection and isolation (FDI) at each node. A systematic method is then proposed to determine the information to be exchanged between nodes to achieve FDI specifications while limiting the computation complexity and communication cost.

1. Introduction

Wireless sensor networks (WSNs) can provide new methods for information gathering for a variety of applications, like environment monitoring and human health monitoring. They are working based on the collaboration of a large number of sensor nodes which integrate the computation, communication and sensing capabilities. In order to ensure the network quality of service, the quality of the measurements has to be guaranteed. Distributed fault detection and isolation schemes are preferred to centralized solutions to diagnose faulty sensors in WSNs. Indeed the first approach avoids the need for a central node that collects information from every sensor node, and hence it limits complexity and energy cost while improving safety.

A taxonomy for classification of faults in sensor network is introduced in [1]. Two common fault classes correspond to dead sensors, namely sensors not reporting any value, on the one hand and sensor bias, on the other hand. The diagnosis of permanent faults for example caused by battery depletion in WSNs are considered in [2] and [3]. A spinning tree fault diagnosis protocol is proposed to identify the crash faults in [2] while considering the time efficiency and energy consumption. An external operator can be used to replace the faulty nodes based on the diagnostic information. In [3], a decentralized method is proposed to detect the permanent faults in the cluster heads which are the leaders of the grouped nodes. According to the energy and radio range, a cluster manager is chosen to be responsible to detect the fault in the cluster head by confirming with 'alive' message, to notify all the other cluster members and to perform the recovery by selecting a new cluster head.

Rather than hard faults like complete breakdown of a node, small and slowly developing malfunctions should also be taken care of. Distributed fault detection algorithms with voting scheme for sensor faults in wireless sensor networks have been proposed in [4, 5]. In these papers, all the sensor nodes in the sensor network measure the same quantity. The sensor readings are compared with their neighbor's to determine whether they are faulty. Time redundancy is used in [5] to tolerate transient faults in sensing and communication, which corresponds the errors caused by some temporary conditions like connectivity issues and will disappear automatically. Distributed fault tolerant methods are used for event region detection in WSNs. The problem is to distinguish the specific events of interest from abnormal measurements. To achieve this goal, spatial correlation between the measurements is exploited by a distributed fault-tolerant Bayesian recognition algorithm in [6] and a clustering approach in [7]. Model-based FDI methods have been developed for distributed fault diagnosis in [8] and in [9], in which the dynamic model of the monitored system is considered.

In order to diagnose an incipient sensor fault, a distributed FDI system in WSNs can be designed with the aid of a measure named distinguishability degree. Distinguishability measures have been used in sensor placement problem for fault diagnosis in [11] [12] [13]. Qualitative detectability and isolability have been defined with a structural method in [11] and with an analytical method in [12]. They are used to decide which sensors to include to achieve fault diagnosis. A quantitative distinguishability measure is defined in a stochastic framework in [14] and it is used as a constraint for a greedy optimization algorithm to find globally optimal sensor sets for diagnosis [13]. The considered cost function is the the number of the selected sensors.

In this paper, we propose a distributed method based on the parity space approach to generate residuals for sensor fault detection and isolation in a WSN. A distinguishability measure is defined in a deterministic framework and a diagnosibility degree is defined for indicating the performance for FDI at each node. Furthermore, a systematic method is presented to determine the information to be exchanged between different nodes in order to achieve, when possible, a given performance level for FDI while the computation complexity and communication cost are kept as small as possible.

The paper is organized as follows. The problem statement is given in section 2. In section 3, the proposed approach for distributed FDI system design in a deterministic framework is described. The distinguishability and diagnosibility degree at each node are defined to characterize the property of the distributed FDI system. The selecting criteria among the neighboring nodes to improve the FDI system property are deduced. Then the proposed approach for distributed FDI system design is described. Next, a simulation example is used to illustrate how to improve the property of fault detection and isolation at a given node with the aid of the neighboring node in section 4.

2. Problem statement

A sensor network of M nodes, like the one illustrated in figure 1, can be spatially placed over some region to monitor some target quantities of a system. This network can be described by a graph which is characterized by its adjacency matrix W, a $M \times M$ matrix with Boolean entries. The $(i, j)^{th}$ entry is equal to 1 if nodes i and j are connected, namely if they can communicate directly with each other. Otherwise, the entry is equal to 0. The connections between the nodes are called the edges. The neighborhood of a given node, say i, is denoted by $\mathcal{N}_i$ and it is defined as the set of nodes that are directly connected to node i. The number of neighbors of node i is called the degree of node i and is denoted by m_i. These nodes can be called one-hop neighbors, for example, in figure 1, node 2 and node 3 are the one-hop neighbors of node 1. The nodes which are not connected with node i but are connected with the neighboring nodes of node i are called proximity neighbors of node i, or two-hop neighbors. Node 1's proximity neighbors are node 4 and node 5. Likewise, the three-hop neighbors, are node 6 and node 7 [10]. The following measurement equation is considered at node i:

$$y_i(k) = C_i x(k) + f_i(k) \tag{1}$$

where $x(k) \in R^n$ is a vector that represents the target quantities, $y_i(k) \in R^{p_i}$ is the measurement vector at node i, C_i is the measurement matrix at node i that indicates which quantities are measured at node i and $f_i(k) \in R^{p_i}$ is the fault vector. In order to get a variety of measurements

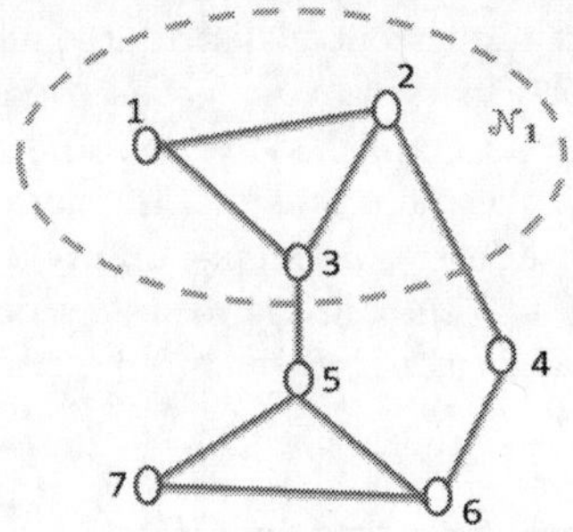

Figure 1. Topology of a sensor network with 7 sensor nodes

with a limited set of sensors, we assume that there are no redundant sensors at each node. The sensor nodes measure one or several of the target quantities, thus $p_i \leqslant n$ and C_i has full row rank. Only single faults are considered here.

A local sensing model can be defined at each node by considering the measurements available in the neighborhood, namely:

$$y_{i\_loc}(k) = C_{i\_loc}x(k) + f_{i\_loc}(k) \tag{2}$$

where $y_{i\_loc}(k) \in R^{p_{i\_loc}}$, $C_{i\_loc} \in R^{p_{i\_loc} \times n}$ and $f_{i\_loc}(k) \in R^{p_{i\_loc}}$ are the local measurement vector, local measurement matrix and local fault vector over the network respectively. The following notations will be needed below. $\mathcal{N}_i = \{i_1, \cdots, i_j \cdots, i_{m_i}\}$ indicates the neighborhood of node i. $y_{i\_loc}(k) = [y_{i_1}^T(k), \cdots, y_{i_j}^T(k), \cdots, y_{i_{m_i}}^T(k)]^T$, $C_{i\_loc} = [C_{i_1}^T, \cdots, C_{i_j}^T, \cdots, C_{i_{m_i}}^T]^T$, $f_{i\_loc}(k) = [f_{i_1}^T(k), \cdots, f_{i_j}^T(k), \cdots, f_{i_{m_i}}^T(k)]^T$ and $p_{i\_loc} = \sum_{i_j \in \mathcal{N}_i} p_{i_j}$. Such notations are illustrated in example 1.

Example 1. *A sensor network made of 7 nodes is considered as an example. The topology of the network is illustrated in figure 1 and the adjacency matrix of this network is thus* $W = \begin{bmatrix} 1 & 1 & 1 & 0 & 0 & 0 & 0 \\ 1 & 1 & 1 & 1 & 0 & 0 & 0 \\ 1 & 1 & 1 & 0 & 1 & 0 & 0 \\ 0 & 1 & 0 & 1 & 0 & 1 & 0 \\ 0 & 0 & 1 & 0 & 1 & 1 & 1 \\ 0 & 0 & 0 & 1 & 1 & 1 & 1 \\ 0 & 0 & 0 & 0 & 1 & 1 & 1 \end{bmatrix}$.

The network measures three different quantities which are represented by $x(k) \in R^3$. *The measurement equation at each node can be described as:*

$$y_i(k) = C_i x(k) + f_i(k) \tag{3}$$

where $C_1 = \begin{bmatrix} 1 & 0 & 0 \\ 0 & 1 & 0 \\ 0 & 0 & 1 \end{bmatrix}$, $C_2 = \begin{bmatrix} 1 & 0 & 0 \\ 0 & 0 & 1 \end{bmatrix}$, $C_3 = \begin{bmatrix} 0 & 1 & 0 \\ 0 & 0 & 1 \end{bmatrix}$, $C_4 = \begin{bmatrix} 1 & 0 & 0 \\ 0 & 1 & 0 \end{bmatrix}$, $C_5 = \begin{bmatrix} 1 & 0 & 0 \\ 0 & 1 & 0 \\ 0 & 0 & 1 \end{bmatrix}$, $C_6 = \begin{bmatrix} 1 & 0 & 0 \\ 0 & 0 & 1 \end{bmatrix}$, $C_7 = \begin{bmatrix} 1 & 0 & 0 \\ 0 & 1 & 0 \\ 0 & 0 & 1 \end{bmatrix}$.

The local sensing model at node 1 is generated based on the adjacency matrix:

$$y_{1\_loc}(k) = C_{1\_loc}x(k) + f_{1\_loc}(k)$$

where $y_{1\_loc}(k) = [y_1^T(k), y_2^T(k), y_3^T(k)]^T$, $C_{1\_loc}(k) = [C_1^T, C_2^T, C_3^T,]^T$, $f_{1\_loc}(k) = [f_1^T(k), f_2^T(k), f_3^T(k)]^T$.

Our objective is to design and implement a distributed monitoring algorithm in order to detect and isolate sensor faults for each sensor node based on the local sensing model. If the FDI system at each node can not accomplish detection and isolation, it can be extended to include information exchange with the other nodes in the network. The process of selecting an extra node is performed with the aid of a distinguishability measure of sensors at each node. The distinguishability is presented and the criteria for selecting a suitable node are analyzed in the next section. Then an on-line algorithm is introduced for the distributed deterministic fault diagnosis system.

3. Distributed sensor fault detection and isolation

The aim of a fault detection and isolation system is to generate fault indicators (also called residuals) and to determine the most likely faulty component through the analysis of the fault signature on these residuals. Since the property of the residual generator can reflect the relation between the residual and faults, detectability and isolability can be defined based on residual generator according to [15] notably. A distinguishability measure which includes detectability and isolability can be used to check whether the residuals can fulfill the FDI tasks. A natural idea for fault detection and isolation which has been investigated for a long time is to design residuals with directional responses namely residuals constrained in a fixed and different direction for each fault vector [16], [17]. Considering the parity space method in the deterministic situation, a fault can be detected by comparing the norm of the residual to a preset threshold, and the fault can be isolated by comparing the orientation of the residual to that of the potential fault directions [18]. These notions are applied to the local model (2), next the performance of the local FDI system is analyzed.

3.1. Residual Generation

A residual can be generated based on the local sensing model (2) with the parity space approach. The condition of existence for physical redundancy requires the existence of a non zero left null space of the measurement matrix, i.e. $rank(C_{i\_loc}) < p_{i\_loc}$. Here we assume it is true that a non zero left null space of the local measurement matrix exists. A basis of the left null space of the measurement matrix at each node can be derived according to the definition: $\Omega_{i\_loc}C_{i\_loc} = 0$. Multiplying the model equation (2) on the left by Ω_{i_loc} yields the residual $r_{i\_loc}(k)$.

$$r_{i\_loc}(k) = \Omega_{i\_loc}y_{i\_loc}(k) \quad (4)$$
$$= \Omega_{i\_loc}f_{i\_loc}(k) \quad (5)$$

The computational form (4) corresponds to the implementation step and it is used to generate the residual on line. The internal form (5) expresses the residual as a function of the fault vector. So in the following, the internal form is used to analyze how the faults affect the residual. Then based on the analysis, the criteria for selecting a suitable neighboring node when necessary are deduced based on detectability and isolability conditions.

3.2. Distinguishability measure in deterministic context based on parity space

The directions of the residual vector upon occurrence of fault l can be defined with the lth column of $\Omega_{i\_loc} = [\omega^1_{i\_loc}, \omega^2_{i\_loc}, ..., \omega^{m_i}_{i\_loc}]$, where $\omega^l_{i\_loc}$ is the lth column of matrix Ω_{i_loc}. Define the set $F_{i\_loc} = \{f_{i1}, f_{i2}, ..., f_{im_i}\}$ which includes all the sensor faults in the neighborhood of node i. The sensor faults are detectable if

$$\forall f_k \in F_{i\_loc}, \| \omega^l_{i\_loc} \| > 0$$

The lth sensor fault $f_l \in F_{i\_loc}$ is isolable if:

$$\forall f_q \in F_{i\_loc}\ with\ f_q \neq f_l, \left\{\theta_{lq} | \theta_{lq}\ is\ the\ angle\ between\ \omega^l_{i\_loc}\ and\ \omega^q_{i\_loc}\right\} \cap \{0, \pi\} = \varnothing$$

Table 1. Deterministic distinguishability at node i

D	NF	f_{i1}	f_{i2}	$\cdots$	f_{im_i}
f_{i1}	$1, if \parallel \omega^1_{i\_loc} \parallel > 0$	0	$\arccos\theta_{12}$		$\arccos\theta_{1m_i}$
f_{i2}	$1, if \parallel \omega^2_{i\_loc} \parallel > 0$	$\arccos\theta_{21}$	0		
$\vdots$	$\vdots$	$\vdots$	$\vdots$	$\ddots$	
f_{im_i}	$1, if \parallel \omega^{m_i}_{i\_loc} \parallel > 0$	$\arccos\theta_{m_i1}$	$\arccos\theta_{m_i2}$		0

Table 2. Distinguishability matrix at node 1

D	NF	f_{11}	f_{12}	f_{13}	f_{21}	f_{22}	f_{31}	f_{32}
f_{11}	1	0	1.5708	1.5708	3.1416	1.5708	1.5708	1.5708
f_{12}	1	1.5708	0	1.5708	1.5708	1.5708	3.1416	1.5708
f_{13}	1	1.5708	1.5708	0	1.5708	2.0944	1.5708	2.0944
f_{21}	1	3.1416	1.5708	1.5708	0	1.5708	1.5708	1.5708
f_{22}	1	1.5708	1.5708	2.0944	1.5708	0	1.5708	2.0944
f_{31}	1	1.5708	3.1416	1.5708	1.5708	1.5708	0	1.5708
f_{32}	1	1.5708	1.5708	2.0944	1.5708	2.0944	1.5708	0

So the detectability property is characterized by a binary value. The isolability can be calculated through the above defined angles: the isolability between fault f_l and fault f_q is related to the angles between their corresponding column vectors in Ω_{i_loc}. It can be determined from $\cos\theta_{lq} = \frac{\omega^{lT}_{i\_loc}\omega^{q}_{i\_loc}}{\|\omega^l_{i\_loc}\|\|\omega^q_{i\_loc}\|}$ [19]. Moreover, the closer the angles between the two fault directions is to 90°, the more easily the two faults can be isolated.

A distinguishability matrix can be derived to analyze the detectability and isolability property of the sensor faults in the neighborhood of a given node. This matrix is illustrated in table 1 for node i. The column NF represents the detectability of each sensor fault and any $(l,q)^{th}$ element from the other columns characterizes the isolability between fault l and q. Unlike in [14], the distinguishability matrix here is symmetric because of the considered definition in the deterministic framework. A notion of diagnosibility degree can be introduced on the basis of the distinguishability matrix. To this end, the indices of the sensors whose faults are detectable and isolable are stored in a set denoted DI_i (the so-called diagnosable set), the indices of the sensors whose faults can not be detected are stored in a set denoted ND_i and the indices of the ones whose faults can not be isolated are stored in NI_i.

Definition 1. *(Diagnosibility degree) The diagnosibility degree d_i at node i can be defined as $d_i = p_{d_i}/p_{i\_loc}$ where p_{d_i} is the cardinality of the set DI_i.*

If $d_i = 1$, all the sensor faults in the neighborhood can be detected and isolated. Otherwise, if $d_i < 1$, node i needs to exchange more information with other nodes to improve the detectability and isolability.

The following example illustrates the distinguishability matrix and the diagnosibility degree.

Example 2. *The distinguishability matrix at node 1 in the previous example can be generated* *based on the local sensing model (1) and the matrix Ω_{1_loc}. It is shown in table 2. f_{11}, f_{12} and f_{13} represent the three sensor faults at node 1, f_{21} and f_{22}, f_{31} and f_{32} represent the sensor faults at node 2 and at node 3 respectively.*

From the distinguishability matrix, we can deduce the set $ND_1 = \varnothing$ which means all the sensor faults are detectable at node 1. The non isolable set is $NI_1 = \{f_{11}, f_{21}, f_{12}, f_{31}\}$. Therefore, the diagnosibility degree is $d_1 = \frac{3}{7}$ at node 1.

3.3. Selection process based on detectability and isolability

If the diagnosibility degree is less than 1, like for node 1 in the above example, one can determine whether there is suitable information from the other nodes in the network that can improve diagnosibility. Since the nodes can communicate with their neighbors directly, the selection process starts with the neighboring nodes. The scheme is illustrated below and it is proven that the distinguishability can be improved by exchanging information with a suitable node.

Consider two parity space residuals at two neighboring nodes, say node i and node j:

$$\begin{aligned} r_{i\_loc}(k) &= \Omega_{i\_loc} y_{i\_loc}(k) = \Omega_{i\_loc} C_{i\_loc} x(k) + \Omega_{i\_loc} f_{i\_loc}(k) \\ r_{j\_loc}(k) &= \Omega_{j\_loc} y_{j\_loc}(k) = \Omega_{j\_loc} C_{j\_loc} x(k) + \Omega_{j\_loc} f_{j\_loc}(k) \end{aligned}$$

As defined in the previous section, the sets F_{i_loc} and F_{j_loc} represent all the sensor faults in the neighborhood of node i and node j. Since the local sensing model at each node is formed based on the measurements from their neighborhood, the neighboring nodes always have a set of common faults. Let F_{ij} denote the set of faults that are common to nodes i and j. The set of fault specific to node i (node j) are respectively denoted as $\bar{F}_{ij} = F_{i\_loc} \backslash F_{ij}$ and $\bar{F}_{ji} = F_{j\_loc} \backslash F_{ij}$. The respective fault vectors associated to these sets are defined as: $f_{ij} \in F_{ij}$, $\bar{f}_{ij} \in \bar{F}_{ij}$, $\bar{f}_{ji} \in \bar{F}_{ji}$. The corresponding measurement vectors are denoted y_{ij}, $\bar{y}_{ij}$ and $\bar{y}_{ji}$. The measurement matrix C_{i_loc} can also be divided into two parts: the part associated to the common faults C_{ij}, and the part associated to the non common faults $\bar{C}_{ij}$.

So the residual at node i, assuming that r_{j_loc} is transmitted to node i, can be written as:

$$\begin{aligned} r_{i\_new} &= \begin{bmatrix} r_{i\_loc}(k) \\ r_{j\_loc}(k) \end{bmatrix} \\ &= \begin{bmatrix} \Omega_{ij} & \bar{\Omega}_{ij} & 0 \\ \Omega_{ji} & 0 & \bar{\Omega}_{ji} \end{bmatrix} \begin{bmatrix} y_{ij} \\ \bar{y}_{ij} \\ \bar{y}_{ji} \end{bmatrix} \\ &= \begin{bmatrix} \Omega_{ij} & \bar{\Omega}_{ij} & 0 \\ \Omega_{ji} & 0 & \bar{\Omega}_{ji} \end{bmatrix} \begin{bmatrix} C_{ij} \\ \bar{C}_{ij} \\ \bar{C}_{ji} \end{bmatrix} x(k) + \begin{bmatrix} \Omega_{ij} & \bar{\Omega}_{ij} & 0 \\ \Omega_{ji} & 0 & \bar{\Omega}_{ji} \end{bmatrix} \begin{bmatrix} f_{ij} \\ \bar{f}_{ij} \\ \bar{f}_{ji} \end{bmatrix} \end{aligned} \tag{6}$$

where $\Omega_{ij} = [\omega_{ij}^1, \omega_{ij}^2, \cdots, \omega_{ij}^{n_i}]$ represents the part of Ω_{i_loc} corresponding to the common faults from F_{ij} and $\bar{\Omega}_{ij} = [\bar{\omega}_{ij}^1, \bar{\omega}_{ij}^2, \cdots, \bar{\omega}_{ij}^{p_{i\_loc} - n_i}]$ is the part of Ω_{i_loc} associated to the faults from $\bar{F}_{ij}$. The following result can be stated on the basis of (6).

Proposition 1. *A non detectable fault at node i, say fault l, which is common to nodes i and j ($f_{ij}^l \in F_{ij}$), can be made detectable by transmitting $r_{j,loc}$ to node i if and only if the fault f_{ij}^l can be detected at node j.*

Proof. According to the definition of detectability, fault f_l is detectable if $\| \omega^l \| > 0$ where ω^l is the lth column of Ω. Hence, if a sensor fault f_{ij}^l that belongs to the common set at node i can not be detected, it means $f_{ij}^l \in F_{ij}$, $f_{ij}^l \neq 0$ and $\| \omega_{ij}^l \| = 0$. Then the detectability can be improved if and only if $\| \omega_{ji}^l \| \neq 0$ because $\left\| \begin{matrix} \omega_{ij}^l \\ \omega_{ji}^l \end{matrix} \right\| \neq 0$ after combining $r_{i,loc}$ and $r_{j,loc}$. □

Proposition 2. *Consider a detectable fault $f^l_{ij} \in F_{ij}$ that cannot be isolated from $f^q_{ij} \in F_{ij}$. Then isolability of f^l_{ij} from f^q_{ij} can be achieved by transmitting $r_{j,loc}$ to node i if and only if the two faults can be isolated at node j.*

Proof. According to the definition of isolability, fault f_l is isolable from fault f_q if $-1 < \cos\theta_{lq} < 1$, where θ_{lq} is the angle between column vector ω^l and any other column ω^q of matrix Ω. If a sensor fault f^l_{ij} can be detected but can not be isolated from f^q_{ij} at node i, ω^l_{ij} and ω^q_{ij} in matrix Ω_{i_loc} are linearly dependent, namely $\omega^l_{ij} = \alpha\omega^q_{ij}$. The isolability between fault f^l_{ij} and f^q_{ij} can be improved if and only if the two faults can be isolated at node j, that is, $\nexists\gamma$ *s.t.* $\omega^l_{ji} = \gamma\omega^q_{ji}$. Thus the angle between the two columns is changed such that $-1 < cos\left\langle \begin{bmatrix} \omega^l_{ij} \\ \omega^l_{ji} \end{bmatrix}, \begin{bmatrix} \omega^q_{ij} \\ \omega^q_{ji} \end{bmatrix} \right\rangle < 1$. It indicates the isolability can be ensured by adding the residual information from node j. □

Proposition 3. *Consider a detectable fault $f^l_{ij} \in F_{ij}$ that cannot be isolated from $\bar{f}^q_{ij} \in \bar{F}_{ij}$ at node i, then fault isolation can be achieved by transmitting $r_{j,loc}$ to node i if and only if f^l_{ij} is detectable at node j.*

Proof. In the considered case $\omega^l_{ij} = \alpha\bar{\omega}^q_{ij}$, the isolability can be ensured if sensor fault f^l_{ij} can be detected at node j namely $\omega^l_{ji} \neq 0$. Indeed $-1 < cos\left\langle \begin{bmatrix} \omega^l_{ij} \\ \omega^l_{ji} \end{bmatrix}, \begin{bmatrix} \bar{\omega}^q_{ij} \\ 0 \end{bmatrix} \right\rangle < 1$, hence the angle between these two fault directions is neither 0° nor 180°any more. □

We can explain the propostions from the physical point of view in the following. Here a special case is used in which each sensor only measures a single component of x.

Detectability If a sensor fault $f^l_{ij} \in F_{ij}$ can not be detected, it means there is only one sensor measuring the quantity. Then the detectability can be improved by adding the residual information from node j which includes the information from the sensors measuring the same quantity.

Isolability If a sensor fault f^l_{ij} can be detected but can not be isolated from f^q_{ij} at node i, we can infer that there are only two sensors measuring the same quantity and they can not be distinguished from each other. Then the isolability of the lth fault can be improved if there is another sensor in the neighborhood measuring the same quantity.

Criteria to improve certain diagnosiblity of faults According to propositions 1 to 3 and the above analysis , the residual information from node j is helpful only when it can fulfill some conditions. The criteria for selecting a suitable node to improve the diagnosibility of the fault f^l_{ij} at node i can be deduced:

Criterion 1 The sensor fault f^l_{ij} is detectable at node j. Adding the residual information from node j ensures the detectability of $f^l_{ij} \in F_{ij} \cap ND_i$. Moreover, it also ensures the isolability of $f^l_{ij} \in F_{ij} \cap NI_i$ from $\bar{f}^q_{ij} \in \bar{F}_{ij} \cap NI_i$.

Criterion 2 The sensor f^l_{ij} is isolable at node j. Adding the residual information from node j ensures the isolability of $f^l_{ij} \in F_{ij} \cap NI_i$ from $f^q_{ij} \in F_{ij} \cap NI_i$.

Criterion 3 In order to avoid bringing other undiagnosable faults, we have to set a constraint on node j that the fault $\bar{f}^t_{ji} \in \bar{F}_{ji}$ should be detectable and isolable at node j.

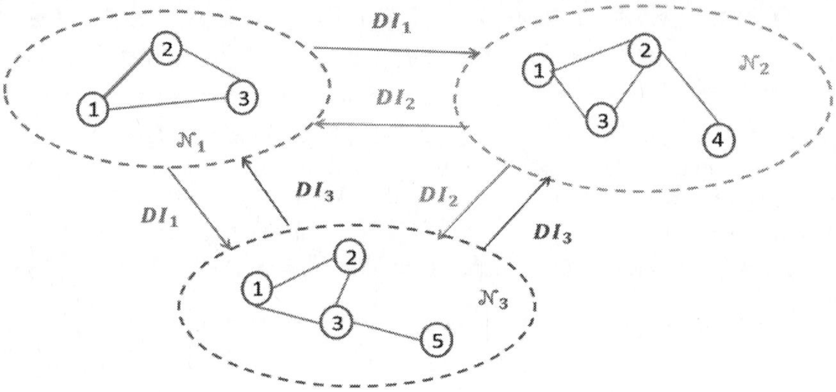

Figure 2. On-line exchanging scheme for diagnosing sensor fault in mobile network

Now we can summarize the steps of computing the distinguishability, checking the diagnosibility and selecting a suitable node if necessary:

1. generate the parity space Ω_{i_loc} based on the local sensing model;
2. compute the distinguishability matrix according to Table 1 on the basis of the columns of Ω_{i_loc};
3. classify the faults into the detectable and isolable set DI_i, the non detectable set ND_i and the non isolable set NI_i, then compute the diagnosibility degree d_i;
4. if $d_i < 1$, one has to associate a neighboring node to node i to ensure fault detection and isolation at node i according to the above mentioned criteria;

There are two different schemes to accomplish the selection, either off-line or on-line. Moreover, an iterative algorithm can be exploited for choosing a suitable neighboring node as described in the following section.

Off-line If the topology of the network is fixed and static, based on the global information on the network, the above 4-step procedure can be realized for each node, in order to determine the residual(s) to be exchanged with the neighboring node, as well as the corresponding decision logic.

On-line If the topology of the network is mobile, each node can know its identifier and the identifiers of its neighbors [2]. Assuming that the global measurement matrix $C_G = [C_{i_1}^T, \cdots, C_j^T, \cdots, C_M^T]^T$ is known by each node, the nodes can form the local measurement matrix from the global measurement matrix and the identifiers of its neighbors, like firstly forming the circles $\mathcal{N}_1, \mathcal{N}_2, \mathcal{N}_3$ in figure 2. Then the diagnosable sets DI_i can be exchanged between node and its neighboring nodes to determine a suitable node which can improve the diagnosibility according to the criteria.

3.4. Selection optimization

If $d_i < 1$, there could be more than one suitable node able to increase the diagnosibility degree at node i. Besides more than one node might be required to improve the diagnosibility. Therefore a criterion for node selection is needed, as well as a methodology for searching among the neighboring nodes.

To describe such a methodology, let us introduce some notations. Let $\mathcal{O}_i^l$ denote the set of l hop-nieghbours of node i which fulfills the criteria to improve the distinguishability. Let $\bar{d}_i^l$ denote the highest achievable value of the diagnosibility degree for a diagnosis system based on the available measurements in the set of nodes $\mathcal{S}_i^l = \mathcal{O}_i^1 \cup \mathcal{O}_i^2 \cup \cdots \cup \mathcal{O}_i^l$.

Furthermore, to characterize the communication and computation costs associated to the introduction of a given node, say node j, in the diagnosis system at node i, let us introduce the following cost function:

$$\mathcal{C}_{ij} = Eg_{ij} + Dim_{ij} \tag{7}$$

where Eg_{ij} represents the number of edges between node i and node j, and Dim_{ij} denotes the dimension of the residual r_{j_loc}. The first term can be seen as a communication cost, while the second one characterizes a computation cost.
Algorithm 1 can be considered for the selection of a relevant set of nodes to build a diagnosis system at node i. This allows one to retain the solution with the lowest computational and communication costs.

Algorithm 1 Selection algorithm

Compute $\bar{d}_i^0 = d_i$,(where d_i is the distinguishability at node i for a diagnosis system based on the local model at node i)
Set $n = 0$, (n stands for n hop neighbors)
while $n < n_{max}$ **do**
  $n = n + 1$
  solve $\max_{s_{i_l} \in \mathcal{S}_i^n} d_i^n$
end while
Let $\mathcal{S}_{i,max}^n = \{s_{i_1}, \cdots, s_{i_q}\}$ denote the set of all solutions that reach the highest diagnosibility degree d_i^{max}.
To determine the set of nodes to be retained for the diagnosis system at node i,
solve $\min_{s_i \in \mathcal{S}_{i,max}^n} \sum_{j \in s_i} \mathcal{C}_{ij}$, (where $\mathcal{C}_{ij}$ is defined in (7))

Remark 1. *Depending on different design requirements, constraints can be added to specify the faults with respect to which detectability and/or isolability should be ensured. For example, increasing one of the elements in the distinguishability matrix could be an objective.*

3.5. Implementation of the distributed fault diagnosis system

Algorithm 2 summarizes the implementation steps in the simple situation where one single additional residual is necessary for achieving a diagnosibility degree equal to 1 at node i.

The decision rules include thresholds h_d and h_{isol} that are introduced to account for uncertainties in the measurement model (1).

4. Simulation

Here we use the same example as in section 2 to illustrate the process of selecting the suitable neighboring node and the improved performance of the fault detection and isolation system at node 1.

According to the topology, node 1 has two neighbors node 2 and 3 and their distinguishability matrices are illustrated in table (3) and (4) respectively. The analysis of these tables yields: $ND_2 = \varnothing$, $NI_2 = \varnothing$, $ND_3 = \varnothing$ and $NI_3 = \varnothing$ which means all the sensor faults can be detected and isolated at node 2 and node 3.

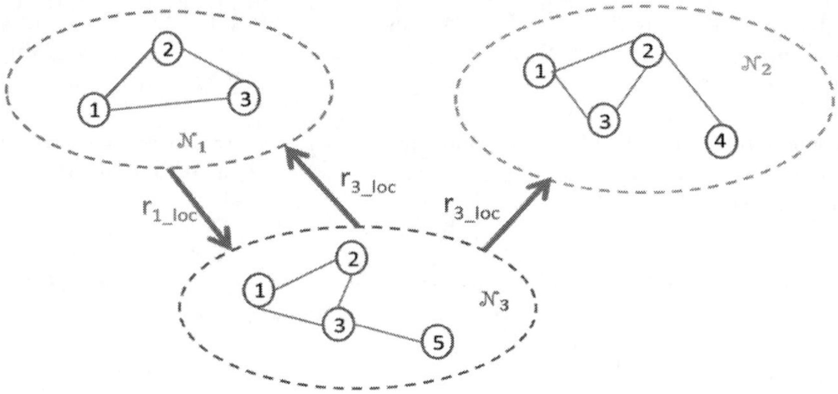

Figure 3. Improving the diagnosibility by exchanging residual vectors with neighboring nodes

Algorithm 2 on-line diagnosis algorithm in a deterministic framework

1. residual generation: $r_{i\_loc}(k) = \Omega_{i\_loc} y_{i\_loc}(k)$;
2. residual exchanging: $r_{i\_new} = \begin{bmatrix} r_{i\_loc}(k) \\ r_{j\_loc}(k) \end{bmatrix}$;
3. fault detection: if$\| r_{i_new} \| > h_d$ indicate that there is a fault, where h_d is the detection threshold;
4. if a fault is detected, fault isolation can be implemented: if $\arccos(r_{i\_new}, \omega^l_{i\_new}) < h_{isol}$ decide that fault l has occurred, where $\omega^l_{i\_new}$ is the lth column of matrix $\Omega_{i\_new} = [\omega^1_{i\_new}, \omega^2_{i\_new}, ..., \omega^{n_{i\_new}}_{i\_new}]$ and h_{isol} is the isolation threshold;

D	NF	f_{11}	f_{12}	f_{13}	f_{21}	f_{22}	f_{31}	f_{32}	f_{41}	f_{42}
f_{11}	1	0	1.5708	1.5708	2.0944	1.5708	1.5708	1.5708	2.0944	1.5708
f_{12}	1	1.5708	0	1.5708	1.5708	1.5708	2.0944	1.5708	1.5708	2.0944
f_{13}	1	1.5708	1.5708	0	1.5708	2.0944	1.5708	2.0944	1.5708	1.5708
f_{21}	1	2.0944	1.5708	1.5708	0	1.5708	1.5708	1.5708	2.0944	1.5708
f_{22}	1	1.5708	1.5708	2.0944	1.5708	0	1.5708	2.0944	1.5708	1.5708
f_{31}	1	1.5708	2.0944	1.5708	1.5708	1.5708	0	1.5708	1.5708	2.0944
f_{32}	1	1.5708	1.5708	2.0944	1.5708	2.0944	1.5708	0	1.5708	1.5708
f_{41}	1	2.0944	1.5708	1.5708	2.0944	1.5708	1.5708	1.5708	0	1.5708
f_{42}	1	1.5708	2.0944	1.5708	1.5708	1.5708	2.0944	1.5708	1.5708	0

Table 3. Distinguishability matrix at node 2

Given the non isolable set $NI_1 = \{f_{11}, f_{21}, f_{12}, f_{31}\}$, node 2 and node 3 can both be selected based on the criteria stated in section 3.3. Node 1 node 2 and node 3 have the common faults set $\{f_{11}, f_{12}, f_{13}, f_{21}, f_{22}, f_{31}, f_{32}, \}$. Fault f_{11} can be isolated from fault f_{21} and fault f_{12} can be isolated from f_{31} at node 2 and node 3. Besides the uncommon faults $\{f_{41}, f_{42}\}$ and $\{f_{51}, f_{52}, f_{53}\}$ can be detected and isolated at node 2 and node 3 respectively.

Since node 2 and node 3 both fulfill the distinguishability specification, the optimization step in algorithm 2 is considered, node 2 and node 3 both satisfy $d^a_1 > d_1$ and $d^2_1 = d^3_1 = 1$. Besides,

D	NF	f_{11}	f_{12}	f_{13}	f_{21}	f_{22}	f_{31}	f_{32}	f_{51}	f_{52}	f_{53}
f_{11}	1	0	1.5708	1.5708	2.0944	1.5708	1.5708	1.5708	2.0944	1.5708	1.5708
f_{12}	1	1.5708	0	1.5708	1.5708	1.5708	2.0944	1.5708	1.5708	2.0944	1.5708
f_{13}	1	1.5708	1.5708	0	1.5708	1.9106	1.5708	1.9106	1.5708	1.5708	1.9106
f_{21}	1	2.0944	1.5708	1.5708	0	1.5708	1.5708	1.5708	2.0944	1.5708	1.5708
f_{22}	1	1.5708	1.5708	1.9106	1.5708	0	1.5708	1.9106	1.5708	1.5708	1.9106
f_{31}	1	1.5708	2.0944	1.5708	1.5708	1.5708	0	1.5708	1.5708	2.0944	1.5708
f_{32}	1	1.5708	1.5708	1.9106	1.5708	1.9106	1.5708	0	1.5708	1.5708	1.9106
f_{51}	1	2.0944	1.5708	1.5708	2.0944	1.5708	1.5708	1.5708	0	1.5708	1.5708
f_{52}	1	1.5708	2.0944	1.5708	1.5708	1.5708	2.0944	1.5708	1.5708	0	1.5708
f_{53}	1	1.5708	1.5708	1.9106	1.5708	1.9106	1.5708	1.9106	1.5708	1.5708	0

Table 4. Distinguishability matrix at node 3.

D	NF	f_{11}	f_{12}	f_{13}	f_{21}	f_{22}	f_{31}	f_{32}	f_{41}	f_{42}
f_{11}	1	0	1.5708	1.5708	2.3664	1.5708	1.5708	1.5708	1.9584	1.5708
f_{12}	1	1.5708	0	1.5708	1.5708	1.5708	2.3664	1.5708	1.5708	1.9584
f_{13}	1	1.5708	1.5708	0	1.5708	2.0944	1.5708	2.0944	1.5708	1.5708
f_{21}	1	2.3664	1.5708	1.5708	0	1.5708	1.5708	1.5708	1.9584	1.5708
f_{22}	1	1.5708	1.5708	2.0944	1.5708	0	1.5708	2.0944	1.5708	1.5708
f_{31}	1	1.5708	2.3664	1.5708	1.5708	1.5708	0	1.5708	1.5708	1.9584
f_{32}	1	1.5708	1.5708	2.0944	1.5708	2.0944	1.5708	0	1.5708	1.5708
f_{41}	1	1.9584	1.5708	1.5708	1.9584	1.5708	1.5708	1.5708	0	1.5708
f_{42}	1	1.5708	1.9584	1.5708	1.5708	1.5708	1.9584	1.5708	1.5708	0

Table 5. Distinguishability matrix at node 1 with extra information from node 2

in (7) $Eg_1^2 = Eg_1^3 = 1$. The residual generated at node 2 is $r_{2\_loc} \in R^6$ and the residual at node 3 is $r_{3\_loc} \in R^7$, so $Dim_1^2 = 6$ and $Dim_1^3 = 7$. The cost function for node 1 associated to node 2 and node 3 are $C_1^2 = 7$ and $C_1^3 = 8$ respectively. Therefore, node 2 corresponds to the optimal selection.

After adding the information from node 2, the distinguishability matrix at node 1 is improved. The angles between f_{11} and f_{21}, f_{12} and f_{31} which indicate the isolability are closer to 90° (1.5708) than before as can be seen from the table (5). All the sensor faults can be diagnosed with the aid of node 2. The diagnosibility at node 1 is $d_1 = 1$.

5. Conclusion

A systematic way to design and implement a local FDI system at each node of a sensor network has been presented. It resorts to a distinguishability measure to indicate the performance of the FDI system and to select a suitable node in the network. The specified fault detection and isolation performance can be achieved while the communication and computation cost is taken into account.

In the presented work, measurement noise is not accounted for. Should this noise be significant, the scheme proposed in this paper can be developed in a stochastic framework, by resorting to a distance measure based on the Kullback-Leibler divergence.

References

[1] F. Koushanfar, M. Potkonjak and A. Sangiovanni-Vincentelli 2003 On-line fault detection of sensor measurements *Sensors, Proceedings of IEEE* **2003** 974-979

[2] Stefano Chessa and Paolo Santi 2002 Crash faults identification in wireless sensor networks *Computer Communications* **25** 1273-1282

[3] Shahram Babaie and Ali Ranjideh Rezaie 2011 DFDM: Decentralized fault detection mechanism to improving fault management in wireless sensor networks *9th International Conference on Reliability, Maintainability and Safety (ICRMS)* 1026-1029

[4] Jinran Chen, Shubha Kher and Arun Somani 2006 Distributed fault detection of wireless sensor networks *Proc. of the 2006 workshop on Dependability issues in wireless ad hoc networks and sensor networks* 65-72

[5] Myeong-Hyeon Lee and Yoon-Hwa Choi 2008 Fault detection of wireless sensor networks *Computer Communications* **31** 3469-3475

[6] Bhaskar Krishnamachari and Sitharama Lyengar 2004 Distributed Bayesian algorithms for fault-tolerantolerent event region detection in wireless sensor networks *IEEE Transactions on Computers* **53** 241-250

[7] Dyi-Rong Duh, Ssu-Pei Li and Victor W. Cheng 2013 Distributed fault-tolerantofault-tevent region detection of wireless sensor networks *International Journal of Distributed Sensor Networks* **2013**

[8] Vasso Reppa, Marios M. Polycarpou and Christos G. Panayiotou 2012 A Distributed architecture for sensor fault detection and isolation using adaptive approximation *IEEE World Congress on Computational Intelligence* **2012**

[9] Michel Kinnaert and Jingjing Hao 2013 Distributed Sensor Fault Detection and Isolation over Network *IFAC Proceedings Volumes* **47** 11458 - 11463

[10] Iman. Shames, Andre M. H. Teixeira, Henrik Sandberg and Karl H. Johansson 2012 Distributed fault detection and isolation with imprecise network models *American Control Conference (ACC)* 5906-5911

[11] Mattias Krysander and Erik Frisk 2008 Sensor placement for fault diagnosis *IEEE Trans. Syst. Man, Cybern. A, Syst., Humans* **38** 1398-1410

[12] Erik Frisk, Mattias Krysander and Jan Aslund 2009 Sensor placement for fault isolation in linear differential-algebraic systems *Automatica* **45** 364-371

[13] Daniel Eriksson, Erik Frisk and Mattias Krysander 2012 Using quantitative diagnosability analysis for optimal sensor placement *Proc of the 8th IFAC Safe Process, Mexico city, Mexico* **2012**

[14] Daniel Eriksson, Erik Frisk and Mattias Krysander 2013 A method for quantitative fault diagnosability analysis of stochastic linear descriptor models *Automatica* **49** 1591-1600

[15] J. Chen and R.J.Patton 1999 *Robust model-based fault diagnosis for dynamic systems* (Springer US)

[16] M. Basseville 2001 On fault detectability and isolability' *European Journal of Control* **7** 625-637

[17] Janos Gertler and Moid Kunwer 1995 Optimal residual decoupling for robust fault diagnosis *International Journal of Control* **61** 395-421

[18] Asok Ray and Rogelio Luck 1991 An introduction to sensor signal validation in redundant measurement systems *IEEE. Control Systems* **11** 44-49

[19] F Gustafsson 2002 Stochastic fault diagnosability in parity spaces 15th Triennial World Congress of the International Federation of Automatic Control **2012**

Sensor Placement for Fault Diagnosis Using Graph of a Process

Anna Sztyber

Institute of Automatic Control and Robotics, Faculty of Mechatronics, Warsaw University of Technology, Poland

E-mail: a.sztyber@mchtr.pw.edu.pl

Abstract. The quality of a diagnostic system strongly depends on the availability of appropriate measurements. In this work sensor placement method using Graph of a Process is introduced. Graph of a Process is a formalization of a causal graph useful in fault diagnosis. Faults are directly incorporated into the model. The necessary and sufficient conditions for fault detectability and isolability are formulated. The analysis is divided into acyclic graph search and calculations within strongly connected components. This method is applicable to the design of the instrumentation of diagnostic systems, when the analytical process model is unavailable.

1. Introduction

Proper fault detection and isolation in large scale industrial plants is important because of safety, environmental protection, and economical reasons. Due to the complexity of the systems to be diagnosed there is a need for tools supporting the design of the diagnostic system. One of the important problems is sensor placement to meet diagnostic specifications. This problem may have more than one solution, so the selected set should be optimal with respect to the chosen objective, like minimal cost or minimal number of sensors.

A review of the methods and criteria for sensor set selection was presented in [1]. The main task of measuring instruments is to estimate the values of process variables, which is important for control, monitoring, and optimization purposes. It should be noted that not every process value needs to be directly measured. Some variables can be estimated by the use of other measurements.

Often the set of measuring instruments installed for control is insufficient for fault diagnosis purposes. Therefore, the set of sensors used by the diagnostic system should be selected taking into account one or more of the above criteria and be sufficient to:

- detect the set of considered faults;
- distinguish the specified pairs of faults.

The problem of sensor placement for fault diagnosis has been an active research area in recent years. This is a complicated problem and finding a solution in the general case is intractable, so some simplifying assumptions are needed. The most popular approach is to use a simplified, qualitative model of a process. There are sensor placement algorithm using structural analysis [2, 3, 4, 5, 6], bond-graphs [7, 8], and SDG graphs [9, 10, 11]. GP approach, presented in this paper, differs from the existing approaches because of the form of the process description and assumptions underlying the diagnostic system.

2. Graph of a Process and Model Structures

A causal graph shows relations between variables. Vertices of the graph represent process variables, disturbances, and control signals. Graph of a Process is a kind of a causal graph with faults included directly in a model as additional vertices. Directed edges depict relations between the variables. The GP is a simple, intuitive, and easily understandable way of describing diagnostic knowledge.

The GP was chosen as a tool for sensor placement because of the following features:

- The GP can be constructed using process description, process instrumentation diagrams or with cooperation with process engineers. Mathematical description is not needed, only basic knowledge about the physics of a process. This is an important feature because sensor placement analysis should be performed at the design stage and a detailed process model can be unavailable.
- We assume diagnosis using residuals in which values are the differences between the measured value of the process variable and the value calculated by a local multi input single output (MISO) model. The models can be black-box models identified using process data (fuzzy, neural, or parametric models). The GP was designed to support finding structures of these models. This approach is justified when mathematical description of the system is unavailable, instead archival measurements can be used to train local models for fault detection. The proposed approach gives better fault isolation than Signed Directed Graph (SDG) approach.

In this part definitions of GP and model structure (MS) will be recalled. For further details and detailed comparison with other qualitative models, interested reader is referred to [12].

2.1. Graph of a Process Definition

Control variables, faults, disturbances, measurements, and physical system variables are included as graph vertices. The set of all physical variables characterizing the system is denoted by X. This set can be divided into the following subsets:

$$X = X_U \cup X_D \cup X_X \cup X_Y,$$

where: X_U - set of control physical variables, X_D - set of input physical variables of unknown values also called disturbances, X_X - set of internal physical variables, and X_Y - set of output physical variables, which is a representation of measured variables in the set X.

The control system generates control signals $u \in U$, which, in the case of a fault free state, are equal to control physical variables $x \in X_U$. All variables $x \in X_Y$ are measured. Therefore, a set of measured variables Y is as numerous as X_Y. If faults are absent, values of variables in both sets are the same within the accuracy of measurement.

The following subsets can be distinguished in the set of faults F: faults of controls F_U, faults of installation components F_C, and faults of sensors F_Y:

$$F = F_U \cup F_C \cup F_Y.$$

In this notation, faults of actuators are also regarded as faults of installation components. The set V describing the object of diagnosis can be divided into the following disjoint sets:

$$V = U \cup Y \cup X \cup F = K \cup X \cup F,$$

where K is a set of known variables, which are control variables and measured variables.

The equivalent of the set K in the set X is a set of physical known variables X_K, the set of physical variables that have equivalents in the sets Y or U.

$$X_K = X_U \cup X_Y$$

GP is a simple directed graph:

$$GP = (V, A),$$

where: $V = V(GP)$ - vertices set, $A = A(GP) \in V \times V$ - binary relation defined on the Cartesian product of the vertices set V, a set of ordered pairs $< v, w > \in V \times V$ representing edges of the graph. Edges represent the direction of influence. The vertices subset X of the graph GP form a subgraph GP_X representing the relation between physical variables.

2.2. Example of a Three Tank System

The ideas described will be illustrated using a three tank system (TTS) shown in Fig. 1. The level in tank $T3$ is controlled by a flow valve to tank $T1$. The GP for the TTS is shown in Fig. 2. Graph GP_X, containing only vertices from the set X, is presented in Fig. 3. Symbols are described in Table 1. The last column in Table 1 shows the classification of the vertices of GP. Variables p_{ZP} and n are classified as disturbances because during normal work of a process they have almost constant values and cannot be measured. It is assumed that no measurements are available. It is an artificial assumption because the controlled level must be measured, but the example is meant as an illustration of the sensor placement algorithm. It should be noted that if some variables are already measured then some of the conditions from the Section 3.1 are already fulfilled and should be excluded from the search. Therefore, without the lost of generality, we assume that there are no available measurements. Faults are listed in Table 2.

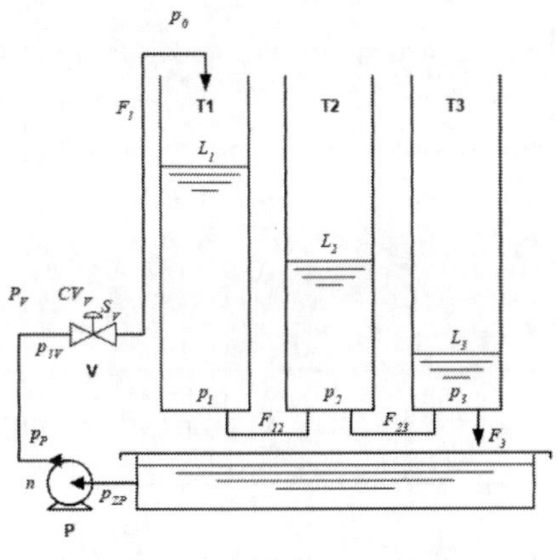

Figure 1: Scheme of TTS.

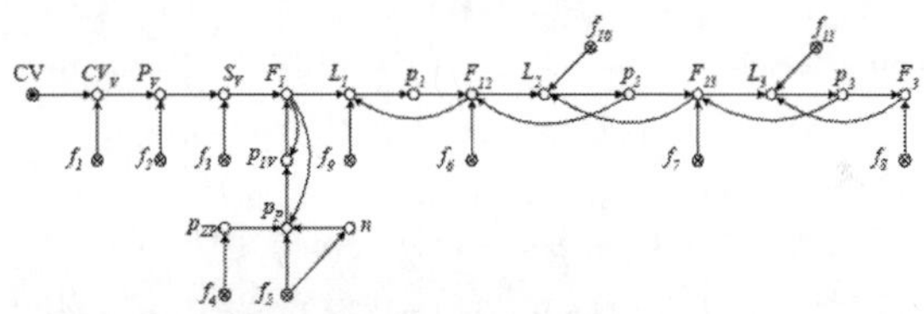

Figure 2: GP of TTS.

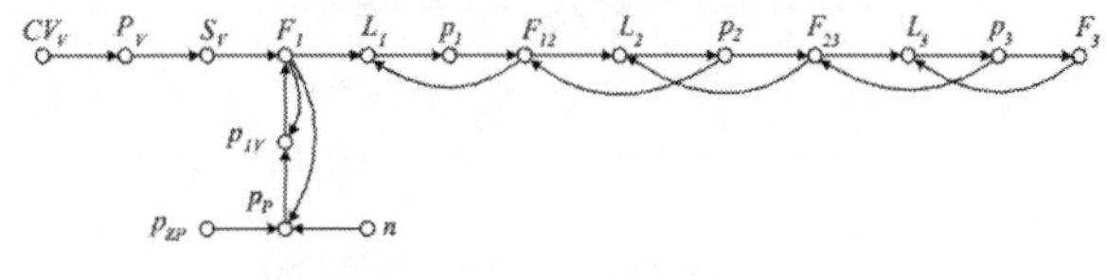

Figure 3: GP_X graph of TTS.

2.3. Model Structures

In this paper model structure (MS) is understood as a pair: a modeled variable and a set of model inputs. Models for fault detection are assumed to be black-box models (e.g., parametric models, neural or fuzzy models). All MSs for fault detection can be defined using GP. The modeled variable is always an element of the set X_Y and the input variables are defined by the following subsets of X:

$$X_I = X_U \cup X_D \cup X_Y.$$

In the model structure definition, only the subgraph GP_X is considered.

Definition 1. ***Model structure*** *is defined as a pair:*

$$M = (o, I),$$

Table 1: Symbols

Symbol	Description	Type
CV	control signal steering a position of the valve V	U
p_{ZP}	pressure of a pump supply P	X_D
n	rotational speed of the pump P	X_D
p_P	pressure after the pump P	X_X
p_{1V}	pressure before the valve V	X_X
P_V	position of the valve V	X_X
CV_V	input state variable steering valve position	X_U
S_V	open cross-section of the valve V	X_X
F_1	flow to the tank $T1$	X_Y
L_1, L_2, L_3	level in the tanks $T1, T2$, and $T3$	X_Y
p_1, p_2, p_3	hydrostatic pressure at the bottom of the tanks $T1, T2$, and $T3$	X_X
F_{12}	flow between the tanks $T1$ and $T2$	X_X
F_{23}	flow between the tanks $T2$ and $T3$	X_X
F_3	outflow from the tank $T3$	X_X

Table 2: Faults in TTS

Fault	Description
f_1	fault in control circuit
f_2	valve actuator fault
f_3	clogging of the valve
f_4	no liquid before the pump
f_5	pump fault
f_6	clogging of a pipe between $T1$ and $T2$
f_7	clogging of a pipe between $T2$ and $T3$
f_8	clogging of an outlet pipe from $T3$
f_9	leakage in $T1$
f_{10}	leakage in $T2$
f_{11}	leakage in $T3$

where: $o \in X_Y$ is the modeled variable, $I \subset X_I$ is the set of input variables, fulfilling the following conditions:

(i) $\forall i \in I$ in GP_X exists a path from i to o, not containing any other vertex from the set I,

(ii) $\forall v \in GP_X$ if there exists a path from v to o, not containing any vertex from the set I, then there exists a path from some $i \in I$ to v.

The method for finding all MSs was presented in [12].

2.4. Faults-symptoms Relation

If a fault causes a difference between the measured value of a variable and its estimation calculated from the model, then we say that MS is sensitive to that fault, which leads to the following:

Definition 2. *Model structure $M = (o, I)$ is* ***sensitive*** *to faults $F = F_U \cup F_X \cup F_Y$, where $\forall f \in F_X$ in a graph there exists a path from f to o, not containing any vertex from the set $I \cap X_Y$. The set F_Y contains sensor faults of measurements of variables from the set $I \cup o$. The set F_U contains faults of controls related to the physical control variables in $I \cap X_U$.*

The algorithm for determining MS sensitivity to faults was presented in [12].

3. Sensor placement

When using GP in diagnosability analysis, we assume that a fault is detectable if there exists at least one MS sensitive to this fault. Two faults f_1 and f_2 are distinguishable if their signatures are different. With binary residual evaluation it means that there must exists at least one MS sensitive to f_1 and insensitive to f_2 or vice versa. Therefore, the proposed solution will be based on the following definitions.

Definition 3. *The fault is* ***detectable within the meaning of GP*** *if there exists a MS sensitive to this fault.*

Definition 4. *Two faults are* ***distinguishable within the meaning of GP*** *if both are detectable and there exists MS sensitive to only one of them.*

We want to find the set of sensors that allows the diagnostic system to fulfill the diagnostic specification. The specification is a set of faults D, which should be detectable and the set of pairs of faults I, which should be distinguishable. Only single faults are considered. This problem has many solutions; therefore, the chosen solution should minimize the selected objective. We chose a minimal number of sensors, but the algorithms to be presented can be easily modified to use some other cost function. These considerations lead to the following problem formulation:

Given GP, find the minimal set of sensors that allows detection all faults in D and isolation of all pairs of faults in I.

3.1. Conditions for Faults Detectability and Isolability

In this Section, the sufficient and necessary conditions for faults detectability and isolability will be analyzed. The formulation of these conditions is a first step of a sensor placement algorithm. Moreover, such an analysis can be used for answering the following questions:

- Which measurements are useful for detection of the fault?
- Which measurements can help isolate indistinguishable faults?

3.1.1. Fault detection

Necessary condition The necessary conditions for detection of the fault is the existence of at least one measured process variable influenced by this fault. In the context of GP influence is identified with the existence of the directed causal path from the fault to the variable.

Let V_{Df} denote the set of variables influenced by the fault f. This set can be obtained by the depth-first search started at f; $V_{Df} = DFS(G, f)$. Therefore, the necessary condition for detectability of f is:

$$\exists v \in V_{Df} : v \in X_Y.$$

This reasoning is similar to SDG approach [9] but there is one crucial difference. In the case of SDG graphs, fault detection is based on the deviations of the variables from the normal state, and the normal state is assumed to be known. In that case, the necessary condition is also sufficient. In case of GP, it is not.

Figure 4: Exemplary graph G_1

Let us consider the exemplary graph G_1 shown in Fig. 4 containing only two vertices: $a \in X_Y$ and $f \in F$. The necessary condition of detectability of f is fulfilled; $V_{Df1} = \{a\}$ and $a \in X_Y$. In the diagnostic system using the control of the alarm thresholds the fault can be detectable, because the influence of f can cause crossing of the alarm limits by the value of a. While diagnosing using MSs the fault is not detectable because, for this graph, no MS can be built (the system does not contain any redundancy).

Necessary condition The necessary condition for the detectability of the fault, in the meaning of GP, is an existence at least one MS sensitive to this fault.

3.1.2. Fault isolation In the isolability analysis, we will use the concept of a strongly connected component. A strongly connected component in a graph is a set of vertices that, for each pair of vertices u and v, there exist paths from u to v and from v to u. One strongly connected component contains a set of variables influencing each other. Being in a strongly connected component is an equivalency relation defined on the set of vertices, so division into components splits the graph into disjointed sets, which can be considered separately.

The considerations concerning fault isolability should be divided into two parts. We should consider separately the faults affecting different strongly connected components of a graph and the faults affecting the same strongly connected component.

Acyclic graph The first step in the analysis is finding and merging strongly connected components of GP. This can be obtained using an algorithm based on a depth-first search [13]. The faults influencing one strongly connected component are identified. After merging strongly connected components we obtain an acyclic graph.

The necessary condition to isolate two faults f_1 and f_2 is that they influence different sets of variables:

$$\exists v \in V_{Df1} \Delta V_{Df2} : v \in X_Y,$$

where Δ denotes a symmetric difference of the sets. The set $V_{Df1} \Delta V_{Df2}$ will be denoted by V_{I12}.

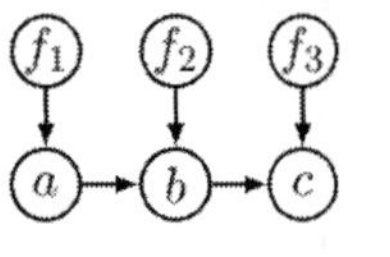

Figure 5: Exemplary graph G_2

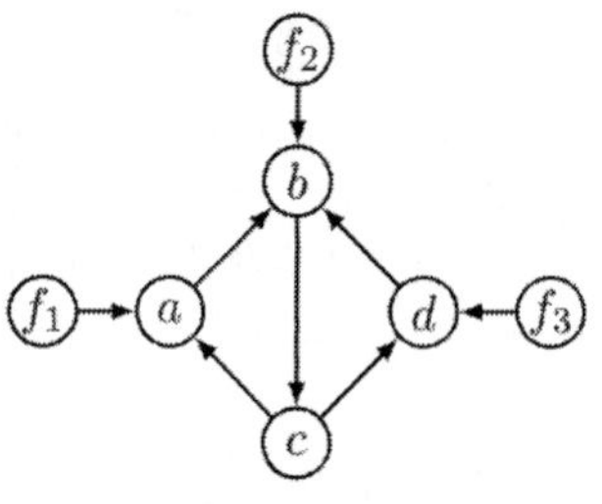

Figure 6: Exemplary graph G_3

Let us consider the exemplary graph G_2 shown in Fig. 5. The detectability sets for this graph are as follows:

$$V_{Df1} = \{a, b, c\} \qquad V_{Df2} = \{b, c\} \qquad V_{Df3} = \{c\}.$$

Considering these sets, we obtain that for detectability of faults, measurement of c is necessary. However this measurement is not sufficient because with only one sensor, no model can be built.

It should be also noted that detection of f_1 is impossible because there is not a MS sensitive to this fault (according to Definitions 1 and 2).

The isolability sets are as follows:

$$V_{I12} = \{a\} \qquad V_{I23} = \{b\} \qquad V_{I13} = \{a, b\}.$$

The fault f_1 is undetectable; therefore, only the pair $\{f_2, f_3\}$ can be distinguished. The measurement of b is necessary for this purpose. The isolability can be obtained using the following MS: $(c, \{b\})$ and $(b, \{a\})$, where the first one is sensitive to f_3 and the second one is sensitive to f_2. Without the measurement of b only one MS is possible - $(c, \{a\})$, which is sensitive to both faults.

It should be noted that, in the case of faults, influencing one strongly connected component, the sets of variables influenced by the faults are equal. This means that in the case of cycles in the graph, another approach must be proposed.

Strongly connected components We assume that, at this stage, the analysis for the acyclic graph was finished and that the set of the sensors was selected and that the faults influencing the strongly connected component are detectable and distinguishable from the other faults.

In the strongly connected component, each fault influences all vertices. For the isolability of faults, according to Definition 4, we should find MS sensitive to only one of the considered faults. For this purpose, we should search for appropriate model inputs, which will cut the influence of one of the faults (and only one of them) on the modeled variable. This leads us to the definition of the cutting path vertex.

Definition 5. ***Cutting path vertex*** *between the faults f_1 and f_2 is a vertex that afterwards:*

(i) it is deleted from the graph,

(ii) direction of the edges incident with the fault f_2 is reversed

there is no path in the graph between f_1 and f_2.

Let us consider G_3 shown in Fig. 6. It contains one strongly connected component containing the vertices a, b, c, and d, and which is influenced by the faults f_1, f_2, and f_3. The vertices b, c, and d cut the path from the fault f_2 to f_3. Vertices b and d cut the path from f_3 to f_2. The cutting path vertices can be used as an inputs of the model used for the fault isolation. For example measurement of d can be used in MS - $(a, \{d\})$ is sensitive to f_2 and insensitive to f_3.

Theorem 1. *The necessary condition for isolation of two faults f_1 and f_2, influencing vertices of the same strongly connected component is measurement of at least one variable cutting path from f_1 to f_2 or from f_2 to f_1.*

The proof is omitted due to space limitations.

The condition in Theorem 1 is not a sufficient condition. Let us consider again graph G_3 shown in Fig. 6. Vertex b cuts the path between the faults f_2 and f_3. The model $(a, \{b\})$ is sensitive to the fault f_1 and insensitive to any of the faults f_2 and f_3. To assure isolability, appropriate selection of the modeled variable is necessary.

If there exists measured vertex v cutting path form f_1 to f_2 or vice versa, then the modeled variable o should be selected in a way that there exists a path from one of the faults to o not containing v (and such a path does not exists from the other fault). An example of a suitable model is $(d, \{b\})$, which is sensitive to f_3 but is not sensitive to f_2 (see Fig. 6).

The condition that v is a vertex cutting path between f_1 and f_2 assures the existence of suitable model outputs o. Finally, to assure isolability of f_1 and f_2, the MS with o as an output and v as an input must exist (measurements of other inputs are needed). If the previous step of the sensor placement algorithm was completed, then the faults are detectable and such a model exists.

The appropriate set of vertices are calculated by the following algorithms. The subgraph of GP containing only one strongly connected component is denoted by G, $DFS(G, f)$ denotes all vertices met in the depth-first search of G started at f. The Algorithm 1 $Faults(G, f)$ finds the sets of faults $f(v)$ that the vertex v is a vertex cutting path from f to the faults in $f(v)$. The Algorithm 2 $PathCut(f_1, f_2, G)$ finds the sets $pc(f_1, f_2)$ of vertices cutting path from f_1 to f_2 or from f_2 to f_1.

The Algorithm 3 $PossibleModelOutput(f_1, f_2, G)$ for the pair of the faults f_1 and f_2 and the cutting path vertex v finds the set of the model outputs $pmo(f_1, f_2, v)$, where if v is an input then the model will be sensitive to only one of the faults from the pair $\{f_1, f_2\}$.

The algorithms will be explained for the graph shown in Fig. 6. We will analyze the pair $\{f_2, f_3\}$.

Algorithm 1 $Faults(G, f)$

Reverse direction of the edges incident with the other faults in G
for all $v \in G \cap X_X$ **do**
  delete v from G
  $mf(v) \leftarrow DFS(G, f) \cap F$
  $f(v) \leftarrow F \backslash \{mf(v) \cup f\}$
  add v to G
end for

Algorithm 2 $PathCut(f_1, f_2, G)$

$pc(f_1, f_2) \leftarrow \emptyset$
for all $v \in G \cap X_X$ **do**
  if $f_2 \in f_1(v)$ or $f_1 \in f_2(v)$ **then**
    $pc(f_1, f_2) \leftarrow pc(f_1, f_2) \cup v$
  end if
end for

Algorithm 3 $PossibleModelOutput(f_1, f_2, G)$

for all $v \in pc(f_1, f_2)$ **do**
  usuñ v
  $V_{f1} \leftarrow DFS(G, f_1)$
  $V_{f2} \leftarrow DFS(G, f_2)$
  $pmo(f_1, f_2, v) \leftarrow V_1 \Delta V_2$
end for

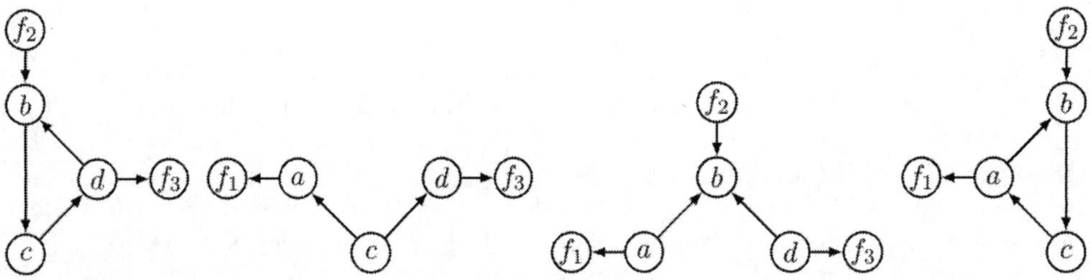

(a) (b) (c) (d)

Figure 7: Work of the Algorithm 1 $Faults(G_3, f_2)$

The work of the Algorithm 1 $Faults(G_3, f_2)$ is shown in Fig. 7. In the subsequent steps, the vertices a, b, c and d are deleted and the edges incident with the other faults (f_1 and f_3) are reversed. We obtain the following sets:

$$mf_2(a) = \{f_3\} \quad mf_2(b) = \emptyset \quad mf_2(c) = \emptyset \quad mf_2(d) = \{f_1\}.$$

$$f_2(a) = \{f_1\} \quad f_2(b) = \{f_1, f_3\} \quad f_2(c) = \{f_1, f_3\} \quad f_2(d) = \{f_3\}.$$

In a similar way, we obtain the sets for the fault f_3:

$$mf_3(a) = \{f_2\} \quad mf_3(b) = \emptyset \quad mf_3(c) = \{f_1, f_2\} \quad mf_3(d) = \emptyset,$$

$$f_3(a) = \{f_1\} \quad f_3(b) = \{f_1, f_2\} \quad f_3(c) = \emptyset \quad f_3(d) = \{f_1, f_2\}.$$

On the basis of the above sets, the Algorithm 2 $PathCut(f_2, f_3, G_3)$ finds the vertices cutting path from f_2 to f_3 or from f_3 to f_2:

$$pc(f_2, f_3) = \{b, c, d\}.$$

The last step is the search for the possible model outputs for each of the cutting path vertices (Algorithm 3 $PossibleModelOutput(f_2, f_3, G_3)$):

$$\begin{array}{llll} b: & V_{f2} = \emptyset & V_{f3} = \{d\} & pmo(f_2, f_3) = \{d\}, \\ c: & V_{f2} = \{b\} & V_{f3} = \{b, d\} & pmo(f_2, f_3) = \{d\}, \\ d: & V_{f2} = \{a, b, c\} & V_{f3} = \emptyset & pmo(f_2, f_3) = \{a, b, c\}. \end{array}$$

Therefore the isolability of the faults f_2 and f_3 can be obtained using the following MS:

$$(d, \{b\}) \quad (d, \{c\}) \quad (a, \{d\}) \quad (b, \{d\}) \quad (c, \{d\}).$$

3.2. Sensor Placement Algorithm

In the previous sections, the considerations concerning sufficient and necessary conditions for fulfillment of the diagnostic specifications were presented. Theses conditions let us formulate the general scheme of the sensor placement algorithm. The analysis should contain the following steps:

(i) Adding all possible sensors, finding all MS and verification of the diagnostic specification. Unrealizable requirements must be deleted from the specification (undetectable faults and indistinguishable pairs of faults).
(ii) Merging strongly connected components of a graph;
(iii) Finding the sets V_{Df} for each fault that should be detectable;
(iv) Finding the sets V_{I12} for each pair of faults $\{f_1, f_2\}$ that should be distinguishable;
(v) Finding the hitting set of the sets from the Steps 3 and 4;
(vi) Verification of the solution for the acyclic graph. The solution is checked by finding all MS for the obtained set of sensors.
(vii) Analysis within each strongly connected component of a graph.

The number of all MS in the Step 1 can be exponential in the number of measurement (but it is often much smaller than the set of analytical redundancy relations (ARRs) due to the causality limitations). If the calculation of all the set of MS is a problem, the method of calculation of only local MS was proposed in [12]. The set of local MS has the same detectability and isolability properties as a full set with exclusion of sensor faults, which is sufficient in that case.

The conditions formulated in Points 3 and 4 are simple and can be obtained in the linear time for each fault (linear in the sense of a size of GP counted as a sum of the number of vertices and edges). Finding the hitting set is known hard algorithmic problem. Analysis in each strongly connected component is carried out locally using only the subset of variables. Both problems can be handled with standard discrete optimization techniques like branch and bound.

3.3. Example

The above conditions and algorithms will be illustrated using the three tank system example. The simple example was chosen to demonstrate the differences between the proposed and the existing approaches. The three tank system is shown in Fig. 1, the lists of faults and symbols are presented in Table 1 and 2, respectively.

At the beginning, there are no assumptions concerning measured variables, only the control signal CV is known. We assume that all process variables can be measured, excluding p_{ZP} and n, which belong to X_D. We consider maximal diagnostic specification, and all faults should be detectable and distinguishable pairwise.

The first step of the analysis is verification if the diagnostic specification can be fulfilled using all possible measurements. For this purpose, sensors for all variables are added, in that case it is the set: $\{CV_V, P_V, S_V, F_1, p_{1V}, p_P, L_1, L_2, L_3, p_1, p_2, p_3, F_{12}, F_{23}, F_3\}$ and all MS are calculated for diagnosability analysis.

For the three tank system we obtain that all faults are detectable. The pair of faults $\{f_4, f_5\}$ cannot be distinguised. These faults denote the lack of supply water and the fault in the pump. They can be distinguised using the sensor of the level in the buffer tank, but this sensor is not considered in the model. The other faults are pairwise distinguishable.

3.3.1. Fault detectability The first step of the fault detectability analysis is merging all strongly connected components of GP. The graph of a three tank system with merged strongly connected components is shown in Fig. 8. The vertices p_P, p_{1V}, and F_1 were merged into SN_1, and process variables describing the tanks were merged into SN_0. All the faults influencing the component SN_0 were merged into one fault denoted by f_Σ.

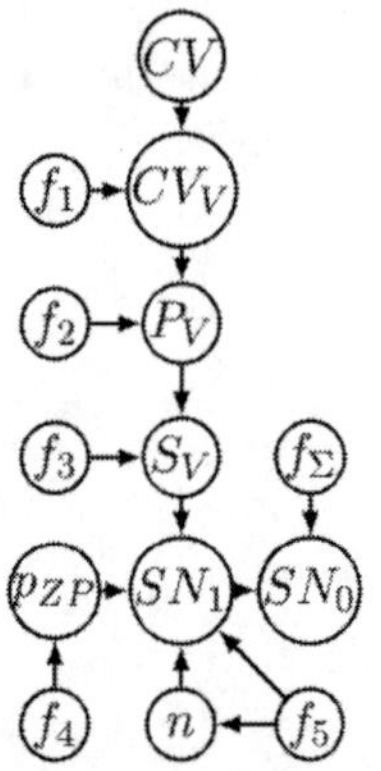

Figure 8: Graph of three tank system with merged strongly connected components

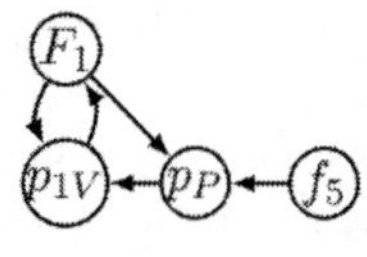

Figure 9: Component SN_1

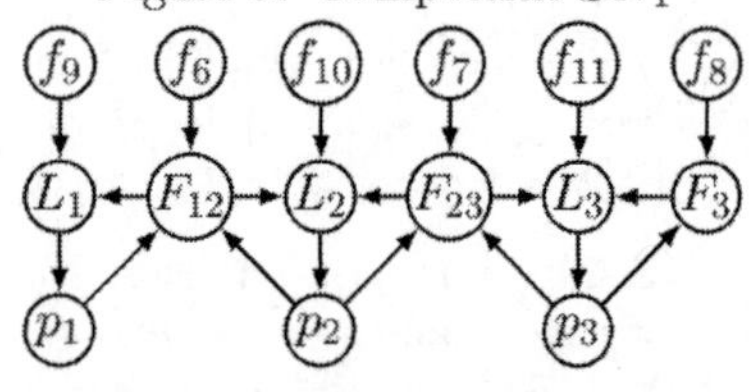

Figure 10: Component SN_0

The sets V_{Df} of variables influenced by the particular faults are listed in Table 3.

Therefore, the necessary condition for the detectability of all the faults is measurement of at least one variable in each set. It should be noted that, in this case, placing a sensor at SN_0 is enough. The measurement of a vertex denoting a strongly connected component should be understood as a measurement of at least one variable in that component.

3.3.2. Fault isolability For each pair of faults, we calculate the symmetric differences of the V_{Df} sets. The obtained sets are shown in Table 6. The necessary condition for the isolabilty of the faults is a measurement of at least one variable from each set. In should be noted that, for the isolation of the pair $\{f_4, f_5\}$, there is a need for measurement of n or p_{ZP}, which were assumed unmeasurable. It confirms the previous isolability analysis.

The nest step is to find the hitting set of the sets listed in Table 3 and Table 6. The hitting set should contain only possible sensor locations; therefore, the variables n and p_{ZP} are not considered. This problem has only one solution containing variables: SN_0, P_V, CV_V, S_V, SN_1.

This solution fulfills all the necessary conditions for detectability and isolability. The next step is a verification by building all MS with their sensitivity to faults. These model structures are listed in Table 4.

Table 3: Detectability sets

f	V_{Df}
f_1	SN_0, P_V, CV_V, SN_1, S_V
f_2	SN_0, P_V, SN_1, S_V
f_3	SN_0, SN_1, S_V
f_4	SN_0, SN_1, p_{ZP}
f_5	SN_0, SN_1, n
f_Σ	SN_0

Table 4: Model structures

Output	Inputs	Faults
SN_0	SN_1	f_Σ
P_V	CV_V	f_2
P_V	CV	f_1, f_2
CV_V	CV	f_1
S_V	P_V	f_3
S_V	P_V	f_3
S_V	CV_V	f_2, f_3
S_V	CV	f_1, f_2, f_3
SN_1	n, p_{ZP}, S_V	f_4, f_5
SN_1	n, p_{ZP}, P_V	f_3, f_4, f_5
SN_1	n, p_{ZP}, CV_V	f_2, f_3, f_4, f_5
SN_1	n, p_{ZP}, CV	f_1, f_2, f_3, f_4, f_5
SN_0	n, p_{ZP}, S_V	f_4, f_5, f_Σ
SN_0	n, p_{ZP}, P_V	f_3, f_4, f_5, f_Σ
SN_0	n, p_{ZP}, CV_V	$f_2, f_3, f_4, f_5, f_\Sigma$
SN_0	n, p_{ZP}, CV	$f_1, f_2, f_3, f_4, f_5, f_\Sigma$

Table 5: Possible model outputs for each vertex cutting path between $\{f_6, f_7\}$

Faults	Input	Possible model outputs
f_6, f_7	p_2	$L_2, p_1, L_1, F_3, p_3, F_{12}, F_{23}, L_3$
	L_2	$p_1, L_1, F_3, p_3, F_{12}, F_{23}, L_3$
	F_{12}	$p_2, L_2, F_3, p_3, F_{23}, L_3$
	F_{23}	$p_2, L_2, p_1, L_1, F_{12}$

Table 6: Isolability sets

Faults	V_I
f_1, f_2	CV_V
f_1, f_3	CV_V, P_V
f_1, f_4	p_{ZP}, P_V, CV_V, S_V
f_1, f_5	n, P_V, CV_V, S_V
f_1, f_Σ	SN_1, P_V, CV_V, S_V
f_2, f_3	P_V
f_2, f_4	p_{ZP}, P_V, S_V
f_2, f_5	n, P_V, S_V
f_2, f_Σ	SN_1, P_V, S_V
f_3, f_4	p_{ZP}, S_V
f_3, f_5	n, S_V
f_3, f_Σ	SN_1, S_V
f_4, f_Σ	p_{ZP}, SN_1
f_5, f_Σ	n, SN_1
f_4, f_5	p_{ZP}, n

Table 7: Cutting path vertices

Faults	Cutting path vertices
f_6, f_7	F_{23}, p_2, F_{12}, L_2
f_6, f_8	$L_3, F_{23}, p_2, F_3, p_3, F_{12}, L_2$
f_6, f_9	p_1, L_1, F_{12}
f_6, f_{10}	p_2, F_{12}, L_2
f_6, f_{11}	$L_3, F_{23}, p_2, p_3, F_{12}, L_2$
f_7, f_8	L_3, F_{23}, F_3, p_3
f_7, f_9	$p_1, F_{23}, p_2, L_1, F_{12}, L_2$
f_7, f_{10}	F_{23}, p_2, L_2
f_7, f_{11}	L_3, F_{23}, p_3
f_8, f_9	$L_3, p_1, F_{23}, p_2, F_3, L_1, p_3, F_{12}, L_2$
f_8, f_{10}	$L_3, F_{23}, p_2, F_3, p_3, L_2$
f_8, f_{11}	L_3, F_3, p_3
f_9, f_{10}	$p_1, p_2, L_1, F_{12}, L_2$
f_{10}, f_{11}	$L_3, F_{23}, p_2, p_3, L_2$

The faults f_1, f_2, f_3, f_Σ are detectable and distinguishable pairwise. The faults f_4, f_5 are detectable and distinguishable from the other faults. In can also be noted that deletion of any sensor changes diagnosability properties of the system.

3.3.3. Strongly connected components The further analysis will be carried out for each of the strongly connected components separately.

The strongly connected components with faults are shown in Figs. 9 and 10. Only one fault influences the component SN_1; therefore, only the faults in SN_0 should be considered.

The list of the vertices cutting the path between each pair of faults in the component SN_0 is shown in Table 7. These sets were calculated using Algorithm 2.

The cutting path vertices can be inputs of a model used for fault isolation. For each of the cutting path vertices, the set of possible model outputs is calculated. These sets are obtained by Algorithm 3. The results for the pair $\{f_6, f_7\}$ are shown in Table 5.

Exemplary minimal solutions for fault isolability in SN_0 are as follows: $\{L_2, F_3, p_1, p_3\}$, $\{F_{23}, p_2, L_1, F_3\}$, $\{L_3, p_2, L_1, F_3\}$, $\{F_{23}, F_3, L_1, L_2\}$, $\{p_1, F_{23}, F_3, F_{12}\}$, $\{p_1, F_{23}, F_3, F_{12}\}$, $\{p_1, p_2, F_3, p_3\}$, $\{L_3, F_3, L_1, L_2\}$. For each of these the faults within SN_0 are distinguishable pairwise.

4. Conclusions

In this paper, the sensor placement method was proposed with use of GP. The sensor placement problem for the fault diagnosis was discussed earlier, but the proposed solution has the following features differing it from the existing approaches:

- The GP is used as a qualitative model of a process to be diagnosed. This model can be obtained at the design stage, without knowledge about the process equations.
- This solution is intended for diagnosis systems using residuals comparing the measured value of a variable with a value calculated by a local black-box model and the results are consistent with that assumption.
- This paper shows how to deal with cycles in a graph within the SDG approach.
- The part of isolability analysis is carried out separately within each strongly connected component, which can limit computational complexity. Moreover, the solutions are provided separately for each strongly connected component, which can be suitable when the number of the solutions is large.

The proposed solution can be extended by including the signs of the influence in the isolability analysis or by analyzing signatures of multiple faults (like in [9]).

Acknowledgments

The work was partially financed from the grant of Dean of Faculty of Mechatronics.

References

[1] Bagajewicz M J 2001 *Process Plant Instrumentation: Design and Upgrade* (Lancaster, Pennsylvania: Technomic Publishing Company, Inc.)
[2] Krysander M and Frisk E 2008 *Systems, Man and Cybernetics, Part A: Systems and Humans, IEEE Transactions on* **38** 1398–1410 ISSN 1083-4427
[3] Rosich A 2012 *IFAC Safeprocess* pp 391–396
[4] Rosich A, Frisk E, Åslund J, Sarrate R and Nejjari F 2012 *IEEE Transactions on Systems, Man, and Cybernetics, Part A: Systems and Humans* **42** 371 – 381
[5] Travé-Massuyès L, Escobet T and Milne R 2001 *17th International Joint Conference on Artificial Intelligence*
[6] Yassine A A, Ploix S and Flaus J M 2008 *International Journal of Applied Mathematics and Computer Science* **18** 497–512
[7] Narasimhan S, Mosterman P and Biswas G 1998 *International Workshop on Principles of Diagnosis* pp 94–101
[8] Khemliche M, Bouamama B O and Haffaf H 2006 *Sensors and Actuators A: Physical* **132** 547–556
[9] Raghuraj R, Bhushan M and Rengaswamy R 1999 *AIChE Journal* **45** 310–322 ISSN 1547-5905 URL `http://dx.doi.org/10.1002/aic.690450212`
[10] Bhushan M, Narasimhan S and Rengaswamyb R 2008 *Computers & Chemical Engineering* **32** 1067–1084
[11] Yang F, Xiao D and Shah S L 2009 *Sensors* **9** 8579–92
[12] Sztyber A, Ostasz A and Kościelny J 2015 *Systems, Man, and Cybernetics: Systems, IEEE Transactions on* **45** 1004–1017 ISSN 2168-2216
[13] Tarjan R E 1972 *SIAM Journal on Computing* **1** 146Ű160

IOP Conf. Series: Journal of Physics: Conf. Series **783** (2017) 012008 doi:10.1088/1742-6596/783/1/012008

Influence of multi-valued diagnostic signals on optimal sensor placement

Kornel Rostek
Warsaw University of Technology, Faculty of Mechatronics, św. A. Boboli 8, Warsaw, Poland
E-mail: k.rostek@mchtr.pw.edu.pl

Abstract. In this paper the comparison of results of optimal sensor placement with the use of binary and multi-valued diagnostic signals is given. The exoneration assumption was introduced and its effects were discussed. The influence of multi-valued diagnostic signals on several fault isolability metrics were discussed. The optimal sensor placement problem under budgetary constraints is formulated. A branch-and-bound algorithm solving this problem is described. It is used in context of three tanks system with 25 possible diagnostic signals.

1. Introduction

The fault isolability is one of the elementary features characterizing the quality of a diagnosis. Therefore, it is used for a comparison of effectiveness of different methods of Fault Detection and Isolation (FDI).

The performance of a FDI system for a given industrial process is strongly dependent on available measurements. A FDI system performance figure can often be improved by installing additional sensors providing additional information about a process. However, it is vital to achieve the best possible FDI system performance without additional costs. The problem of optimal sensor selection can be understood as a combinatorial problem of selecting the optimal set of measurements.

In recent years, numerous papers discussed different problems of the optimal sensor placement. They usually consider the required minimum fault isolability of the diagnostic system [1, 2]. Some of proposed methods also maximize designed fault isolability using heuristic methods, e.g. genetic algorithms [3].

The model-based FDI considers faults as deviations from nominal values of process parameters or as unknown process inputs. Faults are detected when system model and measurements behave differently. In [4], a method for searching for the optimal sensor set based on Analytical Redundancy Relations (ARRs) is proposed. First, all ARRs are found under the assumption that all sensor candidates are installed. Then, a sensor set is selected that minimizes the cost while satisfying detectability and isolability requirements. However, this solution is computationally expensive. A modified, incremental approach, using Minimal Structurally Overdetermined (MSO) sets, was proposed in [5]. In [6] the Binary Integer Programming is used to find the optimal sensor set using the set of all possible MSO sets. FDI requirements were ensured by means of non-linear constraints. The resulting problem is computationally difficult to solve. This method was further improved in [7] and [8]. There, FDI requirements were specified as linear constraints. As the cost function is linear, the problem falls into Binary Integer

Linear Programming (BILP). It can be efficiently solved with branch-and-bound algorithm with standard Linear Programming (LP) solver. Those methods were thoroughly compared in [9]. Budgetary constraints were analyzed in [10]. Branch-and-bound algorithm is used to obtain the optimal solution. Regardless of chosen method, simple, qualitative methods of analysis of fault isolability are insufficient. Generalized, quantitative method of fault isolability analysis is required.

This paper presents the analysis of different approaches to the optimal sensor placement problem with multi-valued diagnostic signals. The main contribution of this paper is assessment of the influence of multi-valued diagnostic signals in optimal sensor placement problem. Both resulting sensor sets and computational complexity of optimization procedure were taken into account. Application of multi-valued signals in fault diagnosis often allows to obtain more accurate diagnosis. Nevertheless, designing such FDI system is usually more time consuming and therefore expensive. An efficient methodology of designing multi-valued FDI systems is required. Different metrics of isolability were used as a performance indices in order to analyze their effect on the optimization results. The model of three tank system was used as a benchmark.

The paper is organized as follows. In Section 2, the preliminary definitions used in this work are given. Section 3 defines different metrics of fault isolability. Section 4 presents the proposed optimization procedure. Section 5 describes the example of a three tank system. In Section 6 the achieved results are presented and discussed. Conclusions and final remarks section finalizes this paper.

2. Preliminaries

A signal sensitive to faults is considered as a diagnostic signal in FDI. A symptom is a value of a diagnostic signal which indicate fault or faults. In the case of multi-valued diagnostic signals, one fault type may generate different values of each diagnostic signal. A fault signature is a vector of diagnostic signal values associated with a particular fault [11]. The specific vector of values of diagnostic signals is called an alternative signature [11]. In case of binary diagnostic signals each fault has exclusively one alternative signature, while in case of mutli-valued diagnostic signals there might be multiple alternative signatures.

The Binary Diagnostic Matrix (BDM) or Incidence Matrix is a widely used form of notation of relation between binary diagnostic signals and faults. It has a form of a column matrix. Each fault signature constitutes one column. Each row displays sensitivity of a given diagnostic signal to each individual fault. The analogous form of notation in the case of multi-valued diagnostic signals is Fault Information System (FIS) defined in [12].

There are different definitions of fault isolability. Generally, faults are considered isolable when at least some of their signatures are different [13]. In the case of multi-valued diagnostic signals, it is possible that a pair of faults is conditionally isolable i.e. isolable for some alternative signatures and unisolable for others.

Commonly, in FDI an exoneration assumption is accepted. It states that a lack of a symptoms exonerates a fault. It means that all symptoms must appear for a fault isolation. This assumption is not always valid. Due to dynamics of symptoms and different sensitivity to faults, they may not appear simultaneously or may even not appear at all.

3. Metrics of fault isolability

There are known quantitative methods of determining fault isolability for pairs of faults. In the simplest case, Hamming distance between binary fault signatures can be used. Examples of other, more advanced metrics, using an internal form of residuals, can be found in [14] and [15]. Unfortunately, those measures can be only used when analyzing pairs of faults. To analyze a whole diagnostic system other methods are required.

Table 1. An example of Binary Diagnostic Matrix.

	f_1	f_2	f_3	f_4
s_1	1	1	1	
s_2		1	1	1
s_3				1

Let us consider a simple example of a BDM given in the Tab. 1. It consists of three binary diagnosis signals detecting four possible faults. This example will be further used in this paper.

3.1. Diagnosability degree

One of the most commonly used measures of fault isolability is diagnosability degree [1, 2, 3, 16, 17]. It was defined in [17]. The value of this measure is calculated in two steps.

(1) The set of all considered faults is divided into disjoint subsets of unisolable faults. Those sets are called D-classes.

(2) The number of D-classes denoted as D_c, is divided by the number of all considered faults. The acquired ratio is called diagnosability degree:

$$\frac{D_c}{card(F)}. \qquad (1)$$

In the special case when all fault are isolable, the number of D-classes is equal to the number of all considered faults. Diagnosability degree is then equal to 1.

If all faults are unisolable then diagnosability degree is equal to $^1/_{card(F)}$.

Diagnosability degree is a simple measure and easy to calculate and to interpret. Unfortunately, it has some disadvantages. Fault diagnosability for each pair of faults is defined only for binary diagnostic matrices and does not apply to multi-valued diagnostic matrices, such as Fault Information System FIS [18].

Considering the example given in the Tab. 1, there are three D-classes of unisolable faults: $\{f_1\}$, $\{f_2, f_3\}$, $\{f_4\}$. Therefore diagnosability degree is equal to 3/4.

3.2. Diagnosis accuracy

In [19], a simple fault isolability measure was defined. It is called a diagnosis accuracy. It is calculated as the reciprocal of an average number of faults in a diagnosis.

$$\left[\frac{\sum_{d_i \in D} card(d_i)}{card(D)}\right]^{-1}, \qquad (2)$$

where d_i denotes the ith diagnosis from the set of all diagnoses D. Only single faults will be analyzed in this work .

When all diagnoses point out only single faults, then value of diagnosis accuracy is equal to 1. In general case value of diagnosis accuracy is in $(0, 1\rangle$.

The value of this measure depends on the chosen diagnosis approach. For example, in the case of binary diagnostic matrix, diagnosis accuracy delivers different results depending whether the exoneration assumption is used or not. This results from ignoring the part of information regarding lack of some symptoms. Nevertheless, in some cases such approach is justified, because after occurrence of a fault, symptoms do not appear simultaneously. Different interim diagnoses are considered with appearance of successive symptoms.

Average number of faults in a diagnosis can be difficult to calculate in case of complex diagnostic systems with multi-valued or continuous diagnostic signals or taking into account sequences of appearance of symptoms. In this case the number of possible diagnoses, grows exponentially with a number of faults.

The value of diagnosis accuracy depends on the exoneration assumption. Returning to the example given in the Tab. 1, with exoneration assumption there are three possible diagnoses: $\{f_1\}$, $\{f_2, f_3\}$, $\{f_4\}$. Average number of faults in a diagnosis equals 4/3 and diagnosis accuracy is equal to 3/4. In case of binary diagnostic signals, under exoneration assumption, the value of diagnosis accuracy is always equal to diagnosability degree, because possible diagnoses are identical with D-classes.

Without exoneration assumption there are also possible three diagnoses: $\{f_1, f_2, f_3\}$, $\{f_2, f_3\}$, $\{f_4\}$. The diagnosis accuracy is then equal to $[6/3]^{-1} = 1/2$.

3.3. Isolability index

Isolability index is often used, as the fault isolability measure [20, 10]. It is defined as a number of ordered pairs of isolable faults. Maximum number of isolability index depends on a number of considered faults. If isolability relation is asymmetric, then maximal isolability index is equal to $n\,(n-1)$, where n is the number of considered faults. Similarly to diagnosability degree, the isolability index is defined for binary diagnostic signal.

Following the example given in the Tab. 1, with exoneration assumption there are ten ordered pairs of faults if the first fault is isolable from the second: (f_1, f_2), (f_1, f_3), (f_1, f_4), (f_2, f_1), (f_2, f_4), (f_3, f_1), (f_3, f_4), (f_4, f_1), (f_4, f_2), (f_4, f_3). It is worth noticing that with exoneration assumption this relation is symmetric.

Without exoneration assumption there are eight such pairs: (f_1, f_2), (f_1, f_3), (f_1, f_4), (f_2, f_4), (f_3, f_4), (f_4, f_1), (f_4, f_2), (f_4, f_3).

3.4. Isolability measure

In [21] a new measure of fault isolability for multi-valued diagnostic systems was proposed. It is calculated in two steps.

(1) Calculate for each ordered pair of faults:

$$D(f_i, f_j) = \frac{card(S_{i,j})}{card(S_i)} \tag{3}$$

where S_i is the set of all possible alternative signatures of f_i, and $S_{i,j}$ is the set of alternative signatures of f_i excluding f_j. If faults are unconditionally isolable [18] then $D(f_i, f_j) = 1$, if they are unconditionally unisolable then $D(f_i, f_j) = 0$, otherwise $D(f_i, f_j) \in (0, 1)$.

(2) Calculate the measure as:

$$\frac{1}{(n-1)n} \sum_{i=1}^{n} \sum_{\substack{j=1 \\ j \neq i}}^{n} D(f_i, f_j). \tag{4}$$

Value of this measure has a practical interpretation. It is an average fraction of all single faults that are excluded during each diagnosis. There are $n+1$ possible diagnoses: a faultless state and n single fault states.

This approach for calculation of isolability measure can be used without modifications for any bi-valued form of notation of diagnostic relation, such as directions in residual space or a sequences of symptoms.

Table 2. Values of $D(f_i, f_j)$.

	f_1	f_2	f_3	f_4
f_1	0	0	0	1
f_2	1	0	0	1
f_3	1	0	0	1
f_4	1	1	1	0

The isolability measure can be calculated for the example given in the Tab. 1 The values of $D(f_i, f_j)$ are presented in the Tab. 2. Using those values we can calculate isolability measure as $\frac{1}{(n-1)n}\sum_{i=1}^{n}\sum_{j=1}^{n} D(f_i, f_j) = 2/3$.

4. Optimization Problem Formulation

The optimal sensor set is determined using following conditions.

(1) A solution that allow all faults to be detectable is preferred i.e. for each fault there is at least one diagnostic signal sensitive to this fault.

(2) If multiple solutions offer the same detectability, the solution with higher isolability metric is chosen.

(3) If there are multiple solutions with the same detectability and isolability metric, the cheapest one is selected.

The optimization problem described in this section is non-linear and computationally difficult to solve. There are generalized techniques that will allow to find a solution. The optimization problem was solved with the branch-and-bound algorithm [10]. For each sensor there was defined a discrete decision variable with three possible values: positive, negative and unknown. In the beginning, all decision variables are set to undecided.

If a current solution is infeasible, i.e. the cost of chosen sensors exceeds available budget, the solution is rejected. Current best fully decided solution is retained by algorithm. The upper bound of currently analyzed solution is compared to current best solution according to criteria given above. If the achieved result is worse, then analyzed solution can be rejected, because it will not lead to solution better than current best. This operation is called bounding.

During each step of the algorithm, after bounding, one undecided variable is chosen and two new intermediate solutions are created, one with positive value of this variable and the other with negative. This operation is called branching. Those solutions are then added to solution queue.

5. Three Tank System example

The proposed sensor placement optimization approach was tested on the example of a FIS for a system of three tanks arranged in series (TTS) as in Fig. 1. In total 16 faults were considered (Tab. 3). Seven proposed new sensor locations with cost estimation are presented in (Tab. 4). The cost of CV_v is 0, because this signal is already available in the diagnosed system.

By using method presented in [23], the 25 possible diagnostic signals were generated, using sensors depicted in the Tab. 4. Tri-valued diagnostic signals were considered. The method proposed in this paper was tested for all budget values lower or equal to the total cost of all proposed sensors.

Table 3. Considered faults.

Fault	Description
f_1	measurement chain $FI102$ fault
f_2	measurement chain $LI103$ fault
f_3	measurement chain $LI104$ fault
f_4	measurement chain $LIC105$ fault
f_5	measurement chain $XI101$ fault
f_6	control signal fault
f_7	valve actuator fault
f_8	valve fault
f_9	pump fault
f_{10}	low water level fault
f_{11}	clogging between tanks $T1$ and $T2$
f_{12}	clogging between tanks $T2$ and $T3$
f_{13}	clogged output flow from tank $T3$
f_{14}	leak from tank $T1$
f_{15}	leak from tank $T2$
f_{16}	leak from tank $T3$

Table 4. Sensor locations and costs.

Symbol	Measurement	Cost
F_1	Input flow	5
P_v	Valve position	2
L_1	Level in T1	1
L_2	Level in T2	1
L_3	Level in T3	1
CV_v	Control signal	0
p_{zp}	Pressure on pump inlet	1
n	Pump rotational speed	2

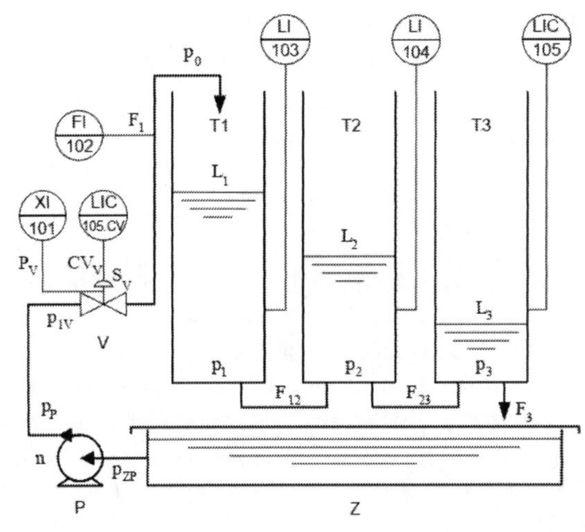

Figure 1. Three tank system [22].

6. Results

Two main cases were analyzed. In the first one, the BDM obtained from Tab. 5 was used. In the second case, three-valued diagnostic signals were used as presented in Tab. 5. In each case, the optimization process was repeated for each appropriate isolability metric and for each reasonable budget value. The optimization results from the first case are presented in the Tab. 7. Similarly, the results from the second case are presented in the Tab. 6. For each combination of metrics and available budgets the optimal sensor set is presented with the corresponding metric value.

The achieved results are very close despite the type of the isolability metrics. In each case, if a budget is greater than 7, the results are identical for all considered metrics. This is due to the fact that the sensor set $\{CV_v, L_1, L_2, L_3, P_v, n, p_{zp}\}$ is the cheapest set providing detectability of all considered faults and detectability had highest priority in the optimization procedure. With

Table 5. Diagnostic signals and the expected directions of its change caused by faults.

Signals (sensor sets)	f_1	f_2	f_3	f_4	f_5	f_6	f_7	f_8	f_9	f_{10}	f_{11}	f_{12}	f_{13}	f_{14}	f_{15}	f_{16}
$s_1(F_1, CV_v, p_{zp}, n)$	±					±	±	−	−	−						
$s_2(F_1, P_v, p_{zp}, n)$	±				±			−	−	−						
$s_3(P_v, CV_v)$					±	±	±									
$s_4(F_1, L_3)$	±			±							−	−	+	−	−	−
$s_5(P_v, L_3, p_{zp}, n)$				±	±			−	−	−	−	−	+	−	−	−
$s_6(L_3, CV_v, p_{zp}, n)$				±		±	±	−	−	−	−	−	+	−	−	−
$s_7(L_2, L_3)$			±	±								−	+			−
$s_8(L_1, L_3)$		±		±							−	−	+		−	−
$s_9(F_1, L_2)$	±		±								−	+	+	−	−	−
$s_{10}(P_v, L_2, p_{zp}, n)$			±		±			−	−	−	−	+	+	−	−	−
$s_{11}(L_2, CV_v, p_{zp}, n)$			±			±	±	−	−	−	−	+	+	−	−	−
$s_{12}(F_1, L_2, L_3)$	±		±	±							−	+		−	−	
$s_{13}(L_1, L_2, L_3)$		±	±	±							−	+		−		
$s_{14}(L_2, L_3, CV_v, p_{zp}, n)$			±	±		±	±	−	−	−	−	+		−	−	
$s_{15}(P_v, L_2, L_3, p_{zp}, n)$			±	±	±			−	−	−	−	+		−	−	
$s_{16}(L_1, L_2)$	±		±								−	+	+		−	−
$s_{17}(F_1, L_1)$	±	±									+	+	+	−	−	−
$s_{18}(F_1, L_1, L_2)$	±	±	±								+			−		
$s_{19}(P_v, L_1, L_2, p_{zp}, n)$		±	±		±			−	−	−	+			−		
$s_{20}(L_1, CV_v, p_{zp}, n)$		±				±	±	−	−	−	+	+	+	−	−	−
$s_{21}(P_v, L_1, p_{zp}, n)$		±			±			−	−	−	+	+	+	−	−	−
$s_{22}(F_1, L_1, L_3)$	±	±		±							+	+		−	−	
$s_{23}(L_1, L_3, CV_v, p_{zp}, n)$		±		±		±	±	−	−	−	+	+		−	−	
$s_{24}(P_v, L_1, L_3, p_{zp}, n)$		±		±	±			−	−	−	+	+		−	−	
$s_{25}(L_1, L_2, CV_v, p_{zp}, n)$		±	±			±	±	−	−	−	+			−		

the available budget equal 13 all proposed sensors are selected in each case.

There are 4 differences in optimization results obtained form multi-valued and binary diagnostic signals. They were marked bold in the Tab. 6. The sensor sets from the Tab. 7 that are different than those from the Tab. 6 were analyzed. Their corresponding values of isolability measures obtained with FIS were calculated. Diagnosis accuracy with exoneration for the set $\{CV_v, P_v\}$ equals 0.13. Diagnosis accuracy without exoneration for the same set is 0.11. Both values are identical as in the case of BDM because the signal s_3 does not provide new isolability options in multi-valued case. The set $\{CV_v, L_2, L_3, P_v, n, p_{zp}\}$ with diagnosis accuracy without exoneration gives value 0.32. Finally, isolability measure of the set $\{CV_v, L_1, L_2, L_3, P_v\}$ is 0.69.

The results obtained with BDM are close to optimal results obtained with FIS. This shows that result of binary optimization can be chosen as reasonable initial guess for optimization algorithm used for multi-valued diagnostic signals. It is especially important because computational complexity of analyzed metrics of isolability is directly proportional to an average number of alternative fault signatures. In analyzed TTS example, the optimization procedure for binary diagnostic signals is on average 2.1 times faster than for multi-valued ones.

It is also worth noticing that in all the cases the value of isolability measure obtained for FIS is greater than for BDM. Similarly, greater value of isolability is obtained where exoneration

Table 6. Results of optimization with various measures of isolability for the set of multi-valued diagnostic signals.

Available budget						
2:	from 3 to 4:	5:	6:	7:	from 8 to 12:	13:
			Diagnosis accuracy with exoneration:			
0.17: $\mathbf{L_1\ L_2}$	0.56: $L_1\ L_2\ L_3$	0.63: $CV_v\ L_1\ L_2\ L_3\ P_v$	0.67: $CV_v\ L_1\ L_2\ L_3$ n p_{zp}	0.67: $CV_v\ L_1\ L_2\ L_3\ n\ p_{zp}$	0.75: $CV_v\ L_1\ L_2\ L_3\ P_v$ n p_{zp}	0.81: $CV_v\ F_1\ L_1\ L_2\ L_3\ P_v$ n p_{zp}
			Diagnosis accuracy without exoneration:			
0.13: $\mathbf{L_2\ L_3}$	0.31: $L_1\ L_2\ L_3$	0.31: $CV_v\ L_1\ L_2\ L_3\ P_v$	0.33: $CV_v\ L_1\ L_2\ L_3$ n p_{zp}	0.34: $\mathbf{F_1\ L_2\ L_3}$	0.54: $CV_v\ L_1\ L_2\ L_3\ P_v$ n p_{zp}	0.75: $CV_v\ F_1\ L_1\ L_2\ L_3\ P_v$ n p_{zp}
			Isolability Measure:			
0.33: $L_1\ L_2$	0.53: $L_1\ L_2\ L_3$	0.73: $\mathbf{CV_v\ L_1\ L_2}$ **n** $\mathbf{p_{zp}}$	0.83: $CV_v\ L_1\ L_2\ L_3$ n p_{zp}	0.85: $CV_v\ L_1\ L_2\ P_v\ n\ p_{zp}$	0.93: $CV_v\ L_1\ L_2\ L_3\ P_v$ n p_{zp}	0.97: $CV_v\ F_1\ L_1\ L_2\ L_3\ P_v$ n p_{zp}

is taken into consideration. The optimization results for budget value 7 with exoneration assumption are in all cases identical to those with available budget equal 6, whereas without that assumption they are different. This implies that smaller differences between solution are taken into consideration without adoption of exoneration assumption.

7. Conclusion

In this work the influence of multi-valued diagnostic signals on optimal sensor placement problem was discussed on an example. The results for both types of diagnostic signal were similar in most cases. Therefore, using results from optimal sensor placement is a good initial guess for optimal sensor placement with multi-valued diagnostic signals. It would allow to early ignore groups of possible solutions using branch-and-bound algorithm and result in significant decrease in time needed to solve optimization problem. Following algorithm can be used to efficiently solve optimal sensor placement problem for FIS:

(1) Transform FIS into BDM by replacing nonzero values with 1.

(2) Solve the optimal sensor placement problem for BDM.

(3) Solve the original problem using the solution from step (2) as a initial guess.

The exoneration assumption affects results of optimal sensor placement. When designing diagnostic system it is important to analyze if this assumption can be met. If the expected time between occurrence of the first and last symptoms is much greater than the time needed to formulate diagnosis, then assumption of exoneration may generate false diagnoses.

Acknowledgments

This work was partially supported by the Mechatronics Faculty Deans Grant 504/02141.

References

[1] Yassine A, Ploix S and Flaus J M 2008 *International Journal of Applied Mathematics and Computer Science* **18** 497–512

[2] Krysander M, Aslund J and Nyberg M 2008 *Systems, Man and Cybernetics, Part A: Systems and Humans, IEEE Transactions on* **38** 197–206

[3] Spanache S, Escobet T and Trave-Massuyes L 2004 Sensor placement optimisation using genetic algorithms *Proceedings of the 15th International Workshop on Principles of Diagnosis, DX-04*

Table 7. Results of optimization with various measures of isolability for the set of binary diagnostic signals.

Available budget						
2:	from 3 to 4:	5:	6:	7:	from 8 to 12:	13:
Diagnosis accuracy with exoneration:						
0.13: CV_v P_v	0.50: L_1 L_2 L_3	0.56: CV_v L_1 L_2 L_3 P_v	0.63: CV_v L_1 L_2 L_3 n p_{zp}	0.63: CV_v L_1 L_2 L_3 n p_{zp}	0.69: CV_v L_1 L_2 L_3 P_v n p_{zp}	0.75: CV_v F_1 L_1 L_2 L_3 P_v n p_{zp}
Diagnosis accuracy without exoneration:						
0.11: CV_v P_v	0.205: L_1 L_2 L_3	0.214: CV_v L_1 L_2 L_3 P_v	0.23: CV_v L_1 L_2 L_3 n p_{zp}	0.24: CV_v L_2 L_3 P_v n p_{zp}	0.39: CV_v L_1 L_2 L_3 P_v n p_{zp}	0.55: CV_v F_1 L_1 L_2 L_3 P_v n p_{zp}
Isolability Measure:						
0.26: L_1 L_2	0.48: L_1 L_2 L_3	0.64: CV_v L_1 L_2 L_3 P_v	0.73: CV_v L_1 L_2 L_3 n p_{zp}	0.75: CV_v L_1 L_2 P_v n p_{zp}	0.87: CV_v L_1 L_2 L_3 P_v n p_{zp}	0.93: CV_v F_1 L_1 L_2 L_3 P_v n p_{zp}
Diagnosability degree:						
0.13: CV_v P_v	0.50: L_1 L_2 L_3	0.56: CV_v L_1 L_2 L_3 P_v	0.63: CV_v L_1 L_2 L_3 n p_{zp}	0.63: CV_v L_1 L_2 L_3 n p_{zp}	0.69: CV_v L_1 L_2 L_3 P_v n p_{zp}	0.75: CV_v F_1 L_1 L_2 L_3 P_v n p_{zp}
Isolability Index with exoneration:						
126: L_1 L_2	194: L_1 L_2 L_3	218: CV_v L_1 L_2 L_3 P_v	218: CV_v L_1 L_2 L_3 P_v	218: CV_v L_1 L_2 L_3 P_v	224: CV_v L_1 L_2 L_3 P_v n p_{zp}	230: CV_v F_1 L_1 L_2 L_3 P_v n p_{zp}
Isolability Index without exoneration:						
63: L_1 L_2	115: L_1 L_2 L_3	154: CV_v L_1 L_2 L_3 P_v	175: CV_v L_1 L_2 L_3 n p_{zp}	179: CV_v L_1 L_2 P_v n p_{zp}	208: CV_v L_1 L_2 L_3 P_v n p_{zp}	223: CV_v F_1 L_1 L_2 L_3 P_v n p_{zp}

[4] Travé-Massuyes L, Escobet T and Olive X 2006 *Systems, Man and Cybernetics, Part A: Systems and Humans, IEEE Transactions on* **36** 1146–1160

[5] Rosich A, Sarrate R, Puig V and Escobet T 2007 Efficient optimal sensor placement for model-based fdi using an incremental algorithm *Decision and Control, 2007 46th IEEE Conference on* (IEEE) pp 2590–2595

[6] Sarrate R, Puig V, Escobet T and Rosich A 2007 Optimal sensor placement for model-based fault detection and isolation *Decision and Control, 2007 46th IEEE Conference on* (IEEE) pp 2584–2589

[7] Nejjari F, Sarrate R and Rosich A 2010 Optimal sensor placement for fuel cell system diagnosis using bilp formulation *Control & Automation (MED), 2010 18th Mediterranean Conference on* (IEEE) pp 1296–1301

[8] Rosich A, Sarrate R and Nejjari F 2009 Optimal sensor placement for fdi using binary integer linear programming *20th International Workshop on Principles of Diagnosis*

[9] Sarrate R, Nejjari F and Rosich A 2012 *Fault Detection, Supervision and Safety of Technical Processes, Volume# 8— Part# 1*

[10] Sarrate R, Nejjari F and Rosich A 2012 Sensor placement for fault diagnosis performance maximization under budgetary constraints *2nd International Conference on Systems and Control*

[11] Bartyś M 2013 *International Journal of Applied Mathematics and Computer Science* **23** 407–417

[12] Korbicz J, Kościelny J, Kowalczuk Z and Cholewa W 2004 *Fault diagnosis: models, artificial intelligence, applications* (Springer Science & Business Media)

[13] Gertler J 1997 *Control Engineering Practice* **5** 653 – 661 ISSN 0967-0661 URL `http://www.sciencedirect.com/science/article/pii/S0967066197000476`

[14] Khorasgani H, Eriksson D, Biswas G, Frisk E and Krysander M 2014 Off-line robust residual selection using sensitivity analysis *Proceedings of 25th International Workshop on Principles of Diagnosis, Graz*

[15] Jung D, Frisk E and Krysander M 2015 Quantitative isolability analysis of different fault modes *IFAC-PapersOnLine, Paris*

[16] Rouissi F and Hoblos G 2013 Fault tolerance in wind turbine sensor systems for diagnosability properties guarantee *Conference on Control and Fault-Tolerant Systems (SysTol)* pp 523–28

[17] Travé-Massuyès L, Escobet T and Milne R 2001 Model-based diagnosability and ensor placement application to frame 6 gas turbine subsystem *In 12th Int, Workshop on principles of diagnosis* pp 205–12

[18] Kościelny J M and Zakroczymski K 2001 *Bulletin of the Polish Academy of Sciences. Technical Sciences* **49** 323–36

[19] Bartyś M, Patton R, Syfert M, de las Heras S and Quevedo J 2006 *Control Engineering Practice* **14** 577 – 596 ISSN 0967-0661 a Benchmark Study of Fault Diagnosis for an Industrial ActuatorA Benchmark Study of Fault Diagnosis for an Industrial Actuator

[20] Sarrate R, Blesa J, Nejjari F and Quevedo J 2014 *Water Science and Technology: Water Supply, 795–803*

[21] Rostek K 2014 Measure of fault isolability of diagnostic system *25th International Workshop on Principles of Diagnosis*

[22] Ostasz A 2006 *Graf przyczynowo-skutkowy procesu i jego zastosowanie w wyznaczaniu zbioru residuow i relacji diagnostycznej* Ph.D. thesis Warsaw University of Technology

[23] Sztyber A, Ostasz A and Koscielny J 2015 *Systems, Man, and Cybernetics: Systems, IEEE Transactions on* **45** 1004–1017

Intelligent Case Based Decision Support System for Online Diagnosis of Automated Production System

N Ben Rabah[1, 2], R Saddem[1], F Ben Hmida[2], V Carre-Menetrier[1] and M Tagina[2]

[1] Centre de Recherche en STIC (CReSTIC), Reims, France
[2] National School of Computer Sciences University of Manouba 2010 Manouba, Tunisia
nourhene.ben-rabah@etudiant.univ-reims.fr, ramla.saddem@univ-reims.fr, faten.benhmida@ensi.rnu.tn, veronique.carre@univ-reims.fr, moncef.tagina@ensi.rnu.tn

Abstract. Diagnosis of Automated Production System (APS) is a decision-making process designed to detect, locate and identify a particular failure caused by the control law. In the literature, there are three major types of reasoning for industrial diagnosis: the first is model-based, the second is rule-based and the third is case-based. The common and major limitation of the first and the second reasonings is that they do not have automated learning ability. This paper presents an interactive and effective Case Based Decision Support System for online Diagnosis (CB-DSSD) of an APS. It offers a synergy between the Case Based Reasoning (CBR) and the Decision Support System (DSS) in order to support and assist Human Operator of Supervision (HOS) in his/her decision process. Indeed, the experimental evaluation performed on an Interactive Training System for PLC (ITS PLC) that allows the control of a Programmable Logic Controller (PLC), simulating sensors or/and actuators failures and validating the control algorithm through a real time interactive experience, showed the efficiency of our approach.

1. Introduction

Recent technological revolutions in the industrial field result in the appearance of more and more complex APS. This complexity is accompanied by the growing needs in terms of operation and security. Indeed, it is essential to implement systematic diagnosis approaches to detect, locate and identify possible failures in real time. In this work, our goal is to support and assist HOS in his/her decision process in order to accomplish his/her diagnosis task. To reach such a goal, it is essential to define the HOS and determine his/her tasks. So, who is a HOS?

HOS is a person responsible for the driving of an APS. He/she has to assume essentially three different tasks: acquiring knowledge from data of sensors and actuators, analyzing knowledge and making the diagnosis decision. Every time, he/she tries to remember past experiences which represent observed failures in similar situations and which can lead to similar results. In fact, it is difficult to perform these tasks, essentially the analysis of a large amount of binary data and the making of the relevant diagnosis decision in real time, notably when there is insufficient data.

In this paper, we present a synergy between the CBR approach and the DSS to propose an Intelligent CB-DSSD of an APS. The proposed system includes two phases: An offline phase in which data, describing the system behavior as a set of cases, are collected and represented. Each instance of case is labeled as a normal or faulty behavior of the system to diagnose. The subset of cases describing normal behavior of the system is listed in normal case base. However, the subset of cases, describing faulty behavior of the system, is stored in abnormal case base. An online phase takes as input the diagnosis request of a HOS and as output the diagnosis response (faulty or normal behavior).

The paper is structured as follows. In section 2, we provide an overview of failures to diagnose, diagnosis definitions and diagnosis methods of APS. In section 3, we present some recent case-based systems for industrial diagnosis and we enumerate the benefits of using CB-DSSD of APS. The functional model of the proposed system is detailed in section 4. In section 5, we illustrate our proposal with an application on the sorting system of ITS PLC Collection. Conclusion and possible futures works are addressed in the last section.

2. Diagnosis of Automated Production System

Before proposing a diagnosis method for an APS, it is very important to specify first the failures to diagnose, the diagnosis definition and the diagnosis method to employ.

2.1 Failures to Diagnose

An APS is composed of two parts: Control Part (CP) and Operative Part (OP). The CP is a set of software means and information concerning the control and the conduct of the process. It sends orders to OP in order to obtain the desired effects. It allows the communication with humans and other systems. Indeed, the OP performs the commands, sent by the CP, using actuators. It also transmits the reports to the CP (information collected from sensors). The APS study can be carried out from the point of view of either the OP or the CP. Failures, on an APS, are bound to an unanticipated event which comes modifying the operation of the system. It can be either external, caused by customer, environment or raw material, or internal caused by hardware of the control system, OP or CP.

In this work, we treat only internal failures, especially those caused by the OP, such as the stuck-off to 1 or 0 of a sensor or an actuator.

2.2 Diagnosis Definition

In Discrete Event System (DES) community, there are several definitions of the diagnosis function. Some works did not consider the detection as a diagnosis entity. They rather dealt it as a monitoring entity. For example, [1] defined the diagnosis as three entities:

- The location determining the sub-system responsible for the failure.
- The identification specifying the causes behind this failure.
- The explanation which justifies the diagnosis conclusions.

Other works, like [2] and [3], considered the detection as a diagnosis entity. They defined the diagnosis as a decision-making process involving three entities:

- The detection which allows detecting any deviation of the process behavior from normal behavior and alert HOS in a failure case.
- The location which allows specifying the failure origin (area and affected component).
- The identification which allows determining the failure occurrence time, its duration and its severity.

Our work is based on the second definition of diagnosis.

2.3 Diagnosis Methods

Literature distinguishes three industrial diagnosis methods: Rule-based diagnosis, Model-based diagnosis and CBR-based diagnosis.

Rule-based diagnosis [4] requires the translation of the diagnosis problem in boolean expressions. It is easily and efficiently applied in simple industrial systems. However, it needs a lot of time in order that experts convert their knowledge in rules. Besides, it is a method having no learning ability (it cannot handle a failure not planned in its rule base).

Model-based diagnosis [2], [3] is a problem-solving technique based on a model representing the system to be diagnosed. Indeed, the model is a description of real system. It is used to detect and locate multi-failures. However, model creation is a long process and the method has no automated learning ability.

CBR-based diagnosis [5] is a method used to solve new diagnosis problems by using previous experiences stored in a case base. It is the only approach which offers a learning capacity. The knowledge acquisition is more economical than that in the rule-based method and updating cases is much easier. For these reasons, we adopted in our work this diagnosis method.

3. Case Based Decision Support System for Industrial Diagnosis

A DSS is an interactive information system, flexible, adaptable and specifically developed to quickly provide decision makers with relevant information that will help them to make the best decisions. The human-machine cooperation is important in the decision support. It assists the human decision-maker, and therefore guarantees the sharing/distribution of tasks between human-machine [6]. According to [7] and [8], there are several types of DSS:

- *DSS guided by models*: It is based on the assumptions of analysis models type « If ... then ... else » or similar.
- *Data-driven DSS*: It analyzes large amounts of stored data to produce associations, sequences, classifications, segments and forecasts.
- *Web-based DSS*: It supports the web client decision-making process by providing more information before deciding to purchase an item or a service.
- *Communication-driven DSS*: It supports more than one person working on a shared task.
- *Knowledge-driven DSS*: It provides the expertise to solve problems formalized in terms of rules, procedures, recommendations, etc...

In our work, we choose a Knowledge-driven DSS where expertise is formalized as cases and in order to produce new knowledge, we use the CBR approach.

This section provides first an overview of CBR. Secondly, it presents some recent case-based systems for industrial diagnosis that aim at classifying our work and making our choices. Finally, the section ends with enumerating the benefits of using a CB-DSSD of APS.

3.1 Overview of CBR

CBR is a methodology for solving problems, by using past experiences to understand and deal with new problems. Psychologists and cognitive scientists proved that humans routinely use it in their decision processes [9]. Past experiences are organized and stored in a case base. Each solved problem is represented as a case. The latter can be defined as a "contextualized piece of knowledge representing an experience that teaches a lesson fundamental to achieving the goals of the reasoner" [10].

According to [11], a case is a set formed by Problem (P) with its solution (Sol (P)): Case= ((P, Sol (P)). It can be structured in various ways. The old cases, stored in the case base, are called Source Cases (SCs), while each solved problem is named Target Case (TC). The CBR cycle (figure 1) consists of four essential steps [12]:

- *Retrieve*: It allows searching the previous cases most likely to satisfy the objective of reasoning (new problem).
- *Reuse*: The desired solution is extracted from the most relevant case for reuse.
- *Revise*: The proposed solution is revised and adapted in order to become compatible with the current problem. Adaptation can also repair a solution causing negative impact.
- *Retain*: Confirmed solution is stored in the case base for future use.

In CBR, there are three general areas that have to be considered when creating a case base [13]: (a) The structure and representation of the cases themselves. (b) The memory model used to organize the entire case base. (c) The selection of indices used to identify each case.

3.2 Recent Case-Based Systems for Industrial Diagnosis

Many research works use the CBR approach to solve problems in the industrial systems. We study some recent case-based systems that aim at solving diagnosis, maintenance or reconfiguration problems while representing the diagnosis phase in their systems.

We notice that the efficiency of a CBR system is related to the case representation step. Indeed, it can influence all other CBR cycle phases. In the works of [14] and [15], the case representation stage was defined according to the domain-related knowledge through an ontology. In fact, in complex systems, the construction of domain ontology remains a difficult task. [16] proposed a very rich case representation. It is based on 15 indexes showing the problem main features. Nevertheless, the choice of these indexes is done manually. Besides, [20] introduced case representation that considers the different types of knowledge (general and contextual). In their study, the case was represented by a huge number of variables (110 variables). Moreover, in their approach, there is no extraction step of the most important features modeling the problem. [17] suggested a typical case representation based on the observed symptoms, their dates and the information about vehicles. [5] is the only work that considered the temporal aspect to carry out an in-depth analysis through the extraction of sensor signals features using temporal abstraction.

Generally, in the stage of searching resource cases that are more similar to the target cases, the measure of similarity or distance depends on the type of data and information stored in the case base. In the research studies of [16], [14], [18] and [15], the authors used similarity measures that take into account the descriptor relevance degree.

The adaptation phase is a very important stage either because the systems do not treat it as it is the case in [16], [18] and [19], or because the systems merge the selected solutions [5], use the dependency relationships of a domain ontology or conduct substitutional adaptation at the case descriptors. Finally, we note that none of the studied system dealt with the maintenance step.

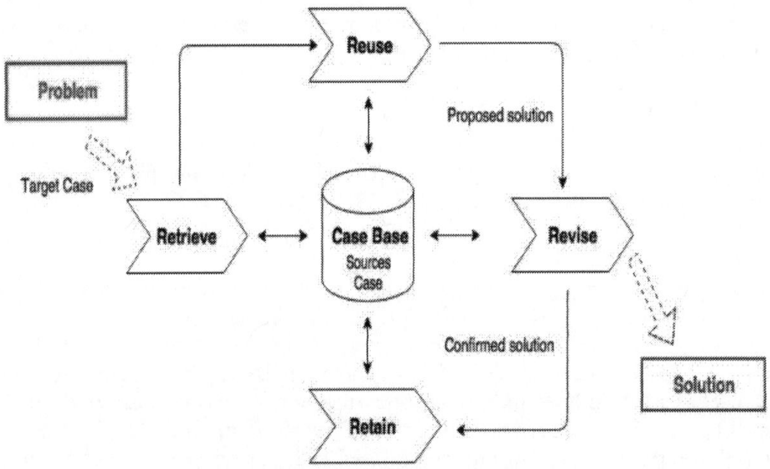

Figure 1. CBR cycle.

The presented study allows us to classify our work and make choices while trying to highlight some limitations of the afore-mentioned research studies. We choose first to diagnose the APS. The diagnosis is carried out from the point of view of OP. Second, our proposed system is based on a case representation according to the structural model. Such representation considers the temporal aspect in order to carry out a thorough analysis through the extraction of sensor and actuator signals to diagnose its current and previous PLC cycles. In addition, the APS sensors and actuators to be diagnosed are of AON types (ALL or Nothing). For this reason, our attributes values are binary (0 or 1), and the applied similarity metrics is a symmetric binary measure. Finally, our introduced system allows automatically formulating a new case when a new diagnosis request is made by the HOS. The new case inherits the solution from the selected cases.

3.3 The benefits of the proposed system

The proposed CB-DSSD of APS has many benefits:

- Simple and easy implementation.
- Achievable in real time: It is able to provide help and assistance to the HOS in order to perform his/her diagnosis task in real time.
- Incremental learning: The system can be created with a small or limited amount of incrementally-developed experiences, adding more cases to the case base.

- Offers several solutions for the same problem: The system can provide more than one alternative for a similar problem, and the HOS can choose the right solution that seems most suitable.
- The system does not need any rules, complete knowledge of the domain or any models: It does not need a complete knowledge of the domain or rules to reproduce new knowledge. In other words, it requires less knowledge acquisition. In fact, it consists in collecting a set of past experiences without extracting a formal domain model from these cases.

4. Functional model of the proposed system

The generic functional model of our proposal (figure 2) consists essentially of a graphical user interface that allows the HOS to interact with the CB-DSSD and express his/her choices and decisions, an offline phase to collect and represent data which describes the APS behaviors as a set of cases and an online phase to extract for each new case the similar or nearest case using a similarity metric. Both phases are developed in the following subsections.

4.1 Offline phase

The objective of this phase is to extract knowledge necessary as cases. Indeed, to achieve such goal, the offline phase is based on the expert knowledge, observations and specifications of the system to be diagnosed. But, firstly, it is necessary to choose the good structure of case representation. We have already mentioned in the third section of this work that the choice of this structure influence the CBR cycle.

Figure 2. Generic functional model of our proposal.

Our proposed case is composed of two parts: problem representation part and solution representation part (see figure 3).

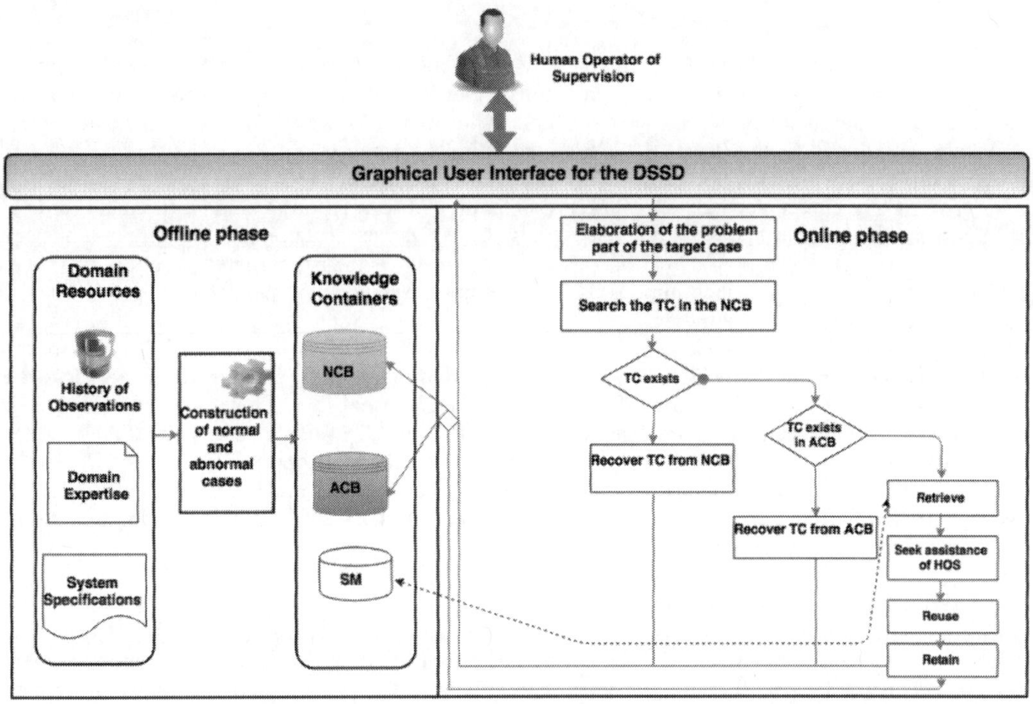

4.1.1 Problem Representation

During this part, the system state is defined to understand the situation to be diagnosed. The problem representation relies on the information obtained from sensors (inputs) and actuators (outputs) for the current PLC cycle (t) and the previous PLC cycle (t-1). This information is modeled by a set of descriptors. Each descriptor is a triplet:

$$<Co_i, V_t, V_{t-1}>$$

Where Co_i is a sensor or a system actuator, V_t is the value of Co_i at t, and V_{t-1} is its value at t-1, with t: the current PLC cycle, and t-1: the previous PLC cycle.

Problem representation		Solution representation
Representation of the current situation for each I/O	Representation of the previous situation for each I/O	Case Type

Figure 3. Case generic format.

4.1.2 Solution Representation

This part describes the problem solution. It also indicates the case type (normal or abnormal behavior) (figure 3). It is composed of three sub-parts: detection, location and identification. The normal cases, representing normal behaviors of APS, are listed in Normal Case Base (NCB). However, a set of abnormal cases, representing faulty behaviors of APS, is stored in Abnormal Case Base (ACB). Figure 4 shows the case detailed format for an APS composed of 2 actuators and 4 sensors.

Actuator 0		Actuator 1		Sensor 0		Sensor 1		Sensor 2		Sensor 3		Normal or Abnormal
Val_t	Val_{t-1}	Val_t	Val_{t-1}	Val_t	Val_{t-1}	Val_t	Val_{t-1}	Val_t	Val_{t-1}	Val_t	Val_{t-1}	

Figure 4. Case detailed format.

4.2 Online phase

The online phase of the CB-DSSD is triggered by a diagnosis request of the HOS. It corresponds to a new problem which is formalized as a problem part of a new case called Target Case (TC). The formalization is based on sensors and actuators data at the current PLC cycle time t (the time of request) and at the previous PLC cycle time t-1(the time preceding the request).

After this step, we choose the most appropriate Similarity Metric (SM) among a set of similarity metrics [20]. Our choice depends on data representation and data type to compare (nominal, ordinal, continuous or binary). In this work, we use APS with sensors and actuators having only two possible values: 0 or 1 (All or nothing). For this reason, we employ a binary symmetric similarity (binary attributes are symmetrical, which implies that there is no preference that the attribute is coded 0 or 1) called the simple matching coefficient [21].

After applying the metric selection, we search the TC in the NCB. If it exists in NCB, we are in a normal situation. So, the CB-DSSD shows that there is no problem in the process. If it is not located in NCB, we search it in ACB. If it is found, then we are in an abnormal situation known as the CB-DSSD. Otherwise, CB-DSSD calculates the distance between TC, all NCB source cases and all ACB source cases. After that, it asks for the HOS help in order to choose the most appropriate solution. Once HOS intervenes, the new case inherits the solution from the selected cases. Finally, it will be stored in the NCB or ACB for future use. These treatments show that the CB-DSSD is able to use the obtained experience to promote learning.

5. Application on the sorting system of ITS PLC

To test the effectiveness of our proposal and describe its performance, we implemented our approach under the beta version of the Interactive Training System for PLC (ITS PLC) proposed by the Portuguese company Real Games [22]. In fact, it is a set of simulation software dedicated to automation training, writing and using scripts written in IronPyton [23]. It allows notably simulating failures in sensors and actuators.

With the failures panel (figure. 5), the user is able to induce malfunctions in the system, presenting new challenges and increasing the realism of the simulation.

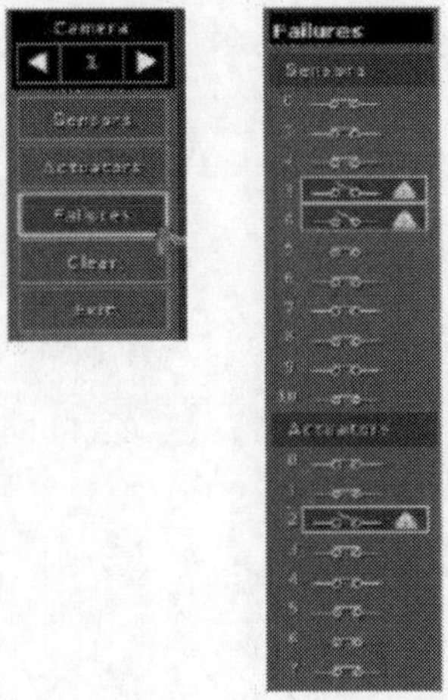

Figure 5. Failures panel of ITS PLC.

In our work, we developed two scripts. The first one allows controlling the APS without the need for a real API. However, in the second script, we implement our approach. In the following section, we operate the virtual system "Sorting System". Then, we present the offline and online phases of the CB-DSSD.

5.1 Description of the sorting system

The objective of the sorting system (figure 6) is to bring boxes of entry conveyor to exit conveyor by sorting them according to their heights. The system has eleven sensors to determine boxes size (small or large) and the box entry or exit in different conveyors (feeding, intermediate, and evacuation) or turntable. It has also seven actuators to activate the various conveyors and the turntable.

Figure 6. The sorting system from ITS PLC.

5.2 Offline phase and online phase of CB-DSSD

From a set of cyclically- recorded raw data (cycle time is 16 ms), representing signals of various sensors and actuators of the sorting system, a regrouping of these data is made, so that each 2 instances will represent a case. Then, an expert accompanies the construction phase of initial cases and expresses his/her opinion with regard to each case (normal behavior or failing behavior). In this stage, we obtain 52 normal cases and 10 abnormal cases.

During online phase and at every 60 ms, the CB-DSSD elaborates the problem part of the TC (new problem). After that, it searches the TC first at NCB. If the TC is already in NCB, then diagnosis result will be normal operation (case type=N). An example of this situation is shown in figure 7.

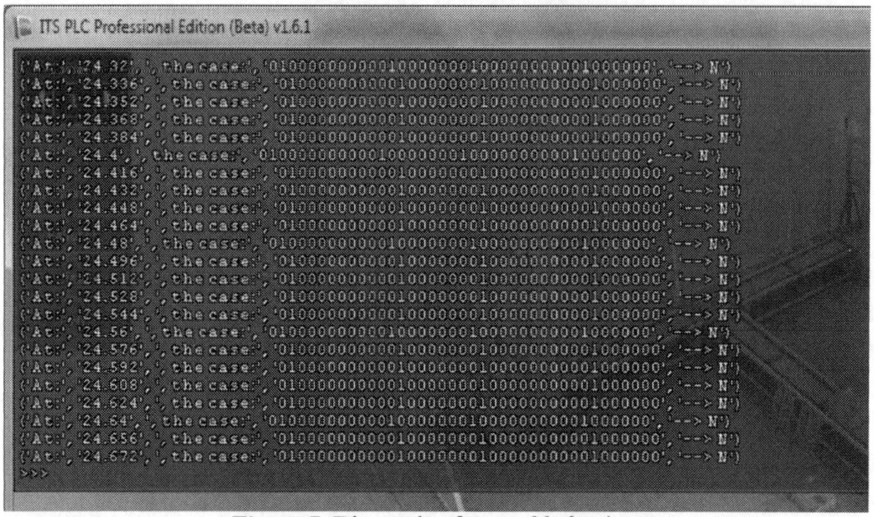

Figure 7. Diagnosis of normal behavior.

Then, we trigger a failure which corresponds to an existing case in ACB. The CB-DSSD is able to detect and locate this failure (figure 8).

Figure 8. Diagnosis of faulty operation.

Figure 9 shows a new case which corresponds to an inexistent case in NCB and ACB. The CB-DSSD begins by calculating similarity between the new case, all abnormal sources cases and all normal sources cases. Then, it asks the help of HOS to choose the most appropriate solution. The new case inherits the behavior (normal or failed) of the selected case and will be stored in the NCB or ACB.

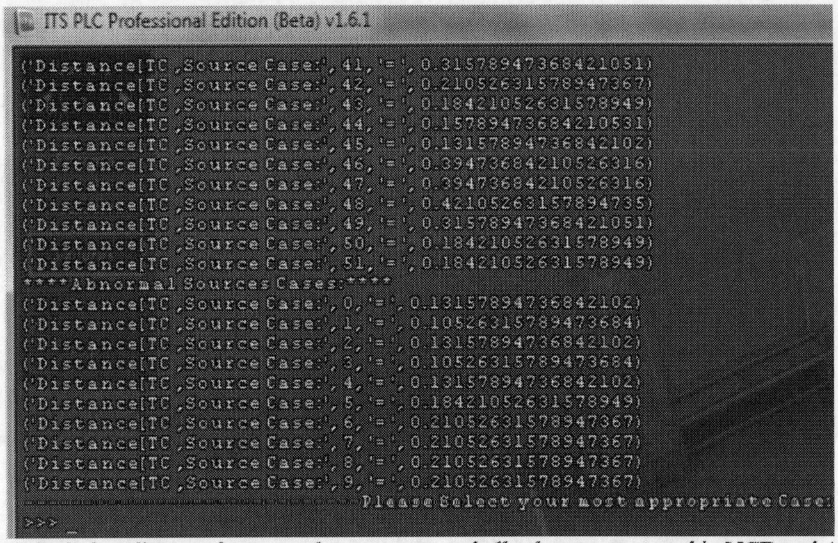

Figure 9. Calculating distance between the new case and all other cases stored in NCB and ACB.

After the learning of this experience, we trigger the same failure. Thus, the DSSD ise able to detect and locate it.

5.3 Difficulties:

The negative point of our proposal is described through this example:

Example:

At the triggering of stuck-off to 1 of actuator 6, the CB-DSSD forms the TC and firstly seeks it in the NCB. The TC exists in the NCB.

According to our CB-DSSD, the TC represents a normal behavior of the sorting system based on sensors and actuators data at the current PLC cycle time t and at the previous PLC cycle time t-1. Hence, to describe a stuck-off to 1 of an actuator or a sensor, these data are not sufficient. They only help to diagnose unexpected move of actuators or sensors from 0 to 1 or from 1 to 0.

To solve this problem, we should add temporal constraints of the occurrences of the observables events to our case representation.

6. Conclusion

Conventional CBR systems for industrial diagnosis were designed as automated problem solvers. Most of them were not efficiently used to diagnose real industrial systems. This paper outlined a case-based decision support system which resulted from a synergy between CBR and DSS. Its main purpose is to support and assist the HOS in his decision process in order to accomplish its diagnosis task.

From a diagnosis request of the HOS, the CB-DSSD begins by formalizing it as a problem part of a target case based on sensors and actuators signals at the current and the previous PLC cycle. Second, it searches the TC in the NCB. If it exists in NCB, the process will be normal. If it is not located in NCB, it will search it in the ACB. If it is found, the process will be falty. Otherwise, the CB-DSSD asks the HOS to choose the appropriate similarity metric and it calculates distance between the TC, all source cases of NCB and all source cases of ACB. Finally, it asks for the HOS help to choose the appropriate solution. The new case inherits the solution from the selected cases and it is stored in the NCB or ACB for future use.

An application on the virtual sorting system from ITS PLC Collection showed the feasibility of our approach created with a small amount of incrementally-developed experiences, which adds more cases to the two case bases based on HOS decisions.

In future works, we will improve our case representation by taking into account the temporal constraints of the occurrences of the observables events. To diagnose complex system, we will use multi-expert knowledge in an aggregate approach in order to facilitate collaborative decision making. In fact, the approach will be based on the multi-agent systems for their adaptation to the efficient management of complex systems.

7. References

[1] Combacau M, Berruet P, Charbonnaud K A and Zamai E 2000 Supervision and monitoring of production systems *IFAC 2nd Conference on Management and Control of Production and Logistics* Grenoble France

[2] Darkhovski B and Staroswiecki M 2003 Theoretic Approach to Decision in FDI *Transactions On Automatic Control* IEEE 48(5) pp 292-297

[3] Phillippot A, Sayed-Mouchaweh M, Carré-Ménétrier V, Riera B 2006 Decentralized approach to diagnose manufacturing systems *Computational Engineering in Systems Applications* IEEE Vol 1 pp 912-918

[4] Saddem R and Philippot A 2014 A Causal Temporal Signature from diagnoser model for online diagnosis of Discrete Event Systems *Control, Decision and Information Technologies* IEEE pp 551-556

[5] Xiong N, Funk P and Olsson T 2012 Case-based reasoning supports fault diagnosis using sensor information *2nd International Workshop and Congress on eMaintenance*

[6] Millot P and Lemoine M P 1998 An attempt for generic concepts toward human-machine cooperation *Man and Cybernetics* IEEE pp 1044-1049

[7] Turban E, Aronson J, Liang T P 2005 *Decision Support Systems and Intelligent Systems* Pearson Prentice Hall

[8] Power D J 2008 Decision support systems: a historical overview *Decision Support Systems* Vol 1 pp 121-140

[9] Kolodner J L 1991 Improving human decision making through case-based decision aiding *AI Magazin* 12(2) pp 52-68

[10] Kolodner J L 1992 An introduction to case-based reasoning *Artificial Intelligence Review* 6(1) pp 3-34

[11] Napoli A, Lieber J and Curien R 1996 Classification-based problem-solving in case based reasoning *European Workshop on Advances in Case-Based Reasoning* Springer Berlin Heidelberg pp 295-308

[12] Aamodt A and Plaza E 1994 Case-based reasoning: Foundational issues, methodological variations, and system approaches *AI communications* 7(1) pp 39-59

[13] Main J, Dillon T S and Shiu S C 2001 A tutorial on case based reasoning *Soft computing in case based reasoning* Springer London pp1-28

[14] Dendani N, Khadir M and Guessoum S 2012 Hybrid approach for fault diagnosis based on CBR and ontology: using jCOLIBRI framework *Int. Conf. on Complex Systems* IEEE pp 1-8

[15] Ruiz P A P, Kamsu-Foguem B and Noyes D 2013 Knowledge reuse integrating the collaboration from experts in industrial maintenance management *Knowledge-Based Systems* Vol 50 pp 171-186

[16] Behbahani M, Saghaee A and Noorossana R 2012 A case-based reasoning system development for statistical process control: Case representation and retrieval *Computers and Industrial Engineering* 63(4) pp 1107-1117

[17] Chougule R, Rajpathak D and Bandyopadhyay P 2011 An integrated framework for effective service and repair in the automotive domain: An application of association mining and case-based-reasoning *Computers in Industry* 62(7) pp 742-754

[18] Lejri O and Tagina M 2012 Representation in case-based reasoning applied to control reconfiguration *Advances in Data Mining* pp 113-120

[19] Vareilles E, Aldanondo M, De Boisse A C, Coudert T, Gaborit P and Geneste L 2012 How to take into account general and contextual knowledge for interactive aiding design: Towards the coupling of CSP and CBR approaches *Engineering Applications of Artificial Intelligence* 25(1) pp 31-47

[20] Cha S H 2007 *Comprehensive survey on distance/similarity measures between probability density*

functions City 1(2) pp 1
[21] Legendre P and Legendre L 1998 *Numerical Ecology* 2nd English edition Elsevier Science BV Amsterdam 852p
[22] www.realgames.pt
[23] http://ironpython.net

Distributed Configuration of Sensor Network for Fault Detection in Spatio-Temporal Systems

Maciej Patan and Damian Kowalów

Institute of Control and Computation Engineering,
University of Zielona Góra,
ul. Podgórna 50, 65–246 Zielona Góra, Poland

E-mail: {m.patan,d.kowalow}@issi.uz.zgora.pl

Abstract. The problem of fault detection in spatio-temporal systems is formulated as that of maximizing the power of a parametric hypothesis test verifying the nominal state of the process under consideration. Then, adopting a pairwise communication schemes, a computational procedure is developed for the spatial configuration of the observation locations for sensor network which monitor changes in the underlying parameters of a distributed parameter system. As a result, the problem of planning the percentage of experimental effort spent at given sensor locations can be solved in a fully decentralized fashion. The approach is verified on a numerical example involving sensor selection for a convective diffusion process.

1. Introduction

Recently, one can observe an extremely fast development of methods of Fault Detection and Isolation (FDI) for dynamical systems. Although a vast amount of existing approaches (for surveys reader can be referred to [3, 7, 9, 24, 39]), most of contributions focus on the methods for lumped parameter systems with just a few approaches dedicated to spatio-temporal systems also known as distributed-parameter systems (DPSs). Since in DPS it is usually impossible to observe the system states over the entire spatial domain, the problem of sensor location and observation scheduling for proper system monitoring become of vital importance. But within the framework of FDI systems for DPSs, the optimization of measurement process which increases the reliability of the diagnosis is usually neglected and there is a substantial lack of techniques dedicated to this area.

Technological advances in communication systems and the development of inexpensive mobile measurement systems made it possible to deploy a group of networked devices in a number of environments, cf. [13, 4]. A cooperated and scalable network of sensors has the potential to substantially enhance performance of the observation systems. Although the number of sensor placement techniques to manage the configuration for sensor networks of practical scale is very limited, some effective approaches has been proposed to cover a number of different experimental settings, including stationary [12, 34, 22, 26, 17, 10, 21, 11], scanning [22, 33, 15] or moving observations [29, 27, 34, 19, 36]. Their adaptation to the sensor-network setting was reported in [32, 31].

However, most techniques communicated by various authors rely on centralized techniques, which assume an existence of some superior entity to maintain the whole network and responsible for optimization of the observation strategy. The distributed nature of the design problem is taken into account very occasionally. Recent advancements in sensor networks necessitate effective, distributed and fault-tolerant algorithms for computation and information exchange.

In the background of diagnostic problems, another difficulty is a development of proper relation between the performance of diagnostic system and the observational strategy. For this purpose, a classical hypothesis testing is used for model-based diagnosis in DPSs. This allows us to construct a qualitative criterion of sensor allocation using the notion of D-optimum designs originated in optimum experiment design theory. Then, proper fault detection and localization schemes are developed based on the results reported by Patan [22] for stationary sensors and further extended for scanning and mobile sensor networks [16, 22, 37, 23, 14, 30].

In this paper we wish to focus on certain computational aspects of sensor placement problems for fault detection in DPSs. Given a finite set of stationary located sensors, the observation horizon is divided into consecutive stages. Then, the non-negative weights are assigned to each sensor at given stage so as to maximize the determinant of the Fisher Information Matrix (FIM) related to the parameters to be identified. The weight assigned to a measurement point can be interpreted as the proportion of observations performed at this point, or the percentage of experimental effort spent at this point. The potential solutions are of considerable interest while assessing which sensors are more informative than the others and allow for complexity reduction of the measurement system. The solution proposed here is close in spirit to the classical optimum experimental design theory for lumped systems [1, 38]. Its novelty relies on a generalization of a known numerical algorithm for computation of a D-optimum design on a finite set together with its efficient distributed implementation dedicated to sensor network.

As a result, an extremely simple computational scheme is developed to determine best sensor configurations. Its idea is based on the class of so-called *gossip* algorithms in which each node communicates with no more than one neighbor at each time instant [40, 5, 6]. The different nodes of the net-work independently calculate and store the desired quantities and their final estimates are obtained in a fully decentralized fashion. The advantage of such information exchange is clear as it is robust with respect to individual sensor faults and the the global estimate of information matrix is stored at all sensor nodes so easily can be recovered.

2. Model-based fault detection in DPS

Parameter estimation is one of the fundamental approaches in the framework of analytical techniques of fault detection [9]. It is especially useful in situations when the faulty system state does not only changes system response but leads also to perturbation in values of model parameters. In practice, this is quite common if only parameters have a physical interpretation built upon proper analysis and modeling of monitored process variables.

Let $y = y(x;t;\theta)$ denote the scalar state of a given DPS at a spatial point $x \in \Omega \subset \mathbb{R}^d$ and time instant $t \in T = [0;t_f], t_f < \infty$. Here θ represents an unknown constant m-dimensional parameter vector which must be estimated using observations of the system.

The state y is observed by N scanning sensors, which possibly change their positions at time instants $0 < t_0 < t_1 < \cdots < t_K = t_f$ and will be remaining stationary for the duration of each subinterval $T_k = [t_{k-1}, t_k]$, $k = 1, \dots, K$. The observations can be formally represented as [20, 15]

$$z^j(t) = y(x_k^j, t; \theta) + \varepsilon(x_k^j, t), \quad x_k^j \in X, \quad t \in T_k, \tag{1}$$

for $j = 1, \dots, N$ and $k = 1, \dots,$ where x_k^j is element selected from defined *a-priori* set of sensor locations $X = \{x^1, \dots, x^N\}$ and $\varepsilon(x_k^j, t)$ is measurement noise which is assumed that is zero-mean, Gaussian, spatial uncorrelated and white [14, 20], i.e. $\mathrm{E}[\varepsilon(x_k^j, t)] = 0$ and $\mathrm{var}(\varepsilon(x_k^j, t)) = \sigma^2$ where $\sigma > 0$ denotes standard deviation of the measurement noise.

A further assumption is that the estimation of the unknown parameter vector θ is performed via minimization of the least-squares criterion

$$\mathcal{J}(\theta) = \sum_{j=1}^{N}\sum_{k=1}^{K}\int_{T_k} |z^j(t) - \hat{y}(x_k^j, t; \theta)|^2 \, \mathrm{d}t, \tag{2}$$

with $\theta \in \Theta_{\mathrm{ad}}$ and Θ_{ad} is the set of admissible parameters and $\hat{y}(\cdot, \cdot; \theta)$ denotes the system model response suitable to θ. The estimate of the true value of $\theta^\star$ is a vector $\hat{\theta}$ which minimize $\mathcal{J}(\theta)$.

An elementary idea of fault diagnosis is to compare the resulting parameter estimates with the corresponding known nominal values, treating possible differences as residuals which contain information about potential faults. Then using thresholding techniques, the appropriate decision making system could be constructed to detect abnormal situations in system functioning, see [22, 18, 37, 23]. Based on the observations, it is possible to test the simple parametric statistical hypothesis $H^0 : \theta = \theta^\star$, where $\theta^\star$ is the nominal value of the vector θ corresponding to the normal system performance.

Introducing the generalization of the likelihood function for the considered experiment[8]:

$$\mathcal{L}(z|\theta) = \left(\frac{1}{2\pi\sigma^2}\right)^{N/2} \exp\left(-\frac{1}{2\sigma}\mathcal{J}(\theta)\right), \tag{3}$$

and setting $\Theta_0 = \{\theta \in \Theta_{\mathrm{ad}} : \theta = \theta^\star\}$ it could be define the following generalized log-likelihood ratio:

$$\lambda(z) = 2\ln\frac{\sup_{\theta\in\Theta_{\mathrm{ad}}} \mathcal{L}(z|\theta)}{\sup_{\theta\in\Theta_0} \mathcal{L}(z|\theta)} = \mathcal{J}(\tilde{\theta}) - \mathcal{J}(\hat{\theta}), \tag{4}$$

where

$$\hat{\theta} = \arg\min_{\theta\in\Theta_{\mathrm{ad}}} \mathcal{J}(\theta), \qquad \tilde{\theta} = \arg\min_{\theta\in\Theta_0} \mathcal{J}(\theta). \tag{5}$$

The generalized log-likelihood ratio test is widely used in statistics, because it can be shown that assuming the validity of the null hypothesis H^0 the sequence $\lambda(z)$ for $N \to \infty$ is weakly convergent to x^2 random variable on m degrees of freedom [8, Thm. 3.6.1, p 55].he meaning of this fact is that for the given significance level γ representing a fixed range of model uncertainty we can compare the observed value of $\lambda(z)$ with some threshold k_γ obtained from the cumulative chi-squared distribution with m degrees of freedom, where k_λ is 'λ-quantile' of the distribution. Then, the decision rule for fault detection takes the following form:

$$S = \begin{cases} S^1 & \text{if} \quad 2\ln\lambda(y) \geq k_\gamma \quad (\text{reject } H^0), \\ S^0 & \text{if} \quad 2\ln\lambda(y) < k_\gamma \quad (\text{accept } H^0). \end{cases} \tag{6}$$

The potential rejection of H^0 indicates a significant deviation of the vector from the nominal value of this parameter and forms a base for proper detection of abnormal states in the system. When the null hypothesis H^0 is true, we do not want to reject it. The mistake in rejecting null hypothesis H^0 when it is true is called the Type I error. Similarly, accepting null hypothesis when the alternative hypothesis of the form $H^1 : \theta \neq \theta^\star$ is true is called the Type II error [8] In fault diagnosis, there is a direct connection between these errors and the probability of a false alarm and missed detection, respectively.

3. Optimal measurement problem

In order to achieve a reliable diagnosis on the DPS state we have to keep the probabilities of false alarms and missed detection low. In practice, in order to avoid the multi-criteria optimization, a typical approach is to minimize the Type II error with a constraint imposed on the maximum level of the Type I error.

Based on the parameter vector θ it can be shown that the power of the presented hypothesis test (i.e. the probability of not committing the Type II error) can be increased by taking a large number of measurements N or, alternatively, by maximizing the D-optimality criterion (see [18] for details):

$$\Psi(M) = -\log\det(M), \tag{7}$$

defined on the so-called *average* Fisher Information Matrix given by [15]

$$M = \frac{1}{Nt_f}\sum_{j=1}^{N}\sum_{k=1}^{K} g(x_k^j, t)g^{\mathsf{T}}(x_k^j, t)\,\mathrm{d}t, \tag{8}$$

where

$$g(x,t) = \left(\frac{\partial y(x,t;\theta)}{\partial\theta}\right)^{\mathsf{T}}_{\theta=\theta^0}, \tag{9}$$

stands for the so-called *sensitivity vector*, θ^0 being a prior estimate to the unknown parameter vector θ. The D-optimality leads to minimization of the volume of the uncertainty ellipsoid for the parameter estimates. Since in real world problems, the number of available observations has to be always limited, thus in practice, such strategy is important alternative to obtain the high powered test which leads directly to the low probability of missed detection.

The introduction of an optimality criterion renders it possible to formulate the sensor location problem as an optimization problem

$$\Psi[M] \to \min, \tag{10}$$

with respect to x_k^j, $j = 1, \ldots, N$, $k = 1, \ldots, K$ belonging to the admissible set X.

Owing to assumption on independent measurement noise, we admit of replicated measurements, i.e. some values x_k^j may appear several times in the optimal solution (we can use multiple sensors located at the same position or we can repeat the process). Consequently, we introduce r_k^j as the numbers of replicated measurements corresponding to the sensor nodes locations $x^1, \ldots, x^N$ at consecutive time subinterval T_k. The collection of variables

$$\xi_k = \{(x_k^1, p_{k1}),\quad (x_k^2, p_{k2}),\quad (x_k^N, p_{kN})\}, \tag{11}$$

where $p_{kj} = r_k^j/N$, $N = \sum_{j=1}^{N} r_k^j$ is called the *exact design* of experiment for subinterval T_k. The proportion p_{kj} of observations performed at x_k^j can be considered as the percentage of experimental effort spent at that point.

Introducing the $\xi = (\xi_1, \xi_2, \ldots, \xi_K)$ as the total experimental design, we rewrite the FIM in the form

$$M(\xi) = \frac{1}{K}\sum_{j=1}^{N}\sum_{k=1}^{K} p_{kj}\int_{T_k} g(x_k^j, t)g^{\mathsf{T}}(x_k^j, t)\,\mathrm{d}t, \tag{12}$$

Here the p_{kj}'s are rational numbers, since both r_k^j's and N are integers. This discrete nature of N-observation exact designs causes serious difficulties and cannot be solved easily by standard optimization techniques, particularly when N is large. A reasonable approach is to relax the definition of the design. When N is large, the weights p_{kj} can be approximated with a real numbers in the interval [0, 1], not necessarily integer multiples of $1/N$. Obviously, we must have $\sum_{j=1}^{N} p_{kj} = 1$, for $k = 1, \dots, K$, so we may think of the designs as probability distributions on X. This leads to the so-called *continuous* designs which constitute the basis of the modern theory of optimal experiments [2, 28]. It turns out that such an approach drastically simplifies the design. Thus, we shall operate on designs which concentrates approximately Np_{kj} measurements at each location x_k^js.

Then we may redefine an optimal design as a solution to the optimization problem

$$\xi^\star = \arg \min_{\xi \in \Xi(X)} \Psi[M(\xi)], \tag{13}$$

where $\Xi(X)$denotes the set of all probability distributions on $X = \{x^1, \dots, x^N\}$.

4. Characterization of the optimal solutions

In the remainder of this paper we shall assume that $g \in C(X \times T; \mathbb{R}^m)$. A number of characterizations of the optimal design $\xi^\star$ can be derived in a rather straightforward manner from the general results given in [35, 16].

Lemma 1. *For any $\xi \in \Xi(\mathrm{X})$ the information matrix $M(\xi)$ is symmetric and non-negative definite. Let us introduce the notation $\mathfrak{M}(X)$ for the set of all admissible information matrices, i.e. $\mathfrak{M}(X) = \{M(\xi) : \xi \in \Xi(X)\}$.*

Lemma 2. *$\mathfrak{M}(X)$ is compact and convex.*

The next result constitutes the necessary and sufficient condition for the optimality of designs. It is usually called an *equivalence theorem*[35, 16].

Theorem 1. *The following conditions are equivalent:*

- *the design $\xi^\star$ minimizes $\Psi[M(\xi)]$,*
- *the design $\xi^\star$ minimizes $\max_{x \in X} \phi_k(x, \xi^\star) = \varsigma_k(\xi^\star)$, $k = 1, \dots, K$, where*

$$\phi_k(x, \xi) = \frac{1}{t_f} \int_{T_k} g^\mathsf{T}(x, t) M^{-1}(\xi) g(x, t)\, \mathrm{d}t, \tag{14}$$

$$\varsigma_k(x, \xi) = \frac{1}{t_f} \sum_{j=1}^{N} \int_{T_k} g^\mathsf{T}(x, t) M^{-1}(\xi) g(x, t)\, \mathrm{d}t. \tag{15}$$

The sensitivity function $\phi_k(x, \xi)$ is of vital importance here, providing us with a simple test for the optimality of designs. In particular, if it takes the values less than or equal to $\varsigma_k(\xi)$ for all $x \in X$ then ξ is optimal, otherwise ξ is not optimal.

5. Distributed algorithm

Analytical determination of optimal designs is possible only in simple situations and for general systems it is usually the case that some iterative design procedure will be required. In the case considered in the paper, i.e. the design for fixed sensor locations, a computational algorithm can be derived based on the mapping $\mathcal{T} : \Xi(X) \to \Xi(X)$ defined by

$$\mathcal{T}\xi = \left\{ \left(x_k^1, p_{k1} \frac{\phi(x_k^1, \xi)}{\varsigma_k(\xi)} \right), \quad \dots, \quad \left(x_k^N, p_{kN} \frac{\phi_k(x_k^N, \xi)}{\varsigma_k(\xi)} \right) \right\}, \quad k = 1, \dots, K. \tag{16}$$

From Theorem 1 it follows that a design $\xi^\star$ is D-optimal if it is a fixed point of the mapping $\mathcal{T}$, i.e.

$$\mathcal{T}\xi^\star = \xi^\star. \tag{17}$$

Also, we have the valuable property that the sum of weights p_{kj} is invariant with respect to $\mathcal{T}$ for any $k = 1, \dots, K$ (i.e. it is always equal to 1). As for interpretation of this mapping, let us note that in case of ξ not being the optimal design, it increases the weights for those sensor locations where we observe the high values of sensitivity function at the cost of decreasing the weights of non-informative locations with small values of ϕ. Using this simple idea, it is possible to develop a decentralized configuration algorithm for sensor network being a distributed generalization of that proposed in [25, p. 139] for the classical optimum experimental design problem consisting in iterative computation of a D-optimum design on a finite set (and further extended in [35, 16]).

In the following we assume the asynchronous time model for the configuration process. Let $r = 0, 1, 2\dots$ be the discrete time index, which partition the continuous configuration time axis into time slots $Z_r = [z_{r-1}, z_r)$.

Owing to Theorem 1 we can use the mapping $\mathcal{T}$ to iteratively improve the experimental effort distribution. Unfortunately $\phi_k(\cdot, \xi)$ and $\varsigma_k(\xi)$ computed according to (14) in a centralized fashion are not very useful in our setting. But the only component which cannot be calculated independently of other network nodes is the global information matrix $M(\xi)$. Moreover, from (12) it is clearly seen the information matrix is a weighted average of the subinterval information matrices

$$M_k = \frac{1}{t_f} \sum_{j=1}^{N} p_{kj} \int_{T_k} g(x_k^j, t) g^\top(x_k^j, t)\, \mathrm{d}t. \tag{18}$$

Thus, our task is closely related to the problem of distributed averaging on a network which appears in many applications and has been a subject of extensive studies [40, 5, 6].

Distributed averaging can be achieved in many ways. One of straightforward techniques is a pairwise communication flooding, also known as a *gossip* scheme, which in its classic version assumes that at the k-th reconfiguration time slot the i-th sensor contacts some neighboring node j with probability Q_{ij} i.e. a pair $(i - j)$ is randomly and independently selected. At this time, both nodes set their values equal to the average of their current values. Denoting the estimate of (18) for subinterval T_k maintained by j-th sensor at configuration time slot Z_r as $M_k^j(\xi^{(r)})$ we have ($\leftarrow$ is an update operator).

$$M_k^\ell(\xi^{(r)}) \leftarrow \frac{p_{ki} M_k^i(\xi^{(r)}) + p_{kj} M_k^j(\xi^{(r)})}{p_{ki} + p_{kj}}, \quad \ell \in \{i, j\}, \quad k = 1, \dots, K. \tag{19}$$

However, in our setting the averaging problem is not typical as the contribution of the nodes with weights tending to zero as r increases should be eliminated from the averaging. This can be achieved by increasing the contribution of the local information matrix via the concept of so-called running consensus [6]

$$M_k^i(\xi^{(r)}) \leftarrow \frac{r-1}{r} M_k^i(\xi^{(r)}) + \frac{1}{r} \Upsilon_k(x_k^i), \tag{20}$$

where $\Upsilon_k(x_k^i) = \int_{T_k} g(x_k^j, t) g^\top(x_k^j, t)$. The first term enforces consensus among the nodes (represents the average information from the rest of network), while the second accounts for the increase in the total contribution of the local node. Note, that there is no need to record the whole design structure of $\xi^{(r)}$ at each node as the information about the current solution is propagated via local estimates of global FIM, i.e. $M_k^i(\xi^{(r)})$. The whole idea of the communication process is given by the Algorithm 1.

IOP Conf. Series: Journal of Physics: Conf. Series **783** (2017) 012010 doi:10.1088/1742-6596/783/1/012010

Algorithm 1 Distributed optimization of experimental effort. Indexes i and j denote, respectively, data from local repository and obtained from neighbor, r is an index of current configuration slot Z_r. Function *EXCHANGE* is responsible for both sending and receiving data to/from connector neighbor (order descending on who initiated communication)

```
procedure EXCHANGE_PROTOCOL(r)
    for k ← 1, K do                                          ▷ for all time subintervals T_k
        EXCHANGE(M_k^j(ξ^(r)), M_k^i(ξ^(r)))                 ▷ sends and receives
        EXCHANGE(p_ki and p_kj)                              ▷ FIM weights
        p ← p_ki + p_kj                                      ▷ store old weights for normalization
        Update M_k^i(ξ^(r)) ← (p_ki M_k^i(ξ^(r)) + p_kj M_k^j(ξ^(r)))/p
        Calculate φ_k(x_k^ℓ, ξ^(r)) and ς_k(ξ^(r)),  ℓ ∈ {i, j}
        Update p_ki ← p_ki φ_k(x_k^i, ξ^(r))/ς_k(ξ^(r))
        p_ki ← p · p_i(p_ki + p_kj)                          ▷ normalization p_ki's to sum up to 1
        Update M_i(ξ) ← ((r−1)/r) M_i(ξ) + (1/r) M_i
    end for
    r ← r + 1
end procedure
```

For all subintervals T_k data are calculated separately to reach most informative configuration for mentioned part of experiment. At $r = 0$ each network node starts with a global FIM estimate $M_k^i(\xi^{(r)})$ initialized with its local information matrix $\Upsilon_k(x_k^i)$ and starting with non-zero weights (it is important since once the weight attains zero, a multiplicative update forces it to remain zero). Then at each of asynchronous configuration time slot Z_r an appropriate pair of nodes exchanges information according to Algortithm 1. It can be shown that algorithm converges to the optimal solution, what is a consequence of independent weight update for each network node and the stochastic convergence of gossip algorithms[40].

Obviously, the sensor network topology and the choice of proper communication scheme significantly influence on the convergence rate. In general, under some assumptions on the network connectivity graph, a suitable gossip algorithm can be provided via the semidefinite programming with an exponential convergence rate. Since discussion on those very important issues is far beyond the scope of this work, therefore we refer the reader to the seminal papers [40, 5] for details.

6. Simulation example

As an illustration of the presented approach, consider the problem of sensor configuration for parameter estimation in the process of air pollutant transport over a given urban area. Within this domain, which has been normalized to the unit square $\Omega = (0,1)^2$, an active source of pollution is present, which yields the pollutant spatial concentration $y = y(x,t)$. The evolution of y over the normalized observation interval $T = (0,1]$ is described by the following advection-diffusion equation:

$$\frac{\partial y(x,t)}{\partial t} + \nabla \cdot \big(v(x,t)y(x,t)\big) = \nabla \cdot \big(\alpha(x)\nabla y(x,t)\big) + f(x), \quad x \in \Omega \tag{21}$$

subject to the boundary and initial conditions:

$$\frac{\partial y(x,t)}{\partial n} = 0, \text{ on } \Gamma \times T, \tag{22}$$

$$y(x,0) = y_0, \text{ in } \Omega, \tag{23}$$

where term $f(x) = 50\exp(-50\|x - c\|^2)$ represents the pollutant source located at the point $c = (0.3, 0.3)$ and $\partial y/\partial n$ stands for the partial derivative of y with respect to the outward normal to the boundary Γ. The mean spatiotemporal changes of the wind velocity field over the area were approximated by $v(x,t) = (2(x_1 + x_2 - t), x_2 - x_1 + t)$. The assumed functional form of the spatial-varying diffusion coefficient is $a(x) = \theta_1 + \theta_2 x_1 x_2 + \theta_3 x_1^2 + \theta_4 x_2^2$. The subject of interest here is a detection of significant increase in the intensity of the pollutant emission from the source. As the symptom of this abrupt fault we can observe the excessive deviation of the parameters θ from their nominal values. Therefore, these parameters need estimation based on measurement data from monitoring stations.

A MATLAB program was written using a PC equipped with Intel Core i5 processor (1.9GHz, 12 GB RAM) running Windows 10. The nominal values of the systems parameters were assumed to be $\theta_1^0 = 0.02$, $\theta_2^0 = 0.01$, $\theta_3^0 = 0.005$. Calculations were performed with a finite element method for a spatial mesh composed of 682 triangles, 378 nodes and an evenly partitioned time interval (3 subintervals $T_k = ((k-1)/3, k/3]$, $k = 1, 2, 3$). The sensitivity coefficients were then linearly interpolated and stored within the sensor nodes repositories. In our scenario, the observation grid was assumed to be created at locations selected from among those elements of above-mentioned 378-point triangulation mesh which do not lie on the outer boundary (there were 312 such nodes, which are indicated with dots in Fig. 2). Given $N = 312$ prospective sites in $\Omega \cup \Gamma$ we aim at finding the locations at which the measurements made by sensors would lead to D-optimum least-squares estimates of the parameters θ. In simulation it was assumed that the network is fully connected with uniform probability distribution for the connection between selected pair of nodes. Sensor configurations at different stages of algorithm are shown in Fig. 2. The sensors tend to form the pattern reflecting the areas of greatest changes in the pollutant concentration but the observations are averaged over time and it is not trivial to follow the dynamics of the observation strategy. It can be observed that during each time subinterval the system dynamics enforce the distribution of experimental effort. In the first stage (T_1), the pollutant concentration takes significant values only in the very close vicinity of the source, thus all measurement effort is spent at this location (cf. Figs. 2(a)-(c)). As time elapses the pollutant spreads over the whole spatial domain, therefore the sensors tends to measure the system state in other regions, where we observe the greatest changes in concentration and gather the rich information about the system dynamics (Figs. 2(d)-(f)). Note, that algorithm approximate the pattern of final design very fast after reaching relative small number of r. Then, the algorithm slows down and achieving a very high accuracy of the solution becomes expensive in the sense of the number of pairwise communications. An application of a more efficient communication scheme at the latter stage of configuration process would be helpful here to overcome this effect.

7. Concluding remarks

The problem of selecting optimal observation points in view of reliable detection of abnormal state of DPS has been addressed. The main contribution of this work is a proper characterization of problem relaxation and its adaptation to a distributed scheme of computation and information exchange. The advantage of introducing continuous designs lies in the fact that the problem dimensionality is dramatically reduced. A decided majority of existing approaches to the sensor location problem for DPSs is designed only for centralized monitoring systems, and in the context of modern sensor networks often this comes out impractical. On the other hand, numerous effective routines for distributed information fusion designed for sensor, peer-to-peer or wireless ad hoc networks are not directly applicable for the purpose of parameter estimation in DPSs. The decided advantage of the proposed distributed calculation scheme is a great simplicity and the ability to produce solutions reasonably fast.

Acknowledgement

This work is supported by National Science Center in Poland under grant 2014/15/B/ST7/03208

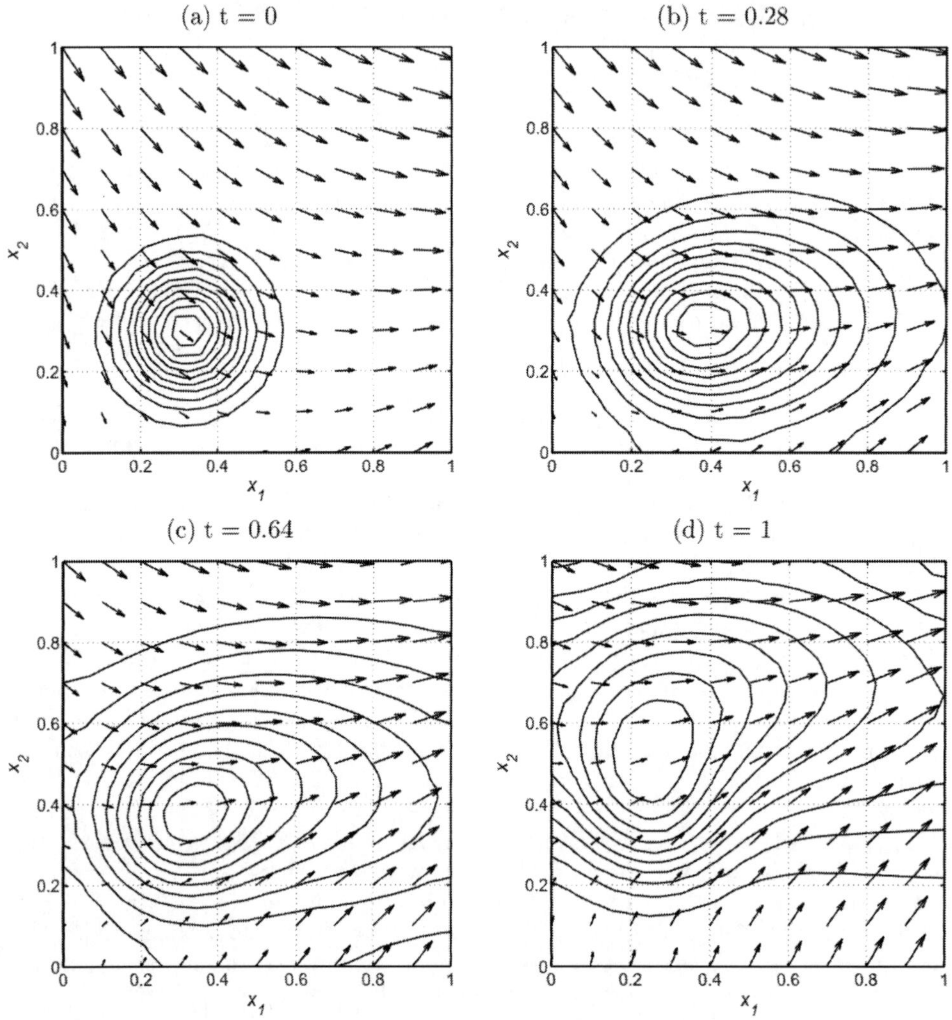

Figure 1. Temporal changes in the wind velocity field and pollutant concentration

References

[1] A. Atkinson, A. Donev, and R. Tobias. *Optimum experimental designs, with SAS*, volume 34. Oxford University Press, 2007.

[2] A. C. Atkinson, A. N. Donev, and R. D. Tobias. Optimum experimental designs, with sas. 2007.

[3] M. Basseville, I. V. Nikiforov, et al. *Detection of abrupt changes: theory and application*, volume 104. Prentice Hall Englewood Cliffs, 1993.

[4] P. H. Bauer. New challenges in dynamical systems: The networked case. 18(3):271–277, 2008.

[5] S. Boyd, A. Ghosh, B. Prabhakar, and D. Shah. Randomized gossip algorithms. *IEEE/ACM Transactions on Networking (TON)*, 14(SI):2508–2530, 2006.

[6] P. Braca, S. Marano, and V. Matta. Enforcing consensus while monitoring the environment in wireless sensor networks. *IEEE Transactions on Signal Processing*, 56(7):3375–3380, 2008.

[7] L. H. Chiang, R. D. Braatz, and E. L. Russell. *Fault detection and diagnosis in industrial systems*. Springer Science & Business Media, 2001.

[8] G. C. Goodwin and R. L. Payne. Dynamic system identification: experiment design and data analysis. 1977.

[9] J. Korbicz and J. M. Kościelny. *Modeling, diagnostics and process control: implementation in the diaster*

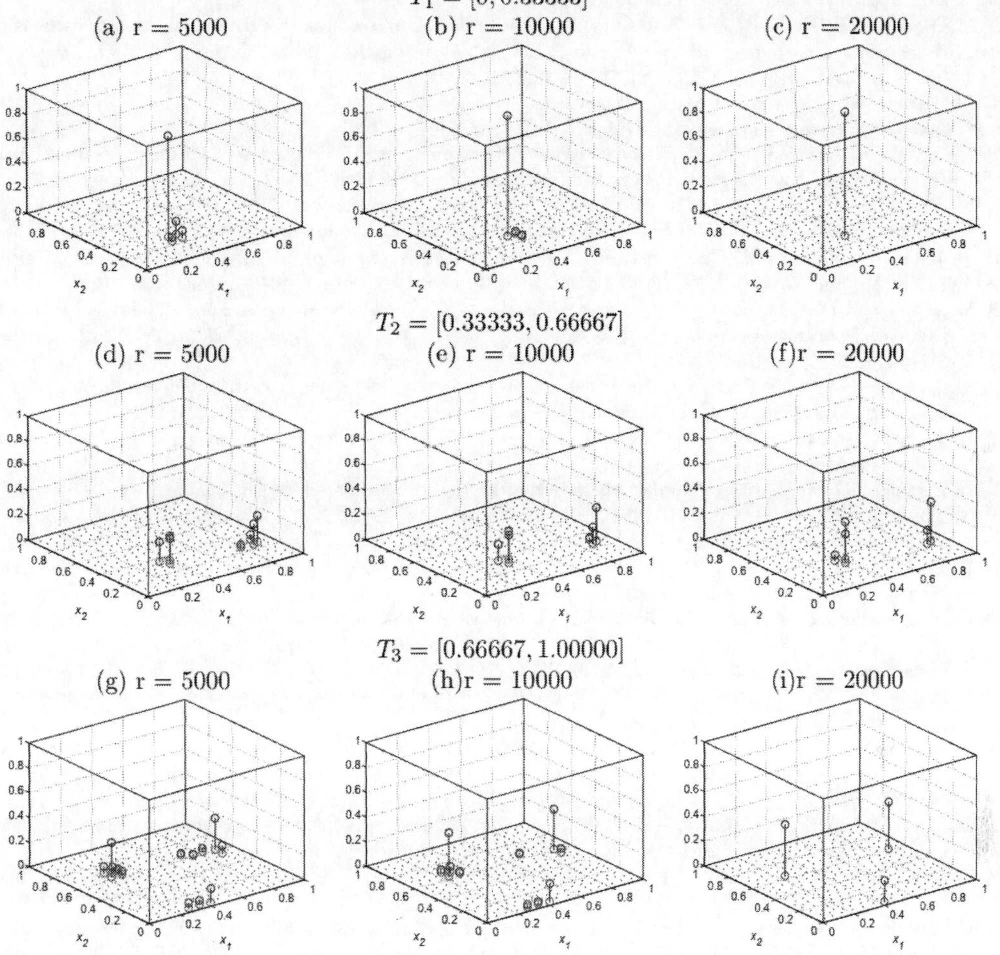

Figure 2. Allocation of active sensors (red circles) with experimental effort (blue stem plot) in consecutive time subintervals for different number of communications r.

system. Springer Science & Business Media, 2010.

[10] D. Kowalów, M. Patan, W. Paszke, and A. Romanek. Sequential design for model calibration in iterative learning control of dc motor. In *Methods and Models in Automation and Robotics (MMAR), 2015 20th International Conference on*, pages 794–799. IEEE, 2015.

[11] D. Kowalw and M. Patan. Optimal sensor selection for model identification in iterative learning control of spatio-temporal systems. In *Methods and Models in Automation and Robotics - MMAR 2016 : 21 international conference; ISBN: 9781509018673*, pages 70–75, Midzyzdroje, Polska, 2016. [B. m.], [B. w.].

[12] A. Nehorai, B. Porat, and E. Paldi. Detection and localization of vapor-emitting sources. *IEEE Transactions on Signal Processing*, 43(1):243–253, 1995.

[13] P. Ögren, E. Fiorelli, and N. E. Leonard. Cooperative control of mobile sensor networks: Adaptive gradient climbing in a distributed environment. 49(8):1292–1302, 2004.

[14] M. Patan. A parallel sensor scheduling technique for fault detection in distributed parameter systems. In *Euro-Par 2008–Parallel Processing*, pages 833–843. Springer, 2008.

[15] M. Patan. Distributed scheduling of sensor networks for identification of spatio-temporal processes. *International Journal of Applied Mathematics and Computer Science*, 22(2):299–311, 2012.

[16] M. Patan. *Optimal sensor networks scheduling in identification of distributed parameter systems*, volume

425. Springer Science & Business Media, 2012.

[17] M. Patan and D. Kowalów. Robust sensor scheduling via iterative design for parameter estimation of distributed systems. In *Methods and Models in Automation and Robotics (MMAR), 2014 19th International Conference On*, pages 618–623. IEEE, 2014.

[18] M. Patan and K. Patan. Optimal observation strategies for model-based fault detection in distributed systems. *International Journal of Control*, 78(18):1497–1510, 2005.

[19] M. Patan, C. Tricaud, and Y. Chen. Resource-constrained sensor routing for parameter estimation of distributed systems. In *Proc. 17th IFAC World Congress*, 2008.

[20] M. Patan and D. Uciński. Time-constrained sensor scheduling for parameter estimation of distributed systems. In *Decision and Control (CDC), 2010 49th IEEE Conference on*, pages 7–12. IEEE, 2010.

[21] M. Patan and D. Uciński. Cost-constrained d-optimum node activation for large-scale monitoring networks. In *Decision and Control (CDC), 2016 55th IEEE Conference on*, pages 1643–1648. IEEE, 2016.

[22] M. Patan and D. Ucinski. Optimal activation strategy of discrete scanning sensors for fault detection in distributed-parameter systems. In *Proceedings of the 16th IFAC world congress, Prague, Czech Republic*, pages 4–8, 2005.

[23] M. Patan and D. Uciński. Configuring a sensor network for fault detection in distributed parameter systems. *International Journal of Applied Mathematics and Computer Science*, 18(4):513–524, 2008.

[24] R. J. Patton, P. M. Frank, and R. N. Clark. *Issues of fault diagnosis for dynamic systems.* Springer Science & Business Media, 2013.

[25] A. Pázman. *Foundations of optimum experimental design*, volume 14. Springer, 1986.

[26] N. Point, A. V. Wouwer, and M. Remy. Practical issues in distributed parameter estimation: Gradient computation and optimal experiment design. *Control Engineering Practice*, 4(11):1553–1562, 1996.

[27] B. Porat and A. Nehorai. Localizing vapor-emitting sources by moving sensors. *Signal Processing, IEEE Transactions on*, 44(4):1018–1021, 1996.

[28] L. Pronzalo and E. Walter. Identification of parametric models from experimental data. *Springer, Berlin Wentzell PD, Andrews DT, Hamilton DC, Faber K, Kowalski BR (1997) Maximum likelihood principal component analysis. J Chemom*, 11:339366Wong, 1997.

[29] E. Rafajłowicz. Optimum choice of moving sensor trajectories for distributed-parameter system identification. *International Journal of Control*, 43(5):1441–1451, 1986.

[30] A. Romanek, M. Patan, and D. Kowalów. Decentralized scheduling of sensor networks for parameter estimation of spatio-temporal processes. *Advanced and Intelligent Computations in Diagnosis and Control*, 386:145, 2015.

[31] Z. Song, Y. Chen, C. R. Sastry, and N. C. Tas. *Optimal observation for cyber-physical systems: a fisher-information-matrix-based approach.* Springer Science & Business Media, 2009.

[32] C. Tricaud and Y. Chen. *Optimal mobile sensing and actuation policies in cyber-physical systems.* Springer Science & Business Media, 2011.

[33] C. Tricaud, M. P. Dariusz, U. Yang, and Q. Chen. D-optimal trajectory design of heterogeneous mobile sensors for parameter estimation of distributed systems. In *2008 American Control Conference*, pages 663–668. IEEE, 2008.

[34] D. Uciński. Optimal selection of measurement locations for parameter estimation in distributed processes. *International Journal of Applied Mathematics and Computer Science*, 10(2):357–379, 2000.

[35] D. Uciński. *Optimal measurement methods for distributed parameter system identification.* CRC Press, 2004.

[36] D. Uciński. Sensor network scheduling for identification of spatially distributed processes. *International Journal of Applied Mathematics and Computer Science*, 22(1):25–40, 2012.

[37] D. Uciński and M. Patan. Sensor network design for the estimation of spatially distributed processes. *International Journal of Applied Mathematics and Computer Science*, 20(3):459–481, 2010.

[38] É. Walter and L. Pronzato. Qualitative and quantitative experiment design for phenomenological modelsa survey. *Automatica*, 26(2):195–213, 1990.

[39] M. Witczak, M. Buciakowski, V. Puig, D. Rotondo, and F. Nejjari. An lmi approach to robust fault estimation for a class of nonlinear systems. *International Journal of Robust and Nonlinear Control*, 2015.

[40] L. Xiao and S. Boyd. Fast linear iterations for distributed averaging. *Systems & Control Letters*, 53(1):65–78, 2004.

IOP Conf. Series: Journal of Physics: Conf. Series **783** (2017) 012011 doi:10.1088/1742-6596/783/1/012011

Fault tolerant multi-sensor fusion based on the information gain

Joelle Al Hage, Maan E.El Najjar and Denis Pomorski
University of Lille, laboratory CRIStAL, UMR CNRS 9189, Villeneuve d'Ascq, France
E-mail: joelle.al-hage@ed.univ-lille1.fr

Abstract. In the last decade, multi-robot systems are used in several applications like for example, the army, the intervention areas presenting danger to human life, the management of natural disasters, the environmental monitoring, exploration and agriculture. The integrity of localization of the robots must be ensured in order to achieve their mission in the best conditions. Robots are equipped with proprioceptive (encoders, gyroscope) and exteroceptive sensors (Kinect). However, these sensors could be affected by various faults types that can be assimilated to erroneous measurements, bias, outliers, drifts,... In absence of a sensor fault diagnosis step, the integrity and the continuity of the localization are affected. In this work, we present a muti-sensors fusion approach with Fault Detection and Exclusion (FDE) based on the information theory. In this context, we are interested by the information gain given by an observation which may be relevant when dealing with the fault tolerance aspect. Moreover, threshold optimization based on the quantity of information given by a decision on the true hypothesis is highlighted.

1. Introduction

In literature, several research deal with the localization of multi-robot system. Initially, localizing a group of robots was done by estimating independently the N robots poses without taking into account useful contributions from the other robots. This approach, based on the odometer measurements, suffers from errors accumulations. To deal with this limitation, collaborative localization could be applied.

In [1], the collaborative localization is formulated through a distributed architecture based on the use of the Kalman Filter (KF). This approach takes advantage of all the information in the robots team. The estimation procedure is distributed among N filters, one for each robot. However, during the update step, when one robot observes another one using its exteroceptive sensors, relative position is measured and all robots should communicate and update their covariance matrices. The localization accuracy is achieved at the cost of high computational requirement.

Collaborative localization could be done through in a decentralized architecture as the method presented in [2] where the covariance intersection algorithm is used to fuse estimates with unknown correlations. This method avoid the calculation of cross correlations terms between robot pose estimates but at cost of lower precision. Other decentralized architectures ignore the cross correlations terms which may lead to inconsistent estimation as in [3]. In [4] synchronous communication between the robots must be ensured.

In literature, the analytical redundancy FDE approaches are divided into model based method (KF, parity relation...), data driven method (artificial neural network, Bayesian belief networks...) and knowledge based expert system [5]. Moreover, endogenous fault detection [6] refers to the case when robot is able to detect faults in itself and exogenous fault detection [7] refers to the case when robot detects faults in another one leading to a more robust system.

Fault detection could be done in a centralized architecture as in [8] where information is sent to a central FDE unit or in a distributed architecture as in [9]. The limitations of the centralized architecture appear in the sensitivity of the system to a single point-failure (of the central unit), the high computational requirement on the server and the problem of scalability to a larger team size.

In [10], a layered fault detection and exclusion is presented where the faults are divided into two types: those that could be monitored on the robots (this type of default should be modeled) and those that could be detected using the collaborative team after taking into account redundant measurements.

In this work, the aim is to develop a method able simultaneously to localize accurately the group of robots and to detect and to exclude the faulty sensors measurements from the team. The method is based on the sate estimation using the Information Filer (IF) which is the informational form of the KF. The main advantage of this filter appears in the update step modeled as a simple summation between the information contributions of different observations. Therefore, a distributed architecture for the data fusion and for the FDE step can be developed through this filter. On the other hand, residuals tests based on the divergence between the priori and the posteriori distributions of the IF are generated to detect and to exclude the faulty sensors measurements. This divergence is calculated in term of the Kullback-Leibler divergence and it includes two terms which allow to achieve two tests: one acts on the mean and the other on the covariance matrix of the data distributions of the IF.

Threshold optimization is raised in the literature in many research. For example, in some works they propose to define a threshold by fixing the probability of false alarm as in [11] or by fixing the value of the costs of errors as in Bayesian optimization [12]. In [13], the threshold is obtained by minimizing or maximizing particular distances. However, these methods do not lead to an optimal threshold because they don't take into account the probability of default and the costs of errors in a rigorous manner. In this work, optimal thresholding method based on the quantity of information is proposed. This method leads to the same rule as in Bayesian optimization but with variable costs values which are function of the prior probability and the probabilities of false alarm and detection.

For the proposed method, we state the following assumptions:

(i) N robots are equipped with proprioceptive and exteroceptive sensors,

(ii) each robot is able to observe at least one another robot and each robot is at least observed by another one,

(iii) communication between the robots is assured (Wi-Fi),

(iv) sensors defaults can be assimilated to: erroneous measurement, conflictual measurement, bias, outliers, drifts, sensors freezing...

This paper is organized as follows: section 2 provides the collaborative localization using the information filter. The fault detection and exclusion approach based on the Kullback-Leibler divergence is presented in section 3, in addition to the proposed thresholding method. Experimental results are detailed in section 4 followed by a conclusion in section 5.

2. Collaborative Localization for multi-robot system

The IF is used to estimate the positions and the orientations of the robots. It consists of two steps:

- the prediction step using the odometer model
- the update step using the relative observations between the robots.

The rest of the paper deals with the case of three robots. Note that the method could be generalized to any number of robots.

2.1. Prediction step

The state vector of the robots team consists of the state vector of each robot:

$$X_k = \left[\begin{array}{ccc} X_k^1 & X_k^2 & X_k^3 \end{array} \right]^T \tag{1}$$

where:
k is the sampling instant

$$X^i = \left[\begin{array}{ccc} x^i & y^i & \theta^i \end{array} \right]^T \tag{2}$$

is the position and the cap of robot i with respect to a fixed frame.

The odometer model is:

$$X^i_{k+1/k} = X^i_{k/k} + A^i_k u^i_k + w^i_k \tag{3}$$

$$= f(X^i_{k/k}, u^i_k) \tag{4}$$

where:
A^i_k is the input matrix:

$$A^i_k = \begin{pmatrix} \cos(\theta^i_{k/k} + \frac{\omega^i_k}{2}) & 0 \\ \sin(\theta^i_{k/k} + \frac{\omega^i_k}{2}) & 0 \\ 0 & 1 \end{pmatrix} \tag{5}$$

u^i_k is the input vector, it consists of the elementary displacement and rotation of robot i obtained from the encoders measurements: $u^i_k = [\Delta^i_k, \omega^i_k]$.

w^i_k is the process state noise modeled as a white Gaussian noise with zero-mean and covariance matrix Q^i_k

The model is non linear, therefore the Extended Information Filter (EIF) is used after linearizing around an estimated value. Consequently, the Jacobian matrix are calculated [14]:

$$F^i_k = \frac{\partial f}{\partial X}|_{X=X^i_{k/k}} \tag{6}$$

$$B^i_k = \frac{\partial f}{\partial u}|_{u=u^i_k} \tag{7}$$

The covariance matrix associated to the robot i is:

$$P^{ii}_{k+1/k} = F^i_k \, P^{ii}_{k/k} \, (F^i_k)^T + B^i_k \, (Q_u)^i_k \, (B^i_k)^T + Q^i_k \tag{8}$$

$(Q_u)^i_k$ is the covariance matrix associated to u^i_k.
Initially, the covariance matrix of the group of robots is bloc diagonal. When two robots met, they exchange information about their states. Therefore, the cross correlations terms appear in the covariance matrix and they are obtained as follows:

$$P^{ij}_{k+1/k} = F^i_k \, P^{ij}_{k/k} (F^j_k)^T \tag{9}$$

The general form of the covariance matrix of the multi-robot system is:

$$P = \begin{pmatrix} P^{11} & P^{12} & P^{13} \\ P^{21} & P^{22} & P^{23} \\ P^{31} & P^{32} & P^{33} \end{pmatrix} \tag{10}$$

In order to use the EIF, the information matrix and the information vector should be calculated:

$$Y_{k+1/k} = P_{k+1/k}{}^{-1} \tag{11}$$

$$y_{k+1/k} = Y_{k+1/k}\, X_{k+1/k} \tag{12}$$

2.2. Update step

When robot i observes robot j using its Kinect, the position of robot j relatively to the frame of robot i is measured: $\begin{bmatrix} x^{ji} & y^{ji} \end{bmatrix}$. Likewise, using the gyroscope of robots i and j, and the communication network (Wi-Fi), the orientation of robot j relatively to the frame of robot i is obtained: θ^{ji}.

The observation vector is then:

$$Z_k^{ji} = \begin{bmatrix} x_k^{ji} & y_k^{ji} & \theta_k^{ji} \end{bmatrix}^T \tag{13}$$

$$= \gamma^T(\theta_k^i)(X_k^j - X_k^i) \tag{14}$$

where:

$$\gamma\left(\theta_k^i\right) = \begin{pmatrix} \cos\theta_k^i & -\sin\theta_k^i & 0 \\ \sin\theta_k^i & \cos\theta_k^i & 0 \\ 0 & 0 & 1 \end{pmatrix} \tag{15}$$

In general case, the observations vector of the multi-robot system can be written as:

$$Z = \begin{bmatrix} Z^{21} & ... & Z^{ji} & ... \end{bmatrix}^T \tag{16}$$

The dimension of this vector may varies over time. R^{ij} is the covariance matrix associated to the observation Z^{ij}.

The observation model is non linear, its linearizion around the predicted estimation yields the Jacobian:

$$H_k^{ji} = \frac{\partial Z_k^{ji}}{\partial X}|_{X_{k/k-1}} = \gamma^T\left(\theta_{k/k-1}^i\right) h_k^{ji} \tag{17}$$

In the case of a state vector $X = \begin{bmatrix} X_k^1 & ... & X_k^i & ... & X_k^j & ... \end{bmatrix}^T$, the general form of h_k^{ji} is:

$$h_k^{ji} = \begin{pmatrix} & & -1 & 0 & y_{k/k-1}^j - y_{k/k-1}^i & & & & \\ 0_{3\times3} & ... & 0 & -1 & -x_{k/k-1}^j + x_{k/k-1}^i & ... & I_{3\times3} & ... & 0_{3\times3} \\ & & 0 & 0 & -1 & & & & \end{pmatrix} \tag{18}$$

The correction of the prediction is done as a simple summation of the information contributions of each observations:

$$Y_{k/k} = Y_{k/k-1} + \sum_{l=1}^{n} I_l(k) \tag{19}$$

$$y_{k/k} = y_{k/k-1} + \sum_{l=1}^{n} i_l(k) \tag{20}$$

where:

n is the number of observations

$I_l(k)$ and $i_l(k)$ are the informations contributions of the observation Z^{ij} on the information matrix and the information vector respectively:

$$I_l(k) = I^{ij}(k) = (H_k^{ij})^T (R_k^{ij})^{-1} H_k^{ij} \tag{21}$$

$$i_l(k) = i^{ij}(k) = (H_k^{ij})^T (R_k^{ij})^{-1} [(Z_k^{ij} - \widehat{Z}_k^{ij}) + H_k^{ij} X_{k/k-1}] \tag{22}$$

$\widehat{Z}_k^{ij}$ is the estimated measurement obtained from the predicted state estimation.

3. Sensors fault tolerance based on the Kullback-Leibler Divergence

In order to ensure the integrity of localization in presence of faulty sensors, a step of FDE should be added.

3.1. Fault detection

Consider two probability distributions $p(x)$ and $q(x)$, the Kullback-Leibler Divergence is defined as:

$$KL\,(p||q) = \int_x p\,(x) \log\left(\frac{p\,(x)}{q\,(x)}\right) \tag{23}$$

log is the natural logarithm

The divergence between the Gaussian distributions obtained in the predicted step of the EIF and the Gaussian distributions obtained in the corrected step is called the Global Kullback-Leibler Divergence ($GKLD$) and is given by [14]:

$$\begin{aligned} GKLD\ (g\,(k/k-1)\,||g\,(k/k)) =& \frac{1}{2} trace\left(Y_{k/k}\ Y_{k/k-1}^{-1}\right) + \frac{1}{2}\log\frac{|Y_{k/k-1}|}{|Y_{k/k}|} - \frac{1}{2}M \\ &+ \frac{1}{2}\left(X_{k/k} - X_{k/k-1}\right)^T Y_{k/k}\left(X_{k/k} - X_{k/k-1}\right) \end{aligned} \tag{24}$$

M is the dimension of the state vector: in the case of three robots, $M = 9$.

Equation 24 could be interpreted as the summation of two terms:

(i) $(X_{k/k} - X_{k/k-1})^T Y_{k/k} (X_{k/k} - X_{k/k-1})$ assimilated to the Mahalanobis distance, it allows a test on the mean

(ii) $\log \frac{|Y_{k/k-1}|}{|Y_{k/k}|} + trace(Y_{k/k} Y_{k/k-1}^{-1}) - M$ assimilated to the Burg matrix divergence [15], it allows a test on the covariance matrices.

The $GKLD$ measures the divergence between the distribution obtained in the predicted step of the EIF and the one obtained in the update step. It could also be interpreted as the information surprise given by the observations. After the convergence of the posteriori distribution (obtained from the update step) to the priori distribution (obtained from the predicted step), this surprise should not exceed a given value.

Therefore, this residual is sensitive to sensors defaults occurring in the predicted or the update step of the EIF.

3.2. Threshod optimization

The comparison of the $GKLD$ to a threshold value is a vital step in order to take the decision about the presence of faulty sensors in the robots team. In this work, an optimized thresholding method based on the quantity of information given by a decision is proposed.

The obtained threshold benefits from several advantages:

- it deals with the problem from an informational point of view: the optimal threshold maximizes the quantity of information provided by a decision on the true hypothesis
- fixing a probability of false alarm isn't required
- the calculation of the costs of errors (costs of choosing hypothesis i when hypothesis j is true) is done in an optimized manner contrary to the case of Bayesian optimization.

Figure 1 presents the probabilities associated with different decisions that could be taken by the detector (u_i) given the true hypothesis (H_i).

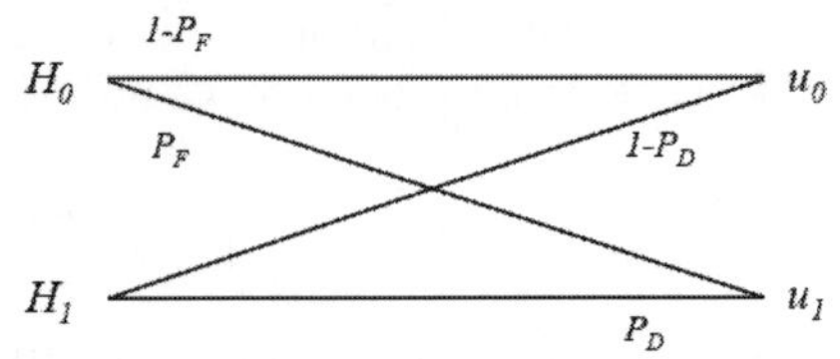

Figure 1. Relation decision/true hypothesis

The decision problem deals with two hypotheses: H_0 represents the null hypothesis (absence of faulty sensors), H_1 represents the alternative hypothesis (presence of faulty sensor). Therefore, we define:

- The detection probability: is the probability of choosing H_1 when H_1 is true:

$$P_D = p(u_1/H_1) \tag{25}$$

- The false alarm probability: is the probability of choosing H_1 when H_0 is true:

$$P_F = p(u_1/H_0) \tag{26}$$

- The miss detection probability: is the probability of choosing H_0 when H_1 is true:

$$P_{mD} = p(u_0/H_1) = 1 - P_D \tag{27}$$

Consider $f(GKLD/H_i)(i = 0, 1)$ to be the $p$robability $d$ensity $f$unction (pdf) of the $GKLD$ under the hypothesis H_i, the detection problem is reduced to:

$$\begin{aligned} H_0 &: GKLD \sim f(GKLD/H_0) \\ H_1 &: GKLD \sim f(GKLD/H_1) \end{aligned} \tag{28}$$

An example of these data distributions is given in figure 2 where:

$$P_F = \int_{\lambda}^{\infty} f\left(GKLD/H_0\right)\, d(GKLD) \tag{29}$$

$$P_D = \int_{\lambda}^{\infty} f\left(GKLD/H_1\right)\, d(GKLD) \tag{30}$$

λ is the detection threshold.

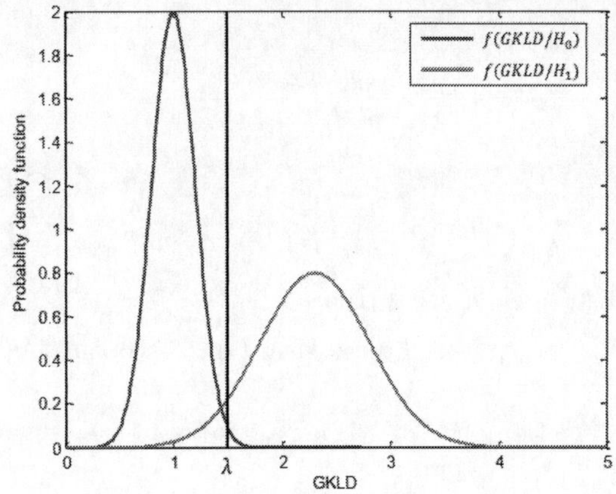

Figure 2. Illustrative example of the probability density functions of the $GKLD$

In Bayes detection theory, the choice of threshold is based on minimizing a risk function [16]. The costs of choosing hypothesis H_i $(i = 0, 1)$ when H_j $(j = 0, 1)$ is true are supposed to be known and constant. In this work, the threshold optimization is formulated using the mutual information between the decision and the hypothesis $I(H, u)$ [12]:

$$I(H,u) = \sum_{H}\sum_{u} p(H,u) \log \frac{p(u/H)}{p(u)} \tag{31}$$

$$\begin{aligned} I(H,u) =& \alpha_0 \log \frac{\alpha_0}{P_0(\alpha_0+\beta_0)} + \beta_0 \log \frac{\beta_0}{(1-P_0)(\alpha_0+\beta_0)} \\ &+ \alpha_1 \log \frac{\alpha_1}{P_0(\alpha_1+\beta_1)} + \beta_1 \log \frac{\beta_1}{(1-P_0)(\alpha_1+\beta_1)} \end{aligned} \tag{32}$$

with:

$$\begin{aligned} \alpha_0 &= P_0\,(1-P_F) \qquad & \beta_0 &= (1-P_0)(1-P_D) \\ \alpha_1 &= P_0 P_F & \beta_1 &= (1-P_0)P_D \end{aligned} \tag{33}$$

$$p(u=0) = P_0(1-P_F) + (1-P_0)(1-P_D) \tag{34}$$

$$p(u=1) = P_0 P_F + (1-P_0)P_D \tag{35}$$

The minimum of $I(H, u)$ is achieved when $P_F = P_D$, indicating that the decision does not provide any information about the true hypothesis. However, maximizing the mutual information is equivalent to maximizing the information gain given by the decision on the true hypothesis [17].

Consequently, the maximization of $I(H, u)$ leads to a likelihood ratio test expressed in function of the prior probability $(P(H_i))$, the probability of false alarm and the probability of detection:

$$\frac{f(GKLD/H_1)}{f(GKLD/H_0)} \underset{H_0}{\overset{H_1}{\gtrless}} \frac{P_0}{1-P_0}\,\frac{C_{10}-C_{00}}{C_{01}-C_{11}} \tag{36}$$

where:

$$p(H_0) = P_0 \tag{37}$$
$$p(H_1) = 1 - P_0 \tag{38}$$

and

$$C_{ij} = \log\left(\frac{p(u_i, H_j)}{p(u_i)p(H_j)}\right) = \log\left(\frac{p(u_i/H_j)}{p(u_i)}\right) \tag{39}$$

is the cost of choosing H_i when H_j is true.

The detection test could be expressed directly in function of the $GKLD$:

$$GKLD \underset{H_0}{\overset{H_1}{\gtrless}} \lambda = \text{function of}(C_{ij}, C_{ii}, P_0, \text{parameter of the } pdf \text{ of the } GKLD) \tag{40}$$

For a given value of P_0, the optimal threshold λ is obtained at the values of P_F and P_D that maximize the mutual information.

3.3. Fault exclusion

After the detection step, the faulty sensors should be excluded from the fusion procedure.

In order to ensure the exclusion of faults in the encoders, the gyroscope or the Kinect, we propose to divide the observation vector into two parts: $Z^{ji,xy} = \begin{bmatrix} x^{ji} & y^{ji} \end{bmatrix}^T$ and $Z^{ji,\theta} = \theta^{ji}$.

Using H^{ji}, we can conclude the observations matrix associated to the observations $Z^{ji,xy}$ ($H^{ji,xy}$) and the one associated to the observations $Z^{ji,\theta}$ ($H^{ji,\theta}$):

$$H_k^{ji,xy} = \left(\gamma^{xy}(\theta^i_{k/k-1})\right)^T h_k^{ji,xy} \tag{41}$$

$$\gamma^{xy}\left(\theta^i_k\right) = \begin{pmatrix} \cos\theta^i_k & -\sin\theta^i_k \\ \sin\theta^i_k & \cos\theta^i_k \end{pmatrix} \tag{42}$$

$$h_k^{ji,xy} = \begin{pmatrix} 0_{2\times3} & \dots & \widetilde{h}_k^{ji,xy} & \dots & I_{2\times3} & \dots \end{pmatrix} \tag{43}$$

$$\widetilde{h}_k^{ji,xy} = \begin{pmatrix} -1 & 0 & y^j - y^i \\ 0 & -1 & -x^j + x^i \end{pmatrix}_{k/k-1} \tag{44}$$

And:

$$H_k^{ji,\theta} = \begin{pmatrix} 0 & 0 & 0 & \dots & 0 & 0 & -1 & \dots & 0 & 0 & 1 & \dots \end{pmatrix} \tag{45}$$

When robot i observes robot j, two residuals ($KL^{ji,xy}$ and $KL^{ji,\theta}$) using the state vector of the observed robot are introduced. $KL^{ji,xy}$ is obtained from a filter denoted $EIF^{ji,xy}$ in such way that only the observation $Z^{ji,xy}$ is used in the update step of the EIF. $KL^{ji,\theta}$ is obtained from a filter denoted $EIF^{ji,\theta}$ designed in such way that only the observation $Z^{ji,\theta}$ is used in the update step:

$$\begin{aligned} KL^{ji,xy} =& \frac{1}{2} trace\left(Y^{j,xy}_{k/k}\left(Y^j_{k/k-1}\right)^{-1}\right) + \frac{1}{2}\log\frac{\left|Y^j_{k/k-1}\right|}{\left|Y^{j,xy}_{k/k}\right|} - \frac{1}{2}m \\ &+ \frac{1}{2}\left(X^{j,xy}_{k/k} - X^j_{k/k-1}\right)^T Y^{j,xy}_{k/k}\left(X^{j,xy}_{k/k} - X^j_{k/k-1}\right) \end{aligned} \tag{46}$$

where

$$Y^j = (P^{jj})^{-1} \tag{47}$$

m is the dimension of the corresponding state vector: $m = 3$

$P^{jj,xy}_{k/k}$, $X^{j,xy}_{k/k}$ are respectively the covariance matrix and the state vector of robot j obtained when the observation $Z^{ji,xy}$ is used in the update step.

Similarly:

$$\begin{aligned} KL^{ji,\theta} =& \frac{1}{2} trace \left(Y^{j,\theta}_{k/k} \left(Y^{j}_{k/k-1} \right)^{-1} \right) + \frac{1}{2} \log \frac{\left| Y^{j}_{k/k-1} \right|}{\left| Y^{j,\theta}_{k/k} \right|} - \frac{1}{2} m \\ &+ \frac{1}{2} \left(X^{j,\theta}_{k/k} - X^{j}_{k/k-1} \right)^T Y^{j,\theta}_{k/k} \left(X^{j,\theta}_{k/k} - X^{j}_{k/k-1} \right) \end{aligned} \tag{48}$$

$KL^{ji,xy}$ is influenced by a default in the encoders of robot i and j, in addition to a default in the Kinect of robot i. $KL^{ji,\theta}$ is influenced by a default in the encoders or the gyroscopes of robots i and j. Consequently, the isolation step becomes feasible after creating a set of residuals $[KL^{ij,xy}, KL^{ij,\theta}]$.

The threshold optimization defined for the $GKLD$ can be applied to the set of $KL^{ij,xy}$ and $KL^{ij,\theta}$ in order to obtain a binary vector of 0 and 1. This vector, compared to an appropriate signature matrix, allows the isolation and the exclusion of the faulty sensors from the fusion procedure.

Table 1 presents the corresponding signature matrix which links the residuals to the potential sensors errors, in the case of three robots. As it can be noticed, each default influences a different set of residuals.

Residual	O1	O2	O3	K1	K2	K3	G1	G2	G3
$KL^{21,xy}$	1	1	0	1	0	0	0	0	0
$KL^{12,xy}$	1	1	0	0	1	0	0	0	0
$KL^{31,xy}$	1	0	1	1	0	0	0	0	0
$KL^{13,xy}$	1	0	1	0	0	1	0	0	0
$KL^{32,xy}$	0	1	1	0	1	0	0	0	0
$KL^{23,xy}$	0	1	1	0	0	1	0	0	0
$KL^{21,\theta}$	1	1	0	0	0	0	1	1	0
$KL^{12,\theta}$	1	1	0	0	0	0	1	1	0
$KL^{31,\theta}$	1	0	1	0	0	0	1	0	1
$KL^{13,\theta}$	1	0	1	0	0	0	1	0	1
$KL^{32,\theta}$	0	1	1	0	0	0	0	1	1
$KL^{23,\theta}$	0	1	1	0	0	0	0	1	1

Table 1. The signatures matrix: O for odometer, K for Kinect et G for gyroscope.

4. Results and discussions

The validation of the proposed method is applied in indoor environment using three Turtlebots equipped with encoders, Kinect and gyroscope (figure 3).

The divergence of the estimated trajectories relatively to the reference one can be noticed in figure 4. These trajectories are obtained after incorporation of different defaults in the encoders, gyroscopes and Kinect.

Figure 3. The Turtlebots in real experimentation

Figure 5 shows the $GKLD$ used for the fault detection. The threshold λ is calculated after maximizing the mutual information applied on the $GKLD$ distributions in the faulty and non faulty cases. Therefore, $f(GKLD/H_0)$ is obtained after using the different sensors in their normal behavior. However, $f(GKLD/H_1)$ is concluded from available data obtained while trying to cover different types of sensors defaults. Consequently, for the calculation of $f(GKLD/H_1)$, we simulate faults in the encoders, the gyroscope and the Kinect measurements in the form of bias and drifts. The maximum of the mutual information is obtained at $P_F = 0.0528$ and $P_D = 0.9405$ corresponding to a threshold: $\lambda = 0.093$.

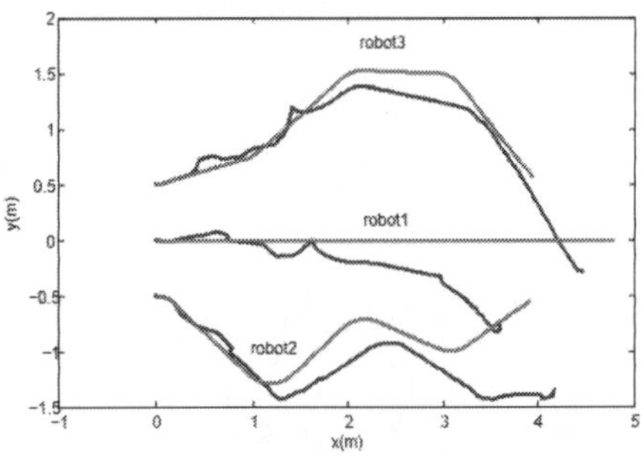

Figure 4. The trajectories of the three robots, before the FDE step (in blue) compared to the reference trajectories (in red).

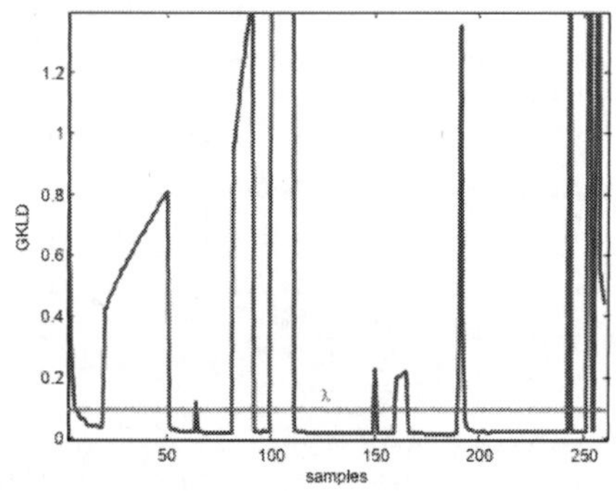

Figure 5. The $GKLD$ for the fault detection (in blue) with the optimal threshold (in red).

After the fault detection, the faulty sensors should be excluded from the fusion procedure. Therefore, the sets of residuals $[KL^{ji;xy}]$ and $[KL^{ji;\theta}]$ are calculated as shown in figure 6 and figure 7 respectively.

After finding the distributions of the $KL^{ij,xy}$ ($KL^{ij,\theta}$) in the faulty and non faulty cases, the maximization of the mutual information applied to each KL^{ij} leads to the optimal threshold value: $\lambda = 1.9924 \times 10^{-5}$ corresponding to $P_F = 0.0121$ and $P_D = 0.9667$.

From $k = 20$ to 50, $KL^{21,xy}$ and $KL^{31,xy}$ exceed the threshold value. After comparison with the signature matrix, Kinect 1 is declared faulty. In the same way, we can conclude that the kinect 2 is faulty from $k = 100$ to 110 and kinect 3 from $k = 82$ to 90.

From $k = 190$ to 192, $KL^{31,xy}$, $KL^{31,xy}$, $KL^{32,xy}$, $KL^{23,xy}$, $KL^{31,\theta}$, $KL^{31,\theta}$, $KL^{32,\theta}$, $KL^{23,\theta}$ exceed the threshold value indicating that the encoders of robot 3 are faulty.

From $k = 160$ to 165 gyroscope 3 is faulty. We note that the gyroscope 2 is faulty at $k = 150$.

Figure 8 proves the effectiveness of the proposed approach after the exclusion of the faulty measurements from the fusion procedure. Indeed, the estimated trajectories of the three robots

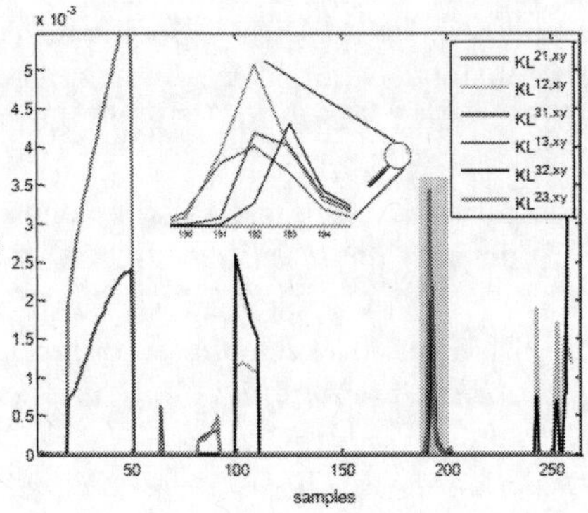

Figure 6. $KL^{ij,xy}$ residuals set.

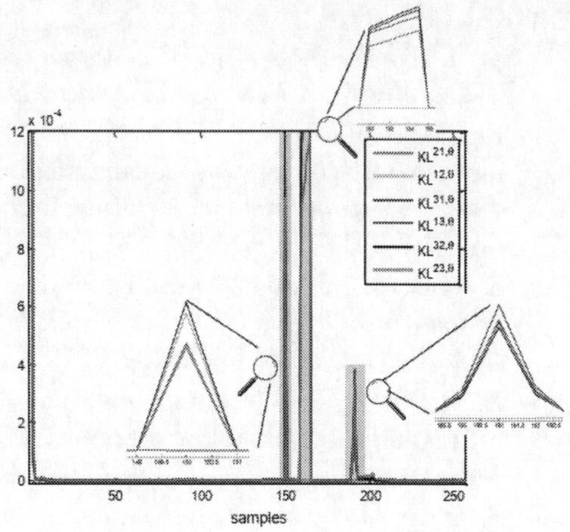

Figure 7. $KL^{ij,\theta}$ residuals set.

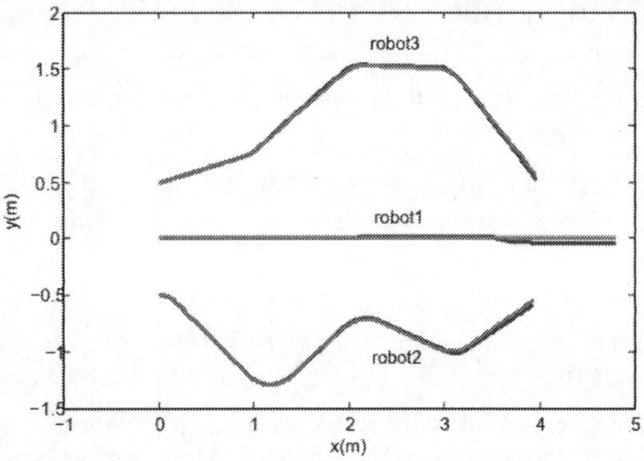

Figure 8. Trajectories after the FDE step (in blue) compared to the reference trajectories (in red).

are roughly similar to the references.

5. Conclusion

In this paper, we propose a method able to localize accurately a group of robots with a sensors faults detection and exclusion step.

The method is based on a informational framework: information filter and tools of the information theory. The global Kullback-Leibler divergence is used for the fault detection and a set of residuals $[KL^{ij,xy}, KL^{ij,\theta}]$ defined at each time one robot sees another one, are used for the faults exclusion.

Threshold optimization based on the maximization of the quantity of information given by a decision on the true hypothesis is proposed. This method does not require to fix the probability of false alarm nor the costs of errors.

The method was able to localize accurately the group of robots while excluding the faulty sensors measurements.

References

[1] S. I. Roumeliotis and G. A. Bekey, "Distributed multirobot localization," *IEEE Transactions on Robotics and Automation*, vol. 18, no. 5, pp. 781–795, 2002.

[2] L. C. Carrillo-Arce, E. D. Nerurkar, J. L. Gordillo, and S. I. Roumeliotis, "Decentralized multi-robot cooperative localization using covariance intersection," in *International Conference on Intelligent Robots and Systems (IROS), 2013 IEEE/RSJ*, IEEE, 2013, pp. 1412–1417.

[3] S. Panzieri, F. Pascucci, and R. Setola, "Multirobot localisation using interlaced extended kalman filter," in *2006 IEEE/RSJ International Conference on Intelligent Robots and Systems*, Oct. 2006, pp. 2816–2821.

[4] E. D. Nerurkar, S. Roumeliotis, and A. Martinelli, "Distributed maximum a posteriori estimation for multi-robot cooperative localization," in *IEEE International Conference on Robotics and Automation, 2009. ICRA '09*, May 2009, pp. 1402–1409.

[5] S. Ding, *Model-based fault diagnosis techniques: Design schemes, algorithms, and tools.* Springer Science & Business Media, 2008.

[6] S. I. Roumeliotis, G. Sukhatme, and G. A. Bekey, "Sensor fault detection and identification in a mobile robot," in *Proceedings in International Conference on Intelligent Robots and Systems, 1998.*, vol. 3, IEEE, 1998, pp. 1383–1388.

[7] A. Christensen, R. O'Grady, and M. Dorigo, "From fireflies to fault-tolerant swarms of robots," *IEEE Transactions on Evolutionary Computation*, vol. 13, no. 4, pp. 754–766, Aug. 2009, ISSN: 1089-778X.

[8] X. Li and L. Parker, "Sensor analysis for fault detection in tightly-coupled multi-robot team tasks," in *2007 IEEE International Conference on Robotics and Automation*, Apr. 2007, pp. 3269–3276.

[9] ——, "Distributed sensor analysis for fault detection in tightly-coupled multi-robot team tasks," in *IEEE International Conference on Robotics and Automation, 2009. ICRA '09*, May 2009, pp. 3103–3110.

[10] R. A. Carrasco, F. Núñez, and A. Cipriano, "Fault detection and isolation in cooperative mobile robots using multilayer architecture and dynamic observers," *Robotica*, vol. 29, no. 4, pp. 555–562, Jul. 2011, ISSN: 1469-8668.

[11] A. Martineau, "Etude de la performance du contrôle autonome d'intégrité pour les approches à guidage vertical," PhD thesis, Université de Toulouse, Nov. 14, 2008.

[12] P. K. Varshney, *Distributed detection and data fusion.* Springer Science & Business Media, 2012.

[13] R. J. Irwin and T. C. Irwin, "A principled approach to setting optimal diagnostic thresholds: Where ROC and indifference curves meet," *European Journal of Internal Medicine*, vol. 22, no. 3, pp. 230–234, Jun. 2011, ISSN: 0953-6205.

[14] J. Al Hage, M. E. El Najjar, and D. Pomorski, "Fault tolerant collaborative localization for multi-robot system," in *2016 24th Mediterranean Conference on Control and Automation (MED)*, Jun. 2016, pp. 907–913.

[15] J. V. Davis and I. S. Dhillon, "Differential entropic clustering of multivariate gaussians," in *Neural information processing systems (NIPS)*, 2006, pp. 337–344.

[16] C. Jutten, *Détection, Estimation, Information.* Polytech' Grenoble: Université Joseph Fourier, 2007.

[17] I. Y. Hoballah and P. K. Varshney, "An information theoretic approach to the distributed detection problem," *IEEE Transactions on Information Theory*, vol. 35, no. 5, pp. 988–994, 1989.

Leak Isolation in Pressurized Pipelines using an Interpolation Function to approximate the Fitting Losses

A. Badillo-Olvera, O. Begovich, A. Peréz-González

CINVESTAV-IPN, Unit Guadalajara, Av. Del Bosque No. 1145, Col. El Bajío, Zapopan, Jalisco, México.

E-mail: [ambadillo, obegovi, aperez]@gdl.cinvestav.mx

Abstract. The present paper is motivated by the purpose of detection and isolation of a single leak considering the Fault Model Approach (FMA) focused on pipelines with changes in their geometry. These changes generate a different pressure drop that those produced by the friction, this phenomenon is a common scenario in real pipeline systems. The problem arises, since the dynamical model of the fluid in a pipeline only considers straight geometries without fittings. In order to address this situation, several papers work with a virtual model of a pipeline that generates a equivalent straight length, thus, friction produced by the fittings is taking into account. However, when this method is applied, the leak is isolated in a virtual length, which for practical reasons does not represent a complete solution. This research proposes as a solution to the problem of leak isolation in a virtual length, the use of a polynomial interpolation function in order to approximate the conversion of the virtual position to a real-coordinates value. Experimental results in a real prototype are shown, concluding that the proposed methodology has a good performance.

1. Introduction

The hydraulic efficiency of a pipeline system, from a physical perspective, is associated to the capacity of the system to provide the injected fluid to its final destination. In order to preserve this efficiency, it is necessary to point out the importance of early detection, isolation and reparation of leaks. A neglectful treatment of these activities can causes fluid losses, discontinuities in the services and low pressures in the pipeline, which produce an excessive energy consumption in the pumping systems and consequently, the operating costs rise[1].

In the last years, different analytical methods have been proposed, based on Fault Model Approach (FMA) and Fault Sensitive Approach (FSA) algorithms, for instance [2], [3], [4], [5], [6], [7], to cite a few. Such techniques use, in first instance, sensors to monitor certain internal quantities (flow, pressure, temperature, etc.), and in second instance, nonlinear mathematical models [8] deduced from a pair of partial differential equations that describe the fluid dynamics in closed conduits, these equations are known as Water Hammer Equations (WHE).

The nonlinear models are generally used to design an observer that estimates the unmeasurable states using input and output measurements from a real system. In the case where the Extended Kalman Filter is used as an observer, when a leak occurs the leak isolation algorithm begins to estimate its location and magnitude.

The WHE are deduced for straight pipelines without fittings, which in most of real systems is not satisfied. Generally, pipeline systems are composed by different accessories, in order to adapt the installation to topography features. Several devices are also included, as valves for flow control, for example. All of the mentioned above cause energy losses, known as *local losses*, which are different from those produced by the friction of the pipes. In the actual literature exist an increasing number of studies focused on leak detection and isolation based on FMA and FSA, these studies takes into account the presence of fittings in the analysis performing the leak isolation on a virtual straight pipeline. In order to do that, each fitting of the original pipeline is replaced by a section of straight pipe that presents the same pressure loss that the respective fitting [2], [9], [10]. Under the mentioned conditions, it is said that the leak location is performed in *Equivalent Straight Length* coordinates (ESL). Several other works say nothing about the coordinates that represent the leak position, even though there is presence of accessories in the pipeline in which the analysis was performed. In the consulted literature, only [2] presents a leak isolation in real length coordinates, however, there is not an explanation of how to arrive to the final result of the research. In another hand, [11] gives a methodology to achieve the leak isolation in pipelines with fittings, in real coordinates, but it introduces proportionality relationships for coefficients of lost owned by each accessory that not always can be reached. Considering the importance of the real-coordinates representation, this paper proposes a new alternative method to isolate a leak in real coordinates, unlike the other approaches found in the literature, where the leak is isolated in equivalent coordinates. This method is performed through the calculation of an interpolation function obtained from the estimation of the pressure at different points of the pipeline. Once the interpolation function is formulated, it is used then as an equivalent-to-real coordinate converter. Results are tested in a prototype and presented in this work, showing a good agreement with the real pipeline conditions.

The paper is organized as follows: Section 2 provides a mathematical model for pipelines and the algorithm of the Extended Kalman Filter (EKF), which is implemented as an observer for Leak Detection and Isolation (LDI) problem. Section 3 presents our interpolation method considering the pressure drops in each pipeline fitting. Section 4 shows the experiments and results obtained. Finally, Section 5 presents some relevant conclusions and discusses the future work.

2. Leak modeling and isolation

The equations that describe the dynamics of the flow in a pipeline, in transient state, are known as the Water Hammer Equations (1) and (2). Considering the *principles of mass and momentum*, let the pipeline be straight, without slope, and with duct wall slightly deformable; also consider that the fluid is slightly compressible, the convective velocity changes are negligible and let the pipeline cross section area and fluid density be constants [12]. Then, the continuity equation is defined as follows:

$$\frac{\partial H(z,t)}{\partial t} + \frac{b^2}{gA}\frac{\partial Q(z,t)}{\partial z} = 0, \tag{1}$$

and the momentum equation takes the form:

$$\frac{\partial Q(z,t)}{\partial t} + gA\frac{\partial H(z,t)}{\partial z} + f\frac{Q(z,t)|Q(z,t)|}{2DA} = 0, \tag{2}$$

where, H is the pressure head $[m]$, Q is the flow rate $[m^3/s]$, z is the length coordinate $[m]$, t is the time coordinate $[s]$, g is the gravity acceleration $[m/s^2]$, A is the cross section area $[m^2]$, b is the pressure wave speed in the fluid $[m/s]$, D is the internal diameter $[m]$ and f is the Darcy-Weisbach friction factor $[dimensionless]$.

The pressure wave speed in the fluid is given by:

$$b = \sqrt{\frac{\frac{\kappa}{\rho}}{1 + \frac{D\kappa}{Ee}}}, \tag{3}$$

where E is the Young's modulus of elasticity for the conduit walls $[Pa]$, κ and ρ are the bulk modulus $[Pa]$ and the density of the fluid $[kg/m^3]$, respectively, and finally, e is the thickness of the pipe wall $[m]$.

The boundary conditions in this work are taken as the pressure heads at the extremes of the pipeline, which are measurable parameters denoted by:

$$H(x = 0, t) = H_{in}(t), \quad H(x = L, t) \quad = H_{out}(t). \tag{4}$$

2.1. *Friction factor*

One of the most commonly used equations, due to its property to give good values of the friction factor f, is the Coolebrok-White equation:

$$\frac{1}{\sqrt{f}} = -2.0 \log\left(\frac{\epsilon/D}{3.71} + \frac{2.51}{R_e\sqrt{f}}\right), \tag{5}$$

where, D is the internal diameter $[m]$, ϵ is the roughness coefficient of the material $[m]$ and Re is the Reynolds number calculated as follows: $Re = QD/\nu A$, where ν is the kinematic viscosity $[m^2/s]$, Q is the flow $[m^3/s]$ and A is the cross section area $[m^2]$.

The general form of the Coolebrok-White equation requires iterative calculations, which can be hard to implement or computationally expensive, therefore several authors have proposed explicit equations for the friction factor based on the Coolebrok-White equation. A set of the most accurate and efficient in the zone of complete turbulence [13], are presented below:

- *Buzzelli.* High accuracy, relative percentage error 0.005:

$$\frac{1}{\sqrt{f}} = B_1 - \left[\frac{B_1 + 2\log(\frac{B_2}{R_e})}{1 + \frac{2.18}{B_2}}\right], \tag{6}$$

where:

$$B_1 = \frac{[0.774\ln(R_e)] - 1.41}{(1 + 1.32\sqrt{\frac{\epsilon}{D}})}, \qquad B_2 = \frac{\epsilon}{3.7D}R_e + 2.51B_1.$$

- *Haaland.* Medium accuracy, relative percentage error 0.0373 and applicable range of $4 \times 10^3 \leq R_e \leq 1 \times 10^8$, $1 \times 10^{-6} \leq \epsilon/D \leq 5 \times 10^{-2}$:

$$\frac{1}{\sqrt{f}} = -3.6 \log\left[\frac{6.9}{R_e} + \left(\frac{\epsilon}{3.7D}\right)^{1.11}\right]. \tag{7}$$

- *Swamee and Jain.* Medium accuracy, relative percentage error 0.478 and applicable range of $5 \times 10^3 \leq R_e \leq 1 \times 10^8$, $1 \times 10^{-6} \leq \epsilon/D \leq 5 \times 10^{-2}$:

$$f = \frac{0.25}{\left[\log_{10}\left(\frac{5.74}{Re^{0.9}} + \frac{\epsilon/D}{3.7}\right)\right]^2}. \tag{8}$$

2.2. Leak model

The equation that describes the behavior of a leak located in an arbitrary point z_L, as seen in the Figure 1, can be formulated using the orifice equation:

$$Q_L = C_d A \sqrt{2g H_L}, \tag{9}$$

where C_d is the discharge coefficient [*dimensionless*] and A is the leak cross section area $[m^2]$. Now, defining $\lambda = C_d A \sqrt{2g}$, the flow in the leak can be expressed as:

$$Q_L = \lambda \sqrt{H_L}. \tag{10}$$

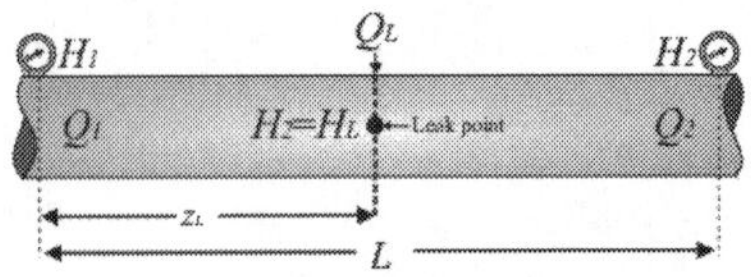

Figure 1. Discretization of the pipeline with a leak located at position z_L.

In order to obtain a representation in the space of states, the equations (1) and (2) are discretized with respect to the spatial variable z, [12], [14], [15] using the following relationships:

$$\frac{\partial H}{\partial z} \approx \frac{H_{i+1} - H_i}{\Delta z} \quad \forall i = 1, ..., n, \tag{11}$$

$$\frac{\partial Q}{\partial z} \approx \frac{Q_i - Q_{i-1}}{\Delta z} \quad \forall i = 1, ..., n. \tag{12}$$

The pipeline is discretized in two sections, as shown in Figure 1, where Δz is the distance step, $z_i, i = 1, 2$ are the distances from the start of the pipeline to the leak position (z_L) and from leak position to the end of the pipeline ($L - z_L$), respectively.

Using approximations (11) and (12), and considering z_L and λ as constant values, the following dynamical system representation is obtained:

$$\begin{bmatrix} \dot{Q}_1 \\ \dot{H}_2 \\ \dot{Q}_2 \\ \dot{\lambda} \\ \dot{z}_L \end{bmatrix} = \begin{bmatrix} \frac{-gA}{z_L}(H_2 - u_1) - \frac{f(Q_1)}{2DA} Q_1 |Q_1| \\ \frac{-b^2}{gAz_L}(Q_2 - Q_1 - \lambda\sqrt{H_2}) \\ \frac{-gA}{L-z_L}(u_2 - H_2) - \frac{f(Q_2)}{2DA} Q_2 |Q_2| \\ 0 \\ 0 \end{bmatrix}, \tag{13}$$

where $[u_1 \quad u_2]^T = [H_1 \quad H_3]^T$ is the input vector and $y = [Q_1 \quad Q_2]^T$ is the output vector.

The model (13) can be written in compact form as:

$$\dot{x} = \xi(x, u), \tag{14}$$

where $x \triangleq [Q_1 \quad H_2, Q_2 \quad \lambda \quad z_L]$ and $\xi(.)$ is a nonlinear function.

2.3. Leak isolation scheme

In order to develop a leak isolation scheme, it is necessary in first place to obtain a discrete representation for the model (14). According to [9], the Heun's method is suitable to perform the discretization.

Defining the initial value problem:

$$\dot{x}(t) = \xi(x(t), u(t)),\ x(t_0) = x_0, \tag{15}$$

then, the Heun's method equation takes the form:

$$x^{i+1} = x^i + \frac{\xi(x^i, u^i) + \xi(x^i + \Delta t \xi(u^i, x^i), u^{i+1})}{2}\Delta t, \tag{16}$$

where Δt is the time step.

The model (14) discretized by equation (16) can be written in compact form as:

$$\begin{aligned} x^{i+1} &= \xi(x^i, u^{i+1}, u^i), \\ y &= Hx^i, \end{aligned} \tag{17}$$

where $x^i \triangleq \begin{bmatrix} Q_1^i & H_2^i & Q_2^i & z_L^i & \lambda^i \end{bmatrix}^T$, $\xi(.)$ is a nonlinear function and H is fixed as:

$$H = \begin{bmatrix} 1 & 0 & 0 & 0 & 0 \\ 0 & 0 & 1 & 0 & 0 \end{bmatrix},$$

Once the discretized model is obtained, an Extended Kalman Filter is implemented as a state observer, taking into account the expressions in Table 1 [16]:

$\widehat{x}^{i^-}$ is the a-priori estimate of x^i: $\widehat{x}^i = \widehat{x}^{i^-} + \mathbb{K}^i(y^i - H\widehat{x}^{i^-})$	P^{i^-} is the a-priori covariance matrix: $P^{i^-} = J^i P^{i-1^-}(J^i)^T + Q$
$\mathbb{K}^i$ is the Kalman gain for the observer: $\mathbb{K}^i = P^{i^-} - H^T(HP^{i^-}H^T + R)^{-1}$	P^i is the posterior covariance matrix: $P^i = (I - \mathbb{K}^i H)P^{i^-}$
J^i is the Jacobian matrix: $J^i = \left.\frac{\partial \xi(x,u)}{\partial x}\right\|_{x=\hat{x}}$	

Table 1. Kalman Filter Equations

R and Q are known as the covariance matrices of measurements and process noises, respectively. Notice that:

$$P^0 = (P^0)^T > 0,\ R = R^T > 0, \text{ and } Q = Q^T > 0.$$

3. Locating leaks in pipelines with fittings

Fittings such as elbows, valves, unions and contractions present resistance to the flow, known as *local* or *minor losses*; the total of these pressure drops in a pipeline system usually has a value between 5% to 20% of the total pressure drop. In the case of the LDI problem, such minor losses must be known with high precision in order to achieve a good leak isolation.

Pressure losses due to friction in a straight pipeline can be obtained through the Darcy-Weisbach equation:

$$h_f = f\frac{\Delta z}{D}\frac{Q(t)^2}{2gA^2}, \tag{18}$$

where h_f is a friction loss $[m]$, f is the friction factor $[dimensionless]$, Δz is the length $[m]$ of a straight pipe section and $Q(t)^2/2gA^2$ is the head velocity in terms of flow $[m]$.

Pipelines with fittings have additional local losses due to the accessories, calculated as:

$$h_l = K_f\frac{Q(t)^2}{2gA^2}, \tag{19}$$

where h_l is the local loss $[m]$, K_f is a dimensionless coefficient, known as *Coefficient of Loss*, which depends on: the type of fitting, the Reynolds number and the roughness of fitting material.

For a pipeline with accessories, the total pressure drop is calculated as:

$$H = h_f - \sum_{i=1}^{N} h_{l(i)}, \tag{20}$$

where $h_{l(i)}$ represents the local loss in the i-th accessory. As an example, let L_a be an arbitrary distance in the pipeline, such that L_1, L_2 are straight pipes and F_1 and F_2 are fittings that belong to the $[0, L_a]$ section. Then, to calculate the pressure drop in the point HL_a of the $[0, L_a]$ section due to each individual component, the Equation (20) takes the form:

$$H_{L_a} = H_{in} - f\frac{L_1}{D}\frac{Q^2}{2gA^2} - f\frac{L_2}{D}\frac{Q^2}{2gA^2} - K_{F_1}\frac{Q^2}{2gA^2} - K_{F_2}\frac{Q^2}{2gA^2},$$

where, H_{in} represents the pressure at the beginning of the $[0, L_a]$ section, provided by the pump, $f\frac{L_1}{D}\frac{Q^2}{2gA^2}$ is the pressure drop produced by the L_1 straight pipe, $f\frac{L_2}{D}\frac{Q^2}{2gA^2}$ is the pressure drop produced by the L_2 straight pipe, and finally, $K_{F_1}\frac{Q^2}{2gA^2}$ and $K_{F_2}\frac{Q^2}{2gA^2}$ represent the pressure drop added by the fittings F_1 and F_2.

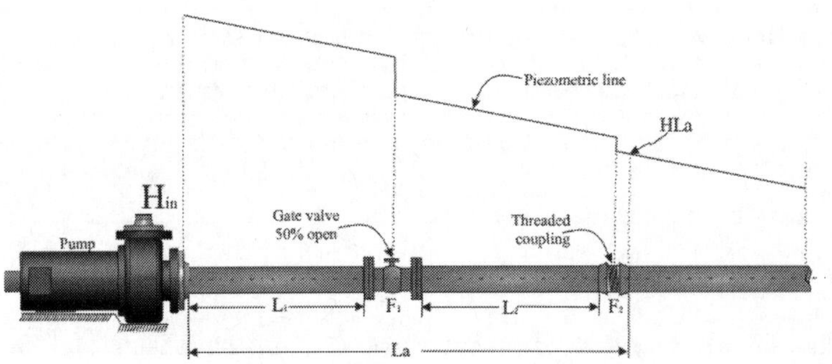

Figure 2. Piezometric line of a pepeline.

Figure 2 shows the piezometric profile of the pipeline described at the previous example. The local losses generate a sharper drop, thus, the piezometric profile is not completely linear.

3.1. Equivalent straight length to real length

To tackle the problem of leak isolation using the FMA in pipelines with fittings, different works as ([2], [9] and [10]) propose the calculation of a virtual equivalent straight pipeline, in which a longitudinal compensation is performed for each fitting, in order to obtain a straight pipeline with losses that are equivalent to the losses in the original pipeline with fittings. The equivalent straight length L_{eq} is calculated and introduced in the mathematical model.

Using the Darcy-Weisbach equation (18) and the pressure head measurements, L_{eq} is calculated as follows:

$$L_{eq} = \frac{\Delta H(t) D^5 \pi^2 g}{8 f(t) Q(t)^2}. \tag{21}$$

Using the Equation (21), the position of a leak is isolated in an equivalent straight length, but from a practical point of view, this result does not represent a complete solution: in fact, an equivalent-to-real coordinates conversion is necessary to locate the leak in the real pipeline. Further more, there does not exist a direct conversion from equivalent to real coordinates, since it can be possible to obtain an infinite number of pipelines with different structures and fittings that present the same equivalent length.

One way, simple but not very accurate, to return to real length coordinates is to establish a linear relationship between the real and the equivalent length, as is shown in (22) (for instance, see Figure 5 in Section 4); however, if the leak occurs near to a fitting, the isolation will be wrong due to the abrupt pressure drop caused by the mentioned accessory. The linear relation is expressed as follows:

$$z_r = \frac{L_t}{L_{eq}} z_{eq}, \tag{22}$$

where z_r is a position in real length, L_{eq} is the total equivalent length of the pipeline, L_t is the total real length of the pipeline and z_{eq} is the leak position in equivalent length, all expressed in m.

For the sake of understanding, when a leak is isolated on a equivalent length, it will be noted that the leak is *isolated in equivalent coordinates*, otherwise, the leak is *isolated in real coordinates.*

3.2. Interpolation function for pressure drop

Due to the existence of an infinite number of pipes with different structures and fittings that presents the same equivalent length, it is necessary to know the structure of the pipeline under study in order to achieve the leak isolation in real coordinates. To known the structure of the pipeline is, in general, an easy task, since usually there are design plans with this information and the standard length of pipes is known.

Using the EKF algorithm to isolate a leak, the position in equivalent coordinates (z_{Leq}) and the value of the pressure at the leak point (H_2) are obtained [9],[2]; this pressure value is the same in both coordinates systems, the equivalent and real one. This correspondence of pressure values can be used to find the leak position in real coordinates (z_{Lr}).

The pressure drop along the length in a pipeline with fittings is not completely linear, however, the pressure behaviour can be approximated by a smooth curve. A simple way to obtain a suitable approximation is to calculate an interpolation function with the Least Square (LS) technique [17].

In order to use the LS technique, a set of measurements is required to approximate the behavior of the pressure throughout the pipeline. The most common scenario, as is taken in this research, is the one in which the measurements of pressure are only available at both extreme points of the pipeline. Given such restriction, the measurements are complemented by a set of calculated pressure values, obtained from (20) in different points of the pipeline. Each fitting produces an inflection in the pressure profile, therefore, to include the local loses caused by a fitting in a pipe, it is required to estimate the pressure with Equation (20). It is necessary to perform this estimation in the two connection points of each fitting. Thus, using the LS technique and having as dependent variable the length (z) and as independent variable the pressure (H), the interpolation function is expressed as follows:

$$\hat{z} = \beta_1 + \beta_2 H + \beta_3 H^2 + \dots + \beta_n H^n, \tag{23}$$

where $\beta_j, j = 1, 2, ..., n$ are the interpolation function coefficients calculated with the LS technique, H is the pressure at a specific point of the pipeline, z is it corresponding length and n is the degree of the polynomial equation. Note that, for n data points, a polynomial of order $n-1$ can be used to obtain a good adjustment.

Using a set of coordinated pairs of the form (z_i, H_i), it is possible to find the coefficients $\beta_j, j = 1, 2, ..., n$ that fit to the Equation (23) in the best way, through the solution of a quadratic minimization problem, where the objective function J_c is given by:

$$J_c = \|z - \mathbf{H}\beta\|^2. \tag{24}$$

The polynomial interpolation must be calculated after the establishment of a constant inflow value, that is, after the transient effect caused by the occurrence of a leak has been vanished. Once the β_j are determined, Equation (23) allows to know a position in the pipeline for a given pressure. In particular, using the pressure at the leak point given by the EKF (H_2) is possible to determine the leak position in real coordinates (z_r).

The next flow chart illustrates the above procedure in the grey boxes:

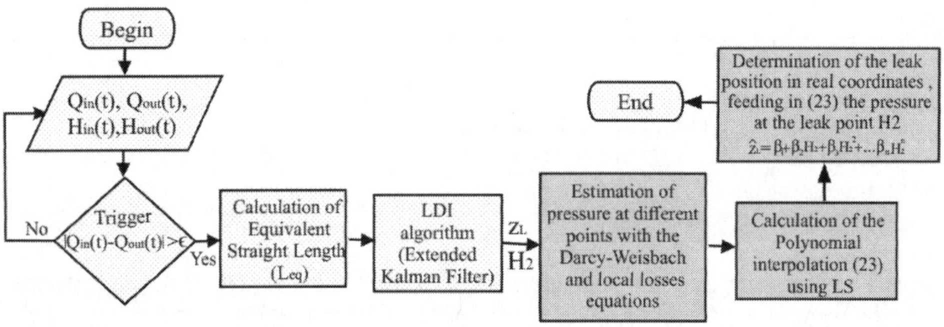

Figure 3. Algorithm to isolate a leak in a pipeline with fittings in real coordinates using the proposed interpolation function.

4. Experimental results

In order to compare the relationship of proportionality (22) against the interpolation function (23), two experiments are performed, using real data acquired from a pipeline prototype.

4.1. Prototype description

The experimental scenarios are implemented in a pipeline prototype built at the Center for Research and Advanced Studies (CINVESTAV) Guadalajara, Mexico. This pipeline prototype is composed by two pressure sensor *Promag Propiline 10P* and two flow sensor *PMP 41*, both of them *Endress Hauser*™. These sensors are placed at the end points of the pipeline. Besides, a temperature sensor *PT100* is mounted at the interior of the water supply tank. To distribute the water, the prototype includes a centrifugal pump of 5HP from *Siemens*™. The pipeline total length is 68.147m and it has three valves located at 17.045m (valve 1), 33.47m (valve2) and 49.895m (valve 3) that allow to emulate leaks. Valves 1 and 3 also contain pressure sensors from *Winters*™. These sensors have the objective of validate the head pressure estimations in the leak points. The data logging for sensors is performed by a DAQ module *NI US-6229* produced by *National Instruments*™. Finally, the user interface, which interacts with hardware devices, is developed in Labview™ and Matlab™. The main flow line parameters are shown in the Table 2. For more technical information about the pipeline prototype consult [18].

The architecture of the pipeline prototype is composed by eight joints with metal thread, eleven plastic joins, two plastic elbows, five metal tees and $64.93m$ of straight plastic pipe. The coefficients of loss of each fitting are shown in the Table 3. Using the coefficients from Table 3, a pressure in kgf/m^2 units is obtained, so, it is convenient to convert the result to a pressure in meters of water column $[mH_2O]$ units.

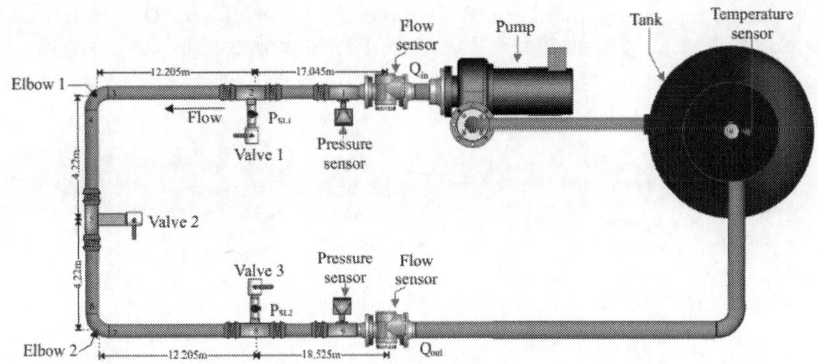

Figure 4. Prototype scheme.

Parameter	Symbol	Value	Units
Total Length	L_r	68.147	m
Internal diameter	D	6.271×10^{-3}	m
Wall thickness	e	13.095×10^{-3}	m
Roughness	ϵ	7×10^{-6}	m
Slope	s	0	%

Table 2. Pipeline prototype parameters.

Type of fitting	K_r
Elbow	2.0
Plastic join	0.25
Metal join	0.4
metal tee	0.7

Table 3. Local loss coefficients.

4.2. Experimental results

Each of the tests presented in this section has a duration of $250s$, with sampling time of $\Delta t = 0.1s$.

For the first experiment, the pressure drop is estimated at nine different points of the pipeline, and the coefficients β_j of the interpolation function (23) are determinate using the LS technique. Then, the results obtained using Equation (20) are compared against the measurements obtained at the points where sensors are available (see Figure 5 (a)). The Mean Squared Error (MSE) between these points is 29.68×10^{-2} units. Figure 5 (a) shows the result obtained using the equivalent straight length and its conversion to real coordinates through the relationship of proportionality (22). It is easy to note that near to the fittings, the pressure drop significantly differs from the real pressure profile. The MSE between the relation of proportionality and the pressure drop profile has a magnitude of 37.21 units. In Figure 5 (b), the comparison between the pressure profile and the interpolation function (23) is shown. For this case, the MSE has a value of 2.4 units.

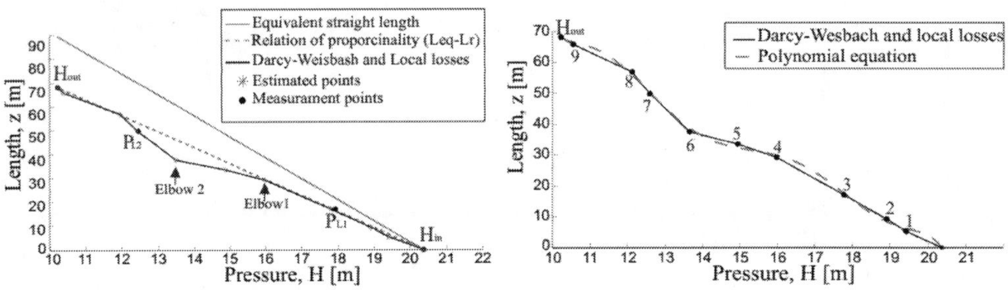

(a) Darcy-weisbach and equivalent straight length comparison.

(b) Polynomial equation.

Figure 5. Pressure drop.

For the second experiment, three different leaks coming from valves 1, 2 and 3 are isolated, one at a time. Each leak is localized in equivalent coordinates and then converted to real coordinates with the relationship of proportionality (22) and using the interpolation function (23). For the leaks coming from valves 1 and 3 (Figure 6 and 8, respectively), results are similar for both methods. Nevertheless, for the leak at valve 2, (7), which is the closest to an elbow, a better result is obtained with the interpolation equation, reaching a MSE of 2.009 units against 9.4 units obtained with the relationship of proportionality. The results would be more evident if there where fittings that produce greater pressure drops, such as elbows and sharp contractions, near to the valves.

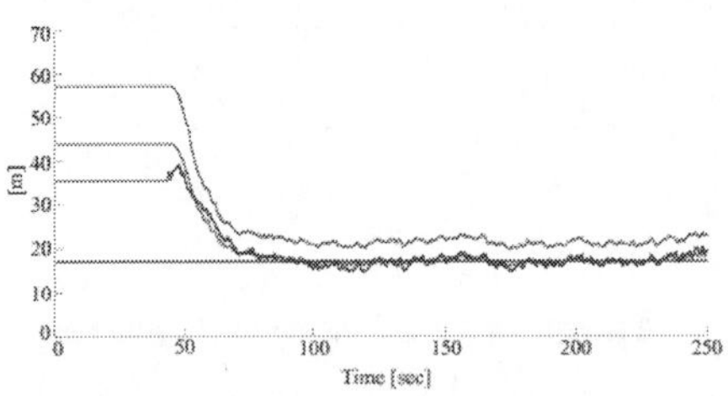

Figure 6. Leak isolation emulated by valve 1. Real leak position (black), estimation of leak position using polynomial function (blue), estimation of leak position in equivalent coordinates (green) and leak position estimated by proportionality relationship (red).

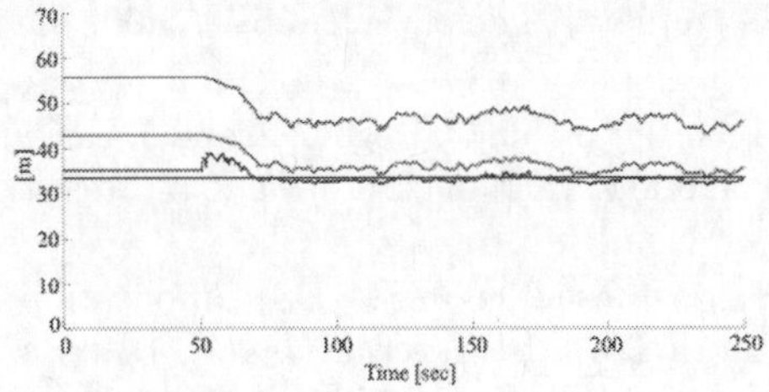

Figure 7. Leak isolation emulated by valve 2. Real leak position (black), estimation of leak position using polynomial function (blue), estimation of leak position in equivalent coordinates (green) and leak position estimated by proportionality relationship (red).

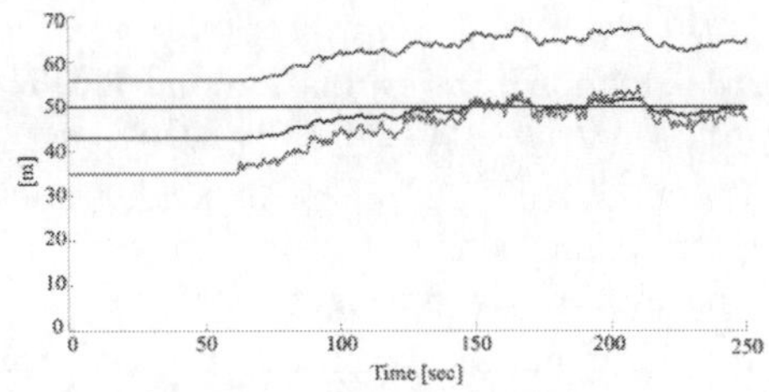

Figure 8. Leak isolation emulated by valve 3. Real leak position (black), estimation of leak position using polynomial function (blue), estimation of leak position in equivalent coordinates (green) and leak position estimated by proportionality relationship (red).

5. Conclusions

In this paper, an interpolation function that allows to estimate a leak position in real length coordinates is proposed, considering as an approach of low practical value the expression of the leak position in equivalent length coordinates, which is a common scenario in the methods based on Fault Model Approach found in the actual literature.

The interpolation function method presented in this work can be used in real situations and not only in theoretical or controlled scenarios. The proposed methodology shows a good performance thought experimental results.

Based on the performed analysis and the obtained results, it is possible to emphasize the following conclusions: (i) It is necessary to know the structure of the pipeline and the characteristics of each fitting in order to isolate a leak in a pipeline with fittings with a high degree of exactitude; (ii) the values of local losses coefficients are a fundamental part of the methodology in order to ensure a good level of accuracy for the isolation of a leak using the polynomial interpolation.

As future work, the interpolation function will be tested in different prototypes.

Acknowledgments

The authors thank the financial support for the project provided by CONACyT: CB-2012 No. 177656.

References

[1] CONAGUA. Manual de incremento de eficiencia física, hidráulica y energética en sistemas de agua potable., 2012.

[2] J. A. Delgado-Aguiñaga, G. Besançon, O. Begovich, and J. E. Carvajal. Multi-leak diagnosis in pipelines based on extended Kalman filter. *Control Engineering Practice*, 49:139–148, 2016.

[3] O. Begovich and G. Valdovinos-Villalobos. DSP application of a water-leak detection and isolation algorithm. In *2010 7th International Conference on Electrical Engineering Computing Science and Automatic Control (CCE)*, pages 93–98. IEEE, 2010.

[4] Lizeth Torres, Gildas Besançon, Adrian Navarro, Ofelia Begovich, and Didier Georges. Examples of pipeline monitoring with nonlinear observers and real-data validation. In *8th IEEE International Multi-Conf on Signals Systems and Devices, Sousse, Tunisia*, 2011.

[5] Tiantian Zhang, Yufei Tan, Xuedan Zhang, and Jinhui Zhao. A novel hybrid technique for leak detection and location in straight pipelines. *Journal of Loss Prevention in the Process Industries*, 35:157–168, 2015.

[6] Ole Morten Aamo. Leak detection, size estimation and localization in pipe flows. *IEEE Transactions on Automatic Control*, 61(1):246–251, 2016.

[7] Ignacio Barradas, Luis E. Garza, Ruben Morales-Menendez, and Adriana Vargas-Martínez. Leaks detection in a pipeline using artificial neural networks. In *Iberoamerican Congress on Pattern Recognition*, pages 637–644. Springer, 2009.

[8] Zdzisław Kowalczuk and Keerthi Gunawickrama. Detecting and locating leaks in transmission pipelines. In *Fault Diagnosis*, pages 821–864. Springer, 2004.

[9] Jorge Delgado-Aguiñaga, Gildas Besançon, and Ofelia Begovich. Leak isolation based on extended Kalman filter in a plastic pipeline under temperature variations with real-data validation. In *2015 23th Mediterranean Conference on Control and Automation (MED)*, pages 316–321. IEEE, 2015.

[10] G. Espinoza-Moreno, O. Begovich, and J. Sanchez-Torres. Real time leak detection and isolation in pipelines: A comparison between sliding mode observer and algebraic steady state method. In *2014 World Automation Congress (WAC)*, pages 748–753. IEEE, 2014.

[11] Adrian Navarro, Ofelia Begovich, and Gildas Besançon. Real-time leak isolation based on state estimation with fitting loss coefficient calibration in plastic pipeline. *Asian Journal of Control*, 19(1):1–11, 2017 (in press).

[12] M Hanif Chaudhry. Applied hydraulic transients. Technical report, Springer, 1979.

[13] Herbert Keith Winning and Tim Coole. Explicit friction factor accuracy and computational efficiency for turbulent flow in pipes. *Flow, turbulence and combustion*, 90(1):1–27, 2013.

[14] C. Verde. Multi-leak detection and isolation in fluid pipelines. *Control Engineering Practice*, 9(6):673–682, 2001.

[15] M. Elena Vázquez-Cendón. Numerical resolution of one-dimensional hyperbolic linear systems. In *Solving Hyperbolic Equations with Finite Volume Methods*, pages 57–72. Springer, 2015.

[16] Dan Simon. *Optimal state estimation: Kalman, H infinity, and nonlinear approaches*. John Wiley & Sons, 2006.

[17] John R. Hauser. *Numerical methods for nonlinear engineering models*. Springer Science & Business Media, 2009.

[18] Ofelia Begovich, Alejandro Pizano, and Gildas Besançon. Online implementation of a leak isolation algorithm in a plastic pipeline prototype. *Latin American applied research*, 57(6):131–140, 2012.

IOP Conf. Series: Journal of Physics: Conf. Series **783** (2017) 012013 doi:10.1088/1742-6596/783/1/012013

Linear Discrete-time State Space Realization of a Modified Quadruple Tank System with State Estimation using Kalman Filter

Sazuan N M Azam

Department of Applied Mathematics and Computer Science, Technical University of Denmark, DK-2800 Kgs. Lyngby, Denmark

E-mail: snaz@dtu.dk

Abstract. In this paper, we used the modified quadruple tank system that represents a multi-input-multi-output (MIMO) system as an example to present the realization of a linear discrete-time state space model and to obtain the state estimation using Kalman filter in a methodical mannered. First, an existing dynamics of the system of stochastic differential equations is linearized to produce the deterministic-stochastic linear transfer function. Then the linear transfer function is discretized to produce a linear discrete-time state space model that has a deterministic and a stochastic component. The filtered part of the Kalman filter is used to estimates the current state, based on the model and the measurements. The static and dynamic Kalman filter is compared and all results is demonstrated through simulations.

1. Introduction

A dynamic model of a system can be described in a various way. From the derivation of the mathematical model it is possible to obtain the underlying information of the system and implement a control algorithm to the system. Most of the systems or processes are usually described by state-space system and by investigating the state of a system at certain time and its present and future inputs, it is possible to predict the output in the future [1]. State space models can be either non-linear or linear form and usually a real system or process is described by a non-linear models whereas in order to estimate and control the system, most mathematical tools are more accessible to a linear models. Therefore, in this paper we want to demonstrate the transformation of a non-linear continuous model of a modified quadruple tank system described as deterministic-stochastic differential equations into a linear discrete-time state space model.

Throughout this work, we will fully utilize the Modified Quadruple Tank System, based on [2] to assimilate the fundamental theory of model realization and state estimation to an exemplification of MIMO system, illustration of the real-world complex system applications which is widely used for education in modeling and demonstrating advanced control strategies [3], [4].

Several works have been done on four tanks system regarding the modeling the dynamic of the system. A full description of linearization of the model for four tank system is presented in [5], [6] and [7]. In [7] the linearization is described in detail using the Jacobian matrix formation to represent the system in state space model while in [8] establishes the linearized model based on the non-linear mechanism of the system. Another method is shown in [9]where the model is

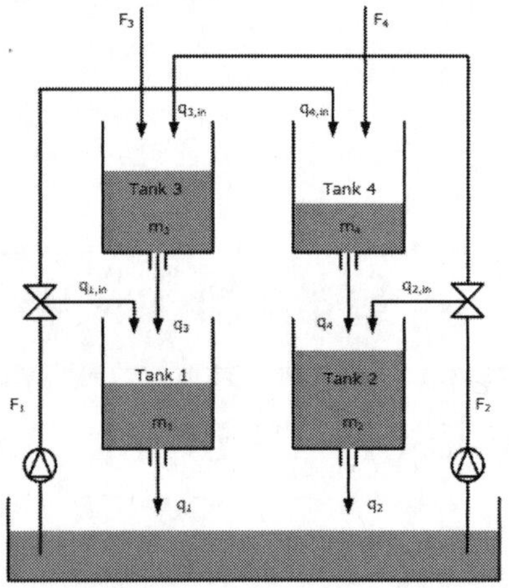

Figure 1. Schematic diagram of the modified quadruple tank process

developed from input and output data resulting an empirical linear state space model using a sub-space identification and can be used effectively for a non-linear system.

As for the state estimation, we want to estimates all the variables which represents the internal condition or the status of the system at a specific given time [1] so as to allow for future output prediction and to design the control algorithms. A Kalman's state estimator for a non-linear multivariable process such as the four tanks system is shown in [10] using a linear state space model solved by the algebraic Ricatti equation. Meanwhile the usage of the estimation of the Kalman filter with full derivation can also be found in [11]. In this work we use the Kalman filter in order to estimates the current state of the modified quadruple tank system and evaluate the response of dynamic and static Kalman filter.

This paper is structured as follows. A brief description of modified quadruple tank system is presented and the realization of the linear discrete-time state space model is shown in detail in Section 2. Then the state estimation using Kalman filter is discussed in Section 3. The following section is where all the results is discussed in Section 4. Finally, we conclude this work in the last section.

2. Linear Discrete-time State Space Model Realization

The first part of this paper is to transform the non-linear continuous state space model of a modified quadruple tank system to a linear discrete state space model through linearization and discretization. A brief description of the system is presented below and followed by the linearization and discretization.

2.1. The Modified Quadruple Tank System

The modified quadruple tank system is a simple process, consist of four identical tanks and two pumping system as shown in Figure 1 but yet illustrates a system that is non-linear with multiple inputs and outputs (MIMO) and complicated interactions between manipulated and controlled variables.

The main objective of this system is to control the level of the water in the lower tanks (Tank 1 and 2) by manipulating the flow rates F_1 and F_2 which are distributed across all four tanks, represents the dynamics of multivariable interaction since each manipulated variables influences the outputs. The height of the water level in these two tanks, h_1 and h_2 is measured and controlled. The flows denoted F_3 and F_4 are unmeasured unknown disturbances.

The dynamic of the process is described in Stochastic Nonlinear Model (SDE) given as:

$$d\mathbf{x}(t) = f(\mathbf{x}(t), u(t), d(t), p)dt + \sigma d\mathbf{w}(t) \tag{1a}$$

$$\mathbf{y}(t) = c(\mathbf{x}(t)) + \mathbf{v}(t) \tag{1b}$$

$$\mathbf{z}(t) = c(\mathbf{x}(t)) \tag{1c}$$

where $\mathbf{w}(t)$ is the process noise and normally distributed, $\mathbf{w}(t) \sim N(0, R_w)$, due to the unknown information regarding the distribution and $\mathbf{v}(t)$ is the measurement noise from the sensors in each tank and it is normally distributed, $\mathbf{v}(t) \sim N(0, R_v)$. $\mathbf{v}(t)$ is being added to the measured variables, (1b). For full description of the modeling part, see [12].

2.2. Linear System Realization

Linearization is required to find the linear approximation to analyze the behaviour of the nonlinear function, given a desired operating point. We apply the first-order term only of Taylor expansion by truncation around the steady state of the non-linear differential equations, $f(x(t), u(t), d(t))$ and consider the derivative of the state variable, x. This derivative is defined as a function, f,

$$f(x(t), u(t), d(t), p) = \begin{pmatrix} \rho\gamma_1 u_1(t) + \alpha_3 - \alpha_1 \\ \rho\gamma_2 u_2(t) + \alpha_4 - \alpha_2 \\ \rho(1-\gamma_2)u_2(t) + \rho d_1(t) - \alpha_3 \\ \rho(1-\gamma_1)u_1(t) + \rho d_2(t) - \alpha_4 \end{pmatrix} \tag{2}$$

where α_i is given by

$$\alpha_i = \rho a_i \sqrt{x_i(t)} \sqrt{\frac{2g}{\rho A_i}} \qquad i = 1, 2, 3, 4 \tag{3}$$

and p denote the vector containing all the parameters of the system, for full description of the parameter see [12]. The Jacobian of f with respect to the state-variables are

$$J_x(x(t), u(t), d(t), p) = \begin{pmatrix} -\beta_1 & 0 & \beta_3 & 0 \\ 0 & -\beta_2 & 0 & \beta_4 \\ 0 & 0 & -\beta_3 & 0 \\ 0 & 0 & 0 & -\beta_4 \end{pmatrix} \tag{4}$$

where β_i is given by

$$\beta_i = \frac{1}{\sqrt{x_i(t)}} \sqrt{\frac{a_i^2 g \rho}{2A_i}} \qquad i = 1, 2, 3, 4 \tag{5}$$

Similarly the Jacobian of f with respect to the manipulated variables giving

$$J_u(x(t), u(t), d(t), p) = \begin{pmatrix} \rho\gamma_1 & 0 \\ 0 & \rho\gamma_2 \\ 0 & \rho(1-\gamma_2) \\ \rho(1-\gamma_1) & 0 \end{pmatrix} \tag{6}$$

and lastly the Jacobian of f with respect to the disturbance variables are

$$J_d(x(t), u(t), d(t), p) = \begin{pmatrix} 0 & 0 \\ 0 & 0 \\ \rho & 0 \\ 0 & \rho \end{pmatrix} \tag{7}$$

In this case, we introduced the deviation variables as

$$X(t) = x(t) - x_s \quad U(t) = u(t) - u_s \quad D(t) = d(t) - d_s \tag{8}$$

and defined the Jacobian matrices evaluated around a stationary point x_s, u_s, d_s to be

$$A_c = J_x(x_s, u_s, d_s, p) \quad B_c = J_u(x_s, u_s, d_s, p) \quad E_c = J_d(x_s, u_s, d_s, p) \tag{9}$$

With these matrices the first order Taylor approximation around the steady state point are given as

$$\begin{aligned} f(x(t), u(t), d(t), p) &\approx f(x_s, u_s, d_s, p) + A_c X(t) \\ &\quad + B_c X(t) + E_c D(t) \\ &= A_c X(t) + B_c X(t) + E_c D(t) \end{aligned} \tag{10}$$

For the measurement and controlled variables we introduced $Y(t)$ and $Z(t)$ respectively and the linearized system of the modified quadruple tank system as

$$\dot{X}(t) = A_c X(t) + B_c X(t) + E_c D(t) \qquad X(t_0) = 0 \tag{11a}$$
$$Y(t) = CX(t) \tag{11b}$$
$$Z(t) = C_z X(t) \tag{11c}$$

where the C matrices are defined as

$$C = C_z = \begin{pmatrix} \frac{1}{\rho A_1} & 0 & 0 & 0 \\ 0 & \frac{1}{\rho A_2} & 0 & 0 \end{pmatrix} \tag{12}$$

2.3. Discretization of a Linear System

The dynamics of the modified quadruple tank system is now described as (11) and to use this linear continuous model of the system to be subjected to MPC, the model needs to be discretized by assuming zero-order-hold (ZOH) of the variables at specified sampling points, that is assuming the exogenous variables are constant between sampling points. The aim is to have a linear discrete-time state space model with piecewise constant u_k, d_k in a form of

$$x_{k+1} = A_d x_k + B_d u_k + E_d d_k \tag{13a}$$
$$y_k = C_d x_k + D_d u_k \tag{13b}$$

with discrete-time consideration

$$\begin{aligned} t_k &= t_0 + kTs, \qquad k = 0, 1, 2... \\ x_k &= x(t_k) \end{aligned}$$

and assuming the inputs on the ZOH is

$$u(t) = u_k, \qquad t_k \leq t \leq t_{k+1}$$

then the solution of (13) with respect to u is given as

$$x_{k+1} = x(t_{k+1}) \tag{14a}$$
$$= e^{A(t_{k+1}-t_k)} x_k + \int_{t_k}^{t_{k+1}} e^{A(t_{k+1}-\tau)} Bu(\tau) d\tau \tag{14b}$$
$$= \left[e^{AT_s}\right] x_k + \left[\int_0^{T_s} e^{A\eta} B d\eta\right] u_k \tag{14c}$$

By comparing both equations (11) and (14) and similar result can be obtained for disturbances variable $d(t)$ giving

$$\begin{array}{lll} A_d = e^{AT_s} & B_d = \int_0^{T_s} e^{A\tau} B d\tau & E_d = \int_0^{T_s} e^{A\tau} E d\tau \\ C_d = C & D_d = D & \end{array} \tag{15}$$

where A_d, B_d, E_d can be computed with

$$\begin{gathered} \begin{bmatrix} A_d & B_d \\ 0 & I \end{bmatrix} = \exp\left(\begin{bmatrix} A & B \\ 0 & I \end{bmatrix} T_s \right) \\ \begin{bmatrix} A_d & E_d \\ 0 & I \end{bmatrix} = \exp\left(\begin{bmatrix} A & E \\ 0 & I \end{bmatrix} T_s \right) \end{gathered} \tag{16}$$

For this particular work, the continuous state space representation matrices were discretized with $T_s = 30s$ assuming ZOH.

Considering the stochastic part of the model, a piecewise constant process noise w, measurement noise v and uncertainty of the initial state x_0 to the process is added. The linear discrete model from (13) is expanded into stochastic version as in the equation below

$$x_{k+1} = A_d x_k + B_d u_k + E_d(d_k + w_k) \tag{17a}$$
$$y_k = C_d x_k + v_k \tag{17b}$$
$$z_k = C_{dz} x_k + v_k \tag{17c}$$

subject to

$$x_0 \sim N(\bar{x}_0, P_p), \quad w_k \sim N(0, Q), \quad v_k \sim N(0, R) \tag{18}$$

where Q, R is given by

$$Q = \begin{bmatrix} 12.5^2 & 0 \\ 0 & 12.5^2 \end{bmatrix} \quad R = \begin{bmatrix} 2^2 & 0 & 0 & 0 \\ 0 & 2^2 & 0 & 0 \end{bmatrix}$$

and P_p is given by

$$P_p = \begin{bmatrix} 0.1^2 & 0 & 0 & 0 \\ 0 & 0.1^2 & 0 & 0 \\ 0 & 0 & 0.1^2 & 0 \\ 0 & 0 & 0 & 0.1^2 \end{bmatrix}$$

2.4. Linear Discrete-time State Space Representation

In order to rewrite the difference equation system representation (13) in a more structured form, the Markov parameters is introduced. It is a discrete impulse coefficients of a discrete state space model. The Markov parameters are calculated to avoid making iterative simulations to keep only the matrix-vector multiplications. In doing so, a significant time saving is introduced to the control algorithm and to have an observer canonical form with minimal realization. Let H_i denote the Markov parameters at the $i'th$ sampling time after an unit-impulse, then to obtain the Markov parameters from u to y is given as

$$H_i = \begin{cases} 0 & i = 0 \\ CA_d^{i-1}B & i = 1, 2, \ldots N \end{cases} \tag{19}$$

N is assigned value to be sufficiently large so that the impulse response can reach the steady state. The Markov parameters for u to z, d to y and d to z is computed the same way and by replacing the appropriate matrices accordingly. With all the information being gathered, it can be re-written in a matrix form of

$$Y = \Phi x_0 + \Gamma U \tag{20}$$

where Y, Φ, U are

$$Y = \begin{bmatrix} y_1 \\ y_2 \\ y_3 \\ \vdots \\ y_i \end{bmatrix} \quad \Phi = \begin{bmatrix} CA_d \\ CA_d{}^2 \\ CA_d{}^3 \\ \vdots \\ CA_d{}^i \end{bmatrix} \quad U = \begin{bmatrix} u_1 \\ u_2 \\ u_3 \\ \vdots \\ u_i \end{bmatrix}$$

while Γ is obtained from the calculated Markov parameters, H_i, $i = 1, 2, ...N$

$$\Gamma = \begin{bmatrix} H_1 & 0 & 0 & \dots & 0 \\ H_2 & H_1 & 0 & \dots & 0 \\ H_3 & H_2 & H_1 & \dots & 0 \\ \vdots & \vdots & \vdots & & \vdots \\ H_N & H_{N-1} & H_{N-2} & \vdots & H_1 \end{bmatrix}$$

As for the system with disturbances, the state space model can be represented as

$$Y = \Phi x_0 + \Gamma_u U + \Gamma_d D \tag{21}$$

where

$$D = \begin{bmatrix} d_1 & d_2 & d_3 & \dots & d_i \end{bmatrix}^T$$

From equations (20) and (21), Φ and Γ can be used for the prediction part from the Kalman filter for a model predictive control strategy.

3. State Estimation for the Discrete-Time Linear System

From the previous section, the discrete-time state space model is a linearized model from the non-linear model. We want to extract information from the measurements of the real system in order to limit the discrepancy between the model and the real system but since the models assume measurement error, the signals need to be filtered. This can be done by using Kalman filter where it is used to filter the measurement [1]. The Kalman filter consists of two parts, filtering part and prediction part. The filtered part is to estimates current state based on the model and the measurements whilst the prediction part is used by the constrained regulator to predict the future output trajectory, given an input trajectory. This is illustrated in the block diagram as in Figure 2. In this paper we focus on the filtering part for the state estimation only and design both dynamic and static filter to evaluate their estimation.

3.1. Dynamic Kalman Filter

From [12] the model is linear time invariant (LTI) discrete-time stochastic difference equations, in the form of

$$x_{k+1} = A_d x_k + B_d u_k + E_d d_k + E d w_k \tag{22a}$$
$$y_k = C_d x_k + v_k \tag{22b}$$

subject to

$$w_k \sim N(0, Q), v_k \sim (0, R) \tag{23}$$

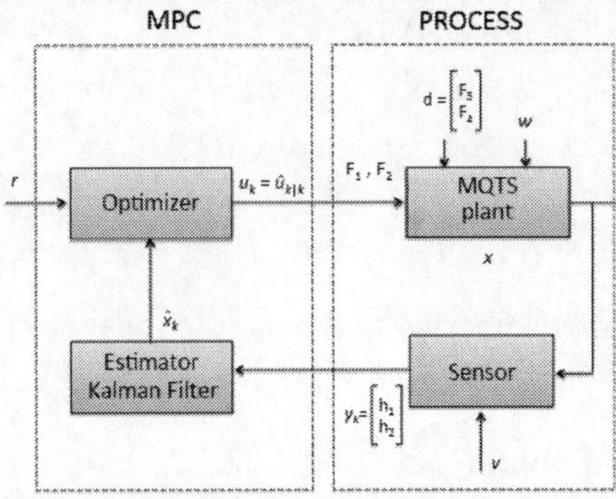

Figure 2. Block diagram of MPC for modified quadruple tank process

where the process noise w_k and measurement noise v_k are distributed as

$$\begin{bmatrix} w_k \\ v_k \end{bmatrix} \sim N_{iid}\left(\begin{bmatrix} 0 \\ 0 \end{bmatrix}, \begin{bmatrix} Q & S \\ S^T & R \end{bmatrix} \right) \tag{24}$$

where R and Q is the covariance matrix of measurement error and disturbances variable accordingly, S is the covariance matrix between disturbance variable and measurement error and the distribution of the initial state is given by

$$x_{0|-1} \sim N(\hat{x}_{0|-1}, P_{0|-1}) \tag{25}$$

Assuming at stationary point $t = t_k$ and the measurement $y_k = y(t_k)$, the filtering part can be performed by calculating

$$\hat{y}_{k|k-1} = C\hat{x}_{k|k-1} \tag{26a}$$
$$e_k = y_k - \hat{y}_{k|k-1} \tag{26b}$$
$$\hat{x}_{k|k} = \hat{x}_{k|k-1} + K_{fx,k}e_k \tag{26c}$$
$$\hat{w}_{k|k} = K_{fw}e_k \tag{26d}$$
$$\hat{x}_{k+1|k} = A\hat{x}_{k|k} + Bu_k + \hat{w}_{k|k} \tag{26e}$$

By using the coefficients

$$R_{e,k} = CP_{k|k-1}C^T + R \qquad K_{fx,k} = P_{k|k-1}C^T R_{e,k}^{-1} \qquad K_{fw} = SR_{e,k}^{-1} \tag{27}$$

and the following expression can be achieved

$$P_{k+1|k} = AP_{k|k}A^T + Q_{k|k} - AK_{fx,k}S^T - SK_{fx,k}^T A^T \tag{28}$$

3.2. Static Kalman Filter
From equation (28) it can be re-written into a difference equation form as

$$P_{k+1|k} = AP_{k|k-1}A^T + Q - (AP_{k|k-1}C^T + S)(CP_{k|k-1}C^T + R)^{-1}(AP_{k|k-1}C^T + S)^T \tag{29}$$

P signifies the stationary one-step ahead state error covariance matrix obtained from the Discrete-time Algebraic Riccati Equation (DARE).

$$P = APA^T + Q - (APC^T + S)(CPC^T + R)^{-1}(APC^T + S)^T \tag{30}$$

and the coefficients in equations (27) can be simplify

$$R_e = CPC^T + R \qquad K_{fx} = PC^T R_e^{-1} \qquad K_{fw} = SR_e^{-1} \tag{31}$$

Since by using this limit as an approximation to the one-step matrix and the Kalman gains K_{fx} and K_{fw} becomes constant matrices, it will lighten the computations of the controller.

4. Results and Simulation of the system

The first part of this work is the linear discrete-time state space realization and computations which is re-written in a more structured form where we introduced the use of Markov parameters. Then the second part is the implementation of Kalman filter and the predictive controller strategy. In this sections, all results from the computations and simulations will be shown.

4.1. Linear Discrete-time State Space Realization

The linearized continuous system matrices are obtained as in equation (11) to ensure that the theoretical linear estimation of the system are almost identical to the non-linear system, it is

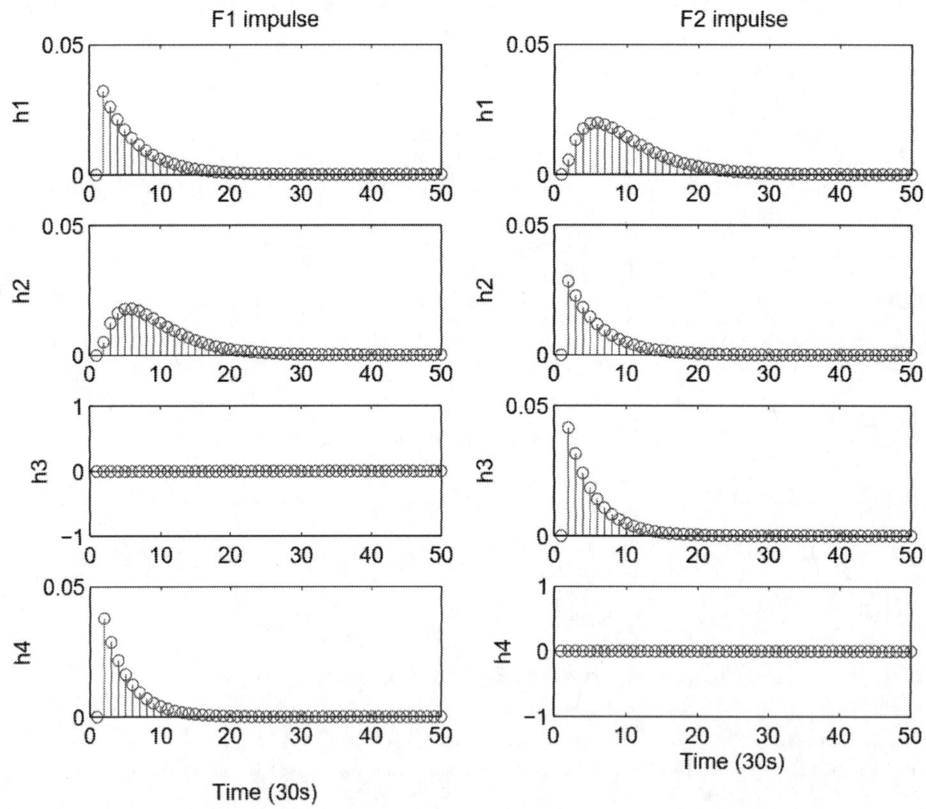

Figure 3. Markov Parameters for the Discrete-time State Space Model Experiments

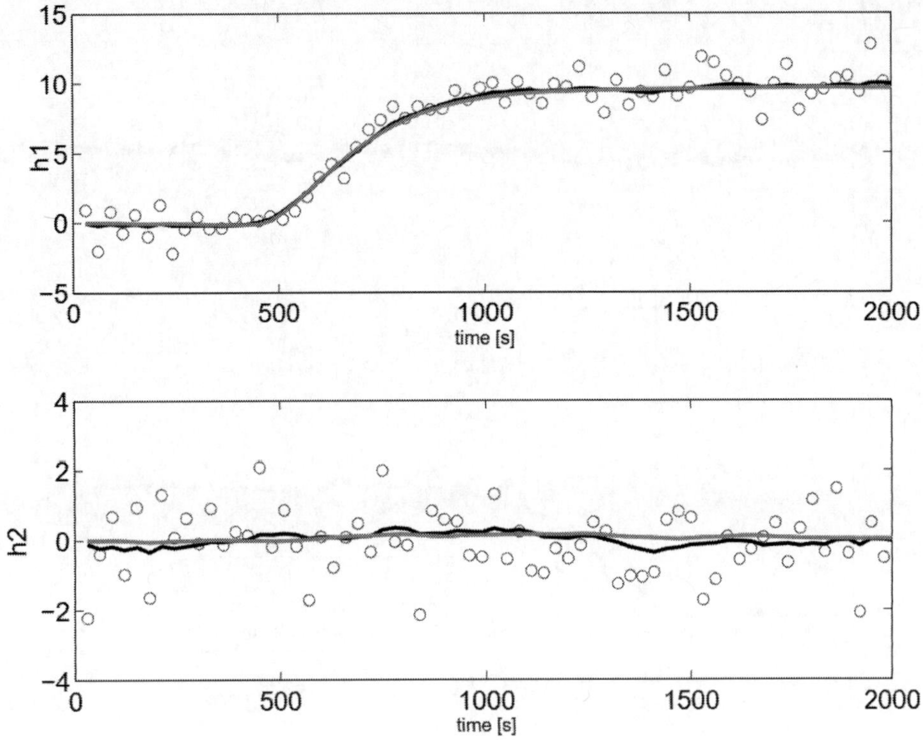

Figure 4. Kalman filter for noisy F_3 and 10% step changes

possible to compare the estimated gains and time-constants from the non linear transfer function. Then, by assuming zero-order hold, the continuous state space representation matrices above were discretized with $T_s = 30s$, hence the discrete state space system matrices is also obtained as in equation (13).

Considering a sampling time of $T_s = 30s$ the Markov parameters has been calculated for the discrete-time state space model, from Figure 3 the plots can be compared to the model from the step response experiment in our previous work [12] and shows that in most cases the plots are very similar. Therefore, the calculated Markov parameters, H_i are reliable and usable for other purposes such as in designing the predictive controller.

4.2. State Estimation

In this experiments, both dynamic and static Kalman filter were tested as a state estimator where the disturbance is an unknown stochastic variable, then after approximately 450s we introduced a 10% step changes and for this simulations, the linear model is used to create the measurements. Figure 4 and Figure 5 were plotted to compare the estimated current states with and without Kalman filter and between dynamic and static Kalman filter with a step change of F_3 and F_4 accordingly. It can be clearly seen that in general, the filter is well performed tracking the output trajectory from the noisy measurements and also it can cope well dealing with an impact of the unknown disturbance step. Although the difference between dynamic and static Kalman filter is not apparent, the dynamic filter is able to even further reduce the noise giving a smoother and more stable response particularly in tanks h_1 in figure 4 and tank h_2 in figure 5, noticeable for both tanks that is directly affected by the given step disturbance.

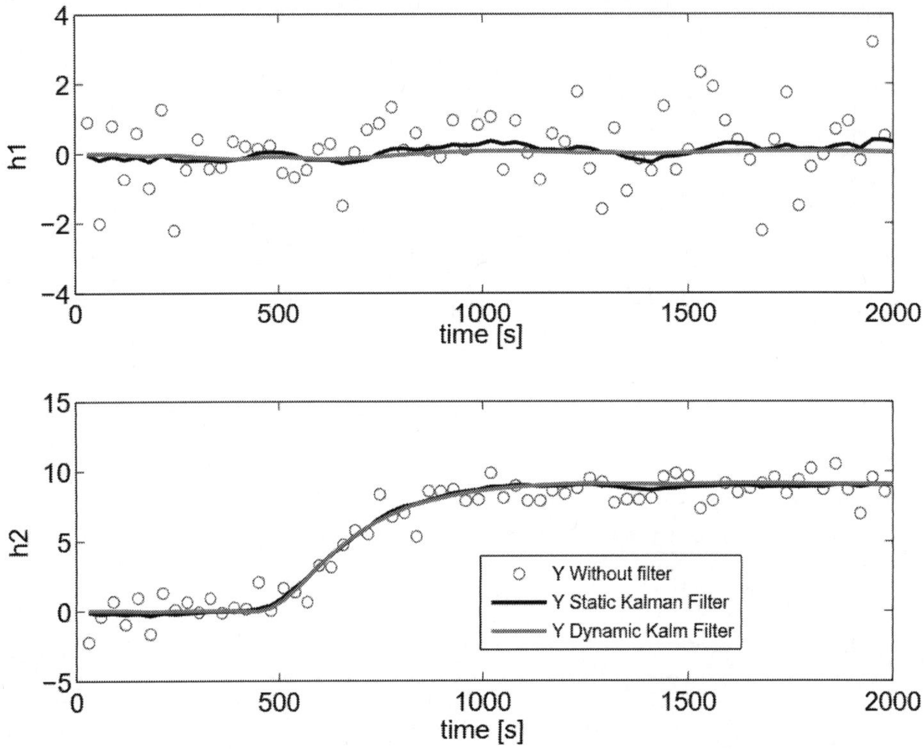

Figure 5. Kalman filter for noisy F_4 and 10% step changes

5. Conclusion

This paper has described comprehensively an outline to obtained a discrete-time state space model for linear system on a modified quadruple tank system in a simple and constructive method. This lab scale system represents a MIMO system which has complicated variables interactions and complex control problems. The dynamics of the system is described by an existing simulation models in terms of deterministic and stochastic non-linear continuous time models. These models were linearized and discretized in order to form a discrete-time linear time-invariant difference equations, the form that is used in the Kalman Filter for estimations. Based on the model and measurements, the current state of the system was estimated and in additional, the comparison between dynamic and static Kalman filter was also presented.

Acknowledgement

This work was supported by Faculty of Electrical Engineering, Universiti Teknikal Malaysia Melaka, Durian Tunggal, 76100, Malaysia and the Ministry of Higher Education, Malaysia.

References

[1] Simon D 2006 *Optimal State Estimation* (John Wiley & Sons)
[2] Johansson K H 2000 *IEEE Transactions on Control Systems Technology* **8** 456–465
[3] Johansson K H, Horch A, Wijk O and Hansson A 1999 *Proceedings of the IEEE Conference on Decision and Control* **1** 807–812
[4] Dormido S and Esquembre F 2003 *European Control Conference (ECC), 2003* (IEEE) pp 3267–3272

[5] Askari M, Moghavvemi M, Almurib H A F and Muttaqi K M 2016 *IEEE Transactions on Industry Applications* **52** 1882–1890
[6] Almurib H A F, Askari M and Moghavvemi M 2011 *Proceedings of the Sice Annual Conference* 6060581, 87–91
[7] J J, T S and Babu T H 2014 *International Journal of Advanced Research in Electrical, Electronics and Instrumentation Engineering, IJAREEIE* **03** 11552–11565
[8] Prusty S B, Pati U C and Mahapatra K K 2015 *Computer, Communication, Control and Information Technology (C3IT), 2015 Third International Conference on* (IEEE) pp 1–6
[9] Gatzke E P, Meadows E S, Wang C and Doyle F J 2000 *Computers & Chemical Engineering* **24** 1503–1509
[10] Vijula D A and Devarajan N 2013 *International Journal of Electrical Engineering and Technology* **6** 149–163
[11] Jorgensen J B and Jorgensen S B 2007 *2007 American Control Conference* (IEEE) pp 128–133
[12] Mohd Azam S N and Jørgensen J B 2015 *Proceedings of the 5th International Conference on Control Systems, Computing and Engineering (ICCSCE 2015)*

Tire Force Estimation using a Proportional Integral Observer

Ahmad Farhat, Damien Koenig

Gipsa-Lab INPG
11 rue des Mathématiques, Grenoble Campus BP 46,
38,402 Saint Martin d'Hères Cedex - France

E-mail: { ahmad.farhat, damien.koenig }@gipsa-lab.grenoble-inp.fr

Diana Hernandez-Alcantara, Ruben Morales-Menendez

Tecnológico de Monterrey, School of Engineering and Sciences
Av Eugenio Garza Sada # 2501, Col. Tecnológico
64,849 Monterrey, Nuevo León, México

E-mail: { d.hernandez.phd.mty, rmm }@itesm.mx

Abstract. This paper addresses a method for detecting critical stability situations in the lateral vehicle dynamics by estimating the non-linear part of the tire forces. These forces indicate the road holding performance of the vehicle. The estimation method is based on a robust fault detection and estimation approach which minimize the disturbance and uncertainties to residual sensitivity. It consists in the design of a *Proportional Integral Observer* (*PIO*), while minimizing the well known H_∞ norm for the worst case uncertainties and disturbance attenuation, and combining a transient response specification. This multi-objective problem is formulated as a *Linear Matrix Inequalities* (*LMI*) feasibility problem where a cost function subject to *LMI* constraints is minimized. This approach is employed to generate a set of switched robust observers for uncertain switched systems, where the convergence of the observer is ensured using a *Multiple Lyapunov Function* (*MLF*). Whilst the forces to be estimated can not be physically measured, a simulation scenario with *CarSim*^{TM} is presented to illustrate the developed method.
Keywords: Proportional Integral Observer, Robust fault detection, Estimation, switched systems, H_∞, Linear Matrix Inequalities, Multiple Lyapunov Function.

1. Introduction

Fully autonomous vehicles are unlikely to be commercially available before 2020. *Advanced Driver-Assistance Systems* (*ADAS*) will play a crucial role in preparing regulators, consumers, and corporations for the medium-term reality of cars taking over control from drivers .

The market introduction of *ADAS* has shown that the primary challenges preventing faster market penetration are pricing, consumer understanding and safety/security issues.The technological challenges are not insignificant, and will likely drive the delay between conditionally autonomous cars which allow the driver to cede control in certain situations (Level 3 according to the *National Highway Traffic Safety Administration* (*NHTSA*)) and fully autonomous cars, which require no driver intervention for the entire trip (Level 4 *NHTSA*). Tech players and startups will likely play an important role in achieving this level of technical complexity. Regulation and consumer acceptance may represent additional hurdles for autonomous vehicles. A progressive scenario would see fully autonomous cars accounting for up to 15 % of passenger vehicles sold worldwide in 2030 [6].

Many safety mechanisms rely on the knowledge of vehicle dynamics states and parameters, that can be measured. Tire forces that act during maneuvers are nonlinear with respect to wheel slip and wheel slip angles and cause a wide range of stability and handling properties. The force depend on external factors, such as road properties. However, some of these parameters can not be directly measured because the expensive cost of the sensors, physical restrictions or simply they are correlated with other vehicle data. Example of this are the tire forces, involved in lateral vehicle dynamics. The tire force can be used as an indicator for road holding of the vehicle. It is well known that the lateral tire force has a nonlinear relationship with respect to the slip angle. During normal driving situations, the slip angle is low and the tire force can be approximated to a linear function of the slip angle. Conversely, in critical situations, where the tire is approaching its adherence limits, the lateral force is highly nonlinear. Therefore, by using the difference between the linear tire model and the actual tire force it is possible to discriminate the vehicle situation as normal or critical. Since measuring the actual tire force is difficult in practice, this paper deals with the estimation of the nonlinear part of the tire force, which is the difference between the actual tire force and the linear model of the force.

To maintain stability during emergencies follow rules based on intensive tests, rather than analytic control laws based on actual forces and accurate state estimates. The model of tire-road contact forces is complex because a wide variety of parameters including environmental factors and pneumatic properties that impact the tire-road contact interface. Many tire-force models have been proposed.

From the theoretical aspects, different longitudinal tire/road friction models can be classified into two types: empirical and analytical models. Empirical models are based on curve-fitting techniques. Empirical models can accurately represent the steady-state characteristics of the tire/road friction phenomena. But, they cannot describe several significant dynamic behaviors such as hysteresis. Analytical models are popular recently; they use differential equations to describe tire/road friction properties. The obtained results were based in Magic Formula proposed by Pacejka; however, there are options such as Piecewise Linear Model, Burckhardt Model [2], Rill Model [13], Dahl Model [3], and LuGre Model [5] that provide different results. Intensive testing is needed to determine the parameters of analytical models. It is very difficult to determine all of them in real-time for every tire, tire pressure and wear state.

Some related work concerning the nonlinear lateral force estimation can be found in the literature. In [11] an *Extended Kalman Filter* (*EKF*) designed using a 9 *DoF* vehicle model and an analytic tire force model is used to simulate the vehicle motion (i.e. maneuvers combining steering, braking, and varying road conditions), and a 5 *DoF* was proposed to estimate the nonlinear tire force. No prior information of the tire force characteristics or external factors is needed for the nonlinear estimation. Some simulations validated the effectiveness of the proposal that suggest the *EKF* could be implemented in real time. In [12] an *Extended Kalman Bucy Filter* and Bayesian hypothesis selection were also designed using a 8 *DoF* longitudinal-lateral vehicle model to estimate morion, tire forces, and road coefficient of friction of vehicles. In both cases simulation results are very promising; however, the approaches were just validated in a mild scenario, where the force remains close to the linear zone.

In [16] a comparison of the lateral force estimated by four observers one based on a linear vehicle model and three based on non linear model (an extended Luenberger observer, *EKF* and Sliding-Mode Observer), was carried out. *EKF* using a yaw rate sensor only gives less accurate estimations than the other two nonlinear observers; however, all non-linear observers have similar results. All observers had satisfactory performance when lateral acceleration was low. However, at larger acceleration values, some assumptions made for models are not totally rigth (i.e. linearity), yielding worse estimations; better models are needed.

Employing the bicycle model dynamical equations, in [17] the tire forces are estimated using algebraic techniques for filtering and estimating derivatives of noisy signals. The drawback of this approach is that the estimation is highly dependent on an accurate identification of the vehicle parameters and on the quality of the yaw an lateral acceleration measurements.

The *Proportional Integral Observer* (*PIO*), [1], was introduced by [7] for *SISO* systems, and extended for *MIMO* systems [18]. The sysnthesis with robustness performances toward parametric uncertainties and unknown inputs was proposed by [14]. Later, they were used for fault estimation [15] [9].

A hybrid process consists of several modes of behavior that has a different dynamic. One class of hybrid system is switched system. Transition between these modes might be state-dependent, time-dependent, input-dependent, event-dependent or a combination of these conditions, [4]. The transitions between modes is defined by a switching signal [8]. A design of *Switched Proportional Integral Observer* (*SPIO*) for fault estimation in discrete time is proposed to the estimate the non-linear tire force. The integral part of this observer provides an additional degree of freedom in comparison to a traditional observer that is used to improve the accuracy of the steady-state fault estimation and the robustness to disturbance inputs.

The outline of this paper is as follows. A lateral vehicle model as a switched system is developed in section 2. In section 3, the method for the synthesis of fault estimation *SPIO* is presented. *CarSim*^{TM} simulations are carried out to validate of the *SPIO* design in section 4. Finally, section 5 concludes the research and shows the possible future work. Table 1 the used variables of this research paper.

2. Switched vehicle model

The well known bicycle model, Fig. 1, is used to model the lateral vehicle dynamics. The reference point *Center of Gravity* (*CoG*) is chosen to represent the center of gravity for the vehicle body, where the vehicle velocity is defined. Symbols A and B denote the positions of the front and rear tire/road interfaces. The heading angle ψ is the angle from the guideline to the longitudinal axis of vehicle body AB. The side-slip angle β is the angle from the longitudinal axis of the vehicle body to the direction of the vehicle velocity, δ is the steering angle, $\dot{\psi}$ is the yaw rate. The involved forces in this model are the lateral tire forces at the front F_{ty_f} and the rear F_{ty_r} of the vehicle, the load transfer forces F_d and the wind forces F_w.

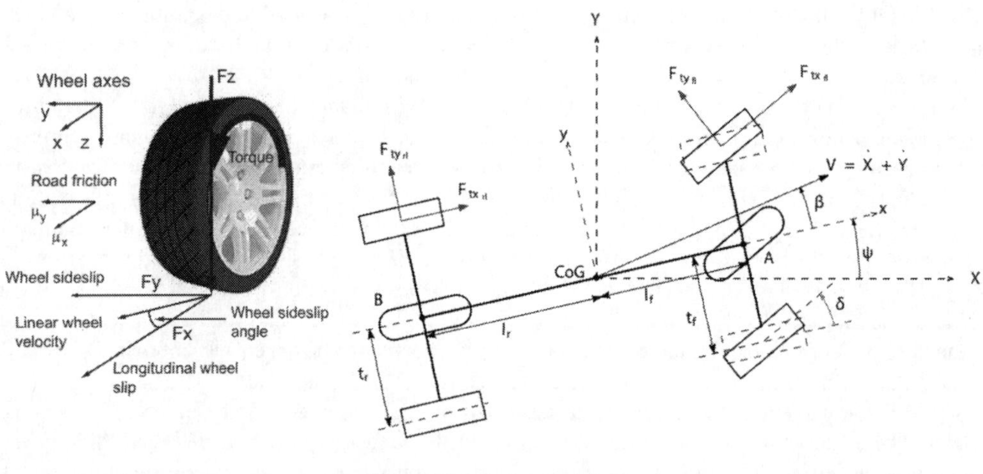

Figure 1. Bicycle Model

Applying the second law of Newton to translational and rotational movements yields:

$$mv(\dot{\beta}+\dot{\psi}) = F_{ty_f} + F_{ty_r} + F_{dy} + F_w \tag{1a}$$

$$I_z\ddot{\psi} = l_f F_{ty_f} - l_r F_{ty_r} + M_{dz} \tag{1b}$$

where m, I_z, l_f and l_r are the vehicle parameters, Table 1.

Table 1. D-Class Sedan parameters for bicycle model

Symbol	Variable	Value	Unit
c_{yf}	Front linear cornering coefficient	66,257.00	[N/rad]
c_{yr}	Rear linear cornering coefficient	46,545.00	[N/rad]
d	Disturbance vector		$\mathbb{R}^{n_d}$
e_x, e_f	State and fault estimation error		
f	Faults vector		$\mathbb{R}^{n_f}$
F_d	Load transfer force		N
F_{tyf}	Lateral force at the front		N
F_{tyr}	Lateral force at the rear		N
F_w	Wind force		N
$L_{I,\alpha}$	Integral gain		
$L_{P,\alpha}$	Proportional gain		
I_z	Inertia around z-axis	2,315.30	[kg.m^2]
l_f	Distance from CoG to front axle	1.11	[m]
l_r	Distance from CoG to rear axle	1.67	[m]
m	Vehicle total mass	1,530.00	[kg]
m_s	Sprung mass		[kg]
m_{us}	Unsprung mass		[kg]
r	Residual error $(y-\hat{y})$		
x	State vector		$\mathbb{R}^n_x$
y	Measurement output vector		$\mathbb{R}^n_y$
α	Switching signal		
δ	Wheel steer angle	-	[rad]
β	Side slip angle	-	[rad]
Γ_{ij}	Non-linear part of the tire force		N
v	Longitudinal speed of the vehicle	-	[m/s]
$\dot{\psi}$	Yaw rate	-	[rad/s]
n_d	Number of disturbances		
n_f	Number of faults		
n_x	Number of states		
n_y	Number of outputs		
N	Number of subsystems		

The lateral tire forces F_{ty_ij} are function of the wheel load, slip angle and camber. Fig 2 shows the lateral force versus slip-angle curves for the tire considered in this paper. As observed, the tire characteristic is highly nonlinear.

One of the most used models for tire forces was proposed by Pacejka [10], often called the *magic formula*. In this model, lateral tire force is computed as:

$$F_{ty_{ij}} = D_y \sin\left[C_y \tan^{-1}\left(B_y\beta_{y_{ij}} - E_y(B_y\beta_{y_{ij}} - \tan^{-1}\left(B_y\beta_{y_{ij}}\right))\right)\right] \tag{2}$$

where the parameters are function of the wheel load. D_y represents the maximum value of the characteristic, B_y and C_y are the slope and the shape of the curve and E_y adjusts the x-coordinate (side-slip angle). This model can also be described as the sum of a linear function $\bar{F}_{ty_{ij}}$ for small slip angles

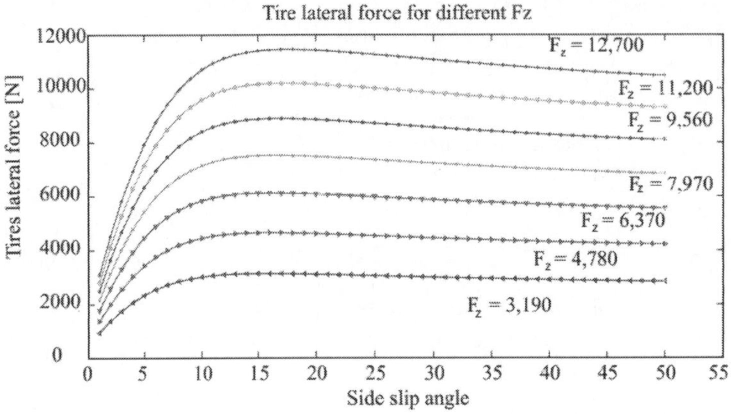

Figure 2. Lateral tire force for different normal efforts

and a non-linear part Γ_{ij}:

$$F_{ty_{ij}} = \bar{F}_{ty_{ij}} + \Gamma_{ij} \tag{3}$$

where

$$\bar{F}_{ty_f} = c_{yf}\beta_{yf} = c_{yf}\left(\delta - \beta - \frac{l_f\dot{\psi}}{v(t)}\right) \quad \bar{F}_{ty_r} = c_{yr}\beta_{yr} = -c_{yr}\left(\beta - \frac{l_r\dot{\psi}}{v(t)}\right) \tag{4}$$

The state space representation of the linear vehicle model is:

$$\begin{bmatrix}\dot{\beta}(t)\\ \ddot{\psi}(t)\end{bmatrix} = \begin{bmatrix} -\frac{c_r+c_f}{mv(t)} & \frac{c_rl_r-c_fl_f}{mv^2(t)} - 1 \\ \frac{c_rl_r-c_fl_f}{I_z} & -\frac{c_rl_r^2+c_fl_f^2}{I_zv(t)}\end{bmatrix}\begin{bmatrix}\beta(t)\\ \dot{\psi}(t)\end{bmatrix} + \begin{bmatrix}\frac{c_f}{mv(t)}\\ \frac{c_rl_f}{I_z}\end{bmatrix}u_L(t) + \begin{bmatrix}\frac{1}{mv} & \frac{1}{mv}\\ \frac{l_r}{I_z} & -\frac{l_f}{I_z}\end{bmatrix}\begin{bmatrix}\Gamma_{yf}\\ \Gamma_{yr}\end{bmatrix} + \begin{bmatrix}\frac{1}{mv}\\ \frac{l_w}{I_z}\end{bmatrix}F_w \tag{5}$$

with $u_L(t)$ is the steering wheel angle, it is assumed equal to the steering angle at the front wheels, if the transmission is considered as a linear constant function $u_L(t) = \delta(t)$, and the yaw rate as the measured output.

The lateral tire force $F_{ty_{ij}}$ can be divided in three operating regions according to the magnitude of the tire slip angle, Fig. 3:

- Normal zone, where $F_{ty_f} = \bar{F}_{ty_f}$ and $\Gamma_{ij} \cong 0$. In this region the road holding performance is satisfied.
- Critical zone, where $\Gamma_{ij} \neq 0$; the vehicle starts loosing its road holding, approaching risk zone,
- Skidding zone, where F_{ty_f} becomes constant. In this condition the tire looses its grip, compromising the stability and handling of the vehicle.

The problem of detecting critical driving situation can be traduced into the detection and estimation of the non-linear tire forces (Γ_f, Γ_r), by considering them as faults. A dedicated residual generator scheme is adopted. Two observers are designed to estimate Γ_f and Γ_r. The state-space representation used to synthesize the residual that estimates Γ_f is given by:

$$\begin{bmatrix}\dot{\beta}(t)\\ \ddot{\psi}(t)\end{bmatrix} = \begin{bmatrix} -\frac{c_r+c_f}{mv(t)} & \frac{c_rl_r-c_fl_f}{mv^2(t)} - 1 \\ \frac{c_rl_r-c_fl_f}{I_z} & -\frac{c_rl_r^2+c_fl_f^2}{I_zv(t)}\end{bmatrix}\begin{bmatrix}\beta(t)\\ \dot{\psi}(t)\end{bmatrix} + \begin{bmatrix}\frac{c_f}{mv(t)}\\ \frac{c_rl_f}{I_z}\end{bmatrix}u_L(t) + \begin{bmatrix}\frac{1}{mv} & \frac{1}{mv}\\ \frac{l_w}{I_z} & \frac{l_r}{I_z}\end{bmatrix}\begin{bmatrix}F_w\\ \Gamma_{yr}\end{bmatrix} + \begin{bmatrix}\frac{1}{mv}\\ -\frac{l_f}{I_z}\end{bmatrix}\Gamma_{yf} \tag{6}$$

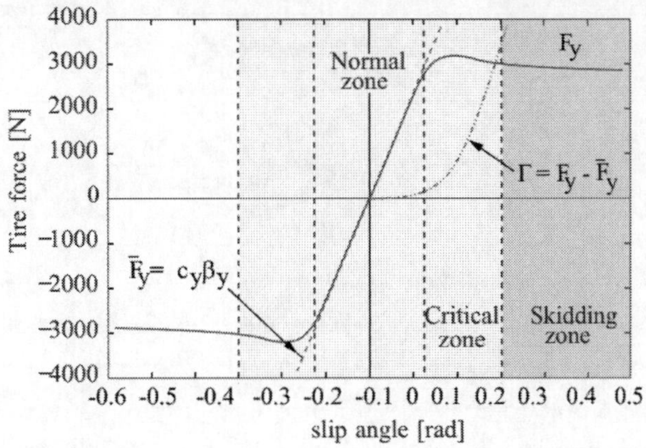

Figure 3. Lateral tire force and working regions

where $\left[F_w\ \Gamma_{yr}\right]^T$ is considered as a perturbation signal to be rejected. Conversely, the representation used to synthesize the residual for Γ_r is:

$$\begin{bmatrix}\dot{\beta}(t)\\ \ddot{\psi}(t)\end{bmatrix} = \begin{bmatrix} -\frac{c_r+c_f}{mv(t)} & \frac{c_rl_r-c_fl_f}{mv^2(t)}-1 \\ \frac{c_rl_r-c_fl_f}{I_z} & -\frac{c_rl_r^2+c_fl_f^2}{I_zv(t)}\end{bmatrix}\begin{bmatrix}\beta(t)\\ \dot{\psi}(t)\end{bmatrix} + \begin{bmatrix}\frac{c_f}{mv(t)}\\ \frac{c_rl_f}{I_z}\end{bmatrix}u_L(t) + \begin{bmatrix}\frac{1}{mv} & \frac{1}{mv}\\ \frac{l_w}{I_z} & \frac{l_r}{I_z}\end{bmatrix}\begin{bmatrix}F_w\\ \Gamma_{yf}\end{bmatrix} + \begin{bmatrix}\frac{1}{mv}\\ -\frac{l_f}{I_z}\end{bmatrix}\Gamma_{yr} \quad (7)$$

with the perturbation signal $\left[F_w\ \Gamma_{yf}\right]^T$.

Clearly, the bicycle model is non-linear because of the terms $v^{-1}(t)$ and $v^{-2}(t)$. This model is linearized around multiple velocities v_α. A switching rule dependent on the measured velocity is deduced, and the switching signal is then calculated as the integer part of $v(t)/\Delta$, where Δ is the half distance between two successive operating points. The non-linear bicycle model is transformed into a switched uncertain model, using a Taylor expansion around v_α:

$$\frac{1}{v}|_{v=v_\alpha} = \frac{1}{v_\alpha} - \frac{1}{v_\alpha^2}(v-v_\alpha) + \mathcal{O}\left(\frac{1}{v^2}\right) \qquad \frac{1}{v^2}|_{v=v_\alpha} = \frac{1}{v_\alpha^2} - \frac{2}{v_\alpha^3}(v-v_\alpha) + \mathcal{O}\left(\frac{1}{v^3}\right) \quad (8)$$

For velocities $v > 10$ km/h, $\mathcal{O}(\frac{1}{v^3}) << \mathcal{O}(\frac{1}{v^2}) << 1$, then:

$$A_\alpha = \underbrace{A_0 + \frac{1}{v_\alpha}A_1 + \frac{1}{v_\alpha^2}A_2}_{A_\alpha} + \underbrace{\left(-\frac{1}{v_\alpha^2}A_1 - \frac{2}{v_\alpha^3}A_2\right)}_{H_{A,\alpha}}\underbrace{(v-v_\alpha)}_{\Delta_{x,\alpha}} \qquad B_\alpha = \underbrace{B_0 + \frac{1}{v_\alpha}B_1}_{B_\alpha} + \underbrace{-\frac{1}{v_\alpha^2}B_1}_{H_{B,\alpha}}\underbrace{(v-v_\alpha)}_{\Delta_{u,\alpha}}$$

$$E_{d_\alpha} = \underbrace{E_d0 + \frac{1}{v_\alpha}E_d1}_{E_{d\alpha}} + \underbrace{-\frac{1}{v_\alpha^2}E_d1}_{H_{C,\alpha}}\underbrace{(v-v_\alpha)}_{\Delta_{d,\alpha}} \qquad E_{f_\alpha} = \underbrace{E_f0 + \frac{1}{v_\alpha}E_f1}_{E_{f\alpha}} + \underbrace{-\frac{1}{v_\alpha^2}E_f1}_{H_{D,\alpha}}\underbrace{(v-v_\alpha)}_{\Delta_{f,\alpha}} \quad (9)$$

Speeds higher than 40 km/h ($v_{min} > 10$ m/s) are considered. In this case, the matrices $H_{A,\alpha}, H_{B,\alpha}, H_{C,\alpha}, H_{D,\alpha}$ can be neglected. The transformed linear switched system is discretized with a sampling time of 20 ms:

$$x_{k+1} = A_\alpha x_k + B_\alpha u_k + E_{d_\alpha} d_k + E_{f_\alpha} f_k \quad (10)$$

where d_k is the disturbance signal and f_k denotes the non-linear tire force to be estimated. An example of a velocity profile and the deduced switching signal is illustrated in Fig. 4

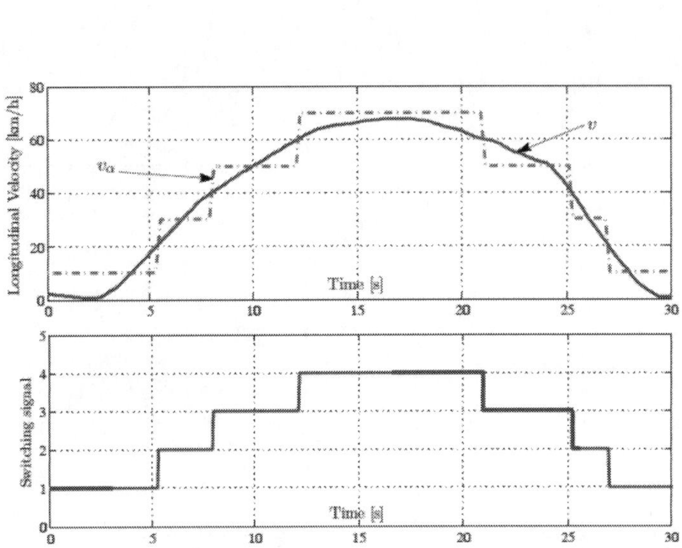

Figure 4. Velocity profile and the corresponding switching signal

3. Observer Design

Consider this discrete-time switched linear system subject to faults and disturbances:

$$\begin{cases} x_{k+1} &= A_{\alpha(k)}x_k + B_{\alpha(k)}u_k + E_{d,\alpha(k)}d_k + E_{f,\alpha(k)}f_k \\ y_k &= C_{\alpha(k)}x_k + D_{\alpha(k)}u_k + F_{d,\alpha(k)}d_k + F_{f,\alpha(k)}f_k \end{cases} \tag{11}$$

where $\alpha(k) : [0,\infty) \rightarrow \mathscr{I} = \{1,2,..,N\}$ is the switching signal. The fault detection problem can be solved by *LMI* optimization. The *SPIO* for the estimation of the faults has the following form:

$$\begin{cases} \dot{\hat{x}} &= A_\alpha \hat{x} + B_\alpha u + L_{P,\alpha}(y-\hat{y}) + E_{f,\alpha}\hat{f} \\ \dot{\hat{f}} &= L_{I,\alpha}(y-\hat{y}) \\ \hat{y} &= C_\alpha \hat{x} + D_\alpha u + F_{f,\alpha}\hat{f} \end{cases} \tag{12}$$

where $L_{P,\alpha}$ and $L_{I,\alpha}$ are the Proportional and Integral gains. The state and fault estimation errors, $e_x = x - \hat{x}$ and $e_f = f - \hat{f}$, respectively, are:

$$\begin{aligned} e_{x k+1} &= (A_\alpha - L_{P,\alpha}C_\alpha)e_{x k} + (E_{f,\alpha} - L_{P,\alpha}F_{f,\alpha})e_{f k} + (E_{d,\alpha} - L_{P,\alpha}F_{d,\alpha})d_k \\ e_{f k+1} &= -L_{I,\alpha}C_\alpha e_{x k} - L_{I,\alpha}F_{f,\alpha}e_{f k} - L_{I,\alpha}F_{d,\alpha}d_k \end{aligned} \tag{13}$$

This can be written in the matrix form:

$$\begin{bmatrix} e_{x k+1} \\ e_{f k+1} \end{bmatrix} = \left(\begin{bmatrix} A_\alpha & E_{f,\alpha} \\ 0 & 0 \end{bmatrix} - \begin{bmatrix} L_{P,\alpha} \\ L_{I,\alpha} \end{bmatrix} \begin{bmatrix} C_\alpha & F_{f,\alpha} \end{bmatrix} \right) \begin{bmatrix} e_{x k} \\ e_{f k} \end{bmatrix} + \begin{bmatrix} E_{d,\alpha} - L_{P,\alpha}F_{d,\alpha} \\ -L_{I,\alpha}F_{d,\alpha} \end{bmatrix} d_k \tag{14}$$

or equivalently:

$$\tilde{x}_{k+1} = (\tilde{A}_\alpha - L_\alpha \tilde{C}_\alpha)\tilde{x}_k + (\tilde{E}_{d,\alpha} - L_\alpha F_{d,\alpha})d_k \tag{15}$$

where $\tilde{x} = \begin{bmatrix} e_x^T & e_f^T \end{bmatrix}^T$ and

$$\tilde{A}_\alpha = \begin{bmatrix} A_\alpha & E_{f,\alpha} \\ 0 & 0 \end{bmatrix},\ \tilde{B}_\alpha = \begin{bmatrix} B_\alpha \\ 0 \end{bmatrix},\ \tilde{E}_{d,\alpha} = \begin{bmatrix} E_{d,\alpha} \\ 0 \end{bmatrix},\ \tilde{C}_\alpha = \begin{bmatrix} C_\alpha F_{f,\alpha} \end{bmatrix},\ L_\alpha = \begin{bmatrix} L_{P,\alpha} \\ L_{I,\alpha} \end{bmatrix} \tag{16}$$

To ensure the robustness of the observer, eqn (12), to unknown inputs and perturbation, a residual signal to be minimized is $r = y - \hat{y}$, where $A^a_\alpha = \tilde{A}_\alpha - L_\alpha \tilde{C}_\alpha$, $E^a_{d,\alpha} = \tilde{E}_{d,\alpha} - L_\alpha F_{d,\alpha}$; then, a switched residual generator is:

$$\begin{cases} \tilde{x}_{k+1} &= A^a_\alpha \tilde{x}_k + E^a_{d,\alpha} d_k \\ r_k &= \tilde{C}_\alpha \tilde{x}_k + F_{d,\alpha} d_k \end{cases} \tag{17}$$

Defining the sensitivity functions of disturbance to the residual as:

$$T_{rd_\alpha}(z) = \tilde{C}_\alpha (zI - A^a_\alpha)^{-1} E^a_{d,\alpha} + F_{d,\alpha} \tag{18}$$

The goal of the *SPIO* in the $\mathscr{H}_\infty$ formulation is:

$$\|T_{rd_\alpha}\|_\infty < \gamma_\alpha \tag{19}$$

The problem is: *Find the gain matrices L_α that minimize γ_α such that the SPIO is stable.* The *SPIO* poles can be arbitrary assigned using an H_∞/pole placement technique. Consider the switched system, eqn. (11), if there exists matrices $P_i > 0$, $P_j > 0$, U_i and U_j and positive scalars γ_i and ξ_i $\forall (i,j) \in \mathscr{I} \times \mathscr{I}$ such that the following *LMI*s are satisfied:

$$\begin{bmatrix} P_j - 2P_i & P_i\tilde{A}_i + U_i\tilde{C}_i & P_i\tilde{E}_{d,i} + U_iF_{d,i} & 0 \\ \star & -P_i & 0 & \tilde{C}_i^T \\ \star & \star & -\gamma_i^2 I & F_{d,i}^T \\ \star & \star & \star & -I \end{bmatrix} < 0 \tag{20a}$$

$$\begin{bmatrix} \xi_i(P_j - 2P_i) & P_i\tilde{A}_i + U_i\tilde{C}_i \\ \star & \xi_i(P_j - 2P_i) \end{bmatrix} \tag{20b}$$

$$P_\alpha > 0 \tag{20c}$$

then, there exist a *SPIO*, eqn (12), such that the $\mathscr{H}_\infty$ and design objectives are satisfied. The *SPIO* gains are computed as:

$$L_i = P_i^{-1} U_i \qquad L_{P_i} = \begin{bmatrix} I_p & 0_{p \times n_f} \end{bmatrix} L_i \qquad L_{I_i} = \begin{bmatrix} 0_{n_f \times p} & I_{n_f} \end{bmatrix} L_i \tag{21}$$

Consider the switched Lyapunov function candidate:

$$V_k = x_k^{aT} P_{\alpha(k)} x_k^a > 0 \tag{22}$$

where $P_{\alpha(k)} > 0$ is a positive definite matrix. If such Lyapunov function exists and $\Delta V = V_{\alpha(k+1)} - V_{\alpha(k)}$ is negative definite along system trajectories of eqn (17), then the origin of the system is globally asymptotically stable. The $\mathscr{H}_\infty$ feasibility problem formulated in eqn (19) with the sufficient Lyapunov-stability condition can be expressed with the following inequality:

$$\mathbb{V}_k = \Delta V_k + r_k^T r_k - \gamma_i^2 w_k^T w_k < 0 \tag{23}$$

The *SPIO* for the fault estimation is implemented using the following structure:

$$\begin{bmatrix} \hat{x}_{k+1} \\ \hat{f}_{k+1} \end{bmatrix} = \begin{bmatrix} A_\alpha - L_{P,\alpha}C_\alpha & E_{f,\alpha} - L_{P,\alpha}F_{f,\alpha} \\ -L_{I,\alpha}C_\alpha & -L_{I,\alpha}F_{f,\alpha} \end{bmatrix} \begin{bmatrix} \hat{x}_k \\ \hat{f}_k \end{bmatrix} + \begin{bmatrix} B_\alpha - L_{P,\alpha}D_\alpha & L_{P,\alpha} \\ -L_{I,\alpha}D_\alpha & L_{I,\alpha} \end{bmatrix} \begin{bmatrix} u_k \\ y_k \end{bmatrix} \tag{24}$$

In this study, a non-conventional modelling of vehicle dynamics has been used. Since the velocity is a continuous varying parameter, the *LPV* method is more commonly used. In fact, within the *LPV* framework, the construction of the gain matrices and the stability analysis at each vertex will consider a common Lyapunov fuction, i.e. a unique P matrix. However, within the result of switching modeling, a switched Lyapunov function in considered, and thus the *LPV* result can be considered as one case of a general solution of the switched observer.

4. Results

The synthesis of *SPIO* is given as the solution of the *LMI* optimization problem. The non-linear tire forces to be estimated are computed as:

$$\Gamma_i = \sum_{j=r}^{j=l} F_{ty_{ij}} - \sum_{j=r}^{j=l} c_{y_{ij}} \beta_{y_i} \tag{25}$$

and it is considered that the force is the same in the left and right side. As Fig. 2 shows, c_{yi} denotes the slope of the curve $F_y(\beta)$ in the linear zone.

In order to obtain c_{yi} for each tire, the parameters of the Pacejka model, eqn. (26), were identified using the data from the tires curves of Fig. 2. As described in eqn (2), the lateral tire forces according to the Pacejka model depend on C_y, which is constant. B_y, D_y and E_y are functions of F_{z_i}:

$$D_y(F_{z_{ij}}) = a_1 F_{z_{ij}}^2 + a_2 F_{z_{ij}} \quad B_y(F_{z_{ij}}) = a_3 \frac{\sin(a_4 \tan^{-1}(a_5 F_{z_{ij}})))}{C_y D_y(F_{z_{ij}})} \quad E_y(F_{z_{ij}}) = a_6 F_{z_{ij}}^2 + a_7 F_{z_{ij}} + a_8 \tag{26}$$

The non-linear least squares algorithm was used to fit the coefficients to the experimental tire data vectors $(F_{ty_{ij}}, F_{z_i}, \beta_{y_{ij}})$. Once these coefficients are estimated, the identification of the model is validated with *CarSim*^{TM} data, Fig. 5, using the *Error to Signal Ratio* (*ESR*) as performance index. An *ESR* of 0.245 % was obtained, thus this model accurately represents the tire lateral force.

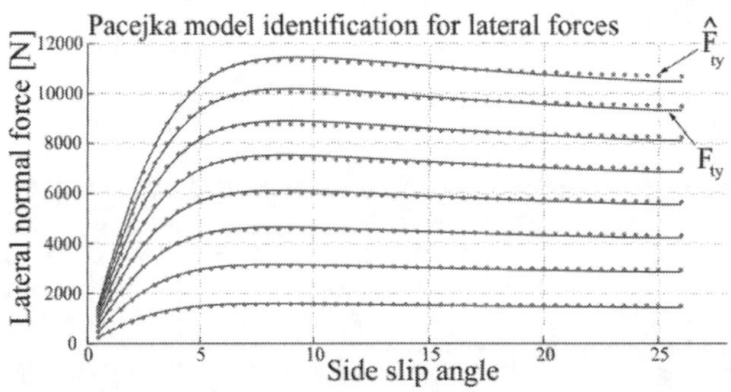

Figure 5. Validation of Pacejka model for lateral pneumatic efforts

On th other hand, the vertical force on the tire (i, j) is calculated as the vehicle weight divided by 2 and weighted by the front-rear load distribution:

$$F_{z_{rj}} = \frac{1}{2}(m_s + m_{us}) g \frac{l_r}{l_r + l_f} \quad F_{z_{fj}} = \frac{1}{2}(m_s + m_{us}) g \frac{l_f}{l_r + l_f} \tag{27}$$

The normal force $F_{z_{ij}}$, eqn (27), was used to obtain the corresponding coefficients c_{y_i}, which are computed as $c_{y_i} = B_y C_y D_y$, [10]. A gyroscope gives the yaw rate as the measured output, the output state matrices are:

$$C_\alpha = \begin{bmatrix} 0 & 1 \end{bmatrix}, \qquad D_\alpha = 0, \qquad F_{d,\alpha} = 0, \qquad F_{f,\alpha} = 0 \tag{28}$$

The driving scenario was the *Double Lane Change* (*DLC*) test, at high velocities ($> 70 km/h$). This test starts with the vehicle driving in straigth line at a constant target speed. After 60 m of longitudinal displacement the driver performs an avoidance steering manoeuvre to have a lateral displacement of 3.4

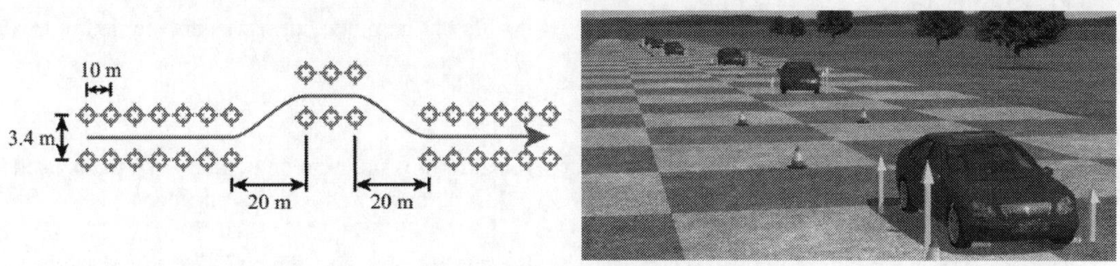

Figure 6. Double Lane Change test

m widthin the next 20 m of longitudinal displacement; and finally, after another 20 m the driver performs a steering manoeuvre in the opposite sense to return to the original lane, Fig. 6.

Figure 7 shows a comparison of the nonlinear lateral force for the front and rear tires (solid blue) with the estimated ones with the *SPIO* (dashed green). It can be observed that at the beginning of the test, when the vehicle runs in straight line and thus the side-slip angle is close to zero, the tires operate in the linear region and the estimation is close to zero. When the vehicle perform the lane change maneuver, the side-slip increase and the tires operate in the nonlinear region. In this scenario, the estimation presents some sign error, however, it has similar dynamical characteristics of the actual nonlinear force. Therefore, even if the estimation is imperfect, this signal can be used for detection purposes, *i.e* bounds in the nonlinear force estimation can be established to determine if the vehicle is entering in a critical condition.

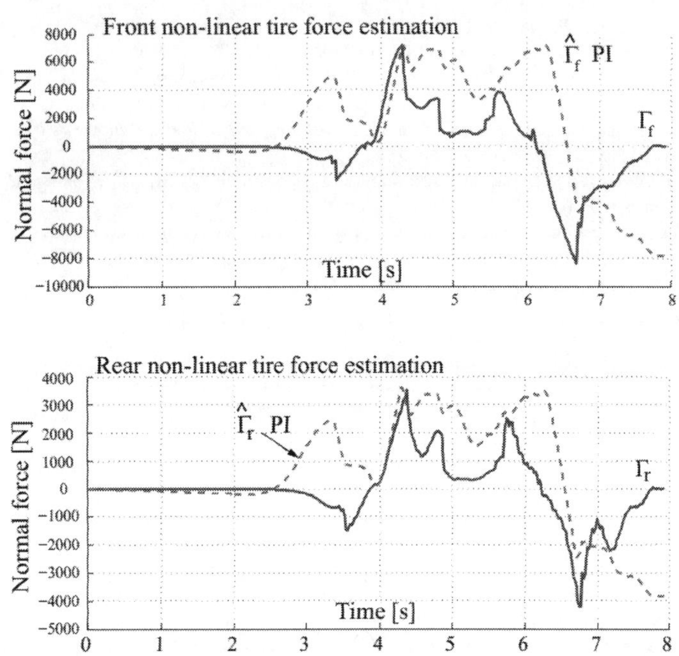

Figure 7. Estimation of non-linear tire forces using *SPIO*

As shows the results in the different figures, the estimated signal roughly follows the actual signal, and this is due to the modeling of the tire dynamics. In fact, the tire stiffness coefficient is not constant, it depends on the tire normal force that varies due to load transfer. A robustness study on this parameter could be performed to overcome these errors, that it is visible on the plots, but it was not studied in this paper.

5. Conclusions

An adaptation of fault estimation problem is used for critical driving situation detection. A signal is used as an indicator of the loss of lateral stability. This signal to be estimated as the non-linear part of the lateral force of the tires.

A model with uncertainties on the tire stiffness coefficients was obtained. This model is then used for the synthesis of *SPIO* to estimate the non-linear force of the tires.

A comparison of the estimated nonlinear normal force by *SPIO* is given with simulation results in $CarSim^{TM}$. Results show that even if the nonlinear force estimation has error, this signal can be used to detect critical situations since the dynamical characteristics of the estimation are similar to the actual ones.

Possible future work should consists on better tire force modeling in order to take into consideration load transfers and the road properties, and extend the estimation of the zone for a coupling with existing driving assistance strategies (*ESP*, *ABS*).

Acknowledgment

Authors thank *Tecnológico de Monterrey*, *Gipsa-Lab INPG* and *CONACyT* because bilateral PCP 03/10 and 06/13 (México-France) projects.

References

[1] F. Bakhshande and D. Söffker. Proportional-Integral-Observer: A Brief Survey with Special Attention to the Actual Methods using ACC Benchmark. In *American Control Conf*, pages 532–537, 2015.
[2] M. Burckhardt. *Fahrwerktechnik: Radschlupfregelsysteme*. Vogel-Verlag, Wüzburg, Germany, 1993.
[3] P. Dahl. A Solid Friction Model. Technical report, The Aerospace Corporation, El Segundo, CA, USA, 1976.
[4] N. De Freitas, Dearden R, F. Hutter, R. Morales-Menendez, J. Mutch, and D. Poole. Diagnosis by a Waiter and a Mars Explorer. *Proc of the IEEE*, 92(3):456–468, Mar 2004.
[5] C. Canudas de Wit, H. Olson, and K.J. Astrom. A New Model for Control of Systems with Friction. *IEEE Trans. Autom. Control*, 40(3):419–425, March 1995.
[6] P. Gao, H.W Kaas, D. Mohr, and D. Wee. Disruptive Trends That will Transform the Auto Industry. Technical report, McKinsey & Company, 2016.
[7] T. Kaczorek. Proportional-Integral Observers for Linear Multivariable Time-Varying Systems. *Regelungstechnik*, 27:359–362, 1979.
[8] Liberzon. *Switching in Systems and Control*. Birkhäuser, 2003.
[9] B. Marx, D. Koenig, and D. Georges. Robust Fault Diagnosis for Linear Descriptor Systems using Proportional Integral Observers. *Proc of the* 42th *IEEE Conf on Decision and Control*, pages 661–670, 2003.
[10] H.B. Pacejka and E. Bakker. The Magic Formula Tyre Model. In 1st *Int Colloquio on Tyre Models for Vehicle Dynamics Analysis*, pages 1–18, 1991.
[11] L. R. Ray. Nonlinear State and Tire Force Estimation for Advanced Vehicle Control. *IEEE Trans on Control Systems Technology*, 3(1):117–124, Mar 1995.
[12] L.R. Ray. Nonlinear Tire Force Estimation and Road Friction Identification: Simulation and Experiments. *Automatica*, 33(10):1819–1833, 1997.
[13] G. Rill. *Simulation von Kraftfahrzeugen*. Vieweg, Wiesbaden, Germany, 1994.
[14] B. Shafai and R. Carroll. Design of Proportional-Integral Observer for Linear Time-Varying Multivariable Systems. *IEEE Conf on Decision and Control*, 24:597–599, 1985.
[15] B. Shafai, C. Pi, and S. Nork. Simultaneous Disturbance Attenuation and Fault Detection using Proportional Integral Observers. *American Control Conf*, 2:1647–1649, 2002.
[16] J. Stephant, A. Charara, and D. Meizel. Virtual Sensor: Application to Vehicle Sideslip Angle and Transversal Forces. *IEEE Trans on Industrial Electronics*, 51(2):278–289, April 2004.
[17] J. Villagra, B. dAndrea-Novel, M. Fliess, and H. Mounier. A Diagnosis-based Approach for TireRoad Forces and Maximum Friction Estimation . *Control Eng Practice*, 19(2):174–184, 2011.
[18] B. Wojciechowski. *Analysis and Synthesis of Proportional-Integral Observers for Single-Input-Single-Output Time-Invariant Continuous Systems*. PhD thesis, Gliwice, Poland, 1978.

A Hybrid Extended Kalman Filter as an Observer for a Pot-Electro-Magnetic Actuator

Simon Schmidt, Paolo Mercorelli

Institute of Product and Process Innovation, Leuphana University of Lueneburg, Volgershall 1, D-21339 Lueneburg, Germany

E-mail: {simon.schmidt, mercorelli}@uni.leuphana.de

Abstract. This paper deals with an application in which a *hybrid* extended Kalman Filter (HEKF) is used to estimate state variables in a U-shaped electro-magnetic actuator to be used in mechanical systems. In this context a hybrid Kalman Filter is the one which switches between different models. The paper proposes a hybrid model for an extended Kalman Filter to be used as an observer to estimate the state and to control the force of the actuator. Applications include position, velocity and force control in automotive, engine and manufacturing systems. This work is focused on the estimation of state variables of the actuator. Simulated results show the effectiveness of the proposed approach.

1. Introduction

Dynamic systems are usually classified as continuous or discrete dynamic ones. However, real systems cannot often be clearly classified into one of these categories. Most real dynamic systems contain continuous and discrete dynamics as well. This mixture of continuous and discrete dynamics is called a hybrid dynamic system (or short: hybrid system). Therefore, hybrid dynamic systems exhibit continuous and instantaneous changes, having features of continuous-time and discrete-time dynamical systems. Many examples of hybrid systems are given not only from physics, but also from automotive and industrial applications. Hybrid systems can be not only found in modern technologies in which continuous variables are mixed with non-continuous ones. An overview on this literature can be found in [1] and in [2]. Very often just the necessity of reconfigurations of the controller as well as the observer generate many difficulties as shown in [3]. Nevertheless, in electro-magnetic mechanical actuators, observers as in [4], [5] and in particular Kalman Filter (KF) as an observer such as in [6] and [7] are very often applied. In [8] an electromagnetic mechanical actuator for engines is considered. In this application the kernel of the actuator is represented by a U-shaped electromagnetic actuator, which controls the intake and exhaust valve for engines. In contrast to this work, the contribution of this paper is oriented to the control of an actuator to be used in manufacturing systems for the problem of precise positioning. In applications it is often desirable to implement control without a position sensor due to the restricted mounting space and the expense. In this case, an estimator could be the basis for sensorless control in which position and velocity signals are reconstructed from electrical measurements [9, 10]. Another challenging problem, for reasons of safety is represented by the soft landing of the positioning. Achieving a soft landing is a difficult job that requires precise state estimation. To achieve sensorless control, in the literature, there are several different approaches propose which are based on observers. The Kalman Filter is one of the most used algorithms in this context, with the extended version utilized when nonlinearities are present in the system. The extended Kalman Filter (EKF) is a heuristic method for nonlinear filtering and estimation problems. When it is tuned well, it is often effective and is widely used in practice. Nevertheless, the EKF needs to be enhanced and adapted to the particular application; in this case, the EKF is a viable and computationally efficient candidate for general applications. In general, applications of Kalman Filters offer the possibility of implementing a state-based controller in any field with automatic controls. An adaptive extended Kalman Filter has been used to improve the performance of industrial drive control by estimating some important loads of the industrial drive [11]. A novel sensorless force control approach for robot-assisted motion of the human arm based on the Kalman Filter has been presented in [12] as well as an intelligent system that incorporates Kalman Filters and a fuzzy expert system to track the tip of a fastening tool and to identify the fastened bolt [13]. In general, because these control structures avoid bulky and complicated measurement systems, they are easily applicable to real world problems. The proposed method presented in this paper is general and can be applied to all types of problems in which position and velocity measurements are not possible or are too expensive to be acquired.

2. Contribution of the paper

The proposed method presented in this paper is quite general and could be applied to all types of problems in which position and velocity measurements are not possible or are too expensive to be acquired. The main contributions of this paper are as follows:

- the introduction of an observer for a sensorless control scheme
- the use of a hybrid extended Kalman Filter based on five switching models

This paper provides an extensive description of an application combining a hysteresis hybrid observer with an extended Kalman Filter and hysteresis automation, resulting in sensorless control. The hybrid observer scheme is designed to expand upon the continuous extended Kalman Filter. The proposed structure consists of five models. Between these models there are five switching conditions to make sure that the observer works inside the correct region of the state space. An artificial hysteresis between the switching states is introduced to avoid chattering problems when the velocity is close to zero. The proposal considers already consolidated structures on extended Kalman Filter and uses theory of hybrid systems to get a hybrid structure of this filter. Using this combination of well known approaches yields to a new structure of Kalman Filter, which shows a very accurate and fast observation of states. The proposed observer works in the way that the current is measured, and the position and velocity of the electromagnet are estimated. The behavior of the system is a hybrid one, because it has points in state space where it switches its behavior. These switching points are met e.g. on upper or lower limits. The paper is organized as follows. In Section 4, a hybrid model of the actuator is presented. In Section 3 fundamental aspects of KF are recalled together with the description of the structure of the proposed EKF which is devoted to estimate the state variables of the actuator. In particular Section 5 presents the validations of the proposed approach though simulations and also shows the performance of the proposed observer. Conclusion closes the paper.

3. Tracking embedded hybrid behaviors using an Extended Kalman Filter

Our state estimation architecture tracks the nominal system dynamics using a robust observer scheme implemented as a combination of an Extended Kalman Filter (EKF) and an automated switching between several models for the magnet. The hybrid observer has to track (i) continuous behavior in individual modes of operation and (ii) discrete mode changes. At mode changes, the new state space model and the initial state of the system are recomputed, and the error covariance matrix is updated. The hybrid observer scheme is designed to expand upon the continuous extended Kalman Filter. Hysteresis is introduced to avoid chattering problems when the velocity is close to zero. Model uncertainty and measurement noise are implemented as white noise, using uncorrelated Gaussian distributions with zero mean. To be able to estimate the operation mode, the model discussed in section 4 is extended by an additional state and then it is as follows:

$$\dot{\mathbf{x}} = \mathbf{f}_q(\mathbf{x}(t)) + \mathbf{g}(t)u(t) + \mathbf{w}(t), \tag{1}$$

$$\mathbf{x}(t) = \begin{bmatrix} i(t) \\ s(t) \\ v(t) \\ q(t) \end{bmatrix}, \quad \mathbf{g}(t) = \begin{bmatrix} \frac{1}{L(s(t))} \\ 0 \\ 0 \\ 0 \end{bmatrix}, \tag{2}$$

where function $\mathbf{f}$ represents the nonlinearities, and $\mathbf{w}(t)$ is the model uncertainty. The measured output can be written as follows:

$$y(t) = \mathbf{H}_m\mathbf{x}(t) + \varsigma(t), \tag{3}$$

where $\mathbf{H}_m = [1 \ \ 0 \ \ 0 \ \ 0]$, and $\varsigma(t)$ is the measurement noise variance. An EKF is based on linearized dynamics, output functions at the current estimate and on propagating an approximation of the conditional expectation and covariance. In the prediction mode q, the predicted state is

$$\mathbf{x}(k/k-1) = \mathbf{x}(k-1/k-1) + T_s\big(\mathbf{f}_q(\mathbf{x}(k-1/k-1)) + \mathbf{B}_m\mathbf{u}(k-1)\big), \tag{4}$$

and the predicted error covariance matrix is

$$\begin{aligned}\mathbf{P}_q(k/k-1) = \mathbf{P}_q(k-1/k-1) + T_s\big(&\mathbf{F}_q(x(k-1/k-1))\mathbf{P}_q(k-1/k-1) \\ &+ \mathbf{P}_q(k-1/k-1)\mathbf{F}_q(x(k-1/k-1))\big) + \mathbf{R}_w,\end{aligned} \tag{5}$$

where $k \in \mathbb{Z}$, $\mathbf{R}_w$ is the process noise covariance matrix, $\mathbf{P}_q(k-1/k-1)$ is the error covariance matrix, $\mathbf{F}_q(x(k-1/k-1))$ is the Jacobian matrix of the system calculated at the estimated state, and T_s is the sampling time.
In particular, the Jacobian matrix $\mathbf{F}_q(\mathbf{x}(k-1/k-1))$ and $\mathbf{P}_q(k-1/k-1)$ with variable $q \in \mathbb{N}$, where $q = 1, 2, ..., 5$ are related to the five dynamic modes of the actuator. In the correction mode,

$$\mathbf{K}_q(k) = \mathbf{P}_q(k/k-1)\mathbf{H}_m^T\big(\mathbf{H}_m\mathbf{P}(k/k-1)\mathbf{H}_m^T + \mathbf{R}_\varsigma\big)^{-1} \tag{6}$$

$$\mathbf{x}(k/k) = \mathbf{x}(k/k-1) + \mathbf{K}_q(k)\big(y(k) - \mathbf{H}_m\mathbf{x}(k/k-1)\big) \tag{7}$$

$$\mathbf{P}_q(k/k) = \mathbf{P}_q(k/k-1) - \mathbf{K}_q(k)\mathbf{H}_m\mathbf{P}_q(k/k-1), \tag{8}$$

where $\mathbf{K}_q(k)$ is the Kalman gain and $\mathbf{R}_\varsigma$ is the measurement noise matrix. The mode change calculations are based on the system mode at time step k, $q(k)$, and on the state of the system $\mathbf{x}(k)$. In contrast to the standard extended Kalman Filter there are discontinuities because of the switching state $q(t)$. To make sure that the observer works in a robust way, all states has to be recalculated at switching points and the Kalman Filter has to use different models for the differnt modes of operation.

4. Description of the actuator and its hybrid mathematical model

Figures 1 and 2 represent a simplified scheme of the U-shaped Electromagnetic actuator which is proposed to be observed. The general mathematical model can be found in the literature such as [14]. To test the proposed algorithm, an actuator is considered. The actuator consists only of

Figure 1. Simplified scheme of pot-magnet

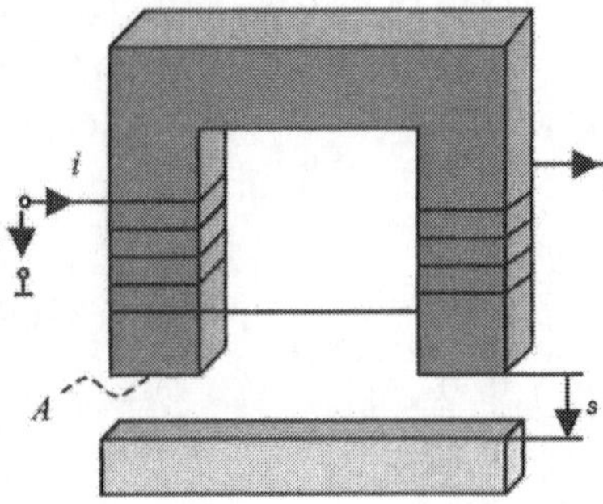

Figure 2. Free-body diagram of magnet anchor

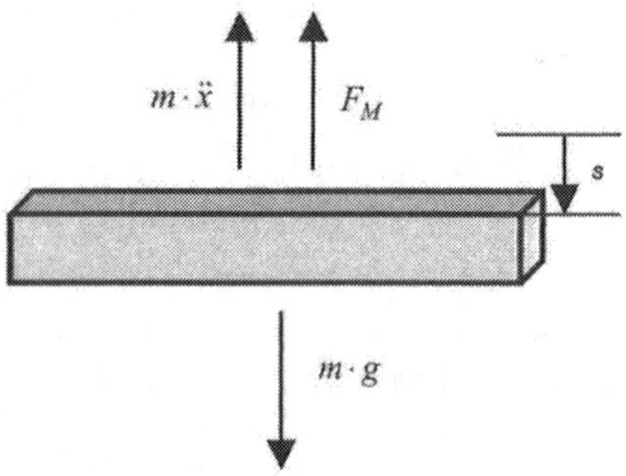

a half magnet system. The data for the physical dimensions and characteristics are taken from a real magnet. A state vector $\mathbf{x(t)} = [i(t), s(t), v(t)]$ and an input voltage $u(t)$ are considered for the system in Fig. 1. The current is represented by $i(t)$, the position of the armature by $s(t)$, the velocity by $v(t)$ and the input voltage by $u(t)$. The general system can be represented in the following way. The electrical dynamics are governed by the equation

$$\frac{di(t)}{dt} = \frac{1}{L(s(t))}\Big(u(t) - \Big(R + \frac{dL(s(t))}{dt}\Big)i(t)\Big), \tag{9}$$

where $i(t)$ indicates the current, u indicates the supply voltage, R indicates the resistance and L indicates the inductance. The inductance L is a function of the current $i(t)$, through the magnetic permeability, and of the distance $s(t)$. In particular,

$$L(s(t)) = \frac{N^2\mu_r\mu_o A}{c_1 + 2\mu_r s(t)}, \tag{10}$$

where N represents the number of windings, μ_r and μ_0 represent the relative permeability and the vacuum permeability respectively, A is the surface involved in the magnetic flux and c_1 is a physical constant that depends on the current position of the magnet. From (10) the following expression can be derived:

$$\frac{dL(s(t))}{ds} = -\frac{2N^2\mu_r^2\mu_o A}{(c_1 + 2\mu_r s(t))^2}. \tag{11}$$

Since there is no equation for $\frac{dL(s(t))}{dt}$ but one for $\frac{dL(s(t))}{ds}$ the following equation is needed:

$$\frac{dL(s(t))}{dt} = \frac{dL(s(t))}{ds}\frac{ds(t)}{dt} = \frac{dL(s(t))}{ds}\frac{ds(t)}{dt} = \frac{dL(s(t))}{ds}v(t). \tag{12}$$

The mechanical dynamic equation is the following:

$$\frac{d^2s(t)}{dt^2} = \frac{F_m(t) - F_g(t)}{m}, \tag{13}$$

where $F_m(t)$ represents the magnetic force and is:

$$F_m(t) = \frac{N^2 \mu_r^2 \mu_o A i^2(t)}{(c_1 + 2\mu_r s(t))^2}, \tag{14}$$

F_g is the gravitational force. According to the range of the velocity $v(t)$ in which the actuator works, it can be considered a linear affine function of the velocity. This model can be summarized as follows:

$$\underbrace{\begin{bmatrix} \frac{di(t)}{dt} \\ \frac{ds(t)}{dt} \\ \frac{dv(t)}{dt} \end{bmatrix}}_{\dot{\mathbf{x}}(t)} = \underbrace{\begin{bmatrix} \frac{-(Ri(t) + \frac{dL(s(t))}{ds} v(t) i(t))}{L(s(t))} \\ v(t) \\ \frac{F_m(t) - F_g(t)}{m} \end{bmatrix}}_{\mathbf{f}(t)} + \underbrace{\begin{bmatrix} \frac{1}{L(s(t))} \\ 0 \\ 0 \end{bmatrix}}_{\mathbf{g}(t)} u(t). \tag{15}$$

The model described above now can be used as basis to build a more accurate model as described in the upcoming part. According to the description of the dynamic modes proposed in [14] it is possible to individuate five operation modes with five switching variables for the proposed actuator in the context of the application presented above. The state diagram represented in Fig. 3 shows the principle of the switching strategy. If variable $q \in \mathbb{N}$, where $q = 0, 1, 2, 3, 4$ are related to the five dynamic modes of the actuator, then the five phases can be described as follows:

- i. $q = 0$, Latent phase lower limit (as long as $i(t) < i_{on}$): initial conditions $\rightarrow i(0) = 0$; $s(0) = 0$; $v(0) = 0$.
- ii. $q = 1$, Lifting phase (as long as $0 < s(t) < s_{max}0$, $v(t) > 0$): initial conditions $i(0) = i_{on}$; $s(0) = 0$; $v(0) = 0$.
- iii. $q = 2$, Contact phase (as long as $s(t) = s_{max}$ and $\frac{di(t)}{dt} = 0$): initial conditions $i(0) = i(t)$; $s(t) = s_{max}$; $v(t) = 0$.
- iv. $q = 3$, Latent phase upper limit (as long as $i(t) > i_{off}$): initial conditions $\rightarrow i(0) = i(t)$; $s(0) = s_{max}$; $v(0) = 0$.
- v. $q = 4$, Back phase (as long as $v(t) < 0$ and $s(t) > 0$): initial conditions $i(0) = i_{off}$; $s(t) = s_{max}$; $v(t) = 0$.

i_{on}, i_{off} are two constant currents, s_{max} is the position in which the actuator establishes the contact. Moreover, it is possible to introduce another switching variable to identify s_{max} in which the contact phase starts and the time in which the back phase starts. To get a clear representation the following constants are used:

$$K_1 = \frac{N^2 \mu_r \mu_0 A}{l_1}, \quad K_2 = l_1, \quad K_3 = 2\mu_r, \quad K_4 = N^2 \mu_r^2 \mu_0 A$$
$$K_5 = N^2 \mu_r \mu_0 A, \quad K_6 = R_i + R_m, \quad K_7 = R_d + R_m.$$

The resistance R is changing at the position limits. At the lower limit $s(t) = 0$ the resistance is $R = K_6$. At the upper limit it is $R = K_7$.
$q = 0$: *Latent phase lower limit (as long as $i(t) < i_{on}$): initial condition* $\rightarrow i(0) = 0$; $s(0) = 0$; $v(0) = 0$. In this phase the inductance $L(s(t))$ is constant because the position is constant $s = 0$. Because of this, the derivative of the inductance gets zero and the system can be simplified to:

$$\underbrace{\begin{bmatrix} \frac{di(t)}{dt} \\ \frac{ds(t)}{dt} \\ \frac{dv(t)}{dt} \end{bmatrix}}_{\dot{\mathbf{x}}(t)} = \underbrace{\begin{bmatrix} \frac{-(K_6 i(t))}{K_1} \\ 0 \\ 0 \end{bmatrix}}_{\mathbf{f}(t)} + \underbrace{\begin{bmatrix} \frac{1}{K_1} \\ 0 \\ 0 \end{bmatrix}}_{\mathbf{g}(t)} u(t), \tag{16}$$

IOP Conf. Series: Journal of Physics: Conf. Series **783** (2017) 012015 doi:10.1088/1742-6596/783/1/012015

Figure 3. State diagram

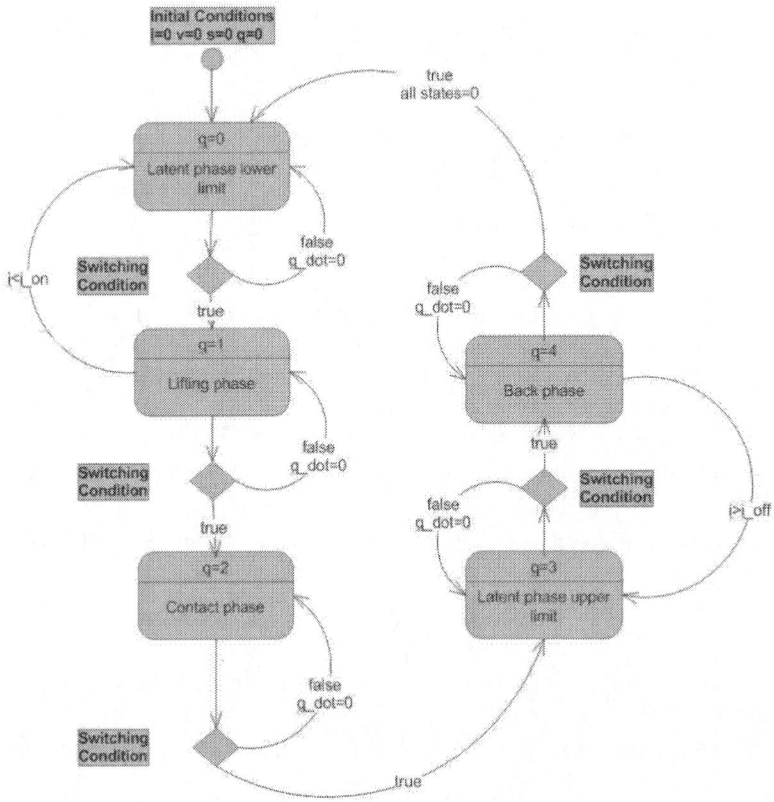

with $L(s(t)) = K_1$.
$q = 1$: *Lifting phase starting with* $i(0) = i_{on}$ *(as long as* $0 < s(t) < s_{max}$, $v(t) > 0$*): initial conditions* $i(0) = i_{on}$; $s(0) = 0$; $v(0) = 0$.

$$\underbrace{\begin{bmatrix} \frac{di(t)}{dt} \\ \frac{ds(t)}{dt} \\ \frac{dv(t)}{dt} \end{bmatrix}}_{\dot{\mathbf{x}}(t)} = \underbrace{\begin{bmatrix} \frac{-(K_6 i(t) + \frac{dL(s(t))}{ds} v(t) i(t))}{L(s(t))} \\ v(t) \\ \frac{F_m(t) - F_g(t)}{m} \end{bmatrix}}_{\mathbf{f}(t)} + \underbrace{\begin{bmatrix} \frac{1}{L(s(t))} \\ 0 \\ 0 \end{bmatrix}}_{\mathbf{g}(t)} u(t), \tag{17}$$

with $L(s(t)) = \frac{K_5}{(K_2 + K_3 s(t))}$, $\frac{dL(s(t))}{ds} = \frac{-2K_4}{(K_2 + K_3 s(t))^2}$ and $F_m = \frac{K_4 i(t)^2}{(K_2 + K_3 s(t))^2}$.
$q = 2$: *Contact phase (as long as* $s(t) = s_{max}$ *and* $\frac{di(t)}{dt} = 0$*): initial conditions* $i(0) = i(t)$;

$s(t) = s_{max}$; $v(t) = 0$.

$$\underbrace{\begin{bmatrix} \frac{di(t)}{dt} \\ \frac{ds(t)}{dt} \\ \frac{dv(t)}{dt} \end{bmatrix}}_{\dot{\mathbf{x}}(t)} = \underbrace{\begin{bmatrix} \frac{-(K_6 i(t)}{L(s(t))} \\ 0 \\ 0 \end{bmatrix}}_{\mathbf{f}(t)} + \underbrace{\begin{bmatrix} \frac{1}{L(s(t))} \\ 0 \\ 0 \end{bmatrix}}_{\mathbf{g}(t)} u(t), \tag{18}$$

with $L(s(t)) = \frac{K_5}{(K_2+K_3 s(t))}$.

$q = 3$: *Latent phase upper limit (as long as $i(t) > i_{off}$): initial conditions $i(0) = i(t)$; $s(0) = s_{max}$; $v(0) = 0$.*

$$\underbrace{\begin{bmatrix} \frac{di(t)}{dt} \\ \frac{ds(t)}{dt} \\ \frac{dv(t)}{dt} \end{bmatrix}}_{\dot{\mathbf{x}}(t)} = \underbrace{\begin{bmatrix} \frac{-(K_7 i(t)}{K8} \\ 0 \\ 0 \end{bmatrix}}_{\mathbf{f}(t)} + \underbrace{\begin{bmatrix} \frac{1}{K_8} \\ 0 \\ 0 \end{bmatrix}}_{\mathbf{g}(t)} u(t). \tag{19}$$

$q = 4$: *Back Phase starting for $i(t) < i_{off}$ (as long as $v(t) < 0$ and $s(t) > 0$): initial condition $i(0) = i_{off}$; $s(t) = s_{max}$; $v(t) = 0$.*

$$\underbrace{\begin{bmatrix} \frac{di(t)}{dt} \\ \frac{ds(t)}{dt} \\ \frac{dv(t)}{dt} \end{bmatrix}}_{\dot{\mathbf{x}}(t)} = \underbrace{\begin{bmatrix} \frac{-(K_7 i(t) + \frac{dL(s(t))}{ds} v(t) i(t))}{L(s(t))} \\ v(t) \\ \frac{F_m(t) - F_g(t)}{m} \end{bmatrix}}_{\mathbf{f}(t)} + \underbrace{\begin{bmatrix} \frac{1}{L(s(t))} \\ 0 \\ 0 \end{bmatrix}}_{\mathbf{g}(t)} u(t), \tag{20}$$

with $L(s(t)) = \frac{K_5}{(K_2+K_3 s(t))}$, $\frac{dL(s(t))}{ds} = \frac{-2K_4}{(K_2+K_3 s(t))^2}$ and $F_m = \frac{K_4 i(t)^2}{(K_2+K_3 s(t))^2}$. All these equations describe the whole hybrid system of a magnet.

5. Validation through simulations

The simulation is done with Matlab/Simulink using the 'Hybrid Equations Toolbox' developed by [15]. The constants used for the simulation are shown in Table 1. To show how the hybrid Kalman Filter is estimating the state, a simulation of the system discussed in Section 4 runs using the Hybrid Equations Toolbox. In parallel, the hybrid Extended Kalman Filter uses the current (plus added white noise) of this simulation as measurement to estimate the position $s(t)$ and the velocity $v(t)$. To show the efficiency and accuracy of this estimation several steps excite the simulation. The simulated state (black line) and estimated state (dashed/dotted red line) states are always shown in the same diagram to give a good possibility of comparison. If the magnet is already lifted and the excitation gets zero, it shows the following behavior. Now the excitation is inverted to a step from 24V to 0V at 0.5 seconds. Now the current shows the behavior as shown in Fig.4. The estimation here is nearly exact and without any delay. This is because the current is the measured state with just a small uncertainty. In Fig.6 the velocity for this kind of step is represented. The velocity is negative for a short time and then zero is reached again because the magnet reaches the lower limit. Finally Fig. 7 shows the switching variable for a step for 0V to 24V. Here the passed operation modes beginning at 0.5 seconds are 2 (Contact Phase), 3 (Latent phase upper limit) and 4 (Back Phase). Afterwards, the system gets back to operation mode 1 and is 'waiting' for an excitation. Concerning the setup parameters of the Kalman Filter, the process noise covariance matrix $\mathbf{R}_w$, the measurement noise variance

Table 1. Constants used for simulation

Constant	Description	Value
g	gravity constant	$9.81\frac{m}{s^2}$
N	number of windings	964
μ_0	vacuum permeability	$1.257e^{-3}\frac{Vs}{Am}$
μ_r	relative permeability	200
A	surface involved in the magnetic flux	$7.6341e^{-5}m^2$
s_{max}	maximum position	$0.5e^{-3}m$
m	mass of magnet armature	$0.34kg$
i_{on}	turn-on current	$0.1A$
i_{off}	cut-off current	$0.02A$
$R = R_i + R_m$	resistance on lower level	114.3Ω
$R = R_d + R_m$	resistance on upper level	120Ω
l_1	average field line length	$22e^{-6}m$
T_s	sample time	$1e^{-4}s$

Figure 4. Current for a step from 24V to 0V at 0.5 seconds

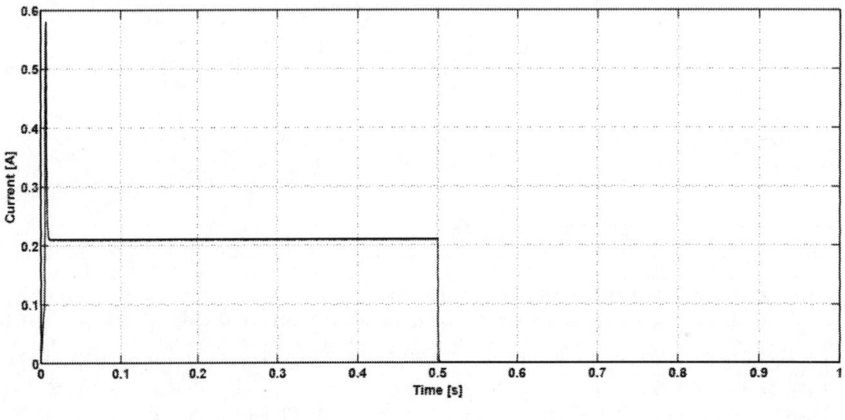

$\mathbf{R}_\zeta$ is and the initial values of the covariance matrix $\mathbf{P}_q(\mathbf{0})$ are chosen equal for each q-model and their values are as follows:

$$\mathbf{R}_w = \begin{bmatrix} 1 & 0 & 0 & 0 \\ 0 & 10^{-10} & 0 & 0 \\ 0 & 0 & 10^{-10} & 0 \\ 0 & 0 & 0 & 10^{-10} \end{bmatrix}, \tag{21}$$

$\mathbf{R}_\zeta = 0.01$ and $\mathbf{P}_q(\mathbf{0}) = \mathbf{0}$. It is possible to observe that matrix $\mathbf{R}_w$ states the precision which which the model is known. In particular, the uncertainty is localised in the first equation in which the nonlinearity is present.

Figure 5. Position for a step from 24V to 0V at 0.5 seconds

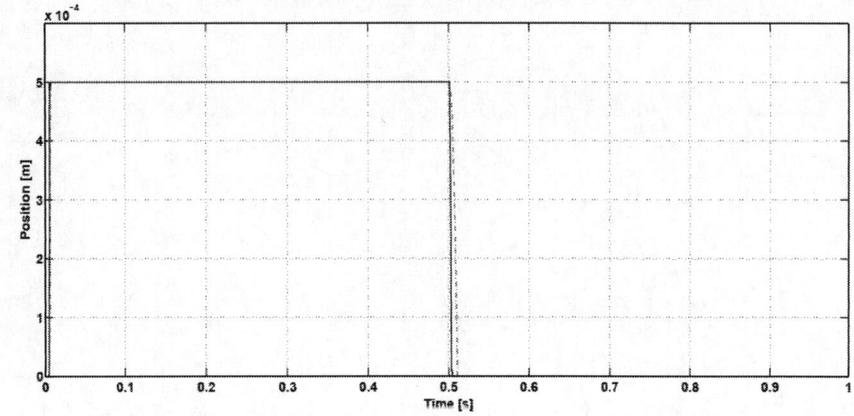

Figure 6. Velocity for a step from 24V to 0V at 0.5 seconds

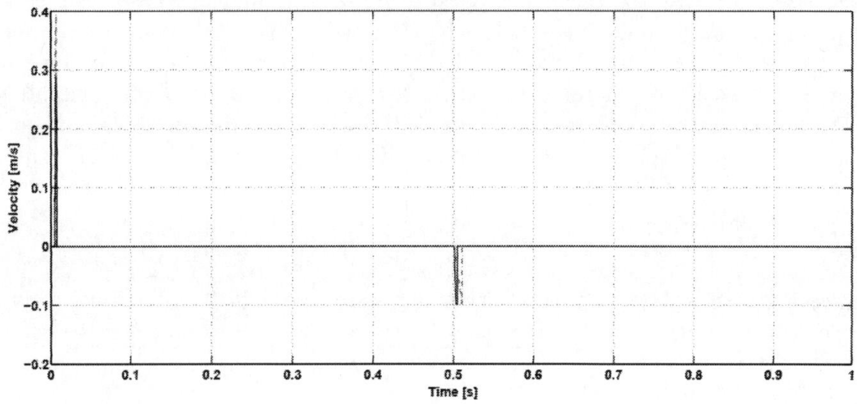

6. Performance of the proposed Observer

A performance evaluation concerning the estimation is provided using a Root-Mean-Square (RMS) Deviation (RMSD). The following RMSD of an estimator $\hat{\theta}$ with respect to an estimated parameter θ is defined as the square root of the mean square error. In case of an unbiased estimator, the square root of the mean square error corresponds to the square root of the variance, known as the standard deviation.

$$RMSD = \sqrt{MSE(\hat{\theta})} = \sqrt{E(\hat{\theta} - \theta)} \tag{22}$$

In a Kalman Filter the presence of a bias is shown to be caused by a correlation between the gain and innovation sequences which are due to the nonlinearity on the measurements, [16]. Using a linearisation, this correlation remains and to obtain an unbiased filter, a modification of the EKF is considered in [17]. The variation proposed in [17] uses a modification of the nonlinear measurement function and is claimed to give unbiased estimates. In our case, the proposed EKF does not consist of a nonlinear output function. In this case indicator RMSD is appropriately applied. In fact, the RMS Deviation (RMSD) or Root-Mean-Square Error (RMSE) is very often

Figure 7. Switching variable q for a step from 24V to 0V at 0.5 seconds

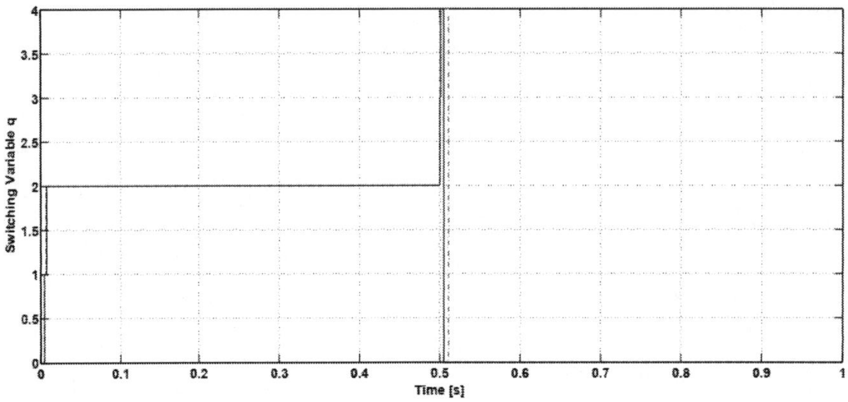

used, in particular in unbiased cases, as a measure of the differences between values predicted by a model or an estimator and the values actually observed. The RMSD represents the sample standard deviation of the differences between predicted values and observed values. These individual differences are called residuals when the calculations are performed over the data sample that was used for estimation, and are called prediction errors when computed out-of-sample. The RMSD serves to aggregate the magnitudes of the errors in predictions for various times into a single measure of predictive power. RMSD is a good measure of accuracy, but only to compare forecasting errors of different models for a particular variable and not between variables, as it is scale-dependent.

Figure 8. RMS Deviation of the Current for a step from 24V to 0V at 0.5 seconds

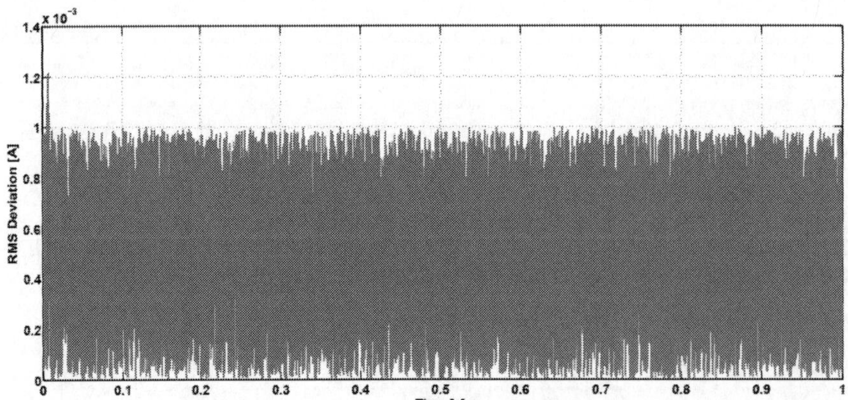

7. Conclusion

The paper deals with an application in which a hybrid extended Kalman Filter is used to estimate state variables in a Pot-Magnet to be used in mechanical systems. The paper proposes a hybrid model to be used in the extended Kalman Filter to monitor the state and to control the

Figure 9. RMS Deviation of the Position for a step from 24V to 0V at 0.5 seconds

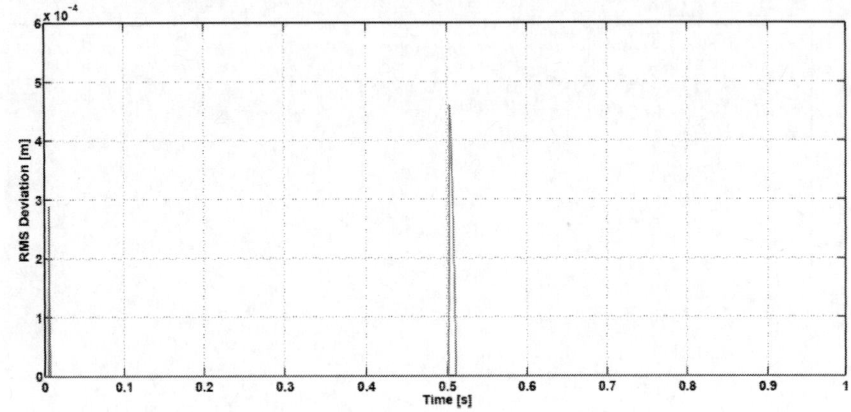

Figure 10. RMS Deviation of the Velocity for a step from 24V to 0V at 0.5 seconds

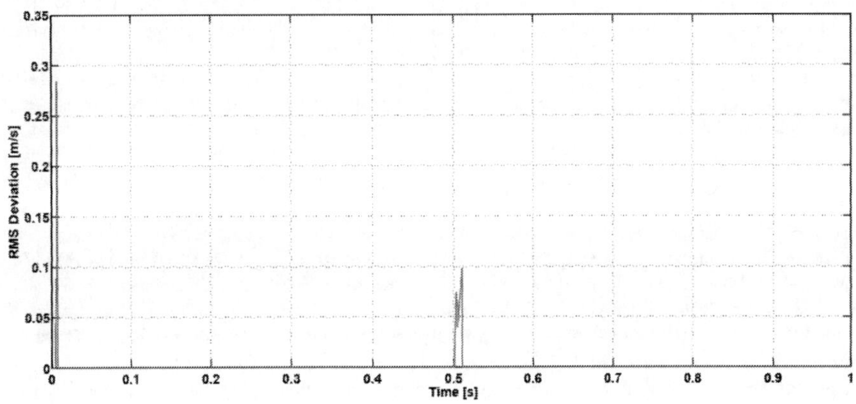

force of the actuator to be applied in manufacturing systems. Validations using simulations are shown and discussed to demonstrate the validity of the approach which proposed five dynamic models to be integrated and coordinated with a switching logic structure. The proposed observer shows a good performance and estimates the states correctly. A possible future work consists of an implementation into a real system and in designing a controller for this system using the proposed hybrid Kalman Filter.

References

[1] Goebel R., Sanfelice R.G. and Teel A 2009 Hybrid dynamical systems, *IEEE Control Systems Magazine*, **29**(2), 28–93.

[2] Sanfelice R.G. 2013 Control of Hybrid Dynamical Systems: An Overview of Recent Advances, *Control of Hybrid Dynamical Systems: An Overview of Recent Advances*, (Wiley).

[3] Mercorelli P. 2012 A Hysteresis Hybrid Extended Kalman Filter as an Observer for Sensorless Valve Control in Camless Internal Combustion Engines, *IEEE Transactions on Industry Applications*, **48**(6), 1940–1949.

[4] Mercorelli P. 2014 An Adaptive and Optimized Switching Observer for Sensorless Control of an Electromagnetic Valve Actuator in Camless Internal Combustion Engines, Robust Adaptive Soft Landing

Figure 11. RMS Deviation of the Switching variable q for a step from 24V to 0V at 0.5 seconds

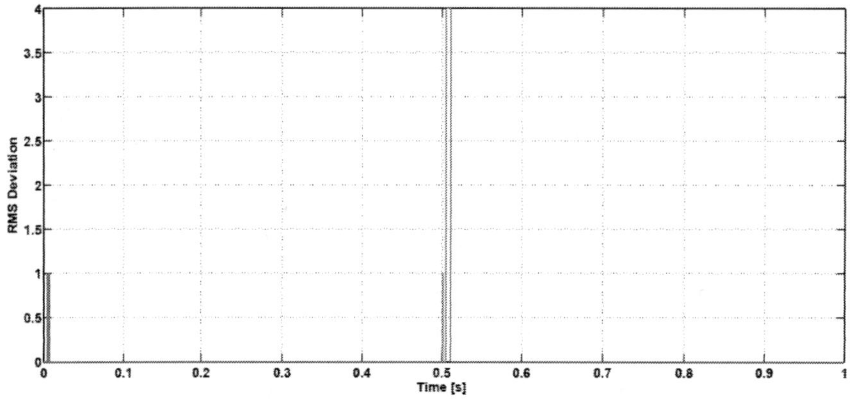

Control of an Electromagnetic Valve Actuator for Camless Engines, *Asian Journal of Control*, **16**(4), 959–973.

[5] Mercorelli P. 2016 Robust Adaptive Soft Landing Control of an Electromagnetic Valve Actuator for Camless Engines, *Asian Journal of Control*, **18**(5), 1299–1312.

[6] Mercorelli P. 2015 A Two-Stage Sliding-Mode High-Gain Observer to Reduce Uncertainties and Disturbances Effects for Sensorless Control in Automotive Applications, *IEEE Transactions on Industrial Electronics* **62**(9), 5929–5940.

[7] Mercorelli P. 2012 A Two-Stage Augmented Extended Kalman Filter as an Observer for Sensorless Valve Control in Camless Internal Combustion Engines, *IEEE Transactions on Industrial Electronics* **59**(11), 4236–4247.

[8] Mercorelli P. 2012 An Anti-Saturating Adaptive Pre-action and a Slide Surface to Achieve Soft Landing Control for Electromagnetic Actuators, *IEEE/ASME Transactions on Mechatronics*, **17**(1), 76–85.

[9] Butzmann S., Melbert J. and Koch A 2000 Sensorless control of electromagnetic actuators for variable valve train, *Proc. SAE Technical Paper 2000-01-1225*.

[10] Tai C. and Tsao T. 2003 Control of an electromechanical actuator for camless engines, *Proceedings of the American control conference* vol 4, 3113–3118.

[11] Szabat K. and Orlowska-Kowalska T. 2008 Performance Improvement of Industrial Drives With Mechanical Elasticity Using Nonlinear Adaptive Kalman Filter, *IEEE Transactions on Industrial Electronics*, **55**(3), 1075–1084.

[12] Mitsantisuk C., Katsura S. and Ohishi K. 2009 Kalman-Filter-Based Sensor Integration of Variable Power Assist Control Based on Human Stiffness Estimation, *IEEE Transactions on Industrial Electronics*, **56**(10), 3897–3905

[13] Won S., Golnaraghi F. and Melek W. 2009 A Fastening Tool Tracking System Using an IMU and a Position Sensor with Kalman Filters and a Fuzzy Expert System, *IEEE Transactions on Industrial Electronics*, **56**(5), 1782–1792.

[14] Kallenbach E., Eick R., Quendt P., Ströhla T, Feindt K, Kallenbach M and Radler O 2012 *Das dynamische Verhalten von Elektromagneten* (Wiesbaden, Germany: Vieweg+Teubner Verlag (Springer)) ISBN 9783834882974-9783834809681.

[15] Sanfelice R.G., Copp D.A. and Nanez P 2014 Hybrid equations (hyeq) toolbox v2.02 URL `https://hybrid.soe.ucsc.edu/software`

[16] Moorman M.J. and Bullock T.E. 1991 Mathematical analysis of bias in the extended kalman filter, *Proceedings of the 30th IEEE Conference on Decision and Control, 1991*, 2733–2737.

[17] Song T. and Speyer J. 1985 A stochastic analysis of a modified gain extended Kalman filter with applications to estimation with bearings only measurements, *IEEE Transactions on Automatic Control*, **30**(10), 940–949, ISSN 0018-9286.

Joint State and Parameter Estimation for Discrete-Time Takagi-Sugeno Model

Krishnan Srinivasarengan, José Ragot, Didier Maquin, Christophe Aubrun

CRAN UMR 7039, Université de Lorraine, Vandoeuvre-lès-Nancy, Cedex, France
CRAN UMR 7039, CNRS, France

E-mail: krishnan.srinivasarengan@univ-lorraine.fr, jose.ragot@univ-lorraine.fr, didier.maquin@univ-lorraine.fr, christophe.aubrun@univ-lorraine.fr

Abstract. Takagi-Sugeno (T-S) models have been popular in analyzing nonlinear systems. Observer designs for T-S models have focused on both cases with measured and unmeasured premise variables. However, the unmeasured premise variables have to be one of the states of the system. If one of the inputs is a premise variable, these approaches are not applicable. A recent literature proposed a joint state and parameter estimation for time-varying unknown parameters in systems of a special type [1]. A discrete-time version of the results is developed in this paper. The modeling approach is described and the stability analysis of the estimation is given using the Lyapunov method with the $\mathbb{L}_2$ approach handling the uncertainties.

1. Introduction

Takagi-Sugeno (T-S) type of systems is a class of quasi-LPV (Linear Parameter Varying) systems whose model structure could depend on inputs, outputs, and states of the system. Observer design for T-S models has been popular for a long time. The sector nonlinearity (SNL) approach is typically used to obtain the T-S equivalent model of a nonlinear system that are an exact representation within a sector. However, the use of this approach gives rise to unmeasured states as premise variables. Since the work of [2], there has been significant interest in developing observers for systems with unmeasured premise variable. These works have been extended to cascaded [3], distributed [4] and other such scenarios. However, there are limited works when it comes to dealing with the case when premise variables contain an unknown input. In this circumstance, the unknown input observer formulations such as [5] wouldn't be applicable.

In [6], an interesting approach was proposed to represent time varying unknown parameters using the sector nonlinearity approach. The estimation of the unknown parameters followed, in spirit, the approaches proposed for observer design with unmeasured premise variables. In [1], this was combined with a T-S system with unmeasured premise variables to develop a joint state and parameter estimation observer. The unknown parameters are represented using the sector nonlinearity approach, assuming that their bounds are known. This leads to a set of weighting functions that depend on estimated parameters and forms a product of estimated weighting functions. Various approaches assume the knowledge of Lipschitz constant that bound the difference between the estimated and actual weighting function. However, the convex sum property along with matrix theory results are used to bound the uncertain components. Further, an $\mathbb{L}_2$ approach is used to guarantee a bounded error on the estimations.

In this work, the discrete-time version of this result is solved. On the outset, the derivation follows similar steps, however, the discrete-time version of some preliminary results (especially the Bounded Real Lemma) makes the task of adapting the existing results to discrete-time a significant challenge. The key contributions are in formulating the problem so as to avoid non-linear matrix inequalities, resolving nonlinear matrix inequalities by simplifying them to linear matrix inequalities (LMIs), developing bounds on the time varying uncertainties and providing the necessary LMI conditions for the convergence of the estimation. The organization is as follows: The Sec.2 provides the preliminaries necessary for the results developed in this work including the definition of model structures. The main results are presented in Sec.3 and an example to illustrate the results are given in 4. The Sec.5 provides some concluding remarks with comments on the way forward.

2. Preliminaries and Problem Formulation

Takagi-Sugeno models are of the form,

$$\begin{aligned} x(k+1) &= \sum_{i=1}^{r} \mu_i(z(k)) \left[A_i x(k) + B_i u(k)\right] \\ y(k) &= Cx(k) \end{aligned} \tag{1}$$

In this work, the output equation is considered to be linear. However, the results can be readily extended to cases with the output equation as,

$$y(k) = \sum_{i=1}^{r} \mu_i(z(k)) \left[C_i x(k) + D_i u(k)\right] \tag{2}$$

Here, $r = 2^{n_p}$, where n_p is the number of premise variables. Further,

$$x \in \mathbb{R}^n, \quad u \in \mathbb{R}^{n_u}, \quad z \in \mathbb{R}^{n_p}, \quad y \in \mathbb{R}^{n_y}$$

and

$$A_i \in \mathbb{R}^{n_x \times n_x}, \quad B_i \in \mathbb{R}^{n_x \times n_u}, \quad C \in \mathbb{R}^{n_x \times n_y}, \forall i \tag{3}$$

$\mu_i(z(k))$ corresponds to the weighting function obtained from the product of corresponding membership functions of the premise variables (see for e.g., [7] for more details). The weighting functions satisfy the convex sum property, such that,

$$\sum_i \mu_i(.) = 1 \qquad \text{and} \qquad 0 \leq \mu_i(.) \leq 1, \ \forall i \tag{4}$$

2.1. Preliminary Results

The following results would be used during the proof of the results in this work,

Lemma 2.1 *[8] The following LMIs, for appropriately suitable dimensions, are equivalent:*

$$A^T P A - Q < 0, \ P > 0 \tag{5}$$

$$\begin{bmatrix} -Q & A^T P \\ PA & -P \end{bmatrix} < 0, \ P > 0 \tag{6}$$

Lemma 2.2 *[9] Consider two matrices X and Y with appropriate dimensions, a time varying matrix $\Delta(k)$ and a positive scalar λ. The following property is verified:*

$$X^T\Delta^T(k)Y + Y^T\Delta(k)X \leq \lambda X^TX + \lambda^{-1}Y^TY \tag{7}$$

for $\Delta^T(k)\Delta(k) \leq I$

Lemma 2.3 *(**Discrete-time Bounded Real Lemma** [10]) For a discrete-time system, of the form,*

$$\begin{aligned} x(k+1) &= Ax(k) + Bu(k) \\ y(k) &= Cx(k) + Du(k) \end{aligned} \tag{8}$$

the Bounded Real Lemma equivalent LMI condition for stability with an $\mathfrak{L}_2$ gain Γ is,

$$P > 0, \begin{bmatrix} A^TPA - P + C^TC & A^TPB + C^TD \\ B^TPA + D^TC & D^TD + B^TPB - \Gamma \end{bmatrix} \leq 0 \tag{9}$$

Lemma 2.4 *(**Schur's Complement**) For a symmetric matrix M, given by,*

$$M = \begin{bmatrix} A & B \\ B^T & C \end{bmatrix}$$

if C is invertible, then the following properties hold:
(i) $M \succ 0$ iff $C \succ 0$ and $A - BC^{-1}B^T \succ 0$
(ii) if $C \succ 0$, then $M \succeq 0$ iff $A - BC^{-1}B^T \succeq 0$.

2.2. System Model Structure
Consider discrete-time Takagi Sugeno type systems of the form,

$$\begin{aligned} x(k+1) &= \sum_{i=1}^{r} \mu_i(x)[A_i(\theta(k))x(k) + B_i(\theta(k))u(k)] \\ y(k) &= Cx(k) \end{aligned} \tag{10}$$

where, $\theta \in \mathbb{R}^{n_\theta}$. In this paper, it is considered that θ is in a bilinear form with the premise variables, which could be the states, outputs, and the inputs of the system, that is:

$$\begin{aligned} A_i(\theta(k)) &= A_i + \sum_{j=1}^{n_\theta} \bar{A}_{ij}\theta_j(k) \\ B_i(\theta(k)) &= B_i + \sum_{j=1}^{n_\theta} \bar{B}_{ij}\theta_j(k) \end{aligned} \tag{11}$$

These matrices could be rewritten by applying sector nonlinearity on the unknown parameter, assuming that the extremum values are known. With this assumption, the time varying matrix is then rewritten as,

$$\begin{aligned} A_i(\theta(k)) &= A_i + \sum_{j=1}^{n_\theta}\sum_{m=1}^{2} \tilde{\mu}_j^m(\theta_j(k))\bar{A}_{ij}\theta_j^m \\ B_i(\theta(k)) &= B_i + \sum_{j=1}^{n_\theta}\sum_{m=1}^{2} \tilde{\mu}_j^m(\theta_j(k))\bar{B}_{ij}\theta_j^m \end{aligned} \tag{12}$$

where, $\tilde{\mu}_j^m(\theta_j(k))$ refers to the weighting function corresponding to mth submodel of θ_j Referring to further steps in [1], the overall system could be rewritten as,

$$\begin{aligned} x(k+1) &= \sum_{i=1}^{r}\sum_{j=1}^{s} \mu_i(x)\tilde{\mu}_j(\theta)[A_{ij}x(k) + B_{ij}u(k)] \\ y(k) &= Cx(k) \end{aligned} \tag{13}$$

where $s \triangleq 2^{n_\theta}$,

$$\forall\ i,j,\ \ A_{ij} = A_i + \sum_{j=1}^{n_\theta} \theta_j^m \bar{A}_j \qquad B_{ij} = B_i + \sum_{j=1}^{n_\theta} \theta_j^m \bar{B}_j \tag{14}$$

where, θ_j^k is the minimum or maximum value of θ_j depending upon the submodel under consideration $(m = 1, 2)$.

2.3. Observer Model Structure

The structure of the observer envisaged to design has a Luenberger like structure for the state estimation part and a first order form for the parameter estimation part, it is given by,

$$\hat{x}(k+1) = \sum_{i=1}^{r}\sum_{j=1}^{s} \mu_i(\hat{x})\tilde{\mu}_j(\hat{\theta})[A_{ij}\hat{x}(k) + B_{ij}u(k) + L_{ij}(y(k) - \hat{y}(k))] \tag{15}$$

$$\hat{\theta}(k+1) = \sum_{i=1}^{r}\sum_{j=1}^{s} \mu_i(\hat{x})\tilde{\mu}_j(\hat{\theta})[K_{ij}(y(k) - \hat{y}(k)) - \eta\theta(k)] \tag{16}$$

The gains $L_{ij} \in \mathbb{R}^{n_x \times n_y}$ and $K_{ij} \in \mathbb{R}^{n_\theta \times n_y}$ are to be estimated with the gain $\eta \in \mathbb{R}^{n_\theta \times n_\theta}$ is chosen. The choice of η shall typically be in the form of a diagonal matrix. The reason for this choice is to avoid the presence of nonlinear matrix inequality terms that arise. In the initial work [1], this was introduced to avoid a marginal stability condition for the error dynamics. As discussed in [11], choosing this reduces the number of variables in the final LMI to solve and hence allows for a computational tractable problem.

3. Results

The main result in this work is stated as the theorem below.

Theorem 3.1 *Given the system model of the form* (13)*, there exists an observer of the form* (15)-(16)*, if there exists* P_0, P_1, R_{ij}, F_{ij}, λ_1, λ_2, λ_3, λ_4, Γ_2^k *(*$\forall i, \forall j, \forall k \in (0,1,2,3)$*), such that,*

$$P_0 > 0 \tag{17}$$

$$P_1 > 0 \tag{18}$$

$$\begin{pmatrix} -P+Q_e & Q_A & \Phi_{ij}^T P & 0 & P_A\mathcal{A} & P_A\mathcal{B} & 0 & 0 \\ Q_A^T & T_{22} & 0 & Q_B^T & 0 & 0 & 0 & 0 \\ P\Phi_{ij} & 0 & -P & 0 & 0 & 0 & 0 & 0 \\ 0 & Q_B^T & 0 & -P & 0 & 0 & P_0\mathcal{A} & P_0\mathcal{B} \\ \mathcal{A}^T P_A & 0 & 0 & 0 & -\lambda_1 I & 0 & 0 & 0 \\ \mathcal{B}^T P_A & 0 & 0 & 0 & 0 & -\lambda_3 I & 0 & 0 \\ 0 & 0 & 0 & \mathcal{A}^T P_0 & 0 & 0 & -\lambda_2 I & 0 \\ 0 & 0 & 0 & \mathcal{B}^T P_0 & 0 & 0 & 0 & -\lambda_4 I \end{pmatrix} < 0 \tag{19}$$

Here, $Q_e \triangleq diag(Q_{e0}, Q_{e1})$ *in line with* $P \triangleq diag(P_0, P_1)$ *and*

$$P_A \triangleq A_{ij}^T P_0 - C^T R_{ij}^T \tag{20}$$

$$T_{22} = \begin{bmatrix} -\Gamma_2^x + (\lambda_1+\lambda_2)E_A^T E_A & 0 & 0 & 0 \\ 0 & -\Gamma_2^x + (\lambda_3+\lambda_4)E_B^T E_B & 0 & 0 \\ 0 & 0 & -\Gamma_2^\theta & 0 \\ 0 & 0 & 0 & -\Gamma_2^{\Delta\theta} \end{bmatrix} \tag{21}$$

$$Q_A = \begin{bmatrix} 0 & 0 & -C^T F_{ij}^T (I+\eta) & -C^T F_{ij}^T \\ 0 & 0 & -\eta^T P_1 (I+\eta) & -\eta^T P_1 \end{bmatrix} \qquad Q_B = \begin{bmatrix} 0 & 0 \\ 0 & 0 \\ 0 & (I+\eta)^T P_1 \\ 0 & P_1 \end{bmatrix}$$

$$\Phi_{ij}^T P = \begin{bmatrix} P_A & -C^T F_{ij}^T \\ 0 & -\eta^T P_1 \end{bmatrix}$$

and,

$$\mathcal{A} = \begin{bmatrix} A_{11} & \dots & A_{rs} \end{bmatrix}, \quad \mathcal{B} = \begin{bmatrix} B_{11} & \dots & B_{rs} \end{bmatrix},$$
$$E_A = \begin{bmatrix} I_{n_x} & .. & I_{n_x} \end{bmatrix}, \quad E_B = \begin{bmatrix} I_{n_u} & .. & I_{n_u} \end{bmatrix}$$
$$\Sigma_A(k) = diag(\delta_{11}(k) I_{n_x}, .., \delta_{rs}(k) I_{n_x}), \ \Sigma_B(k) = diag(\delta_{11}(k) I_{n_u}, .., \delta_{rs}(k) I_{n_u}),$$
$$\delta_{ij}(k) = \mu_i(x)\tilde{\mu}_j(\theta) - \mu_i(\hat{x})\tilde{\mu}_j(\hat{\theta}) \tag{22}$$

The observer gains are given by,

$$K_{ij} = P_1^{-1} F_{ij}, \qquad L_{ij} = P_0^{-1} R_{ij} \tag{23}$$

Proof Given that the system model and observer are not in same form, comparing the error dynamics would be difficult. Hence the original system (13) is represented in an uncertain-like form,

$$x(k+1) = \sum_{i=1}^{r}\sum_{j=1}^{s} \mu_i(\hat{x})\tilde{\mu}_j(\hat{\theta})[(A_{ij} + \Delta A(k))x(k) + (B_{ij} + \Delta B(k))u(k)]$$
$$y(k) = Cx(k) \tag{24}$$

with,

$$\Delta A(k) = \mathcal{A}\Sigma_B(k)E_A, \qquad \Delta B(k) = \mathcal{B}\Sigma_B(k)E_B \tag{25}$$

where, the components are as given in (22). Due to the convex sum property of the weighting functions, we also have,

$$-1 \le \delta_{ij}(k) \le 1, \qquad \Sigma_A^T(k)\Sigma_A(k) \le I \qquad \Sigma_B^T(k)\Sigma_B(k) \le I \tag{26}$$

Defining $e_\theta(k) \triangleq \theta(k) - \hat{\theta}(k)$ and $\Delta\theta(k) \triangleq \theta(k+1) - \theta(k)$, the estimation errors are given as below:

$$e_x(k+1) = \sum_{i=1}^{r}\sum_{j=1}^{s} \mu_i(\hat{x})\tilde{\mu}_j(\hat{\theta})[(A_{ij} - L_{ij}C]e_x(k) + \Delta A(k)x(k) + \Delta B(k)u(k)] \tag{27}$$

$$e_\theta(k+1) = \sum_{i=1}^{r}\sum_{j=1}^{s} \mu_i(\hat{x})\tilde{\mu}_j(\hat{\theta})[\theta(k+1) - K_{ij}Ce_x(k) + \eta\hat{\theta}]$$
$$= \sum_{i=1}^{r}\sum_{j=1}^{s} \mu_i(\hat{x})\tilde{\mu}_j(\hat{\theta})[\Delta\theta(k) + (I+\eta)\theta(k) - K_{ij}Ce_x(k) - \eta e_\theta(k)] \tag{28}$$

The augmented error equation is given by,

$$e_a(k+1) = \sum_{i=1}^{r}\sum_{j=1}^{s} \mu_i(\hat{x})\tilde{\mu}_j(\hat{\theta})[\Phi_{ij}e_a(k) + \Psi_{ij}(k)\tilde{u}] \tag{29}$$

where,

$$\Phi_{ij} = \begin{bmatrix} A_{ij} - L_{ij}C & 0 \\ -K_{ij}C & -\eta \end{bmatrix} \quad \text{and} \quad \Psi_{ij}(k) = \begin{bmatrix} \Delta A(k) & \Delta B(k) & 0 & 0 \\ 0 & 0 & I+\eta & I \end{bmatrix} \tag{30}$$

with $e_a \triangleq [e_x \;\; e_\theta]^T$ and $\tilde{u}(k) \triangleq [x(k) \; u(k) \; \theta(k) \; \Delta\theta(k)]^T$.

Consider a quadratic Lyapunov candidate $V(k) = e_a^T(k)Pe_a(k)$ and an extended stability condition such that,

$$V(k+1) - V(k) + e_a^T(k)Q_e e_a(k) - \tilde{u}^T(k)\Gamma_2\tilde{u}(k) < 0 \tag{31}$$

where Q_e is a scaling factor for the errors that is chosen and $\Gamma_2 = \text{block}(\Gamma_2^x, \Gamma_2^u, \Gamma_2^\theta, \Gamma_2^{\Delta\theta})$ is the $\mathbb{L}_2$ gain value under which the uncertainties ($\tilde{u}(t)$) are expected to be bounded. By applying the discrete-time Bounded Real Lemma (BRL) (Lemma 2.3) we get,

$$\begin{bmatrix} \Phi_{ij}^T P\Phi_{ij} - P + Q_e & \Phi_{ij}^T P\Psi_{ij}(k) \\ (\Phi_{ij}^T P\Psi_{ij}(k))^T & \Psi_{ij}^T(k)P\Psi_{ij}(k) - \Gamma_2 \end{bmatrix} < 0 \tag{32}$$

There are nonlinear terms in this matrix condition. Specifically, the term $\Phi_{ij}P\Phi_{ij}$ is not quadratic as it appears, but it includes cross terms where the index for the first Φ is different from that of the second Φ. However, based on the illustration in Theorem 17 in [12] on non-symmetric terms in the discrete-time models, it could be shown that ignoring the cross terms would not affect the negativity of the inequality condition.

Given the structure of Φ and $\Psi(k)$ in (30), there are likely nonlinear terms arising due to the quadratic structure in (32). At first however, applying the Lemma 2.1 on the $(1,1)$ and $(2,2)$ blocks of (32) leads to,

$$\begin{aligned} \Phi_{ij}^T P\Phi_{ij} - P + Q_e &\equiv \begin{bmatrix} -P + Q_e & \Phi_{ij}^T P \\ P\Phi_{ij} & -P \end{bmatrix} < 0 \\ \Psi_{ij}^T(k)P\Psi_{ij} - \Gamma_2 &\equiv \begin{bmatrix} -\Gamma_2 & \Psi_{ij}^T(k)P \\ P\Psi_{ij}(k) & -P \end{bmatrix} < 0 \end{aligned} \tag{33}$$

Inserting these into (32), we get,

$$\begin{bmatrix} -P + Q_e & \Phi_{ij}^T P\Psi_{ij}(k) & \Phi_{ij}^T P & 0 \\ (\Phi_{ij}^T P\Psi_{ij}(k))^T & -\Gamma_2 & 0 & \Psi_{ij}^T(k)P \\ P\Phi_{ij} & 0 & -P & 0 \\ 0 & P\Psi_{ij}(k) & 0 & -P \end{bmatrix} < 0 \tag{34}$$

In what follows, the aim is to reduce the matrix inequality in (34), which is both time varying and possibly nonlinear, to an LMI. The key steps are as follows:

- Handling the nonlinearity in the matrix entries
  - Enforce a structure on the positive definite matrix P such that, $P = diag(P_0, P_1)$.

  - Rewrite quadratic terms in variables into new variables.
- Handling the time varying terms in the matrix entries
  - Partition the time varying terms so that the Lemma 2.2 could be applied

The diagonal structure enforcing of the Lyapunov function is followed by the following change of variables:

$$R_{ij} = P_0 L_{ij}, \qquad F_{ij} = P_1 K_{ij} \tag{35}$$

This leads to the terms in $(1,2)$, $(1,3)$, and $(2,4)$ blocks in (34) to become,

$$\Phi_{ij}^T P \Psi_{ij}(k) = \begin{bmatrix} (A_{ij}^T P_0 - C^T R_{ij}^T)\Delta A(k) & (A_{ij}^T P_0 - C^T R_{ij}^T)\Delta B(k) & -(C^T F_{ij}^T)(I+\eta) & -C^T F_{ij}^T \\ 0 & 0 & -\eta^T P_1 (I+\eta) & -\eta^T P_1 \end{bmatrix} \tag{36}$$

and,

$$\Phi_{ij}^T P = \begin{bmatrix} A_{ij}^T P_0 - C^T R_{ij}^T & -C^T F_{ij}^T \\ 0 & -\eta^T P_1 \end{bmatrix}$$

$$\Psi_{ij}^T(k) P = \begin{bmatrix} \Delta A^T(k) P_0 & 0 \\ \Delta B^T(k) P_0 & 0 \\ 0 & (I+\eta)^T P_1 \\ 0 & P_1 \end{bmatrix} \tag{37}$$

The terms $\Psi_{ij}(k)$ and $\Phi_{ij}^T P \Psi_{ij}(k)$ in (34) have time varying terms which need to be bounded. To do it, the original error dynamics equation in (29) is split into terms with and without time varying components, that is,

$$\sum_{i=1}^{r} \sum_{j=1}^{s} \mu_i(\hat{x}) \tilde{\mu}_j(\hat{\theta}) \left(Q_{ij} + \mathcal{L}_{ij}(k) + \mathcal{L}_{ij}^T(k) \right) < 0 \tag{38}$$

where,

$$Q_{ij} = \begin{bmatrix} -P + Q_e & Q_A & \Phi_{ij}^T P & 0 \\ Q_A^T & -\Gamma_2 & 0 & Q_B \\ P\Phi_{ij} & 0 & -P & 0 \\ 0 & Q_B^T & 0 & -P \end{bmatrix} \tag{39}$$

with Q_A and Q_B given by,

$$Q_A = \begin{bmatrix} 0 & 0 & -C^T F_{ij}^T (I+\eta) & -C^T F_{ij}^T \\ 0 & 0 & -\eta^T P_1 (I+\eta) & -\eta^T P_1 \end{bmatrix} \qquad Q_B = \begin{bmatrix} 0 & 0 \\ 0 & 0 \\ 0 & (I+\eta)^T P_1 \\ 0 & P_1 \end{bmatrix} \tag{40}$$

And the time varying terms are gathered as,

$$\mathcal{L}_{ij}(k) = \begin{bmatrix} 0 & \mathcal{L}_A(k) & 0 & 0 \\ 0 & 0 & 0 & 0 \\ 0 & 0 & 0 & 0 \\ 0 & \mathcal{L}_B(k) & 0 & 0 \end{bmatrix}$$

$$\mathcal{L}_A(k) = \begin{bmatrix} (A_{ij}^T P_0 - C^T R_{ij}^T)\Delta A(k) & (A_{ij}^T P_0 - C^T R_{ij}^T)\Delta B(k) & 0 & 0 \\ 0 & 0 & 0 & 0 \end{bmatrix}$$

$$\mathcal{L}_B(k) = \begin{bmatrix} P_0 \Delta A(k) & P_0 \Delta B(k) & 0 & 0 \\ 0 & 0 & 0 & 0 \end{bmatrix} \tag{41}$$

and their corresponding transposes. To bound the time varying matrices, the Lemma 2.2 is used. Each of the four time varying components (two in $\mathcal{L}_{ij}(k)$ and two in its transpose) would be bounded by applying the Lemma individually.

We define, for simplified notation purposes, $P_A \triangleq A_{ij}^T P_0 - C^T R_{ij}^T$. We can write each of the summation element in an expanded form by splitting $\mathcal{L}_{ij}(k)$ as follows:

$$\mathcal{L}_{ij}(k) = \begin{bmatrix} 0 & \begin{bmatrix} P_A \Delta A(k) & 0 \\ 0 & 0 \end{bmatrix} & 0 & 0 \\ 0 & 0 & 0 & 0 \\ 0 & 0 & 0 & 0 \\ 0 & 0 & 0 & 0 \end{bmatrix} + \begin{bmatrix} 0 & 0 & 0 & 0 \\ 0 & 0 & 0 & 0 \\ 0 & 0 & 0 & 0 \\ 0 & \begin{bmatrix} P_0 \Delta A(k) & 0 \\ 0 & 0 \end{bmatrix} & 0 & 0 \end{bmatrix}$$
$$+ \begin{bmatrix} 0 & \begin{bmatrix} 0 & P_A \Delta B(k) & 0 \\ 0 & 0 & 0 \end{bmatrix} & 0 & 0 \\ 0 & 0 & 0 & 0 \\ 0 & 0 & 0 & 0 \\ 0 & 0 & 0 & 0 \end{bmatrix} + \begin{bmatrix} 0 & 0 & 0 & 0 \\ 0 & 0 & 0 & 0 \\ 0 & 0 & 0 & 0 \\ 0 & \begin{bmatrix} 0 & P_0 \Delta B(k) & 0 \\ 0 & 0 & 0 \end{bmatrix} & 0 & 0 \end{bmatrix} \tag{42}$$
$$= \mathcal{L}_{A1}(k) + \mathcal{L}_{A2}(k) + \mathcal{L}_{B1}(k) + \mathcal{L}_{B2}(k) \tag{43}$$

The zeros are not of the same size, but have appropriate dimensions. The $\Delta A(k)$ and $\Delta B(k)$ contain the time varying terms $\Sigma_A(k)$ and $\Sigma_B(k)$ as given in (26). To do this, the matrices need to be split making use of the components as in (25), so that we get,

$$\mathcal{L}_{A1} = \begin{pmatrix} P_A \mathcal{A} \\ 0 \end{pmatrix} \Sigma_A(k) \begin{pmatrix} 0 & E_A & 0 \end{pmatrix} \qquad \mathcal{L}_{B1} = \begin{pmatrix} P_A \mathcal{B} \\ 0 \end{pmatrix} \Sigma_B(k) \begin{pmatrix} 0 & E_B & 0 \end{pmatrix} \tag{44}$$

$$\mathcal{L}_{A2} = \begin{pmatrix} 0 \\ P_0 \mathcal{A} \\ 0 \end{pmatrix} \Sigma_A(k) \begin{pmatrix} 0 & E_A & 0 \end{pmatrix} \qquad \mathcal{L}_{B2} = \begin{pmatrix} 0 \\ \mathcal{B}^T P_0 \\ 0 \end{pmatrix} \Sigma_B(k) \begin{pmatrix} 0 & E_B & 0 \end{pmatrix} \tag{45}$$

And this could be further extended to their corresponding transposes. For each of the matrix and transpose pair, we can apply the Lemma 2.2, so that,

$$\mathcal{L}_{A1}(k) + \mathcal{L}_{A1}^T(k) \leq \lambda_1^{-1} \begin{pmatrix} P_A \mathcal{A} \\ 0 \end{pmatrix} \begin{pmatrix} \mathcal{A}^T P_A & 0 \end{pmatrix} + \lambda_1 \begin{pmatrix} 0 \\ E_A^T \\ 0 \end{pmatrix} \begin{pmatrix} 0 & E_A & 0 \end{pmatrix}$$
$$\leq \begin{pmatrix} \begin{bmatrix} \lambda_1^{-1} P_A \mathcal{A} \mathcal{A}^T P_A & 0 \\ 0 & 0 \end{bmatrix} & 0 & 0 & 0 \\ 0 & 0 & 0 & 0 \\ 0 & 0 & 0 & 0 \\ 0 & 0 & 0 & 0 \end{pmatrix} + \begin{pmatrix} 0 & 0 & 0 & 0 \\ 0 & \begin{bmatrix} \lambda_1 E_A^T E_A & 0 \\ 0 & 0 \end{bmatrix} & 0 & 0 \\ 0 & 0 & 0 & 0 \\ 0 & 0 & 0 & 0 \end{pmatrix} \tag{46}$$

$$\mathcal{L}_{B1}(k) + \mathcal{L}_{B1}^T(k) \leq \begin{pmatrix} \begin{bmatrix} \lambda_3^{-1} P_A \mathcal{B} \mathcal{B}^T P_A & 0 \\ 0 & 0 \end{bmatrix} & 0 & 0 & 0 \\ 0 & 0 & 0 & 0 \\ 0 & 0 & 0 & 0 \\ 0 & 0 & 0 & 0 \end{pmatrix} + \begin{pmatrix} 0 & 0 & 0 & 0 \\ 0 & \begin{bmatrix} 0 & 0 & 0 \\ 0 & \lambda_3 E_B^T E_B & 0 \\ 0 & 0 & 0 \end{bmatrix} & 0 & 0 \\ 0 & 0 & 0 & 0 \\ 0 & 0 & 0 & 0 \end{pmatrix} \tag{47}$$

Similarly,

$$\mathcal{L}_{A2}(k) + \mathcal{L}_{A2}^T(k) \leq \begin{pmatrix} 0 & 0 & 0 & 0 \\ 0 & 0 & 0 & 0 \\ 0 & 0 & 0 & 0 \\ 0 & 0 & 0 & \begin{bmatrix} \lambda_2^{-1} P_0 \mathcal{A} \mathcal{A}^T P_0 & 0 \\ 0 & 0 \end{bmatrix} \end{pmatrix} + \begin{pmatrix} 0 & 0 & 0 & 0 \\ 0 & \begin{bmatrix} \lambda_2 E_A^T E_A & 0 \\ 0 & 0 \end{bmatrix} & 0 & 0 \\ 0 & 0 & 0 & 0 \\ 0 & 0 & 0 & 0 \end{pmatrix} \tag{48}$$

$$\mathcal{L}_{B2}(k) + \mathcal{L}_{B2}^T(k) \leq \begin{pmatrix} 0 & 0 & 0 & 0 \\ 0 & 0 & 0 & 0 \\ 0 & 0 & 0 & 0 \\ 0 & 0 & 0 & \begin{bmatrix} \lambda_4^{-1} P_0 \mathcal{B} \mathcal{B}^T P_0 & 0 \\ 0 & 0 \end{bmatrix} \end{pmatrix} + \begin{pmatrix} 0 & 0 & 0 & 0 \\ 0 & \begin{bmatrix} 0 & 0 & 0 \\ 0 & \lambda_4 E_B^T E_B & 0 \\ 0 & 0 & 0 \end{bmatrix} & 0 & 0 \\ 0 & 0 & 0 & 0 \\ 0 & 0 & 0 & 0 \end{pmatrix} \tag{49}$$

The zeros inside the matrices are of different sizes and so the position of the nonzero elements are not identical as visible. The overall bound time varying would be,

$$\mathcal{L}_{ij}(k) + \mathcal{L}_{ij}^T(k) \leq \begin{pmatrix} \begin{bmatrix} \mathcal{L}^1 & 0 \\ 0 & 0 \end{bmatrix} & 0 & 0 & 0 \\ 0 & \mathcal{L}^2 & 0 & 0 \\ 0 & 0 & 0 & 0 \\ 0 & 0 & 0 & \begin{bmatrix} \mathcal{L}^3 & 0 \\ 0 & 0 \end{bmatrix} \end{pmatrix} \tag{50}$$

with,

$$\mathcal{L}^1 = \lambda_1^{-1} P_A \mathcal{A} \mathcal{A}^T P_A + \lambda_3^{-1} P_A \mathcal{B} \mathcal{B}^T P_A, \tag{51}$$

$$\mathcal{L}^2 = \begin{bmatrix} (\lambda_1 + \lambda_2) E_A^T E_A & 0 & 0 & 0 \\ 0 & (\lambda_3 + \lambda_4) E_B^T E_B & 0 & 0 \\ 0 & 0 & 0 & 0 \\ 0 & 0 & 0 & 0 \end{bmatrix} \tag{52}$$

$$\mathcal{L}^3 = \lambda_2^{-1} P_0 \mathcal{A} \mathcal{A}^T P_0 + \lambda_4^{-1} P_0 \mathcal{B} \mathcal{B}^T P_0 \tag{53}$$

This would mean, we could rewrite the matrices of (38) with bounds as ($\forall\ i, j$),

$$Q_{ij} + \mathcal{L}_{ij}(k) + \mathcal{L}_{ij}^T(k) \leq Q_m \tag{54}$$

$$Q_m = \begin{bmatrix} \begin{pmatrix} -P_0 + Q_{e0} + \mathcal{L}^1 & 0 \\ 0 & -P_1 + Q_{e1} \end{pmatrix} & Q_A & \Phi_{ij}^T P & 0 \\ Q_A^T & T_{22} & 0 & Q_B^T \\ P\Phi_{ij} & 0 & \begin{pmatrix} -P_0 & 0 \\ 0 & -P_1 \end{pmatrix} & 0 \\ 0 & Q_B^T & 0 & \begin{pmatrix} -P_0 + \mathcal{L}^3 & 0 \\ 0 & -P_1 \end{pmatrix} \end{bmatrix} \tag{55}$$

Here, $Q_e \triangleq \text{diag}(Q_{e0}, Q_{e1})$ and

$$T_{22} = \begin{bmatrix} -\Gamma_2^x + (\lambda_1 + \lambda_2) E_A^T E_A & 0 & 0 & 0 \\ 0 & -\Gamma_2^u + (\lambda_3 + \lambda_4) E_B^T E_B & 0 & 0 \\ 0 & 0 & -\Gamma_2^\theta & 0 \\ 0 & 0 & 0 & -\Gamma_2^{\Delta\theta} \end{bmatrix} \tag{56}$$

The nonlinear terms of $\mathcal{L}^1$ and $\mathcal{L}^3$ could be converted to linear terms using Schur's complement (Lemma 2.4). The two terms in each of these two entries would contribute to four matrices, that is,

$$\lambda_1^{-1} P_A \mathcal{A} \mathcal{A}^T P_A < 0 \Rightarrow 0 - (P_A \mathcal{A}) \lambda_1^{-1} (\mathcal{A}^T P_A) < 0 \Rightarrow \begin{bmatrix} 0 & P_A \mathcal{A} \\ \mathcal{A}^T P_A & -\lambda_1 I \end{bmatrix} \tag{57}$$

The same steps would lead to the other three nonlinear terms to give an overall matrix condition such that,

$$\left[\begin{array}{c|c} \begin{pmatrix} \begin{bmatrix} -P_0 + Q_{e0} & 0 \\ 0 & -P_1 + Q_{e1} \end{bmatrix} & Q_A & \Phi_{ij}^T P & 0 \\ Q_A^T & T_{22} & 0 & Q_B^T \\ P\Phi_{ij} & 0 & \begin{bmatrix} -P_0 & 0 \\ 0 & -P_1 \end{bmatrix} & 0 \\ 0 & Q_B^T & 0 & \begin{bmatrix} -P_0 & 0 \\ 0 & -P_1 \end{bmatrix} \end{pmatrix} & \begin{pmatrix} P_A \mathcal{A} & P_A \mathcal{B} & 0 & 0 \\ 0 & 0 & 0 & 0 \\ 0 & 0 & 0 & 0 \\ 0 & 0 & P_0 \mathcal{A} & P_0 \mathcal{B} \end{pmatrix} \\ \hline \begin{pmatrix} \mathcal{A}^T P_A & 0 & 0 & 0 \\ \mathcal{B}^T P_A & 0 & 0 & 0 \\ 0 & 0 & 0 & \mathcal{A}^T P_0 \\ 0 & 0 & 0 & \mathcal{B}^T P_0 \end{pmatrix} & \begin{pmatrix} -\lambda_1 I & 0 & 0 & 0 \\ 0 & -\lambda_3 I & 0 & 0 \\ 0 & 0 & -\lambda_2 I & 0 \\ 0 & 0 & 0 & -\lambda_4 I \end{pmatrix} \end{array}\right] \tag{58}$$

or in a simplified form, it leads to (19). Hence the proof. $\blacksquare$

4. Simulation Example

Consider a simple second order nonlinear system,

$$\begin{aligned} x_1(k+1) &= -0.5x_1^2(k) + 0.5x_2(k) + x_1(k)\theta(k) \\ x_2(k+1) &= 0.8x_1(k) + (1 - 0.5\theta)u(k) \\ y(k) &= x_1(k) \end{aligned} \tag{59}$$

Considering that the state $x_1(k) \in [0, 1]$ and the $\theta(k) \in [-1, 1]$ and consider $z \triangleq x_1$ as the premise variables, the system matrices for (13) are given by,

$$A_{11} = \begin{bmatrix} -1 & 0.5 \\ 0.8 & 0 \end{bmatrix} \; A_{12} = \begin{bmatrix} 1 & 0.5 \\ 0.8 & 0 \end{bmatrix} \; A_{21} = \begin{bmatrix} -1.5 & 0.5 \\ 0.8 & 0 \end{bmatrix} \; A_{11} = \begin{bmatrix} 0.5 & 0.5 \\ 0.8 & 0 \end{bmatrix} \tag{60}$$

and with

$$B_{11} = B_{13} = \begin{bmatrix} 0 \\ 1 \end{bmatrix} \qquad B_{21} = B_{22} = \begin{bmatrix} 0 \\ 0.5 \end{bmatrix} \tag{61}$$

The weighting functions are given by,

$$\mu_1(\hat{x}) = 1 - \hat{x}, \; \mu_2(\hat{x}) = \hat{x}, \; \tilde{\mu}_1(\hat{\theta}) = \frac{1 - \hat{\theta}}{2}, \; \tilde{\mu}_2(\hat{\theta}) = \frac{\hat{\theta} + 1}{2} \tag{62}$$

For developing the observer, the aspects were considered:

- The value of η was fixed at -0.995. Apart from avoiding the nonlinear matrix term, this would also avoid Φ_{ij} to become marginal stable in (30).
- A condition to ensure that the value of K_{ij} are sufficiently larger than that of η as discussed in [13], the following LMI condition was introduced

$$F_{ij} > P_1 \eta$$

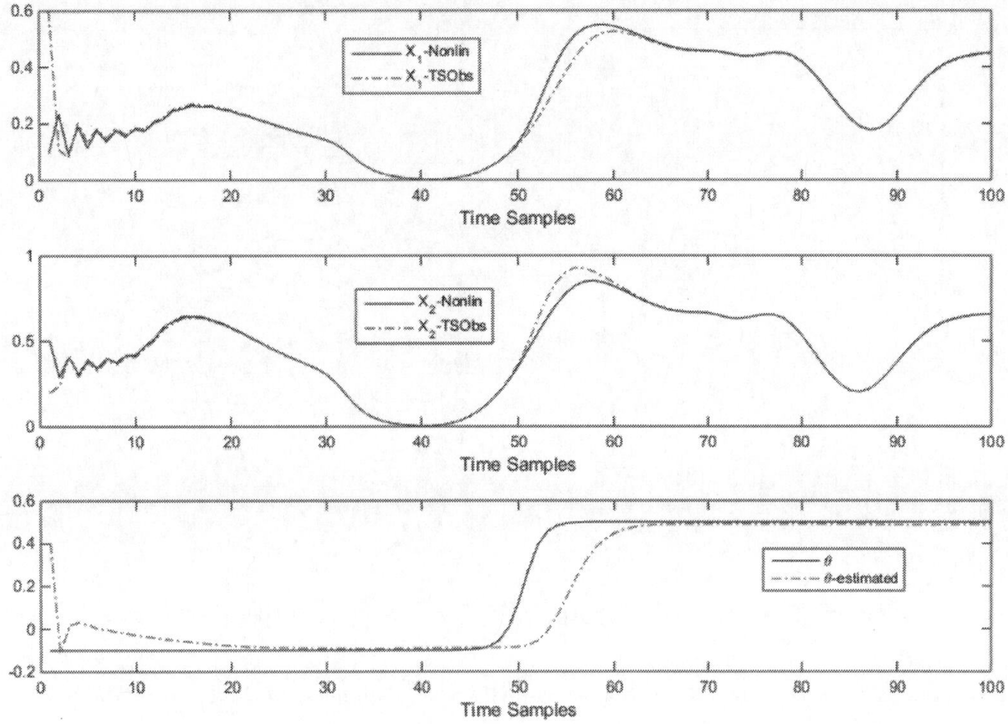

Figure 1. States and the parameter of the system and their estimates

- The chosen values for $Q_e = diag(Q_{e0}Q_{e1})$ as in (34) are given by

$$Q_{e0} = \begin{bmatrix} 0.001 & 0 \\ 0 & 0.001 \end{bmatrix}, \; Q_{e1} = 0.001$$

The resultant observer gains were as follows:

$$L_{11} = \begin{bmatrix} -1 \\ 0.8 \end{bmatrix} \; L_{12} = \begin{bmatrix} 1 \\ 0.8 \end{bmatrix} \; L_{21} = \begin{bmatrix} -1.5 \\ 0.8 \end{bmatrix} \; L_{22} = \begin{bmatrix} 0.5 \\ 0.8 \end{bmatrix} \quad \text{and} \;\; K_{ij} = 1.014 \;\forall i, j \qquad (63)$$

The state and the parameter estimation results are shown in the Fig. 1. As could be seen, when there is an error in the state estimates, the parameter estimation component does not take corrective actions. Only when the state estimates settle down does the parameter estimation component kicks in, which is visible in the form of the delay in the parameter estimation. The input used for the simulation as well as the variation of the four weighting functions during the simulation are shown in the Fig.2.

5. Concluding Remarks

This paper provided the results for a joint state and parameter estimation observer for a discrete time Takagi Sugeno models. The techniques illustrated to handle the nonlinear matrix inequalities and time varying terms could be useful in similar discrete-time version implementation of other results. A proof for the results were given as well and a simple academic example illustrated the results. In future, one exploration could be to use piece-wise Lyapunov

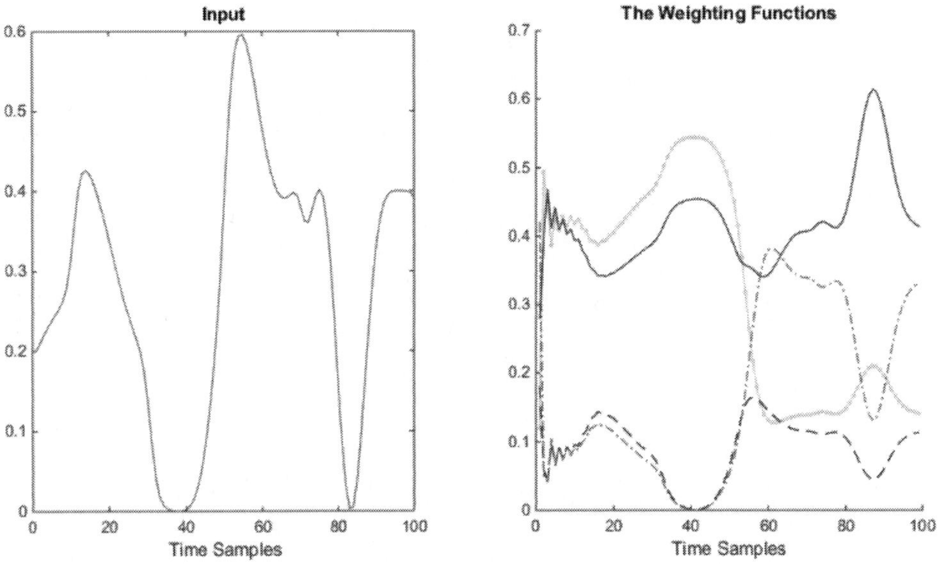

Figure 2. States and the parameter of the system and their estimates

function to avoid the problems with the nonlinearity of matrix inequalities. Further, these results could be applied to some system models of practical systems.

References

[1] Bezzaoucha S, Marx B, Maquin D and Ragot J 2013 State and parameter estimation for nonlinear systems: a Takagi-Sugeno approach *American Control Conference, ACC 2013, Washington, DC, USA, June 17-19*
[2] Bergsten P, Palm R and Driankov D 2001 Fuzzy observers. *FUZZ-IEEE* pp 700–703
[3] Lendek Z, Babuška R, De Schutter B *et al.* 2009 *IEEE Transactions on Fuzzy Systems* **17** 641–653
[4] Lendek Z, Babuška R and De Schutter B 2011 *Fuzzy Sets and Systems* **174** 1–30
[5] Ichalal D, Marx B, Ragot J and Maquin D 2009 State and unknown input estimation for nonlinear systems described by Takagi-Sugeno models with unmeasurable premise variables *17th Mediterranean Conference on Control and Automation MED* (IEEE) pp 217–222
[6] Bezzaoucha S, Marx B, Maquin D and Ragot J 2013 State and parameter estimation for time-varying systems: a Takagi-Sugeno approach *5th Symposium on System Structure and Control, IFAC Joint Conference 2013 SSSC, TDS, FDA*
[7] Tanaka K and Wang H O 2004 *Fuzzy control systems design and analysis: a linear matrix inequality approach* (John Wiley & Sons)
[8] Kruszewski A 2006 *Lois de commande pour une classe de modèles non linéaires sous la forme Takagi-Sugeno: Mise sous forme LMI* Ph.D. thesis Université de Valenciennes et du Hainaut-Cambresis
[9] Zhou K and Khargonekar P P 1988 *Systems & Control Letters* **10** 17–20
[10] de Souza C E and Xie L 1992 *Systems & Control Letters* **18** 61–71
[11] Srinivasarengan K, Ragot J, Maquin D and Aubrun C 2016 Nonlinear joint state-parameter observer for vav damper position estimation *3rd Conference on Control and Fault-Tolerant Systems, SysTol 2016*
[12] Blanco Y 2001 *STABILISATION DES MODELES TAKAGI-SUGENO ET LEUR USAGE POUR LA COMMANDE DES SYSTEMES NON LINEAIRES* Ph.D. thesis Université des Sciences et Technologies de Lille
[13] Srinivasarengan K, Ragot J, Maquin D and Aubrun C 2016 Takagi-sugeno model based nonlinear parameter estimation in air handling units *4th IFAC International Conference on Intelligent Control and Automation Sciences, ICONS 2016*

Performance comparison of attitude determination, attitude estimation, and nonlinear observers algorithms

M.A. SI MOHAMMED[1], **H. BOUSSADIA**[2], **A. BELLAR**[1], **A. ADNANE**[1]

[1]Département de recherche en Mécanique Spatiale, Centre de Développement des Satellites (CDS), BP 4065 Ibn Rochd USTO Oran, Algeria

[2]USTO-MB University, Dept of Electronic, LARESI Laboratory, BP 1505 El Mnaouar Bir El Djir Oran, Algeria.

masimohammed@cds.asal.dz, halima.boussadia@univ-usto.dz, abellar@cds.asal.dz, aadnane@cds.asal.dz.

Abstract. This paper presents a brief synthesis and useful performance analysis of different attitude filtering algorithms (attitude determination algorithms, attitude estimation algorithms, and nonlinear observers) applied to Low Earth Orbit Satellite in terms of accuracy, convergence time, amount of memory, and computation time. This latter is calculated in two ways, using a personal computer and also using On-board computer 750 (OBC 750) that is being used in many SSTL Earth observation missions. The use of this comparative study could be an aided design tool to the designer to choose from an attitude determination or attitude estimation or attitude observer algorithms. The simulation results clearly indicate that the nonlinear Observer is the more logical choice.

1. Introduction

The objective of the attitude filtering algorithms is to calculate the satellite orientation, using measurement data of satellite sensors such as star sensor, sun sensor, infrared horizon sensor, gyroscope and the magnetometer, etc. Among these, Sun sensors or star trackers are very used in small satellites because of the limitations of small satellite missions in terms of power and structure [1].

There are many algorithms for determining the attitude of the satellite. These algorithms are divided into three families: attitude determination algorithms (such as TRIAD, Q-Method, Quest, Least Squares, etc.), attitude estimation algorithms (Kalman Filter, H-infinity Filter, Particle Filter, etc.), and nonlinear observers (Sliding Mode Observer, Luenberger observer, Backstepping observer, etc.). Each of these algorithms is able to determinate the attitude, and it has an advantages and disadvantages.

The attitude determination algorithms have an advantage of low execution time. This is due to its independence of dynamic models, but they are able to estimate only the attitude but not other system states such as angular rate, biases and noise measurement (not presented here). So, what do we do if the angular rates are also calculated from angular sensor measurements? Because we can't use the attitude determination algorithms to estimate this system state, to solve this problem we use the attitude estimation algorithms.

The attitude estimation algorithms are characterized by good performances in term of accuracy, but they have a high time consuming because its structure uses the dynamic models. This point presents a disadvantage in comparison with attitude determination algorithms. In addition, the attitude estimation algorithms are able to estimate the attitude and the angular rate using the sensors installed in the

IOP Conf. Series: Journal of Physics: Conf. Series **783** (2017) 012017 doi:10.1088/1742-6596/783/1/012017

satellite body. Among these sensors, the gyroscope is a sensor which provides the angular rate measurements but it is costly and heavy. Therefore, this sensor presents a problem in small satellites. To eliminate the use of this sensor, and to solve the problem of the time consuming, we use nonlinear observers such as sliding mode observer.

The sliding mode observers are helpful for micro satellite when gyro is unavailable or gyro fails, and they have robustness proprieties similar to sliding mode controllers. They are characterized by high performances in terms of time consuming and accuracy at the same time with respect to attitude determination and attitude estimation algorithms.

The attitude determination and control system (ADCS) is the more important subsystems of satellite. The main aim of the designer of this subsystem is the choice of the best algorithm to extract the full attitude of spacecraft (Euler angles or quaternions or Rodriguez parameters and angular velocities) in term of accuracy or execution time or both, which depends on the mission requirement during the satellite design. Therefore, many questions will be asked, among the algorithms that are presented at the previous paragraph, which algorithm must be used? how does the designer choose this algorithm? and why do we use nonlinear observers while it is possible to use attitude estimation or attitude determination algorithms? Therefore, it is important to give a comparative study between the three filtering algorithms and help the ADCS designer to choose the best algorithm.

[2-3] gave a comparative study between attitude determination and estimation algorithms and did not discuss of nonlinear observers.

In this paper, we give a comparative performance analysis between attitude determination algorithms (Least Squares, Quest method and Optimized TRIAD), attitude estimation algorithms (Extended Kalman Filter and Unscented Kalman Filter), and nonlinear observer (Sliding Mode Observer based) to be applied to Low Earth Orbit satellite, we choose those particular algorithms because they are the most applied in spatial field. The selected algorithms are compared to find the required accuracy, computation time, convergence time, and amount memory, the software implementation of these algorithms is executed in a personal computer and in the On-board Computer 750 (OBC 750) that was developed in SSTL to function as the primary flight computer for application requiring processing On-board.

The presented study will help the designer to determine which type of algorithms should be used to extract spacecraft attitude.

The paper is organized as follows. Section 2 presents the dynamic and kinematic models used for the satellite. Section 3 presents the observability test of our system. In the section 4, 5, and 6, we describe the design of the different algorithms which are presented at the beginning. In next section, we present the simulation results. Finally, the conclusion of this paper is presented in section 8.

2. Spacecraft attitude model

The dynamic and the kinematic of the satellite are governed by the following equation system as [4-5],

$$\dot{\mathbf{x}} = \mathbf{f}(\mathbf{x}, \mathbf{t}) + \mathbf{B}\,\mathbf{u} \tag{1.1}$$
$$\mathbf{y} = \mathbf{H}\,\mathbf{x} \tag{1.2}$$

where,

$$\mathbf{x}(\mathbf{t}) = [q_1, q_2, q_3, q_4, \omega_x, \omega_y, \omega_z]^T \tag{2}$$

$$\mathbf{f(x,h)} = \begin{bmatrix} 0.5(\omega_{oz}q_2 - \omega_{oy}q_3 + \omega_{ox}q_4) \\ 0.5(-\omega_{oz}q_1 + \omega_{ox}q_3 + \omega_{oy}q_4) \\ 0.5(\omega_{oy}q_1 - \omega_{ox}q_2 + \omega_{oz}q_4) \\ 0.5(-\omega_{ox}q_1 - \omega_{oy}q_2 - \omega_{oz}q_3) \\ I_x^{-1}(N_x - (I_z - I_y)\omega_y\omega_z - \omega_y h_z + \omega_z h_y) \\ I_y^{-1}(N_y - (I_x - I_z)\omega_x\omega_z + \omega_x h_z - \omega_z h_x) \\ I_y^{-1}(N_z - (I_y - I_x)\omega_x\omega_y - \omega_x h_y + \omega_y h_x) \end{bmatrix} \quad (3)$$

$\mathbf{I} = [I_x \quad I_y \quad I_z]$ moment of inertia of spacecraft;
$\boldsymbol{\omega}_B^I = [\omega_x \quad \omega_y \quad \omega_z]^T$ angular velocity vector in the inertial frame;
$\mathbf{q} = [q_1 q_2 q_3 q_4]$ quaternion;
$\boldsymbol{\omega}_S^O = [\omega_{ox} \quad \omega_{oy} \quad \omega_{oz}]^T$ angular velocity vector in the orbital reference frame;
$\mathbf{h} = [h_x \quad h_y \quad h_z]^T$angular moment vector;
$\mathbf{N_{ext}} = [N_x \quad N_y \quad N_z]^T$external disturbance torque vector;
$\mathbf{B} = [\mathbf{0}^{4\times3} \ \mathbf{I}^{3\times3}]^T$ control matrix;
$\mathbf{U} = -\dot{\mathbf{h}}$ control input torque;
It assumed that the quaternion measurements are directly available $\mathbf{H} = [\mathbf{I}^{4\times4} \ \mathbf{0}^{4\times3}]$.

3. Observability test

The observability concept is related to dynamic systems. It has a fundamental importance when studying attitude determination and estimation algorithms that we will presented in the following sections. For that reason we introduce this concept in what follows.
The nonlinear observability is tied to the Lie derivative. The observability test of our system is described as follows.
Firstly, we compute the matrix **G** of Lie derivative,

$$\mathbf{G} = \begin{bmatrix} L_f^0(h_1) & \dots & L_f^0(h_p) \\ \dots & \dots & \dots \\ L_f^{n-1}(h_1) & \dots & L_f^{n-1}(h_p) \end{bmatrix} \quad (4)$$

Where, $\mathbf{h(x)} = [q_1 \quad q_2 \quad q_3 \quad q_4]^T$ is the measurement equation, L_f is the Lie derivative, p is the size of the measurement equation, and n is the size of the state vector.
Then, the gradients operator on matrix **G** is calculated as follow,

$$\mathbf{dG} = \begin{bmatrix} dL_f^0(h_1) & \dots & dL_f^0(h_p) \\ \dots & \dots & \dots \\ dL_f^{n-1}(h_1) & \dots & dL_f^{n-1}(h_p) \end{bmatrix} \quad (5)$$

The system is observable if the matrix **dG** has rank n (seven).

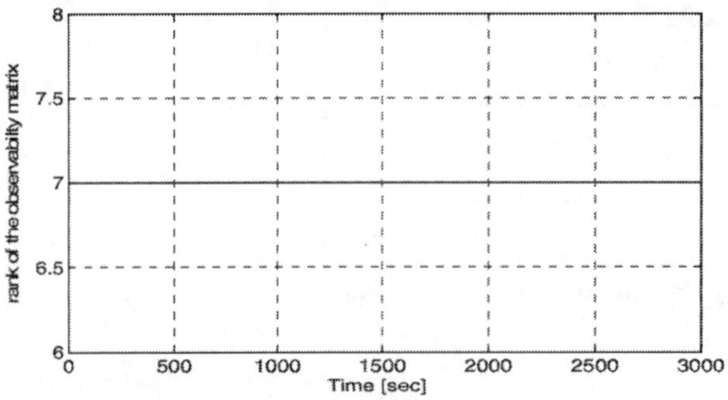

Figure 1. Rank of the observability matrix**dG**.

Figure 1 shows that the rank of the matrix **dG** is equal to seven. Therefore, our system is observable.

4. Attitude determination algorithms

The objective of the attitude determination algorithms is to calculate the space orientation of the satellite based on sensors that can give measurements of known quantity (magnetic field, direction of the sun, direction of the star, etc) in the form of three dimensional vectors [6].
The three known algorithms are: Least Squares, Quest method and Optimized TRIAD.

4.1. Least Squares

Consider the following nonlinear system

$$\dot{\mathbf{x}} = \mathbf{f}(\mathbf{x},\mathbf{t}) + \mathbf{B}\,\mathbf{u}(\mathbf{t}) \tag{6.1}$$

$$\mathbf{y} = \mathbf{H}\mathbf{x} + \mathbf{v} \tag{6.2}$$

Now define $\boldsymbol{\epsilon}_{\mathbf{y}}$ as the difference between the noisy measurements and the vector $\mathbf{H}\hat{\mathbf{x}}$ [7]

$$\boldsymbol{\epsilon}_{\mathbf{y}} = \mathbf{y} - \mathbf{H}\hat{\mathbf{x}} \tag{7}$$

$\hat{\mathbf{x}}$ represents the state estimates.
$\boldsymbol{\epsilon}_{\mathbf{y}}$ is called the measurement residual. The most probable value of the vector **x** is the vector $\hat{\mathbf{x}}$ that minimizes the sum of squares between the observed values **y** and the vector $\mathbf{H}\hat{\mathbf{x}}$. So we will try to compute the $\hat{\mathbf{x}}$ that minimizes the cost function J, where J is given as [7],

$$\mathrm{J} = \boldsymbol{\epsilon}_{\mathbf{y}}^{\mathbf{T}}\,\boldsymbol{\epsilon}_{\mathbf{y}} \tag{8}$$

The algorithm of LS is written in the following form [7]

$$\mathbf{K}_{\mathrm{k}} = \mathbf{P}_{\mathrm{k}-1}\mathbf{H}_{\mathrm{k}}^{\mathrm{T}}\left[\mathbf{H}_{\mathrm{k}}\mathbf{P}_{\mathrm{k}-1}\mathbf{H}_{\mathrm{k}}^{\mathrm{T}} + \mathbf{R}\right]^{-1} \tag{9}$$

$$\hat{\mathbf{X}}_{\mathrm{k}} = \hat{\mathbf{X}}_{\mathrm{k}-1} + \mathbf{K}_{\mathrm{k}+1}\left[\mathbf{Y}_{\mathrm{k}+1} - \mathbf{H}_{\mathrm{k}}\,\hat{\mathbf{X}}_{\mathrm{k}-1}\right] \tag{10}$$

$$\mathbf{P}_{\mathrm{k}} = [\mathbf{I} - \mathbf{K}_{\mathrm{k}}\mathbf{H}_{\mathrm{k}}]\mathbf{P}_{\mathrm{k}-1}[\mathbf{I} - \mathbf{K}_{\mathrm{k}}\mathbf{H}_{\mathrm{k}}]^{\mathbf{T}} + \mathbf{K}_{\mathrm{k}}\,\mathbf{R}\,\mathbf{K}_{\mathrm{k}}^{\mathrm{T}} \tag{11}$$

4.2. Quest-Method

The principle of the Quest algorithm can be done by solving Wahba's problem, this latter can be described as follows.
Considering n vectors $\mathbf{b}_{\mathrm{i}}$, $\mathrm{i} = 1 \ldots \mathrm{n}$, where n corresponds to the number of independent sensors installed in the satellite body and $\mathbf{b}_{\mathrm{i}}$ is the observation vectors. For each observed vector, a reference vector $\mathbf{r}_{\mathrm{i}}$ is necessary to be used in the attitude computation [6]. This description can be described as follows,

$$\mathbf{b}_{\mathrm{i}} = \mathbf{A}\,\mathbf{r}_{\mathrm{i}} + \mathbf{n}_{\mathrm{i}} \tag{12}$$

In mathematical terms, the Wahba's problem consists of finding the matrix **A** which minimizes the following cost function [6]

$$\mathrm{L}(\mathbf{A}) = \frac{1}{2}\sum_{\mathrm{i}=1}^{\mathrm{n}} \mathrm{a}_{\mathrm{i}}\,|\,\mathbf{b}_{\mathrm{i}} - \mathbf{A}\,\mathbf{r}_{\mathrm{i}}|^{2} \tag{13}$$

This problem is formulated as an eigenvector problem. The Quest algorithm estimates the optimal eigenvalue and eigenvector for the Wahba problem.
It is equivalent to a research problem of the maximum eigenvalue λ_{max} of the matrix **K**

$$\mathbf{K}=\begin{bmatrix}\mathbf{S}-\sigma\mathbf{I} & \mathbf{z}\\ \mathbf{z}^T & \sigma\end{bmatrix} \tag{14}$$

where, σ, **S**, and **Z** are defined as follows,

$$\mathbf{B}=\sum_{i=1}^{n} a_i\, \mathbf{b}_i \mathbf{r}_i^T \qquad \sigma=\text{trace}\,\mathbf{B} \qquad \mathbf{S}=\mathbf{B}+\mathbf{B}^T \qquad \mathbf{Z}=\sum_{i=1}^{n} a_i\, \mathbf{b}_i \times \mathbf{r}_i \tag{15}$$

The largest eigenvalue is obtained by solving numerically the equation [6],

$$\det(\mathbf{K}-\lambda\mathbf{I}_{4\times4})=\mathbf{0} \tag{16}$$

The optimal value λ_{opt} is given as,

$$\lambda_{opt} \approx \sum_{i=1}^{n} a_i - L(\mathbf{A}) \tag{17}$$

The values of L(**A**) are quite smalls, an approximate solution is given as,

$$\lambda_{opt} \approx \sum_{i=1}^{n} a_i \tag{18}$$

Using this eigenvalue, the respective eigenvector is calculated as follows,

$$\bar{\mathbf{q}}=\frac{1}{\sqrt{1+\mathbf{p}^T\mathbf{p}}}\begin{bmatrix}\mathbf{p}\\ 1\end{bmatrix} \qquad \mathbf{p}=\left[(\lambda_{opt})\mathbf{I}-\mathbf{S}\right]^{-1}\mathbf{Z} \tag{19}$$

4.3. Optimized TRIAD

The optimized TRIAD is a linear combination between two TRIAD algorithms, called TRIAD-I and TRIAD-II. The first algorithm generates the matrix $\boldsymbol{A}_1$, and the second generates the matrix $\boldsymbol{A}_2$.
It consists of the construction of two vectors triads using two pairs of vector measurements: two in the orbital reference frame, noted $\mathbf{v}_1$ and $\mathbf{v}_2$, and two in the body reference frame, noted $\mathbf{w}_1$ and $\mathbf{w}_2$. Let $\mathbf{n}_1$ and $\mathbf{n}_2$ be the measured noises of $\mathbf{w}_1$ and $\mathbf{w}_2$, respectively [8].
The optimized TRIAD algorithm requires the computation of the matrices $\mathbf{r}_i$ in body coordinate, and the corresponding column matrices $\mathbf{s}_i$ in the reference system.

For the TRIAD-I algorithm, we define

$$\mathbf{r}_1=\mathbf{w}_1/|\mathbf{w}_1| \qquad \mathbf{r}_2=(\mathbf{r}_1\times\mathbf{w}_2)/|\mathbf{r}_1\times\mathbf{w}_2| \qquad \mathbf{r}_3=\mathbf{r}_1\times\mathbf{r}_2 \tag{20}$$

$$\mathbf{s}_1=\mathbf{v}_1/|\mathbf{v}_1| \qquad \mathbf{s}_2=(\mathbf{s}_1\times\mathbf{v}_2)/|\mathbf{s}_1\times\mathbf{v}_2| \qquad \mathbf{s}_3=\mathbf{s}_1\times\mathbf{s}_2 \tag{21}$$

The matrix $\mathbf{A}_1$ is expressed as,

$$\mathbf{A}_1=\mathbf{r}_1.\mathbf{s}_1^T+\mathbf{r}_2.\mathbf{s}_2^T+\mathbf{r}_3.\mathbf{s}_3^T \tag{22}$$

For the TRIAD-II algorithm, we define

$$\mathbf{r}_5=\mathbf{w}_2/|\mathbf{w}_2| \qquad \mathbf{r}_2=(\mathbf{r}_1\times\mathbf{w}_2)/|\mathbf{r}_1\times\mathbf{w}_2| \qquad \mathbf{r}_4=\mathbf{r}_5\times\mathbf{r}_2 \tag{23}$$

$$\mathbf{s}_5=\mathbf{v}_2/|\mathbf{v}_2| \qquad \mathbf{s}_2=(\mathbf{s}_1\times\mathbf{v}_2)/|\mathbf{s}_1\times\mathbf{v}_2| \qquad \mathbf{s}_4=\mathbf{s}_5\times\mathbf{s}_2 \tag{24}$$

And the matrix $\mathbf{A}_2$ is expressed as,

$$\mathbf{A}_2=\mathbf{r}_5.\mathbf{s}_5^T+\mathbf{r}_2.\mathbf{s}_2^T+\mathbf{r}_4.\mathbf{s}_4^T \tag{25}$$

Then, the optimization attitude matrix $\widehat{\mathbf{A}}'$ is calculated as,

$$\widehat{\mathbf{A}}'=\frac{\sigma_2^2}{\sigma_1^2+\sigma_2^2}\mathbf{A}_1+\frac{\sigma_1^2}{\sigma_1^2+\sigma_2^2}\mathbf{A}_2 \tag{26}$$

Finally, the attitude matrix of optimized TRIAD is expressed as,

$$\mathbf{A} = 0.5\left[\widehat{\mathbf{A}}' + \left(\widehat{\mathbf{A}}'^{-1}\right)^{\mathrm{T}}\right] \tag{27}$$

5. Attitude estimation algorithms

The attitude estimation theory is very useful in the space field, the attitude estimation algorithms estimate the full elements of the state vector (attitude and angular rate) using the dynamic and kinematic models to propagate the estimated state vector.
The two known algorithms are: Extended Kalman Filter and Unscented Kalman Filter.

5.1. Extended Kalman Filter

The Extended Kalman filter (EKF) [9] is a recursive state estimator. Its cost function is given as [7],

$$J = \lim_{n\to\infty} \sum_{k=0}^{N} \mathbf{E}(\|\mathbf{x}_k - \hat{\mathbf{x}}_k\|) \tag{28}$$

where, $\mathbf{E}$: the expectation; $\mathbf{x}_k$: true state vector; $\hat{\mathbf{x}}_k$: estimated state vector.
The EKF algorithm consists of two parts: propagation and correction cycles, it is written in the following form [10-11].

5.1.1. Propagation cycle.

$$\hat{\mathbf{X}}_{k+1/k} = \mathbf{f}\,(\hat{\mathbf{X}}_{k/k}, \boldsymbol{u}_k) \tag{29}$$

$$\mathbf{P}_{k+1} = \boldsymbol{\Phi}_k \mathbf{P}_k \boldsymbol{\Phi}_k^{\mathrm{T}} + \mathbf{Q}_k \tag{30}$$

5.1.2. Correction cycle.

$$\mathbf{K}_{k+1} = \mathbf{P}_{k+1}\mathbf{H}_k^{\mathrm{T}}\left[\mathbf{H}_k\mathbf{P}_{k+1}\mathbf{H}_k^{\mathrm{T}} + \mathbf{R}\right]^{-1} \tag{31}$$

$$\mathbf{P}_{k+1/k+1} = [\mathbf{I} - \mathbf{K}_{k+1}\mathbf{H}_k]\mathbf{P}_{k+1/k} \tag{32}$$

$$\hat{\mathbf{X}}_{k+1/k+1} = \hat{\mathbf{X}}_{k+1/k} + \mathbf{K}_{k+1}\left[\mathbf{Y}_{k+1} - \mathbf{H}_k\hat{\mathbf{X}}_{k+1/k}\right] \tag{33}$$

where,
$\mathbf{P}$: Covariance matrix (7x7);
$\boldsymbol{\Phi}$: State transformation matrix (7x7) ;
$\mathbf{Q}$: Process noise covariance matrix (7x7) ;
$\mathbf{K}$: Kalman gain matrix (7x4);
$\mathbf{R}$: Measurement noise covariance matrix (4x4).

5.2. Unscented Kalman Filter

The Unscented Kalman Filter was proposed by Julier and Uhlmann [12]. It is an estimator which applies to the nonlinear systems without passing by the linearization required by the EKF.
The principle of the UKF is to perform a deterministic sampling of the state vector using a nonlinear transformation called "Unscented transformation". The state vector is represented by a set of samples (sigma points), these samples are stored in a matrix of size $n \times (2n + 1)$. The algorithm of this filter is described as follows [13],

5.2.1. Calculation of the sigma point. The columns of the sigma point matrix are calculated as follows

$$\boldsymbol{\mathcal{X}}_{k-1} = \left[\hat{\mathbf{X}}_{k-1} \quad \hat{\mathbf{X}}_{k-1} + \sqrt{(n+\lambda)\left(\hat{\mathbf{P}}_{k-1}\right)_i} \quad \hat{\mathbf{X}}_{k-1} - \sqrt{(n+\lambda)\left(\hat{\mathbf{P}}_{k-1}\right)_i}\right] \tag{34}$$

The square root of the $(n + \lambda)\hat{\mathbf{P}}_{k-1}$ matrix is computed using Cholesky decomposition, and the parameter λ is defined as,

$$\lambda = \alpha^2(n + k) - n \tag{35}$$

where, n is the dimension of the state vector.

5.2.2. Propagation. If the $\mathcal{X}_{k-1}$ matrix is calculated, we propagate each column of this matrix using the mathematical model.

$$\mathcal{X}^{i}_{k/k-1} = f(\mathcal{X}^{i}_{k-1}) \tag{36}$$

The mean and covariance of the state are calculated as follows

$$\hat{X}_{k/k-1} = \sum_{i=0}^{2n} W_i^{(m)}\, \mathcal{X}^{i}_{k/k-1} \tag{37}$$

$$\hat{P}_{k/k-1} = \sum_{i=0}^{2n} W_i^{(c)}\, (\mathcal{X}^{i}_{k/k-1} - \hat{X}_{k/k-1})(\mathcal{X}^{i}_{k/k-1} - \hat{X}_{k/k-1})^T + Q_k \tag{38}$$

where,

$$W_0^{(m)} = \frac{\lambda}{n+\lambda} \qquad W_0^{(c)} = \frac{\lambda}{n+\lambda} + (1 - \alpha^2 + \beta) \qquad W_i^{(m)} = W_i^{(c)} = \frac{1}{2(n+\lambda)}\ ,i = 1,2 \ldots n \tag{39}$$

5.2.3. Update. The sigma points for the measurements are calculated as follows,

$$Y^{i}_{k/k-1} = h(\mathcal{X}^{i}_{k-1}) \tag{40}$$

The mean and covariance of the sigma points for the measurements $Y^{i}_{k/k-1}$

$$\hat{Y}_{k/k-1} = \sum_{i=0}^{2n} W_i^{(m)}\, Y^{i}_{k/k-1} \tag{41}$$

$$\hat{P}_{yy,k} = \sum_{i=0}^{2n} W_i^{(c)}\, (Y^{i}_{k/k-1} - \hat{Y}_{k/k-1})(Y^{i}_{k/k-1} - \hat{Y}_{k/k-1})^T + R_k \tag{42}$$

The covariance between state and measurement

$$\hat{P}_{xy,k} = \sum_{i=0}^{2n} W_i^{(c)}\, (\mathcal{X}^{i}_{k/k-1} - \hat{X}_{k/k-1})\,(Y^{i}_{k/k-1} - \hat{Y}_{k/k-1})^T \tag{43}$$

Finally, the gain, the state vector, and the covariance matrix are calculated as follows,

$$K_k = \hat{P}_{xy,k}\, \hat{P}^{-1}_{yy,k} \qquad \hat{X}_k = \hat{X}_{k/k-1} + K_k[Y_k - \hat{Y}_{k/k-1}] \qquad \hat{P}_k = \hat{P}_{k/k-1} - K_k\, \hat{P}_{yy,k}\, K_k^T \tag{44}$$

6. Nonlinear observers

An observer is a dynamic system that can be called "software sensor". It estimates the state vector using the system input, the system output, and knowledge of the mathematic model.

Among the most robust observers, we have the sliding mode observers. The sliding mode observers are a nonlinear state observers based on the theory of variable structure systems [14].

Consider the following nonlinear system

$$\dot{x} = f(x,t) + B\,u(t) \tag{45.1}$$

$$y = C\,x + v \tag{45.2}$$

where, v presents the measurement noise.

The Sliding Mode Observer is chosen as the following form [14]

$$\dot{\hat{x}} = f(\hat{x},t) + B\,u(t) + K\,\mathrm{sat}\left(\frac{S}{\Phi}\right) \tag{46.1}$$

$$\hat{y} = C\,\hat{x} \tag{46.2}$$

where, S is the sliding surface function, it is defined as,

$$S = y - C\,\hat{x} \tag{47}$$

$\hat{x}$ represents the state estimates;

$f(\hat{x},t)$is the model of the nonlinear system;

$K = \begin{bmatrix} K_\omega \\ K_q \end{bmatrix}, H = \begin{bmatrix} H_\omega \\ H_q \end{bmatrix}, \Phi = \begin{bmatrix} \Phi_\omega \\ \Phi_q \end{bmatrix}$ are gain matrices to be determined.

To accelerate the estimation convergence, the linear correction term of Luenberger $H(y - \hat{y})$ is included in the observer.

$$\dot{\hat{x}} = f(\hat{x},t) + B\,u(t) + H(y - \hat{y}) + K\,\mathrm{sat}\left(\frac{S}{\Phi}\right) \tag{48.1}$$

$$\hat{y} = C\,\hat{x} \tag{48.2}$$

where, $\mathbf{H} = \begin{bmatrix} \mathbf{H}_\omega \\ \mathbf{H}_q \end{bmatrix}$ is gain matrix of Luenberger.

7. Simulation results

In this section, we present the simulation results obtained for the different algorithms exposed previously. The satellite is placed at an average altitude of 686 km on a circular orbit inclined 98° to the equator. The inertia matrix is **I**=[12 0 0;0 14 0;0 0 10] $Kg.m^2$. And the initial conditions are [5 10 -10] deg and [0 -0.06 0] rad/sec for the Euler angles and the attitude rate respectively.
The presented algorithms are simulated using the same simulation method, and they are initialized in the same way, using the same initialization values. The initial parameters are detailed in what follows.

Table 1. Extended Kalman Filter parameters.

Parameter	Value
$\mathbf{X_0}$	$[-0.22 \quad -0.16 \quad 0.21 \quad 0.93 \quad 0 \quad -0.01 \quad 0]$
$\mathbf{P_0}$	$\begin{bmatrix} (0.1)^2\mathbf{I}_{4\times4} & \mathbf{0}_{4\times3} \\ \mathbf{0}_{3\times4} & 10^{-14}\mathbf{I}_{3\times3} \end{bmatrix}$
Measurement	$(10^{-6})^2$
Process Noise	$(10^{-5})^2$

Table 2. Unscented Kalman Filter parameters.

Parameter	Value
$\mathbf{X_0}$	$[-0.22 \quad -0.16 \quad 0.21 \quad 0.93 \quad 0 \quad -0.01 \quad 0]$
$\mathbf{P_0}$	$\begin{bmatrix} (0.1)^2\mathbf{I}_{4\times4} & \mathbf{0}_{4\times3} \\ \mathbf{0}_{3\times4} & 10^{-14}\mathbf{I}_{3\times3} \end{bmatrix}$
Measurement	$(10^{-6})^2$
Process Noise	$(10^{-5})^2$
α	10^{-4}
β	2
k	0

Table 3. Sliding Mode Observer parameters.

Parameter	Value
$\mathbf{K}_\omega$	diag([0.077 0.009 0.009])
$\mathbf{K}_q$	diag([0.035 0.099 0.0987 0.098])
$\mathbf{H}_\omega$	diag([0.05 0.09 0.098])
$\mathbf{H}_q$	diag([0.28 0.28 0.28 0.28])
ϕ_ω	diag([0.57 0.03 0.03])
ϕ_q	diag([0.52 0.027 0.03 0.03])
$\mathbf{I}_{SMO}[Kg.m^2]$	[14 13 12]

Due to a lack of obvious physical interpretation of the quaternion, the Euler angles will normally be used to present the attitude during simulation tests.

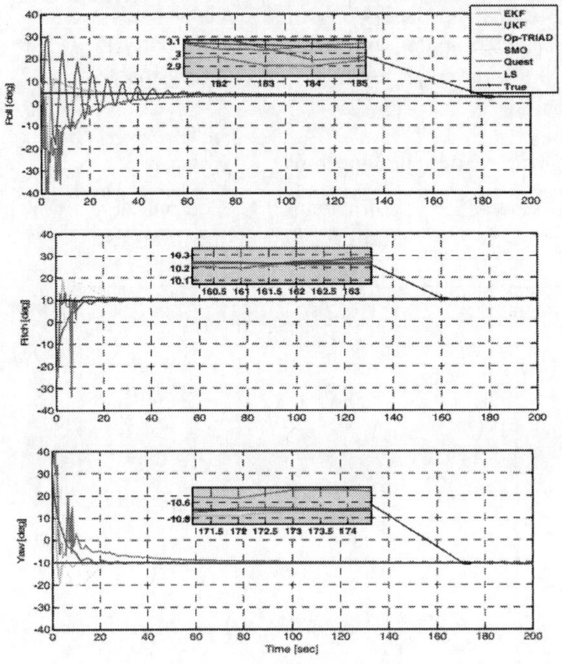

Figure 2. Actual and estimated attitude.

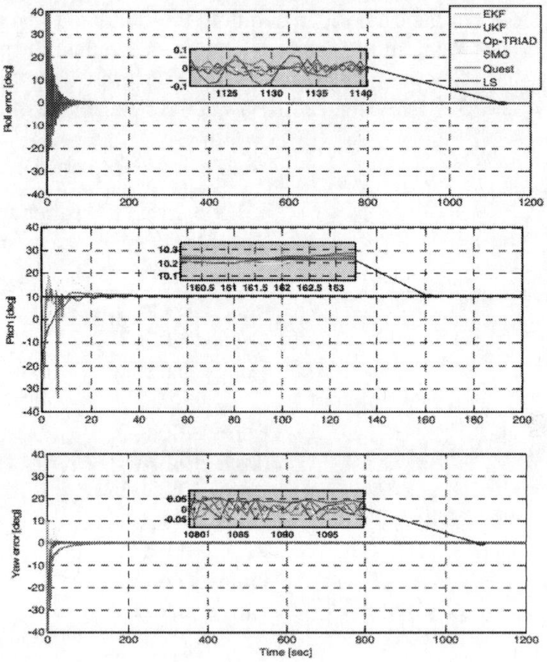

Figure 3. Errors of the estimated attitude.

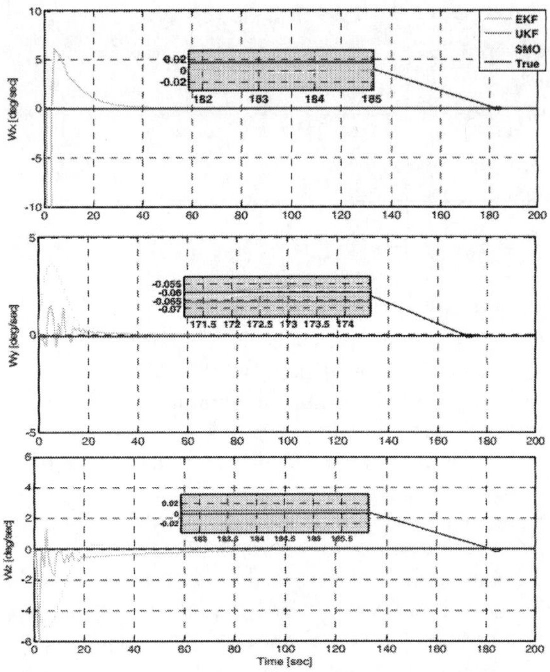

Figure 4. Estimated angular rates.

For more detailed analysis, the root mean square (RMS) values of error results for the time period 35000-36000 sec, the amount of memory and the convergence time are computed; they are presented in the Table 5. In addition, the estimated computation time along the 3000 sec is calculated with two ways. In the fourth column, it is computed using a personal computer with i54590S CPU 3.00 GHz, 4 GByte of RAM, and using On-board Computer (OBC 750) in the last column, this OBC is capable of 500 Dhrystone 2 MIPS and has 512 MByte of RAM.

Table 4. Performance evaluation of the different algorithms.

	RMS error [deg]	Amount of memory [M Bytes]	Convergence time [sec]	Computation time for PC [sec]	Computation time for OBC 750 [sec]
Least Squares	Roll 0.0726, Pitch 0.0563, Yaw 0.0250 0.0951	0.765	60	0.56	12.30
Quest Method	Roll 0.0402, Pitch 0.0278, Yaw 0.0271 0.0559	0.6484	/	0.31	6.81
Optimized TRIAD	Roll 0.0415, Pitch 0.0332, Yaw 0.0312 0.0616	0.8867	/	0.49	10.76
EKF	Roll 0.0294, Pitch 0.0294, Yaw 0.0294 0.0509	0.9492	100	0.74	16.25
UKF	Roll 0.0297, Pitch 0.0316, Yaw 0.0295 0.0524	10.218	200	2.72	59.75
SMO	Roll 0.0245, Pitch 0.0192, Yaw 0.0195 0.0367	1.4063	30	0.63	13.84

The figures 2-4 prsent the attitude, the errors of the estimated attitude and the angular rates respectively of the different algorithms presented in this paper. These figures show that the attitude determination algorithms (Least Sequares, Quest method and optimized TRIAD) are able to estimate only the attitude. However, the attitude estimation algorithms (EKF and UKF) and the sliding mode observer estimate the attitude and the angular rates.

The table 4 presents a comparison between the different algorithms in terms of accuracy, amount of memory convergence time, and computation time. From this table, it is seen that the attitude determination algorithms are characterized by fast computation time (0.56 sec, 0.31 sec, and 0.49 sec) for personal computer and (12.30 sec, 6.81 sec, and 10.76 sec) for OBC750, and they require less amount of memory but they provide less accuracy in comparaison with the attitude estimation algorithms and the Sliding mode observer. In addition, the Quest and Optimized TRIAD algorithms have not a convergence time because they do not need the initial parameters. The attitude estimation algorithms have a good performance in term of accuracy but they are more expensive in computation time, 16.25 sec for the EKF and 59.75 for the UKF. the computation time of this latter is

approximately nine times the computation time of Quest-Method, it is seen that this filter has the slowest convergence. In addition, its requirement of memory is very largest (10.21 M Bytes) due to calculation of the sigma points .
The table 4 results indicate also that the SMO has a good performance in terms of accuracy, computation time and convergence time (it is charactirized by accuracy two times better than the Optimized TRIAD and the Quest-Method).

8. Conclusion

The primary focus of this paper is to present a performance analysis of the attitude determination algorithms, attitude estimation algorithms, and nonlinear observer for application on low earth orbit satellites. The selected algorithms were compared to find the required accuracy, computation time, convergence time, and amount of memory. The softwere implementation if these algorithms was executed in a personal computer and in the On-board Computer 750 that was developped in SSTL to function as the primary flight computer for applications requiring processing On-board. The presented study can be a help tool for the ADCS designer.
The results of this work indicate that each of the attitude determination, attitude estimation, and nonlinear observer algorithms is able to determinate the attitude. However, there are some differences between these algorithms.
Attitude determination algorithms are deterministic methods. They are characterized by low computational efforts, less accuracy and amount of memory in comparison with attitude estimation algorithms. This is due to its independence of dynamic equation. However, this latter is used by attitude estimation algorithms, the use of this equation increases the accuracy of these algorithms, but they have a high time consuming (the execution time of the UKF is approximately nine times the execution time of the Quest-Method) because they need much calculation (calculation of the covariance matrix) and its structure uses the dynamic models. In addition, the memory requirement of these algorithms is very largest.
On the other hand, the nonlinear observers have high performances in terms of time consuming and accuracy, it is characterized by accuracy two times better than the Optimized TRIAD and the Quest-Method.
From all the above, we can give a solution to the problem posed in the beginning of this paper, we should use the nonlinear observer to extract the attitude of spacecraft because its accuracy is more high than the others algorithms. In addition, it is helpful for micro satellite not equipped with gyroscope (which is costly and heavy).

References

[1] Shuster M D 1993 A survey of attitude representations *the Astronautical Science* vol **41** pp 439-517.
[2] Tamer Mekky A H A 2013 comparative study of spacecraft attitude determination and estimation algorithms (a cost-benefit approach) *Aerospace Science and Technology (Elsevier)* vol **26** pp 211-215.
[3] Mohd zamri H and all 2016 Review on attitude estimation algorithm of attitude determination system *ARPN journal of engineering and applied sciences* vol **11** pp 4455-4460.
[4] Wertz J R 1986 *Spacecraft Attitude Determination and Control* D Reidel Publishing Company Boston U.S.A Reprint.
[5] Sidi M1997 *Spacecraft Dynamics and Control* Cambridge Univ press.
[6] Shuster M D and S D Oh 1981 Three axis attitude determination from vector observations *Guid. Control* vol **4** pp 70-77.
[7] Simon D 2006 *Optimal State Estimation* John Wiley & Sons Inc Hoboken New Jersey.
[8] Itzhack Y Bar-Itzhack, Richard R Harman1996 Optimized Triad Algorithm for Attitude Determination *Guidance Control and Dynamics* pp 208-211.
[9] Jazwinski 1970 *A Stochastic Processes and Filtering Theory* New York Academic Press.

[10] Gelb M 1974 *Applied Optimal Estimation* MIT press.
[11] Brown R G and Hwang 1996 *Introduction to Random Signals and Applied Kalman Filtering* Tohn Wiley New york.
[12] Julier S J Uhlmann J K and Durrant-Whyte H 1995 A new approach for filtering nonlinear system *Proceedings of the American Control Conference* pp1628-1632.
[13] Julier S J Uhlmann J K A 1996 general method for approximating nonlinear transformation of probability distributions *Technical report RRG Dept of Engineering Science University of Oxford.*
[14] Slotine J J E Hedrick J K Misawa E A 1987 On Sliding Observers for nonlinear Systems *Dynamic Systems Measurement and Control* vol **109** pp245-252.

IOP Conf. Series: Journal of Physics: Conf. Series **783** (2017) 012018 doi:10.1088/1742-6596/783/1/012018

Decentralized Fault-Tolerant Control of Inland Navigation Networks: a Challenge

P. Segovia[*,], L. Rajaoarisoa[*], F. Nejjari[**], J. Blesa[+], V. Puig[**,+], E. Duviella[*]**
[*] Mines-Douai IA, France.
[**] Automatic Control Department, Technical University of Catalonia (UPC), Spain.
[+] Institut de Robòtica i Informàtica Industrial (CSIC-UPC), Spain.

E-mail: {pablo.segovia,lala.rajaoarisoa,eric.duviella}@mines-douai.fr, {fatiha.nejjari,joaquim.blesa,vicenc.puig}@upc.edu

Abstract. Inland waterways are large-scale networks used principally for navigation. Even if the transport planning is an important issue, the water resource management is a crucial point. Indeed, navigation is not possible when there is too little or too much water inside the waterways. Hence, the water resource management of waterways has to be particularly efficient in a context of climate change and increase of water demand. This management has to be done by considering different time and space scales and still requires the development of new methodologies and tools in the topics of the Control and Informatics communities. This work addresses the problem of waterways management in terms of modeling, control, diagnosis and fault-tolerant control by focusing in the inland waterways of the north of France. A review of proposed tools and the ongoing research topics are provided in this paper.

1. Introduction

Inland navigation networks cover more than 38,000 km in Europe and more than 8,500 km in France. They are used to transport billions of tons of water per year with a demand increase that should reach +35% by 2050 in Europe [6]. Inland navigation takes part in the Trans-European network program (TEN-T[1]), which promotes the development of transport infrastructure policies to close the gaps between Member States' transport networks and to guarantee seamless transport chains for passengers and freight. As an example of this, the new Seine-Nord canal between the French towns of Compiègne and Creil will connect Paris to the Belgian and Dutch waterways. In the north of France, the Nord-Pas de Calais region is seeking to shift 20% of transport volume from road to rail and navigation by 2050[2]. An intensification in the use of inland waterways is expected in the near future, mainly due to the fact that navigation transport offers an environmental-friendly and economical alternative in comparison to road transport [9]. This increase of navigation demand will be in the context of climate change, since it is now recognized that inland waterways will be impacted by global change [2,3,16,29,64]. Thus, the constraints on inland waterway management, and more exactly on the available water resources dedicated to the navigation, will be more severe.

Inland waterways management is a synonym of water resources management, as it is necessary to allocate the water resources amongst the networks [12, 14, 43, 44] and to control the water levels

[1] http://ec.europa.eu/transport/themes/infrastructure/ten-t-guidelines/index_en.htm
[2] https://www.nordpasdecalais.fr/upload/docs/application/pdf/2014-09/schema_climat-web-170914.pdf

everywhere. By focusing on the water level control, the main objectives can be efficiently reached by the design of modeling approaches, control techniques, fault detection and isolation (FDI) methods and finally fault-tolerant control (FTC) strategies. The modeling approaches are useful for the design of control, FDI and FTC methods, even if some model-free based control techniques are proposed in the literature for hydroelectric power plants [31], model-free based FDI for a gate of waterways [33] or a dam-gallery open channel system [1]. The control algorithms have to take into account the complexity and the scale of the inland waterways by considering centralized [61] or decentralized approaches [17, 46, 50, 51]. FDI techniques can be based on models of the systems [5, 30, 42], on the measured data [15] or on both [26]. Finally, FTC strategies aim at allowing the system to continue operating properly with fault occurrences [25, 67]. To the knowledge of the authors, the proposal of FTC strategies for inland waterways is an open issue. The development of these approaches is still an ongoing research topic if all the complexity of the waterway networks has to be taken into account.

After describing the characteristics of inland waterways and their management objectives in Section 2, the application of FTC to inland waterways is presented in Section 3 by proposing an architecture for interconnected systems. This architecture is based on active FTC strategy that aims at reconfiguring control actions so that the stability and an acceptable performance of the entire system can be maintained. The proposed scheme is built on modeling approaches that are described in Section 4. The decentralized control techniques are discussed in Section 5, and Section 6 presents the FDI methods. For each section, proposed contributions in the literature are discussed and the still existing challenges are presented.

2. Inland waterways management objectives

Inland waterways are large-scale networks composed of several interconnected navigation reaches (NR) (*see* Fig. 1). The main management goal is to maintain the water level of each reach around a setpoint: the Normal Navigation Level (NNL). Two boundaries are considered around the NNL and define the navigation rectangle: when the water level crosses one of these boundaries, the navigation has to stop. Due to the increase of navigation demand and the accommodation of the biggest ships, the navigation rectangle is restricted to several centimeters around the NNL.

To simplify the study of open-flow water networks, they are usually divided into reaches by considering initial and final ends, which usually correspond to hydraulic human-made devices such as gates or weirs. Three different kinds of reaches can be considered according to whether or not there are water streams leaving or entering such reach in an intermediate point: a reach with a tributary, when an external water stream flows into the considered navigation reach; a reach with a distributary, when a stream of water branches off from the navigation reach; and finally a *simple* reach, when there is neither a tributary nor a distributary. Fig. 2 depicts these three possibilities. Due to the configuration of the network, a navigation reach is a Multiple-Input-Multiple-Output (MIMO) system.

Each reach is linked with other reaches thanks to controlled gates and locks. In this way, the control of a gate or a lock located on a reach has a direct influence on the upstream and downstream reaches. The dynamics of the gates mainly depend on the water levels of the upstream and downstream reaches. They are located in the center of watersheds, and controlled or uncontrolled water exchanges occur between natural rivers and them. As an example, the Cuinchy-Fontinettes reach, located in the north of France, has more than 300 uncontrolled inputs or outputs that link it with the watershed. Finally, reaches are generally equipped with sensors that provide water level or discharge measurements. Thus, a waterway network is a strong interconnected MIMO system subject to uncertainties.

A reach is a large-scale system with complex nonlinear dynamics and variable time delays. It can be affected by resonance phenomena that causes waves from downstream to upstream (resp. upstream to downstream) with a magnitude of several centimeters. Depending on the magnitude of the waves, the attenuation of this phenomena might require several hours.

It is also known that leaks are present between two reaches due to the water-tightness problem of locks and gates. In addition, sensors and actuators are vulnerable to faults. The occurrence of faults can lead to an inefficient water level control and a waste of water resources, which is not acceptable in

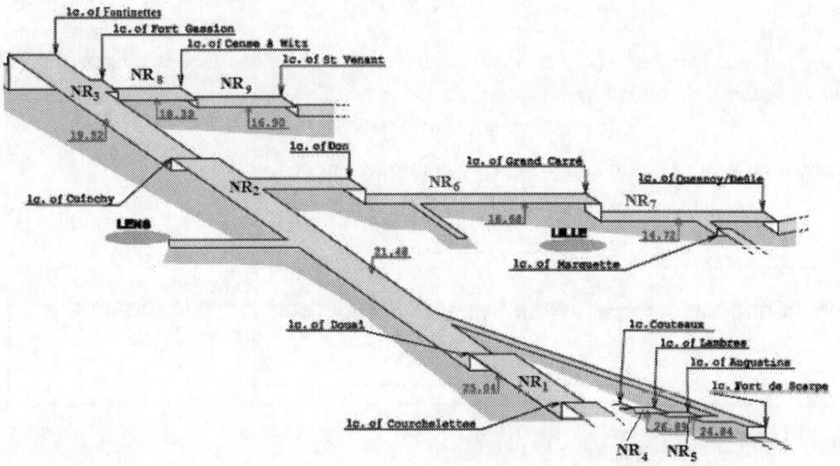

Figure 1. Part of the waterways in the north of France composed of 9 NR.

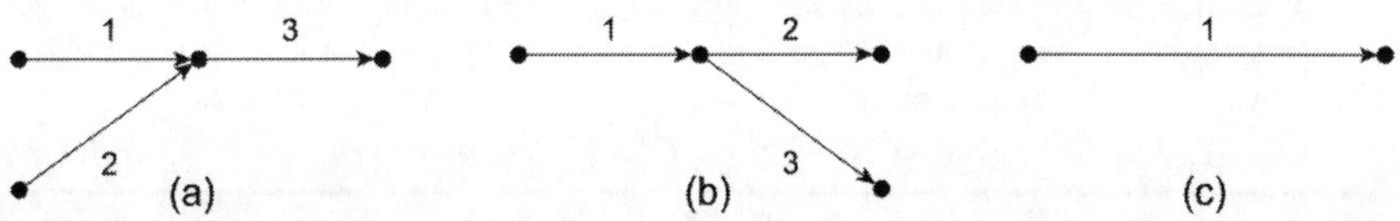

Figure 2. (a) reach with a tributary, (b) reach with a distributary, (c) *simple* reach.

a context of climate change and increase of navigation demand. A decentralized FTC architecture is proposed to deal with this challenge.

3. Decentralized fault tolerant control architecture

The fundamental characteristics of the complexity of the presented case study are dimensionality and uncertainty, according to the author's research results in modeling and control. These two main features will be the overriding concern and will serve as major motivating factors for the development of decentralized control for this kind of studied systems.

This work only addresses the study of sensor faults. However, two other main kinds of faults can affect the system: actuator faults, when there is a mismatch between the specified control action and the action performed by the actuator; and system faults, when a change in the system dynamics occur. These two other types of faults are not considered in this work but will be addressed in the future.

Controllers aim at canceling the error between the measurements and their reference inputs. But when a sensor fault occurs, the faulty measurements directly corrupt the closed-loop behavior and may drive the system to its physical limitations or even to instability. To avoid or limit this behavior, FTC strategies that do not require any modification of the control law are proposed in [47, 66]. They are based on the conventional fault-tolerant configuration depicted in Fig. 3. The estimator provides y^{corr}, which is used to estimate the fault-free sensor magnitude such as:

$$y^{corr}(s) = y(s) - f_y, \tag{1}$$

where s is the Laplace variable, $y(s)$ the output of the sensor and f_y an adaptive estimation of the fault magnitude that can be determined by:

$$f_y = g(u(s), y(s)), \tag{2}$$

where g is any nonlinear function and $u(s)$ is the controlled input.

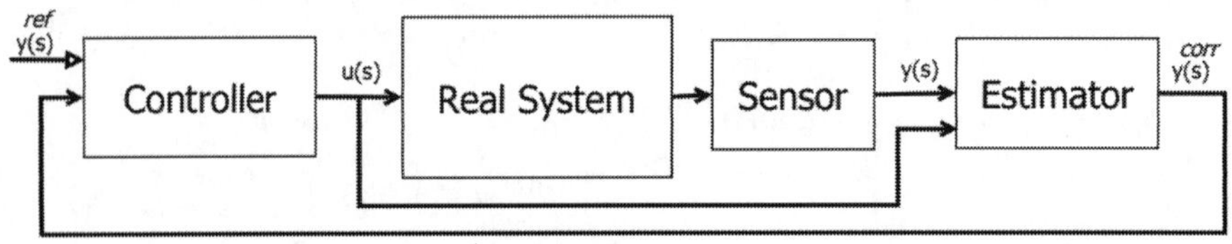

Figure 3. Conventional Fault Tolerant Control configuration.

The estimation y^{corr} is then considered in the computation of the new control law, minimizing the effects of the faults.

Based on this principle, a decentralized FTC architecture is proposed in Fig. 4, whose purpose is to take into account large-scale interconnected systems. The real system is equipped with m actuators and n sensors, and a setpoint y^{ref} is attributed to each of the m controllers. According to the configuration of the network, consider that the global interconnected system can be described by (3).

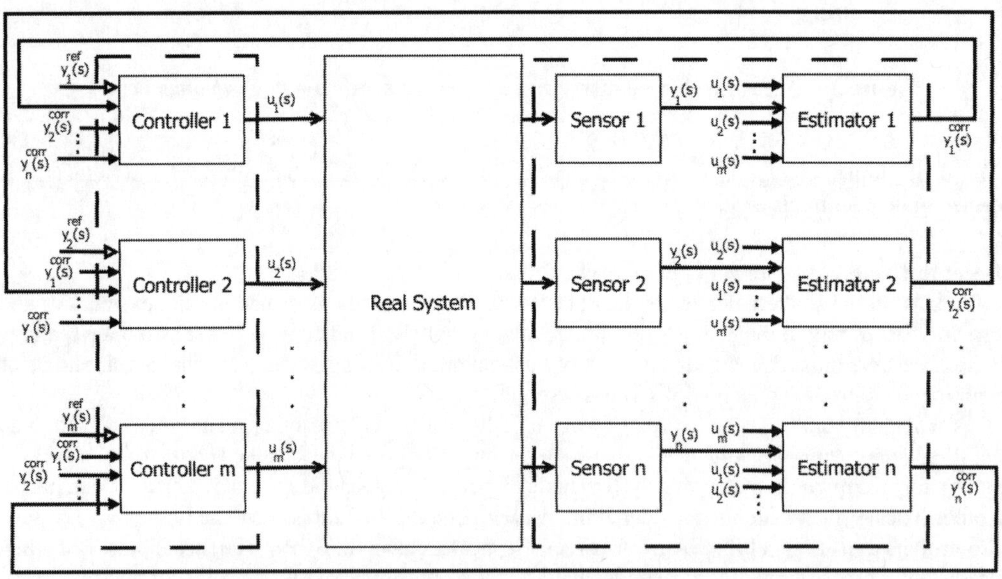

Figure 4. Decentralized fault tolerant control architecture.

$$\Sigma: \ \mathbf{Y}(s) = \ \mathbf{P}(s)\mathbf{U}(s), \tag{3}$$

where $\mathbf{P}(s)$ is the transfer matrix defined as follows:

$$\mathbf{P}(s) = \begin{bmatrix} p_{1,1}(s) & p_{1,2}(s) & \cdots & p_{1,m}(s) \\ p_{2,1}(s) & p_{2,2}(s) & \cdots & \vdots \\ \vdots & \vdots & \ddots & \vdots \\ p_{n,1}(s) & p_{n,2}(s) & \cdots & p_{n,m}(s) \end{bmatrix} \tag{4}$$

The structure of the elements of $\mathbf{P}(s)$ is given in the next section.

Therefore, the feedback system Σ_{cl} consists of:

$$\Sigma_{cl}: \begin{cases} \text{Plant } \Sigma: \mathbf{Y}(s) = & \mathbf{P}(s)\mathbf{U}(s) \\ \text{Controller } \Lambda: \mathbf{U}(s) = & \mathbf{K}(s)\mathbf{E}(s) \\ \text{Feedback } \Pi: \mathbf{E}(s) = & \mathbf{Y}^{ref}(s) - \mathbf{Y}^{corr}(s) \end{cases}, \tag{5}$$

where matrices $\mathbf{P}(s)$ and $\mathbf{K}(s) \in \mathbb{R}(s)$ have dimensions $n \times m$ and $m \times n$, respectively, and belong to the set of real rational matrices, and $\mathbf{Y}^{ref}$ is the set of the objectives.

The global interconnected system can be decomposed as:

$$\Sigma_{cl}: \begin{cases} \Sigma_i: \ y_i(s) = & p_{ii}(s)u_i(s) + \sum_{j=1}^{m} p_{ji}(s)u_j(s) \\ \Lambda_j: u_j(s) = & K_{jj}(s)e_j(s) + \sum_{i=1}^{n} K_{ij}(s)e_i(s) \\ \Pi_i: e_i(s) = & y_i^{ref}(s) - y_i^{corr}(s) \end{cases}, \tag{6}$$
$$\text{for } i = 1,2,...,n, \ \ j = 1,2,...,m, \text{ and } j \neq i$$

where $p_{ii}(s)$ and $K_{ij}(s)$ are sub-elements of $\mathbf{P}(s)$ and $\mathbf{K}(s)$ respectively.

The ith term $\sum_{j=1}^{m}(p_{ji}(s)u_j(s))$ models the effect of all inputs on the measurement point i, with $i \neq j$. It represents the weighting interconnection values used to model the strength of interconnections. The jth term $\sum_{i=1}^{n}(K_{ij}(s)e_i(s))$ is the contribution of all errors between $y_i^{ref}(s)$ and $y_i^{corr}(s)$ in the computation of the control law for the actuator j, with $j \neq i$. In both cases, the elements that have to be taken into account depend on the configuration of the network. In Fig. 4, all elements ($y_i^{corr}(s)$ resp. $u_j(s)$) are considered as inputs of the blocks *Controller* and *Estimator* by default, but some of them might not be considered, depending on the configuration of the network.

Hence, it is necessary to determine these interconnections according to the dynamics and the configuration of the network. A modeling approach has to be proposed to obtain this information as well as the transfer matrix $\mathbf{P}(s)$ defined in (4). Therefore, it is necessary to develop decentralized control algorithms. Finally, the adaptive estimators have to be designed according to FDI methods. The ongoing work on each of these three topics is described hereinafter.

4. Modeling of inland navigation networks

As water networks are large-scale systems, they are usually decomposed into reaches to simplify the study. Reaches are generally defined as parts of a water stream that are comprised between a beginning and an ending point. These points generally correspond to the location of hydraulic structures that clearly delimit the water course, but they might as well be chosen due to topological reasons. Thus, water networks are regarded as the connection of individual reaches.

Following this idea, the first question that arises is which is the most suitable way to obtain an accurate mathematical representation of the open-flow canal system. The real behavior of the reaches can be accurately described by the Saint-Venant equations [11]. However, as there is no known analytical solution for these equations and they are strongly sensitive to errors in the geometry and other unmodeled dynamics, simplified models are used instead [8]. Some examples are the Integrator Resonance model (IR) [23,62], the Integrator Delay model [58], the Integrator Delay Zero model (IDZ) [34] and Gray-box model [13,22].

The IDZ model is chosen due to its capability to characterize the dynamics of the system in all regimes with great accuracy. Its structure is as follows:

$$p_{ij}(s) = \frac{\alpha s + 1}{\mathscr{A} s} e^{-\tau s}, \tag{7}$$

where α represents the inverse of the zero, $\mathscr{A}$ the integrator gain and τ the propagation time delay. The exact values of these parameters cannot be computed; however, they can be accurately approximated as shown in [34]. Since the used parameters are an approximation of the theoretical ones, the notation $\hat{p}_{ij}$ replaces p_{ij}.

In low frequencies, the canal is close to a tank-like behavior, with the integrator term reproducing this phenomenon, which can be approximated by the backwater area inverse. On the other hand, the delay represents the minimum time that a wave requires to travel from its origin to the measurement point. Assuming that the wave velocity V and the wave celerity C_w are constant, τ_d (perturbation generated upstream and measured downstream) and τ_u (perturbation generated downstream and measured upstream) can be computed in the following fashion:

$$\tau_u = \frac{L}{C_w - V} \tag{8a}$$

$$\tau_d = \frac{L}{C_w + V} \tag{8b}$$

where $C_w = \sqrt{g\mathscr{A}/T}$ and $V = q/\mathscr{A}$, with L the total length of the reach (m), g the gravitational acceleration (m/s^2), $\mathscr{A}$ the section (m^2), T the top width of the canal section (m) and q the discharge (m^3/s).

On the other hand, the zero, whose calculation procedure is omitted here but can be found in detail in [34] along with the other parameters, accounts for the high frequency dynamics. When the three elements are merged, a model that provides fitting accuracy in the whole frequency range is obtained.

With all that, (7) is used to model the different reaches that compose the overall system. By considering a reach with measured depths $y(0,s)$ and $y(L,s)$ and discharges $q(0,s)$ and $q(L,s)$—where 0 and L are the upstream and downstream ends of the reach, respectively—, the following transfer function matrix $\mathbf{P}_r(s)$ links these variables for the rth reach of the network in the following way:

$$\begin{bmatrix} y(0,s) \\ y(L,s) \end{bmatrix} = \underbrace{\begin{bmatrix} \hat{p}_{11}(s) & \hat{p}_{12}(s) \\ \hat{p}_{21}(s) & \hat{p}_{22}(s) \end{bmatrix}}_{\mathbf{P}_r(s)} \begin{bmatrix} q(0,s) \\ q(L,s) \end{bmatrix}, \tag{9}$$

where the $\hat{p}_{ij}(s)$ elements of matrix $\mathbf{P}_r(s)$ follow the IDZ model introduced in (7). Considering that the total number of reaches in the whole network is $\mathscr{R}$, $\mathbf{P}_r(s)$ can be computed, with $r = 1, ..., \mathscr{R}$.

Individual matrices $\mathbf{P}_r(s)$ are embedded into matrix $\mathbf{P}(s)$ defined in (4) in the following fashion: each $\hat{p}_{ij}(s)$ term of matrices $\mathbf{P}_r(s)$ is carefully placed inside $\mathbf{P}(s)$ according to the input-output (I/O) pair (u,y) that it links. Of course, not all possible (u,y) pairs are linked, hence $\mathbf{P}_r(s)$ contains null elements that account for such reason.

The final goal is to check the accuracy of such IDZ models by comparing them with the real behavior of the canal. To do so, two main options are considered. The first one consists in using an hydraulic software that enables the modeling and simulation of waterways: for this purpose, SIC^2 [3], a commercial

[3] `http://sic.g-eau.net/`

software, is chosen. The second approach is based on using available data from the real system. However, these data need to be previously treated or filtered due to the presence of a large number of uncertainties in the network.

5. Decentralized control approaches

Control approaches for free-surface water systems have been widely proposed in literature [36,39]. Most of them are dedicated to irrigation canals [28, 37, 49, 53, 63]. In [21, 55, 56], decentralized control techniques have been conceived for interconnected irrigation pools. The possible destabilization of decentralized controllers is addressed in [57], and the use of decouplers is suggested. When analyzing the coupling of the subsystems, [38] also studied the coupling effects and the possible decouplers. Decentralized control using decouplers is implemented in [40]. In [46, 50], a distributed Model-based Predictive Control based on multi-agent systems applied to interconnected irrigation pools is proposed with the main objective to overcome the computational effort of centralized control techniques.

More recently, control algorithms have been proposed for inland navigation reaches by taking into account their characteristics such as resonance, no significant slope or locks [24,27]. These contributions are focused on a single navigation reach. A decentralized control algorithm has been proposed in [51] considering interconnected inland navigation reaches.

In decentralized control, there are two main approaches to deal with the problem of interconnections [35]. The first one is the passive approach, where a decentralized control is designed for each isolated subsystem for certain desirable performance and the design is independent of the knowledge of the interconnections; the control is then applied to the overall system [59]. The second one is the synthetic approach: in this case, interconnections are explicitly taken into account when designing the controllers [20]. There is a large number of survey papers [10,54] and books [19,32] that provide further information on decentralized control theory and practice.

With regard to the global interconnected system given by (6), consider Γ as a weighting matrix of positive values used to model the strength of the interconnection between subsystems i and j. This interconnection matrix can be defined in the following manner:

$$\Gamma_{ij} = K_{ij}(s)e_i(s). \tag{10}$$

Considering that the gains of the controllers have a strong interaction, the mathematical relationship of input/output pair that describes the control problem is the following:

$$u_j = -(K_{j,j}(s)e_j(s) + \sum_{i=1}^{n} \Gamma_{ij}) \text{ with } j \neq i \text{ for } j = 1,...,m \tag{11}$$

such that controller K exists and its diagonal blocks are m separate controllers defined by:

$$K = diag(K_{jj}), \ \forall j = 1,...,m \tag{12}$$

In other terms, the problem consists in computing the controller K which stabilizes the global system (3) according to any specific feature. Indeed, the main idea is to design a feedback controller in such a way that the stability of the closed-loop system can be guaranteed for a wide range of uncertainties in the structure and subsystems dynamics. Thus, a special attention should be paid to the domain of uncertain systems.

On the other hand, the decentralized control strategy is naturally appended with the global control, whose role is to enhance the dominance of subsystems' stability by reducing the effect of the interconnections. Then, a condition for stability of the overall system can be proposed by using aggregated information about subsystems and interconnections. This condition requires that the system remains stable under any structural perturbation. So, in a few words, this work should lead to the generation of an algorithm capable of designing decentralized control laws for robust stabilization of uncertain interconnected systems by using and extending the works in [45,65].

6. Fault detection and isolation methods

Due to the fact that waterways are large-scale systems, a large number of sensors and actuators are placed throughout the whole system in order to collect data and to operate the hydraulic equipment, respectively. However, these devices are not perfect but are subject to measurement errors and faults. This fact justifies the need for tools or procedures that are able to detect the occurrence of a fault. But just knowing that the system is in a faulty situation is usually not enough, since the faulty element needs to be addressed in order to go back to the nominal behavior.

FDI allows the user, under certain circumstances, to detect whether or not a fault appears in the system and to be able to isolate it from the rest of possible faults, that is, to identify in which component the fault occurs. Many different diagnosis methods exist, but they all follow the principle of consistency [7]. First, the nominal behavior of the system is characterized by means of a plant model. Next, it is checked whether the measurement information (U,Y) (I/O pair) and the nominal behavior are consistent: if so, no inconsistency is detected, even if a certain fault exists. Moreover, it is possible that two (or more) different faults cause the same I/O pair, and thus an ambiguity in the diagnostic result exists. Hence, it is possible to detect a fault with only information about the nominal behavior of the system.

Conditions about the detection of a fault have just been given, but fault isolation and identification are two other crucial aspects. To address them, information about the faults and their impact on the system are needed, which means that fault models are required. But of course, this information is not always enough to isolate a fault, since more than one fault may generate the same effect, which means that, again, an ambiguity is present.

Many application examples of FDI can be found for water systems. A brief survey of these works include FDI using model-based set-membership applied to open-flow channels [25], decentralized FDI of a water transport network [48], an application to drinking water networks [52], the development of a new FI architecture for irrigation canals [41], FDI for irrigation canals for time-varying delayed systems [4] and the development of algorithms for leak detection in open-flow water channels [5].

There are two main approaches used to solve the fault detection and isolation problem [60]: model-based and data-driven methods. The model-based FDI approach compares the measured data with the estimated data from the knowledge of the nominal behavior of the system. This comparison generates a residual, which is an expression that is close to zero when no fault is present in the system and that diverts from zero when a fault occurs. However, since it is necessary to compute a model that represents the normal behavior of the plant in order to contrast the measured data, a deep knowledge of the physical principles that take part in the system is required.

On the other hand, data-driven methods rely on the usage of experimental data, and no previous physical knowledge of the system is needed. The main idea is to build a classifier that is able to localize in which component the fault occurs [18]. However, the main drawback of this approach is that the applicability of this problem is restricted by the available data. This means that only past faults may be detected, as they are the only ones that have been experimented by the system.

The selected approach to this problem is a blend between both model-based and data-driven methods. The idea is to obtain a fairly accurate model built from the physical equations that govern the system. With this model, residuals can be generated by comparing the nominal value to the measured data. Afterwards, a classification problem can be solved by considering these residuals and a set of labels. Therefore, by combining the two methods, it is possible, to some extent, to take advantage of the best features and to avoid the main drawbacks of each methodology.

7. Conclusion

This paper presented the ongoing research in the topic of decentralized fault-tolerant control applied to inland navigation networks in the framework of an increase in the water demand. The main problems for water management in the topics of modeling, control, FDI and FTC were discussed, and a survey of tools and references in the literature was detailed. The inland waterways in the north of France served to illustrate the work.

The mid-term work plan consists in comparing the theoretical IDZ models with the results provided by an hydraulic simulation software and by a computed model from real data to check the accuracy of the former. This verification seeks to obtain reliable models that will later be used for control strategies such as Model Predictive Control (MPC), FDI and FTC. For FTC, only sensor faults are considered at the present time, but actuator and system faults will also be considered and addressed.

References

[1] A. Akhenak, E. Duviella, L. Bako, and S. Lecoeuche. On-line fault diagnosis using recursive subspace identification algorithm of a dam-gallery open channel system. *Control Engineering Practice*, 21:797–806, 2013.

[2] B.P. Arkell and G.J.C. Darch. Impact of climate change on London's transport network. *Proceedings of the ICE - Municipal Engineer*, 159:231–237, 2006.

[3] B. Bates, Z. Kundzewicz, S. Wu, and J. Palutikof. Climate change and water. *Technical repport, Intergovernmental Panel on Climate Change, Geneva*, 2008.

[4] N. Bedjaoui, X. Litrico, D. Koenig, and P.O. Malaterre. H-infinity observer design for time-delay systems. Application to FDI for irrigation canals. *45th IEEE CDC, San Diego, CA, USA, December 13-16*, 2006.

[5] N. Bedjaoui and E. Weyer. Algorithms for leak detection, estimation, isolation and localization in open water channels. *Control Engineering Practice*, 19(6):564–573, 2011.

[6] M. Beuthe, B. Jourquin, N. Urbain, F. Bruinsma, I. Lingemann, B. Ubbels, and E. Van Heumen. Estimating the impacts of water depth and new infrastructures on transport by inland navigation: A multimodal approach for the rhine corridor. *Procedia - Social and Behavioral Sciences - Proceedings of EWGT2012 - 15th Meeting of the EURO Working Group on Transportation*, 54:387 – 401, 2012.

[7] M. Blanke, M. Kinnaert, J. Lunze, and M. Staroswiecki. Diagnosis and fault-tolerant control. *Springer-Verlag*, 2003.

[8] Y. Bolea, V. Puig, and J. Blesa. Linear parameter varying modeling and identification for real-time control of open-flow irrigation canals. *Environmental modelling & software*, 53:87–97, 2014.

[9] C. Brand, M. Tran, and J. Anable. The UK transport carbon model: An integrated life cycle approach to explore low carbon futures. *Energy Policy*, 41:107–124, 2012.

[10] S. Chae and Z. Bien. Techniques for decentralized control for interconnected systems. *in Control and Dynamic Systems*, 41:273–315, 1991.

[11] V. T. Chow. Open-channel hydraulics. *McGraw-Hill. New York.*, 1959.

[12] G. Desquesnes, G. Lozenguez, A. Doniec, and E. Duviella. Dealing with large MDPs, case study of waterway networks supervision. *in Advances in Practical Applications of Scalable Multi-agent Systems, the PAAMS Collection, Springer International Publishing*, 5124:48–59, 2016.

[13] E. Duviella, L. Bako, M. Sayed-Mouchaweh, J. Blesa, Y. Bolea, V. Puig, and K. Chuquet. Inland navigation channel model: Application to the Cuinchy-Fontinettes reach. In *Networking, Sensing and Control (ICNSC), 2013 10th IEEE International Conference on*, pages 164–169. IEEE, 2013.

[14] E. Duviella, H. Nouasse, A. Doniec, and K. Chuquet. Dynamic optimization approaches for resource allocation planning in inland navigation networks. *ETFA2016, Berlin, Allemagne, September 6-9*, 2016.

[15] E. Duviella, L. Rajaoarisoa, J. Blesa, and K. Chuquet. Fault detection and isolation of inland navigation channel: Application to the Cuinchy-Fontinettes reach. In *52nd IEEE Conference on Decision and Control, Florence, Firenze, Italy, December 10-13*, pages 4877–4882, 2013.

[16] EnviCom. Climate change and navigation - waterborne transport, ports and waterways: A review of climate change drivers, impacts, responses and mitigation. *EnviCom - Task Group 3*, 2008.

[17] F. Fele, J.M. Maestre, S. Mehdy Hashemy, D. Muñoz de la Peña, and E.F. Camacho. Coalitional model predictive control of an irrigation canal. *Journal of Process Control*, 24(4):314 – 325, 2014.

[18] L. Ferrandez-Gamot, P. Busson, J. Blesa, S. Tornil-Sin, V. Puig, E. Duviella, and A. Soldevila. Leak localization in water distribution networks using pressure residuals and classifiers. *IFAC-PapersOnLine*, 48(21):220 – 225, 2015.

[19] Z. Gaji and X. Shen. Parallel algorithms for optimal control of large-scale linear systems. *in springer-Verlag*, 1993.

[20] D.T. Gavel and D.D. Siljak. Decentralized adaptive control: structural conditions for stability. *in IEEE Transactions on Automatic Control*, 34:413–426, 1989.

[21] M. Gómez, J. Rodellar, and J. Mantecón. Predictive control method for decentralized operation of irrigation canals. *in Applied Mathematical Modelling*, 26:1039–1056, 2002.

[22] K. Horvàth, E. Duviella, J. Blesa, L. Rajaoarisoa, Y. Bolea, V. Puig, and K. Chuquet. Gray-box model of inland navigation channel: application to the Cuinchy-Fontinettes reach. *Journal of Intelligent Systems*, 23(2):183–199, 2014.

[23] K. Horvàth, E. Duviella, M. Petreczky, L. Rajaoarisoa, and K. Chuquet. Model predictive control of water levels in a navigation canal affected by resonance waves. *HIC 2014, New York, USA, 17-21 August*, 2014.

[24] K. Horvàth, E. Duviella, L. Rajaoarisoa, and K. Chuquet. Modelling of a navigation reach with unknown inputs: the Cuinchy-Fontinettes case study. *Simhydro, Sofia Antipolis, 11-13 June*, 2014.

[25] K. Horvàth, J. Blesa J., E. Duviella, and K. Chuquet. Fault-tolerant model predictive control of open channels. *Planning,*

Operation and Automation of Irrigation Delivery Systems: USCID Water Management, Phoenix, Arizona, December 2-5, 2014.

[26] K. Horvàth, J. Blesa J., E. Duviella, L. Rajaoarisoa, V. Puig, and K. Chuquet. Sensor fault diagnosis of inland navigation system using physical model and pattern recognition approach. *IFAC Proceedings Volumes*, 47(3):5309–5314, 2014. 19th IFAC World Congress.

[27] K. Horvàth, M. Petrecsky, L. Rajaoarisoa, E. Duviella, and K. Chuquet. MPC of water level in a navigation canal - The Cuinchy-Fontinettes case study. *European Control Conference, Strasbourg, France, June 24-27*, 2014.

[28] K. Horvàth, L. Rajaoarisoa, E. Duviella, J. Blesa, M. Petreczky, and K. Chuquet. Enhancing inland navigation by model predictive control of water level the Cuinchy-Fontinettes case. *Operations Research/ Computer Science Interfaces Series*, 58:211–234, 2015.

[29] IWAC. Climate change mitigation and adaptation. implications for inland waterways in england and wales. *Report*, 2009.

[30] J. Blesa J., K. Horvàth, E. Duviella, V. Puig, Y. Bolea, L. Rajaoarisoa, and K. Chuquet. Model-based sensor supervision in inland navigation networks: Cuinchy-Fontinettes case study. *Journal Maritime Research*, 11(2):81–88, 2014.

[31] C. Join, G. Robert, and M. Fliess. Model-free based water level control for hydroelectric power plants. *IFAC CMTEE, Vilamoura, Portugal*, 2010.

[32] C.T. Leondes. Control and dynamic systems. *Academic Press, Orlando, Fl*, pages 22–24, 1985.

[33] O. LePocher, E. Duviella, L. Bako, and K. Chuquet. Sensor fault detection of a real undershot/overshot gate based on physical and nonlinear black-box models. *Safeprocess'12, Mexico, Mexico, 29-31 August*, 2012.

[34] X. Litrico and V. Fromion. Simplified modeling of irrigation canals for controller design. *Journal of Irrigation and Drainage Engineering*, pages 373–383, 2004.

[35] J. Lunze. Feedback control of large-scale systems. *in Prentice-Hall*, 1992.

[36] P.-O. Malaterre. Classification of canal control algorithms. *Journal of Irrigation and Drainage Engineering*, 124(1):310, 1998.

[37] P.-O. Malaterre, D. Dorchies, and J.-P. Baume. Automatic tuning of robust PI controllers for a cascade of rivers or irrigation canals pools. *European Control Conference, Strasbourg, France, June 24-27*, 2014.

[38] P.O. Malaterre and J. Rodellar. Optimal choice of control action variables and linked algorithms: Comparison of different alternatives. *in Proceedings of USCID Workshop on modernization of irrigation water delivery systems*, pages 387–405, 1999.

[39] I. Mareels, E. Weyer, S. Ooi, M. Cantoni, Y. Li, and G. Nair. Systems engineering for irrigation systems: Successes and challenges. *Annual reviews in control*, 29(2):191204, 2005.

[40] A. Montazar, P.J. van Overloop, and R. Brouwer. Centralized controller for the Narmada main canal. *in Irrigation and drainage*, 54:79–89, 2005.

[41] J.L. Nabais, L.F. Mendonça, and M. A. Botto. New fault isolation architecture for irrigation canals. *8th IFAC Symposium on Fault Detection, Supervision and Safety of Technical Processes (SAFEPROCESS), Mexico City, Mexico, August 29-31*, 2012.

[42] J.L. Nabais, L.F. Mendonça, and M. A. Botto. Sensor fault tolerant architecture for irrigation canals. *10th Portuguese Conference on Automatic Control 16-18 July, Funchal, Portugal*, 2012.

[43] H. Nouasse, A. Doniec, G. Lozenguez, E. Duviella, P. Chiron, B. Archimde, and K. Chuquet. Constraint satisfaction problem based on flow transport graph to study the resilience of inland navigation networks in a climate change context. *IFAC Conference MIM, Troyes, France, 28-30 June*, 2016.

[44] H. Nouasse, L. Rajaoarisoa, A. Doniec, P. Chiron, E. Duviella, B. Archimde, and K. Chuquet. Study of drought impact on inland navigation systems based on a flow network model. *ICAT, Sarajevo, Bosnie Herzegovia*, 2015.

[45] J.H. Park. Robust decentralized stabilization of uncertain large-scale discrete-time systems with delays. *Journal of Optimization Theory and Applications*, 113(1):105–119, 2002.

[46] V. T. Pham, L. Lefèvre, D. Georges, and G. Besançon. Decentralized predictive control for 1D cascaded systems of conservation laws. *IFAC Proceedings Volumes*, 47(3):5163–5168, 2014. 19th IFAC World Congress.

[47] J.-C. Ponsart, D. Theilliol, and C. Aubrun. Virtual sensors design for active fault tolerant control system applied to a winding machine. *Control Engineering Practice*, 18:10371044, 2010.

[48] V. Puig and C.A. Ocampo-Martínez. Decentralised fault diagnosis of large-scale systems: Application to water transport networks. *26th International Workshop on Principles of Diagnosis (DX-2015), Paris, France, August 31- September 3*, 2015.

[49] V. Puig, J. Romera, J. Quevedo, C. M. Cardona, A. Salterain, E. Ayesa, I. Irizar, A. Castro, M. Lujan, P. Charbonnaud, P. Chiron, and J.-L. Trouvat. Optimal predictive control of water transport systems: Arret-Darre/Arros case study. *Water science and technology*, 60(8):21252133, 2009.

[50] C. Raievsky, V. T. Pham, J.-P. Jamont, and L. Lefèvre. Irrigation canals distributed model-based predictive control using multi-agent systems. *MOSIM 2014, Nancy, France, November*, 2014.

[51] L. Rajaoarisoa, K. Horvàth, E. Duviella, and K. Chuquet. Large-scale system control based on decentralized design. application to Cuinchy-Fontinettes reach. *IFAC Proceedings Volumes*, 47(3):11105–11110, 2014. 19th IFAC World Congress.

[52] D. Robles, V. Puig, C.A. Ocampo-Martínez, and L.E. Garza. Reliable fault-tolerant model predictive control of drinking water networks. *Control Engineering Practice (Special Issue on Industrial Diagnosis and Control)*, 55:197–211, 2016.

[53] J. Rodellar, M. Gómez, and J. P. Martín Vide. Stable predictive control of open-channel flow. *Journal of Irrigation and Drainage Engineering*, 115(4):701713, 1989.

[54] N.R. Sandell, P. Varaiya, M. Athans, and M.G. Safonov. Survey of decentralized control methods for large-scale systems. *in IEEE Transactions on Automatic Control*, 23:108–128, 1978.

[55] S. Sawadogo, R. Faye, A. Benhammou, and K. Akouz. Decentralized adaptive control of multi-reach irrigation canal. *IEEE International Conference on Systems, Man and Cybernetics*, page 34373442, 2000.

[56] S. Sawadogo, R.M. Faye, P.O. Malaterre, and F. Mora-Camino. Decentralized predictive controller for delivery canals. *IEEE International Conference on Systems, Man and Cybernetics*, pages 3880–3884, 1998.

[57] J. Schuurmans. Control of water levels in open channels. *PhD Thesis*, 1997.

[58] J. Schuurmans, A.J. Clemmens, S. Dijkstra, A. Hof, and R. Brouwer. Modeling of irrigation and drainage canals for controller design. *Journal of Irrigation and Drainage Engineering, December*, 125(6), 1999.

[59] D.D. Siljak. Decentralized control of complex systems. *in Control and Dynamic Systems*, 1991.

[60] A. Soldevila, J. Blesa, S. Tornil-Sin, E. Duviella, R. M. Fernandez-Canti, and V. Puig. Leak localization in water distribution networks using a mixed model-based/data-driven approach. *Control Engineering Practice*, 55:162 – 173, 2016.

[61] J. Toyoda, N. Toriumi, and Y. Inoue. An adaptive predictor of river flow for on-line control of water resource systems. *Automatica*, 5(2):175 – 181, 1969.

[62] P.J. van Overloop, I.J. Miltenburg, X. Bombois, A.J. Clemmens, R.J. Strand, N.C. van de Giesen, and R. Hut. Identification of resonance waves in open water channels. *Control Engineering Practice, Volume 18, Issue 8, August*, pages 863–872, 2010.

[63] B. T. Wahlin. Performance of model predictive control on ASCE test canal 1. *Journal of Irrigation and Drainage Engineering*, 130(3):227238, 2004.

[64] N. Wanders and Y. Wada. Human and climate impacts on the 21st century hydrological drought. *Journal of Hydrology*, 526:208–220, 2015.

[65] Y. Wang, L. Xie, and C.E. de Souza. Robust decentralized control of interconnected uncertain linear systems. In *Decision and Control, 1995., Proceedings of the 34th IEEE Conference on*, volume 3, pages 2653–2658 vol.3, 1995.

[66] E.N. Wu, S. Thavamani, Y. Zhang, and M. Blanke. Sensor fault masking of a ship propulsion. *Control Engineering Practice*, 14:13371345, 2006.

[67] Y. Zhang and J. Jiang. Bibliographical review on reconfigurable fault-tolerant control systems. *Annual Reviews in Control*, 32:229252, 2008.

IOP Conf. Series: Journal of Physics: Conf. Series **783** (2017) 012019 doi:10.1088/1742-6596/783/1/012019

Cascade structures of fault-tolerant control schemes with the static and dynamic output controllers

Dušan Krokavec and Anna Filasová

Department of Cybernetics and Artificial Intelligence, Faculty of Electrical Engineering and Informatics, Technical University of Košice, Letná 9, 042 00 Košice, Slovakia

E-mail: dusan.krokavec@tuke.sk; anna.filasova@tuke.sk

Abstract. An enhanced approach to fault-tolerant control design is proposed in the paper for linear systems subject to cascade control strategy, while static and dynamic output controllers are employed to maintain the stability of the overall interconnected control structure. The controller gains are solved simultaneously using two-step linear matrix inequality formulation, conditioned by linear matrix equalities. A simulation example, subject to a system matrix parameter fault, demonstrates the effectiveness of the proposed method of design and cascade control technique.

1. Introduction

Operating conditions in modern engineering systems are still exposed to possibility of system failure. Any failure of the sensors, actuators or other system components can drastically change the system behavior. Fault tolerant control (FTC) allows a strategy to improve reliability of the whole system and so many techniques have been proposed especially for sensor and actuator failures with application to a wide range of engineering fields. To recover at least in part the performance of the fault-free control, one of the main ideas to control reconfiguration design is to modify the feedback gain so that the reconfigured system approximates the dynamics of nominal closed-loop system. In the case of complex technological systems, it is possible to predesign controllers for the anticipated fault cases, and switch to the corresponding control law once a fault is detected and isolated.

One of the key reconfigurable control methods is reference model based control, where the main objective is to maintain as much similarity as possible achieved by reassigning the feedback gains to the original designed. Based on modified pseudoinverse methods [6], [7], [8], the lack of stability guarantee limits hardly such use in the design of FTC [15], [16]. Rather than claiming the closed-loop faulty system to have the same model as the nominal system, it can be required to design the control reconfiguration for trajectory following problem in which the reference is the nominal closed-loop system trajectory, or its approximates [2], [4].

Cascade control structures are mainly used to achieve primarily suppression of a disturbance before it propagates to the parts of the plant [13]. To apply in reconfiguration, a cascade state controller structure is proposed in [1], [9], to keep intact the original controller and the nominal model trajectory tracking, while the error between the nominal output and the measurable faulty output variables is proposed as a basis for the correction. Relaxing the existence conditions for reference model state following, the technique proposed in this paper generalizes the design principle in accordance to the model-reference control tenets subject to a system matrix parameter fault [5], [11], [12]. To achieve the desired control objective, the bi-proper dynamic output controller in combination with the static output control part is realized, as a principal novelty, within the reconfiguration cascade structure. A characteristic feature of such structure is that even after the fault occurrence is maintained the inner loop with the nominal control law parameters. The approach utilizes the measurable input and output vector variables, design conditions are based on linear matrix inequality (LMI) technique combined with regularization of bilinear forms by linear matrix inequality approach, which give an effective way to calculate the controller parameters.

The paper is organized as follows. Ensuing the introduction given in Sec. 1, Sec. 2 presents the problem formulation focusing on assumptions about the controller structure properties. In Sec. 3 the main properties of the method exploiting the cascade structure for linear systems with system dynamic faults is presented and the separation principle is proven. Subsequently, Sec. 4 - 6 derive new design results when using the bi-proper dynamic output controller, static output control and static decoupling principle in the fault-tolerant cascade control structure, all in the framework of LMIs and LMEs. Conforming the results, Sec. 7 follows with an illustrative example and simulations and, finally, some concluding remarks are reached in Sec. 8.

Throughout the paper, the following notation was used: $\boldsymbol{x}^T$, $\boldsymbol{X}^T$ denotes the transpose of the vector $\boldsymbol{x}$ and the matrix $\boldsymbol{X}$, respectively, rank$(\cdot)$ remits the rank of a matrix, for a square matrix $\boldsymbol{X} < 0$ means that $\boldsymbol{X}$ is a symmetric negative definite matrix, the symbol $\boldsymbol{I}_n$ indicates the n-th order unit matrix, $I\!R$ notes the set of real numbers, and $I\!R^n$, $I\!R^{n\times r}$ refer to the set of all n-dimensional real vectors and $n \times r$ real matrices, respectively.

2. Problem formulation

In the paper, there are taken into account square linear dynamic systems described in the fault-free conditions as

$$\dot{\boldsymbol{q}}(t) = \boldsymbol{A}\boldsymbol{q}(t) + \boldsymbol{B}\boldsymbol{u}(t)\,, \tag{1}$$

$$\boldsymbol{y}(t) = \boldsymbol{C}\boldsymbol{q}(t)\,, \tag{2}$$

where $\boldsymbol{q}(t) \in I\!R^n$ stands up for the system state vector, $\boldsymbol{u}(t) \in I\!R^r$ denotes the input vector, $\boldsymbol{y}(t) \in I\!R^m$ is the measurable output vector, $\boldsymbol{A} \in I\!R^{n\times n}$, $\boldsymbol{B} \in I\!R^{n\times r}$ and $\boldsymbol{C} \in I\!R^{m\times n}$ are real matrices and $r = m$. It is supposed that the system (1), (2) is stabilizable by the the bi-proper dynamic output controller (DOC)

$$\dot{\boldsymbol{p}}(t) = \boldsymbol{J}\boldsymbol{p}(t) + \boldsymbol{L}\boldsymbol{y}(t)\,, \tag{3}$$

$$\boldsymbol{u}(t) = \boldsymbol{M}\boldsymbol{p}(t) + \boldsymbol{N}\boldsymbol{y}(t) - \boldsymbol{W}\boldsymbol{w}(t) \tag{4}$$

of an order p, where it can be accepted $1 \leq p < n$ (reduced order), $p = n$ (full order) and $n < p \leq p_m$ (upgraded order) and where $\boldsymbol{p}(t) \in I\!R^p$ is the vector of the controller state variables, $\boldsymbol{w}(t) \in I\!R^m$ is desired output signal vector and $\boldsymbol{W} \in I\!R^{r\times m}$ is the signal gain matrix. With respect to the real matrices $\boldsymbol{J} \in I\!R^{p\times p}$, $\boldsymbol{L} \in I\!R^{p\times m}$, $\boldsymbol{M} \in I\!R^{r\times p}$, $\boldsymbol{N} \in I\!R^{r\times m}$, the controller parameter notation takes for $\boldsymbol{K}^\circ \in I\!R^{(p+r)\times(p+m)}$ the following prescribed structure

$$\boldsymbol{K}^\circ = \begin{bmatrix} \boldsymbol{J} & \boldsymbol{L} \\ \boldsymbol{M} & \boldsymbol{N} \end{bmatrix}. \tag{5}$$

The model of the system with a faulted dynamics is taken as [11]

$$\dot{\boldsymbol{q}}_f(t) = \boldsymbol{A}_f\boldsymbol{q}_f(t) + \boldsymbol{B}\boldsymbol{u}^\circ(t)\,, \tag{6}$$

$$\boldsymbol{y}_f(t) = \boldsymbol{C}\boldsymbol{q}_f(t)\,, \tag{7}$$

where $\boldsymbol{q}_f(t) \in I\!R^n$, $\boldsymbol{u}_f(t) \in I\!R^r$, $\boldsymbol{y}_f(t) \in I\!R^m$ denote the faulty system state variables vector, the vector of the input variables and the vector of output variables, respectively, $\boldsymbol{u}^{\circ}(t) \in I\!R^r$ is a composed reference input signal to the cascade structure and $\boldsymbol{A}_f \in I\!R^{n\times n}$ is the faulty system dynamics matrix. That the problem of the cascade structure design is solvable, it is assumed the pair $(\boldsymbol{A}_f, \boldsymbol{B})$ is stabilizable [12].

3. Separation principle

Considering the static output control (SOC) principle so that

$$\boldsymbol{u}^{\circ}(t) = \boldsymbol{K}\boldsymbol{y}_f(t) + \boldsymbol{u}^{\diamond}(t) = \boldsymbol{K}\boldsymbol{C}\boldsymbol{q}_f(t) + \boldsymbol{u}^{\diamond}(t)\,, \tag{8}$$

such control corresponds to the closed-loop state-space representation of the faulty system

$$\dot{\boldsymbol{q}}_f(t) = \boldsymbol{A}_{cr}\boldsymbol{q}_f(t) + \boldsymbol{B}\boldsymbol{u}^{\diamond}(t)\,, \tag{9}$$

$$\boldsymbol{y}_f(t) = \boldsymbol{C}\boldsymbol{q}_f(t)\,, \tag{10}$$

where

$$\boldsymbol{A}_{cr} = \boldsymbol{A}_f + \boldsymbol{B}\boldsymbol{K}\boldsymbol{C}\,, \tag{11}$$

while $\boldsymbol{u}^{\diamond}(t)$ is the signal acting on the input of the faulty system.

Theorem 1 *The state of the faulty system (9), (10) under the control policy*

$$\boldsymbol{u}(t) = \boldsymbol{M}\boldsymbol{p}(t) + \boldsymbol{N}\boldsymbol{y}(t) - \boldsymbol{W}\boldsymbol{w}(t)\,, \tag{12}$$

$$\boldsymbol{u}^{\circ}(t) = \boldsymbol{M}\boldsymbol{p}_f(t) + \boldsymbol{N}\boldsymbol{y}_f(t) - \boldsymbol{K}\boldsymbol{y}_r(t)\,, \tag{13}$$

asymptotically converge to the state of the reference model if $\boldsymbol{A}_{cn}$, $\boldsymbol{A}_{cr}$, $\boldsymbol{J}$ are Hurwitz, where

$$\boldsymbol{A}_{cn} = \boldsymbol{A} + \boldsymbol{B}\boldsymbol{N}\boldsymbol{C}\,, \quad \boldsymbol{A}_{cr} = \boldsymbol{A}_s + \boldsymbol{B}\boldsymbol{K}\boldsymbol{C}\,, \quad \boldsymbol{A}_s = \boldsymbol{A}_{cn} + \boldsymbol{A}_f - \boldsymbol{A}\,, \tag{14}$$

$\boldsymbol{N} \in I\!R^{r\times m}$ is the dynamic controller gain matrix, $\boldsymbol{K} \in I\!R^{r\times r}$ is the static output control gain matrix, $\boldsymbol{W} \in I\!R^{r\times m}$ is the signal gain matrix and $\boldsymbol{J} \in I\!R^{p\times p}$ is the system matrix of the dynamic controller.

If $\boldsymbol{A}_{cn}$, $\boldsymbol{A}_{cr}$, $\boldsymbol{J}$ are Hurwitz and

$$\boldsymbol{e}_q(t) = \boldsymbol{q}(t) - \boldsymbol{q}_f(t)\,, \quad \boldsymbol{e}_p(t) = \boldsymbol{p}(t) - \boldsymbol{p}_f(t)\,, \quad \boldsymbol{e}_y(t) = \boldsymbol{C}\boldsymbol{e}_q(t)\,, \tag{15}$$

then the autonomous equations

$$\dot{\boldsymbol{q}}_f(t) = \boldsymbol{A}_{cr}\boldsymbol{q}_f(t)\,, \quad \dot{\boldsymbol{e}}_q(t) = \boldsymbol{A}_{cn}\boldsymbol{e}_q(t)\,, \quad \dot{\boldsymbol{p}}_f(t) = \boldsymbol{J}\boldsymbol{p}_f(t)\,, \quad \dot{\boldsymbol{e}}_p(t) = \boldsymbol{J}\boldsymbol{e}_p(t) \tag{16}$$

are asymptotically stable.

Proof: Describing the dynamics of the bi-proper dynamic controller in a faulty regime as

$$\dot{\boldsymbol{p}}_f(t) = \boldsymbol{J}\boldsymbol{p}_f(t) + \boldsymbol{L}\boldsymbol{y}_f(t)\,, \tag{17}$$

where $\boldsymbol{p}_f(t) \in I\!R^p$ is the vector of the controller state variables in the faulty regime then, incorporating all state vectors occurring in the control loop, the composed dynamic model becomes

$$\begin{bmatrix} \dot{\boldsymbol{q}}_f(t) \\ \dot{\boldsymbol{q}}(t) \\ \dot{\boldsymbol{p}}_f(t) \\ \dot{\boldsymbol{p}}(t) \end{bmatrix} = \begin{bmatrix} \boldsymbol{A}_{cr} & \boldsymbol{0} & \boldsymbol{0} & \boldsymbol{0} \\ \boldsymbol{0} & \boldsymbol{A} & \boldsymbol{0} & \boldsymbol{0} \\ \boldsymbol{L}\boldsymbol{C} & \boldsymbol{0} & \boldsymbol{J} & \boldsymbol{0} \\ \boldsymbol{0} & \boldsymbol{L}\boldsymbol{C} & \boldsymbol{0} & \boldsymbol{J} \end{bmatrix} \begin{bmatrix} \boldsymbol{q}_f(t) \\ \boldsymbol{q}(t) \\ \boldsymbol{p}_f(t) \\ \boldsymbol{p}(t) \end{bmatrix} + \begin{bmatrix} \boldsymbol{B} & \boldsymbol{0} \\ \boldsymbol{0} & \boldsymbol{B} \\ \boldsymbol{0} & \boldsymbol{0} \\ \boldsymbol{0} & \boldsymbol{0} \end{bmatrix} \begin{bmatrix} \boldsymbol{u}^{\diamond}(t) \\ \boldsymbol{u}(t) \end{bmatrix}. \tag{18}$$

To apply the separation principle, the transform matrix $\boldsymbol{T}$ is defined as

$$\boldsymbol{T}=\begin{bmatrix}\boldsymbol{I}_n&\boldsymbol{0}&\boldsymbol{0}&\boldsymbol{0}\\-\boldsymbol{I}_n&\boldsymbol{I}_n&\boldsymbol{0}&\boldsymbol{0}\\\boldsymbol{0}&\boldsymbol{0}&\boldsymbol{I}_p&\boldsymbol{0}\\\boldsymbol{0}&\boldsymbol{0}&-\boldsymbol{I}_p&\boldsymbol{I}_p\end{bmatrix},\quad \boldsymbol{T}^{-1}=\begin{bmatrix}\boldsymbol{I}_n&\boldsymbol{0}&\boldsymbol{0}&\boldsymbol{0}\\\boldsymbol{I}_n&\boldsymbol{I}_n&\boldsymbol{0}&\boldsymbol{0}\\\boldsymbol{0}&\boldsymbol{0}&\boldsymbol{I}_p&\boldsymbol{0}\\\boldsymbol{0}&\boldsymbol{0}&\boldsymbol{I}_p&\boldsymbol{I}_p\end{bmatrix},\tag{19}$$

where, carrying out the indicated separation, it yields

$$\boldsymbol{T}\begin{bmatrix}\boldsymbol{q}_f(t)\\\boldsymbol{q}(t)\\\boldsymbol{p}_f(t)\\\boldsymbol{p}(t)\end{bmatrix}=\begin{bmatrix}\boldsymbol{I}_n&\boldsymbol{0}&\boldsymbol{0}&\boldsymbol{0}\\-\boldsymbol{I}_n&\boldsymbol{I}_n&\boldsymbol{0}&\boldsymbol{0}\\\boldsymbol{0}&\boldsymbol{0}&\boldsymbol{I}_p&\boldsymbol{0}\\\boldsymbol{0}&\boldsymbol{0}&-\boldsymbol{I}_p&\boldsymbol{I}_p\end{bmatrix}\begin{bmatrix}\boldsymbol{q}_f(t)\\\boldsymbol{q}(t)\\\boldsymbol{p}_f(t)\\\boldsymbol{p}(t)\end{bmatrix}=\begin{bmatrix}\boldsymbol{q}_f(t)\\\boldsymbol{e}_q(t)\\\boldsymbol{p}_f(t)\\\boldsymbol{e}_p(t)\end{bmatrix},\tag{20}$$

$$\boldsymbol{T}\begin{bmatrix}\boldsymbol{A}_{cr}&\boldsymbol{0}&\boldsymbol{0}&\boldsymbol{0}\\\boldsymbol{0}&\boldsymbol{A}&\boldsymbol{0}&\boldsymbol{0}\\\boldsymbol{LC}&\boldsymbol{0}&\boldsymbol{J}&\boldsymbol{0}\\\boldsymbol{0}&\boldsymbol{LC}&\boldsymbol{0}&\boldsymbol{J}\end{bmatrix}\boldsymbol{T}^{-1}=\begin{bmatrix}\boldsymbol{A}_{cr}&\boldsymbol{0}&\boldsymbol{0}&\boldsymbol{0}\\\boldsymbol{A}-\boldsymbol{A}_{cr}&\boldsymbol{A}&\boldsymbol{0}&\boldsymbol{0}\\\boldsymbol{LC}&\boldsymbol{0}&\boldsymbol{J}&\boldsymbol{0}\\\boldsymbol{0}&\boldsymbol{LC}&\boldsymbol{0}&\boldsymbol{J}\end{bmatrix},\tag{21}$$

$$\begin{bmatrix}\boldsymbol{I}_n&\boldsymbol{0}&\boldsymbol{0}&\boldsymbol{0}\\-\boldsymbol{I}_n&\boldsymbol{I}_n&\boldsymbol{0}&\boldsymbol{0}\\\boldsymbol{0}&\boldsymbol{0}&\boldsymbol{I}_p&\boldsymbol{0}\\\boldsymbol{0}&\boldsymbol{0}&-\boldsymbol{I}_p&\boldsymbol{I}_p\end{bmatrix}\begin{bmatrix}\boldsymbol{B}&\boldsymbol{0}\\\boldsymbol{0}&\boldsymbol{B}\\\boldsymbol{0}&\boldsymbol{0}\\\boldsymbol{0}&\boldsymbol{0}\end{bmatrix}=\begin{bmatrix}\boldsymbol{B}&\boldsymbol{0}\\-\boldsymbol{B}&\boldsymbol{B}\\\boldsymbol{0}&\boldsymbol{0}\\\boldsymbol{0}&\boldsymbol{0}\end{bmatrix},\tag{22}$$

that is (18) can be rewritten in an equivalent form

$$\begin{bmatrix}\dot{\boldsymbol{q}}_f(t)\\\dot{\boldsymbol{e}}_q(t)\\\dot{\boldsymbol{p}}_f(t)\\\dot{\boldsymbol{e}}_p(t)\end{bmatrix}=\begin{bmatrix}\boldsymbol{A}_{cr}&\boldsymbol{0}&\boldsymbol{0}&\boldsymbol{0}\\\boldsymbol{A}-\boldsymbol{A}_{cr}&\boldsymbol{A}&\boldsymbol{0}&\boldsymbol{0}\\\boldsymbol{LC}&\boldsymbol{0}&\boldsymbol{J}&\boldsymbol{0}\\\boldsymbol{0}&\boldsymbol{LC}&\boldsymbol{0}&\boldsymbol{J}\end{bmatrix}\begin{bmatrix}\boldsymbol{q}_f(t)\\\boldsymbol{e}_q(t)\\\boldsymbol{p}_f(t)\\\boldsymbol{e}_p(t)\end{bmatrix}+\begin{bmatrix}\boldsymbol{B}&\boldsymbol{0}\\-\boldsymbol{B}&\boldsymbol{B}\\\boldsymbol{0}&\boldsymbol{0}\\\boldsymbol{0}&\boldsymbol{0}\end{bmatrix}\begin{bmatrix}\boldsymbol{u}^{\circ}(t)\\\boldsymbol{u}(t)\end{bmatrix}.\tag{23}$$

Considering the composed cascade structure input signals as (12), (13), it is easily verified that

$$\begin{aligned}&\begin{bmatrix}\boldsymbol{B}&\boldsymbol{0}\\-\boldsymbol{B}&\boldsymbol{B}\\\boldsymbol{0}&\boldsymbol{0}\\\boldsymbol{0}&\boldsymbol{0}\end{bmatrix}\begin{bmatrix}\boldsymbol{Mp}_f(t)+\boldsymbol{Ny}_f(t)-\boldsymbol{Ky}_r(t)\\\boldsymbol{Mp}(t)+\boldsymbol{Ny}(t)-\boldsymbol{Ww}(t)\end{bmatrix}\\&=\begin{bmatrix}\boldsymbol{BMp}_f(t)+\boldsymbol{BNy}_f(t)-\boldsymbol{BKy}_r(t)\\\boldsymbol{BMe}_p(t)+\boldsymbol{BNCe}_q(t)-\boldsymbol{BWw}(t)+\boldsymbol{BKy}_r(t)\end{bmatrix}\end{aligned}\tag{24}$$

and, consequently,

$$\begin{aligned}\begin{bmatrix}\dot{\boldsymbol{q}}_f(t)\\\dot{\boldsymbol{e}}_q(t)\\\dot{\boldsymbol{p}}_f(t)\\\dot{\boldsymbol{e}}_p(t)\end{bmatrix}&=\begin{bmatrix}\boldsymbol{A}_{cr}+\boldsymbol{BNC}&\boldsymbol{0}&\boldsymbol{BM}&\boldsymbol{0}\\\boldsymbol{A}-\boldsymbol{A}_{cr}&\boldsymbol{A}+\boldsymbol{BNC}&\boldsymbol{0}&\boldsymbol{BM}\\\boldsymbol{LC}&\boldsymbol{0}&\boldsymbol{J}&\boldsymbol{0}\\\boldsymbol{0}&\boldsymbol{LC}&\boldsymbol{0}&\boldsymbol{J}\end{bmatrix}\begin{bmatrix}\boldsymbol{q}_f(t)\\\boldsymbol{e}_q(t)\\\boldsymbol{p}_f(t)\\\boldsymbol{e}_p(t)\end{bmatrix}\\&+\begin{bmatrix}\boldsymbol{0}&-\boldsymbol{BK}\\-\boldsymbol{BW}&\boldsymbol{BK}\\\boldsymbol{0}&\boldsymbol{0}\\\boldsymbol{0}&\boldsymbol{0}\end{bmatrix}\begin{bmatrix}\boldsymbol{w}(t)\\\boldsymbol{y}_r(t)\end{bmatrix}.\end{aligned}\tag{25}$$

Because of the block structure of the connected system matrix in (25), it is evident that the separation principle holds.

Using the terms given above, it can write

$$\boldsymbol{A}_{cr} + \boldsymbol{BNC} = \boldsymbol{A}_f + \boldsymbol{BKC} + \boldsymbol{BNC} = \boldsymbol{A} + \boldsymbol{BNC} + \boldsymbol{A}_f - \boldsymbol{A} + \boldsymbol{BKC}, \tag{26}$$

$$\boldsymbol{A}_{cr} + \boldsymbol{BNC} = \boldsymbol{A}_{cn} + \boldsymbol{A}_f - \boldsymbol{A} + \boldsymbol{BKC} = \boldsymbol{A}_s + \boldsymbol{BKC} = \boldsymbol{A}_{sc}, \tag{27}$$

respectively, and (27) gives (14).

Since the separation principle yields, to obtain an asymptotic stable solution of (25), the matrices $\boldsymbol{A}_{cn}$, $\boldsymbol{A}_{cr}$ of the structures (14) and the matrix $\boldsymbol{J}$ have to be Hurwitz. This concludes the proof. ∎

4. Dynamic controller parameter design

Exploiting the separation principle, it can write for (15) and the error subsystem autonomous block in (25) that

$$\begin{bmatrix} \dot{\boldsymbol{e}}_q(t) \\ \dot{\boldsymbol{e}}_p(t) \end{bmatrix} = \begin{bmatrix} \boldsymbol{A} + \boldsymbol{BNC} & \boldsymbol{BM} \\ \boldsymbol{LC} & \boldsymbol{J} \end{bmatrix} \begin{bmatrix} \boldsymbol{e}_q(t) \\ \boldsymbol{e}_p(t) \end{bmatrix}, \tag{28}$$

$$\boldsymbol{e}_y(t) = \begin{bmatrix} \boldsymbol{0} & \boldsymbol{I}_m \end{bmatrix} \begin{bmatrix} \boldsymbol{0} & \boldsymbol{I}_p \\ \boldsymbol{C} & \boldsymbol{0} \end{bmatrix} \begin{bmatrix} \boldsymbol{e}_q(t) \\ \boldsymbol{e}_p(t) \end{bmatrix}. \tag{29}$$

Introducing the notations

$$\boldsymbol{e}^{\circ T}(t) = \begin{bmatrix} \boldsymbol{e}_q^T(t) & \boldsymbol{e}_p^T(t) \end{bmatrix}, \tag{30}$$

$$\boldsymbol{A}^\circ = \begin{bmatrix} \boldsymbol{A} & \boldsymbol{0} \\ \boldsymbol{0} & \boldsymbol{0} \end{bmatrix}, \ \boldsymbol{B}^\circ = \begin{bmatrix} \boldsymbol{0} & \boldsymbol{B} \\ \boldsymbol{I}_p & \boldsymbol{0} \end{bmatrix}, \ \boldsymbol{C}^\circ = \begin{bmatrix} \boldsymbol{0} & \boldsymbol{I}_p \\ \boldsymbol{C} & \boldsymbol{0} \end{bmatrix}, \ \boldsymbol{I}^\circ = \begin{bmatrix} \boldsymbol{0} & \boldsymbol{I}_m \end{bmatrix}, \tag{31}$$

where $\boldsymbol{A}^\circ \in I\!\!R^{(n+p)\times(n+p)}$, $\boldsymbol{B}^\circ \in I\!\!R^{(n+p)\times(p+r)}$, $\boldsymbol{C}^\circ \in I\!\!R^{(p+m)\times(n+p)}$, $\boldsymbol{I}^\circ \in I\!\!R^{m\times(p+m)}$, than the state-space equations (28), (29) take the forms

$$\dot{\boldsymbol{e}}^\circ(t) = \boldsymbol{A}_c^\circ \boldsymbol{e}^\circ(t), \tag{32}$$

$$\boldsymbol{y}^\circ(t) = \boldsymbol{I}^\circ \boldsymbol{C}^\circ \boldsymbol{e}^\circ(t), \tag{33}$$

where, with $\boldsymbol{K}^\circ$ given in (5),

$$\boldsymbol{A}_c^\circ = \boldsymbol{A}^\circ + \boldsymbol{B}^\circ \boldsymbol{K}^\circ \boldsymbol{C}^\circ. \tag{34}$$

In the sequel, it is supposed that the pair $(\boldsymbol{A}^\circ, \boldsymbol{B}^\circ)$ is stabilizable and the couple $(\boldsymbol{A}^\circ, \boldsymbol{C}^\circ)$ is detectable [3].

Theorem 2 *The error equation (32) converges asymptotically to their equilibrium if there exist a symmetric positive definite matrix $\boldsymbol{Q}^\circ \in I\!\!R^{(n+p)\times(n+p)}$, a regular matrix $\boldsymbol{H}^\circ \in I\!\!R^{(p+m)\times(p+m)}$ and a matrix $\boldsymbol{Y}^\circ \in I\!\!R^{(p+r)\times(p+m)}$ such that*

$$\boldsymbol{Q}^\circ = \boldsymbol{Q}^{\circ T} > 0, \tag{35}$$

$$\boldsymbol{A}^\circ \boldsymbol{Q}^\circ + \boldsymbol{Q}^\circ \boldsymbol{A}^{\circ T} + \boldsymbol{B}^\circ \boldsymbol{Y}^\circ \boldsymbol{C}^\circ + \boldsymbol{C}^{\circ T} \boldsymbol{Y}^{\circ T} \boldsymbol{B}^{\circ T} < 0, \tag{36}$$

$$\boldsymbol{C}^\circ \boldsymbol{Q}^\circ = \boldsymbol{H}^\circ \boldsymbol{C}^\circ. \tag{37}$$

When the above conditions hold then the composed gain matrix of the bi-proper dynamic controller is given as

$$\boldsymbol{K}^\circ = \boldsymbol{Y}^\circ (\boldsymbol{H}^\circ)^{-1} = \begin{bmatrix} \boldsymbol{J} & \boldsymbol{L} \\ \boldsymbol{M} & \boldsymbol{N} \end{bmatrix}. \tag{38}$$

Proof: Defining the Lyapunov function candidate as follows

$$v(\boldsymbol{e}^{\circ}(t)) = \boldsymbol{e}^{\circ T}(t)\boldsymbol{P}^{\circ}\boldsymbol{e}^{\circ}(t) > 0\,, \tag{39}$$

where $\boldsymbol{P}^{\circ} > 0$ is a symmetric positive definite matrix, then it has to be satisfied

$$\dot{v}(\boldsymbol{e}^{\circ}(t)) = \dot{\boldsymbol{e}}^{\circ T}(t)\boldsymbol{P}^{\circ}\boldsymbol{e}^{\circ}(t) + \boldsymbol{e}^{\circ T}(t)\boldsymbol{P}^{\circ}\dot{\boldsymbol{e}}^{\circ}(t) < 0\,. \tag{40}$$

Substituting (32) into (40) results

$$\dot{v}(\boldsymbol{e}^{\circ}(t)) = \boldsymbol{e}^{\circ T}(t)(\boldsymbol{A}_c^{\circ T}\boldsymbol{P}^{\circ} + \boldsymbol{P}^{\circ}\boldsymbol{A}_c^{\circ})\boldsymbol{e}^{\circ}(t) < 0 \tag{41}$$

and, evidently, (41) implies

$$\boldsymbol{P}^{\circ}\boldsymbol{A}_c^{\circ} + \boldsymbol{A}_c^{\circ T}\boldsymbol{P}^{\circ} < 0\,. \tag{42}$$

Using (34), the open structure of (42) takes the form of the bilinear matrix inequality

$$\boldsymbol{P}^{\circ}(\boldsymbol{A}^{\circ} + \boldsymbol{B}^{\circ}\boldsymbol{K}^{\circ}\boldsymbol{C}^{\circ}) + (\boldsymbol{A}^{\circ} + \boldsymbol{B}^{\circ}\boldsymbol{K}^{\circ}\boldsymbol{C}^{\circ})^{T}\boldsymbol{P}^{\circ} < 0\,, \tag{43}$$

which can be rewritten as

$$(\boldsymbol{A}^{\circ} + \boldsymbol{B}^{\circ}\boldsymbol{K}^{\circ}\boldsymbol{C}^{\circ})\boldsymbol{Q}^{\circ} + \boldsymbol{Q}^{\circ}(\boldsymbol{A}^{\circ} + \boldsymbol{B}^{\circ}\boldsymbol{K}^{\circ}\boldsymbol{C}^{\circ})^{T} < 0\,, \tag{44}$$

where $\boldsymbol{Q}^{\circ} = (\boldsymbol{P}^{\circ})^{-1}$. Writing now

$$\boldsymbol{B}^{\circ}\boldsymbol{K}^{\circ}\boldsymbol{C}^{\circ}\boldsymbol{Q}^{\circ} = \boldsymbol{B}^{\circ}\boldsymbol{K}^{\circ}\boldsymbol{H}^{\circ}(\boldsymbol{H}^{\circ})^{-1}\boldsymbol{C}^{\circ}\boldsymbol{Q}^{\circ}, \tag{45}$$

where $\boldsymbol{H}^{\circ}$ is a regular square matrix, and defining the equality

$$(\boldsymbol{H}^{\circ})^{-1}\boldsymbol{C}^{\circ} = \boldsymbol{C}^{\circ}(\boldsymbol{Q}^{\circ})^{-1}, \tag{46}$$

then it yields

$$\boldsymbol{B}^{\circ}\boldsymbol{K}^{\circ}\boldsymbol{C}^{\circ}\boldsymbol{Q}^{\circ} = \boldsymbol{B}^{\circ}\boldsymbol{K}^{\circ}\boldsymbol{H}^{\circ}\boldsymbol{C}^{\circ} = \boldsymbol{B}^{\circ}\boldsymbol{Y}^{\circ}\boldsymbol{C}^{\circ}, \tag{47}$$

where

$$\boldsymbol{Y}^{\circ} = \boldsymbol{K}^{\circ}\boldsymbol{H}^{\circ}. \tag{48}$$

Thus, with (47) then (44) imply (36) and the equality (46) gives (37). This concludes the proof. ■

5. Static output control parameter design

Theorem 3 *The controlled state in the closed-loop structure with faulty system is stable if there exist a positive definite symmetric matrix* $\boldsymbol{Q} \in \mathbb{R}^{n \times n}$, *a regular matrix* $\boldsymbol{H} \in \mathbb{R}^{m \times m}$ *and matrix* $\boldsymbol{Y} \in \mathbb{R}^{m \times m}$ *such that*

$$\boldsymbol{Q} = \boldsymbol{Q}^{T} > 0\,, \tag{49}$$

$$\boldsymbol{A}_s\boldsymbol{Q} + \boldsymbol{Q}\boldsymbol{A}_s^{T} + \boldsymbol{B}\boldsymbol{Y}\boldsymbol{C} + \boldsymbol{C}^{T}\boldsymbol{Y}^{T}\boldsymbol{B}^{T} < 0\,, \tag{50}$$

$$\boldsymbol{C}\boldsymbol{Q} = \boldsymbol{H}\boldsymbol{C}\,, \tag{51}$$

where

$$\boldsymbol{A}_s = \boldsymbol{A}_{cn} + \boldsymbol{A}_f - \boldsymbol{A}\,, \quad \boldsymbol{A}_{cn} = \boldsymbol{A} + \boldsymbol{B}\boldsymbol{N}\boldsymbol{C}\,. \tag{52}$$

When the above conditions hold, the static output control gain matrix is computed as

$$\boldsymbol{K} = \boldsymbol{Y}\boldsymbol{H}^{-1}. \tag{53}$$

Proof: Using the separation principle, (25), (27) imply the following equation, describing the autonomous regime of the closed-loop system state with the faulty system matrix $\boldsymbol{A}_f$,

$$\dot{\boldsymbol{q}}_f(t) = (\boldsymbol{A}_s + \boldsymbol{BKC})\boldsymbol{q}_f(t)\,, \tag{54}$$

where $\boldsymbol{A}_s$ is prescribed in (52).

Considering the Lyapunov function candidate in the form

$$v(\boldsymbol{q}_f(t)) = \boldsymbol{q}_f^T(t)\boldsymbol{P}\boldsymbol{q}_f(t) > 0\,, \tag{55}$$

where $\boldsymbol{P} \in I\!R^{n\times n}$ is a square, symmetric, positive definite matrix, then the time derivative of (55) along a trajectory of (54) takes the form

$$\dot{v}(\boldsymbol{q}(t)) = \dot{\boldsymbol{q}}_f^T(t)\boldsymbol{P}\boldsymbol{q}_f(t) + \boldsymbol{q}_f^T(t)\boldsymbol{P}\dot{\boldsymbol{q}}_f(t) < 0\,. \tag{56}$$

Therefore, using (54), the inequality (56) can be written as

$$\dot{v}(\boldsymbol{q}_f(t)) = \boldsymbol{q}_f^T(t)\boldsymbol{P}_s\boldsymbol{q}_f(t) < 0\,, \tag{57}$$

where

$$\boldsymbol{P}_s = (\boldsymbol{A}_s + \boldsymbol{BKC})^T\boldsymbol{P} + \boldsymbol{P}(\boldsymbol{A}_s + \boldsymbol{BKC}) < 0\,. \tag{58}$$

Pre-multiplying the left-hand side and post-multiplying the right-hand side of (58) by the matrix $\boldsymbol{Q} = \boldsymbol{P}^{-1}$ leads to the bilinear matrix inequality

$$(\boldsymbol{A}_s + \boldsymbol{BKC})\boldsymbol{Q} + \boldsymbol{Q}(\boldsymbol{A}_s + \boldsymbol{BKC})^T < 0\,. \tag{59}$$

Writing, analogously as above,

$$\boldsymbol{BKCQ} = \boldsymbol{BKHH}^{-1}\boldsymbol{CQ} = \boldsymbol{BYC}\,, \tag{60}$$

where $\boldsymbol{H}$ is a regular square matrix, and

$$\boldsymbol{H}^{-1}\boldsymbol{C} = \boldsymbol{CQ}^{-1}, \quad \boldsymbol{Y} = \boldsymbol{KH}\,, \tag{61}$$

then (59) implies (50) and the equalities in (61) gives (51), (53). This concludes the proof. ∎

6. Signal gain matrix design

Naturally, the control law (12) inherently defines a forced regime, where $\boldsymbol{w}(t) \in I\!R^m$ is desired output signal vector and $\boldsymbol{W} \in I\!R^{m\times m}$ is the signal gain matrix. Using the static decoupling principle, the conditions to design the signal gain matrices $\boldsymbol{W}_n$, $\boldsymbol{W}_f$ can be derived considering that $\boldsymbol{y}_r(t) = \boldsymbol{0}$ by using the principle of superposition.

Theorem 4 *If the system (1), (2) is stabilizable by the control policy (3), (4), and [17]*

$$rank\begin{bmatrix} \boldsymbol{A} & \boldsymbol{B} \\ \boldsymbol{C} & \boldsymbol{0} \end{bmatrix} = rank\begin{bmatrix} \boldsymbol{A}_f & \boldsymbol{B} \\ \boldsymbol{C} & \boldsymbol{0} \end{bmatrix} = n + m\,, \tag{62}$$

then the matrix $\boldsymbol{W}$ *takes, in the dependency on the system regime, the forms*

$$\boldsymbol{W}_n = \Big(\boldsymbol{C}(\boldsymbol{A} - \boldsymbol{BMJ}^{-1}\boldsymbol{LC} + \boldsymbol{BNC})^{-1}\boldsymbol{B}\Big)^{-1}. \tag{63}$$

$$\boldsymbol{W}_f = \Big(\boldsymbol{C}(\boldsymbol{A}_f - \boldsymbol{BMJ}^{-1}\boldsymbol{LC} + \boldsymbol{BNC})^{-1}\boldsymbol{B}\Big)^{-1}. \tag{64}$$

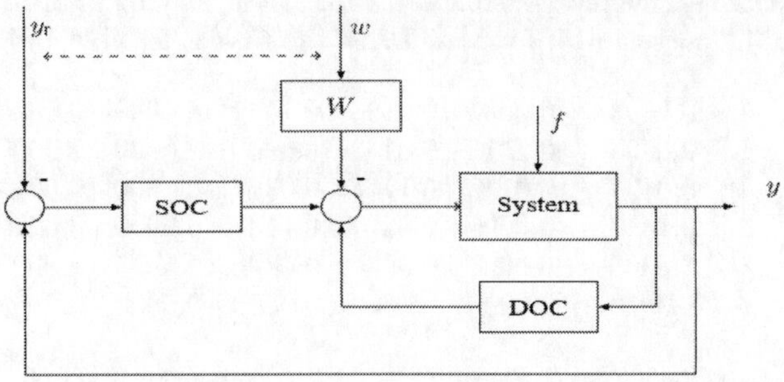

Figure 1. Cascade control reconfiguration scheme

Proof: In a steady-state, the system equations (1), (2) and the control (3), (4) imply

$$\mathbf{0} = \boldsymbol{A}\boldsymbol{q}_o + \boldsymbol{B}\boldsymbol{u}_o, \tag{65}$$

$$\mathbf{0} = \boldsymbol{J}\boldsymbol{p}_o + \boldsymbol{L}\boldsymbol{C}\boldsymbol{q}_o, \tag{66}$$

$$\boldsymbol{u}_o = \boldsymbol{M}\boldsymbol{p}_o + \boldsymbol{N}\boldsymbol{C}\boldsymbol{q}_o - \boldsymbol{W}\boldsymbol{w}_o\,, \tag{67}$$

where $\boldsymbol{q}_o$, $\boldsymbol{u}_o$, $\boldsymbol{p}_o$, $\boldsymbol{w}_o$ are the steady-state values of the vectors $\boldsymbol{q}(t)$, $\boldsymbol{u}(t)$, $\boldsymbol{p}(t)$, $\boldsymbol{w}(t)$, respectively.

Since (66), (67) in the steady-state give

$$\boldsymbol{u}_o = (-\boldsymbol{M}\boldsymbol{J}^{-1}\boldsymbol{L}\boldsymbol{C} + \boldsymbol{N}\boldsymbol{C})\boldsymbol{q}_o - \boldsymbol{W}\boldsymbol{w}_o\,, \tag{68}$$

then, substituting (68) into (65), it yields

$$\mathbf{0} = (\boldsymbol{A} - \boldsymbol{B}\boldsymbol{M}\boldsymbol{J}^{-1}\boldsymbol{L}\boldsymbol{C} + \boldsymbol{B}\boldsymbol{N}\boldsymbol{C})\boldsymbol{q}_o - \boldsymbol{B}\boldsymbol{W}\boldsymbol{w}_o\,, \tag{69}$$

$$\boldsymbol{q}_o = (\boldsymbol{A} - \boldsymbol{B}\boldsymbol{M}\boldsymbol{J}^{-1}\boldsymbol{L}\boldsymbol{C} + \boldsymbol{B}\boldsymbol{N}\boldsymbol{C})^{-1}\boldsymbol{B}\boldsymbol{W}\boldsymbol{w}_o\,, \tag{70}$$

respectively, and

$$\boldsymbol{y}_o = \boldsymbol{C}(\boldsymbol{A} - \boldsymbol{B}\boldsymbol{M}\boldsymbol{J}^{-1}\boldsymbol{L}\boldsymbol{C} + \boldsymbol{B}\boldsymbol{N}\boldsymbol{C})^{-1}\boldsymbol{B}\boldsymbol{W}\boldsymbol{w}_o\,. \tag{71}$$

Thus, considering $\boldsymbol{y}_o = \boldsymbol{w}_o$, then (71) implies (63).

Analogously can be derived (64). This concludes the proof. ■

The cascade reconfiguration control scheme is presented in Fig. 1, where $\boldsymbol{W} = \boldsymbol{W}_n$ under nominal control and $\boldsymbol{W} = \boldsymbol{W}_f$ after reconfiguration activation. To obtain a forced mode structure, the reference input signal $\boldsymbol{y}_r(t)$ has to be connected to $\boldsymbol{w}(t)$.

7. Illustrative example

In the example, there is considered the system (1), (2) in the state-space representation, where the system matrices are [10]

$$\boldsymbol{A} = \begin{bmatrix} 1.380 & -0.208 & 6.715 & -5.676 \\ -0.581 & -4.290 & 0.000 & 0.675 \\ 1.067 & 4.273 & -6.654 & 5.893 \\ 0.048 & 4.273 & 1.343 & -2.104 \end{bmatrix}, \quad \boldsymbol{B} = \begin{bmatrix} 0.000 & 0.000 \\ 5.679 & 0.000 \\ 1.136 & -3.146 \\ 1.136 & 0.000 \end{bmatrix},$$

$$\boldsymbol{A}_f = \begin{bmatrix} 1.380 & -0.208 & 6.715 & -5.676 \\ -0.581 & -4.290 & 0.000 & 0.675 \\ 2.134 & 8.546 & -13.308 & 11.786 \\ 0.048 & 4.273 & 1.343 & -2.104 \end{bmatrix}, \quad \boldsymbol{C} = \begin{bmatrix} 1 & 2 & 1 & 1 \\ 1 & 1 & 0 & 2 \end{bmatrix}.$$

Considering the order of dynamic controller $p = 2$ then, solving (35)-(37) using Self-Dual-Minimization (SeDuMi) package [14] for MATLAB, the task is feasible and

$$\boldsymbol{Q}^{\circ} = \begin{bmatrix} 0.2498 & -0.0036 & 0.0193 & 0.1547 & 0.0000 & 0.0000 \\ -0.0036 & 0.2515 & 0.0142 & 0.0000 & 0.0000 & 0.0000 \\ 0.0193 & 0.0142 & 0.1773 & 0.0387 & 0.0000 & 0.0000 \\ 0.1547 & 0.0000 & 0.0387 & 0.1437 & 0.0000 & 0.0000 \\ 0.0000 & 0.0000 & 0.0000 & 0.0000 & 0.5865 & 0.0000 \\ 0.0000 & 0.0000 & 0.0000 & 0.0000 & 0.0000 & 0.5865 \end{bmatrix},$$

$$\boldsymbol{Y}^{\circ} = \begin{bmatrix} -0.4337 & 0.0000 & 0.0000 & 0.0000 \\ 0.0000 & -0.4337 & 0.0000 & 0.0000 \\ 0.0000 & 0.0000 & 0.0033 & -0.2246 \\ 0.0000 & 0.0000 & 0.1453 & 0.1304 \end{bmatrix}, \quad \boldsymbol{H}^{\circ} = \begin{bmatrix} 0.5865 & 0.0000 & 0.0000 & 0.0000 \\ 0.0000 & 0.5865 & 0.0000 & 0.0000 \\ 0.0000 & 0.0000 & 0.2547 & 0.6573 \\ 0.0000 & 0.0000 & 0.0387 & 0.1437 \end{bmatrix}.$$

The dynamic controller gain matrices are computed using (38) as follows

$$\boldsymbol{J} = \begin{bmatrix} -0.7395 & 0.0000 \\ 0.0000 & -0.7394 \end{bmatrix}, \quad \boldsymbol{L} = 10^{-3}\begin{bmatrix} -0.0131 & 0.1031 \\ 0.0369 & -0.3260 \end{bmatrix},$$

$$\boldsymbol{M} = 10^{-4}\begin{bmatrix} -0.1101 & -0.0180 \\ -0.1553 & -0.1631 \end{bmatrix}, \quad \boldsymbol{N} = \begin{bmatrix} 0.8193 & -5.3100 \\ 1.4158 & -5.5686 \end{bmatrix},$$

and guaranties the stable closed-loop state convergence, where the system matrix eigenvalues spectrum is

$$\rho(\boldsymbol{A}_c^{\circ}) = \{-0.7394 \ -0.7395 \ -3.6642 \pm 3.1351\,\mathrm{i} \ -6.9476 \pm 11.6631\,\mathrm{i}\}.$$

Moreover, evidently, the structure of $\boldsymbol{J}$ implies that the bi-proper dynamic controller is stable.

To obtain the parameters for the static output controllers, (49)-(51) is solved with the result

$$\boldsymbol{Q} = \begin{bmatrix} 0.2249 & -0.0143 & 0.0248 & 0.1534 \\ -0.0143 & 0.2553 & 0.0571 & 0.0000 \\ 0.0248 & 0.0571 & 0.1317 & 0.0384 \\ 0.1534 & 0.0000 & 0.0384 & 0.1455 \end{bmatrix},$$

$$\boldsymbol{Y} = \begin{bmatrix} 0.0023 & -0.0128 \\ 0.0650 & 0.2597 \end{bmatrix}, \quad \boldsymbol{H} = \begin{bmatrix} 0.2311 & 0.6521 \\ 0.0384 & 0.1455 \end{bmatrix}.$$

The obtained gain matrix $\boldsymbol{K}$ is

$$\boldsymbol{K} = \begin{bmatrix} 0.0956 & -0.5161 \\ -0.0590 & 2.0498 \end{bmatrix}$$

which gives that all eigenvalues of $\boldsymbol{A}_{sc} = \boldsymbol{A}_s + \boldsymbol{BKC}$ are stable, where

$$\rho(\boldsymbol{A}_{sc}) = \{\ -2.8442 \ -11.9300 \ -6.6978 \pm 9.5406\,\mathrm{i}\ \}.$$

One can verify the stable eigenvalue spectrum of the system matrix in (25), where

$$\rho(\boldsymbol{A}_{cs}^{\circ}) = \left\{ \begin{array}{l} -0.7394 \ -0.7394 \ -0.7395 \ -0.7395 \ -2.8442 \ -11.9300 \\ -3.6642 \pm 9.5406\,\mathrm{i} \ -6.6978 \pm 9.5406\,\mathrm{i} \ -6.9476 \pm 11.6631\,\mathrm{i} \end{array} \right\},$$

which certify stability of the cascade control structure.

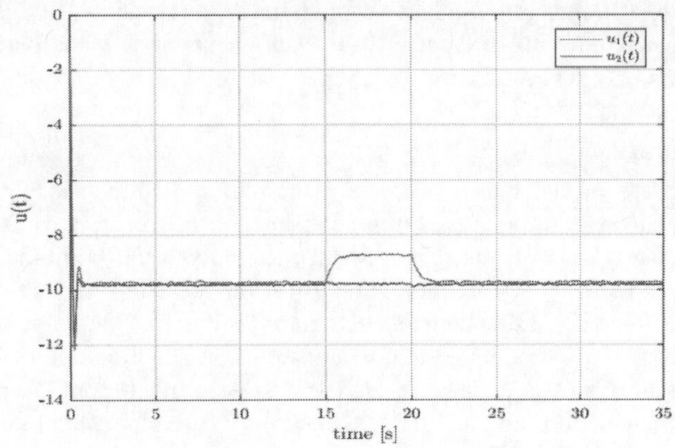

Figure 2. System input signals

Simulations are realized in the forced mode and the noise environment where, to set up the equal working point of the closed-loop system in the forced mode for nominal and faulty regimes, the signal gains matrices were computed, and the system noise input is given so that

$$\boldsymbol{W}_n = \begin{bmatrix} 0.7880 & -5.3257 \\ 1.1933 & -5.7555 \end{bmatrix}, \ \boldsymbol{W}_f = \begin{bmatrix} 0.7880 & -5.3257 \\ 0.9820 & -5.9368 \end{bmatrix},$$

$$\boldsymbol{V}_d^T = \begin{bmatrix} 1.136 & 1.136 & 5.679 & 3.146 \end{bmatrix}, \ \sigma_d^2 = 0.08\,.$$

The desired output values are prescribed by the vector $\boldsymbol{w}(t)$ as $\boldsymbol{w}^T(t) = [\,1\ 2\,]$ (in simulation set as $-\boldsymbol{w}(t)$ because the cascade structure is in a dual form - see Fig. 1), the system initial state is $\boldsymbol{q}(0) = \boldsymbol{0}$ and the dynamic controller initial state is $\boldsymbol{p}(0) = \boldsymbol{0}$.

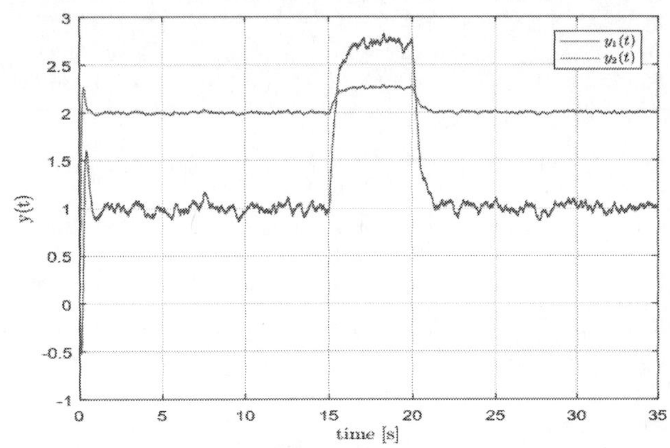

Figure 3. System output response

As results, Fig. 2 and Fig. 3 show the input and output of the closed-loop system in the cascade reconfiguration structure and forced mode, when the fault occurs in the matrix $\boldsymbol{A}$ at the time instant $t = 15$s. The control reconfiguration law applied to the cascade structure with the faulty system at the time instant $t = 20$s (reflecting the fault detection and isolation time delay lasted approximately 5 s). Practically, it means that the control law parameters of the cascade structure are still the same and, with unchanged $\boldsymbol{w}(t)$, there $\boldsymbol{W}_n$ is switched to $\boldsymbol{W}_f$. It is obvious that the system fault impact on the system output was compensated by using the cascade reconfiguration control structure.

8. Concluding remarks

Exploiting the reference model control design principle, the influence of the system matrix parameter faults on the system output is analyzed for control reconfiguration. To achieve the desired control objective, the bi-proper dynamic output controller, in combination with the static output control part, is realized within the reconfiguration cascade structure. A specific feature of this is that even after the fault occurrence is maintained the inner loop with the nominal control loop parameters, so it was necessary to use as a reference model a state form based on the faulty system description. Because dynamic and static output controllers are exploited, in terms of synthesis it has proven advantageous to use dual cascade control structure.

To optimize the cascade reconfiguration structure properties, the gain matrix parameters design conditions are designed to reflect the stability of both regimes, verifying them separately in the sense of Lyapunov asymptotic stability. In consequence, the design requires solution of two set of consistent linear matrix inequalities, every one combined with an associated matrix equality.

Adapting to the faulty plant and the fault detection subsystem with a time delay, the reconfigured gain matrix parameter is changed after a fault occurrence, keeping the rated control intact. The reconfiguration is so conveyed as an autonomous formula that may be designed and organized online. Because the faulty mode is optimized and bi-proper dynamic output control is used, large system input and output signal peeks, occurring at the starting interval of control when the standard cascade structure is used, are substantially reduced.

Presented method for linear continuous-time systems provides useful and easily implementable structure in process of control reconfiguration for the system matrix parameter faults. A simulation example, subject to given type of failures, demonstrates the effectiveness of the proposed form of the fault tolerant control design technique.

Acknowledgment

The work presented in the paper was supported by VEGA, the Grant Agency of the Ministry of Education and the Academy of Science of Slovak Republic, under Grant No. 1/0348/14. This support is very gratefully acknowledged.

References

[1] Acosta-Santana P Rodriguez-Alfaro L H and Alcorta-Garcia E 2013 A cascade structure approach to control reconfiguration *Proc. 2nd Int Conf Control and Fault-Tolerant Systems SysTol'13* Nice France 347-52
[2] Bezzaoucha S Marx B Maquin D and Ragot J 2015 State constrained tracking control for nonlinear systems *J Franklin Institute* **352**(7) 2866-86
[3] Doyle J C Francis B A and Tannenbaum A R 2009 *Feedback Control Theory* (Mineola:Dover Publications)
[4] Erzberger H 1968 Analysis and design of model following control systems by state space techniques *Prepr 9th Joint Automatic Control Conference JACC 1968* Ann Arbor MI USA 572-581
[5] Filasová A and Krokavec D 2016 One approach to cascade framework design within control structure reconfiguration in faulty systems *Proc 16th Int Carpathian Control Conference ICCC 2016* Tatranská Lomnica Slovakia 188-93
[6] Gao Z 1996 Techniques in reconfigurable control system design *Digital Control Systems Implementation and Computational Techniques* (San Diego:Academic Press) 102-15
[7] Gao Z and Antsaklis P J 1989 On the stability of the left pseudo-inverse method for reconfigurable control systems *Proc IEEE National Aerospace and Electronics Conf NAECON 1989* Dayton OH USA 333-7
[8] Gao Z and Antsaklis P J 1991 Reconfigurable control systems desing via perfect model-following *Proc 1991 AIAA Guidance Navigation and Control Conf* New Orleans LA USA 239-46

[9] Huang C Bai Y and Liu X 2010 H-infinity state feedback control for a class of networked cascade control systems with uncertain delay *IEEE Tran Industrial Informatics* **6**(1) 62-72
[10] Kautsky J Nichols N K and Van Dooren P 1985 Robust pole assignment in linear state feedback *Int J Control* **41**(5) 1129-55
[11] Krokavec D Filasová A and Liščinský P 2016 D-stable condition formulation for cascade reconfiguration control design *Proc 3rd Int Conf Control Decision and Information Technologies CoDiT'16* St Pauls Bay Malta 1-6
[12] Krokavec D Filasová A and Liščinský P 2016 LMI-based design of cascade reconfiguration control structures *Proc 3rd IEEE Int Conf Control and Fault-Tolerant Systems SysTol 2016* Barcelona Spain 269-74
[13] Maffezzoni C Schiavoni N and Ferretti G 1990 Robust design of cascade control *IEEE Control Systems Magazine* **10**(1) 21-5
[14] Peaucelle D Henrion D Labit Y and Taitz K 2002 *User's Guide for SeDuMi Interface 1.04* (Toulouse: LAAS-CNRS)
[15] Staroswiecki M 2005 Fault tolerant control The pseudo-inverse method revisited *IFAC Proceedings Volumes* **38**(1) 418-23
[16] Staroswiecki M and Cazaurang F 2004 Fault recovery by nominal trajectory tracking *Proc 2008 American Control Conference* Seattle WA USA 1070-5
[17] Wang Q G 2003 *Decoupling Control* (Berlin:Springer Verlag)

Adaptive backstepping control for three axis microsatellite attitude pointing under actuator faults

M.A. SI MOHAMMED[1], H. BOUSSADIA[2], A. BELLAR[1], A. ADNANE[1]

[1]Département de recherche en Mécanique Spatiale, Centre de Développement des Satellites (CDS), BP 4065 Ibn Rochd USTO Oran, Algeria
[2]USTO-MB University, Dept of Electronic, LARESI Laboratory, BP 1505 El Mnaouar Bir El Djir Oran, Algeria.

masimohammed@cds.asal.dz, halima.boussadia@univ-usto.dz, abellar@cds.asal.dz, aadnane@cds.asal.dz

Abstract. This paper presents the design of Low Earth Orbit (LEO) micro-satellite attitude controller using reaction wheels, and under actuator faults. Firstly, a backstepping controller is developed when the actuator is fault-free. Then, a fault tolerant controller is designed to compensate the actuator fault. Two types of this latter are considered (additive and multiplicative faults). The presented control strategy is based on adaptive backstepping technique. The simulation results clearly demonstrate the effectiveness of the presented technique.

1. Introduction

The attitude is the orientation of satellite in the space. In absence of control, it evolves naturally under effect of external disturbances. The attitude control has the role to compensate this disturbing torques.

The attitude control is a very important field of research in space technology. It attracted much attention in the recent years because of many types of space missions.

From automatic point of view, the satellite is a nonlinear dynamic system, it is highly coupled. Therefore, the design of the attitude controllers is usually difficult. Various nonlinear controllers have been proposed to solve this problem. These controllers include sliding mode control [1-2], fuzzy control [3], backstepping control and feedback control [4]. Among these techniques, backstepping control is a recursive method, it is based on Lyapunov theory which ensures the stabilization of each step of synthesis.

The backstepping has been very useful in the space field over the past years. In [5], this technique was used with the inverse optimal control to stabilize spacecraft attitude. In [6], the backstepping was based on similar skew-symmetric structure. [7] presented control by integrator backstepping with internal stabilization. All these controllers are based on the knowledge of the system parameters. However, if faults appear in satellite subsystems such as actuators, the control laws [5-7] induce no desired behaviours, and the consequences of these faults can be catastrophic. Therefore, it is important to develop other control strategies that can tolerate faults.

In [8], a combination between backstepping control, adaptive control and new matrix product (which called the semitensor product) was used to control the spacecraft attitude with unknown external disturbances, but no faults were present. In [9-11] adaptive control was developed in the presence of unknown inertia parameters and also with no faults. In [12] fault tolerant attitude control was designed using adaptive neural network.

In this paper, a controller design for Low Earth Orbit microsatellite is presented in the presence of the combination of two actuator faults: additive and multiplicative faults. The presented controller is developed using the adaptive backstepping technique, which is based on adaptive design of Lyapunov. The paper is organized as follows. Section 2 presents the dynamic and kinematic models used for the satellite. Section 3 presents the controllability test of our system. In the section 4, 5, and 6, we describe the design of the control laws which are presented in this work. In next section, we present the simulation results. Finally, the conclusion of this paper is presented in section 8.

2. Spacecraft attitude model

The dynamics of the spacecraft in inertial space governed by Euler's equations of motion can be expressed as follows in vector form as [13-14],

$$\mathbf{I}\dot{\boldsymbol{\omega}}_s^I = \mathbf{C}_{gg} + \mathbf{C}_{ext} + \mathbf{I}\boldsymbol{\omega}_s^I \times (\mathbf{I}\boldsymbol{\omega}_s^I + \mathbf{h}) - \dot{\mathbf{h}} \tag{1}$$

where $\boldsymbol{\omega}_s^I$, $\mathbf{I}$, $\mathbf{C}_{gg}$ and $\mathbf{C}_{ext}$ are respectively the inertial referenced body angular velocity vector, moment of inertia of spacecraft, gravity gradient torque vector, and external disturbance torque vector. The expression of the gravity gradient vector in body coordinate is expressed as,

$$C_{ggx} = 3\omega_0^2\left[(I_{zz} - I_{yy})A_{23}A_{33} + I_{yx}A_{13}A_{33} + I_{yz}A_{33}^2 - I_{zx}A_{13}A_{23} - I_{zy}A_{23}^2\right] \tag{2.a}$$

$$C_{ggy} = 3\omega_0^2\left[(I_{xx} - I_{zz})A_{13}A_{33} - I_{xy}A_{23}A_{33} - I_{xz}A_{33}^2 + I_{zx}A_{13}^2 + I_{zy}A_{13}A_{23}\right] \tag{2.b}$$

$$C_{ggz} = 3\omega_0^2\left[(I_{yy} - I_{xx})A_{13}A_{23} + I_{xy}A_{23}^2 + I_{xz}A_{23}A_{33} - I_{yx}A_{13}^2 - I_{yz}A_{13}A_{33}\right] \tag{2.c}$$

where,
ω_0 mean orbital angular velocity of the satellite;
A_{ij} attitude matrix elements.

The kinematic quaternion is expressed as,

$$\dot{\mathbf{q}} = \frac{1}{2}\,\boldsymbol{\Omega}\,\mathbf{q} = \frac{1}{2}\Lambda(\mathbf{q})\boldsymbol{\omega}_s^o \tag{3}$$

where,

$$\boldsymbol{\Omega} = \begin{bmatrix} 0 & \omega_{oz} & -\omega_{oy} & \omega_{ox} \\ -\omega_{oz} & 0 & \omega_{ox} & \omega_{oy} \\ \omega_{oy} & -\omega_{ox} & 0 & \omega_{oz} \\ -\omega_{ox} & -\omega_{oy} & -\omega_{oz} & 0 \end{bmatrix} \tag{4}$$

and,

$$\Lambda(\mathbf{q}) = \begin{bmatrix} q_4 & -q_3 & q_2 \\ q_3 & q_4 & -q_1 \\ -q_2 & q_1 & q_4 \\ -q_1 & -q_2 & -q_3 \end{bmatrix} \tag{5}$$

$\boldsymbol{\omega}_s^o = [\omega_{ox} \quad \omega_{oy} \quad \omega_{oz}]^T$: body angular velocity vector referenced to orbital coordinates.
The angular body rates referenced to the orbit coordinates can be obtained from the inertial referenced body rates by using the transformation matrix $\mathbf{A}$ as [13],

$$\boldsymbol{\omega}_s^o = \boldsymbol{\omega}_s^I - \mathbf{A}\boldsymbol{\omega}_0 \tag{6}$$

From equation (1) and equation (3), the satellite mathematical model can be written as,

$$\dot{\mathbf{x}} = \mathbf{f}(\mathbf{x}, \mathbf{h}) + \mathbf{B\,U} \tag{7.1}$$
$$\mathbf{y} = \mathbf{H\,x} \tag{7.2}$$

where,

$$\mathbf{f}(\mathbf{x},\mathbf{h}) = \begin{bmatrix} 0.5(\omega_{oz}q_2 - \omega_{oy}q_3 + \omega_{ox}q_4) \\ 0.5(-\omega_{oz}q_1 + \omega_{ox}q_3 + \omega_{oy}q_4) \\ 0.5(\omega_{oy}q_1 - \omega_{ox}q_2 + \omega_{oz}q_4) \\ 0.5(-\omega_{ox}q_1 - \omega_{oy}q_2 - \omega_{oz}q_3) \\ I_x^{-1}\left(C_x - (I_z - I_y)\omega_y\omega_z - \omega_y h_z + \omega_z h_y\right) \\ I_y^{-1}\left(C_y - (I_x - I_z)\omega_x\omega_z + \omega_x h_z - \omega_z h_x\right) \\ I_y^{-1}\left(C_z - (I_y - I_x)\omega_x\omega_y - \omega_x h_y + \omega_y h_x\right) \end{bmatrix} \tag{8}$$

$\mathbf{x} = [q_1, q_2, q_3, q_4, \omega_x, \omega_y, \omega_z]^T$ state vector;
$\mathbf{I} = [I_x \quad I_y \quad I_z]$ moment of inertia of spacecraft;
$\boldsymbol{\omega}_B^I = [\omega_x \quad \omega_y \quad \omega_z]^T$ angular velocity vector in the inertial frame;
$\mathbf{q} = [q_1 q_2 q_3 q_4]$ quaternion;
$\mathbf{h} = [h_x \quad h_y \quad h_z]^T$ angular moment vector;
$\mathbf{C}_{ext} = [C_x \quad C_y \quad C_z]^T$ external disturbance torque vector;
$\mathbf{B} = [\mathbf{0}^{4\times3}\ \mathbf{I}^{3\times3}]^T$ control matrix;
$\mathbf{U} = -\dot{\mathbf{h}}$ control input torque;
It assumed that the quaternion measurements are directly available $\mathbf{H} = [\mathbf{I}^{4\times4} \quad \mathbf{0}^{4\times3}]$.

3. Controllability test

The controllability concept is related to dynamic systems. It has a fundamental importance when studying attitude control algorithms that we will presented in the following sections. For that reason we introduce this concept in what follows.
The controllability test is based on the rank test of the controllability matrix **CO** which is defined as,

$$\mathbf{CO} = [\mathbf{B} \quad \mathbf{FB} \quad \mathbf{F^2B} \quad \mathbf{F^3B} \quad \mathbf{F^4B} \quad \mathbf{F^5B} \quad \mathbf{F^6B}] \tag{9}$$

where,

$$\mathbf{F} = \frac{\partial \mathbf{f}}{\partial \mathbf{x}} \qquad \mathbf{B} = \begin{bmatrix} 1 & 0 & 0 \\ 0 & 1 & 0 \\ 0 & 0 & 1 \\ 0 & 0 & 0 \\ 0 & 0 & 0 \\ 0 & 0 & 0 \\ 0 & 0 & 0 \end{bmatrix} \tag{10}$$

The system is controllable if the rank of the matrix **CO** is equal to seven.

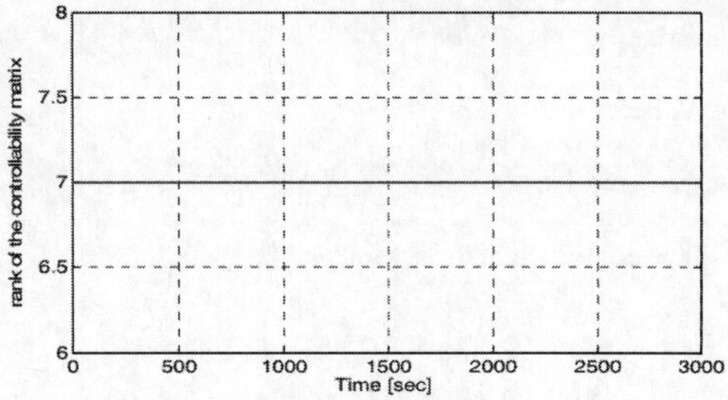

Figure 1. Rank of the controllability matrix.

Figure 1 shows that the rank of the controllability matrix is equal to seven. Therefore, our system is controllable.

4. Backstepping control design

In this section, backstepping control law is developed. This technique is inspired by [15].
The backstepping algorithm is described in two steps as follow.

4.1. Step 1

We define the first and the second variable of backstepping

$$\mathbf{z}_1 = \mathbf{x}_1 = \mathbf{q}_e = \mathbf{q}_c\,\mathbf{q} \tag{11}$$
$$\mathbf{z}_2 = \mathbf{x}_2 - \boldsymbol{\alpha}_1 \tag{12}$$

where,
$\mathbf{x}_2 = \boldsymbol{\omega}_s^o$;
$\boldsymbol{\alpha}_1$ is a virtual control law.
The time derivative of $\mathbf{z}_1$ is expressed as,

$$\dot{\mathbf{z}}_1 = \dot{\mathbf{x}}_1 = \dot{\mathbf{q}}_e = \mathbf{q}_c \left(\frac{1}{2}\boldsymbol{\Lambda}(\mathbf{q})\boldsymbol{\omega}_s^o\right) \tag{13}$$

The first Lyapunov function is defined as,

$$V_1(\mathbf{z}_1) = \mathbf{z}_1^T\,\mathbf{z}_1 \tag{14}$$

Its time derivative is expressed as,

$$\dot{V}_1 = 2\mathbf{z}_1^T\dot{\mathbf{z}}_1 = \mathbf{z}_1^T\mathbf{G}(\mathbf{q})\mathbf{z}_2 + \mathbf{z}_1^T\mathbf{G}(\mathbf{q})\boldsymbol{\alpha}_1 \tag{15}$$

where,

$$\mathbf{G}(\mathbf{q}) = \mathbf{q}_c\;\boldsymbol{\Lambda}(\mathbf{q}) \tag{16}$$

To make $\dot{V}_1$ negative, $\boldsymbol{\alpha}_1$ is chosen as,

$$\boldsymbol{\alpha}_1 = -\mathbf{k}_1\mathbf{G}(\mathbf{q})^T\mathbf{z}_1 \tag{17}$$

where,
$\mathbf{k}_1$: is a positive gain matrix.

The time derivative of V_1 becomes

$$\dot{V}_1 = -\mathbf{z}_1^T\mathbf{G}(\mathbf{q})\mathbf{k}_1\mathbf{G}(\mathbf{q})^T\mathbf{z}_1 + \mathbf{z}_1^T\mathbf{G}(\mathbf{q})\mathbf{z}_2 \tag{18}$$

The term $\mathbf{z}_1^T\mathbf{G}(\mathbf{q})\mathbf{z}_2$ will be eliminated in the next step.

4.2. Step 2
The time derivative of $\mathbf{z}_2$ is expressed as,

$$\dot{\mathbf{z}}_2 = \dot{\boldsymbol{\omega}}_s^o - \dot{\boldsymbol{\alpha}}_1 \tag{19}$$

where,

$$\boldsymbol{\omega}_s^o = \boldsymbol{\omega}_s^I - \mathbf{A}\boldsymbol{\omega}_0 \tag{20}$$

We replace the equation (20) in (19), we obtain

$$\dot{\mathbf{z}}_2 = \dot{\boldsymbol{\omega}}_s^I - \dot{\mathbf{A}}\boldsymbol{\omega}_0 - \dot{\boldsymbol{\alpha}}_1 \tag{21}$$
$$\mathbf{I}\dot{\mathbf{z}}_2 = \mathbf{I}\dot{\boldsymbol{\omega}}_s^I - \mathbf{I}\dot{\mathbf{A}}\boldsymbol{\omega}_0 - \mathbf{I}\dot{\boldsymbol{\alpha}}_1 \tag{22}$$

The second Lyapunov function is defined as,

$$V_2(\mathbf{z}_1, \mathbf{z}_2) = V_1(\mathbf{z}_1) + \frac{1}{2}\mathbf{z}_2^T\mathbf{I}\,\mathbf{z}_2 \tag{23}$$

Its time derivative is expressed as,

$$\dot{V}_2 = \dot{V}_1 + \mathbf{z}_2^T\mathbf{I}\,\dot{\mathbf{z}}_2 \tag{24}$$
$$\dot{V}_2 = -\mathbf{z}_1^T\mathbf{G}(\mathbf{q})\mathbf{k}_1\mathbf{G}(\mathbf{q})^T\mathbf{z}_1 + \mathbf{z}_1^T\mathbf{G}(\mathbf{q})\mathbf{z}_2$$
$$+\mathbf{z}_2^T\left[\left(\mathbf{C}_{gg} + \mathbf{C}_{ext} - \boldsymbol{\omega}_s^I \times \left(\mathbf{I}\boldsymbol{\omega}_s^I + \mathbf{h}\right) - \dot{\mathbf{h}}\right) - \mathbf{I}\dot{\mathbf{A}}\boldsymbol{\omega}_0 - \mathbf{I}\dot{\boldsymbol{\alpha}}_1\right] \tag{25}$$

To make $\dot{V}_2$ negative, the control law $\dot{\mathbf{h}}$ is chosen as,

$$\dot{\mathbf{h}} = \mathbf{k}_2\mathbf{z}_2 + \mathbf{G}(\mathbf{q})^T\mathbf{z}_1 + \left(\mathbf{C}_{gg} + \mathbf{C}_{ext} - \boldsymbol{\omega}_s^I \times \left(\mathbf{I}\boldsymbol{\omega}_s^I + \mathbf{h}\right)\right) - \mathbf{I}\dot{\mathbf{A}}\boldsymbol{\omega}_0 - \mathbf{I}\dot{\boldsymbol{\alpha}}_1 \tag{26}$$

5. Adaptive control based on Lyapunov theory
The adaptive design of Lyapunov is presented in this section as an introduction of the adaptive backstepping method.
Consider the following nonlinear system

$$\dot{\mathbf{x}} = \mathbf{u} + \boldsymbol{\varphi}(\mathbf{x})^T\boldsymbol{\theta} \tag{27}$$

where,
$\boldsymbol{\theta}$ is a vector of unknown parameters.

We seek to find the control law $\mathbf{u}(\mathbf{x},\boldsymbol{\theta})$ which ensures the stability of the system (27). Therefore, we choose the following Lyapunov function

$$V_1(\mathbf{x}) = \frac{1}{2}\mathbf{x}^T\mathbf{x} \tag{28}$$

which is definite positive, its derivative is expressed as,

$$\dot{V}_1 = \mathbf{x}^T\dot{\mathbf{x}} = \mathbf{x}^T[\mathbf{u} + \boldsymbol{\varphi}(\mathbf{x})^T\boldsymbol{\theta}] \tag{29}$$

The control law is chosen as,

$$\mathbf{u}(\mathbf{x},\boldsymbol{\theta}) = -\boldsymbol{\varphi}(\mathbf{x})^T\boldsymbol{\theta} - \mathbf{k}_1\mathbf{x} \tag{30}$$

where, $\mathbf{k}_1 > \mathbf{0}$
Therefore, two cases arise:

- $\boldsymbol{\theta}$ is known: the control law (30) can be realized, which makes the system (25) stable.
- $\boldsymbol{\theta}$ is unknown : the controller law (30) cannot be realized. We propose to replace it by its equivalent $\widehat{\boldsymbol{\theta}}$.

$$\mathbf{u} = -\boldsymbol{\varphi}(\mathbf{x})^T\widehat{\boldsymbol{\theta}} - \mathbf{k}_1\mathbf{x} \tag{31}$$

The derivative of the Lyapunov function will be expressed as,

$$\dot{V}_1 = \mathbf{x}^T\left[-\mathbf{k}_1\mathbf{x} + \boldsymbol{\varphi}(\mathbf{x})^T\widetilde{\boldsymbol{\theta}}\right] = -\mathbf{x}^T\mathbf{k}_1\mathbf{x} + \mathbf{x}^T\boldsymbol{\varphi}(\mathbf{x})^T\widetilde{\boldsymbol{\theta}} \tag{32}$$

where, $\widetilde{\boldsymbol{\theta}}$ represents the estimation error $(\boldsymbol{\theta} - \widehat{\boldsymbol{\theta}})$.
This expression contains an unknown term $\widetilde{\boldsymbol{\theta}}$, its sign is indefinite, and no conclusion can be drawn for the stability of the system. Therefore, we define a new Lyapunov function by adding to the initial function (28) a quadratic term according to the estimation error $\widetilde{\boldsymbol{\theta}}$.

$$V_2(\mathbf{x},\widetilde{\boldsymbol{\theta}}) = \frac{1}{2}\mathbf{x}^T\mathbf{x} + \frac{1}{2}\widetilde{\boldsymbol{\theta}}^T\boldsymbol{\Gamma}^{-1}\widetilde{\boldsymbol{\theta}} \tag{33}$$

where, $\boldsymbol{\Gamma}$ is a gain matrix (definite positive) which represents the adaptation gain.
The derivative of this function becomes:

$$\dot{V}_2 = \mathbf{x}^T\dot{\mathbf{x}} + \widetilde{\boldsymbol{\theta}}^T\boldsymbol{\Gamma}^{-1}\dot{\widetilde{\boldsymbol{\theta}}} = -\mathbf{x}^T\mathbf{k}_1\mathbf{x} + \widetilde{\boldsymbol{\theta}}^T\boldsymbol{\Gamma}^{-1}\left(\dot{\widetilde{\boldsymbol{\theta}}} + \boldsymbol{\tau}\right) \tag{34}$$

where,

$$\boldsymbol{\tau} = \boldsymbol{\Gamma}\boldsymbol{\varphi}(\mathbf{x})\,\mathbf{x} \tag{35}$$

This derivative remains always indefinite, but this time the choice of the update law can cancel the second term of the equation.

$$\dot{\widehat{\boldsymbol{\theta}}} = -\dot{\widetilde{\boldsymbol{\theta}}} = -\boldsymbol{\Gamma}\boldsymbol{\varphi}(\mathbf{x})\,\mathbf{x} \tag{36}$$

6. Fault tolerant adaptive backstepping control design

The adaptive backstepping is designed from the fusion of the adaptive design of Lyapunov presented in section (5) and the no adaptive backstepping technique presented in section (4). The direct combination of these two methods gives a triplet (Lyapunov function, control law, adaptation law). Figure 2 presents the adaptive backstepping schematic diagram.

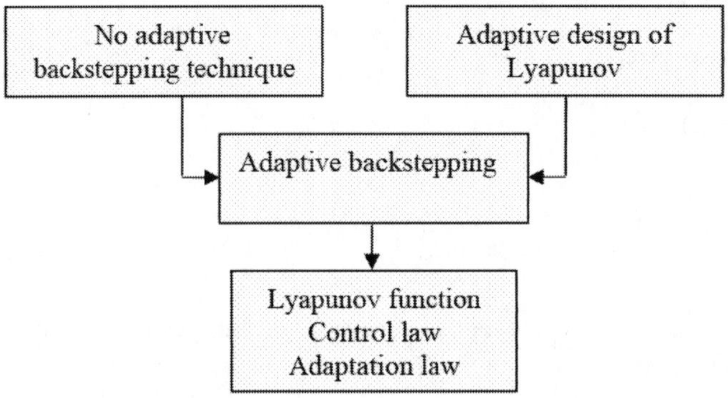

Figure 2. Adaptive backstepping schematic diagram.

This technique is used to develop a controller which can tolerate actuator faults. The design of this controller is described as follow [16],
In the presence of actuator faults, the dynamic equation model is rewritten as,

$$\mathbf{I}\dot{\boldsymbol{\omega}}_s^I = \mathbf{C}_{gg} + \mathbf{C}_{ext} + \mathbf{I}\boldsymbol{\omega}_s^I \times \left(\mathbf{I}\boldsymbol{\omega}_s^I + \mathbf{h}\right) - \left(\mathbf{f}_m\dot{\mathbf{h}} + \mathbf{f}_a\right)$$

where,
$\mathbf{f}_a$: additive fault;
$\mathbf{f}_m$: multiplicative fault.

6.1. Step 1

We define the first variable of backstepping $\mathbf{z}_1 = \mathbf{x}_1 = \mathbf{q}_e$, and the estimated fault errors $\tilde{\mathbf{f}}_a = \hat{\mathbf{f}}_a - \mathbf{f}_a$, $\tilde{\mathbf{f}}_m = \hat{\mathbf{f}}_m - \mathbf{f}_m$.
where,
$\hat{\mathbf{f}}_a$: estimated additive fault;
$\hat{\mathbf{f}}_m$: estimated multiplicative fault;
$\tilde{\mathbf{f}}_a$: additive fault error;
$\tilde{\mathbf{f}}_m$: multiplicative fault error.

The second variable of backstepping is:

$$\mathbf{z}_2 = \mathbf{x}_2 - \boldsymbol{\alpha}_1 = \boldsymbol{\omega}_s^o - \boldsymbol{\alpha}_1 \tag{37}$$

where, $\boldsymbol{\alpha}_1$ is a virtual control law.

The first subsystem of the mathematic model is independent of faults. Therefore, the design of the first step of the adaptive backstepping technique is the same as the traditional backstepping and the derivative of the first Lyapunov function does not change.

$$\dot{V}_1 = -\mathbf{z}_1^T\mathbf{G}(\mathbf{q})\mathbf{k}_1\mathbf{G}(\mathbf{q})^T\mathbf{z}_1 + \mathbf{z}_1^T\mathbf{G}(\mathbf{q})\mathbf{z}_2 \tag{38}$$

The term $\mathbf{z}_1^T\mathbf{G}(\mathbf{q})\mathbf{z}_2$ will be eliminated in the next step.

6.2. Step 2

$$\mathbf{z}_2 = \boldsymbol{\omega}_s^o - \boldsymbol{\alpha}_1 \tag{39}$$

Its time derivative is expressed as,

$$\dot{\mathbf{z}}_2 = \dot{\boldsymbol{\omega}}_s^o - \dot{\boldsymbol{\alpha}}_1 \tag{40}$$
$$\boldsymbol{\omega}_s^o = \boldsymbol{\omega}_s^I - \mathbf{A}\boldsymbol{\omega}_0 \tag{41}$$

We replace the equation (41) in (40), we obtain

$$\dot{\mathbf{z}}_2 = \dot{\boldsymbol{\omega}}_s^I - \dot{\mathbf{A}}\boldsymbol{\omega}_0 - \dot{\boldsymbol{\alpha}}_1 \tag{42}$$
$$\mathbf{I}\dot{\mathbf{z}}_2 = \mathbf{I}\dot{\boldsymbol{\omega}}_s^I - \mathbf{I}\dot{\mathbf{A}}\boldsymbol{\omega}_0 - \mathbf{I}\dot{\boldsymbol{\alpha}}_1 \tag{43}$$

We choose the following Lyapunov function

$$V_2(\mathbf{z}_1, \mathbf{z}_2, \tilde{\mathbf{f}}) = V_1(\mathbf{z}_1) + \frac{1}{2}\mathbf{z}_2^T\mathbf{I}\,\mathbf{z}_2 + \frac{1}{2}\tilde{\mathbf{f}}_a^T\,\boldsymbol{\Gamma}^{-1}\tilde{\mathbf{f}}_a + \frac{1}{2}\tilde{\mathbf{f}}_m^T\,\boldsymbol{\Gamma}^{-1}\,\tilde{\mathbf{f}}_m \tag{44}$$

where,
$\boldsymbol{\Gamma}$ is a design parameter which must be positive.

The time derivative of the Lyapunov function is expressed as,

$$\dot{V}_2 = \dot{V}_1 + \mathbf{z}_2^T\mathbf{I}\,\dot{\mathbf{z}}_2 + \tilde{\mathbf{f}}_a^T\boldsymbol{\Gamma}^{-1}\dot{\tilde{\mathbf{f}}}_a + \tilde{\mathbf{f}}_m^T\boldsymbol{\Gamma}^{-1}\dot{\tilde{\mathbf{f}}}_m \tag{45}$$

$$\dot{V}_2 = \mathbf{z}_2^T\left[\left(\mathbf{C}_{gg} + \mathbf{C}_{ext} - \boldsymbol{\omega}_s^I \times (\mathbf{I}\boldsymbol{\omega}_s^I + \mathbf{h}) - \mathbf{f}_m\dot{\mathbf{h}} - \mathbf{f}_a\right) - \mathbf{I}\dot{\mathbf{A}}\boldsymbol{\omega}_0 - \mathbf{I}\dot{\boldsymbol{\alpha}}_1\right] - \mathbf{z}_1^T\mathbf{G}(\mathbf{q})\mathbf{k}_1\mathbf{G}(\mathbf{q})^T\mathbf{z}_1 + \mathbf{z}_1^T\mathbf{G}(\mathbf{q})\mathbf{z}_2 + \tilde{\mathbf{f}}_a^T\boldsymbol{\Gamma}^{-1}\dot{\tilde{\mathbf{f}}}_a + \tilde{\mathbf{f}}_m^T\boldsymbol{\Gamma}^{-1}\dot{\tilde{\mathbf{f}}}_m \tag{46}$$

$$\begin{aligned}\dot{V}_2 = {} & -\mathbf{z}_1^T\mathbf{G}(\mathbf{q})\mathbf{k}_1\mathbf{G}(\mathbf{q})^T\mathbf{z}_1 + \mathbf{z}_1^T\mathbf{G}(\mathbf{q})\mathbf{z}_2 \\ & + \mathbf{z}_2^T\left[\left(\mathbf{C}_{gg} + \mathbf{C}_{ext} - \boldsymbol{\omega}_s^I \times (\mathbf{I}\boldsymbol{\omega}_s^I + \mathbf{h}) - \hat{\mathbf{f}}_m\dot{\mathbf{h}} - \hat{\mathbf{f}}_a\right) - \mathbf{I}\dot{\mathbf{A}}\boldsymbol{\omega}_0 - \mathbf{I}\dot{\boldsymbol{\alpha}}_1\right] + \tilde{\mathbf{f}}_a^T\left[\mathbf{z}_2 + \boldsymbol{\Gamma}^{-1}\dot{\tilde{\mathbf{f}}}_a\right] \\ & + \tilde{\mathbf{f}}_m^T\left[\dot{\mathbf{h}}\mathbf{z}_2 + \boldsymbol{\Gamma}^{-1}\dot{\tilde{\mathbf{f}}}_m\right]\end{aligned} \tag{47}$$

To make this derivative negative, we choose the control law $\dot{\mathbf{h}}$ as follow,

$$\dot{\mathbf{h}} = \hat{\mathbf{f}}_m^{-1}\left[\mathbf{k}_2\mathbf{z}_2 + \mathbf{G}(\mathbf{q})^T\mathbf{z}_1 + \left(\mathbf{C}_{gg} + \mathbf{C}_{ext} - \boldsymbol{\omega}_s^I \times (\mathbf{I}\boldsymbol{\omega}_s^I + \mathbf{h})\right) - \hat{\mathbf{f}}_a - \mathbf{I}\dot{\mathbf{A}}\boldsymbol{\omega}_0 - \mathbf{I}\dot{\boldsymbol{\alpha}}_1\right] \tag{48}$$

The update laws of the estimated faults are expressed as,

$$\dot{\hat{\mathbf{f}}}_a = -\boldsymbol{\Gamma}\,\mathbf{z}_2 \tag{49}$$
$$\dot{\hat{\mathbf{f}}}_m = -\boldsymbol{\Gamma}\,\dot{\mathbf{h}}\,\mathbf{z}_2 \tag{50}$$

7. Simulation results

In this section, we present the simulation results obtained for the controllers exposed previously. These results are obtained using the following parameters.

Table1. Satellite simulation parameters.

Parameter	Value
Inertia [kg.m^2]	$\begin{bmatrix} 12 & 0 & 0 \\ 0 & 14 & 0 \\ 0 & 0 & 10 \end{bmatrix}$
Orbit [km]	686
Inclination [deg]	98
Initial attitude [deg]	$[5 \quad 10 \quad -10]$
Initial attitude rate	$[0 \quad -0.06 \quad 0]$
External Torques [N.m]	$\begin{bmatrix} 10^{-7}(5\cos(\omega_0 t)+1) \\ 10^{-7}(5\cos(\omega_0 t)+2\sin(\omega_0 t)) \\ 10^{-7}(5\cos(\omega_0 t)+1) \end{bmatrix}$

Table 2. Desired attitude for the Euler angles.

Desired attitude [deg]
$[10 \quad 30 \quad 20]$

Table 3. Backstepping parameters.

$\mathbf{k_1}$	$\mathbf{k_2}$
$0.01 * \text{eye}(3)$	$8 * \text{eye}(3)$

Table 4. Adaptive backstepping parameters.

$\mathbf{k_1}$	$\mathbf{k_2}$	$\mathbf{\Gamma}$
$0.01 * \text{eye}(3)$	$8 * \text{eye}(3)$	$50 * \text{eye}(3)$

Two actuator faults are considered simultaneously. We introduce at the time t=400 sec an additive fault $\boldsymbol{f_a(t)}$.

$$f_a(t) = \begin{cases} 0 & t < 400\ sec \\ 2 + 0.01\cos(0.5\pi t) & t \geq 400 \text{ sec} \end{cases}$$

Then, at the time t=600 sec a multiplicative fault $\boldsymbol{f_m(t)}$.

$$f_m(t) = \begin{cases} 1 & t < 600\ sec \\ 0.5 + 0.07\cos(0.5\pi t) & t \geq 600 \text{ sec} \end{cases}$$

The simulation results are presented as follow.

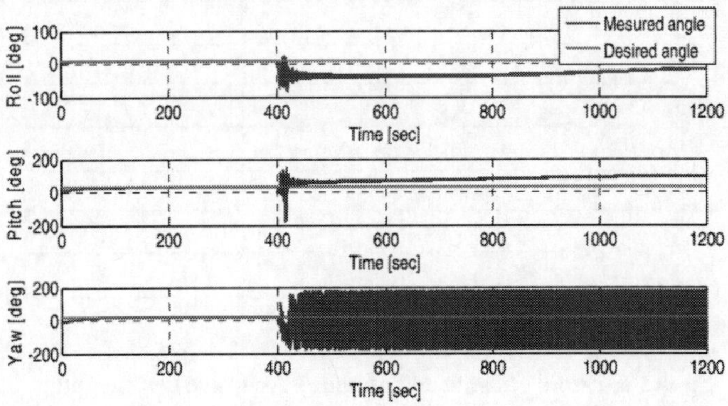

Figure 3. Actual and estimated attitude - No adaptive backstepping.

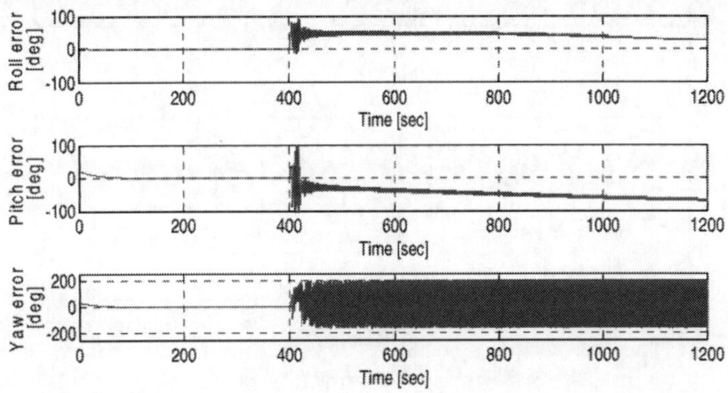

Figure 4. Errors of estimated attitude-No adaptive backstepping.

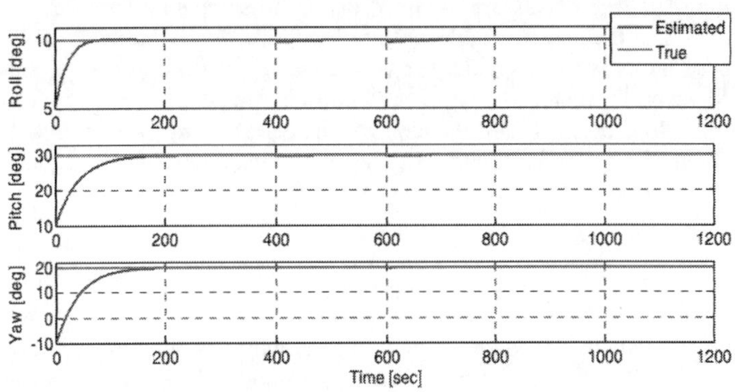

Figure 5. Actual and estimated attitude - Adaptive backstepping

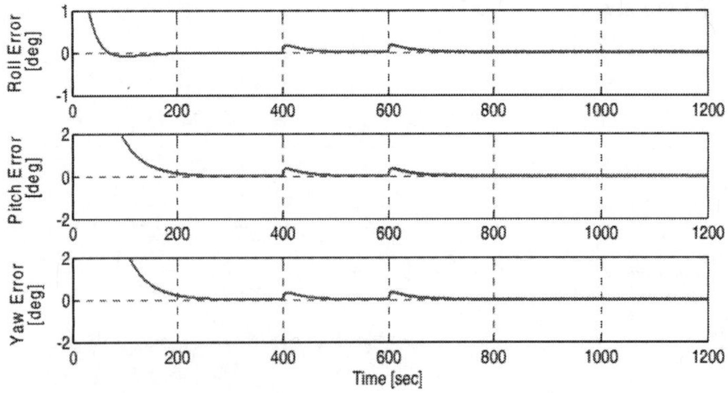

Figure 6. Errors of estimated attitude - Adaptive backstepping.

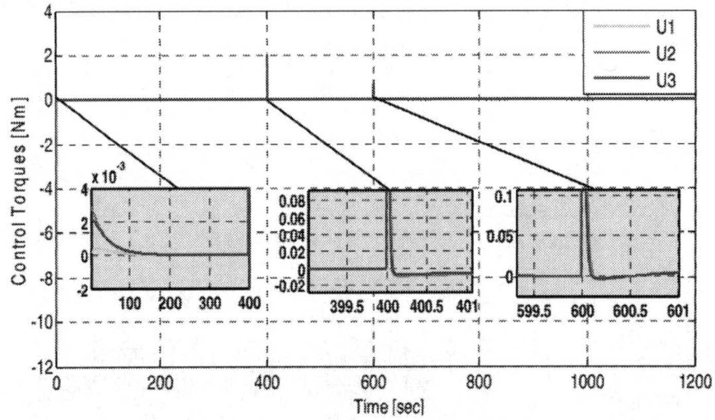

Figure 7. Control torques.

The figures 3-6 show the estimated attitude and its error of the two controllers (no adaptive and adaptive backstepping). These results are obtained using the same simulation conditions. We clearly observe in figures 2-3 that the actuator faults are not compensated by the backstepping controller, the consequence of these faults is catastrophic, the attitude diverges and the Euler angle errors are very large (200 deg). However, in figures 5-6 the desired attitude tracking is assured even the occurrence of the actuator faults at t=400 sec and t=600. In addition, the control torques converges to zero.
From all of the above, it is clearly found that the presented controller can guarantees the control performance, it succeed despite the combination of these faults.

8. Conclusion

In this work, we presented a synthesis of control laws for Low Earth Orbit (LEO) micro-satellite attitude stabilization using three axis controls by reaction wheels, and under actuator faults. All these laws are based on the Lyapunov theorem. Firstly, a nominal backstepping controller was developed when the actuator is fault-free. Then, a fault tolerant controller is designed to compensate the actuator faults. Two types of this latter were considered (additive and multiplicative faults). The presented control strategy is based on adaptive backstepping technique. This latter is designed by combining two methods (the adaptive design of Lyapunov and the no adaptive backstepping technique).
The results of this work demonstrate the effectiveness of the presented technique. It is found that the presented controller can guarantees the control performance despite the combination of these faults.

As a future research, we will develop other control techniques to compensate the actuator faults such as the super twisting algorithm.

References

[1] Boskovic J D Li SM and Mehra M K 2001Robust adaptive variable structure control of spacecraft under input saturation *Guidance Control and Dynamics* vol **24** pp14-22.

[2] Lu K Xia Y Zhu Z and Basin M V 2012 Sliding mode attitude tracking of rigid spacecraft with disturbance *Journal of the Franklin Institue* vol **349** pp 413-440.

[3] Zou A and Kumar K D 2011 Adaptive fuzzy fault-tolerant attitude control of spacecraft Control Engineering Practice vol **19** pp10-21.

[4] Wong H de M S Queiroz and Kapila V 2001 Adaptive tracking control using synthesized velocity from attitude measurements *Automatica* vol **37** pp 947-953.

[5] Krstié M Tsiotras P 1999 Inverse optimale stabilization of a rigid spacecraft *IEEE Transactions on Automatic Control* vol **44** pp1014-1027.

[6] Liu Y C, Zhang T Song J Y 2009 Attitude controller design for larger angle maneuver based on similar skew-symmetric structure *Journal of Astronautics* vol **30** pp 1017-1023.

[7] Dork R Ö Kocaoglan E 2008 Satellite attitude control by integrator backstepping with internal stabilization *Aircraft Engineering and Aerospace Technology* vol **80** pp 3-10.

[8] Zhang X 2012 Semitensor product based adaptive control for attitude tracking of spacecraft with unknown external disturbances *Control Theory Applications* vol **10** pp 292-296.

[9] Ahmed J Coppola V T Bernstein D S 1998 Adaptive asymptotic tracking of spacecraft attitude motion with inertia identification *Guidance Control and Dynamics* vol **21** pp 684-690.

[10] Schaub H Akella M R Junkins J L 2001 Adaptive control of nonlinear attitude motions realizing linear closed loop dynamics *Guidance Control and Dynamics* vol **24** pp 95-100.

[11] Seo D Akella M R 2008 High-performance spacecraft adaptive attitude-tracking control through attractingmanifold design Guidance Control and Dynamics vol **31** pp 884-891.

[12] Zhang M Shen X and Li T 2016 Fault tolerant control for cubesats with input saturation based on dynamic adaptive neural network *Innovative Computing Information and Control* Vol **12** pp 651-663.

[13] Wertz J R 1986 *Spacecraft Attitude Determination and Control* D Reidel Publishing Company Boston U.S.A Reprint.

[14] Sidi M 1997 *Spacecraft Dynamics and Control* Cambridge Univ press.

[15] Kopoulos K, Kokotovic I and all 1991Systematic Design of Adaptive Controllers for Feedback Linearizable Systems *IEEE Transactions on Automatic Control* vol **36**.

[16] Krstic M Kokotovié P 1995 Adaptive nonlinear design with controller-identifier separation and swapping *IEEE Transactions On Automatic Control* vol **40** pp 426-440.

IOP Conf. Series: Journal of Physics: Conf. Series **783** (2017) 012021 doi:10.1088/1742-6596/783/1/012021

An adaptive actuator failure compensation scheme for two linked 2WD mobile robots

Yajie Ma, Ayad Al-Dujaili, Vincent Cocquempot and Maan El Badaoui El Najjar

University of Lille, CNRS, Centrale Lille, UMR 9189-CRIStAL-Centre de Recherche en Informatique, Signal et Automatique de Lille, F-59000 Lille, France

E-mail: `vincent.cocquempot@univ-lille1.fr`.

Abstract. This paper develops a new adaptive compensation control scheme for two linked mobile robots with actuator failurs. A configuration with two linked two-wheel drive (2WD) mobile robots is proposed, and the modelling of its kinematics and dynamics are given. An adaptive failure compensation scheme is developed to compensate actuator failures, consisting of a kinematic controller and a multi-design integration based dynamic controller. The kinematic controller is a virtual one, and based on which, multiple adaptive dynamic control signals are designed which covers all possible failure cases. By combing these dynamic control signals, the dynamic controller is designed, which ensures system stability and asymptotic tracking properties. Simulation results verify the effectiveness of the proposed adaptive failure compensation scheme.

1. Introduction

Due to simplicity, efficiency and flexibility, two-wheel drive (2WD) mobile robots are widely used [1]. In some harsh conditions, resulting from a terrorist attack, a nuclear accident or nature disasters, 2WD rescue robots carrying instruments are employed to help rescuers. These adverse conditions may increase the probability of actuator (wheel motor) failures which will leads to the loss of the failed robots and the important instruments. To deal with this situation, several physical links may be used to connect the robots. For example, if one robot fails, then another robot carrying a mechanical arm can be sent to link the failed robot and to help it continue moving. This robot architecture provides actuator and sensor redundancies which improve the fault tolerance of the system. This paper is concerned with the fault tolerant control design for such a configuration with two linked 2WD robots.

There are a lot of research results on the motion control of 2WD mobile robots [2, 3] but without the consideration of actuator faults. For fault tolerant control, various effective design methods are proposed for different applications [4, 5]. As for the application to wheeled robots, a sensor fault accommodation scheme is presented in [6]; a fault-tolerant control design method is given in [7] for four-wheel drive (4WD) mobile robots; and a hybrid adaptive fault tolerant control scheme is proposed in [8] to accommodate partial faults and degradation for 2WD robots. Unlike 4WD robots, a 2WD robot has no redundant actuator, so that if one motor is lost, then the 2WD robot becomes uncontrollable.

To our knowledge, there is no research result on the fault tolerant control of 2WD robots with the motors (actuators) being totally faulty. To deal with such actuator failures, physical links are used to connect several 2WD mobile robots as shown in Fig. 1 for two robots. For this configuration, if one or two motors are lost, the remaining motors can continue moving the two robots. In this paper, a multi-design integration based adaptive control scheme is developed. The main contributions are as follows.

(i) The kinematic and dynamic models of the two 2WD mobile robots with fixed link are proposed.
(ii) A new adaptive actuator failure compensation scheme is developed using a multi-design integration method, which ensures system stability and asymptotic tracking properties, despite the presence of actuator failures including simultaneous multi-failures.

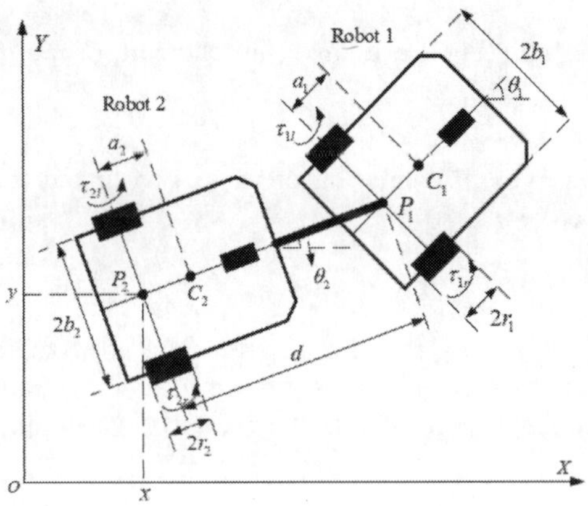

Figure 1: Two linked 2WD mobile robots

The rest of this paper is as follows. In Section 2, system modelling is given and the actuator failure compensation problem is formulated. In Section 3, a multi-design integration based adaptive actuator failure compensation scheme is developed. In Section 4, a simulation study is presented to demonstrate the effectiveness of the proposed adaptive control scheme. Conclusions follow in Section 5.

2. System Modeling and Problem Formulation

In this section, the system models are given for the two linked 2WD mobile robots as shown in Fig. 1, and its actuator failure compensation problem is formulated.

As shown in Fig. 1, the orientation of robot 2 is consistent with the one of the physical link, but the orientation of robot 1 is independent, and for each robot: the front wheel is passive and the two rear wheels are actuated; P_i $(i = 1, 2)$ is the center between two actuated wheels, C_i is the center of mass, a_i is the distance between P_i and C_i, b_i is half of the distance between two actuated wheels, r_i is the radius of wheels, θ_i is the orientation of the robot, and τ_{il} and τ_{ir} are the control torques applied to the left and right actuated wheels, respectively. Moveover, d is the distance between P_1 and P_2, OXY is the inertial frame, and (x, y) denotes the position of P_2 in frame OXY.

Kinematic model. Let $q = [x, y, \theta_2, \theta_1]^T$ and $\eta = [v_2, \omega_1]^T$, where v_2 is the linear velocity of robot 2, and ω_1 is the angular velocity of robot 1. The kinematic model of the two linked 2WD mobile robots in Fig. 1 is given by

$$\dot{q} = S(q)\eta. \tag{1}$$

where

$$S(q) = \begin{bmatrix} \cos\theta_2 & \sin\theta_2 & \frac{1}{d}\tan(\theta_1-\theta_2) & 0 \\ 0 & 0 & 0 & 1 \end{bmatrix}^T. \tag{2}$$

Moreover, the system constraints may be expressed as

$$A(q)\dot{q} = 0, \tag{3}$$

where

$$A(q) = \begin{bmatrix} \sin\theta_1 & -\cos\theta_1 & -d\cos(\theta_1-\theta_2) & 0 \\ \sin\theta_2 & -\cos\theta_2 & 0 & 0 \end{bmatrix} \tag{4}$$

is the system constraint matrix. Then, we have the following property:

$$S^T(q)A^T(q) = 0. \tag{5}$$

Dynamic model. The dynamic equation of these two linked 2WD robots is obtained by using the Lagrange method as:

$$M(q)\ddot{q} + E(q,\dot{q}) = B(q)\tau + A^T(q)\lambda, \tag{6}$$

where $M(q) \in \Re^{4\times4}$ is the inertia matrix that is symmetric positive definite, $E(q,\dot{q}) \in \Re^4$ is the centripetal and coriolis vector, $B(q) \in \Re^{4\times4}$ is the input injection matrix, $\tau = [\tau_{1r}, \tau_{1l}, \tau_{2r}, \tau_{2l}]^T$ is the vector of control torques, and $\lambda \in \Re^2$ is the vector of constraint forces. The matrices and vectors in (6) are given by

$$M(q) = \begin{bmatrix} m_1+m_2 & 0 & -(a_2m_2+dm_1)\sin\theta_2 & -a_1m_1\sin\theta_1 \\ 0 & m_1+m_2 & (a_2m_2+dm_1)\cos\theta_2 & a_1m_1\cos\theta_1 \\ -(a_2m_2+dm_1)\sin\theta_2 & (a_2m_2+dm_1)\cos\theta_2 & m_2a_2^2+m_1d^2+I_{m2} & a_1dm_1\cos(\theta_1-\theta_2) \\ -a_1m_1\sin\theta_1 & a_1m_1\cos\theta_1 & a_1dm_1\cos(\theta_1-\theta_2) & m_1a_1^2+I_{m1} \end{bmatrix},$$

$$E(q,\dot{q}) = \begin{bmatrix} -a_1m_1\dot{\theta}_1^2\cos\theta_1 - (a_2m_2+dm_1)\dot{\theta}_2^2\cos\theta_2 \\ -a_1m_1\dot{\theta}_1^2\sin\theta_1 - (a_2m_2+dm_1)\dot{\theta}_2^2\sin\theta_2 \\ -a_1dm_1\dot{\theta}_1^2\sin(\theta_1-\theta_2) \\ a_1dm_1\dot{\theta}_2^2\sin(\theta_1-\theta_2) \end{bmatrix},$$

$$B(q) = \begin{bmatrix} \frac{\cos\theta_1}{r_1} & \frac{\cos\theta_1}{r_1} & \frac{\cos\theta_2}{r_2} & \frac{\cos\theta_2}{r_2} \\ \frac{\sin\theta_1}{r_1} & \frac{\sin\theta_1}{r_1} & \frac{\sin\theta_2}{r_2} & \frac{\sin\theta_2}{r_2} \\ \frac{d\sin(\theta_1-\theta_2)}{r_1} & \frac{d\sin(\theta_1-\theta_2)}{r_1} & \frac{b_2}{r_2} & -\frac{b_2}{r_2} \\ \frac{b_1}{r_1} & -\frac{b_1}{r_1} & 0 & 0 \end{bmatrix}, \tag{7}$$

where m_1 and m_2 are the masses of the robot 1 and robot 2, and I_{m1} and I_{m2} are the corresponding inertia parameters.

Substituting (1) into (6), and multiplying by $S^T(q)$, $A^T(q)\lambda$ is eliminated with (5). Then equation (6) becomes

$$\bar{M}_1(q)\dot{\eta} + \bar{M}_2(q)\eta + \bar{E}(q,\dot{q}) = \bar{B}(q)\tau, \tag{8}$$

where $\eta = [v_2, \omega_1]^T$, and

$$\bar{M}_1(q) = S^T(q)M(q)S(q), \;\; \bar{M}_2(q) = S^T(q)M(q)\dot{S}(q),$$
$$\bar{E}(q,\dot{q}) = S^T(q)E(q,\dot{q}), \qquad \bar{B}(q) = S^T(q)B(q). \tag{9}$$

Note that $\bar{M}_1(q)$ is also symmetric positive definite as well as $M(q)$.

Actuator failure model. The considered actuator failure is that some motors totally lose power or are stuck which will introduce additional frictions. This type of actuator failure for one motor is modeled as

$$\tau_j(t) = \bar{u}_j, \qquad t \geq t_j \tag{10}$$

where $j = 1r, 1l, 2r, 2l$, $\bar{u}_j$ is the friction value that is unknown but constant, and $\bar{u}_j = 0$ means that the motor loses its power but can rotate freely and $\bar{u}_j > 0$ means that the motor is stuck or can not rotate freely caused by some frictions in the bearing, and t_j is the unknown failure occurring time instant.

Consider that all actuators may be faulty. The control torque in (8) becomes

$$\tau(t) = \sigma(t)u(t) + (I_4 - \sigma(t))\bar{u}, \tag{11}$$

where $\tau = [\tau_{1r}, \tau_{1l}, \tau_{2r}, \tau_{2l}]^T$ is the vector of control torques generated by the wheel motors, $u = [u_{1r}, u_{1l}, u_{2r}, u_{2l}]^T$ is the control signal vector to be designed, $\bar{u} = [\bar{u}_{1r}, \bar{u}_{1l}, \bar{u}_{2r}, \bar{u}_{2l}]^T$ is the vector of constant frictions, I_4 is the identity matrix, and $\sigma = \text{diag}\{\sigma_{1r}, \sigma_{1l}, \sigma_{2r}, \sigma_{2l}\}$ is the uncertain failure pattern matrix with

$$\sigma_j(t) = \begin{cases} 0, & \text{if the } j\text{th motor fails,} \\ 1, & \text{otherwise.} \end{cases} \tag{12}$$

Actuation redundancy. For the two linked 2WD mobile robots in Fig. 1, the following actuation redundancy condition needs to be satisfied:

$$\text{rank}(\bar{B}\sigma) = 2. \tag{13}$$

for all possible failure pattern matrices σ. This condition means that there are enough actuated motors to control $\eta = [v_2, \omega_1]^T$.

Remark 1 *With $\bar{B}(q)$ in (9), the compensable failure cases satisfying this redundancy condition are: 1) fault free case with four actuated wheels; 2) one actuator fails with three remaining actuated wheels; 3) two actuators fail with two remaining actuated wheels. However, for case 3), if the two failed actuators are in robot 2, then the system is similar with a tractor-trailer system [9], and the failures may be tolerated; if each robot has one faulty actuator, then the failures are also compensable; but if the two failed actuators are in robot 1, then the failures are not compensable, because in this case,* $\text{rank}(\bar{B}\sigma) = 1$ *with* $\sigma = \text{diag}\{0, 0, 1, 1\}$ *and ω_1 is uncontrollable.* □

Control objective. The control objective is to develop an actuator failure compensation scheme for two linked 2WD mobile robots in Fig. 1 to asymptotically track a reference trajectory, that is, to design a control signal $u(t)$ to guarantee that all closed-loop system signals are bounded, and $\lim_{t\to\infty}(x(t) - x_r(t)) = 0$, $\lim_{t\to\infty}(y(t) - y_r(t)) = 0$ and $\lim_{t\to\infty}(\theta_2(t) - \theta_r(t)) = 0$ for system (3) and (8), in the presence of some actuator failures modeled as (10)-(12) that satisfies the actuation redundancy condition in (13), where x_r, y_r, θ_r are the reference trajectories.

Remark 2 *In Fig. 1, the position of P_1 is determined by x, y and θ_2, in this sense, the control objective contains the position for the whole system. In addition, this objective is implemented by the control of ω_1 with the dynamic equation (8), as $\dot{\theta}_1 = \omega_1$, we can see that the orientation angle θ_1 is an intermediate variable to be controlled.* □

A virtual robot is employed to generate the reference trajectories as follows:

$$\dot{x}_r = v_r \cos\theta_r,\ \dot{y}_r = v_r \sin\theta_r,\ \dot{\theta}_r = \omega_r. \tag{14}$$

where v_r and ω_r are the linear velocity and angular velocity. By choosing appropriate v_r, ω_r and initial values $x_r(0)$, $y_r(0)$ and $\theta_r(0)$, the reference trajectories x_r, y_r and θ_r are determined. In this paper, we consider the tracking problem of a two-robot system. Then the following assumption is given for the reference trajectories.

Assumption 1: The reference trajectories x_r, y_r and θ_r, the velocities v_r and ω_r, and their derivatives are continuous and uniformly bounded, moreover $v_r \neq 0$.

Design issues. The structure of the proposed actuator failure compensation scheme is shown in Fig. 2. To design such a control scheme, the following technical issues need to be solved:

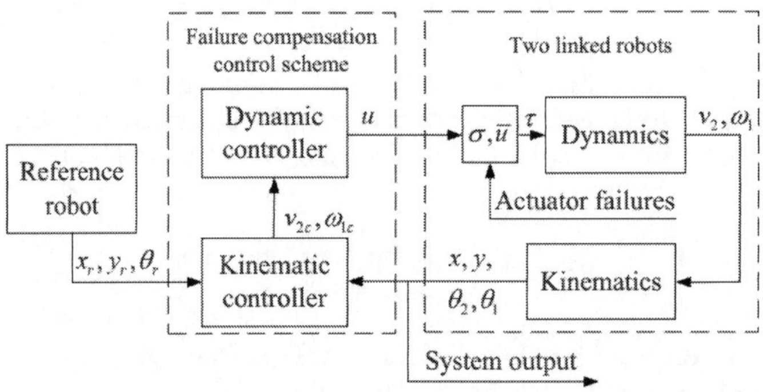

Figure 2: Actuator failure compensation control scheme.

- to design a kinematic control law $\eta_c = [v_{2c}, \omega_{1c}]^T$, such that when it is applied, the desired control performance can be ensured;
- to design a dynamic control law u using the designed kinematic control law, to achieve the control objective;
- to handle the failure uncertainties, for which, a multi-design integration based adaptive method will be employed;
- to evaluate the control performance.

3. Multi-Design Integration based Adaptive Failure Compensation Scheme

In this section, a multi-design integration based adaptive actuator failure compensation scheme is developed including a kinematic controller and a dynamic controller.

3.1. Kinematic Controller Design

3.1.1. Kinematic control law Define the output tracking error vector as

$$\tilde{e} = [\tilde{e}_x, \tilde{e}_y, \tilde{e}_\theta]^T = [x - x_r, y - y_r, \theta_2 - \theta_r]^T, \tag{15}$$

and a transformation matrix as

$$T_e = \begin{bmatrix} \cos\theta_r & \sin\theta_r & 0 \\ -\sin\theta_r & \cos\theta_r & 0 \\ 0 & 0 & 1 \end{bmatrix}. \tag{16}$$

Then, a new error vector is defined as

$$e = [e_x, e_y, e_\theta]^T = T_e \tilde{e}. \tag{17}$$

Since T_e is nonsingular with $\det[T_e] = 1$, if $\lim_{t\to\infty} e(t) = 0$, then $\lim_{t\to\infty} \tilde{e}(t) = 0$. It follows that

$$\dot{e}_x = \omega_r e_y + v_2 \cos e_\theta - v_r, \tag{18}$$
$$\dot{e}_y = -\omega_r e_x + v_2 \sin e_\theta, \tag{19}$$
$$\dot{e}_\theta = \frac{v_2}{d}\tan(\theta_1 - \theta_2) - \omega_r. \tag{20}$$

Introduce the diffeomorphism:

$$z_1 = e_x, \tag{21}$$
$$z_2 = e_y, \tag{22}$$
$$z_3 = \tan e_\theta, \tag{23}$$

an additional signal:

$$z_4 = \frac{\tan(\theta_1 - \theta_2)}{d\cos^3 e_\theta} - \frac{\omega_r}{v_r \cos^2 e_\theta} + e_y, \tag{24}$$

and an input transformation:

$$\alpha = \begin{bmatrix} \alpha_1 \\ \alpha_2 \end{bmatrix} = \begin{bmatrix} v_2 \cos e_\theta - v_r \\ \dot{z}_4 \end{bmatrix}. \tag{25}$$

Then, the derivatives of z_1, z_2, z_3 and z_4 are

$$\dot{z}_1 = \omega_r z_2 + \alpha_1, \tag{26}$$
$$\dot{z}_2 = -\omega_r z_1 + (v_r + \alpha_1) z_3, \tag{27}$$
$$\dot{z}_3 = v_r(z_4 - z_2) + \alpha_1\left(z_4 - z_2 + \frac{\omega_r}{v_r}(1 + z_3^2)\right), \tag{28}$$
$$\dot{z}_4 = \alpha_2. \tag{29}$$

From (24) and (25), we can obtain

$$\alpha = T_\alpha \eta + f_\alpha, \tag{30}$$

where

$$T_\alpha = \begin{bmatrix} T_{\alpha 11} & T_{\alpha 12} \\ T_{\alpha 21} & T_{\alpha 22} \end{bmatrix}, \quad f_\alpha = \begin{bmatrix} f_{\alpha 1} \\ f_{\alpha 2} \end{bmatrix}, \tag{31}$$

with $T_{\alpha 11} = \cos e_\theta$, $T_{\alpha 12} = 0$,

$$T_{\alpha 21} = \frac{3\tan^2(\theta_1 - \theta_2)\sin e_\theta}{d^2 \cos^4 e_\theta} - \frac{\tan(\theta_1 - \theta_2)}{d^2 \cos^3 e_\theta \cos^2(\theta_1 - \theta_2)} - \frac{2\omega_r \tan(\theta_1 - \theta_2)\sin e_\theta}{d v_r \cos^3 e_\theta} + \sin e_\theta,$$
$$T_{\alpha 22} = \frac{1}{d\cos^3 e_\theta \cos^2(\theta_1 - \theta_2)}, \quad f_{\alpha 1} = -v_r,$$
$$f_{\alpha 2} = -\frac{3\omega_r \tan(\theta_1 - \theta_2)\sin e_\theta}{d\cos^4 e_\theta} - \frac{\dot{\omega}_r}{v_r \cos^2 e_\theta} + \frac{\omega_r \dot{v}_r}{v_r^2 \cos^2 e_\theta} + \frac{2\omega_r^2 \sin e_\theta}{v_r \cos^3 e_\theta} - \omega_r e_x.$$

Define a virtual kinematic control signal as

$$\alpha_c = T_\alpha \eta_c + f_\alpha, \tag{32}$$

and design it as

$$\alpha_{c1} = -k_1(z_1 + z_3(z_4 + \frac{\omega_r}{v_r}(1 + z_3^2)), \tag{33}$$

$$\alpha_{c2} = -k_2 v_r z_3 - k_3 z_4, \tag{34}$$

where $k_1 > 0$, $k_2 > 0$ and $k_3 > 0$ are chosen to be constant. Then, with (32), the kinematic control law is

$$\eta_c = T_\alpha^{-1}(\alpha_c - f_\alpha). \tag{35}$$

3.1.2. Performance analysis Define the velocity tracking error as

$$\eta_e = \eta - \eta_c. \tag{36}$$

Then, we have

$$\alpha_e = [\alpha_{e1}, \alpha_{e2}]^T = \alpha - \alpha_c = T_\alpha \eta_e. \tag{37}$$

For the preliminary analysis of the designed kinematic control law, a positive-definite function is chosen as

$$V_1 = \frac{1}{2}(z_1^2 + z_2^2 + z_3^2 + \frac{1}{k_2} z_4^2). \tag{38}$$

With (26)-(29), the time derivative of V_1 is

$$\begin{aligned} \dot{V}_1 =& (z_1 + z_3(z_4 + \frac{\omega_r}{v_r}(1 + z_3^2)))\alpha_{c1} + v_r z_3 z_4 + \frac{z_4}{k_2}\alpha_{c2} \\ &+ (z_1 + z_3(z_4 + \frac{\omega_r}{v_r}(1 + z_3^2)))\alpha_{e1} + \frac{z_4}{k_2}\alpha_{e2}. \end{aligned} \tag{39}$$

Letting $f_\eta = [z_1 + z_3(z_4 + \frac{\omega_r}{v_r}(1 + z_3^2)), \frac{z_4}{k_2}]^T$ and substituting (33), (34) and (37) into (39), we have

$$\dot{V}_1 = -k_1(z_1 + z_3(z_4 + \frac{\omega_r}{v_r}(1 + z_3^2)))^2 - \frac{k_3}{k_2} z_4^2 + f_\eta^T T_\alpha \eta_e. \tag{40}$$

If there is no $f_\eta^T T_\alpha \eta_e$, then $\dot{V}_1$ is nonpositive, which means the system is stable. To eliminate it and ensure desired system performance, a dynamic controller is designed in the next section.

3.2. Dynamic Controller Design

3.2.1. Multi-design integration Substituting (11) into (8), we have the dynamic equations with actuator failures as follows:

$$\begin{aligned} \dot{\eta} &= -\bar{M}_1^{-1}\bar{M}_2\eta - \bar{M}_1^{-1}\bar{E} + \bar{M}_1^{-1}\bar{B}\sigma u + \bar{M}_1^{-1}\bar{B}(I_4 - \sigma)\bar{u} \\ &= -\bar{M}_1^{-1}\bar{M}_2\eta - \bar{M}_1^{-1}\bar{E} + \bar{M}_1^{-1}\bar{B}\sigma u + \bar{M}_1^{-1}\bar{B}\bar{u}_f, \end{aligned} \tag{41}$$

where $\bar{u}_f = (I_4 - \sigma)\bar{u}$, and $u = [u_{1r}, u_{1l}, u_{2r}, u_{2l}]^T$ is the applied control signal to be designed.

Let $\sigma_{(k)}, k = 1, 2, \ldots, N$ denote the kth possible failure pattern matrix satisfying actuation redundancy condition (13), where N is the number of all possible failure pattern matrices that are under consideration. For each $\sigma_{(k)}$, a corresponding dynamic control signal $u_{(k)}$ will be designed such that the desired system performance can be ensured if $u = u_{(k)}$ and $\sigma = \sigma_{(k)}$. Then, to cover all possible failure patterns, a nominal multi-design integration based dynamic control signal is constructed as

$$u^* = \sum_{k=1}^{N} \chi_{(k)} u_{(k)} = \sum_{k=1}^{N} \text{diag}\{\chi_{(k)}, \chi_{(k)} \chi_{(k)}, \chi_{(k)}\} u_{(k)}, \tag{42}$$

where

$$\chi_{(k)} = \begin{cases} 1 & \text{if } \sigma = \sigma_{(k)} \\ 0 & \text{otherwise.} \end{cases} \tag{43}$$

3.2.2. Dynamic control law Since the actual failure pattern σ is uncertain, $\chi_{(k)}$ is also uncertain for $k = 1, 2, \ldots, N$. To deal with such uncertainties, in this paper, the dynamic control law is designed as

$$u = \sum_{k=1}^{N} \hat{\chi}_{(k)} u_{(k)}, \tag{44}$$

where $\hat{\chi}_{(k)} = \text{diag}\{\hat{\chi}_{(k)1r}, \hat{\chi}_{(k)1l}, \hat{\chi}_{(k)2r}, \hat{\chi}_{(k)2l}\}$ is the estimate matrix that will be described in the following.

With (41), the time derivative of the velocity tracking error $\eta_e = \eta - \eta_c$ in (36) is

$$\dot{\eta}_e = -\bar{M}_1^{-1}\bar{M}_2\eta - \bar{M}_1^{-1}\bar{E} + \bar{M}_1^{-1}\bar{B}\sigma u + \bar{M}_1^{-1}\bar{B}\bar{u}_f - \dot{\eta}_c. \tag{45}$$

Let $\hat{\bar{u}}_f$ denote the estimate of $\bar{u}_f$. Then, the dynamic control signal $u_{(k)} = [u_{(k)1r}, u_{(k)1l}, u_{(k)2r}, u_{(k)2l}]^T$ is designed as

$$u_{(k)} = (\bar{M}_1^{-1}\bar{B}\sigma_{(k)})^+(-k_4\eta_e - T_\alpha^T f_\eta + \bar{M}_1^{-1}\bar{M}_2\eta + \bar{M}_1^{-1}\bar{E} - \bar{M}_1^{-1}\bar{B}\hat{\bar{u}}_f + \dot{\eta}_c) \tag{46}$$

for $\sigma_{(k)}, k = 1, 2, \ldots, N$, where $k_4 > 0$ is a chosen constant, and $(\bar{M}_1^{-1}\bar{B}\sigma_{(k)})^+$ is a generalized inverse matrix satisfying $\bar{M}_1^{-1}\bar{B}\sigma(k)(\bar{M}_1^{-1}\bar{B}\sigma_{(k)})^+ = I_2$.

Remark 3 *Some elements of some $u_{(k)}$ may be zero. For example, if $\sigma_{(k)} = \text{diag}\{0,1,1,1\}$ and $(\bar{M}_1^{-1}\bar{B}\sigma_{(k)})^+ = \sigma_{(k)}(\bar{M}_1^{-1}\bar{B})^T(\bar{M}_1^{-1}\bar{B}\sigma_{(k)}(\bar{M}_1^{-1}\bar{B})^T)^{-1}$, then $u_{(k)1r} = 0$. In this sense, the estimate matrix $\hat{\chi}_{(k)}$ in (44) can be simplified for some $\sigma_{(k)}$. For instance, for $\sigma_{(k)} = \text{diag}\{0,1,1,1\}$, $\hat{\chi}_{(k)} = \text{diag}\{0, \hat{\chi}_{(k)1l}, \hat{\chi}_{(k)2r}, \hat{\chi}_{(k)2l}\}$; and for $\sigma_{(k)} = \text{diag}\{0,1,1,0\}$, $\hat{\chi}_{(k)} = \text{diag}\{0, \hat{\chi}_{(k)1l}, \hat{\chi}_{(k)2r}, 0\}$. This can reduce the number of estimates.* □

Define the estimation errors as

$$\begin{aligned} &\tilde{\bar{u}}_f = \bar{u}_f - \hat{\bar{u}}_f, \\ &\tilde{\chi}_{(k)1r} = \chi_{(k)} - \hat{\chi}_{(k)1r}, \qquad \tilde{\chi}_{(k)1l} = \chi_{(k)} - \hat{\chi}_{(k)1l}, \\ &\tilde{\chi}_{(k)2r} = \chi_{(k)} - \hat{\chi}_{(k)2r}, \qquad \tilde{\chi}_{(k)2l} = \chi_{(k)} - \hat{\chi}_{(k)2l}, \end{aligned} \tag{47}$$

for $k = 1, 2, \ldots, N$. Then, substituting (42), (44) and (46) into (45), we have

$$\begin{aligned}\dot{\eta}_e =& -\bar{M}_1^{-1}\bar{M}_2\eta - \bar{M}_1^{-1}\bar{E} + \bar{M}_1^{-1}\bar{B}\sigma u^* + \bar{M}_1^{-1}\bar{B}\bar{u}_f - \dot{\eta}_c + \bar{M}_1^{-1}\bar{B}\sigma(u - u^*)\\ =& -k_4\eta_e - T_\alpha^T f_\eta + \bar{M}_1^{-1}\bar{B}\tilde{\bar{u}}_f\\ &-\sum_{k=1}^{N}(\bar{M}_1^{-1}\bar{B})_{c1}\sigma_{1r}\tilde{\chi}_{(k)1r}u_{(k)1r} - \sum_{k=1}^{N}(\bar{M}_1^{-1}\bar{B})_{c2}\sigma_{1l}\tilde{\chi}_{(k)1l}u_{(k)1l}\\ &-\sum_{k=1}^{N}(\bar{M}_1^{-1}\bar{B})_{c3}\sigma_{2r}\tilde{\chi}_{(k)2r}u_{(k)2r} - \sum_{k=1}^{N}(\bar{M}_1^{-1}\bar{B})_{c4}\sigma_{2l}\tilde{\chi}_{(k)2l}u_{(k)2l},\end{aligned} \tag{48}$$

where $(\bar{M}_1^{-1}\bar{B})_{ci}$ $(i = 1, 2, 3, 4)$ denotes the ith column vector of matrix $\bar{M}_1^{-1}\bar{B}$.

3.2.3. Adaptive laws To construct the control signal $u_{(k)}$ in (46), the adaptive law of $\hat{\bar{u}}_f$ is chosen as

$$\dot{\hat{\bar{u}}}_f = \Gamma_f(\bar{M}_1^{-1}\bar{B})^T\eta_e, \tag{49}$$

where $\Gamma_f = \Gamma_f^T \in \Re^{4\times4}$ is the positive-definite adaptation gain matrix that is chosen to be constant.

To construct the dynamic control law u in (44), the adaptive laws of $\hat{\chi}_{(k)1r}$, $\hat{\chi}_{(k)1l}$, $\hat{\chi}_{(k)2r}$ and $\hat{\chi}_{(k)2l}$ are chosen as

$$\begin{aligned}\dot{\hat{\chi}}_{(k)1r} &= -\gamma_{k1r}u_{(k)1r}\eta_e^T(\bar{M}_1^{-1}\bar{B})_{c1} + f_{k1r},\\ \dot{\hat{\chi}}_{(k)1l} &= -\gamma_{k1l}u_{(k)1l}\eta_e^T(\bar{M}_1^{-1}\bar{B})_{c2} + f_{k1l},\\ \dot{\hat{\chi}}_{(k)2r} &= -\gamma_{k2r}u_{(k)2r}\eta_e^T(\bar{M}_1^{-1}\bar{B})_{c3} + f_{k2r},\\ \dot{\hat{\chi}}_{(k)2l} &= -\gamma_{k2l}u_{(k)2l}\eta_e^T(\bar{M}_1^{-1}\bar{B})_{c4} + f_{k2l},\end{aligned} \tag{50}$$

for $k = 1, 2, \ldots, N$, where $\gamma_{k1r} > 0$, $\gamma_{k1l} > 0$, $\gamma_{k2r} > 0$ and $\gamma_{k2l} > 0$ are chosen to be constant, and f_{k1r}, f_{k1l}, f_{k2r} and f_{k2l} are the projection signals.

Let $p_{k1r} = -\gamma_{k1r}u_{(k)1r}\eta_e^T(\bar{M}_1^{-1}\bar{B})_{c1}$, $p_{k1l} = -\gamma_{k1l}u_{(k)1l}\eta_e^T(\bar{M}_1^{-1}\bar{B})_{c2}$, $p_{k2r} = -\gamma_{k2r}u_{(k)2r}\eta_e^T(\bar{M}_1^{-1}\bar{B})_{c3}$ and $p_{k2l} = -\gamma_{k2l}u_{(k)2l}\eta_e^T(\bar{M}_1^{-1}\bar{B})_{c4}$. The projection signals are given as follows:

$$f_{kj} = \begin{cases} 0, & \text{if } \hat{\chi}_{(k)j} \in (0,1), \text{or}\\ & \text{if } \hat{\chi}_{(k)j} = 0 \text{ and } p_{kj} \geq 0, \text{or}\\ & \text{if } \hat{\chi}_{(k)j} = 1 \text{ and } p_{kj} \leq 0,\\ -p_{kj}, & \text{otherwise,}\end{cases} \tag{51}$$

for $j = 1r, 1l, 2r, 2l$.

Lemma 1 The adaptive laws in (50) with the projection signals in (51) for $k = 1, 2, \ldots, N$ guarantee that: i) $\hat{\chi}_{(k)1r}, \hat{\chi}_{(k)1l}, \hat{\chi}_{(k)2r}, \hat{\chi}_{(k)2l} \in [0,1]$; and ii) $\tilde{\chi}_{(k)1r}f_{k1r} \geq 0$, $\tilde{\chi}_{(k)1l}f_{k1l} \geq 0$, $\tilde{\chi}_{(k)2r}f_{k2r} \geq 0$ and $\tilde{\chi}_{(k)2l}f_{k2l} \geq 0$.

Proof: The proofs for the four adaptive laws are similar. Here the details for $\hat{\chi}_{(k)1r}$ are given as an example.

Choose the initial estimate as $\hat{\chi}_{(k)1r}(0) \in [0,1]$. The projection signal in (51) with $j = 1r$ ensures $\hat{\chi}_{(k)1r}(t) \in [0,1]$, and

$$(\chi_{(k)1r} - \hat{\chi}_{(k)1r})f_{k1r} = \tilde{\chi}_{(k)1r}f_{k1r} \geq 0 \tag{52}$$

that is analyzed for the following three cases: (i) if $\hat{\chi}_{(k)1r} = 0$ and $p_{k1r} < 0$, then $\chi_{(k)1r} - \hat{\chi}_{(k)1r} \geq 0$ and $f_{k1r} = -p_{k1r} > 0$, which means (52) is ensured; (ii) if $\hat{\chi}_{(k)1r} = 1$ and $p_{k1r} > 0$, then $\chi_{(k)1r} - \hat{\chi}_{(k)1r} \leq 0$ and $f_{k1r} = -p_{k1r} < 0$, which also means (52) is ensured; and (iii) otherwise, we have $f_{k1r} = 0$, then (52) is ensured.

Similarly, we can also obtain the same properties for the other three adaptive laws. The proof is completed. ∇

3.3. System performance

Choose the global Lyapunov function candidate as

$$V_2 = V_1 + \frac{1}{2}\eta_e^T\eta_e + \frac{1}{2}\tilde{\bar{u}}_f^T\Gamma_f^{-1}\tilde{\bar{u}}_f + \frac{1}{2}\sum_{k=1}^{N}\sigma_{1r}\gamma_{k1r}^{-1}\tilde{\chi}_{(k)1r}^2 + \frac{1}{2}\sum_{k=1}^{N}\sigma_{1l}\gamma_{k1l}^{-1}\tilde{\chi}_{(k)1l}^2$$
$$+ \frac{1}{2}\sum_{k=1}^{N}\sigma_{2r}\gamma_{k2r}^{-1}\tilde{\chi}_{(k)2r}^2 + \frac{1}{2}\sum_{k=1}^{N}\sigma_{2l}\gamma_{k2l}^{-1}\tilde{\chi}_{(k)2l}^2. \tag{53}$$

Then, the derivative of V_2 is

$$\dot{V}_2 \leq -k_1(z_1 + z_3(z_4 + \frac{\omega_r}{v_r}(1 + z_3^2)))^2 - \frac{k_3}{k_2}z_4^2 - k_4\eta_e^T\eta_e \leq 0, \tag{54}$$

which indicates: $z_1, z_2, z_3, z_4, \eta_e, z_1 + z_3(z_4 + \frac{\omega_r}{v_r}(1 + z_3^2)) \in L^\infty$ and all estimates are bounded, and $z_4, \eta_e, z_1 + z_3(z_4 + \frac{\omega_r}{v_r}(1 + z_3^2)) \in L^2$. It follows from (23) and (24) that $\cos e_\theta \neq 0$ and $\tan(\theta_1 - \theta_2) \in L^\infty$ meaning $\cos(\theta_1 - \theta_2) \neq 0$. Then, with (18)-(35) and (41)-(46), we have: T_α is nonsingular and bounded, and $f_\alpha, \alpha_c, \alpha_e, \alpha, \eta_c, \eta, \dot{z}_1, \dot{z}_2, \dot{z}_3, \dot{z}_4, \dot{\alpha}_c, \dot{\eta}_c, u_{(k)}, \dot{\eta}, \dot{\eta}_e, \dot{\alpha}_e, \dot{\alpha} \in L^\infty$, which also implies that the derivative of $z_1 + z_3(z_4 + \frac{\omega_r}{v_r}(1 + z_3^2))$ is bounded. According to Barbalat's lemma, we can conclude that all closed-loop signals are bounded, and $\lim_{t\to\infty}(z_1 + z_3(z_4 + \frac{\omega_r}{v_r}(1 + z_3^2))) = 0$, $\lim_{t\to\infty} z_4 = 0$ and $\lim_{t\to\infty} \eta_e = 0$, which also means $\lim_{t\to\infty} \alpha_e = 0$, $\lim_{t\to\infty} \alpha_{c1} = 0$, and $\lim_{t\to\infty} \alpha_1 = 0$ with (37) and (33).

From (29), $\ddot{z}_4 = \dot{\alpha}_2 = \dot{\alpha}_{c2} + \dot{\alpha}_{e2} \in L^\infty$ with $\dot{\alpha}_{c2}, \dot{\alpha}_{e2} \in L^\infty$, which means that $\dot{z}_4$ is uniformly continuous, together with $\lim_{t\to\infty}\int_0^t \dot{z}_4(\tau)d\tau = z_4(\infty) - z_4(0) = -z_4(0)$, we have $\lim_{t\to\infty} \dot{z}_4 = \lim_{t\to\infty} \alpha_2 = \lim_{t\to\infty}(\alpha_{c2} + \alpha_{e2}) = 0$ according to Barbalat's lemma. Then, with $\lim_{t\to\infty} \alpha_e = 0$, $\lim_{t\to\infty} z_4 = 0$, $\alpha_{c2} = -k_2v_rz_3 - k_3z_4$ and $v_r \neq 0$, we have $\lim_{t\to\infty} \alpha_{c2} = 0$ and $\lim_{t\to\infty} z_3 = 0$, it follows that $\lim_{t\to\infty} z_1 = 0$ with $\lim_{t\to\infty}(z_1 + z_3(z_4 + \frac{\omega_r}{v_r}(1 + z_3^2))) = 0$. On the other hand, from $\dot{z}_3 = v_r(z_4 - z_2) + \alpha_1(z_4 - z_2 + \frac{\omega_r}{v_r}(1 + z_3^2))$ in (28), we have $\ddot{z}_3 \in L^\infty$. Similarly, $\lim_{t\to\infty} \dot{z}_3 = 0$ is ensured according to Barbalat's lemma. Then, we obtain $\lim_{t\to\infty} z_2 = 0$ with $\lim_{t\to\infty} z_4 = 0$, $\lim_{t\to\infty} \alpha_1 = 0$ and $v_r \neq 0$.

Finally, we can conclude that: all closed-loop signals are bounded, and $\lim_{t\to\infty} z_i(t) = 0$ $(i = 1, 2, 3, 4)$ and $\lim_{t\to\infty}(\eta(t) - \eta_c(t)) = 0$, which also means $\lim_{t\to\infty}(x(t) - x_r(t)) = 0$, $\lim_{t\to\infty}(y(t) - y_r(t)) = 0$ and $\lim_{t\to\infty}(\theta_2(t) - \theta_r(t)) = 0$ with the diffeomorphism in (21)-(23) and the transformation in (17).

In summary, we have the following theorem.

Theorem 1 The developed multi-design integration based adaptive actuator failure compensation control scheme, constituted by the kinematic control law in (35), and dynamic control law in (44) with multiple control signals in (46) and updated by the projection adaptive laws in (49) and (50), applied to two linked 2WD mobile robots modeled as (1) and (8), guarantees that all closed-loop signals are bounded, and $\lim_{t\to\infty}(x(t) - x_r(t)) = 0$, $\lim_{t\to\infty}(y(t) - y_r(t)) = 0$ and $\lim_{t\to\infty}(\theta_2(t) - \theta_r(t)) = 0$, despite the presence of actuator failures modeled as (10)-(12).

4. Simulation Study

To verify the effectiveness of the developed multi-design integration based adaptive actuator failure compensation scheme for two linked 2WD robots shown in Fig. 1, the following simulation study is presented.

4.1. Simulation Conditions

In this simulation, the physical parameters of each robot are $a_1 = a_2 = 0.3$ m, $b_1 = b_2 = 0.75$ m, $r_1 = r_2 = 0.15$ m, $m_1 = m_2 = 30$ kg, $I_{m1} = I_{m2} = 15.625$ kg m^2. The length of the link is assumed to be $d = 1.7$ m.

To verify the system tracking property, a circle reference trajectory is considered. To generate such reference a trajectory, the velocities v_r and ω_r are chosen as: $v_r = 0.5$ m/s and $\omega_r = 0.5$ rad/s. Then, x_r, y_r and θ_r are generated by (14) with $x_r(0) = y_r(0) = 0$ and $\theta_r(0) = 45$ deg.

In order to verify the failure compensation effectiveness of the developed adaptive control scheme, the following failure cases are simulated:

$$\begin{array}{lll}
\text{no failure,} & \sigma_{(1)} = \text{diag}\{1,1,1,1\}, & 0 \le t < 10\text{s}. \\
\tau_{1r} \text{ fails,} & \sigma_{(2)} = \text{diag}\{0,1,1,1\}, \tau_{1r} = 0, & 10\text{s} \le t < 20\text{s}; \\
\tau_{1r}, \tau_{2l} \text{ fail,} & \sigma_{(3)} = \text{diag}\{0,1,1,0\}, \tau_{1r} = 0, \tau_{2l} = -2 \text{ Nm}, & 20\text{s} \le t < 30\text{s}; \\
\tau_{2l} \text{ fails,} & \sigma_{(4)} = \text{diag}\{1,1,1,0\}, \tau_{2l} = -2 \text{ Nm}, & 30\text{s} \le t < 40\text{s}; \\
\tau_{2r}, \tau_{2l} \text{ fail,} & \sigma_{(5)} = \text{diag}\{1,1,0,0\}, \tau_{2r} = -1 \text{ Nm}, \tau_{2l} = -2 \text{ Nm}, & t \ge 40\text{s}.
\end{array}$$

There are 5 failure pattern matrices which satisfy the actuation redundancy condition in (13), covering the cases of fault free, one actuator fails, both two actuators of robot 2 fail, and one actuator of each robot fails.

The initial conditions are chosen as: $x(0) = 0$, $y(0) = 1$ m, $\theta_2(0) = 10$ deg, $\theta_1(0) = 0$, $v_2(0) = 0$, and $\omega_1(0) = 0$. The adaptation gains are chosen as: $\Gamma_f = I_4$, and $\gamma_{k1r} = \gamma_{k1l} = \gamma_{k2r} = \gamma_{k2l} = 1$ for $k = 1,2,3,4,5$. The control gains are chosen as: $k_1 = 1$, $k_2 = 1$, $k_3 = 0.2$ and $k_4 = 20$.

4.2. Simulation Results

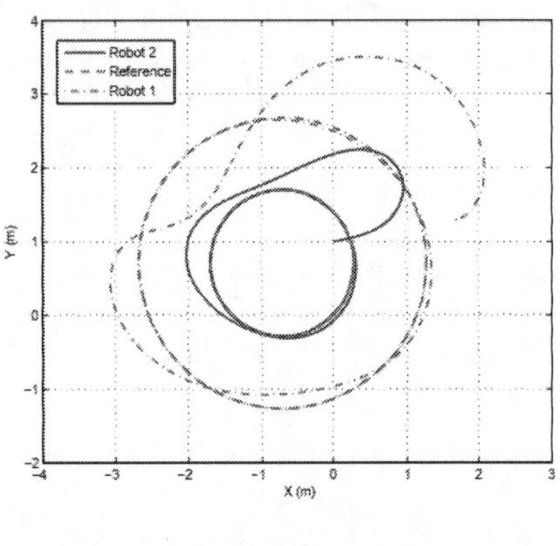

Figure 3: Robot trajectories in (X, Y) plane.

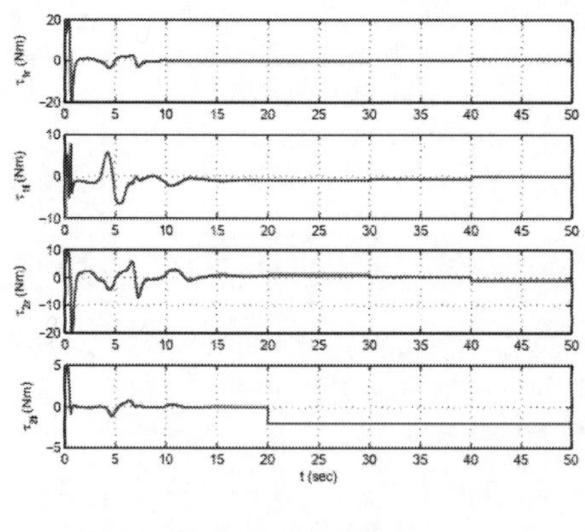

Figure 4: Control torques.

Fig. 3 shows the positions of the robot 2, the reference robot and the robot 1. Fig. 4 shows the control torques generated by the four wheels, which are consistent with the faulty cases in simulation conditions. Fig. 5 shows the tracking errors. Fig. 6 shows the orientation error

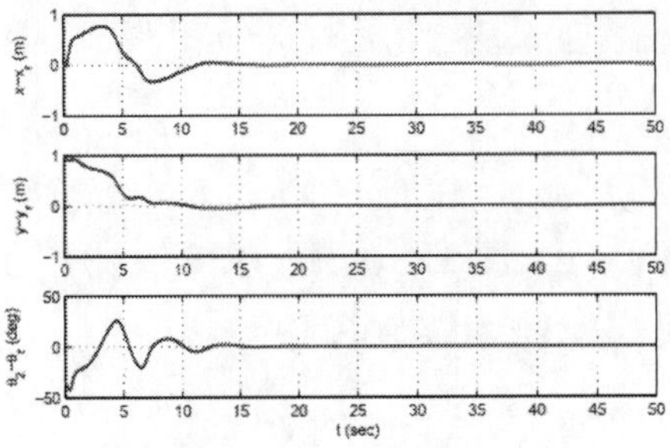

Figure 5: Tracking errors.

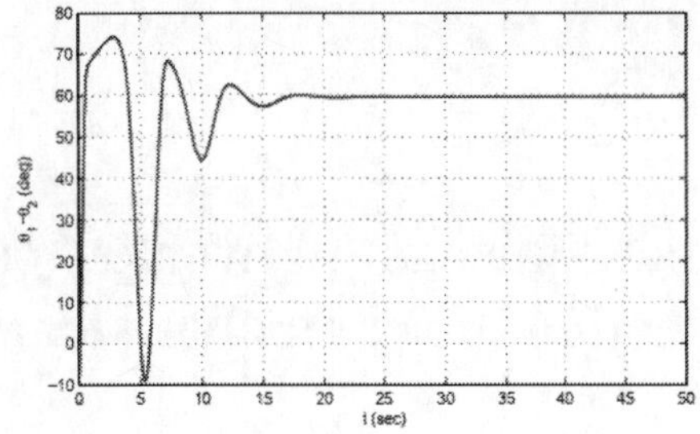

Figure 6: Orientation error of two robots.

between two robots. From them, we can see that the desired system stability and asymptotic tracking properties are ensured by the developed multi-design integration based adaptive failure compensation scheme, despite the presence of some actuator failures.

5. Conclusions

This paper developed a multi-design integration based adaptive actuator failure compensation scheme for two linked 2WD robots. The kinematics and dynamics of this robot configuration were modeled, for which, the proposed adaptive failure compensation scheme was designed. The developed control scheme ensures desired system stability and asymptotic tracking properties, which was verified by simulation results. Extending the proposed method for $n(n > 2)$ linked 2WD mobile robots is our interest in the future work.

ACKNOWLEDGMENT

This work was partially supported by the Regional project SUCRé (Sûreté de fonctionnement et résilience pour la gestion et le contrôle coopératifs des systèmes sociotechniques: Coopération Homme(s)-Robot(s) en milieu hostile) of the Hauts de France region.

References

[1] Dixon W, D. Dawson, Zergeroglu E and Behal A. 2001 Adaptive tracking control of a wheeled mobile robot via an uncalibrated camera system *IEEE T. Syst., Man, Cybern., Cybern.* **31** 341-52,

[2] Fierro R and Lewis F 1998 Control of a nonholonomic mobile robot using neural networks *IEEE T. Neural Network* **9** 589-600

[3] Huang J, Wen C, Wang W and Jiang Z 2014 Adaptive output feedback tracking control of a nonholonomic mobile robot *Automatica* **50** 821-31

[4] Zhang Y and Jiang J, 2008 Bibliographical review on reconfigurable fault-tolerant control systems *Annu. Rev. Control* **32** 229-52

[5] Blanke M, Kinnaert M, Lunze J and Staroswiecki M 2006 *Diagnosis and Fault-Tolerant Control* Springer-Verlag Berlin Heidelberg

[6] Ji M and Sarkar N 2007 Supervisory fault adaptive control of a mobile robot and its application in sensor-fault accommodation *IEEE T. Robotics* **23** 174-8

[7] Rotondo D, Puig V, Nejjari F and Romera J 2014 A fault-hiding approach for the switching Quasi-LPV fault-tolerant control of a four-wheeled omnidirectional mobile robot *IEEE T. Ind. Electron.* **62** 3932-44

[8] Ji M, Zhang Z, Biswas G and Sarkar N 2003 Hybrid fault adaptive control of a wheeled mobile robot *IEEE-ASME T. Mech.* **8** 226-33

[9] Khalaji A and Moosavian S 2014 Robust adaptive controller for a trackor-trailer mobile robot *IEEE-ASME T. Mech.* **19** 943-53

IOP Conf. Series: Journal of Physics: Conf. Series **783** (2017) 012022 doi:10.1088/1742-6596/783/1/012022

Attack-tolerant networked control system: an approach for detection the controller stealthy hijacking attack

Amer Atta Yaseen, Mireille Bayart

Univ. Lille, CNRS UMR 9189 CRIStAL – Centre de Recherche en Informatique Signal et Automatique de Lille, Bât Polytech-Lille, 59655 Villeneuve d'Ascq Cedex, France

Email : amer.yaseen@polytech-lille.fr, mireille.bayart@univ-lille1.fr

Abstract. In this work, a new approach will be introduced as a development for the attack-tolerant scheme in the Networked Control System (NCS). The objective is to be able to detect an attack such as the Stuxnet case where the controller is reprogrammed and hijacked. Besides the ability to detect the stealthy controller hijacking attack, the advantage of this approach is that there is no need for a priori mathematical model of the controller. In order to implement the proposed scheme, a specific detector for the controller hijacking attack is designed. The performance of this scheme is evaluated be connected the detector to NCS with basic security elements such as Data Encryption Standard (DES), Message Digest (MD5), and timestamp. The detector is tested along with networked PI controller under stealthy hijacking attack. The test results of the proposed method show that the hijacked controller can be significantly detected and recovered.

1. Introduction

Any network medium is susceptible to be easily intercepted. Research in NCS were initialized as regards the safety and convenience in hazardous environments such as power plants, nuclear reactor space projects, military applications, etc. In all these applications, security is of utmost concern [1]. The cyber-attacks can have heavy consequences on the plant, but can also have wide-ranging effects on environment and the individuals. The cyber-attacks include the data modification, deceptive sender identity, and data replay [2]. Security mechanisms of NCS related to the cyber-attacks have been addressed in several directions.

Some works on network security algorithms: In [3], DES (Data Encryption Standard), 3DES (Triple Data Encryption Standard), and AES (Advanced Encryption Standard) are integrated with the application to secure the sensor as well as control data flow on the network, then 1-D gain scheduler was designed and implemented to alleviate the adverse effect due to security. The encryption of the data transmitted on the network is also realized by DES in [4], and they add message digest 5 (MD5) to detect the data integrity. In [5], the DES was also adopted as security solutions for the DC motor networked control system in TrueTime platform algorithm. In [6], to prevent most of the attacks in which the IP systems are vulnerable, the authors proposed a Network management of Named Data.

Others interest on communication and control with real times constraints: The trade-off between NCS security and its real-time performance was demonstrated in [7]. In [8], a detection module based on implementation of DES algorithm is given, furthermore, to protect NCSs from getting out of control,

the authors design also a response module. In [9], the authors described a quick detection approach against data-injection attack in the smart grid.
However, all the methods [3]- [9] are handled only with the attack cases, which are stated in [2].
From The British Columbia Institute of Technology Industrial Security Incident Database [1] we can note other attacks on process control and industrial networked systems (e.g. Stuxnet case) which are not mentioned in [2].
This paper deals with the effects of special attack scenario such as Stuxnet case [10] where a controller was reprogrammed and hijacked.
The Stuxnet case needs to the implementation of special attack detectors along with backup controller. Until now, a typical approach to implement such attack detectors is based on system models e.g. [11], [12] and it is inspired by fault diagnosis detectors [13], [14]. The Stuxnet attack detector, which not based on controller model was proposed in [15]. However, the restriction of [15] was represented by assuming the value of set point (operator's desired value) is always equal to zero.
The attack-tolerant scheme for NCS was introduced in [16]; this scheme considered three types of attacks in NCS, these types are represented by injection, modification, and replication of the sent and received data in the networked control loop. A modification for the scheme in [16] was adopted in [17] to handle the Stuxnet case. However, the approach of [17] is still need to develop the sensitivity of detection. In this paper, we specially focus on the controller hijacking attack with the injection of a destructive control signal into the networked control loop. The stealthy hijacking attack will be considering rather than the rough attack. The architecture of the attack-tolerant system developed will be detailed in the next sections.

2. The proposed secure NCS

Fig.1 illustrates the general block diagrams of the proposed secure NCS. Due to the real time nature of the NCS, UDP is preferred for use in NCS over IP network. The Data Encryption Standard (DES), Message Digest (MD5), and timestamp will be used along with the controller and plant sides' security mechanisms; the details of these mechanisms were described in [16]

In this paper, the controller hijacking attack detector will be introduced as a new enhancement to the attack tolerant scheme of [16].

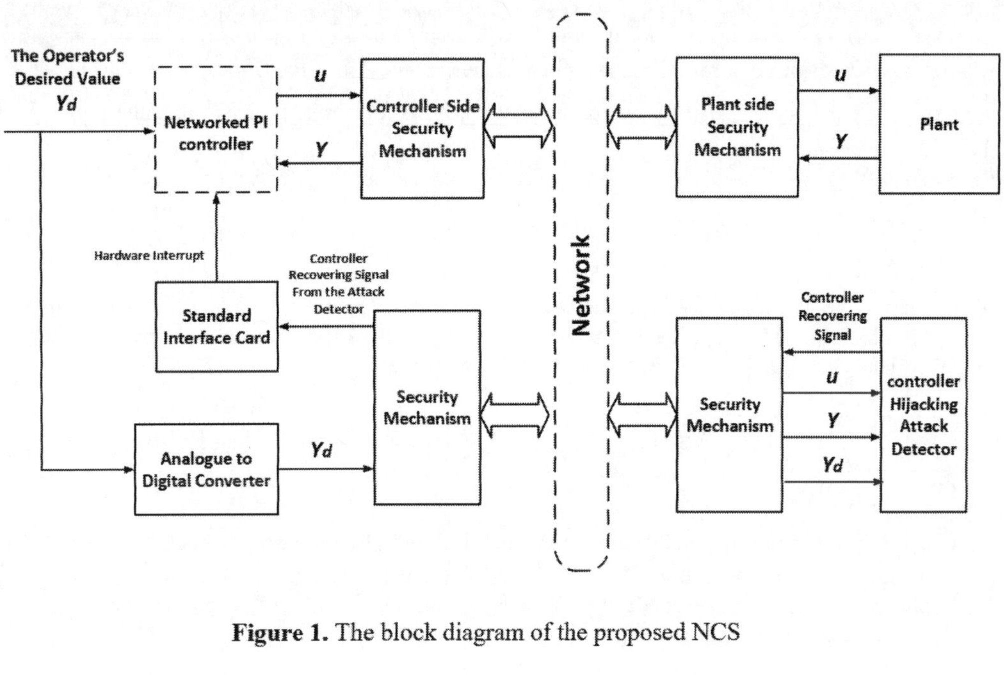

Figure 1. The block diagram of the proposed NCS

IOP Conf. Series: Journal of Physics: Conf. Series **783** (2017) 012022 doi:10.1088/1742-6596/783/1/012022

3. Detection of the controller hijacking attack

The block diagram of the NCS is illustrated in Fig. 1. For detecting the attacks on the controller, a new approach is proposed. The advantage of this proposed approach is that there is no need for a priori mathematical model of the controller. The detector is connected separately from the controller and plant sides to the NCS with basic security elements such as Data Encryption Standard (DES), Message Digest (MD5), and timestamp, which are used for the secure communication over NCS. Assume we have a controller (see Fig. 2) and that one wants to monitor the controller behaviour by mean of the controller input and output. The controller with a single input variable e and a single output variable u is selected.

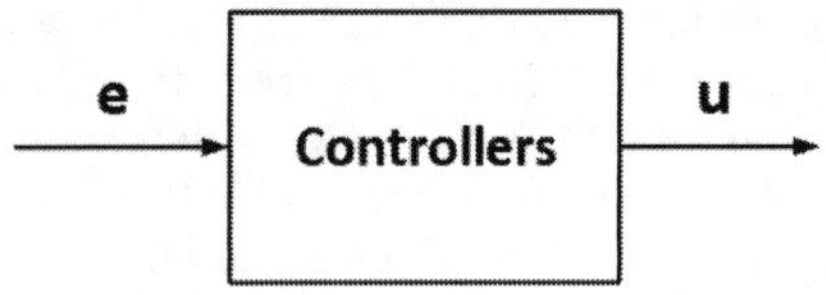

Figure 2. SISO Controller

The attack on the plant was expressed in [18] as follows

$$ur_k = u_k(1-\psi) \quad (1)$$

where, ψ, $0 \leq \psi \leq 1$, is the loss of the response efficiency of the plant due to control signal which is sent from the attacker. ur_k, is the true control variable.

Rewrite (1) for the attack on the controller (e.g. controller hijacking attack) or,

$$er_k = e_k(1-\phi) \quad (2)$$

where, er_k is the true error variable which is measured at the controller input during all cases including the presence of an attack on controller side, e_k is the nominal error value and ϕ, $0 \leq \phi \leq 1$, is the controller performance index.

Rewrite (1),

$$\phi = 1 - \frac{er_k}{e_k} \quad (3)$$

In this paper, the controller performance index will be used to detect the controller hijacking attack. Now, the objective is to estimate the nominal error as well as the true error. Recall the ultra-local model which is given in [19] and introduced for the controller in [17]

$$u^{(v)} = F + \alpha e \quad (4)$$

where,

- $\alpha \in R$ is a non-physical constant parameter. It is chosen by the designer such that αe and $u^{(v)}$ are of the same magnitude. It should be therefore clear that its numerical value, which is obtained by trials and errors during attack free condition, is not a priori precisely defined.
- F represents all the unusual input-output behavior of the controller;

- The order v is also a design parameter of the numerical model of (4) that can be arbitrarily chosen by the designer. In this paper, first order derivative $\dot{u}$ will be chosen.
- $e = y_d - y$, y_d. is the operator's desired set point and y is the plant response.

For the attack free controller, the values of $\dot{u}$ and αe are equal in magnitude [17] or,

$$\left| \left\langle \left\langle e \right\rangle \right\rangle_T^c (t) \right| = \left| \frac{\left\langle \left\langle \dot{u} \right\rangle \right\rangle_T^c (t)}{\alpha} \right| \tag{5}$$

where, $\left\langle \left\langle \dot{u} \right\rangle \right\rangle_T^c (t)$ is continuous-time estimation of the first derivative of the controller output.

Due to the network effects e.g. noise [20], packet drop, and the out of sequence [16], the estimated values for the first derivative of the controller output and the true error will be used in this paper.

By applying the derivative estimation method which introduced in [21], the first derivative of the controller output can be estimated as follows:

Let *um* be a measured value of the controller output *u*, and *um* is the image of *u* and we consider that *u* is distorted by some of added noise Ω, therefore we have: $um = u + \Omega$.

The objective is to estimate the derivatives of the controller output *u*, up to a finite order of derivation, from its measurement *um* observed on a given time interval.

The Taylor expansion of the controller output around *0* is given by

$$u(\tau) = \sum_{n=0}^{\infty} u^{(n)}(0) \frac{\tau^n}{n!} \tag{6}$$

By the polynomial we can approximate *u(t)* for the interval *[0,T], T>0*.

$$u_N(\tau) = \sum_{n=0}^{N} u^{(n)}(0) \frac{\tau^n}{n!} \tag{7}$$

For *N* degree, ΦN is the operational analogue of *uN* and can be written as

$$\Phi_N(s) = \frac{u(0)}{s} + \frac{\dot{u}(0)}{s^2} + \ldots + \frac{u^{(N)}(0)}{s^{N+1}} \tag{8}$$

By applying a convenient operator to $\Phi N(s)$, we can separate each coefficient $u^{(n)}(0)$ appearing in previous expression. Thus,

$$\forall i = 0, \ldots, N$$

$$\frac{u^{(i)}(0)}{s^{2N+1}} = \frac{(-1)^i}{N!(N-i)!} \cdot \frac{1}{s^{N+1}} \cdot \frac{d^i}{ds^i} \cdot \frac{1}{s} \cdot \frac{d^{N-i}}{ds^{N-i}} \left(s^{N+i} \Phi N(s) \right) \tag{9}$$

The expression of $u^{(i)}(0)$ in the time domain can be written as

$$u^{(i)}(0) = \int_{0}^{T} P(\delta;T)uN(\delta)\,d\delta \tag{10}$$

where $P(\delta;T)$ is polynomial in δ and T. Notice that (11) gives the calculation of $u^{(i)}(0)$ from an integral on the time interval [0; T] for a given small $T > 0$.
As $\left.\frac{d^i u(t-\delta)}{d\delta^i}\right|_{\delta=0} = (-1)^i u^{(i)}(t)$ it is possible to express $u^{(i)}(t)$ as an integral, which includes values of uN on the time interval $[t - T, t]$:

$$u^{(i)}(t) = (-1)^{(i)} \int_{0}^{T} P(\delta;T)uN(t-\delta)\,d\delta \tag{11}$$

By using the noisy signal *um,* a simple estimator of the derivative $u^{(i)}(t)$ can be expressed as follows:

$$\left\langle\left\langle u^{(i)}\right\rangle\right\rangle_T^C (t) = (-1)^{(i)} \int_{0}^{T} P(\delta;T)um(t-\delta)\,d\delta \tag{12}$$

(13) is realized from (12) by changing *uN* by *um*. Noting that the integral operation acting the rule of low-pass filter and reduced the noise that distorts *um*. The choice of *T* results in a trade-off: small value of *T* leads to the effect of the noise; the large value of *T* is leads to better integrals low pass filtering but there is error due to truncation.
In practice, the integral expressed in (13) is obtained by a numerical integration method; therefore, the estimator $\left\langle\left\langle u^{(i)}\right\rangle\right\rangle_T^C (t)$ will be performed at each sample of *k*.

Let *Ts* is the sampling period, then the discretization of any continuous time function f will be denoted by *f[k]* i.e.

$$f[k] = f(k,Ts), \ k \in Z$$

With these notations, the discrete-time approximation of the first derivative estimation of the controller output is simply a discrete sum that can be written as:

$$\left\langle\left\langle \dot{u}\right\rangle\right\rangle_{Ts,ns}^d [k] = \sum_{j=0}^{ns} w(j)P(jT_s;n_sT_s)Um[k-j] \tag{13}$$

Where *ns* is the number of samples used in the time window $T = nsTs$ and the $w(j)$ is the weight related to the used numerical integration method.
To eliminate the effects of network which are mentioned above, the true error function values will be integrated as a tabular data. These values are taken at certain discrete points during the same time window of the derivative output estimation.
First is to fit a curve through the true error data, and then integrate the resulting curve. It is possible to use any integration methods, but the most common approach is to use a piecewise polynomial such as a spline as follows:

- Suppose we are given as set of true error samples point *(t, er(t))*

- Fit a spline *E(t),* through this samples, keep in mind that *E(t)* is a piecewise polynomial on several intervals in particular.

$$Er(t)=\begin{cases} erp_1(t) & t_0 \leq t \leq t_1 \\ erp_2(t) & t_1 \leq t \leq t_2 \\ \vdots & \\ erp_{ns}(t) & t_{ns-1} \leq t \leq t_{ns} \end{cases} \tag{14}$$

Then

$$\left\langle\left\langle e_r \right\rangle\right\rangle_T^c(t) = \int_{t-nsT}^{t} e(t)\, dt \approx \int_{t-nsT}^{t} E(t)\, dt = \sum_{j=0}^{ns} \int_{t_{j-1}}^{t_j} erp_j(t)\, dt \tag{15}$$

It is possible to use a cubic spline interpolant. In this case, each $erp_j(t)$ is a cubic polynomial, which can be written as

$$erp_j(t) = a_{j,1}(t-t_{j-1})^3 + a_{j,2}(t-t_{j-1})^2 + a_{j,3}(t-t_{j-1}) + a_{j,4} \tag{16}$$

Solving for $a_{j,1}, a_{j,2}, a_{j,3},$ and $a_{j,4}$ using Gauss elimination [22], then we can easily integrate $erp_j(t)$

$$\int_{t_{j-1}}^{t_j} erp_j(t)\, dt = \frac{a_{j,1}}{4}(t_j - t_{j-1})^4 + \frac{a_{j,2}}{3}(t_j - t_{j-1})^3 + \frac{a_{j,3}}{2}(t_j - t_{j-1})^2 + a_{j,4}(t_j - t_{j-1}) \tag{17}$$

Let, $T = t_j - t_{j-1}$ then,

$$\int_{t_{j-1}}^{t_j} erp_j(t)\, dt = \frac{a_{j,1}}{4}T^4 + \frac{a_{j,2}}{3}T^3 + \frac{a_{j,3}}{2}T^2 + a_{j,4}T \tag{18}$$

Thus, the estimated true error is given by

$$\left\langle\left\langle e_r \right\rangle\right\rangle_{Ts,ns}^d[k] = \sum_{j=0}^{ns} \left(\frac{a_{j,1}}{4}Ts^4 + \frac{a_{j,2}}{3}Ts^3 + \frac{a_{j,3}}{2}Ts^2 + a_{j,4}Ts \right) \tag{19}$$

Note that a direct implantation of this summation would require 7*ns* multiplications, 3*ns* additions, and 3*ns* exponentiations. An algebraic rearrangement results in a nested approach of this computation [22],

$$\left\langle\left\langle e_r\right\rangle\right\rangle_{Ts,ns}^{d}[k]=\sum_{j=0}^{ns}Ts\left(Ts\left(Ts\left(\frac{a_{j,1}}{4}Ts+\frac{a_{j,2}}{3}\right)+\frac{a_{j,3}}{2}\right)+a_{j,4}\right) \tag{20}$$

(20) requires only $7ns$ multiplications, $3ns$ additions, and no exponentiations.
Assuming nominal and true error are in the same sign (i.e. stealthy controller hijacking attack) then, rewrite (3) in term of estimated values

$$\phi k=1-\left|\frac{\alpha\left\langle\left\langle er\right\rangle\right\rangle_{Ts,ns}^{d}[k]}{\left\langle\left\langle \dot{u}\right\rangle\right\rangle_{Ts,ns}^{d}[k]}\right| \tag{21}$$

Or,

$$\phi k=1-\frac{\alpha\sum_{j=0}^{ns}Ts\left(Ts\left(Ts\left(\frac{a_{j,1}}{4}Ts+\frac{a_{j,2}}{3}\right)+\frac{a_{j,3}}{2}\right)+a_{j,4}\right)}{\sum_{j=0}^{ns}w(j)P(jTs;nsTs)um[k-j]} \tag{22}$$

If $\phi=0$ means that there is no controller hijacking attack.
$0<\phi<1$ means that there is stealthy controller hijacking attack.
If $\phi=1$ implies that the controller is completely hijacked.

4. Controller recovering

On referring to Fig.1, at each sample time k, the controller hijacking attack detector receives yd, y, and u over NCS and calculates ϕ according to the (22). If ϕ increased, a recovering signal will be sent over NCS to the interface circuit. The interface circuit converts the recovering signal to hardware interrupt, which restores the controller software to its design parameters.

5. Simulation results

The proposed method is developed as per the scheme mentioned with a simulation of the local downstream controller at the Partiteur cross-regulator in the Gignac canal which located 40 km north-west of Montpellier, in the south of France.
The Integrator Delay (ID) model is an approximate representation of the dynamics of canal pool for low frequencies, this model and its parameters were given in [23] as follows,

$$y=\frac{1}{A_d s}\left(e^{-\tau_d s}u-p\right) \tag{23}$$

where,

- y is the downstream water elevation.
- τ_d is the delay of the canal pool.
- p the downstream perturbation.

When the control input u is a discharge, Ad represents the backwater area, and when u is the upstream gate opening, Ad is the inverse integrator gain.

For the Partiteur Left Bank cross regulator, the Ad = 25.2113 and τ_d =30 seconds. The PI controller will be used in this paper as the downstream controller with *Kp=0.60* and *Ti=96* seconds [23].
Initially, the system was tested without any attack; the test result is illustrated in Fig.3. After that, the simulation is carried out by randomly reprogramming the PI controller, in this case the original settings of the controller will be lost and its behaviour will be unknown (i.e. hijacked by the attacker). The detection of the controller hijacking attack and the controller recovering method which described in sections 2, 3 and 4 are applied to the NCS.
In order to evaluate the proposed detector, the controller reprogramed for three times (attack A, attack B, and attack C), this is made by change the parameters (*kp* and *Ki*) of *PI* controller in stealthy way. The test results are shown in Fig 4., Fig.5., Fig.6 and Fig.7. The attack is successfully detected and the controller recovered to its normal operation.

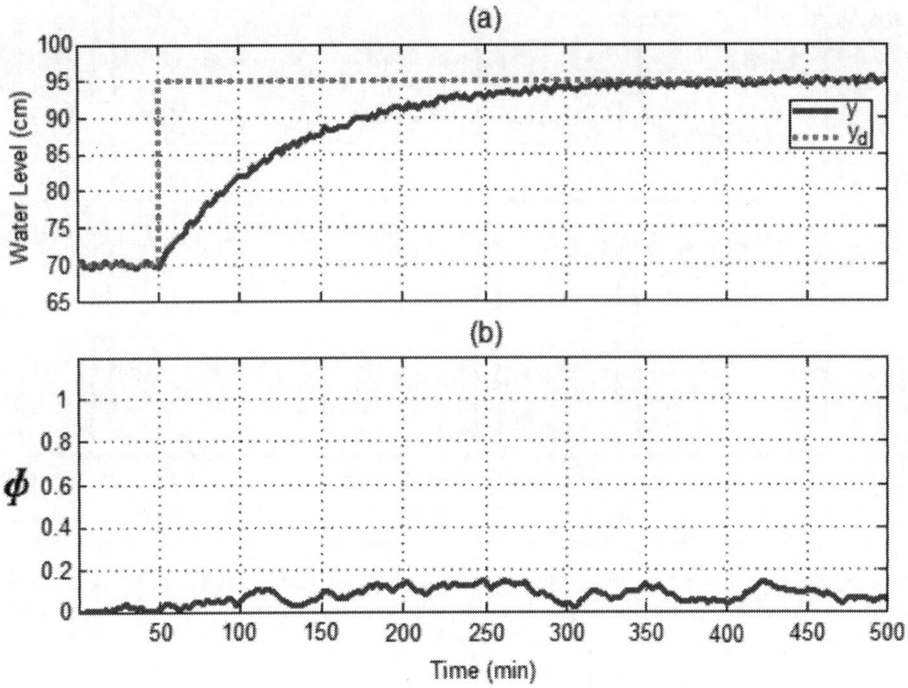

Figure 3. The NCS response without any attack

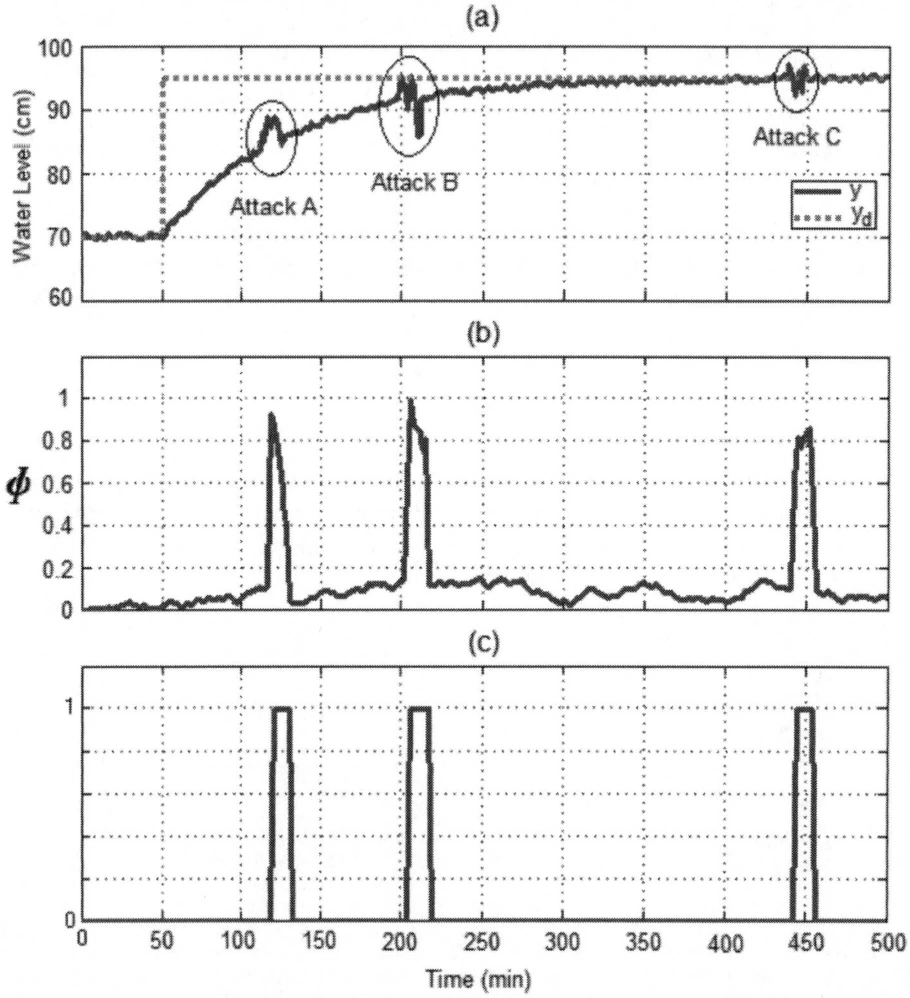

Figure 4. (a) The response of the plant when the attack is (or has been) detected and the controller is recovered. (b) The effect of the controller hijacking attack on the value of ϕ (c) controller recovering signal

From Fig.5.b, Fig.6.b, and Fig.7.b, one can note the effect of controller hijacking attack on the value of ϕ The abnormal change of ϕ will be used to generate controller recovering signal as illustrate in Fig.4.c.

IOP Conf. Series: Journal of Physics: Conf. Series **783** (2017) 012022 doi:10.1088/1742-6596/783/1/012022

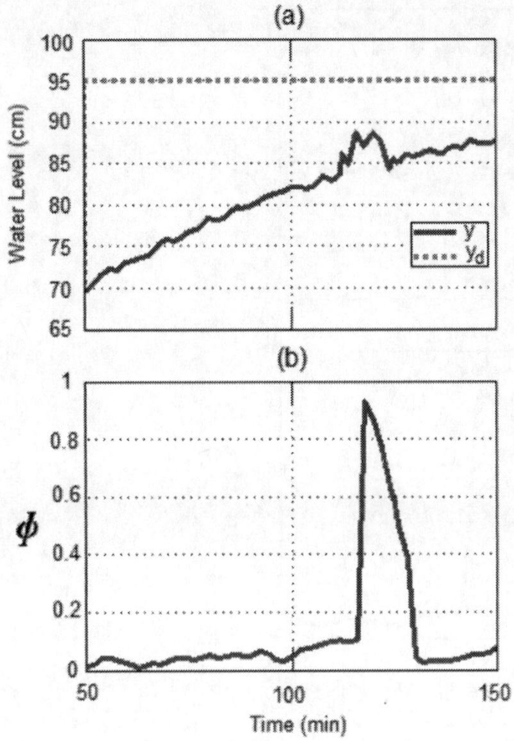

Figure 5. (a) The response of the plant when the ***attack A*** is detected and the controller is recovered, (b) The effect of the controller hijacking attack on the value of ϕ (with zoom)

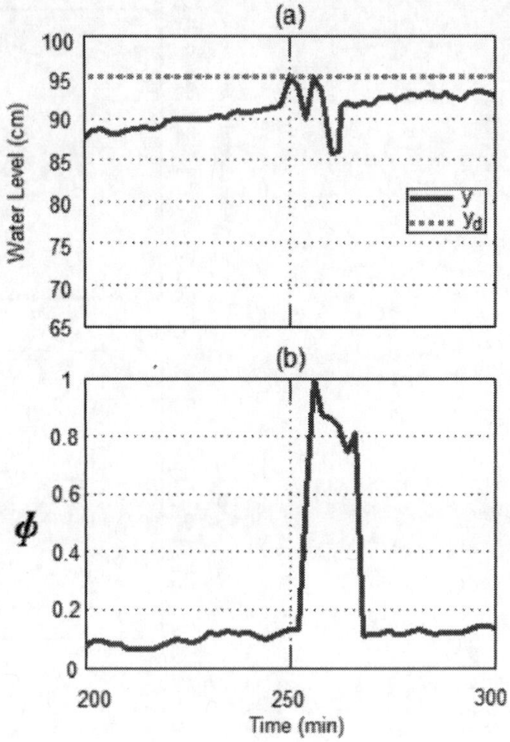

Figure 6. (a) The response of the plant when the ***attack B*** is detected and the controller is recovered, (b) The effect of the controller hijacking attack on the value of ϕ (with zoom)

6. Conclusion and perspectives

In this paper, an attack-tolerant networked control system in presence of the controller hijacking attack is introduced.

The facility of this method is that no a priori controller mathematical model is required to detect this type of attacks.

The proposed detector is connected to the NCS separately from the controller and plant sides. The secure mechanism is used for the secure communication of the proposed detector over NCS.

Even though only one type of controllers is used in this paper, the proposed attack detector is suitable enough to be considered as basis for the future development with other types of the controllers.

The investigation reveals that the proposed attack tolerant scheme can be used successfully for NCS in presence of the controller hijacking attack such as Stuxnet case.

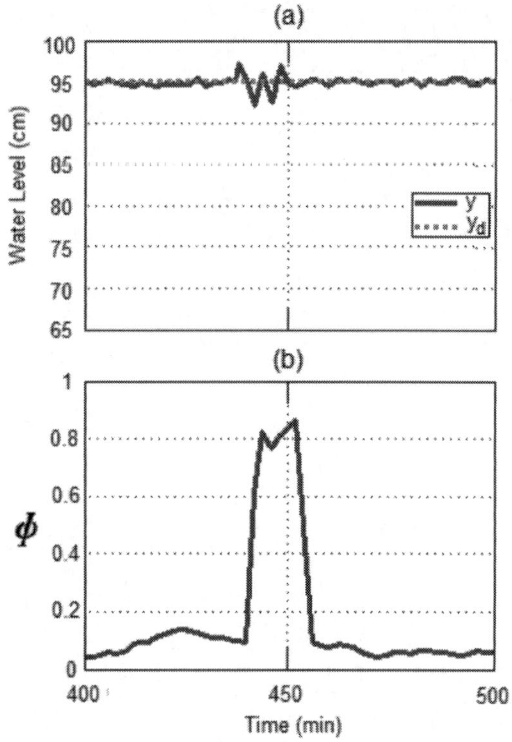

Figure 7. (a) The response of the system when the ***attack C*** is detected and the controller is recovered, (b) The effect of the controller hijacking attack on the value of ϕ (with zoom)

7. References

[1] Gupta, R.A.; Mo-Yuen Chow, *Networked Control System: Overview and Research Trends, IEEE Transactions on Industrial Electronics,* , vol.57, no.7, pp.2527-2535, July 2010.

[2] W. Stallings, *Cryptography and Network Security: Principles and Practice*, 4th ed. Englewood Cliffs, NJ: Pearson/Prentice-Hall, 2006.

[3] Gupta, R.A.; Mo-Yuen Chow, *Performance assessment and compensation for secure networked control systems, 34th Annual Conference of IEEE Industrial Electronics (IECON 2008)*, Orlando, Florid, pp.2929-2934, 10-13 Nov. 2008

[4] Pang Zhong hua; Liu Guoping, *Secure networked control systems under data integrity attacks, 29th Chinese Control Conference (CCC)*, Beijing, China, pp.5765-5771, 29-31 July 2010

[5] Liying Zhang, Lun Xie, Weize Li, Zhiliang Wang, *Security Solutions for Networked Control Systems Based on DES Algorithm and Improved Grey Prediction Model* , IJCNIS, vol.6, no.1, pp.78-85, 2014.

[6] Perez, V.; Garip, M.T.; Lam, S.; Lixia Zhang, *Security evaluation of a control system using Named Data Networking, 21st IEEE International Conference on Network Protocols (ICNP)*, Goettingen, Germany, pp.1-6, 7-10 Oct. 2013

[7] Wente Zeng; Mo-Yuen Chow, *A trade-off model for performance and security in secured Networked Control Systems, 2011 IEEE International Symposium on Industrial Electronics (ISIE)*, Gdansk, Poland, pp. 1997- 2002, 27-30 June 2011.

[8] Zhang Liying; Xie Lun; Li Weize; Wang Zhiliang, *A secure mechanism for networked control systems based on TrueTime, International Conference on Cyberspace Technology (CCT 2013)*, Beijing, China, pp.44-49, 23-23 Nov. 2013.

[9] Shuguang Cui; Zhu Han; Kar, S.; Kim, T.T.; Poor, H.V.; Tajer, A., *Coordinated data-injection attack and detection in the smart grid: A detailed look at enriching detection solutions,IEEE Signal Processing Magazine*, vol.29, no.5, pp. 106-115, Sept. 2012

[10] Nourian, A.; Madnick, S., *A Systems Theoretic Approach to the Security Threats in Cyber Physical Systems Applied to Stuxnet, IEEE Transactions on Dependable and Secure Computing*, vol.PP, no.99, pp.1-1. 2015.

[11] Y. Mo and B. Sinopoli, *False data injection attacks in control systems, in First Workshop on Secure Control Systems*, Cyber Physical Systems Week 2010, April 2010.

[12] I. Shames, A. M. Teixeira, H. Sandberg, and K. H. Johansson, *Distributed fault detection for interconnected second-order systems, Automatica*, vol. 47, no. 12, pp. 2757 – 2764, 2011.

[13] M. Blanke, M. Kinnaert, J. Lunze, and M. Staroswiecki, *Diagnosis and Fault-Tolerant Control, 2nd ed. Springer*, 2006

[14] J. Gertler, *Fault Detection and Diagnosis in Engineering Systems*. New York: Marcel Dekker, Inc., 1998.

[15] Rosich, A.; Voos, H.; Yumei Li; Darouach, M., *A model predictive approach for cyber-attack detection and mitigation in control systems, 52nd IEEE Conference on Decision and Control (CDC)*, Florence, Italy, pp.6621-6626, 10-13 Dec. 2013.

[16] A. Yaseen, Mireille Bayart, *Attack-Tolerant Networked Control System Based on the Deception for the Cyber-Attacks, World Congress on Industrial Control Systems Security (WCICSS-2015)*, London, UK , pp.37-44, 14-16 Dec. 2015

[17] A. Yaseen, Mireille Bayart, *Attack-tolerant networked control system in presence of the controller hijacking attack, 2016 International Conference on Military Communications and Information Systems (ICMCIS)*, Brussels, Belgium, pp. 1-8, 2016

[18] A. Yaseen, Mireille Bayart, *Intelligent Generalized Predictive Control Strategy for Networked Control System with an Internal Cyber Attack Detector, 21st IEEE International Conference on Emerging Technologies and Factory Automation (IEEE ETFA'2016)*, Berlin, Germany , 6 - 9 Sep. 2016.

[19] M. Fliess, C. Join, *Model-Free Control*, *International Journal of Control*, Vol. 86, Issue 12, pp. 2228–2252, 2013.

[20] Xi-Sheng Zhanb, a, Jie Wua,Tao Jiangb, Xiao-Wei Jiangb, *Optimal performance of networked control systems under the packet dropouts and channel noise, ISA Transactions*,Volume 58, September 2015, Pages 214–221.

[21] E. Delaleau, *A proof of stability of model-free control, 2014 IEEE Conference on Norbert Wiener in the 21st Century (21CW)*, Boston, MA, pp. 1-7, 2014.

[22] Joe D. Hoffman, *Numerical Methods for Engineers and Scientists*, Marcel Dekker, Inc, second edition,2001

[23] Litrico, X., Malaterre, P., Baume, J., Vion, P., and Ribot-Bruno, J., *Automatic Tuning of PI Controllers for an Irrigation Canal Pool*, *Journal of Irrigation and Drainage Engineering*, 2007 133:1, 27-37.

Decentralized scenario-based plug and play MPC for linear systems with multiplicative uncertainties

Anna Nehring, Ping Zhang
Institute of Automatic Control, University of Kaiserslautern, Germany
E-mail: nehring@eit.uni-kl.de, pzhang@eit.uni-kl.de

Abstract. This paper proposes a novel approach to decentralized control with Plug and Play capabilities. Plug and Play allows a flexible addition and removal of subsystems to an existing plant. The proposed approach guarantees stability and robustness to a certain degree for systems with additive disturbances and multiplicative uncertainties. For the controller design, the plant is decomposed into subsystems. The algorithms for adding and removing subsystems are given. By applying the basic idea of scenario-based methods, the proposed approach is less conservative and can handle complex multiplicative uncertainties.

1. Introduction

Plug and Play applied to control systems is a principle where new subsystems can be added to or removed from a system without the need of redetermining all controller parameters for the system [1], [2]. This principle can be very useful in large-scale systems and distributed systems, as they usually consist of a number of subsystems. The class of studied systems varies, from cooperative systems, where all information is shared, to decentralized systems, where only static informations or no information is shared. Most approaches use a form of cooperative design. Some cooperative designs are given below.

In [3], a Plug and Play MPC scheme is introduced for distributed LTI systems. A Lyapunov approach is used, redefining the constraints of each subsystem on a Plug and Play operation, to guarantee boundedness of the states. To guarantee input-to-state stability, in [4] an approach is proposed where each subsystem communicates to other subsystem it's desired trajectory and the guaranteed set it will lie in. In [5], a Plug and Play method is proposed using identification and parameter estimation for Plug and Play operations. When a system changes, the new system is identified and the controller is changed accordingly. In [6], a Plug and Play algorithm is presented, where network communication concepts are used to notify other subsystems of new players in the network. The overall cost function is minimized with only minimal communication between the subsystems. This method has been applied to microgrids. In [7], a Plug and Play procedure is proposed for adding and removing subsystems in a running system for an agent based architecture, where the Nyquist plots of the subsystems are used to prove the input-output stability of the whole system. A decentralized approach is proposed in [8], which utilizes only the information of constraints of neighbouring subsystems to guarantee the input-to-state stability.

To deal with additive disturbances and multiplicative uncertainties, robust MPC methods have been studied [9], [10], [11]. Robust MPC requires the state and input constraints to be satisfied for all realizations of the uncertainty and usually considers the worst case. Most of the robust MPC approaches modify the constraint sets [12], by using offline information such as the bounds of the disturbance. One example for this is tube-based MPC which is referenced later in this paper [12]. Its basic idea is to restrict the state trajectories to a tube which always lies within the constraints regardless of a (bounded) disturbance, which makes the control scheme rather conservative in some cases. The $\mathcal{H}_\infty$-optimization based robust MPC approach takes the $\mathcal{H}_\infty$-norm as the cost function to be minimized under constraints [10]. Mainly used for nonlinear systems, the input-to-state stability methods [10] impose stricter conditions for the Lyapunov stability. This results in new terminal costs in the cost function and time-varying state constraints.

In order to reduce the conservatism, stochastic MPC has been developed, which does not aim at the satisfaction of all constraints [13], [14], [15], [16]. Basic ideas include the reformulation of deterministic constraints into chance constraints (which only have to be satisfied most of the time) paired with affine disturbance feedback [14] and the prediction of state trajectories via samples of the uncertainties [17], so called scenario-based approaches. For scenario-based approaches, two basic methods exist. One is based on random convex programming [18], which guarantees feasibility of the optimization problem and constraint satisfaction by requiring a minimum number of used scenarios [17], [19], [20]. The other method is based on stochastic programming [21], which utilizes a stochastic Lyapunov function to achieve feasibility and a scenario tree to generate and weight scenarios based on probability to achieve constraint satisfaction [22], [23], [24]. In [25] and [26], distributed cooperative scenario-based MPC algorithms are presented, which use information from other subsystems to determine their output sequence.

Recently, based on the tube-based MPC, a Plug and Play variant for decentralized systems has been developed [8]. In order to reduce the conservativeness of the design, we propose another way to take into account the knowledge about uncertainties and disturbances in the design. The scenario-based approach provides an interesting alternative [27]. We propose to apply the basic idea of the scenario-based approach, proposed in [27], to enhance the capabilities of the Plug and Play MPC controller to deal with uncertainties.

The main contribution of this paper is the enhancement of the decentralized control scheme proposed in [8] with the capabilities of scenario-based control as proposed in [27]. Compared with [8], the proposed approach is capable of considering complex model uncertainties and moreover, it reduces the conservatism of the solution.

This paper is organized as follows. First, in Section 2 a description of the uncertain system under consideration will be given and the basic idea of Plug and Play control will be explained. In Section 3, the decentralized scenario-based MPC will be introduced and the stability of the decentralized system will be discussed as well as Plug and Play operations. In Section 4, the proposed approach will be compared with the existing approach (i.e. the tube-based Plug and Play MPC approach) by means of an example.

2. Problem formulation

2.1. System description

In this paper, we consider the following linear system with additive disturbances and multiplicative uncertainties described by

$$x(k+1) = A(\theta)x(k) + B(\theta)u(k) + B_\gamma(\theta)\gamma(k) \tag{1}$$

where $x \in \mathbb{X} \subset \mathbb{R}^n$ is the state vector, $u \in \mathbb{U} \subset \mathbb{R}^m$ is the control input vector, $\gamma \in \Gamma \subset \mathbb{R}^{m_\gamma}$ is the additive disturbance vector and $\theta \in \Theta \subset \mathbb{R}^{r_\theta}$ denotes the multiplicative uncertainties. The system matrices $A(\theta), B(\theta), B_\gamma(\theta)$ and the set Γ are bounded. It is assumed that $(A(\theta), B(\theta))$ is stabilizable for all θ and

$$[A(\theta)\ B(\theta)\ B_\gamma(\theta)] = \sum_{\xi=1}^{r_\theta} \theta_\xi [A_\xi\ B_\xi\ B_{\gamma\xi}], \quad 0 \le \theta_\xi \le 1, \ \sum_{\xi=1}^{r_\theta} \theta_\xi = 1. \tag{2}$$

The additive disturbances γ may have time-varying distributions and may be correlated in time [28]. The constraints on x and u are characterized as zonotopes [29], i.e. centrally symmetric convex polytopes which are represented by

$$\mathbb{X} = \{x \in \mathbb{R}^n | g_r x \leq 1, r = 1, \cdots, r_x\} = \{x \in \mathbb{R}^n | \mathcal{G}x \leq \mathbf{1}_{r_x}\} \tag{3}$$

$$\mathbb{U} = \{u \in \mathbb{R}^m | h_r u \leq 1, r = 1, \cdots, r_u\} = \{u \in \mathbb{R}^m | \mathcal{H}u \leq \mathbf{1}_{r_u}\} \tag{4}$$

where r_x, r_u represent the number of constraints, $g_r \in \mathbb{R}^{1\times n}$, $h_r \in \mathbb{R}^{1\times m}$ are row vectors with only one non-zero entry, $\mathcal{G} = [g_1^T \cdots g_{r_x}^T]^T \in \mathbb{R}^{r_x \times n}$, $\mathcal{H} = [h_1^T \cdots h_{r_u}^T]^T \in \mathbb{R}^{r_u \times m}$, $\mathbf{1}_{r_x}$ denotes a column vector of ones of length r_x.

2.2. System decomposition

Assume that system (1) can be partitioned into M subsystems as described in [30] (see Chapter 3.3 therein). Partition the state vector and the control input vector as $x = [x_1^T\ x_2^T\ \cdots\ x_M^T]^T$, $u = [u_1^T\ u_2^T\ \cdots\ u_M^T]^T$, where $x_i \in \mathbb{R}^{n_i}$, $u_i \in \mathbb{R}^{m_i}$, $i = 1, \cdots, M$, $\sum_{i=1}^M n_i = n$, $\sum_{i=1}^M m_i = m$. Correspondingly, partition the matrix $A(\theta)$ into

$$A(\theta) = \begin{bmatrix} A_{11}(\theta) & \cdots & A_{1M}(\theta) \\ \vdots & \ddots & \vdots \\ A_{M1}(\theta) & \cdots & A_{MM}(\theta) \end{bmatrix} \tag{5}$$

where $A_{ij} \in \mathbb{R}^{n_i \times n_j}$, $i = 1, \cdots, M$, $j = 1, \cdots, M$. Assume that the subsystems are input decoupled (i.e. the input u_i only affects subsystem S_i). Therefore, $B(\theta)$ and $B_\gamma(\theta)$ can be partitioned into

$$B(\theta) = \begin{bmatrix} B_1(\theta) & \cdots & 0 \\ \vdots & \ddots & \vdots \\ 0 & \cdots & B_M(\theta) \end{bmatrix}, \quad B_\gamma(\theta) = \begin{bmatrix} B_{\gamma 1}(\theta) \\ \vdots \\ B_{\gamma M}(\theta) \end{bmatrix}$$
$$B_i(\theta) \in \mathbb{R}^{n_i \times m_i},\ B_{\gamma i}(\theta) \in \mathbb{R}^{n_i \times m_\gamma},\ i = 1, \cdots, M \tag{6}$$

The i-th subsystem S_i, $i = 1, \cdots, M$, is given by

$$\begin{aligned} S_i:\ x_i(k+1) &= A_{ii}(\theta)x_i(k) + B_i(\theta)u_i(k) + w_i(k) \\ w_i(k) &= B_{\gamma i}(\theta)\gamma(k) + \sum_{j\in\mathcal{N}_i} A_{ij}(\theta)x_j(k) \end{aligned} \tag{7}$$

where $\mathcal{N}_i = \{j | A_{ij}(\theta) \neq 0\}$ is the set of neighbouring subsystems to the subsystem S_i. Assume that if the subsystems are decoupled, they stay decoupled independent of the multiplicative uncertainties, i.e. if the i-th subsystem and the j-th subsystem are decoupled, then $A_{ij}(\theta) = 0$, $\forall\,\theta$.

For the i-th subsystem, the coupling between it and other subsystems described by $\sum_{j\in\mathcal{N}_i} A_{ij}(\theta)x_j$ is treated as an additional additive disturbance and added to the disturbance w_i.

The constraint sets $\mathbb{X}$ and $\mathbb{U}$ are also partitioned into

$$\mathbb{X}_i = \{x_i \in \mathbb{R}^{n_i} | g_{r_i} x_i \leq 1, r_i = 1, \cdots, r_{x_i}\} = \{x_i \in \mathbb{R}^{n_i} | \mathcal{G}_i x_i \leq \mathbf{1}_{r_{x_i}}\},\ \sum_{i=1}^M r_{x_i} = r_x \tag{8}$$

$$\mathbb{U}_i = \{u_i \in \mathbb{R}^{m_i} | h_{r_i} u_i \leq 1, r_i = 1, \cdots, r_{u_i}\} = \{u_i \in \mathbb{R}^{m_i} | \mathcal{H}_i u_i \leq \mathbf{1}_{r_{u_i}}\},\ \sum_{i=1}^M r_{u_i} = r_u \tag{9}$$

where g_{r_i} are the vectors g_r corresponding to the constraints of x_i and h_{r_i} are the vectors h_r corresponding to the constraints of u_i.

2.3. Control objective

The goal is to drive the state vector x to the target x_r with a small control effort u while being robust against additive disturbances and multiplicative uncertainties. Moreover, the computational effort should be kept relatively small for the control scheme to be applicable in real systems, which do not necessarily have much computing power. Plug and Play enables a flexible system structure allowing for changes in subsystems as well as in the number of subsystems. Changes such as the removal and the addition have to be executed with relatively easy operations. After a change in the system structure, the local controllers as well as the overall system should still be stable and perform well.

3. The new approach

In this section, an approach is proposed to combine scenario-based MPC with a Plug and Play scheme.

3.1. Controller structure

Each subsystem S_i, $i = 1, \cdots, m$, is equipped with a controller

$$C_i: \; u_i(k) = K_i x_i(k) + v_i(k) \tag{10}$$

where K_i is the local feedback gain matrix, v_i is a correction term that will be calculated later by the local MPC scheme. The main task of K_i is to stabilize the subsystem S_i, while the correction term v_i is used to compensate for disturbances. The basic idea of the proposed approach is to use a Plug and Play framework to guarantee input-to-state stability and then ensure constraint satisfaction with a local scenario-based MPC scheme.

3.2. Design of the local feedback gain matrix K_i

Lemma 1 *Given a linear system (1) that can be decomposed into M subsystems as (5)-(7) with the state constraints (8) and a controller given by (10). Assume that the local feedback gain matrices $K_1, K_2, \cdots, K_M$ of the subsystems are selected so that the matrices $F_i = A_{ii}(\theta) + B_i(\theta)K_i$, $i = 1, \cdots, M$ are Schur. If the following condition is fulfilled*

$$\alpha_i = \sum_{j \in N_i} \sum_{k=0}^{\infty} ||\mathcal{G}_i F_i^k A_{ij} \mathcal{G}_j^+||_\infty < 1, \quad \forall i = 1, \cdots, M \tag{11}$$

where $\|\cdot\|_\infty$ denotes the largest absolute row sum of a matrix [31] and $\mathcal{G}_j^+$ is the pseudo-inverse of $\mathcal{G}_j$, then

(i) the matrix $F = A(\theta) + B(\theta)K$ is Schur, where $K = diag\{K_1, \cdots, K_M\}$.

(ii) the whole system (1) is asymptotically stable.

Lemma 1 tells us that if the gain of subsystem S_i caused by the other subsystems S_j, $j \in \mathcal{N}_i$, is small enough and the disturbance γ is bounded, then the overall system (1) is asymptotic stable if (11) holds. In the control design, K_i is the only parameter which affects α_i, therefore this condition has to hold when designing K_i. Lemma 1 can be proven in a way similar to [8] while taking into account the influence of multiplicative uncertainties. For the sake of clarity, the proof is put in the appendix.

The local feedback gain matrix $K_i, i = 1, \cdots, M$, can be selected with the help of the LMI technique proposed by [32]. The state feedback matrix K_i is given by $K_i = Y_i G_i^{-1}$, where the square matrix G_i, the rectangular matrix Y_i and the symmetric matrices $Q_{1\xi}$, $Q_{2\xi}$, $\cdots$, $Q_{r_\theta\xi}$ satisfy the following LMI

$$\begin{bmatrix} G_i + G_i^T - Q_{i\xi} & G_i^T A_\xi^T + Y_i^T B_\xi^T \\ A_\xi G_i + B_\xi Y_i & Q_{il} \end{bmatrix} > 0, \quad \forall \xi = 1, \cdots, r_\theta, \ \forall l = 1, \cdots, r_\theta. \tag{12}$$

The solution of this LMI is not unique and has to satisfy (11). If no solution can be found, the subsystem can not be plugged in using the proposed method. Equation (12) represents Lyapunov functions for the vertices of the uncertainty (2).

3.3. MPC scheme

Scenario-based MPC is used in this approach, as it is a way of dealing with additive and multiplicative uncertainties. The randomized finite horizon control problem is formulated here. The technical details of scenario-based MPC theory are largely omitted here due to limitations of space. For further informations, the reader is referred to [28], [33], [34] and the references therein. The reliability of the solution, the number of scenarios and the basic setup are discussed below.

For each subsystem a local MPC scheme is used. Let N_i be the chosen prediction horizon length. In the proposed approach, the disturbance is dealt with by considering a discrete set of state trajectories x_i obtained for a number L_i of randomly extracted samples of θ, γ and x_j at the sampling time k, so called scenarios. These random parameters are collected in $\delta = (\theta, \gamma, x_j)$. These samples can be generated based on the probability distribution if it is known. Otherwise, the samples can be obtained from empirical data or by sampling a stochastic model [28]. It is assumed that enough samples of δ can be generated. By substituting (10) into the system model (7), the dynamics of the subsystem S_i is governed by

$$x_i(k+1) = F_i(\theta)x_i(k) + B_i(\theta)v_i(k) + w_i(k, \delta) \tag{13}$$

Based on the random selected samples $\delta^l = (\theta^l, \gamma^l, x_j^l)$, $l = 1, 2, \cdots, L_i$, different state and input predictions over a prediction horizon of length N_i can be obtained from (13) as

$$\begin{aligned} x_i^l(k+1) &= F_i(\theta^l)x_i^l(k) + B_i(\theta^l)v_i(k) + w_i(k, \delta^l) \\ x_i^l(0) &= x_i(0) \\ u_i^l(k) &= K_i x_i^l(k) + v_i(k), \ k = 0, \cdots, N_i - 1 \end{aligned} \tag{14}$$

Let the cost function be

$$J_i(v_i) = \sum_{l=1}^{L_i} \sum_{k=0}^{N_i-1} \{(x_i^l(k) - x_{i,r}(k))^T Q_i (x_i^l(k) - x_{i,r}(k)) + v_i(k)^T R_i v_i(k)\} \tag{15}$$

where $Q_i \in \mathbb{R}^{n_1 \times n_1} > 0$ and $R_i \in \mathbb{R}^{m_i \times m_i} > 0$ are weighting matrices chosen by the designer. The optimization problem is

$$\min_{v_i} J_i(v_i) \tag{16a}$$

subject to

$$x_i^l(k+1) = F_i(\theta^l)x_i^l(k) + B_i(\theta^l)v_i(k) + w_i(k, \delta^l), \tag{16b}$$
$$x_i^l(0) = x_i(0), \tag{16c}$$
$$u_i^l(k) = v_i(k) + K_i x_i^l(k), \tag{16d}$$
$$\mathcal{G}_i x_i^l(k) \leq \mathbf{1}_{r_{x_i}}, \tag{16e}$$
$$\mathcal{H}_i u_i^l(k) \leq \mathbf{1}_{r_{u_i}}, \tag{16f}$$
$$k = 0, \cdots, N_i - 1, \tag{16g}$$
$$l = 1, \cdots, L_i \tag{16h}$$

The optimization problem (16a) can be solved with quadratic programming [35]. In comparison with [8], the scenario-based optimization can deal with the time varying parameters $A_{ii}(\theta)$, $B_i(\theta)$ and $B_{\gamma i}(\theta)$ of the uncertain system (7).

Since scenario-based MPC does not aim at satisfying all constraints at all times, the hard constraints are transformed into chance constraints which only have to be satisfied with a probability R_i

$$R_i = Prob\{\mathcal{H}_i u_i(k) \leq \mathbf{1} \text{ and } \mathcal{G}_i x_i(k) \leq \mathbf{1}\}, \tag{17}$$

where $Prob$ refers to the probability of the event. The reliability of the optimal solution v_i^* to the optimization problem (16a) is defined as the probability R_i that the optimal solution satisfies the constraints under all possible realizations of the disturbances and uncertainties. Note that $R_i \in [0,1]$ is itself a random variable, since it depends on the extracted scenarios δ, i.e. $R_i = R_i(\delta)$. In [36] a connection between the number of scenarios L_i and the probability of constraint violations (17) for the optimal solution v_i^* is given.

Lemma 2 ([36]) *Given a system described by (7) and assume that* $(\theta^1,\ \gamma^1,\ x_j^1), \cdots,$ $(\theta^L,\ \gamma^L,\ x_j^L)$ *are* L_i *samples of the multiplicative uncertainties* θ*, the additive disturbance* γ *and the states of the neighbouring system* x_j*,* $j \in \mathcal{N}_j$*. Let* N_i *be the chosen horizon length in an MPC scheme and* $d_i = m_i N_i + q_i$ *represent the decision complexity of the system, where* q_i *is the number of soft constraints. Let* p_i *be the acceptable reliability level for* R_i*. If the number of scenarios* L_i *is chosen such that*

$$\sum_{s=0}^{d_i-1} \binom{L_i}{s} (1-p_i)^s p_i^{L-s} \leq \beta_i \tag{18}$$

where $\binom{L_i}{s}$ *is the binomial coefficient, then it holds that*

$$Prob\{R_i(\delta) \geq p_i\} \geq 1 - \beta_i. \tag{19}$$

Also, the optimal solution v_i^* *of (16a) is a feasible solution with a confidence of at least* $1-\beta_i$*.*

Lemma 2 tells us that the probability that the reliability R_i of the solution is higher than p_i is at least $1-\beta_i$. Note that β_i can be chosen by the designer to be very small, for instance, in the order of 10^{-12}. In [36] it has been proven that

$$L_i \geq \frac{2}{1-p_i}(ln\beta_i^{-1} + d_i) \tag{20}$$

implies the fulfilment of condition (18).

As it can be seen in (18), the number of scenarios L_i depends on the reliability of constraint fulfilment p_i, the confidence of a feasible solution β_i and the decision complexity d_i. For larger systems this leads to a large number of scenarios that quickly becomes unmanageable due to high computational effort. In [17], scenario removal techniques are proposed to remove unlikely scenarios for a quicker optimization. However, a smaller number of scenarios L_i than given in (20) leads to more constraint violations.

Each subsystem S_i, $i \in M$. may have its own reliability level p_i and confidence level β_i. The subsystems S_i and S_j, $j \neq i$, $j \in \{1,\ 2,\ \cdots,\ M\}$ are decoupled from each other as the states of the other systems are treated as an unknown disturbance. Therefore, the probabilities $Prob(R_i \geq p_i)$ and $Prob(R_j \geq p_j)$ are independent. The probability of constraint satisfaction for the whole system (1) is then [37]

$$Prob\left(\bigcap_{i=1}^{M} R_i \geq p_i\right) = \prod_{i=1}^{M} Prob(R_i \geq p_i). \tag{21}$$

3.4. Plug and Play operations

In this subsection, we shall give the details of the Plug and Play aspects of the proposed approach. The Plug and Play operations are done in offline mode.

3.4.1. Generating samples of neighbouring systems To generate the samples of the subsystem S_i, a description of S_i has to be provided to each coupled subsystem $S_j,\ j \in \mathcal{N}_i$. We assume that no information is shared between the subsystems other than static constraints $\mathcal{G}_i,\ \mathcal{H}_i$ and system matrices $A_{ii}(\theta),\ B_i(\theta),\ B_{\gamma i}(\theta)$ during Plug and Play operations. Therefore no information about setpoints or actual values of the other subsystems can be used to generate these samples. For simplicity we generate these samples of x_i as uniformly distributed in $\mathbb{X}_i$, independent of the other subsystem S_j [38].

3.4.2. Plug in Operation Let S_{M+1} be a new subsystem described by $x_{M+1}(k+1) = A_{M+1,M+1}(\theta)x_{M+1}(k) + B_{M+1}(\theta)u_{M+1}(k) + w_{M+1}(k)$ with the bounds $\mathbb{X}_{M+1},\ \mathbb{U}_{M+1}$ and the neighbouring systems $\mathcal{N}_{M+1} = \{j | A_{M+1,j}(\theta) \neq 0,\ j = 1, \cdots, M\}$. To get the new local feedback gain matrix K_{M+1}, the LMI (12) has to be solved under condition (11) for $i = M+1$. Since S_{M+1} is now a new neighbour to the subsystems $S_j, j \in \mathcal{N}_{M+1}$, the sets $\mathcal{N}_j$ change accordingly and α_j may change. When $\mathcal{N}_j$ grows in size, α_j can only increase and therefore the controller $K_j,\ j \in \mathcal{N}_{M+1}$ may have to be redesigned with (12) so that condition (11) is still satisfied. The state of the new system S_{M+1} has to be added to the disturbance vectors w_j of the neighbouring subsystems $S_j,\ j \in \mathcal{N}_{M+1}$ and the samples of x_{M+1} can be drawn as described in Section 3.4.1. If no feedback gain matrix K_i can be found satisfying (11) and (12), the new system S_{M+1} can not be plugged in. This procedure is summarized in Algorithm 1.

Algorithm 1: Plug in operation of a new subsystem S_{M+1}

(i) Determine $A_{M+1,M+1}(\theta),\ B_{M+1}(\theta),\ B_{\gamma,M+1}(\theta)$ and the coupling terms $A_{M+1,j}(\theta)$ as well as the constraint sets $\mathbb{X}_{M+1},\ \mathbb{U}_{M+1}$

(ii) Determine the neighbouring set $\mathcal{N}_{M+1} = \{j | A_{M+1,j}(\theta) \neq 0,\ j = 1, \cdots, M\}$

(iii) Add $M+1$ to the neighbouring sets $\mathcal{N}_j$ for all $j \in \mathcal{N}_{M+1}$

(iv) Determine K_{M+1} such that (11) and (12) hold

(v) For each $\alpha_j,\ j \in \mathcal{N}_{M+1}$, check (11) and redesign K_j if necessary

(vi) Add x_{M+1} to the disturbance vectors of S_j and x_j to the disturbance vector of S_{M+1} as described in Subsection 3.4.1

(vii) Choose $p_{M+1},\ \beta_{M+1}$ (see Subsection 3.4.4)

(viii) Optional: change $p_i,\ i = 1, \cdots, M+1$, so that the reliability level for the whole system p holds its prespecified value (see Subsection 3.4.4)

Algorithm 2: Plug out operation of subsystem S_λ

(i) Remove λ from $\mathcal{N}_j$ for all $j \in \mathcal{N}_\lambda$

(ii) Remove x_λ from the disturbance vectors of $S_j,\ j \in \mathcal{N}_\lambda$

(iii) Optional: redesign $K_j,\ j \in \mathcal{N}_\lambda$ for better performance

(iv) Optional: change $p_i, i = 1, \cdots, M$, such the reliability level for the whole system p holds its prespecified value.

Algorithm 3: Design of a distributed scenario-based Plug and Play MPC controller for the system described by (1)

(i) Decompose the system (1) into M subsystems in the form of (7)
(ii) For each subsystem S_i, $i = 1, \cdots, M$, execute the plug in operation detailed in Algorithm 1 and obtain the local feedback controller gain K_i
(iii) For each subsystem S_i, generate L_i samples for γ, θ and each x_j, $j \in \mathcal{N}_i$ coupled to S_i
(iv) Solve the optimization problem (16a) for each subsystem, e.g. with quadratic programming
(v) For each subsystem S_i, apply the first element of $v_i(k)$ from the solution of (16a) to $u_i = K_i x_i + v_i$
(vi) To add a new subsystem S_{M+1}: execute Algorithm 1
(vii) To remove a subsystem S_λ: execute Algorithm 2
(viii) To change a subsystem S_λ: execute Algorithm 2 and then Algorithm 1

3.4.3. Plug out operation The unplugging of subsystem S_λ, $\lambda \in \{1, \cdots, M\}$ is comparatively easy. In this case, the state x_λ has to be removed from all neighbouring disturbance vectors w_j. Since for each neighbouring subsystem $S_j \in \mathcal{N}_\lambda$ the set $\mathcal{N}_j$ gets smaller with the disconnection of subsystem S_λ, α_j can only decrease and the inequality (11) still holds. Therefore, the controllers $K_j, j \in \mathcal{N}_\lambda$ do not have to be redesigned upon removal of S_λ. Algorithm 2 summarizes this procedure.

Remark 1 *A redesign of K_j, $j \in \mathcal{N}_\lambda$, might improve the local control performance as it might be less conservative.*

3.4.4. Reliability of the subsystems The designer has to make a choice between keeping the reliability level of the whole system p at a certain value and keeping the reliability level of the subsystems p_i constant. In the first option, the reliability p of the whole system is fixed. When a subsystem is plugged in, the reliability of all subsystems p_i has to change according to (21) such that p is still above the specified value. Here the number of scenarios changes with the number of subsystems and grows larger with each new subsystem. The second option fixes the reliability p_i for each subsystem. Here the number of scenarios L_i is constant, but the reliability of the whole system p changes with the number of subsystems.

The design procedure for the proposed controller of (1) is summarized in Algorithm 3.

4. Example

In this section, an example of building climate control is given to illustrate the proposed approach. Consider a system described by

$$x(k+1) = A(\theta)x(k) + B(\theta)u(k) + B_\gamma(\theta)\gamma(k)$$
$$A = \begin{pmatrix} 0.200+\theta & 0.001+\theta & 0.001+\theta \\ 0.002+\theta & 0.300+\theta & 0 \\ 0.002+\theta & 0 & 0.400+\theta \end{pmatrix},$$
$$B = \begin{pmatrix} 0.010+\theta & 0 & 0 \\ 0 & 0.010+\theta & 0 \\ 0 & 0 & 0.010+\theta \end{pmatrix}, \quad B_\gamma = \begin{pmatrix} 0.020+\theta \\ 0.020+\theta \\ 0.020+\theta \end{pmatrix} \tag{22}$$

where the uncertainty θ is truncated normally distributed with mean 0, variance 1 and a maximal magnitude of 0.03, i.e. θ varies between -0.03 and 0.03, the additive disturbance γ is a sine signal

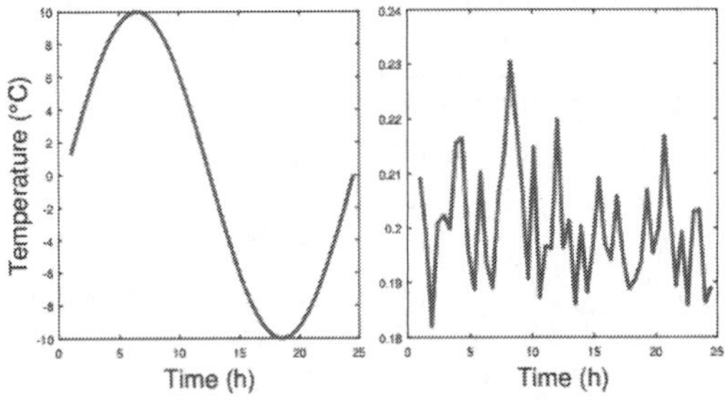

(a) disturbance γ (b) uncertainity $A_{11}(\theta)$

Figure 1. The additive disturbance γ (a) and the uncertain system matrix $A_{11}(\theta)$ (b) in the simulation

with a amplitude of 10 and an unknown offset of maximal 10%. In the simulation, the samples are generated using random number generators for known distributions. The system matrices are indeed a simplified model of a three-room-building. In this context the states x represent the room temperature, the uncertainty θ can be anything from open windows to modelling errors and γ can be, for instance, the outside temperature of a winter day influencing the room temperature of the building. The input u represents a climate control unit which can provide heating as well as cooling.

The initial state is $x(0) = [15\ 15\ 15]^T$ and the target constraints are $[21\ 21\ 21]^T \leq x \leq [25\ 25\ 25]^T$, $[-1500\ -1500\ -1500]^T \leq u \leq [1500\ 1500\ 1500]^T$ due to actuator saturation.

The goal of the controller design is to drive all states to $x_r = [23\ 23\ 23]^T$ while being robust to the external disturbance γ as well as the uncertainty θ.

The system can be decomposed into three subsystems. For instance, the first subsystem is

$$S_1 :\ x_1(k+1) = A_{11}(\theta)x_1(k) + B_1(\theta)u_1(k) + w_1(k) \tag{23}$$

where $A_{11}(\theta) = 0.2 + \theta$, $B_1(\theta) = 0.01 + \theta$ and $w_1 = (0.001 + \theta)x_2 + (0.001 + \theta)x_3 + (0.02 + \theta)\gamma$. Since the constraints have to be zonotopes, the state x_1 has to be shifted to $x_1 - 23$. The $\mathbb{X}_i$ can be written in the form given in (3):

$$\mathbb{X}_1 = \left\{ x_1 \in \mathbb{R}^1 \middle| \begin{bmatrix} -0.5 \\ 0.5 \end{bmatrix} x_1 \leq \begin{bmatrix} 1 \\ 1 \end{bmatrix} \right\} \tag{24}$$

In Figure 1 the disturbance γ (i.e. the outside temperature) is shown, which is simplified as a sine signal. In Figure 1(b) the changes of the system matrix $A_{11}(\theta)$ due to the existence of model uncertainty θ are shown over the time.

At first only the first subsystem is examined. The reliability level of subsystem S_1 is set to $p_1 = 0.9$ and the confidence level to $\beta_1 = 10^{-9}$. According to the formula (20), in the optimization 855 scenarios are used. By solving (12), it gives $Y_1 = 0.58$, $G_1 = 1$, $Q_{11} = 1$, $Q_{12} = 1$. As a result $K_1 = Y_1 G_1^{-1} = 0.58$. As $\mathcal{N}_1 = \emptyset$, $\alpha 1 = 0 < 1$. At each time instant, v_1 is determined by solving the optimization problem (16a). The controller designed with the proposed approach is compared with the Plug and Play tube-based MPC controller, which is constructed as shown in [8].

Figure 2(a) shows the state trajectory of x_1 with $\theta = 0$. It can be seen that both trajectories of the tube-based control and the new approach show little difference. In Figure 2(b) the state trajectories are shown with $|\theta| \leq 0.03$ as described above. In this simulation for the validation of the performance, the states of the other subsystems are assumed to stay at $x_j = 0$ so that they have no influence on x_1. In the following we compare x_1 with x_{1r} to determine the robustness of the controller [30], [31]. The maximum control error is defined as $e = |x - x_r|$ and is $e_{x1,tube} = 1.52$ and $e_{x1,scenario} = 0.1$. Here the proposed approach clearly outperforms the tube-based MPC control, as the tube-based controlled state is rather erratic.

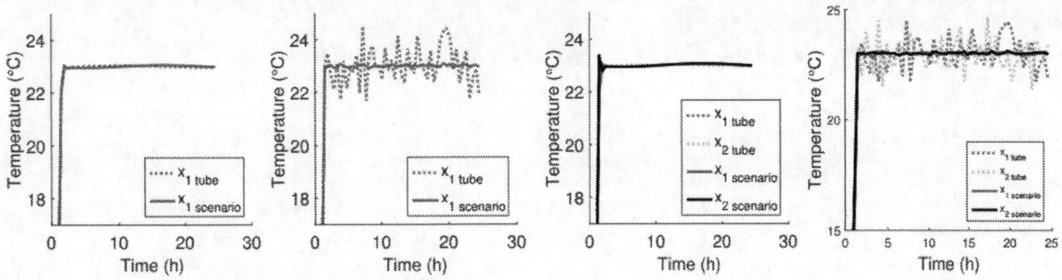

(a) $\theta = 0$ (b) $\theta \neq 0$ (c) $\theta = 0$ (d) $\theta \neq 0$

Figure 2. The controlled state trajectories x_1 and x_2 got by the tube-based Plug and Play MPC controller (dotted line) and the scenario-based Plug and Play MPC controller (solid line). In (a) and (b), only the first subsystem is considered, in (c) and (d) the first and the second subsystem are considered.

Now the second subsystem S_2 is added by applying Algorithm 1. The new local controller gain K_2 is $K_2 = 0.8$. As both $\alpha_1 = 0.16 < 1$ and $\alpha_2 = 0.25 < 1$, K_1 does not have to be redesigned. x_1 is added to w_2 and x_2 is added to w_1 to take the interaction of the subsystems into account. In Figure 2(c) the trajectories for x_1 and x_2 with no multiplicative uncertainties $\theta = 0$ are shown and in Figure 2(d) the trajectories are shown with an uncertainty $|\theta| \leq 0.03$ present. The maximum error with scenario-based Plug and Play MPC here is $e_{x_1,scenario} = 0.11, e_{x_2,scenario} = 0,11$. Note that with the second subsystem added, the maximum error e_{x_1} is increased by 10%. For the Plug and Play tube-based MPC controller, the control error is $e_{x_1,tube} = 1.53$ and $e_{x_2,tube} = 1.54$. All state trajectories lie within the constraints.

The addition of the third subsystem would only affect subsystem S_1 as it is physically decoupled from subsystem S_2. Due to the limitations of space, the Plug in operation of the third subsystem is omitted here.

5. Conclusions

In this paper, a decentralized scenario-based Plug and Play MPC approach is proposed. In comparison with the existing tube-based Plug and Play MPC approach, the new approach leads to smaller control errors if there are uncertainties in the system matrices. The application of the approach is demonstrated by an example in the field of building climate control. In future work, the stochastic description of the neighbouring states x_j as well as scenario reduction methods will be further investigated.

Appendix A. Proof of Lemma 1

First we prove that an expansion of the original system is input-to-state stable which implies asymptotic stability for linear systems [39]. Suppose that a linear transformation matrix T satisfies the conditions in [30] (see Theorem 3.3 of [30]). Then the linear transformation of the original system with respect to T results in an expansion of this system. Then, the inclusion principle given in [30] (see Theorem 3.4 of [30]) is utilized. It shows that if the expansion of a system is asymptotically stable, the original system is also asymptotically stable and vice versa. Using $\mathcal{G}_i$ as the transformation matrix and denoting $\tilde{x}_i(k) = \mathcal{G}_i x_i(k)$, we can define the expanded system

$$\tilde{x}_i(k+1) = (\tilde{A}_{ii}(\theta) + \tilde{B}_i(\theta)\tilde{K}_i)\tilde{x}_i(k) + \tilde{B}_i(\theta)v_i(k) + \sum_{j \in \mathcal{N}_j} \tilde{A}_{ij}(\theta)\tilde{x}_j(k) + \tilde{B}_{\gamma,i}(\theta)\gamma_i(k) \tag{A.1}$$

$$\tilde{A}_{ii}(\theta) = \mathcal{G}_i A_{ii}(\theta)\mathcal{G}_i^+, \ \tilde{B}_i(\theta) = \mathcal{G}_i B_i(\theta), \ \tilde{K}_i = K_i \mathcal{G}_i^+, \ \tilde{A}_{ij}(\theta) = \mathcal{G}_i A_{ij}(\theta)\mathcal{G}_j^+, \ \tilde{B}_{\gamma,i}(\theta) = \mathcal{G}_i B_{\gamma,i}(\theta)$$

where $\mathcal{G}_i^+$ denotes the pseudo-inverse of $\mathcal{G}_i$.

Let $\tilde{F}_i(\theta) = \tilde{A}_{ii}(\theta) + \tilde{B}_i(\theta)\tilde{K}_i$. The state equation can be rewritten as

$$\tilde{x}_i(k) = \tilde{F}_i^k(\theta)\tilde{x}_i(0) + \sum_{t=0}^{k-1} \tilde{F}_i^t(\theta) \sum_{j\in\mathcal{N}_i} \tilde{A}_{ij}(\theta)\tilde{x}_j(k-t-1) + \sum_{t=0}^{k-1} \tilde{F}_i^t(\theta)[\tilde{B}_i(\theta)\ \tilde{B}_{\gamma,i}(\theta)] \begin{bmatrix} v_i(k-t-1) \\ \gamma_i(k-t-1) \end{bmatrix}$$

which satisfies the following:

$$\|\mathcal{G}_i x_i(k)\|_\infty \leq \|\mathcal{G}_i F_i^k(\theta)\mathcal{G}_i^+\|_\infty \|\mathcal{G}_i x_i(0)\|_\infty + \sum_{j\in\mathcal{N}_i} \underbrace{\sum_{t=0}^{\infty} \|\mathcal{G}_i F_i^t(\theta) A_{ij}(\theta)\mathcal{G}_j^+\|_\infty}_{\psi_{ij}} \max_{t\leq k} \|\mathcal{G}_j x_j(k)\|_\infty$$

$$+ \sum_{t=0}^{\infty} \|\mathcal{G}_i F_i^t(\theta)[B_i(\theta) B_{\gamma,i}(\theta)] \max_{t\leq k} \left\| \begin{bmatrix} v_i \\ \gamma \end{bmatrix} \right\|_\infty \tag{A.2}$$

Corollary 16 in [40] states that the overall system $\tilde{x}(k+1) = (\tilde{A} + \tilde{B}\tilde{K})\tilde{x}(k) + \tilde{B}v(k) + \tilde{B}_\gamma\gamma(k)$ with $\tilde{x}(k) = \tilde{\mathcal{G}}x(k)$, $\tilde{\mathcal{G}} = diag(\mathcal{G}_1, \cdots, \mathcal{G}_M)$, is input-to-state stable as long as the spectral radius of the gain matrix $\rho(\Psi) < 1$ with ψ_{ij} being the entries of Ψ and

$$\psi_{ij} = \begin{cases} 0, & \text{if } i = j \\ \sum_{t=0}^{\infty} \|\mathcal{G}_i F_i(\theta)^t A_{ij}(\theta)\mathcal{G}_j^+\|_\infty, & \text{if } i \neq j. \end{cases} \tag{A.3}$$

Since $A(\theta), B(\theta), B_\gamma(\theta)$ are bounded the norm can be found via algebraic means in spite of the uncertainty.

According to Gershgorin circle theorem [41], all eigenvalues of Ψ lie within a circle centred at the origin with a radius equal to the largest row sum of Ψ. Since (11) represents the largest row sum of Ψ, if (11) holds, all eigenvalues of Ψ lie within the unit circle. According to Corollary 16 in [40], the overall system is input-to-state stable. Hence, the expanded system (A.1) is asymptotically stable. Based on the inclusion principle, it can be concluded that the original system (1) is asymptotically stable.

References

[1] Stoustrup J Plug & play control: Control technology towards new challenges 2009 *European Journal of Control* **15** 311–330
[2] Kambhampati C, Patton R J and Uppal F J 2006 Reconfiguration in networked control systems: Fault tolerant control and plug-and-play In *Proc. of 6th IFAC Symposium SAFEPROCESS* pp 126–131
[3] Zeilinger M N, Pu Y, Riverso S, Ferrari-Trecate G and Jones C N 2013 Plug and play distributed model predictive control based on distributed invariance and optimization In *Proc. of the IEEE 52nd Conference on Decision and Control* pp 5770–5776
[4] Lucia S, Kögel M and Findeisen R Contract-based predictive control of distributed systems with plug and play capabilities 2015 *IFAC-PapersOnLine* **48** 205–211
[5] Trangbaek K, Knudsen T and Kallesoe C S 2009 Plug and play process control of a district heating system In *Proc. of the 2009 European Control Conference* pp 4084–4089
[6] Crisostomi E, Liu M, Raugi M and Shorten R Plug-and-play distributed algorithms for optimized power generation in a microgrid 2014 *IEEE Transactions on Smart Grid* **5** 2145–2154
[7] Bodenburg S and Lunze J 2015 Plug-and-play control of interconnected systems with a changing number of subsystems In *Proc. of the 2015 European Control Conference* pp 3520–3527
[8] Riverso S 2014 *Distributed and plug-and-play control for constrained systems* doctoral dissertation (Università degli studi di Pavia)
[9] Bemporad A and Morari M Robust model predictive control: A survey In *Robustness in identification and control, Lecture Notes in Control and Information Sciences* (Springer) pp 207–226
[10] Raimondo D M, Limon D, Lazar M, Magni L and Fernndez Camacho E F Min-max model predictive control of nonlinear systems: A unifying overview on stability 2009 *European Journal of Control* **15** 5–21

[11] Mayne D Q Model predictive control: Recent developments and future promise 2014 *Automatica* **50** 2967–2986
[12] Mayne D Q, Kerrigan E C, van Wyk E J and Falugi P Tube-based robust nonlinear model predictive control 2011 *International Journal of Robust and Nonlinear Control* **21** 1341–1353
[13] Tempo R, Calafiore G and Dabbene F 2012 *Randomized algorithms for analysis and control of uncertain systems: with applications* (Springer Science & Business Media)
[14] Oldewurtel F, Jones C N, Parisio A and Morari M Stochastic model predictive control for building climate control 2014 *IEEE Transactions on Control Systems Technology* **22** 1198–1205
[15] Couchman P, Cannon M and Kouvaritakis B Stochastic MPC with inequality stability constraints 2006 *Automatica* **42** 2169–2174
[16] Bernardini D and Bemporad A Stabilizing model predictive control of stochastic constrained linear systems 2012 *IEEE Transactions on Automatic Control* **57** 1468–1480
[17] Schildbach G, Fagiano L, Frei C and Morari M The scenario approach for stochastic model predictive control with bounds on closed-loop constraint violations 2014 *Automatica* **50** 3009–3018
[18] Calafiore G and Campi M C Uncertain convex programs: randomized solutions and confidence levels 2005 *Mathematical Programming* **102** 25–46
[19] Parisio A, Varagnolo D, Molinari M, Pattarello G, Fabietti L and Johansson K H 2014 Implementation of a scenario-based mpc for hvac systems: an experimental case study In *Proc. of the 19th IFAC World Congress*
[20] Zhang X, Schildbach G, Sturzenegger D and Morari M 2013 Scenario-based mpc for energy-efficient building scenario-based mpc for energy-efficient building climate control under weather and occupancy uncertainty In *Proc. of the 2013 European Control Conference*
[21] de la Penad, D M, Bemporad A and Alamo T 2005 Stochastic programming applied to model predictive control In *Proc. of the 2005 European Control Conference* pp 1361–1366
[22] Di Cairano S, Bernardini D, Bemporad A and Kolmanovsky I V Stochastic MPC with learning for driver-predictive vehicle control and its application to HEV energy management 2014 *IEEE Transactions on Control Systems Technology* **22** 1018–1031
[23] Bernardini D and Bemporad A Scenario-based model predictive control of stochastic constrained linear systems 2009 *In Proc. of the 2009 Conference on Decision and Control* 6333–6338
[24] Calafiore G C and Fagiano L Robust model predictive control via scenario optimization 2013 *IEEE Transactions on Automatic Control* **58** 219–224
[25] Maestre J M, Muñoz de la Peña, D and Camacho E F Distributed model predictive control based on a cooperative game 2011 *Optimal Control Applications and Methods* **32** 153–176
[26] Long Y, Liu S, Xie L and Johansson K H A scenario-based distributed stochastic mpc for building temperature regulation 2014 *Proc. of 2014 IEEE International Conference on Automation Science and Engineering*
[27] Schildbach G, Calafiore G C, Fagiano L and Morari M 2012 Randomized model predictive control for stochastic linear systems In *Proc. of the 2012 American Control Conference* pp 417–422
[28] Schildbach G and Morari M 2015 Scenario MPC for linear time-varying systems with individual chance constraints In *Proc. of the 2015 American Control Conference* pp 415–421
[29] McMullen P On zonotopes 1971 *Transactions of the American Mathematical Society* **159** 91–109
[30] Lunze J 1992 *Feedback Control of Large-Scale Systems* (New Jersey: Prentice Hall)
[31] Skogestad S and Postlethwaite I 2007 *Multivariable Feedback Control: Analysis and Design* (Wiley New York)
[32] Mao W J Robust stabilization of uncertain time-varying discrete systems and comments on "an improved approach for constrained robust model predictive control" 2003 *Automatica* **39** 1109–1112
[33] Calafiore G C Random convex programs 2010 *SIAM Journal on Optimization* **20** 3427–3464
[34] Campi M C and Garatti S The exact feasibility of randomized solutions of uncertain convex programs 2008 *SIAM Journal on Optimization* **19** 1211–1230
[35] Rawlings J B and Mayne D Q 2009 *Model Predictive Control: theory and design* (Madison, Wisconsin: Nob Hill Publishing)
[36] Schildbach G, Fagiano L and Morari M Randomized solutions to convex programs with multiple chance constraints 2013 *SIAM Journal on Optimization* **23** 2479–2501
[37] Papoulis A and Unnikrishna Pillai S 1985 *Probability, Random variables, and Stochastic processes* (McGraw-Hill)
[38] Barmish B R and Lagoa C M The uniform distribution: A rigorous justification for its use in robustness analysis 1997 *Mathematics of Control, Signals and Systems* **10** 203–222
[39] Sontag E D On the input-to-state stability property 1995 *European Journal of Control* **1** 24–36
[40] Dashkovskiy S, Rüffer B S and Wirth F R An ISS small gain theorem for general networks 2007 *Mathematics of Control, Signals, and Systems* **19** 93–122
[41] Varga R S 2010 *Geršgorin and his circles* (Springer Science & Business Media)

Constrained time-optimal control of double-integrator system and its application in MPC

Marek Fehér, Ondřej Straka

Department of Cybernetics & European Center of Excellence: New Technologies for Information Society University of West Bohemia, Univerzitní 8, 30614 Pilsen, Czech Republic

E-mail: mfeher@ntis.zcu.cz, straka30@ntis.zcu.cz

Václav Šmídl

Regional Innovation Centre for Electrical Engineering (RICE), University of West Bohemia, Pilsen, Czech Republic

E-mail: vsmidl@rice.zcu.cz

Abstract. The paper deals with the design of a time-optimal controller for systems subject to both state and control constraints. The focus is laid on a double-integrator system, for which the time-to-go function is calculated. The function is then used as a part of a model predictive control criterion where it represents the long-horizon part. The designed model predictive control algorithm is then used in a constrained control problem of permanent magnet synchronous motor model, which behavior can be approximated by a double integrator model. Accomplishments of the control goals are illustrated in a numerical example.

1. Introduction

Model predictive control (MPC) has gained a significant attention in the field of automatic control mainly because of its possibilities in real-world industry applications. Increasing interest in the MPC practically in the entire industry can be observed. It is a giant leap since the 1980s when the first applications of the MPC have been developed to meet the specialized control needs of power plants and petroleum rafineries [1]. The current state-of-the-art of the MPC is well-presented in [2], where its future potential is outlined as well. It is also important to note that there already exist several excellent books and papers that create a quite strong background for those who want to understand the issues of the MPC, e.g. [3], [4], or [5].

In simple terms, the MPC can be characterized as a class of computer control algorithms that utilize an explicit plant model to predict the future response of a process. At each time step an MPC algorithm attempts to optimize the future plant behavior by computing a sequence of future manipulated variable adjustments. The first control action in the optimal sequence is then sent into the plant, and the entire calculation is repeated at subsequent control intervals [1]. This process is time consuming due to the optimization carried out at each time step while only the first control is utilized. To cope with such problem, the lookahead policy can be used. The lookahead policies were proposed to simplify a solution to the Bellman functional equation for a finite-time horizon problem and are based on a separation of the control horizon into a short-horizon and a long-horizon part [6]. The optimization is then simplified by approximating the long-horizon part by a precomputed value function. This paper focuses on an alternative solution. The information about the long horizon is provided by a time-to-go function [7] which contains at least a piece of the information about the future behavior of the controlled system. It was motivated by the fact that even obtaining a reasonable approximation of the long-horizon value function can be an uneasy task.

Many physical systems are also subject to constraints [7, 8] on both the control and the state variables of the system state-space model, which should be taken into account for both the short-horizon and the long-horizon parts of the MPC. In the case of continuous-time linear systems, the time-optimal control subject to the control constraints has been treated in the literature leading to the bang-bang control strategy [9]. If additionally the state constraints are considered, finding the optimal control strategy can be challenging. As the treatment of this problem for a general system is difficult, the attention will be focused on the double-integrator system. This is motivated by the fact that many physical systems possess the constrained double-integrator dynamics. It means that the relation between two quantities of the system can be investigated as the relation, for instance, between the position and the velocity. The similar idea has been presented in [7], where the fast finite-set MPC for permanent-magnet synchronous motor (PMSM) was proposed owing to such approximation.

Once the optimal control strategy is known, the time-to-go function of the double-integrator system with the control and the state constraints can be obtained. The goal of the paper is thus to obtain analytically the time-to-go function of the double-integrator system subject to both control and state constraints. Then, the calculated time-to-go function will be applied in the MPC for the long-horizon part. Application of such MPC will be illustrated in a numerical example involving the PMSM, which exhibits characteristics of the double-integrator system.

The paper is structured as follows. Section 2 is devoted to the time-optimal control of a constrained continuous-time double-integrator system. Then, the MPC is introduced in Section 3. Section 4 deals with the application of the MPC to PMSM which represents the complex system with hard constraints. The paper is concluded by final considerations in Section 5.

2. Constrained minimum-time control of double-integrator system

2.1. Problem formulation

Consider a continuous-time double-integrator system with the following state dynamics

$$\frac{dx_1(t)}{dt} = K_1 u(t), \tag{1a}$$

$$\frac{dx_2(t)}{dt} = K_2 x_1(t), \tag{1b}$$

where the state variables $x_1(t)$, $x_2(t)$, the control $u(t)$ and the constants K_1, K_2 meet $x_1(t) \in [-\overline{x}_1, \overline{x}_1]$, $x_2(t) \in \mathbb{R}$, $u(t) \in [-1, 1]$, $K_1 \in \mathbb{R}_{>0}$, and $K_2 \in \mathbb{R}_{>0}$. It means that the state $x_1(t)$ is supposed to be constrained by a constant parameter $\overline{x}_1 \in \mathbb{R}_{>0}$ and the control $u(t)$ is constrained by the interval $[-1, 1]$. As can be seen, both state and control constraints are chosen as symmetrical, which is quite usual situation in many physical systems (e.g. PMSM). Nevertheless, the solution for general non-symmetrical constraints can be obtained easily. The initial state of the double-integrator system $\mathbf{x}_0 = [x_1(0), x_2(0)]^T$ satisfies the state constraint and is assumed to be known.

Let a T be the time, in which the control system starting from $\mathbf{x}_0$ reaches a required state $\mathbf{x}_r = [x_{1,r}, x_{2,r}]^T$, i.e. $x_{1,r} = x_1(T)$ and $x_{2,r} = x_2(T)$, where again $x_{1,r} \in [-\overline{x}_1, \overline{x}_1]$ and $x_{2,r} \in \mathbb{R}$. Then the problem of seeking a minimum time, in which the required state is reachable from the initial state, can be formally expressed as

$$t_f(\mathbf{x}_0, \overline{x}_1) = \min T \tag{2}$$

$$\text{subject to} \quad (1),$$

$$x_1(t) \in [-\overline{x}_1, \overline{x}_1], \tag{2a}$$

$$u(t) \in [-1, 1], \tag{2b}$$

where $t_f(\mathbf{x}_0, \overline{x}_1)$ is called time-to-go function. The form of the time-to-go function can be derived if the optimal control strategy $u^*(t) = \gamma(\mathbf{x}(t))$, which governs the system from its initial state to the required state in minimum time, is known.

2.2. *Time-optimal control strategy*

The control strategy ensuring time optimality can be determined via Pontryagin's minimum principle [10]. The solution can be divided into two special cases. In the first special case, the optimization problem can be solved regardless the state constraint. Then, the bang-bang control strategy is obtained [9]. The second case leads to the bang-off-bang control strategy because in this case the state reaches the region, in which its constraint has to be taken into account.

Both the bang-bang and bang-off-bang control strategies require specification of a switching curve, which governs switching between the minimum and the maximum admissible control strategies. For both control strategies, the switching curve has the form [9]

$$x_2(t) = \frac{K_2}{2K_1}\text{sign}\left(-x_1(t) + x_{1,r}\right)\left(x_1^2(t) - x_{1,r}^2\right) + x_{2,r}. \tag{3}$$

If the initial state lies on the switching curve and meets the state constraint, it is obvious that the switching is not necessary and only the optimum admissible control depending on the $x_{1,r}$ and $x_{1,0}$ locations is applied. If the initial state does not lie on the switching curve, in the first part of the problem the switching between the minimum and the maximum of admissible controls or vice versa has to be performed to get the optimal solution. The switching is performed when the state of the controlled system reaches the switching curve. As the first control, the minimum admissible control ($u(t) = -1$) is applied if the state is above the switching curve and the maximum admissible control ($u(t) = 1$) is used if the state is below the switching curve. The special bang-bang control case is illustrated in Figure 1.

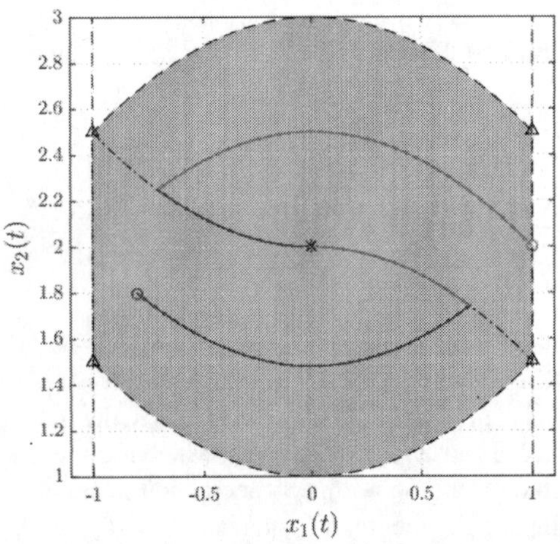

Figure 1: The time-optimal bang-bang control of the double-integrator system ($K_1 = 1$, $K_2 = 1$) from the initial states $\mathbf{x}_0^1 = [-0.8, 1.8]^T$ (blue circle) and $\mathbf{x}_0^2 = [1, 2]^T$ (red circle) to the required state $\mathbf{x}_r = [0, 2]^T$ (black asterisk). In this case, the state constraints given by $\overline{x}_1 = 1$ cannot be reached. Red (blue) lines illustrate whether $u(t) = -1$ or $u(t) = 1$ control was used. Dash-dotted line denotes the switching curve. Shading delineates the region in which the initial condition has to lie to use the bang-bang control.

As can be seen, if the minimum or the maximum admissible control is applied, the state trajectory in the phase plane can be described using the following parabolic curves

$$x_2(t) = \begin{cases} -\frac{K_2}{2K_1}\left(x_1^2(t) - x_{1,0}^2\right) + x_{2,0}, & \text{if } u(t) = -1, \\ \frac{K_2}{2K_1}\left(x_1^2(t) - x_{1,0}^2\right) + x_{2,0}, & \text{if } u(t) = 1. \end{cases} \tag{4}$$

Because of the state constraint, there are initial conditions, for which the switching curve cannot be reached by the bang-bang control strategy. From the intersections of the switching curve and the state constraint (denoted by triangles in Figure 1), it can be easily calculated that the region, in which the bang-bang control is sufficient, is given by

$$\begin{aligned}\mathcal{R} = \Big\{\mathbf{x}(t) \in \mathrm{R}^2 : x_1(t) \in [-\overline{x}_1, \overline{x}_1] \wedge \frac{K_2}{2K_1}\left(x_1(t)^2 + x_{1,r}^2\right) - \frac{K_2}{K_1}\overline{x}_1^2 + x_{2,r} \leq x_2(t) \\ \wedge x_2(t) \leq -\frac{K_2}{2K_1}\left(x_1(t)^2 + x_{1,r}^2\right) + \frac{K_2}{K_1}\overline{x}_1^2 + x_{2,r}\Big\}.\end{aligned} \tag{5}$$

For the initial state lying outside the region (5), the bang-off-bang control strategy for time-optimal control has to be applied. The strategy consists in bringing the state to its constraint by the same control as in the first part of the solution. Once the state reaches the constraint, the zero control is being applied till the switching curve is reached. Then, again the opposite value of admissible control is used to bring the system to the required state. An example of the bang-off-bang control is illustrated in Figure 2.

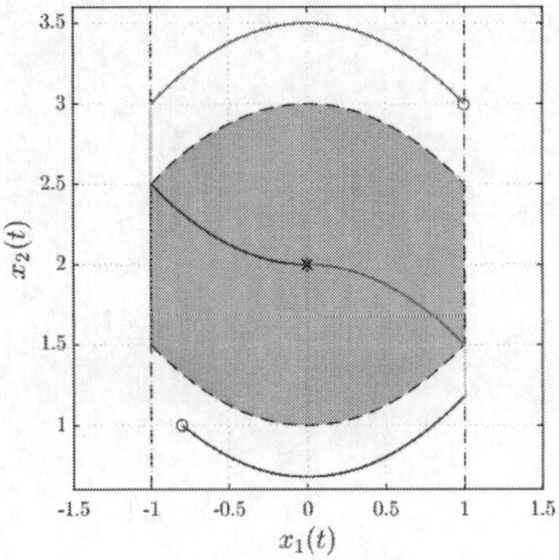

Figure 2: The time-optimal bang-off-bang control of the double-integrator system ($K_1 = 1$, $K_2 = 1$) from the initial states $\mathbf{x}_0^1 = [-0.8, 1]^T$ (blue circle) and $\mathbf{x}_0^2 = [1, 3]^T$ (red circle) to the required state $\mathbf{x}_r = [0, 2]^T$ (black asterisk). In this case, the zero control (green line) is applied when the constraint given by $\overline{x}_1 = 1$ is reached. Red (blue) lines illustrate whether $u(t) = -1$ or $u(t) = 1$ control was used. Dash-dotted line denotes the switching curve. Shading delineates the region $\mathcal{R}$.

Finally, the complete time-optimal control strategy for the constrained double-integrator system (1) can be expressed as

$$u^*(t) = \begin{cases} 1, & \text{if } \left[\underbrace{\left(x_1(t) \in (-\overline{x}_1, \overline{x}_1)\right)}_{\text{from (2a)}} \wedge \underbrace{\left(x_2(t) < \frac{K_2}{2K_1} \operatorname{sign}\left(-x_1(t) + x_{1,r}\right)\left(x_1^2(t) - x_{1,r}^2\right) + x_{2,r}\right)}_{\text{below the switching curve (3)}} \right] \\ & \vee \left[\underbrace{\left(x_1(t) \in [-\overline{x}_1, \overline{x}_1]\right)}_{\text{from (2a)}} \wedge \underbrace{\left(x_1(t) < x_{1,r}\right) \wedge \left(x_2(t) = \frac{K_2}{2K_1}\left(x_1^2(t) - x_{1,r}^2\right) + x_{2,r}\right)}_{\text{on the corresponding part of switching curve (3)}} \right] \\ 0, & \text{if } \underbrace{\left[\left(x_1(t) = -\overline{x}_1\right) \wedge \left(x_2(t) \neq \frac{K_2}{2K_1}\left(x_1^2(t) - x_{1,r}^2\right) + x_{2,r}\right)\right]}_{\text{on the lower constraint and not on the corresponding part of switching curve (3)}} \\ & \vee \underbrace{\left[\left(x_1(t) = x_{1,r}\right) \wedge \left(x_2(t) = x_{2,r}\right)\right]}_{\text{at the required state}} \\ & \vee \underbrace{\left[\left(x_1(t) = \overline{x}_1\right) \wedge \left(x_2(t) \neq -\frac{K_2}{2K_1}\left(x_1^2(t) - x_{1,r}^2\right) + x_{2,r}\right)\right]}_{\text{on the upper constraint and not on the corresponding part of switching curve (3)}} \\ -1, & \text{if } \left[\underbrace{\left(x_1(t) \in (-\overline{x}_1, \overline{x}_1))\right)}_{\text{from (2a)}} \wedge \underbrace{\left(x_2(t) > \frac{K_2}{2K_1} \operatorname{sign}\left(-x_1(t) + x_{1,r}\right)\left(x_1^2(t) - x_{1,r}^2\right) + x_{2,r}\right)}_{\text{above the switching curve (3)}} \right] \\ & \vee \left[\underbrace{\left(x_1(t) \in [-\overline{x}_1, \overline{x}_1]\right)}_{\text{from (2a)}} \wedge \underbrace{\left(x_1(t) > x_{1,r}\right) \wedge \left(x_2(t) = -\frac{K_2}{2K_1}\left(x_1^2(t) - x_{1,r}^2\right) + x_{2,r}\right)}_{\text{on the corresponding part of switching curve (3)}} \right]. \end{cases} \tag{6}$$

2.3. Time-to-go function

Once the strategy ensuring the minimum time to achieve the desired state $\mathbf{x}_r$ is known, the time-to-go function t_f in the sense of (2) can be determined. In simple terms, the determination of the time-to-go function is based on the computation of switching time instants. For this purpose, in the argument of the function (2) the initial conditions are substituted by the state $\mathbf{x} = [x_1, x_2]^T$. This leads to a general specification of the problem with arbitrary initial state $\mathbf{x}$, which enables calculation of the time-to-go function.

Based on the obtained control strategy, the time-to-go function can be described as follows

$$t_f(\mathbf{x}, \overline{x}_1) = \begin{cases} \frac{-K_2(x_1+x_{1,r})+2\sqrt{K_2^2x_1^2-K_1K_2\left(\frac{K_2}{2K_1}\left(x_1^2-x_{1,r}^2\right)+x_2-x_{2,r}\right)}}{K_1K_2}, & \text{if A} \wedge (x_2 \leq \text{B}) \wedge (x_2 \geq \text{C}) \\ \frac{\overline{x}_1-x_1-x_{1,r}}{K_1} + \frac{x_1^2+x_{1,r}^2}{2K_1\overline{x}_1} + \frac{x_{2,r}-x_2}{K_2\overline{x}_1}, & \text{if A} \wedge (x_2 \leq \text{B}) \wedge (x_2 < \text{C}) \\ \frac{K_2(x_1+x_{1,r})+2\sqrt{K_2^2x_1^2+K_1K_2\left(\frac{K_2}{2K_1}\left(-x_1^2+x_{1,r}^2\right)+x_2-x_{2,r}\right)}}{K_1K_2}, & \text{if A} \wedge (x_2 > \text{B}) \wedge (x_2 \leq \text{D}) \\ \frac{\overline{x}_1+x_1+x_{1,r}}{K_1} + \frac{x_1^2+x_{1,r}^2}{2K_1\overline{x}_1} + \frac{-x_{2,r}+x_2}{K_2\overline{x}_1}, & \text{if A} \wedge (x_2 > \text{B}) \wedge (x_2 > \text{D}) \\ p, & \text{otherwise} \end{cases} \tag{7}$$

where

A: $x_{1,r} \in [-\overline{x}_1, \overline{x}_1] \wedge x_{2,r} \in \mathbb{R} \wedge x_1 \in [-\overline{x}_1, \overline{x}_1]$,
B $= \frac{K_2}{2K_1}\text{sign}\left(-x_1 + x_{1,r}\right)\left(x_1^2 - x_{1,r}^2\right) + x_{2,r}$,
C $= \frac{K_2}{2K_1}\left(x_1^2 + x_{1,r}^2\right) - \frac{K_2}{K_1}\overline{x}_1^2 + x_{2,r}$,
D $= -\frac{K_2}{2K_1}\left(x_1^2 + x_{1,r}^2\right) + \frac{K_2}{K_1}\overline{x}_1^2 + x_{2,r}$,

and p is a penalty for violating the constraints. Therefore, it should be chosen according to requirements of the problem that is treated. In the paper $p = \infty$ will be considered. An example of the time-to-go function is illustrated in Figure 3.

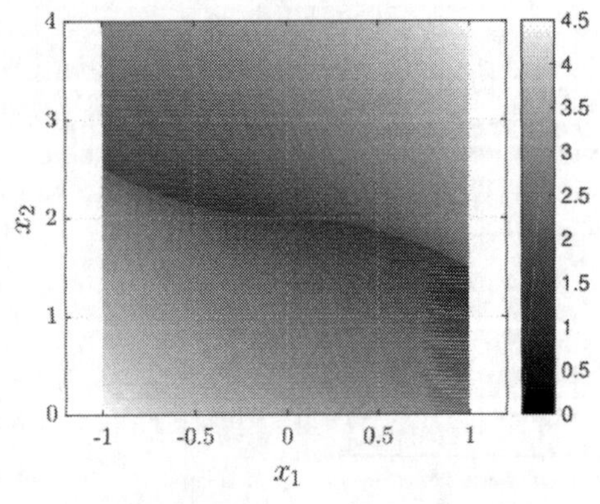

Figure 3: The example of the time-to-go function $t_f(\mathbf{x}, 1)$, where the required state was $\mathbf{x}_r = [0, 2]^T$ and the parameters were chosen as $K_1 = 1$ and $K_2 = 1$.

2.4. Vehicle moving along a single axis

To demonstrate possibilities of the obtained time-to-go function for time-optimal control problems, a simple illustration example will be presented.

Consider the problem of control of a vehicle moving only along single axis and its state being given by its velocity $x_1(t)$ and its position $x_2(t)$. The state is supposed to be measured with the constant sampling period $T_s = 0.01$ [s]. For simplicity, the mass of the vehicle is considered to be $m = 1$ [kg]. Then, the control $u(t)$ represents directly an actuation, a force driving the vehicle. The dynamics of the vehicle is described by the following continuous-time model

$$\frac{dx_1(t)}{dt} = u(t), \tag{8a}$$

$$\frac{dx_2(t)}{dt} = x_1(t). \tag{8b}$$

A discrete-time model of the system (8) based on the zero order hold is

$$\mathbf{x}(k+1) = \mathbf{f}(\mathbf{x}(k), u(k)) = \begin{bmatrix} 1 & 0 \\ T_s & 1 \end{bmatrix} \mathbf{x}(k) + \begin{bmatrix} T_s \\ \frac{1}{2}T_s^2 \end{bmatrix} u(k), \tag{9}$$

where $k \in \mathbb{N}_0$ represents the time index.

Let us further assume that the control satisfies $u(k) \in \{-1, 0, 1\}$. It is also assumed that the speed limit of the vehicle is given by $1\left[m \cdot s^{-1}\right]$. At the initial time, the vehicle is located at the position $x_2(0) = -3\,[m]$ and moves with the maximal admissible velocity $x_1(0) = -1\left[m \cdot s^{-1}\right]$. The task is to park $x_1(t_f) = 0\left[m \cdot s^{-1}\right]$ the vehicle at the position $x_2(t_f) = 1\,[m]$ in minimum time (the situation is illustrated in Figure 4).

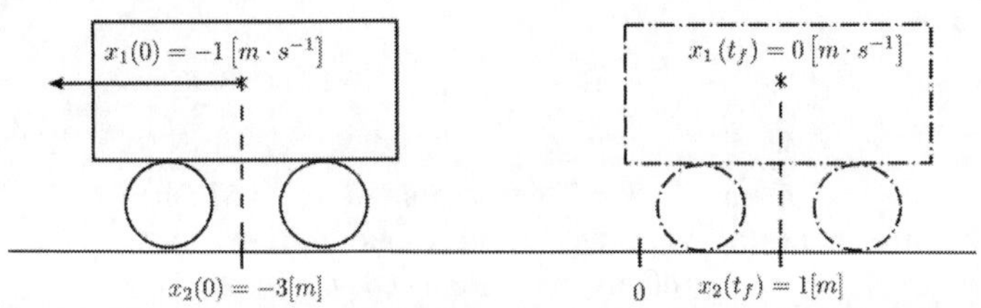

Figure 4: The initial state (solid vehicle) and the required state (dash-dotted vehicle) of the system.

The model of the considered system corresponds to the double-integrator system (1) where $K_1 = K_2 = 1$ and the constraint of the state representing the speed limit is $\overline{x}_1 = 1$. The obtained time-to-go function (7) with the corresponding parameters can be seen as the value function V defined as

$$V\left(\mathbf{x}(k)\right) \triangleq t_f\left(\mathbf{x}(k), 1\right). \tag{10}$$

Thus, the control action in time step k is chosen as

$$u^*(k) = \underset{u(k)\in\{-1,0,1\}}{\arg\min}\; V\left(\mathbf{f}(\mathbf{x}(k), u(k))\right), \tag{11}$$

The results of the control (11) applied to the system (9) are illustrated in Figures 5 and 6. As can be seen, the required state was successfully reached at time $t_f = 6.54\,[s]$ by meeting all requirements. The process of the proposed control was a little bit slower than the pre-calculated continuous-time control strategy (6) (6.50 $[s]$). It is mainly caused by discretization of the problem i.e. the control can be changed only at sampling times. For instance, if the actual state lies close to its constraint, the predictions for some control actions fall behind the constraint at the next time step and thus the control algorithm would rather choose the control ensuring non-exceeding of the constraint, however, these control actions would be possible in the continuous case. It is obvious that this property can be adjusted by the choice of the sampling period.

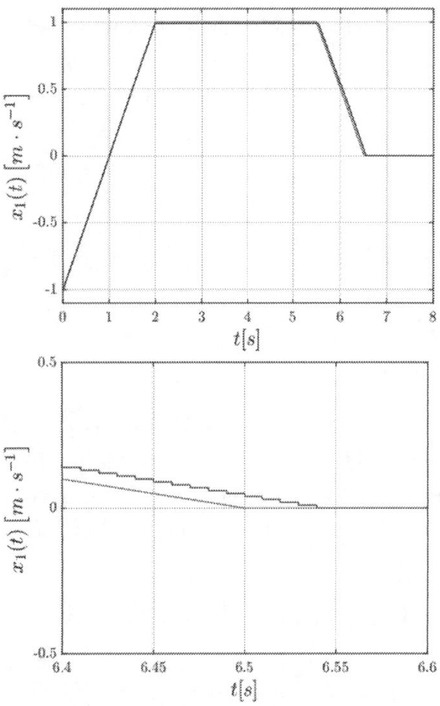

Figure 5: Time evolution of velocity (top) with the detail (bottom) (red line - continuous pre-calculated time-optimal control, blue stair-step line - proposed discrete-time controller (11)).

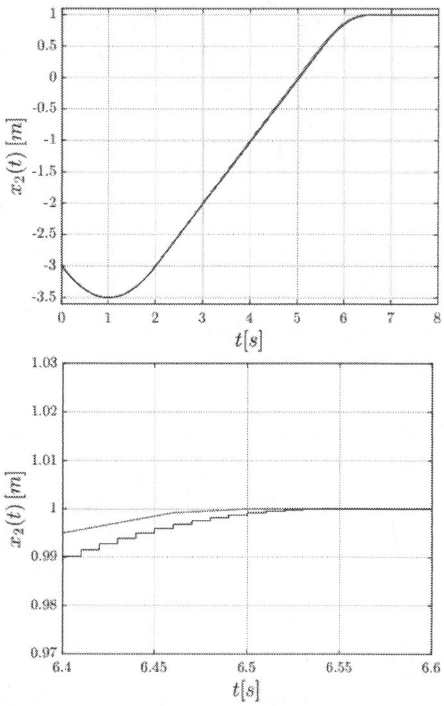

Figure 6: Time evolution of position (top) with the detail (bottom) (red line - continuous pre-calculated time-optimal control, blue stair-step line - proposed discrete-time controller (11)).

3. MPC with constraints

In this section, the general problem of the MPC subjected to constraints will be formulated. The theoretical relations will be later applied in a practical problem of the PMSM control.

Consider the general problem of control of a system given by the following state-space model

$$\mathbf{x}(k+1) \;=\; \mathbf{f}\left(\mathbf{x}(k), \mathbf{u}(k)\right), \tag{12}$$

where $\mathbf{x} \in \mathcal{X} \subset \mathbb{R}^n$, $\mathbf{u} \in \mathcal{U} \subset \mathbb{R}^m$, and $\mathbf{f}\colon \mathcal{X} \times \mathcal{U} \to \mathcal{X}$, which means that the state and control are constrained by $\mathcal{X}$ and $\mathcal{U}$, respectively. Basically, the idea of the MPC is to find a sequence of inputs $\{\mathbf{u}(j)\}_{j=k}^{N+k-1} \in \mathcal{U}^N$ that minimizes the criterion

$$J\left(\mathbf{x}(k), \{\mathbf{u}(j)\}_{j=k}^{N+k-1}\right) = g_N\left(\mathbf{x}(N+k)\right) + \sum_{i=k}^{N+k-1} g_{i-k}\left(\mathbf{x}(i), \mathbf{u}(i)\right), \tag{13}$$

where $g_N\colon \mathcal{X} \to \mathbb{R}$ and $g_i\colon \mathcal{X} \times \mathcal{U} \to \mathbb{R}$ are given cost functions, subject to the constraints. Only the first control action $\mathbf{u}^*(k)$ is used once the optimal control strategy $\{\mathbf{u}^*(i)\}_{i=k}^{N+k-1}$ is obtained. Then, at all subsequent time steps $k+1, k+2, \ldots$ the MPC repeats the optimization and application of the first control action.

It is important to mention that the minimization problem can be separated into several stages in the sense of

$$J^*\left(\mathbf{x}(k)\right) = \min_{\mathbf{u}(k)}\Big(g_0\left(\mathbf{x}(k), \mathbf{u}(k)\right) + \min_{\mathbf{u}(k+1)}\left(g_1\left(\mathbf{x}(k+1), \mathbf{u}(k+1)\right) + \ldots + V_{k+l}\left(\mathbf{x}(k+l)\right)\right)\Big) \tag{14}$$

subject to the constraints. Then, the first l, $l < N$, stages are called short horizon and the rest represented by the value function V in eq. (14), defined as

$$V_{k+l}\left(\mathbf{x}(k+l)\right) = \min_{\{\boldsymbol{\gamma}_i(\mathbf{x}(i))\}_{i=k}^{N+k-1}} \sum_{i=k+l}^{N+k-1} g_{i-k}\left(\mathbf{x}\left(i\right), \boldsymbol{\gamma}_{i-k}\left(\mathbf{x}(i)\right)\right) + g_{N+k}\left(\mathbf{x}(N+k)\right) \tag{15}$$

subject to the constraints, are called long horizon. The reason why the separation is mentioned here is because the optimization with respect to the whole control strategy $\{\mathbf{u}(i)\}_{i=k}^{N+k-1}$ at each time instant can be computationally very intensive depending on N and functions $\mathbf{f}$ and $\mathbf{g}$. This is the major reason why the approximation (15) is used. Unfortunately, for the complex systems even obtaining the long horizon can be problematic. Therefore, in the example introduced in the next section, the long-horizon part of the criterion will be substituted by the time-to-go function derived in Section 2.

4. Application of the MPC with constraints to the PMSM

The control of the PMSM is a very challenging task because the system has to satisfy hard constraints. A strong emphasis is also placed on the speed of control, since the sampling period has to be small and a control algorithm is usually implemented in a hardware with limited computational power. For this reason, the control of the PMSM will serve as an ideal example for employing the proposed MPC to a system with a complex dynamics.

The conventional model of the PMSM in the d-q reference frame [11] has the form

$$\frac{di_d(t)}{dt} = -\frac{R_s}{L_d} i_d(t) + \frac{L_q}{L_d} i_q(t)\omega(t) + \frac{1}{L_d} u_d(t), \tag{16a}$$

$$\frac{di_q(t)}{dt} = -\frac{R_s}{L_q} i_q(t) - \frac{\Psi_{PM}}{L_q}\omega(t) - \frac{L_d}{L_q} i_d(t)\omega(t) + \frac{1}{L_q} u_q(t), \tag{16b}$$

$$\frac{d\omega(t)}{dt} = \frac{p_p}{J}\left(T_e(t) - T_L(t)\right), \tag{16c}$$

$$\frac{d\vartheta(t)}{dt} = \omega(t), \tag{16d}$$

$$\frac{dT_L(t)}{dt} = 0, \tag{16e}$$

$$T_e(t) = k_p p_p \left(\Psi_{PM} i_q(t) + \left(L_d - L_q\right) i_d(t) i_q(t)\right), \tag{16f}$$

where $i_d(t)$, $i_q(t)$, $u_d(t)$, and $u_q(t)$ represent components of the stator current and voltage vector in the rotating reference frame, respectively. The variable $\omega(t)$ is the electrical rotor speed and $\vartheta(t)$ is the electrical rotor position. The variable $T_L(t)$ represents the load torque. The system parameters are: the components of stator inductance L_d and L_q, the stator resistance R_s, the moment of inertia J, the number of pole pairs p_p, the flux linkage excited by permanent magnets on the rotor Ψ_{PM}, and the Park constant k_p.

For the purpose of the MPC, an approximate discrete-time model has to be obtained. This can be done by using the Taylor series based discretization method [12], which can be for the state $\mathbf{x}_P(t)$ of the PMSM given by $\mathbf{x}_P(t) = \left[i_d(t), i_q(t), \omega(t), \vartheta(t), T_L(t)\right]^T$ and for the input $\mathbf{u}(t)$ represented by the voltage vector $\mathbf{u}(t) = \left[u_d(t), u_q(t)\right]^T$ generally expressed as

$$\mathbf{x}_P(k+1) = \mathbf{x}_P(k) + \sum_{l=1}^{Z} \mathbf{A}_l\left(\mathbf{x}_P(k), \mathbf{u}(k)\right) \frac{T_s^l}{l!}, \tag{17}$$

where $\mathbf{A}_l\left(\mathbf{x}_P, \mathbf{u}\right)$ is determined recursively as

$$\mathbf{A}_1\left(\mathbf{x}_P, \mathbf{u}\right) = \mathbf{f}\left(\mathbf{x}_P\right) + \mathbf{g}\left(\mathbf{x}_P\right)\mathbf{u}, \tag{17a}$$

$$\mathbf{A}_{l+1}\left(\mathbf{x}_P, \mathbf{u}\right) = \frac{\partial \mathbf{A}_l\left(\mathbf{x}_P, \mathbf{u}\right)}{\partial \mathbf{x}_P}\left(\mathbf{f}\left(\mathbf{x}_P\right) + \mathbf{g}\left(\mathbf{x}_P\right)\mathbf{u}\right), \tag{17b}$$

where the form of the functions $\mathbf{f}(\mathbf{x}_P)$ and $\mathbf{g}(\mathbf{x}_P)$ can be determined from the PMSM model (16). It is also important to mention how to choose the truncation of the series given by the parameter Z. For the stator current dynamics, where the input has an immediate effect on the derivative, it is sufficient to set $Z = 1$. On the other hand, for the mechanical variables, higher expansions, at least $Z = 2$, are required [13].

4.1. Constraints of the PMSM

From the control point of view, the biggest challenge is represented by the constraint on currents. This constraint can formally be expressed as

$$i_d^2(t) + i_q^2(t) \leq I_{max}^2, \tag{18}$$

where I_{max} is a given maximum current. Besides current, the input voltage is also constrained. Since a finite set of control actions is required for the computational costs reason and these control actions have to satisfy some other conditions too, the usual way of control generation is to add a voltage source invertor (VSI) to the PMSM [7], [8]. The VSI can generate stepwise six active and two zero input voltage vectors according to the relation

$$\mathbf{u}(k) = \frac{2}{3}U_{dc}\begin{bmatrix} \cos(\vartheta(k)) & \sin(\vartheta(k)) \\ -\sin(\vartheta(k)) & \cos(\vartheta(k)) \end{bmatrix}\begin{bmatrix} 1 & \cos\left(\frac{2}{3}\pi\right) & \cos\left(\frac{4}{3}\pi\right) \\ 0 & \sin\left(\frac{2}{3}\pi\right) & \sin\left(\frac{4}{3}\pi\right) \end{bmatrix}\begin{bmatrix} S_a(k) \\ S_b(k) \\ S_c(k) \end{bmatrix}, \tag{19}$$

where U_{dc} is the DC-link voltage and all possible input actions are generated through all possible switching combinations given by $S_x(k) \in \{0, 1\}$.

4.2. Approximation of the PMSM by the double-integrator system

In this subsection an approximation of the PMSM model will be presented. The approximation will later be used to obtain an approximate time-to-go function for the PMSM model.

The behavior of the PMSM torque and angular speed roughly corresponds to the behavior of the double-integrator system (1), where $x_1(t) \approx T_e(t)$ and $x_2(t) \approx \omega(t)$ [7]. The key question is how to choose the parameters of double integrator K_1, K_2 and how to fit the currents constraint (18) to the state constraint $\overline{x}_1$. Since the torque is approximated by the first state of the double-integrator system $x_1(t)$, the parameter K_1 can be obtained from the time derivative of the torque $T_e(t)$. This derivative is

$$\frac{dT_e(t)}{dt} = k_p p_p\left(\Psi_{PM}\frac{di_q(t)}{dt} + \left(L_d - L_q\right)\left(\frac{di_d(t)}{dt}i_q(t) + \frac{di_q(t)}{dt}i_d(t)\right)\right). \tag{20}$$

In many practical applications the values of the inductances L_d and L_q are close to each other. Therefore, assuming $L_d \approx L_q$ and neglecting the state-variables in (16b) depending only on the last control, the following approximation of K_1

$$K_1 \approx \frac{k_p p_p \Psi_{PM}}{L_q} \tag{21}$$

seems to be reasonable. The approximation of K_2 can be based on neglecting the load torque, so it is assumed as

$$K_2 \approx \frac{p_p}{J}. \tag{22}$$

The constraint for the currents i_d and i_q (18) should be transformed to a constraint for the torque. Unfortunately, the constraint for the torque varies with the currents. Hence, the state constraint appearing in the time-to-go function will also vary with the currents. If the value of current $i_d(k)$ is known, the relation for the current $i_q(k)$ on the boundary of the constraint (18) can be expressed as

$$i_q(k) = \pm\sqrt{\left(I_{max}^2 - i_d^2(k)\right)}. \tag{23}$$

By substituting relation (23) into eq. (16f), the torque constraint at time step k with respect to known current $i_d(k)$ can be described by

$$\overline{x}_1(k) = k_p p_p \left[\left(L_d - L_q\right) i_d(k)\sqrt{I_{max}^2 - i_d^2(k)} + \Psi_{PM}\sqrt{I_{max}^2 - i_d^2(k)}\right]. \tag{24}$$

Such choice of the state constraint ensures that the currents will always satisfy the constraint (18).

4.3. Control strategy

For the MPC of the PMSM, only the horizon $N = 1$ in criterion (13) is considered. The reason is that the control of PMSM has to be very fast to achieve good performance.

For the short horizon part, the cost function g_0 in criterion (13) was chosen as

$$g_0(\mathbf{x}_P(k), \mathbf{u}(k)) = c_1\left(\omega'(k+1) - \omega_r\right)^2 + c_2\left(i_d'(k+1) - i_{d,r}\right)^2 \tag{25}$$

saying that, in addition to reach the required speed, the current i_d has to be choke, which is a reasonable requirement [8]. The weights c_1 and c_2 in the cost function (25) allow to set compromise between the requirements. Apostrophes in the cost function (25) denote predictions of the state variables based on the discretized model of system (16).

However, even when considering $N = 1$, it is desirable to include some information about the long horizon at least in some way. Thus, the following choice of cost function $g_1\left(\mathbf{x}_p'(k+1)\right)$ in criterion (13) employing the obtained time-to-go function for constrained double-integrator system is assumed

$$g_1\left(\mathbf{x}_P'(k+1)\right) = t_f\left(\left[T_e'(k+1), \omega'(k+1)\right]^T, \overline{x}_1(k+1)\right), \tag{26}$$

where the parameters K_1 and K_2 are given by relations (21) and (22), respectively. Such choice of the cost function utilizes the knowledge of dynamics between the torque and speed in the sense of providing the rough information about regulation time. Thus, this information can represent, to a certain extent, the long horizon. The time-to-go function also contains the transversality conditions within, so there is a huge emphasis on reaching the required state. All is in compliance with the constraints.

The complete algorithm of the MPC for the PMSM based on the approximation by the double-integrator system can be summarized as follows.

Step 1: Obtain the value of the actual state variables of the controlled PMSM $\mathbf{x}_P(k)$.

Step 2: Compute the set of seven possible control actions generated by the VSI (19) and make predictions of the state for the next time step $\mathbf{x}_p'(k+1)$ based on the approximate discrete-time model.

Step 3: From the set of the control actions find the action $\mathbf{u}^*(k)$ minimizing

$$J\left(\mathbf{x}_P(k), \mathbf{u}(k)\right) = c \cdot g_1\left(\mathbf{x}_p'(k+1)\right) + g_0\left(\mathbf{x}(k), \mathbf{u}(k)\right), \tag{27}$$

where the approximation of g_1 is done as in eq. (26) and $c \in \mathbb{R}_+$ is a constant allowing eventually tune the influence of the long-horizon information represented by the time-to-go function.

Step 4: Apply the optimal control $\mathbf{u}^*(k)$.

Step 5: Set $k = k + 1$ and return to Step 1.

Table 1: The numerical example parameters.

Parameter	Value	Parameter	Value
R_s	$0.28\,[\Omega]$	p_p	4
L_d	$0.003465\,[H]$	J	$0.0098\,\left[kg \cdot m^2\right]$
L_q	$0.004465\,[H]$	U_{dc}	$200\,[V]$
Ψ_{PM}	$0.1989\,[Wb]$	I_{max}	$\sqrt{10}\,[A]$
k_p	1.5	T_s	$50 \cdot 10^{-6}\,[s]$

4.4. Numerical results

The proposed control algorithm is illustrated using a PMSM model (16) with parameters listed in Table 1. It is supposed that the PMSM starts from its standstill mode ($\mathbf{x}_P(0) = \mathbf{0}$) and that the required state is given by speed $100\left[rad \cdot s^{-1}\right]$ ($\mathbf{x}_r = [0, 0, 100, 0, 0]^T$). The tune constants in criterion (27) are chosen as $c = 1$, $c_1 = 0.1$, and $c_2 = 0.005$.

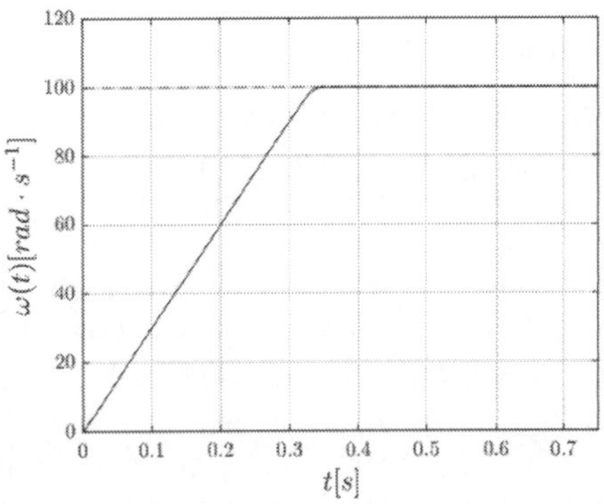

Figure 7: The time evolution of angular speed (red dashed line - required speed, blue line - true speed).

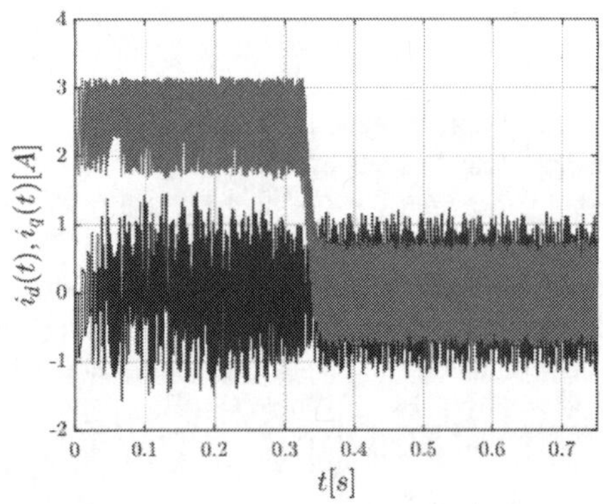

Figure 8: The time evolution of currents $i_d(t)$ and $i_q(t)$ during the process of speed control (blue line - $i_d(t)$, red line - $i_q(t)$).

It can be seen that the required speed is successfully reached (see Figure 7) while the constraints are satisfied (see Figure 9). This demonstrates satisfactory control quality of the MPC of the PMSM with the long-horizon part of the criterion approximated by the time-optimal control obtained for a double-integrator dynamics. The time evolution of the currents i_d and i_q and the used control actions are illustrated in Figure 8 and Figure 10, respectively.

Simulations (not provided here for space reasons) also showed better properties of the designed MPC compared to the simple MPC with criterion (27) where the cost function g_1 is equal to zero. This simple MPC was not able to reach the required speed at all and the only way to reach the required speed was to increase the constant c_1 penalizing the square of deviation of the speeds in loss function (25). However, this leads to an undesirable increase in the current i_d. For this reason, the designed MPC represents the good compromise between the performance and the demands. The use of the time-to-go function directly as the criterion for MPC (assuming the function g_0 in (27) is equal to zero) is also interesting. In such a case, the controller is unable to reach high speed requirements ($10^2\left[rad \cdot s^{-1}\right]$). It is caused by the fact that the constraint (24) and consequently the time-to-go function vary with each state prediction.

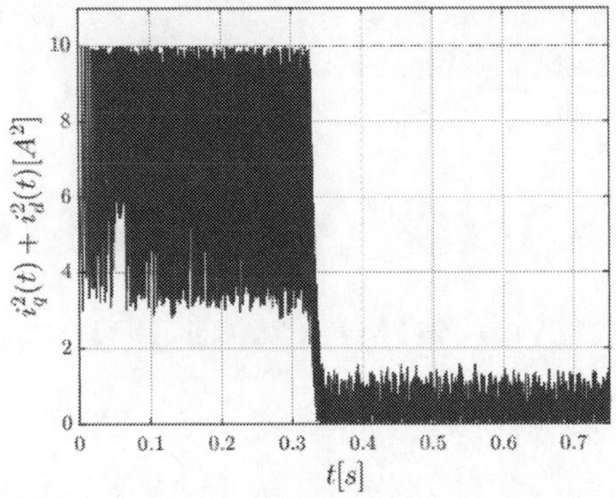

Figure 9: The fulfillment of the currents constraint (red dashed line - I_{max}^2, blue line - $i_q^2(t) + i_d^2(t)$).

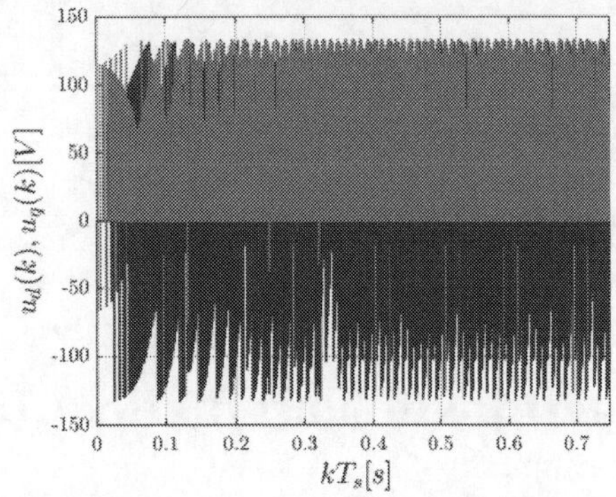

Figure 10: The voltages generated by the VSI (blue line - $u_d(k)$, red line - $u_q(k)$).

5. Conclusion

The paper focused on a design of a time-optimal controller of a double-integrator system subject to both state and control constraints. The derived time-to-go function was then utilized as a long-horizon part of a criterion for the model predictive control problem. The obtained controller was then applied to a control problem involving a permanent magnet synchronous motor model. As this model exhibits the double-integrator behavior, the derived time-to-go function was convenient for it. Satisfactory performance and achievement of the control goals were illustrated using a numerical example.

Acknowledgement

This work was supported by the project LO1506 of the Czech Ministry of Education, Youth and Sports.

References

[1] S. J. Quin and T. A. Badgwell, "A survey of industrial model predictive control technology," *Control Engineering Practice*, vol. 11, p. 733.

[2] D. Q. Mayne, "Model predictive control: Recent developments and future promise," *Automatica*, vol. 50, p. 2967.

[3] E. F. Camacho and C. Bordons, *Model Predictive Control*. New York: Springer, 2004.

[4] L. Grüne and J. Pannek, *Nonlinear model predictive control: theory and algorithms*. New York: Springer, 2011.

[5] J. B. Rawlings and D. Q. Mayne, *Model predictive control: theory and design*. Madison, Wis.: Nob Hill Pub., 2009.

[6] D. P. Bertsekas, *Dynamic programming and optimal control*. Massachusetts: Athena Scientific, 2000.

[7] E. Fuentes, D. Kalise, and J. Rodrígez, "Cascade-free predictive speed control for electrical drives," *IEEE Transactions on Industrial Electronics*, vol. 61, p. 2176.

[8] M. Preindl and S. Bolognani, "Model predictive direct speed control with finite control set of PMSM drive systems," *IEEE Transactions on Power Electronics*, vol. 28, p. 1007.

[9] A. E. Bryson and Y.-C. Ho, *Applied optimal control*. Washington: John Wiley & Sons, 1975.

[10] L. S. Pontryagin, *The mathematical theory of optimal processes*. New York: Gordon and Breach, 1986.

[11] P. Pillay and R. Krishnan, "Modeling, simulation, and analysis of permanent-magnet motor drives, part I: the permanent-magnet synchronous motor drive," *IEEE Transactions on industry applications*, vol. 25, p. 265.

[12] Z. Zhang, D. U. An, H. Kim, and K. T. Chong, "Comparative study of matrix exponential and Taylor series discretization methods for nonlinear odes," *Simulation Modelling Practice and Theory*, vol. 17, p. 471.

[13] E. J. Fuentes, C. A. Silva, and J. I. Yuz, "Predictive speed control of a two-mass system driven by a permanent magnet synchronous motor," *IEEE Transactions on industrial electronics*, vol. 59, p. 2840.

IOP Conf. Series: Journal of Physics: Conf. Series **783** (2017) 012025 doi:10.1088/1742-6596/783/1/012025

Nonlinear Model Predictive Control with Constraint Satisfactions for a Quadcopter

Ye Wang[1], Andres Ramirez-Jaime[2], Feng Xu[3] and Vicenç Puig[1]

[1] Advanced Control Systems (SAC) Research Group at Institut de Robòtica i Informàtica Industrial, CSIC-UPC, Universitat Politècnica de Catalunya-BarcelonaTech (UPC), C/. Llorens i Artigas 4-6, 08028 Barcelona, Spain

[2] Departamento de Ingeniería Eléctrica y Electrónica, Universidad de los Andes, Carrera 1, No. 18A-10, Colombia

[3] Center of Intelligent Control and Telescience, Graduate School at Shenzhen, Tsinghua University, 518055 Shenzhen, P.R. China

E-mail: ywang@iri.upc.edu, af.ramirez236@uniandes.edu.co, xu.feng@sz.tsinghua.edu.cn, vicenc.puig@upc.edu

Abstract. This paper presents a nonlinear model predictive control (NMPC) strategy combined with constraint satisfactions for a quadcopter. The full dynamics of the quadcopter describing the attitude and position are nonlinear, which are quite sensitive to changes of inputs and disturbances. By means of constraint satisfactions, partial nonlinearities and modeling errors of the control-oriented model of full dynamics can be transformed into the inequality constraints. Subsequently, the quadcopter can be controlled by an NMPC controller with the updated constraints generated by constraint satisfactions. Finally, the simulation results applied to a quadcopter simulator are provided to show the effectiveness of the proposed strategy.

1. Introduction

The quadcopter has become one of the most popular unmanned aerial vehicles (UAVs), which attracts considerable attention in both academia and industry. Due to its simple structure and flexible flight ability, the quadcopter is playing a significant role in applications such as video surveillance, agricultural service, mapping, goods delivery and other military applications [1, 2, 3].

The motion of a quadcopter relies on the rotation of four propellers to generate thrusts and to control its attitude and position. In general, a simultaneous increasing/decreasing of angular speed of four propellers will make the quadcopter fly upward/downward. When keeping the vertical component of total thrust from four propellers equal to the gravity of quadcopter, the quadcopter can be controlled to fly forward/backward or /leftward/rightward by increasing the angular speed of one propeller and simultaneous decreasing the angular speed of the propeller on its centrosymmetric position, respectively. Additionally, simultaneous increasing the speed of one pair of propellers centrosymmetric to each other and decreasing the speed of the other pair can control the quadcopter to rotate in horizontal plane. However, due to the nonlinear dynamics and high degrees of couplings between different control channels, to achieve high-performance control of quadcopter is still an issue full of challenges.

The control problem of quadcopter has been widely investigated by using a variety of approaches such as proportional-integral-derivative (PID) control, backstepping control, feedback linearization, linear quadratic regulator and model predictive control (MPC). In [4, 5, 6], the PID control was used to control the quadcopter with the objective of achieving robustness and fault tolerance. The application of backstepping control and feedback linearization to the quadcopter could be found in [7, 8, 9, 10]. Related to the topic of the current paper, we can also find MPC control of quadcopter in [2, 11, 12], where a common feature of these works is that they are based on the linear MPC techniques.

Commonly, the quadcopter is normally considered as a system with high nonlinearities and fast dynamics. Although it is possible to achieve some control objectives by using a linearized model or a group of linearized models to approximate the nonlinear dynamics of quadcopter, we have to face a degree of performance degradation. Motivated by this point, we propose to directly use the nonlinear MPC technique to control the quadcopter. In this case, we can not only simultaneously obtain better performance but also make use of the inherent capability of MPC, i.e., explicit constraint handling, which is very helpful to deal with the constraints on states, inputs and outputs. Additionally, turning to the constraint-handling capability, MPC is quite convenient to be used for tolerance of faults in actuators (i.e., motors of quadcopter) by adjusting its input constraints, which is also of vital important interest of many works [13, 14].

The main contribution of this paper is to present a nonlinear MPC (NMPC) strategy combined with constraint satisfactions for a quadcopter. Paritial nonlinear constraints and modeling errors are updated through the procedure of constraint satisfactions. Then, the NMPC strategy is implemented by solving a nonlinear optimization problem through the nonlinear programming technique. As a result, simulation results by applying the proposed NMPC strategy to a quadcopter simulator implemented in MATLAB environment are shown in order to verify the effectiveness of the proposed MPC strategy. Besides, comparison results show the performance of the proposed MPC strategy.

The remainder of this paper is organized as follows: The nonlinear dynamics and control-oriented model of the quadcopter are briefly introduced in Section 2. The constraint satisfaction for the quadcopter control is addressed in Section 3. In Section 4, the NMPC design for the quadcopter is presented. The simulation results as well as comparisons results with different uncertain cases based on a quadcopter simulator are shown in Section 5. Finally, some conclusions are written in Section 6.

2. Nonlinear Dynamics of the Quadcopter

2.1. Quadcopter Dynamic Description

The motion of the quadcopter is generated by the lift forces produced by the rotating propeller blades, where the translations and rotations are realized by manipulating the difference between velocities of four rotors. The quadcopter has 6 degrees of freedom (attitude and position) according to the earth-fixed frame. The position of this helicopter is described by the position vector $[x, y, z]^T$ and its attitude is defined by three Euler angles (pitch, roll and yaw) as $[\phi, \theta, \psi]^T$, where these three angles are assumed to satisfy the following constraints [15]:

$$-\frac{\pi}{2} \leq \phi \leq \frac{\pi}{2}, \tag{1a}$$

$$-\frac{\pi}{2} \leq \theta \leq \frac{\pi}{2}, \tag{1b}$$

$$-\pi \leq \psi \leq \pi. \tag{1c}$$

Consider the earth-fixed frame as inertial frame and body-fixed frame with the origin in the mass center of the quadcopter. It is further assumed that the body of quadcopter is rigid and symmetric. The dynamics of the quadcopter can be characterized by a set of continuous-time differential equations as follows [2]:

$$\ddot{\phi} = \dot{\theta}\dot{\psi}a_1 + \dot{\theta}a_2\Omega_r + b_1U_2, \tag{2a}$$
$$\ddot{\theta} = \dot{\phi}\dot{\psi}a_3 - \dot{\phi}a_4\Omega_r + b_2U_3, \tag{2b}$$
$$\ddot{\psi} = \dot{\phi}\dot{\theta}a_5 + b_3U_4, \tag{2c}$$
$$\ddot{z} = -g + (\cos\phi\cos\theta)\,U_1/m, \tag{2d}$$
$$\ddot{x} = (\cos\phi\sin\theta\cos\psi + \sin\phi\sin\psi)\,U_1/m, \tag{2e}$$
$$\ddot{y} = (\cos\phi\sin\theta\sin\psi - \sin\phi\cos\psi)\,U_1/m, \tag{2f}$$

with

$$\begin{bmatrix} U_1 \\ U_2 \\ U_3 \\ U_4 \\ \Omega_r \end{bmatrix} = \begin{bmatrix} b\left(\Omega_1^2 + \Omega_2^2 + \Omega_3^2 + \Omega_4^2\right) \\ b\left(\Omega_2^2 - \Omega_4^2\right) \\ b\left(-\Omega_1^2 + \Omega_3^2\right) \\ d\left(-\Omega_1^2 + \Omega_2^2 - \Omega_3^2 + \Omega_4^2\right) \\ -\Omega_1 + \Omega_2 - \Omega_3 + \Omega_4 \end{bmatrix} \tag{3}$$

and the parameters of the quadcopter are defined as

$$a_1 = \frac{I_{yy} - I_{zz}}{I_{xx}},\ a_2 = \frac{J_r}{I_{xx}},\ a_3 = \frac{I_{zz} - I_{xx}}{I_{yy}},$$
$$a_4 = \frac{J_r}{I_{yy}},\ a_5 = \frac{I_{xx} - I_{yy}}{I_{zz}},$$
$$b_1 = \frac{\ell}{I_{xx}},\ b_2 = \frac{\ell}{I_{yy}},\ b_3 = \frac{1}{I_{zz}}.$$

where Ω_1, Ω_2, Ω_3 and Ω_4 denote the speeds of four rotors, respectively, I_{xx}, I_{yy} and I_{zz} are the moments of inertial of the quadcopter around three body axes, respectively, J_r is the propeller rotational moment of inertia, ℓ denotes the arm length of the quadcopter, b and d represent thrust and drag coefficients. Besides, the constraints on the rotor angular speeds can be formulated by

$$0 \leq \Omega_i \leq \Omega_i^{max}, \quad i = 1, 2, \ldots, 4, \tag{4}$$

where Ω_i^{max} denotes the maximum angular speed of the i-th rotor.

It is noticed that the dynamic model in (2) for designing the MPC controller is a simplified one, where some modeling errors are ignored, such as the wind resistances. In the simulation, the uncertain simulator model including some modeling errors will be used.

2.2. Control-oriented Model of the Quadcopter

Considering the dynamic equations of the quadcopter in (2), the variables of system states and inputs are chosen as

$$\mathbf{x} \triangleq \left[\phi\ \dot{\phi}\ \theta\ \dot{\theta}\ \psi\ \dot{\psi}\ z\ \dot{z}\ x\ \dot{x}\ y\ \dot{y}\right]^T, \tag{5a}$$
$$\mathbf{u} \triangleq [\Omega_1\ \Omega_2\ \Omega_3\ \Omega_4]^T. \tag{5b}$$

Then, by means of the Euler discretization method, the discrete-time control-oriented model of the quadcopter can be generally formulated as

$$\mathbf{x}(k+1) = \mathbf{F}(\mathbf{x}(k), \mathbf{u}(k)) + \mathbf{w}(k), \tag{6a}$$
$$\mathbf{y}(k) = \mathbf{G}(\mathbf{x}(k)) + \mathbf{v}(k), \tag{6b}$$

where $\mathbf{x}(k)$ and $\mathbf{u}(k)$ denote vectors of system states and inputs defined in (5) at time instant k, respectively. $\mathbf{F}(\cdot)$ denotes the nonlinear mapping function describing the system dynamics in (2) and (3). $\mathbf{G}(\cdot)$ represents the measurement function based on the characteristics of the available sensors. Besides, $\mathbf{w}(k)$ and $\mathbf{v}(k)$ denote the vectors of the system disturbances of modeling errors and the measurement noise of sensors at time instant k, respectively.

The quadcopter operates under some constraints, which can be written as follows:

$$\mathbf{x}_{\min} \leq \mathbf{x}(k) \leq \mathbf{x}_{\max}, \tag{7}$$

$$\mathbf{0} \leq \mathbf{u}(k) \leq \mathbf{u}_{\max}, \tag{8}$$

where $\mathbf{x}_{\min}$ denotes the vector of minimum values of the system states, $\mathbf{x}_{\max}$ and $\mathbf{u}_{\max}$ denote vectors of maximum values of system states and inputs.

2.3. Partial Nonlinearity Embeddings

Depending on (3), it is clear that the selected input of the quadcopter Ω_i is hidden in $[U_1, U_2, U_3, U_4, \Omega_r]^T$. Hence, the virtual inputs of the quadcopter are chosen as

$$\tilde{\mathbf{u}} \triangleq [U_1, U_2, U_3, U_4, \Omega_r]^T. \tag{9}$$

Therefore, the nominal and uncertain control-oriented model of the quadcopter (6) can be reformulated with the virtual inputs as

$$\mathbf{x}(k+1) = \mathbf{F}(\mathbf{x}(k), \tilde{\mathbf{u}}(k)) + \mathbf{w}(k), \tag{10a}$$

$$\mathbf{y}(k) = \mathbf{G}(\mathbf{x}(k)) + \mathbf{v}(k). \tag{10b}$$

3. Constraint Satisfactions for the Quadcopter Control

In this section, the constraint-satisfaction approach is used to find the suitable bounds of the defined virtual inputs (9) in order to guarantee the consistency (3). The utilized constraint-satisfaction approach relies on an interval-based method, which has been popularly applied in the robotics [16]. The interval-based constraint-satisfaction approach can be implemented by a procedure of the forward and backward propagations to refine the value range from wide intervals to short ones including eliminating the impossible values violated from some defined equality and inequality constraints.

3.1. Constraint-Satisfaction Approach

As introduced in [16], a constraint-satisfaction approach on sets can be formulated as a 3-tuple $\mathcal{H} = (\mathcal{Z}, \mathcal{D}, \mathcal{C})$, where

- $\mathcal{Z} = \{z_1, z_2, \cdots, z_n\}$ is a finite set of variables.
- $\mathcal{D} = \{\mathcal{D}_1, \mathcal{D}_2, \cdots, \mathcal{D}_n\}$ is the set of domains of the variables.
- $\mathcal{C} = \{c_1, c_2, \cdots, c_n\}$ is a finite set of constraints, where each constraint c_i is specified by the pair $(\mathcal{Z}_i, \mathcal{R}_i)$, in which $\mathcal{Z}_i$, that is called as the constraint scope, is a subset of $\mathcal{Z}$ and $\mathcal{R}_i$, that is called as constraint relation, is a relation indicating the allowed combination of values for the variables belonging to $\mathcal{Z}_i$.

The contractors are the useful tools for solving the constraint-satisfaction problem [16], [17]. Solving a constraint-satisfaction problem consists in finding all variable assignments such that all constraints are satisfied. The variable value assignment $(\hat{z}_1, \hat{z}_2, \cdots, \hat{z}_n) \in \mathcal{D}$ is a solution of $\mathcal{H}$ if all constraints in $\mathcal{C}$ are satisfied. The set of all solution points of $\mathcal{H}$ is called the global solution set and denoted by $\mathcal{S}(\mathcal{H})$. The variable $z_i \in \mathcal{Z}$ is consistent in $\mathcal{H}$ if and only if $\forall \hat{z}_i \in \mathcal{D}_i$,

$$\exists(\hat{z}_1 \in \mathcal{D}_1, \cdots, \hat{z}_n \in \mathcal{D}_n)$$

such that $(\hat{z}_i, \cdots, \hat{z}_n) \in \mathcal{S}(\mathcal{H})$.

The solution of a constraint-satisfaction problem is said to be globally consistent if and only if every variable is consistent. A variable is locally consistent if and only if it is consistent with respect to all directly connected constraints. Thus, the solution of the constraint-satisfaction problem is said to be locally consistent if all variables are locally consistent. An algorithm for finding an approximation of the solution set of a constraint-satisfaction problem can be found in [16].

3.2. Constraint Satisfactions for the Quadcopter Control

As aforementioned, partial nonlinearities in (3) are embedded into the control-oriented model of the quadcopter. Therefore, $\mathbf{u}(k)$ is replaced by $\widetilde{\mathbf{u}}(k)$ in (9). The constraints on $\mathbf{u}(k)$ and Equation (3) are transformed into the new constraints on $\widetilde{\mathbf{u}}(k)$. The constraint-satisfaction approach for the quadcopter control along the MPC prediction horizon H_p is implemented by using the following algorithm.

Algorithm 1 Constraint Satisfaction applied to the Quadcopter

for $k := 1$ to H_p **do**
  $\mathcal{X}_k \Leftarrow [\mathbf{0}, \mathbf{x}_{\max}]$
  $\mathcal{U}_{k-1} \Leftarrow [\mathbf{0}, \mathbf{u}_{\max}]$
  $\widetilde{\mathcal{U}}_{k-1} \Leftarrow [\mathbf{0}, \widetilde{\mathbf{u}}_{\max}]$
end for
$\mathcal{V} \Leftarrow \overbrace{\mathbf{x}(1), \mathbf{x}(2), \ldots \mathbf{x}(H_p)}^{X}, \overbrace{\mathbf{u}(0), \mathbf{u}(1), \ldots, \mathbf{u}(H_p-1)}^{U}, \overbrace{\widetilde{\mathbf{u}}(0), \widetilde{\mathbf{u}}(1), \ldots, \widetilde{\mathbf{u}}(H_p-1)}^{\widetilde{U}}$
$\mathcal{D} \Leftarrow \mathcal{X}_1, \mathcal{X}_2, \ldots, \mathcal{X}_{H_p}, \mathcal{U}_0, \mathcal{U}_1, \ldots, \mathcal{U}_{H_p-1}, \widetilde{\mathcal{U}}_0, \widetilde{\mathcal{U}}_1, \ldots, \widetilde{\mathcal{U}}_{H_p-1}$
$\mathcal{C} \Leftarrow$ Equation (3)
$\mathcal{H} \Leftarrow \mathcal{V}, \mathcal{D}, \mathcal{C}$
$\mathcal{S} = solve(\mathcal{H})$
Implement the new constraint $\mathcal{S}$ into the MPC optimization problem.

4. Nonlinear Model Predictive Control of the Quadcopter

4.1. Closed-loop NMPC Scheme

The closed-loop NMPC scheme is shown in **Figure 1**. The constraint-satisfaction approach presented in Algorithm 1 is solved off-line and the updated constraints can be found. A path planner is usually required to generate a reference trajectory in order to let the quadcopter track it, where the reference trajectory contains not only the position vector $[x, y, z]^T$ but also the attitude of the quadcopter ψ. In this paper, the NMPC controller is divided into two layers: In the upper layer, a nonlinear optimization problem is proposed in order to control the quadcopter to track the given references generated by the path planner and subsequently the virtual inputs $\widetilde{\mathbf{u}}(k)$ can be obtained. In the lower layer, the actual input can be computed by solving a set of algebraic equations. Besides, when there are no enough sensors available in the quadcopter to feedback the values of the system states, an observer for the quadcopter is also required, for instance, the zonotopic extended Kalman filter proposed in [18]. Finally, a quadcopter simulator is used to test the effectiveness of the proposed NMPC controller.

4.2. The Upper Level: Nonlinear Optimization Problem

The MPC prediction model of the quadcopter is chosen by the nominal model of (10) without considering system disturbances and measurement noise. The constraints on system states $\mathbf{x}(k)$,

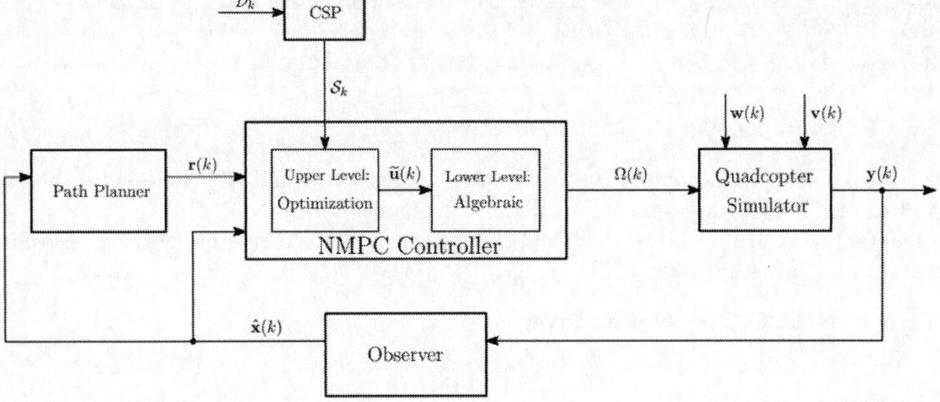

Figure 1. The closed-loop simulation scheme of the NMPC strategy with constraint satisfactions for the quadcopter

the virtual inputs $\widetilde{\mathbf{u}}(k)$ at time instant k can be denoted as

$$\mathbf{0} \leq \mathbf{x}(k+i \mid k) \leq \mathbf{x}_{\max}, \quad i = 1, 2, \ldots, H_p, \tag{11a}$$

$$\mathbf{0} \leq \widetilde{\mathbf{u}}(k+j \mid k) \leq \widetilde{\mathbf{u}}_{\max}, \quad j = 0, 1, \ldots, H_p - 1. \tag{11b}$$

The control objective of the quadcopter is mainly towards controlling the quadcopter to track the given position references and attitude references both generated by the path planner. Therefore, the cost function for the NMPC controller mainly includes two parts: a tracking part of the minimization of the deviations between x, y, z and ψ with their references x^r, y^r, z^r and ψ^r and a smoothness part with the minimization of the slew rate of the virtual inputs, which can be defined as

$$\begin{aligned} L(\mathbf{x}, \widetilde{\mathbf{u}}) \triangleq \sum_{i=1}^{H_p} \Big(& \lambda_x \ell_x \left(x_9(k+i \mid k), x^r(k+i)\right) \\ & + \lambda_y \ell_y \left(x_{11}(k+i \mid k), y^r(k+i)\right) \\ & + \lambda_z \ell_z \left(x_7(k+i \mid k), z^r(k+i)\right) \\ & + \lambda_\psi \ell_\psi \left(x_5(k+i \mid k), \psi^r(k+i)\right) \Big) \\ + \sum_{j=0}^{H_p-1} & \lambda_{\Delta u} \ell_{\Delta u} \left(\widetilde{\mathbf{u}}(k+j \mid k)\right), \end{aligned} \tag{12}$$

with

$$\ell_x \left(x_9(k+i \mid k)\right) = \left\| x_9(k+i \mid k) - x^r(k+i) \right\|_2, \tag{13a}$$

$$\ell_y \left(x_{11}(k+i \mid k)\right) = \left\| x_{11}(k+i \mid k) - y^r(k+i) \right\|_2, \tag{13b}$$

$$\ell_z \left(x_7(k+i \mid k)\right) = \left\| x_7(k+i \mid k) - z^r(k+i) \right\|_2, \tag{13c}$$

$$\ell_\psi \left(x_5(k+i \mid k)\right) = \left\| x_5(k+i \mid k) - \psi^r(k+i) \right\|_2, \tag{13d}$$

and

$$\ell_{\Delta u} \left(\widetilde{\mathbf{u}}(k+j \mid k)\right) = \left\| \Delta\widetilde{\mathbf{u}}(k+j \mid k) \right\|_2, \tag{14a}$$

$$\Delta\widetilde{\mathbf{u}}(k+j \mid k) = \widetilde{\mathbf{u}}(k+j \mid k) - \widetilde{\mathbf{u}}(k+j-1 \mid k), \tag{14b}$$

where λ_x, λ_y, λ_z, λ_ψ and $\lambda_{\Delta u}$ denote the weights for each objective and $\|\cdot\|_2$ denotes the 2-norm operator.

The selected observer, here the zonotopic extended Kalman filter in [18], can be written as

$$\hat{\mathbf{x}}(k) = \chi(\mathbf{y}(k)), \tag{15}$$

where $\hat{\mathbf{x}}(k)$ denotes the estimated full states at time instant k and $\chi(\cdot)$ represents the model-based observer function based on the model of the quadcopter in (2).

In general, the NMPC strategy in the upper layer can be implemented by solving the following nonlinear optimization problem $\mathbb{P}_q(k)$:

Problem 1 (NMPC-Quadcopter Problem)

$$\min_{\widetilde{\mathbf{u}}^*(k|k),\ldots,\widetilde{\mathbf{u}}^*(k+H_p-1|k)} L(\mathbf{x}, \widetilde{\mathbf{u}}), \tag{16a}$$

subject to

$$\begin{aligned}
&\mathbf{x}(k+i+1 \mid k) = F(\mathbf{x}(k+i \mid k), \widetilde{\mathbf{u}}(k \mid k)), && (16b)\\
&\mathbf{0} \leq \mathbf{x}(k+i \mid k) \leq \mathbf{x}_{\max}, && (16c)\\
&\mathbf{0} \leq \widetilde{\mathbf{u}}(k+i \mid k) \leq \widetilde{\mathbf{u}}_{\max}, && (16d)\\
&\mathbf{x}(k \mid k) = \hat{\mathbf{x}}(k), && (16e)\\
&\hat{\mathbf{x}}(k) = \chi(\mathbf{y}(k)). && (16f)
\end{aligned}$$

After solving the nonlinear optimization problem $\mathbb{P}_q(k)$ at time step k, a series of the virtual inputs $\widetilde{\mathbf{u}}^*(k \mid k)$, ... , $\widetilde{\mathbf{u}}^*(k+H_p-1 \mid k)$ can be obtained. By means of the receding horizon approach, the first value $\widetilde{\mathbf{u}}^*(k \mid k)$ is used as the optimal control action $\widetilde{\mathbf{u}}^*(k)$ at this time step.

4.3. The Lower Level: Algebraic Problem

As introduced in the previous subsection, the virtual input can be found by solving the nonlinear optimization problem. Then, the actual input can be computed after the virtual input obtained by means of the algebraic problem.

Depending on (3), the virtual input $\widetilde{\mathbf{u}}(k)$ and actual input $\mathbf{u}(k)$ has the obvious algebraic relationship, which can be written as follows:

$$\widetilde{\mathbf{u}}(k) = \begin{bmatrix} b & b & b & b \\ 0 & -b & 0 & b \\ b & 0 & -b & 0 \\ -d & d & -d & d \end{bmatrix} \mathbf{u}^2(k). \tag{17}$$

And then the actual input at time instant k can be computed by

$$\mathbf{u}(k) = \begin{bmatrix} b & b & b & b \\ 0 & -b & 0 & b \\ b & 0 & -b & 0 \\ -d & d & -d & d \end{bmatrix}^{-\frac{1}{2}} \widetilde{\mathbf{u}}^{\frac{1}{2}}(k). \tag{18}$$

The solution $\mathbf{u}(k)$ of (17) satisfies the nonlinear optimization problem proposed in the upper layer as well as (18). Therefore, the solution $\mathbf{u}(k)$ can be used for the actual input at time instant k.

IOP Conf. Series: Journal of Physics: Conf. Series **783** (2017) 012025 doi:10.1088/1742-6596/783/1/012025

5. Simulation Results

In this section, the proposed NMPC controller combined with constraint satisfactions is tested using continuous-time quadcopter simulator implemented in MATLAB. In order to evaluate the inherent robustness of the NMPC controller, the simulator model includes the uncertainties in (6), which are not explicitly considered in the prediction model in the NMPC controller.

5.1. Simulator Model

Taking into account the uncertainties discussed before, the mathematical model of the simulator can be formulated as follows:

$$\ddot{\phi} = \dot{\theta}\dot{\psi}a_1 + \dot{\theta}a_2\Omega_r + b_1U_2 + \omega_1, \tag{19a}$$

$$\ddot{\theta} = \dot{\phi}\dot{\psi}a_3 - \dot{\phi}a_4\Omega_r + b_2U_3 + \omega_2, \tag{19b}$$

$$\ddot{\psi} = \dot{\phi}\dot{\theta}a_5 + b_3U_4 + \omega_3, \tag{19c}$$

$$\ddot{z} = K_a^z\dot{z} - g + (\cos\phi\cos\theta)\,U_1/m, \tag{19d}$$

$$\ddot{x} = K_a^x\dot{x} + (\cos\phi\sin\theta\cos\psi + \sin\phi\sin\psi)\,U_1/m, \tag{19e}$$

$$\ddot{y} = K_a^y\dot{y} + (\cos\phi\sin\theta\sin\psi - \sin\phi\cos\psi)\,U_1/m, \tag{19f}$$

where $\omega_1, \omega_2, \omega_3 \in \mathbb{R}$ denote the system disturbances that can be regarded as the Gaussian white noise following some certain distributions, respectively. K_a^x, K_a^y and K_a^z denote the aerodynamic friction coefficients for the axes x, y and z, respectively. All the parameters of the quadcopter are obtained from [2].

The measurement outputs of the quadcopter simulator are selected as

$$\mathbf{y} = \begin{bmatrix} \dot{x} & \dot{y} & \dot{z} & \dot{\psi} \end{bmatrix}^T. \tag{20}$$

5.2. Results

The NMPC optimization problem is solved by using the nonlinear programming technique implemented through *YALMIP* toolbox [19] and *IPOPT* solver implemented in *OPTI* toolbox [20]. The constraint-satisfaction algorithm for the quadcopter control is implemented using the *Interval Peeler* software [21]. The continuous-time quadcopter simulator is built by means of the function of *ode* in MATLAB.

In this work, the reference trajectory is assumed to be available generated by a path planner. The weights for each objective are set as $\lambda_x = 10$, $\lambda_y = 10$, $\lambda_z = 10$, $\lambda_\psi = 10$ and $\lambda_{\Delta u} = 0.1$. The sampling time is chosen as $0.1s$. The NMPC prediction horizon H_p is chosen equal to 10.

Besides, three scenarios are provided in the simulations defined as follows:

- *Deterministic scenario*:

$$\omega_1 = \omega_2 = \omega_3 = 0,$$
$$K_a^z = K_a^x = K_a^y = 0,$$

- *Resistant scenario*:

$$\omega_1 = \omega_2 = \omega_3 = 0,$$
$$K_a^z = 4.36, K_a^x = 2.91, K_a^y = 2.91,$$

- *Noisy scenario*:

$$\omega_1, \omega_2, \omega_3 \sim \mathcal{N}(0, 0.05),$$
$$K_a^z = K_a^x = K_a^y = 0,$$

where $\mathcal{N}(0, 0.05)$ denotes the Gaussian distribution with the zero mean and the variance of 0.05.

IOP Conf. Series: Journal of Physics: Conf. Series **783** (2017) 012025 doi:10.1088/1742-6596/783/1/012025

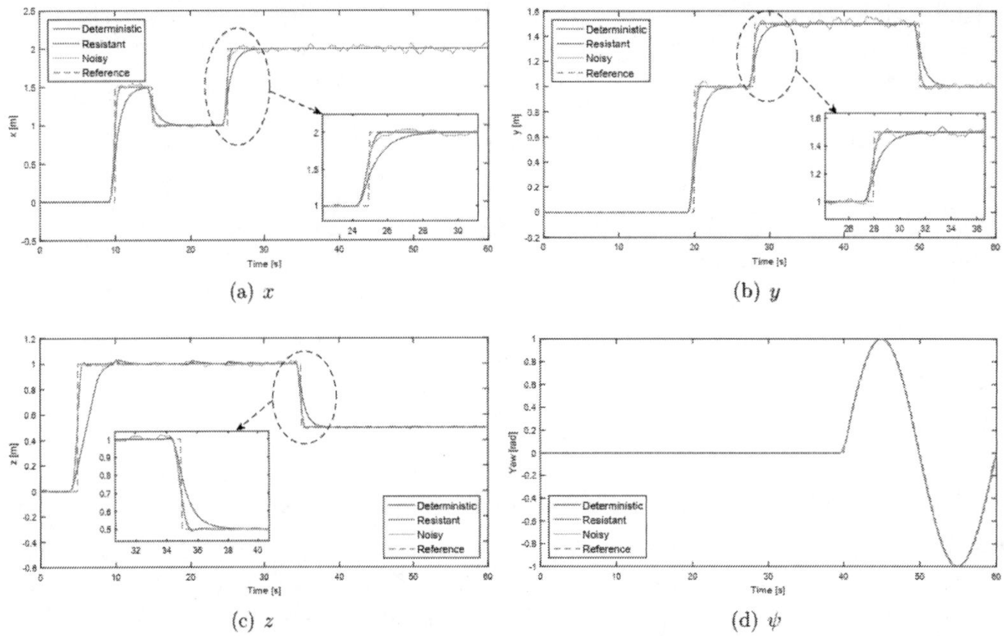

(a) x (b) y (c) z (d) ψ

Figure 2. Simulation results of x, y, z and ψ for three scenarios

The selected simulation results with the three scenarios are shown in **Figure 2** and **Figure 3**. Four references for x, y, z and ψ are generated by the path planner in red dashed lines. It can be seen that the control objective of tracking the given references is well achieved for all the scenarios. The deterministic scenario presents the case that the simulator model is the same as the prediction model in the controller. In this case, the real trajectories on each axis and angle can follow the given references with a fast response, which means that the MPC is able to find feasible solutions with the mathematical model of the quadcopter. For the resistant scenario, the continuous resistant forces are included in order to simulate the quadcopter is operated in an environment with the effect of the wind. Consequently, as shown in the simulation results, the real trajectories of the quadcopter have some delays to reach the references compared to the deterministic scenario. In terms of the noisy scenario, additive noise following a Gaussian distribution with zero mean and variance of 0.05 is assumed, which is mainly generated from sensors located in the quadcopter. Therefore, there exist small deviations around the given the references on each axis. Besides, four actual control inputs Ω_1, Ω_2, Ω_3 and Ω_4 are plots. For the deterministic and resistant cases, these inputs are quite smooth compared to the noisy case.

The real trajectory in 3-dimensional space ($[x, y, z]^T$) is plotted in **Figure 4**. The quadcopter is controlled to do a set of motions including taking off, moving along x, y and z axis and turning the head of the quadcopter through changing ψ. As shown in **Figure 4**, the quadcopter is able to reach the end point from a given start point for the deterministic and resistant cases. For the noisy scenario, the quadcopter arrives at an area around the target place (the end point) since there are small deviations for the each axis. Hence, the tracking accuracy for the quadcopter full-state control is quite significant. As a result, the quadcopter controlled by the NMPC controller is able to track the given references.

In reality, the quadcopter could suffer some unexpected disturbances at a certain time. Hence, it is necessary to test the behavior of the designed NMPC controller under these disturbances. From **Figure 5**, two peak disturbances with the same magnitude of 0.5 in $\dot{z}$ around two different

IOP Conf. Series: Journal of Physics: Conf. Series **783** (2017) 012025 doi:10.1088/1742-6596/783/1/012025

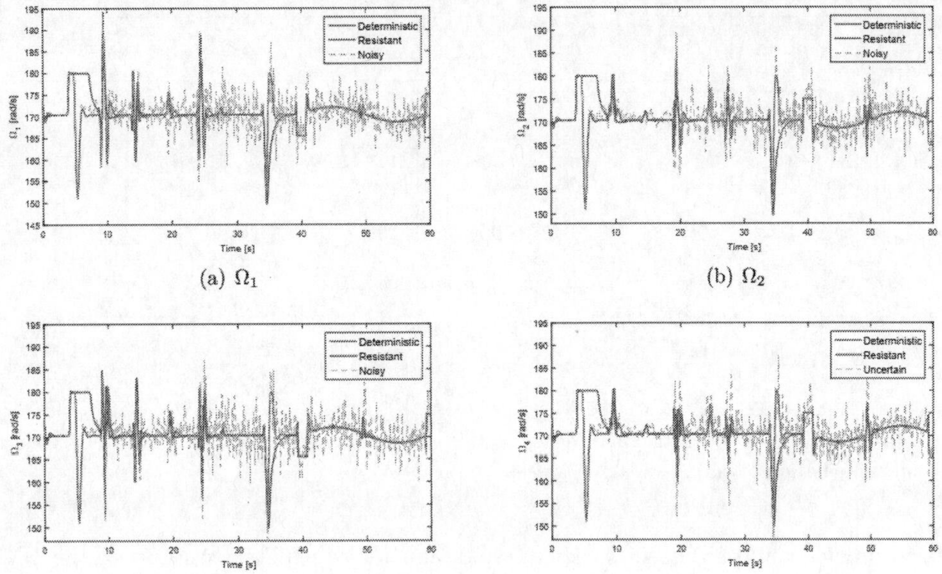

(a) Ω_1 (b) Ω_2

(c) Ω_3 (d) Ω_4

Figure 3. Simulation results of Ω_1, Ω_2, Ω_3 and Ω_4 for three scenarios

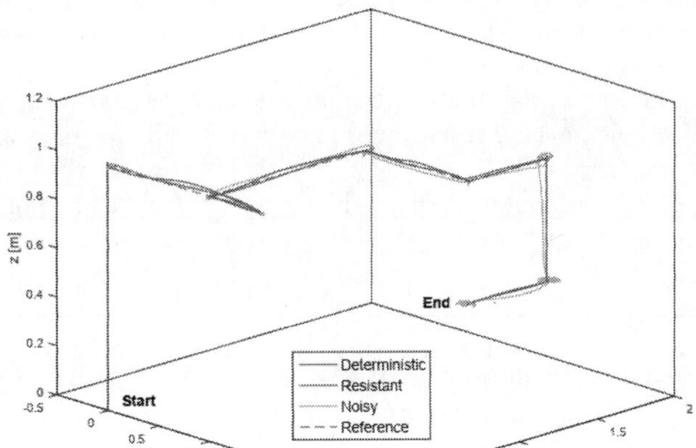

Figure 4. Real trajectory and reference for three scenarios

directions in z axis are tested. The NMPC controller is able to deal with these disturbances within some transient periods. In particular, the tracking objective can be guaranteed by using the proposed NMPC strategy as well.

6. Conclusions

In this paper, an NMPC strategy combined with constraint satisfactions is studied for a quadcopter to track the given references. From the simulation results, the NMPC controller

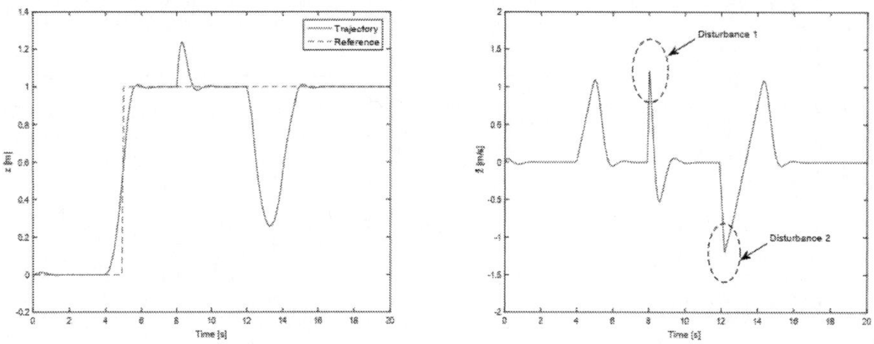

(a) z (b) $\dot{z}$

Figure 5. Result of adding disturbances: k=80, a disturbance towards positive z axis and k=120, a disturbance towards negative z axis

is able to provide quite accurate tracking results and the inherent robustness of the NMPC controller is able to deal with uncertainties.

On the other hand, the nonlinear optimization problem is not easy to be solved compared with the linear one requiring more time to find a solution. Taking into account the continuous hardware and software development in the computer science field, the nonlinear programming technique is expected to find the solution that satisfies the required real-time constraints allowing the real implementation. Thus, as future research, an important direction is to improve nonlinear optimization algorithm by some possible ways, such as properly relaxing the nonlinear constraints and updating the initial guess of the nonlinear solver. Besides, the MPC strategy is normally considered as a flexible control strategy and a quite good framework for fault-tolerant control because the system reconfiguration after faults appeared in the control system is easy to implement in the MPC constraint settings. Hence, as another future work, the fault-tolerant capability of the proposed control strategy will be investigated.

Acknowledgement

This work has been partially funded by the Spanish Government (MINECO) through the project CICYT ECOCIS (ref. DPI2013-48243-C2-1-R), by MINECO and FEDER through the project CICYT HARCRICS (ref. DPI2014-58104-R), by the Natural Science Foundation of Guangdong (No.2015A030313881), by the National Natural Science Foundation of China (No.61673239) and the Shenzhen Key Lab of Space Robotic Technology and Telescience.

References

[1] Alexhs K, Nikolakopoulos G and Tzes A 2011 *International Review of Automatic Control* **4**
[2] Alexis K, Nikolakopoulos G and Tzes A 2012 *IET Control Theory Applications* **6** 1812 – 1827
[3] Valavanis K and Vachtsevanos G 2015 *Handbook of Unmanned Aerial Vehicles* (Springer)
[4] Garciá R, Rubio F and Ortega M 2012 Robust PID control of the quadrotor helicopter *IFAC Conference on Advances in PID Control* (Brescia, Italy)
[5] Lippiello V, Ruggiero F and Serra D 2014 Emergency landing for a quadrotor in case of a propeller failure: A PID based approach *IEEE International Symposium on Safety, Security, and Rescue Robotics (SSRR)* (Montreal, Canada) pp 1 – 7
[6] Sadeghzadeh I, Mehta A, Chamseddine A and Zhang Y 2012 Active fault tolerant control of a quadrotor UAV based on gainscheduled PID control *25th IEEE Canadian Conference on Electrical Computer Engineering (CCECE)* (Hokkaido, Japan) pp 1 – 4

[7] Benallegue A, Mokhtari A and Fridman L 2006 Feedback linearization and high order sliding mode observer for a quadrotor UAV *International Workshop on Variable Structure Systems (VSS)* (Alghero, Italy) pp 365 – 372
[8] Qingtong W, Honglin W, Qingxian W and Mou C 2015 Backstepping-based attitude control for a quadrotor UAV using nonlinear disturbance observer *34th Chinese Control Conference (CCC)* (Hangzhou, China) pp 771 – 776
[9] Rashad R, Aboudonia A and El-Badawy A 2015 Backstepping trajectory tracking control of a quadrotor with disturbance rejection *XXV International Conference on Information, Communication and Automation Technologies (ICAT)* (Sarajevo, Bosnia and Herzegovina) pp 1 – 7
[10] Shulong Z, Honglei A, Daibing Z and Lincheng S 2014 A new feedback linearization LQR control for attitude of quadrotor *13th International Conference on Control Automation Robotics Vision (ICARCV)* (Singapore) pp 1593 – 1597
[11] Bavili R E, Khosrowjerdi M J and Vatankhah R 2015 *Control Engineering and Applied Informatics* **17** 68 – 76
[12] Deori L, Garatti S and Prandini M 2015 A model predictive control approach to aircraft motion control *American Control Conference (ACC)* (Chicago, USA) pp 2299 – 2304
[13] Drak A, Noura H, Hejase M and Younes Y A 2015 Sensor fault diagnostic and fault-tolerant control for the altitude control of a quadrotor UAV *8th IEEE GCC Conference and Exhibition (GCCCE)* pp 1 – 5
[14] López-Estrada F, Ponsart J, Theilliol D, Zhang Y and Astorga-Zaragoza C 2015 *Journal of Intelligent & Robotic Systems* 1 – 15
[15] Derafa L, Madani T and Benallegue A 2006 Dynamic modelling and experimental identification of four rotors helicopter parameters *IEEE International Conference on Industrial Technology (ICIT)* (Mumbai, India) pp 1834 – 1839
[16] Jaulin L, Kieffer M, Didrit O and Walter E 2001 *Applied Interval Analysis, with Examples in Parameter and State Estimation, Robust Control and Robotics* (Springer)
[17] Jaulin L 2012 *Computing* **94** 297 – 311
[18] Wang Y and Puig V 2016 Zonotopic extended Kalman filter and fault detection of discrete-time nonlinear systems applied to a quadrotor helicopter *3rd International Conference on Control and Fault-Tolerant Systems (SysTol)* (Barcelona, Spain) pp 361 – 366
[19] Löfberg J 2004 YALMIP: A toolbox for modeling and optimization in MATLAB URL `http://users.isy.liu.se/johanl/yalmip`
[20] Currie J and Wilson D 2012 *OPTI: Lowering the Barrier Between Open Source Optimizers and the Industrial MATLAB User* URL `http://www.i2c2.aut.ac.nz/Wiki/OPTI`
[21] Baguenard X 2005 *Interval Peeler Solver* URL `https://www.ensta-bretagne.fr/jaulin/demo.html`

Model Predictive Control of Sewer Networks

Einar B. Pedersen[2], Hannes R. Herbertsson[2], Henrik Niemann[1,2], Niels K. Poulsen[3] and Anne K.V. Falk[4]

[2]Dept. of Electrical Engineering and [3]Dept. of Applied Math. and Compures Science Technical University of Denmark, DK 2800 Kgs. Lyngby, Denmark
[4]DHI, Agern Alle 5, DK-2970 Hørsholm, Denmark

E-mail: einarbp@mazterinn.com, hannesh08@gmail.com, hhn@elektro.dtu.dk, nkpo@dtu.dk, akf@dhigroup.com

Abstract. The developments in solutions for management of urban drainage are of vital importance, as the amount of sewer water from urban areas continues to increase due to the increase of the world's population and the change in the climate conditions. How a sewer network is structured, monitored and controlled have thus become essential factors for efficient performance of waste water treatment plants.

This paper examines methods for simplified modelling and controlling a sewer network. A practical approach to the problem is used by analysing simplified design model, which is based on the *Barcelona* benchmark model. Due to the inherent constraints the applied approach is based on *Model Predictive Control*.

Keywords: *Sewer network, modelling, model predictive control, simulations.*

1. Introduction

Controlling sewer networks is a demanding and necessary task in modern society, that prevents property damages, accidents and more importantly reduces the risk of environmental pollution. Therefore, it is important to manage the sewage water to wastewater treatment plants (WWTPs) as efficiently as possible. Inadequate or no treatment of sewage water can have a major impact on biodiversity in rivers, lakes and the sea, where harmful bacteria and toxic chemicals that are carried by the sewage water, can cause serious health problems and even lead to more deadly epidemic outbreaks such as malaria. Many cities have combined sewer systems and the need for efficient control of sewer networks is going to increase, especially according to some predictions stating that by the year 2050, 80% of the world's population will be living in urban areas [14].

Sewer networks are considered complex systems, as they are geographically distributed and structured for the collection of *sewage* and *run-off* water. Usually, sewer systems are divided into two categories: *separate* or *combined*. The *separate* sewer system uses different pipelines for sewage and run-off water, while in the *combined* sewer system, sewage and run-off water are mixed together in the same pipeline. Whereas, the separate sewer system directs sewage

[1] Coorsponding author: Henrik Niemann

water to a WWTP, the run-off water is released into the environment, treated or untreated, then the combined sewer system, sewage and run-off water are both transported to a WWTP. As sewage water is considered to be more contaminated than run-off water, consequences become more severe when an overflow situation occurs in the combined sewer system, as the risk of contaminated water escaping into the environment is higher. *Rainstorms* are considered the main influencing factor that cause combined sewer systems to overflow. Therefore, minimizing overflow situations in combined sewer system is a more demanding procedure than for the separate system, making the overall control problem more difficult to manage [16]. In general, waste water is categorized as originating from households, industry and sewage, while run-off water arise when rainfalls occurs. The main purpose of WWTPs, is to purify as much contaminated water as possible, before releasing it back into the environment. Thus, the number of treatment plants needed for a specific sewer network is dependent on the quantity of waste water in the plants' districts and their maximum inflow capacity. This issue is most commonly addressed by controlling flow rates of sewer pipes to maximize treatment plant performance. Furthermore, pipelines are limited by their dimensions and therefore flow quantities need to be controlled within boundary limits of the pipes, by manipulating the flow rates in the pipes.

The focus in this paper is to study the Barcelona sewer network system with respect to control it. A benchmark model of the Barcelona sewer network has been developed by Ocampo-Martnez [9], [11]. Based on this benchmark model, the system is controlled by using Model Predictive Control (MPC). The main reason for choosing the MPC method as the control strategy for the sewer network, relies on its unique ability to include boundary constraints in the optimization procedures for MIMO systems. Further, it is also possible to include prediction of the water inflow from rainfall in the control optimization. The control objectives for system performance are:

- Minimize flooding to streets.
- Maximize the waste water treatment plant/plants (less pollution to the environment)

For satisfying the the above performance conditions, a MPC problem is formulated with both hard and soft constraints. Hard constraints are used for flows in pipes and levels in the tanks. Soft constraints are used for flooding in streets and flooding to the sea. Simulation with both know and unknown inflow from rainfall is shown in this paper. (The results presented in this paper is based on the work in [13].)

The rest of the paper is organized as follows. In Section 2, the modelling of the single elements in a sewer network is presented. A sewer network model is given in Section 3. The model is based on the virtual tank method. Section 4 show a number of simulation results followed by a conclusion in Section 5.

2. System modelling

Modelling methods are analyzed in this section, in order to find a simple sewer network approach that is applicable to the objectives in this paper. The literature and research material presented in this section is mostly based on Marinaki & Papageorgiou [8], Martnez [9], [11] and in Duran [4]. A suitable modelling method will be selected, which will be used for ongoing analysis.

2.1. The Saint Venant Equations

The most general approach when modelling a sewer network, is to use the *Saint Venant nonlinear partial differential equations*. Assuming that the movement of water is one dimensional, the water flow can be represented as an open channel where the distribution of pressure is *hydrostatic*, the sewer pipes' cross-sectional areas are constant with a uniform velocity over the cross-section, the flow is incompressible and the average slope in the channel is small, [3].

The general Saint Venant equations are given as:

$$\frac{\partial q(x,t)}{\partial x} + \frac{\partial A(x,t)}{\partial t} = 0 \tag{1}$$

$$\underbrace{\frac{\partial q(x,t)}{\partial t}}_{1} + \underbrace{\frac{\partial}{\partial x}\frac{q(x,t)^2}{A(x,t)}}_{2} + \underbrace{g\ A(x,t)\frac{\partial h(x,t)}{\partial x}}_{3} - \underbrace{g\ A(x,t)\ (I_0(x) - I_f(x,t))}_{4} = 0 \tag{2}$$

Eq. (1) expresses mass conservation equation. Moreover, the numbered terms in the momentum equation given in eq. (2), describe the following physical quantities, [12]: 1 - Inertia, 2 - Advection, 3 - Gravitational force and 4 - Friction force. The other symbols in eqs. (1) and (2) are as follows: $q(x,t)$ is the flow, $A(x,t)$ is the cross-sectional area of the pipe, t is the time, x is the spatial variable measured in the direction of the sewage flow, g is the gravitational acceleration, $I_0(x)$ is the sewer pipe slope and $I_f(x,t)$ is the friction slope.

Modelling methods for simplified models of sewer networks/channels that include the theoretical basis of the Saint Venant equations are introduced and examined in this section. There exist a number of different model types. The must used models are: The Virtual Tank Model, [8], that are used in this paper, The Hybrid Linear delayed Model described in [4], The Muskingum Model described in [4, 1], The Integrator Delay Model described in [10] and The Integrator Delay Model with Zero introduced in [6, 5].

2.2. Sewer Network Components

The main elements in a sewer network is introduced in the following.

Virtual Tanks The virtual tanks are often referred to as *catchments*, where volume quantities from numerous pipelines are collected together in certain areas. This simplification reduces the amount of system variables significantly. The schematic of a virtual tank component is provided in Figure 1, showing the characteristics and associated parameters.

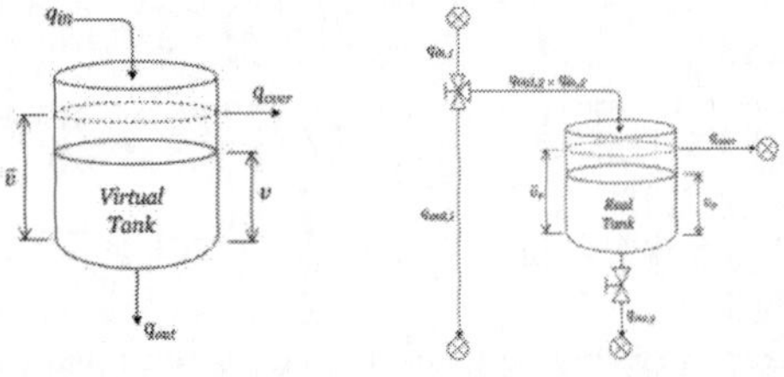

Figure 1. Schematic of virtual tank component (left) and a real tank component (right).

Discretization methods are applied on the Saint Venant equations to obtain a discrete time linear expression for the stored volume in the virtual tanks, [4]. The discrete time difference equation of the virtual tank is presented in eq. (3).

$$v(k+1) = v(k) + T_s(q_{in}(k) - q_{out}(k) - q_{over}(k)) \tag{3}$$

The stored volume is denoted as v, the inflow given as q_{in}, the outflow as $q_{out,}$, the overflow as q_{over} and the sampling time as T_s. The inflow to the virtual tank is calculated as:

$$q_{in,j}(k) = q_{in,1}(k) + q_{in,2}(k) +, \cdots, + q_{in,n}(k), \quad \text{for } j = 1,\ 2, \ldots, n \tag{4}$$

The outflow from the virtual tank is then expressed as:

$$q_{out}(k) = \beta v(k) \tag{5}$$

The outflow equation can also be obtained for the nonlinear system, where the relation between them is: $q_{out}(k) = \bar{\beta}\sqrt{v(k)}$.

Real Tanks Real tanks are *actual water containers*, which serve an important role within a sewer network, where they can be used as buffers to produce a slack behavior within the system. The inflow and outflow of the tanks can be controlled in order to reduce the risk of overflow and to maximize the performance of a WWTP. A schematic of the real tank and associated components, are presented in Figure 1. The discrete time difference equation that describes the volume conservation for the real tank, is presented in eq. (6),

$$v_r(k+1) = v_r(k) + T_s(q_{in,2}(k) - q_{out,3}(k) - q_{over}(k)) \tag{6}$$

where the inflow, $q_{in,2}$, is controlled by using a redirection gate, while the retention gate is used to control the outflow, $q_{out,3}$.

Rainfall A rainfall is regarded as the most influential factor that affects a sewer network, where the estimated run-off water from rainfalls are assumed to be a disturbance factor that enters the system, as presented in eq. (7).

$$d(k) = S\varphi I_t(k) \tag{7}$$

where S, denotes the surface area of the catchment, φ, the ground absorption coefficient and I_t, the rain intensity. The absorption factor needs to be derived for each catchment due to the rainfall water that is absorbed by the ground and is based on a real data measurements, further explained in [15].

Gates Two types of gates are introduced in this section, referred to as the *redirection* and *retention* gates.

Redirection Gates is used to control the inflow rates to connected pipes by changing the direction of the flow through the gate, as eq. (8) illustrates.

$$q_{in}(k) = q_{out,1}(k) + q_{out,2}(k) \tag{8}$$

$$\underline{q}_{out,i} \leq q_{out,i}(k) \leq \bar{q}_{out,i} \quad \text{for } i = 1, 2 \tag{9}$$

Notice that redirection gates can be implemented as binary logic gates or for continuous time. The continous implementation is used in this paper.

Retention Gates are used to control tank outflows and are controlled by using local controllers. The benefit of using retention gates is maintaining wastewater flow at a certain point in the sewer network, as eqs. (10) and (11) show.

$$q_{in}(k) = q_{out}(k) \tag{10}$$

$$\underline{q}_{out} \leq q_{out}(k) \leq \bar{q}_{out} \tag{11}$$

3. Sewer network model

A nine Tank Catchment model is introduced. The augmented model will consist of eight virtual tanks, one real tank and two waste water treatment plants, WWTP 1 and WWTP 2. It will be referred to as the 9TC model. A schematic of the 9TC model is provided in Figure 2. The structure of the model can be related to the *Barcelona model* from Martnez [9].

The recursion of the tanks' geometric linear equations are provided in eqs. (12), (13) and (14).

$$\begin{aligned}
T_{1,V1}(k+1) &= T_{1,V1}(k) + T_s(d_1(k) - q_{out,1}(k))\\
T_{2,V2}(k+1) &= T_{2,V2}(k) + T_s(u_1(k) + d_2(k) - q_{out,2}(k))\\
T_{3,R}(k+1) &= T_{3,R}(k) + T_s(u_2(k) - u_3(k))\\
T_{4,V3}(k+1) &= T_{4,V3}(k) + T_s(u_3(k) + d_3(k) + q_{13}(k) + q_{23}(k) - q_{out,4}(k))\\
T_{5,V4}(k+1) &= T_{5,V4}(k) + T_s(d_4 - q_{out,5}(k))\\
T_{6,V5}(k+1) &= T_{6,V5}(k) + T_s(u_4(k) + d_5(k) - q_{out,6}(k))\\
T_{7,V6}(k+1) &= T_{7,V6}(k) + T_s(d_6(k) + q_{in,6}(k) - q_{out,7}(k))\\
T_{8,V7}(k+1) &= T_{8,V7}(k) + T_s(d_7(k) + q_{out,6}(k) - q_{out,8}(k))\\
T_{9,V8}(k+1) &= T_{9,V8}(k) + T_s(d_8(k) + q_{out,7}(k) - q_{out,9}(k))
\end{aligned} \tag{12}$$

$$q_{13}(k) = q_{out,1}(k) - u_1(k) \tag{13}$$

$$q_{in,6}(k) = q_{out,4}(k) + q_{out,5}(k) - u_4(k) \tag{14}$$

A state space model for the 9TC system is given by eq. (15), where the obtained 9TC model matrices $\mathbf{A}_e$, $\mathbf{B}_e$, $\mathbf{E}_{x,e}$ and $\mathbf{C}_e$ can be found in [13].

$$\begin{aligned}
\mathbf{x}_e(k+1) &= \mathbf{A}_e\mathbf{x}_e(k) + \mathbf{B}_e\mathbf{u}_e(k) + \mathbf{E}_{x,e}\mathbf{d}_e(k)\\
\mathbf{y}_e(k) &= \mathbf{C}_e\mathbf{x}_e(k)
\end{aligned} \tag{15}$$

where

$$\begin{aligned}
\mathbf{x}_e &= \begin{bmatrix} T_{1,V1} & T_{2,V2} & T_{3,R} & T_{4,V3} & T_{5,V4} & T_{6,V5} & T_{7,V6} & T_{8,V7} & T_{9,V8} \end{bmatrix}^T\\
\mathbf{u}_e &= \begin{bmatrix} u_1 & u_2 & u_3 & u_4 \end{bmatrix}^T\\
\mathbf{d}_e &= \begin{bmatrix} d_1 & d_2 & d_3 & d_4 & d_5 & d_6 & d_7 & d_8 \end{bmatrix}^T
\end{aligned} \tag{16}$$

Further, the system parameters can also be found in [13]. Additionally, the sampling time, $T_s =$ 300 sec is used.

4. Simulation Results

First, the MPC setup is shortly described followed by the a number of simulation results for the 9TC model. The simulation is done both when the disturbance is know and when the disturbance is uncertain.

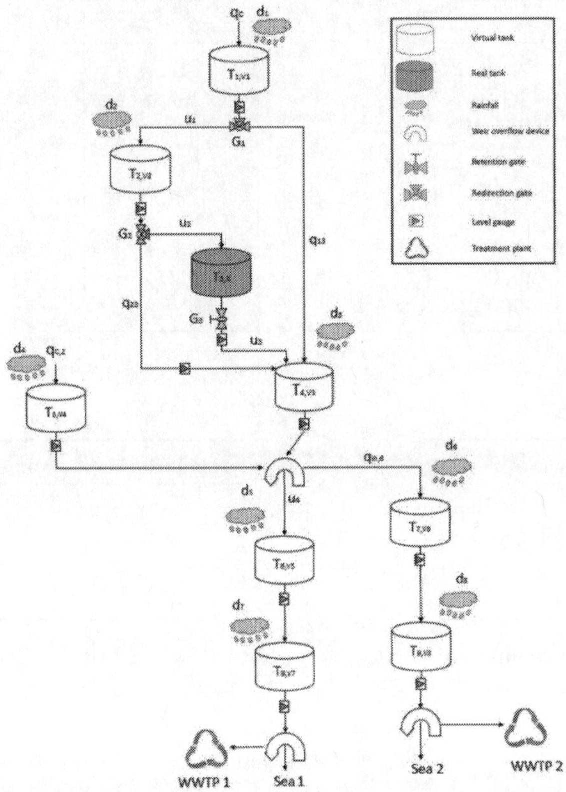

0.35

Figure 2. Schematic of the augmented sewer network model (9TC).

4.1. Model predictive control

The applied controller is a MPC (see eg [2] or [7]) with hard constraints on the inputs and input rates. The outputs are limited using soft constraints. The cost function is a quadratic function in the output derivations from a reference. The cost function also contain a quadratic term penalizing control actions. The constraints are formulated as linear inequalities. The numerical values of the boundary limits of the model (i.e. system volume levels, outputs, control signals, flow rates and pipes) are given in Table 1 through Table 5, respectively.
The linear inequality constraint equations for control signal u_1, u_2, u_3 and u_4 becomes:

$$u_{\min,1} \leq q_{out,1}(k) - \overline{q}_{13} \leq u_1(k) \leq q_{out,1}(k) - \underline{q}_{13} \leq u_{\max,1}$$
$$u_{\min,2} \leq q_{out,2}(k) - \overline{q}_{23} \leq u_2(k) \leq q_{out,2}(k) - \underline{q}_{23} \leq u_{\max,2}$$
$$u_{\min,3} \leq u_3(k) \leq q_{out,3}(k) \leq u_{\max,3}$$
$$u_{\min,4} \leq q_{out,5}(k) + q_{out,6}(k) - \overline{q}_{in,6} \leq u_4(k) \leq q_{out,5}(k) + q_{out,6}(k) - \underline{q}_{in,6} \leq u_{\max,4} \quad (17)$$

4.2. Results

The 9TC model is simulated for different intensity of rainfalls, using normal ISOC (input, input rate and soft output constrained) MPC as control strategy. Pulse signals imitating moderate,

$\mathbf{z}$ $[m^3]$	$\mathbf{z}_{\min}$	$\mathbf{z}_{\max}$
z_1	0	19,718
z_2	0	24,138
z_3	0	40,000
z_4	0	20,000
z_5	0	61,538
z_6	0	31,111
z_7	0	27,778
z_8	0	80,000
z_9	0	90,000

Table 1. Numerical values of output constraints.

$\mathbf{q}_{out}$ $[m^3/sec]$	$\mathbf{q}_{out,\min}$	$\mathbf{q}_{out,\max}$
$q_{out,1}$	0	14
$q_{out,2}$	0	14
$q_{out,3}$	0	8
$q_{out,4}$	0	20
$q_{out,5}$	0	8
$q_{out,6}$	0	14
$q_{out,7}$	0	15
$q_{out,8}$	0	28
$q_{out,9}$	0	45

Table 2. Numerical values of pipes outflow constraints.

$\mathbf{u}$ $[m^3/sec]$	$\mathbf{u}_{\min}$	$\mathbf{u}_{\max}$
u_1	0	12
u_2	0	14
u_3	0	8
u_4	0	14

Table 3. Numerical values of input constraints on control signals in $\mathbf{u}$.

$\Delta\mathbf{u}$ $[m^3/sec]$	$\Delta\mathbf{u}_{\min}$	$\Delta\mathbf{u}_{\max}$
Δu_1	-6	6
Δu_2	-6	6
Δu_3	-5	5
Δu_4	-10	10

Table 4. Numerical values of input rate constraints on control signal rates, $\Delta\mathbf{u}$.

$\mathbf{q}$ $[m^3/sec]$	$\underline{q}$	$\overline{q}$
q_{13}	0	5
q_{23}	0	8
$q_{in,6}$	0	15
$q_{\text{WWTP},1}$	0	7.5
$q_{\text{WWTP},2}$	0	9

Table 5. Numerical values of constraints for pipes.

heavy and violent rainfall episodes are simulated for separate simulations for rain areas d_1, d_2, d_5, d_6 and d_8, for different time intervals, while no rain occurs for other rainfall areas. The simulation time for all closed-loop simulations is 20 hours (1,200 minutes) and time intervals of pulse signals and corresponding magnitudes, are listed in Table 6. Further, the time intervals for simulations are chosen so that overflow situations are avoided in order to analyze the system responses, and controller performance.

Additionally, the system responses from simulations are shown in Figures 3, 4 and 5, where corresponding signals from all three simulations are plotted together in each figure.

A resemblance of the system responses from the three simulations are noticed, as the rainfall intensities rises. From Figure 4 it is seen that in the time intervals where the external inflows d_5, d_6 and d_8 to tanks $T_{6,V5}$, $T_{7,V6}$ and $T_{9,V8}$, are active, corresponding tanks' volume levels start to rise. The advantage of including a real tank in a sewer network, is lowering the chances of overflow situations, and maximizing the WWTP capacity. The controller responds to the disturbance by increasing the inflow, u_2, to the real tank, $T_{3,R}$, while the outflow, u_3, from the real tank is decreased, as further observed in Figure 5. Therefore, the volume level in the real

Dist.	Time int. [min]	Moderate [m^3/sec]	Heavy [m^3/sec]	Violent [m^3/sec]
d_1	$445 \leq t \leq 545$	0.6741	2.6965	5.3929
d_2	$395 \leq t \leq 495$	0.3434	1.3739	2.7478
d_5	$295 \leq t \leq 345$	1.0206	4.0824	8.1649
d_6	$195 \leq t \leq 245$	1.9280	7.7119	15.424
d_8	$95 \leq t \leq 145$	3.9856	15.942	31.884

Table 6. Time intervals for disturbance signals and numerical values of magnitudes for moderate, heavy and violent rainfalls.

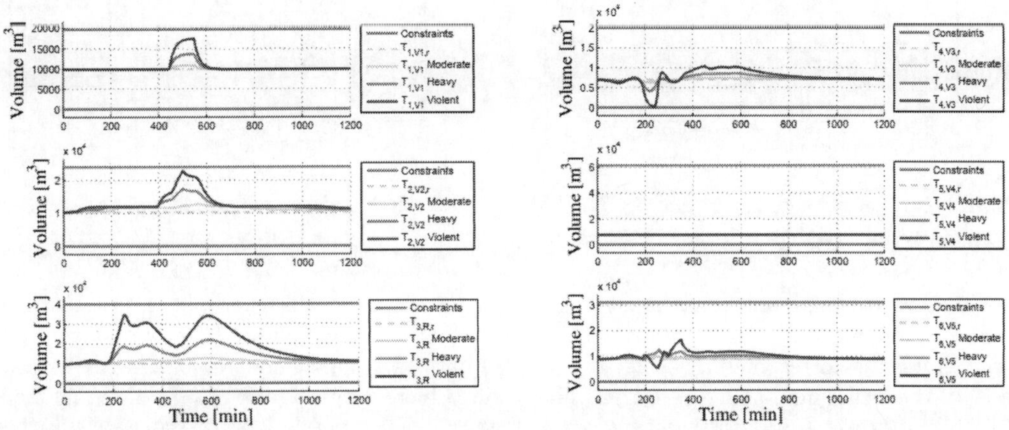

Figure 3. Volume in the tanks $T_{1,V1}$, $T_{2,V2}$ and $T_{3,R}$ (left) and the volume in the tanks $T_{4,V3}$, $T_{5,V4}$ and $T_{6,V5}$ (right).

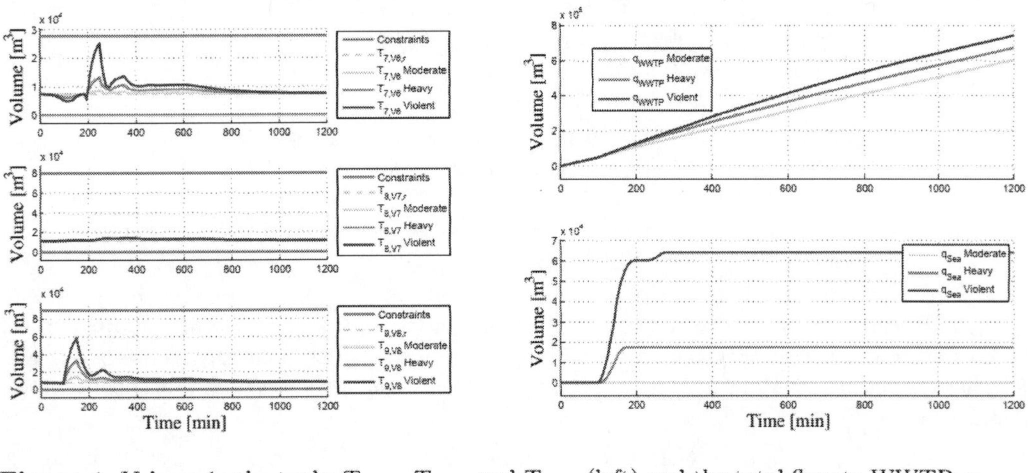

Figure 4. Volume in the tanks $T_{7,V6}$, $T_{8,V7}$ and $T_{9,V8}$ (left) and the total flow to WWTP, q_{wwtp}, and sea, q_{sea} accumulated volumes (right).

tank starts to rise, as seen in Figure 3. Notice as well, that the volume level in tank, $T_{4,V3}$, starts to decrease around minute 180, and control signal, u_4, to increase, thus increasing wastewater flow directed towards treatment plant WWTP 1.

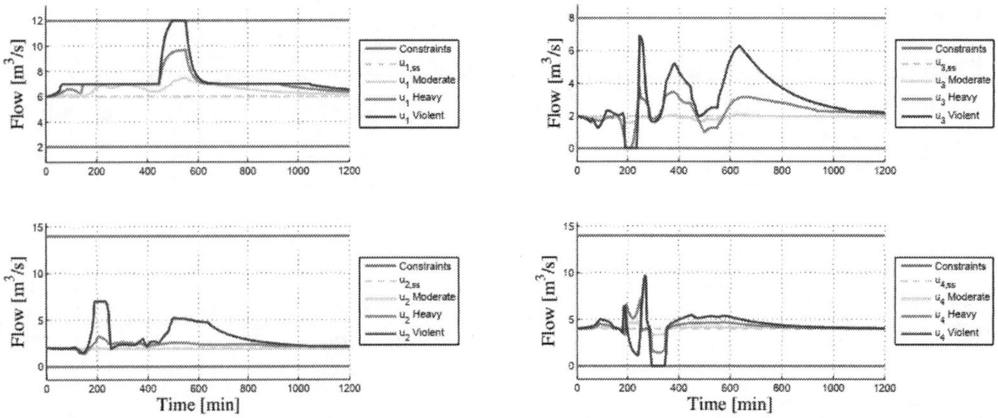

Figure 5. Control signals, u_1 and u_2, the 9TC system (left) and the control signals, u_3 and u_4, the 9TC system (right).

Furthermore, when system behavior is evaluated for the time intervals where the external inflows, d_1 and d_2, are active, it can be seen from Figure 3, that the volume levels rises within the time intervals for the corresponding tanks. It is as well noticed around minute 500, that the volume level in the real tank, $T_{3,R}$, decreases while slightly increasing in tank, $T_{4,V3}$, to counter for the disturbances entering the system, seen in Figure 3. The volume level in the real tank then increases again around minute 600, and when the rainfall episodes have passed, the tanks and pipes settle slowly to their steady state values.

	Moderate [m^3]	Heavy [m^3]	Violent [m^3]
Total Inflow	605,293	693,174	810,348
Sea area 1	0	0	0
Sea area 2	0	17,578	64,144
Sea total	0	17,578	64,144
WWTP 1	295,118	315,518	334,519
WWTP 2	309,412	358,151	408,742
WWTP total	604,530	673,669	743,261

Table 7. Total volume of inflow, and total volume directed to WWTPs and sea.

In Figure 4, the accumulated volume of wastewater that is treated and directed to sea, is compared. For the moderate rain episode, no change is noticed in the graph, where approximately 99% of the wastewater is treated, as seen from the numerical data presented in Tables 7 and 8. Thus, the system performance is considered sufficient, where the result is based on a known disturbance.

Further, for the heavy rain episode, it is clear from Figure 4, that the weir overflow device at the second terminal of the sewer network, is overflowing and wastewater is contaminating the sea. This is considered a small quantity (since the volume of wastewater is only about 2.5% of the

	Moderate [%]	Heavy [%]	Violent [%]
Sea area 1	0	0	0
Sea area 2	0	2.54	7.92
Sea total	0	2.54	7.92
WWTP 1	48.76	45.52	41.28
WWTP 2	51.12	51.67	50.44
WWTP total	99.87	97.19	91.72

Table 8. The difference between corresponding volume parameter, and the difference of the total inflow and overflow volume.

total inflow) and compared to the treated water (which is approximately 97%), the same can be concluded for the system performance during heavy rain.

For the violent rain episode, an even larger volume change is noticed for the quantity of wastewater that flows to sea (only around 8% of the total inflow). Notice, if the difference in total volume directed to sea are compared between simulations of heavy and violent rain episodes, it is approximately 5.5%, since there is a large difference between corresponding magnitudes of disturbances rain consequently resulting in a large difference in total inflows, as seen in Table 6. When taking these factors into consideration, the increase in contaminated water flowing to the sea, is not considered serious since the quantity of treated wastewater is around 92%, as seen in Table 8.

Subsequently, the distribution of wastewater directed to the two treatment plants, WWTP 1 and WWTP 2, is shown in Table 7. The wastewater flowing to the two treatment plants seems to be distributed equally between them. Notice that WWTP 2 receives more wastewater than WWTP 1, as its capacity of treating contaminated water is larger. This behavior is produced by the weighting chosen for the system and can be adjusted further.

As a overall conclusion for the simulations, no abnormal behavior was observed from the closed-loop system responses from simulations. The tanks signals responded to the external inflows, $\mathbf{d}_e$, then as rainfall episodes have ended the tank signals settle slowly to their steady state values.

4.3. Uncertainty Analysis of Disturbance

Now, simulation are shown for uncertain disturbances. For simulations, actual and estimated disturbances entering the system are simulated as pulse signals, and the 9TC system evaluated for the simulation cases described in Table 9. System responses of tank volume signals from simulations are presented, along with plots of the total accumulated volume directed to $WWTPs$ and that received by the sea (q_{sea}). Further, the total quantity of external inflow to the 9TC system, simulated for the actual and estimated disturbances, are close to being equal (with a very small difference of approximately 0.0025% between the actual and estimated disturbances).

Cases / Type of rain	Controller	Process
9TCa	Estimated	Actual rain
9TCb	Estimated	Estimated
9TCc	Actual	Actual

Table 9. Simulations are distinguished by denoted cases: 9TCa, 9TCb and 9TCc.

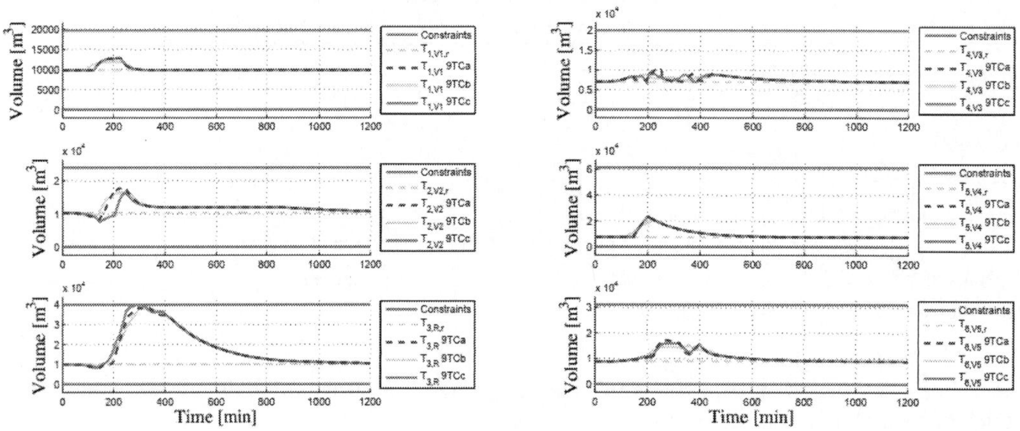

Figure 6. Volume in the tanks $T_{1,V1}$, $T_{2,V2}$ and $T_{3,R}$ (left) and the volume in the tanks $T_{4,V3}$, $T_{5,V4}$ and $T_{6,V5}$ (right).

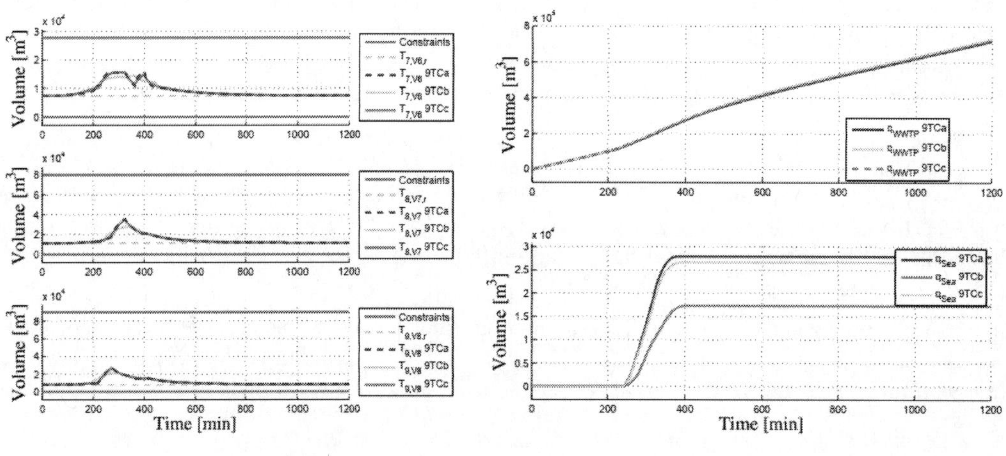

Figure 7. Volume in the tanks $T_{7,V6}$, $T_{8,V7}$ and $T_{9,V8}$ (left) and the total flow to WWTP, q_{wwtp}, and sea, q_{sea} accumulated volumes (right).

When tank volume signals from Figure 7, are evaluated, only small differences are noticed between simulation cases (9TCa, 9TCb and 9TCc). A considerable difference between volume signal of tank, $T_{2,V2}$, of case 9TCc compared to cases 9TCa and 9TCb, is noticed in Figure 6. This behavior is due to the difference between the actual and estimated rain. Whereas the actual rain enters the system for shorter periods but with larger intensities, while the estimated rain enters the system for longer periods but with smaller intensities.

If the system performance in treating wastewater and minimizing that received by the sea, then differences in performances can be compared by using Figure 7, and the numerical data from simulation, seen in Table 10. It is noticed when simulating case 9TCb, better performance in minimizing overflow to the sea is achieved, compared to cases 9TCa and 9TCc. Similar can be concluded for the performance in treating wastewater (WWTP total), when cases are compared. The total differences of treated wastewater is similar between cases 9TCa and 9TCc, and slightly better performance observed from case 9TCb.

	9TCa [%]	9TCb [%]	9TCc [%]
Sea area 1	2.11	1.50	2.01
Sea area 2	1.63	0.82	1.55
Sea total	3.74	2.32	3.56
WWTP 1	46.55	47.17	46.65
WWTP 2	49.50	50.31	49.59
WWTP total	96.06	97.48	96.24

Table 10. The difference between corresponding volume parameter, and the difference of the total inflow and overflow volume.

5. Conclusion

The Barcelona benchmark model for a sewer network has been considered in this paper. The sewer network has been controlled by using MPC, where it is possible to handle system constraints as well as to include prediction in the control optimization. It is shown by simulations, that using MPC, that it is possible to control the waste water with respect to minimizing flooding in the stress as well as minimizing the flow waste water directly to the sea. The simulations shows clearly that the prediction of the inflow from rainfall is important to optimize the control of the sewer network. The MPC methodology allow to use the capacity in the network to minimize the flooding as well as minimizing the flow of waste water directly to the sea. In this case, the MPC methodology of controlling a sewer network, showed sufficient results and can be used for ongoing study of sewer networks.

References

[1] Yolanda Bolea, Vicenç Puig, and Antoni Grau. Discussion on muskingum versus integrator-delay models for control objectives. *Applied Mathematics*, 2014.

[2] E.F Camacho and C.B. Alba. *Model Predictive Control.* Springer-Verlag London, 2nd edition, 2007.

[3] Mathias Fui Lin FOO. *Modelling and Control Design of River Systems.* PhD thesis, National ICT Australia, NICTA Department of Electrical and Electronic Engineering The University of Melbourne, 2012.

[4] Bernat Joseph i Duran. *Hybrid Modelling and Receding Horizon Control of Combined Sewer Networks.* PhD thesis, Universitat Polit'ecnica de Catalunya, 2014.

[5] X. Litrico and V. Fromion. Simplified modeling of irrigation canals for controller design. *Journal of Irrigation and Drainage Engineering*, 130:373–383, 2004.

[6] X. Litrico and V. Fromion. Tuning of robust distant downstream PI controllers for an irrigation canal pool. *Journal of Irrigation and Drainage Enginnering*, 133:27–37, 2007.

[7] Jan Marian Maciejowski. *Predictive Control with Constraints.* Prentice Hall, 2002.

[8] Markos Papageorgiou Magdalene Marinaki. *Optimal Real-time Control of Sewer Networks.* Springer, 2005.

[9] C. Ocampo Martínez. *Model Predictive Control of Complex Systems including Fault Tolerance Capabilities: Application to Sewer Networks.* PhD thesis, Technical University of Catalonia, April 2007.

[10] Clemmens M.ASCE, X. Litrico P.J. van Overloop, A.M.ASCE, and R. J. Strand. Estimating canal pool resonance with auto tune variation. *Journal of Irrigation and Drainage Engineering*, 138:9–15, 2012.

[11] C. Ocampo-Martinez. *Model Predictive Control of Wastewater Systems.* Springer-Verlag London, 2010.

[12] Peter-Jules VAN OVERLOOP. *Model Predictive Control on Open Water Systems.* PhD thesis, Delft University of Technology, juni 2006.

[13] E.B. Pedersen and H.R. Herbertsson. Modelling and control of sewer network. Master's thesis, Dept. of Electrical Engineering, Technical University of Denmark, 2015.

[14] Roland K. Price. Hydroinformatics and urban drainage: an agenda for the beginning of the 21st century. *Journal of Hydroinformatics*, 2:133–147, 2000.

[15] V. Puig, J. Romera G. Cembrano, J. Quevedo, B. Aznar, G. Ramón, and J. Cabot. Predictive optimal control of sewer networks using coral tool: application to riera blanca catchment in barcelona. *Water Science and Technology*, 60.4:869 – 878, 2009.

[16] Sara De Toffol. *Sewer system performance assessment -an indicators based methodology.* PhD thesis, University of Innsbruck, 2006.

Analysis and Application of Advanced Control Strategies to a Heating Element Nonlinear Model

C. Turhan[1], S. Simani[2], I. Zajic[3], G. Gokcen Akkurt[1]

[1]Izmir Institute of Technology, Department of Mechanical Engineering, Gulbahce Campus, Urla – 35430 Izmir, Turkey. [2]Department of Engineering, University of Ferrara. Via Saragat 1E, 44122 Ferrara (FE), Italy. [3]Control Theory and Applications Centre, Coventry University, Coventry, United Kingdom.

E-mail: cihanturhan@iyte.edu.tr, silvio.simani@unife.it, zajici@uni.coventry.ac.uk

Abstract. This paper presents the design of different control strategies applied to a heating element nonlinear model. The description of this heating element was obtained exploiting a data–driven and physically meaningful nonlinear continuous–time model, which represents a test–bed used in passive air conditioning for sustainable housing applications. This model has low complexity while achieving high simulation performance. The physical meaningfulness of the model provides an enhanced insight into the performance and functionality of the system. In return, this information can be used during the system simulation and improved model–based and data–driven control designs for tight temperature regulation. The main purpose of this study is thus to give several examples of viable and practical designs of control schemes with application to this heating element model. Moreover, extensive simulations and Monte-Carlo analysis are the tools for assessing experimentally the main features of the proposed control schemes, in the presence of modelling and measurement errors. These developed control methods are also compared in order to evaluate advantages and drawbacks of the considered solutions. Finally, the exploited simulation tools can serve to highlight the potential application of the proposed control strategies to real air conditioning systems.

1. Introduction

Buildings are responsible for approximately 40% of the total energy consumption in the entire world [1, 2]. In developed countries, 50% of this energy is used for air conditioning purposes in buildings [3].

Energy control of the air conditioning systems is crucial to satisfy thermal comfort including relative humidity and temperature which is the primary variable in controlling air conditioners. Heating Elements (HEs) play an essential role for supplying treated air with specified temperature to the conditioned space in the buildings. Building control strategies require extensive computational requirements and necessity of a mathematical model of the HE system. Since the mathematical model is a description of the system behaviour, accurate modelling for a complex nonlinear system is very difficult to achieve in practice.

The HE system is a nonlinear system in the control of view, therefore, most of temperature control systems use conventional controllers like on–off controllers, including standard PID or PI regulators for its relative simplicity. However, these controllers do not always produce fast response and suffer the problem of overshoot, large settling time. Moreover, the tuning of the conventional controllers is difficult [4]. For example, when HE systems include bilinear terms depending on temperature and mass flow rate, advanced control systems can be needed to optimise the design problem. To this aim, Artificial Intelligence (AI) has been applied to temperature control due to substantial advantages of applicability to nonlinear systems with unknown or partially known dynamics [5].

In the last years, AI based methods, namely, Fuzzy Logic (FL), Adaptive Neuro–Fuzzy Inference System (ANFIS), Artificial Neural Network (ANN) and Model Predictive Controllers (MPC) have been considered for trying to achieve more comfortable environments in buildings, see *e.g.* [6, 7, 4, 8, 9, 10, 11, 12]. These controllers have better advantages compared to the conventional ones. For instance, the paper [11] addressed the implementation of a MPC for the temperature control of a building. The controller achieved a saving of $17-24\%$ compared to a weather–compensated control technique. Similarly, the work [3] shows the development of a MPC algorithm to maintain the thermal comfort resulting in 28% of heating energy savings. In addition to the MPC, FL controllers were studied to control the temperature in heating systems by some authors [10, 9]. The authors demonstrated that the nonlinearity of the system can be easily managed with a FL controller, thus obtaining good performances. The paper [4] compared PID and ANN controllers. The authors designed an ANN controller feeding–back the air conditioning system. The ANN controllers allowed for shorter rise time and zero overshoot of the controller output. Although ANNs are widely used in the literature [8], some researchers studied more efficient control techniques based on the ANFIS tool, with the advantages of both ANN and FL [10, 5, 13]. In particular, the paper [13] used ANFIS controller effectively to model nonlinear functions to obtain thermal comfort conditions including temperature and relative humidity. The authors observed that maximum absolute error was less than $0.015\,^oC$ for temperature control. Adaptive controllers help the heating systems to maintain thermal comfort, while enhancing energy savings, as shown *e.g.* in [3]. Note finally that the increased popularity of AI control techniques for building temperature control is incontrovertible; however most of them do not give a comparison with extensive simulations.

This paper is focussed on the design of different control strategies with application to the HE nonlinear model developed by one of the same authors in [14]. The simulations and the comparisons with the achieved performances have been implemented in the Matlab and Simulink environments.

The rest of the paper is organised as follows. The heating element system is summarised in Section 2. Section 3 provides the control methods. The achieved results are summarised in Section 4. Comparisons among the different control methods and their performances with respect to measurements and modelling errors are also investigated and discussed. Finally, Section 5 concludes the paper by summarising the main achievements of the work, and providing some suggestions for further research topics.

2. Heating Element Model Description

The HE is a large test–bed system for testing and modelling of Phase Change Materials (PCM) used in the passive air conditioning for sustainable housing applications. The test–bed is installed in a large indoor space with approximately constant ambient temperature, as shown in Fig. 1.

The inputs of the HE system are the speed and temperature of air, v $[m\,s^{-1}]$ and T_i $[K]$, respectively, passing through the tested PCM. The output is the outlet temperature T_o $[K]$. The air is conditioned by passing through the considered HE, which is connected in a downstream series connection with a cooling unit. The schematic diagram of the heating element installed in the supply duct is represented in Fig. 2. The measurements of v, T_i, and T_o are taken from the centre of the cross sectional area of the supply duct.

The HE system is mainly described by 3 nonlinear differential relations. The energy balance equations for the HE, the control volume of air surrounding the HE and the adjacent duct walls

IOP Conf. Series: Journal of Physics: Conf. Series **783** (2017) 012027 doi:10.1088/1742-6596/783/1/012027

are represented by Eqs. (1), (2), and (3), respectively:

$$C_h \frac{dT_h(t)}{dt} = q(t) - (UA)_h \left[T_h(t) - T_o(t)\right] \quad (1)$$

$$\begin{aligned} 0 = {} & (UA)_h \left[T_h(t) - T_o(t)\right] + \\ & -v(t)\, \rho_a\, A_a\, c_a \left[T_o(t) - T_i(t)\right] + \\ & - (UA)_{int} \left[T_o(t) - T_w(t)\right] \end{aligned} \quad (2)$$

$$\begin{aligned} C_w \frac{dT_w(t)}{dt} = {} & (UA)_{int} \left[T_o(t) - T_w(t)\right] + \\ & - (UA)_{ext} \left[T_w(t) - T_a(t)\right] \end{aligned} \quad (3)$$

where C_h $[J/K]$ is the thermal capacity of the heating element, C_w $[J/K]$ the thermal wall capacity (insulated plywood), c_a [J/kgK] is the air specific heat capacity, A_a $[m^2]$ denotes the cross sectional area of the duct, ρ_a $[kg/m^3]$ is the air density.

Figure 1. The HE test–bed.

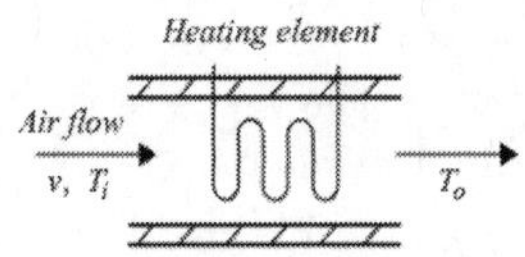

Figure 2. HE schematic diagram.

The heat transfer coefficient is denoted by U $[J/m^2K]$, whilst the products of the heat transfer coefficient by the efficient surface area, A $[m^2]$, through which the heat is transmitted are denoted with $(UA)_h$ $[J/K]$ (when referred to the HE), $(UA)_{int}$ $[J/K]$ (with reference to the inner duct wall), and $(UA)_{ext}$ $[J/K]$ (for the the outer duct wall). The mean temperature of the heating element and wall temperature are represented by $T_h(t)$ $[K]$ and $T_w(t)$ $[K]$, respectively. $q(t)$ is the power supplied by the heating element. Eq. (2) assumes that the passing air has negligible thermal capacity, hence the left hand side of Eq. (2) equals to zero. Note finally that the HE model was described in more detail in [14].

3. Control Scheme Designs

The general description of the dynamic model of the HE can be represented by the nonlinear dynamic function P:

$$y = P(u, t) \quad (4)$$

where y is the process output, u is its input, and t is the time. The control strategy applied to the HE should determine the control input such that the controlled process is able to track a given reference $r(t)$ [14].

It can be shown that the overall dynamic behaviour of the HE process (4) can be approximated with very good accuracy by a product of 2 second–order continuous–time transfer functions, as shown in [14]. On the other hand, the continuous–time linearised state–space description of the HE nonlinear process (4) is described by the system (5):

$$\begin{cases} \dot{x}(t) &= A\,x(t) + B\,u(t) \\ y(t) &= C\,x(t) \end{cases} \tag{5}$$

with $x \in \Re^4$. The control input is represented by the mass flow rate speed $u(t) = v$, and the measured output is $y(t) = T_o$. In general, Heating, Ventilation, and Air Conditioning (HVAC) systems have a second input T_i, *i.e.* the air flow input temperature, which is considered as a measurable disturbance, d.

This paper recalls different control strategies including the standard PID controller as well as AI techniques, such as fuzzy logic, adaptive, model predictive controllers, which are used for the regulation of the outlet temperature of the HE system. These methodologies are briefly outlined in the following subsections.

3.1. PID Controller Design

Standard PID regulators are the most commonly used feedback controllers for industrial processes. The control logic is based on the computation of the error $e(t)$ between the desired and the measured values of the output, *i.e.* $e(t) = r(t) - y(t)$, which is fed back to the system after proportional, integral and derivative operations [15]. In this way, the continuous–time control law of the PID regulator is described by Eq. (6):

$$u(t) = K_p\,e(t) + K_i \int_0^t e(\tau)\,d\tau + K_d \frac{de(t)}{dt} \tag{6}$$

where K_p, K_i, K_d are the PID proportional, integral, and derivative gains. The optimal selection of this gains is performed by using the automatic tuning algorithm in the Simulink environment that balances the performance (response time) and the robustness (stability margins) of the controlled system [16]. The PID automatic tuning Simulink toolbox uses the linearised model (5) of HE system.

3.2. Fuzzy Controller Design

Fuzzy Logic Controllers (FLCs) are extensively used in processes where the system dynamics are either very complex or exhibit highly nonlinear characteristics [10]. The controller design approach relies on the identification of transparent rule–based Takagi–Sugeno (TS) fuzzy models using an Adaptive Neuro–Fuzzy Inference System (ANFIS) tool implemented in the Simulink toolbox [17].

The TS fuzzy model consists of a set of rules R_i, where the consequents are deterministic functions f_i:

$$R_i : IF\, x\, is\, A_i\, THEN\, u_i = f_i(x) \tag{7}$$

with $i = 1, 2, \ldots, K$ and K is the number of clusters or rules of the rule–based system, x represents the vector of the *antecedent* variables, and u_i describes the *consequent* output. A_i represents the antecedent fuzzy set of the i–th rule, defined by its (multivariable) membership function $\mu_{A_i}(x) \to [0, 1]$ The function f_i is represented by suitable parametric models, whose structure remains equal in all rules, whilst the parameters can vary. A parametrisation in affine form is usually exploited, and described by Eq. (8):

$$u_i = a_i\,x + b_i \tag{8}$$

where the vector a_i and the scalar b_i are the i–th submodel parameters. The vector x contains a suitable number n of delayed samples of the model inputs and output. In this way, the product $a_i\, x$ represents an Auto–Regressive eXogenous (ARX) parametric dynamic model of order n.

The final output u of the TS fuzzy model is the weighted average of all rule outputs, computed as:

$$u = \frac{\sum_{i=1}^{K} \mu_{A_i}(x)\, y_i(x)}{\sum_{i=1}^{K} \mu_{A_i}(x)} \tag{9}$$

The modelling approach used by ANFIS is similar to many system identification techniques. First, the TS fuzzy model structure described by the its order n, the form of the membership functions μ_{A_i} and the number of clusters K are hypothesised. Next, the input–output data are used by ANFIS for training the TS model according to a chosen error criterion, thus determining the optimal values of the controller parameters a_i and b_i [17].

The paper considers also an alternative approach to ANFIS for the derivation of the controller fuzzy model, which is represented by the Fuzzy Modelling and Identification (FMID) toolbox developed in the Matlab environment [18]. Also in this case, the estimation of the controller prototype relies on the identification of rule–based fuzzy models and using the input–output data acquired from the controlled process. This method exploits Takagi–Sugeno fuzzy models and employs the Gustafson–Kessel clustering method to divide the data into subsets with a common local linear (affine) behaviour [18].

The identified fuzzy controller is thus obtained by selecting an proper model structure n and a number of clusters K. The FMID toolbox provides the parameters a_i, b_i and the estimation of the membership functions μ_{A_i} of the optimal controller minimising the tracking error $e(t)$, *i.e.* the difference between the reference signal $r(t)$ and the measured output $y(t)$.

Note finally that the fuzzy controller in the form of Eq. (9) is described by a discrete–time input–output model, which is connected to the controlled continuous–time nonlinear system of Eq. (4) using Digital–to–Analog (D/A) and Analog–to–Digital (A/D) converters.

3.3. Adaptive Control Design

The adaptive control method exploited in this paper is based on the on–line identification of a second order discrete–time transfer function of an ARX time–varying model in the form:

$$G(z) = \frac{b_1\, z^{-1} + b_2\, z^{-2}}{1 + a_1\, z^{-1} + a_2\, z^{-2}} \tag{10}$$

whose parameters are recursively estimated at each sampling time $t_k = k T$, with $k = 1, 2, \ldots, N$, N the number of samples, and T the sampling interval. z represents the unit advance operator. The parameters in (10) are estimated using the Recursive Least–Square Method (RLSM) with directional forgetting factor [19].

The synthesis of the adaptive control law is derived using a modified Ziegler–Nichols criterion, in the form of Eq. (11):

$$\begin{aligned} u_k &= q_0\, e_k + q_1\, e_{k-1} + q_2\, e_{k-2} + \\ &\quad +(1-\gamma)\, u_{k-1} + \gamma\, u_{k-2} \end{aligned} \tag{11}$$

where e_k is the tracking error $e(t)$ at the sampling time t_k, u_k the control signal $u(t)$ at the sampling time t_k, whilst q_0, q_1, q_2, and γ are the time–varying controller parameters, which are calculated by solving a Diophantine equation that leads to the following relations [19]:

$$\begin{aligned} q_0 &= \tfrac{1}{b_1}\left(d_1 + 1 - a_1 - \gamma\right), & \gamma &= q_2\, \tfrac{b_2}{a_2} \\ q_1 &= \tfrac{a_2}{b_2} - q_2\left(\tfrac{b_1}{b_2} - \tfrac{a_1}{a_2} + 1\right), & q_2 &= \tfrac{s_1}{r_1} \end{aligned} \tag{12}$$

where $r_1 = (b_1 + b_2)\left(a_1 b_2 b_1 - a_2 b_1^2 - b_2^2\right)$ and $s_1 = a_2\left((b_1 + b_2)(a_1 b_2 - a_2 b_1) + \right.$ $\left. + b_2\,(b_1 d_2 - b_2 d_1 - b_2)\right)$. It is assumed that the final closed loop model has a behaviour similar to a second–order continuous time system with characteristic polynomial $s^2 + 2\,\delta\,\omega\,s + \omega^2$, where δ and ω represent its damping factor and natural frequency, respectively. In this case, if $\delta \leq 1$, $d_1 = -2\,e^{-\delta\,\omega\,T}\cos\left(\omega\,T\sqrt{1-\delta^2}\right)$ and $d_2 = e^{-2\,\delta\,\omega\,T}$.

Both the on–line identification procedure and the adaptive controller parameter computation are implemented in the self–tuning controller Simulink library [19]. In this way, the sampled output y_k of the time–varying ARX model (10) should follow the sampled reference signal r_k when regulated by the control law (11). Note finally that, also in this case, the adaptive controller (11) is connected to the continuous–time nonlinear system (4) using the D/A and A/D converters.

3.4. Model Predictive Control Design

Model Predictive Control (MPC) relies on dynamic models of the process, most often linear models obtained by system identification or linearisation of a nonlinear plant. The main advantage of MPC is the fact that it allows the current sampling time to be optimised, while keeping future sampling times in account. This is achieved by optimising a finite time–horizon, but only implementing the current sampling time. MPC has the ability to anticipate future events and can take control actions accordingly. PID controllers do not have this predictive ability. MPC is nearly universally implemented as a digital control.

MPC is based on iterative, finite–horizon optimisation of the plant model. At the sample k ($k = 1, 2, \ldots, N$) the current plant output is sampled and a cost minimising control strategy is computed (via a numerical minimisation algorithm) for a relatively short time horizon in the future: $[k, k + N_p]$. Specifically, an online calculation is used to explore state trajectories that emanate from the current state and find (via the solution of Euler–Lagrange equations) a cost–minimising control strategy until time $k + N_p$. Only the first step of the control strategy is implemented, then the plant state is sampled again and the calculations are repeated starting from the new current state, yielding a new control and new predicted state path. The prediction horizon keeps being shifted forward and for this reason MPC is also called receding horizon control. Although this approach is not optimal, in practice it has given very good results [20].

An example of a cost function J for optimisation is given by:

$$J = \sum_{k}^{k+N_p} w_{y_k}\,(r_k - y_k)^2 + \sum_{k}^{k+N_c} w_{u_k}\,\Delta u_k^2 \tag{13}$$

where w_{y_k} the weighting coefficient reflecting the relative importance of the monitored output, and w_{u_k} the weighting coefficient penalising relative big changes in u_k, with $\Delta u_k = u_k - u_{k-1}$. N_p represents the prediction horizon, whilst N_c the control horizon.

Note finally that the discrete–time MPC design is performed by using the MPC toolbox in the Simulink environment, which computes a linearisation of the HE nonlinear model (5). The discrete–time MPC is thus connected to the continuous–time nonlinear system of Eq. (4) using the D/A and the A/D converters. The MPC Simulink toolbox uses the linearised model (5) of HE system.

4. Simulation Results

This study exploits the 5 different control methods described in the previous subsections to regulate the outlet temperature of the HE nonlinear system. The simulations are performed in the Simulink environment and the toolboxes described above. The achieved results are compared in terms of settling time T_s, maximum overshoot $S\%$ and Mean Sum of Squared Errors (MSSE), defined as:

$$MSSE\% = 100\sqrt{\frac{\sum_{k=1}^{N}(r_k - y_k)^2}{\sum_{k=1}^{N} r_k^2}} \tag{14}$$

The standard control strategy for HVAC systems uses a classic PI regulator, whose complete structure is recalled in Section 3.1 [14]. The optimal proportional and integral gains are determined using the automatic PID tuning procedure and settled to $K_p = 22$ and $K_i = 36$, respectively.

Fig. 3 shows the reference signal T_o (blue continuous line) of test–bed in Fig. 1 and the process output (red dashed line) controlled by the PI regulator (6), as shown in Fig. 4. The settling time achieved by the PI controller is $T_s = 2.92s.$, with an overshoot of $S\% = 40.19\%$, which are computed by applying a step change in the reference output temperature from 39^oC to 40^oC. Using the performance index (14), the tracking error is $MSSE\% = 0.51\%$.

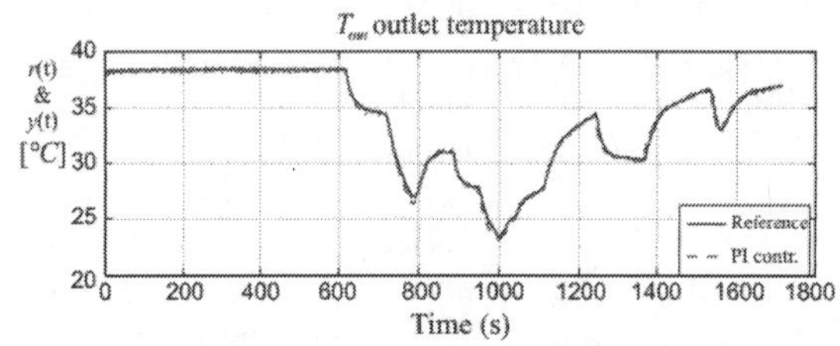

Figure 3. Outlet temperature controlled by the continuous–time PI regulator.

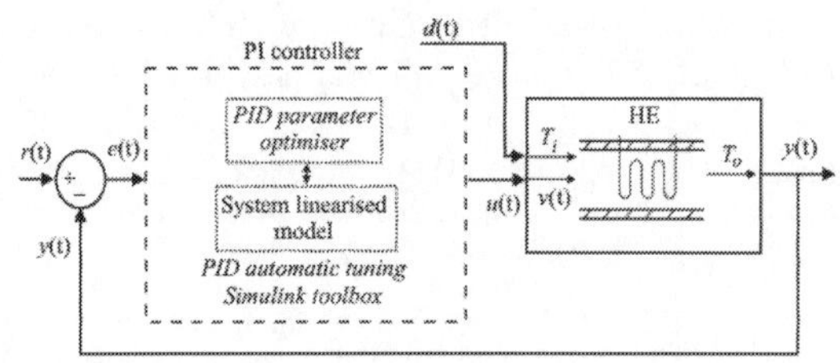

Figure 4. Block diagram of the HE system controlled by the standard PI regulator.

Standard industrial controllers, such the PID of Section 3.1, are quite simple and have the benefit of quite straightforward implementation. However, when applied to the control of HE systems, the control laws are not efficient, thus leading to possible high maintenance costs. Therefore, advanced controllers are exploited to reduce energy consumption and enhance thermal comfort.

With reference to the strategies described in Section 3.2, fuzzy identification is used to derive the models of the controllers by exploiting the so–called model reference control approach [21]. For this purpose, the PI regulator of Fig. 1 represents the reference controller for the generation of the data used by the identification strategy proposed described in Section 3.2. In this way, the fuzzy controller parameters are identified such that the performances in terms of tracking error $e(t)$ are optimised.

In particular, with reference to the TS fuzzy controller derived with the ANFIS tool, a sampling interval $T = 0.1s.$ is exploited. Moreover, the fuzzy controller (9) uses a number

$K = 3$ of Gaussian membership functions, with a number of delayed inputs and output $n = 1$. The antecedent vector is thus $x = [e_k, e_{k-1}, u_{k-1}]$. The achieved performances of the controller obtained with the ANFIS tool are shown in Fig. 5, whilst its implementation is sketched in Fig. 6.

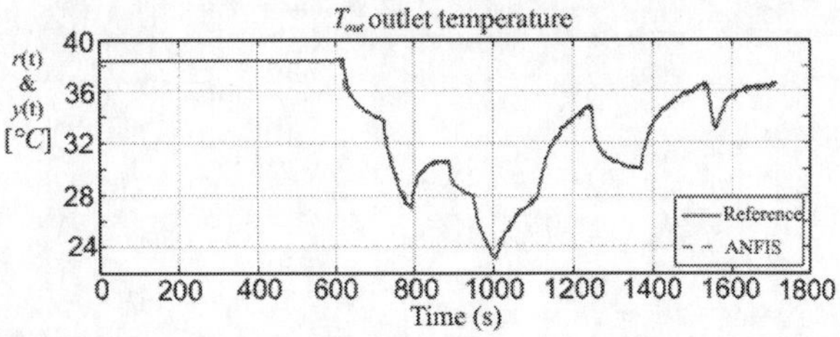

Figure 5. Outlet temperature controlled by the fuzzy controller obtained via the ANFIS tool.

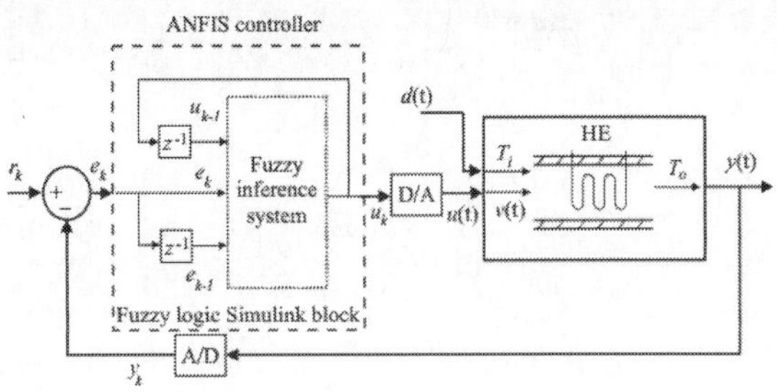

Figure 6. Block diagram of the HE system controlled by the ANFIS fuzzy regulator.

Fig. 5 highlights that fuzzy regulator performs slightly better than the standard PI controller. In this case, the settling time is $T_s = 3.05s.$, with an overshoot $S\% = 41.66\%$, and an $MSSE\% = 0.47\%$. .

Using the same data from the PI reference regulator, a second fuzzy controller (9) has been estimated using the FMID tool, with a number of clusters $K = 3$, a number of delays $n = 2$, and the antecedent vector $x = [u_{k-1}, u_{k-2}, r_k, r_{k-1}, y_k, y_{k-1},]$. The FMID tool provides also the optimal estimate of the shapes of the membership functions μ_{A_i}. The implementation scheme is represented in Fig. 7, and the results of this fuzzy controller are shown in Fig. 8. Note that the high overshoot at the beginning of the simulation in Fig. 8 is due to the initial conditions of the delay blocks of the fuzzy controller represented in Fig. 7 that are zero.

Fig. 8 highlights that the reference signal is tracked with good accuracy, with a $MSSE\% = 0.09\%$. In this case, the settling time is $T_s = 3.15s.$ with a maximum overshoot $S\% = 41.25\%$.

On the other hand, by considering the on–line procedure recalled in Section 3.3, Fig. 9 shows the tracking capabilities of the adaptive controller (11). Its time–varying parameters have been obtained via the relations (12) with the damping factor and the natural frequency $\delta = \omega = 1$. The adaptive controller implementation using the Self Tuning controller Simulink Library (STCSL) is represented in Fig. 10.

The settling time of the HE system output with the adaptive controller is $T_s = 3.16s.$, and the overshoot is $S\% = 42.87\%$. The tracking error is $MSSE\% = 5.97\%$.

Finally, with reference to the MPC strategy recalled in Section 3.4, the reference and the monitored output signals are depicted in Fig. 11. Its implementation is sketched in Fig. 12.

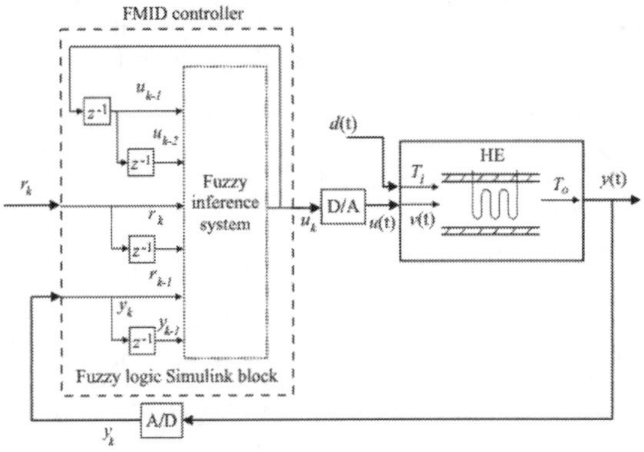

Figure 7. Block diagram of the HE system controlled by the FMID fuzzy regulator.

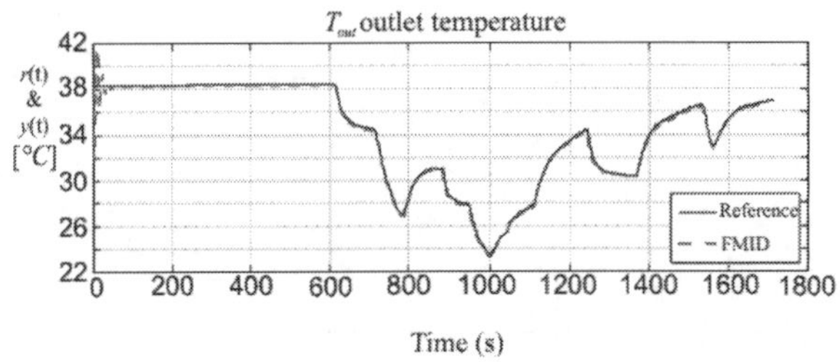

Figure 8. Outlet temperature controlled by the fuzzy controller estimated by the FMID tool.

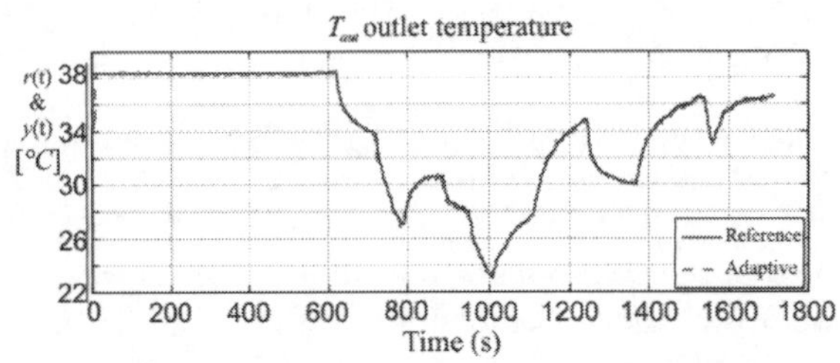

Figure 9. Outlet temperature controlled by the adaptive controller.

The results shown in Fig. 11 have been achieved by using a prediction horizon $N_p = 10$ and a control horizon $N_c = 2$. The weighting parameters have been settled to $w_{y_k} = 0.1$ and $w_{u_k} = 1$, in order to reduce abrupt changes of the control input that would lead to higher energy consumption. In this case, the settling time is $T_s = 3.49s$. and the overshoot is 40.68%. The tracking error corresponds to $MSSE\% = 3.2\%$. The achieved results in terms of settling time, maximum overshoot and nominal $MSSE\%$ values are summarised in Table 1.

Note that the standard PI regulator leads to the best values of settling time and maximum overshoot, as its parameters are automatically tuned in the Simulink environment in order to optimise these indices, as recalled in Section 3.1. On the other hand, the best performance in terms of tracking error is obtained with the fuzzy controller estimated via the FMID tool recalled Section 3.2.

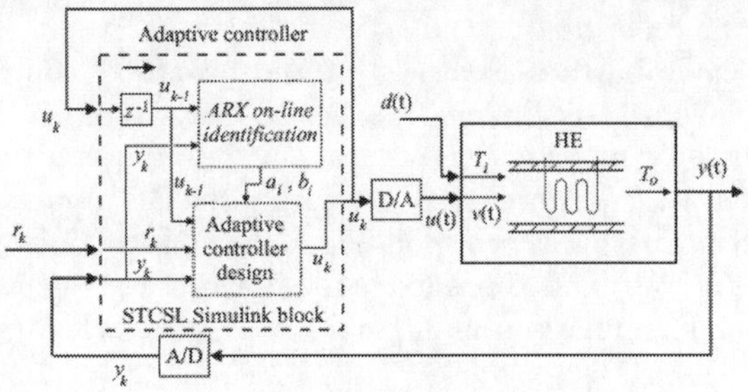

Figure 10. Block diagram of the HE system controlled by the adaptive regulator.

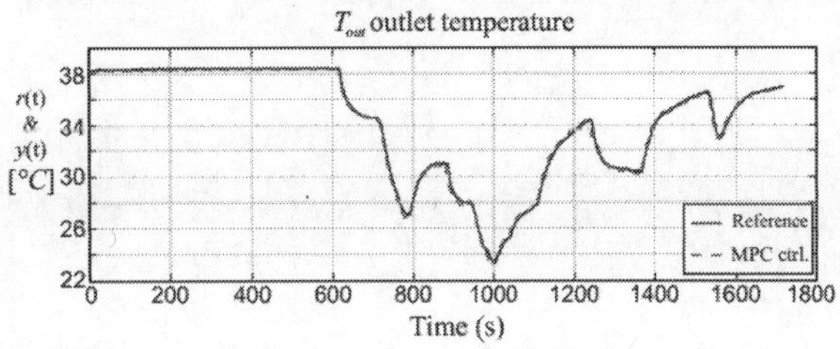

Figure 11. Outlet temperature controlled by the MPC.

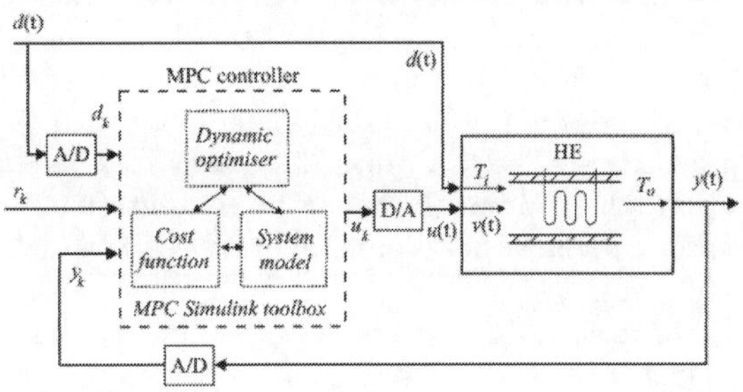

Figure 12. Block diagram of the HE system controlled by the MPC strategy.

Table 1. Comparison of the achieved performance in terms of $MSSE\%$.

Controller type	Settling time T_s	Overshoot $S\%$	$MSSE\%$ nominal value
PI	$2.92s.$	40.19%	0.51%
ANFIS	$3.05s.$	41.66%	0.47%
FMID	$3.15s.$	41.25%	0.094%
Adaptive	$3.16s.$	42.87%	5.97%
MPC	$3.49s.$	40.68%	3.19%

The next section provides further results with some final comments regarding the robustness features of the proposed controllers applied to the HE nonlinear model.

4.1. Parameter Sensitivity Analysis

This section analyses the robustness features of the proposed controllers with respect to parameter variations. This analysis exploits the Monte–Carlo tool, as the control performance depends on the model–reality mismatch as well as on the input–output measurement errors. Therefore, the analysis has been performed by describing the HE model parameters as Gaussian stochastic processes with mean values corresponding to the nominal ones and standard deviations of ±20%. The average, the best and the worst $MSSE$ index values have been computed with 100 Monte–Carlo runs, and summarised in Table 2.

Table 2. Comparison of the achieved performance in terms of $MSSE\%$.

Controller type	$MSSE\%$ best case	$MSSE\%$ average case	$MSSE\%$ worst case
PI	0.19%	0.51%	1.45%
ANFIS	0.12%	0.47%	3.75%
FMID	0.093%	0.094%	0.26%
Adaptive	4.22%	5.97%	6.95%
MPC	0.13%	3.19%	3.57%

A few comments can be drawn here. When the modelling of the dynamic system can be taken into account, the MPC scheme is preferred, even if an optimisation procedure is required. However, in the case of a system with modelling errors, after a certain amount of off–line learning, the fuzzy–based estimation error can fall below the value of the MPC–based scheme, as shown for the controller estimated via the FMID tool. On the other hand, the FMID controller achieves the best control capabilities. The adaptive approach takes advantage of its improved features, as it is able to track possible variations of the controlled system, but with quite complicated and not straightforward design procedures. The fuzzy–based schemes rely on the learning accumulated from off–line simulations, but the training stage can be computationally heavy. Regarding the developed PI control strategy, it is rather simple and straightforward, even if the achievable performances are quite limited.

Finally, the results demonstrate also that the Monte–Carlo analysis is an effective tool for testing the suggested control methods. Moreover, in presence of uncertainty and modelling errors, this simulation tool seems to facilitate the validation of the considered control schemes for application to real HVAC cases.

5. Conclusion

In this study, different temperature control techniques were applied to a heating element nonlinear model, which can represent a part of a larger system used also for modelling phase change materials in passive air conditioning for sustainable housing applications. Simulations on the heating element nonlinear model and the Monte–Carlo analysis were the tools for assessing experimentally the properties of the designed control schemes, in the presence of modelling and measurement errors. The proposed control solutions relied on standard PID, fuzzy, adaptive, and model–predictive control strategies. The achieved results showed that these control schemes can be successfully used for the regulation of the temperature of the heating element system. Moreover, the proposed comparisons among the different control strategies highlighted also that the suggested control solutions can be promising for the real building control applications. Further studies will try to extend the developed schemes to the control of more advanced thermal comfort parameters, such as the relative humidity and the air flow.

References

[1] N. Payam, J. Fatemeh, M. T. Mohammad, G. Mohammad, and M. A. Muhd Zaimi, "A global review of energy consumption, CO_2 emissions and policy in the residential sector (with an overview of the top ten CO_2 emitting countries)," *Renewable and Sustainable Energy Reviews*, vol. 43, pp. 843–862, March 2015. DOI: 10.1016/j.rser.2014.11.066.

[2] Y. Xing, N. Hewitt, and P. Griffiths, "Zero carbon buildings refurbishment – A hierarchical pathway," *Renewable and Sustainable Energy Reviews*, vol. 15, pp. 3229–3236, August 2011. DOI: 10.1016/j.rser.2011.04.020.

[3] D. Lindelof, H. Afshari, M. Alisafaee, J. Biswas, M. Caban, X. Mocellin, and J. Viaene, "Field tests of an adaptive, model–predictive heating controller for residential buildings," *Energy and Buildings*, vol. 99, pp. 292–302, May 2015. DOI: 10.1016/j.enbuild.2015.04.029.

[4] P. Kumar and K. P. Singh, "Comparative analysis of air conditional system using PID and neural network controller," *International Journal of Scientific and Research Publications*, vol. 3, pp. 1–6, August 2013. ISSN: 2250–3153. DOI: 10.1.1.415.2205.

[5] J. Moon, S. K. Jung, Y. Kim, and S. H. Han, "Comparative study of artificial intelligence–based building thermal control – Application of fuzzy, adaptive neuro–fuzzy inference system and artificial neural network," *Applied Thermal Engineering*, vol. 31, pp. 2422–2429, October 2011. DOI: 10.1016/j.applthermaleng.2011.04.006.

[6] A. I. Dounis and C. Caraiscos, "Advanced control systems engineering for energy and comfort management in a building environment – a review," *Renewable and Sustainable Energy Reviews*, vol. 13, pp. 1246–1261, August–September 2009. DOI: 10.1016/j.rser.2008.09.015.

[7] M. Krarti, "An overview of artificial intelligence–based methods for building energy systems," *Journal of Solar Energy Engineering*, vol. 125, pp. 331–342, August 2003. DOI: 10.1115/1.1592186.

[8] P. M. Ferreira, A. E. Ruano, S. Silva, and E. Z. E. Conceicao, "Neural networks based predictive control for thermal comfort and energy savings in buildings," *Energy and Buildings*, vol. 55, pp. 238–251, December 2012. DOI: 10.1016/j.enbuild.2012.08.002.

[9] M. He, W. J. Cai, and S. Y. Li, "Multiple fuzzy model–based temperature predictive control for HVAC systems," *Information Science*, vol. 169, pp. 155–174, January 2005. DOI: 10.1016/j.ins.2004.02.016.

[10] S. Soyguder and H. Alli, "An expert system for the humidity and temperature control in HVAC systems using ANFIS and optimization with Fuzzy Modelling Approach," *Energy and Buildings*, vol. 41, pp. 814–822, August 2009. DOI: 10.1016/j.enbuild.2009.03.003.

[11] S. Privera, J. Siroky, L. Ferkl, and J. Cigler, "Model predictive control of a building heating system: The first experience," *Energy and Buildings*, vol. 43, pp. 564–572, February–March 2011. DOI: 10.1016/j.enbuild.2010.10.022.

[12] J. Siroky, F. Oldewurtel, J. Cigler, and S. Privara, "Experimental analysis of model predictive control for an energy efficient building heating system," *Applied Energy*, vol. 88, pp. 3079–3087, September 2011. DOI: 10.1016/j.apenergy.2011.03.009.

[13] K. L. Ku, J. S. Liaw, M. Y. Tsai, and T. Liu, "Automatic control system for thermal comfort based on predicted mean vote and energy saving," *IEEE Transactions on Automation Science and Engineering*, vol. 12, pp. 378–383, Jan. 2015. DOI: 10.1109/TASE.2014.2366206.

[14] I. Zajic, M. Iten, and K. J. Burnham, "Modelling and data–based identification of heating element in continuous–time domain," *Journal of Physics: Conference Series*, vol. 570, no. 1, p. 012003, 2014. DOI: 10.1088/1742-6596/570/1/012003.

[15] K. J. Åström and B. Wittenmark, *Computer Controlled Systems: Theory and Design*. Englewood Cliffs, N.J. 07632: Prentice-Hall, third ed., 1990.

[16] K. J. Åström and T. Hägglund, *Advanced PID Control*. Research Triangle Park, NC 27709: ISA - The Instrumentation, Systems, and Automation Society, 2006. ISBN: 978–1–55617–942–6.

[17] J. S. R. Jang, "ANFIS: Adaptive–Network–based Fuzzy Inference System," *IEEE Transactions on Systems, Man., & Cybernetics*, vol. 23, pp. 665–684, May 1993.

[18] R. Babuška, *Fuzzy Modeling for Control*. Boston, USA: Kluwer Academic Publishers, 1998.

[19] V. Bobál, J. Böhm, J. Fessl, and J. Machácek, *Digital Self–Tuning Controllers: Algorithms, Implementation and Applications*. Advanced Textbooks in Control and Signal Processing, Springer, 1st ed., 2005.

[20] M. Nikolaou, *Advances in Chemical Engineering*, vol. 26, ch. Model predictive controllers: A critical synthesis of theory and industrial needs, pp. 131–204. Academic Press, 2001.

[21] M. Brown and C. Harris, *Neurofuzzy Adaptive Modelling and Control*. Prentice Hall, 1994.

IOP Conf. Series: Journal of Physics: Conf. Series **783** (2017) 012028 doi:10.1088/1742-6596/783/1/012028

Improved LTVMPC design for steering control of autonomous vehicle

Shridhar Velhal
M.Tech Student, Department of Electrical Engineering, National Institute of Technology, Calicut, India
E-mail: shridharbvelhal@gmail.com

Susy Thomas
Professor, Department of Electrical Engineering, National Institute of Technology, Calicut, India
E-mail: susy@nitc.ac.in

Abstract. An improved linear time varying model predictive control for steering control of autonomous vehicle running on slippery road is presented. Control strategy is designed such that the vehicle will follow the predefined trajectory with highest possible entry speed. In linear time varying model predictive control, nonlinear vehicle model is successively linearized at each sampling instant. This linear time varying model is used to design MPC which will predict the future horizon. By incorporating predicted input horizon in each successive linearization the effectiveness of controller has been improved. The tracking performance using steering with front wheel and braking at four wheels are presented to illustrate the effectiveness of the proposed method.

1. Introduction

In this paper, the problem being addressed is the steering control of autonomous vehicle for effective tracking of predefined trajectory. A double lane change scenario on slippery road is considered as trajectory. The assumption is that a trajectory planning system is available for designing the path to be followed. The problem involves yaw and lateral vehicle dynamics stabilization via steering control while tracking a desired path as close as possible while fulfilling the physical constraints. In [1][2][3][4] this problem has been addressed using front wheel steering control where the authors rely on linear time varying model predictive control (LTVMPC) for the controller design. In [4] the authors have improved the design by incorporating braking at four wheels to the steering control. In the work reported in this paper this design is further improved by modifying the linearization process involved in controller design.

Model predictive control (MPC) qualifies to be an adequate control strategy by virtue of its capacity to predict the future states and to handle constraints. In MPC a model of the plant is used to predict the future horizon of states of system. Based on this prediction horizon at each sampling time t, a performance index is optimized under operating constraints. Optimization gives sequence of future input moves in order to best follow a given trajectory. The first of the optimal moves is the control action applied to the plant at time t. At time t+1, a new optimization is defined and solved over a shifted prediction horizon [5] [6].

The nonlinear vehicle model discussed in [1] is being considered for the LTVMPC design. For nonlinear model, predictions of horizon based on nonlinear equations are very complex involving difficult computation. For such systems a sub-optimal controller using successive linearization is often used. In [1][2][3] the authors depend on such a methodology for controller design using front wheel steering.

The reported LTVMPC [3], uses a linearized state space model to predict the future horizon. For a time t, predictions are calculated using linearized model obtained by successive linearization at each sampling instant. This model is then used for the calculation of next input horizon. In next sampling instant this procedure of modeling and computing control horizon is repeated. In the methodology used in [3], at every instant successive linearization makes use of the previous input. Since predicted input is available at every instant, the accuracy of linearization process can be improved by making use of this predicted value rather than previous input as this will reduce the approximation involved in jacobian matrix of linearized model. This improvisation in successive linearization of system leads to better tracking performance compared to the methodology in [3]. This has been verified by applying this proposed method to the same autonomous vehicle being used in [3] [4]

The control inputs being used are front steering angle and braking at each wheel. Steering angle and its rate of change has physical limitations which has been taken into consideration in design. Also brakes are being applied within limits so that stability is maintained.

The paper is structured as follows: Section II describes the system model which includes vehicle model and tyre model in detail. Since the proposed improved LTVMPC method relies on LTVMPC methodology reported in [3], section III describes this methodology first and then deals with the proposed method. Simulation results are presented in section IV to illustrate the effectiveness of this method. Section V concludes the paper.

2. System model

The vehicle dynamic model as given in [1][2][3][4] is presented in this section. Fig 1 shows schematic of four wheel autonomous vehicle.

The two front and two rear wheels of car are denoted by $\star \in \{f, r\}$ and the left and right wheels by $\bullet \in \{l, r\}$. $F_{l\star,\bullet}$, $F_{c\star,\bullet}$ denotes longitudinal and lateral tire forces receptively. $F_{x\star,\bullet}$, $F_{y\star,\bullet}$ are forces in car body frame, F_z is normal tire load. I denotes the inertia of car, a, b are distance of front and rear wheels from center of gravity of car, c is the distance of vehicle longitudinal axis from wheels, g is gravitational constant, m is car mass, r is wheel radius, ψ is heading angle, X, Y are absolute car position coordinates (global reference), $s_{\star,\bullet}$ are slip ratios for four wheels, $\dot{x}$ is the longitudinal speed of vehicle, $\dot{y}$ is the lateral speed of vehicle, $\alpha_{\star,\bullet}$ are slip angles at four wheels, $\delta_{\star,\bullet}$ are wheel steering angle at each wheel and $\mu_{\star,\bullet}$ is road friction coefficient at the four wheels.

2.1. Vehicle model

Assuming the mass of vehicle to be constant, the normal tyre force (i.e. vertical force on tire) F_z is constant, the dynamics of four wheel model as shown in fig 1 is represented by following equations :

$$m\ddot{x} = m\dot{y}\dot{\psi} + F_{x_{f,l}} + F_{x_{f,r}} + F_{x_{r,l}} + F_{x_{r,r}} \quad (1a)$$

$$m\ddot{y} = -m\dot{x}\dot{\psi} + F_{y_{f,l}} + F_{y_{f,r}} + F_{y_{r,l}} + F_{y_{r,r}} \quad (1b)$$

$$I\ddot{\psi} = a(F_{y_{f,l}} + F_{y_{f,r}}) - b(F_{y_{r,l}} + F_{y_{r,r}}) + c(-F_{x_{f,l}} + F_{x_{f,r}} - F_{x_{r,l}} + F_{x_{r,r}}) \quad (1c)$$

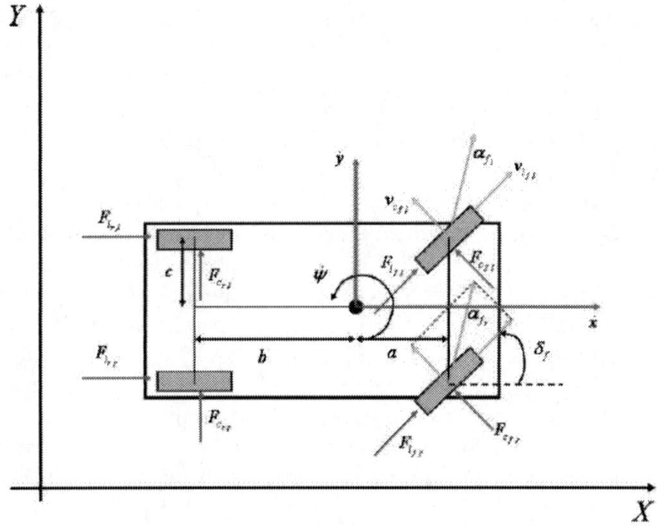

Figure 1: Vehicle model

The equations of motion in an absolute inertial frame are :

$$\dot{X} = \dot{x}\cos\psi - \dot{y}\sin\psi \quad (2a)$$
$$\dot{Y} = \dot{x}\sin\psi + \dot{y}\cos\psi \quad (2b)$$

Longitudinal and lateral tire forces lead to the following forces acting on the center of gravity :

$$F_{y_{\star,\bullet}} = F_{l_{\star,\bullet}}\sin\delta_{\star,\bullet} + F_{c_{\star,\bullet}}\cos\delta_{\star,\bullet} \quad (3a)$$
$$F_{x_{\star,\bullet}} = F_{l_{\star,\bullet}}\cos\delta_{\star,\bullet} - F_{c_{\star,\bullet}}\sin\delta_{\star,\bullet} \quad (3b)$$

Tire forces for each tire are given by [7] :

$$F_{l_{\star,\bullet}} = f_l(\alpha_{\star,\bullet}, s_{\star,\bullet}, \mu_{\star,\bullet}, F_z) \quad (4a)$$
$$F_{c_{\star,\bullet}} = f_c(\alpha_{\star,\bullet}, s_{\star,\bullet}, \mu_{\star,\bullet}, F_z) \quad (4b)$$

The tire slip angle represents the angle between the wheel velocity and the direction of the wheel itself and can be expressed as :

$$\alpha_{\star,\bullet} = \tan^{-1}\frac{v_{c_{\star,\bullet}}}{v_{l_{\star,\bullet}}} \quad (5)$$

The slip ratio formula is defined as :

$$s = \frac{r\omega}{v_l} - 1, \qquad if\ v_l > r\omega,\ v \neq 0 \qquad for\ braking \quad (6a)$$
$$s = 1 - \frac{v_l}{r\omega}, \qquad if\ v_l < r\omega,\ \omega \neq 0 \qquad for\ driving \quad (6b)$$

The lateral (or cornering) and longitudinal wheel velocities are :

$$v_{l_{\star,\bullet}} = v_{y_{\star,\bullet}}\sin\delta_{\star,\bullet} + v_{x_{\star,\bullet}}\cos\delta_{\star,\bullet} \quad (7a)$$
$$v_{c_{\star,\bullet}} = v_{y_{\star,\bullet}}\cos\delta_{\star,\bullet} - v_{x_{\star,\bullet}}\sin\delta_{\star,\bullet} \quad (7b)$$

and

$$v_{y_{f,l}} = \dot{y} + a\dot{\psi} \qquad v_{x_{f,l}} = \dot{x} - c\dot{\psi} \tag{8a}$$
$$v_{y_{f,r}} = \dot{y} + a\dot{\psi} \qquad v_{x_{f,l}} = \dot{x} + c\dot{\psi} \tag{8b}$$
$$v_{y_{r,l}} = \dot{y} - b\dot{\psi} \qquad v_{x_{f,l}} = \dot{x} - c\dot{\psi} \tag{8c}$$
$$v_{y_{r,r}} = \dot{y} - b\dot{\psi} \qquad v_{x_{f,l}} = \dot{x} + c\dot{\psi} \tag{8d}$$

The parameter $\mu_{\star,\bullet}$ denotes the road frictional coefficients and are assumed equal for all the four wheels. F_z is the total vertical load of the vehicle and is distributed between the front and rear wheels based on the geometry of the car model (described by the parameters a and b) is given by :

$$F_{z_f} = \frac{\text{bmg}}{2(a+b)}, \qquad F_{z_r} = \frac{\text{amg}}{2(a+b)} \tag{9}$$

Equations (1)-(9) describes the nonlinear vehicle dynamics. For simplicity of representation this is expressed as :

$$\frac{d\xi}{dt} = f_{s(t),\mu(t)}(\xi(t), u(t)) \tag{10}$$

The state and input vectors are given by,

$$\xi = [\dot{y} \;\; \dot{x} \;\; \psi \;\; \dot{\psi} \;\; Y \;\; X]^T \tag{11}$$
$$u = [\delta_{fl} \;\; \delta_{fr} \;\; \delta_{rl} \;\; \delta_{rr} \;\; s_{fl} \;\; s_{fr} \;\; s_{rl} \;\; s_{rr}]^T \tag{12}$$

The front wheel steering model is obtained by letting $\delta_{fl} = \delta_{fr} = \delta_f$ and $\delta_{rl} = \delta_{rr} = \delta_r = 0$.

2.2. Tyre model

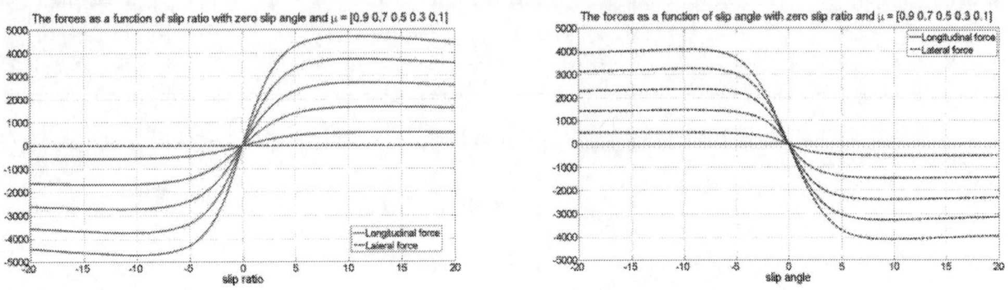

(a) longitudinal force (b) lateral force

Figure 2: Longitudinal and lateral tire forces with different frictional coefficient (μ) values. [7]

The model for tyre tractive and cornering forces presented in [3] are based on magic formula described by Pacejka model [7]. The same tyre model has been used in this paper.

The Pacejka tire model calculates lateral force and longitudinal force based on percent longitudinal slip (slip ratio), slip angle, frictional coefficient and vertical normal force. Figure 2 shows the longitudinal and lateral forces versus slip ratio and slip angle respectively for constant friction coefficient and vertical load.

3. Controller design

In MPC at each sampling time, an optimal control problem is solved over a finite horizon. The optimal control problem minimizes the performance index, subject to operating constraints. The resulting optimal command signal is applied to the vehicle system for that sampling interval. At the next instant, a new optimal control problem is solved for new measurements of the state over the shifted horizon. MPC is a step by step procedure where design ensures that states and control input are satisfying their constraints. As the problem being considered is steering control with braking and all outputs are to be tracked output constraints does not figure in problem. The continuous time model given in eq (10) is discretized with T_s as sampling time. This discretized model is represented by :

$$\xi(k+1) = f_{s(k),\mu(k)}(\xi(k), \Delta u(k)) \tag{13}$$

$$u(k) = u(k-1) + \Delta u(k) \tag{14}$$

3.1. LTV model

Controller design for nonlinear systems are made simple by linearizing the system model around its operating point. By successive linearization over each shifting operating point the accuracy of the MPC design for autonomous vehicle can be improved. Such a model is referred to as linear time varying (LTV) model.

With the initial state $\xi_0 \in X$ and initial control input $u_0 \in U$, the one step ahead dynamics of system is :

$$\begin{aligned} \xi(k+1) &= f(k, \hat{\xi}_0(k), \Delta u(k)) \\ u(0) &= u_0 \\ \hat{\xi}(0) &= \xi_0 \end{aligned} \tag{15}$$

Linearization of eq (13) about its initial condition will lead to :

$$\xi(k+1) = A_{k,0}\xi(k) + B_{k,0}u(k) + Bw_{k,0}W(k) \tag{16}$$

where,

$$A_{k,0} = \left.\frac{\partial f}{\partial \xi}\right|_{\hat{\xi}_0(k),u_0} \tag{17a}$$

$$B_{k,0} = \left.\frac{\partial f}{\partial u}\right|_{\hat{\xi}_0(k),u_0} \tag{17b}$$

$$\delta\xi(k) = \xi(k) - \hat{\xi}_0(k) \tag{17c}$$

Bw is the input matrix for disturbance and $W(k)$ is the disturbance to system.

3.2. LTVMPC controller

The MPC designed for linear model in eq (16) is referred to as LTVMPC [1] [2] [3] [4]. The fact that, the model in eq (16) is recalculated at every sampling instant justifies the term LTV. At each sampling instant t, the system dynamics can be represented as :

$$\xi(k+1) = A_{k,t}\xi(k) + B_{k,t}u(k) + Bw_{k,0}W(k) \tag{18}$$

The matrices $A_{k,t}, B_{k,t}$ are as defined in eq (17), where fixed initial index of zero is replaced by current time t.

$$A_{k,t} = \left.\frac{\partial f}{\partial \xi}\right|_{\hat{\xi}(k,t),u_t} \qquad B_{k,t} = \left.\frac{\partial f}{\partial u}\right|_{\hat{\xi}(k,t),u_t} \tag{19}$$

and,

$$\hat{\xi}(k+1,t) = f(\hat{\xi}(k,t), \Delta u_t, t), \quad k = t, t+1, ...t+N-1 \tag{20a}$$

$$\hat{\xi}(t,t) = \xi_t \tag{20b}$$

$$u_t = u_{t-1} \tag{20c}$$

For the above system, control horizon is generated by minimizing cost function J_N given by :

$$J_N(\xi_t, \Delta U_t, \Xi_{reft}) = \sum_{i=1}^{n_y} \|(\eta_{t+i,t} - \eta ref_{t+i,t})\|_Q^2 + \sum_{i=0}^{n_u-1} \|\Delta u_{t+i,t}\|_R^2 + \sum_{i=0}^{n_u-1} \|u_{t+i,t}\|_\lambda^2 + \rho\epsilon \tag{21}$$

where $\eta = [\ velocity\ \ \dot{\psi}\ \ \psi\ \ Y\ ]$ are the output variables of the system and ηref denote the corresponding reference signal. n_u is size of input prediction horizon and n_y is size of output prediction horizon. $\Delta U_t = [\ \Delta u_t\ \ \Delta u_{t+1}\ \ .\ \ .\ \ \Delta u_{t+n_u-1}\ ]$ *and* $\Xi_{ref}(t) = [\eta_{ref}(t+1)\ \ \eta_{ref}(t+2)\ \ .\ \ \eta_{ref}(t+n_y)]$ are the sequence of input and reference trajectory over time horizon n_u and n_y respectively.

Eq (21) indicates that the cost function involves trajectory tracking with weight Q while minimizing both control effort and its increment with weights λ and R respectively.

At each sampling instant t, the cost function given in eq (21) is minimized. The optimization problem can be stated as follows :

$$\min_{\Delta U_t} J(\xi_t, \Delta U_t, \Xi_{ref_t})$$

$$\begin{aligned} \text{s.t.}\ \ \xi_{k+1,t} &= A_{k,t}\xi_{k,t} + B_{k,t}u_{k,t} + d_{k,t} \\ \alpha_{k,t} &= \mathcal{C}_t\xi_{k,t} + \mathcal{D}_t u_{k,t} \\ \Delta u_{\min} \le \Delta u_{k,t} &\le \Delta u_{\max} \\ u_{\min} \le u_{k,t} &\le u_{\max} \\ u_{k,t} &= u_{k-1,t} + \Delta u_{k,t} \\ \alpha_{\min} - \epsilon \le \alpha_{k,t} &\le \alpha_{\max} + \epsilon \end{aligned} \tag{22}$$

In eq (22) α is the slip angle which is the constrained output. The constraint on α is soft constraint defined using slack variable ε [9] [10]. The soft constraint are those which can be violated. However the weight factor on the violation ensures that violation penalizes the cost function as indicated by the term $\rho\xi$ in eq (21).

The optimization problem stated in eq (22) can be reformulated as a quadratic problem (QP). The details of quadratic programming and MPC control law has been discussed in [5]-[6]. The sequence of optimal input deviations are computed at every sampling time t by solving (22). The optimal control horizon is

$$\Delta U_{t,t}^* = [\ \Delta u_{t,t}^*\ \ \Delta u_{t+1,t}^*\ \ .\ \ .\ \ .\ \ \Delta u_{t+n_u-1,t}^*\ ] \tag{23}$$

Only the first deviation value is used to obtain optimal control at instant t

$$u_t = u_{t-1} + \Delta u_{t,t}^* \tag{24}$$

$$\hat{u}_{t+1} = u_t + \Delta u_{t+1,t}^* \tag{25}$$

As already explained, at each sampling time t, the linear model is computed by linearization of (13) around state ξ_t and input u_t.

In [1][3][4] for linearization the input used is $u_t = u_{t-1}$ as indicated by (20c), which is not correct in the strict sense. Since MPC is capable of generating the predicted input at instant t which is a better approximation to the actual u_t than u_{t-1}, the linearized model with predicted $\hat{u}_t$ will reduce the error involved in the approximation. For incorporating this improvement, u_{t-1} in (20c) is replaced by $\hat{u}_t$ computed as follows:

$$\hat{u}_t = u_{t-1} + \Delta u^*_{t,t-1} \tag{26}$$

As ideal MPC control law always follows the tail of previous input horizon, the above predicted control is a logical improvisation to equating u_t to u_{t-1}.

In LTV model, over a prediction horizon state space model is assumed to be constant. At next sampling instant $(t+1)$, the horizon moves one step and the system is linearized and optimization problem defined by eq (22) is solved based on the state ξ_{t+1} and input $\hat{u}_{t+1}$.

Use of eq (20c) rather than eq (26) in each successive linearization steps is bound to add to the tracking error because of the cumulative effect of the approximation error.

The improvised model which makes use of eq (26) is seen to fetch better tracking. This is presented in simulation results.

4. Simulation results

As in [1][2][3][4], it is assumed that vehicle is entering a double lane change maneuver on very low frictional road with a given initial speed. The control input is the steering angle and the goal is to follow the trajectory as close as possible by minimizing the vehicle deviation from the target path with highest possible entry speed. In [1][2] only steering control has been used for trajectory tracking. In [3][4] both steering and braking control has been used for this purpose. The simulation results presented in this section are based on both steering and braking. As in [4] a vehicle speed of $21m/sec$ is considered for simulation. As in [1][2][3][4] The reference model for MPC used is same as the system model at starting point. The simulation results for the improved LTVMPC are in accordance with the expected improvement in tracking error.

The control inputs are the front wheel steering angle and slip ratios at the four wheels. Following has been used for tuning of controller :

- Sample time T = 0.05 sec.
- Constraints on steering input and input increments : $\delta_{fmin} = -10°, \delta_{fmax} = 10°, \Delta\delta_{fmin} = -0.85°, \Delta\delta_{fmax} = 0.85°$
- The upper and lower bounds on four slip ratio are 0% and -2.7% respectively. The upper and lower bounds on braking rate are 10 % and -10% respectively.
- Weights on tracking errors: $Q_{\dot{x}} = 0.5, Q_\psi = 12000, Q_{\dot{\psi}} = 3000, Q_Y = 600$
- Weights on input rates: $R_{\delta_f} = 5000, R_{s_{\star,\bullet}} = 50$
- Weights on inputs: $\lambda_{\delta_f} = 1000, \lambda_{s_{\star,\bullet}} = 16$
- Horizon $n_y = 21, n_u = 10$
- Constraints on the tyre slip angles: $\alpha_{min} = -3°, \alpha_{max} = 3°$
- Weight on soft constraint $\rho = 28$
- Entry velocity of vehicle $v = 21m/sec$
- Road friction $\mu_{\star,\bullet} = 0.3$

In figures 3 - 5 simulation results obtained by using the LTVMPC design as reported in [4] are shown. The entry speed of $21m/sec$ is reduced to $18.75m/sec$ at the end of double lane change as seen in fig 3(a). The tracking performance with LTVMPC for ψ, Y, $\dot{\psi}$ are shown in fig 3(b), 3(c), 3(d) respectively.

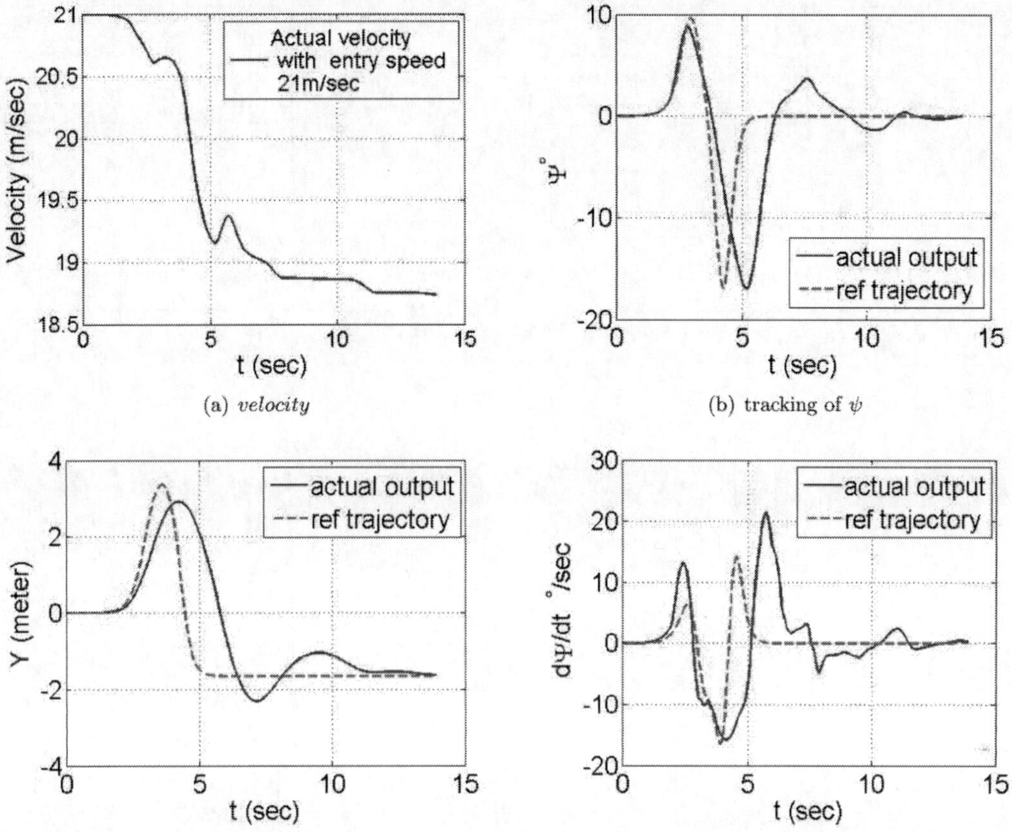

(a) *velocity* (b) tracking of ψ

(c) tracking of Y (d) tracking of $\dot{\psi}$

Figure 3: Tracking with LTVMPC.

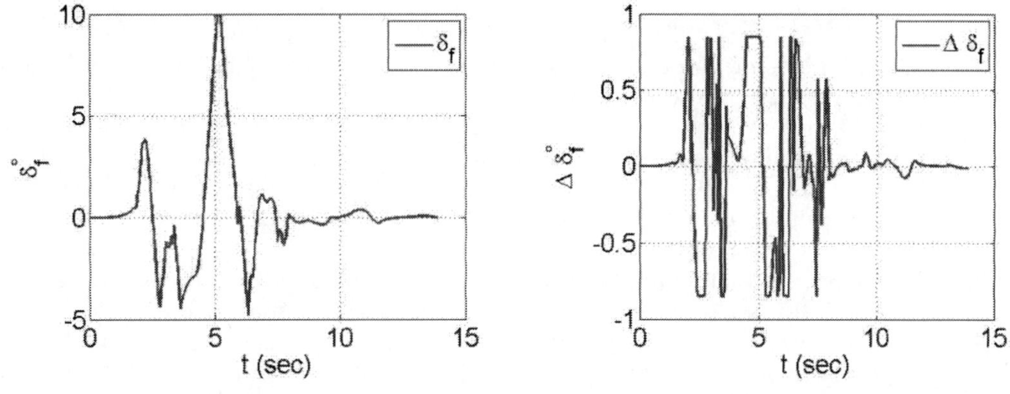

(a) input δ_f (b) input $\Delta\delta_f$

Figure 4: Input and input increment of front wheel steering angle with LTVMPC.

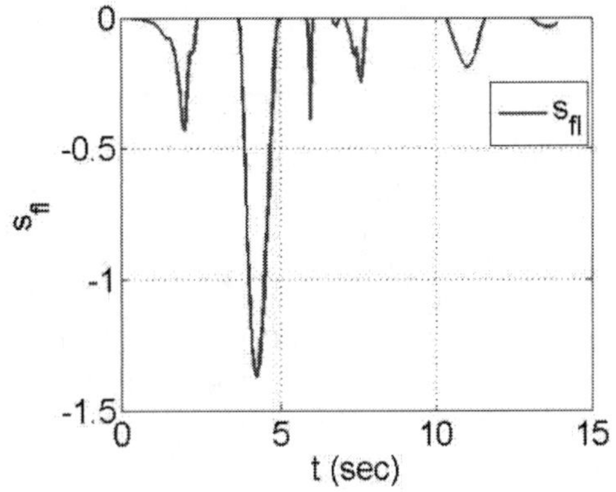

(a) input s_{fl}

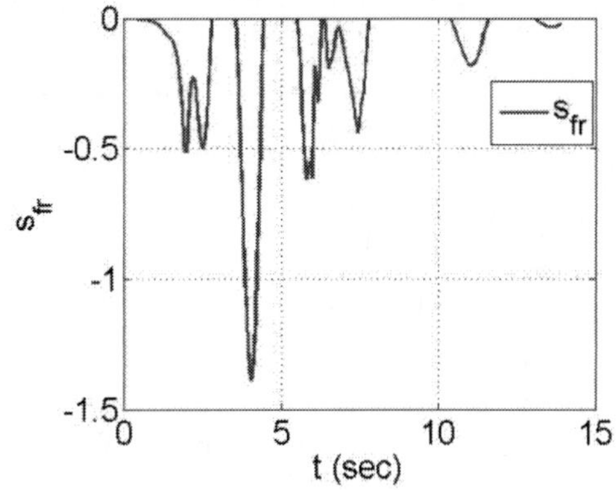

(b) input s_{fr}

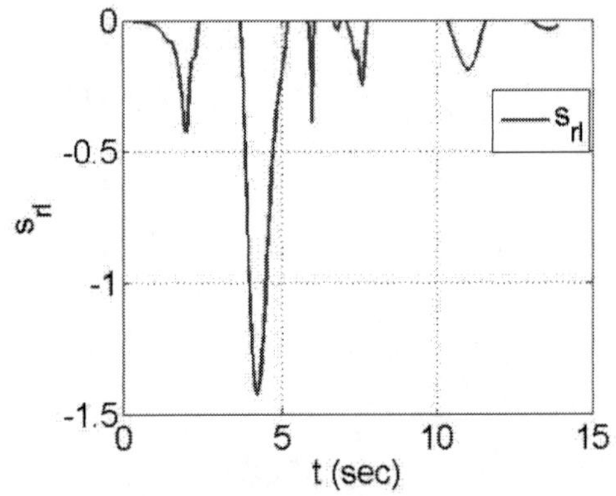

(c) input s_{rl}

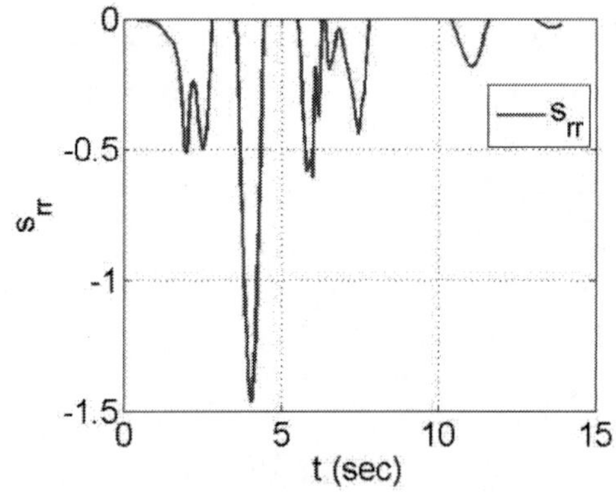

(d) input s_{rr}

Figure 5: Slip ratios at four wheels with LTVMPC.

Fig 4 shows the steering angle and increments of steering angle for LTVMPC. Both these values are bounded within the constraints.

Fig 5 shows the slip ratio at the four wheels for LTVMPC. It can be seen that the slip ratio of right side front and rear wheels are similar to each other. Also for left side front and rear slip ratio are similar. More braking is applied to right side wheels. This input provides differential braking which helps to turn right. This shows that braking is not only applied to reduce speed of vehicle but also to help the steering.

In figures 6 - 8 simulation results of the proposed improved LTVMPC controller are shown. The entry speed of $21m/sec$ is reduced to $19.10m/sec$ which is an improvisation on the achieved speed of $18.75m/sec$ with LTVMPC. Comparison of the velocity profile in fig 3(a) and fig 6(a) shows an improvement in velocity profile for the proposed method towards the latter half of lane change where it is seen that velocity profile continuously maintains a higher velocity than the velocity observed in fig 3(a).

The tracking performance with improved LTVMPC for ψ, Y, $\dot{\psi}$ are shown in fig 6(b), 6(c), 6(d) respectively. An improvement in tracking of Y for improved LTVMPC can be seen in fig

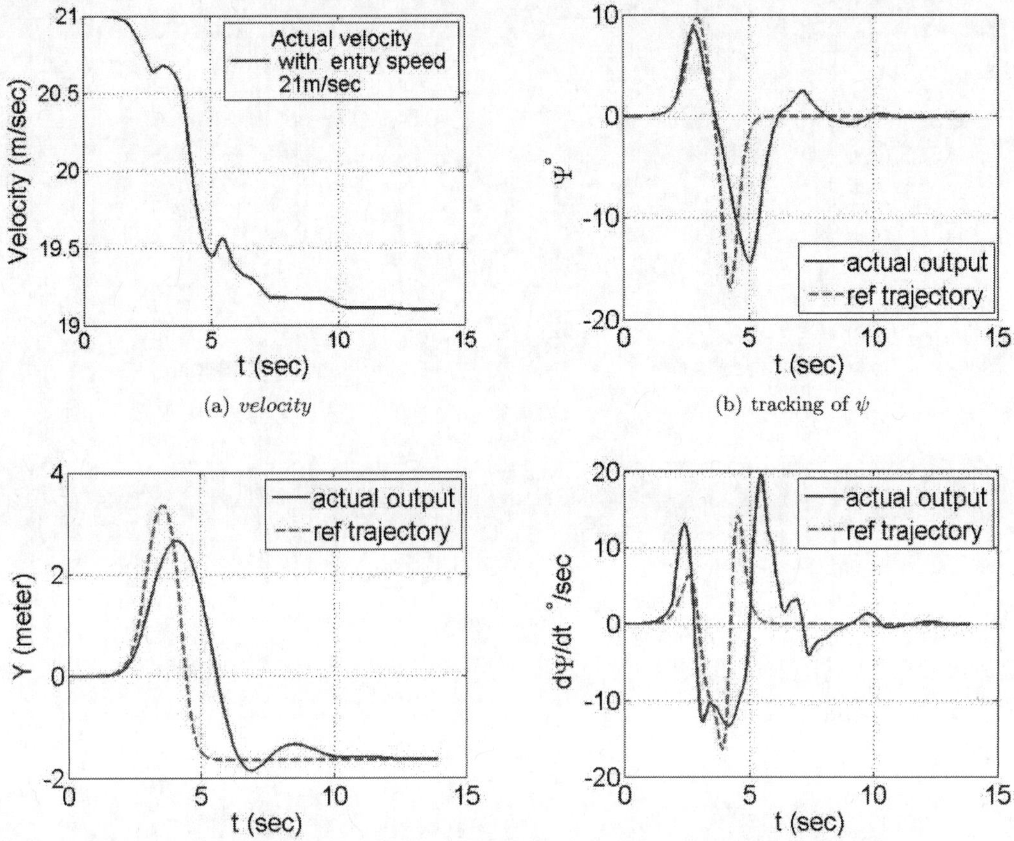

(a) *velocity* (b) tracking of ψ

(c) tracking of Y (d) tracking of $\dot{\psi}$

Figure 6: Tracking with improved LTVMPC.

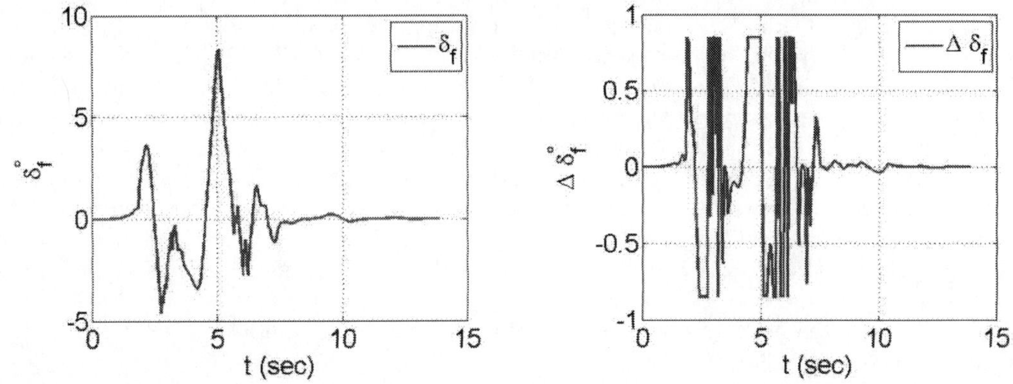

(a) input δ_f (b) input $\Delta\delta_f$

Figure 7: Input and input increment of front wheel steering angle with improved LTVMPC.

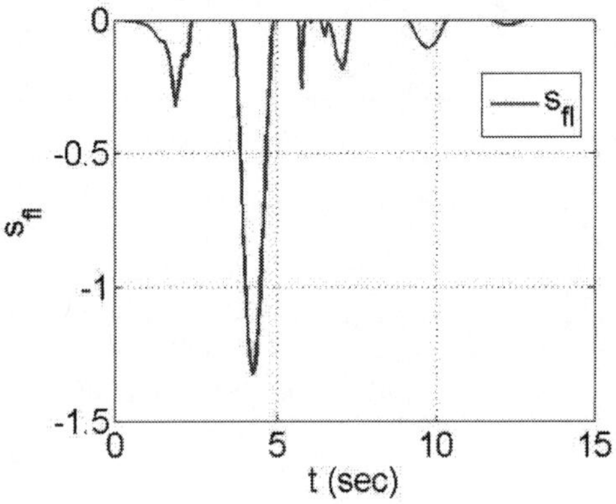

(a) input s_{fl}

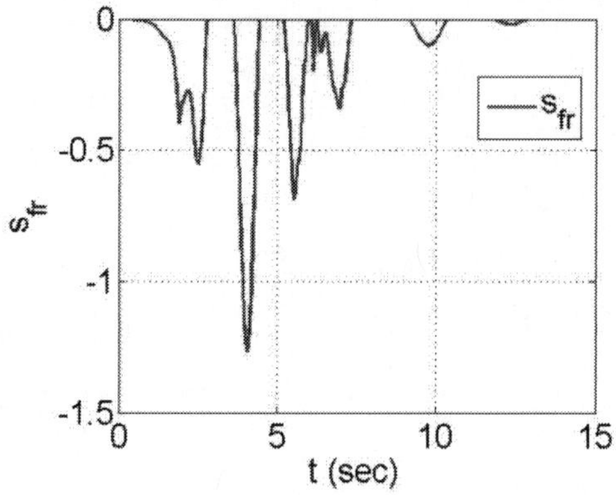

(b) input s_{fr}

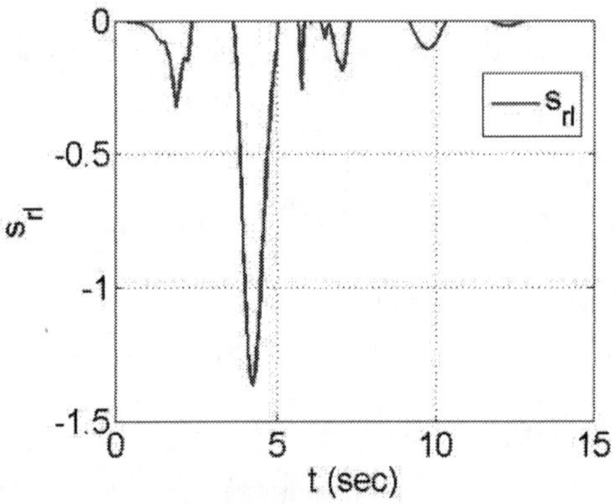

(c) input s_{rl}

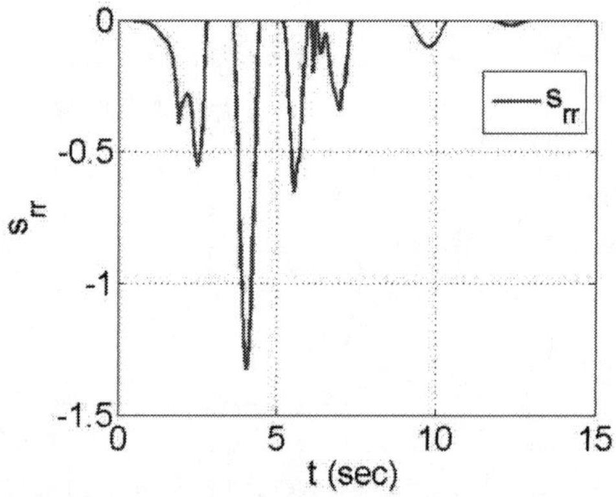

(d) input s_{rr}

Figure 8: Slip ratios at four wheels with improved LTVMPC.

6(c). Also improved LTVMPC reaches steady state faster than that of LTVMPC which can be seen by comparing fig 6(b) and 6(d) with 3(b) and 3(d) respectively.

Fig 7 shows the steering angle and its increment for improved LTVMPC. Both of these values are within the constraints. By comparing input steering angle for LTVMPC shown in fig 4(a) with the input steering angle for improved LTVMPC in fig 7(a), it can be seen that input for LTVMPC reaches the boundary value of 10° and for improved LTVMPC this value is well within its bounds.

Fig 8 shows the slip ratio at four wheels with the improved LTVMPC. As with the LTVMPC here also the braking shown is differential braking where more brakes has been applied to right wheels so that vehicle will turn right. Comparison of fig 8 with fig 5 clearly shows that the braking applied with the improved controller is less than the braking with LTVMPC. This is in accordance with the velocity profiles in fig 3(a) and fig 6(a) where the effect of less braking as in fig 8 is reflected on the higher velocity being maintained in fig 6(a) in comparison to the velocity observed in fig 3(a).

The effect of the improved LTVMPC on its tracking performance will be directly reflected in the rms values of error. This is shown in table 1. In the table, ψ is in degree, $\dot{\psi}$ is in degree/sec and Y is in meter.

Table 1: Comparison of improved LTVMPC with LTVMPC

Controller		e_{avg}	e_{min}	e_{max}	e_{rms}
LTVMPC	ψ	1.0836	−3.6500	16.2548	4.6595
	$\dot{\psi}$	−1.7172	−21.3342	10.74186	5.9206
	Y	−0.3821	−3.5790	0.6580	0.9294
Improved LTVMPC	ψ	0.7907	−2.4357	13.0706	3.1355
	$\dot{\psi}$	−1.4964	−19.0279	4.1650	4.7914
	Y	−0.2862	−3.0738	0.2113	0.7114

From comparison it is seen that rms value of error is substantially less with the improved LTVMPC for all the 3 output variables; this reduction in error being 32.70%, 17.07% and 23.45% for ψ, $\dot{\psi}$ and Y respectively for the entry speed of $21m/sec$.

5. Conclusion

An improved linear time varying model predictive control (LTVMPC) for combined front wheel steering and four wheel braking of autonomous vehicle has been discussed. The performance of the controller for double lane change trajectory on low frictional road has been presented. The LTVMPC controller as reported earlier in literature has been modified by improvising on the linearization of the nonlinear model for the autonomous vehicle which is used for the design of steering control with braking. The accuracy in the linearization is directly reflected on the tracking performance with high entry speed during double lane change. Simulation results are presented to show the efficacy of the methodology. Comparison of the simulation results from LTVMPC and improved LTVMPC shows that the proposed methodology leads to better traits in tracking performance.

References

[1] Falcone P, Borrelli F, Asgari J, Tseng H E, and Hrovat D 2007 Predictive active steering control for autonomous vehicle systems, *IEEE Trans. Control System Technology.* **15** 566-580.

[2] Borrelli F, Falcone P, Keviczky T, Asgari J, and Hrovat D 2005 MPC-based approach to active steering for autonomous vehicle systems *Int. J. Veh. Autonomous Syst.* **3** 265-291.

[3] Falcone P, Tufo M, Borrelli F, Asgari J and Tseng H E 2007 A linear time varying model predictive control approach to the integrated vehicle dynamics control problem in autonomous systems *Proc. of the 46th IEEE Conf. on Decision and Control (New Orleans)* 2980-2985.

[4] Falcone P, Borrelli F, Asgari J, Tseng H E, and Hrovat D 2007 A model predictive control approach for combined braking and steering in autonomous vehicles *15th Mediterranean Conf. on Control and Automation.*

[5] Mayne D, Rawlings J, Rao C and Scokaert P 2000 Constrained model predictive control: Stability and optimality *Automatica* **36** 789-814.

[6] Mayne D 2001 Control of constrained dynamic systems *European J. of Control* **7** 87-99.

[7] Bakker E, Nyborg L and Pacejka H B 1987 Tyre modeling for use in vehicle dynamics studies *SAE paper No. 870421.*

[8] Palmieri G, Barbarisi O, Scala S and Glielmo L 2010 On low complexity predictive approaches to control of autonomous vehicles *Springer book* 195-210.

[9] Kerrigan, Eric C and Jan M 2000 Soft constraints and exact penalty functions in model predictive control *Control Conf. (Cambridge).*

[10] Scokaert, Pierre O and James B 1999 Feasibility issues in linear model predictive control *AIChE J.* **45** 1649-1659.

Output-Feedback Model Predictive Control of a Pasteurization Pilot Plant based on an LPV model

Fatemeh Karimi Pour, Carlos Ocampo-Martinez and Vicenç Puig

Automatic Control Department of the Universitat Politècnica de Catalunya, Institut de Robòtica i Informàtica Industrial (CSIC-UPC). Llorens i Artigas, 4-6, 08028 Barcelona, Spain.

E-mail: {fkarimi, cocampo, vpuig}@iri.upc.edu

Abstract. This paper presents a model predictive control (MPC) of a pasteurization pilot plant based on an LPV model. Since not all the states are measured, an observer is also designed, which allows implementing an output-feedback MPC scheme. However, the model of the plant is not completely observable when augmented with the disturbance models. In order to solve this problem, the following strategies are used: (i) the whole system is decoupled into two subsystems, (ii) an inner state-feedback controller is implemented into the MPC control scheme. A real-time example based on the pasteurization pilot plant is simulated as a case study for testing the behavior of the approaches.

1. Introduction

One of the important food preservation techniques is pasteurization, which is widely used in food industries. The pasteurization process implies applying heat to some products such as milk, cream, beer and others at a specified temperature for a specified period of time. This process is distributed into three sections that are heating, cooling and regeneration sections. The most important section is heating, which involves heat exchanger equipment to heat up the temperature of the product at the desired setpoint, and then maintaining this temperature during a constant time [11].

Among many pasteurization approaches, the High-Temperature Short-Time (HTST) approach is generally accepted as the industry standard [2]. In this process, the time temperature compound can change depending on the product and some of its properties as viscosity, fat percentage, solid residues, etc. Controlling and maintaining the temperature of the process is an important key in pasteurization. Accordingly, a suitable control system to control product temperature needs to be designed for keeping the desired product quality. The necessity of a significant control for the process arises from the saving in energy, product and time if an accurate tracking of the setpoint is performed, and if disturbance influences are rapidly and productively rejected.

Recently, model-based predictive control (MPC) technology gained its industrial position in the process industry [13], because of its ability to deal with multivariable control, delays and constraints on system variables [5]. The basic idea behind MPC is to repeatedly solve an optimal control problem on-line to find an optimal input to the controlled system. Nonetheless, continuous technological advances during the last years make possible the real-time implementation of the MPC controllers based on more complex and large-scale dynamical models. Therefore, several possibilities arise when implementing real-time predictive controllers for industrial processes.

Furthermore, a pasteurization system includes typical behaviors of relevant processes, such as complex dynamical models with nonlinearities, which imply important challenges when a suitable controller should be designed [14]. In the literature, there are different modeling approaches and controller strategies for pasteurization systems. In [6], the complete process model leads to a multivariable first order with pure delay transfer functions with variable parameters after decomposing the system into functional subsystems. Other models from the whole system and/or some subsystems are obtained in [1, 11]. Regarding its control, in [7] a Dynamic Matrix Control (DMC) is designed and implemented by using the system models proposed in [6]. In [11], it is proposed a scheme based on PID controllers with different tuning methods. The regulation of both water and milk temperatures by using MPC is reported in [12], where transient behaviors have been suitably handled with respect to other control techniques such as cascade generic model control (GMC) strategy. A recent study by [14] involved three different control topologies based on MPC that aim to the minimization of the energy consumption. However, after revising the literature the pasteurization plant have never been applied controller based on Linear Parameters Varying (LPV) models of the pasteurization plants. In addition, theses studies used MPC controller with full knowledge of the state, whereas, in many control problems, not all the states can be measured exactly. In this situation, the output-feedback MPC is more practical than the state-feedback one. Thus, an observer can be employed for estimating the states if the model of the system has the observability condition. One of the main problems that can be found for designing a state observer and estimating the states is to have an unobservable model. In this paper, some states of the system can not be measured exactly, therefore an observer is needed but the system model including disturbances has observability problem.

This paper proposes the design and simulation of an observer-based output-feedback MPC controller to regulate the temperature of the output product in a small-scale pasteurization plant, taking into account the complete model from the pasteurization system has unobservable states. The LPV model of the pasteurization plant is obtained from experimental data according to [7]. Then, the identified models obtained as transfer functions are suitably formulated by their equivalent controllable realizations in state-space with varying parameters. Owing to this, the main contribution of this paper is not only the suitable design of an output-feedback MPC controller with a state observer based on an LPV model of the pasteurization plant in order to satisfy the control objectives fulfilling system constraints, but also to address the observability problem of the pasteurization system when considering the disturbance model. Moreover, in this paper, the MPC controller is computed by using the LPV model during the simulation scenario and Linear Time Varying (LTI) along the prediction horizon.

The remainder of the paper is organized as follows. In Section 2, the pasteurization plant and pasteurization process are described. Besides, the mathematical models and statement of the control problem are presented. In Section 3, approaches for solving problems described in Section 3 are proposed. Section 4 presents and discusses the results. Finally, in Section 5, the conclusions of this work are drawn and some research lines for future work are proposed.

2. PROBLEM STATEMENT

2.1. System Description

The considered pasteurization process is the small-scale plant PCT-23 MKII, manufactured by Armfield (UK) [3]. This laboratory plant is the small version (1.2m, 0.6m, 0.6m) of several real-time industrial pasteurization process. The High-Temperature Short-Time (HTST) approach is generally accepted as the industry standard for the pasteurization process [6]. In this process, the goal is to heat and preserve the product, which is typically a liquid, at a predetermined temperature for a minimum time. This is achieved by circulating the heated liquid through a holding tube that delays the product stream [14]. Throughout the pasteurization process, the liquid is pumped at a predetermined flow rate from one of two storage tanks to the heat exchanger indirectly. The water heat is transferred to the product inside the first phase of the heat exchanger, which is called regenerator. The raw product is heated to an intermediate temperature by using missed energy from the pasteurized product. Then, this product is heated from that middle temperature to the full pasteurization temperature, in the second phase, while using a hot-water flow (F_h) coming from a closed circuit with a heater. The temperature (T_{past}) at the output of the holding

IOP Conf. Series: Journal of Physics: Conf. Series **783** (2017) 012029 doi:10.1088/1742-6596/783/1/012029

tube is used to monitor the product temperature after the pasteurization process. Finally, the product is cooled in the third phase of the heat exchanger, where remaining heat is removed to the inlet product. This last phase does not add anything new to the model, and so it was not considered in this paper. In summary, from a control point of view, the pasteurization plant can be seen as a multiple-input and multiple-output (MIMO) system with electric heater power, P, and hot/cold flow ratio ($R = Fh/Fc$) as inputs, and temperatures water heater T_{ow}, and T_{past} as outputs. In addition, T_{iw} and T_{ic} are disturbances of the system, where T_{iw} is the input of water heater and T_{ic} is the cold temperature input of the heat exchanger.

2.2. Mathematical Model

The block diagram of the model used in this article has been established in [7], where the hot /cold water flow ratio ($R = F_h/F_c$) was modified to make possible its control. The complete scheme of plant is shown in Figure 1.

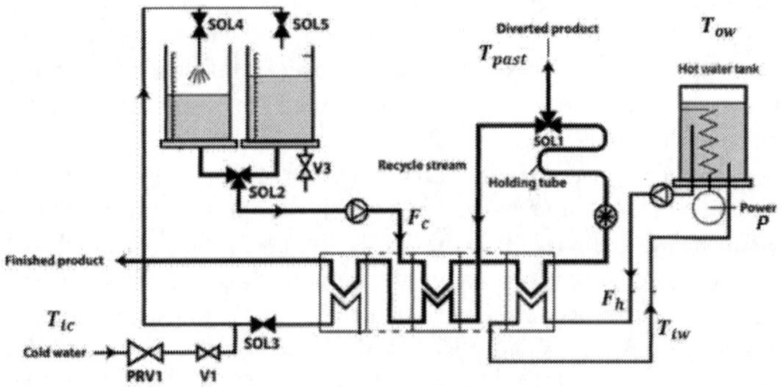

Figure 1. Plant scheme.

2.2.1. *Hot water tank model:*

The heater subsystem is an electrically heated reservoir that is covered in order to minimize heat losses. In the hot water tank, the water is warmed by means of the power resistor, whereas a water pump with an upper limit of 700 ml/min moves the heated water. This flow, defined as hot flow F_h, transfers heat to the pasteurization product in a heat exchanger, before returning it to the hot water tank. Generally, this hot flow changes along the time in order to keep the pasteurization process. According to the dotted red box in Figure 2, the input temperature T_{iw} and the input power P are considered as two inputs for the hot water tank subsystem, while the temperature of the water heater T_{ow} is the system output.

2.2.2. *Heat exchanger model:*

This section includes two effective phases, the former is based on the regeneration phase and the lattes is related to the heating phase. The heating phase (second phase) is more complex than the regeneration phase (first phase), also the first phase is obtained as a simplification of the second one, hence, the heating phase is studied first.

Heating phase (second phase): The block diagram obtained for the heat exchanger is shown in Figure 2 (green box). As reported in [7], the transfer functions related to

$$G_{ij} = \frac{K_{ij}(R)}{1 + \tau_{ij}(F_c, F_h)s} e^{-\theta_{ij}(F_c, F_h)s}. \quad (1)$$

The transfer functions G_{11} and G_{21} generally correspond to the hot water flow, while G_{12} and G_{22} describe the variations caused by the cold water flow. The pasteurization time (or holding time) is fixed

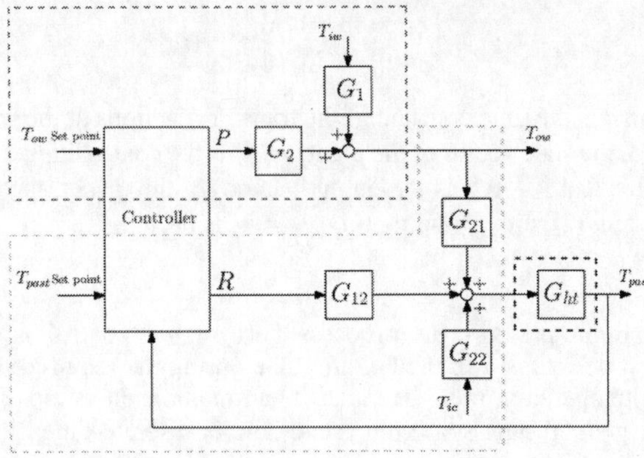

Figure 2. Block diagram of the control loop.

during the real pasteurization process. Hence, the cold water flow must be kept as a constant value. Therefore, G_{12} and G_{22} will show fixed dynamics, but with varying gain, since the static gain depends on R according to (1). Note that the parameters of (1) were obtained experimentally in [7].

Regeneration phase (first phase): In this phase of the heat exchanger, the cold water and the hot water flows are the same, thus the relation $R = F_h/F_c$ is constant and of unitary value.

2.2.3. Holding tube model The model of the holding tube is not purely a transport delay because there were some heat losses. The holding tube can be modeled as a single-input single-output system. According to [7], the estimation parameters for transfer function G_{ht} (see inside dotted black line in Figure 2) are considered as static gain $K_{ht} = 0.91$ and time constant $\tau_{ht} = 21$s.

Accordingly, identified models obtained as transfer functions are suitably stated by their equivalent controllable realizations in stat-space, with varying parameters according to the hot water flow, F_h, as exogenous input and the hot/cold water flow ratio, R, as the input of the system. As discussed before, the state-space model of this plant has four inputs: P and R are the manipulated inputs, while T_{iw} and T_{ic} are disturbances. Moreover, T_{ow} and T_{past} the temperature of the hot water tank and the pasteurization temperature of the process are the first and second outputs of this system, respectively. Therefore, the continuous time state-space model of the pasteurization plant can be represented as follows:

$$x(k+1) = \begin{bmatrix} \frac{-1}{\tau_1(F_h(t))} & 0 & 0 & 0 & 0 & 0 \\ 0 & \frac{-1}{\tau_2(F_h(t))} & 0 & 0 & 0 & 0 \\ \frac{K_{21}(R(t))}{\tau_{21}(R,(t))} & \frac{K_{21}(R(t))}{\tau_{21}(R(t))} & \frac{-1}{\tau_{21}(R(t))} & 0 & 0 & 0 \\ 0 & 0 & 0 & \frac{-1}{\tau_{12}(R(t))} & 0 & 0 \\ 0 & 0 & 0 & 0 & \frac{-1}{\tau_{22}(R(t))} & 0 \\ 0 & 0 & \frac{K_{ht}}{\tau_{ht}} & \frac{K_{ht}}{\tau_{ht}} & \frac{K_{ht}}{\tau_{ht}} & \frac{-1}{\tau_{ht}} \end{bmatrix} x(k)$$

$$+ \begin{bmatrix} 0 & 0 \\ \frac{K_2(F_h(t))}{\tau_2(F_h(t))} & 0 \\ 0 & 0 \\ 0 & \frac{K_{12}(R(t))}{\tau_{12}(R(t))} \\ 0 & 0 \\ 0 & 0 \end{bmatrix} u(k) + \begin{bmatrix} \frac{K_1(F_h(t))}{\tau_1(F_h(t))} & 0 \\ 0 & 0 \\ 0 & 0 \\ 0 & 0 \\ 0 & \frac{K_{22}(R(t))}{\tau_{22}(R(t))} \\ 0 & 0 \end{bmatrix} w(k),$$

$$y(k) = \begin{bmatrix} 1 & 1 & 0 & 0 & 0 & 0 \\ 0 & 0 & 0 & 0 & 0 & 1 \end{bmatrix} x(k), \tag{2}$$

where K is a static gain and τ is the time constant of the transfer functions of the subsystems. In addition, the variable $x \in \mathbb{R}^{n_x}$ denotes the state vector of the plant, $u \in \mathbb{R}^{n_u}$ the control input vector, $w \in \mathbb{R}^{n_d}$ denotes the disturbance input vector and $y \in \mathbb{R}^{n_y}$ is the output vector. In this paper, the discretized state-space model of the complete system (2) with sampling time $T_s = 4$s is used.

2.3. Control Problem

As previously mentioned, in the pasteurization process there are two control objectives to be achieved. First, the raw product must be retained in the holding tube during the required pasteurization time and second the pasteurization temperature must be reached and maintained as close as possible to the set-point values. Consider the general state-space representation for discrete-time LPV systems

$$x(k+1) = A(\theta(k))x(k) + B(\theta(k))u(k) + E(\theta(k))w(k), \tag{3a}$$
$$y(k) = C(\theta(k))x(k), \tag{3b}$$

where the discrete time is denoted by $k \in \mathbb{Z}$, the system matrices $A(\theta(k)) \in \mathbb{R}^{n_x \times n_x}$, $B(\theta(k)) \in \mathbb{R}^{n_x \times n_u}$, $E(\theta(k)) \in \mathbb{R}^{n_x \times n_d}$ and $C(\theta(k)) \in \mathbb{R}^{n_y \times n_x}$ are assumed to depend linearly on the parameter vector $\theta(k) := [\theta_{1,k}, \theta_{2,k}, ..., \theta_{N,k}]^T \in \mathbb{R}^N$, which belongs to a convex polytope Θ defined by

$$\Theta := \left\{ \theta(k) \in \mathbb{R}^N | \sum_{j=1}^{N} \theta_{j,k} = 1, \theta_{j,k} \geq 0 \right\}, \tag{4}$$

where N is the number of vertices. Clearly, as $\theta(k)$ varies inside the convex polytope Θ, the matrices of the system (3) vary inside a corresponding polytope Ψ, which is defined by the convex hull (Co) of N local matrix vertices $[A_i, B_i, E_i, C_i]$, $i \in [1, .., N]$,

$$\Psi := Co\{[A_1 \quad B_1 \quad E_1 \quad C_1], [A_2 \quad B_2 \quad E_2 \quad C_2], ..., [A_N \quad B_N \quad E_N \quad C_N]\}, \tag{5}$$

and the matrices of the system (3) can be rewritten as

$$A(\theta(k)) = \sum_{j=1}^{N} \theta_{j,k} A_j, \quad B(\theta(k)) = \sum_{j=1}^{N} \theta_{j,k} B_j, \quad E(\theta(k)) = \sum_{j=1}^{N} \theta_{j,k} E_j, \quad C(\theta(k)) = \sum_{j=1}^{N} \theta_{j,k} C_j. \tag{6}$$

Moreover, the hot water flow F_h, and hot/cold water flow ratio R are the scheduled variables of the pasteurization model, where F_h and R are exogenous input and endogenous input of the system, respectively. The process of pasteurization changes so fast. Therefore, the signal at the current time step is equal to the previous step and next step ahead. In this situation, the equal scheduled variables of models are assumed along prediction horizon N_p in the LPV model. Thus, $\theta_{(k|k)} \simeq \theta_{(k+1|k)} \simeq \theta_{(k+2|k)} \simeq ... \simeq \theta_{(k+N_p-1|k)}$ can be considered in a short N_p. Based on the fact that one of the scheduled varietals is the manipulated input of the system, the following condition can be written in a few N_p:

$$u_{(k|k)} \simeq u_{(k+1|k)} \simeq u_{(k+2|k)} \simeq ... \simeq u_{(k+N_p-1|k)},$$

it means that current input $u_{k|k}$ can be used for the next system matrix $A(\theta_{k+1|k})$. Consequently, the model of pasteurization plant during the simulation scenario is considered as an LPV model. In addition, the LTI model is used along N_p into the MPC controller. Taking into account the control objectives described before, an MPC controller can be achieved for the pasteurization system with the following cost function:

$$\min_u J(k) = \sum_{i=0}^{N_p-1} \|Cx(k+i|k) - r(k+i+1|k)\|_{W_1}^p + \|u(k+i|k)\|_{W_2}^p, \tag{7a}$$

subject to

$$x(k+i+1) = A(\theta(k+i))x(k+i) + B(\theta(k+i))u(k+i) + E(\theta(k+i))w(k+i), \quad \forall i \in \{0,...,N_p - 1\} \tag{7b}$$

$$y(k+i) = Cx(k+i), \quad \forall i \in \{0,...,N_p\} \tag{7c}$$

$$\underline{x} \le x(k+i) \le \overline{x}, \quad \forall i \in \{0,...,N_p\} \tag{7d}$$

$$\underline{u} \le u(k+i) \le \overline{u}, \quad \forall i \in \{0,...,N_p - 1\} \tag{7e}$$

$$\theta(k+i) \in \Theta, \quad \forall i \in \{0,...,N_p\} \tag{7f}$$

where p denotes the norm used, also $W_1 \in \mathbb{R}^{n_x \times n_x}$ and $W_2 \in \mathbb{R}^{n_u \times n_u}$ are positive definite weighting matrices needed to prioritize the different control goals within the multi-objective cost function for state and the control input, respectively. The vector $r(k)$ is the reference. Vectors $\underline{x}$ and $\overline{x}$ determine the minimum and maximum possible state values of the system, respectively. Also $\underline{u}$ and $\overline{u}$ determine the minimum and maximum possible control inputs, respectively. To distinguish actual state and predicted trajectories, in what follows $k+i|k$ denotes future values at time $k+i$ predicted at time k. According to the model presented in this paper, not all states of the system can be measured while only the output $y(k)$ is available for measurement. Thus, an observer to estimate the system state is required for closing the feedback control loop. However, in (2), the pair (A,C) is not observable because of the inclusion of the disturbance models and it has caused the state related to the disturbance input generates an observability problem for the complete model. The following sections present and discuss two specific approaches for solving this problem.

3. PROPOSED APPROACH

3.1. Problem formulation

The pasteurization plant described in Section 2 can be controlled with a holistic approach by means of one single MPC controller that directly operates the actuators from the states measurement. A completed model of the pasteurization plant is needed for this approach. In order to design the corresponding MPC, the constraints and cost function in (7) should be defined. As already discussed, some states of the system can not be measured exactly and hence an observer should be employed to estimate the states. In this paper, the topology of the observer is based on the classical Luenberger structure [9]. The observer equations can be written as

$$\hat{x}(k+1) = A(\theta(k))\hat{x}(k) + B(\theta(k))u(k) + L(\theta(k))(y(k) - \hat{y}(k)) + E(\theta(k))w(k), \tag{8a}$$

$$\hat{y}(k) = C(\theta(k))\hat{x}(k), \tag{8b}$$

where $\hat{x} \in \mathbb{R}^{n_x}$ is the current observer state, $u \in \mathbb{R}^{n_u}$ is the current control input, $\hat{x}(k+1)$ is the successor state of the observer system, $\hat{y} \in \mathbb{R}^{n_y}$ is the current observer output and $L \in \mathbb{R}^{n_x \times n_y}$ denotes the observer gain. The design of the LPV observer can be achieved by considering the LMI pole placement technique [4], which allows to locate the poles of the observer in a subregion of the unit circle using an LMI region [10]. The motivation for seeking pole clustering in specific regions inside the unitary circle is to obtain fast observer dynamics for all considered operating points.

As mentioned in Section 2.3, the considered model in (2) becomes unobservable when the disturbance is added into the matrix system. In the following, two ways of solving this issue will be considered.

3.2. Decouple the complete system into two separate subsystems

In particular, while the complete model of pasteurization plant is unobservable, it becomes possible that the decomposed system becomes observable. In this way, the complete system can be decomposed into two separate subsystems: (i) hot water tank, (ii) heat exchanger and holding tube. Then, the two subsystems become observable. In this case, two independent controllers are needed. The first control loop (see inside the dotted black line in Figure 3) is dedicated to control T_{past} with the controller output R and considering the other inputs to the heat exchanger as measurable perturbations. The second controller will maintain T_{ow} constant in spite of changes both in R (and therefore in F_h) and in T_{iw}, with the power P as the controller output (see inside dotted red line in Figure 3).

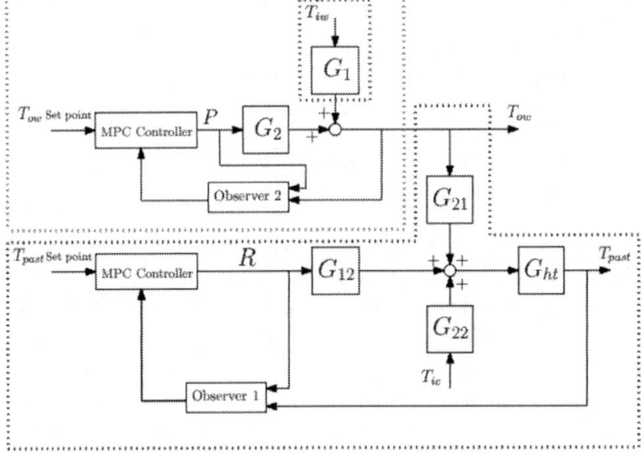

Figure 3. Scheme of split the complete system into two separate subsystems.

3.2.1. Hot water tank: According to Section 2, the hot water tank transfer function presents as dependent on the exogenous variable F_h, while the input temperature of water heater T_{iw} and power P are disturbance input and manipulated input of the hot water tank, respectively. Furthermore, the state-space model of this subsystem can be represent as

$$x_{hot}(k+1) = \begin{bmatrix} \frac{-1}{\tau_1(F_h(t))} & 0 \\ 0 & \frac{-1}{\tau_2(F_h(t))} \end{bmatrix} x_{hot}(k) + \begin{bmatrix} 0 & \frac{K_1(F_h(t))}{\tau_1(F_h(t))} \\ \frac{K_2(F_h(t))}{\tau_2(F_h(t))} & 0 \end{bmatrix} u_{hot}(k), \tag{9a}$$

$$y_{hot}(k) = \begin{bmatrix} 1 & 1 \end{bmatrix} x_{hot}(k), \tag{9b}$$

where x_{hot}, u_{hot} and y_{hot} are the states, input and output of the hot water tank subsystem, respectively. Note that the pair (A,C) is not observable based on the existence of the disturbance in this subsystem. In this case, to avoid this problem, one option is to consider the unobservable part (see inside dotted blue line in Figure 3) of the system as a measurement disturbance. Therefore, the model (9) can be reduced by removing the unobservable part dynamic (blue line in Figure 3). Then, the evaluation of the disturbance in the future can be anticipated. Thus, the state-space of hot water model can be rewritten as

$$x_{hot}(k+1) = A(\theta(k))x(k) + B(\theta(k))u(k), \tag{10a}$$

$$y_{hot}(k) = C(\theta(k))x(k) + y_{unobs}(k), \tag{10b}$$

and

$$x_{unobs}(k+1) = A_{unobs}(\theta(k))x_{unobs}(k) + B_{unobs}(\theta(k))u(k), \tag{11a}$$

$$y_{unobs} = C_{unobs}(\theta(k))x_{unobs}(k), \tag{11b}$$

where x_{unobs} and y_{unobs} are states and output of the unobservable part dynamic of the subsystem, respectively. In addition, A_{unobs} and B_{unobs} are matrices system of this part. After all, the pair (A,C) is observable in the system (10). Therefore, the Luenberger observer is used for estimating the states of hot water tank subsystem.

3.2.2. Heat exchanger and Holding tube: As mentioned in Section 2, the transfer function of the heat exchanger second phase introduces a dependence on R, and the holding tube is a transfer function with constant values of both the static gain and time constant. In addition, the hot/cold flow R, hot water tank temperature $T_o w$ and input temperature of cold water, T_{ic} are inputs of this subsystem. Therefore, the state-space model of this part is represented as

$$x_{heat}(k+1) = \begin{bmatrix} \frac{K_{21}(R(t))}{\tau_{21}(R(t))} & 0 & 0 & 0 \\ 0 & \frac{K_{12}(R(t))}{\tau_{12}(R(t))} & 0 & 0 \\ 0 & 0 & \frac{K_{22}(R(t))}{\tau_{22}(R(t))} & 0 \\ \frac{K_{ht}}{\tau_{ht}} & \frac{K_{ht}}{\tau_{ht}} & \frac{K_{ht}}{\tau_{ht}} & \frac{-1}{\tau_{ht}} \end{bmatrix} x_{heat}(k) + \begin{bmatrix} 0 & \frac{K_{21}(R(t))}{\tau_{21}(R(t))} & 0 \\ \frac{K_{12}(R(t))}{\tau_{12}(R(t))} & 0 & 0 \\ 0 & 0 & \frac{K_{22}(R(t))}{\tau_{22}(R(t))} \\ 0 & 0 & 0 \end{bmatrix} u_{heat}(k) \quad (12a)$$

$$y_{heat}(k) = \begin{bmatrix} 0 & 0 & 0 & 1 \end{bmatrix} x_{heat}(k), \quad (12b)$$

where x_{heat}, u_{heat} and y_{heat} are the states, input and output of the second subsystem (heat exchanger and holding tube), respectively. According to observability rule, the state-space model in (12) is observable.

3.3. Complete system with local state-feedback controller

Another possibility to solve the unobservability of the complete model for pasteurization plant is to use another control loop inside the control scheme. The control scheme for this approach is presented in Figure 4.

In other words, making a closed-loop model of a complete dynamic model of the pasteurization plant by using the state-feedback controller before the MPC controller and in this case, the observability of the complete model of the pasteurization system is determined. It is assumed that a model of the plant is controllable, also as in the first approach, the control objective remains the same, forcing the controlled output to track the references whilst satisfying the constraints. As a matter of fact, just some states of the pasteurization system are measurable and these can be used into inner controller for determining the static gain (see Figure 4). Consider a state-feedback control law for inner control loop as

$$u^*(k) = K(\theta(k))x^*(k), \quad (13)$$

where x^* includes some measurable states and $K(\theta(k))$ is the scheduled control gain. A suitable gain for the whole LPV model is obtained via the weighted average of the control law designed for each vertex, i.e.,

$$K(\theta(k)) = \sum_{j=1}^{N} \theta_{j,k} K_j, \quad (14)$$

where $K_j \in \mathbb{R}^{n_u \times n_x}$ is a time-invariant feedback gain associated with the $j-th$ vertex system. Moreover, the stack of gains for the scheduling control law K_j should be found by solving an LQR problem [8] at each vertex. For the complete system in (2), by using the control law (13), the closed-loop model of system representation is written as

$$x(k+1) = A_{cl}(\theta(k))x(k), \quad (15a)$$

$$y(k) = C(\theta(k))x(k), \quad (15b)$$

where

$$A_{cl} = \sum_{i=1}^{N} \sum_{j=1}^{N} \theta_{i,k}\theta_{j,k}(A_i + B_i K_j), \quad (16)$$

and A_i, B_i are local matrix vertices with $[A_i, B_i]$, $i \in [1,..,N]$. In this case, analyzing the observability of the closed-loop model yields that the complete model with local controller is observable.

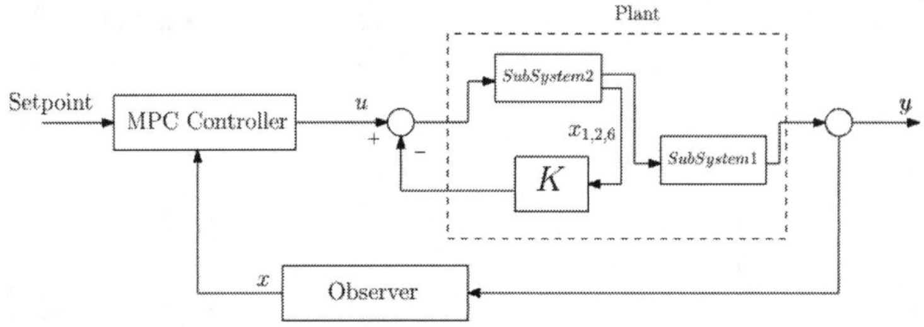

Figure 4. Control scheme of pasteurization process with inner state-feedback controller.

4. SIMULATION RESULTS

In this section, results of two of the proposed approaches are presented and analyzed in detail.

4.1. Scenarios

4.1.1. Scenario 1: As discussed in Section 3, the complete model of the pasteurization plant is separated in two parts: hot water tank and heat exchanger with holding tube. For this experiment, the water heater input temperature T_{iw} was maintained constant at $\simeq 30\,°\mathrm{C}$. Therefore, the behavior of the subsystem inside the dotted blue box in Figure 3 can be computed separately. Then, the value of the response is added to the subsystem as measurement disturbance in output. In addition, the LPV state observer is designed assuming that the eigenvalues are placed in an LMI region that is the intersection between the disk of radius $r = 0.6$ and center $q = 0$. The control objective of MPC controller for the hot water tank subsystem is temperature T_{ow} should be track the setpoint where their references change from 40 °C to 50 °C. The MPC controller has been implemented in simulation, where the weight matrices are $Q = 1$, $R = 0.01$. The prediction horizon is chosen $N_p = 5$ with $T_s = 1$ min. In second part of this approach, the pasteurization temperature T_{past} is the variable to be controlled and R will be used as the controller output. The first stage is supposed to be disconnected, so that the input temperature to the heat exchanger T_{ic} is the raw temperature and it will be maintained at $\simeq 20\,°\mathrm{C}$.

The control objective of MPC controller for the second subsystem is the temperature T_{past} and it should follow a predefined profile that is tracked the setpoint where their references change from 25 °C to 27 °C. The prediction horizon is chosen $N_p = 7$ where the weight matrices are $Q = 1$, $R = 0.04$ with $T_s = 4$s. Moreover, the state observer is designed by using the LMI method, where $r = 0.4$ and $q = 0$.

4.1.2. Scenario 2: The second approach involves the controller and observer of the complete system with inner state-feedback controller. The control gain for the state-feedback controller is computed by using the LQR method, while the measurable states as x_1,x_2 and x_6 are used also the control gain is computed at every iteration. Similar to the first scenario, the state observer is designed by LMI considering that the eigenvalues are located in an LMI region that is the intersection between the disk of radius $r = 0.6$ and $q = 0$. In order to improve the convergence of the observer, the following initial condition on the pasteurization plant and the observer states have been considered: $x(0) = \begin{bmatrix}0 & 0 & 0 & 0 & 0 & 0\end{bmatrix}^T$, $\hat{x}(0) = \begin{bmatrix}5 & 5 & 5 & 5 & 5 & 5\end{bmatrix}^T$. Now, the MPC controller was implemented with the weighting matrices are $Q = 1$, $R = 1$ and the prediction horizon $N_p = 7$. Furthermore, the references of T_{past} and T_{ow} are considered in the same way as introduced in Scenario 1 when the T_{ic} and T_{iw} are maintained constant at $\simeq 30\,°\mathrm{C}$ and 20 °C as measurement input disturbances. In addition, the first input heating power of resistor P can take value in the range $[0,2]$ KW and the range value for the second input, the hot/cold flow ratio R, is $[0.2,2]$.

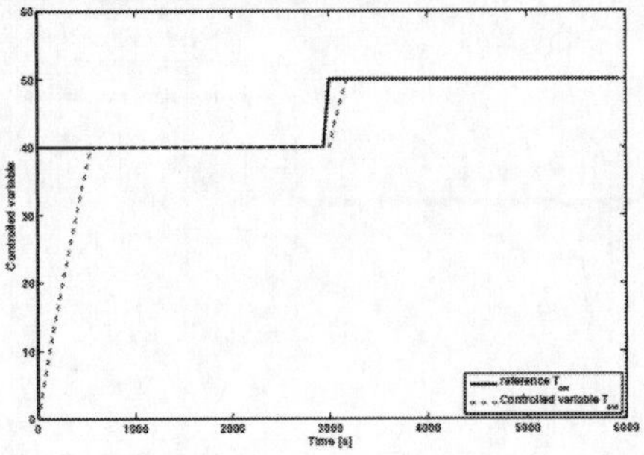

(a) T_{ow} with two MPC controller

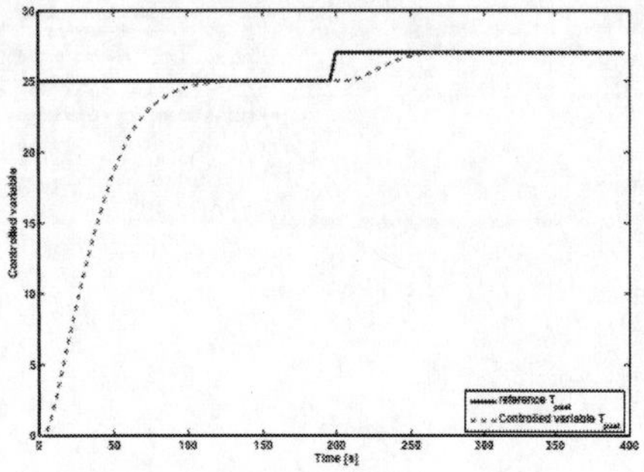

(b) T_{past} with two MPC controller

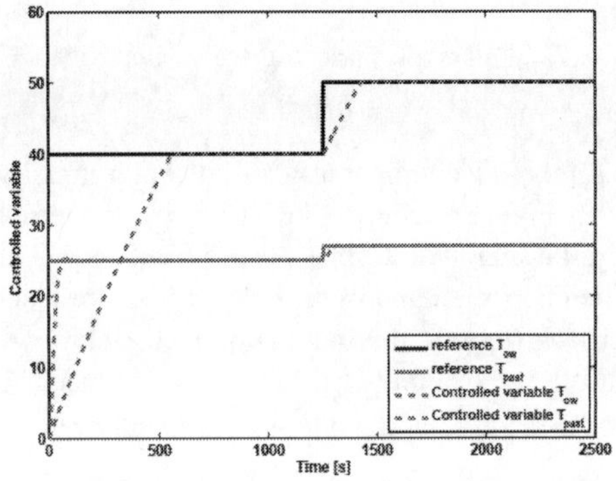

(c) T_{ow} and T_{past} with a single MPC controller

Figure 5. Controlled temperature of pasteurization process with single MPC and two independent MPC.

4.2. *Discussion*

In order to test the behavior of the proposed control scheme regarding tracking and disturbance rejection, several simulations were carried out and the results obtained are presented in this part.

In Figure 5, the temperatures T_{ow} and T_{past} from the pasteurization plant under the MPC controller of the split model (two subsystems) and complete system without the observer are shown.

Figure 5a shows the output temperature of the hot water tank subsystem. In addition, the first reference point is assumed 40°C from time instant 0s until 3000s and the second reference point is considered 50°C from time instant 3000s until 6000s in Figure 5a. In Figure 5b, from time instant 0s until 200s, the first reference point is 25°C and 27°C is considered from time instant 200s until 400s for the second reference point in simulation of output temperature of heat exchanger subsystem. The behavior of the controlled variable of both output temperatures T_{ow} and T_{past} can be observed in Figures 5a and 5b, while the controlled variables track the references. The behavior of controlled temperatures, T_{ow} and T_{past}, from the pasteurization plant with one MPC controller are presented in Figure 5c together with their corresponding references. The output temperatures of hot water tank and heat exchanger in Figures 5c have the similar behaviors in comparison with the same variables in Figures 5a and 5b. As it can be seen in Figure 5c the settling times of hot water tank and heat exchanger are $\sim$ 500s and $\sim$ 120s, respectively. Besides the settling times of hot water tank and heat exchanger are $\sim$ 500s and $\sim$ 120s, in Figure 5a and 5b, respectively. Results obtained by applying first approach are depicted in Figure 6 and Figure 7. These figures shows a comparison between the measurement response of controlled variables

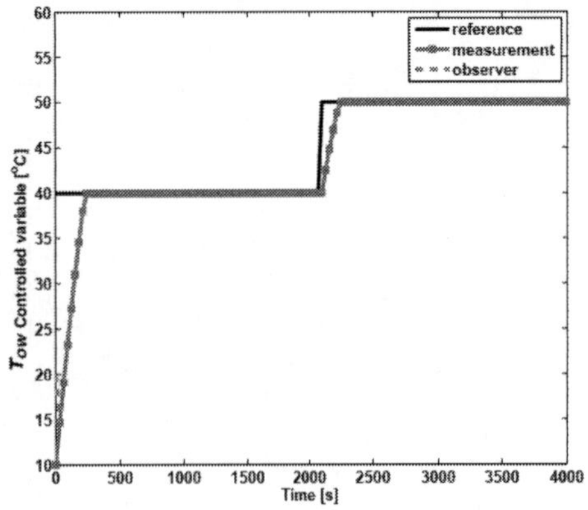

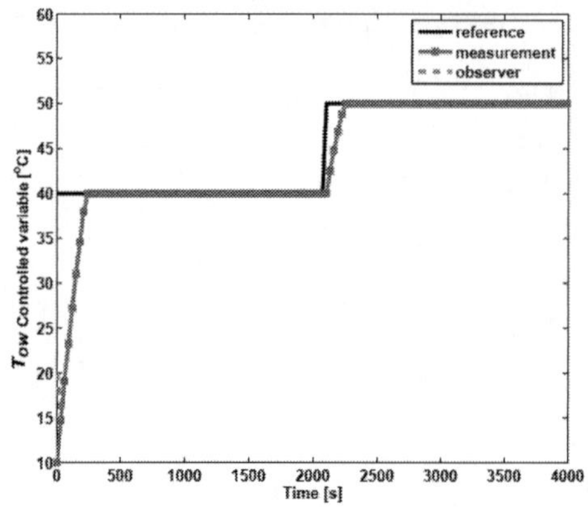

Figure 6. Controlled temperature (T_{ow}) in first Scenario.**Figure 7.** Controlled temperature T_{past} in first Scenario.

and observer output. Figure 6 refers to the temperature of hot water tank T_{ow}, where the complete system is separated into two subsystems. Figure 6 relate to the pasteurization temperature T_{past}, where one of the inputs is T_{ow}. In all cases, it can be seen that despite the difference in the initial conditions, the observer converges to the real state. The controlled and observed temperature responses from the pasteurization process using the second approach (that is, the inner state-feedback controller) are shown in Figure 8. The measurement outputs and observer outputs tracking the references also, the observer converges to measurement outputs after the effect of the initial conditions disappears.

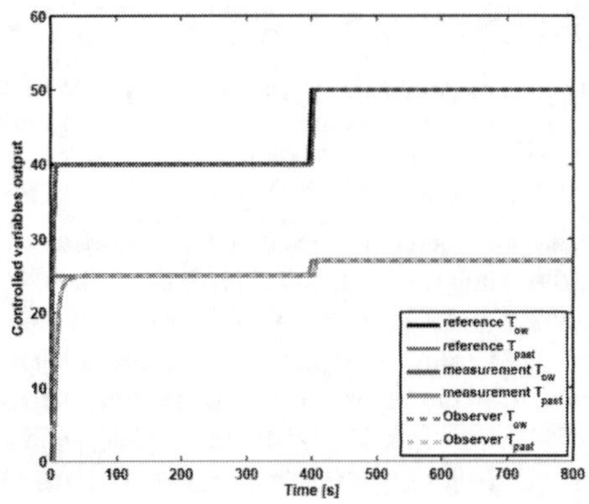

Figure 8. Controlled temperatures T_{ow} and T_{pas} with state-feedback controller (second Scenario).

5. CONCLUSIONS

In this paper, an output-feedback MPC controller has been designed for a pasteurization process, when an LPV mathematical model of the pasteurization plant obtained from experimental data is considered. However, the complete model of this process is not totally observable. Two different ways are proposed for solving this issue. First, the complete model is decomposed into two subsystems. Second, an inner state-feedback controller is used. Then, the output-feedback MPC controller based on the observer is implemented, while the control objective is tracked setpoint with different reference points.

According to the results discussion, the objectives of the pasteurization process are kept close to the reference points by using any of the both approaches proposed. This implies savings in energy because the accuracy and performance are achieved with output-feedback MPC controller and it is not necessary to create a setpoint with a safety differences above the accurate pasteurization temperature. As future research, the proposed approaches will be tested in the real pilot pasteurization plant.

References

[1] P. Aadaleesan, N. Miglan, R. Sharma, and P. Saha. Nonlinear system identification using Wiener type Laguerre–Wavelet network model. *Chemical Engineering Science*, 63(15):3932–3941, 2008.

[2] C. F. Alastruey, M. De la Sen, and M. García-Sanz. Modelling and identification of a high temperature short time pasteurization process including delays. In *Proceedings of the 7th Mediterranean Conference on Control and Automation*, pages 28–30, Israel, 1999.

[3] Armfield. *Process Plant trainer PTC23-MKII, Instruction Manual*. 2015.

[4] P. Gahinet, P. Apkarian, and M Chilali. Affine parameter-dependent Lyapunov functions and real parametric uncertaint. *IEEE Transactions on Automatic Control*, 41(3):436–442, 1996.

[5] D. Hrovat, S. Di Cairano, H. E. Tseng, and I. V. Kolmanovsky. The development of model predictive control in automotive industry: A survey. In *IEEE International Conference on Control Applications (CCA)*, pages 295–302. IEEE, 2012.

[6] J. Ibarrola, J. Guillén, J. Sandoval, and M. García-Sanz. Modelling of a high temperature short time pasteurization process. *Food Control*, 9(5):267–277, 1998.

[7] J. Ibarrola, J. Sandoval, M. Garcıa-Sanz, and M. Pinzolas. Predictive control of a high temperature–short time pasteurisation process. *Control Engineering Practice*, 10(7):713–725, 2002.

[8] H. Kwakernaak and R. Sivan. *Linear optimal control systems*. Wiley-interscience New York, 1972.

[9] D. G. Luenberger. Observing the state of a linear system. *IEEE Transactions on Military Electronics*, 8(2):74–80, 1964.

[10] J. Luna, C. Ocampo-Martinez, and M. Serra. Nonlinear predictive control for the concentrations profile regulation under unknown reaction disturbances in a fuel cell anode gas channel. *Journal of Power Sources*, 282:129–139, 2015.

[11] W. Mokhtar, F. S. Taip, N. Aziz, and S. Noor. Process control of pink guava puree pasteurization process: Simulation and validation by experiment. *International Journal on Advanced Science, Engineering and Information Technology*, 2(4):302–305, 2012.

[12] S. Niamsuwan, P. Kittisupakorn, and I. M. Mujtaba. Control of milk pasteurization process using model predictive approach. *Computers & Chemical Engineering*, 66:2–11, 2014.

[13] S. Qin and T. A. Badgwell. A survey of industrial model predictive control technology. *Control Engineering Practice*, 11(7):733–764, 2003.

[14] A. Rosich and C. Ocampo-Martinez. Real-time experimental implementation of predictive control schemes in a small-scale pasteurization plant. In *Developments in Model-Based Optimization and Control*, pages 255–273. Springer, 2015.

Health-aware Model Predictive Control of Pasteurization Plant

Fatemeh Karimi Pour, Vicenç Puig and Carlos Ocampo-Martinez
Automatic Control Department of the Universitat Politècnica de Catalunya, Institut de Robòtica i Informàtica Industrial (CSIC-UPC). Llorens i Artigas, 4-6, 08028 Barcelona, Spain.

E-mail: {fkarimi, cocampo, vpuig}@iri.upc.edu

Abstract. In order to optimize the trade-off between components life and energy consumption, the integration of a system health management and control modules is required. This paper proposes the integration of model predictive control (MPC) with a fatigue estimation approach that minimizes the damage of the components of a pasteurization plant. The fatigue estimation is assessed with the rainflow counting algorithm. Using data from this algorithm, a simplified model that characterizes the health of the system is developed and integrated with MPC. The MPC controller objective is modified by adding an extra criterion that takes into account the accumulated damage. But, a steady-state offset is created by adding this extra criterion. Finally, by including an integral action in the MPC controller, the steady-state error for regulation purpose is eliminated. The proposed control scheme is validated in simulation using a simulator of a utility-scale pasteurization plant.

1. Introduction

Pasteurization is a quite common procedure in the food industry. The pasteurization process implies applying heat to some products such as milk, cream, beer and others. Mainly, pasteurization implies that a food product is exposed to some temperature profile during a predetermined period of time, in order to reduce the proportion of microorganisms. This process is distributed into three sections that are heating, cooling and regeneration sections. The most important section is heating, which involves heat exchanger equipment to heat up the temperature of the product at requested setpoint, and then maintaining this temperature during a constant time. An important key in pasteurization plant is controlling and maintaining the temperature of the process. Particularly, in the case of milk pasteurization the heating temperature is in general $72-74$ °C during $15-20$ seconds followed by cooling [9]. Therefore, a suitable control system needs to be designed to control the product temperature for keeping the desired product quality. The necessity of a significant control of the process arises from the saving in energy, product and time if an accurate tracking of the setpoint is performed. Because of the high number of load cycles that occur during the life of the pasteurization's pump, fatigue measurement is of particular importance in pasteurization control. For this reason, interest in the integration of control with a fatigue-based prognosis of components has increased in recent years. Fatigue can be taken as a breakdown of the material subject to stress, especially when repeated series of stresses are applied. It has been widely and exhaustively studied from different perspectives [15]. In this paper, the damage rule used to perform fatigue analysis is the Palmgren-Miner linear damage rule [14]. This rule commonly called the Miners rule is being currently used throughout the industry and in academia [12].

Along the last decades, model-based predictive control (MPC) gained its relevant position in the process industry, because of its ability to deal with multivariable control, delays and constraints on system variables and actuators. Also, MPC has started to attract the attention of academia and industry due to the possibility of dealing with the conflicting power optimization and fatigue load reduction. A data-based MPC strategy that incorporates fatigue estimation was presented in [2]. In the work of [18], an approach including dynamic inflow into the MPC controller is proposed to decrease fatigue load.

This paper propose a health-aware control (HAC) that considers the information about the system health to adapt the objectives of the control law to extend the system remaining useful life (RUL) [5]. Thus, the control actions are generated to fulfill the control objectives/constraints but at the same time to extend the life of the system components. In case that the controller is implemented using MPC, the trade-off is based on modifying the control objective including new terms that take care of the system health. This leads to solving a multi-objective optimization problem where a trade-off between system health and performance should be established [23].

The main contribution of this paper is the integration of MPC with fatigue-based prognosis to minimize the damage of pasteurization components while still, the pasteurization temperature tracks the suitable references related to the products. The integration of a system health management module with MPC control is done by including a fatigue-based model using the rain-flow counting approach and adding an extra criterion in the control objective function that takes into account the accumulated damage. In practice, the extra criterion can lead to steady-state offset unless precautions are taken in the control design. For this reason, the integral action for eliminating steady-state offset involves augmenting the process model to include a constant step disturbance is used in MPC controller. The control scheme is implemented using a high fidelity simulator of a utility-scale pasteurization plant.

The remainder of the paper is organized as follows. In Section 2, the pasteurization plant and process of this system are represented. Besides, the general statement of the MPC problems is described. Section 3 presents the fatigue analysis applied to pasteurization system and show how to implement health-aware control using MPC. In Section 4, approaches for eliminating steady-state offset in MPC controller are presented. Section 5 describes the case study based on the pasteurization benchmark, where the proposed approach is assessed and the results are analyzed. Finally, in Section 6, the conclusions of this work are drawn and some research lines for future work are proposed.

2. PROBLEM STATEMENT

2.1. Pasteurization Model

The considered pasteurization process is the small-scale plant PCT-23 MKII, manufactured by Armfield (UK) [1]. The system emulates an industrial high-temperature short-time (HTST) pasteurization process. The goal of this process is to heat and keep the product, which is typically a liquid, at a predetermined temperature for a minimum time. Throughout the pasteurization process, the liquid is pumped at a predetermined flow rate from one of two storage tanks to an indirect plate heat exchanger. The water heat is transferred to the product inside the first phase of the heat exchanger. The raw product is heated to an intermediate temperature by using missed energy from the pasteurized product. Then, this product is heated from that middle temperature to the full pasteurization temperature, in the second phase, while using a hot-water flow (F_h) coming from a closed circuit with a heater. The temperature (T_{past}) at the output of the holding tube is used to monitor the product temperature after the pasteurization process. Finally, the product is cooled in the third phase of the heat exchanger, where remaining heat is removed to the entrance product. This last phase does not add anything new to the model, and so it was not considered in this paper.

In order to have a model that can be used with an MPC controller, suitable models of the pasteurization plant are needed. The control-oriented pasteurization model is represented in terms of behavior equations of each subsystem, consisting of power, water pump, heat exchanger and hot water tank. Accordingly, identified models obtained as transfer functions are suitably stated by their equivalent controllable realizations in state space. The controlled inputs are the power of the electrical heater, P, and water pump speed, N, respectively. The input temperature of the water heater, T_{iw}, and temperature of cold water, T_{ic}, are measured non-controlled inputs (disturbance). Therefore, the continuous time state-space

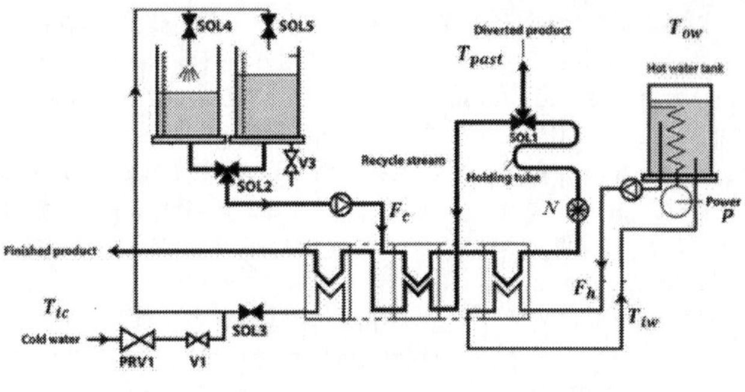

Figure 1. Plant scheme.

model of the pasteurization system can be expressed in the following form:

$$A = \begin{bmatrix} \frac{-1}{\tau_1} & 0 & 0 & 0 & 0 & 0 & 0 \\ 0 & \frac{-1}{\tau_2} & 0 & 0 & 0 & 0 & 0 \\ \frac{K_{21}}{\tau_{21}} & \frac{K_{21}}{\tau_{21}} & \frac{-1}{\tau_{21}} & 0 & 0 & 0 & 0 \\ 0 & 0 & 0 & \frac{-1}{\tau_{21}} & 0 & 0 & 0 \\ 0 & 0 & 0 & 0 & \frac{-1}{\tau_{22}} & 0 & 0 \\ 0 & 0 & 0 & 0 & 0 & \frac{-1}{\tau_f} & 0 \\ 0 & 0 & 0 & \frac{K_{ht}}{\tau_{ht}} & \frac{K_{ht}}{\tau_{ht}} & \frac{K_{ht}}{\tau_{ht}} & \frac{-1}{\tau_{ht}} \end{bmatrix}, \quad B = \begin{bmatrix} 0 & 0 \\ \frac{K_2}{\tau_2} & 0 \\ 0 & 0 \\ 0 & 0 \\ 0 & \frac{K_{12}}{\tau_{12}} \\ 0 & 0 \\ 0 & 0 \end{bmatrix},$$

$$E = \begin{bmatrix} \frac{K_1}{\tau_1} & 0 & 0 & 0 & 0 & 0 & 0 \\ 0 & 0 & 0 & 0 & \frac{K_{22}}{\tau_{22}} & 0 & 0 \end{bmatrix}^T, \quad C = \begin{bmatrix} 1 & 1 & 0 & 0 & 0 & 0 & 0 \\ 0 & 0 & 0 & 0 & 0 & 0 & 1 \end{bmatrix}, \tag{1}$$

where K is static gain and τ is the time constant of the transfer functions of the subsystems. In spite of this, the discretized state space-model form a complete system of pasteurization plant (1) with sampling time T_s = 120 s is used here.

2.2. *Operational Control*

As previously mentioned, the main goal of pasteurization process is to guarantee that the pasteurization temperature is reached and maintained as close as possible to the set-point value. At the same time, reduction of energy consumption and health management of the system expressed as a multi-objective problem. Hence, MPC is suitable technique to control a pasteurization system due to its capability to deal efficiently with dynamic constrained systems and predict the proper action to achieve the optimal performance according to a user defined cost function. In this paper, it is assumed that the system behavior can be described at each time instant $k \in \mathbb{Z}$ by the following discrete-time difference equation:

$$x(k+1) = Ax(k) + Bu(k) + Ed(k), \tag{2a}$$

$$y(k) = Cx(k), \tag{2b}$$

where $x \in \mathbb{R}^{n_x}$ is the state of system, $u \in \mathbb{R}^{n_u}$ is the vector of manipulated variable, $d \in \mathbb{R}^{n_d}$ is vector of measurable disturbances, and $y \in \mathbb{R}^{n_y}$ is vector of controlled variables. Moreover, A, B, E and C are time-invariant matrices of adequate dimensions. It also considered that the system is subject to the hard state and input constraints, which can be posed as

$$x(k) \in \mathbb{X} \triangleq \{x(k) \in \mathbb{R}^{n_x} | \underline{x} \leq x(k) \leq \overline{x}, \quad \forall k\}, \tag{3a}$$

$$u(k) \in \mathbb{U} \triangleq \{u(k) \in \mathbb{R}^{n_u} | \underline{u} \leq u(k) \leq \overline{u}, \quad \forall k\}. \tag{3b}$$

The control goal is to minimize a convex multi-objective cost function $J(x,u) : \mathbb{X} \times \mathbb{U} \longrightarrow \mathbb{R}$, which might tolerate any functional relationship to the economic and safety of the system operation. To do so, the MPC controller design is based on the solution of the following finite-time horizon optimization problem (FHOP):

$$\min_{\mathbf{u}_k} \sum_{i=0}^{N_p-1} [\|x(k+i|k)\|_{w_1}^p + \|u(k+i|k)\|_{w_2}^p + \|\Delta u(k+i|k)\|_{w_3}^p], \tag{4a}$$

subject to:

$$x(k+i+1|k) = Ax(k+i|k) + Bu(k+i|k) + Ed(k+i|k), \tag{4b}$$
$$\Delta u(k+i|k) = u(k+i|k) - u(k+i-1|k), \tag{4c}$$
$$u(k+i|k) \in \mathbb{U}, \tag{4d}$$
$$x(k+i|k) \in \mathbb{X}, \tag{4e}$$

for all $i \in \mathbb{Z}_{[0,N_p-1]}$, where $\mathbf{u}_k = \{u_{k+i|k}\}_{i\in\mathbb{Z}_{[0,N_p-1]}}$ is the decision variable, with $\mathbf{u}_k$ being the sequence of controlled inputs. Furthermore, weighting matrices $w_1 \in \mathbb{R}^{n_x \times n_x}$, $w_2 \in \mathbb{R}^{n_u \times n_u}$ and $w_3 \in \mathbb{R}^{n_u \times n_u}$ are used to establish the antecedence of the different control objectives. Also, p denotes the norm used.

Denote with $\mathbf{u}_k^\star$ the optimal solution of (4) at time step k. Then, following the MPC receding horizon philosophy, only the first optimal control action is applied, i.e., $u_k = u_{k|k}^\star$. Then, the new measurements are accumulated to initialize initial conditions and then the optimization problem (4) is solved again. This procedure is repeated at each time step k.

3. Health-aware MPC Controller

In this section, the rainflow counting algorithm (RFC) is represented and later on it is formulated and integrated by MPC controller for the pasteurization plant as a case study.

3.1. Rainflow counting algorithm (RFC)

The damage accumulation process on a component produced by cyclic loading is called as *fatigue*. Exposing a material to cyclic loading of constant amplitude will cause fatigue failure after a certain number of cycles. A usual method to quantify the fatigue impact of fluctuating loads is the combination of a rainflow counting algorithm and a damage equivalent load approach, enabling the relative comparison of different load samples [13]. RFC method, first introduced by [4], has a complex sequential and nonlinear structure in order to decompose ideal sequences of loads into cycles. Usually, to compute a lifetime estimate from a given structural stress input signal, the RFC method is applied by counting cycles and maxima, jointly with the Palmgren-Miner rule to calculate the expected damage. The input signal is obtained from time history of the loading parameter of interest, such as force, torque, stress, strain, acceleration, or deflection [11].

There are several types of RFC algorithms have been proposed such as [22] and [3], with different rules but providing the same results. The RFC algorithm applied in this paper is presented in [17]. This algorithm computes the stress for each rainflow cycle in four steps [23]:

- the stress history is converted to an extremum sequence of alternating maxima and minima;
- for each local maximum M_j, the left (m_j^-) and right (m_j^+) region where all stress values are below M_j is identified;
- the minimum stress value is processed being $m_j = min\{m_j^-, m_j^+\}$;
- the equivalent stress per rainflow cycle s_j related to the M_j is given by the amplitude $s_j = M_j - m_j$ or the mean value $s_j = \frac{M_j - m_j}{2}$.

The damage, D, at each stress cycle is calculated by using the $S-N$ curve [7]. In fact, for many materials there is an explicit relation between the number of cycles to failure and cycle amplitude, which is known as $S-N$ or Wöhler curves, given as a line in a log-log scale as

$$s^{c_w}N = K, \tag{5}$$

where c_w and K are material specific parameters, being the c_w Wöhler-coefficient, and N is the number of cycles to failure at a given stress amplitude s. The damage imposed by a stress cycle with a ranges_j is computed as

$$D_j \equiv \frac{1}{N_j} = \frac{1}{K}s_j^{c_w}. \tag{6}$$

Then, for a time history, the total damage under the linear accumulation damage (Palmgren-Miner) rule is given as

$$D_{ac} = \sum_{j=1}^{\lambda} \frac{1}{N_j} = \sum_{j=1}^{\lambda} \frac{1}{K}s_j^{c_w}, \tag{7}$$

for damage increments D_{ac} associated to each counted cycle, N_j the number of cycles to failure associated to stress amplitude s_j, and the number of all counted cycles λ. These sequences are presented in Figure 2. On the top the input stress is shown, and in the bottom part the instantaneous damage and the accumulated damage are shown.

For real-time applications, applying the customary rainflow counting algorithm is quite challenging and computationally heavy. Considerable amounts of data must be stored and processed periodically to establish a magnitude of the data in equivalent regular cycles. Besides the algorithm must be applied to a stored set of data. One of the objectives of this paper is to analyze the fatigue due to pump load of the pasteurization system. Loads in the pump structure arise from several factors, being the primary cause of failure in the pump is bearing failure. While there are two influential and important factors such as pressure (PSI) and speed (RPM) which are related to the pump, and in this case, the speed of the pump is used as fatigue stress.

Utilizing the RFC algorithm the accumulated damage is obtained as a function of the cycles of the pump speed stress signal. In order to have available an accumulated damage variable that can be integrated with a linear MPC model, a simplified approach to compute fatigue on time series signal is proposed based on RFC theory explained before. The outcome of this approach is that the accumulated damage is obtained as a function of time instead of the number of cycles.

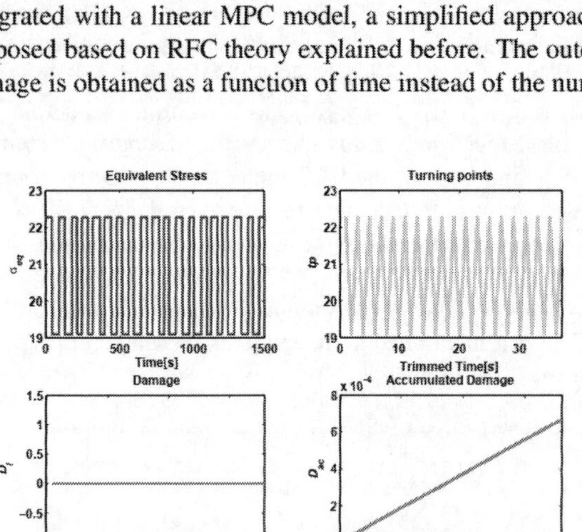

Figure 2. Rainflow counting damage estimation.

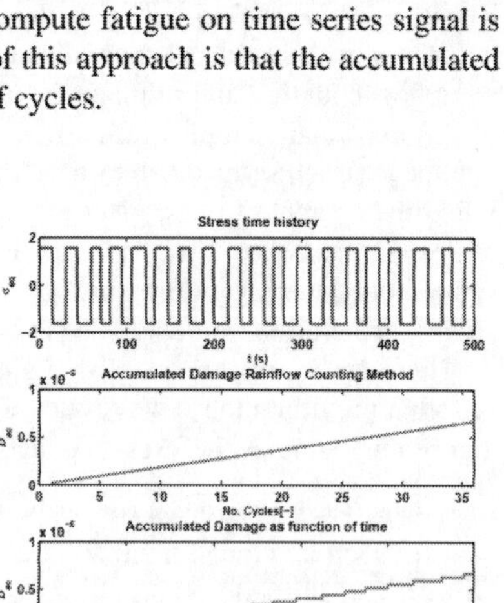

Figure 3. Accumulated Damage Comparison.

The proposed approach finds the changes of the sign which corresponds to a cycle in the stress time signal. The obtained function at each sample step k is the following:

$$D(k) = \begin{cases} 0 & \text{if} \quad I(k) = I(k-1), \\ \frac{1}{K}(s(k))^{c_w} & \text{if} \quad I(k) \neq I(k-1), \end{cases} \tag{8}$$

where $s(k)$ is the stress at time k

$$s(k) = \frac{1}{L} \sum_{q=k-L}^{k} N_s(q), \tag{9}$$

and L is the number of samples per cycle, N_s is the pump speed moment, q is difference between the the number of samples per cycle and sample step and $I(k)$ is the signal adapted to indicate cycle (10)

$$I(k) = N_s(k) - s(k). \tag{10}$$

Then, the accumulated damage is calculated by

$$D_{acc}(k) = D_{acc}(k-1) + D(k). \tag{11}$$

Notice that, at the end of the scenario, the accumulated damage is practically the same. The difference, as explained before, depends on the fact that the damage obtained by RFC technique is represented as a function of the cycles count while the other is a function of time.

3.2. *MPC with health-aware objective*

As shown in Section 3.1, the degradation process of the heat pump of pasteurization plant can be assessed using the water pump speed sensor information. In order to decrease the accumulated damage, a new objective is considered in the MPC controller that the rainflow counting model is approximated by means of a linear model in this objective.

As first approximation, later on observing that the proposed approach gives a quite close approximation of the accumulated damage obtained by the RFC algorithm (see Figure 3), the slope m of the accumulated damage curve in function of time is calculated and then employed as one of the parameters in the linear fatigue-damage model proposed in this paper. According to the [21], an experimental model that relates the values of the pump speed and flow signals in steady state was proposed. Therefore, it can be shown the relation between the pump speed moment as a function of the flow. The proposed model for pump speed dynamics is a first order pump speed model with an slope of α_1 plus a constant value α_0, i.e.,

$$N_s(k) = \alpha_1 F(k) + \alpha_0. \tag{12}$$

Briefly, a linear fatigue damage model is suggested as a function of the flow value signal while there is a relation between the accumulated damage of the pump and the control signal expressed as:

$$z(k+1) = z(k) + \frac{m}{L}(\alpha_1 F(k) + \alpha_0), \tag{13}$$

where $z(k+1)$ is the accumulated damage of the pump. This damage model (13) can be included into the MPC as a new state of model and additional objective is increased in the MPC cost function (4) to minimize the accumulated damage. Taking into account (13), the MPC problem (4) can be represented as follows:

$$\min_{\mathbf{u}_k} \sum_{i=0}^{N_p-1} [\|x(k+i|k)\|_{w_1}^{p} + \|u(k+i|k)\|_{w_2}^{p} + \|\Delta u(k+i|k)\|_{w_3}^{p} + z(k+i|k)\|_{w_4}^{p}], \tag{14a}$$

subject to:

$$x(k+i+1|k) = Ax(k+i|k) + Bu(k+i|k) + Ed(k+i|k), \tag{14b}$$
$$\Delta u(k+i|k) = u(k+i|k) - u(k+i-1|k), \tag{14c}$$
$$z(k+1) = z(k) + \frac{m}{L}(\alpha_1 F(k) + \alpha_0), \tag{14d}$$
$$u(k+i|k) \in \mathbb{U}, \tag{14e}$$
$$x(k+i|k) \in \mathbb{X}, \tag{14f}$$

where the new objective with the corresponding matrix weight w_4 is is appended in the MPC cost function to minimize the accumulated damage. The degree of fitting between the RFC approximation as a function of time presented in Section 3.1 and the linear z state damage model are described in (13) can be compared in Figure 4.

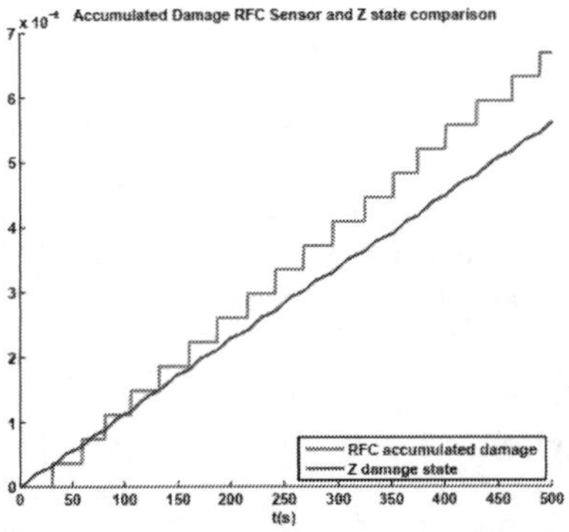

Figure 4. Accumulated damage RFC as a function of time and z fatigue damage state.

4. Condition for Offset Elimination in MPC Controller

As it described before, the new objective in cost function of MPC controller can be lead to steady-state offset unless precautions are taken into control design. Elimination of steady-state offset is carried out in two basic manners. The first method involves modifying the controller objective to include integration of the tracking error [16]. The tracking error state and process model are augmented and incorporated as an integral term into the MPC structure. But, this augmentation can significantly increase the computational cost of the dynamic optimization which grows in proportion to the cube of the state dimension for some complex system. Also, this method is needed an anti-windup algorithm for the integral term, and sometimes costly, performance penalty [20]. A second general method, by augmenting the process model that is included a constant step disturbance can be eliminated the steady-state offset, whereas this disturbance, which is estimated from the measured process variables, is generally assumed to remain constant in the future and its effect on the controlled variables is eliminated by shifting the steady-state target for the controller. In this paper, by using the disturbance model remove the steady-state offset which is created with the new state in the model.

Now, it is considered the general type in which the particular recent MPC controllers eliminate the offset by means of disturbance models [19], [16], [6]. Hence, a general way to include disturbance sub-model given from the state-space model is as follows [16]:

$$\begin{bmatrix} x(k+1) \\ d(k+1) \\ g(k+1) \end{bmatrix} = \begin{bmatrix} A & G_d & 0 \\ 0 & I & 0 \\ 0 & 0 & I \end{bmatrix} \begin{bmatrix} x(k) \\ d(k) \\ g(k) \end{bmatrix} + \begin{bmatrix} B \\ 0 \\ 0 \end{bmatrix} u(k), \qquad y(k) = \begin{bmatrix} C & 0 & G_g \end{bmatrix} \begin{bmatrix} x(k) \\ d(k) \\ g(k) \end{bmatrix}, \tag{15}$$

where the output and state disturbance are $g(k) \in \mathbb{R}^{s_g}$ and $d(k) \in \mathbb{R}^{s_d}$, respectively; and s_g and s_d are the numbers of augmented output and state disturbance. In addition, G_g and G_d are matrices that determine the effect of the disturbance on the output and the states. In the general form of considering the disturbance, if $G_d = 0$ and $G_g = I$ model (15) have an output disturbance, on the other hand, if $G_d = B$ and $G_g = 0$ model (15) have an input disturbance. According to the complete model of pasteurization system (1), the matrix E is utilized as disturbance states in this paper. The states of the process model and the unmeasured disturbance model must be estimated simultaneously using the augmented system model. The state estimation can be performed within a stochastic framework using a Kalman filter or a deterministic framework using a Luenberger observer. In both cases, a stable filter gains L can be determined using standard methods provided the augmented system is detectable [10]. Thus, the steady-state observer equation is given by

$$\hat{x}(k) = A\hat{x}_{\bar{k}} + Bu_{\bar{k}} + G_d\hat{d}_{\bar{k}} + L_x[y(k) - C(A\hat{x}_{\bar{k}} + Bu_{\bar{k}} + G_d\hat{d}_{\bar{k}})], \tag{16}$$

$$\begin{bmatrix} \hat{d}_{\bar{k}} \\ \hat{g}_{\bar{k}} \end{bmatrix} = \begin{bmatrix} \hat{d}_{\bar{k}} \\ \hat{g}_{\bar{k}} \end{bmatrix} + \begin{bmatrix} L_d \\ L_g \end{bmatrix} \left[y(k) - C(A\hat{x}_{\bar{k}} + Bu_{\bar{k}} + G_d\hat{d}_{\bar{k}}) - G_g\hat{g}_{\bar{k}}\right], \tag{17}$$

where L_x, L_d and L_g are the observer gains that corresponds to the states estimation $\hat{x}_{\bar{k}}$, and the disturbance estimation $\hat{d}_{\bar{k}}$ and $\hat{g}_{\bar{k}}$, respectively.

The states from the augmented system (15) which corresponding to the disturbances, that cannot be controlled by means of the input u. In that case, the MPC regulator makes use of the original model and shifts the steady-state targets in order to remove the estimated disturbance effect. Moreover, the dynamic control is separated from the stationary computation: the target calculus block is completely devoted to obtain the stationary values, while the MPC regulator block is dedicated to guiding the states $\hat{x}(k)$ to its corresponding targets. Then, the optimization problem that must be solved in the regulator block is represented by

$$\min_{\tilde{\mathbf{u}}_k} \sum_{i=0}^{N_p-1} [\|C\tilde{x}(k+i|k)\|_{w_1}^p + \|\tilde{u}(k+i|k)\|_{w_2}^p + \|\Delta\tilde{u}(k+i|k)\|_{w_3}^p + z(k+i|k)\|_{w_4}^p], \tag{18a}$$

subject to:

$$\tilde{x}(k+i+1|k) = A\tilde{x}(k+i|k) + B\tilde{u}(k+i|k) + Ed(k+i|k), \tag{18b}$$

$$\Delta\tilde{u}(k+i|k) = \tilde{u}(k+i|k) - \tilde{u}(k+i-1|k), \tag{18c}$$

$$z(k+1) = z(k) + \frac{m}{L}(\alpha_1 F(k) + \alpha_0), \tag{18d}$$

$$\tilde{u}(k-1) = u(k-1) - u_s \tag{18e}$$

$$\tilde{u}(k+i) = 0 \tag{18f}$$

$$\tilde{x}(k) = \hat{x}(k) - x_s \tag{18g}$$

$$u_{min} \le \tilde{u}(k+i) + u_s \le u_{max} \qquad u(k+i|k) \in \mathbb{U}, \tag{18h}$$

$$\Delta u_{min} \le \Delta\tilde{u}(k+i) \le \Delta u_{max} \qquad x(k+i|k) \in \mathbb{X}, \tag{18i}$$

where it can be seen that the state and input are led to the targets x_s and u_s. The role of this part is to steer the shifted states and input to zero, which means that the ability of this strategy to eliminate the offset depends on the computation of the targets x_s and u_s. For obtaining this operation, the following optimization problem is solved:

$$\min_{u_s, x_s} V_t(k) \triangleq \{(y_{sp} - y_t^a)^T Q(y_{sp} - y_t^a) + (u_s - u_{sp})^T R(u_s - u_{sp})\}, \tag{19a}$$

subject to:

$$x_s = Ax_s + Bu_s + G_d\hat{d}(k), \tag{19b}$$

$$y_t^a \triangleq Cx_s + G_g\hat{g}(k), \tag{19c}$$

$$u_{min} \le u_s \le u_{max}, \tag{19d}$$

where y_{sp} stands for the output set points, and y_t^a is the achievable stationary output. Also, Q and R are positive definite weighting matrices.

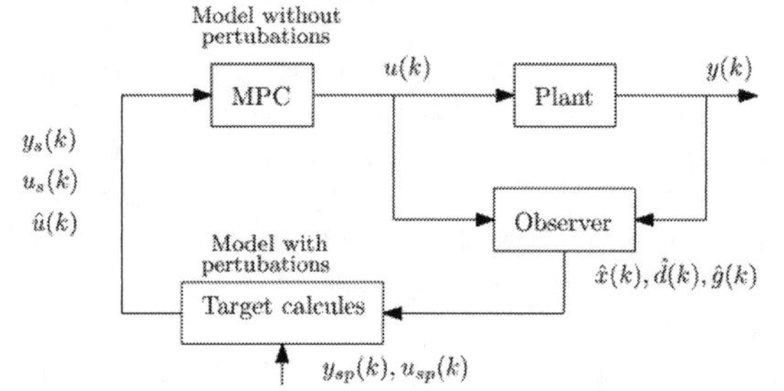

Figure 5. Representative block diagram of the predictive control system with linear regulator.

Following the strategy describe in Figure 5, two optimization problems must be solved at each sample time k. Assume that constraints (19b) and (19c) hold true for the stationary disturbances $\hat{d}$ and $\hat{d}$ presented by the observer and in that case, for a time $\bar{k}$ large enough, the states and input target satisfies,

$$x_s = Ax_s + Bu_s + G_d\hat{d}_{\bar{k}}, \tag{20a}$$

$$y_{sp} \triangleq Cx_s + G_g\hat{g}_{\bar{k}}. \tag{20b}$$

On the contrary, from (17), the states and disturbances provided by the observer satisfy

$$\hat{x}_{\bar{k}} = A\hat{x}_{\bar{k}} + B\hat{u}_{\bar{k}} + G_d\hat{d}_{\bar{k}}, \tag{21a}$$

$$y_{\bar{k}} = C\hat{x}_{\bar{k}} + G_g\hat{g}_{\bar{k}}. \tag{21b}$$

Afterward, by subtracting (20a) from (21a) can be driven as

$$(\hat{x}_{\bar{k}} - x_s) = A(\hat{x}_{\bar{k}} - x_s) + B(\hat{u}_{\bar{k}} - u_s), \tag{22}$$

which relates to the original system considered by the target tracking optimization. If the original model given by A, B and C can be stabilized, then the regulator will lead the states $\tilde{x}(k)$ to zero, that is,

$$(\hat{x}_{\bar{k}} - x_s) = 0. \tag{23}$$

Now, subtracting (20b) from (21b) can be written as

$$(y_{\bar{k}} - y_{sp}) = C(\hat{x}_{\bar{k}} - x_s), \tag{24}$$

and, from (23) and (24), implies

$$y_{\bar{k}} = y_{sp}. \tag{25}$$

To sum up, that the use of an augmented model requires the simultaneous solution of both optimization problems. This method can increase the computational cost of the algorithm, especially if the system has a large number of inputs and outputs.

5. Case Study

5.1. Benchmark description

As discussed in Section 2, the pasteurization system is used as a case study which the considered pasteurization process is the benchmark PCT-23 MKII, manufactured by Armfield (UK). This laboratory plant is the small version (1.2 m, 0.6 m, 0.6 m) of several real-time industrial pasteurization processes. Moreover, the model of the pasteurization plant are obtained from experimental data according to [8] also the identified models obtained as transfer functions are suitably formulated by their equivalent controllable realizations in state-space model.

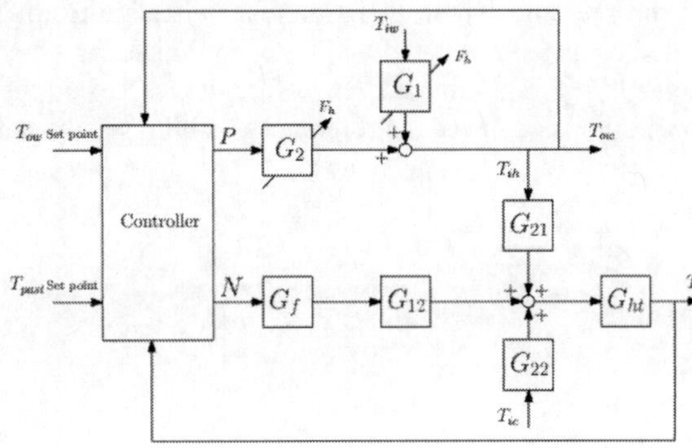

Figure 6. Block diagram of the control loop.

Besides, the size of handle tube of pasteurization pumps is 1.6 mm with wall thickness from 0.5 mm to 8.0 mm bore. The rotor speed is 400 rpm while the flow rate is 2000 ml/min. Figure 6 introduce a block diagram of the pasteurization simulation model, provided with the benchmark, including the feedback loops corresponding to the flow and power. Also, the MPC controller, fatigue monitoring and integral action block that will be designed in an integrated manner are presented in this Figure 7.

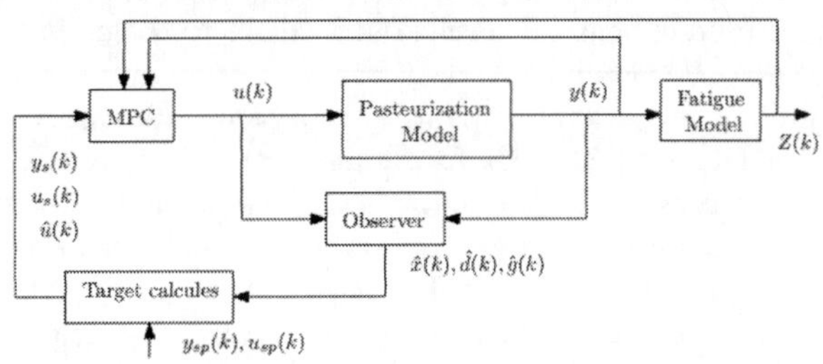

Figure 7. Representative block diagram of the predictive control system with linear regulator and fatigue model.

5.2. Implementation of MPC with health-aware objective and integral action

The health-aware MPC, designed as described in Section 3, is obtained adding the accumulated damage model presented before, as a new state introduced in (13). In addition, for removing steady-state offset that is obtained with the new objective in the health-aware MPC cost function is used disturbance model.

In brief then, the MPC with the new augmented model uses the following model:

$$x(k+1) = A_n x(k) + B_n u(k) + E_n d(k), \quad (26a)$$
$$y(k) = C_n x(k), \quad (26b)$$

where A_n, B_n, C_n, and E_n are state matrices of proper dimensions with included accumulated damage and disturbance model given by (13) and (15), respectively. The MPC controller has been implemented in simulation where the weight matrices are $Q = 1$, $R = 0.01$. The prediction horizon is chosen $N_p = 400$.

The control objective of the MPC controller for the pasteurization system is pasteurization temperature, T_{past}, tracking the setpoint, where their references change from 70 °C to 40 °C and in the same time, the accumulated damaged evaluated as (13) is minimized and the power of the electrical heater, P, and water pump speed, N, are minimized. Therefore, it can be said the cost and degradations are decreased while the temperature is achieved the references without steady-state offset.

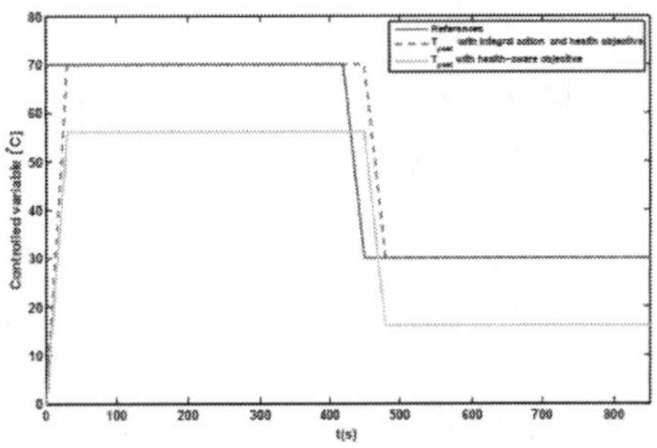

Figure 8. Evolution of pasteurization temperature with (red) and without (green) integral action in the MPC.

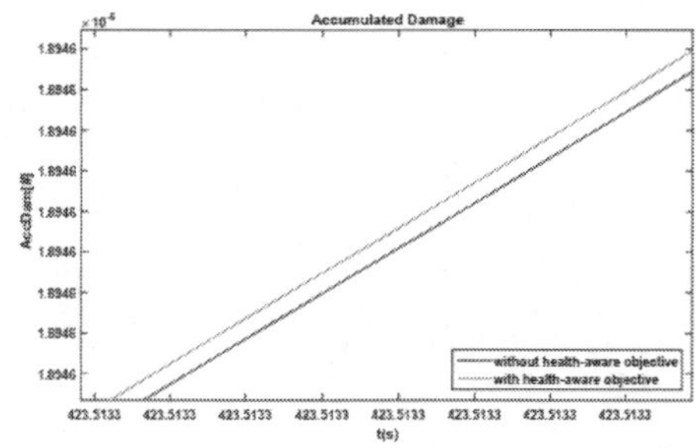

Figure 9. Evolution of accumulated damages with (yellow) and without (red) health-aware objective in the MPC.

5.3. Results

In order to test the behavior of the proposed health-aware MPC control scheme regarding tracking, several simulations were carried out and the results obtained are presented in this part. The operating point for the flow is chosen as 200 ml/min.

The behavior of controlled temperature T_{past} from the pasteurization plant under the health-aware MPC controller with and without integral action for elimination steady-state offset is presented in Figure 8 with its corresponding references, while the controlled variables track the references. As it can be seen from the Figure 8, the temperature of the pasteurization process T_{past} (green line) wants to follow the references but, due to the adding the accumulated damage state as new states in the model generated same steady state error between the reference and output . However, by including an integral action In the MPC controller, the steady-state is eliminated (red line).

In Figure 9, it is provided the evolution of the accumulated damages obtained with (yellow) and without a (red) health-aware objective in the MPC controller. From Figure 9, it can be seen that the inclusion of the fatigue objective relieves 0.11 of the accumulated damage. At the same time, the pasteurization temperature, T_{past}, tracking the setpoint, where their references change from 70 °C to 40 °C is shown in Figure 8.

6. Conclusion

This paper has presented the integration of MPC with fatigue-based prognosis to minimize the damage of the components of a pasteurization plant. The integration of a system health management module with MPC control has provided the pasteurization plant with a mechanism to operate safely and optimize the trade-off between components life and savings energy, product if an accurate following of the setpoint is performed. The MPC controller objective has been modified by adding an extra criterion that takes into account the accumulated damage. Finlay, by including an integral action into the MPC controller design, the steady-state offset has been eliminated. The control scheme has been satisfactorily implemented using simulator of a utility-scale pasteurization plant. The results obtained show that there exists a trade-off between the minimization of the accumulated damage and tracking setpoint that is for saving energy. As future research, the proposed MPC control scheme to reduce the accumulated damage more than this approach and designing the health-aware nonlinear MPC based on the real model of the pasteurization system .

References

[1] Armfield. *Process Plant trainer PTC23-MKII, Instruction Manual*. 2015.

[2] J. de Jesus Barradas-Berglind, R. Wisniewski, and M. Soltani. Fatigue damage estimation and data-based control for wind turbines. *IET Control Theory & Applications*, 9(7):1042–1050, 2015.

[3] S. D. Downing and D. F. Socie. Simple rainflow counting algorithms. *International journal of fatigue*, 4(1):31–40, 1982.

[4] T. Endo, K. Mitsunaga, and H. Nakagawa. Fatigue of metals subjected to varying stressprediction of fatigue lives. In *Preliminary Proceedings of the Chugoku-Shikoku District Meeting, Japanese Society of Mechanical Engineers, Tokyo*, pages 41–44, 1967.

[5] T. Escobet, J. Quevedo, V. Puig, and F. Nejjari. Combining health monitoring and control. *Diagnostics and Prognostics of Engineering Systems: Methods and Techniques: Methods and Techniques*, page 230, 2012.

[6] AH. González, EJ. Adam, and JL. Marchetti. Conditions for offset elimination in state space receding horizon controllers: A tutorial analysis. *Chemical Engineering and Processing: Process Intensification*, 47(12):2184–2194, 2008.

[7] K. Hammerum, P. Brath, and N. K. Poulsen. A fatigue approach to wind turbine control. In *Journal of Physics: Conference Series*, volume 75, page 012081. IOP Publishing, 2007.

[8] J. Ibarrola, J. Sandoval, M. Garcia-Sanz, and M. Pinzolas. Predictive control of a high temperature–short time pasteurisation process. *Control Engineering Practice*, 10(7):713–725, 2002.

[9] M. T. Khadir. Enthalpy predictive functional control of a pasteurisation plant based on a plate heat exchanger. *International Journal of Modelling, Identification and Control*, 13(1-2):78–87, 2011.

[10] H. Kwakernaak and R. Sivan. *Linear optimal control systems*. Wiley-interscience New York, 1972.

[11] Y. L. Lee. *Fatigue testing and analysis: theory and practice*, volume 13. Butterworth-Heinemann, 2005.

[12] J. C. Marin, A. Barroso, F. Paris, and J. Canas. Study of damage and repair of blades of a 300kW wind turbine. *Energy*, 33(7):1068–1083, 2008.

[13] S. Martinen, I. Carlén, K. Nilsson, S. P. Breton, and S. Ivanell. Analysis of the effect of curtailment on power and fatigue loads of two aligned wind turbines using an actuator disc approach. In *Journal of Physics: Conference Series*, volume 524, page 012182. IOP Publishing, 2014.

[14] M. A Miner et al. Cumulative damage in fatigue. *Journal of applied mechanics*, 12(3):159–164, 1945.

[15] M. Musallam and C. M. Johnson. An efficient implementation of the rainflow counting algorithm for life consumption estimation. *IEEE Transactions on reliability*, 61(4):978–986, 2012.

[16] K. R. Muske and T. A. Badgwell. Disturbance modeling for offset-free linear model predictive control. *Journal of Process Control*, 12(5):617–632, 2002.

[17] A. Niesłony. Determination of fragments of multiaxial service loading strongly influencing the fatigue of machine components. *Mechanical Systems and Signal Processing*, 23(8):2712–2721, 2009.

[18] P. F. Odgaaard, T. Knudsen, A. Overgaard, H. Steffensen, and M. Jørgensen. Importance of dynamic inflow in model predictive control of wind turbines. *IFAC-Congress*, 48(30):90–95, 2015.

[19] G. Pannocchia and J. B. Rawlings. Disturbance models for offset-free model-predictive control. *AIChE journal*, 49(2):426–437, 2003.

[20] C. V. Rao, S. J. Wright, and J. B. Rawlings. Application of interior-point methods to model predictive control. *Journal of optimization theory and applications*, 99(3):723–757, 1998.

[21] A. Rosich and C. Ocampo-Martinez. Real-time experimental implementation of predictive control schemes in a small-scale pasteurization plant. In *Developments in Model-Based Optimization and Control*, pages 255–273. Springer, 2015.

[22] I. Rychlik. A new definition of the rainflow cycle counting method. *International journal of fatigue*, 9(2):119–121, 1987.

[23] H. Sanchez, T. Escobet, V. Puig, and P. F. Odgaard. Health-aware model predictive control of wind turbines using fatigue prognosis. *IFAC-Congress*, 48(21):1363–1368, 2015.

IOP Conf. Series: Journal of Physics: Conf. Series **783** (2017) 012031 doi:10.1088/1742-6596/783/1/012031

Robust stabilization using LMI techniques of neutral time-delay systems subject to input saturation

Nabil El Fezazi[1], **Fatima El Haoussi**[1,2], **El Houssaine Tissir**[1], **Teresa Alvarez**[3] and **Fernando Tadeo**[3]

[1]University of Sidi Mohammed Ben Abdellah, Department of Physics, Faculty of Sciences Dhar El Mehraz, LESSI, BP 1796, Fes-Atlas, Morocco
[2]Polydisciplinary Faculty, BP 300, Selouane 62700, Nador, Morocco
[3]University of Valladolid, Industrial Engineering School, 47005, Valladolid, Spain

E-mail: fizazi.99@gmail.com, elhaous_fati@yahoo.fr, elhoussaine.tissir@usmba.ac.ma, tere@autom.uva.es, fernando@autom.uva.es

Abstract. The robust stabilization of uncertain saturated neutral systems with state delay is solved in this paper: based on a free weighting matrix approach, sufficient conditions are obtained via an LMI formulation. From these conditions, state feedback gains that ensure stability for the largest set of admissible initial conditions can be calculated solving optimization problems with LMI constraints. Some applications of this methodology to feedback control are then presented and compared with previous results in the literature.

1. Introduction

The stability analysis of time-delay systems is a topic of theoretical and practical importance, because delays appear in many areas: mechanics, physics, biology, economy, epidemics, population dynamic models, large-scale systems, automatic control systems, neural networks, chaotic systems, practical control system, and so on. It is well known that delays often lead to poor system performance, even instabilities. Therefore, stability analysis of time-delay systems has been extensively studied by many researchers (see, for example, [3, 18, 19, 28, 29]).

Moreover, almost all practical control systems are subject to input saturation, because of the existence of physical, technological or even safety constraints (see [5, 6, 7, 8, 13]). Input saturation is a also a source of performance degradation, generating also limit cycles, multiple equilibrium points and even instability. For time-delay systems, some works addressing the stability analysis and stabilization in the presence of saturating control signals can be found in the literature. In [27], conditions for stability or stabilization are proposed with state feedback. However, in that paper, the set of admissible initial conditions, for which the asymptotic stability is ensured (i.e. the domain of attraction) in the presence of control saturation, is not mentioned or explicitly defined. In [1], method for computing stabilizing state feedback control laws aiming at enlarging well defined estimates of the domain of attraction of the closed-loop system have been proposed. This method is based on the use of polytopic differential inclusions for describing the behavior of the closed-loop system with saturating inputs. In [25], the synthesis of stabilizing static anti-windup loops is addressed for the case of retarded systems presenting fixed delays. On the other hand, considering neutral systems, we can cite [11, 12, 19, 26]. In that paper, using a polytopic approach for modelling saturation effects, a method for computing stabilizing state feedback controls with the aim of maximizing the set of admissible initial conditions is proposed.

Thus, in this paper the stabilization problem for neutral systems with time-varying delay and actuator saturation subject to uncertainty is solved. By incorporating Lyapunov-Krasovskii (L-K) functional theory, free-weighting matrix technique, integral inequalities and function that corresponds to a decentralized dead-zone nonlinearity, then efficient stabilization conditions are obtained in terms of LMIs. Compared with the existing results, an optimization problem is formulated with the aim of computing stabilizing state feedback control laws. This optimization problem search the maximal delay bound for which a stabilizing control law can be found. On the other hand, when the open-loop system is unstable, the optimization objective consists in finding a control law that maximizes an estimate of the domain of attraction, or alternatively that ensures the stability for a given set of admissible initial states.

The rest of this paper is organized as follows: in Section 2, we formulate the problem of stabilization of uncertain neutral time-delay system with saturating actuator. The main results are presented in Section 3. Finally, three numerical examples are included to illustrate the results developed in this paper.

Notation: The Banach space of continuous vector functions mapping the interval $[-h_m, 0]$ into $\Re^n$ with the norm $\|\phi\|_c = \sup_{-h_m \leq t \leq 0} \|\phi(t)\|$ is denoted by $\mathcal{C}_{h_m} = \mathcal{C}([-h_m, 0], \Re^n)$. Additionally, $\overline{\lambda}(P)$ denotes the maximal eigenvalue of matrix P.

2. Problem Formulation and Preliminaries

Consider the following uncertain neutral delayed system with saturating actuators

$$\begin{aligned} &\dot{x}(t) - (C + \Delta C(t))\dot{x}(t - \tau(t)) = (A + \Delta A(t))x(t) + (A_d + \Delta A_d(t))x(t - \tau(t)) + Bu(t) \\ &x(t) = \phi(t), \ \ \forall t \in [-h_m, 0], \ \ \phi(t) \in \mathcal{C}_{h_m} \end{aligned} \tag{1}$$

where $x(t) \in \Re^n$ and $u(t) \in \Re^m$ are respectively the state and the control vectors. The matrices C, A, A_d and B are real constants of appropriate dimensions. the delay $\tau(t)$ is assumed to be unknown but bounded function of time, continuously differentiable, with their rate of change bounded as follows

$$0 \leq \tau(t) \leq h_m \ , \ \ \dot{\tau}(t) \leq d \tag{2}$$

we assume $0 < d < 1$ to ensure causality (see [3]).

In this paper, the uncertainties can be described as follows

$$[\Delta A(t) \quad \Delta A_d(t) \quad \Delta C(t)] = DF(t)[E_0 \quad E_1 \quad E_2] \tag{3}$$

where D, E_0, E_1 and E_2 are known real matrices of appropriate dimensions, and $F(t)$ denotes the time-varying parameter uncertainties with Lebesgue-measurable elements satisfying $F^T(t)F(t) \leq I, \forall t \geq 0$.

We suppose that the input vector u is subject to amplitude limitations defined as follows

$$|u_i| \leq u_{0i}, \ \ u_{0i} > 0; \ \ i = 1, \ldots, m \tag{4}$$

The control input is $u(t) = Kx(t)$. Due to the control bounds defined in (4), the effective control signal to be applied to the system is $u(t) = sat(Kx(t))$ where $u_i(t) = sat(K_i x(t)) = sign(K_i x(t)) min\{u_{0i}, |K_i x(t)|\}$.

Hence, the closed-loop system (1) reads

$$\begin{aligned} \dot{x}(t) - (C + \Delta C(t))\dot{x}(t - \tau(t)) \ = \ & (A + \Delta A(t))x(t) + (A_d + \Delta A_d(t))x(t - \tau(t)) \\ & + Bsat(Kx(t)) \end{aligned} \tag{5}$$

In this paper, it is assumed that $\phi(t)$ is continuously differentiable over $[-h_m, 0]$, and one of our interests is to estimate the domain of attraction of the following form

$$\Xi = \{\phi(t) \in \mathcal{C}_{h_m} : \|\phi\|_c \leq \delta_1, \ \|\dot{\phi}\|_c \leq \delta_2\}$$

where δ_1 and δ_2 are some scalars to be determined.

Define the following function $\psi(Kx(t)) = Kx(t) - sat(Kx(t))$ where $\psi(Kx(t))$ corresponds to a decentralized dead-zone nonlinearity. Thus, the closed-loop system (5) can be written as

$$\begin{aligned} \dot{x}(t) - (C + \Delta C(t))\dot{x}(t - \tau(t)) &= (A + \Delta A(t) + BK)x(t) + (A_d + \Delta A_d(t))x(t - \tau(t)) \\ &\quad -B\psi(Kx(t)) \end{aligned} \tag{6}$$

Furthermore, the following useful lemmas are used in this paper

Lemma 2.1: [24] Let Ω, D, and E be real matrices of appropriate dimensions. Then for any $F(t)$

$$\Omega + DF(t)E + E^T F^T(t) D^T < 0$$

if and only if there exists some $\varepsilon > 0$ such that

$$\Omega + \varepsilon DD^T + \varepsilon^{-1} E^T E < 0.$$

Lemma 2.2: [18] Jensen Inequality: For any scalar $b > a$, the following inequality holds

$$(b-a)\int_a^b x^T(s)Rx(s)ds \geq \Big(\int_a^b x(s)ds\Big)^T R\Big(\int_a^b x(s)ds\Big)$$

Considering a matrix G and defining the following polyhedral set

$$\mathcal{S} = \{x(t) \in \Re^n; |(K_{(i)} - G_{(i)})x(t)| \leq u_{0(i)}\}$$

Lemma 2.3: [25] If $x(t) \in \mathcal{S}$, then the following relation

$$\psi^T(Kx(t))T_0\Big[\psi(Kx(t)) - Gx(t)\Big] \leq 0$$

is verified for any diagonal positive matrix $T_0 \in \Re^{m\times m}$.

Consider that $x(t) \in \mathcal{S}$ and the following L-K functional which will be used throughout the paper

$$\begin{aligned} V(t) &= x^T(t)Px(t) + \int_{t-\tau(t)}^t x^T(s)Qx(s)ds + \int_{-h_m}^0 \int_{t+\theta}^t \dot{x}^T(s)R\dot{x}(s)dsd\theta \\ &\quad + \int_{t-\tau(t)}^t \dot{x}^T(s)W\dot{x}(s)ds \end{aligned} \tag{7}$$

where $P, Q, R, W > 0$ need to be determined.

Finally, for a positive scalar β, the ellipsoid D_e is defined as follows

$$D_e = \{x(t) \in \Re^n; \ x^T(t)Px(t) \leq \beta^{-1}\}$$

The result in Lemma 2.3 can be seen as a generalized sector condition. As will be seen in the sequel, differently from the classical sector condition (used for instance in [25]), this condition will allow to obtain stability conditions directly in an LMI form.

The stabilization problems that we are interested in studying can be summarized as follows

Problem 2.1: Given d, maximize h_m in order to ensure the robust stability of the closed-loop system for some set of admissible initial conditions.

Problem 2.2: Given h_m and d, find K and a set of admissible initial conditions, as large as possible, for which the stability of the closed-loop system is ensured.

3. Main Results

Some results are derived to ensure the robust stabilization of neutral system with saturating actuators. These results will be applied to a real system control problem.

3.1. Stability results

Some results are established to ensure the asymptotic stabilization of the nominal system ($DF(t)E_0 = DF(t)E_1 = DF(t)E_2 = 0$). Next the robust stabilization of the uncertain system is derived for any initial conditions in an estimated domain of attraction.

Lemma 3.1: If there exist symmetric positive definite matrices P, Q, R, W, appropriately sized matrices T_1, T_2, Y_1, Y_2, T_1, T_2 and a positive definite diagonal matrix T_0 satisfying

$$\Psi = \begin{bmatrix} \Psi_{11} & * & * & * & * & * \\ \Psi_{21} & \Psi_{22} & * & * & * & * \\ \Psi_{31} & \Psi_{32} & \Psi_{33} & * & * & * \\ \Psi_{41} & \Psi_{42} & 0 & \Psi_{44} & * & * \\ \Psi_{51} & \Psi_{52} & 0 & 0 & \Psi_{55} & * \\ \Psi_{61} & \Psi_{62} & 0 & 0 & 0 & \Psi_{66} \end{bmatrix} < 0 \tag{8}$$

where

$$\begin{aligned} \Psi_{11} &= T_1A + A^TT_1^T + T_1BK + K^TB^TT_1^T + Y_1 + Y_1^T + Q, \ \Psi_{31} = A_d^TT_1^T - Y_1^T \\ \Psi_{21} &= -T_1^T + T_2A + T_2BK + Y_2 + P, \ \Psi_{22} = -T_2 - T_2^T + h_mR + W, \ \Psi_{32} = A_d^TT_2^T - Y_2^T \\ \Psi_{33} &= -(1-d)Q, \ \Psi_{41} = -Y_1^T, \ \Psi_{42} = -Y_2^T, \ \Psi_{44} = -\frac{R}{h_m}, \ \Psi_{51} = C^TT_1^T, \ \Psi_{52} = C^TT_2^T \\ \Psi_{55} &= -(1-d)W, \ \Psi_{61} = T_0G - B^TT_1^T, \ \Psi_{62} = -B^TT_2^T, \ \Psi_{66} = -2T_0 \end{aligned}$$

Then, the nominal system (6) is asymptotically stable.

Proof 1: Using (2), the time derivative of the functional (7) along the trajectory of the system (6) is given by

$$\begin{aligned} \dot{V}(t) &= 2x^T(t)P\dot{x}(t) + x^T(t)Qx(t) - (1-d)x^T(t-\tau(t))Qx(t-\tau(t)) + h_m\dot{x}^T(t)R\dot{x}(t) \\ &\quad - \int_{t-\tau(t)}^{t} \dot{x}^T(s)R\dot{x}(s)ds + \dot{x}^T(t)W\dot{x}(t) - (1-d)\dot{x}^T(t-\tau(t))W\dot{x}(t-\tau(t)) \end{aligned}$$

Using the free weighting matrix approach, for appropriately dimensioned matrices T_1, T_2, Y_1 and Y_2, we have

$$\begin{aligned} &2\Big[x^T(t)T_1 + \dot{x}^T(t)T_2\Big]\Big[-\dot{x}(t) + C\dot{x}(t-\tau(t)) + (A+BK)x(t) + A_dx(t-\tau(t)) \\ &\qquad -B\psi(Kx(t))\Big] = 0 \\ &2\Big[x^T(t)Y_1 + \dot{x}^T(t)Y_2\Big]\Big[x(t) - x(t-\tau(t)) - \int_{t-\tau(t)}^{t} \dot{x}(s)ds\Big] = 0 \end{aligned} \tag{9}$$

From Lemma 2.3, it follows that

$$\dot{V}(t) \le \dot{V}(t) - 2\psi^T(Kx(t))T_0[\psi(Kx(t)) - Gx(t)] \tag{10}$$

Then, applying Lemma 2.2, adding the terms on the left of (9) and (10) to $\dot{V}(t)$ allows us to express $\dot{V}(t)$ as

$$\begin{aligned}\dot{V}(t) \;\le\; & 2x^T(t)P\dot{x}(t) + x^T(t)Qx(t) - (1-d)\Big(x^T(t-\tau(t))Qx(t-\tau(t)) + \dot{x}^T(t-\tau(t))\\ & \times W\dot{x}(t-\tau(t))\Big) + \dot{x}^T(t)\Big(h_m R + W\Big)\dot{x}(t) - \Big(\int_{t-\tau(t)}^{t}\dot{x}(s)ds\Big)^T \frac{R}{h_m}\Big(\int_{t-\tau(t)}^{t}\dot{x}(s)ds\Big)\\ & +2\Big[x^T(t)T_1 + \dot{x}^T(t)T_2\Big]\Big[-\dot{x}(t) + C\dot{x}(t-\tau(t)) + (A+BK)x(t) + A_d x(t-\tau(t))\\ & -B\psi(Kx(t))\Big] + 2\Big[x^T(t)Y_1 + \dot{x}^T(t)Y_2\Big]\Big[x(t) - x(t-\tau(t)) - \int_{t-\tau(t)}^{t}\dot{x}(s)ds\Big]\\ & -2\psi^T(Kx(t))T_0\Big[\psi(Kx(t)) - Gx(t)\Big] \end{aligned} \qquad (11)$$

By simple manipulation, (11) can be rewritten as

$$\dot{V}(t) \le \eta^T(t)\Psi\eta(t)$$

where Ψ is defined in (8) and

$$\eta^T(t) = \Big[\; x^T(t) \;\; \dot{x}^T(t) \;\; x^T(t-\tau(t)) \;\; \int_{t-\tau(t)}^{t}\dot{x}^T(s)ds \;\; \dot{x}^T(t-\tau(t)) \;\; \psi^T(Kx(t)) \;\Big]$$

The condition (8) holds implies that $\dot{V}(t) < 0$. It follows that the trajectories of nominal system (6) converge asymptotically to the origin. This result gives a general solution for testing stability. In the following, we provide a new result that permits a robust stabilizing controller to be calculated. Thus, based on Lemma 3.1, some results are now derived to ensure robust stabilization of saturated neutral state-delayed system subject to uncertainty (3).

Theorem 3.1: If there exist symmetric positive definite matrices $\overline{P}$, $\overline{Q}$, $\overline{R}$, $\overline{W}$, appropriately sized matrices X, $\overline{Y_1}$, $\overline{Y_2}$, M, U, a diagonal matrix S of appropriate dimension, a real scalar α and positive scalars $\varepsilon, \beta, \delta$ satisfying the conditions (12)-(14)

$$\begin{bmatrix} \Omega_{11} + \varepsilon DD^T & * & * & * & * & * & * \\ \Omega_{21} + \alpha\varepsilon DD^T & \Omega_{22} + \alpha^2\varepsilon DD^T & * & * & * & * & * \\ \Omega_{31} & \Omega_{32} & \Omega_{33} & * & * & * & * \\ \Omega_{41} & \Omega_{42} & 0 & \Omega_{44} & * & * & * \\ \Omega_{51} & \Omega_{52} & 0 & 0 & \Omega_{55} & * & * \\ \Omega_{61} & \Omega_{62} & 0 & 0 & 0 & \Omega_{66} & * \\ E_0X^T & 0 & E_1X^T & 0 & E_2X^T & 0 & -\varepsilon I \end{bmatrix} < 0 \qquad (12)$$

$$\begin{bmatrix} \overline{P} & * \\ U_i - M_i & \beta u_{0i}^2 \end{bmatrix} \ge 0,\; i = 1, ..., m \qquad (13)$$

$$\begin{aligned} &\Big(\overline{\lambda}(X^{-1}\overline{P}X^{-T}) + h_m\overline{\lambda}(X^{-1}\overline{Q}X^{-T})\Big)\|\phi\|_c^2 \\ &+\Big(\frac{h_m^2}{2}\overline{\lambda}(X^{-1}\overline{R}X^{-T}) + h_m\overline{\lambda}(X^{-1}\overline{W}X^{-T})\Big)\|\dot{\phi}\|_c^2 \le \beta^{-1} \end{aligned} \qquad (14)$$

where

$$\begin{aligned} \Omega_{11} &= AX^T + XA^T + BU + U^TB^T + \overline{Y_1} + \overline{Y_1}^T + \overline{Q},\; \Omega_{21} = -X + \alpha AX^T + \alpha BU + \overline{P} + \overline{Y_2} \\ \Omega_{22} &= -\alpha X - \alpha X^T + h_m\overline{R} + \overline{W},\; \Omega_{31} = XA_d^T - \overline{Y_1}^T,\; \Omega_{32} = \alpha XA_d^T - \overline{Y_2}^T \\ \Omega_{33} &= -(1-d)\overline{Q},\; \Omega_{41} = -\overline{Y_1}^T,\; \Omega_{42} = -\overline{Y_2}^T,\; \Omega_{44} = -\frac{\overline{R}}{h_m},\; \Omega_{51} = XC^T,\; \Omega_{52} = \alpha XC^T \\ \Omega_{55} &= -(1-d)\overline{W},\; \Omega_{61} = M - SB^T,\; \Omega_{62} = -\alpha SB^T,\; \Omega_{66} = -2S^T \end{aligned}$$

then, the uncertain saturated delayed system is robustly stable and the trajectories of $x(t)$ remain within the ellipsoid D_e when the state feedback control law is used, with $K = UX^{-T}$.

Proof 2: Set $T_2 = \alpha T_1$, $\alpha > 0$. From (8), it is easily seen that T_2 is nonsingular and consequently T_1 is inversible. Multiplying (8) by $diag\{T_1^{-1}, T_1^{-1}, T_1^{-1}, T_1^{-1}, T_1^{-1}, T_0^{-1}\}$ on the left and by its transpose on the right. Then, introduce a new change of variables such that $X = T_1^{-1}$, $U = KX^T$, $M = GX^T$ $S = T_0^{-1}$ and $\bar{\Pi} = X\Pi X^T$ where $\Pi = P, Q, R, W, Y_1, Y_2$. Thus, we obtain the following equation

$$\Omega = \begin{bmatrix} \Omega_{11} & * & * & * & * & * \\ \Omega_{21} & \Omega_{22} & * & * & * & * \\ \Omega_{31} & \Omega_{32} & \Omega_{33} & * & * & * \\ \Omega_{41} & \Omega_{42} & 0 & \Omega_{44} & * & * \\ \Omega_{51} & \Omega_{52} & 0 & 0 & \Omega_{55} & * \\ \Omega_{61} & \Omega_{62} & 0 & 0 & 0 & \Omega_{66} \end{bmatrix} < 0 \tag{15}$$

Now, let us replace A, A_d and C by $A + DF(t)E_0$, $A_d + DF(t)E_1$ and $C + DF(t)E_2$, respectively. We find that equation (15) is equivalent to the following condition

$$\Omega + \begin{bmatrix} D \\ \alpha D \\ 0 \\ 0 \\ 0 \\ 0 \end{bmatrix} F(t) \begin{bmatrix} E_0X^T & 0 & E_1X^T & 0 & E_2X^T & 0 \end{bmatrix}$$
$$+ \begin{bmatrix} XE_0^T \\ 0 \\ XE_1^T \\ 0 \\ XE_2^T \\ 0 \end{bmatrix} F^T(t) \begin{bmatrix} D^T & \alpha D^T & 0 & 0 & 0 & 0 \end{bmatrix} < 0$$

According to Lemma 2.1 and by the Schur complement we obtain LMI (12). On other hand, the satisfaction of (13) guarantees that $\forall x \in D_e$, $x \in \mathcal{S}$. In fact, $D_e \subset \mathcal{S}$ is verified by the following conditions

$$\begin{bmatrix} P & * \\ K_i - G_i & \beta u_{0i}^2 \end{bmatrix} \geq 0 \tag{16}$$

pre and post multiplying (16) by $\Delta = diag\{X, I\}$ and its transpose will result in the LMI (13).

Moreover, the satisfaction of (14) can be proven as follows. From the L-K functional defined in (7) we have

$$\begin{aligned} V(0) &\leq x^T(0)Px(0) + \int_{-h_m}^{0} x^T(s)Qx(s)ds + \int_{-h_m}^{0}\int_{\theta}^{0} \dot{x}^T(s)R\dot{x}(s)dsd\theta + \int_{-h_m}^{0} \dot{x}^T(s)W\dot{x}(s)ds \\ &\leq \left(\bar{\lambda}(P) + h_m\bar{\lambda}(Q)\right)\|\phi(\theta)\|_c^2 + \left(\frac{h_m^2}{2}\bar{\lambda}(R) + h_m\bar{\lambda}(W)\right)\|\dot{\phi}(\theta)\|_c^2 = \delta \end{aligned}$$

Therefore, we have $x^T(t)Px(t) \leq V(t) \leq V(0) \leq \delta \leq \beta^{-1}$, that is for all $t \geq 0$ the trajectories of the system do not leave the set D_e for any initial functions $\phi(\theta)$ in D_e which ensures that $x(t) \in \mathcal{S}$. This completes the proof.

The above results can be applied to the real system control problem by adopting a fourth-order linearized model of the match number in a wind tunnel.

3.2. Application to a feedback control of Mach Number in a Wind Tunnel

In steady-state operating conditions (some constant fan speed, liquid nitrogen injection rate, and gaseous-nitrogen vent rate), the dynamic response of the Mach Number perturbations δM to small perturbations in the guide vane angle actuator $\delta\theta_a$ in a driving fan is described by the following equations [22, 23]

$$\begin{aligned} &\frac{1}{a}\delta\dot{M}(t) + \delta M(t) = k\delta\theta(t-\tau(t)) \\ &\delta\ddot{\theta}(t) + 2\xi w\delta\dot{\theta}(t) + w^2\delta\theta(t) = w^2\delta\theta_a(t) \end{aligned} \tag{17}$$

where $\delta\theta$ is the guide vane angle, a, k, ξ, w are parameters depending on the operating point which are presumed constant when the perturbation δM, $\delta\theta$, $\delta\theta_a$ are small and the delay $\tau(t)$ represents the time of the transport between the fan and the test section.

Rewriting (17) in state space form yields

$$\begin{aligned} &\dot{x}(t) = Ax(t) + A_d x(t-\tau(t)) + Bsat(u(t)), \ t>0 \\ &x(\theta) = \phi(\theta), \ \forall\theta \in [-h_m, 0], \ \phi(\theta) \in C^v_{h_m}, \end{aligned} \tag{18}$$

where

$$x = \begin{bmatrix} \delta M \\ \delta\theta \\ \delta\theta_a \end{bmatrix}, \ A = \begin{bmatrix} -a & 0 & 0 \\ 0 & 0 & 1 \\ 0 & -w^2 & -2\xi w \end{bmatrix}, \ A_\tau = \begin{bmatrix} 0 & ka & 0 \\ 0 & 0 & 0 \\ 0 & 0 & 0 \end{bmatrix}, \ B = \begin{bmatrix} 0 \\ 0 \\ w^2 \end{bmatrix}$$

The control $u(t)$ represents $\delta\theta_a$. This system can be seen as a particular case of nominal system (5) when $C = \Delta A(t) = \Delta A_d(t) = \Delta C(t) = 0$. Then, the following theorem gives a condition to stabilize system (18).

Theorem 3.2: If there exist symmetric positive definite matrices $\overline{P}$, $\overline{Q}$, $\overline{R}$, appropriately sized matrices X, $\overline{Y_1}$, $\overline{Y_2}$, M, U, a diagonal matrix S of appropriate dimension and a real scalar α satisfying the condition (19)-(21)

$$\begin{bmatrix} \Omega_{11} & * & * & * & * \\ \Omega_{21} & \Omega_{22} - \overline{W} & * & * & * \\ \Omega_{31} & \Omega_{32} & \Omega_{33} & * & * \\ \Omega_{41} & \Omega_{42} & 0 & \Omega_{44} & * \\ \Omega_{61} & \Omega_{62} & 0 & 0 & \Omega_{66} \end{bmatrix} < 0 \tag{19}$$

$$\begin{bmatrix} \overline{P} & * \\ U_i - M_i & \beta u^2_{0i} \end{bmatrix} \geq 0 \tag{20}$$

$$\Big(\overline{\lambda}(X^{-1}\overline{P}X^{-T}) + h_m\overline{\lambda}(X^{-1}\overline{Q}X^{-T})\Big)\|\phi(\theta)\|^2_c + \Big(\frac{h^2_m}{2}\overline{\lambda}(X^{-1}\overline{R}X^{-T})\Big)\|\dot{\phi}(\theta)\|^2_c \leq \beta^{-1} \tag{21}$$

then, the saturated state-delayed system (18) with the state feedback control law and $K = UX^{-T}$ is robustly stable and the trajectories of $x(t)$ remain within the ellipsoid D_e.

Proof 3: It suffices to follow the same steps of the Proof 1 and 2 considering $W = 0$, $C = 0$ and $\Delta A(t) = \Delta A_d(t) = \Delta C(t) = 0$.

Remark 3.1: In deriving Lemma 3.1, the slack variable T_1, T_2, Y_1, Y_2 are introduced in order to reduce the conservatism of the asymptotic stability conditions. It can be seen from the Proof 1 that $\dot{V}(t)$ remains unaffected by the introduction equation (9) i.e. the slack variables T_1, T_2,

Y_1, Y_2. So these matrices leads to a more flexible LMI condition in (8) and consequently reduce the conservatism of Lemma 3.1 and consequently that of the next results. This advantage can be seen from the numerical examples.

Remark 3.2: In the above proof, the free-weighting matrices T_1, T_2, Y_1, Y_2 are employed to offer more flexibility to our results which may lead to some improvements. Specifically, the first equation of the (9) which is equal to zero, was added to the derivative of the L-K functional. Moreover, the Leibniz-Newton formula was employed to obtain a delay-dependent criterion, and the relationships between those terms was also taken into account. That is, the second equation of the (9) which is equal to zero, was added to the derivative of the L-K functional, as well.

Remark 3.3: The tuning parameter α is designed to improve results and gives a better performance to system. An adequate choice of this parameter gives improved result for which the system is robustly stable.

4. Optimization Problems

In this section we show how the theoretical conditions can be casted into LMI-based optimization problems to determine a suitable stabilizing gain K and a domain of attraction which ensure that the state trajectory of the closed-loop system (6) starting from any initial functions $\phi(\theta)$ in D_e will remain within D_e for all $t > 0$.

The idea is to develop a methodology to estimate the largest possible domain of initial conditions for which it can be ensured that the closed-loop system trajectories remain bounded. As in [5], we impose the following conditions

$$\begin{bmatrix} \sigma_1 I & \widetilde{X} \\ \widetilde{X}^T & \widetilde{P} \end{bmatrix} \geq 0, \begin{bmatrix} \sigma_2 I & \widetilde{X} \\ \widetilde{X}^T & \widetilde{Q} \end{bmatrix} \geq 0, \begin{bmatrix} \sigma_3 I & \widetilde{X} \\ \widetilde{X}^T & \widetilde{R} \end{bmatrix} \geq 0, \begin{bmatrix} \sigma_4 I & \widetilde{X} \\ \widetilde{X}^T & \widetilde{W} \end{bmatrix} \geq 0 \tag{22}$$

It follows that condition (14) is satisfied if the following LMI holds

$$\left[\sigma_1 + h_m\sigma_2 + \frac{h_m^2}{2}\sigma_3 + h_m\sigma_4\right]\delta^2 \leq \beta^{-1} \tag{23}$$

where $\delta^2 = \max(\|\phi(\theta)\|_c^2, \|\dot{\phi}(\theta)\|_c^2)$, $X^{-1} = \widetilde{X}$ and $\overline{\Pi}^{-1} = \widetilde{\Pi}$ with $\Pi = P, Q, R, W$ and the *stability radius* $\delta > 0$ a scalar to be determined.

Therefore, let us construct a feasibility problem as follows

$$\begin{aligned} &\text{Minimize Trace}\Big(\overline{P}\widetilde{P} + \overline{Q}\widetilde{Q} + \overline{R}\widetilde{R} + \overline{W}\widetilde{W} + (X + X^T)(\widetilde{X} + \widetilde{X}^T)\Big) \\ &\text{subject to } \overline{\Pi} > 0, \widetilde{\Pi} > 0, \beta > 0, \delta > 0, \sigma_{i=1,\ldots,4} > 0, (12), (13), (22), (23), \\ &\begin{bmatrix} \overline{P} & * \\ I & \widetilde{P} \end{bmatrix} \geq 0, \begin{bmatrix} \overline{Q} & * \\ I & \widetilde{Q} \end{bmatrix} \geq 0, \begin{bmatrix} \overline{R} & * \\ I & \widetilde{R} \end{bmatrix} \geq 0, \begin{bmatrix} \overline{W} & * \\ I & \widetilde{W} \end{bmatrix} \geq 0, \\ &\begin{bmatrix} X + X^T & * \\ I & \widetilde{X} + \widetilde{X}^T \end{bmatrix} \geq 0. \end{aligned} \tag{24}$$

The new LMIs problem can be solved by using the following like cone complementarity algorithm

Step 1: Given h_m, β, fix initial values $\alpha = \alpha_0$ and choose a sufficiently large initial δ such that there exits a feasible solution to LMI conditions in (24). Set $\delta_0 = \delta$ and $\alpha_0 = \alpha$.

Step 2: Find a set of feasible matrices $\left(\overline{\Pi}, X, \widetilde{\Pi}, \widetilde{X}, \sigma_{i=1,\ldots,4}\right)_0$ that satisfies (24).

IOP Conf. Series: Journal of Physics: Conf. Series **783** (2017) 012031 doi:10.1088/1742-6596/783/1/012031

Step 3: Solve the following LMI minimization problem

$$\text{Minimize Trace}\Big(\overline{P}\widetilde{P}_0+\overline{Q}\widetilde{Q}_0+\overline{R}\widetilde{R}_0+\overline{W}\widetilde{W}_0+(X+X^T)(\widetilde{X}_0+\widetilde{X}_0^T)+\overline{P}_0\widetilde{P}+\overline{Q}_0\widetilde{Q}+\overline{R}_0\widetilde{R}$$
$$+\overline{W}_0\widetilde{W}+(X_0+X_0^T)(\widetilde{X}+\widetilde{X}^T)\Big)$$

subject to LMIs in (24)

Step 4: Substitute the new matrix variables from the previous step into (24). If the result is feasible, then set $\delta_0=\delta$ and $\alpha_0=\alpha$. If it is not feasible, then set the new matrices to be $\left(\overline{\Pi}, X, \widetilde{\Pi}, \widetilde{X}, \sigma_{i=1,\ldots,4}\right)_0$ and go to step 3.

5. Illustrative Examples

In this section we illustrate our methodology over three examples borrowed from [2, 4, 20].
Example 1 [2]. Consider the nominal system of (1), where

$$A=\begin{bmatrix}0.5 & -1\\ 0.5 & -0.5\end{bmatrix},\ A_d=\begin{bmatrix}0.6 & 0.4\\ 0 & -0.5\end{bmatrix},\ B=\begin{bmatrix}1\\ 1\end{bmatrix},\ u_0=5,\ d=0,\ C=0,\ W=0$$

We can apply the stability results presented in Theorem 3.1 with $\Delta A(t)=\Delta A_d(t)=\Delta C(t)=0$. Taking $\beta=1$, we obtain a stability radius $\delta=271$ when a tuning parameter is $\alpha=143$. For the comparison with other approaches we give the Table 1

Table 1. Comparison of h_m with obtained K.

Approach	h_m	K
[1]	0.35	Not Reported
[15]	1.854	$(-25.8809 \quad -4.9315)$
[30]	2.248	Not Reported
[10]	2	$(-\ 5.7702 \quad -0.9754)$
[9]	2	$(-\ 5.6104 \quad -0.9147)$
Th. 3 [4]	1.854	$(-\ 1.7008 \quad 0.2776)$
[2]	1.854	$(-\ 2.2346 \quad 0.0580)$
Th. 3.1	3.262	$(-\ 3.9263 \quad -2.5783)\times 10^3$

Note that as pointed in Section 2, the effective control signal applied to the system is $u(t)=sat(Kx(t))$. in this case, $u(t)$ is a scalar so we have $u(t)=sat(Kx(t))=sign(Kx(t))min\{u_0,|Kx(t)|\}$ and the amplitude limitation of the input is always satisfied.

From the numerical example it is clear that our approach is less conservative in stabilizing the system with larger time-delay bound than those of [1, 15, 30, 10, 9, 4, 2].
Example 2 [4]. The system is described by (1) with

$$A=\begin{bmatrix}1 & 1.5\\ 0.3 & -2\end{bmatrix},\ A_d=\begin{bmatrix}0 & -1\\ 0 & 0\end{bmatrix},\ B=\begin{bmatrix}10\\ 1\end{bmatrix},\ C=cI,\ u_0=15,\ h_m=1,\ D=E_{0,1,2}=0$$

Taking $\beta=1$ and applying the proposed algorithm, we obtain a stability radius of $\delta=344$ and $\delta=404$. Their results are listed in Table 2 and Table 3 along with the results obtained by Theorem 3.1 for $\alpha=111$ and $\alpha=94$, respectively.

The domain of initial condition obtained for neutral system in which $c = 0.2$ and $d = 0.1$ is given by $\delta = 344$, while for retarded system (i.e. $c = 0$), it is given by $\delta = 404$. For the comparison with other approaches, we give Table 2 for neutral system and Table 3 for time delay system.

It is clear that the obtained δ is significantly larger than those obtained in [17, 16, 2, 1, 15, 4].

To see graphically the improvements of our approach, we represent the domains of attraction of Table 2 and Table 3 in Figure 1 and Figure 2.

Table 2. Comparison of δ and K for $c = 0.2$ and $d = 0.1$.

Approach	δ	K
[17]	12.88	$(-0.2780 \quad -0.1390)$
[16]	70.74	$(-0.1325 \quad 0.0153)$
[2]	76.2262	$(-0.2359 \quad -0.0453)$
Th. 3.1	344	$(-0.1032 \quad -0.0272)$

Table 3. Comparison of δ and K for $c = d = 0$.

Approach	δ	K
[1]	67.0618	Not Reported
[15]	79.43	$(-7.9130 \quad 0.7323)$
[16]	83.55	$(-0.1950 \quad 0.0649)$
[2]	84.6074	$(-0.2223 \quad -0.0246)$
Th. 3 [4]	106.2856	$(-0.6646 \quad -0.0239)$
Th. 3.1	404	$(-0.1013 \quad -0.0495)$

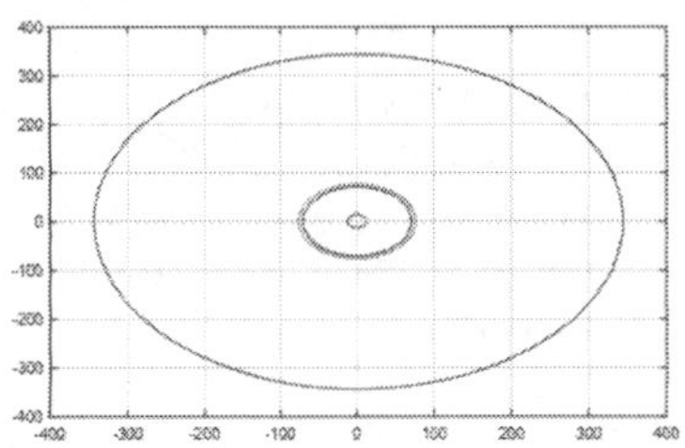

Figure 1. The domain of attraction in Table 2.

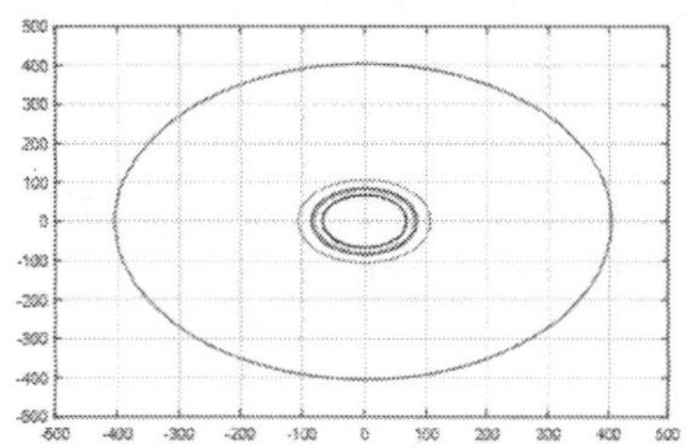

Figure 2. The domain of attraction in Table 3.

Example 3 In this example, we consider the linearized model of the Mach Number in a Wind Tunnel described by (17)-(18), all the numerical values of the parameters are borrowed from [22]. The same parameters have been used in [20] and are as follows

$$\frac{1}{a} = 1.964s,\ k = -0.0117deg^{-1},\ \xi = 0.8\, and\, w = 6rad/s$$

IOP Conf. Series: Journal of Physics: Conf. Series **783** (2017) 012031 doi:10.1088/1742-6596/783/1/012031

Using Theorem 3.2, letting, $u_0 = 1$ and $d = 0$. The following state feedback control law

$$K = \begin{pmatrix} 0.0000 & 0.7308 & 0.1398 \end{pmatrix}$$

stabilizes the system (18), hence the system is stabilizable for any time delay less than the upper bound 26, and the stability radius is 389 where $\alpha = 2$ and $\beta = 1$. For comparison with the results of others papers see Table 4.

Table 4. Comparison of h_m.

method	[22]	[20]	this paper
h_m	0.33	0.9685	26

From Table 4, the proposed method gives much larger time delay bound than the results of [22, 20].

On other hand, the values of the control parameters are adjusted to 0.75, 1 and 0.5, respectively. Thus, our results are less conservative than those of [21].

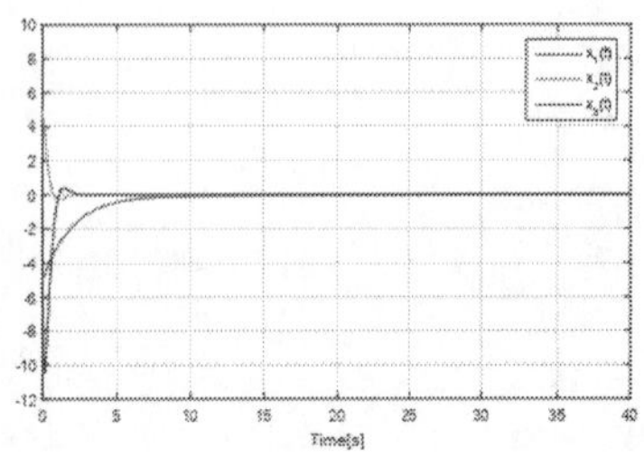

Figure 3. The response of the state variables using Runge-Kutta Method.

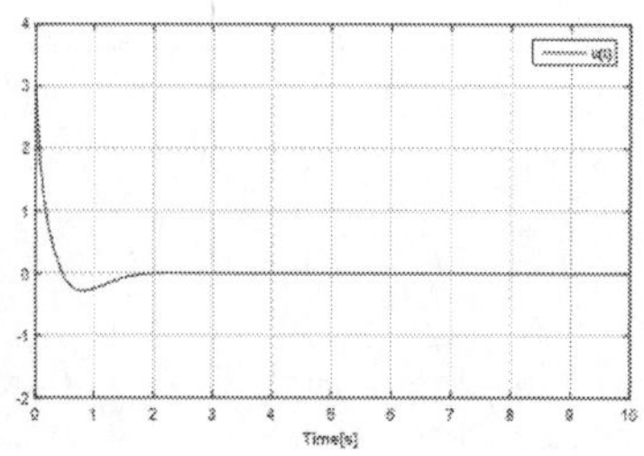

Figure 4. The response of the control input variable using Runge-Kutta Method.

Note that even though the control input is initially saturated, the states, in due course, are driven to the origin with $x_0(t) = [-5 \quad 5 \quad -5]^T$. The numerical simulation presented in this example shows that very good transient responses can be obtained by using control law compared to [21], which can be implemented in a controlling microprocessor as a simple feedback, using state variable $x(t)$ and delayed state variable $x(t - \tau(t))$ and control with a saturation at its input.

6. Conclusions

We have presented a methodology for robust stabilization of uncertain neutral time-delay system with saturating actuator, with the derived conditions given as LMIs that depend of the maximum value of the delay. These conditions then guarantee the stability of the closed loop system when the initial states are taken within a calculated region of attraction. The proposed conditions have being shown to be less conservative than those previously proposed in the literature by numerical examples, that have also illustrated the feasibility of the proposed approach.

Since the class of systems investigated in this paper appear in many process control applications, one can expect that the feedback design method presented here may have a wide range of applicability. The results were applied to the feedback control of Mach Number in a Wind Tunnel for illustration.

Acknowledgment
This work was partly funded by MiCInn project DPI2014-54530-R.

References

[1] Cao Y-Y, Lin Z and Hu T 2002 Stability analysis of linear time-delay systems subject to input saturation, *IEEE Trans. on Circ. and Sys.* 49 233–40.
[2] Chen Y, Fei S and Li Y 2015 Stabilization of neutral time-delay systems with actuator saturation via auxiliary time-delay feedback, *Automatica* 52 242–7.
[3] Delgado E and Barreiro A 2007 Stability of teleoperation systems for time varying delays by LMI techniques, *In Proc. of the Eur. Conf. Contr. Kos Greece* 4214–21.
[4] Dey R, Ghosh S, Ray G and Rakshit A 2014 Improved delay-dependent stabilization of time-delay systems with actuator saturation, *Int. J. of Robust and Nonlin. Control* 24 902–17.
[5] El Fezazi N, El Haoussi F, Tissir E H and Tadeo F 2015 Delay dependent anti-windup synthesis for time-varying delay systems with saturating actuators, *Int. J. of Comp. Appl. (0975 8887)* 111 1–6.
[6] El Fezazi N, Tissir E H and El Haoussi F 2015 Delay-dependent anti-windup design to address challenges in AQM congestion control for TCP/IP networks, *In Proc. of the Int. Conf. AIT2S Settat Morocco* 309–13.
[7] El Fezazi N, Lamrabet O, El Haoussi F, Tissir E H, Alvarez T and Tadeo F 2016 Robust Controller Design for Congestion Control in TCP/IP Routers, *In Proc. of the Int. Conf. ICSC Marrakesh Morocco* 243–9.
[8] El Fezazi N, El Haoussi F, Tissir E H, Husain A R and Zakaria M I 2016 Delay-dependent Robust Anti-windup Synthesis Approach to AQM in TCP/IP Networks, *In Proc. of the Int. Conf. ICEIT Tanger Morocco* 294–9.
[9] El Haoussi F, Tissir E H, Hmamed A and Tadeo F 2012 Stabilization of neutral systems with saturating actuators, *J. of Contr. Sci. and Eng.* 2012 1–9.
[10] El Haoussi F and Tissir E H 2010 Delay and its time-derivative dependent robust stability of uncertain neutral systems with saturating actuators, *Int. J. of Autom. and Compu.* 7 455–62.
[11] El Haoussi F, Tissir E H and Tadeo F 2014 Advances in the robust stabilization of neutral systems with saturating actuators, *Int. J. of Ecol. and Devel.* 28 49–62.
[12] El Haoussi F, Tissir E H, Tadeo F and Hmamed A 2013 Robust stabilization with saturating actuators of neutral and state delayed systems, *Int. J. of Sci. and Tech. of Autom. Contr. and Compu. eng.* 7 1878–89.
[13] El Haoussi F and Tissir E H 2007 An LMI-based approach for robust stabilization of time delay systems containing saturating actuators, *J. of Math. Contr. and Inf.* 24 347–56.
[14] Flores J V, Gomes da Silva J M Jr and Seuret A 2011 Static anti-windup synthesis for linear systems with time-varying input delays, *In Proc. of the 18th IFAC World Congress Milan Italy* 44 14483–8.
[15] Fridman E, Pila A and Shaked U 2003 Regional stabilization and H_∞ control of time-delay systems with saturating actuators, *Int. J. of Rob. and Nonlin. Contr.* 13 885–907.
[16] Gomes da Silva J M Jr, Seuret A, Fridman E and Richard J P 2011 Stabilization of neutral systems with saturating control inputs, *Int. J. of Sys. Sci.* 42 1093–103.
[17] Gomes da Silva J M Jr, Fridman E, Seuret A and Richard J P 2005 Stabilization of neutral systems with saturating inputs, *In Proc. of the 16th IFAC World Congress Praga Czech Republic* 38 448–53.
[18] Gu K, Chen J and Kharitonov V L 2003 Stability of time-delay systems, *Springer Sci. and Bus. Med.*.
[19] He Y., Wu M and She J-H 2008 Delay-dependent robust stability and stabilization of uncertain neutral systems, *Asian J. of Contr.* 10 376–83.
[20] Liu P L 2011 Delay-dependent stabilization for linear time-delay uncertain systems with saturating actuators, *Int. J. of Gene. Sys.* 40 301–12.
[21] Loukianov A G, Espinosa-Guerra O and Castillo-Toledo B 2005 Deadtime block compensation sliding mode control of linear system with delay, *In Proc. of the 16th IFAC World Congress Praga Czech Republic* 38 712–7.
[22] Manitius A Z 1984 Feedback controllers for a wind tunnel model involving a delay: analytical design and numerical simulation, *IEEE Trans. on Autom. Contr.* 29 1058–68.
[23] Manitius A and Tran H 1986 Numerical simulation of a non-linear feedback controller for a wind tunnel model involving a time delay, *Opt. Contr. Appl. and Meth.* 7 19–39.
[24] Peterson I R 1987 A stabilization algorithm for a class of uncertain linear systems, *Sys. and Contr. Lett.* 8 351–7.
[25] Tarbouriech S, Gomes da Silva J M Jr and Garcia G 2003 Delay-dependent anti-windup loops for enlarging the stability region of time-delay systems with saturating inputs, *J. of dyn. sys., meas., and contr.* 125 265–7.

[26] Tarbouriech S, Gomes da Silva J M Jr and Garcia G 1999 Stabilization of neutral linear time-delay systems with saturating actuators, *In Proc. of the 38th IEEE Conf. on Dec. and Contr.* 2 2011–6.

[27] Tissir E H and Hmamed A 1992 Further results on the stabilization of time delay systems containing saturating actuators, *Int. J. of Sys. Sci.* 23 615–22.

[28] Tissir E H 2007 Delay-dependent robust stability of linear systems with non-commensurate time varying delays, *Int. J. of Sys. Sci.* 38 749–57.

[29] Tissir E H 2010 Delay-dependent robust stabilization for systems with uncertain time varying delay, *J. of Dyn. Sys., Meas., and Contr.* 132 054504.

[30] Zhang L, Boukas E-K and Haidar A 2008 Delay-range-dependent control synthesis for time-delay systems with actuator saturation, *Automatica* 44 2691–5.

Design of robust iterative learning control schemes for systems with polytopic uncertainties and sector-bounded nonlinearities

Marcin Boski, Wojciech Paszke

Institute of Control and Computation Engineering, University of Zielona Góra, ul. Szafrana 2, 65-516 Zielona Góra, Poland.

E-mail: m.boski@issi.uz.zgora.pl, w.paszke@issi.uz.zgora.pl

Abstract. This paper deals with designing of iterative learning control schemes for uncertain systems with static nonlinearities. More specifically, the nonlinear part is supposed to be sector bounded and system matrices are assumed to range in the polytope of matrices. For systems with such nonlinearities and uncertainties the repetitive process setting is exploited to develop a linear matrix inequality based conditions for computing the feedback and feedforward (learning) controllers. These controllers guarantee acceptable dynamics along the trials and ensure convergence of the trial-to-trial error dynamics, respectively. Numerical examples illustrate the theoretical results and confirm effectiveness of the designed control scheme.

1. Introduction

Iterative learning control (ILC) is an effective method for providing a significant improvements of the tracking performance for systems that operate in a repetitive mode. The novel feature of this control scheme is to take a benefit of learning of the feedforward signals for subsequent task executions by iteratively updating them based on the control error accumulated in the previous task executions. As a result of successive repetitions or trials, a high performance feedforward signal is provided and hence a small tracking error is obtained by removing the repetitive part of the error. Specifically, the objective in ILC design is to construct the input such that the corresponding output precisely tracks a reference signal that is specified over a finite time interval. A particular feature is the availability to the control law of temporal information that would be non-causal outside the ILC setting, e.g., at sample instant p information at $p+\lambda$, $\lambda > 0$, can be used provided it has been generated on a previous trial.

Since the original work of Arimoto [2], ILC has remained as a significant area of control systems research with many algorithms experimentally verified in the research laboratory and applied in many industrial applications - see [1, 5] for many practical systems and processes where ILC has been successfully applied. Particular application examples include industrial robotics, see, e.g., [9], where the pick and place operation common in many mass manufacturing processes is an immediate fit to ILC, and wafer stage motion systems, see, e.g., [8]. Also, the survey paper [18] focuses on run-to-run control, as often found in the chemical process industries.

A literature review, see, e.g., [1, 5], shows that a popular approach used for ILC design is to first apply a feedback control law to produce acceptable along the trial dynamics and then

IOP Conf. Series: Journal of Physics: Conf. Series **783** (2017) 012032 doi:10.1088/1742-6596/783/1/012032

apply ILC to force trial-to-trial error convergence of the resulting system. This means that the objective of stabilization of the system in the time domain and convergence of the ILC scheme in the trial domain can be written as convergence conditions on the tracking error and control input

$$\lim_{k\to\infty} \|e_k\| = 0, \quad \lim_{k\to\infty} ||u_k - u_\infty|| = 0,$$

where $\|\cdot\|$ is an appropriate vector norm and u_∞ is termed the learned control. The reason for the condition on the control input sequence is to prevent large increases in the control vector from one trial to the next.

The main drawback of the classical synthesis procedure is that it does not lead to an optimal combination of the feedback and feedforward control actions. In particular, since the convergence condition includes terms related to the feedback controller, it should be treated simultaneously with the learning controller. Therefore, an obvious alternative to classical approach is to use a repetitive process setting for analysis [12–14]. This seems to be a natural setting for ILC analysis and design since the dynamics evolve in two independent directions and information in the temporal domain is limited to a finite duration. The main advantage of this approach is that gives a systematic way to simultaneously consider behaviour along the time axis and from trial-to-trial [15].

The analysis of linear repetitive processes has received considerable attention in the literature, see, e.g. [14]. Although this approach is relatively well known in the ILC area, there remain open problems to be solved. In this paper, the repetitive setting is applied and the results on the inclusion of nonlinear dynamics for uncertain systems are developed and hence the results of [11] are significantly extended.

As it is well known, an important class of nonlinear systems is a feedback system whose forward path contains a linear time invariant subsystem and the feedback path contains a memoryless nonlinearity, see [16] or [17] for some further details on this class of systems. In particular, systems where the nonlinearity satisfies certain sector condition (i.e. Lur'e systems) are treated. Such a description can be used to model many nonlinear dynamics arising in applications, such as systems with saturation and deadzone nonlinearities that can be treated as a linear system in connection with a nonlinearity satisfying some sector conditions. Hence, based on sector nonlinearity properties, the paper objective is to extend an approach of [11] to ILC design for uncertain systems with sector bounded nonlinearities in the repetitive process setting. Obviously, such results are required since practical control specifications require to involve the size of perturbation which affect system matrices.

This paper uses a repetitive setting to develop robust ILC design conditions for uncertain discrete linear systems with sector-bounded nonlinearities. It has to be emphasized that design is performed in discrete domain only, since practical implementation requires the control and error signals to be stored in a digital memory. Furthermore, a set of transformations is used to express the design conditions in terms of linear matrix inequalities (LMIs) for calculation purposes. A numerical example is also given to demonstrate the effectiveness of the design.

This paper is organized as follows. In Section 2, repetitive processes will be addressed in more detail as they are the basis for the main result in this paper and Section 3, an ILC scheme for uncertain systems with vector-valued nonlinear dynamics is embedded into the repetitive process setting. The main result of the paper on design in the presence of polytopic uncertainties and sector-valued nonlinearities is developed in Section 4. A numerical example is given in Section 5 to illustrate the effectiveness of the new design algorithm and is based on the model of an inverted pendulum with nonlinearities. The main results are summarized in Section 6 together with some possible areas for further research.

The notation used throughout the paper is fairly standard. The null and identity matrices with the required dimensions are denoted by 0 and I, respectively, and the notation $X \succ Y$

means that the matrix $X - Y$ is positive definite. Also sym{M} is used to denote the symmetric matrix $M + M^T$ and $\rho(\cdot)$ denotes the spectral radius of its matrix argument, i.e., if $h_i, 1 \le i \le q$, denotes the eigenvalues of a $q \times q$ matrix, say H, $\rho(H) = \max_{1 \le i \le q} |h_i|$. The symbol $(\star)$ denotes block entries in symmetric matrices.

Use will also be made of the following results, which are essential for later developments.

Lemma 1. *[7] Given a symmetric matrix $\Gamma \in \mathbb{R}^{p \times p}$ and two matrices Λ, Σ of column dimension p, there exists a matrix $\mathcal{W}$ such that the following inequality holds*

$$\Gamma + \text{sym}\{\Lambda^T \mathcal{W} \Sigma\} \prec 0,$$

if, and only if the following two projection inequalities with respect to $\mathcal{W}$ are satisfied

$$\Lambda^{\perp^T} \Gamma \Lambda^{\perp} \prec 0, \ \Sigma^{\perp^T} \Gamma \Sigma^{\perp} \prec 0,$$

where $\Lambda^{\perp}$ and $\Sigma^{\perp}$ denote the orthogonal complement of Λ and Σ, respectively; i.e., $\Lambda\Lambda^{\perp} = 0$ and $\Sigma\Sigma^{\perp} = 0$.

Lemma 2. *[4] Let $\Omega_1(\varpi)$ and $\Omega_2(\varpi)$ be two arbitrary quadratic forms of $\varpi \in \mathbb{R}^n$, then the following two conditions are equivalent*

(i) $\varpi^T \Omega_1(\varpi) \varpi \le 0$, $\forall \varpi \ne 0$ such that $\varpi^T \Omega_2(\varpi) \varpi \le 0$.

(ii) $\exists \, \tau \geqslant 0$ such that $\Omega_1(\varpi) - \tau \Omega_2(\varpi) \prec 0$.

Before formulating the design problem, we introduce some useful preliminaries on linear repetitive processes needed to develop the proposed main contributions.

2. Discrete Linear Repetitive Processes

Linear repetitive processes are one of the most important classes of two-dimensional (2D) linear systems of both industrial and algorithmic interest [14, 15]. The essential unique characteristic of such processes is a series of sweeps, termed passes, through a set of dynamics defined over a fixed finite duration known as the pass length and here denoted by α. On each pass an output, termed the pass profile, is produced which acts as a forcing function on, and hence contributes to, the next pass profile. This characteristic shows that linear repetitive processes have clear 2D system structure and hence it is natural to exploit links between these processes and ILC schemes.

Following [14], the state-space model of a discrete linear repetitive process has the following form over $0 \leqslant p \leqslant \alpha - 1$, $k \geqslant 0$, $\alpha < +\infty$

$$\begin{aligned} x_{k+1}(p+1) &= \mathcal{A} x_{k+1}(p) + \mathcal{B} u_{k+1}(p) + \mathcal{B}_0 y_k(p), \\ y_{k+1}(p) &= \mathcal{C} x_{k+1}(p) + \mathcal{D} u_{k+1}(p) + \mathcal{D}_0 y_k(p), \end{aligned} \tag{1}$$

where on pass k, $x_k(p) \in \mathbb{R}^n$ is the state vector, $y_k(p) \in \mathbb{R}^m$ is the output (pass profile) vector and $u_k(p) \in \mathbb{R}^l$ is the control input vector. In addition, the terms $\mathcal{B}_0 y_k(p)$ and $\mathcal{D}_0 y_k(p)$ represent the contribution of the previous pass profile to the current pass state and output vectors respectively.

Also, it is required to specify the boundary conditions, i.e., the state initial vector on each pass and the initial pass profile (i.e., on pass 0). Without loss of generality, it is assumed that the state initial vector at the start of each new pass is zero and that the entries in initial pass profile vector $y_0(p)$ are known functions of p over the pass length.

It is important to note that several sets of necessary and sufficient conditions for stability of discrete linear repetitive processes of the form considered here are known [14]. From the practical point of view, the most important are those which guarantee asymptotic stability or stability along the pass. Asymptotic stability guarantees a bounded sequence of pass profiles (output signals) for a bounded initial pass profile over the finite and fixed pass length α, whereas stability along the pass is stronger since it requires this property uniformly, that is, for all possible values of the pass length and hence, it is not surprising that asymptotic stability is a necessary condition for stability along the pass.

The following result characterizes stability along the pass of processes described by (1).

Lemma 3. *[14] A discrete linear repetitive process represented by (1) is stable along the pass if and only if*

i) $\rho(\mathcal{D}_0) < 1$,

ii) $\rho(\mathcal{A}) < 1$,

iii) all eigenvalues of $\mathcal{G}(z) = \mathcal{C}(zI - \mathcal{A})^{-1}\mathcal{B}_0 + \mathcal{D}_0$, $\forall |z| = 1$ *have modulus strictly less than unity.*

Obviously, the main problem to solve is to find the efficient condition for checking the third condition only. More precisely, the third condition requires to check if all eigenvalues of $\mathcal{G}(e^{j\omega})$, i.e., the characteristic loci, lie inside the unit circle in the complex plane. In addition, transformation of this condition into an LMI based condition is not trivial. However, as shown in [10,13], by making extensive use of the Kalman-Yakubovich-Popov lemma the third condition in Lemma 3 can be converted into the LMIs, leading to the following result that provides the basis for the control law design results developed in this paper.

Lemma 4. *[10] A discrete linear repetitive process described by (1) is stable along the pass if there exist* $Q \succ 0$ *and* $P \succ 0$ *such that the following LMI*

$$\begin{bmatrix} -Q & Q\mathcal{A}^T & Q\mathcal{C}^T & 0 \\ \mathcal{A}Q & -Q & 0 & \mathcal{B}_0 P \\ \mathcal{C}Q & 0 & -P & \mathcal{D}_0 P \\ 0 & P\mathcal{B}_0^T & P\mathcal{D}_0^T & -P \end{bmatrix} \prec 0, \tag{2}$$

is feasible. Clearly, if the above inequality is feasible, then

$$\begin{bmatrix} -Q & Q\mathcal{A}^T \\ \mathcal{A}Q & -Q \end{bmatrix} \prec 0, \text{ and } \begin{bmatrix} -P & \mathcal{D}_0 P \\ P\mathcal{D}_0^T & -P \end{bmatrix} \prec 0,$$

and it is straightforward to conclude that conditions *i)* and *ii)* of Lemma 3 hold. Furthermore, by the Schur's complement formula, it follows that (2) can be converted to the form

$$\begin{bmatrix} \mathcal{A} & I \\ \mathcal{C} & 0 \end{bmatrix} \begin{bmatrix} Q & 0 \\ 0 & -Q \end{bmatrix} \begin{bmatrix} \mathcal{A} & I \\ \mathcal{C} & 0 \end{bmatrix}^T + \begin{bmatrix} \mathcal{B}_0 & 0 \\ \mathcal{D}_0 & I \end{bmatrix} \begin{bmatrix} P & 0 \\ 0 & -P \end{bmatrix} \begin{bmatrix} \mathcal{B}_0 & 0 \\ \mathcal{D}_0 & I \end{bmatrix}^T \prec 0.$$

Using the result of Kalman-Yakubovich-Popov lemma (see [10, 13] for details of required transformations) it follows that there exists $P \succ 0$ such that

$$\begin{bmatrix} \mathcal{C}(e^{j\omega}I - \mathcal{A})^{-1} & I \end{bmatrix} \begin{bmatrix} \mathcal{B}_0 & 0 \\ \mathcal{D}_0 & I \end{bmatrix} \begin{bmatrix} P & 0 \\ 0 & -P \end{bmatrix} \begin{bmatrix} \mathcal{B}_0^T & \mathcal{D}_0^T \\ 0 & I \end{bmatrix} \begin{bmatrix} (e^{j\omega}I - \mathcal{A})^{-T}\mathcal{C}^T \\ I \end{bmatrix} \prec 0,$$

is satisfied $\forall \omega \in [-\pi, \pi]$. This last inequality implies that condition *iii)* of Lemma 3 must hold. On the other side, the LMI of (2) can be derived by applying the Lyapunov theory. This approach, see [14] for more details, uses the following candidate Lyapunov function for processes described by (1)

$$V(k,p) = V_1(k,p) + V_2(k,p) = x_{k+1}^T(p)Qx_{k+1}(p) + y_k^T(p)Py_k(p), \tag{3}$$

for some $Q \succ 0$ and $P \succ 0$. This function is a combination of two independent functions due to the two-dimensional character of the systems considered. Also

$$\Delta V_1(k,p) = x_{k+1}^T(p+1)Qx_{k+1}(p+1) - x_k^T(p)Qx_k(p),$$
$$\Delta V_2(k,p) = y_{k+1}^T(p)Py_{k+1}(p) - y_k^T(p)Py_k(p),$$

and the associated increment for (3) is calculated as

$$\Delta V(k,p) = \Delta V_1(k,p) + \Delta V_2(k,p). \tag{4}$$

Using this approach, a version of (2) is obtained by using the below result.

Lemma 5. *[14] A discrete linear repetitive process described by (1) is stable along the pass if*

$$\Delta V(k,p) < 0. \tag{5}$$

Clearly, based on the above inequality one can obtain the following condition for checking of stability along the pass that allows us to handle the considered control problem

$$\begin{bmatrix} x_{k+1}(p) \\ y_{k+1}(p) \end{bmatrix}^T \left(\begin{bmatrix} \mathcal{A} & \mathcal{B}_0 \\ \mathcal{C} & \mathcal{D}_0 \end{bmatrix}^T \begin{bmatrix} Q & 0 \\ 0 & P \end{bmatrix} \begin{bmatrix} \mathcal{A} & \mathcal{B}_0 \\ \mathcal{C} & \mathcal{D}_0 \end{bmatrix} - \begin{bmatrix} Q & 0 \\ 0 & P \end{bmatrix} \right) \begin{bmatrix} x_{k+1}(p) \\ y_{k+1}(p) \end{bmatrix} \prec 0. \tag{6}$$

3. Problem formulation

Consider a class of discrete-time systems with sector-bounded nonlinearities and time-invariant polytopic uncertainties represented by the following state-space model in the ILC setting as

$$\begin{cases} x_{k+1}(p+1) = A(v)x_{k+1}(p) + B(v)u_{k+1}(p) + f(x_{k+1}(p), p), \\ y_{k+1}(p) = C(v)x_{k+1}(p), \end{cases} \tag{7}$$

where k is the trial number, $x_k(p) \in \mathbb{R}^n$ is the state vector, $y_k(p) \in \mathbb{R}^m$ is the output vector and $u_k(p) \in \mathbb{R}^l$ is the control input vector. $A(v)$, $B(v)$ and $C(v)$ are constant matrices of appropriate dimensions and belong to the following uncertainty polytope of the form

$$\left\{ \left[A(v)\ B(v)\ C(v) \right] = \sum_{i=1}^{N} v_i \left[A_i\ B_i\ C_i \right], v_i \geq 0, \sum_{i=1}^{N} v_i = 1 \right\}, \tag{8}$$

In the above description, A_i, B_i, C_i define N^{th} vertices of a polytope in which it is assumed that the actual matrices A, B and C lie.

Remark 1. *It has to be emphasized that known results on designing of ILC schemes for polytopic systems require to impose some constraints on the system matrices. In particular it is assumed that $C(v)$ is fixed, i.e., $C(v) = C$. The result developed in this paper avoids this drawback by application of the approach presented in [6].*

Furthermore, it is assumed that $f(x_{k+1}(p), p) \in \mathbb{R}^n$ is a given nonlinear function that satisfy the following condition

$$\|f(x_{k+1}(p), p)\| \leqslant \zeta \|Mx_{k+1}(p)\|,$$

where $\zeta > 0$ is the bounding parameter on the uncertain disturbances function f, $M \in \mathbb{R}^{n \times n}$ is known constant matrix and $\|\cdot\|$ is Euclidean norm - see [16] or [17] for details on this condition. Maximize parameter ζ leads to an increased the level of robustness of the system for non-linear disturbances, so the parameter ζ can be defined as a degree of robustness. A given function $f(x_{k+1}(p),p)$ satisfies the following constrains

$$f^T(x_{k+1}(p),p)f(x_{k+1}(p),p)\leqslant\zeta^2x_{k+1}^T(p)M^TMx_{k+1}(p). \tag{9}$$

The above quadratic constraints can be written in the following matrix form

$$\begin{bmatrix} x_{k+1}(p) \\ f(x_{k+1}(p),p) \end{bmatrix}^T \begin{bmatrix} -\zeta^2M^TM & 0 \\ 0 & I \end{bmatrix} \begin{bmatrix} x_{k+1}(p) \\ f(x_{k+1}(p),p) \end{bmatrix} \leqslant 0. \tag{10}$$

Linear repetitive process with the nonlinear disturbances is stable along the pass with $\zeta > 0$ if and only if all three conditions of Lemma 3 are satisfied for all $f(x_{k+1}(p),p)$ such that inequality (9) holds.

The primary problem here is to find a practical guideline for design of ILC schemes for systems represented by (7). To proceed assume that $y_d(p)$ denotes the reference trajectory and then the error on trial k is

$$e_k(p) = y_d(p) - y_k(p).$$

The control signal can be written as the sum of current input signal $u_k(p)$ and $\Delta u_{k+1}(p)$

$$u_{k+1}(p) = u_k(p)+\Delta u_{k+1}(p),$$

where $\Delta u_{k+1}(p)$ denotes the modification term added to the previous pass input.

To formulate the design problem in a discrete linear repetitive process setting assume that the modification term in the ILC law takes the form

$$\Delta u_{k+1}(p) = K_1\eta_{k+1}(p) + K_2e_k(p). \tag{11}$$

In the above equation K_1 is the controller in the feedback loop and K_2 is the learning controller in the feedforward loop. Clearly, the main problem is to provide the procedure for computing both K_1 and K_2. To proceed, suppose that the vectors $\eta_{k+1}(p)$ and $\psi_{k+1}(p)$ are defined as

$$\begin{aligned} \eta_{k+1}(p) &= x_{k+1}(p-1)-x_k(p-1), \\ \psi_{k+1}(p) &= f(x_{k+1}(p-1),p-1)-f(x_k(p-1),p-1). \end{aligned} \tag{12}$$

The above vectors $\eta_{k+1}(p)$ and $\psi_{k+1}(p)$ represent the the difference between the current state x_{k+1} and the state in the previous iteration x_k, and the difference between the non-linear distortion in the current $f(x_{k+1})$ and previous iterations $f(x_k)$.Based on previous equations the controlled dynamics can be written as

$$\begin{aligned} \eta_{k+1}(p+1) &= \mathcal{A}(v)\eta_{k+1}(p) + \mathcal{B}_0(v)e_k(p) + \psi_{k+1}(p), \\ e_{k+1}(p) &= \mathcal{C}(v)\eta_{k+1}(p) + \mathcal{D}_0(v)e_k(p), \end{aligned} \tag{13}$$

where

$$\begin{aligned} \mathcal{A}(v) &= A(v)+B(v)K_1,\ \mathcal{B}_0(v) = B(v)K_2, \\ \mathcal{C}(v) &= -C(v)A(v)-C(v)B(v)K_1,\ \mathcal{D}_0(v) = I-C(v)B(v)K_2. \end{aligned} \tag{14}$$

The obtained model (13) in the state space form based on the discrete linear repetitive model given by equation (1). The state vector is given as $\eta_{k+1}(p)$, pass profile vector as $e_k(p)$ and $\psi_{k+1}(p)$ is a nonlinear term. Treated as a repetitive process, the stability along the pass property will force the trial-to-trial error to converge to zero, see the next section.

4. Main result

In this section, the stability theory for discrete linear repetitive processes is used to develop a new ILC design algorithm for systems with polytopic uncertainties for systems with sector-bounded nonlinearities. To proceed, observe that the result of (9) and equation (12) yields

$$\psi_{k+1}^T(p)\psi_{k+1}(p) \le \zeta^2 \eta_{k+1}^T(p) M^T M \eta_{k+1}(p), \tag{15}$$

and then the above inequality which can be written in the matrix form as

$$\begin{bmatrix} \eta_{k+1}(p) \\ e_{k+1}(p) \\ \psi_{k+1}(p) \end{bmatrix}^T \begin{bmatrix} -\zeta^2 M^T M & 0 & 0 \\ 0 & 0 & 0 \\ 0 & 0 & I \end{bmatrix} \begin{bmatrix} \eta_{k+1}(p) \\ e_{k+1}(p) \\ \psi_{k+1}(p) \end{bmatrix} \leqslant 0. \tag{16}$$

Hence, based on presented transformations, we have the following theorem that provides the sufficient condition for stability along the pass (with degree $\zeta > 0$) of ILC scheme represented by (13).

Theorem 1. *Consider an ILC scheme described as a discrete linear repetitive process of the form (13). Then stability along the pass holds with degree $\zeta > 0$ and hence trial-to-trial error convergence occurs if there exist matrices $Q \succ 0$, $P \succ 0$ of appropriate dimensions and a positive scalar τ such that the following inequality is feasible*

$$\Omega_1 - \tau\Omega_2 \prec 0, \tag{17}$$

where

$$\Omega_1 = \begin{bmatrix} \mathcal{A}^T(v)Q\mathcal{A}(v) - Q + \mathcal{C}^T(v)P\mathcal{C}(v) & \mathcal{A}^T(v)Q\mathcal{B}_0(v) + \mathcal{D}_0^T(v)P\mathcal{C}(v) & \mathcal{A}^T(v)Q \\ \mathcal{B}_0^T(v)Q\mathcal{A}(v) + \mathcal{C}^T(v)P\mathcal{D}_0(v) & \mathcal{B}_0^T(v)Q\mathcal{B}_0(v) + \mathcal{D}_0^T(v)P\mathcal{D}_0(v) - I & \mathcal{B}_0^T(v)Q \\ Q\mathcal{A}(v) & Q\mathcal{B}_0(v) & Q \end{bmatrix}, \Omega_2 = \begin{bmatrix} -\zeta^2 M^T M & 0 & 0 \\ 0 & 0 & 0 \\ 0 & 0 & I \end{bmatrix}.$$

Proof. Assume that the inequality of (17) is feasible for some $Q \succ 0$, $P \succ 0$ and $\tau > 0$. Then, by the result of Lemma 2, the inequality (17) holds if

$$\varpi^T \Omega_1 \varpi \le 0, \forall \varpi \ne 0 \text{ such that } \varpi^T \Omega_2 \varpi \le 0.$$

Next, set $\varpi = \begin{bmatrix} \eta_{k+1}^T(p) & e_{k+1}^T(p) & \zeta_{k+1}^T(p) \end{bmatrix}^T$ and then $\varpi^T \Omega_2 \varpi \le 0$ is equivalent to (16). Furthermore, observe that $\varpi^T \Omega_1 \varpi \le 0$ holds when

$$\begin{bmatrix} \eta_{k+1}(p+1) \\ e_{k+1}(p) \end{bmatrix}^T \begin{bmatrix} Q & 0 \\ 0 & P \end{bmatrix} \begin{bmatrix} \eta_{k+1}(p+1) \\ e_{k+1}(p) \end{bmatrix} - \begin{bmatrix} \eta_{k+1}(p) \\ e_k(p) \end{bmatrix}^T \begin{bmatrix} Q & 0 \\ 0 & P \end{bmatrix} \begin{bmatrix} \eta_{k+1}(p) \\ e_k(p) \end{bmatrix} \prec 0.$$

Obviously, by application of Lemma 5 to a system represented by (13) we can establish that feasibility of the above inequality guarantees stability along the pass and the proof is complete. □

Unfortunately, the condition of Theorem 1 cannot be implemented due to their infinite-dimensional nature in the parameter v. Moreover, the inequality (17) is not LMI as it is bilinear in Q, P and the matrices defining the controller (11). This means no computationally effective method exists for checking (17). However, by using a particular set of transformations, the inequality (17) can be converted into an LMI form where the matrix variables Q and P are separated from the linear repetitive process matrices. Therefore, the affine parameter dependent matrix variables of the following form can be introduced

$$Q(v) = \sum_{i=1}^{N} v_i Q_i, \; P(v) = \sum_{i=1}^{N} v_i P_i. \tag{18}$$

The above matrix variables can take N different values corresponding to the vertices of the polytope (8). As the result, less conservatism can be attained.

Theorem 2. *Consider an ILC scheme described as a discrete linear repetitive process of the form (11) and (13). Then stability along the pass holds for a given $\tau > 0$ with degree $\zeta > 0$ and hence trial-to-trial error convergence occurs if there exist matrices $Q_i \succ 0$, $P_i \succ 0$, N_1, N_2, G_1 and G_2 of appropriate dimensions such that the following LMIs are feasible*

$$\Pi_{ii} \prec 0, \quad (i = 1, 2, ..., N), \tag{19}$$

$$\Pi_{ij} + \Pi_{ji} \prec 0, \quad i < j, \ (i, j = 1, 2, ..., N), \tag{20}$$

where

$$\Pi_{ij} = \begin{bmatrix} -\widehat{Q}_j & 0 & 0 & 0 & \widehat{Q}_j M^T & 0 & \widehat{Q}_j & 0 \\ (\star) & -\frac{1}{\tau}\widehat{P}_j & 0 & 0 & 0 & 0 & 0 & \widehat{P}_j \\ (\star) & (\star) & -I & I & 0 & 0 & 0 & 0 \\ (\star) & (\star) & (\star) & -\widehat{Q}_j & 0 & 0 & A_iG_1 + B_iN_1 & B_iN_2 \\ (\star) & (\star) & (\star) & (\star) & -\beta I & 0 & 0 & 0 \\ (\star) & (\star) & (\star) & (\star) & (\star) & -\widehat{P}_j & -C_iA_iG_1 - C_iB_iN_1 & G_2 - C_iB_iN_2 \\ (\star) & (\star) & (\star) & (\star) & (\star) & (\star) & -G_1 - G_1^T & 0 \\ (\star) & (\star) & (\star) & (\star) & (\star) & (\star) & (\star) & -G_2 - G_2^T \end{bmatrix},$$

and $\beta = 1/\sqrt{\zeta}$. Moreover, when the above LMIs hold then the required controller gains K_1 and K_2 of (11) are computed as

$$K_1 = N_1 G_1^{-1}, K_2 = N_2 G_2^{-1}.$$

Proof. Suppose that the LMIs (19) and (20) are feasible. Then $G_1 + G_1^T \succ 0$, and $G_2 + G_2^T \succ 0$, which implies that the matrices G_1 and G_2 are non-singular and hence invertible. Furthermore, from (8), (19) and (20)

$$\sum_{i=1}^{N} \upsilon_i^2 \Pi_{ii} + \sum_{i=1}^{N}\sum_{i<j}^{N} \upsilon_i\upsilon_j(\Pi_{ij} + \Pi_{ji}) = \sum_{i=1}^{N}\sum_{j=1}^{N} \upsilon_i\upsilon_j\Pi_{ij}$$

$$= \begin{bmatrix} -\widehat{Q}(\upsilon) & 0 & 0 & 0 & \widehat{Q}(\upsilon)M^T & 0 & \widehat{Q}(\upsilon) & 0 \\ (\star) & -\frac{1}{\tau}\widehat{P}(\upsilon) & 0 & 0 & 0 & 0 & 0 & \widehat{P}(\upsilon) \\ (\star) & (\star) & -I & I & 0 & 0 & 0 & 0 \\ (\star) & (\star) & (\star) & -\widehat{Q}(\upsilon) & 0 & 0 & \Upsilon_2(\upsilon) & B(\upsilon)N_2 \\ (\star) & (\star) & (\star) & (\star) & -\beta I & 0 & 0 & 0 \\ (\star) & (\star) & (\star) & (\star) & (\star) & -\widehat{P}(\upsilon) & \Upsilon_1(\upsilon) & G_2 - C(\upsilon)B(\upsilon)N_2 \\ (\star) & (\star) & (\star) & (\star) & (\star) & (\star) & -G_1 - G_1^T & 0 \\ (\star) & (\star) & (\star) & (\star) & (\star) & (\star) & (\star) & -G_2 - G_2^T \end{bmatrix}, \tag{21}$$

where $\Upsilon_1(\upsilon) = -C(\upsilon)(A(\upsilon)G_1 + B(\upsilon)N_1)$, $\Upsilon_2(\upsilon) = A(\upsilon)G_1 + B(\upsilon)N_1$. Next, observe that the resulting matrix in (21) can be rewritten as

$$\begin{bmatrix} \Phi(\upsilon) & J(\upsilon) + R(\upsilon)G \\ J^T(\upsilon) + G^T R^T(\upsilon) & -G - G^T \end{bmatrix} \prec 0,$$

where $G = \text{diag}(G_1, G_2)$ and

$$\Phi(\upsilon) = \begin{bmatrix} -\widehat{Q}(\upsilon) & 0 & 0 & 0 & \widehat{Q}(\upsilon)M^T & 0 \\ (\star) & -\frac{1}{\tau}\widehat{P}(\upsilon) & 0 & 0 & 0 & 0 \\ (\star) & (\star) & -I & I & 0 & 0 \\ (\star) & (\star) & (\star) & -\widehat{Q}(\upsilon)_j & 0 & 0 \\ (\star) & (\star) & (\star) & (\star) & -\beta I & 0 \\ (\star) & (\star) & (\star) & (\star) & (\star) & -\widehat{P}(\upsilon) \end{bmatrix},$$

IOP Conf. Series: Journal of Physics: Conf. Series **783** (2017) 012032 doi:10.1088/1742-6596/783/1/012032

$$R(v)=\begin{bmatrix} 0 & 0 \\ 0 & 0 \\ 0 & 0 \\ A(v)+B(v)K_1 & B(v)K_2 \\ 0 & 0 \\ -C(v)(A(v)+B(v)K_1) & I-C(v)B(v)K_2 \end{bmatrix},\ J(v)=\begin{bmatrix} \widehat{Q}(v) & 0 \\ 0 & \widehat{P}(v) \\ 0 & 0 \\ 0 & 0 \\ 0 & 0 \\ 0 & 0 \end{bmatrix}.$$

Since $\Phi(v) \prec 0$, a direct application of Lemma 1 yields

$$\Phi(v) + R(v)J^T(v) + J(v)R^T(v) \prec 0.$$

Also, the above inequality can be rewritten as

$$\begin{bmatrix} -\widehat{Q}(v) & 0 & 0 & \widehat{Q}(v)\mathcal{A}^T & \widehat{Q}(v)M^T & \widehat{Q}(v)\mathcal{C}^T \\ 0 & -\frac{1}{\tau}\widehat{P}(v) & 0 & \widehat{P}(v)\mathcal{B}_0^T & 0 & \widehat{P}(v)\mathcal{D}_0^T \\ 0 & 0 & -I & I & 0 & 0 \\ \mathcal{A}\widehat{Q}(v) & \mathcal{B}_0 P(v) & I & -\widehat{Q}(v) & 0 & 0 \\ M\widehat{Q}(v) & 0 & 0 & 0 & -\beta I & 0 \\ \mathcal{C}\widehat{Q}(v) & \mathcal{D}_0\widehat{P}(v) & 0 & 0 & 0 & -\widehat{P}(v) \end{bmatrix} \prec 0,$$

where $\mathcal{A}$, $\mathcal{B}_0$, $\mathcal{C}$ and $\mathcal{D}_0$ are defined in (14). Next, post- and pre-multiply the above inequality by diag $\left\{\widehat{Q}^{-1}(v), \widehat{P}^{-1}(v), I, I, I, I\right\}$ and its transpose, respectively. Set $\widehat{Q}(v) = \tau Q(v)$, $\widehat{P}(v) = \tau P(v)$ and apply the Schur's complement formula to obtain

$$\frac{1}{\tau}\begin{bmatrix} \mathcal{A}^T Q(v)\mathcal{A}-Q(v)+\mathcal{C}^T P(v)\mathcal{C} & \mathcal{A}^T Q(v)\mathcal{B}_0+\mathcal{D}_0^T P(v)\mathcal{C} & \mathcal{A}^T Q(v) \\ \mathcal{B}_0^T Q(v)\mathcal{A}+\mathcal{C}^T P(v)\mathcal{D}_0 & \mathcal{B}_0^T Q(v)\mathcal{B}_0+\mathcal{D}_0^T P(v)\mathcal{D}_0-P(v) & \mathcal{B}_0^T Q(v) \\ Q(v)\mathcal{A} & Q(v)\mathcal{B}_0 & Q(v) \end{bmatrix} + \begin{bmatrix} -\zeta^2 M^T M & 0 & 0 \\ 0 & 0 & 0 \\ 0 & 0 & I \end{bmatrix} \prec 0. \quad (22)$$

Observe that the above inequality is just a version of (17) where the affine parameter dependent matrix variables of (18) are used. In view of Theorem 1 the resulting repetitive process is stable along the pass and hence ILC scheme is convergent and the proof is complete. □

5. Simulation results

In this section, an example is given to illustrate the validity of our theoretical results on design of ILC schemes for uncertain systems with nonlinearities. The considered system, taken from [3], is an inverted pendulum system which consists of a pendulum rod with a fixed pivot point attached to a motor-driven shaft - see Figure 1. Based on motion equations of the pendulum, a simplified

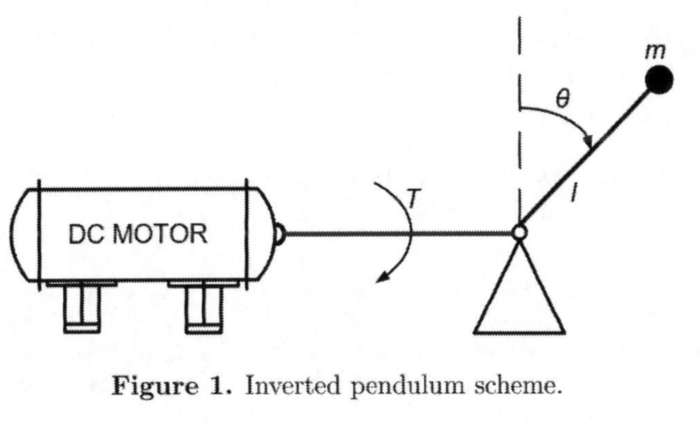

Figure 1. Inverted pendulum scheme.

model of an inverted pendulum system can be provided, see [3] for more details, and state space model is

$$\begin{aligned} x_{k+1}(p+1) =& A(v)x_{k+1}(p) + B(v)(u_{k+1}(p) + \zeta \sin(y_{k+1}(p))), \\ y_{k+1}(p) =& C(v)x_{k+1}(p), \end{aligned} \tag{23}$$

where $x(p) = [x_1^T(p)\ x_2^T(p)]^T$, $x_1(p)$ is the angular displacement, $x_2(p)$ is the velocity and $u(p)$ is the field current of the DC motor. The matrices $A(v), B(v)$ and $C(v)$ are assumed to be given as

$$\begin{aligned} A(v) =& v_1 \begin{bmatrix} 1 & 0.0953 \\ 0 & 0.9048 \end{bmatrix} + v_2 \begin{bmatrix} 1 & 0.0951 \\ 0 & 0.9048 \end{bmatrix}, \quad B(v) = v_1 \begin{bmatrix} 0.0481 \\ 0.952 \end{bmatrix} + v_2 \begin{bmatrix} 0.0479 \\ 0.952 \end{bmatrix}, \\ C(v) =& v_1 \begin{bmatrix} 1.0001 & 0 \end{bmatrix} + v_2 \begin{bmatrix} 0.999 & 0 \end{bmatrix}, \end{aligned}$$

where v_1, v_2 are uncertainty parameters, satisfying $v_1 + v_2 = 1$. Additionally, it is assumed that the nonlinear function $f(x_{k+1}(p), p)$ is given by the following equation

$$f(x_{k+1}(p), p) = \zeta B \sin(y_{k+1}(p)) = \zeta B \sin(x_1(p)),$$

and satisfies the below constraints

$$f^T(x_{k+1}(p), p) f(x_{k+1}(p), p) = \zeta^2 B^T B \sin^2(x_1(p)) \leqslant \zeta^2 \begin{bmatrix} x_1(p) \\ x_2(p) \end{bmatrix}^T \begin{bmatrix} 0.9487 & 0 \\ 0 & 0 \end{bmatrix} \begin{bmatrix} x_1(p) \\ x_2(p) \end{bmatrix}.$$

Based on the above inequalities one can find that the matrix M of (9) is given as

$$M = \begin{bmatrix} 0.9487 & 0 \\ 0 & 0 \end{bmatrix}.$$

Executing the design procedure given in Theorem 2 for $\tau = 1.5$ gives the feedback (K_1) and learning (K_2) controllers as

$$K_1 = \begin{bmatrix} -20.8012 & -1.9818 \end{bmatrix}, K_2 = 17.1929.$$

The designed ILC algorithm was simulated over 15 trials. As the reference signal we took a little change (approx. 1 degree) in the angular position of the pendulum during 2 seconds - see Figure 2. Since the sampling frequency is 100 Hz, the trial length is 200 points, i.e. $\alpha = 200$. The simulation results are depicted in Figure 2. The reference signal is marked with a solid line. The dashed lines marked the signal after the first and the second iteration. Obviously, fast convergence is observed. Furthermore, on completion of each trial the corresponding root mean square (RMS) value of the tracking error was computed using RMS $= \sqrt{\frac{1}{N}\sum_{p=1}^{N} e(p)^2}$, where $N = 200$ is the number of sampled data in the trial time. The result of these computations are depicted in Figure 3. Clearly, Figure 3 shows the convergence of error signal in subsequent repetitions of the process. Also noise (RMS ranges between 10^{-4} and 10^{-3}) is included and hence the convergence curve stays at noise level after 5 trials.

6. Conclusions

In this paper a systematic procedure for design of ILC schemes with polytopic uncertainties for systems with sector-bounded nonlinearities using the repetitive process setting has been investigated. Developed results are based on the transformation to the problem of robust stability along the pass for linear repetitive processes. Then, the set of transformations is

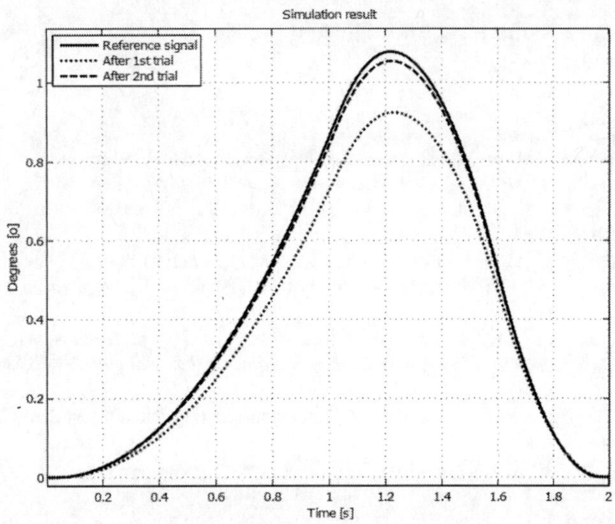

Figure 2. The response of the control system for a given signal after 1st and 2nd trial.

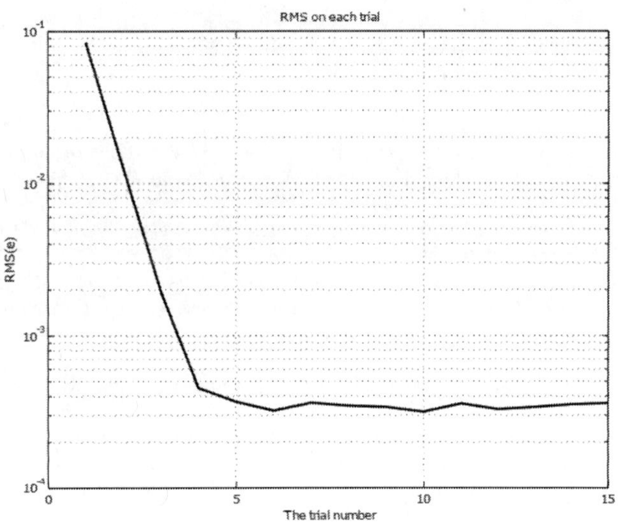

Figure 3. The error convergence.

provided to allow the formulation of the problem in the form of linear matrix inequalities. Provided conditions allows to use the affine parameter dependent matrix variables which often introduce less conservatism. The theoretical findings have been illustrated by simulation results from an inverted pendulum system model and demonstrate the benefits of the developed design procedure. Future work will focus on imposing additional design specifications such as constraints imposed on control and output signals. Additionally, future work directions include design of ILC schemes for time-delay and non-minimum phase systems.

Acknowledgments
This work is partially supported by National Science Centre in Poland, grant No. 2014/15/B/ST7/03208.

References
[1] H.-S. Ahn, Y.-Q. Chen, and K. L. Moore. Iterative learning control: brief survey and categorization. *IEEE Transactions on Systems, Man and Cybernetics, Part C*, 37(6):1109–1121, 2007.
[2] S. Arimoto, S. Kawamura, and F. Miyazaki. Bettering operation of robots by learning. *Journal of Robotic Systems*, 1(2):123–140, 1984.
[3] N. Bedioui, S. Salhi, and M. Ksouri. Robust stabilization approach and $\mathcal{H}_\infty$ performance via static output feedback for a class of nonlinear systems. *Mathematical Problems in Engineering*, 2009, 2009. Article ID 486470.
[4] S. Boyd, L. E. Ghaoui, E. Feron, and V. Balakrishnan. *Linear Matrix Inequalities in System and Control Theory*, volume 15 of *SIAM Studies in Applied and Numerical Mathematics*. SIAM, Philadelphia, USA, 1994.
[5] D. A. Bristow, M. Tharayil, and A.G. Alleyne. A survey of iterative learning control. *IEEE Control Systems Magazine*, 26(3):96–114, 2006.
[6] J. Dong and G.-H. Yang. Robust static output feedback control synthesis for linear continuous systems with polytopic uncertainties. *Automatica*, 49(6):1821–1829, 2013.
[7] P. Gahinet and P. Apkarian. A linear matrix inequality approach to $\mathcal{H}_\infty$ control. *International Journal of Robust and Nonlinear Control*, 4:421–448, 1994.
[8] M. Heertjes and T. Tso. Nonlinear iterative learning control with applications to lithographic machinery. *Control Engineering Practice*, 15(12):1545–1555, 2007.
[9] M. Norrlöf. An adaptive iterative learning control algorithm with experiments on an industrial robot. *IEEE Transactions on Robotics and Automation*, 18(2):245–251, 2002.
[10] W. Paszke and O. Bachelier. Robust control with finite frequency specification for uncertain discrete linear repetitive processes. *Multidimensional Systems and Signal Processing*, 24(4):727–745, 2013.
[11] W. Paszke, M. Boski, E. Rogers, and K. Gałkowski. Design of iterative learning control schemes for systems with sector-bounded nonlinearities. In *Proc. Int. Conf. Methods and Models in Automation and Robotics (MMAR)*, pages 76–81, Miedzyzdroje, Poland, 2016.
[12] W. Paszke, R. Merry, and R. van de Molengraft. Iterative learning control by two-dimensional system theory applied to a motion system. In *Proc. American Control Conference (ACC)*, pages 5484–5489, New York, USA, 9-13 July 2007.
[13] W. Paszke, E. Rogers, and K. Gałkowski. Experimentally verified generalized KYP lemma based iterative learning control design. *Control Engineering Practice*, 53(10):57–67, 2016.
[14] E. Rogers, K. Gałkowski, and D. H. Owens. *Control Systems Theory and Applications for Linear Repetitive Processes*, volume 349 of *Lecture Notes in Control and Information Sciences*. Springer-Verlag, Berlin, Germany, 2007.
[15] E. Rogers, K. Gałkowski, W. Paszke, K. L. Moore, P. H. Bauer, L. Hładowski, and P. Dabkowski. Multidimensional control systems: case studies in design and evaluation. *Multidimensional Systems and Signal Processing*, 26(4):895–939, 2015.
[16] D.M. Stipanović and D.D. Šiljak. Robust stability and stabilization of discrete-time nonlinear systems: the LMI approach. *International Journal of Control*, 74(9):873–879, 2001.
[17] D.D. Šiljak and D.M. Stipanović. Robust stabilization of nonlinear systems: the LMI approach. *Mathematical Problems in Engineering*, 6:461–493, 2000.
[18] Y. Wang, F. Gao, and F.J. Doyle. Survey on iterative learning control, repetitive control, and run-to-run control. *Journal of Process Control*, 19(10):1589–1600, 2009.

Robust optimization based energy dispatch in smart grids considering demand uncertainty

M Nassourou[1,2], V Puig[1,2], J Blesa[2]

[1] Automatic Control Department, Technical University of Catalonia (UPC), Carrer Pau Gargallo, 5, 08028 Barcelona (Spain).

[2] Institut de Robòtica i Informàtica Industrial (CSIC-UPC), Carrer Llorens Artigas, 4-6, 08028 Barcelona (Spain)

Abstract: In this study we discuss the application of robust optimization to the problem of economic energy dispatch in smart grids. Robust optimization based MPC strategies for tackling uncertain load demands are developed. Unexpected additive disturbances are modelled by defining an affine dependence between the control inputs and the uncertain load demands. The developed strategies were applied to a hybrid power system connected to an electrical power grid. Furthermore, to demonstrate the superiority of the standard Economic MPC over the MPC tracking, a comparison (e.g average daily cost) between the standard MPC tracking, the standard Economic MPC, and the integration of both in one-layer and two-layer approaches was carried out. The goal of this research is to design a controller based on Economic MPC strategies, that tackles uncertainties, in order to minimise economic costs and guarantee service reliability of the system.

*Keywords:*Smart Grid, Model Predictive Control, Demand Uncertainty, Robust Optimization

1. Introduction

Due to model uncertainty and noise, the development of robust MPC strategies is not a trivial task [2,10]. The problem of model uncertainty and noise in linear time-invariant (LTI) systems can be solved through enforcement of computational constraints as reported in [2,11,12]. However, this approach is usually raising the issue of tractability.

There are other techniques based on optimization techniques such as Minimax MPC [12,16], that can be used to tackle model uncertainties.

In [13,14,16] Adjustable Robust Solutions have been proposed, which assume that adjustable control inputs can be made to depend affinely on the uncertain parameters of the problem. This approach is more flexible, and is most of the time expected to result in a computationally tractable problem [15].

In this paper, we follow the technique of affine dependence to solve the problem of demand uncertainty in a smart micro-grid.

In [3] uncertainties were ignored, we showed that the standard Economic MPC (EMPC) was not only superior to the standard MPC tracking, but also to the integration of both in a two-layer approach.

In this paper, we repeat the study in [3] by considering unexpected additive load demands in the development of control strategies. Additionally we have also analysed the integration of EMPC and MPC tracking into a single layer.

Several studies [1,3,4] have dealt with the issue of tackling uncertainties using stochastic approaches.

In this work, a deterministic approach namely robust optimization is used. Each uncertain variable in the system is splitted into a non-adjustable one and an adjustable one. After that, using affine dependence method, some relationship between the control inputs (desired input) and the unexpected inputs (additive disturbances) are established.

By setting some bounds on the uncertain load demand, it is easier, realistic and robust to develop a

controller capable of guaranteeing satisfaction of load demands. We have explicitly included the uncertainty information into the optimization problem, by replacing uncertain constraints of states and control inputs with robust counterpart ones, and by substituting the uncertain variables with their robust counterparts in the objective function.
We consider a smart grid consisting of ***G*** fossil fuel generators, ***S*** clean energy generators, and ***B*** storage devices integrated through a DC Bus into a power grid.
The DC Bus collects the energy generated by the subsystems and delivers it to the load, and if necessary to the storage devices.
We have developed some MPC control strategies by using both single-layer and two-layer approaches. In the one-layer approach, we started by applying MPC tracking and EMPC strategies separately to the hybrid system. After that, we integrated both the economic optimisation and the tracking formulation in a single layer.
In the two-layer approach, the upper layer consists of an EMPC controller viewed as the supervisory unit, which is in charge of scheduling the operation of the subsystems, and computing their power references. At the lower layer, standard MPC tracking controllers responsible for implementing the computed reference values for each subsystem were used.
The main goal of this work is the development of a robust optimization based Economic MPC strategy capable of satisfying uncertain load demands in smart grids.

2. Control-oriented modelling

A typical smart grid architecture can be represented with a directed graph consisting of n_s energy sources, n_x storage elements, n_q intersection Buses, and n_d loads. Smart grids could be viewed as instances of generalized flow-based networks.

2.1. Component modeling

Smart grids could be considered as examples of generalized flow-based networks. Basically, every flow-based network is made up of some components e.g: flow sources, links, nodes, storage, flow handling, and sink elements. Considering [1, 5, 7, 26], we can model the components as follows:

2.1.1. Flow source elements:

Flow sources consist of the generators and the storage elements, which supply the required energy to the load. Since everything has got its limits, flow sources should also be constrained. Some lower and upper limits can be defined.
For any given generator we can write: $P^{min} \leq P(k) \leq P^{max}$; where $P(k)$ the supplied energy at time k, and P^{min} and P^{max} are the lower and upper limits respectively.

2.1.2. Flow handling elements:

Inverters are used to convert direct current (DC) power into alternating current (AC) power. They usually connect the batteries to an AC Bus. Inverters being a type of rectifiers have the characteristic of modifying the nature of an input signal. Hence, they can be considered as flow handling elements. They can amplify or decrease an input power if necessary. Therefore, we can define the maximum and minimum power that they can handle. $P^{min} \leq P(k) \leq P^{max}$
Moreover, inverters could even be considered as "*Active Nodes*". To some extend flow handling elements can basically be considered as links that can modify the nature of the flowing substance i.e current.

2.1.3. Link elements:

Links are basically made up of converters and inverters that connect the generators to the AC/DC Bus. Contrary to inverters, converters do not change the nature of an input power i.e (AC to AC, or DC to DC). They could be termed as "*Passive nodes*". They can also be constrained by setting the maximum and minimum energy allowed to flow through those links. $P^{min} \leq P(k) \leq P^{max}$

2.1.4. Node elements:

Node elements are generally the buses or other elements (e.g transformers), which interconnects the subsystems and the load. They are also known as junctions, where energy flows are either merged or propagated [26] to other elements of a system. For any given node, we can write: $\sum_i P_{\text{in},i}(k)=\sum_i P_{\text{out},i}(k)$

2.1.5. Storage elements:

Storage elements are the batteries that store energy. Their storage capacity should also be constrained as follows:

$SOC^{min} \leq SOC(k) \leq SOC^{max}$

where SOC^{min} and SOC^{max} are the maximum and minimum values of the stored energy, and SOC(k) the stored energy at time k.

Moreover we can describe the dynamic process of storing the energy with the following equation:

$SOC(k+1) = SOC(k) + \Delta t(P_{in}(k) - P_{out}(k))$

2.1.6. Sink elements:

Loads are the sink elements, because they consume the flowing energy. Loads should not consume more than what is produced: $P_L(k) \leq Tp^{max}(k)$, where $P_L(k)$ is the load demand, and $Tp^{max}(k)$ is the total energy flow at time k.

2.2. Control-oriented model

We consider a smart grid consisting of n_x storage elements, n_u energy flow handling and sources elements, n_d sinks and n_q intersection nodes. The n_s source elemets are considered as inflows.

2.2.1. State space model

The hybrid power system is an example of a MIMO (multiple-input multiple-output) system, whose linear state space modeling is given by the following equations:

$\mathbf{x}(k+1) = \mathbf{A}\mathbf{x}(k) + \mathbf{B}\mathbf{u}(k) + \mathbf{B}_p\mathbf{d}(k)$ with $k = 0,1,2,...,N-1$ (1a)

$\mathbf{E}_u\mathbf{u}(k) + \mathbf{E}_d\mathbf{d}(k)=0$ with $k = 0,1,2,...,N$ (1b)

where:

$\mathbf{x} \in \mathbb{R}^{nx}$ is the state vector, $\mathbf{u} \in \mathbb{R}^{nu}$ stands for the vector of control inputs, $\mathbf{d} \in \mathbb{R}^{nd}$ denotes the disturbances vector. $\mathbf{A} \in \mathbb{R}^{nx \times nx}$, $\mathbf{B} \in \mathbb{R}^{nx \times nu}$, $\mathbf{B}_p \in \mathbb{R}^{nx \times nd}$ are system matrices.

$\mathbf{E}_u \in \mathbb{R}^{nq \times nu}$ and $\mathbf{E}_d \in \mathbb{R}^{nq \times nd}$ are matrices of suitable dimensions relating the supply and the load demand on the DC Bus(ses).

2.2.2. Constraints

Control inputs are subject to some bounds (upper and lower limits):

$\mathbf{u}_i^{min}(k) \leq \mathbf{u}_i(k) \leq \mathbf{u}_i^{max}(k), \quad \forall k$ (2)

where i denotes a subsystem, and ($\mathbf{u}_i^{min}(k)$ is in this case always zero, because energy flow from the generators is positive).

The state of charge (SOC) of each storage element is subject to the following constraint:

$\mathbf{x}^{min} \leq \mathbf{x}(k) \leq \mathbf{x}^{max}, \quad \forall k$ (3)

where $\mathbf{x}^{min}$ and $\mathbf{x}^{max}$ are the lower and upper limit values of the state of charge respectively.

To guarantee availability of energy in the batteries we set: $\mathbf{x}^{min} \geq \boldsymbol{\delta}$ (4)

where $\boldsymbol{\delta}$ is the minimum quantity of energy that should always be available in the batteries.

3. Robustness in MPC strategy

Instead of using probabilistic methods, we opted for a deterministic approach namely robust optimization, which offers some possibilities of bounding uncertain load demands. The adjustable

robust optimization (ARO) method [13,14] is used.
Let's assume that, load demands can be subject to the following constraint:

$$\mathbf{d}_i^{min}(k) \le \mathbf{d}(k) \le \mathbf{d}_i^{max}(k) \qquad \text{where } i=1,\dots,n_d \tag{5}$$

where $\mathbf{d}_i^{min}$ *and* $\mathbf{d}_i^{max}$ are the lower and upper bounds of load demands respectively.

3.1. *Affine dependence*

The main objective of MPC is to generate some optimal control inputs, which should be fed to the plant in order to produce desired outputs.
If the control inputs could be adjusted according to the uncertain variables, the MPC strategy would be robust. To achieve this aim, an affine dependence between the control input variables and the load demand is developed. The dependence is able to adjust the control inputs according to past measured disturbances.
In fact, the adjustable robust optimization method [13,14] relies on affine dependence technique, because the decision variables are made to be "wait and see" variables, which depend on the past measured disturbances.
Mathematically, the dependence can be derived as follows:

$$\mathbf{u}(k) = \mathbf{u}^k(k)$$
$$\mathbf{u}(k+1) = \mathbf{u}^{k+1}(k+1) + \mathbf{L}_{10}\mathbf{d}(k)$$
$$\mathbf{u}(k+2) = \mathbf{u}^{k+2}(k+2) + \mathbf{L}_{21}\mathbf{d}(k+1) + \mathbf{L}_{20}\mathbf{d}(k)$$
$$\mathbf{u}(k+3) = \mathbf{u}^{k+3}(k+3) + \mathbf{L}_{32}\mathbf{d}(k+2) + \mathbf{L}_{31}\mathbf{d}(k+1) + \mathbf{L}_{30}\mathbf{d}(k)$$
$$\dots$$
$$\mathbf{u}(k+N) = \mathbf{u}^{k+N}(k+N) + \mathbf{L}_{N(N-1)}\mathbf{d}(k+N-1) + \dots + \mathbf{L}_{N0}\mathbf{d}(k)$$

where
$\mathbf{u}^k(k)$ is the computed control input at time k and $\mathbf{L}_{ij} \in \mathbb{R}^{nu \times nd}$.

In compact matricial form:

$$\underbrace{\begin{pmatrix} \mathbf{u}(k) \\ \mathbf{u}(k+1) \\ \mathbf{u}(k+2) \\ \mathbf{u}(k+3) \\ \vdots \\ \mathbf{u}(k+N) \end{pmatrix}}_{\underline{\mathbf{u}}} = \underbrace{\begin{pmatrix} \mathbf{u}^k(k) \\ \mathbf{u}^{k+1}(k+1) \\ \mathbf{u}^{k+2}(k+2) \\ \mathbf{u}^{k+3}(k+3) \\ \vdots \\ \mathbf{u}^{k+N}(k+N) \end{pmatrix}}_{\underline{\mathbf{v}}} + \underbrace{\begin{pmatrix} 0 & 0 & 0 & \dots & 0 \\ \mathbf{L}_{10} & 0 & 0 & \dots & 0 \\ \mathbf{L}_{20} & \mathbf{L}_{21} & 0 & \dots & 0 \\ \mathbf{L}_{30} & \mathbf{L}_{31} & \mathbf{L}_{32} & \dots & 0 \\ \vdots & \vdots & \vdots & \dots & 0 \\ \mathbf{L}_{N0} & \dots & & \dots & \mathbf{L}_{N(N-1)} \end{pmatrix}}_{\mathbf{W}} \underbrace{\begin{pmatrix} \mathbf{d}(k) \\ \mathbf{d}(k+1) \\ \mathbf{d}(k+2) \\ \vdots \\ \mathbf{d}(k+N-1) \end{pmatrix}}_{\underline{\mathbf{d}}}$$

Then, equations can be compacted as follows:

$$\underline{\mathbf{u}} = \underline{\mathbf{v}} + \mathbf{W}\underline{\mathbf{d}} \tag{6}$$

The new optimization variables are contained in the vector $\underline{\mathbf{v}}$ and in the matrix $\mathbf{W}$.
The adjusted input vector $\underline{\mathbf{u}}$ can now be used in the formulation of the MPC problem. It can also be plugged into (1), as well as into the objective function.

It should be noticed that, the vectors of control input **u** and disturbances **d** might not have the same size. Hence, elements of vector **d** must be combined.

3.2. *Problem formulation*

All the problems are formulated using the worst-case robust optimization approach, namely the minimax format of the Wald's maximin model. The goal is to minimize the costs of energy production in the presence of uncertain load demand.

The optimization problem is formulated as follows:

$$\min_{\mathbf{x},\mathbf{u}} \max_{\mathbf{d}} f(\mathbf{x},\mathbf{u},\mathbf{d})$$

$$\text{s.t } \mathbf{g}(\mathbf{x},\mathbf{u},\mathbf{d}) \preceq 0 \quad \forall \mathbf{d} \in D = \{\mathbf{d}: \mathbf{d}^{min} \leq \mathbf{d} \leq \mathbf{d}^{max}\}, \tag{7}$$

where:
$f(\mathbf{x},\mathbf{u},\mathbf{d})$ is the cost function, D is the uncertainty set defined as a box−bounded set,
and $\mathbf{g}(\mathbf{x},\mathbf{u},\mathbf{d})$ represents the constraints

3.2.1. MPC tracking

MPC tracking operates by following some reference trajectories from time k over a span of N steps. The objective function of MPC tracking is usually given in Quadratic Program (QP) form.
In this study, we use a QP of the following form:

$$J_{MPC} = (\mathbf{x}-\mathbf{x}^r)^T\mathbf{Q}(\mathbf{x}-\mathbf{x}^r) + (\underline{\mathbf{u}}-\mathbf{u}^r)^T\mathbf{R}(\underline{\mathbf{u}}-\mathbf{u}^r) + (\mathbf{x}_{Np}-\mathbf{x}_{Np}{}^r)^T\mathbf{S}_x(\mathbf{x}_{Np}-\mathbf{x}_{Np}{}^r) \tag{8}$$

where:
$\mathbf{x}=[x(k+1), x(k+2), ..., x(k+N)]^T$ is the state vector,
$\mathbf{x}^r=[x^r(k+1), x^r(k+2), ..., x^r(k+N)]^T$ is the reference trajectory vector of the states,
$\underline{\mathbf{u}}$ is the adjusted control input vector,
$\mathbf{u}^r=[u^r(k+1), u^r(k+2), ..., u^r(k+N)]^T$ is the reference trajectory vector of the control inputs,
$\mathbf{x}_{Np}$ is the vector of terminal state and $x_{Np}{}^r$ its reference trajectory,
Q, R and $\mathbf{S}_x$ are weights on the states, control inputs, and terminal state respectively.

$$\mathbf{Q}=\begin{bmatrix}1 & 0\\ 0 & 1\end{bmatrix}$$

$$\mathbf{R}=\text{diag}(c_b, c_b, c_h, c_h, c_g, c_g, c_d, c_{hy}, c_w, c_{pv})$$

$$\mathbf{Sx}=W_c.\mathbf{Q}; \text{ where } W_c \text{ is a positive scalar}$$

c_b, c_h, c_g, c_d c_d, c_{pv}, c_w, c_{hy} , are positive weight coefficient ($\leq$ 1) for the lead-acid battery, hydrogen battery, grid connection, diesel, solar, wind, and hydroelectric generators respectively.

The reference trajectory for the input controls were computed using the following formula:

$$\mathbf{u}^r(k) = \rho \cdot \sigma(k) \tag{9}$$

where:
ρ is the nominal maximum power of each generator.
$\sigma(k)$ is the effective generated power at time k. It is measured directly from the generators.
This formula is a result of some trial and error simulations based on the idea that, the reference trajectory is expected to be proportional to the control inputs.

For instance, the updating of the states can be done by substituting the adjusted control input in (1a), and in the objective function as follows:

$$\mathbf{x}(k+1)=\mathbf{Ax}(k) + \mathbf{B}\underline{\mathbf{u}}(k) + \mathbf{B}_p\mathbf{d}(k) = \mathbf{Ax}(k) + \mathbf{B}(\underline{\mathbf{v}} + \mathbf{W}\underline{\mathbf{d}})(k) + \mathbf{B}_p\mathbf{d}(k)$$

$$J_{MPC} = (\mathbf{x}-\mathbf{x}^r)^T\mathbf{Q}(\mathbf{x}-\mathbf{x}^r) + ((\underline{\mathbf{v}} + \mathbf{W}\underline{\mathbf{d}})-\mathbf{u}^r)^T\mathbf{R}((\underline{\mathbf{v}} + \mathbf{W}\underline{\mathbf{d}})-\mathbf{u}^r) + (\mathbf{x}_{Np}-\mathbf{x}^r_{Np})^T\mathbf{Sx}(\mathbf{x}_{Np}-\mathbf{x}^r_{Np})$$

The MPC tracking problem can be formulated as follows:

$$\min_{\mathbf{u}(k)\mathbf{x}(k)} \max_{\mathbf{d}(k)} [\sum_{k=0}^{N-1} J_{MPC}(k)]$$

$$\text{s.t } (1),(2),(3),(4), \ \mathbf{d}(k) \in D = \{\mathbf{d}: \mathbf{d}^{min} \leq \mathbf{d} \leq \mathbf{d}^{max}\} \ \forall k \tag{10}$$

3.2.2. Economic MPC

Economic MPC does not require a reference trajectory. Three operational goals have been considered: economic, safety, and smoothness as described in [3,5].

Economic cost:
The total cost is given by: $f_E(k) = (\boldsymbol{\alpha}_1 + \boldsymbol{\alpha}_2(k))\mathbf{u}(k)\Delta t$
where $\mathbf{u}(k)$ is a vector of control actions at time k;
$\boldsymbol{\alpha}_1$ is a known vector related to economic costs of maintenance of generators and its accessories; $\boldsymbol{\alpha}_2(k)$ is a known time-varying vector associated to the economic cost of power flows related to transmission and distribution. The time dependence of $\boldsymbol{\alpha}_2$ is given by the power distribution, which varies along the time. Δt is the sampling time.
Safety Storage Measures:
The safety measures are defined as: $f_S(k) = \varepsilon(k)^T\boldsymbol{\varepsilon}(k)$
where $\varepsilon(k)$ is the amount of soft constraint violation. $\boldsymbol{\varepsilon} = 0$ means there is no violation.
Smoothness/Stability of the control action:
The rate of change of the control action must be made smooth, in order to ensure that, consecutive control inputs are either continuously increasing or decreasing.
$f_{\Delta u}(k) = \Delta\mathbf{u}(k)^T\Delta\mathbf{u}(k)$
where $\Delta\mathbf{u}(k)$ is the rate of change of control signal, defined as $\Delta\mathbf{u}(k)=\mathbf{u}(k) - \mathbf{u}(k\text{-}1)$.
Let $F_1 = f_E(k)$, $F_2 = f_S(k)$, $F_3 = f_{\Delta u}(k)$; substituting the control input $\mathbf{u}$ with (6), we get:
$\Delta\mathbf{u}(k)=\mathbf{u}(k) - \mathbf{u}(k\text{-}1) = \underline{\mathbf{u}}(k) - \underline{\mathbf{u}}(k\text{-}1)$
The objective function is given by:

$$J_{EMPC} = \lambda_1 F_1 + \lambda_2 F_2 + \lambda_3 F_3 = \lambda_1(\boldsymbol{\alpha}_1 + \boldsymbol{\alpha}_2(k))\underline{\mathbf{u}}(k)\Delta t + \lambda_2\varepsilon(k)^T\boldsymbol{\varepsilon}(k) + \lambda_3[\underline{\mathbf{u}}(k) - \underline{\mathbf{u}}(k\text{-}1)]^T[\underline{\mathbf{u}}(k) - \underline{\mathbf{u}}(k\text{-}1)] \quad (11)$$

The Economic MPC optimization problem is formulated as follows:

$$\min_{\mathbf{u}(k)\mathbf{x}(k)} \max_{\mathbf{d}(k)} \left[\sum_{k=0}^{N-1} J_{EMPC}(k)\right] \quad (12)$$

$$\text{s.t } (1),(2),(3),(4),\ \mathbf{d}(k) \in \mathrm{D} = \{\mathbf{d}:\ \mathbf{d}^{\min} \le \mathbf{d} \le \mathbf{d}^{\max}\}\ \forall k$$

3.2.3. Two-layer approach: EMPC and MPC tracking
The main idea behind the use of a two-layer approach is to overcome the problem of non-reachable reference trajectories. We integrate the EMPC and the MPC tracking in a cascaded fashion.

a) Upper layer: Economic MPC
This layer comprises the EMPC described in (*3.2.2*). The problem to be solved is expressed in equation (12).
In this layer the prediction horizon is set to $2N$-1, in order to be able to track N time-varying set points with the receding horizon strategy of the lower layer.

b) Lower layer: MPC tracking
The lower layer consists of the MPC tracking described in (*3.2.1*). Instead of using manually selected reference trajectories, the computed states and control inputs by the upper layer are used.
The problem to be solved is expressed in equation (11).

3.2.4. Single-layer approach: EMPC and MPC tracking
The main idea behind the use of a two-layer approach is to overcome the problem of non-reachable reference trajectories (feasibility).
Contrary to the two-layer approach as defined above, the economic optimisation (EMPC) and the tracking formulation (MPC tracking) are integrated in a single layer.
The problem to be solved is given as follows:

$$\min_{\mathbf{u}(k)\mathbf{x}(k)} \max_{\mathbf{d}(k)} \left[\sum_{k=0}^{N-1} J_{EMPC}(k) + J_{MPC}(k)\right] \quad (13)$$

$$\text{s.t } (1),(2),(3),(4),\ \mathbf{d}(k) \in \mathrm{D} = \{\mathbf{d}:\ \mathbf{d}^{\min} \le \mathbf{d} \le \mathbf{d}^{\max}\}\ \forall k$$

4. Stability analysis

Weights on the terminal state (where MPC tracking is involved) are used to improve (guarantee) stability of the system. However, in the works ahead we will use asymptotic stability method to analyse stability conditions of the controllers. As for the EMPC strategy, the stability of the system around a feasible region can be guaranteed following the results in [18] for periodic systems, whereby a Lyapunov function was developed.

The infeasibilty issue was overcome using the following two strategies:

a) reusing previously predicted solutions,

b) online adjustment of the upper and lower limits of the control inputs and states by considering the highest computed control input and state respectively.

5. Case study

5.1. Description

This subsection presents a smart micro-grid that consists of: two storage elements (batteries), three sinks (loads) and one virtual sink (external grid connection), one node (DC Bus), four sources (PV, Wind, Hydroelectric, and Diesel generators), and one virtual source (external grid connection). A grid connection is a sink when buying energy, and a source when selling energy. Since all the components (excluding sinks) are connected to a single node (DC Bus) through flow handling elements, they are all considered as manipulated inputs. The state of charge of the storage elements are the states of the micro-grid. A block diagram of the micro-grid is shown in Figure 1.

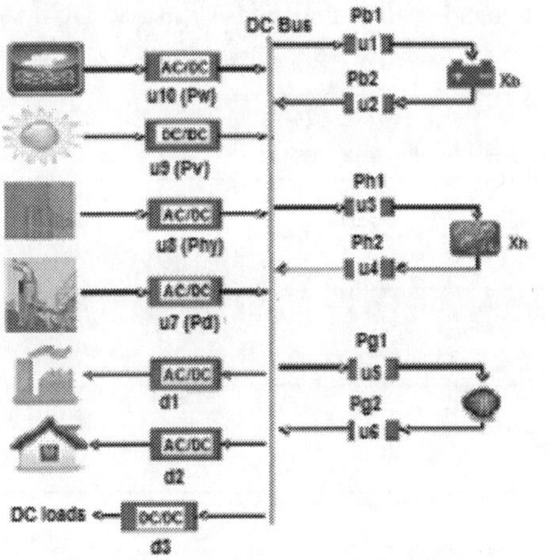

Figure 1. block diagram of the micro-grid

5.2. Control-oriented model

State variables:

x_b and x_h are the state of charge of the batteries (lead-acid and hydrogen respectively).

$\mathbf{x}(k) \triangleq [x_b(k), x_h(k)]^T$

Control input variables:

P_{b1} and P_{b2} are charged power and discharged power of the lead-acid battery;

P_{h1} and P_{h2} are the charged and discharged power of the hydrogen battery;

P_{g1} and P_{g2} are the exported and imported power into/from the external grid;

P_d, P_{hy}, P_{pv} and P_w stand for the power supplied to the DC Bus by the diesel, hydroelectric, wind, and photovoltaic generators respectively;

$\mathbf{u}(k) \triangleq [P_{b1}(k), P_{b2}(k), P_{h1}(k), P_{h2}(k), P_{g1}(k), P_{g2}(k), P_d(k), P_{hy}(k), P_w(k), P_{pv}(k)]^T$

Disturbance variables:
d_1, d_2, d_3 are the industrial load, the residential load, and the DC-load respectively. The disturbance vector **d** comprises the three loads. $\mathbf{d}(k) \triangleq [d_1(k), d_2(k), d_3(k)]^T$
The matrices and vectors that define the system and theconstraints are given as follows:

$$\boldsymbol{A}=\begin{bmatrix}1 & 0\\ 0 & 1\end{bmatrix} \quad \boldsymbol{B}=\begin{bmatrix}\eta_{bc} & -\eta_{bd} & 0 & 0 & 0 & 0 & 0 & 0 & 0 & 0\\ 0 & 0 & \eta_{hc} & -\eta_{hd} & 0 & 0 & 0 & 0 & 0 & 0\end{bmatrix} \quad \boldsymbol{B}_p=\begin{bmatrix}-1 & -1 & -1\\ -1 & -1 & -1\end{bmatrix}$$

$\mathbf{x}^{min} = [0\ 0]^T$, $\mathbf{x}^{max} = [100\ 100]^T$
$\mathbf{u}^{min} = [0\ 0\ 0\ 0\ 0\ 0\ 0\ 0\ 0\ 0]^T$, $\mathbf{u}^{max} = [35\ 10\ 35\ 18\ 18\ 18\ 18\ 18\ 18\ 18]^T$
$\mathbf{E_u} = [-1\ 1\ -1\ 1\ -1\ 1\ 1\ 1\ 1\ 1]$, $\mathbf{E_d} = [-1\ -1\ -1]$

We have considered a scenario (days with bad weather conditions) where all the generators are required to be in operation, in order to satisfy load demands. If they cannot, then the batteries, the external grid, and the diesel generator will compensate the shortage one after the other. Initial values of the subsystems as well as the state of charge of the batteries are set to zero. The simulations were made for 96 hours (4 days).
The diesel generator (1-1.3 kWh) was operated in summer during the first six hours of the day, and in winter six hours in the afternoon. It supplied 1.3 kWh in the first hour, and 1 kWh in the remaining hours. The batteries were used during the first two hours of the day. They delivered 2 kWh in the first hour and 1 kWh in the second hour.
1 kWh was bought from the external grid during the second hour of the day.
Each additive uncertain load demand (residential, industrial and DC-load) is bounded between -1 and +1 kWh. Figure 2. shows the profiles of the load demands. The additive uncertain demand is represented with the shadowed area.

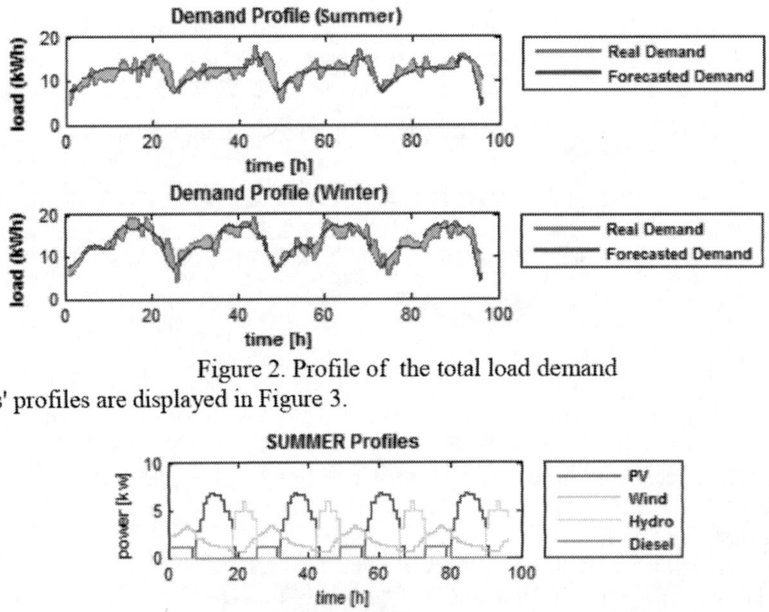

Figure 2. Profile of the total load demand

Generators' profiles are displayed in Figure 3.

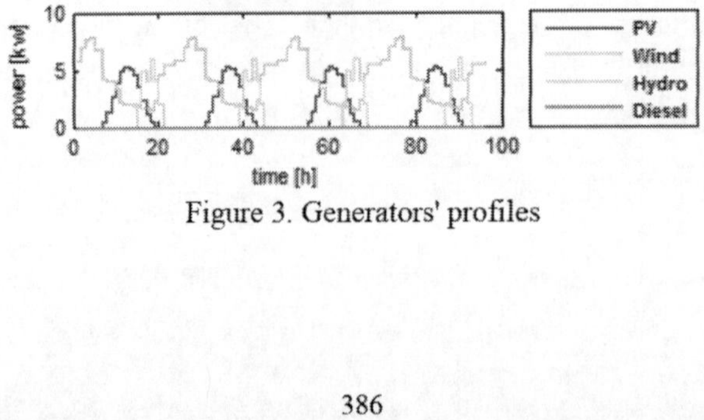

Figure 3. Generators' profiles

IOP Conf. Series: Journal of Physics: Conf. Series **783** (2017) 012033 doi:10.1088/1742-6596/783/1/012033

System parameters		**Control parameters**		**Energy prices**	
Parameters	Values in kW	Parameters	Values	Lead-acid battery charging	2.2
P^{max}_{pv}	15	N_p	24	Lead-acid battery discharging	2.2
P^{max}_{w}	15	N_c	24	Hydrogen battery charging	2.2
P^{max}_{hy}	15	c_{pv}	0.2	Hydrogen battery discharging	2.2
P^{max}_{d}	15	c_w	0.3	External grid selling	3
P^{max}_{b1}	35	c_{hy}	0.4	External grid buying	3
P^{max}_{b2}	15	c_b	0.75	Diesel	3.3
P^{max}_{h1}	35	c_h	0.75	Hydroelectric	2.1
P^{max}_{h2}	15	c_d	1	Wind	2.1
P^{max}_{g1}	15	c_g	0.75	Solar	2
P^{max}_{g2}	15	$\mathbf{Q}$	as defined previously		
η_{bc}	0.95	$\mathbf{R}$	as defined previously		
η_{bd}	1				
η_{hc}	0.85				
η_{hd}	1.0				
$\boldsymbol{\delta}$	$[35\ 35]^T$				

Table 2. System and control parameters, and energy prices

5.3. *Simulation results*

All the plots have been made using YALMIP (CPLEX and QuadProg solvers) [16] within the Matlab environment.

The purpose of including all the simulations figures is to enable a visual comparison of the four MPC strategies.

5.3.1. *One-layer: MPC tracking*

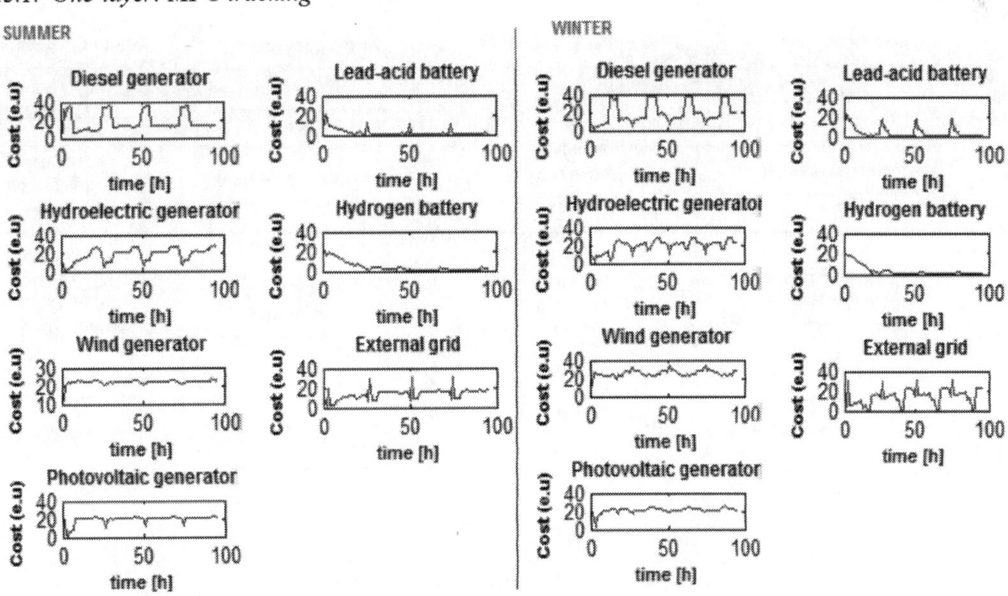

Figure 4. Plots of the economic costs

5.3.2. One-layer: EMPC

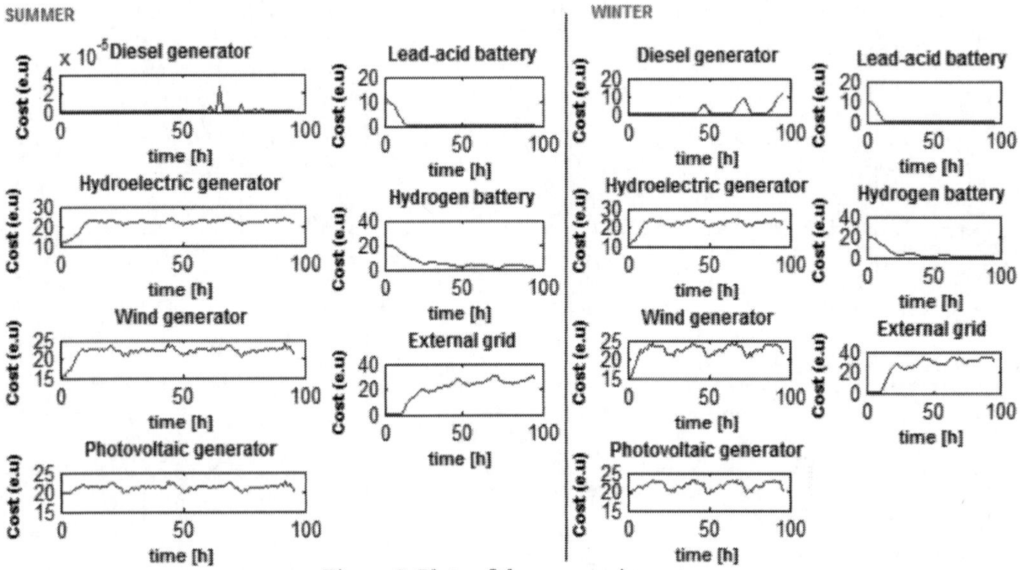

Figure 5. Plots of the economic costs

5.3.3. Two-layer approach: EMPC and MPC tracking

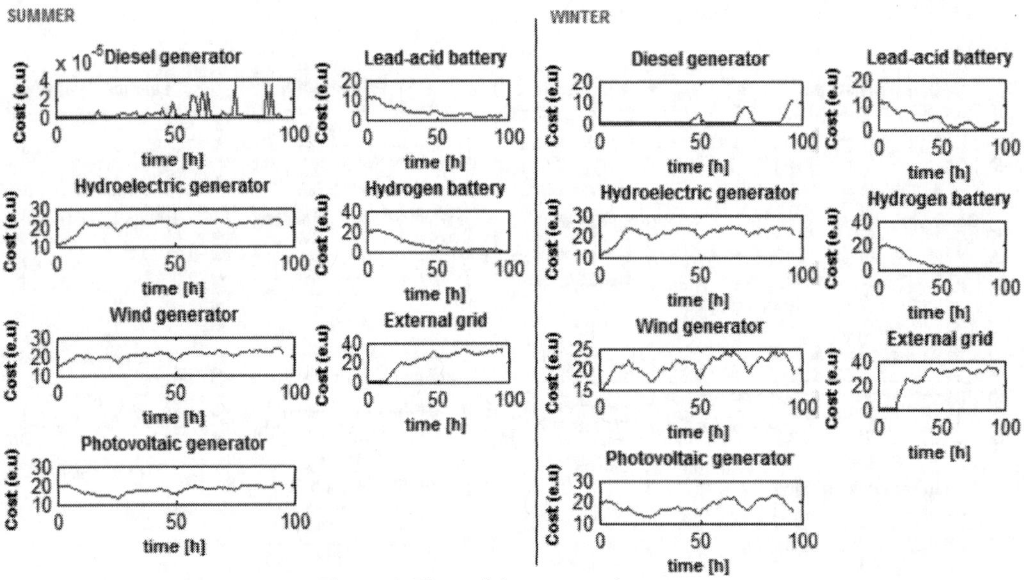

Figure 6. Plots of the economic costs

5.3.4. One-layer: EMPC and MPC tracking

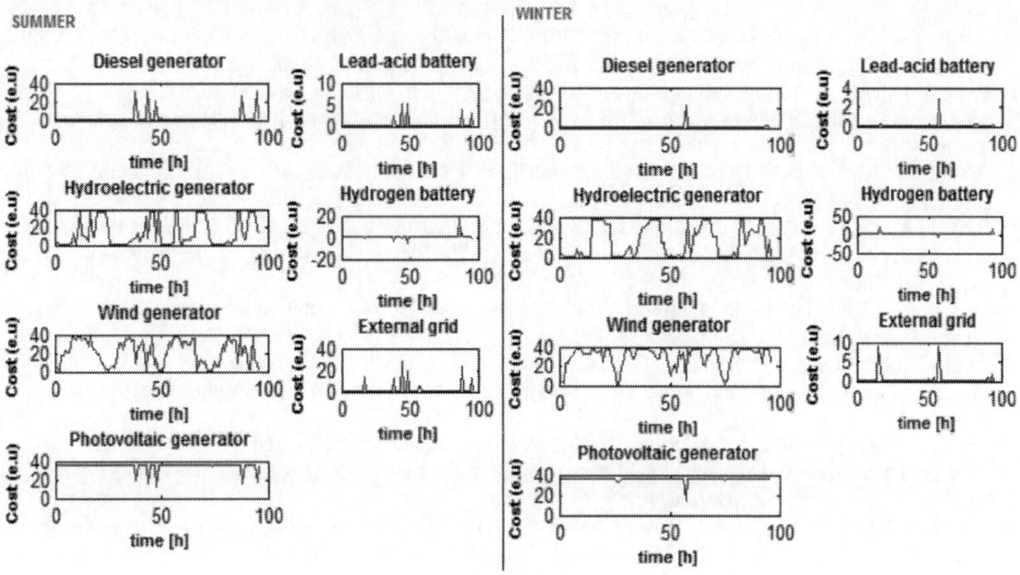

Figure 7. Plots of the economic costs

Remark:

It has been found that, one-layer approach is economically superior to the two-layer hierarchical scheme. Similar result was obtained in [1,4]. Moreover, it can also be seen (table 2) that, the EMPC produces the lowest economic costs, thereby proving its superiority to the MPC tracking.

Table 2. presents a comparison of the four MPC approaches' economic costs in economic unit (e.u).

	EMPC	EMPC + MPC tracking (single-layer)	EMPC + MPC tracking (two-layer)	MPC tracking
Summer economic cost	2934.4	2963.2	3761.5	4175.1
Winter economic cost	3369.0	3378.7	3802.0	4451.5

Table 2. Quantitative comparison of the economic costs

6. Conclusion and future work

In this study, we have presented the application of four variations of robust optimization based MPC strategies for controlling energy dispatch in a grid connected smart micro-grid. An affine dependence between the control inputs and the load demand has been derived. The standard EMPC, the MPC tracking, and their combination in a single and two-layer approaches have been discussed and compared. It has been found that, the standard Economic MPC yields a better economic result, because the energy consumption is the lowest. The next steps for completing this research will be devoted to tackling uncertainties of energy prices and renewable energy sources.

Acknowledgments

This work was supported by Spanish Government (MINISTERIO ECNONOMIA Y COMPETITIVIDAD) and FEDER under project DPI2014-58104-R (HARCRICS).

References

[1] Grosso J M, Ocampo-Martinez C, and Puig V 2012 Chance-constrained model predictive control for large-scale Networks, *Transactions on Control Systems Technology,* UPC (Barcelona, Spain)
[2] Abate A, Laurent El G 2004 Robust model predictive control through adjustable variables: an application to path planning *43rd IEEE Conference on Decision and Control*, (Atlantis, Paradise Island, Bahamas)
[3] Nassourou M 2016 Model Predictive Control Strategies for Energy Dispatch of a Smart Micro-grid System UPC (Barcelona, Spain)
[4] Ocampo-Martinez C, Puig V, Cembrano G, Creus R, and Minoves M 2009 Improving water management efficiency by using optimization-based control strategies: The Barcelona case study *Water Science & Technology: Water supply*, vol. **9**, no. 5, pp. 565–575
[5] Grosso J M 2014 On model predictive control for economic and robust operation of generalised flow-based networks, *PhD thesis UPC* (Barcelona, Spain)
[6] Rosero J L 2012 Estrategia de control de modelo predictivo para el despacho y almacenamiento de energia renovable en sistemas hibridos *Master thesis*, *Instituto Tecnologico y de Estudios Superiores de Monterrey*
[7] Grosso G, Ocampo-Martinez C, and Puig V 2012 A service reliability model predictive control with dynamic safety stocks and actuators health monitoring for drinking water networks *in Proceedings of the IEEE Conference on Decision and Control*
[8] Pereira M, Limon D, Muñoz de la Peña D, Valverde L, Alamo T 2015 Periodic economic control of a non-isolated micro-grid, *Transactions on Industrial Electronics 62*, Institute of Electrical and Electronics Engineers, 5247-5255
[9] Limon D, Pereira M, Muñoz de la Peña D, Alamo T, Grosso J.2014 Single-layer economic model predictive control for periodic operation *Journal of Process Control* **24**, 1207–1224
[10] Pereira M, Limon D, Alamo T, Valverde L 2015 Application of periodic economic MPC to a Grid-connected micro-grid, *conference poster, 5th IFAC Nonlinear Model Predictive Control Conference International Federation of Automatic control (*Sevilla, España)
[11] Bemporad A and Morari M 1999 Robust model predictive control: A survey hybrid systems *Computation and Control, F.W. Vaandrager and J.H. van Schuppen,* vol. **1569**, pp. 31–45,1999, lecture Notes in Computer Science.
[12] Loefberg J 2003 Minimax approaches to robust model predictive control *Ph.D. dissertation, Univ. of Linkoping*
[13] Ben-Tal A, Goryashko A, Guslitzer E, Nemirovski A 2004 Adjustable solutions of uncertain linear programs *Mathematical Programming*, **Vol. 99**, 2004, pp. 351–376
[14] Ben-Tal A and Nemirovski 1998 A Robust convex optimization *Mathematics of Operations Research*, Vol. **23**, No. 4, Nov. 1998, pp. 769–805
[15] Vandenberghe L and Boyd S 1996 Semidefinite programming *SIAM Rev.*, Vol. **38(1)**, 49–95
[16] Löfberg J 2012 Automatic robust convex programming *Optimization methods and software*, **27(1)**:115-129, 2012
[17] Prodan I, Enrico Z 2013 Predictive control for reliable microgrid energy management under uncertainties *22nd European Safety and Reliability (ESREL 2013) (*Amsterdam, Netherlands) <hal-00912003>
[18] Zanon M, Gros S, and Diehl M 2013 A Lyapunov function for periodic economic optimizing model predictive control *In Proc. 52nd IEEE Annual Conference on Decision and Control (CDC),* pages 5107–5112, (Firenze, Italy)

Real Time & Power Efficient Adaptive - Robust Control

Lavinius Ioan Gliga, Cosmin Constantin Mihai, Ciprian Lupu, Dumitru Popescu

gliga.l.ioan@hotmail.com, mihaicosmin2007@hotmail.com, ciprian.lupu@acse.pub.ro, dumitru.popescu@acse.pub.ro

gliga.l.ioan@hotmail.com

Abstract. A design procedure for a control system suited for dynamic variable processes is presented in this paper. The proposed adaptive - robust control strategy considers both adaptive control advantages and robust control benefits. It estimates the degradation of the system's performances due to the dynamic variation in the process and it then utilizes it to determine when the system must be adapted with a redesign of the robust controller. A single integral criterion is used for the identification of the process, and for the design of the control algorithm, which is expressed in direct form, through a cost function defined in the space of the parameters of both the process and the controller. For the minimization of this nonlinear function, an adequate mathematical programming minimization method is used. The theoretical approach presented in this paper was validated for a closed loop control system, simulated in an application developed in C. Because of the reduced number of operations, this method is suitable for implementation on fast processes. Due to its effectiveness, it increases the idle time of the CPU, thereby saving electrical energy.

1. Introduction

When dealing with processes with dynamic variables, classical control designs are no longer suitable. If a process evolves rapidly, or is subject to structural disturbances, it may quickly go into states that were not accounted for when designing the controller, thus making it impossible to obtain the desired performance or even stability. Several design strategies exist to deal with these kind of systems, the most important ones being adaptive and robust control. Both involve the presence of a high-level supervisor in the control system, which will handle the evolutions of the process. In the case of adaptive control, the supervisor computes a new algorithm at every sampling moment, the strategy being very potent when used on processes that evolve quickly over time. Robust control is used for processes with structural disturbances: a single control algorithm is computed in the design phase, which only works if the disturbances do not change the process' behaviour too much. Obviously, robust control does not work for highly dynamic processes, but it is cost effective when implemented on predictable systems (like train controls).

To design control system, one must first identify the dynamic model M of the process P, and then calculate the control algorithm C. The pair (C, M) defines the nominal system in fig. 1.

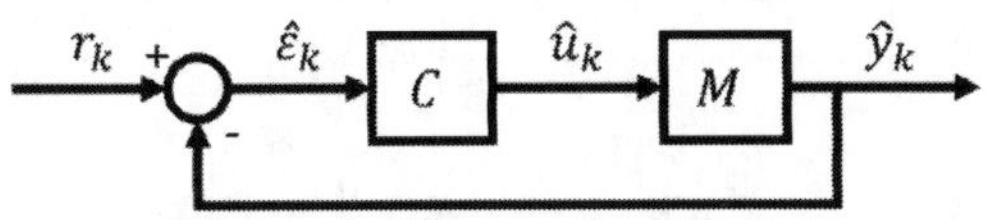

Figure 1. The nominal System

The mathematical model M is identified by solving the optimization problem [1][2]

$$\min J_I(C_{fixed}, M) \quad (1)$$

where J_I is the identification criterion built using the steady state error and the constant controller.

The calculation of the control algorithm is based on the identified model and is obtained as the solution to the minimization problem

$$\min J_C(C, M_{fixed}) \tag{2}$$

with M fixed and J_C being the optimal criterion built using the steady state error of the close loop system. The control algorithm design for the nominal system implies the successive solving of these two different optimization problems.

However, if the integral criterion would be identical for the two problems, they could be solved in a unified manner, using the same method, as will be detailed in subsequent sections. Compared with other system identification and controller design methodologies, where two different procedures are used: one for identification and another one for computing the controller, this new method has multiple advantages: easier implementation (only one function can achieve multiple ends), easier code maintenance and lower learning curve for system and/or design engineers who need to maintain and/or design such a system.

The main objective is to ensure that the real system represented by the pair (C, P), from fig. 2, has optimal performance at any given time.

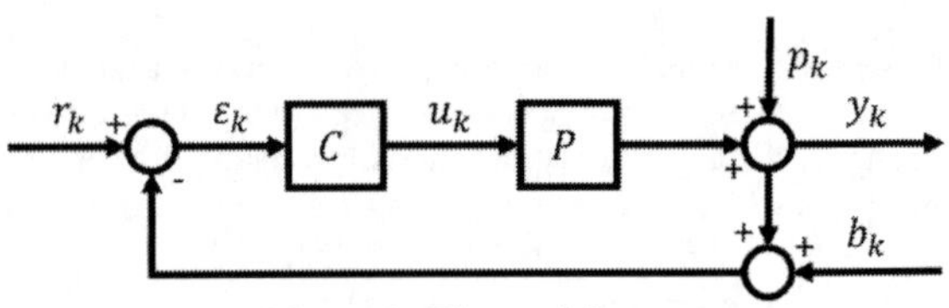

Figure 2. The real System

For the real process P, with dynamic variable parameters, either the adaptive strategy or the robust one are recommended for implementation. Adaptive control [3][4] implies the (re)identification of the process' model and the (re)design of the control algorithm at every sampling moment: for a new model, only a new controller can preserve the system's performances. From the (k) to the (k+1) sampling moments, adaptive control strategy can be summarized as

$$(C,\ M)_k \rightarrow (C,\ M)_{k+1} \tag{3}$$

Robust control [5] is recommended for processes with parametric or structural disturbances. It requires a single controller to ensure the desired performances. It is suitable for a class of models that must reside near the process [6], which has the same parameters values as the ones used for designing the controller. The performance degradation of the resulting closed loop system is formally expressed by equation (4), with distance δ (being the performance degradation threshold) small enough

$$\left|\frac{CP}{1-CP} - \frac{CM}{1-CM}\right| < \delta \tag{4}$$

If the degradation becomes too great, the controller will no longer ensure the needed performance, making robust control inappropriate for system with highly dynamic variable.

We propose a combined adaptive – robust [7] algorithm design methodology, built on the advantages of the two previous strategies, which also eliminates their disadvantages. Compared to classical adaptive strategy, the new procedure preserves nominal performance for the physical system by (re)identifying only in case of significant performance degradation, not at each sampling time. This clearly reduces the numerical calculations, thus increasing the idle time of the processor, saving electrical energy. The robust strategy is used to compute a single control algorithm, tolerant to disturbances. This controller is suited for a whole range of process models. The control algorithm remains unchanged if the degradation condition – inequality (4) remains satisfied. The controller is then recalculated after the identification of the dynamic model. This eliminates the computation of the control algorithm at every sampling moment, by imposing a reserve of robustness.

2. Adaptive – Robust Control

An optimal criterion [8] (integral criterion for optimal control) is considered and we propose an iterative design procedure, where the two problems of identification and control are intertwined (playing the main role, one after another):

- Using the current available controller, the next model M_{i+1} is determined, if it is necessary
- Using the obtained model, the subsequent controller C_{i+1} is computed

The connection between the two problems, identification and control, is illustrated by the recursive mechanism expressed in (1) and (2).

The integral optimality criterion J and its absolute numerical value are considered. For process P, model M and controller C, (4) it can be written

$$|J(C,P) - J(C,M)| < d \tag{5}$$

where

- $J(C,P)$ is the measurement of the real system's performance
- $J(C,M)$ is the measurement of the nominal system's performance
- $|J(C,P) - J(C,M)|$ is the measurement of the performances' degradation, due to changes in the process
- d is the performance degradation threshold, but scaled for the difference in integral criterions

Considering (5), the following necessary condition must be satisfied for the nominal performance is achieved to be achieved by the real system

$$|J(C,P) - J(C,M)| \leq |J(C,M)| \tag{6}$$

where $|J(C,M)| < \delta$. $|J(C,M)|$ has larger values when a change occurs in the process, therefore behaving like a classical robustness margin, but converges to 0 when the parameters are constant, thus providing more sensitivity to future unexpected variations.

The restriction (6) expresses the robust behaviour of the algorithm, since the degradation is bordered by the low values provided by $|J(C,M)|$. In addition, if inequality (8) is unsatisfied, the performance of the closed loop system has degraded so much that the process needs to be identified again and a new controller must be computed.

The C_i is considered, available for the current model M_i, and the next replica is computed [9]

$$M_{i+1} = \underset{M}{\operatorname{argmin}}[J(C_i, M) + J(C,P)] \tag{7}$$

For model M_{i+1} the following controller is computed

$$C_{i+1} = \underset{C}{\operatorname{argmin}}\, J(C, M_{i+1}) \tag{8}$$

(7) and (8) are solved if (6), which is being checked at each sampling moment, is unsatisfied.

The system needs to be adapted (i.e. the controller must be recalculated) when (4) is no longer satisfied, which means that the parameters of the process have changed so much, that the closed loop performance of the real system is too different from the nominal one's. Thus, a new model of the process is identified and used to compute a robust controller. This regulator should be able to passively accommodate changes in the process, in the vicinity of the current one for which it was designed [6], thus minimizing the number of necessary adaptations.

3. The adaptive – robust control approach

The proposed approach is presented for a first order process controlled by a PI regulator. While a multitude of different processes can be described by such transfer functions, there are many more which cannot. However, the design methodology remains the same, just the number of mathematical calculations increases with the grade of the process' transfer function and of the controller's. The authors did not find including long mathematical calculations and formulas necessary or useful for understanding the proposed approach. While they may help others integrate the solution more easily into their different application, it also makes the paper more difficult to understand. The authors are always ready to help anyone with using the proposed methodology on more complicated systems with different controllers. The mathematical formulations for the general case may be presented in a future paper.

A process expressed using a first order transfer function is considered [10][11]

$$P = \frac{K}{Ts + 1} \tag{9}$$

and the PI control algorithm which will be used for the process

$$C = K_r(1 + \frac{1}{T_i s}) \tag{10}$$

The integral criterion J used in (7) and (8) is

$$J = \int_0^\infty \varepsilon^2(t) + T_e^2 \dot{\varepsilon}^2(t) dt \tag{11}$$

where the first part of the criterion is used to minimize the response time, while the second one is needed to control the overshoot – these two performance indices being very important when designing a control system. While others, such as frequency specifications or command signal constraints, may appear, they are not usually as important as the ones considered by this criterion. The criterion will also be minimized, thus imposing a zero value of the steady state error, the most important parameter in performance specifications.

The criterion is, however, useless in its current form, as it is impossible to implement in an effective way on actual microcontrollers, so it has to be rewritten using algorithm from [10] to its direct form, expressed in the space of the model and controller parameters.

First, the terms of the integral are separated, and each will be computed separately.

$$J = \int_0^\infty \varepsilon^2(t) + T_e^2 \dot{\varepsilon}^2(t) dt = \int_0^\infty \varepsilon^2(t) dt + T_e^2 \int_0^\infty \dot{\varepsilon}^2(t) dt = J_1 + T_e^2 J_2 \tag{12}$$

The Laplace transform is applied

$$J_1 = \int_0^{\infty} \varepsilon^2(t)dt = \int_0^{\infty} \varepsilon(t)\varepsilon(t)dt = \int_0^{\infty} \varepsilon(t)\{L^{-1}[\varepsilon(s)]\}dt$$

$$J_1 = \int_0^{\infty} \varepsilon(t)\left[\frac{1}{2\pi j}\int_{-j\infty}^{j\infty} \varepsilon(s)e^{st}ds\right]dt \tag{13}$$

$$J_1 = \frac{1}{2\pi j}\int_{-j\infty}^{j\infty} \varepsilon(s)\left[\int_0^{\infty} \varepsilon(t)e^{st}dt\right]ds = \frac{1}{2\pi j}\int_{-j\infty}^{j\infty} \varepsilon(s)\varepsilon(-s)ds$$

The steady state error is written as

$$\varepsilon(s) = \frac{1}{1 + H_d(s)}R(s) = [1 - H_o(s)]R(s) \tag{14}$$

Therefore, the first part of the criterion can be written as

$$J_1 = \frac{1}{2\pi j}\int_{-j\infty}^{j\infty} [1 - H_o(s)][1 - H_o(-s)]R(s)R(-s)ds \tag{15}$$

Before formulating the direct expression of J_1, the following condition must be met: the difference between the grades of the denominator and the numerator of the steady state error's transfer function must the at least 1. This condition will be tested for a first order process controlled with a PI algorithm. The open loop transfer function of the resulting system is

$$H_d = \frac{K_r(T_i s + 1)}{T_i s} \cdot \frac{K}{Ts + 1} = \frac{KK_r(T_i s + 1)}{sT_i(Ts + 1)} = \frac{M(s)}{sN(s)} \tag{16}$$

and the steady state error is

$$\varepsilon(s) = \frac{N(s)}{sN(s) + M(s)} = \frac{T_i(Ts + 1)}{sT_i(Ts + 1) + KK_r(T_i s + 1)} = \frac{T_i Ts + T_i}{T_i Ts^2 + (1 + KK_r)T_i s + KK_r}$$

$$\varepsilon(s) = \frac{b_1 s + b_0}{a_2 s^2 + a_1 s + a_0} \tag{17}$$

The direct form of the criterion's first term is, per [10],

$$J_1 = \frac{T_i + K_r KT}{2K_r K(1 + K_r K)} \tag{18}$$

The second part of the criterion follows roughly the same steps, but the condition applied to the transfer function of the steady state error is checked upon the error's derivative

$$L[\dot{\varepsilon}(t)] = \frac{-M(s)}{sN(s) + M(s)} = \frac{-KK_r(T_i s + 1)}{sT_i(Ts + 1) + KK_r(T_i s + 1)} = \frac{-KK_r T_i s - KK_r}{T_i Ts^2 + (1 + KK_r)T_i s + KK_r}$$

$$L[\dot{\varepsilon}(t)] = \frac{c_1 s + c_0}{a_2 s^2 + a_1 s + a_0} \tag{19}$$

The direct expression for the second part of the integral criterion, according to [10], is

$$J_2 = \frac{K_r K(1 + K_r K \frac{T_i}{T})}{2T_i(1 + K_r K)} \tag{20}$$

Overall, the J integral criterion is equal to

$$J = J_1 + J_2 = \frac{T_i + K_r KT}{2K_r K(1 + K_r K)} + T_e^2 \frac{K_r K(1 + K_r K \frac{T_i}{T})}{2T_i(1 + K_r K)} \tag{21}$$

The minimization of the above non-linear criterion is accomplished using an appropriate nonlinear direct method [12][13]. The Nelder – Mead [14] procedure was chosen because of its good performance for this class of problems. It is based on the SIMPLEX [15] algorithm.

As it can be seen the controller is designed in continuous time. For an implementation, it must be discretized, using one of the known methods: Euler, Tustin, etc.

The robustness of the controller refers to the different changes in the process that it can accommodate, while first maintaining the stability of the real system and then the attained performances. The same principle which is described in [6] will be used here. It will be detailed in the next section.

By knowing the parameters of the process, this nonlinear optimization method can be used to minimize the criterion J to compute the parameters of the controller, and the value of the integral criterion.

The opposite is also applicable: by knowing the parameters of the controller and the integral criterion's value, the parameters of the process can be obtained. It is impossible to know the exact value of the integral criterion, when considering a real application for a real process; therefore, it has to be estimated. The same estimation will be used for the degradation condition. Using a Riemann sum [16], the integral criterion will be rewritten as

$$J = \int_0^\infty \varepsilon(t) + T_e^2 \dot{\varepsilon}(t) dt \cong \hat{J} = T_e^2 \cdot \varepsilon(N) \cdot h + \sum_{k=1}^{\infty} \varepsilon(k) \cdot h \tag{22}$$

where N is the current sampling moment, translating the criterion from the continuous time domain to the discrete one. As the discrete criterion depends only on the steady state error, its value can be computed fast and easy, at each sampling moment.

To sum it all up, the proposed control algorithm is:

1. At each sampling moment, compute the integral criterion's value

$$\hat{J} = T_e^2 \cdot \varepsilon(N) \cdot h + \sum_{k=a}^{b} \varepsilon(k) \cdot h \tag{23}$$

where "a" is the sampling moment when the last identification of the process was performed, and "b" is the current sampling moment.

2. If the degradation condition

$$|\hat{J} - \hat{J}^*| < \hat{J}^* \tag{24}$$

where $\hat{J}^*$ is the integral criterion's value computed using the last identified parameters of the process and computed parameters of the controller, is unsatisfied, then estimate the current model of the process

$$M_{new} = arg\min_{M}[J(C_{old}, M) + \hat{J}] \quad (25)$$

and compute the new controller

$$C_{new} = arg\min_{C} J(C, M_{new}) \quad (26)$$

using the Nelder - Mead method for minimizing the criterion.

At every sampling moment, the performance of the real system is estimated using the discretized version of the integral criterion. Also, the performance degradation (the difference between the estimated performance of the real system and known one of the nominal system) is checked. If it is too great, then a new model of the process is reidentified using the current parameters of the controller and the degradation's value, by minimizing the integral criterion. After the model is obtained, its parameters are utilized to compute the new parameters of the controller, that is suited for that model. The performance value of the nominal system is remembered (it being computed together with the new regulator) and this will be used to check the performance degradation of the real system. Thus, the number of adaptations is reduced, a re-computation being performed only when necessary.

4. Implementation

To test the proposed control system, we developed a software process simulator. The application, developed in C [17] using the LabWindws/CVI [18] IDE and the Nlopt library [19], validated the theoretical approach proposed in this paper. The name of the program is CAR ("Control Adaptiv – Robust" = "Adaptive – Robust Control") and it simulates the response of a closed loop system, being composed of a first order process and a PI controller, at a given sampling period.

The user can modify any of the following parameters whenever he/she desires:

- The parameters (K, T) of the process P (they are the real parameters of the real process). Whenever this pair of parameters is changed, the real value of the integral criterion "J4" is (re)evaluated, using the current parameters of the controller. The parameters correspond to the gain and time constant of a first order transfer function. The process' discrete equivalent is computed automatically, as is for the controller.
- The sampling period
- The weighting of the overshoot
- The robustness (degradation) coefficient
- The reference

The UI informs the operator about the estimated parameters of the process („K_e" and „T_e"), the integral criterion's values estimated for the process („J4") and computed for the model („J4_e"), the current steady state error and the evolution of the process in the last 100 sampling moments, via the graph in the upper right side of the UI.

The workflow of the program is the following:

1. The user sets the initial values, then starts the simulation by pressing the button located in the lower middle part of the UI.

2. The process is assumed to be properly identified, so the estimated parameters (K_e, T_e) are set to be equal with the real parameters of the process. The PI controller is computed (K_r, T_i) using the estimated values of the parameters of the process, and the simulation starts, using the sampling period provided by the user.
3. At each sampling moment, the algorithm estimates the value of the integral criterion and checks the degradation condition. If it is not satisfied:
    a. Use the estimated value of the integral criterion, and the current controller to compute the estimated parameters of the process
    b. Calculate a new controller, using the previously estimated parameters, and record the resulting value of the integral criterion ("J4_e" in the UI)
4. Compute the command to the process and the output of the process. Display the steady state error. The desired (red line) and the real (yellow line) output of the process are displayed on the graph, in the upper right side of the UI.

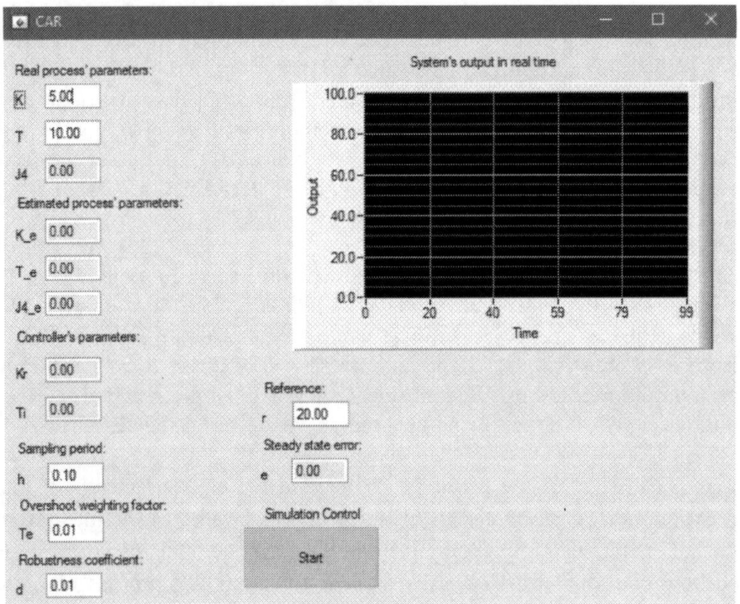

Figure 3. Application UI

Multiple tests were performed on multiple processes. The user can input any parameters for a first order transfer function, subject to certain reasonable constraints (which in no way diminish the generality of the proposed approach), which will be presented later.

The user can input the parameters of a continuous first order transfer function, and the indicators will display the estimated parameters of the model and the controller's parameters in continuous time. However, the processes and controller are discretized for the implementation. After the user modifies a parameter of the process, the discrete time equivalent is computed. When a new controller is recomputed, it is immediately discretized. These are not time consuming operations, because the discrete equivalents are already hard coded, and the new parameters are directly used in them, as constants.

Many real-world processes can be described using a first order transfer function. Among these are the evacuation of liquid from a reservoir, the flow of a liquid through short and long pipes, etc. In the case of Figure 4, we chose, for testing, an air heater. It's transfer function is

$$H(s) = \frac{5}{10s + 1} \quad (27)$$

In the left most image of Figure 4, the air heater is turned on, and the temperature of the air exiting it, after a time, stabilizes at 20° C, the desired value. However, at some point (the centre image in Figure 4) more electrical power is fed into the heater, thus modifying its heating capabilities, the new transfer function being

$$H(s) = \frac{30}{10s + 1} \quad (28)$$

The change in the output error is detected and the process' performance is estimated using the discretized integral criterion. As its value becomes too different from the one of the nominal system, a new controller is computed. Although it is computed very fast after the change in the process has taken place (just a few sampling moments later), the heating rods have a certain inertia and it also takes a while for them to cool off, although the power to them has been lowered. When their temperature has been sufficiently lowered, the exiting air again has the desired temperature (the right-most image from Figure 4).

The sampling period of the system was 0.05 seconds, to test how low the sampling time can be. As the code implementation was non-optimal, lower sampling moments can still be achieved.

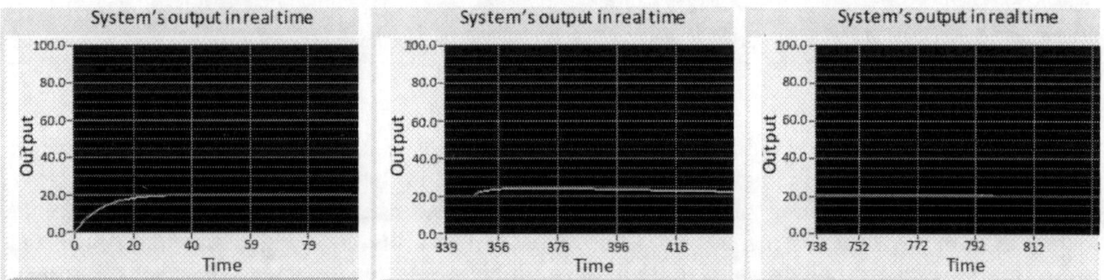

Figure 4. Program running (only the graph from the UI): start of simulation (left), change of parameters – gain changed from "5" to "30" (centre), performance after the change (right)

The robustness of this controller, the range of changes in the parameters of the process which can be accommodated, is described as in [6]. If the process' parameters change, however both still reside inside the circle of robustness which can be defined in the space of the parameters of the process, a circle centred on the model which was used to compute the current controller, the system does not need to be adapted. The zone of robustness for the exemplified process is shown in Figure 5, for the initial transfer function.

After thoroughly testing the simulator for different pairs of parameters representing different processes, from the intervals $K = [0.25, 100]$ and $T = [0.25, 100]$, the following results were obtained:

- The identification procedure's precision is less than 10^{-2}, in the absence of noises. It is not perfect, due to the estimation of the integral criterion.
- If the sampling period is properly chosen for the given process, the stability is maintained, and the desired performance is always reached.

- Any change of the processes' amplification can be accommodated. The changes in the time constant can negatively influence the performances and even the stability, if the sampling period is not corrected accordingly.
- This method is not suitable for unstable processes, or when the amplification is negative.
- The lowest sampling time attained is 50ms, without using any parallel programming techniques [20][21].

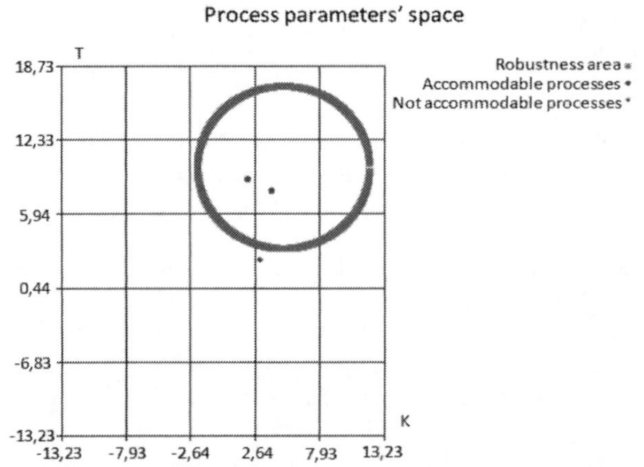

Figure 5. The controller's zone of robustness in the process parameters' space

5. Conclusion

An adaptive – robust method for designing a control system was presented. It combines the advantages of both robust and adaptive control, while minimizing their disadvantages.

The process identification problem is treated implicitly, using an indicator to measure the degradation of the system's performances, in case of dynamically changing parameters. When this estimation is greater than a threshold, the robust controller needs to be adapted. The adaptive - robust strategy computes the controller using the direct expression of an integral criterion, which depends only on the parameters of the process and the controller. It minimizes this criterion using the Nelder – Mead method. This approach can also be used for slow processes or processes with little to no variance in parameters. In such cases, it will behave like a classical robust control algorithm, with the bonus of being very sensitive to unexpected sudden changes in the system behaviour.

The theoretical approach presented in this work is validated after implementing an adaptive – robust algorithm for a closed loop control system, simulated in an application developed in C. The minimum sampling period attained is 50 ms. Thus, not only is it practical to use for fast processes, but it also increases the CPU's idle time when being used for slower ones. Therefore, electrical energy is being saved, making the microcontroller/PLC more power efficient, or saving computational load which can be used elsewhere.

The proposed adaptive – robust approach has the significant advantage of eliminating the traditional identification procedure and assuring robust qualities for the numerical control.

As future work, the proposed design methodology can be extended for:

- Processes which have such a dynamic behaviour, that the order of the transfer function changes. Therefore, different forms of the integral criterion have to be used.

- Testing the identification method in real conditions, where noises are present. The estimation works well in the absence of perturbations, but in a real environment, noises are present.
- Development of an automatic procedure which changes the sample time of the adative – robust system accordingly to the change detected in the process' time constant

References

[1] P. Borne, F. Gh. Filip, D. Popescu, M. Benrejeb, "Automatique Avancée et Informatique Appliquée", Editura Academiei Romane, Bucureşti, România, 2009
[2] F. L. Lewis, D. Vrabie, V. L. Syrmos, "Optimal Control", John Wiley and Sons, 2012
[3] K. J. Åström, B. Wittenmark, "Adaptive Control", Courier Corporation, 2008
[4] N. Doelman, R. Fraanje, I. Houtzager, M. Verhaegen, "Adaptive and real-time optimal control for adaptive optics systems", European Control Conference Proceedings, 2009
[5] M. Green, D. J. N. Limebeer, "Linear Robust Control", Courier Corporation, 2012
[6] L. I. Gliga, C. C. Mihai, C. Lupu, D. Popescu, "Adaptive – Robust control a computational efficient real time implementation", 13th International Conference on Engineering of Modern Electric Systems, 2015
[7] P. Ioannou, J. Sun, "Robust Adaptive Control", Courier Corporation, 2013
[8] S. S. Ge, C. Wang, Y. Li, T. H. Lee, M. H. Ang, "Adaptive optimal control for linear discrete time-varying systems", IEEE Conference on Cybernetics and Intelligent Systems Proceedings, 2013
[9] A. Locatelli, "Optimal Control: An Introduction", Springer Science & Business Media, 2001
[10] S. Călin, M. Tertişco, I. Dumitrache, C. Popeea, D. Popescu, "Optimizări în automatizări industrial", Editura Tehnică, 1979
[11] F. Li, "Robust Control Design: An Optimal Control Approach", John Wiley & Sons, 2007
[12] P. Borne, D. Popescu, F. Gh. Filip, D. Ştefănoiu, "Optimization in Engineering Sciences – Exact Methods", ISTE – Wiley, 2012
[13] P. Borne, D. Popescu, F. Gh. Filip, D. Ştefănoiu, "Optimization in Engineering Sciences – Metaheuristics, Stochastic Methods and Decision Support", ISTE – Wiley, 2012
[14] V. Sinlapakun, W. Assawinchaichote, "Optimized PID controller design for electric furnace temperature systems with Nelder Mead Algorithm", 12th Internation Conference on Electrical Engineering/Electronics, Computer, Telecommunications and Information Technology (ECTI-CON) Proceedings, 2015
[15] I. Maros, "Computational Techniques for the Simplex Method", Springer Science and Business Media, 2012
[16] S. Waner, S. Costenoble, "Applied Calculus", Cengage Learning, 2013
[17] K. Jamsa, L. Klander, "Totul despre C şi C++ - Manualul fundamental de programare în C şi C++", Teora, 2000
[18] National Instruments, "Getting Started with LabWindows/CVI", 2015
[19] S. G. Johnson, "The NLopt nonlinear-optimization package", http://ab-initio.mit.edu/nlopt, 2014
[20] M. McCool, J. Reinders, A. Robison, "Structured Parallel Programming: Patterns for Efficient Compuration", Elsevier, 2012
[21] Microsoft Corporation, "Multithreading with C and Win32", https://msdn.microsoft.com/en-us/library/y6h8hye8.aspx, 2016
[22] D. Popescu, C. Lupu, C. Petrescu, A. Monica, M. Mihaela, "Sisteme de Conducere a Proceselor Industriale", Printech, 2004
[23] D. Ştefănoiu, J. Culiţa, P. Stoica, "Fundamentele identificării şi modelării sistemelor", Printech 2005
[24] O. Nelles, "Nonlinear Systems Identification: From Classical Approaches to Neural Networks and Fuzzy Models", Springer Science and Business Media, 2013

Robust Power Management Control for Stand-Alone Hybrid Power Generation System

Elkhatib Kamal

Industrial Electronics and Control Department, Faculty of Electronic Engineering, Menoufia University, Menouf, Egypt and Institut Pascal / IMobS3, Blaise Pascal University–UMR CNRS 6602, Clermont-Ferrand, France.
(email: elkhatib.ibrahim@univ-bpclermont.fr).

Lounis Adouane

Institut Pascal / IMobS3, Blaise Pascal University – UMR CNRS 6602, Clermont-Ferrand, France (email: Lounis.Adouane@univ-bpclermont.fr).

Abdel Aitouche

Laboratory CRISTAL (Center of Research in Informatics, Signal and Automatic in Lille), University of Lille 1, France (email: abdel.aitouche@hei.fr).

Walaa Mohammed

Industrial Electronics and Control Dep., Faculty of Electronic Engineering, Menoufia University, Menouf, Egypt (email: walaa.mohammed57@el-eng.menofia.edu.eg).

Abstract. This paper presents a new robust fuzzy control of energy management strategy for the stand-alone hybrid power systems. It consists of two levels named centralized fuzzy supervisory control which generates the power references for each decentralized robust fuzzy control. Hybrid power systems comprises: a photovoltaic panel and wind turbine as renewable sources, a micro turbine generator and a battery storage system. The proposed control strategy is able to satisfy the load requirements based on a fuzzy supervisor controller and manage power flows between the different energy sources and the storage unit by respecting the state of charge and the variation of wind speed and irradiance. Centralized controller is designed based on If-Then fuzzy rules to manage and optimize the hybrid power system production by generating the reference power for photovoltaic panel and wind turbine. Decentralized controller is based on the Takagi-Sugeno fuzzy model and permits us to stabilize each photovoltaic panel and wind turbine in presence of disturbances and parametric uncertainties and to optimize the tracking reference which is given by the centralized controller level. The sufficient conditions stability are formulated in the format of linear matrix inequalities using the Lyapunov stability theory. The effectiveness of the proposed Strategy is finally demonstrated through a SAHPS (stand-alone hybrid power systems) to illustrate the effectiveness of the overall proposed method.

Keywords— Hybrid power system, Lyapunov function, Takagi-Sugeno fuzzy control system, Power management, Wind and photovoltaic energy system.

1. Introduction

The consumption of fuels as natural gas, coal and oil threat the energy spare of our generations [1], due to a lot of problems like a high cost of operation, transportation of fuel and difficult maintenance, renewable energy is one of the most solutions to be used [2]. The hybrid system is mainly composed of renewable energy sources (i.e., wind power and solar power), and a storage battery and, a micro turbine generator may be integrated into the system as well. Because of their discontinuous and unstable production, hybrid systems expand a very level of energy security with a great efficiency by the mixture of generation systems with a permanent, and storage energy system to ensure maximum efficiency of power supply [3].

Energy management is a term that collects all the systematic procedures to control and minimize the quantity and the cost of energy used to provide a certain application with its requirements. Some of the power management studies used linear programming to implement energy management, and others used intelligent techniques [4]-[8]. Ref. [4] presented the performance of two power management strategies that use the hysteresis band in operating the hybrid power system that contains Photovoltaic Panel (PV), Wind Turbine (WT), and hydrogen as energy storage. Ref. [5] proposed a power management control strategy for a standalone PV/battery power system. In this strategy, the PV system provides the steady-state energy, while the battery compensates for the dynamic energy. The goal of the strategy is to control the unidirectional and bi-directional DC–DC converter to operate in suitable modes based on the battery SOC and weather conditions. A hybrid system consisting of PV panels, a battery system, and a micro turbine as a backup source are introduced in [6]. Ref. [7] developed an algorithm for the optimal power management of a standalone hybrid system that contains a PV, wind, diesel generator, and battery energy storage. In this strategy, the PV subsystem operates at maximum power with the aid of a maximum power point tracking algorithm to supply the load as a priority. In [8] the authors presented a method for determining the optimum power management strategy of hybrid power systems consisting of various sources of energy and storage systems. In this research, the dispatch strategy employed gives priority to the effective use of renewable energy sources (PV and wind) to meet the load demand, while other sources have no predefined priorities except for the priorities determined by the power management optimization process.

Batteries are employed to store excessive energy from renewable energy. Many of researchers proposed several State of Charge (SOC) estimation methods for the batteries based on fuzzy logic, neural network and Kalman extended filters [9]. Micro turbine with a hybrid power generation system is presented to serve the load demands in remote areas because micro turbine instead of diesel generator because of more fuel flexibility and to less: weight, maintenance, noise, and pollution. The micro turbine and the battery storage are used as a support for the hybrid system of the WT and the solar array.

Stability conditions is studied based on Lyapunov stability theory is proposed in [10], fuzzy scheduling controllers is given in [11], switching controllers is presented in [12] and other methods studied the robust fuzzy control are given in [13]-[17]. Control and maximize the wind power for variable speed wind turbines based on TS fuzzy model and Linear Matrix Inequalities (LMI) are given in [15], [17].

This approach is an extension of the work proposed in [15] and [16]. This paper proposes a new Robust Fuzzy Control of Energy Management Strategy (RFCEMS), by introducing Centralized Fuzzy Supervisory Control (CFSC) to manage the power based on TS fuzzy control for the whole stand-alone hybrid power systems (SAHPS) and generate the power references for each Decentralized Robust Fuzzy Control (DRFC) which based on Takagi-Sugeno (TS) fuzzy control as given in [15]-[17]. DRFC maximize power extraction PV and WT and stabilize a nonlinear system in the presence of the parameter uncertainties. The derived stability conditions for DRFC are used to analyze the stability of the TS fuzzy control systems with uncertainty. Sufficient design conditions are derived for robust asymptotic tracking in terms LMIs.
Based on the aforementioned works, therefore the proposed algorithm combines the merits.

- The SAHPS control model is built-up with a hierarchical modular structure. This modular control structure has the advantage that, if it is desired, is easily extendable by adding new control functions either at the centralized and/or at the decentralized level.
- The modifications approach is designed such that to maximizes the produced power by PV and WT unit and is able to maintain the system stable during the parameter uncertainties and disturbance; in addition, it optimizes and manages the total power produced by the SAHPS.
- The simulation results for SAHPS system are demonstrated to visualize the feasibility of the proposed method.
- The battery maximum and minimum SOC level are chosen so as to increase its lifetime. The dump load added to the system aims to dissipate the power product excess.

The rest of the paper is organized as follows. TS fuzzy model with parameter uncertainty is presented in section 2. Section 3 shows the proposed RFCEMS algorithm. Stability and robustness analysis of proposed DRFC control systems is provided in section 4. In Section 5, the description of the power generation system is introduced. Simulation results are given in section 6. Finally, conclusion is deduced in section 7.

2. Parameters uncertainty matrices based on TS fuzzy model

The overall theoretical development will be introduced in this section and its application will be applied on a hybrid system which will be presented in section 5. To study the stability, the TS fuzzy dynamic model with parameter uncertainties is given by [17],

$$\dot{x}(t)=[\sum_{i=1}^{q}\mu_i(Z(t))A_i+\Delta A]x(t)+[\sum_{i=1}^{q}\mu_i(Z(t))B_i+\Delta B]u(t)]$$

$$y(t)=\sum_{i=1}^{q}\mu_i(Z(t))C_i\,x(t) \tag{1}$$

Where $Z_1(t),\dots,Z_\psi(t)$ are the premise variables; q is the number fuzzy rules; μ_i is a fuzzy strength, $i=1,2,\dots,q$; $x(t)$, $y(t)$ and $u(t)$ are respectively the state vector, output vector and input control signal; A_i, B_i, C_i are system matrices and output matrix, respectively, ΔA_i, and ΔB_i are the uncertainties parameters for A_i and B_i. The regenerator uncertainty based on TS fuzzy is given by the following format:

$$\Delta A=\sum_{j=1}^{n}h_j(\Delta A,\Delta B)\Delta\bar{A}_j\,,\qquad \Delta B=\sum_{j=1}^{n}h_j(\Delta A,\Delta B)\Delta\bar{B}_j \tag{2}$$

$n=2^f$ is the number of the parameter uncertainty regenerator fuzzy rules; f is the number of elements represented by uncertainty in ΔA and ΔB, which are given by $\Delta\bar{A}_j$ and $\Delta\bar{B}_j$. To express the TS fuzzy plant model of (1) as a weighted sum of the corner fuzzy systems, we use the property $\sum_{i=1}^{q}\mu_i=\sum_{j=1}^{n}h_j=\sum_{i=1}^{q}\sum_{j=1}^{n}\mu_i h_j=1$, from (1) and (2) the TS fuzzy plant model becomes

$$\dot{x}(t)=\sum_{i=1}^{q}\sum_{j=1}^{s}\mu_i h_j[(A_i+\Delta\bar{A}_j)x(t)+(B_i+\Delta\bar{B}_j)u(t)] \tag{3}$$

3. Proposed hierarchical supervisory control algorithm

3.1. Control system structure

Based on Matlab/Simulink and the analysis of the mathematical model of the SAHPS, a simulation model of the control system is established in this paper, with the block diagram of the control system being shown in Fig.1. The goal is to design a centralized controller to management and optimization the active power injected by the whole SAHPS. A DRFC for a PV and WT will be designed and will be tested in simulations. The proposed control system is based on a complex hierarchical control

IOP Conf. Series: Journal of Physics: Conf. Series **783** (2017) 012035 doi:10.1088/1742-6596/783/1/012035

architecture. A CFSC manages the power production of the whole SAHPS by sending out reference power signals to PV and WT, while the DRFC ensures that this reference power signal is reached. In addition, this strategy has the ability to keep the PV and WT system states stable in the presence of parameter uncertainties, wind disturbance and irradiance. First of all, we need to calculate the reference and the total power which are given by PV and WT. Error (*e*) is given by $e(t)=P_{Load} - P_{total}$ and Δe its derivative, where P_{Load} is the load power and P_{total} is given by $P_{total} = P_{PV} + P_{WT}$, where P_{PV} and P_{WT} are the PV and WT power at maximum point of the operation, respectively. In order to obtain optimality for the WT, the $r(t)=\Omega_{gref}=\Omega_{g(opt)}=\lambda_{opt}V/R$ profile is chosen in such a way as to follow the optimal tip speed ratio (λ_{opt}), V is the wind speed, R is the rotor-plane radius, and $\Omega_{g(opt)}$ is the optimal mechanical generator speed at λ_{opt}.

3.2. Proposed decentralized controller

The objective in this sub-section is to design a DRFC for WT-generator unit and PV unit. The control is performed such that the power is maximized and gives good tracking of the reference power generated by the centralized CFSC control.

3.2.1. Nonlinear fuzzy controller

Final output of the modified fuzzy controller for the fuzzy model (3) becomes [17]

$$u(t) = \sum_{k=1}^{c} \mu_k(Z(t))[-G_k x(t) + r(t)] \quad (4)$$

Where $G_k \in \Re^{mxn}$ are local feedback gains for DRFC of rule *k*, *r*(*t*) is the reference input and *c* is the number of the fuzzy rules (*k=1,2,...c*).

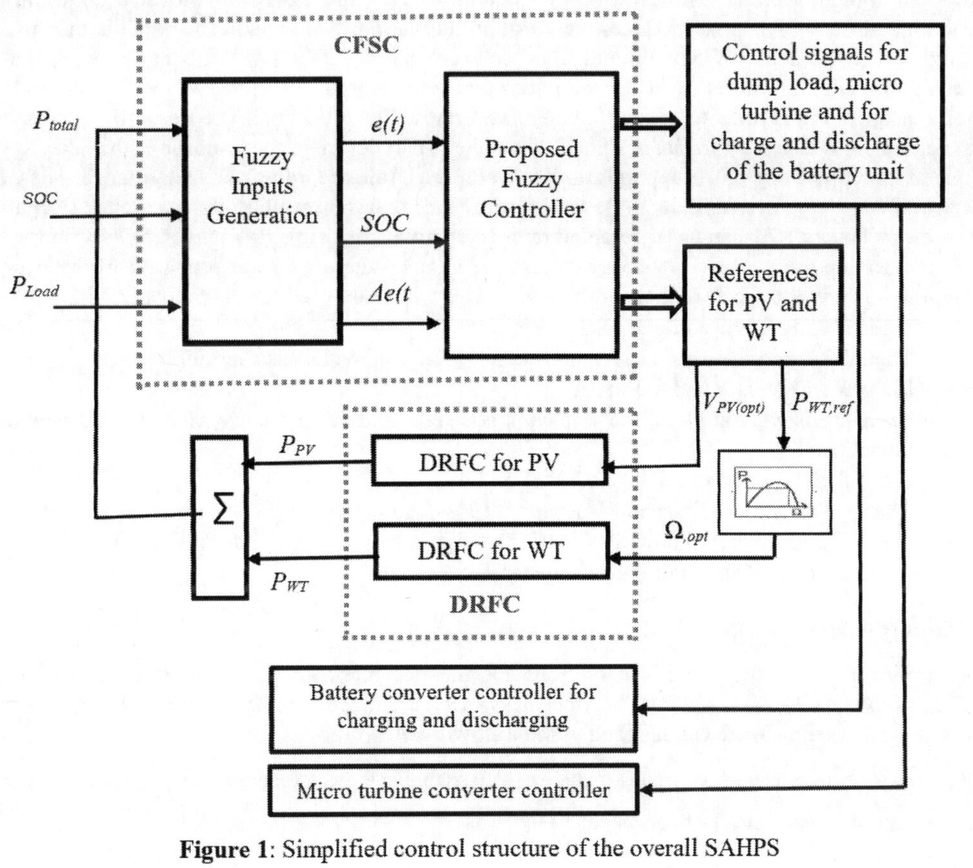

Figure 1: Simplified control structure of the overall SAHPS

3.2.2. Decentralized robust fuzzy controllers

The number of rules and the antecedents of the fuzzy scheduler are the same as those of the fuzzy uncertainty rules. The inferred output of the DRFC is given by:

$$u(t) = \sum_{k=1}^{c} \sum_{j=1}^{n} \mu_i h_j [-G_{kj}\, x(t) + r(t)] \tag{5}$$

With the DRFC (5) employed, the TS fuzzy system (3) has the following closed-loop:

$$\dot{x}(t) = \sum_{i=1}^{q} \sum_{k=1}^{c} \sum_{j=1}^{n} \mu_i \mu_k h_j [((A_i - B_i G_{kj}) + (\Delta\bar{A}_j - \Delta\bar{B}_j G_{kj}))x(t) + (B_i + \Delta\bar{B}_j) r(t)] \tag{6}$$

3.3. Proposed centralized supervisory controller

SAHPS is energy generation systems which are autonomous and controllable generating units. Given the current technological status, turbine-generator-power converter and PV converter set should be considered as a unit. These units work as an individual system for variable speed systems and solar irradiance. Consequently, through the controller and conditions of different wind regimes and solar irradiance, it is possible to establish an optimum point of operation for each generation system. The control method is based in the aim of its control on a fuzzy controller as shown in Fig. 1. After we design the DRFC for WT and PV unit, the CFSC is developed as the following. In the first step summing of the entire active power produced (P_{total}) by each unit is done. Then, with this value and with the load power (P_{Load}), the error and its derivative are calculated. Next, these two variables with SOC variable are the inputs of a CFSC that produces a control output variable. This output variable is then used to determine the optimal reference input for each subsystem PV and WT control unit, battery charging/discharging and set point for the micro turbine operation. Finally, PV and WT control unit adjusts its operation in order to obtain the reference imposed by the CFSC. The rules 'IF' part describes the situation for which the rules are designed. The "Then" part describes the response of the fuzzy system in this situation. The degree of support is used to weigh each rule according to its importance. The CFSC is constituted by 27 rules. With the following significance: e has 3 membership function N (Negative), Z (Zero), P (Postive). Δe can be represented by N (Negative), Z (Zero), P (Postive). SOC is introduced by 3 membership function L (Low), M (Medium), H(High). Where Δe is the derivative of the error. As examples, it is presented below some of the fuzzy rules which are used for the proposed CFSC.

- If e is N and Δe is N and SOC is L
  Then WT is work at Ω_{opt} and PV is work at $V_{PV(opt)}$ and battery is charging.
- If e is N and Δe is N and SOC is H
  Then WT is work at Ω_{opt} and PV is work at $V_{PV(opt)}$ and battery is natural and e is dissipated in dump load.
- If e is P and Δe is P and SOC is H
  Then WT is work at Ω_{opt} and PV is work at $V_{PV(opt)}$ and battery is discharging.
- If e is P and Δe is P and SOC is low Then WT is work at Ω_{opt} and PV is work at $V_{PV(opt)}$ and battery is natural and micro turbine generator operates.

4. Stability analysis of DRFSC

This section presents the stability conditions for the decentralized controlled system (6) and the calculation of the DRFSC gains (5). The robustness analysis and the stability conditions for the uncertain fuzzy control system (6) are summarized in the following theorem.

Theorem: The fuzzy control system as given by (6) is stable if the controllers set to $G_k = M^{-1} Y_k$ with the matrices M, and Y_k satisfying the following LMIs.

$$MA_i^T + A_i M - (BY_{i\,kj})^T - (BY_{i\,kj}) < 0 \quad (7)$$

The proof can be given directly from [17].

5. Description of the proposed hybrid controlled system

The proposed hybrid energy generation system is depicted in Fig.2. This system consists of variable speed WT, a solar array as renewable sources, a micro turbine which has a permanent source power to provide the load of the overall required energy, a battery storage storing the power that exceeded from the load, and a dump load which saves the battery from over charging. The HRFCSEMS is used for obtaining maximum power from renewable sources, managing between different power sources and stabilize the system in the presence of the parameters uncertainties.

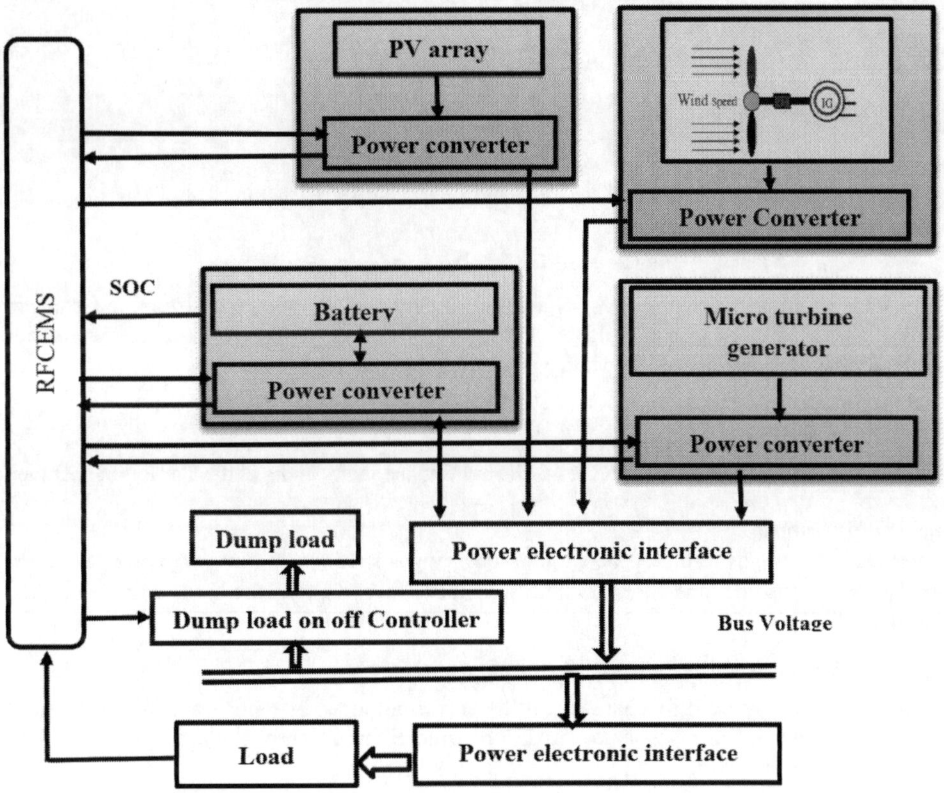

Figure 2. The schematic of the proposed hybrid system.

5.1. PV model system

To show the effectiveness of the proposed controller design techniques, PV model System [18] are simulated. The current-voltage characteristic of a PV array conducts as a function of solar irradiance and cell temperature is described as follows:

$$P_{pv} = I_{PV} V_{PV} = n_p I_{ph} V_{PV} - n_p I_{rs} V_{PV} (\exp(q V_{PV} / n_s \varphi KT) - 1) \quad (8)$$

where n_p and n_s are the number of the parallel and series cells, respectively, K is the Boltzmanns' constant, T is the cell temperature, φ is the p-n junction characteristic, q is the electron charge, I_{PV} and

V_{PV} are the output current and the PV array voltage on the capacitance C_1, I_{ph} and I_{rs} are the photocurrent and reverse saturation current, respectively. When the power slope $dP_{PV}/dV_{PV}=0$, the system operates at the maximum power generation. The dynamic equation can be expressed by the following differential equations given in [18].

5.2. Dynamic model of wind turbine generator unit

If the air density is ρ, R is turbine radius, power coefficient is $C_p(\lambda,\beta)$, and tip speed ratio (λ) is given by $\lambda=\Omega_l R/V$, the WT mechanical power is given by [19].

$$P_m = 0.5\rho\pi R^2 V^3 C_p(\lambda, \beta) \tag{9}$$

From (16) if we maximized $C_p(\lambda,\beta)$, the maximum power from the WT will obtain. $C_{p(max)}(\lambda_{opt},\beta)$ *can obtained at* $\lambda=\lambda_{opt}$. The state space of the WT using Doubly-Fed Induction Generator (DFIG) is given by [19]. The WT system parameters are defined in the appendix. The stator active (P_s) and reactive (Q_s) power of the WT with DFIG are given by [19].

$$P_s = -\frac{L_m V_s}{L_s} i_{qr} \qquad , Q_s = \frac{V_s^2}{\omega_s L_s} - \frac{L_m V_s}{L_s} i_{rd} \tag{10}$$

Where V_s is the stator voltage

5.3. Battery model

In this paper, we use the battery model presented in [20]. During the loaded condition, the electrical model of the battery can be expressed as

$$V_{bat} = V_{oc} + IR_o + IR_1(1-\exp(-t/C_1R_1)) + IR_2(1-\exp(-t/C_2R_2)) \tag{11}$$

where I is the current and t is time, R_1, R_2 and C_1, C_2 are the RC branch resistors and capacitors, respectively, V_{oc} is the open circuit voltage. The quantity that describes the ratio of the remaining capacity to the nominal capacity of the battery is known SOC and is computed by

$$SOC = SOC_i + \frac{\int \eta I dt}{C_n} , \tag{12}$$

where SOC_i is the initial value of the SOC, C_n is the nominal capacity and η is the Coulombic efficiency.

5.4. Micro turbine model

Micro turbine is small gas turbine produce high energy gas stream that turns an electrical generator based on burns gaseous or liquid fuels [21], [22]. In this paper, we will use the Micro turbine model presented in [21],[22].

5.5. Dump load

When the storage battery are fully charged and the power produced by the PV and WT is greater that the required load power, in this case we need to dissipate the difference through the dissipative resistance which known as the dump load.

6. Simulation and results

The simulation model of the hybrid energy system has been developed using MATLAB/Simulink. The performance of the proposed supervision and energy management has been verified by simulation studies for 50 seconds under various scenarios using the wind speed, irradiance, and temperature profiles as shown in Figures 3, 4 and 5, respectively. The control objective of this paper is to design a robust fuzzy control law for the PV and WT system to ensure that all signals in the closed-loop system are bounded and the parametric uncertainties are considered within 40% of their nominal values for WT and PV, which are not time varying.

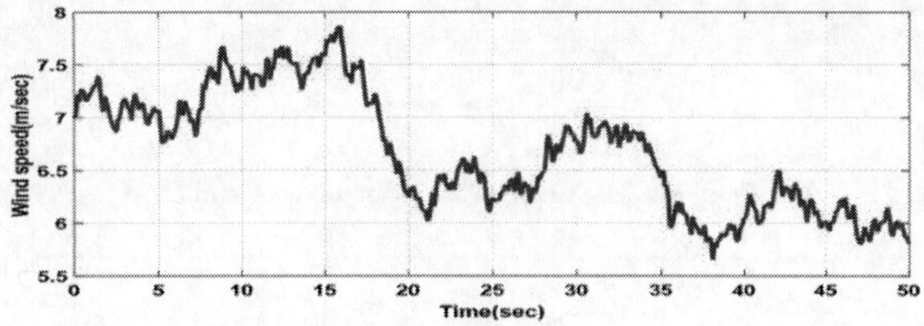

Figure 3. Wind speed profile.

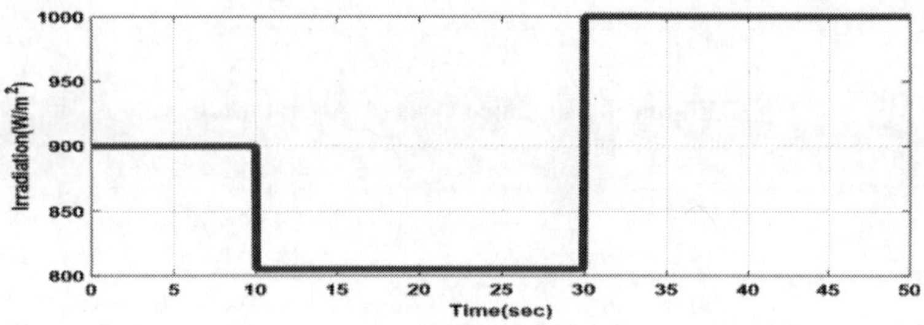

Figure 4. Irradiation profile.

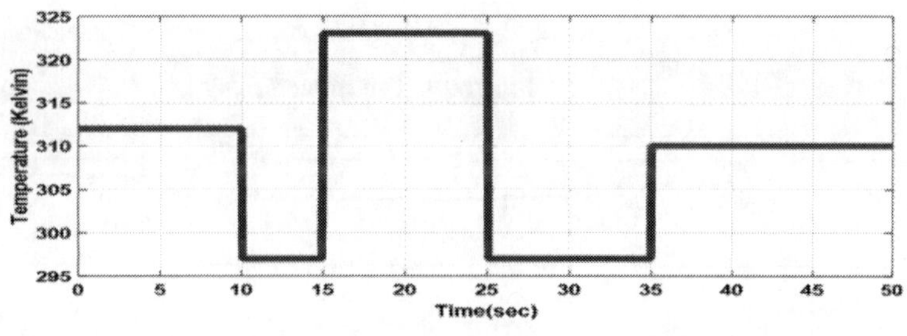

Figure 5. Temperature profile.

The WT, generator rotational speed and the active power P_s are shown in Figures 6, 7 and 8, respectively, in the presence of parametric uncertainties Fig. 9 and 10 show the PV array voltage and power, respectively. From the simulation results, it can be observe that, the outputs tracking of the PV and WT subject to parameter uncertainties are bounded and good performance for the proposed control strategy.

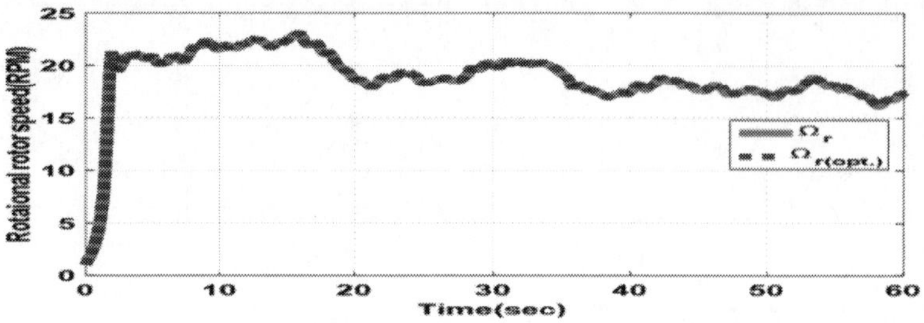

Figure 6. The trajectories of rotational speed Ω_r.

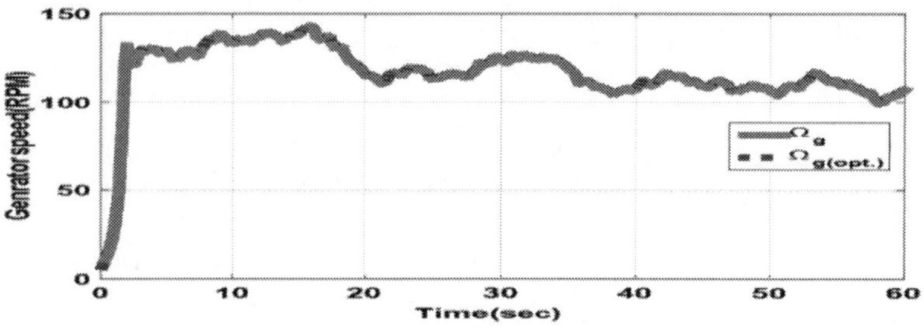

Figure 7. The trajectories of generator Ω_g

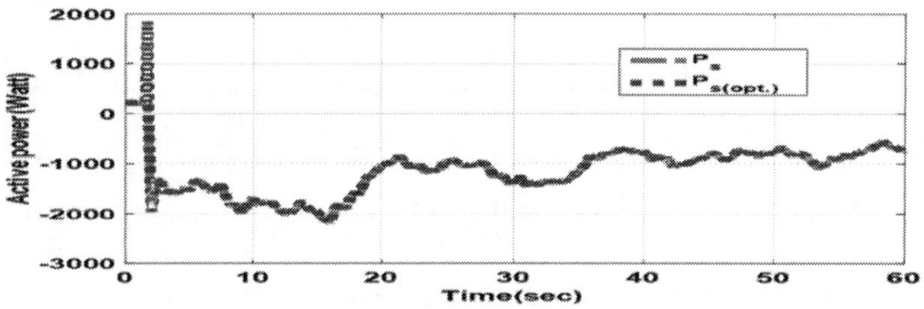

Figure 8. The trajectories of P_s

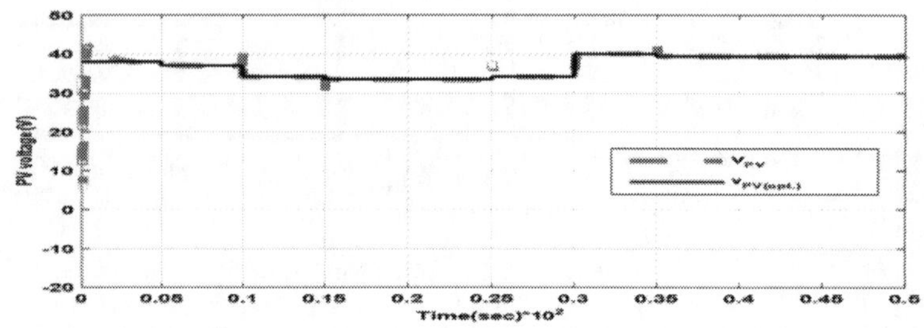

Figure 9. PV array output voltage (V)

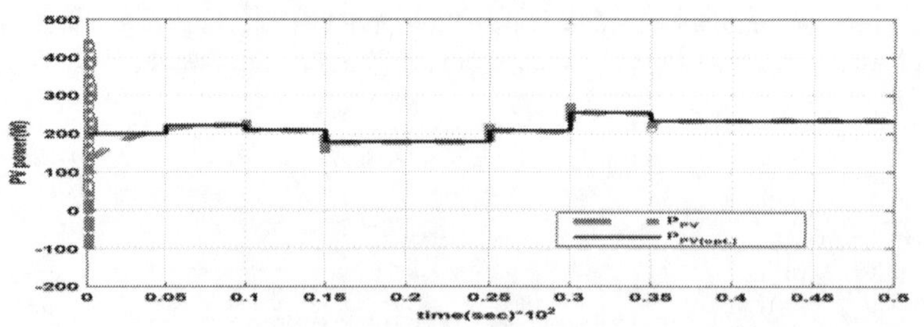

Figure 10. PV array output power (W)

Load power profile is shown in Fig.11. It is variable to ensure that the hybrid energy system can serve the load in any case. In Fig.12, SOC of battery changed from charging and discharging depending on the need of load. Fig.13 is presented the dump load which is used to dissipate the exceeded power to save the battery and the load. Fig. 14 is clarified the operation of micro turbine generator which operates when the load need power more than (WT+PV) power and SOC is low.

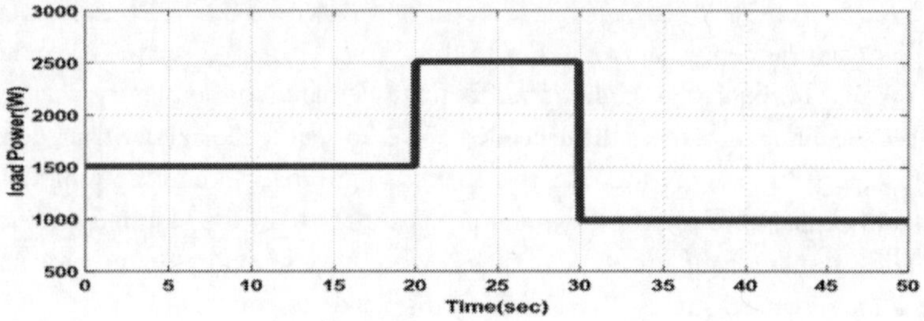

Figure 11. Load power profile.

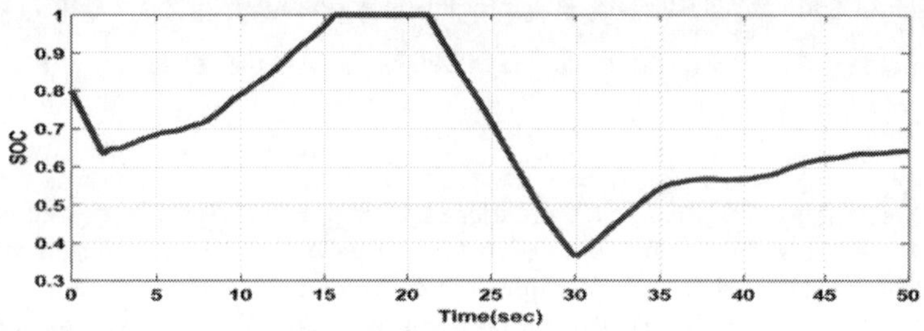

Figure 12. SOC for battery.

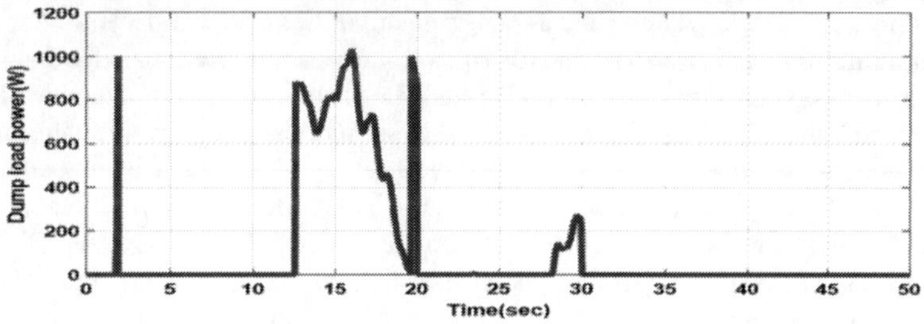

Figure 13. Dump load profile.

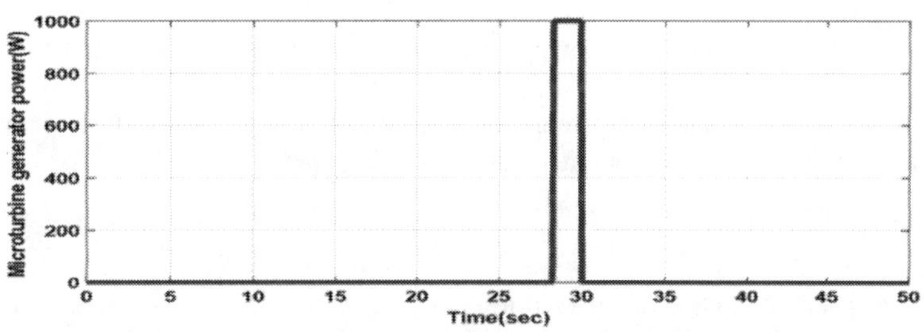

Figure 14. Micro turbine generator power

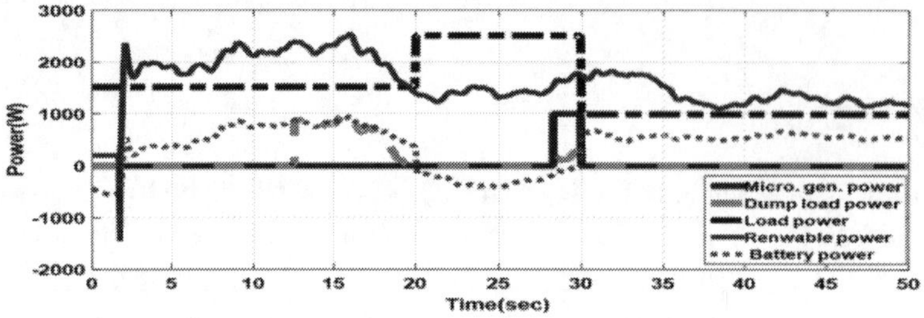

Figure 15. Power generated for each element.

Power generated for each element is shown in Fig.15. In Fig.15, from t = 2 sec to 20 sec the total power from renewable (WT+ PV) energy is higher than the load demand so the battery charged even max. SOC, at t = 12.5 sec the dump load takes the exceeded power to secure the battery from damaging. After that the battery discharged to supply the required energy to the load when renewable energy has not ability to serve the load, then micro turbine generator gives the load of the required power.

From the simulation results of the proposed strategy, with those given by the previous algorithms, it can be noticed that the proposed strategy has the following features:

- It maximizes the produced power by PV and WT unit and is able to maintain the system stable even with parameter and disturbance uncertainties up to 40% compared with [15]-[17]. In addition, it optimizes and manages the total power produced by the SAHPS.
- The simulation results demonstrate the effectiveness and the reliability of the proposed control approach.

7. Conclusion

New algorithm strategy is applied on a stand-alone hybrid power systems, equipped with: photovoltaic panel and WT (as renewable sources), a micro turbine generator and a battery storage system. The proposed control strategy is based on a complex hierarchical control architecture which consists of two levels named centralized and decentralized fuzzy control. Centralized fuzzy control level derives the active power references for PV and WT based on fuzzy control. A decentralized control level is based on TS fuzzy models, gives a good tracking for the reference power signal sent by the centralized control level and is robust for a wide range of parameter uncertainties and wind disturbances. The battery maximum and minimum SOC level are chosen so as to increase its lifetime. The dump load added to the system aims to dissipate the power product excess. The proposed robust power management strategy satisfy the load and battery bank SOC and it is tested in simulation using theoretical profiles of weather conditions to illustrate the performance of the proposed strategy. A set of sufficient conditions for robust stabilization of the TS fuzzy model are formulated in the LMI format based on Lyapunov stability. The concept of PDC is employed to design fuzzy control from the TS fuzzy models. Simulating control actions imposed on the SAHPS controller has been tested the performance of the control system. An extension of the SAHPS controller with fault tolerant control is in the implementation stage and it will be described in detail in a future paper.

Acknowledgments

The authors would like to thank the support of the ADEME (Agence De l'Environnement et de la Matrise de l'Energie), through BUSINOVA Evolution project. The authors would like also to thank the Industrial Electronics and Control Department, Faculty of Electronic Engineering, Menoufia University (Egypt) for their support which considerably improved this paper.

References

[1] Bajpai, P., & Dash, V., Hybrid renewable energy systems for power generation in stand-alone applications: a review. Renewable and Sustainable Energy Reviews, 16(5), 2926-2939, 2012.

[2] Bortolini, M., Gamberi, M., & Graziani, A., Technical and economic design of photovoltaic and battery energy storage system. Energy Conversion and Management, 86, 81-92, 2014.

[3] Daud, A. K., & Ismail, M. S., Design of isolated hybrid systems minimizing costs and pollutant emissions. Renewable energy, 44, 215-224, 2012.
[4] Ipsakis, D., Voutetakis, S., Seferlis, P., Stergiopoulos, F., Papadopoulou, S., & Elmasides, C., The effect of the hysteresis band on power management strategies in a stand-alone power system. Energy, 33(10), 1537-1550, 2008.
[5] Liao, Z., & Ruan, X., A novel power management control strategy for stand-alone photovoltaic power system. In Power Electronics and Motion Control Conference, 2009. IPEMC'09. IEEE 6th International Power Electronics and Motion Control Conference, 445-449, 2009.
[6] Ismail, M. S., Moghavvemi, M., & Mahlia, T. M. I., Design of an optimized photovoltaic and microturbine hybrid power system for a remote small community: case study of Palestine. Energy Conversion and Management, 75, 271-281, 2013.
[7] Dahmane, M., Bosche, J., El-Hajjaji, A., & Dafarivar, M., Renewable energy management algorithm for stand-alone system, 2013 International Conference on In Renewable Energy Research and Applications (ICRERA), 621-626, 2013.
[8] Abedi, S., Alimardani, A., Gharehpetian, G. B., Riahy, G. H., & Hosseinian, S. H., A comprehensive method for optimal power management and design of hybrid RES-based autonomous energy systems. Renewable and Sustainable Energy Reviews, 16(3), 1577-1587, 2012.
[9] Lin, C., Mu, H., Xiong, R., & Shen, W., A novel multi-model probability battery state of charge estimation approach for electric vehicles using H-infinity algorithm. Applied Energy, 166, 76–83, 2016.
[10] Athanasius, G. X., & Zhu, J. G., Design of robust controller for wind turbines. In 2009 Second International Conference on Emerging Trends in Engineering & Technology, 7-12, 2009.
[11] Cao, Y. Y., & Frank, P. M., Robust H∞ disturbance attenuation for a class of uncertain discrete-time fuzzy systems. IEEE Transactions on Fuzzy Systems, 8(4), 406-415, 2000.
[12] Kalantar, M., Dynamic behavior of a stand-alone hybrid power generation system of wind turbine, microturbine, solar array and battery storage. Applied energy, 87(10), 3051-3064, 2010.
[13] Bianchi, F. D., De Battista, H., & Mantz, R. J., Robust Multivariable Gain-Scheduled Control of Wind Turbines for Variable Power Production. International Journal of Systems Control, 1(3), 103-112, 2010.
[14] Korba, P., Babuska, R., Verbruggen, H. B., & Frank, P. M., Fuzzy gain scheduling: controller and observer design based on Lyapunov method and convex optimization. IEEE transactions on fuzzy systems, 11(3), 285-298, 2003.
[15] Kamal, E., Koutb, M., Sobaih, A. A., & Kaddah, S., Maximum power control of hybrid wind-diesel-storage system. Advances in Fuzzy Systems, 8(1), 1-8, 2008.
[16] Kamal, E., Koutb, M., Sobaih, A. A., & Abozalam, B. An intelligent maximum power extraction algorithm for hybrid wind–diesel-storage system. International journal of electrical power & energy systems, 32 (3), 170-177, 2010.
[17] Tanaka, K., Ikeda, T., & Wang, H. O., Robust stabilization of a class of uncertain nonlinear systems via fuzzy control: quadratic stabilizability, H∞ control theory, and linear matrix inequalities. IEEE Transactions on Fuzzy systems, 4(1), 1-13, 1996.
[18] Kim, I. S., Kim, M. B., & Youn, M. J., New maximum power point tracker using sliding-mode observer for estimation of solar array current in the grid-connected photovoltaic system. IEEE Transactions on Industrial Electronics, 53(4), 1027-1035, 2006.
[19] Kamal, E., Aitouche, A., Ghorbani, R., & Bayart, M., Robust fuzzy fault-tolerant control of wind energy conversion systems subject to sensor faults. IEEE Transactions on Sustainable Energy, 3(2), 231-241, 2012.
[20] Sepasi, S., Ghorbani, R., & Liaw, B. Y., Improved extended Kalman filter for state of charge estimation of battery pack. Journal of Power Sources, 255, 368-376, 2014.
[21] Xu, X., Li, K., Qi, F., Jia, H., & Deng, J., Identification of microturbine model for long-term dynamic analysis of distribution networks. Applied Energy, 2016.
Balamurugan, S., Janarthanan, N., & Chandrakala, K. V., Small and large signal modeling of heavy duty gas turbine plant for load frequency control. International Journal of Electrical Power & Energy Systems, 79, 84-88, 2016.

IOP Conf. Series: Journal of Physics: Conf. Series **783** (2017) 012036 doi:10.1088/1742-6596/783/1/012036

Locally optimal control under unknown dynamics with learnt cost function: application to industrial robot positioning

Joris Guérin, Olivier Gibaru, Stéphane Thiery and Eric Nyiri

Arts et Métiers ParisTech, 8 Boulervard Louis XIV, 59800, Lille, FRANCE

E-mail: `firstname.lastname@ensam.eu`

Abstract.
Recent methods of Reinforcement Learning have enabled to solve difficult, high dimensional, robotic tasks under unknown dynamics using iterative Linear Quadratic Gaussian control theory. These algorithms are based on building a local time-varying linear model of the dynamics from data gathered through interaction with the environment. In such tasks, the cost function is often expressed directly in terms of the state and control variables so that it can be locally quadratized to run the algorithm. If the cost is expressed in terms of other variables, a model is required to compute the cost function from the variables manipulated. We propose a method to learn the cost function directly from the data, in the same way as for the dynamics. This way, the cost function can be defined in terms of any measurable quantity and thus can be chosen more appropriately for the task to be carried out. With our method, any sensor information can be used to design the cost function. We demonstrate the efficiency of this method through simulating, with the V-REP software, the learning of a Cartesian positioning task on several industrial robots with different characteristics. The robots are controlled in joint space and no model is provided a priori. Our results are compared with another model free technique, consisting in writing the cost function as a state variable.

1. Introduction

Optimal feedback control theory provides an efficient framework for robot manipulators movement generation as it enables to compute both an optimal open-loop trajectory and a feedback controller at once. To carry out such tasks, classical optimal control uses a model of the dynamics and selects the solution that minimizes a properly designed cost function. Under linear dynamics and quadratic cost, an optimal solution can be found analytically thanks to the widely studied theory of Linear-Quadratic-Regulators (LQR) [1]. When the dynamics are more complex and non-linear, the problem becomes more difficult. However, iterative Linear-Quadratic-Regulator algorithm (iLQR) [2], [3] still enables to converge towards a locally optimal solution by iteratively fitting local linear approximations to the dynamics.

On the other hand, autonomous learning of manipulation skills have received much interest over the past decades. Thanks to different Reinforcement Learning (RL) techniques [4], various problems involving robot manipulators have been explored and successfully applied [5], [6], [7]. Among RL techniques, policy search methods seem to be the most appropriate for high dimensional robotics control problems [8]. Recently, several researchers have proposed methods to link these two fields by using iLQR to compute control laws under unknown dynamics [9], [10]. This approach is interesting as it removes the need to model the dynamics, which can be difficult for complex systems and environments.

Besides the dynamics, an optimal control problem is also defined by its cost function. In order to find a good trajectory, we need to design a proper cost function in relation to the task to be carried out. Nevertheless, in most applications, we acknowledge that the cost function is quadratic and expressed explicitly in terms of the state and control variables. Such form of the cost function eases the implementation of the iLQR algorithm and appears to suit many applications quite well. However, in some cases, it can be useful to formulate the cost function in terms of variables that are not explicitly the state and control vectors. For instance, [11] proposes a cost function that is defined in the Cartesian space whereas the robot is controlled in the joint space. In such cases, we need to have geometric and kinematic models of the system (e.g., for robot manipulator, a direct Denavit-Hartenberg (DH) model [12]) to be able to express analytically the quadratic expansion of the cost function, which is necessary to run iLQR. In robotics, the need for a DH-model can be problematic as it requires a lot of efforts to be calibrated precisely [13], [14]. Moreover, the calibration is very specific to a certain robot [15], [16]. It is literally impossible to derive a model taking into account every dynamic phenomena. For example, if the payload on the robot is changed, the model calibration might not be valid anymore as strains might appear. When it comes to autonomous learning, it does not seem appropriate to have to redesign a model after any change in the system.

In this paper, we propose a method to avoid such problem by learning the quadratic approximation of the cost function in the same way as for the dynamics. Thanks to this method, the cost function can be defined from any measurable quantity, which allows to have a more appropriate cost in some cases. Moreover, the cost function can be defined in a more intuitive way and robot programming can become more accessible to non-specialists. The method is validated through simulation using the V-REP software [17]. Robots with very different specifications (number of joints, length of links, ...) are trained to learn a simple Cartesian positioning task using only position sensors information and no DH-model of any robot. The cost function is the distance measured between the end effector and the target Cartesian position while the robot is controlled in joint space. We also compare our results with another method consisting in including the Cartesian distance in the state vector, thus making the cost a function of the dynamics.

The paper is organized as follows. The derivation of the iLQR algorithm is presented in Section 2 and a new first order method to update the controller is proposed. Section 3 introduces a quadratic regression method for the cost function estimation. Simulation validation of the method is presented in Section 4 and Section 6 proposes a discussion on the results and future work.

2. TRAJECTORY OPTIMIZATION USING ILQR

iLQR is a trajectory optimization method to compute iteratively a locally optimal linear-Gaussian controller. In this section, the algorithm is derived (following [3]) in order to ease the understanding of the main contribution in Section 3. In this paper, no noise is included in the equations, yet it does not require much changes to take noise into account.

2.1. Definitions and overview

To understand the concept of iLQR, a few terms need to be introduced.

A trajectory of length T is defined by the set

$$\{x_0, u_0, l_0, x_1, u_1, l_1, ..., u_{T-1}, l_{T-1}, x_T\},$$

where x_t and u_t represent respectively state and control vectors, and l_t represents the value of the cost function. For example, for serial robot control, x can be joint positions and u target angular positions for next step. Such a state vector is easy to obtain with industrial robots and thus it makes sense to consider full-state feedback in the remaining of this section.

The cost and dynamics functions which map actual state and action to cost and new state are represented by

$$l_t = L_t(x_t, u_t), \tag{1}$$

$$x_{t+1} = F_t(x_t, u_t). \tag{2}$$

In the general framework of iLQR, L_t and F_t are complex non-linear functions. In Section 3, it is shown that they can even be unknown functions which values are only measured.

The deterministic controller to be learned is designated by Π. This controller is in the special form of a time-varying linear controller

$$u_t = \Pi(x_t) = K_t x_t + k_t. \tag{3}$$

We underline that the controller Π depends on time, however, the t index is dropped for better clarity.

The basic principle of iLQR is the following: from a nominal trajectory, denoted by

$$\{\bar{x}_0, \bar{u}_0, \bar{l}_0, \ldots, \bar{x}_T\},$$

a new controller, better than the previous one, is derived. From this new controller, a new nominal trajectory is computed. This process is repeated until a satisfactory controller is obtained. In most application, this iterative process turns out to converge rather rapidly towards a locally optimal trajectory.

2.2. Cost-to-go computation

From a nominal trajectory, the goal is to find a new, better controller. To do so, the first step is to compute local approximations of both (1) and (2) around the nominal trajectory. The dynamics are approximated linearly and the cost function quadratically as follows:

$$F_t(\bar{x}_t + \delta x_t, \bar{u}_t + \delta u_t) = \bar{x}_{t+1} + F_{x_t}\delta x_t + F_{u_t}\delta u_t, \tag{4}$$

$$\begin{aligned} L_t(\bar{x}_t + \delta x_t, \bar{u}_t + \delta u_t) &= \bar{l}_t + L_{x_t}\delta x_t + \frac{1}{2}\delta x_t^T L_{x,x_t}\delta x_t \\ &+ L_{u_t}\delta u_t + \frac{1}{2}\delta u_t^T L_{u,u_t}\delta u_t + \delta x_t^T L_{x,u_t}\delta u_t, \end{aligned} \tag{5}$$

where δx_t and δu_t represent variations from the nominal trajectory. Capital letters indexed by one (respectively two) letters represent sub-vectors (respectively sub-matrices) of the appropriate Jacobian (respectively Hessian) matrices.

These two local approximations are the key to all the derivation that follows. One way to compute these approximations is explored in Section 3.

For the sake of clarity, we also define an intermediate controller that is created during each improvement pass:

$$\delta u_t = \pi(\delta x_t) = P_t \delta x_t + p_t. \tag{6}$$

This controller enables to compute improvements on the time-varying linear controller Π. The P_t term improves the controller response to state variations and the p_t term is a shift in the nominal trajectory.

Finally, The two following value functions are defined:

$$Q_t^{\Pi}(x_t, u_t) = \sum_{j=t}^{T}(l_t), \tag{7}$$

$$V_t^{\Pi}(x_t) = \sum_{j=t}^{T}(l_t). \tag{8}$$

They represent the cost-to-go until the end of the trajectory if following Π. In the case of stochastic controllers, we would consider the expected values of the sums. These functions depend on Π, Q is defined with respect to x and u whereas V depends only on x. For clarity purposes, we drop the Π superscripts of Q and V in the remainder of this paper.

In the context of iLQR, quadratic Taylor expansions of both value functions are required to compute an improved controller:

$$\begin{aligned} Q_t(\bar{x}_t + \delta x_t, \bar{u}_t + \delta u_t) &= Q_{0_t} + Q_{x_t}\delta x_t + \frac{1}{2}\delta x_t^T Q_{x,x_t}\delta x_t \\ &+ Q_{u_t}\delta u_t + \frac{1}{2}\delta u_t^T Q_{u,u_t}\delta u_t + \delta x_t^T Q_{x,u_t}\delta u_t, \end{aligned} \tag{9}$$

$$V_t(\bar{x}_t + \delta x_t) = V_{0_t} + V_{x_t}\delta x_t + \frac{1}{2}\delta x_t^T V_{x,x_t}\delta x_t. \tag{10}$$

These quadratic approximations must be computed at each time step. This is done using different formulations for Q and V, widely used in RL:

$$Q_t(x_t, u_t) = L_t(x_t, u_t) + V_{t+1}(x_{t+1}), \tag{11}$$

$$V_t(x_t) = Q_t(x_t, \Pi(x_t)). \tag{12}$$

This point being reached, Q_t can be expressed from V_{t+1}. By plugging (4), (5) and (10) into (11), the coefficients of (9) are easily obtained:

$$\begin{aligned} Q_{0_t} &= \bar{l}_t + V_{0_{t+1}} + V_{x_{t+1}}\bar{x}_{t+1} + \frac{1}{2}\bar{x}_{t+1}^T V_{x,x_{t+1}}\bar{x}_{t+1}, \\ Q_{x_t} &= L_{x_t} + V_{x_{t+1}}F_{x_t} + \bar{x}_{t+1}^T V_{x,x_{t+1}}F_{x_t}, \\ Q_{x,x_t} &= L_{x,x_t} + F_{x_t}^T V_{x,x_{t+1}}F_{x_t}, \\ Q_{u_t} &= L_{u_t} + V_{x_{t+1}}F_{u_t} + \bar{x}_{t+1}^T V_{x,x_{t+1}}F_{u_t}, \\ Q_{u,u_t} &= L_{u,u_t} + F_{u_t}^T V_{x,x_{t+1}}F_{u_t}, \\ Q_{x,u_t} &= L_{x,u_t} + F_{x_t}^T V_{x,x_{t+1}}F_{u_t}. \end{aligned} \tag{13}$$

Then, we get (10) using (12) and (6) as follows,

$$\begin{aligned} V_{0_t} &= Q_{0_t} + Q_{u_t}p_t + \frac{1}{2}p_t^T Q_{u,u_t}p_t, \\ V_{x_t} &= Q_{x_t} + Q_{u_t}P_t + p_t^T Q_{u,u_t}P_t + p_t^T Q_{x,u_t}^T, \\ V_{x,x_t} &= Q_{x,x_t} + P_t^T Q_{u,u_t}P_t + Q_{x,u_t}P_t. \end{aligned} \tag{14}$$

Once all the value functions have been computed backward in time (starting from $V_T = 0$), the idea is to find the controller with the lowest cost-to-go at each time step.

IOP Conf. Series: Journal of Physics: Conf. Series **783** (2017) 012036 doi:10.1088/1742-6596/783/1/012036

2.3. Update the controller

In order to improve the controller, δu_t that minimizes Q_t needs to be computed at each time step t. Hence, we start by isolating the terms of Q_t depending on δu_t, which gives the following function to minimize,

$$f(\delta x_t, \delta u_t) = (Q_{u_t} + \delta x_t^T Q_{x,u_t})\delta u_t + \frac{1}{2}\delta u_t^T Q_{u,u_t}\delta u_t. \tag{15}$$

Then, we derive this function with respect to δu_t:

$$\frac{\partial f}{\partial \delta u_t}(.) = Q_{u_t} + \delta x_t^T Q_{x,u_t} + \delta u_t^T Q_{u,u_t}. \tag{16}$$

Note that this partial derivative is in the Jacobian form (row wise).

To optimize the controller, several methods (first and second order) have been applied [3]. In this paper, it has been chosen to implement gradient descent, which is more intuitive, in order to ease the characterization of the method (see Section 3). Thus, steps towards better controllers are made in the opposite direction of the gradient.

$$\nabla_{\delta u_t} f(\delta x_t, 0) = Q_{u_t}^T + Q_{x,u_t}^T \delta x_t. \tag{17}$$

That is to say that given the same input δx_t, the new chosen δu_t should be

$$\delta u_t = -\epsilon(Q_{u_t}^T + Q_{x,u_t}^T \delta x_t). \tag{18}$$

Obviously, the model being only valid locally, the step ϵ should be chosen small enough not to violate a "proximity to nominal" constraint.

In practice, we have implemented a new method, similar to super-SAB [18]. The variance for exploration (see Section 3) is modified according to the evolution of the partial derivatives signs, and ϵ is tuned to stay inside the variance boundaries. In this way, as we get closer to the optimal solution, the exploration gets tighter and the improvements are more accurate. This controller improvement method appears to converge within a reasonable number of steps for positioning task (Section 4).

From (18) and (6), we identify the temporary controller parameters

$$P_t = -\epsilon Q_{x,u_t}^T, \tag{19}$$

$$p_t = -\epsilon Q_{u_t}^T. \tag{20}$$

From this temporary controller, the global controller can be updated as follows

$$\left.\begin{aligned} u_t &= \hat{u}_t + \delta u_t \\ x_t &= \bar{x}_t + \delta x_t \\ \hat{u}_t &= \hat{K}_t x_t + \hat{k}_t \end{aligned}\right\} u_t = \hat{K}_t x_t + \hat{k}_t + P_t(x_t - \bar{x}_t) + p_t$$

where $\hat{u}_t$ represents the action chosen under the current controller, represented by $\hat{K}_t$ and $\hat{k}_t$.

Finally, the global controller is modified

$$K_t = \hat{K}_t + P_t, \tag{21}$$

$$k_t = \hat{k}_t + p_t - P_t\bar{x}_t. \tag{22}$$

To obtain the new average trajectory, this new controller just need to be applied to the initial state.

3. COMPUTE COST AND DYNAMICS FROM REAL-WORLD INTERACTION

Several authors have already proposed regression methods to compute the dynamics [9], [10]. The idea is to replace the current controller by a stochastic Gaussian controller of same mean and a with certain covariance matrix Σ_t:

$$P(u_t|x_t) = \mathcal{N}(K_t x_t + k_t, \Sigma_t),$$

where $\mathcal{N}$ represents a normal distribution. Then, we generate several trajectories on the system using this controller and record all the x_t's and u_t's. In this way, the environment is explored around the nominal trajectory and (4) can be computed in this area, using linear regression.

In this section, the focus is on a method to compute the cost function, expressed in the output space, from experience. To our knowledge, such approach of combining cost function regression and iLQR has never been proposed yet.

3.1. Usefulness of the method

As shown in Section 2, once the dynamics have been linearized and the cost quadratized, improving the controller is just a few calculations away. In the original paper for iLQR [3], equations (1) and (2) are considered to be known and their partial derivatives computable. Hence, both Taylor expansions are computed analytically. Later on, it was proposed to compute the linear dynamics through regression on observed values [10], [9]. Thus, the need to have a representation of the environment, which can be tricky, is avoided.

However, the cost function still needs to be expressed directly in terms of the state and action variables. If we want to design a cost function from sensor measurements, it might not be straight forward to obtain the partial derivatives with respect to the state and control vectors. A good example is robot positioning task, where the cost function is a Cartesian distance and the variables controlled are joint positions (Section 4). In such case, a geometric model of the robot [12] is required to express the cost with respect to the state variables. Such model can be hard to obtain with enough precision (e.g. humanoid robots with a lot of degrees of freedom). Indeed, building a precise model of a robot manipulator requires a lot of work [13]. Moreover, the need for a model of the robot reduces the generality of the algorithm and its precision is dependant on the model, which varies with the robot used, its loading, ...

One possibility to avoid modelling the robot is to increase the state vector with a measurement directly proportional to the cost function. For example, the Cartesian distance for the positioning task in Section 4, can be considered to be part of the state. Doing so, the cost function can be expressed directly from the state and the dynamics, thus, the derivation can be done. This trick enables to define the cost function from any measurable quantity.

The method detailed in the following section provides a learning technique to compute the cost function. This method uses data to directly compute the quadratic approximation from regression.

Figure 1, illustrates all three methods in the form of block diagrams. This helps to visualize that the first method (Figure 1.a.) relies on a geometric model of the robot, which we do not want for the reasons explained above. In the second method (Figure 1.b.), the cost computation is dependent on a first order regression whereas in the proposed method (Figure 1.c.) the cost function is directly approximated quadratically. These two methods efficiencies are compared in Section 4.

3.2. Technical explanations

The technique described here uses the same idea as it was done for the dynamics. However, the model is no more linear but quadratic. For this reason, we need to use polynomial regression

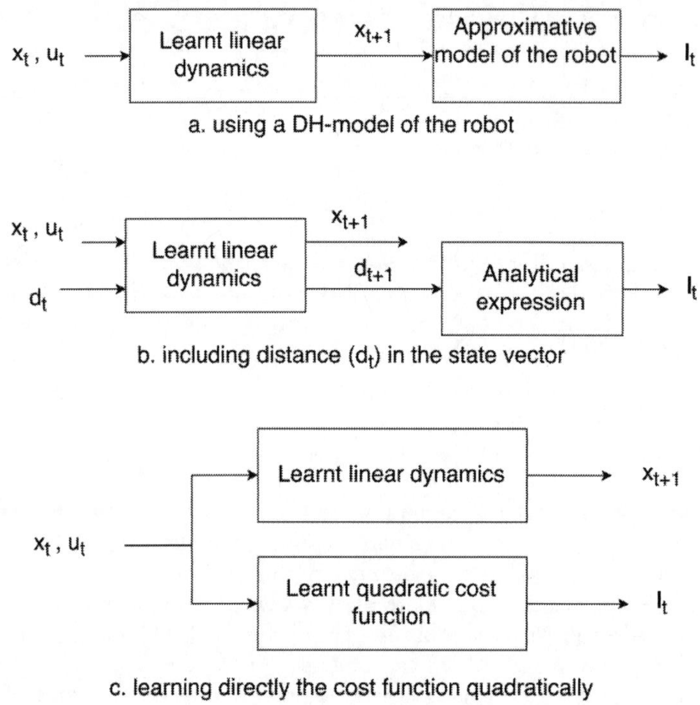

Figure 1. Different ways to compute a Cartesian distance cost function (l_t) from angular state (x_t) and control (u_t) vectors.

[19]. The trick for quadratic regression consists in noticing that a second order polynomial is a linear combination of the squared coefficients. For example, in the two dimensional case, a quadratic regression of

$$f : \mathbb{R}^2 \to \mathbb{R},$$
$$z = f(x, y),$$

is in fact a simple multivariate linear regression with input data of the form

$$v = [1, x, x^2, y, y^2, xy]^T,$$

which provides the coefficients of

$$z = a_0 + a_1x + a_2x^2 + a_3y + a_4y^2 + a_5xy.$$

In the general case, a quadratic expansion of the cost with respect to δx_t and δu_t(respectively n and m entries) is obtained with a training set in the form

$$v = [1, \underbrace{\delta x_1, \delta x_2, ..., \delta x_n}_{n}, \underbrace{\delta x_1^2, \delta x_1 \delta x_2, ..., \delta x_n^2}_{\frac{n(n+1)}{2}}, \underbrace{\delta u_1, ..., \delta u_n}_{m}, \underbrace{\delta u_1^2, ..., \delta u_n^2}_{\frac{m(m+1)}{2}}, \underbrace{\delta x_1 \delta u_1, ..., \delta x_n \delta u_n}_{n*m}]^T.$$

The only drawback is that the new regression is of higher dimension than the one for dynamics, which involves a higher number of samples to be an accurate approximation. However, all the samples found during the dynamics exploration phase work just as well and do not have to

IOP Conf. Series: Journal of Physics: Conf. Series **783** (2017) 012036 doi:10.1088/1742-6596/783/1/012036

be re-explored. All that is required is to record the cost at the same time than the new state observed. Moreover, the controller is learned off-line and even if the learning time is longer, the on-line speed of the controller is not impacted.

Once this regression is done, an easy reshaping of the linear coefficients found provides the matrix coefficients of

$$\begin{aligned} L_t(\bar{x}_t + \delta x_t, \bar{u}_t + \delta u_t) &= \bar{l}_t + L_{x_t}\delta x_t + \frac{1}{2}\delta x_t^T L_{x,x_t}\delta x_t \\ &+ L_{u_t}\delta u_t + \frac{1}{2}\delta u_t^T L_{u,u_t}\delta u_t + \delta x_t^T L_{x,u_t}\delta u_t \end{aligned}$$

which can be used directly for an iLQR update.

4. SIMULATION RESULTS

4.1. Problem definition

In order to show the effectiveness of the method, a robot manipulator positioning task has been implemented using the V-REP simulation software. The robot is controlled in angular position, the state vector is the vector of joint coordinates and the control vector is the desired angular positions for next step. The goal is to reach a target Cartesian point with the end-effector of the robot. The cost function is expressed in the most natural way possible: it is the Cartesian distance between the connector and the target. On Figures 2 and 3, the end-effector is represented by the black connectors at the extremity of the robots. The objective is displayed with the red floating sphere (the target is actually the center of the sphere). The distance (cost function) is extracted directly from the software. In real physical applications, such a cost can be provided by any distance measurement sensor (e.g. laser tracker).

We have chosen this demonstration problem because results can be interpreted intuitively. Distances can be compared easily and are directly related to precision. Such an application, even if not really challenging in robotics, is a good start to evaluate the efficiency of the proposed cost regression method before implementing it for more complex robotics tasks.

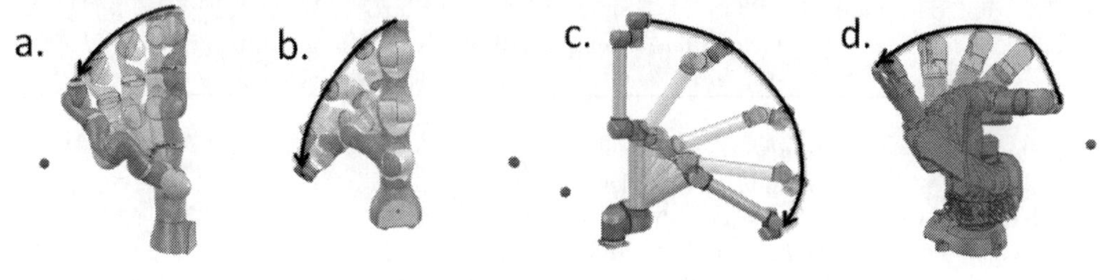

Figure 2. Initial, randomly chosen, nominal trajectory for different industrial robots. a. KUKA LBR iiwa / b. F&P P-ARM / c. UR 10 / d. ABB IRB 140

To solve such a control problem with classical methods, a geometric model of the robot (e.g. DH-model), giving the end-effector position with respect to joint coordinates needs to be built. Indeed, to be able to get the quadratic approximation of the cost, we need to map analytically the joint space to the Cartesian space in order to differentiate it. Obviously, no such model has been derived for the simulations and, using V-REP, we implemented the algorithm on four industrial robots with different specifications (number of links, lengths of links, ...) to illustrate that the method is model-free (Figure 2 and 3). Since the method does not need to explicit a

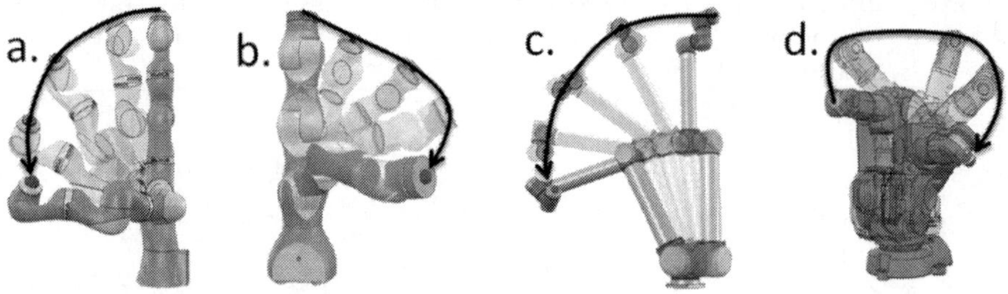

Figure 3. Trajectory learned for positioning task for different industrial robots. a. KUKA LBR iiwa / b. F&P P-ARM / c. UR 10 / d. ABB IRB 140

model, it can be applied directly to any robot. Thanks to this method, we get rid of modelling imprecision and only sensor precision is involved in the learning process.

It is also worth underlining that the task would not have been different if adding orientation constraint to the end-effector. All that is needed is to take distances between three points instead of only one.

4.2. Trajectory found on different robots
The nominal trajectory used to initialize the iLQR algorithm was chosen randomly. We chose to set all desired joint positions to 60° and explore around this initial command. Figure 2 represents the initial nominal trajectory, the distance between the end-effector and the target is 490 mm. Note that the angular point-to-point (PTP) movement between two joints configurations is considered as a black box, part of the V-REP environment. In this way, the dynamics can really be considered unknown.

Figure 3 represents a set of final trajectories found for the positioning task. The algorithm stopped running when a precision of 0.1 mm was reached. As it can be seen, iLQR with no model of the robot and Cartesian cost function succeeded to reach its objective point with various robots. Indeed, they have different lengths, different numbers of links, but the algorithm can still converge towards a solution without being restricted on the cost function definition.

4.3. Number of samples needed
Figure 4 shows how the number of samples influences the rate of convergence on one specific robot (KUKA LBR iiwa). The y-axis scale is too small to see the distances for the first iterations, but this choice is necessary to be able to differentiate curves in the final stages.

The first thing we notice is that for a sample count above 40, the solution does not evolve anymore, which means that 40 samples is enough to characterize with high precision the local cost function. In simulation, we found that for the positioning task, it is possible to find a trajectory with a precision greater than 0.1 mm if we do the off-line learning with at least 40 samples during 20 iterations.

On the other hand, if the number of samples gathered is too small (e.g. 5), the cost and dynamics will not be properly approximated and the algorithm might not converge towards the desired solution, as shown in Figure 4.

4.4. Comparison with state modification method
As mentioned in Section 3, another method to avoid robot modelling consist in modifying the state vector and include the Cartesian distance in it. This technique corresponds to diagram

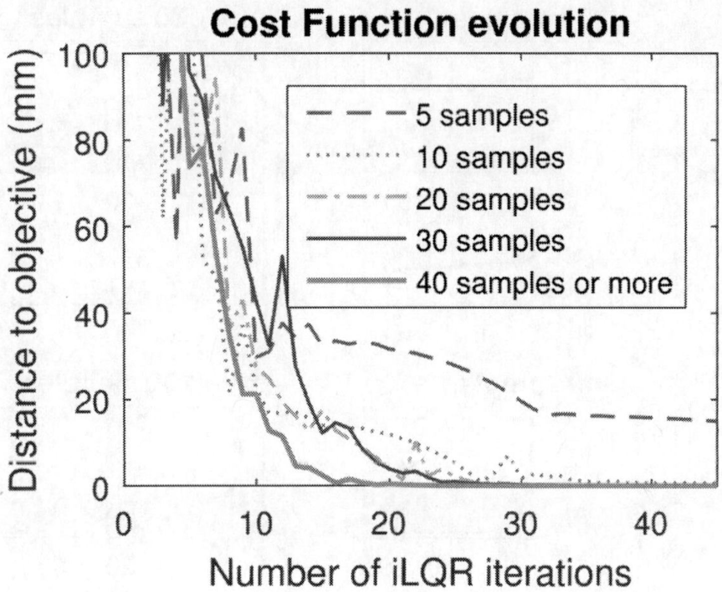

Figure 4. Distance to target point with respect to the number of iLQR passes. Different curves represent different number of samples generated for learning the cost and dynamics. Initial distance was around 650 mm

b. of Figure 1. In this section, our new quadratic regression method is compared with this relatively simple technique. Figure 5 illustrates the learning curves for both methods under different learning sample sizes.

We acknowledge that state modification is less stable and robust. Indeed, in this simulation, the results using 30 samples were better than the ones with 40, which did not happen using quadratic approximation. The state modification method is more sensitive to the samples generated whereas quadratic regression is behaving more regularly for a given number of samples. The state modification does not exhibit the nice property of being independent of the number of samples when this number becomes high (> 40).

On the other hand, with very small number of samples (e.g. 5), it is better to stick to first order approximation. Indeed, as shown on the top left plot of Figure 5, state modification works better for small number of samples because quadratic regression is not accurate enough.

Finally, the quadratic regression appears to be more efficient in the final stages of the algorithm, which might come from the fact that a better model of the cost is required to reach high precision in the positioning.

5. Conclusion

In order to democratize robotics in small companies, there is a need for self-programming robots. An unqualified worker should be able to program a robot only by specifying the task to accomplish. iLQR under unknown dynamics seems a good approach for learning local policies in a short amount of time and has already been used to solve robotics problem by several authors [9],[10]. However, such a learning algorithm should be independent of the robot model, tool and calibration so that it can be applied by any worker. This limitation eliminates approaches based on the geometric model of the robot [11]. The only real model-free method was proposed in [10], and was compared to our polynomial regression learning of the cost function approach.

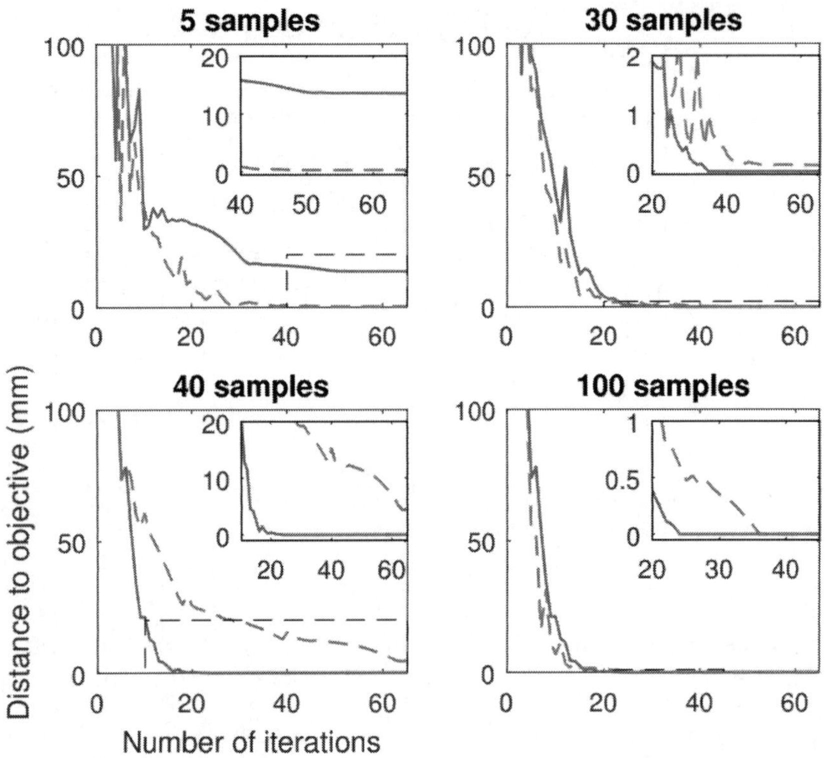

Figure 5. Comparisons of two methods: Quadratic regression (plain blue lines) and modified state (dotted orange lines). Right-upper boxes on each figures are zoom on the final stages of iLQR. Initial distance was around 650 mm

Results obtained on simulation show that our approach converges faster towards high precision behaviors and is more stable and robust if the exploration phase gathers enough samples. Hence, for industrial application, we recommend to learn the quadratic cost function directly from data.

6. Discussion

6.1. Sample count reduction

Despite the number of samples needed can be considered high in the simulation validation, we would like to emphasis that there exists tools, that have already been used in other work [11], [10], to reduce sample count and optimize regressions. In this paper, the idea was to illustrate that it is possible to draw an optimal control problem with a cost function being any measurable quantity, without any geometric model of the system. We did not implement any algorithm optimization, which will be done in future work. Moreover, the learning time does not impact the speed of the final controller as the learning process is realized off-line.

6.2. Link with Policy Search

The process of learning Linear Gaussian Controllers (LGC) is a necessary step for Guided Policy Search (GPS) algorithm [10], [11]. GPS is a very promising method that has proven its ability to succeed for complex RL tasks, particularly in robotics. Indeed, it scales gracefully to high dimensional state spaces and can learn highly general policies for complex tasks thanks to guiding trajectories in the form of LGC. Thus, any improvement on the trajectory optimization methods to learn such controllers benefits directly to the global method. Implementing our method in the framework of GPS is also an axis of future research for our team.

6.3. Handling contact

As explained earlier, we have chosen to control the robot in its angular space in order to ease the results interpretation and the comparison with other methods. Such high level controller has the disadvantage of being less flexible then lower level ones. For example, pure angular position control is not able to deal with contact because there are missing information for feedback. Our plan in the close future being to deal with both grasping and insertion tasks, we need to start including force and torque information in both the control and state vectors. Such future needs explain the use of the KUKA LBR iiwa robot, which is equipped with high accuracy torque sensors in each joint.

References

[1] Lewis F L, Vrabie D and Syrmos V L 2012 *Optimal Control* (John Wiley & Sons)
[2] Li W and Todorov E 2004 Iterative linear quadratic regulator design for nonlinear biological movement systems *Proceedings of the First International Conference on Informatics in Control, Automation and Robotics* pp 222–229
[3] Todorov E and Li W 2005 A generalized iterative lqg method for locally-optimal feedback control of constrained nonlinear stochastic systems *American Control Conference, 2005. Proceedings of the 2005* pp 300–306 vol. 1
[4] Kober J and Peters J 2012 Reinforcement learning in robotics: A survey *Reinforcement Learning* (Springer) pp 579–610
[5] Lin C K 2009 H reinforcement learning control of robot manipulators using fuzzy wavelet networks *Fuzzy Sets and Systems* **160** 1765–1786
[6] Deisenroth M P, Rasmussen C E and Fox D 2011 Learning to control a low-cost manipulator using data-efficient reinforcement learning *MIT Press*
[7] Park J J, Kim J H and Song J B 2007 Path planning for a robot manipulator based on probabilistic roadmap and reinforcement learning *International Journal of Control Automation and Systems* **5** 674
[8] Deisenroth M P, Neumann G, Peters J *et al.* 2013 A survey on policy search for robotics. *Foundations and Trends in Robotics* **2** 1–142
[9] Mitrovic D, Klanke S and Vijayakumar S 2010 Adaptive optimal feedback control with learned internal dynamics models *From Motor Learning to Interaction Learning in Robots* (Springer) pp 65–84
[10] Levine S and Abbeel P 2014 Learning neural network policies with guided policy search under unknown dynamics *Neural Information Processing Systems (NIPS)*
[11] Levine S, Wagener N and Abbeel P 2015 Learning contact-rich manipulation skills with guided policy search *CoRR* **abs/1501.05611**
[12] Dombre E and Khalil W 2013 *Robot manipulators: modeling, performance analysis and control* (John Wiley & Sons)
[13] Majarena A C, Santolaria J, Samper D and Aguilar J J 2010 An overview of kinematic and calibration models using internal/external sensors or constraints to improve the behavior of spatial parallel mechanisms *Sensors* **10** 10256–10297
[14] Elatta A, Gen L P, Zhi F L, Daoyuan Y and Fei L 2004 An overview of robot calibration *Information Technology Journal* **3** 74–78
[15] Nubiola A and Bonev I A 2013 Absolute calibration of an abb irb 1600 robot using a laser tracker *Robotics and Computer-Integrated Manufacturing* **29** 236–245
[16] Jubien A, Gautier M and Janot A 2014 Dynamic identification of the kuka lwr robot using motor torques and joint torque sensors data *World Congress of the International Federation of Automatic Control (IFAC)* pp 1–6
[17] E Rohmer S P N Singh M F 2013 V-rep: a versatile and scalable robot simulation framework *Proc. of The International Conference on Intelligent Robots and Systems (IROS)*
[18] Allard R and Faubert J 2004 Neural networks: different problems require different learning rate adaptive methods *Electronic Imaging 2004* vol 5298 pp 516–527
[19] De Brabanter K, De Brabanter J, De Moor B and Gijbels I 2013 Derivative estimation with local polynomial fitting *The Journal of Machine Learning Research* **14** 281–301

IOP Conf. Series: Journal of Physics: Conf. Series **783** (2017) 012037 doi:10.1088/1742-6596/783/1/012037

Enhanced approach to PD control design for linear time-invariant descriptor systems

Anna Filasová and Dušan Krokavec

Department of Cybernetics and Artificial Intelligence, Faculty of Electrical Engineering and Informatics, Technical University of Košice, Letná 9, 042 00 Košice, Slovakia

E-mail: `anna.filasova@tuke.sk; dusan.krokavec@tuke.sk`

Abstract. Enhanced approaches to PD controller design, adjusted for linear time-invariant descriptor systems, are proposed in the paper. Presented in the sense of the second Lyapunov method, an associated structure of linear matrix inequalities is outlined to possess the regular closed-loop system dynamic properties. A simulation example, subject to the state and output PD control, demonstrates the effectiveness of the proposed form of the design technique.

1. Introduction

Descriptor models are frequently used for modeling of industrial systems such as power system, rigid body mechanisms or chemical processes (see, e.g., [6], [7], [10], [18] and the references therein). The need for application-oriented simulation technique for such systems focused the research activities at first on the conditions of system regularity and observability [1], [3], which pave the basic way for controller and observer design algorithms, relying on the feasibility of a system of linear matrix inequalities (LMI) [12], [15], [16]. In the descriptor system context, novel bounded real lemmas for discrete and continuous-time descriptor systems are adressed in [2], [11], respectively, an extension of the positive real lemma to descriptor systems is given in [4], system output signal contraints are typically considered in the LMI synthesis of H_2 and mixed H_2/H_∞ controllers for singular systems in [19] and the state variables of a singular system can be bounded solving singular LQ problem for descriptor systems by the way presented in [20]. From the stand point of these principles, the main problem is the canonical form, or SVD coordinate form, of the singular system description to be provided the required structural information within design [9], [14]. It seems, for singular systems, to be very attractive in controller and observer design to explore the potential of the Ljapunov matrix parametrization principle [5], [17], to improve the system performances and to simplify the design conditions.

Adapting the idea presented in [17], two methods for proportional-derivative controller (PD) design are proposed in the paper. Following examination of the PD state and output schemes, enhanced algorithms using the Ljapunov matrix parametrization approach are provided. Applied enhanced conditions in the control design develop a general framework for PD control structures for continuous-time singular systems. The approach utilizes the measurable state or output vector variables, the design conditions are based on linear matrix inequality (LMI) technique (applicable conditiond by one matrix equality), which give an effective way to calculate the PD control law parameters.

IOP Conf. Series: Journal of Physics: Conf. Series **783** (2017) 012037 doi:10.1088/1742-6596/783/1/012037

The paper is organized as follows. Ensuing the introduction given in Sec. 1, Sec. 2 presents the problem formulation focusing on assumptions about some dual singular system properties. Thus, in Sec. 3 and Sec. 4, short descriptions of the main properties of the method exploiting the Ljapunov matrix parametrization principle are presented and the enhanced conditions of the PD controller existence are analyzed and proven in Sec. 5. Finaly, Sec. 6 presents the simulation results and some concluding remarks are reached in Sec. 7.

Throughout the paper, the following notation was used: $\boldsymbol{x}^T$, $\boldsymbol{X}^T$ denotes the transpose of the vector $\boldsymbol{x}$ and the matrix $\boldsymbol{X}$, respectively, $\rho(\boldsymbol{X})$ indicates the eigenvalue spectrum of the square matrix $\boldsymbol{X}$, for a square matrix $\boldsymbol{X} \geq 0$ (> 0) means that $\boldsymbol{X}$ is a symmetric positive (semi)definite matrix, the symbol $\boldsymbol{I}_n$ indicates the n-th order unit matrix, $I\!R$ notes the set of real numbers, and $I\!R^n$, $I\!R^{n\times r}$ refer to the set of all n-dimensional real vectors and $n \times r$ real matrices, respectively.

2. Problem formulation

A linear, time-invariant descriptor multi-input, multi-output (MIMO) system in presence of an unknown disturbance can be described by the state-space equations in the following form

$$\boldsymbol{E}\dot{\boldsymbol{x}}(t) = \boldsymbol{A}\boldsymbol{x}(t) + \boldsymbol{B}\boldsymbol{u}(t) + \boldsymbol{D}\boldsymbol{d}(t)\,, \tag{1}$$

$$\boldsymbol{y}(t) = \boldsymbol{C}\boldsymbol{x}(t)\,, \tag{2}$$

where $\boldsymbol{x}(t) \in I\!R^n$, $\boldsymbol{u}(t) \in I\!R^r$, and $\boldsymbol{y}(t) \in I\!R^m$ are vectors of the system, input and output variables, respectively, $\boldsymbol{d}(t) \in I\!R^p$ is the external disturbance vector, $\boldsymbol{A} \in I\!R^{n\times n}$ is the system dynamic matrix, $\boldsymbol{B} \in I\!R^{n\times r}$ is the system input matrix $\boldsymbol{C} \in I\!R^{m\times n}$ is the system output matrix, $\boldsymbol{D} \in I\!R^{n\times p}$ is the disturbance input matrix and $\boldsymbol{E} \in I\!R^{n\times n}$ is a singular matrix.

It is considered that the pair $(\boldsymbol{A}, \boldsymbol{B})$ is controllable and the control law is of the form

$$\boldsymbol{u}(t) = -\boldsymbol{K}\boldsymbol{x}(t)\,, \tag{3}$$

where $\boldsymbol{K} \in I\!R^{r\times n}$ is the control gain matrix. This implies the closed-loop system description in the form

$$\boldsymbol{E}\dot{\boldsymbol{x}}(t) = \boldsymbol{A}_c\boldsymbol{x}(t) + \boldsymbol{D}\boldsymbol{d}(t)\,, \tag{4}$$

$$\boldsymbol{y}(t) = \boldsymbol{C}\boldsymbol{x}(t)\,, \tag{5}$$

where

$$\boldsymbol{A}_c = \boldsymbol{A} - \boldsymbol{B}\boldsymbol{K} \tag{6}$$

is the closed-loop system dynamic matrix.

The transfer function of the closed-loop system (4), (5) is

$$\boldsymbol{G}(s) = \boldsymbol{C}(s\boldsymbol{E} - \boldsymbol{A}_c)^{-1}\boldsymbol{D}\,, \tag{7}$$

where a complex number s is the transform variable (Laplace variable) of the Laplace transform.

Since

$$\|\boldsymbol{G}(s)\|_\infty < \gamma \Leftrightarrow \|\boldsymbol{G}_d(s)\|_\infty < \gamma\,, \tag{8}$$

where $\|\boldsymbol{G}(s)\|_\infty$ is the H_∞ norm of the transfer function (7) and $\|\boldsymbol{G}_d(s)\|_\infty$ is the H_∞ norm of the dual system transfer function

$$\boldsymbol{G}_d(s) = \boldsymbol{D}^T(s\boldsymbol{E}^T - \boldsymbol{A}_c^T)^{-1}\boldsymbol{C}^T\,, \tag{9}$$

it is more suitable in the control law parameter design for descriptor system (4), (5) to work with the dual state-space description of the closed-loop system

$$\boldsymbol{E}^T\dot{\boldsymbol{q}}(t) = \boldsymbol{A}_c^T\boldsymbol{q}(t) + \boldsymbol{C}^T\boldsymbol{g}(t)\,, \tag{10}$$

$$\boldsymbol{z}(t) = \boldsymbol{D}^T\boldsymbol{q}(t)\,, \tag{11}$$

where $\boldsymbol{g}(t) \in I\!R^m$, $\boldsymbol{z}(t) \in I\!R^p$ are associated input and output vector variables.

IOP Conf. Series: Journal of Physics: Conf. Series **783** (2017) 012037 doi:10.1088/1742-6596/783/1/012037

3. Full state feedback control design

These propositions give necessary and sufficient conditions for (1), (2) to be admissible.

Proposition 1 *The closed-loop singular system in (4), (5) is admissible if for given a nonzero square matrix* $\boldsymbol{S} \in I\!\!R^{n \times n}$ *satisfying the condition* $\boldsymbol{SE} = 0$ *there exist a symmetric positive definite matrix* $\boldsymbol{T} \in I\!\!R^{n \times n}$ *and matrices* $\boldsymbol{Q} \in I\!\!R^{n \times n}$, $\boldsymbol{Y} \in I\!\!R^{r \times n}$ *such that*

$$\boldsymbol{T} = \boldsymbol{T}^T > 0\,, \tag{12}$$

$$\boldsymbol{A}(\boldsymbol{TE}^T + \boldsymbol{QS}) + (\boldsymbol{TE}^T + \boldsymbol{QS})^T \boldsymbol{A}^T - \boldsymbol{BY} - \boldsymbol{Y}^T \boldsymbol{B}^T < 0\,. \tag{13}$$

When the above conditions hold, the control law gain matrix is given by

$$\boldsymbol{K} = \boldsymbol{Y}(\boldsymbol{TE}^T + \boldsymbol{QS})^{-1}. \tag{14}$$

Proof: Considering the Lyapunov function candidate in the following form

$$v(\boldsymbol{q}(t)) = \boldsymbol{q}^T(t)\boldsymbol{EPq}(t)\,, \tag{15}$$

$$\boldsymbol{EP} = \boldsymbol{P}^T\boldsymbol{E}^T \geq 0 \tag{16}$$

and $\boldsymbol{P} \in I\!\!R^{n \times n}$ is a square matrix, then the time derivative of (15) for the disturbance free system (10), (11) takes the form

$$\begin{aligned} \dot{v}(\boldsymbol{q}(t)) &= \dot{\boldsymbol{q}}^T(t)\boldsymbol{EPq}(t) + \boldsymbol{q}^T(t)\boldsymbol{P}^T\boldsymbol{E}^T\dot{\boldsymbol{q}}(t) \\ &= \boldsymbol{q}^T(t)(\boldsymbol{A}_c\boldsymbol{P} + \boldsymbol{P}^T\boldsymbol{A}_c^T)\boldsymbol{q}(t) < 0\,. \end{aligned} \tag{17}$$

Evidently, this implies

$$\boldsymbol{A}_c\boldsymbol{P} + \boldsymbol{P}^T\boldsymbol{A}_c^T < 0\,. \tag{18}$$

Inserting in (18) the matrix (6) and the matrix $\boldsymbol{P}$ parametrized as [17]

$$\boldsymbol{P} = \boldsymbol{TE}^T + \boldsymbol{QS}\,, \tag{19}$$

where $\boldsymbol{T} \in I\!\!R^{n \times n}$ is a positive definite matrix, $\boldsymbol{Q} \in I\!\!R^{n \times n}$ is a square matrix and the matrix $\boldsymbol{S} \in I\!\!R^{n \times n}$ is a non-zero square matrix such that

$$\boldsymbol{SE} = 0\,, \tag{20}$$

then (18) gives

$$(\boldsymbol{A} - \boldsymbol{BK})(\boldsymbol{TE}^T + \boldsymbol{QS}) + (\boldsymbol{TE}^T + \boldsymbol{QS})^T(\boldsymbol{A} - \boldsymbol{BK})^T < 0\,. \tag{21}$$

Using the notation

$$\boldsymbol{Y} = \boldsymbol{K}(\boldsymbol{TE}^T + \boldsymbol{QS})\,, \tag{22}$$

then (21) implies (13). This concludes the proof. ■

Proposition 2 *The closed-loop singular system in (4) and (5) is admissible if for given a nonzero square matrix* $\boldsymbol{S} \in I\!\!R^{n \times n}$ *satisfying the condition* $\boldsymbol{SE} = 0$ *there exist a symmetric positive definite matrix* $\boldsymbol{T} \in I\!\!R^{n \times n}$, *matrices* $\boldsymbol{Q} \in I\!\!R^{n \times n}$, $\boldsymbol{Y} \in I\!\!R^{r \times n}$ *and a positive scalar* $\gamma \in I\!\!R$ *such that*

$$\boldsymbol{T} = \boldsymbol{T}^T > 0\,, \quad \gamma > 0\,, \tag{23}$$

$$\begin{bmatrix} \boldsymbol{A}(\boldsymbol{TE}^T + \boldsymbol{QS}) + (\boldsymbol{TE}^T + \boldsymbol{QS})^T\boldsymbol{A}^T - \boldsymbol{BY} - \boldsymbol{Y}^T\boldsymbol{B}^T & * & * \\ \boldsymbol{C}(\boldsymbol{TE}^T + \boldsymbol{QS}) & -\gamma\boldsymbol{I}_m & * \\ \boldsymbol{D}^T & \boldsymbol{0} & -\gamma\boldsymbol{I}_p \end{bmatrix} < 0\,. \tag{24}$$

When the above conditions hold, the control law gain matrix is given by

$$\boldsymbol{K} = \boldsymbol{Y}(\boldsymbol{TE}^T + \boldsymbol{QS})^{-1}. \tag{25}$$

Hereafter, $*$ *denotes the symmetric item in a symmetric matrix.*

Proof: Considering the Lyapunov function candidate in the form

$$v(\boldsymbol{q}(t)) = \boldsymbol{q}^T(t)\boldsymbol{E}\boldsymbol{P}\boldsymbol{q}(t) + \gamma^{-1}\int_0^t (\boldsymbol{z}^T(\tau)\boldsymbol{z}(\tau) - \gamma^2\boldsymbol{g}^T(\tau)\boldsymbol{g}(\tau))\mathrm{d}\tau > 0\,, \tag{26}$$

where

$$\boldsymbol{E}\boldsymbol{P} = \boldsymbol{P}^T\boldsymbol{E}^T \geq 0 \tag{27}$$

and $\boldsymbol{P} \in I\!R^{n\times n}$ is a square matrix, then the time derivative of (26) for the system (10), (11) takes the form

$$\begin{aligned} \dot{v}(\boldsymbol{q}(t)) &= \dot{\boldsymbol{q}}^T(t)\boldsymbol{E}\boldsymbol{P}\boldsymbol{q}(t) + \boldsymbol{q}^T(t)\boldsymbol{P}^T\boldsymbol{E}^T\dot{\boldsymbol{q}}(t) + \gamma^{-1}\boldsymbol{z}^T(t)\boldsymbol{z}(t) - \gamma\boldsymbol{g}^T(t)\boldsymbol{g}(t) \\ &= \boldsymbol{q}^T(t)(\boldsymbol{A}_c\boldsymbol{P} + \boldsymbol{P}^T\boldsymbol{A}_c^T + \gamma^{-1}\boldsymbol{D}\boldsymbol{D}^T)\boldsymbol{q}(t) \\ &+ \boldsymbol{g}^T(t)\boldsymbol{C}\boldsymbol{P}\boldsymbol{q}(t) + \boldsymbol{q}^T(t)\boldsymbol{P}^T\boldsymbol{C}^T\boldsymbol{g}(t) - \gamma\boldsymbol{g}^T(t)\boldsymbol{g}(t) < 0\,. \end{aligned} \tag{28}$$

Defining the composed vector

$$\boldsymbol{q}_c^T(t) = \begin{bmatrix} \boldsymbol{q}^T(t) & \boldsymbol{g}^T(t) \end{bmatrix}, \tag{29}$$

(28) can be written as

$$\dot{v}(\boldsymbol{q}(t)) = \boldsymbol{q}_c^T(t)\boldsymbol{P}_c\boldsymbol{q}_c(t) < 0\,, \tag{30}$$

where

$$\boldsymbol{P}_c = \begin{bmatrix} \boldsymbol{A}_c\boldsymbol{P} + \boldsymbol{P}^T\boldsymbol{A}_c^T + \gamma^{-1}\boldsymbol{D}\boldsymbol{D}^T & \boldsymbol{P}^T\boldsymbol{C}^T \\ \boldsymbol{C}\boldsymbol{P} & -\gamma\boldsymbol{I}_m \end{bmatrix} < 0\,. \tag{31}$$

Then, the Schur complement implies

$$\begin{bmatrix} \boldsymbol{A}_c\boldsymbol{P} + \boldsymbol{P}^T\boldsymbol{A}_c^T & \boldsymbol{P}^T\boldsymbol{C}^T & \boldsymbol{D} \\ \boldsymbol{C}\boldsymbol{P} & -\gamma\boldsymbol{I}_m & \boldsymbol{0} \\ \boldsymbol{D}^T & \boldsymbol{0} & -\gamma\boldsymbol{I}_p \end{bmatrix} < 0 \tag{32}$$

and, substituting in (32) the matrices (6) and (19), the following inequality is obtained

$$\begin{bmatrix} (\boldsymbol{A}-\boldsymbol{B}\boldsymbol{K})(\boldsymbol{T}\boldsymbol{E}^T+\boldsymbol{Q}\boldsymbol{S}) + (\boldsymbol{T}\boldsymbol{E}^T+\boldsymbol{Q}\boldsymbol{S})^T(\boldsymbol{A}-\boldsymbol{B}\boldsymbol{K})^T & (\boldsymbol{T}\boldsymbol{E}^T+\boldsymbol{Q}\boldsymbol{S})^T\boldsymbol{C}^T & \boldsymbol{D} \\ \boldsymbol{C}(\boldsymbol{T}\boldsymbol{E}^T+\boldsymbol{Q}\boldsymbol{S}) & -\gamma\boldsymbol{I}_m & \boldsymbol{0} \\ \boldsymbol{D}^T & \boldsymbol{0} & -\gamma\boldsymbol{I}_p \end{bmatrix} < 0\,. \tag{33}$$

Using the notation (22) then (33) implies (24). This concludes the proof. ■

Note, a pair $(\boldsymbol{E};\boldsymbol{A})$ is admissible if is regular and has neither impulsive modes nor unstable finite modes. In this sense, the proposition can be used to design the gain matrix $\boldsymbol{K}$.

4. Static output control

The following theorems give necessary and sufficient conditions for system (1), (2) under static output control to be admissible.

Theorem 1 *The closed-loop singular system in (4) and (5) is admissible if for given a nonzero square matrix $\boldsymbol{S} \in I\!R^{n\times n}$ satisfying the condition $\boldsymbol{S}\boldsymbol{E} = 0$ there exist a symmetric positive definite matrix $\boldsymbol{T} \in I\!R^{n\times n}$ and matrices $\boldsymbol{H} \in I\!R^{m\times m}$, $\boldsymbol{Q} \in I\!R^{n\times n}$, $\boldsymbol{Y}_o \in I\!R^{r\times m}$ such that*

$$\boldsymbol{T} = \boldsymbol{T}^T > 0\,, \tag{34}$$

$$\boldsymbol{A}(\boldsymbol{T}\boldsymbol{E}^T+\boldsymbol{Q}\boldsymbol{S}) + (\boldsymbol{T}\boldsymbol{E}^T+\boldsymbol{Q}\boldsymbol{S})^T\boldsymbol{A}^T - \boldsymbol{B}\boldsymbol{Y}_o\boldsymbol{C} - \boldsymbol{C}^T\boldsymbol{Y}_o^T\boldsymbol{B}^T < 0\,, \tag{35}$$

$$\boldsymbol{C}(\boldsymbol{T}\boldsymbol{E}^T+\boldsymbol{Q}\boldsymbol{S}) = \boldsymbol{H}\boldsymbol{C}\,. \tag{36}$$

When the above conditions hold, the control law gain matrix is given by

$$\boldsymbol{K}_o = \boldsymbol{Y}_o\boldsymbol{H}^{-1}. \tag{37}$$

Proof: The static output feedback control law is defined as

$$\boldsymbol{u}(t) = -\boldsymbol{K}_o\boldsymbol{y}(t) = -\boldsymbol{K}_o\boldsymbol{C}\boldsymbol{x}(t)\,, \tag{38}$$

where $\boldsymbol{K}_o \in \mathbb{R}^{r\times m}$ is the control law gain matrix. This implies the closed-loop system state-space equations

$$\boldsymbol{E}\dot{\boldsymbol{x}}(t) = \boldsymbol{A}_{co}\boldsymbol{x}(t) + \boldsymbol{D}\boldsymbol{d}(t)\,, \tag{39}$$

$$\boldsymbol{y}(t) = \boldsymbol{C}\boldsymbol{x}(t)\,, \tag{40}$$

where

$$\boldsymbol{A}_{co} = \boldsymbol{A} - \boldsymbol{B}\boldsymbol{K}_o\boldsymbol{C} \tag{41}$$

is the closed-loop system matrix. Thus, the dual state-space description of the closed-loop system takes the form

$$\boldsymbol{E}^T\dot{\boldsymbol{q}}(t) = \boldsymbol{A}_{co}^T\boldsymbol{q}(t) + \boldsymbol{C}^T\boldsymbol{g}(t)\,, \tag{42}$$

$$\boldsymbol{z}(t) = \boldsymbol{D}^T\boldsymbol{q}(t)\,. \tag{43}$$

Considering the Lyapunov function candidate as in (15), (16) then, analogously to (18), it can obtain for the disturbance free system that

$$\boldsymbol{A}_{co}\boldsymbol{P} + \boldsymbol{P}^T\boldsymbol{A}_{co}^T < 0 \tag{44}$$

and substituting (19) and (41) it yields

$$(\boldsymbol{A} - \boldsymbol{B}\boldsymbol{K}_o\boldsymbol{C})(\boldsymbol{T}\boldsymbol{E}^T + \boldsymbol{Q}\boldsymbol{S}) + (\boldsymbol{T}\boldsymbol{E}^T + \boldsymbol{Q}\boldsymbol{S})^T(\boldsymbol{A} - \boldsymbol{B}\boldsymbol{K}_o\boldsymbol{C})^T\,. \tag{45}$$

Writing here

$$\boldsymbol{B}\boldsymbol{K}_o\boldsymbol{C}(\boldsymbol{T}\boldsymbol{E}^T + \boldsymbol{Q}\boldsymbol{S}) = \boldsymbol{B}\boldsymbol{K}_o\boldsymbol{H}\boldsymbol{H}^{-1}\boldsymbol{C}(\boldsymbol{T}\boldsymbol{E}^T + \boldsymbol{Q}\boldsymbol{S}) = \boldsymbol{B}\boldsymbol{K}_o\boldsymbol{H}\boldsymbol{C} = \boldsymbol{B}\boldsymbol{Y}_o\boldsymbol{C}, \tag{46}$$

where $\boldsymbol{H} \in \mathbb{R}^{m\times m}$ is a regular matrix and

$$\boldsymbol{H}^{-1}\boldsymbol{C} = \boldsymbol{C}(\boldsymbol{T}\boldsymbol{E}^T + \boldsymbol{Q}\boldsymbol{S})^{-1}, \tag{47}$$

$$\boldsymbol{Y}_o = \boldsymbol{K}_o\boldsymbol{H}\,, \tag{48}$$

then (45) implies (35) and (47) gives (36), This concludes the proof. ■

Theorem 2 *The closed-loop square singular system in (4) and (5) is admissible if for given a nonzero square matrix $\boldsymbol{S} \in \mathbb{R}^{n\times n}$ satisfying the condition $\boldsymbol{S}\boldsymbol{E} = 0$ there exist a symmetric positive definite matrix $\boldsymbol{T} \in \mathbb{R}^{n\times n}$, matrices $\boldsymbol{H} \in \mathbb{R}^{m\times m}$, $\boldsymbol{Q} \in \mathbb{R}^{n\times n}$, $\boldsymbol{Y}_o \in \mathbb{R}^{r\times m}$ and a positive scalar $\gamma \in \mathbb{R}$ such that*

$$\boldsymbol{T} = \boldsymbol{T}^T > 0\,, \quad \gamma > 0\,, \tag{49}$$

$$\begin{bmatrix} \boldsymbol{A}(\boldsymbol{T}\boldsymbol{E}^T + \boldsymbol{Q}\boldsymbol{S}) + (\boldsymbol{T}\boldsymbol{E}^T + \boldsymbol{Q}\boldsymbol{S})^T\boldsymbol{A}^T - \boldsymbol{B}\boldsymbol{Y}_o\boldsymbol{C} - \boldsymbol{C}^T\boldsymbol{Y}_o^T\boldsymbol{B}^T & * & * \\ \boldsymbol{C}(\boldsymbol{T}\boldsymbol{E}^T + \boldsymbol{Q}\boldsymbol{S}) & -\gamma\boldsymbol{I}_m & * \\ \boldsymbol{D}^T & \boldsymbol{0} & -\gamma\boldsymbol{I}_p \end{bmatrix} < 0\,, \tag{50}$$

$$\boldsymbol{C}(\boldsymbol{T}\boldsymbol{E}^T + \boldsymbol{Q}\boldsymbol{S}) = \boldsymbol{H}\boldsymbol{C}\,. \tag{51}$$

When the above conditions hold, the control law gain matrix is given by

$$\boldsymbol{K}_o = \boldsymbol{Y}_o\boldsymbol{H}^{-1}. \tag{52}$$

Proof: Considering the Lyapunov function candidate in the form (26), (27) then, analogously to (31), it can obtain that

$$\begin{bmatrix} \boldsymbol{A}_{co}\boldsymbol{P} + \boldsymbol{P}^T\boldsymbol{A}_{co}^T & \boldsymbol{P}^T\boldsymbol{C}^T & \boldsymbol{D} \\ \boldsymbol{C}\boldsymbol{P} & -\gamma\boldsymbol{I}_m & \boldsymbol{0} \\ \boldsymbol{D}^T & \boldsymbol{0} & -\gamma\boldsymbol{I}_p \end{bmatrix} < 0\,. \tag{53}$$

Substituting in (53) the matrices (19) and (41), the following inequality is obtained

$$\begin{bmatrix} (\boldsymbol{A}-\boldsymbol{B}\boldsymbol{K}_o\boldsymbol{C})(\boldsymbol{T}\boldsymbol{E}^T+\boldsymbol{Q}\boldsymbol{S})+(\boldsymbol{T}\boldsymbol{E}^T+\boldsymbol{Q}\boldsymbol{S})^T(\boldsymbol{A}-\boldsymbol{B}\boldsymbol{K}_o\boldsymbol{C})^T & (\boldsymbol{T}\boldsymbol{E}^T+\boldsymbol{Q}\boldsymbol{S})^T\boldsymbol{C}^T & \boldsymbol{D} \\ \boldsymbol{C}(\boldsymbol{T}\boldsymbol{E}^T+\boldsymbol{Q}\boldsymbol{S}) & -\gamma\boldsymbol{I}_m & \boldsymbol{0} \\ \boldsymbol{D}^T & \boldsymbol{0} & -\gamma\boldsymbol{I}_p \end{bmatrix} < 0 \tag{54}$$

and using the substitutions (47), (48) then (54) implies (50). This concludes the proof. ■

5. PD feedback control design

The results of the previous sections can be extended to LMI-based control design conditions for linear time-invariant descriptor systems.

Lemma 1 *The generalized dual state-space description of the descriptor system with full state PD feedback control takes the form*

$$\boldsymbol{E}^{\bullet T}\dot{\boldsymbol{q}}^{\bullet}(t) = \boldsymbol{A}^{\bullet T}\boldsymbol{q}^{\bullet}(t) + \boldsymbol{C}^{\bullet T}\boldsymbol{g}(t)\,, \tag{55}$$

$$\boldsymbol{z}(t) = \boldsymbol{D}^{\bullet T}\boldsymbol{q}^{\bullet}(t)\,, \tag{56}$$

where

$$\boldsymbol{q}^{\bullet T}(t) = \begin{bmatrix} \boldsymbol{q}^T(t) & \dot{\boldsymbol{q}}^T(t) \end{bmatrix}, \tag{57}$$

$$\boldsymbol{A}_c = \boldsymbol{A} - \boldsymbol{B}\boldsymbol{K}\,, \quad \boldsymbol{W}_c = \boldsymbol{E} + \boldsymbol{B}\boldsymbol{L}\,, \tag{58}$$

$$\boldsymbol{E}^{\bullet T} = \begin{bmatrix} \boldsymbol{I}_n & \boldsymbol{0} \\ \boldsymbol{0} & \boldsymbol{0} \end{bmatrix}, \ \boldsymbol{A}^{\bullet T} = \begin{bmatrix} \boldsymbol{0} & \boldsymbol{I}_n \\ \boldsymbol{A}_c^T & -\boldsymbol{W}_c^T \end{bmatrix}, \ \boldsymbol{C}^{\bullet T} = \begin{bmatrix} \boldsymbol{0} \\ \boldsymbol{C}^T \end{bmatrix}, \ \boldsymbol{D}^{\bullet T} = \begin{bmatrix} \boldsymbol{D}^T & \boldsymbol{0} \end{bmatrix}. \tag{59}$$

Proof: The full state PD feedback control law is defined as

$$\boldsymbol{u}(t) = -\boldsymbol{K}\boldsymbol{x}(t) - \boldsymbol{L}\dot{\boldsymbol{x}}(t)\,, \tag{60}$$

where $\boldsymbol{K}, \boldsymbol{L} \in I\!R^{r \times n}$ are the control law gain matrices. This implies the closed-loop system state-space equations

$$(\boldsymbol{E} + \boldsymbol{B}\boldsymbol{L})\dot{\boldsymbol{x}}(t) = (\boldsymbol{A} - \boldsymbol{B}\boldsymbol{K})\boldsymbol{x}(t) + \boldsymbol{D}\boldsymbol{d}(t), \tag{61}$$

$$\boldsymbol{y}(t) = \boldsymbol{C}\boldsymbol{x}(t)\,. \tag{62}$$

Using the notations (58), the dual state-space description of the closed-loop system with the full state PD feedback control is

$$\boldsymbol{W}_c^T\dot{\boldsymbol{q}}(t) = \boldsymbol{A}_c^T\boldsymbol{q}(t) + \boldsymbol{C}^T\boldsymbol{g}(t), \tag{63}$$

$$\boldsymbol{z}(t) = \boldsymbol{D}^T\boldsymbol{q}(t). \tag{64}$$

Considering the equality

$$\dot{\boldsymbol{q}}(t) = \dot{\boldsymbol{q}}(t) \tag{65}$$

and rewriting (64) as

$$\boldsymbol{A}_c^T\boldsymbol{q}(t) + \boldsymbol{C}^T\boldsymbol{g}(t) - \boldsymbol{W}_c^T\dot{\boldsymbol{q}}(t) = 0\,, \tag{66}$$

then (65) and (66) can be written compactly as

$$\begin{bmatrix} \boldsymbol{I}_n & \boldsymbol{0} \\ \boldsymbol{0} & \boldsymbol{0} \end{bmatrix} \begin{bmatrix} \dot{\boldsymbol{q}}(t) \\ \boldsymbol{q}(t) \end{bmatrix} = \begin{bmatrix} \boldsymbol{0} & \boldsymbol{I}_n \\ \boldsymbol{A}_c^T & -\boldsymbol{W}_c^T \end{bmatrix} \begin{bmatrix} \boldsymbol{q}(t) \\ \dot{\boldsymbol{q}}(t) \end{bmatrix} + \begin{bmatrix} \boldsymbol{0} \\ \boldsymbol{C}^T \end{bmatrix} \boldsymbol{g}(t), \tag{67}$$

while (65) implies

$$\boldsymbol{z}(t) = \begin{bmatrix} \boldsymbol{D}^T & \boldsymbol{0} \end{bmatrix} \begin{bmatrix} \boldsymbol{q}(t) \\ \dot{\boldsymbol{q}}(t) \end{bmatrix}. \tag{68}$$

Using the notations (57), (59), then (67), (68) implies (55), (56), respectively. This concludes the proof. ■

Lemma 2 *The generalized dual state-space description of the descriptor system with output PD feedback control takes the form*

$$\boldsymbol{E}^{\bullet T} \dot{\boldsymbol{q}}^{\bullet}(t) = \boldsymbol{A}_o^{\bullet T} \boldsymbol{q}^{\bullet}(t) + \boldsymbol{C}^{\bullet T} \boldsymbol{g}(t), \tag{69}$$

$$\boldsymbol{z}(t) = \boldsymbol{D}^{\bullet T} \boldsymbol{q}^{\bullet}(t), \tag{70}$$

where

$$\boldsymbol{q}^{\bullet T}(t) = \begin{bmatrix} \boldsymbol{q}^T(t) & \dot{\boldsymbol{q}}^T(t) \end{bmatrix}, \tag{71}$$

$$\boldsymbol{A}_{co} = \boldsymbol{A} - \boldsymbol{B}\boldsymbol{K}_o\boldsymbol{C}, \quad \boldsymbol{W}_{co} = \boldsymbol{E} + \boldsymbol{B}\boldsymbol{L}_o\boldsymbol{C}, \tag{72}$$

$$\boldsymbol{E}^{\bullet T} = \begin{bmatrix} \boldsymbol{I}_n & \boldsymbol{0} \\ \boldsymbol{0} & \boldsymbol{0} \end{bmatrix}, \ \boldsymbol{A}_o^{\bullet T} = \begin{bmatrix} \boldsymbol{0} & \boldsymbol{I}_n \\ \boldsymbol{A}_{co}^T & -\boldsymbol{W}_{co}^T \end{bmatrix}, \ \boldsymbol{C}^{\bullet T} = \begin{bmatrix} \boldsymbol{0} \\ \boldsymbol{C}^T \end{bmatrix}, \ \boldsymbol{D}^{\bullet T} = \begin{bmatrix} \boldsymbol{D}^T & \boldsymbol{0} \end{bmatrix}. \tag{73}$$

Proof: The output PD feedback control law is defined as

$$\boldsymbol{u}(t) = -\boldsymbol{K}_o\boldsymbol{y}(t) - \boldsymbol{L}_o\dot{\boldsymbol{y}}(t) = -\boldsymbol{K}_o\boldsymbol{C}\boldsymbol{x}(t) - \boldsymbol{L}_o\boldsymbol{C}\dot{\boldsymbol{x}}(t), \tag{74}$$

where $\boldsymbol{K}_o, \boldsymbol{L}_o \in I\!R^{r\times m}$ are the control law gain matrices. This implies the closed-loop system state-space equations

$$(\boldsymbol{E} + \boldsymbol{B}\boldsymbol{L}_o\boldsymbol{C})\dot{\boldsymbol{x}}(t) = (\boldsymbol{A} - \boldsymbol{B}\boldsymbol{K}_o\boldsymbol{C})\boldsymbol{x}(t) + \boldsymbol{D}\boldsymbol{d}(t), \tag{75}$$

$$\boldsymbol{y}(t) = \boldsymbol{C}\boldsymbol{x}(t). \tag{76}$$

Applying the notations (72), the dual state-space description of the closed-loop system with the output PD feedback control is

$$\boldsymbol{W}_{co}^T \dot{\boldsymbol{q}}(t) = \boldsymbol{A}_{co}^T \boldsymbol{q}(t) + \boldsymbol{C}^T \boldsymbol{g}(t), \tag{77}$$

$$\boldsymbol{z}(t) = \boldsymbol{D}^T \boldsymbol{q}(t). \tag{78}$$

Following the same way as before, it can obtain

$$\begin{bmatrix} \boldsymbol{I}_n & \boldsymbol{0} \\ \boldsymbol{0} & \boldsymbol{0} \end{bmatrix} \begin{bmatrix} \dot{\boldsymbol{q}}(t) \\ \boldsymbol{q}(t) \end{bmatrix} = \begin{bmatrix} \boldsymbol{0} & \boldsymbol{I}_n \\ \boldsymbol{A}_{co}^T & -\boldsymbol{W}_{co}^T \end{bmatrix} \begin{bmatrix} \boldsymbol{q}(t) \\ \dot{\boldsymbol{q}}(t) \end{bmatrix} + \begin{bmatrix} \boldsymbol{0} \\ \boldsymbol{C}^T \end{bmatrix} \boldsymbol{g}(t), \tag{79}$$

$$\boldsymbol{z}(t) = \begin{bmatrix} \boldsymbol{D}^T & \boldsymbol{0} \end{bmatrix} \begin{bmatrix} \boldsymbol{q}(t) \\ \dot{\boldsymbol{q}}(t) \end{bmatrix} \tag{80}$$

and using the notations (71), (73), then (79), (80) implies (69), (70), respectively. This concludes the proof. ■

Theorem 3 *The closed-loop singular system in (4) and (5) is admissible if for given positive $\delta \in \mathbb{R}$ and a nonzero square matrix $\boldsymbol{S} \in \mathbb{R}^{n \times n}$ satisfying the condition $\boldsymbol{SE} = 0$ there exist symmetric positive definite matrices $\boldsymbol{P}, \boldsymbol{T} \in \mathbb{R}^{n \times n}$, matrices $\boldsymbol{Q} \in \mathbb{R}^{n \times n}$, $\boldsymbol{Y}, \boldsymbol{Z} \in \mathbb{R}^{r \times n}$ and a positive scalar $\gamma \in \mathbb{R}$ such that*

$$\boldsymbol{P} = \boldsymbol{P}^T > 0\,, \quad \boldsymbol{T} = \boldsymbol{T}^T > 0\,, \quad \gamma > 0\,, \tag{81}$$

$$\begin{bmatrix} \boldsymbol{\Phi}(1,1) & * & * & * \\ \boldsymbol{\Phi}(2,1) & -\boldsymbol{\Phi}(2,2) & * & * \\ \boldsymbol{C}(\boldsymbol{TE}^T + \boldsymbol{QS}) & \delta \boldsymbol{C}(\boldsymbol{TE}^T + \boldsymbol{QS}) & -\gamma \boldsymbol{I}_m & * \\ \boldsymbol{D}^T & \boldsymbol{0} & \boldsymbol{0} & -\gamma \boldsymbol{I}_p \end{bmatrix} < 0\,, \tag{82}$$

$$\boldsymbol{\Phi}(1,1) = \boldsymbol{A}(\boldsymbol{TE}^T + \boldsymbol{QS}) + (\boldsymbol{TE}^T + \boldsymbol{QS})^T \boldsymbol{A}^T - \boldsymbol{BY} - \boldsymbol{Y}^T \boldsymbol{B}^T, \tag{83}$$

$$\boldsymbol{\Phi}(2,1) = \delta(\boldsymbol{TE}^T + \boldsymbol{QS})^T \boldsymbol{A}^T - \delta \boldsymbol{Y}^T \boldsymbol{B}^T + \boldsymbol{P} - \boldsymbol{E}(\boldsymbol{TE}^T + \boldsymbol{QS}) - \boldsymbol{BZ}\,, \tag{84}$$

$$\boldsymbol{\Phi}(2,2) = \delta \boldsymbol{E}(\boldsymbol{TE}^T + \boldsymbol{QS}) + \delta(\boldsymbol{TE}^T + \boldsymbol{QS})^T \boldsymbol{E}^T + \delta \boldsymbol{BZ} + \delta \boldsymbol{Z}^T \boldsymbol{B}^T\,. \tag{85}$$

When the above conditions hold, the control law gain matrices are given by

$$\boldsymbol{K} = \boldsymbol{Y}(\boldsymbol{TE}^T + \boldsymbol{QS})^{-1}, \quad \boldsymbol{L} = \boldsymbol{Z}(\boldsymbol{TE}^T + \boldsymbol{QS})^{-1}. \tag{86}$$

Proof: Considering the Lyapunov function candidate in the form

$$v(\boldsymbol{q}^\bullet(t)) = \boldsymbol{q}^{\bullet T}(t) \boldsymbol{E}^\bullet \boldsymbol{P}^\bullet \boldsymbol{q}(t) + \gamma^{-1} \int_0^t (\boldsymbol{z}^T(\tau)\boldsymbol{z}(\tau) - \gamma^2 \boldsymbol{g}^T(\tau)\boldsymbol{g}(\tau))\mathrm{d}\tau > 0\,, \tag{87}$$

where

$$\boldsymbol{E}^\bullet \boldsymbol{P}^\bullet = \boldsymbol{P}^{\bullet T} \boldsymbol{E}^{\bullet T} \geq 0 \tag{88}$$

and $\boldsymbol{P}^\bullet \in \mathbb{R}^{2n \times 2n}$ is a square matrix, then following the same way as above it can obtain in analogy with (32) that

$$\begin{bmatrix} \boldsymbol{A}^\bullet \boldsymbol{P}^\bullet + \boldsymbol{P}^{\bullet T} \boldsymbol{A}^{\bullet T} & \boldsymbol{P}^{\bullet T} \boldsymbol{C}^{\bullet T} & \boldsymbol{D}^\bullet \\ \boldsymbol{C}^\bullet \boldsymbol{P}^\bullet & -\gamma \boldsymbol{I}_m & \boldsymbol{0} \\ \boldsymbol{D}^{\bullet T} & \boldsymbol{0} & -\gamma \boldsymbol{I}_p \end{bmatrix} < 0\,. \tag{89}$$

Considering the matrix $\boldsymbol{P}^\bullet$ of the following form

$$\boldsymbol{P}^\bullet = \begin{bmatrix} \boldsymbol{P}_1^\circ & \boldsymbol{P}_2^\circ \\ \boldsymbol{P}_3^\circ & \boldsymbol{P}_4^\circ \end{bmatrix}, \tag{90}$$

then with respect to (88)

$$\begin{bmatrix} \boldsymbol{I}_n & \boldsymbol{0} \\ \boldsymbol{0} & \boldsymbol{0} \end{bmatrix} \begin{bmatrix} \boldsymbol{P}_1^\circ & \boldsymbol{P}_2^\circ \\ \boldsymbol{P}_3^\circ & \boldsymbol{P}_4^\circ \end{bmatrix} = \begin{bmatrix} \boldsymbol{P}_1^{\circ T} & \boldsymbol{P}_3^{\circ T} \\ \boldsymbol{P}_2^{\circ T} & \boldsymbol{P}_4^{\circ T} \end{bmatrix} \begin{bmatrix} \boldsymbol{I}_n & \boldsymbol{0} \\ \boldsymbol{0} & \boldsymbol{0} \end{bmatrix} \geq 0\,, \tag{91}$$

which gives

$$\begin{bmatrix} \boldsymbol{P}_1^\circ & \boldsymbol{P}_2^\circ \\ \boldsymbol{0} & \boldsymbol{0} \end{bmatrix} = \begin{bmatrix} \boldsymbol{P}_1^{\circ T} & \boldsymbol{0} \\ \boldsymbol{P}_2^{\circ T} & \boldsymbol{0} \end{bmatrix} \geq 0\,. \tag{92}$$

It is evident that (92) can be satisfied only if

$$\boldsymbol{P} = \boldsymbol{P}_1^\circ = \boldsymbol{P}_1^{\circ T} > 0, \quad \boldsymbol{P}_2^\circ = \boldsymbol{P}_2^{\circ T} = \boldsymbol{0}\,. \tag{93}$$

Thus, using (93), it yields

$$\boldsymbol{A}^{\bullet}\boldsymbol{P}^{\bullet} = \begin{bmatrix} \mathbf{0} & \boldsymbol{A}_c \\ \boldsymbol{I}_n & -\boldsymbol{W}_c \end{bmatrix} \begin{bmatrix} \boldsymbol{P}_1^{\circ} & \mathbf{0} \\ \boldsymbol{P}_3^{\circ} & \boldsymbol{P}_4^{\circ} \end{bmatrix} = \begin{bmatrix} \boldsymbol{A}_c\boldsymbol{P}_3^{\circ} & \boldsymbol{A}_c\boldsymbol{P}_4^{\circ} \\ \boldsymbol{P}_1^{\circ} - \boldsymbol{W}_c\boldsymbol{P}_3^{\circ} & \boldsymbol{W}_c\boldsymbol{P}_4^{\circ} \end{bmatrix}, \tag{94}$$

$$\boldsymbol{C}^{\bullet}\boldsymbol{P}^{\bullet} = \begin{bmatrix} \mathbf{0} & \boldsymbol{C} \end{bmatrix} \begin{bmatrix} \boldsymbol{P}_1^{\circ} & \mathbf{0} \\ \boldsymbol{P}_3^{\circ} & \boldsymbol{P}_4^{\circ} \end{bmatrix} = \begin{bmatrix} \boldsymbol{C}\boldsymbol{P}_3^{\circ} & \boldsymbol{C}\boldsymbol{P}_4^{\circ} \end{bmatrix} \tag{95}$$

and, after some algebraic manipulations, (89) takes the following form

$$\begin{bmatrix} \boldsymbol{A}_c\boldsymbol{P}_3^{\circ} + \boldsymbol{P}_3^{\circ T}\boldsymbol{A}_c^T & \boldsymbol{A}_c\boldsymbol{P}_4^{\circ} + \boldsymbol{P}_1^{\circ} - \boldsymbol{P}_3^{\circ T}\boldsymbol{W}_c^T & \boldsymbol{P}_3^{\circ T}\boldsymbol{C}^T & \boldsymbol{D} \\ \boldsymbol{P}_4^{\circ T}\boldsymbol{A}_c^T + \boldsymbol{P}_1^{\circ} - \boldsymbol{W}_c\boldsymbol{P}_3^{\circ} & -\boldsymbol{W}_c\boldsymbol{P}_4^{\circ} - \boldsymbol{P}_4^{\circ T}\boldsymbol{W}_c^T & \boldsymbol{P}_4^{\circ T}\boldsymbol{C}^T & \mathbf{0} \\ \boldsymbol{C}\boldsymbol{P}_3^{\circ} & \boldsymbol{C}\boldsymbol{P}_4^{\circ} & -\gamma\boldsymbol{I}_m & \mathbf{0} \\ \boldsymbol{D}^T & \mathbf{0} & \mathbf{0} & -\gamma\boldsymbol{I}_p \end{bmatrix} < 0. \tag{96}$$

Applying the substitutions

$$\boldsymbol{P}_3^{\circ} = \boldsymbol{T}\boldsymbol{E}^T + \boldsymbol{Q}\boldsymbol{S}, \quad \boldsymbol{P}_4^{\circ} = \delta\boldsymbol{P}_3^{\circ}, \tag{97}$$

$$\boldsymbol{Y} = \boldsymbol{K}(\boldsymbol{T}\boldsymbol{E}^T + \boldsymbol{Q}\boldsymbol{S}), \quad \boldsymbol{Z} = \boldsymbol{L}(\boldsymbol{T}\boldsymbol{E}^T + \boldsymbol{Q}\boldsymbol{S}), \tag{98}$$

where $\delta \in I\!R$ is a positive scalar, $\boldsymbol{T} \in I\!R^{n\times n}$ is a positive definite matrix, $\boldsymbol{Q} \in I\!R^{n\times n}$ is a square matrix and the matrix $\boldsymbol{S} \in I\!R^{n\times n}$ is a non-zero square matrix such that $\boldsymbol{S}\boldsymbol{E} = 0$, then

$$\boldsymbol{\Phi}(1,1) = \boldsymbol{A}_c\boldsymbol{P}_3^{\circ} + \boldsymbol{P}_3^{\circ T}\boldsymbol{A}_c^T = \boldsymbol{A}(\boldsymbol{T}\boldsymbol{E}^T + \boldsymbol{Q}\boldsymbol{S}) + (\boldsymbol{T}\boldsymbol{E}^T + \boldsymbol{Q}\boldsymbol{S})^T\boldsymbol{A}^T - \boldsymbol{B}\boldsymbol{Y} - \boldsymbol{Y}^T\boldsymbol{B}^T, \tag{99}$$

$$\boldsymbol{\Phi}(2,2) = \boldsymbol{W}_c\boldsymbol{P}_4^{\circ} + \boldsymbol{P}_4^{\circ T}\boldsymbol{W}_c^T = \boldsymbol{E}(\boldsymbol{T}\boldsymbol{E}^T + \boldsymbol{Q}\boldsymbol{S}) + (\boldsymbol{T}\boldsymbol{E}^T + \boldsymbol{Q}\boldsymbol{S})^T\boldsymbol{E}^T + \delta\boldsymbol{B}\boldsymbol{Z} + \delta\boldsymbol{Z}^T\boldsymbol{B}^T, \tag{100}$$

$$\begin{aligned} \boldsymbol{\Phi}(2,1) &= \boldsymbol{P}_4^{\circ T}\boldsymbol{A}_c^T + \boldsymbol{P}_1^{\circ} - \boldsymbol{W}_c\boldsymbol{P}_3^{\circ} \\ &= \delta(\boldsymbol{T}\boldsymbol{E}^T + \boldsymbol{Q}\boldsymbol{S})^T\boldsymbol{A}^T - \delta\boldsymbol{Y}^T\boldsymbol{B}^T + \boldsymbol{P}_1^{\circ} - \boldsymbol{E}(\boldsymbol{T}\boldsymbol{E}^T + \boldsymbol{Q}\boldsymbol{S}) - \boldsymbol{B}\boldsymbol{Z}, \end{aligned} \tag{101}$$

$$\boldsymbol{C}\boldsymbol{P}_3^{\circ} = \boldsymbol{C}(\boldsymbol{T}\boldsymbol{E}^T + \boldsymbol{Q}\boldsymbol{S}), \tag{102}$$

$$\boldsymbol{C}\boldsymbol{P}_4^{\circ} = \delta\boldsymbol{C}(\boldsymbol{T}\boldsymbol{E}^T + \boldsymbol{Q}\boldsymbol{S}). \tag{103}$$

Using the notations given above, the resulting formulas to (96) take the forms (82)-(85). This concludes the proof. ∎

Theorem 4 *The closed-loop square singular system in (4) and (5) is admissible if for given positive $\delta \in I\!R$ and a nonzero square matrix $\boldsymbol{S} \in I\!R^{n\times n}$ satisfying the condition $\boldsymbol{S}\boldsymbol{E} = 0$ there exist symmetric positive definite matrices $\boldsymbol{P}, \boldsymbol{T} \in I\!R^{n\times n}$, matrices $\boldsymbol{Q} \in I\!R^{n\times n}$, $\boldsymbol{Y}_o, \boldsymbol{Z}_o \in I\!R^{r\times m}$, a regular matrix $\boldsymbol{H} \in I\!R^{m\times m}$ and a positive scalar $\gamma \in I\!R$ such that*

$$\boldsymbol{P} = \boldsymbol{P}^T > 0, \quad \boldsymbol{T} = \boldsymbol{T}^T > 0, \quad \gamma > 0, \tag{104}$$

$$\begin{bmatrix} \boldsymbol{\Phi}_o(1,1) & * & * & * \\ \boldsymbol{\Phi}_o(2,1) & -\boldsymbol{\Phi}_o(2,2) & * & * \\ \boldsymbol{C}(\boldsymbol{T}\boldsymbol{E}^T + \boldsymbol{Q}\boldsymbol{S}) & \delta\boldsymbol{C}(\boldsymbol{T}\boldsymbol{E}^T + \boldsymbol{Q}\boldsymbol{S}) & -\gamma\boldsymbol{I}_m & * \\ \boldsymbol{D}^T & \mathbf{0} & \mathbf{0} & -\gamma\boldsymbol{I}_p \end{bmatrix} < 0, \tag{105}$$

$$\boldsymbol{\Phi}_o(1,1) = \boldsymbol{A}(\boldsymbol{T}\boldsymbol{E}^T + \boldsymbol{Q}\boldsymbol{S}) + (\boldsymbol{T}\boldsymbol{E}^T + \boldsymbol{Q}\boldsymbol{S})^T\boldsymbol{A}^T - \boldsymbol{B}\boldsymbol{Y}_o\boldsymbol{C} - \boldsymbol{C}^T\boldsymbol{Y}_o^T\boldsymbol{B}^T, \tag{106}$$

$$\boldsymbol{\Phi}_o(2,1) = \delta(\boldsymbol{T}\boldsymbol{E}^T + \boldsymbol{Q}\boldsymbol{S})^T\boldsymbol{A}^T - \delta\boldsymbol{Y}^T\boldsymbol{B}^T + \boldsymbol{P} - \boldsymbol{E}(\boldsymbol{T}\boldsymbol{E}^T + \boldsymbol{Q}\boldsymbol{S}) - \boldsymbol{B}\boldsymbol{Z}_o\boldsymbol{C}, \tag{107}$$

$$\boldsymbol{\Phi}_o(2,2) = \delta\boldsymbol{E}(\boldsymbol{T}\boldsymbol{E}^T + \boldsymbol{Q}\boldsymbol{S}) + \delta(\boldsymbol{T}\boldsymbol{E}^T + \boldsymbol{Q}\boldsymbol{S})^T\boldsymbol{E}^T + \delta\boldsymbol{B}\boldsymbol{Z}_o\boldsymbol{C} + \delta\boldsymbol{C}^T\boldsymbol{Z}_o^T\boldsymbol{B}^T. \tag{108}$$

$$\boldsymbol{C}(\boldsymbol{T}\boldsymbol{E}^T + \boldsymbol{Q}\boldsymbol{S}) = \boldsymbol{H}\boldsymbol{C}, \tag{109}$$

When the above conditions hold, the control law gain matrices are given by

$$\boldsymbol{K}_o = \boldsymbol{Y}_o\boldsymbol{H}^{-1}, \quad \boldsymbol{L}_o = \boldsymbol{Z}_o\boldsymbol{H}^{-1}. \tag{110}$$

Proof: Considering the Lyapunov function candidate (87), (88) and following the same way as above, after some algebraic manipulations it can be obtained

$$\begin{bmatrix} A_{co}P_3^\circ + P_3^{\circ T}A_{co}^T & A_{co}P_4^\circ + P_1^\circ - P_3^{\circ T}W_{co}^T & P_3^{\circ T}C^T & D \\ P_4^{\circ T}A_{co}^T + P_1^\circ - W_{co}P_3^\circ & -W_{co}P_4^\circ - P_4^{\circ T}W_{co}^T & P_4^{\circ T}C^T & 0 \\ CP_3^\circ & CP_4^\circ & -\gamma I_m & 0 \\ D^T & 0 & 0 & -\gamma I_p \end{bmatrix} < 0. \tag{111}$$

Using the substitutions (47), (97) it yields

$$BK_oC(TE^T + QS) = BK_oHH^{-1}C(TE^T + QS) = BK_oHC = BY_oC, \tag{112}$$

$$BL_oC(TE^T + QS) = BL_oHH^{-1}C(TE^T + QS) = BL_oHC = BZ_oC, \tag{113}$$

and, consequently,

$$\Phi_o(1,1) = A_{co}P_3^\circ + P_3^{\circ T}A_{co}^T = A(TE^T + QS) + (TE^T + QS)^T A^T - DY_oC - C^TY_o^TB^T, \tag{114}$$

$$\Phi(2,2) = W_{co}P_4^\circ + P_4^{\circ T}W_{co}^T = E(TE^T + QS) + (TE^T + QS)^TE^T + \delta BZ_oC + \delta C^TZ_o^TB^T, \tag{115}$$

$$\begin{aligned} \Phi(2,1) &= P_4^{\circ T}A_{co}^T + P_1^\circ - W_{co}P_3^\circ \\ &= \delta(TE^T + QS)^TA^T - \delta C^TY^TB^T + P_1^\circ - E(TE^T + QS) - BZC, \end{aligned} \tag{116}$$

where

$$Y_o = K_oH, \quad Z_o = L_oH. \tag{117}$$

Using (102), (103) together with (114)-(116), the resulting formulas to (111) take the forms (105)-(108). This concludes the proof. ■

6. Illustrative example

In the example, there is considered the system (1), (2) in the state-space representation, where the system matrices are

$$A = \begin{bmatrix} -1.0522 & -1.8666 & 0.5102 \\ -0.4380 & -5.4335 & 0.9205 \\ -0.5522 & 0.1334 & -0.4898 \end{bmatrix}, \ E = \begin{bmatrix} 0 & 0 & 0 \\ 0 & 1 & 0 \\ 0 & 0 & 1 \end{bmatrix}, \ B = \begin{bmatrix} 3 & 1 \\ 1 & -1 \\ 3 & 0 \end{bmatrix}, \ D = \begin{bmatrix} 0.5 \\ 0.0 \\ -0.4 \end{bmatrix},$$

$$S = \begin{bmatrix} 1 & 0 & 0 \\ 0 & 0 & 0 \\ 0 & 0 & 0 \end{bmatrix}, \ C = \begin{bmatrix} 1 & 1 & 0 \\ 0 & 1 & 1 \end{bmatrix}.$$

Solving (81), (82) using Self-Dual-Minimization (SeDuMi) package [13] for MATLAB, the PD control parameter design is feasible for $\delta = 8.5$ and

$$P = \begin{bmatrix} 3.7029 & 0.4317 & -0.2470 \\ 0.4317 & 4.3529 & -0.6160 \\ -0.2470 & -0.6160 & 3.7778 \end{bmatrix}, \ T = \begin{bmatrix} 2.3214 & -0.1005 & 0.0306 \\ -0.1005 & 0.1029 & -0.0287 \\ 0.0306 & -0.0287 & 0.2532 \end{bmatrix},$$

$$Y = \begin{bmatrix} 0.0154 & -0.0040 & 0.1063 \\ 0.3175 & -0.0202 & -0.2021 \end{bmatrix}, \ Z = \begin{bmatrix} 0.0154 & -0.0040 & 0.1063 \\ 0.3175 & -0.0202 & -0.2021 \end{bmatrix},$$

$$Q = \begin{bmatrix} -0.0568 & 0 & 0 \\ 0.0586 & 0 & 0 \\ -0.0638 & 0 & 0 \end{bmatrix}, \ \gamma = 3.1113,$$

The PD state control gain matrices are computed using (86) as follows

$$\boldsymbol{K} = \begin{bmatrix} 42.7410 & 41.6846 & -0.0198 \\ 324.6550 & 315.5865 & -4.2538 \end{bmatrix}, \ \boldsymbol{L} = \begin{bmatrix} 25.6406 & 25.1733 & -0.2372 \\ 125.1027 & 120.9441 & -1.5855 \end{bmatrix},$$

and because the matrix $\boldsymbol{W}_c = \boldsymbol{E} + \boldsymbol{BL}$ is regular, the closed-loop of given singular system is regular and the control law guaranties the stable matrix $\boldsymbol{A}_{cr} = \boldsymbol{W}_c^{-1}\boldsymbol{A}_c$, where the regular closed-loop system matrix eigenvalues spectrum is

$$\rho(\boldsymbol{A}_{cr}) = \{-1.6619 \ -2.1503 \ -3.2236\}.$$

Setting the tuning parameter $\delta = 8.5$ and solving the conditions (104)-(106), the LMI matrix variables are

$$\boldsymbol{P} = \begin{bmatrix} 2.5516 & 1.7656 & -1.2073 \\ 1.7656 & 5.4100 & 0.0585 \\ -1.2073 & 0.0585 & 2.1754 \end{bmatrix}, \ \boldsymbol{T} = \begin{bmatrix} 2.7041 & -0.1202 & 0.0228 \\ -0.1202 & 0.1329 & -0.0131 \\ 0.0228 & -0.0131 & 0.2250 \end{bmatrix},$$

$$\boldsymbol{Y}_o = \begin{bmatrix} -0.0161 & 0.0499 \\ 0.2263 & -0.2072 \end{bmatrix}, \ \boldsymbol{Z}_o = \begin{bmatrix} 0.0284 & 0.0019 \\ 0.0867 & -0.1082 \end{bmatrix}, \ \boldsymbol{H} = \begin{bmatrix} 0.0029 & 0.0097 \\ -0.0921 & 0.2119 \end{bmatrix},$$

$$\boldsymbol{Q} = \begin{bmatrix} -0.0546 & 0 & 0 \\ 0.0575 & 0 & 0 \\ -0.1496 & 0 & 0 \end{bmatrix}, \ \gamma = 2.9852.$$

The obtained output PD controller gain matrices are

$$\boldsymbol{K}_o = \begin{bmatrix} 0.7739 & 0.1999 \\ 19.0529 & -1.8541 \end{bmatrix}, \ \boldsymbol{L}_o = \begin{bmatrix} 4.0836 & -0.1787 \\ 5.5590 & -0.7660 \end{bmatrix}$$

and because the matrix $\boldsymbol{W}_{co} = \boldsymbol{E} + \boldsymbol{BL}_o\boldsymbol{C}$ is regular, the closed-loop with the singular system under the output PD controller is regular and guaranties the stable matrix $\boldsymbol{A}_{cro} = \boldsymbol{W}_{co}^{-1}\boldsymbol{A}_{co}$, where the regular closed-loop system matrix eigenvalues spectrum is

$$\rho(\boldsymbol{A}_{cro}) = \{-0.9701 \ -1.3546 \ -4.5882\}.$$

Thus, one can see that the assymptotic stability is guarantied.

7. Concluding remarks

In the paper, new enhanced formulations of the PD control parameter design for continuous-time descriptor linear systems are given in the strict LMI forms to retain admissibility condition and eliminate the impulsive modes of the plant. Applying linear feedback PD control laws to the system (1), (2), the triple $(\boldsymbol{A},\boldsymbol{E},\boldsymbol{B})$ is finite dynamics stabilizable. The proposed new formulations are fully adapted to the H_∞control problem design by introducing a tuning slack variable. The design approach can be preferable in computations, and simple in a reference model control, since gives the regular nominal closed-loop system dynamic properties. A simulation example, subject to given type of controller, demonstrates the effectiveness of the proposed LMI-based design forms.

Acknowledgment

The work presented in the paper was supported by VEGA, the Grant Agency of the Ministry of Education and the Academy of Science of Slovak Republic, under Grant No. 1/0348/14. This support is very gratefully acknowledged.

References

[1] Campbell S L 1980 *Singular Systems of Differencial Equations* (London:Pitman)
[2] Chadli M and Darouach M 2012 Novel bounded real lemma for discrete-time descriptor systems Application to H_∞ control design *Automatica* **48**(2) 449-53
[3] Dai L 1989 *Singular Control Systems* (Berlin:Springer-Verlag)
[4] Freund R W and Jarre F 2004 An extension of the positive real lemma to descriptor systems *Optimization Methods and Software* **19**(1) 69-87
[5] Gao Z Ding S X 2007 Actuator fault robust estimation and fault-tolerant control for a class of nonlinear descriptor systems *Automatica* **43**(5) 912-20
[6] Guerra T M Delprat S and Labiod S 2005 Stabilization of Fuzzy Takagi-Sugeno descriptor models Application to a double inverted pendulum *Cutting Edge Robotics* Kordic V et al. (Eds) 331-46 (Mammendorf: Pro Literatur Verlag)
[7] Hahn H 2003 *Rigid Body Dynamics of Mechanisms 2. Applications* (Berlin:Springer-Verlag)
[8] Ikeda M Lee T and Uezato E 2000 A strict LMI condition for H_2 control of descriptor systems *Proc 39th IEEE Conf Decision and Control* Sydney Australia 601-4
[9] Kaczorek T and Borawski K 2016 Minimum energy control of descriptor discrete-time linear systems by the use of Weierstrass-Kronecker decomposition *Archives of Control Sciences* **26**(2) 177-87
[10] Li J Liu Y and Li C 2007 Feedback control of nonlinear differential algebraic systems with application to power systems. An extended inverse system method. *Proc 2007 IEEE Int Conf Control and Automation* Guangzhou China 3179-83
[11] Lu R Du W Wang J and Xue A 2009 Robust H_∞ control for descriptor system based on new bounded real lemma *Circuits Syst Signal Process* **28** 869-82
[12] Masubuchi I Kamitane Y Ohara A and Suda N 1997 H_∞ control for descriptor systems A matrix inequalities approach *Automatica* **33**(4) 669-73
[13] Peaucelle D Henrion D Labit Y and Taitz K 2002 *User's Guide for SeDuMi Interface 1.04* (Toulouse: LAAS-CNRS)
[14] Sjöberg J *Descriptor Systems and Control Theory Technical report* (Linköping:Linköpings Universitet)
[15] Takaba K and Katayama T 1998 H_2 output feedback control for descriptor systems *Automatica* **34**(7) 841-50
[16] Takaba K Morihira N and Katayama T 1995 A generalized Lyapunov theorem for descriptor system *Systems & Control Letters* **24**(1) 49-51
[17] Xu S and Lam J 2006 *Robust Control and Filtering of Singular Systems* (Berlin:Springer-Verlag)
[18] Wang G Zhang Q and Yan X 2015 *Analysis and Design of Singular Markovian Jump Systems* (Cham: Springer)
[19] Zhang L Huang B and Lam J 2003 LMI synthesis of H_2 and mixed H_2 H_∞ controllers for singular systems *IEEE Tran Circuits and Systems II Analog and Digital Signal Processing* **50**(9) 615-26
[20] Zhu J Ma S and Cheng Z 2002 Singular LQ problem for nonregular descriptor systems *IEEE Tran Automatic Control* **47**(7) 1128-33

The comparison of network congestion avoidance algorithms in data exchange networks.

S Grzyb, P Orłowski

Faculty of Electrical Engineering, West Pomeranian University of Technology Szczecin, Poland

E-mail: slawomir.grzyb@zut.edu.pl, orzel@zut.edu.pl

Abstract. Effective congestion control strategies allow in maintaining low delay and high throughput in data exchange networks. These requirements seem to be the most desired by network environment participants. Wide range of algorithms are proposed in the literature to approximate to these ideal parameters. All of these approaches are focused on alleviating the results of sudden, unexpected network state changes. This paper discuss a comparison of four control strategies, focused on congestion avoidance. All of them are in charge of queue length control in active network nodes. For research purpose, non-stationary, discrete, dynamical model of communication channel has been used. Research results are presented in table and chart form.

1. Introduction

Modern data exchange networks are facing huge continued traffic increase. This phenomenon appears both in business and consumer sector. This boom cannot be balanced by worldwide hardware efficiency development. It results in network congestion appearance. To keep networks reliable and efficient, some congestion avoidance algorithm must be implemented. Different approaches focused on alleviating unfavorable effects of bottlenecks, like [1], [2] and [3]. In fact, blockage forming is characteristic in wide range of network types like power networks [4], waterworks and discussed data exchange networks [5]. Each network system can be considered as a collection of certain number of nodes and links [5]. Such decomposition facilitates designing, building and troubleshooting in network environment. In fact congestion phenomenon comes from limited hardware resource of active nodes and limited bandwidth between them [6]. Many approaches leading to eliminate or mitigate congestion results are discussed and presented in the literature [7]. Among wide spectrum of algorithms we can distinguish sliding-mode [8], fuzzy logic [9],[10] and predictive methods [11], [12], [13]. But there are also other approaches like [14],[15],[16]. Piecewise affine method is a frequent method of shaping control trajectory [17],[18],[19]. Invaluable input into solving congestion issues in computer networks has been made, using algorithms operating in higher level of ISO Open Systems Interconnection Reference Model [20], [21], [22] and [23].

In this research, a non-stationary, discrete dynamical model of packet exchange network is taken into calculation and discuss. To simplify considerations just single communication channel between two participants is included in taken model. Particular aspects of congestion control in similar models are presented in [24],[25],[26],[27]. In contrast to other approaches, this model is characterized by volatility of delays in time. It reflects the real network environment. Stability of these systems is widely presented in [28] and [29]. The analysis of them with the use of simplified frequency characteristics is described in [28].

The goal of this paper is the comparison of different control strategies of the same plant. In general these strategies can be divided into two groups. One group is based on quasi strict knowledge of disturbance level and the second one is not. This disturbance is an unknown and varying in time bandwidth, available for congested network node. To calculate optimal controller set, PSO algorithm has been used. This method can be also successfully applied to decrease the impact of overloads in other sort of networks, like power distribution grids [30], [31].

First part of the paper is focused on short presentation of discrete, non-stationary and dynamical mathematical model. Consecutive sections discuss different control strategies and they present simulation results in charts and table form.

2. Model description

Model taken into calculations in this research has been introduced in [32], [33] and developed in [34] and [35]. It belongs to non-stationary, discrete dynamical systems group and assumes delays varying in time. It makes a complementary and development to other models, which do not assume delay variation. Considered model presents a fragment of a packet network and consists of source and destination. On the packets path there are certain number of intermediate nodes. These nodes are the source of varying packet delay, because in each time step any node can be experienced by bottleneck. There is also a distinguished node called congested node (CN). Congestion control in this case is focused on maintaining egress buffer occupancy of CN on some assumed or optimal level. The main goal is not to overload or empty the buffer. Graphical illustration is shown in Figure 1.

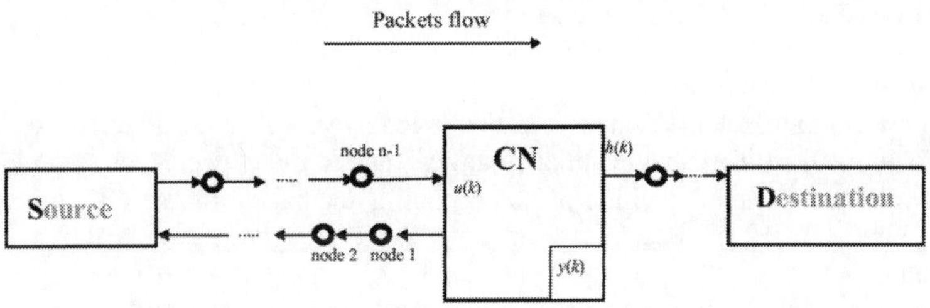

Figure 1. Block diagram of communication channel with time-varying delay.

Taking into calculation following constraints:

$$0 \le h(k) \le d(k) \le d_{max} \tag{1}$$

$$0 \le h(k) \le y(k) \tag{2}$$

$$u(k) \le u_{max} \tag{3}$$

A full model of described communication channel in in vector-matrix notation, can be described as follows:

$$\mathbf{x}(k+1)=\begin{bmatrix}1\\0\\\vdots\\0\\0\end{bmatrix}u(k)+\begin{bmatrix}0\\0\\\vdots\\0\\-1\end{bmatrix}h(k)+\begin{bmatrix}q_1(k) & 0 & 0 & \cdots & 0\\ \bar{q}_1(k) & q_2(k) & 0 & \cdots & 0\\ 0 & \bar{q}_2(k) & q_3(k) & 0 & \vdots\\ \vdots & \ddots & \ddots & \ddots & 0\\ 0 & \cdots & 0 & q_{n-1}(k) & 1\end{bmatrix}\mathbf{x}(k) \tag{4}$$

$$y(k)=\begin{bmatrix}0 & 0 & \cdots & 0 & 1\end{bmatrix}\mathbf{x}(k) \tag{5}$$

where:

$h(k)$ – number of packets sent from CN toward the destination in time *k*.

$d(k)$ – available egress bandwidth for CN in time *k*.

d_{max} – maximum number of packets that can be sent towards destination

$y(k)$ – number of packets in output buffer of CN in time *k*

$u(k)$ – number of packets, that CN orders from the source

$x_j(k)$ – number of packets in node *j* in time *k*

$q_j(k)$ – queuing factor $q_j(k)=\begin{cases}0\text{ - transmission}\\1\text{ - congestion}\end{cases}$

$\bar{q}_j(k)=1-q_j(k)$

In each time step, for every intermediate node a bottleneck can pause further packet transmission. Model simplification assumes, that these intermediate nodes can either do not forward any data, or push their all buffer content in one step to the next node, depending if it's experiencing congestion or not. In contrast, CN is can forward accumulated data in number related to available egress bandwidth. Packets incoming into CN and not forwarded towards the destination, increase the buffer occupancy. The length of the queue in the CN is presented as $y(k)$ and is controlled by the controller. This paper discusses the use of four different controllers. Congested node CN sends the control signal backwards to the source. This is performed to adjust the number of packets sent from source to destination through the congested node to network conditions, which vary in time

Wider description of given model is presented in [32]-[35].

3. Congestion control strategies

One of the most important factor which affects blockage forming in data exchange networks, is the level of nodes buffer utilization. This level cannot float near maximum capacity, or stay close to 0. In these circumstances we are experiencing either packet dropping and packet retransmissions or lack of data to be forwarded. Not wanting to face this disadvantages, a safe margin of buffer utilization must be absolutely maintained.

In order to ensure effective bandwidth utilization and buffer occupancy, four control strategies have been proposed and tested. All of these algorithms have no strict feedback concerning available bandwidth for CN. Figure 2 presents block diagrams of the control system.

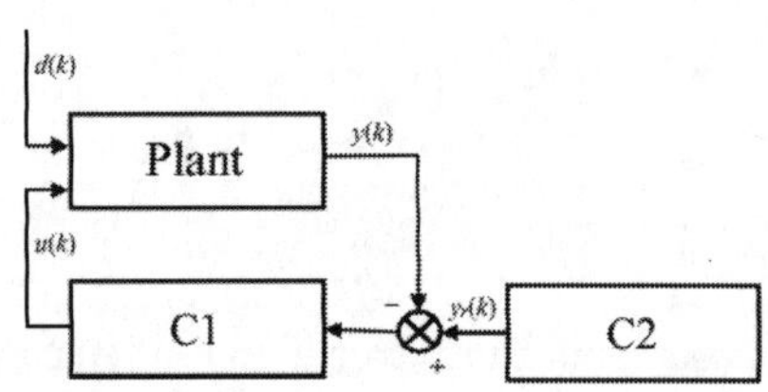

Figure 2. Block diagram of the control system.

Table 1 presents all four numbered control strategies (CS), starting from the simplest with assumed, constant reference que level utilization and proportional gain, and approaching more generalized. Used controllers trajectory is described in each line. In fact either proportional controller or piecewise affine (PWA) could be chosen.

Table 1. Characteristics of researched control systems.

CS	C1 – gain	C2 – constant
1	Optimized proportional gain	Assumed value
2	Optimized proportional gain	Optimized variable
3	Optimized PWA gain	Assumed value
4	Optimized PWA gain	Optimized variable

Particular strategies are presented in following subsections:

3.1. Control strategy 1
The first control strategy is the simplest layout and can be treated as reference trajectory supplier. In this case, wanted vector **z** is single dimensional $\mathbf{z}=[z_1]$. Reference queue length is assumed to be a half of maximum buffer capacity: $y_r = 2500$ and the proportional controller gain is:

$$u(k) = z_1(y_r - y(k)) = z_1(2500 - y(k)) \tag{6}$$

3.2. Control strategy 2
In this strategy, reference value is subjected to be optimized. It follows that wanted vector **z** contains additional component: $\mathbf{z}=[z_1 \quad z_2]$. Reference queue length takes value $y_r = z_2$. It this case, the gain of the controller is proportional too:

$$u(k) = z_1(y_r - y(k)) = z_1(z_2 - y(k)) \tag{7}$$

3.3. Control strategy 3
Like in the first strategy, reference queue length is assumed to be a half of maximum buffer capacity $y_r = 2500$. Control signal $u(k)$ takes value:

$$u(k) = g(y(k))(y_r - y(k)) = g(y(k))(2500 - y(k)) \tag{8}$$

The gain parameter $g(y(k))$ can be related to the current buffer utilization with the use of piecewise affine method:

$$g(y(k)) = \begin{cases} z_3 & y(k) \le z_1 \\ \dfrac{(z_4 - z_3)(y(k) - z_1)}{(z_2 - z_1)} + z_3, & z_1 < y(k) < z_2 \\ z_4 & y(k) \ge z_2 \end{cases} \tag{9}$$

The outline of relations presented in (9) is illustrated in Figure 3.

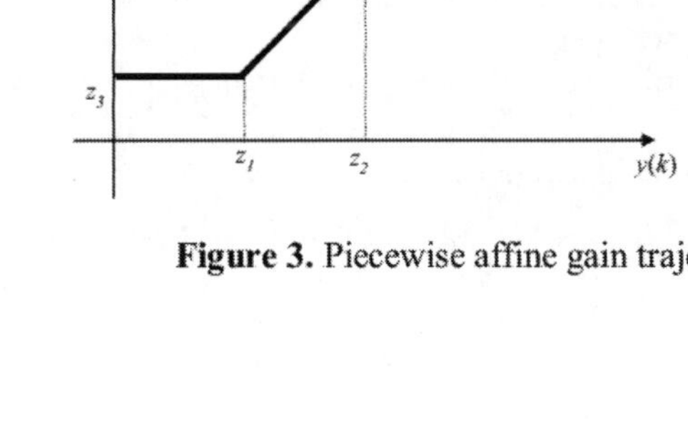

Figure 3. Piecewise affine gain trajectory.

As it can be seen, wanted vector $\mathbf{z}$ is four dimensional $\mathbf{z} = [z_1 \quad z_2 \quad z_3 \quad z_4]$

3.4. Control strategy 4

The most comprehensive control strategy is the development of the third one. Unlike in previous subsection reference queue length is subjected to be optimized $y_r = z_5$. First four coordinates, which reflect gain parameter, are determined exactly like in (9). With these assumptions control signal can be written:

$$u(k) = g(y(k))(y_r - y(k)) = g(y(k))(z_5 - y(k)) \tag{10}$$

Then, wanted vector $\mathbf{z}$ has additional variable $\mathbf{z} = [z_1 \quad z_2 \quad z_3 \quad z_4 \quad z_5]$

3.5. Common assumptions

We define a single cost function to adjust controller's sets to described design requirements. The same cost functions has been used for all seven systems and it is listed below:

$$J(\mathbf{z}) = \sum_{k=1}^{N} \delta(k) \tag{11}$$

$$\delta(k) = \begin{cases} 0 & |2500 - y(k)| \le \text{margin} \\ (2500 - y(k))^2 & ,|2500 - y(k)| > \text{margin} \end{cases} \tag{12}$$

Optimized variables in $\mathbf{z}$ vector can be determined by numerical optimization. A particle swarm optimization (PSO) method is used in these optimization iterations. Dependly on control strategy number, vector $\mathbf{z}$ has from one to five dimensions. We assume, that the total buffer capacity is 5000 packets. Construction of used cost function increases its value in two cases. First case is when the buffer utilization is less then 1/2 of buffer capacity reduced by some margin. Second case is when the buffer utilization is more then 1/2 of buffer capacity enlarged by the same margin.

The optimal controller can for each control strategy can be determined by solving the following optimization problem:

$$\min J(\mathbf{z}) \tag{13}$$

4. Simulation results

Relating to (4) and (5), our discrete, non-stationary and dynamical model takes linear form:

$$\begin{gathered} \mathbf{x}(k+1) = \mathbf{A}(k)\mathbf{x}(k) + \mathbf{B}(k)u(k) + \mathbf{F}(k)h(k) \\ y(k) = \mathbf{C}(k)\mathbf{x}(k) \end{gathered} \tag{14}$$

Vectors **B**, **C** and **F** are constant in all time horizon, whereas matrix A takes different values depending on which intermediate node is congested. Construction of discrete non-stationary linear model taken into research process as well as detailed description of its components is presented in [32] and [33]. For purpose of numerical simulation some necessary assumptions have been made:

a) $u_{max} = d_{max} = 100$ packets per sample
b) maximum buffer capacity of congested node is 5000 packets
c) Sampling period is 10 ms
d) Initial conditions equal: $\mathbf{x}(0) = [0,0,0,\ldots,0]^T$

Assumed bandwidth $d(k)$ actually available for the congested node towards destination is like in [33],[34] and is shown in Figure 4.

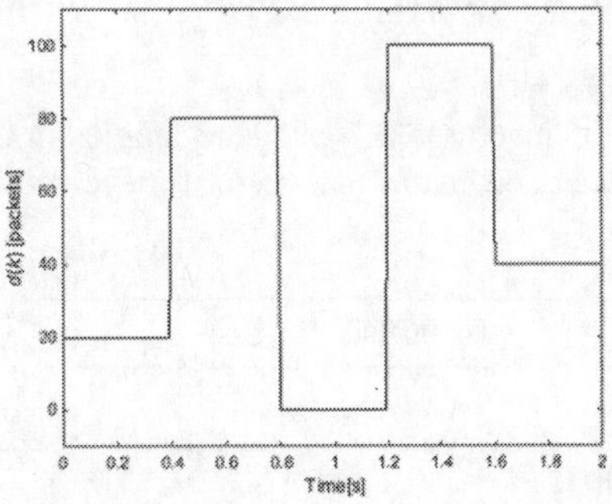

Figure 4. Available outgoing bandwidth for the congested node.

Optimization process has been conducted for each control strategy (CS) with three different margin levels. On the basis of presented strategies, calculated cost functions (J) are printed in table 2.

Table 2. Calculated cost functions.

Margin	CS	J
0	1	552.4236e+006
	2	512.3824e+006
	3	415.4589e+006
	4	335.6319e+006
1000	1	248.0248e+006
	2	150.9136e+006
	3	96.6898e+006
	4	92.9182e+006
2000	1	154.6280e+006
	2	13.3188e+006
	3	9.9518e+006
	4	8.2174e+006

These data are presented in graphical form in Figure 5.

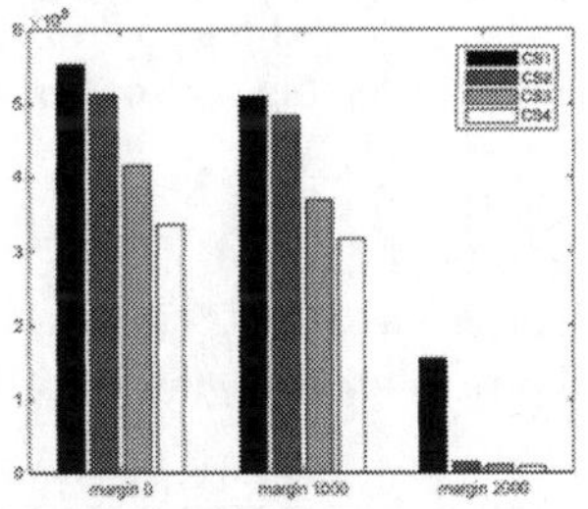

Figure 5. Values of cost functions for different margins and control strategies CS.

Graphical simulation results are divided into three subsections in view of these margin levels.

4.1. First cost function with margin = 0

Assuming margin=0, cost function increases when queue length of CN differs from assumed 2500 packets. Table 3 displays calculated coordinates of **z** vector for each control strategy.

Table 3. Vector **z** values, calculated **z** with margin=0.

Margin	CS	z
0	1	z=[2.5748e+003]
	2	z=[122.7930e+000, 3.3803e+003]
	3	z=[1.8861e+003, 1.9327e+003, 38.4705e-003, 238.1142e+000]
	4	z=[2.5889e+003, 3.1697e+003, 30.1896e-003, 1.1785e+009, 3.3124e+003]

Research results are presented in Figure 5 and 6.

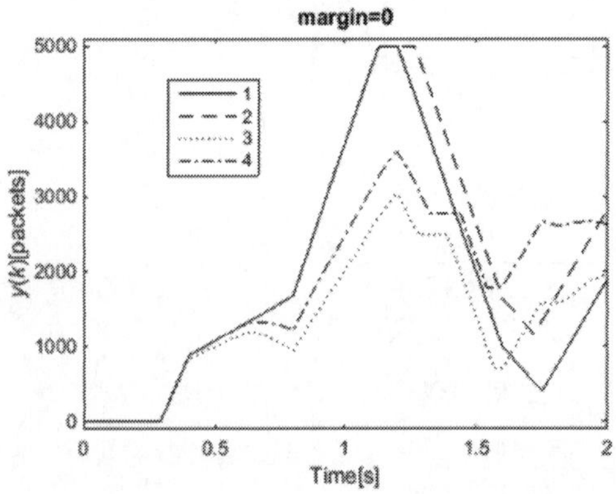

Figure 6. Queue length $y(k)$ of the congested node CN, calculated with margin = 0.

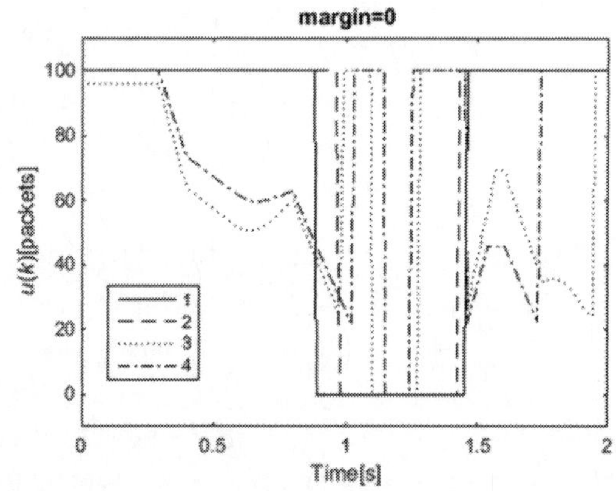

Figure 7. Control signal $u(k)$, calculated with margin = 0.

As it can be seen, CS1 and CS2 lead to total fulfillment of egress buffer. This situation takes place when available outgoing bandwidth shown in Figure 4 equals 0. While stored packets cannot be forwarded, new pieces of data, ordered previously, are incoming and it causes overfill. These two strategies also generate similar control signal. It can be seen, that the controller detects congestion and stops ordering data from the source, but ordered number is too big to avoid packet dropping.

4.2. Second cost function with margin=1000

Second cost function is built on the base of the first one. As it is presented in (12), it accepts buffer utilization in range between 1500 packets and 3500 packets, what reflects half buffer capacity with margin 1000 (2500-1000, 2500+1000). Table 4 displays calculated coordinates of vector **z** for each control strategy.

Table 4. Vector **z** values, calculated **z** with margin=1000.

Margin	CS	z
1000	1	z=[2.9596e+003]
	2	z=[611.0730e+000, 1.2935e+003]
	3	z=[1.9793e+003, 3.7832e+003, 39.9978e-003, 474.2351e+000]
	4	z=[2.1579e+003, 2.3406e+003, 35.2209e-003, 69.1476e+009, 2.8392e+003]

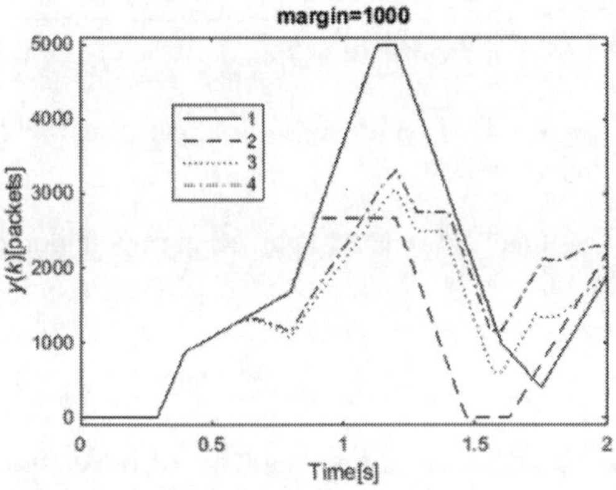

Figure 8. Queue length $y(k)$ of the congested node CN, calculated with margin = 1000.

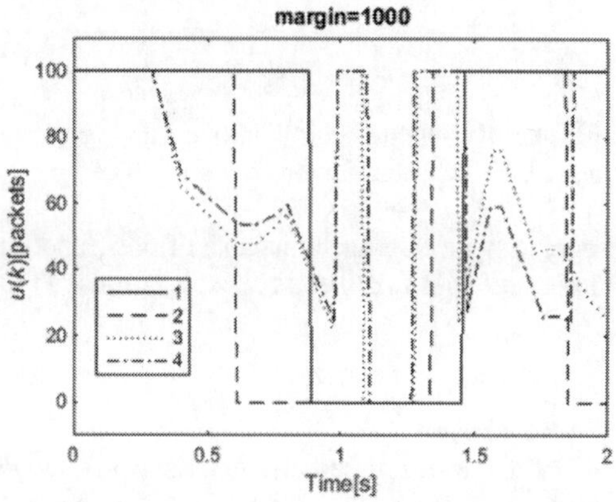

Figure 9. Control signal $u(k)$, calculated with margin = 1000.

We can observe increasing difference between $y(k)$ trajectories. CS1 still leads to total fulfillment of egress buffer, while CS2 leads to total emptiness.

4.3. Third cost function with margin=2000

Third cost function gives the widest spectrum of buffer utilization level acceptance. It can be in range between 500 packets and 4500 packets. Table 5 displays calculated variables of vector for each control strategy.

Table 5. Vector **z** values, calculated **z** with margin=2000.

Margin	CS	z
2000	1	z=[862.9197e+000]
	2	z=[4.7499e+003, 1.2450e+003]
	3	z=[1.7398e+003, 2.1901e+003, 32.1632e-003, 633.3482e-003]
	4	z=[1.4286e+003, 2.3881e+003, 37.4055e-003, 277.3974e+000, 2.2024e+003]

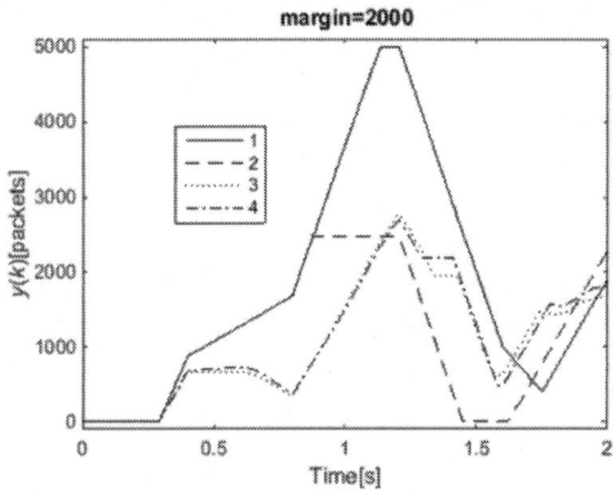

Figure 10. Queue length $y(k)$ of the congested node CN, calculated with margin = 2000.

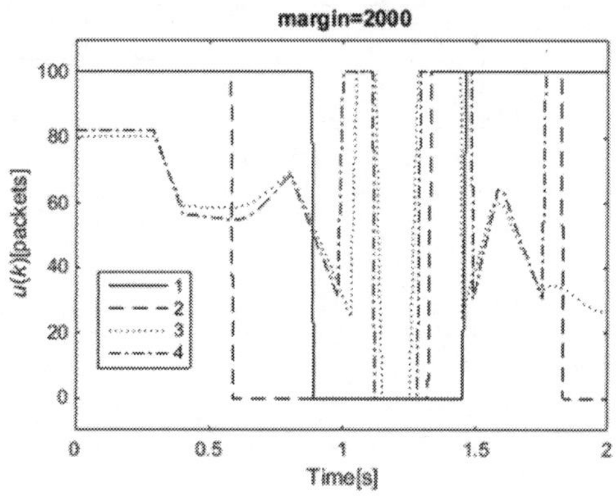

Figure 11. Control signal $u(k)$, calculated with margin = 2000.

Together with tolerance increase of cost functions, CS3 and CS4 runs are approaching to each other. These strategies fulfill control expectations in the best way.

5. Conclusions

The paper discusses a comparison of network congestion avoidance algorithms. Particle swarm optimization method is applied for tuning controllers set for each three values of the margin. Presented four control strategies are focused on queue length control in active network nodes. It can be seen clearly, the improving trend in each group together with control strategy development. The best results were obtained with the most comprehensive algorithm, which takes more parameters under optimization process. In the first approach we define reference queue length and we optimize only the level of constant proportional gain. The second iteration, uses also constant gain value, but allows to optimize also the reference queue utilization level. Second and fourth iteration are modifications first two control strategies. They are depending controller gain level on egress queue utilization.

References

[1] Hazara J, Sinha A 2007 *Congestion management using multiobjective particle swarm optimization*. IEEE Trans. Power Syst., vol. 22, no.4, , pp. 1726–1734

[2] Panigrahi B, Pandi V 2009 *Congestion management using adaptive bacterial foraging algorithm* Energy Conversion and Management, vol. 50, pp. 1202-1209

[3] Ignaciuk P, Bartoszewicz A 2010 *Discrete-time sliding-mode congestion control in multisource communication networks with time-varying delay.* IEEE Trans. on Control Systems Technology 19

[4] Daielson C, Borrelli F, Oliver D, Anderson D, Phillips T 2013 *Constrained flow control in storage networks: capacity maximization and balancing.* Automatica, vol. 49, no. 9, pp. 2612-2621

[5] Grzyb S, Orłowski P 2013 *Mechanisms of forming communication congestion and blockades in networks with time-varying parameters.* Pomiary Automatyka Kontrola, vol. 59, no. 7, pp. 704-707

[6] Sun H, Wu J 2014 *Spatial distribution complexities of traffic congestion and bottlenecks in different network topologies.* Applied Mathematical Modelling, vol 38(2), pp. 496-505

[7] Jiang H 2012 *Network congestion avoidance through speculative reservation.* High Performance Computer Architecture (HPCA). IEEE 18th International Symposium on. IEEE.

[8] Ignaciuk P, Bartoszewicz A 2007 *Linear quadratic optimal sliding mode controllers for a single virtual circuit in a connection-oriented communication network.* Proceedings of the 13th

IEEE/IFAC International Conference on Methods and Models in Automation and Robotics, Szczecin, Poland, pp. 121-128

[9] Nyirenda C, Dawou S 2007 *Fuzzy logic congestion control in IEEE 802.11 wireless local area networks* A performance evaluation. Proc. of AFRICON

[10] Guan X, Yang B, Zhao B 2007 *Adaptive fuzzy sliding mode active queue management algorithms*. Telecommunication Systems, vol.35, no.1-2

[11] Azuma T, Fujita T, Fujita, M 2005 *Congestion control for TCP/AQM networks using state predictive control*. IEEJ Transactions on Electronics Information and Systems, 125(9), 1491.

[12] Azuma, T 2014 *State predictive lqi congestion control for tcp/aqm networks considering effective equilibrium points*. International Journal of Advanced Electrical and Electronics Engineering, 3(4), 32-38.

[13] Chołodowicz E, Orłowski P 2016 *Comparison of a Perpetual and PD Inventory Control System with Smith Predictor and Different Shipping Delays Using Bicriterial Optimization and SPEA2*. Pomiary Automatyka Robotyka, R. 20, no. 3, pp. 5-12.

[14] Masoumzadeh S, Meshgi K, Shiry G 2011 *FQL-RED: an adaptive scalable schema for active queue management*. International Journal of Network Management, vol. 21, no.:2, pp. 147-167

[15] Nyirenda C, Dawoud D 2006 *Multi-objective particle swarm optimization for fuzzy logic based active queue management*. IEEE International Conference on Fuzzy Systems, pp. 2231–2238

[16] Orłowski P 2011 *Convergence of the Discrete-Time Nonlinear Model Predictive Control with Successive Time-Varying Linearization along Predicted Trajectories*. Electronika ir Elektrotechnika, vol. 113, no. 7, pp. 27-31

[17] Grieder P, Kvasnica M, Baotic M, Morari M 2004 *Low Complexity Control of Piecewise Affine Systems with Stability Guarantee*. American Control Conference, Boston, USA

[18] Orłowski P 2012 *Wpływ ilości partycji przybliżonego metodą siecznych modelu odcinkami afinicznego na jakość sterowania układu z nieliniowością zależną od stanu*. Przegląd Elektrotechniczny, vol. 88. no. 10, pp. 69-73.

[19] Orłowski P 2014 *Complexity Analysis of the Piece-Wise Affine Approximation for the Car on the Nonlinear Hill Model Related to Discrete–Time, Minimum Time Control Problem*. Elektronika Ir Elektrotechnika, vol. 20, no. 10, pp.3-6

[20] Hollot C, Misra V, Towsley D, Gong 2001 *A control theoretic analysis of RED*. Twentieth Annual Joint Conference of the IEEE Computer and Communications Societies. Proceedings. IEEE vol. 3, pp. 1510-1519

[21] Hollot C, Chris V 2001 *On designing improved controllers for AQM routers supporting TCP flows*. Twentieth Annual Joint Conference of the IEEE Computer and Communications Societies. Proceedings. IEEE. pp. 1726-173

[22] Han H, Shakkottai S, Hollot C, Srikant R, Towsley D 2006 *Multi-path tcp: a joint congestion control and routing scheme to exploit path diversity in the internet*. IEEE/ACM Transactions on Networking, vol 14(6), pp. 1260-1271.

[23] Alvarez-Flores E, Ramos-Munoz J, Ameigeiras P, Lopez-Soler J 2011 *Selective packet dropping for VoIP and TCP flows*. Telecommunication Systems, vol. 46(1), pp. 1-16.

[24] Chołodowicz E, Orłowski P 2015 *Sterowanie przepływem towarów w magazynie z wykorzystaniem predyktora Smitha*. Pomiary Automatyka Robotyka, vol. 19, no. 6, pp55-60.

[25] Chołodowicz E, Orłowski P 2015 *A periodic inventory control system with adaptive reference stock level for long supply delay*. Measurement Automation Monitoring, vol. 61, no. 12, pp. 568-572

[26] Abrahamowicz E, Orłowski P 2016 *Zastosowanie regulatora ułamkowego rzędu PD do automatycznego sterowania zamówieniami dla magazynu ze zmiennym w czasie opóźnieniem dostaw*. Pomiary Automatyka Robotyka, R. 20, Nr 2/2016, 5-10, DOI: 10.14313/PAR_220/5.

[27] Orłowski P 2016 *Analiza dwukryterialnego problemu optymalizacji w zastosowaniu do automatycznego sterowania systemem zamówień w magazynie z dużymi opóźnieniami dostaw*

z wykorzystaniem regulatora feedback-feedforward z predyktorem Smitha. Przegląd Elektrotechniczny, ISSN 0033-2097

[28] Grzyb S, Orłowski P 2014 *Zastosowanie uproszczonych charakterystyk częstotliwościowych do analizy kanału komunikacyjnego o zmiennych w czasie parametrach.* Pomiary Automatyka Kontrola, vol. 60, no. 5, pp. 317-320

[29] P. Orłowski, Generalized feedback stability for periodic linear time–varying, discrete–time systems. Bulletin of the Polish Academy of Sciences: Technical Sciences – Polish Academy of Sciences, vol. 60, no. 1., 2012, pp. 171–178

[30] Bemporad A, Oliveri A, Poggi T, Storace M 2011 *Ultra-fast stabilizing model predictive control via canonical piecewise affine approximations.* IEEE Trans. Automatic Control, vol. 56(12), 2883-2897.

[31] Yoshida H, Kawata K, Fukuyama Y, Takayama S, Nakanishi Y 2000 *A particle swarm optimization for reactive power and voltage control considering voltage security assessment.* IEEE Trans. Power Syst., vol. 15, no. 4, pp. 1232–1239

[32] Grzyb S, Orłowski P 2013 *Model matematyczny kanału komunikacyjnego z zatorem w sieciach o zmiennych w czasie parametrach.* Pomiary Automatyka Kontrola, vol. 59, no. 11, pp. 1151-1154

[33] Grzyb S, Orłowski P 2014 *Congestion control in computer networks - Application of piece-wise affine controller and particle swarm optimization.* 19th International Conference on Methods and Models in Automation and Robotics (MMAR), Miedzyzdroje, Poland, pp. 834-838.

[34] Grzyb S, Orłowski P 2015 *Congestion feedback control for computer networks with bandwith estimation.* 20th International Conference on Methods and Models in Automation and Robotics (MMAR), Miedzyzdroje, Poland, pp. 1151-1156

[35] Grzyb S, Orłowski P 2016 *Feedback control system with PWA load dependent reference buffer occupancy for congestion control in computer networks.* Przegląd Elektotechniczny, no. 46, pp. 42-45

IOP Conf. Series: Journal of Physics: Conf. Series **783** (2017) 012039 doi:10.1088/1742-6596/783/1/012039

MODELLING AND CONTROL OF THE QBALL X4 QUADROTOR SYSTEM BASED ON PID AND FUZZY LOGIC STRUCTURE

Tolga Bodrumlu, Mehmet Turan Soylemez, Ilhan Mutlu
Istanbul, TURKEY
E-mail: tbodrumlu@itu.edu.tr,soylemezm@itu.edu.tr, mutlui@itu.edu.tr

Abstract. This work focuses on a quadrocopter model, which was developed by Quanser™ and named as Qball X4. First, mathematical model of the Qball X4 is obtained. Then, a conventional PID control technique is presented. This PID control parameters come from Qball user manual. After the presentation of conventional PID control, as an extension of the conventional PID control theory, a different fuzzy controller structure is given. The proposed fuzzy controller structure is based on fuzzy logic and its name is PID type fuzzy controller. All of the simulations are done in MATLAB™ environment.

1. Introduction

Unmanned Aerial Vehicles(UAV) have gained a high level of popularity during the last decade in civilian, military and engineering applications because of the recent advances in sensing, communicating, computing and controlling technologies. UAVs have some advantages than the manned systems like longer endurance, lower cost, maneuverability and less risk to humanity.As an example of UAV systems, the quadrocopter is a relatively simple, affordable and easy to fly system thus it has been widely used to develop, implement and test-fly methods in control. A quadrotor is an aircraft that becomes airborne due to the lift force provided by four rotors usually mounted in cross configuration, hence its name. In this study, Qball-X4 which is constructed and developed by the Quanser is used. This system is a test platform for several unmanned aerial vehicle research applications.

In the literature, there are numerous studies discussing Qball X4 system or fuzzy controller structure. Some of them are given as following: Sadeghzadeh, Chamseddine, Mehta and Zhang [1] proposed a Gain-Scheduled PID control technique which was developed from Active Fault-Tolerant Control (AFTC) and applied to Qball X4 quadrotor system. In this paper, a Fault Detection and Diagnosis (FDD) block is implemented to find and identify the actuator fault. Chamseddine, Zhang, Rabbath, Fulford and Apkarian [2] worked on actuator fault-tolerant control (FTC) for Qball-X4. Their strategy is based on Model Reference Adaptive Control (MRAC). Three different MRAC techniques have been implemented and compared with Linear Quadratic Regulator (LQR) controller. These three different methods are the MIT rule MRAC, the conventional MRAC and the modified MRAC. Nia Maharani Raharja , Iswanto, Muhammad Faris and Adha Imam Cahyadi [3] worked on hover position quadrotor control with fuzzy logic. In this work, fuzzy logic controller algorithm has been proposed for hovering mode. The input of the controller is height and change of height. Membership function of the inputs are triangular. Demet Canpolat Tosun, Yasemin Isik and Hakan Korul works on a Qball X4 system with both PID and LQR controller. The LQR and PID controls applied on height, x and y positions and roll, pitch, yaw angles. [4]Ridvan Ozdemir, Mustafa Kaya, Monier Elfarra, Mehmet Onder Efe[5] work on a Qball X4 through analyzing the object following. Regional surveillance for object tracking is done using micro cameras. Communication between camera and computer are done by using the RF system.Indoor autonomous flight of the aerial vehicle is provided by software which is written in MATLAB^{TM} Simulink^{TM} environment. Moghadam[6] addressed the problem of Fault-Tolerant Control (FTC) of the Qball-X4 quadrotor in his thesis. Both actuator loss of effectiveness and sensor bias faults and their impacts on system response were considered. After Fault Detection and Diagnosis (FDD) of actuators, a leader-follower controller with dynamic reference input was given to counteract the actuator faults effects. Sensor FDD was followed by a new approach of Active Fault-Tolerant Control (AFTC) to correct faulty measured values and feed them back to the controller. Also Two-Stage Kalman Filter (TSKF) is utilized to estimate noisy and unmeasured states and realize the actuator FDD and sensor AFTC.

In our study, first, dynamic model of the Qball-X4 system has been revealed. Then the PID control technique has been discussed for the Qball-X4 system.This PID control parameters adopted from Qball user manual in to the simulink environment. Later, PID type fuzzy controller has been proposed. Finally, results of the simulations of classical PID controller, PID type fuzzy controller are shown and compared to each other.

2. Dynamical Model of Qball X4

In this part of the paper, the dynamic model of the Qball-X4 is described. Nonlinear model of Qball X4 is discussed so that different control algorithms can be applied on it. Roll, pitch and yaw angles of the system are defined as the rotation about x,y and z axis.

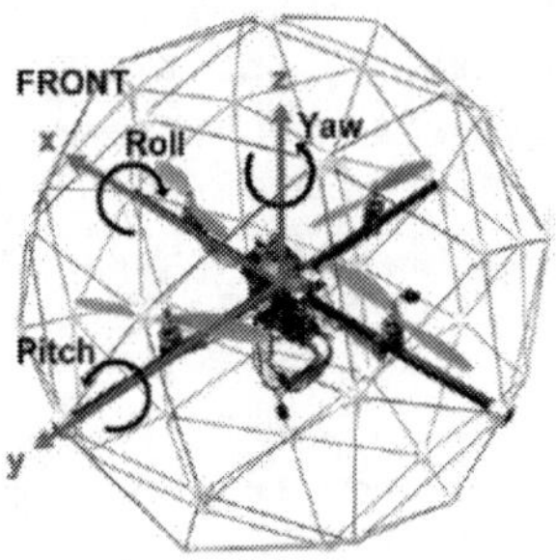

Figure 1. Qball-X4 axes and sign convention.[7]

2.1. Actuator Dynamics

Thrust force is modeled using the following first-order[5] system. That thrust force generated by each of the propeller.

$$F = K\frac{\omega}{S+\omega}u \quad (1)$$

u: PWM input to the actuator
$\omega : Actuator\ \ Bandwidth$
$K : Positive\ \ Gain$

IOP Conf. Series: Journal of Physics: Conf. Series **783** (2017) 012039 doi:10.1088/1742-6596/783/1/012039

v is the state variable and will be used for the representation of the actuator dynamics. And it is defined as follows:

$$v = \frac{\omega}{s+\omega}u \tag{2}$$

2.2. Roll/Pitch Model

When we assume the rotation about x and y axes are decoupled, we can model roll/pitch motion as follows:

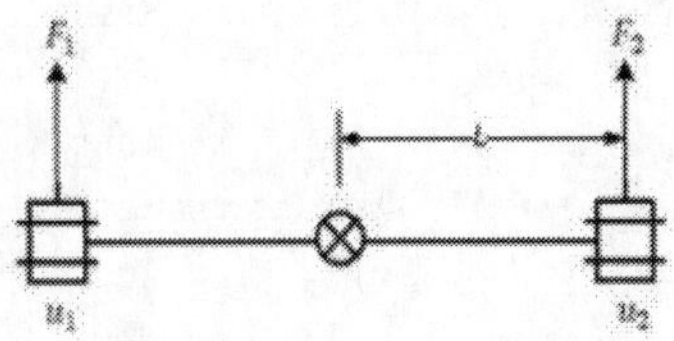

Figure 2. A model of the roll/pitch axis.[7]

Also, roll/pitch angle(θ), can be found from the following equations:

$$J\ddot{\theta} = \Delta FL \tag{3}$$

Where J equals to $J_{roll} = J_{pitch}$. These are the rotational inertia of the device for both roll and pitch axes.
L: distance between the propeller and the center of gravity

$$\Delta F = \Delta F_1 - \Delta F_2 \tag{4}$$

2.3. Height Model

Motion of the Qball-X4 along Z axis is affected by all of the propellers. Thus, the dynamic model of the height could be written as follows:

$$M\ddot{Z} = 4F\cos r\cos p - Mg \tag{5}$$

F: Thrust which is generated by each propeller
M: Total mass of the device
Z: Height
r and p: roll and pitch angles

2.4. X-Y Position Model

Total thrust and the changing roll/pitch angles are the main reason for the motion of the Qball-X4 along the X and Y axes. If the yaw angle is zero then the dynamics of the motion in X and Y axes can be obtain as follows [7]:

$$M\ddot{X} = 4F\sin p \tag{6}$$

$$M\ddot{Y} = -4F\sin r \tag{7}$$

Height, x and y axis model can be connected like that:

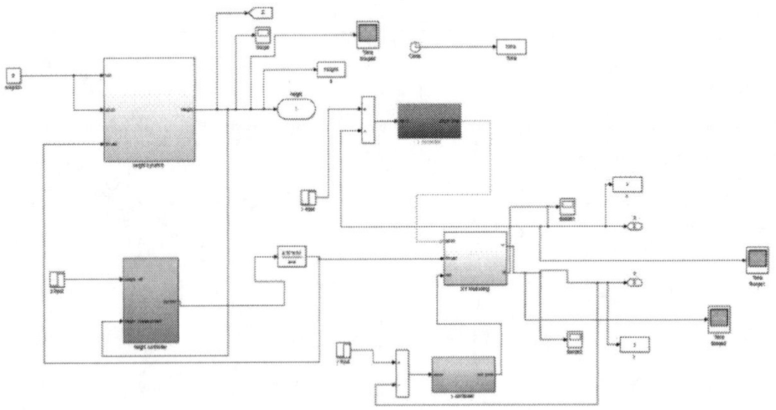

Figure 3. Height, x and y axis model connection.[8]

2.5. Yaw Model

The torque,τ, has the following equation with respect to the PWM input,u

$$\tau = K_y u \tag{8}$$

where K_y is a positive gain. The difference between the torques exerted by the two clockwise and the two counter-clockwise rotating propellers results in the motion of the yaw axis.[7] Motion in the yaw axis could be describe using the following equation:

$$J_y \, \ddot{\theta}_y = \Delta\tau \tag{9}$$

θ_y : yaw angle
J_y rotational inertia about the z axis.

Total torque of the motors, $\Delta\tau$ can be calculated from:

$$\Delta\tau = \tau_1 + \tau_2 - \tau_3 - \tau_4 \tag{10}$$

Table 1. System Parameters.

K	120 N
ω	15 rad/sec
J_{roll}	0.03 $kg.m^2$
J_{pitch}	0.03 $kg.m^2$
M	1.4 kg
K_y	4 N.m
J_y	0.032 $kg.m^2$
L	0.2 m

3. PID Control

This classical PID control and how to apply the dynamical model of the Qball-X4 quadrotor system are shown in [8,9,10]. All of the parameters are adopted from the Qball-X4 user manual(Table 1)[7].For example, for height model:Kp = 0.011 Kd = 0.009 Ki = 0.007. For x and y axis model:Kp = 0.366 Kd = 0.382 Ki = 0.045, for roll and pitch model: Kp = 0.062 Kd = 0.013 Ki = 0.018 and for yaw axis: Kp = 0.013 Kd = 0.009 Ki = 0.010. Although, system is analyzed for different values of parameters such as K, J, M and Ky, Jy and their responses, to be able to understand the system behavior if the parameters are uncertain.

4. Fuzzy Control

In this part, a different method has been applied to our Qball-X4 system. As an extension of the conventional PID control theory, a different fuzzy controller structure is applied. The name of the control structure is PID type fuzzy controller. Fuzzy controller structure is based on a fuzzy logic. Fuzzy logic is a logic, in which the truth variables can take any real number between 0 and 1. It is different than the boolean logic, because in boolean structure truth variables can take only 0 or 1. This control structure has just two inputs, and the rules base ,is two dimensions. Also, compared to fuzzy PI and fuzzy PD in general it is possible to enhance system performance using fuzzy PID. Fuzzy control design includes three important aspects. First one is, knowledge based design, second is control tuning parameters and the last one is membership functions.[8,9,10]

4.1. PID Type Fuzzy Controller Structure

A PID type fuzzy controller structure includes both PD type and PI type fuzzy controllers. And it is known that fuzzy PI type control is more practical than the fuzzy PD type.[8,9,10,11] Removing the steady state error is difficult for the fuzzy PD, and it is the main reason why PI type is more practical. Although, fuzzy PI type control does not provide adequate performance in transient response for high order process because of the internal integration operation. PID type fuzzy controller structure is using the error and the rate of change of error as it inputs.[8,9,10] PI and PD type fuzzy controllers connect in parallel and it shows in the Figure 4.
And the rules are expressed like that, if e is __ and $\dot{e}$ is __ then u is __[6,7].

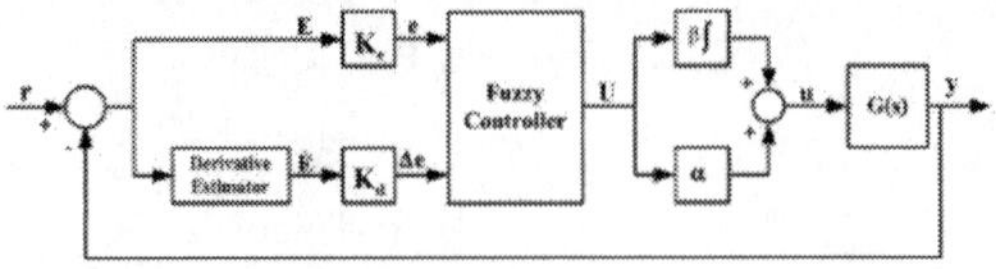

Figure 4. The PID Type fuzzy control structure.

Also, the output of the PID type fuzzy controller as follows:

$$u_c = \alpha u + \beta \int u dt \tag{11}$$

$$u_c = \alpha \left(A + PK_e e + DK_d \dot{e} \right) + \beta \int \left(A + PK_e e + DK_d \dot{e} \right) dt \tag{12}$$

$$u_c = \alpha A + \beta At + (\alpha K_e P + \beta K_d D) e + \beta K_e P \int edt + \alpha DK_d \dot{e} \tag{13}$$

And the control components are like[9,10]:
Proportional: $\alpha K_e P + \beta K_d D$
Integral:βK_e
Derivative:$\alpha D K_d$

These control components show that how difficult and complex of the fuzzy logic control structure. For example α parameter can be in both proportional and derivative terms.
Parameter of α, β, K_e and K_d are obtained as follows. First, using the genetic algorithm to find the parameter of α, β, K_e and K_d, provides a starting point for these parameters. In this method, MATLAB^{TM} genetic algorithm toolbox has been used. ISE function is minimized with using this toolbox. Then using the trial and error method, much better values of the parameters have been found. Membership functions of error, change rate of the error and u are shown in Figure 5.

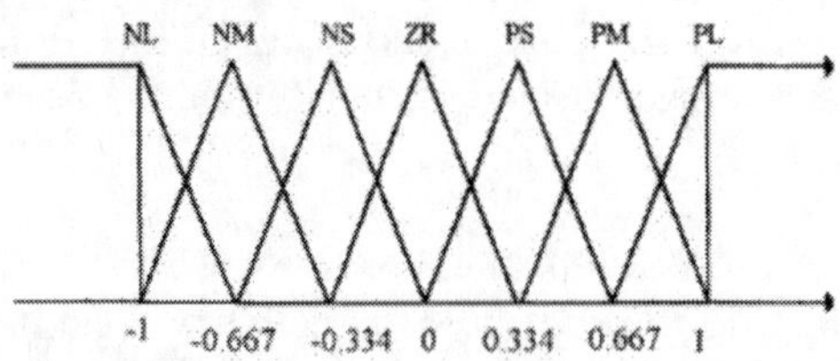

Figure 5. Membership functions of e, $\dot{e}$ and u.

IOP Conf. Series: Journal of Physics: Conf. Series **783** (2017) 012039 doi:10.1088/1742-6596/783/1/012039

The fuzzy PID type control rule is shown in Figure 6[8,9,10]:

E/E	NL	NM	NS	ZR	PS	PM	PL
PL	ZR	ps	pm	pl	pl	pl	pl
PM	ns	ZR	ps	pm	pl	pl	pl
PS	nm	ns	ZR	ps	pm	pl	pl
ZR	nl	nm	ns	ZR	ps	pm	pl
NS	nl	nl	nm	ns	ZR	ps	pm
NM	nl	nl	nl	nm	ns	ZR	ps
NL	nl	nl	nl	nl	nm	ns	ZR

Figure 6. General fuzzy PID type rule base.

The surface of the error, change rate of the error and u are shown in Figure 7.

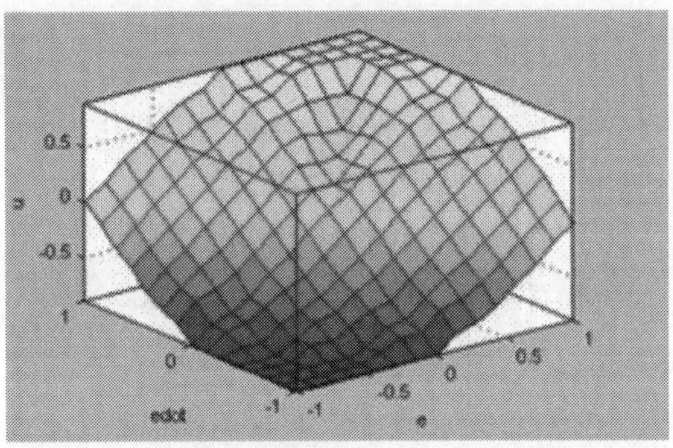

Figure 7. Surface of the e, $\dot{e}$ and u.

The output u is determined by using the method of the center of gravity(in Figure 7).

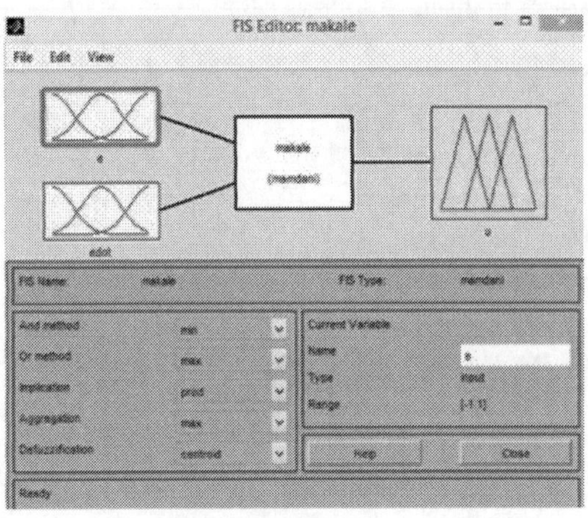

Figure 8. Output u.

5. Comparison

In this part of paper, there is a detailed comparison of the controller structures. Firstly, height response of the system is given.[8,9]

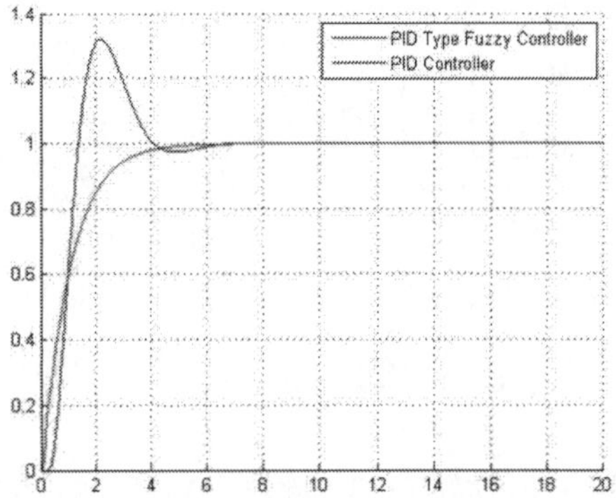

Figure 9. Response of Height for different controllers using K=120 gain.

Table 2. Height Performance Comparison

Time Response	PID	PID type Fuzzy
Overshoot	57.937 %	0,5 %
Rising Time	514.484 ms	2.203 sec
Settling Time	15.447 sec	6.711 sec

According to figure(9) and the table(2), PID type fuzzy controller structure better than classical PID structure. Because, with the PID type Fuzzy Controller, there is a negligible overshoot and the settling time is shortened(Figure 9)[8,9].
X and Y positions responses are given as follows. [8,9]

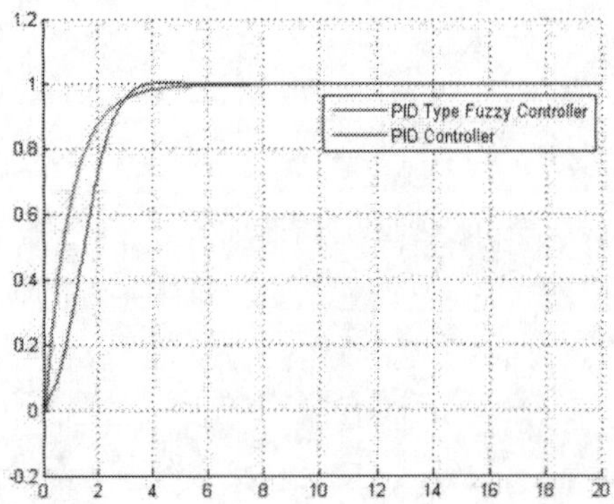

Figure 10. Response of X position for different controllers using K=120 gain.

Table 3. X axis Performance Comparison

Time Response	PID	PID type Fuzzy
Overshoot	0.505 %	0,497 %
Rising Time	2.114 sec	2.036 sec
Settling Time	7.862 sec	6.726 sec

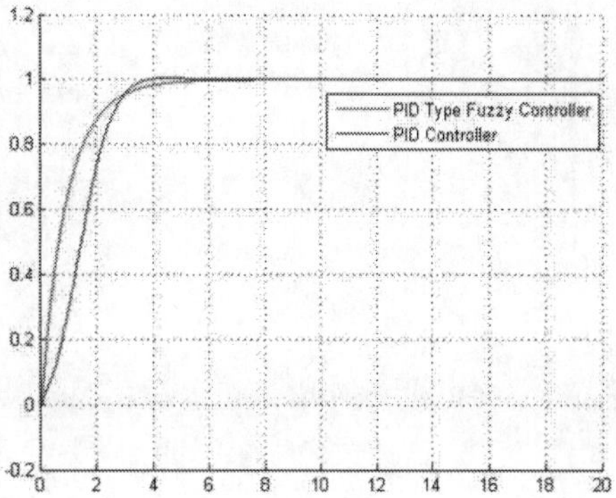

Figure 11. Response of Y position for different controllers using K=120 gain.

Table 4. Y axis Performance Comparison

Time Response	PID	PID type Fuzzy
Overshoot	0.505 %	0,497 %
Rising Time	2.093 sec	2.036 sec
Settling Time	6.649 sec	6.336 sec

With these figures(10,11) and tables(3,4), we can say that PID type fuzzy controller structure better than classical PID structure since with the PID type Fuzzy Controller, both overshoot and settling time are reduced(Figure 10, 11)[8,9].

Pitch and Roll angle responses are given as follows:

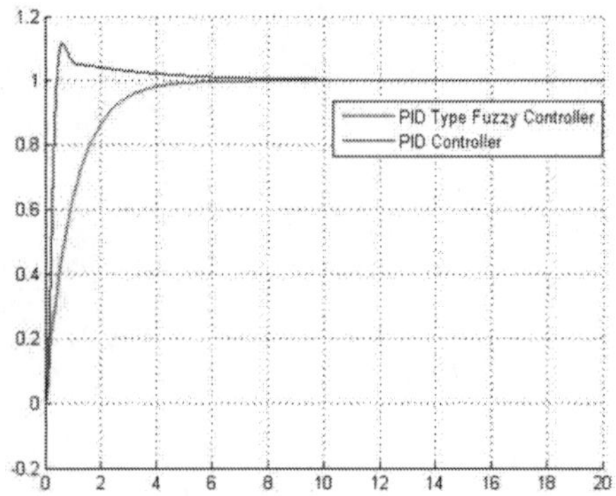

Figure 12. Response of Pitch for different controllers using K=120 gain.

Table 5. Pitch Performance Comparison

Time Response	PID	PID type Fuzzy
Overshoot	11,798 %	0,5 %
Rising Time	242.04 ms	2.143 sec
Settling Time	15.731 sec	8.810 sec

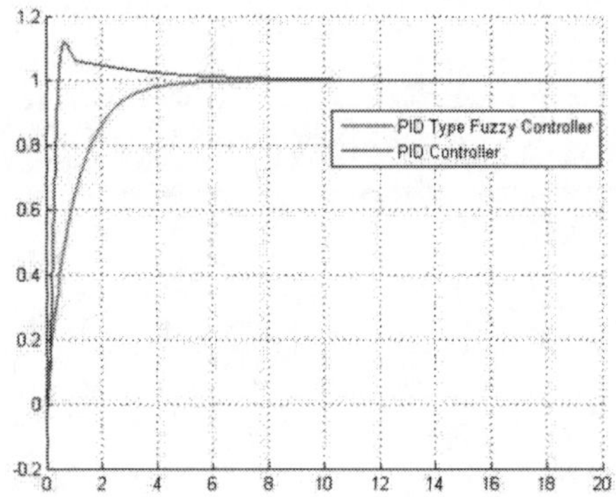

Figure 13. Response of Roll for different controllers using K=120 gain.

Table 6. Roll Performance Comparison

Time Response	PID	PID type Fuzzy
Overshoot	11.798 %	0,5 %
Rising Time	257.211 ms	2.143 sec
Settling Time	16.213 sec	8.779 sec

According to these figures(12,13) and tables(5,6), if the PID type fuzzy controller structure is used, there is a negligible overshoot lin the system. Moreover, the settling time is shortened(Figure 12, 13)[8,9].

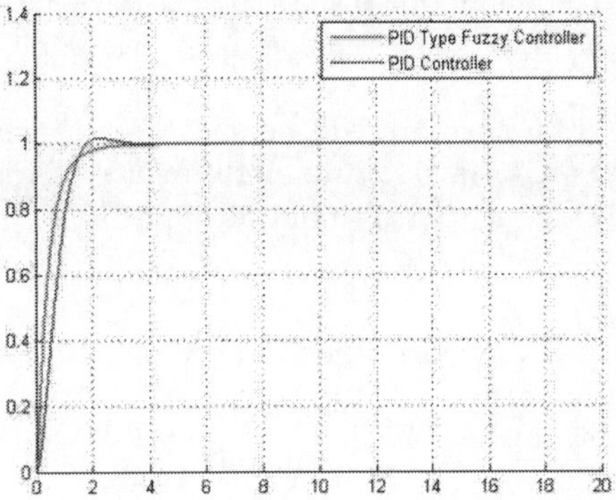

Figure 14. Response of yaw for different Controllers using Jy=0.1 gain.

Table 7. Yaw Performance Comparison

Time Response	PID	PID type Fuzzy
Overshoot	0,505 %	0,486 %
Rising Time	856.946 ms	1.046 sec
Settling Time	5.292 sec	4.569 sec

With this figure(14) and table(7), we can say that PID type fuzzy controller structure better than classical PID structure. Because, with the PID type Fuzzy Controller, both overshoot and settling time are reduced.(Figure 13)[8,9]

6. Conclusion and Future Work

In this paper, firstly the dynamical model of the Qball-X4 system has been described. Non-linear model of the system have been given. After the examination of the dynamical model, PID control has been described. After the PID control part, fuzzy control part is presented. In this part, as an extension of the conventional PID control theory, a different fuzzy controller structure is described. The name of control structure is PID type fuzzy controller. Both of the controllers simulation results' are shown for the position controls along x,y,z axis and roll, pitch yaw angles.

To improve further the performance of classical PID control structure, we proposed a fuzzy controller structure which is based on fuzzy logic and its name is PID type fuzzy controller. With this PID type fuzzy controller x,y,z positions and roll,pitch, yaw angles are controlled in MATLAB^{TM} environment. The simulation results show that with the PID type fuzzy controller, there is not any overshoot and the settling time is shortened.[8,9]

In the future a different kind of structure will be suggested in which the proposed PID type fuzzy controller is developed. Also, a real time performance of the PID type fuzzy controller would be examined or fault tolerant control could be applied on Qball X4 quadrotor system.

References

[1] I. Sadeghzadeh, A. Mehta, A. Chamseddine, and Y. Zhang, "Active fault tolerant control of a quadrotor uav based on gainscheduled pid control," Electrical and Computer Engineering (CCECE), vol. ., pp. 1-4, May 2012.
[2] A. Chamseddine, Y.M. Zhang, C.A. Rabbath, C. Fulford, J. Apkarian, "Model reference adaptive fault tolerant control of a quadrotor UAV," in: AIAA Infotech@Aerospace, St. Louis, Missouri, USA, 29-31 March 2011.
[3] N.M. Raharja , Iswanto, M. Faris, A.I. Cahyadi, "Hover Position Quadrotor Control with Fuzzy Logic",1st International Conference on Infonnation Technology, Computer and Electrical Engineering,2014
[4] D.C. Tosun, Y. Isik, H. Korul, "Comparison of PID and LQR controllers on a quadrotor helicopter", International Journal of Systems Applications, Engineering and Development, Volume 9, 2015
[5] R.Ozdemir, M.Kaya, M.Elfarra, M.O.Efe, "Doner Kanatli Insansiz Hava Araci Kullanarak Bolgesel Gozetim Amacli Kisi veya Nesne Takibi", Turk Hava Kurumu Universitesi, Ankara,2012
[6] Moghadam, M., Fault Tolerant Control of a Quadrotor UAV,2016
[7] Qball Innovate Educate, Quanser Qball-X4: User Manual
[8] T.Bodrumlu Modelling and Control of the Qball X4 Quadrotor System based on PID and Fuzzy Logic Structure, Msc Thesis,Istanbul, 2016.
[9] T.Bodrumlu, M.T. Soylemez, Qball X4 Dort Pervaneli Rotor Sisteminin Modellenmesi ve PID ve Bulanik Mantik Yapisina Dayali Kontrolu, TOK 2016, 29 September - 1 October, Eskisehir
[10] Z.W. Woo, H.Y. Chung , J.J. Lin, A PID type fuzzy controller with self-tuning scaling factors, *Fuzzy Sets and Systems 115*, pp. 321-326, 2000.
[11] Han-Xiong, H.B. Gatland, Conventional fuzzy and its enhancement, IEEE Trans. Systems Man Cybernet. 26 (5), 791-796 ,1996

IOP Conf. Series: Journal of Physics: Conf. Series **783** (2017) 012040 doi:10.1088/1742-6596/783/1/012040

Rigid Body Attitude Control Based on a Manifold Representation of Direction Cosine Matrices

David Nakath, Joachim Clemens, Carsten Rachuy

Cognitive Neuroinformatics, University of Bremen,
Enrique-Schmidt-Str. 5, 28359 Bremen, Germany

E-mail: {dnakath,jclemens,rachuy}@informatik.uni-bremen.de

Abstract. Autonomous systems typically actively observe certain aspects of their surroundings, which makes them dependent on a suitable controller. However, building an attitude controller for three degrees of freedom is a challenging task, mainly due to singularities in the different parametrizations of the three dimensional rotation group $SO(3)$. Thus, we propose an attitude controller based on a manifold representation of direction cosine matrices: In state space, the attitude is globally and uniquely represented as a direction cosine matrix $R \in SO(3)$. However, differences in the state space, i.e., the attitude errors, are exposed to the controller in the vector space $\mathbb{R}^3$. This is achieved by an operator, which integrates the matrix logarithm mapping from $SO(3)$ to $\mathfrak{so}(3)$ and the map from $\mathfrak{so}(3)$ to $\mathbb{R}^3$. Based on this representation, we derive a proportional and derivative feedback controller, whose output has an upper bound to prevent actuator saturation. Additionally, the feedback is preprocessed by a particle filter to account for measurement and state transition noise. We evaluate our approach in a simulator in three different spacecraft maneuver scenarios: (i) stabilizing, (ii) rest-to-rest, and (iii) nadir-pointing. The controller exhibits stable behavior from initial attitudes near and far from the setpoint. Furthermore, it is able to stabilize a spacecraft and can be used for nadir-pointing maneuvers.

1. Introduction

Autonomous systems typically need to keep track of their position and attitude, i.e., their pose, as well as properties of their surroundings for successful operations. The gathered information is mutually beneficial, as – on the one hand – precise pose information improves the ability to describe the surroundings and – on the other hand – a precise description of the latter improves the ability to determine an exact pose. To this end, it is often useful for autonomous systems to actively observe certain aspects of their surroundings, which makes them dependent on a suitable controller.

This especially holds for space missions, where active observation of planetary objects can be used for absolute navigation [11, 18] as well as for nadir-pointing maneuvers for relative navigation in the vicinity of a celestial object, e.g., within a LIDAR-based SLAM approach [6, 20]. However, the scenario may also apply to airplanes, underwater vehicles [9], ground vehicles, and robotic systems, which have to inspect certain aspects of their surroundings by aligning some sensor to a point of interest.

The only global and unique representation of the set of all attitudes a rigid body might assume in three dimensions is the *direction cosine matrix* (DCM) [5]. Parameterizations like

Euler angles have the disadvantage of geometric and kinematic singularities as well as different conventions. The (modified) Rodriguez parameters, which are derived from unit quaternions, suffer from geometric singularities. Even the globally defined unit quaternions can lead to undesired rotations, which is known as controller unwinding [22, 2, 16]. The latter problem stems from the non-unique map between the unit quaternion space $\hat{\mathbb{H}}$ and the rotation space $SO(3)$ [2, 5].

As the computational power and available memory of autonomous systems constantly increases, it has become feasible to directly employ DCMs for pose representations. Thus, the drawbacks of the parametrized attitude representations can be omitted by running control algorithms directly on DCMs [10].

In this paper, we present a filtered proportional derivative saturation preventing rigid body controller. It represents the attitude in $SO(3)$ using a DCM, while exposing the differences between two attitudes in the vector space $\mathbb{R}^3$ employing the $\boxplus$-method [13]. The derivative term is bounded by a nonlinearity, to ensure a bounded control torque in each particular axis. The feedbacks are preprocessed by a particle filter to account for state and measurement noise. The tuning of the controller in the respective maneuver scenarios can be considered as another modest contribution.

This paper is structured as follows. First, in Sect. 2, the problems to be addressed and the corresponding mathematical and geometrical preliminaries will be presented. In Sect. 3, the spacecraft (s/c)-state and a particle filter-based (PF) estimation scheme will briefly be introduced. Based on these foundations, the controller will be derived in Sect. 4. Subsequently, it will be evaluated and discussed in Sect. 5. Finally, this paper concludes in Sect. 6.

2. Preliminaries

In the following, we will briefly cover the mathematical foundations of the derivation of the controller.

2.1. Attitude

Consider a rigid body representing a s/c.[1] An attitude is the rotation of the principal axes of some body-fixed frame F_b relative to some global inertial frame F_a, which is uniquely described by a DCM denoted by R_{ba}. The set of all attitudes is composed of all orthonormal rotation matrices with determinant equal to one, i.e., $\{R \in \mathbb{R}^{3\times3} | R^T R = 1, \det R = +1\}$. This set forms the special orthogonal group $SO(3)$ of rigid body rotations in 3D [17].

2.2. $\boxplus$-Method

We use a $\boxplus$-encapsulation of $SO(3)$ introduced by Hertzberg et al. [13], which is capable of integrating a higher dimensional state representation $\mathcal{S}$ with a local lower dimensional state representation $\mathbb{R}^n$ (cf. Fig. 1). This is achieved by the mappings

$$\begin{aligned} \boxplus &: \mathcal{S} \times \mathbb{R}^n \to \mathcal{S}, \\ \boxminus &: \mathcal{S} \times \mathcal{S} \to \mathbb{R}^n. \end{aligned} \tag{1}$$

The DCM-specific operators encapsulate the manifold by combining the matrix exponential ($\boxplus$-operator) or logarithm ($\boxminus$-operator) respectively with the isomorphism between the Lie-algebra $\mathfrak{so}(3)$ and $\mathbb{R}^3$, in particular

$$\begin{aligned} x \boxplus \delta &= x \exp \delta, \\ y \boxminus x &= \log(x^{-1} y), \end{aligned} \tag{2}$$

[1] We will only take attitude control into account in this paper, as it can be treated independently of translation in our scenario.

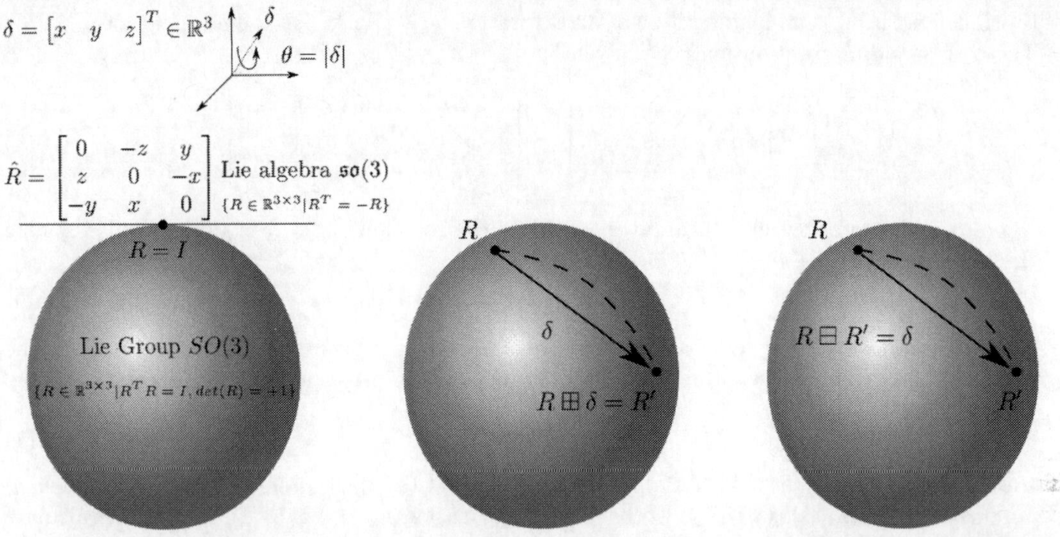

(a) Joint structure of $SO(3)$, $\mathfrak{so}(3)$, and $\mathbb{R}^3$.

(b) $\boxplus$-operator: $\delta \in \mathbb{R}^3$ added to element $R \in SO(3)$, resulting in $R' \in SO(3)$.

(c) $\boxminus$-operator: Distance computation between R and R' on the $SO(3)$-manifold with result $\delta \in \mathbb{R}^3$.

Figure 1: (a) Joint structure of $SO(3)$, $\mathfrak{so}(3)$, and $\mathbb{R}^3$ with corresponding encapsulating operators in (b) and (c). Dashed lines indicate the geodesics on the manifold, while solid arrows indicate the corresponding vectors in $\mathbb{R}^3$. Following [13], the $\boxplus$-principles are shown using the unit sphere S^2 to simplify visualization.

with $x, y \in SO(3)$ and $\delta \in \mathbb{R}^3$. In the case of 3D rotations, the matrix exponential can be computed in closed-form by the Rodriguez-formula [13, 17, 14]. Combined with the hat map from $\mathbb{R}^3$ to $\mathfrak{so}(3)$ this results in

$$\exp \begin{bmatrix} x \\ y \\ z \end{bmatrix} = \begin{bmatrix} \cos\theta + cx^2 & -sz + cxy & sy + cxz \\ sz + cxy & \cos\theta + cy^2 & -sx + cyz \\ -sy + cxz & sx + cyz & \cos\theta + cz^2 \end{bmatrix}, \tag{3}$$

where $\theta = \sqrt{x^2 + y^2 + z^2}$, $s = \frac{\sin\theta}{\theta}$, and $c = \frac{1-\cos\theta}{\theta^2}$. The inverse mapping is a combination of the matrix logarithm and the hat map from $\mathfrak{so}(3)$ to $\mathbb{R}^3$, which is computed according to

$$\log x = \frac{\theta}{2\sin\theta} \begin{bmatrix} x_{32} - x_{23} \\ x_{13} - x_{31} \\ x_{21} - x_{12} \end{bmatrix}, \tag{4}$$

where $\theta = \text{acos}\frac{\text{tr}\, x - 1}{2}$ and $|\theta| < \pi$. Thus, the global state is represented uniquely in $SO(3)$, while state differences δ, i.e., state transitions or errors, are uniquely represented in $\mathbb{R}^3$ for angles $< \pi$.

2.3. Dynamic Setpoint Calculation

To have an autonomous system looking at some observable, we have to rotate the sensor normal into the desired view direction. The rotation of a unit vector A to some desired unit vector B

(both defined in F_a) can be described as a 2D rotation on a plane with the normal vector $A \times B$. The corresponding rotation matrix is given by

$$G = \begin{bmatrix} \cos\phi & -\sin\phi & 0 \\ \sin\phi & \cos\phi & 0 \\ 0 & 0 & 1 \end{bmatrix} = \begin{bmatrix} A \cdot B & -||A \times B|| & 0 \\ ||A \times B|| & A \cdot B & 0 \\ 0 & 0 & 1 \end{bmatrix}, \tag{5}$$

where ϕ is the rotation angle. However, to apply the rotation, we have to temporarily change the basis with

$$F = \begin{bmatrix} A & \frac{B-(A \cdot B)A}{||B-(A \cdot B)A||} & B \times A \end{bmatrix}^{-1}. \tag{6}$$

The final rotation matrix R_e is obtained by a transformation from the inertial frame F_a to the rotation frame, followed by a rotation about G, and a transformation back to the inertial frame. Thus, we get

$$R_e = F^{-1}GF. \tag{7}$$

If we now assume, for the sake of simplicity, that we want to orient a sensor whose sensor normal points along the z_{ba} axis of R_{ba}, we have that $A = z_{ba}$ and $B = z_{da}$. Thus, we can calculate a rotation matrix R_e following the procedure described above. The resulting R_e can also be applied to another rotation matrix, as $SO(3)$ is closed under matrix multiplication. Hence, we can obtain the desired attitude R_{da}, i.e., where z_{ba} is rotated to z_{da}, by rotating the current attitude R_{ba} according to

$$R_{da} = R_e R_{ba}. \tag{8}$$

In order to obtain a suitable representation for the controller, the rotation from R_{ba} onto R_{da} can be retrieved in vector space using Eq. (2). This yields

$$\delta = R_{ba} \boxminus R_{da}, \tag{9}$$

where the $\boxminus$-operator ensures that δ represents the geodesic, i.e., the shortest rotation on $SO(3)$ [13]. The approach described above will be used for time-dependent setpoint (SP) calculation $R_{da}(t)$ within the nadir-pointing maneuver (cf. Sect. 5).

2.4. Attitude Kinematics and Dynamics

Applying this geometric framework to attitude kinematics, we let $\omega \in \mathbb{R}^3$ be the angular velocity relative to F_a given in F_b, then the time rate of change of the rigid body attitude is

$$\dot{R}_{ba} = R_{ba}\hat{\omega}. \tag{10}$$

Here, $\hat{\omega}$ is a skew symmetric matrix, as it is an element of the Lie algebra $\mathfrak{so}(3)$ (cf. Fig. 1a), and $\dot{R}_{ba}$ maps the body coordinates of a point to the spatial velocity of that point. The attitude dynamics are the rate of change of angular velocity expressed in the body frame given by

$$I\dot{\omega} = I\omega \times \omega + u, \tag{11}$$

where I is the inertia tensor and u is the sum of external moments applied to the body given in F_b. Finally, throughout this paper we assume that the s/c is fully actuated regarding attitude, i.e., angular moments can be applied on all three axes.

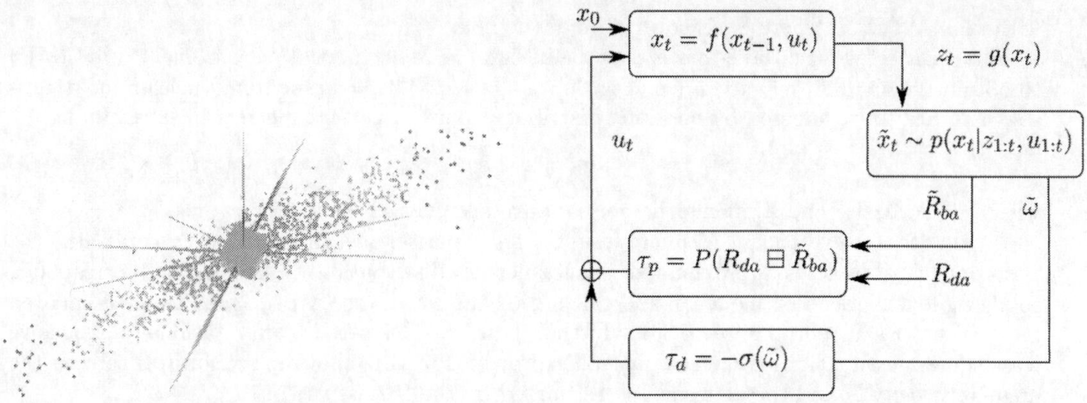

(a) Particle filter (b) Controller structure

Figure 2: (a) shows the s/c visualization with the corresponding particle distribution, where green and red represents a high and low weight respectively. (b) depicts the control diagram of the filtered pd controller.

3. Particle Filtered Measurements

The controller proposed in the next section relies on measurements in the closed loop. To account for noise in state transition as well as sensor measurements we use a particle filter (PF) [8, 21] for state estimation (cf. Fig. 2a). This type of filter is chosen, as it integrates well with the SLAM approach [6] to be used for relative navigation in the vicinity of asteroids. The basic concepts and integration of attitude sensors (star tracker and 3-axis gyroscope), which are relevant in the context of this paper, are described in the following. Further details on our quaternion-based approach including all sensor models can be found in [11, 18]. The state space is defined as $\mathcal{S} = SE(3) \times \mathbb{R}^6$, where $SE(3) = \mathbb{R}^3 \times SO(3)$ represents the space of position and attitude. The remaining $\mathbb{R}^6$ is the space of the corresponding first derivatives, i.e., the linear and angular velocities. The PF estimates a probability distribution of the current state $x_t \in \mathcal{S}$ of the following form

$$\tilde{x}_t \sim p(x_t|z_{1:t}, u_{1:t}), \tag{12}$$

where $\tilde{x}_t \in \mathcal{S}$ is the estimated state, $z_{1:t}$ are all preceding measurements and $u_{1:t}$ are all preceding controls. This probability density function is approximated using a set of particles, where each particle represents a state hypothesis. The particle set is updated recursively over time. In each step, the particles are predicted from the previous to the current time step by sampling from the proposal distribution. This is done by integrating the dynamic model of the s/c, which depends on the previous state estimate and the most recent control u_t, using a fourth order Runge-Kutta integrator and applying additive normally-distributed noise with zero mean. Subsequently, the distribution is corrected using importance resampling. The importance measure is obtained by weighting the particles according to the measurement likelihood given the sampled state. Finally, new particles are sampled with a probability proportional to the weights.

A star tracker is used for determining the attitude R_{ba} of the s/c. It operates by measuring angles to stars (see Fig. 3b) and comparing these to known star positions indexed in a star catalog, e.g., the *HIPPARCOS* [19]. For simplicity, we assume that the resulting attitude measurement z_s corresponds to the true attitude R_{ba} and additive normally-distributed noise with zero mean. Hence, the measurement model is given by

$$z_s = R_{ba} \boxplus \varepsilon_s, \tag{13}$$

with $\varepsilon_s \sim \mathcal{N}(0, \Sigma_s)$, where Σ_s is covariance of the noise.

Furthermore, we use the 3-axis gyroscope integrated in an inertial measurement unit (IMU) to obtain the angular velocity ω (see Fig. 3c) of the s/c. The corresponding measurement z_g is also assumed to be affected by normally-distributed noise with zero mean, which results in

$$z_g = \omega + \varepsilon_g, \tag{14}$$

with $\varepsilon_g \sim \mathcal{N}(0, \Sigma_g)$ and Σ_g being the noise covariance matrix.

In order to retrieve a single estimate from the set of particles as an input for our controller, the posterior distribution is approximated using a normal distribution. This basically corresponds to a weighted average for the mean and calculating the covariance with respect to the estimated mean using the $\boxminus$-operator (see [18] for further details). This avoids computationally expensive least squares estimations [12]. The controller proposed in the following section will rely on the angular velocity $\tilde{\omega}$ and the attitude $\tilde{R}_{ba}$ estimated by the PF as part of $\tilde{x}_t$.

4. Filtered Proportional Derivative Controller

The proportional derivative controller (cf. Fig. 2b) computes the current control u_t as a sum of the proportional τ_p and the derivative τ_d control torque according to

$$u_t = \tau_p + \tau_d. \tag{15}$$

In [4], a feedback controller is presented, which uses the matrix logarithm mapped into vector space to expose the geodesic, i.e., the shortest path on $SO(3)$, as the error term. This approach has been adopted by [1, 15], who additionally show the stability of proportional derivative controllers using such an error term. Analogously, the proportional control term can be computed according to a weighted error function based on the $\boxminus$-operator, resulting in

$$\tau_p = P\,\delta = P(R_{da} \boxminus \tilde{R}_{ba}), \tag{16}$$

where P is the proportional gain matrix. The error function calculates the difference $\delta \in \mathbb{R}^3$ between the desired $R_{da} \in SO(3)$ and the current estimated attitude $\tilde{R}_{ba} \in SO(3)$.

To implement an actuator saturation prevention mechanism [3, 7] based on the proportional term, we observe that δ is guaranteed to never exceed $|\pi|$ in its components. This is asserted, as the error function in Eq. (16) returns the vector representation of the geodesic, whose magnitude can never be greater than π. We initially set $Q = diag(\bar{u}_1/\pi, \bar{u}_2/\pi, \bar{u}_3/\pi)$, where $\bar{u}_i$ indicates the maximum allowed torque per axis, to ensure a bounded maximum value on every axis. For tuning purposes, this value can be scaled with the factor $K = diag([0,1],[0,1],[0,1])$ (cf. Sect. 5). Thus, we let $P = KQ$, where $K_{ii} = 0$ puts all weight on the derivative term, and $K_{ii} = 1$ puts all weight on the proportional term of the respective axis i.

The derivative term is the negative output of the nonlinearity σ, which acts upon the estimated angular velocity $\tilde{\omega}$

$$\tau_d = -\sigma(\tilde{\omega}). \tag{17}$$

The nonlinearity applies upper and lower bounds to the estimated angular velocity $\tilde{\omega}$ [10]

$$\sigma(\tilde{\omega}) = [\sigma_1(\tilde{\omega}_1), \sigma_2(\tilde{\omega}_2), \sigma_3(\tilde{\omega}_3)]^T \tag{18}$$

$$\sigma_i(\tilde{\omega}_i) = \begin{cases} \bar{\omega}_i & \tilde{\omega}_i > \bar{\omega}_i \\ \tilde{\omega}_i & -\bar{\omega}_i \leq \tilde{\omega}_i \leq \bar{\omega}_i, \qquad i = 1,2,3 \\ -\bar{\omega}_i & \tilde{\omega}_i < -\bar{\omega}_i, \end{cases} \tag{19}$$

where the bounding term $\bar{\omega}_i$, $i = 1,2,3$ is computed as the difference between the maximum allowed torque and the maximal value the scaled error term might assume, i.e., $\bar{\omega}_i = \bar{u}_i - P_{ii}\pi$. The indirect dependence on the choice of K ensures that the tuned control u_t never exceeds the maximum allowed torque $\bar{u}$.

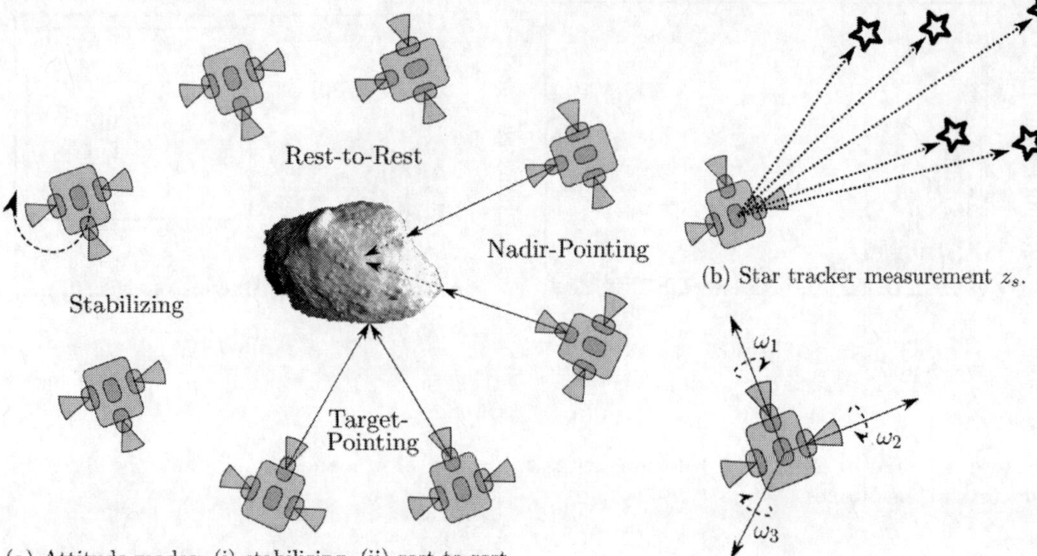

(a) Attitude modes: (i) stabilizing, (ii) rest-to-rest, (iii) nadir-pointing, and (iv) target-pointing.

(b) Star tracker measurement z_s.

(c) Gyroscope measurement z_g.

Figure 3: (a) s/c-attitude modes exemplified around the Itokawa asteroid, (b) gyroscope measurement, and (c) star tracker measurement geometry.

5. Evaluation

We conducted the evaluation in a custom written software, which simulates the solar system bodies as well as rigid body kinematics and dynamics using a fourth order Runge-Kutta integrator. The attitude propagation is based on Eqs. (10) and (11), where the inertia tensor is $I = diag(I_1, I_2, I_3)$, with $I_1 = 3305.49\,kgm^2$, $I_2 = 4653.63\,kgm^2$, and $I_3 = 4814.78\,kgm^2$. For evaluation, we chose an orbit around an asteroid with the gravitational parameter $2.342520648\,m^3s^{-2}$, which is orbiting with a semi-major axis of 3 AU around the Sun. The orbital elements defining the orbit of the s/c around the asteroid are: semi-major axis 0.00000000235 AU $\approx 0.34\,km$, inclination 0.0°, eccentricity 0.1, ascending node 0.0°, periapsis 114.6°, and longitude 57.3°. The simulation operated with a timedelta $dt = 60\,s$ per step. The initial attitude in the evaluation is $R_{ba}(0) = I$ in the **T1/T2** scenarios and varied in the **T3** scenario. Of the possible maneuvers shown in Fig. 3a, we evaluate the following ones under different conditions:

T1 *Rest-to-rest*: From a resting state with $\omega(0) = 0$, we try to reach a constant setpoint R_{da}.
- **T1A** The setpoint (SP) is close to the initial state $R_{da} = I \boxplus [3°, 3°, 3°]^T$, and
- **T1B** the setpoint is exactly 180° rotated with respect to the initial state $R_{da} = diag(-1, -1, 1)$.

T2 *Stabilization*: From an initial angular velocity $\omega(0) = [\pi/2, \pi/2, \pi/2]^T$, we try to reach a constant setpoint R_{da}.
- **T2A** Again, the setpoint is close to the initial state $R_{da} = I \boxplus [3°, 3°, 3°]^T$, and
- **T2B** the setpoint is exactly 180° rotated with respect to the initial state $R_{da} = diag(-1, -1, 1)$.

T3 *Nadir-pointing*: From a resting state $\omega(0) = 0$, we try to follow a dynamic setpoint $R_{da}(t)$, i.e., we want to orient our sensors towards the center of mass of the celestial object to be observed. As the setpoint is determined by the position of the s/c and the asteroid, we now change the initial attitude of the s/c to achieve the desired initial position deltas.

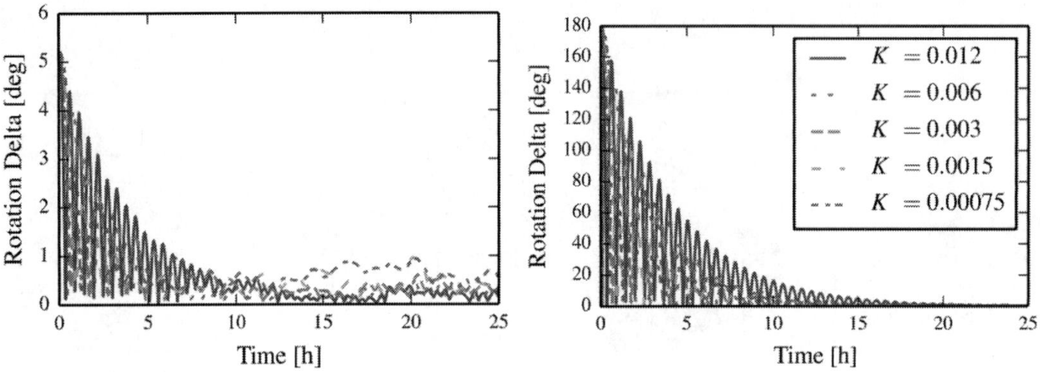

(a) T1A: SP close to the initial attitude.

(b) T1B: SP 180° rotated wrt. to the initial attitude.

Figure 4: T1A/B: 2-norm of rotation deltas for rest-to-rest maneuvers. (Note the different scalings of the Y-axes.)

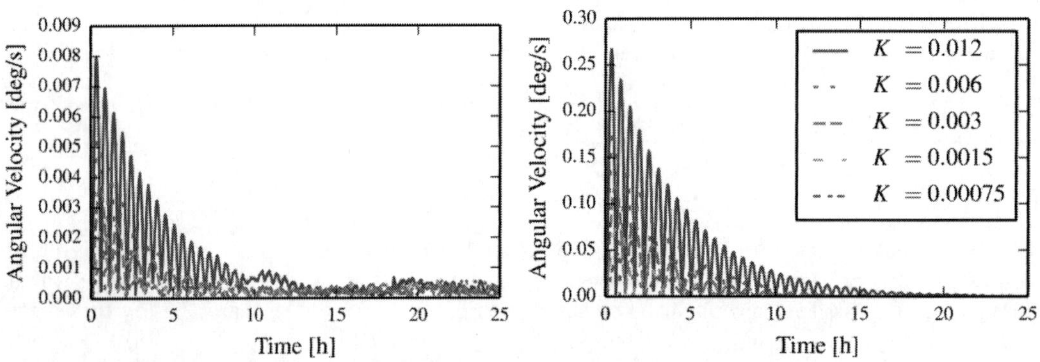

(a) T1A: SP close to the initial attitude.

(b) T1B: SP 180° rotated wrt. to the initial attitude.

Figure 5: T1A/B: 2-norm of angular velocities for rest-to-rest maneuvers. (Note the different scalings of the Y-axes.)

T3A Here, the initial state is modified to differ from the first setpoint setting $R_{ba}(0) = R_{da}(0) \boxplus [3°, 3°, 3°]^T$, and

T3B the initial state is exactly 180° rotated with respect to the initial setpoint $R_{ba}(0) = R_{da}(0)\, diag(-1, -1, 1)$.

We evaluated these scenarios with different gain scaling factors $K_{ii} = 0.00075, 0.0015, 0.003, 0.006, 0.012$, applied uniformly over all axes i. For simplicity, the factor will just be denoted K in the plots in Figs. 4–10. The rotation deltas in Figs. 4–10 are computed in every epoch t as the 2-norm of the current attitude error in vector space

$$\Delta R(t) = \|R_{ba}(t) \boxminus R_{da}(t)\|. \tag{20}$$

Thus, the error states the angle, which has to be carried out about a unique rotation axis to rotate $R_{ba}(t)$ onto $R_{da}(t)$. Similarly, the angular velocities and the control torques are shown in the plots as the 2-norm over all axes.

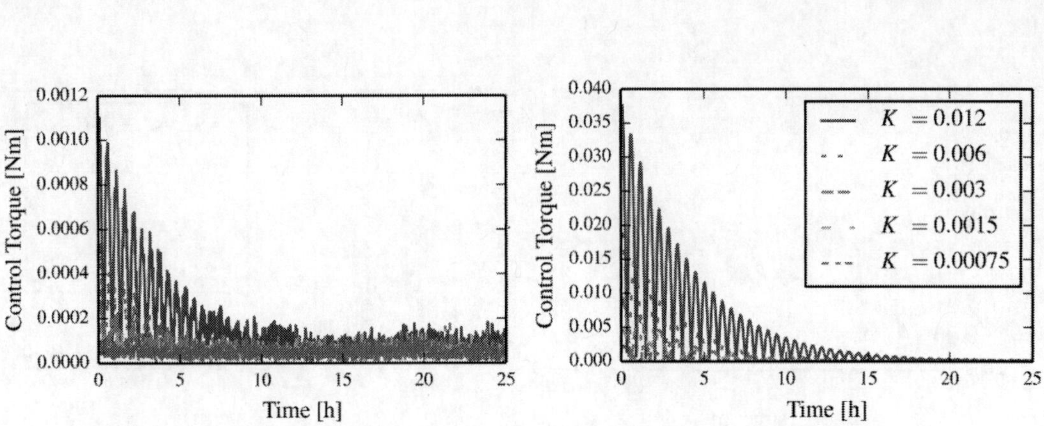

(a) T1A: SP close to the initial attitude.

(b) T1B: SP 180° rotated wrt. to the initial attitude.

Figure 6: T1A/B: 2-norm of spent control torques for rest-to-rest maneuvers. (Note the different scalings of the Y-axes.)

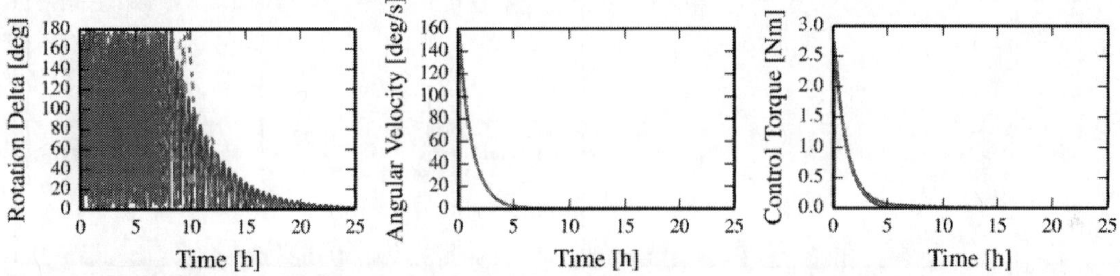

(a) 2-norm of rotation deltas.

(b) 2-norm of angular vel.

(c) 2-norm of spent torques.

Figure 7: T2A: Stabilizing with a setpoint close to the initial attitude. (Note that the error is bounded by 180°.)

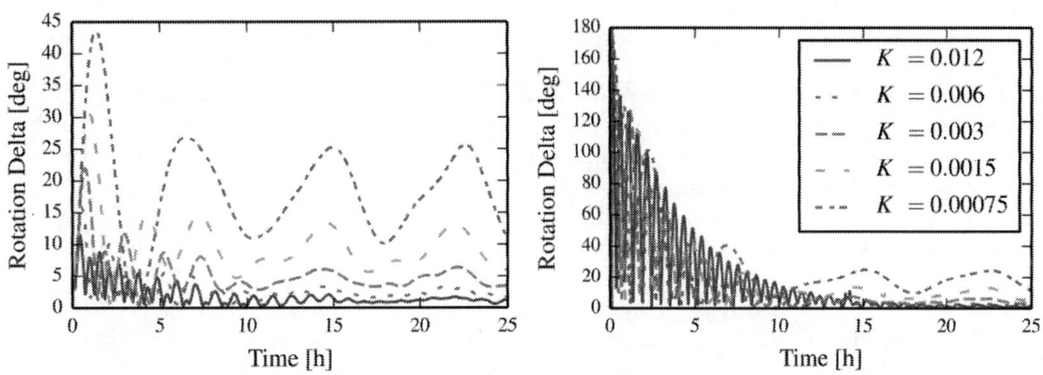

(a) T3A: SP close to the initial attitude.

(b) T3B: SP 180° rotated wrt. to the initial attitude.

Figure 8: T3A/B: 2-norm of rotation deltas for nadir-pointing maneuvers. (Note the different scalings of the Y-axes.)

IOP Conf. Series: Journal of Physics: Conf. Series **783** (2017) 012040 doi:10.1088/1742-6596/783/1/012040

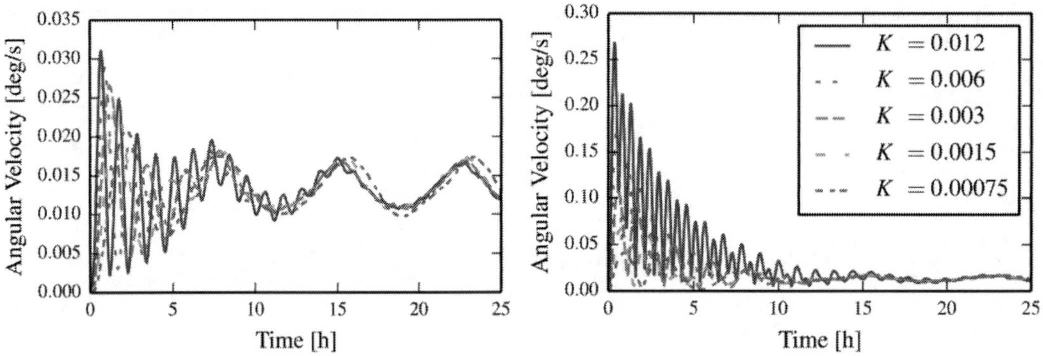

(a) T3A: SP close to the initial attitude.

(b) T3B: SP 180° rotated wrt. to the initial attitude.

Figure 9: T3A/B: 2-norm of angular velocities for nadir-pointing maneuvers. (Note the different scalings of the Y-axes.)

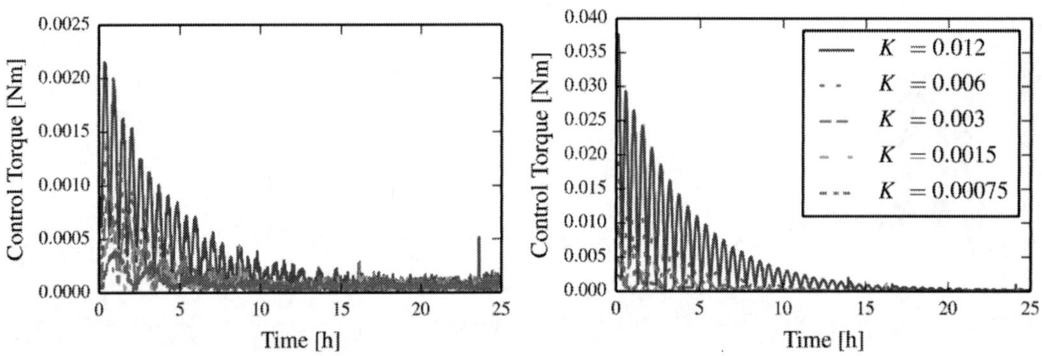

(a) T3A: SP close to the initial attitude.

(b) T3B: SP 180° rotated wrt. to the initial attitude.

Figure 10: T3A/B: 2-norm of spent control torques for nadir-pointing maneuvers. (Note the different scalings of the Y-axes.)

In the first scenario T1A, the controller is able to reach the setpoint within a range $< 1°$ in roughly 6h (cf. Fig. 4). The residual error is due to the state transition and measurement noise, which cannot be entirely smoothed by the filter. As expected, a small K is not able to keep track of the setpoint, while a higher K shows better results at the cost of more spent torque (cf. Fig. 6). In the second scenario T1B, the controller shows its robustness by stabilizing at the setpoint in about 15h (cf. Fig. 5). In T1A and T1B, the gain factor has a similar proportional impact on the control torque although the error starts from a much higher level (cf. Fig. 6).

The scenarios T2A and T2B show that the controller successfully stabilizes the s/c from tumbling after 20h (cf. Fig. 7). Starting from a high error the gain factor K is not that important anymore. The controller neatly decelerates over all axes and decreases the spent torque regardless of the gain factor (cf. Figs. 7b and 7c). We do not show plots of T2B here, as the initial pose does not have a high impact compared to the high initial angular velocity. Thus, the results are qualitatively equal to those of T2A. The gain factor $K = 0.006$ seems to be preferable over $K = 0.012$ for the T1 / T2 scenarios, as it yields a comparable performance while applying significantly less control torque.

In the nadir-pointing scenarios T3A and T3B, the gain factor becomes important again (cf. Figs. 8 and 9). It is obvious that a stiff controller is needed for such a demanding maneuver, as the weaker K are not able to reach the setpoint, neither in condition A (cf. Fig. 8a) nor in condition B (cf. Fig. 8b). In this scenario, we furthermore note a constant residual angular velocity, which is essential to follow the dynamic setpoint (cf. Figs. 9a and 9b). The controllers with a higher gain are able to apply higher control torques (cf. Fig. 10), and thus are able to follow the moving setpoint. Hence, for scenario T3 a gain factor of $K = 0.012$ seems to be preferable over $K = 0.006$ as it (i) approaches the dynamic setpoint faster and (ii) is able to maintain a slightly lower residual error.

6. Conclusion

This paper presented a proportional derivative controller working on a manifold encapsulation of direction cosine matrices. The controller uses particle filtered angular velocity and attitude estimates to compute control torques. The proportional term is calculated from the weighted error between two attitudes on the manifold $SO(3)$ which is mapped to the vector space $\mathbb{R}^3$. The derivative term is bounded by a nonlinearity, which depends on the proportional gain to prevent actuator saturation. In a numerical evaluation, we examined its behavior in different maneuvers: (i) stabilizing, (ii) rest-to-rest, and (iii) nadir-pointing. The controller exhibits stable behavior near the setpoint and from instability-prone initial attitudes. Furthermore, it is able to stabilize a s/c rotating about all three axes and can be used for nadir-pointing maneuvers, which involve a dynamic setpoint.

For future work, a combination with an information driven selection [18] of the setpoint with the control approach presented here is of high interest. Furthermore, the controller can be combined with a SLAM-approach [6] to examine possible landing sites by actively reducing uncertainty. This would result in a highly dynamic setpoint, which is not only determined by decisions as well as the motion of the autonomous system, but also by the motion of the observable itself. However, achieving this would allow autonomous systems to actively select dynamic observables and directly come up with a corresponding control.

Acknowledgments

This work was supported by the German Aerospace Center (DLR) with financial means of the German Federal Ministry for Economic Affairs and Energy (BMWi), project "KaNaRiA" (grant No. 50 NA 1318) and project "EnEx-CAUSE" (grant No. 50 NA 1505).

References

[1] S Bharadwaj, M Osipchuk, K D Mease, and F C Park. Geometry and inverse optimality in global attitude stabilization. *Journal of Guidance, Control, and Dynamics*, 21(6):930–939, 1998.
[2] S P Bhat and D S Bernstein. A topological obstruction to continuous global stabilization of rotational motion and the unwinding phenomenon. *Systems & Control Letters*, 39(1):63–70, 2000.
[3] J D Bošković, S Li, and R K Mehra. Robust adaptive variable structure control of spacecraft under control input saturation. *Journal of Guidance, Control, and Dynamics*, 24(1):14–22, 2001.
[4] F Bullo and R M Murray. Proportional derivative (pd) control on the euclidean group. In *European Control Conference*, volume 2, pages 1091–1097, 1995.
[5] N A Chaturvedi, A K Sanyal, and H N McClamroch. Rigid-body attitude control. *IEEE Control Systems*, 31(3):30–51, 2011.
[6] J Clemens, T Reineking, and T Kluth. An evidential approach to SLAM, path planning, and active exploration. *International Journal of Approximate Reasoning*, 73:1–26, 2016.

[7] A H J De Ruiter. Adaptive spacecraft attitude control with actuator saturation. *Journal of Guidance, Control, and Dynamics*, 33(5):1692–1696, 2010.

[8] A Doucet, N de Freitas, and N Gordon. *Sequential Monte Carlo methods in practice.* Springer Science & Business Media, 2001.

[9] N Fairfield, G Kantor, and D Wettergreen. Real-time SLAM with octree evidence grids for exploration in underwater tunnels. *Journal of Field Robotics*, 24(1):3–22, 2007.

[10] J R Forbes. Direction-cosine-matrix-based attitude control subject to actuator saturation. *IET Control Theory & Applications*, 9(11):1653–1661, 2015.

[11] G González Peytaví, J Clemens, D Nakath, A Probst, R Förstner, K Schill, and B Eissfeller. Autonomous orbit navigation for a mission to the asteroid main belt. In *Proceedings of the 66th International Astronautical Congress, Jerusalem, Israel.* International Astronautical Federation, 2015.

[12] R Hartley, J Trumpf, Y Dai, and H Li. Rotation averaging. *International Journal of Computer Vision*, 103(3):267–305, 2013.

[13] C Hertzberg, R Wagner, U Frese, and L Schröder. Integrating generic sensor fusion algorithms with sound state representations through encapsulation of manifolds. *Information Fusion*, 14(1):57–77, 2013.

[14] A Iserles, H Z Munthe-Kaas, S P Nørsett, and A Zanna. Lie-group methods. *Acta Numerica 2000*, 9:215–365, 2000.

[15] T Lee. Exponential stability of an attitude tracking control system on SO(3) for large-angle rotational maneuvers. *Systems & Control Letters*, 61(1):231–237, 2012.

[16] C G Mayhew, R G Sanfelice, and A R Teel. On quaternion-based attitude control and the unwinding phenomenon. In *Proceedings of the 2011 American Control Conference*, pages 299–304. IEEE, 2011.

[17] R M Murray, Z Li, and S S Sastry. *A Mathematical Introduction to Robotic Manipulation.* CRC Press, 1994.

[18] D Nakath, C Rachuy, J Clemens, and K Schill. Optimal rotation sequences for active perception. In *Proc. SPIE*, volume 9872, pages 987204-1–987204-13, 2016.

[19] M A C Perryman, L Lindegren, J Kovalevsky, E Hoeg, U Bastian, P L Bernacca, M Crézé, F Donati, M Grenon, M Grewing, et al. The HIPPARCOS catalogue. *Astronomy and Astrophysics*, 323, 1997.

[20] A Probst, G González Peytaví, D Nakath, A Schattel, C Rachuy, P Lange, J Clemens, M Echim, V Schwarting, A Srinivas, K Gadzicki, R Förstner, B Eissfeller, K Schill, C Büskens, and G Zachmann. KaNaRiA: Identifying the challenges for cognitive autonomous navigation and guidance for missions to small planetary bodies. In *Proceedings of the 66th International Astronautical Congress, Jerusalem, Israel.* International Astronautical Federation, 2015.

[21] S Thrun, W Burgard, and D Fox. *Probabilistic Robotics.* MIT Press, Cambridge, MA, 2005.

[22] B Wie and P M Barba. Quaternion feedback for spacecraft large angle maneuvers. *Journal of Guidance, Control, and Dynamics*, 8(3):360–365, 1985.

Benchmarking of Advanced Control Strategies for a Simulated Hydroelectric System

S. Finotti, S. Simani, S. Alvisi, M. Venturini

Dipartimento di Ingegneria, Università degli Studi di Ferrara. Via Saragat 1E – 44122 Ferrara (FE), Italy

E-mail: simona.finotti@student.unife.it,
{silvio.simani,stefano.alvisi,mauro.venturini}@unife.it

Abstract. This paper analyses and develops the design of advanced control strategies for a typical hydroelectric plant during unsteady conditions, performed in the Matlab and Simulink environments. The hydraulic system consists of a high water head and a long penstock with upstream and downstream surge tanks, and is equipped with a Francis turbine. The nonlinear characteristics of hydraulic turbine and the inelastic water hammer effects were considered to calculate and simulate the hydraulic transients. With reference to the control solutions addressed in this work, the proposed methodologies rely on data–driven and model–based approaches applied to the system under monitoring. Extensive simulations and comparisons serve to determine the best solution for the development of the most effective, robust and reliable control tool when applied to the considered hydraulic system.

1. Introduction

Hydroelectric plants convert hydraulic energy into useful energy (mainly electric and mechanical energy). Hydropower is, in fact, the most widely adopted form of renewable energy in the world today, accounting for approximately 16% of global energy production, *i.e.* 3673.1 TWh of energy are consumed from hydropower in various countries [1]. With increasing demand for electricity, and concern about reducing fossil fuel consumption, hydropower is likely to continue to play a key role in global energy production. Indeed, changing conditions in the power market have led to an increase in the demand of peak energy generation, short response time and fast frequency changes. Hydroelectric power plants thus need to be operated accounting for different load conditions. More in general, in the operation of hydropower systems the occurrence of variations in the flow is frequently experienced, being true both in routine operation and in accidental or exceptional unforeseen events. The turbine operations such as start–up, load acceptance, load rejection and shutdown may result in hydraulic transients, which can cause large pressure and sub–pressure oscillations in turbine hydraulic systems and must be evaluated to avoid mechanical failures. Matlab and Simulink represent interactive tools for modelling, simulating, and analysing dynamic systems that have been successfully applied also for nonlinear dynamics investigations in hydroelectric processes [2]. Furthermore, power plants are usually equipped with particular control systems to ensure stable operation. The design of proper control systems for hydraulic turbines remains a challenging and important problem.

This paper considers the simulation and the development of different control solutions for a typical hydroelectric power plant, which has a high water head and a long penstock with upstream and downstream surge tanks, and is equipped with a Francis turbine [3]. In the proposed control systems, an electric servomotor is used as a governor. The nonlinear characteristics of hydraulic turbine and the inelastic water hammer effect were considered to calculate and simulate hydraulic transients. The hydraulic system is described by a nonlinear model, therefore, most of compensation schemes use conventional controllers like on–off strategies, including standard PID regulators for their relative simplicity. However, these controllers do not always produce fast response and suffer the problem of high overshoot and large settling time. Moreover, the tuning of the conventional controllers can be difficult [2]. To this aim, Artificial Intelligence (AI) has been applied to control designs due to substantial advantages of applicability to nonlinear systems with unknown or partially known dynamics [4].

In the last years, AI based methods, namely, Fuzzy Logic (FL), Adaptive Neuro–Fuzzy Inference System (ANFIS), Artificial Neural Network (ANN) and Model Predictive Controllers (MPC) have been considered [5, 6, 4]. When used for fault diagnosis and fault tolerant control purpose, some of these controllers have advantages compared to conventional ones, as shown by the same authors *e.g.* in [7, 8, 9].

Regarding classic compensation strategies applied to hydroelectric systems, several methodologies have been proposed in the literature [10]. They can use optimal control theory [11] or intelligent approaches [12].

The contribution of the paper consists of analysing different control designs already considered by the same authors in [8, 9] but for fault diagnosis and fault tolerant control with application to the simulated hydraulic system described in [7]. Moreover, the simulations and the comparisons with the achieved performances that have been implemented in the Matlab and Simulink environments serve to define the most viable and practical control tool when applied to the considered hydroelectric model. On the other hand, the proposed design tools are fundamental for the assessment, the verification and the validation of the considered control strategies in connection with the most important features of the hydroelectric system.

The paper is organised as follows. The mathematical model of the hydroelectric system is summarised in Section 2. Section 3 analyses the proposed control methods. The achieved results are summarised in Section 4. Comparisons among the different control methods and their performances with respect to measurements and modelling errors are also investigated and discussed. Finally, Section 6 concludes the paper by summarising the main achievements of the work, and providing some suggestions for further research topics.

2. Hydroelectric System

The hydroelectric power plant considered in this work is described in detail in [7]. It consists of a reservoir with water constant level, an upstream water tunnel, an upstream surge tank, a penstock, a downstream surge tank, and a downstream tail water tunnel. Finally, a tail water lake has water constant level.

The expressions (1) and (2) represent the non–dimensional flow rate and water pressure in terms of the corresponding relative deviations:

$$\frac{Q}{Q_r} = 1 + q \qquad (1)$$

$$\frac{H}{H_r} = 1 + h \qquad (2)$$

where q is the flow rate relative deviation, h the water pressure relative deviation, Q_r and H_r the rated flow rate and the rated water pressure, respectively. With reference to a pressure water supply system, the Newton's second law for a fluid element inside a tube and the conservation mass law for a control volume, which accounts for water compressibility and tube elasticity, can be written. Under the assumption that the penstock is short or medium in length, water and pipeline can be considered incompressible and rigid, respectively. Therefore, (3) considers only the inelastic water hammer effect:

$$\frac{h}{q} = -T_w\, s - H_f \tag{3}$$

where s is the derivative operator. Under this assumption, the expression (3) represents the flow rate deviation and the water pressure deviation transfer functions for a simple penstock, where H_f is the hydraulic loss and T_w is the water inertia time:

$$T_w = \frac{L\, Q_r}{g\, A\, H_r} \tag{4}$$

depending on the penstock length L, the rated flow rate Q_r, the gravity acceleration g, the cross–section area A, and the rated water pressure H_r.

Regarding the Francis turbine described in [7], the second order polynomial curve (5) relates the non–dimensional water flow rate Q/Q_r to the non–dimensional rotational speed n/n_r. The non–dimensional parameter G (varying in the range between 0 and 100%) represents the turbine wicked gate opening.

$$\frac{Q}{Q_r} = G\left[a_1\left(\frac{n}{n_r}\right)^2 + b_1\left(\frac{n}{n_r}\right) + c_1\right] = f_1\,(n, G) \tag{5}$$

The non–dimensional turbine torque M in (6) is a function of the water flow rate Q, the water pressure H, and the rotational speed n. According to the relation (5), the turbine torque M is a function of the water flow rate Q, the rotational speed n and wicked gate opening G.

$$\frac{M}{M_r} = \frac{\frac{Q}{Q_r}\frac{H}{H_r}}{\frac{n}{n_r}} = f_2\,(Q, n, G) \tag{6}$$

Finally, the relations (7) and (8) express all the non–dimensional parameters for the turbine in terms of the corresponding relative deviations.

$$\frac{n}{n_r} = 1 + x \tag{7}$$

$$G = 1 + y \tag{8}$$

where q_t represents the turbine flow rate relative deviation, h_t the turbine water pressure relative deviation, x the turbine speed relative deviation, and y the wicket gate servomotor stroke relative deviation. Note that, since G varies in the range $0\% - -100\%$, the definition of (8) makes y vary in the range between -1 and 0.

If the generator unit supplies an isolated load, then the dynamic process of the generator unit considering the load characteristic is represented as [7]:

$$\frac{x}{m_t - m_{g0}} = \frac{1}{T_a\, s + e_g} \tag{9}$$

where m_{g0} is the load torque, T_a the generator unit mechanical time, and e_g the load self–regulation factor. The relationship between the control signal u and the wicket gate servomotor stroke y is thus expressed by means of a first–order model:

$$\frac{y}{u} = \frac{1}{T_y\, s + 1} \tag{10}$$

where T_y is the wicket gate servomotor response time.

3. Control Scheme Design

The general description of the dynamic model of the hydroelectric system can be represented by the nonlinear dynamic function $\mathcal{F}$:

$$u(t) = \mathcal{F}\left(x(t)\right) \tag{11}$$

where x is the process output, u represents the control input, and t is the time. The control strategy applied to the system should determine the control input $u(t)$ such that the controlled process output $x(t)$ is able to track a given reference or set–point $r(t)$.

It can be shown that a continuous–time linearised state–space description of the hydroelectric system can be described by the model (12):

$$\begin{cases} \dot{x}_s(t) &= A_s\, x_s(t) + B_s\, u(t) \\ x(t) &= C_s\, x_s(t) \end{cases} \tag{12}$$

where $x_s \in \Re^6$ represents the state–space vector for the hydroelectric model described in Section 2. The matrices A_s, B_s and C_s of the state–space model have appropriate dimensions. With reference to the model of Eq. (12), the monitored output $x(t)$ represents the turbine speed relative deviation, whilst the control variable is $u(t)$ that is applied to the servomechanism for actuating the wicket gate servomotor stroke relative deviation y according to Eq. (10).

This paper recalls different control strategies including the standard PID controller as well as AI techniques, such as fuzzy logic, adaptive, model predictive controllers, which are used for the regulation of the hydroelectric system. These methodologies are briefly outlined in the following subsections.

3.1. Standard PID Controller

Standard PID regulators are the most commonly used feedback controllers for industrial processes. The control logic is based on the computation of the error $e(t)$ between the desired and the measured values of the output, *i.e.* $e(t) = r(t) - x(t)$, which is fed back to the system after proportional, integral and derivative operations. In this way, the continuous–time control law of the PID regulator is described by Eq. (13):

$$u(t) = K_p\, e(t) + K_i \int_0^t e(\tau)\, d\tau + K_d \frac{de(t)}{dt} \tag{13}$$

where K_p, K_i, K_d are the PID proportional, integral, and derivative gains. The optimal selection of this gains is performed by using the automatic tuning algorithm in the Simulink environment that balances the performance (response time) and the robustness (stability margins) of the controlled system [13].

The PID automatic tuning Simulink toolbox uses the linearised model (12) of the hydroelectric system.

3.2. Fuzzy Controller

Fuzzy Logic Controllers (FLCs) are extensively used in processes where the system dynamics are either very complex or exhibit highly nonlinear characteristics. The controller design approach relies on the identification of transparent rule–based Takagi–Sugeno (TS) fuzzy models using an Adaptive Neuro–Fuzzy Inference System (ANFIS) tool implemented in the Simulink toolbox. The same authors have already proposed the use of fuzzy regulators for the problem of the fault tolerant control design as shown in [8, 9].

The TS fuzzy model consists of a set of rules R_i, where the consequents are deterministic functions f_i:

$$R_i : IF\, x\, is\, A_i\, THEN\, u_i = f_i(x) \tag{14}$$

with $i = 1, 2, \ldots, K$ and K is the number of clusters or rules of the rule–based system, x represents the input vector of the *antecedent* variables, and u_i describes the *consequent* output. A_i represents the antecedent fuzzy set of the i–th rule, defined by its (multivariable) membership function $\mu_{A_i}(x) \to [0, 1]$ The function f_i is represented by suitable parametric models, whose structure remains equal in all rules, whilst the parameters can vary. A parametrisation in affine form is usually exploited, and described by Eq. (15):

$$u_i = a_i\, x + b_i \tag{15}$$

where the vector a_i and the scalar b_i are the i–th submodel parameters. The vector x contains a suitable number n of delayed samples of the model inputs and output. In this way, the product $a_i\, x$ represents an Auto–Regressive eXogenous (ARX) parametric dynamic model of order n.

The final output u of the TS fuzzy model is the weighted average of all rule outputs, computed as:

$$u = \frac{\sum_{i=1}^{K} \mu_{A_i}(x)\, y_i(x)}{\sum_{i=1}^{K} \mu_{A_i}(x)} \tag{16}$$

The modelling approach used by ANFIS is similar to many system identification techniques. First, the TS fuzzy model structure described by the its order n, the form of the membership functions μ_{A_i} and the number of clusters K are hypothesised. Next, the input–output data are used by ANFIS for training the TS model according to a chosen error criterion, thus determining the optimal values of the controller parameters a_i and b_i [5].

The paper considers also an alternative approach to ANFIS for the derivation of the controller fuzzy model, which is represented by the Fuzzy Modelling and Identification (FMID) toolbox developed in the Matlab environment [5]. Also in this case, the estimation of the controller prototype relies on the identification of rule–based fuzzy models and using the input–output data acquired from the controlled process. This method exploits Takagi–Sugeno fuzzy models and employs the Gustafson–Kessel clustering method to divide the data into subsets with a common local linear (affine) behaviour [5].

The identified fuzzy controller is thus obtained by selecting an proper model structure n and a number of clusters K. The FMID toolbox provides the parameters a_i, b_i and the estimation of the membership functions μ_{A_i} of the optimal controller minimising the tracking error $e(t)$.

Note finally that the fuzzy controller in the form of Eq. (16) is described by a discrete–time input–output model, which is connected to the controlled continuous–time nonlinear system of Eq. (11) using Digital–to–Analog (D/A) and Analog–to–Digital (A/D) converters.

3.3. Linear Adaptive Controller

The adaptive control method exploited in this paper is based on the on–line identification of a second order discrete–time transfer function of an ARX time–varying model in the form:

$$G(z) = \frac{b_1\, z^{-1} + b_2\, z^{-2}}{1 + a_1\, z^{-1} + a_2\, z^{-2}} \tag{17}$$

whose parameters are recursively estimated at each sampling time $t_k = k\,T$, with $k = 1, 2, \ldots, N$, N the number of samples, and T the sampling interval. z represents the unit advance operator. The parameters in (17) are estimated using the Recursive Least–Square Method (RLSM) with directional forgetting factor [13]. The same authors have proposed similar approaches but for fault tolerant control applications, as shown in [8].

The synthesis of the adaptive control law is derived using a modified Ziegler–Nichols criterion, in the form of Eq. (18):

$$u_k = q_0\, e_k + q_1\, e_{k-1} + q_2\, e_{k-2} + (1 - \gamma)\, u_{k-1} + \gamma\, u_{k-2} \tag{18}$$

where e_k is the tracking error $e(t)$ at the sampling time t_k, u_k the control signal $u(t)$ at the sampling time t_k, whilst q_0, q_1, q_2, and γ are the time–varying controller parameters, which are calculated by solving a Diophantine equation that leads to the following relations [13]:

$$\begin{array}{l} q_0 = \frac{1}{b_1}\left(d_1 + 1 - a_1 - \gamma\right),\ \gamma = q_2 \frac{b_2}{a_2} \\ q_1 = \frac{a_2}{b_2} - q_2\left(\frac{b_1}{b_2} - \frac{a_1}{a_2} + 1\right),\ q_2 = \frac{s_1}{r_1} \end{array} \tag{19}$$

where $r_1 = (b_1 + b_2)\left(a_1 b_2 b_1 - a_2 b_1^2 - b_2^2\right)$ and $s_1 = a_2\left((b_1 + b_2)\left(a_1 b_2 - a_2 b_1\right) + b_2\left(b_1 d_2 - b_2 d_1 - b_2\right)\right)$. It is assumed that the final closed loop model has a behaviour similar to a second–order continuous time system with characteristic polynomial $s^2 + 2\,\delta\,\omega\,s + \omega^2$, where δ and ω represent its damping factor and natural frequency, respectively. In this case, if $\delta \leq 1$, $d_1 = -2\,e^{-\delta\,\omega\,T}\cos\left(\omega\,T\sqrt{1-\delta^2}\right)$ and $d_2 = e^{-2\,\delta\,\omega\,T}$.

Both the on–line identification procedure and the adaptive controller parameter computation are implemented in the self–tuning controller Simulink library [13]. In this way, the sampled output y_k of the time–varying ARX model (17) should follow the sampled reference signal r_k when regulated by the control law (18).

Note finally that, also in this case, the adaptive controller (18) is connected to the continuous–time nonlinear system (11) using the D/A and A/D converters.

3.4. Model Predictive Control

Model Predictive Control (MPC) relies on dynamic models of the process, most often linear models obtained by system identification or linearisation of a nonlinear plant. The main advantage of MPC is the fact that it allows the current sampling time to be optimised, while keeping future sampling times in account. This is achieved by optimising a finite time–horizon, but only implementing the current sampling time. MPC has the ability to anticipate future events and can take control actions accordingly. PID controllers do not have this predictive ability. MPC is nearly universally implemented as a digital control.

MPC is based on iterative, finite–horizon optimisation of the plant model. At the sample k ($k = 1, 2, \ldots, N$) the current plant output is sampled and a cost minimising control strategy is computed (via a numerical minimisation algorithm) for a relatively short time horizon in the future: $[k, k + N_p]$. Specifically, an online calculation is used to explore state trajectories that emanate from the current state and find (via the solution of Euler–Lagrange equations) a cost–minimising control strategy until time $k + N_p$. Only the first step of the control strategy is implemented, then the plant state is sampled again and the calculations are repeated starting from the new current state, yielding a new control and new predicted state path. The prediction horizon keeps being shifted forward and for this reason MPC is also called receding horizon control. The same authors have recently proposed a similar approach, but for the fault tolerant control problem, as shown in [14].

An example of a cost function J for optimisation is given by:

$$J = \sum_{k}^{k+N_p} w_{x_k}\,(r_k - x_k)^2 + \sum_{k}^{k+N_c} w_{u_k}\,\Delta u_k^2 \tag{20}$$

where w_{x_k} the weighting coefficient reflecting the relative importance of the monitored output x_k, and w_{u_k} the weighting coefficient penalising relative big changes in u_k, with $\Delta u_k = u_k - u_{k-1}$. N_p represents the prediction horizon, whilst N_c the control horizon.

Note finally that the discrete–time MPC design is performed by using the MPC toolbox in the Simulink environment, which computes a linearisation of the hydroelectric nonlinear model (12). The discrete–time MPC is thus connected to the continuous–time nonlinear system of Eq. (11) using the D/A and the A/D converters. The MPC Simulink toolbox uses the linearised model (12) of the hydroelectric system.

4. Results

This study exploits the 4 different control methods described in the previous subsections to regulate the output of the hydroelectric nonlinear system. The simulations are performed in the Simulink environment and the toolboxes described above. The achieved results are compared in terms of settling time T_s, maximum overshoot $S\%$ and the percent Normalised Sum of Squared Errors (NSSE), defined as:

$$NSSE\% = 100\sqrt{\frac{\sum_{k=1}^{N}(r_k - x_k)^2}{\sum_{k=1}^{N} r_k^2}} \tag{21}$$

where r_k and x_k represent the samples of the continuous–time signals $r(t)$ and $x(t)$, respectively. The simulation model with two surge tanks and a Francis turbine described in Section 2 allows the simulation of the behavoir of a hydroelectric power plant in the presence of large hydraulic transients after full load rejection m_{g0}. Usually, the most severe hydraulic transients will happen after full load rejection. Therefore, all of these simulations are performed on full load rejection operating conditions.

The standard control strategy for hydroelectric systems can use a classic PID regulator, whose complete structure has been recalled in Section 3.1. The optimal proportional and integral gains are determined using the automatic PID tuning procedure and settled to $K_p = 0.6567$, $K_i = 0.4468$, $K_d = 1.6567$. This issue was already addressed by the same authors in [7]. The turbine speed governor plays a very important role in hydraulic transients caused by load changes. As already remarked, the classic PID controller proposed earlier in [2] required an optimal tuning of its gains, and in this way only the dynamic performance of the generator unit can be improved. Moreover, in order to get the best dynamic performance, it is necessary to set different optimal PID gains on different operating conditions for turbine speed governor.

The controller capabilities have been assessed in simulation by considering different load torque m_{g0} values, which represent the turbine start–up and shutdown conditions. In particular, the start–up phase is assumed to last 300 s (due to the large size of the considered Francis turbine), while the shutdown maneuver takes just 30 s, to simulate an unplanned emergency shutdown.

4.1. PID Controller Results

As an example, the results summarised in Fig. 1 show that the PID governor of Eq. (13), whose parameters were optimised according to the PID autotuning Simulink toolbox, is able to keep the relative deviation of the rotational speed null ($r(t) = 0$, *i.e.* the rotational speed constant) in steady–state conditions. Its performances are better than the standard PID regulator designed in [2].

In the following, the suggested PID controller, with automatic tuning, together with the classic PID governor, have been implemented and compared in the Matlab and Simulink environments. Regarding the standard PID governor parameters in Eq. (13), they were selected as $K_p = 1.0$, $K_i = 0.2$, $K_d = 1.0$, as described in [2]. On the other hand, the PID governor with automatic tuning will be used as reference controller and compared to the remaining control strategies proposed in this work.

Fig. 1 shows the turbine speed relative deviations x when the load torque m_{g0} changes in start–up and shutdown conditions. The hydroelectric system output is controlled by the continuous–time PID regulator, as shown in Fig. 1.

Table 1 summarises the achieved results in terms of $NSSE\%$, the per–cent undershoot $s\%$, overshoot $S\%$ and the settling time T_s for different values of the load torque m_{g0}. According to these simulation results, good properties of the proposed autotuning PID controller are highlighted, and they are better than the PID governor [2]. In fact, the autotuning design

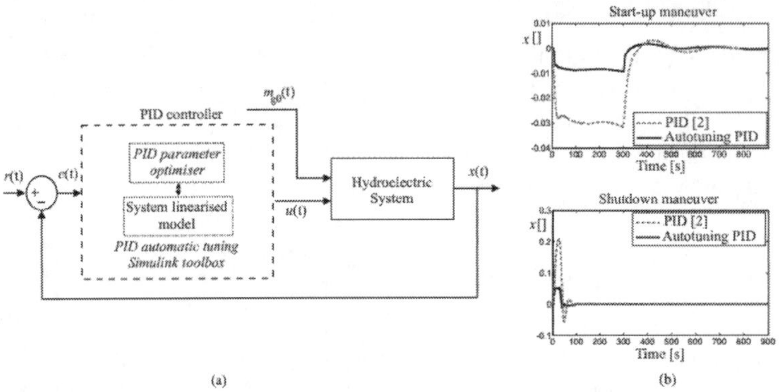

Figure 1. (a) block diagram of the hydroelectric system compensated by the PID regulator with automatic tuning and (b) controlled output.

is able to limit the effect of high–gains for the proportional and the integral contributions of the PID control. Some further comments can be drawn in general regarding the capability of these autotuning solutions. The $NSSE$ values are considerably lower in case of significant transient maneuvers (*i.e.* start–up and shutdown). On the other hand, though always lower, the settling time T_s is not significantly decreased and remains comparable to that obtainable by means of the PID [2]. This is probably due to the inherent dynamics of the simulated hydraulic systems. Similarly, the per–cent $s\%$ and $S\%$ are decreased in all cases by using the autotuning PID controller, and this effect is mainly highlighted when considering the most severe transients (*i.e.* start–up and shutdown).

Note that standard industrial controllers, such the PID recalled in Section 3.1, are quite simple and have the benefit of quite straightforward implementation. However, when applied to the control of hydroelectric systems, the control laws cannot be very efficient. Therefore, the use of more advanced controller solutions can be motivated.

4.2. Fuzzy Controller Results

With reference to the strategies described in Section 3.2, fuzzy identification is used to derive the models of the controllers by exploiting the so–called model reference control approach [15]. For this purpose, the PID regulator of Fig. 1 represents the reference controller for the generation of the data used by the identification strategy proposed described in Section 3.2. In this way, the fuzzy controller parameters are identified such that the performances in terms of tracking error $e(t)$ are optimised.

In particular, with reference to the TS fuzzy controller derived with the ANFIS tool, a sampling interval $T = 0.1s.$ is exploited. Moreover, the fuzzy controller (16) uses a number $K = 3$ of Gaussian membership functions, with a number of delayed inputs and output $n = 1$. The antecedent vector is thus $x = [e_k, e_{k-1}, u_{k-1}]$. The achieved performances of the controller obtained with the ANFIS tool are shown in Fig. 2.

Using the same data from the PID reference regulator, a second fuzzy controller (16) has been estimated using the FMID tool, with a number of clusters $K = 3$, a number of delays $n = 2$, and the antecedent vector $x = [u_{k-1}, u_{k-2}, r_k, r_{k-1}, y_k, y_{k-1},]$. The FMID tool also provides the optimal estimate of the shapes of the membership functions μ_{A_i}. The implementation scheme and the achieved results are represented and compared in Fig. 2.

Fig. 2 highlights that both the fuzzy regulators perform better than the PID controller with autotuning. Also in this case, the settling times T_s, the maximum overshoot $S\%$, and the

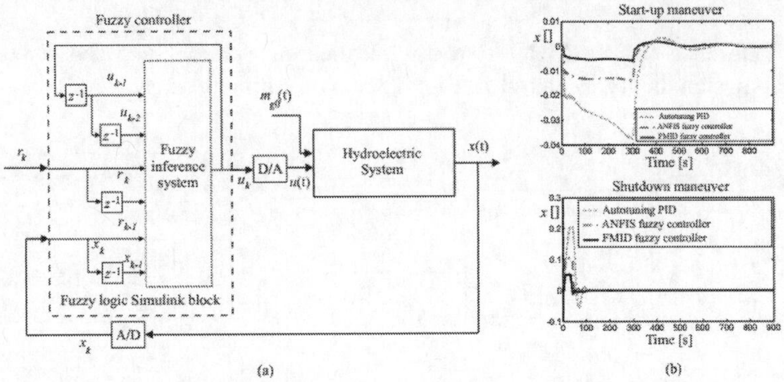

Figure 2. (a) block diagram of the hydroelectric system compensated by the fuzzy regulators and (b) controlled output.

$NSSE\%$ values are reported in Tab. 1, which are computed for the different fuzzy controllers. The simulation results also highlight better properties of the proposed the FMID controller with respect to the ANFIS one, which are motivated by the better capability and flexibility of the FMID tool [5].

4.3. Adaptive Controller Results

On the other hand, by considering the on–line procedure recalled in Section 3.3, Fig. 3 shows the tracking capabilities of the adaptive controller (18). Its time–varying parameters have been obtained via the relations (19) with the damping factor and the natural frequency $\delta = \omega = 1$. The adaptive controller implementation using the Self Tuning controller Simulink Library (STCSL) and the achieved results are reported Fig. 3.

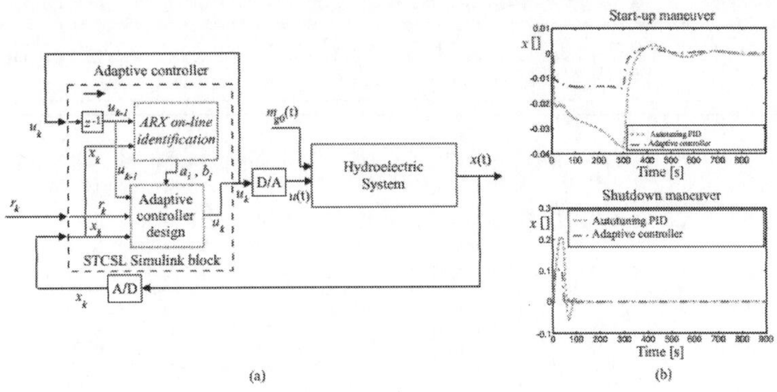

Figure 3. (a) block diagram of the hydroelectric system compensated by the adaptive regulator and (b) controlled output.

Table 1 summarises the achieved results of the adaptive controller in terms of $NSSE\%$, the per–cent undershoot $s\%$, overshoot $S\%$ and the settling time T_s for different values of the load torque m_{g0}. According to these simulation results, good properties of the proposed adaptive controller are highlighted, and they are better than the autotuning PID governor.

IOP Conf. Series: Journal of Physics: Conf. Series **783** (2017) 012041 doi:10.1088/1742-6596/783/1/012041

4.4. MPC Results

Finally, with reference to the MPC strategy recalled in Section 3.4, the reference and the monitored output signals are depicted in Fig. 4 with its implementation.

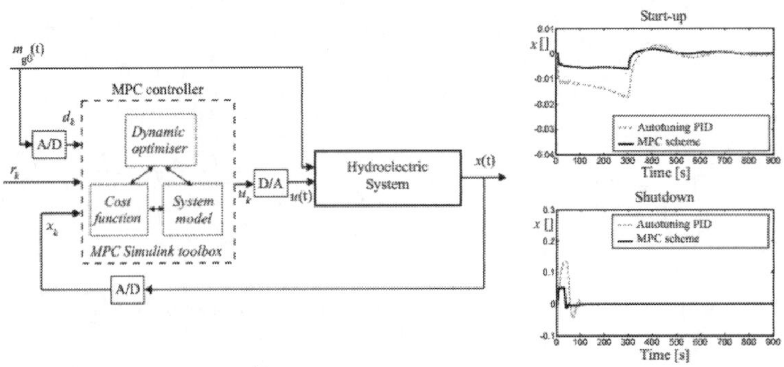

(a) (b)

Figure 4. (a) block diagram of the hydroelectric system compensated by the MPC law and (b) controlled output.

The results shown in Fig. 4 have been achieved by using a prediction horizon $N_p = 10$ and a control horizon $N_c = 2$. The weighting parameters have been settled to $w_{y_k} = 0.1$ and $w_{u_k} = 1$, in order to reduce abrupt changes of the control input. Finally, the MPC strategy performance is summarised in Table 1.

4.5. Quantitative Comparison of the Advanced Control Strategies

In order to provide a comparison of the performances obtained by the considered control solutions, Table 1 summarises the achieved results for different values of the load torque.

Table 1. $NSSE\%$, T_s, $s\%$, and $S\%$ for the considered control solutions.

Controller Type	m_{g0}	$NSSE\%$	T_s	$s\%$	$S\%$
PID [2]	+100%	1.86%	725.59s	3.15%	0.32%
	−100%	3.34%	76.05s	5.66%	20.73%
Autotuning PID	+100%	0.62%	701.36s	0.96%	0.12%
	−100%	1.38%	54.08s	1.19%	5.76%
Fuzzy ANFIS	+100%	0.21%	525.59s	3.15%	0.32%
	−100%	0.18%	76.05s	5.66%	20.73%
Fuzzy FMID	+100%	0.17%	501.36s	0.96%	0.12%
	−100%	0.12%	54.08s	1.19%	5.76%
Adaptive controller	+100%	0.32%	601.36s	0.76%	0.09%
	−100%	0.78%	22.08s	0.58%	2.31%
MPC scheme	+100%	0.29%	581.23s	0.64%	0.08%
	−100%	0.67%	19.98s	0.49%	1.99%

Note that the MPC regulator leads to the best values of settling time and maximum overshoot, as its parameters are automatically tuned in the Simulink environment in order to optimise the MPC cost function of Eq. (20), as recalled in Section 3.4.

5. Parameter Sensitivity Analysis

This section analyses the robustness features of the proposed controllers with respect to parameter variations. This analysis exploits the Monte–Carlo tool, as the control performance depends on the model–reality mismatch as well as on the input–output measurement errors. Therefore, the analysis has been performed by describing the hydroelectric model parameters as Gaussian stochastic processes with mean values corresponding to the nominal ones and standard deviations of $\pm 20\%$. The average $NSSE\%$ index values have been computed with 100 Monte–Carlo runs, and summarised in Table 2.

Table 2. Monte–Carlo analysis for the designed controllers: $NSSE\%$ average values with parameter variations.

Controller type	Torque value m_{g0}	Average $NSSE\%$
PID [2]	+100%	5.67%
	−100%	20.67%
Autotuning PID	+100%	1.35%
	−100%	10.59%
Fuzzy ANFIS	+100%	1.22%
	−100%	0.96%
Fuzzy FMID	+100%	1.01%
	−100%	0.41%
Adaptive controller	+100%	1.18%
	−100%	3.96%
MPC scheme	+100%	0.19%
	−100%	0.22%

It is worth noting that, with reference to the values summarised in Table 2 achieved via the Monte–Carlo analysis, they can serve to compare the overall behaviour of the considered controllers in transient conditions with respect to the standard PID solution. Moreover, the values in Table 2 suggest that when the modelling of the dynamic system can be taken into account, the MPC scheme is preferred, even if an optimisation procedure is required. However, in the case of a system with modelling errors, after a certain amount of off–line learning, the fuzzy–based estimation error can fall below the value of the MPC–based scheme, as shown for the controller estimated via the FMID tool. On the other hand, the FMID controller achieves the best control capabilities. The adaptive approach takes advantage of its improved features, as it is able to track possible variations of the controlled system, but with quite complicated and not straightforward design procedures. The fuzzy–based schemes rely on the learning accumulated from off–line simulations, but the training stage can be computationally heavy. Regarding the standard PID control strategy, it is rather simple and straightforward, even if the achievable performances are quite limited. Similar results, but with application to different dynamic processes, were addressed by the same authors in [16, 17].

6. Conclusion

This paper addressed the design of advanced control strategies for a hydroelectric power plant modelled in the Matlab and Simulink environments. The hydraulic system consisted of a high water head and a penstock with upstream and downstream surge tanks and a Francis turbine. The nonlinear characteristics of hydraulic model were considered to simulate the hydraulic transients and to evaluate the behaviour of the proposed hydraulic turbine regulating systems. The suggested control methodologies were designed using data–driven and model–based approaches applied to the system under monitoring. Extensive simulations showed that the model predictive control scheme has to be preferred when the modelling of the dynamic system can be taken into account. However, in case of modelling errors, artificial intelligence schemes can rely on the learning accumulated from off–line simulations, but the training stage can be computationally heavy. On the other hand, adaptive approaches can track variations of the controlled system, but with complex design procedures. Finally, standard PID control solutions are rather simple and straightforward, but the achievable performances are quite limited.

References

[1] R. R. Singh, T. R. Chelliah, and P. Agarwal, "Power electronics in hydro electric energy systems – A review," *Renewable and Sustainable Energy Reviews*, vol. 32, pp. 944–959, April 2014.

[2] H. Fang, L. Chen, N. Dlakavu, and Z. Shen, "Basic modeling and simulation tool for analysis of hydraulic transients in hydroelectric power plants," *IEEE Trans. Energy Convers.*, vol. 23, pp. 424–434, Sept. 2008.

[3] M. Popescu, D. Arsenie, and P. Vlase, *Applied Hydraulic Transients: For Hydropower Plants and Pumping Stations*. Lisse, The Netherlands: CRC Press, Jan. 2003.

[4] J. Korbicz, J. M. Koscielny, Z. Kowalczuk, and W. Cholewa, eds., *Fault Diagnosis: Models, Artificial Intelligence, Applications*. London, UK: Springer–Verlag, 1st ed., February, 12 2004. ISBN: 3540407677.

[5] R. Babuška, *Fuzzy Modeling for Control*. Boston, USA: Kluwer Academic Publishers, 1998.

[6] J. Maciejowski, "Modelling and predictive control: enabling technologies for reconfiguration," *Annual Reviews in Control*, vol. 23, pp. 13–23, 1999.

[7] S. Simani, S. Alvisi, and M. Venturini, "Study of the Time Response of a Simulated Hydroelectric System," in *Journal of Physics: Conference Series* (H. Schulte and S. Georg, eds.), vol. 570 of *Conference Series*, pp. 1–13, Bristol, United Kingdom: IOP Publishing Limited, Dec. 2014. ISSN: 1742–6596. DOI: 10.1088/1742-6596/570/5/052003.

[8] S. Simani, S. Alvisi, and M. Venturini, "Data–Driven Design of a Fault Tolerant Fuzzy Controller for a Simulated Hydroelectric System," in *Proceedings of the 9th IFAC Symposium on Fault Detection, Supervision and Safety for Technical Processes – SAFEPROCESS'15* (IFAC, ed.), vol. 48, (Paris, France), pp. 1090–1095, IFAC, Sept. 2–4 2015. DOI: 10.1016/j.ifacol.2015.09.672. ISBN: 978–3–642–27644–6. ISSN: 1474–6670. Special session invited paper.

[9] S. Simani, S. Alvisi, and M. Venturini, "Fault Tolerant Control of a Simulated Hydroelectric System," *Control Engineering Practice*, vol. 51, pp. 13–25, June 2016. DOI: http://dx.doi.org/10.1016/j.conengprac.2016.03.010.

[10] J. A. Laghari, H. Mokhlis, A. H. A. Bakar, and H. Mohammad, "A comprehensive overview of new designs in the hydraulic, electrical equipments and controllers of mini hydro power plants making it cost effective technology," *Renewable and Sustainable Energy Reviews*, vol. 20, pp. 279–293, 2013.

[11] P. Schniter and L. Wozniak, "Efficiency based optimal control of kaplan hydro generators," *IEEE Trans. Energy Conver.*, vol. 10, pp. 348–353, June 1995.

[12] M. Djukanovic, M. Novicevic, D. J. Dobrijevic, B. Babic, J. Dejan, and Y. H. P. Sobajic, "Neural–net based coordinated stabilizing control for the exciter and governor loops of low head hydropower plants," *IEEE Trans. Energy Conver.*, vol. 10, pp. 760–767, Dec. 1995.

[13] V. Bobál, J. Böhm, J. Fessl, and J. Machácek, *Digital Self–Tuning Controllers: Algorithms, Implementation and Applications*. Advanced Textbooks in Control and Signal Processing, Springer, 1st ed., 2005.

[14] S. Simani, S. Alvisi, and M. Ventirini in *Proceedings of the* 3rd *International Conference on Control and Fault–Tolerant Systems – SysTol'16* (I. Control Systems Society, ed.), (Barcelona, Spain), pp. 1–6, Research Center for Supervision, Safety and Automatic Control of the Universitat Politècnica de Catalunya in Barcelona, IEEE, 7–9 Sept. 2016. (Accepted). Special session invited paper.

[15] M. Brown and C. Harris, *Neurofuzzy Adaptive Modelling and Control*. Prentice Hall, 1994.

[16] S. Simani, "Overview of Modelling and Advanced Control Strategies for Wind Turbine Systems," *Energies*, vol. 8, pp. 13395–13418, 25 November 2015. Invited paper of the special issue "Wind Turbine 2015". ISSN: 1996-1073. DOI: 10.3390/en81112374.

[17] C. Turhan, S. Simani, I. Zajic, and G. Gokcen Akkurt, "Comparative analysis of thermal unit control methods for sustainable housing applications," in *Proc. of the 12th REHVA World Congress CLIMA 2016*, vol. 8, (Aalborg, Denmark), pp. 1–10, REHVA – Federation of European Heating, Ventilation and Air Conditioning Associations, REHVA, 22–25 May 2016.

Virtual Metrology applied in Run-to-Run Control for a Chemical Mechanical Planarization process*

M.A. Jebri[1], E.M. El Adel[1], G. Graton[1,2], M. Ouladsine[1] and J. Pinaton[3]

[1]Aix-Marseille Université, CNRS, ENSAM, Université de Toulon, LSIS UMR 7296, Av. Normandie-Niemen, 13397 Marseille, France

[2]Ecole Centrale Marseille, Technopôle de Château-Gombert, 38 rue Frédéric Joliot-Curie, F-13451 Marseille Cedex 13, France

[3]STMicroelectronics, ZI Rousset, 13106 Rousset Cedex, France

*This work is part of the European project "INTEGRATE", and carried by STMicroelectronics Fab of Rousset

E-mail: mohamed-ali.jebri@lsis.org

Abstract. This paper deals with missing data in semiconductor manufacturing derived from a measurement sampling strategies. The idea is to construct a virtual metrology module to estimate non measured variables using a new modified Just-In-Time Learning approach (JITL). The aim of this paper is to integrate estimated data into product control loop.

In collaboration with our industrial partner STMicroelectronics Rousset, the accuracy of the proposed method is illustrated by using industrial data-sets derived from Chemical Mechanical Planarization (CMP) process that enables us to compare results obtained with the classical and the modified version of JITL approach. Then, the contribution of the estimated data is shown in product quality improvement.

1. Introduction

In semiconductor manufacturing, a lot which contains usually 25 wafers, can undergo plenty of different treatments (over than 350 according to [1]) to obtain the final product. Among the known and used treatments in semiconductor manufacturing is the Chemical Mechanical Planarization (CMP) process. This latter is a polishing process, which uses a chemical slurry formulation and mechanical polishing process to remove unwanted conductive or dielectric materials on the silicon wafer. In a same CMP equipment, many products characterized by different targets (desired properties) and different recipes, can be processed.

At the end of this step, and to control product quality, the layer thickness of each wafer of the lot can be measured using one measurement strategy (according to [2] and [3]). Depending on whether the measurement equipment is located within or outside the CMP process, two measurement strategies can be considered. The first refers to the Integrated Metrology Module (IMM) and the second is called Stand-Alone Metrology (SAM), where the metrology tool is located outside the process equipment. In the Integrated Metrology Module (IMM) case, all products (lots and wafers) are measured. These physical measurements can then be used for the R2R control, in order that the process operates in the best conditions and improves the product quality.

While this configuration of measuring (IMM) is very expensive, and requires a high cycle time, it provides a good process control and it is applied when the CMP action seems to be critical in the production chain. However, in the Stand Alone Metrology (SAM) case, two strategies are opposed. The first one consists in a control at the end of all treatments. This makes the situation hazardous because in the case of defective lots, the fault is detected too late and lots of resources may be lost. The second strategy consists of measuring the products after each CMP treatment. This is not beneficial because it

requires additional processing time and may slow down the production and reduce the Fab efficiency. It implies also a high cost and introduces an added time delay to the process [4].
To overcome these two main issues, a compromise is achieved by sampling the physical metrology. According to [5], at lot level, the sampling rate is fixed using the criticality of step, machine and the product requirement. However, the main drawback of these measurements sampling is the lack of information. When layer thicknesses $y_j(k)$ are not provided, the recipe parameters $u_j(k)$ are not updated at each run. Then R2R controller is not working properly. Consequently, product quality cannot be well controlled. So, it is not possible to verify whether the variable respects the control limits which are defined as the upper control limit (UCL) and the lower control limit (LCL). Thus, it is necessary to find out an approach that fills in the lack of information. This is the main purpose of virtual metrology (VM).
VM is introduced in semiconductor manufacturing in order to cover the lack of layer thickness measurements by providing estimation $\hat{y}_j(k)$ for non measured products (when $y_j(k)$ is not available). Then, R2R takes into account estimations of missing data and the recipe parameters $u_j(k)$ can be updated at each run. Therefore, VM module with the SAM strategy have the same function of IMM strategy and reacts with the same way with R2R controller with a lower cost.
The interaction between VM module, SAM strategy and R2R controller is described in Figure 1.
The main goal of this paper is to create a VM module for n types of product, manifested by different targets (T_j, where $j \in \{1,\dots,n\}$), using the old sampled measurement $y_j(k)$ with their corresponding old process information $x_j(k)$. Then, to improve a reliable control of product quality, the VM module should provide accurate estimations to R2R control by improving the recipe parameters $u_j(k)$.

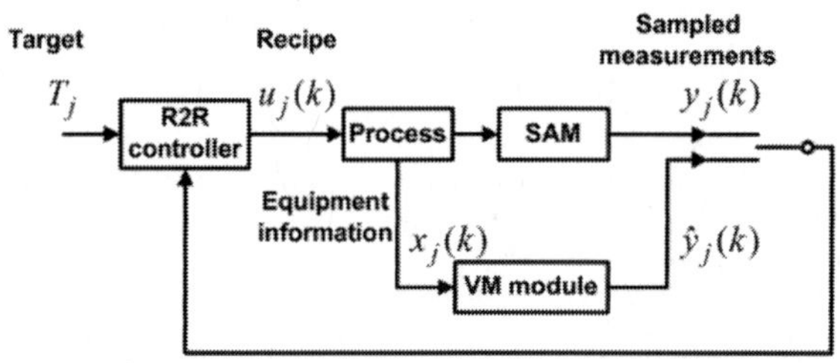

Figure 1. VM and R2R interaction

The paper is organized as follows: The R2R control and the virtual metrology are introduced in section 2. In section 3, a description of the proposed approach to deal with the lack of information and its interaction with R2R control is presented. By considering some data-sets extracted from CMP process and delivered by our industrial partner, the efficiency of the proposed method is given in section 4 followed by concluding remarks. The conclusion fenced by perspectives is given at the end of the paper.

2. Run-to-Run Control and Virtual Metrology

II-A. Virtual Metrology

In the literature, several definitions are given for the virtual metrology (VM). In [6], VM is defined as the use of mathematical models with available data to estimate the variables of interest. According to [7]; VM is presented as the prediction of metrology variables (either measurable or non-measurable) using the process state and the product's information. VM aims also to estimate non-measured variables called difficult-to-measure (DTM) variables [8], from a directly collected equipment data called also easy-to-measure (ETM) variables [8].
Due to the diversity of processes and products in semiconductor manufacturing, many approaches are proposed in the literature to build the VM module. However, there are no standard or generalized methods that can be applied.
Otherwise, the lack of knowledge of physio-chemical relationships in semiconductor manufacturing processes makes the determination of mathematical models describing the behavior of equipment

practically difficult. For this reason the majority of approaches are data-based methods.
There exist two strategies for VM model creation. The first one is based on global modeling strategies which exploits whole databases to provide a single estimation model for all areas of operation. The second one is based on local modeling strategies, also called "modeling on demand", where the model is dynamically built from the similar samples stored in the database to provide a single estimation, when it is needed.
In semiconductor manufacturing, equipment is characterized by many operating domains. Here, local models can provide more precise estimations than the global models which provide more general estimations [1].
The Just-In-Time Learning (JITL) approach is considered as a strategy based on local modeling. JITL approach provides an estimation when it is required [9].
For the two already listed strategies, there are two categories of modeling approaches.
The first strategy is based on linear approaches. The multiple linear regression (MLR) [11], the partial least squares regression (PLSR) defined in [10] which combines a principal components analysis (PCA) and MLR techniques, the recursive partial least squares regression (R-PLSR) [7], the principal component regression (PCR) [4] and Kalman filter [4] are considered as linear based approaches to build VM module.
The second strategy is based on non-linear based approaches. For instance, in [13], the authors adopt a neural network of the radial basis function to build VM model for a CVD process.
In another context, Su *et al.* develop, in [14], a VM module based on neural networks by back-propagation and recurrent neural networks on CVD equipment of 5th generation. Then, they make a comparative study of VM-based models.
In [15], Besnard *et al.* adopt tree ensembles method as a non-linear approach to estimate the oxide thickness.
In another context, in [16], techniques based on data mining are developed on a photo-lithography equipment. Output estimations are used then in R2R control.
To summarize, we notice that in semiconductor manufacturing, CMP equipment features are characterized by their non-linear aspect. We notice also that many product types are processed into CMP machines. So the idea is to use the combination of local models to cover different operation ranges (multiples product types) and linear modeling methods to keep the simplicity of the calculation and make their implantation easier for industrial context. This strategy allows to avoid the high cost of calculation time of techniques based on non-linear modeling.
In this context, JITL approach is proposed as a local modeling technique based on local linear models. This approach is already used for the development of VM modules in chemical processes which are characterized by non-linear aspects. In the next section we aim to introduce JITL algorithm as described by [9].
For our case, JITL, with some contributions, is opted as a VM module in CMP equipment of our industrial partner. Estimation outcomes from VM module are used then in R2R controller.

II-B. Run-to-Run control
The R2R Control module improves essentially product quality, increases yields and reduces defective products by optimizing recipe parameters from lot-to-lot using metrology data or estimations if necessary.
To implement R2R controllers, there exist different types of control laws (recipe parameters): the Exponentially Weighted Moving Average (EWMA) and the double EWMA (dEWMA) are the most widely explored in semiconductor manufacturing process.
The CMP process, for many processed type of product, can be modeled as:

$$y_j(k) = \alpha_j + \beta_j u_j(k) + \Delta(k) + \varepsilon(k) \tag{1}$$

where, $j \in \{1,\dots,n\}$ is the product type, $y_j(k)$ is the output at the end of run k, α_j is the linear process model intercept term, $u_j(k)$ is the recipe parameters at the start of run k and β_j is the system gain. $\varepsilon(k)$

and $\Delta(k)$ are respectively a white noise and systematic disturbance. Note that, these two terms depend only on the considered machine.
First, an offline estimation of the process gain $\hat{\beta}_j$ is required and remains unchanged. Then, at each run k, an estimation of the process model intercept term ($\hat{\alpha}_j(k)$) must be done.
When output is measured (SAM is applied), $\hat{\alpha}_j(k)$ is computed as follows:

$$\hat{\alpha}_j(k) = \lambda_j[y_j(k-1) - \hat{\beta}_j u_j(k-1)] + [1-\lambda_j]\hat{\alpha}_j(k-1) \tag{2}$$

When output is estimated (VM is applied), $\hat{\alpha}_j(k)$ is computed by replacing the measured output $y_j(k-1)$ by its estimation $\hat{y}_j(k-1)$ in (2). $\lambda_j \in [0,1]$ is a discounting factor used to minimize the influence of old data in favor of the new data. The control law for the next run is calculated as follows:

$$\hat{u}_j(k+1) = \frac{T_j - \hat{\alpha}_j(k)}{\hat{\beta}_j} \tag{3}$$

where T_j is the desired target (which reflects product type) value for the process output.
The simple EWMA control method as described above is adequate for controlling equipment with small disturbances and slow changes in output variation. However, the process output results a continuous error from the target if there is a consistent process drift. Therefore, a dEWMA control scheme must be employed. This controller involves two expressions, the first for estimating step-change $\hat{\alpha}_j(k)$ and the second for drift speed $\hat{p}_j(k)$. When the output is measured $\hat{\alpha}_j(k)$ and $\hat{p}_j(k)$ are computed as follows:

$$\begin{cases} \hat{\alpha}_j(k) &= \lambda_{1j}[y_j(k-1) - \hat{\beta}_j u_j(k-1)] + (1-\lambda_{1j})[\hat{\alpha}_j(k-1) + \hat{p}_j(k-1)] \\ \hat{p}_j(k) &= \lambda_{2j}[y_j(k-1) - \hat{\beta}_j u_j(k-1) - \hat{\alpha}_j(k-1)] + (1-\lambda_{2j}) + \hat{p}_j(k-1) \end{cases} \tag{4}$$

where λ_{1j} and $\lambda_{2j} \in [0,1]$, are discounting factors used to minimize the influence of old data in favor of the new data.
Otherwise, when the system output is estimated, $\hat{\alpha}_j(k)$ and $\hat{p}_j(k)$ are computed by replacing the measured output $y_j(k-1)$ by its estimation $\hat{y}_j(k-1)$ in (4).
Then the control law for the next process run is calculated as follows:

$$\hat{u}_j(k+1) = \frac{T_j - \hat{\alpha}_j(k) - \hat{p}_j(k)}{\hat{\beta}_j} \tag{5}$$

3. Proposed Approach

As mentioned before, to deal with the lack of information, and for our industrial case, the JITL approach as described in [9], is used to estimate non measured DTM data ($y_j(k)$). Then, for the R2R controller, and due to the amount of disturbances in CMP process which we have dealt with, the dEWMA is used.
The classical JITL approach, as described in [9], is based on a three-step VM module to the query data. First of all, the relevant data samples in the database are searched to match the query data by some nearest neighborhood criterion. Then, a local model is built based on the most similar data. Finally, the output estimation is calculated based on the local model and the current query data. Right after that, the local model is discarded. When the next estimation is required, a new local model will be built based on the three aforementioned steps.
By observing carefully the classical approach, some contributions can be integrated in the classical JITL version to cope with some drawbacks. In fact, the JITL approach described in [9] has at least three main drawbacks.
For the first step, when some nearest neighborhood criterion is evaluated, there is no fixed similarity threshold. In other words, "most relevant data samples" can be selected while these data are not truly similar to the query data at the run k. Therefore, these selected data are not considered as real neighbors to the query point observation. In this case, local VM model construction is based on dissimilar

observations. Consequently, this model cannot reflect the query point domain. Therefore, current DTM estimation will be biased. In order to face the first drawback, a threshold of similarity must be fixed. Then, data is considered as relevant only if it reaches this threshold.
For the second step of the classical JITL, and just after the threshold definition, it may be that a sufficient number of real neighbors can not be found to build the modeling database. In this case, it is impossible to build a local model. So, the idea is to define minimum number of real similar observations which allows to construct a local model. In this case, the estimation becomes possible.
A third weakness of the classical approach can be listed when using a static sampling measurement. Indeed, a physical measurement must be provided at each time the run of measurement is coming. However, a sufficient number of similar observations could be found before proceeding to physical measurement. Therefore, an estimation of the measured variable could be easily performed instead of physical measurement which increases the manufacturing cost.
The detailed description of the modified JITL approach with its interaction with R2R controller is described in the sequel.

III-A. JITL approach description
As mentioned before, the ETM variables (collected data from equipment), denoted by $x_j(k)$, are measured or collected during the entire processing task (or run). They have also a direct impact on wafer properties which are defined throughout DTM variables denoted by $y_j(k)$. The sequence of the approach is described as follows:

1. To initialize measurement and estimation counter
Measurement and estimation procedures need to be initialized. To do this, let's define two counters k_{est} and k_{mes} which define respectively the number of estimates and measures. At the beginning, both counters are set to zero.

2. To find similar observations
In this step, all anterior ETM variables of the complete database are extracted when the DTMs were measured, which gives a collection $[X_c, Y_c]$ of N_c data. Then, the current ETM $x_j(k)$ is compared to all x_c variables. Such a comparison is achieved by using the similarity criterion (6), defined by [9] as follows:

$$\begin{cases} s_i(x_j(k), x_c(i)) = \gamma\sqrt{e^{-d(x_j(k), x_c(i))^2}} + (1-\gamma)\cos\theta_i \\ \text{with } \cos\theta_i = \dfrac{x_j(k)^T x_c(i)}{||x_j(k)|| \, ||x_c(i)||} \end{cases} \tag{6}$$

where $d(x_j(k), x_c(i))$ denotes the Euclidean distance between the current ETM data $x_j(k)$ and the ith extracted variable $x_c(i)$ from $[X_c, Y_c]$. $\gamma \in [0,1]$ is a weight parameter, i an index with $1 \leq i \leq N_c$ and θ_i is the angle between $x_j(k)$ and $x_c(i)$. Note that from (6), $s_i \in [0,1]$ and s_i tends to unity when $x_j(k)$ is close to $x_c(i)$. Referring to [9], it is also important to note that $s_i(x_j(k), x_c(i))$ will not be computed if $\cos\theta_i$ is negative ($x_j(k)$ and $x_c(i)$ are not collinear and consequently are not similar).
Then, a threshold of similarity s_{min} is defined. This means that the observations $x_j(k)$ and $x_c(i)$ are considered as similar if the following inequality holds:

$$s_i(x_j(k), x^c(i)) \geq s_{min} \tag{7}$$

Then, M data of the database which satisfy the last condition (Eq.7) are selected and noted $[X^M, Y^M]$.

3. To test if the number of similar observations is sufficient
In this step, a minimum number (threshold) of real similar observations noted N_{min} is fixed. This allows

the construction of a local model if the number of nearest ETM neighbors M exceeds the predefined threshold N_{min} ($M \geq N_{min}$). In this case, the estimation becomes possible and k_{est} is incremented ($k_{est} = k_{est} + 1$). Otherwise, if $M \leq N_{min}$, the measurement procedure must be started instead of estimation and k_{mes} counter is incremented ($k_{mes} = k_{mes} + 1$). Physical measurement is then explored in R2R controller to provide control laws $u_j(k)$.

4. To build VM local model
Using $[X^M, Y^M]$ and MLR, the linear local model for each needed estimation can be modeled as follows [17]:

$$y_j^M(l) \;=\; \sum_{p=1}^{N^p} B_p x_j^M p(l) \tag{8}$$

where $l \in \{1,\ldots,M\}$, $x_j(l) = \left[x_{1,j}(l),\cdots,x_{N_p,j}(l)\right]^T$ denotes the vector of ETM variables of the lth run, N_j the number of ETM variables and β_j the regression coefficients.
Let's define $B_j = [B_{1,j}, \ldots, B_{N_p,j}]^T$ the matrix of regression coefficients, $Y_j^M = \left[y_j^M(1), y_j^M(2) \ldots y_j^M(M)\right]^T$ the matrix of M nearest DTM neighbors and $X_j^M = \begin{pmatrix} x_{1,j}^M(1) & \cdots & x_{N_p,j}^M(1) \\ \vdots & & \vdots \\ x_{1,j}^M(M) & \cdots & x_{N_p,j}^M(M) \end{pmatrix}$ the matrix of M nearest ETM neighbors.
Let's define the estimation error $\varepsilon(k) = y_j(k) - \hat{y}_j(k) = y_j(k) - \hat{B}_j^T x_j(k)$
The main objective is to find the best parameter vector B_j such that the estimated output $\hat{y}_j(k)$ is close to the $y_j(k)$. It is well know that this objective is achieved by using the least square minimization [17], and the solution is given as follows:

$$\hat{B}_j = \left(X_j^{M^T} X_j^M\right)^{-1} X_j^{M^T} Y_j^M \tag{9}$$

5. To estimate DTM and to discard local model
Once the model is built, $\hat{y}_j(k)$ can be estimated as follows:

$$\hat{y}_j(k) = \hat{B}_j^T x_j(k) \tag{10}$$

where $x_j(k) = \left[x_{1,j}(k),\cdots,x_{N_p,j}(k)\right]^T$ denotes vector of ETM variables of current run (when estimation is needed).
Once $\hat{y}_j(k)$ is estimated, the local model will be discard.

6. To integrate estimation in R2R control
Using dEWMA strategy and when the estimation $\hat{y}_j(k)$ is provided, the control law $\hat{u}_j(k+1)$ of the next run is calculated using (4) and (5). The whole description of the approach is illustrated by Figure 2.

III-B. Modified JITL approach highlights
Using the new version of the JITL, a new dynamic sampling measurement strategy is newly integrated. This strategy does not take into account the old static sampling measurement strategy already used in semiconductor manufacturing.
For the new proposed version of JITL approach, the sampling frequency of the measurement is not predefined in advance by the user. The number of physical measurements is conditioned by the number of products which cannot be estimated. The number of physical measurements will be reduced using the modified version of the approach. Therefore, more estimations will be provided. Consequently, R2R controller which will find less difficulty to update control laws. Then product quality will be well

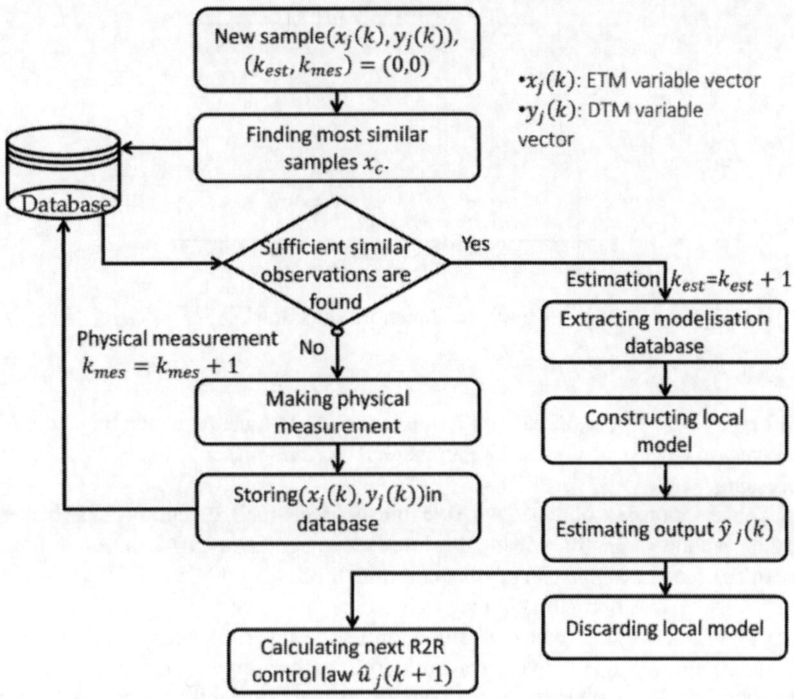

Figure 2. Modified JITL algorithm description

controlled. For example if the sampling measurement rate is prefixed and if accidentally for many runs of measurement (where physical measurement is required), the selected measured product is the same and does not cover all products. The estimation of measurements in other runs will not be possible or the estimation will be very biased.

Another point can be discussed. If physical measurements are forced at these runs, a redundancy of measurements could be made. These redundancy of measurements could be avoided using the modified JITL by improving estimations instead of physical measurements. Therefore, only needed products will be measured. This reduces measurement and production costs.

We note that it is important to include a threshold of non-measured products (threshold of successive estimation) k_{est}^{max}. When the number of successive estimation, k_{est}^{succ} reaches the threshold ($k_{est}^{succ} = k_{est}^{max}$), a physical measurement must be made to guarantee the estimation and the product quality. The next paragraph will be dedicated to a case study using industrial examples which will be used to compare results of the two approaches. The industrial examples will be dedicated also to show how R2R manifests with estimations provided by the second approach.

4. Case study

To test our approach and in collaboration with our industrial partner STMicroelectronics Rousset, 5 historical databases were extracted from 5 different CMP machines. In each equipment, many different products are processed (3 for most machines).

As a first step, a comparison was made between estimation results provided by the classical approach and the modified one.

The first approach is based on a static measurement sampling plan. For this reason, the sampling measurement rate must be defined. For CMP machines, we consider that the sampling rate is equal to 10, i.e. only one product is measured over 10 processed products.

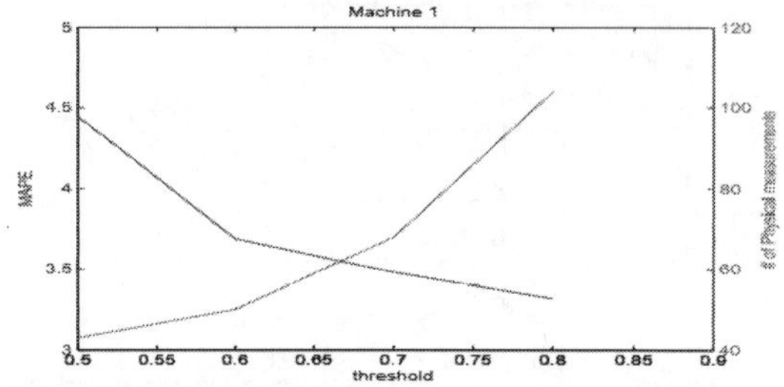

Figure 3. Threshold selection

As mentioned before, the number of real similar observations M must reach the threshold N_{min}. For our example, input vector $x_j(k)$ is made from 3 parameters. For this reason, $N_{min} = N_j = 3$, during second approach processing.

The true similar observations are obtained with the threshold that realize a compromise between the number of physical measurement and the Mean Absolute Percentage Error (MAPE). According to Figure 3, the optimum choice of the threshold can be obtained from the interval $[0.65, 0.7]$. In the present example, the value 0.7 is considered as optimum choice.

According to Figure 4, we note that the estimation of the layer thickness $\hat{y}_j(k)$, using the second approach has been improved qualitatively for the two first represented machines. It is noted that the obtained estimation of $\hat{y}_j(k)$ is filtered i.e. the peaks vanish and variability is reduced. In other words, the estimation error decreases.

Table 1 shows the quality estimation of DTM variables using both approaches, including the proposed new method and the classical approach of the algorithm JITL. One can remark clearly that the number of physical measurements (k_{mes}) is reduced to nearly third when using the second approach for all machines. For example, for the first machine, when the first approach is applied, the total number of estimations is 1331 among 1502. This means that the operation required 171 physical measures to estimate 1331 outputs. However, using the modified version of the approach shows that the number of estimations increases and reaches 1434 among 1502. This means that the operation needed only 68 physical measurements to estimate 1482 outputs. Here, the second approach requires less physical measurements than the first one and contributes in the decrease of the production cost. This intake is also valid for the other machines. This also allows to show the interest of integrating dynamic sampling using the proposed new version of JITL.

Table 1. Results comparison between the classical approach and the modified approach

Machine	Approach	k_{mes}	MAPE% [(1)]	σ_r [(2)]
1	Classical JITL	171	9.49	0.2
	Modified JITL	68	3.5	0.045
2	Classical JITL	157	12.4	1.43
	Modified JITL	55	4.38	0.046
3	Classical JITL	167	7.3	0.3373
	Modified JITL	69	3.8	0.045
4	Classical JITL	164	6.5	0.18
	Modified JITL	62	3.2	0.04
5	Classical JITL	155	7.4	0.13
	Modified JITL	59	3.5	0.044

[(1)] $MAPE\% = 100\frac{1}{n}\sum_{k=1}^{n}\frac{|\hat{y}(k)-y(k)|}{|y(k)|}$, [(2)] $\sigma_r = \sqrt{\frac{1}{n-1}\sum_{k=1}^{n}\frac{|\hat{y}(k)-y(k)|}{|y(k)|}}$

Referring also to Table 1, we note that MAPE (the Mean Absolute Percentage Error) undergoes a slight decrease for all machines. The MAPE is between two times and three times smaller using the modified JITL approach.
The relative standard deviation σ_r of error of estimation also decreases to nearly a third for all machines. So, it can be concluded that the proposed approach contributes to reducing the number of physical measurements by introducing a dynamic sampling measurement plan which contributes in the diminution of the cost of production, and ensures better estimation quality compared to the ones given by the classical JITL.

Table 2. Comparison the MAPE of static error using R2R based on VM and using industrial R2R

Product	R2R type	MAPE %
1	Industrial R2R controller	2.35
	R2R controller based on VM	0.37
2	Industrial R2R controller	1.13
	R2R controller based on VM	0.3
3	Industrial R2R controller	1.6
	R2R controller based on VM	1.4

The second step of the test, consists of showing how the VM based on modified JITL interacts with R2R controller. To highlight the contribution of the estimated outputs, static errors of the R2R controller based on VM and that based on industrial data are compared.Without loss of generality, the comparison is made only for the first machine. In this one, 3 product types are processed.

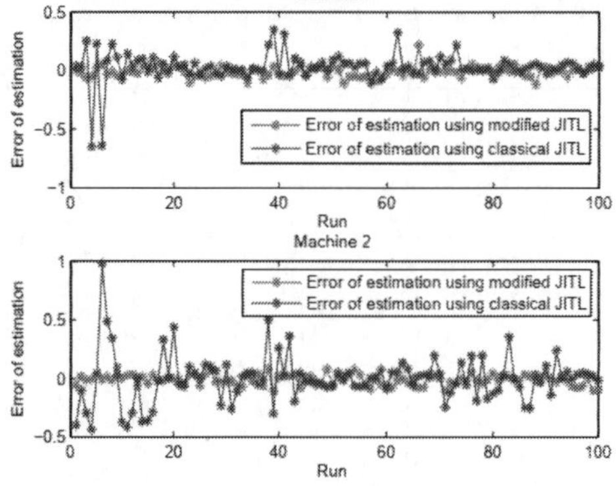

Figure 4. Comparison of error of estimation performance using the two approaches

Referring to Figure 5 and Table 2, we note that static error is improved using R2R controller based on VM for the three products and most noteworthy for the two first ones. In fact, by investigating the MAPE, this latter is 6 and 4 times less for the first and the second products using R2R controller based on VM than using the industrial R2R controller.

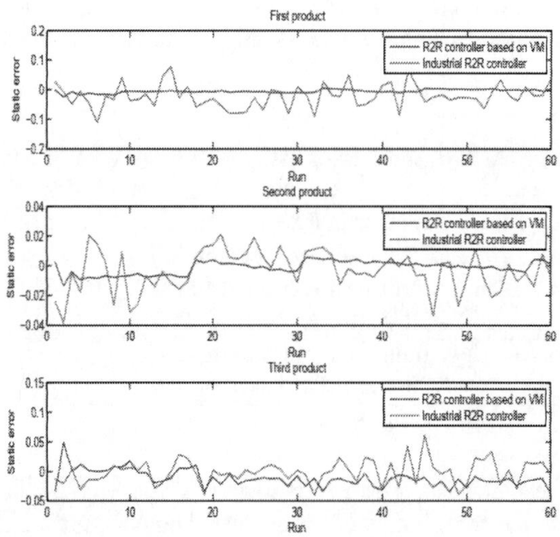

Figure 5. Comparison of static error using R2R based on VM and using industrial R2R

5. Conclusion

In this paper, VM module has been used to estimate non-measured DTM variables (layer thickness) based on current ETM variables. This module is calculated using anterior measured DTM with their corresponding ETM data. We have opted for that a new proposed version of JITL approach which allows to overcome the problem related to static sampling measurement, and the drawbacks of classical JITL version. We have also introduced the interaction and the interest of VM in R2R controller. Finally, a

comparison between the classical JITL approach and the modified one is made on real industrial data-sets derived from CMP machines and delivered by our industrial partner. It has been shown that the contribution of the proposed approach is manifested by its improved results, especially by defining a new dynamic sampling measurement plan reduces the number of physical measurements.
Without loss of generality, we focused the work only on the first machine to see how VM estimations can react with R2R controller. We noted that quality product is improved using R2R controller based on VM.
In partnership with our industrial leader, future works will concern the application of the proposed approach to deal with missing data for the case of interconnected control loops.

References

[1] Lynn, S., MacGearailt, N., and Ringwood, J.V. (2012). Real-time virtual metrology and control for plasma etch. Journal of Process Control, 22(4), 666-676.

[2] Moyne, J., Del Castillo, E., and Hurwitz, A.M. (2000). Run-to-run control in semiconductor manufacturing. CRC press.

[3] Maas, J., Ebert, M., Bhattacharyya, K., Cramer, H., Becht, A., Keij, S., Plug, R., Fuchs, A., Kubis, M., Hoogenboom, T., (2011). Yieldstar: a new metrology platform for advanced lithography control. In 27th European Mask and Lithography Conference, 79850H-79850H. International Society for Optics and Photonics.

[4] Gill, B.S. (2011). Development of virtual metrology in semiconductor manufacturing. Ph.D. thesis, The University of Texas at Austin.

[5] Elliott, R.C., Nurani, R.K., Gudmundsson, D., Preil, M., Nasongkhla, R., and Shanthikumar, J.G. (1999). Critical dimension sample planning for sub-0.25 micron processes. In Advanced Semiconductor Manufacturing Conference and Workshop, 1999 IEEE/SEMI, 139-142. IEEE

[6] Lynn, S. (2011). Virtual metrology for plasma etch processes. Ph.D. thesis, National University of Ireland, Maynooth.

[7] Khan, A.A., Moyne, J.R., and Tilbury, D.M. (2008). Virtual metrology and feedback control for semiconductor manufacturing processes using recursive partial least squares. Journal of Process Control, 18(10), 961-974.

[8] Slišković, D,Grbić, R and Hocenski, Ž (2012). Methods for plant data-based process modeling in soft-sensor development. AUTOMATIKA: časopis za automatiku, mjerenje, elektroniku, računarstvo i komunikacije 52(4), 306-318.

[9] Cheng, C. and Chiu, M.S. (2004). A new data-based methodology for nonlinear process modeling. Chemical Engineering Science, 59(13), 2801-2810.

[10] Abdi, H. (2010). Partial least squares regression and projection on latent structure regression (PLS regression). Wiley Interdisciplinary Reviews: Computational Statistics, 2(1), 97-106.

[11] Pan, J.C.H. and Tai, D.H. (2009). Implementing virtual metrology for in-line quality control in semiconductor manufacturing. International Journal of Systems Science, 40(5), 461-470.

[12] Purwins, H., Barak, B., Nagi, A., Engel, R., Hockele, U., Kyek, A., Cherla, S., Lenz, B., Pfeifer, G., and Weinzierl, K. (2014). Regression methods for virtual metrology of layer thickness in chemical vapor deposition. IEEE/ASME Transactions on Mechatronics, 19(1), 1-8.

[13] Hung, M.H., Lin, T.H., Cheng, F.T., and Lin, R.C. (2007). A novel virtual metrology scheme for predicting CVD thickness in semiconductor manufacturing. IEEE/ASME Transactions on Mechatronics, 12(3), 308-316.

[14] Su, Y.C., Lin, T.H., Cheng, F.T., and Wu, W.M. (2008). Accuracy and real-time considerations for implementing various virtual metrology algorithms. IEEE Transactions on Semiconductor Manufacturing, 21(3), 426-434.

[15] Besnard, J., Gleispach, D., Gris, H., Ferreira, A., Roussy, A., Kernaflen, C., and Hayderer, G. (2012). Virtual metrology modeling for cvd film thickness. International Journal of Control Science and Engineering, 2(3), 26-33.

[16] Kang, P., Kim, D., Lee, H. J., Doh, S., and Cho, S. (2011). Virtual metrology for run-to-run control in semiconductor manufacturing. Expert Systems with Applications, 38(3), 2508-2522.

[17] Ghasemi, J., Saaidpour, S., and Brown, S.D. (2007). Qspr study for estimation of acidity constants of some aromatic acids derivatives using multiple linear regression MLR analysis. Journal of Molecular Structure: THEOCHEM, 805(1), 27-32.

IOP Conf. Series: Journal of Physics: Conf. Series **783** (2017) 012043 doi:10.1088/1742-6596/783/1/012043

Multimodel Control of Diesel Engines

Silviu Cirstoiu*, Dumitru Popescu*, Catalin Dimon*, Severus Olteanu*
*Faculty of Automatic Control and Computer Science, UPB, Bucharest, Romania,
E-mails: silviu.cirstoiu@yahoo.com, popescu_upb@yahoo.com, catalin.dimon@acse.pub.ro, severus.olteanu@gmail.com

Abstract. In this article it is proposed and designed a modern control configuration of the type multicontroler-multimodel (MM) that pilots the nonlinear combustion process of the Diesel engine, needed to adjust the pressure in the intake manifold and the airflow circulating through the compressor. The MM simulator developed by the authors allows the implementation of control systems represented by pairs (Mi, Ci) with the Mi candidate closest to the current operating point of the process and the paired controller Ri, for controlling the key parameters of the combustion process. The proposed configuration is built with robust controllers and thus it is able to ensure superior performance, tolerance to nonlinearities and parametric and structural perturbations in the system.

1. Introduction

Multimodel control strategy for nonlinear systems configuration is a relatively new approach. In recent studies from the literature, increasing interest is shown towards the control of nonlinear processes with multimode operations, as diesel engines, processes that can be approximated only roughly with a single global model. The first papers that have proposed solutions and methods suitable for this type of processes, are those of Balakrishnan and Narenda the 90s [1], relying mainly on the construction of adaptive-robust systems using classical algorithms. Later detailed studies and positive results were presented by Athans [20] who employed the concept of multi model control and demonstrated the effectiveness of this approach, with the disadvantage of additional computation and implementation effort. The principle of building a MM configuration is the same, and is based on known identification procedures of Mi models and design of control algorithms Ci, the differences are mainly due to the selection mode and command algorithms interchange.

Landau and Karimi [19] use the so-called Cloe (Closed Loop Output Error) procedure, adjusting the parameters in a MM control structure [6]. Later studies have appeared on the use of neural networks and fuzzy logic systems, involved in the development of MM structures. Research by the authors have been focused on proposing and testing MM control configurations for nonlinear processes implemented on punctual applications, dedicated control parameters of the combustion process in the Diesel engines.

Following the preliminary results obtained in the paper, the proposed solution for implementing the concept appeals to a control structure (MM) in a robust version, to be tolerant to nonlinearities and disturbed regime of the diesel engine combustion.

The used management configuration contains control loops (feedback) with pairs of models having robust controllers designed for the pre-specified operating points of the nonlinear characteristic of the process, dependent mainly on the variations in the engine load and actuators behaviour.

We fix three possible operating points and for the designed (Mi, Ci), i = 1,2,3, systems, presumably linearized model Mi is disrupted in parameters and/or structure. The choice of three points was made as a consequence of the observed system behaviour. A robustness analysis of systems in closed loop is performed and the nominal command is adjusted for all the preset operating points. An adaptive strategy for MM structure occurs when the engine is about to commute between two operating points. The multi model control structure designed is shown in the following figure:

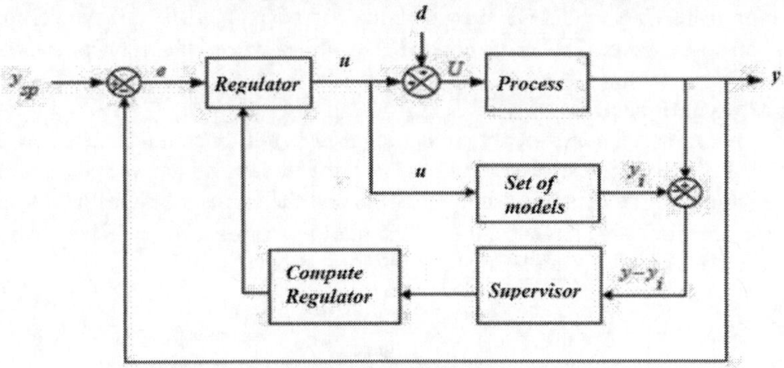

Figure 1. Multi model control structure

where,
y_{sp} – reference measure
e – the error between actual and desired output
u – command
d – disturbance
U – disturbed command
y – the controlled output of the process
y_i – i[th] model output
$y - y_i$ – difference between process output and i model output.

The supervising module must choose the right model and control algorithm for the engine functioning point. Thus we define the model error calculated at each sampling moment, which represents the difference between output y_i and output y for the same u value of the applied command.

$$\varepsilon_i(k) = y(k) - y_i(k) \tag{1}$$

The used criteria for selection of the closest model of the current process operating point is a square criteria built with the help of the model error, as in:

$$J_i(k) = \alpha\varepsilon_i^2(k) + \beta \sum_{j-1}^{k} e^{-\lambda(k-j)}\, e_i^2(j) \tag{2}$$

where $\alpha > 0$ and $\beta > 0$ are the criteria ponderation factors, and $\gamma > 0$ is the forgiveness factor which ensures its action window limitation over the model error $\varepsilon_i(k)$.
The α, β and γ parameters choice depends on the systems' characteristics, being:

- $\alpha = 1$ and $\beta = 0$, for fast response systems:
- better performances in observing the process' changes,
- distrubances' sensibility,
- $\alpha = 0$ and $\beta = 0$, for slow response systems,
- weak performances in observing the process' changes,
- disturbances' insensibility.

We properly configured the MM structure, such that it can act on faster systems, as one can see in the case of controlling the working parameters of the diesel engine.
For each identified Mi model, a Ci regulator (controller) was computed, that meets the objectives and performance requirements; As such, we provided the set of pairs (Mi, Ci), corresponding to the possible nominal operating points Pi. For Ci control algorithms, we applied robust correction.

2. Building the MM configuration

In order to design an effective multi-model control configuration, one should consider the following issues: the inevitability of non-linear processes based on the various points of operation; the choice of the most suitable model for the dynamics of the process and the choice of algorithms respectively; the proper control for the multi-model system's commutation in order to achieve stability [8]. We represent a standard architecture for multi model configuration in Figure 2:

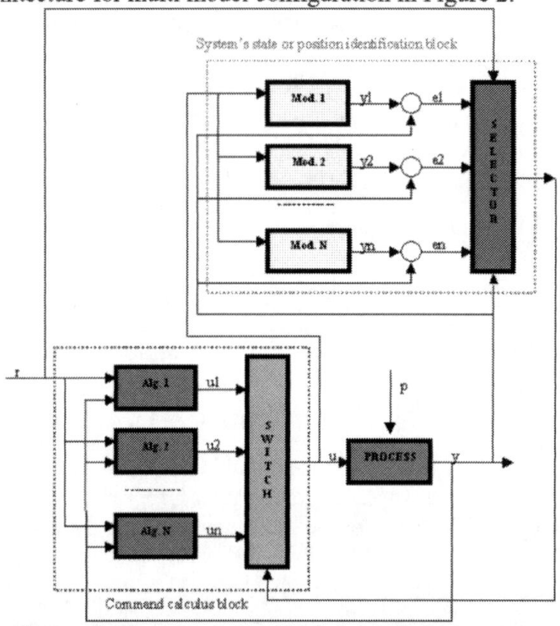

Figure 2. The classic MultiModel (MM) structure

MM framing components in Figure 2 are:
♣ PROCESS - system to be controlled;
♣ BCC - block for calculating the order;
♣ SUPERVISOR System's state - component which gives information about the best matching of operating at a time and pair model - algorithm;
♣ M1, M2, ..., Mn - models for different operating points r;
♣ C1, C2, ..., Cn - control algorithms
♣ SWITCH - switch control law;
♣ BSM - block selection most suitable model for the current state of the system based on predetermined criteria;
♣ y - process output;
♣ y_1, y_2, ..., y_N - models output;
♣ u and u_1, u_2,..., u_N - controls generated by BCC calculation block;
♣ r - reference system;
♣ p - physical disturbance process.

The overall structure can be sized according to its particularities and the system's nonlinearities. The switching is done according to procedures chosen by the reconfiguration mechanism.

After specifying the operating points and the design of the pairs (Mi, Ci), function switching control algorithms play an essential role. The issue is this: in what way and to what extent can we move from one control algorithm obtained, to another so that it preserves the stability offered by the initial performance. The novelty in this paper is given by the method used to calculate the pair (Mi, Ci) and by the approach proposed for obtaining a suitable solution for the switching operation using polynomial control algorithms of RST type.
Therefore, by selecting the most suitable model for a particular process' operating mode and by using pair adjustment mapping algorithms, one can immediately imply the requirement of switching between different selected active algorithms at a given time.
The Diesel Engine disposes of a variable geometry turbine (VGT - Variable Geometry Turbocharger), which allows efficient adaptation to the needs of the operating point of the engine equipped with exhaust gas recirculation valve, EGR (Exhaust Gas Recirculation) directing the gas flow outlet (residual) in the inlet in order to maintain a constant temperature in the combustion process. Although a technologically innovative solution, unfortunately the variable turbine geometry introduces nonlinear behaviour. Therefore, the adopted MM control design structure is justified for the control of the nonlinear regime. Thus we design the MM control solution for the key parameter: Wci airflow.

3. MM System configuration

Regarding the design of the configuration of the MM system for controlling the Wci airflow, we use the I/O model, given by the discrete transfer function:

$$Hd_{VGT-W_{ci}}(z) = \frac{0.03711*z^3-0.0785*z^2+0.09375*z+0.02185}{z^3-2.271*z^2+1.617*z-0.3458} \tag{4}$$

The choice of the operating points was made by taking into account the actuator nonlinearity (VGT turbine xv valve opening position) that can operate through the opening in the range 0-40%, 40-60% and 60-100%. We considered a piecewise linearization around the operating points P1, P2, P3 each corresponding to the selected intervals and for each interval we associated the segment tangent to the corresponding linear model.
Under the reasoning of the previous section, we determine for each operating point, the corresponding pairs(Mi,Ci).
Using the CMMPR identification method and a sampling period Te = 0.1 sec, on the software platform Adaptech / WinPim, we obtained discrete linearized models, so that the input xv can now be adjusted to the WCI output:

M1, given by:

$$Hd_{EGR-p_i}(z) = \frac{280.8*z^3-102.9*z^2-273.6*z-99.22}{z^3-2.271*z^2+1.617*z-0.3458}$$

M2, given by:

$$Hd_{EGR-p_i}(z) = \frac{210*z^3-56.8*z^2-198.45*z-77.28}{z^3-2.271*z^2+1.617*z-0.3458}$$

M3, given by:

$$Hd_{EGR-p_i}(z) = \frac{178.5*z^3-12.7*z^2-102.33*z-24.44}{z^3-2.271*z^2+1.617*z-0.3458}$$

Next it calls for the design platform Adaptech / WinReg with pole assignment methods to calculate the RSTcontrollers.
For the first operating point, corresponding to 0-40% range, we have the following results: the controller computed with override $\xi = 0.95$ and natural pulsation $w_0 = 10.5$ rad / sec for tracking targets, respectively, $\xi = 0.95$, $w_0 = 12.5$ rad/sec for setting targets, has the following components:

$$R(z^{-1}) = -1.38 - 2.98z^{-1} - 0.11z^{-2}$$

$$S(z^{-1}) = 1 - 0.96z^{-1} - 0.04z^{-2}$$
$$T(z^{-1}) = -8.45 + 4.77z^{-1} - 0.79z^{-2}$$

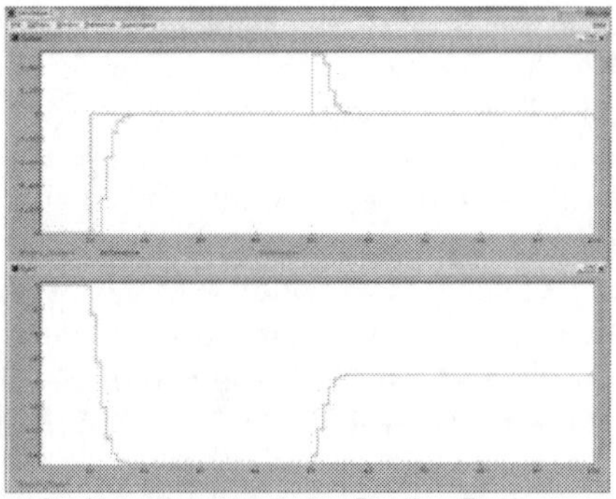

Figure 3. Nominal system response for the first operating point P1

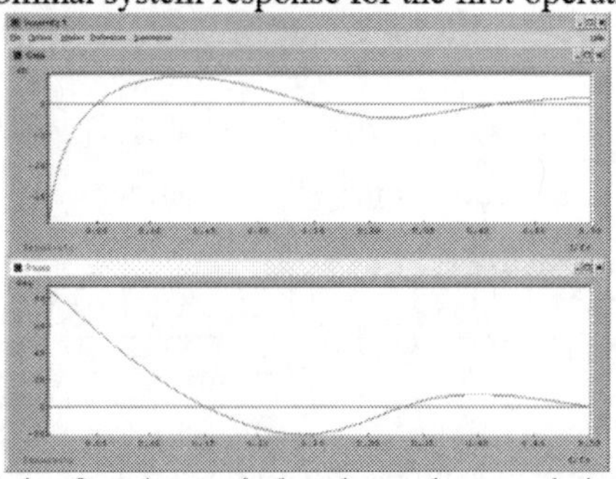

Figure 4. The sensitivity function andnthe phase characteristic of the robust system

New edge module is 0.72, which means that by introducing a pair of zeros in the polynomial S, robustness of the system was improved.

Thus, for the second operating point corresponding to the range 40-60%, we have the following results: the controller calculated similarly with $\xi = 0.95$ and $w_0 = 10.5$ rad/sec for tracking goals and $\xi = 0.95$, $w_0 = 12.5$ rad/sec for setting goals, has the following components:

$$R(z^{-1}) = -1.37 - 2.97z^{-1} - 1.09z^{-2}$$

$$S(z^{-1}) = 1 - 0.96z^{-1} - 0.037z^{-1}$$

$$T(z^{-1}) = -8.42 + 4.75z^{-1} - 0.78z^{-2}$$

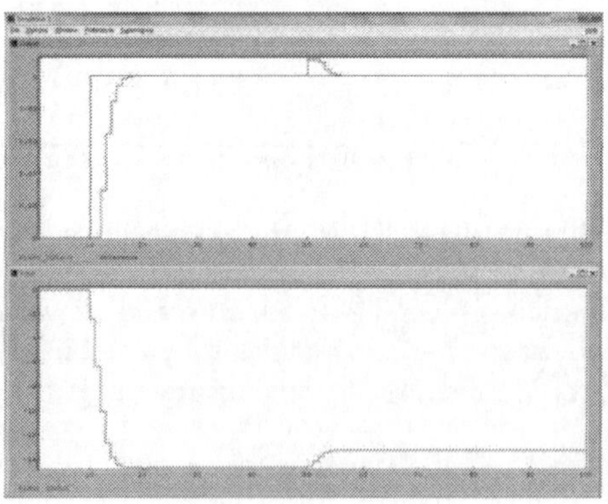

Figure 5. System response for the second functioning point P2

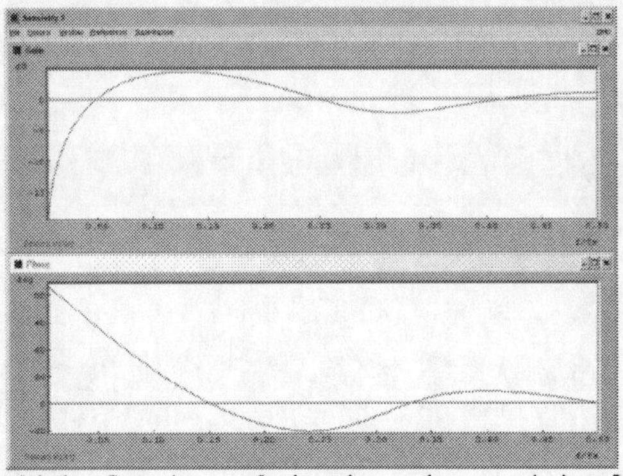

Figure 6. The sensitivity function andnthe phase characteristic of the robust system

New edge module is 0.73, which means that by introducing a pair of zeros in the polynomial S, robustness of the system was improved.

For the third operating point corresponding range 60-100%, we have the following results:
The controller computed with ξ= 0.95 and w_0 = 10.5 rad / sec for tracking goals and w_0 = 12.5 rad / sec with ξ= 0.95 for target setting, has the following components:

$$R(z^{-1}) = -2.17 - 2.02z^{-1} - 0.33z^{-2}$$

$$S(z^{-1}) = 1 - 0.82z^{-1} - 0.17z^{-2}$$

$$T(z^{-1}) = -8.56 + 4.83z^{-1} - 0.80z^{-2}$$

New edge module is 0.71, which means that by introducing a pair of zeros in the polynomial S, robustness of the system was improved.

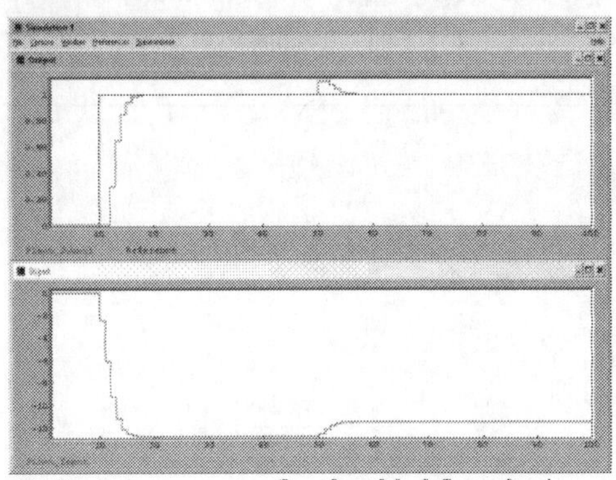

Figure 7. System response for the third functioning point P3

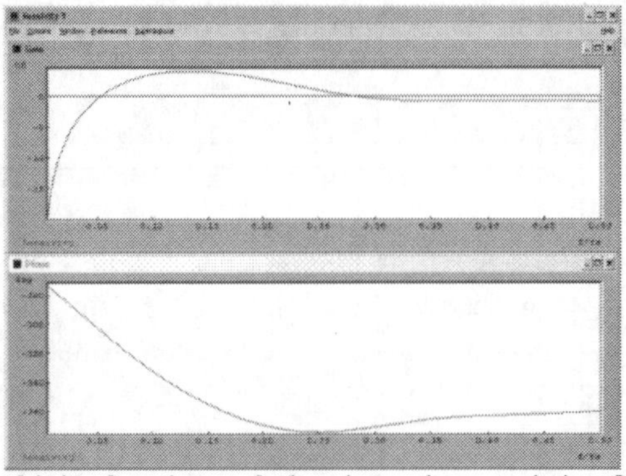

Figure 8. The sensitivity function andnthe phase characteristic of the robust system

4. The MM application and simulation results

After we have selected the operating points of the nonlinear characteristic associated to the motor functioning regime, the dynamic models (Mi-type I / O) were determined as well as the control algorithms (Ci type RST). We performed the simulation in closed loop systems to verify performance and ultimately the MM control structure was configured.

The model' selector was designed to specify the correct position in the real system and to specify the operating conditions for the switching point. It is based on the error model and the performance criteria.

$$\varepsilon_i(k) = y(k) - y_i(k) \tag{4}$$

$$J_i(k) = \alpha \varepsilon_i^{\,2}(k) + \beta \sum_{j=1}^{k} e^{-\lambda(k-j)} e_i^{\,2}(j) \tag{5}$$

Regarding the switch control algorithm, switching is performed based on information received from the pattern selection, which builds a switching signal and calls for the new controller to be activated. The three controllers are brought to active state, *ie.* each controller receives the information as y(k) and reference r(k), and calculate their order without letting u(k) to be applied.

For switching, the reference set will change given the output active control algorithm, and the resulting expression is:

$$y^*(k) = \frac{1}{t_0}\left[\sum_{i=0}^{n_S} s_i u(k-i) + \sum_{i=0}^{n_R} r_i y(k-i) - \sum_{i=1}^{n_T} t_i y^*(k-i)\right] \tag{5}$$

with respect to the equation:

$$y^*(k) = \frac{B_m(q^{-1})}{A_m(q^{-1})} r(k) \tag{6}$$

The switching system will function according to the below schematic:

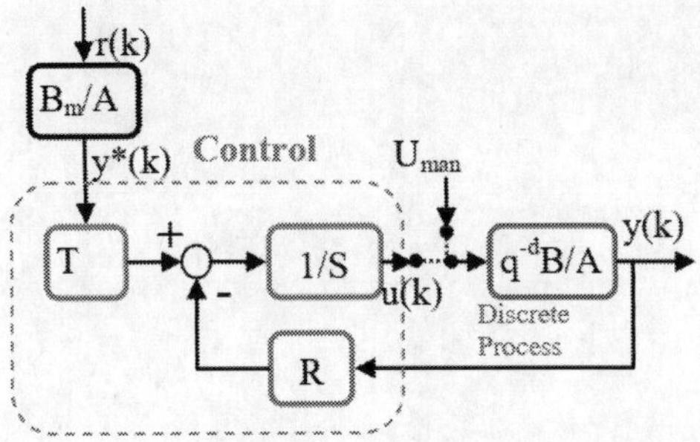

Figure 9. The set point value computation of the active algorithm using U_{man} command

The set point value will be identical to the active control algorithm's command value when switching. By using the solution shown in [9], we can eliminate the possibility of instability due to the probable gaps in the algorithm memory command.

It is therefore important to use the system memory command to always access the history in terms of switching between algorithms.

To evaluate the simulation results, we use the software application CVI - National Instruments, which allows viewing MM configuration dynamics including control algorithms behaviour in switching mode.
Simulation module is user friendly and real-time simulations are performed after all input data has been placed on the platform as shown in Fig. 10, Fig. 11, Fig. 12.

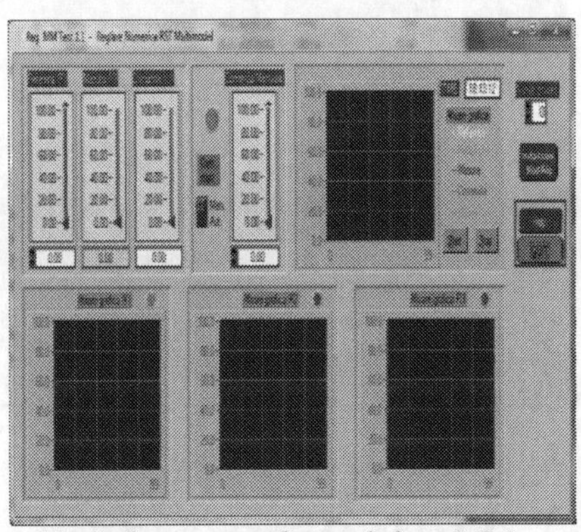

Figure 10. Software app for real time MM command

The application allows quick access to a range of information and commands, such as graphics displaying the time evolution of the process and control algorithms R1, R2, R3, highlighting the active algorithm, automatic or manual control of the system, the choice of the studied process, the possibility of initializing pairs of obtained controlled models, the possibility of selection of references, etc.
The graphical display allows simultaneously viewing the reference measure, the control and response model, each with a corresponding colour code.

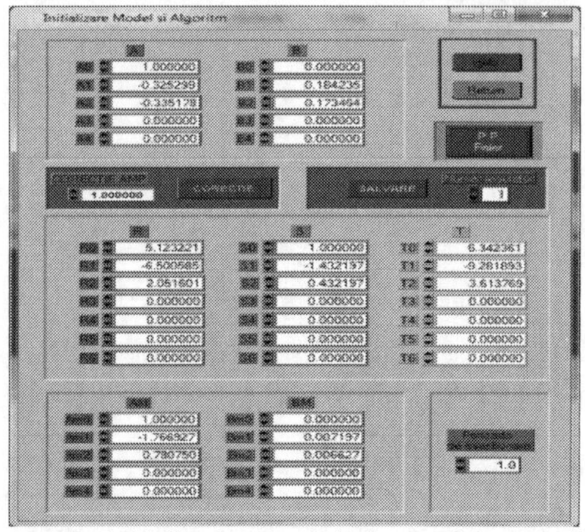

Figure 11. The initialization module for model-controller pairs

In the module initialization model, the control algorithm allows the introduction into the application of polynomial coefficients A and B for models definitions, and polynomials R, S and T for controllers.

After defining corrections, amplification and sampling period, it generates the actual simulation.
For example, we have considered the case of the multi model system for controlling the air flow Wci, in which case we considered three distinct operating points P1, P2 and P3, in accordance to the reasoning from the beginning of the chapter; following the same rationale, we have identified three models: M1 (0-40%), M2 (40-60%) and M3 (60-100%). For these models we have developed robust control algorithms, namely C1, C2, and C3.

Applying the superposition principle for model-algorithms' pairs to determine their control areas we get: M1 (0-55%), M2 (45-90%) and M3 (80-100%).

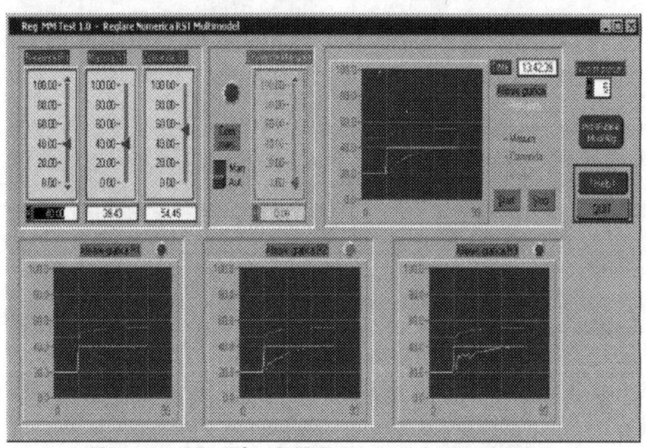

Figure 12. The MM system's response

Two switching tests were performed: the first in the range 40% -60%, and the second in the range 80% -100%, the results can be seen below.

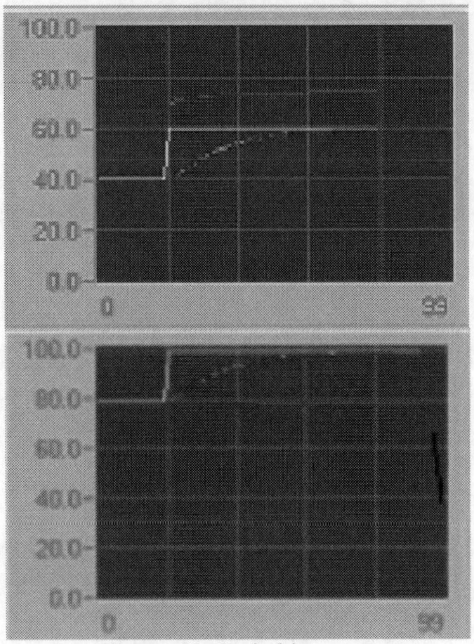

Figure 13. Switching tests

5. Conclusions and perspectives

In this paper, an approach is proposed for modeling and controling a turbocharged Diesel engine with EGR valve. A configuration was designed and validated, in order to steer the multi model MM-multi controller nonlinear combustion process of a diesel engine, available for adjusting the intake manifold pressure and airflow circulating through the compressor.

The I/O Models are obtained by CMMP-recursive identification methods, by processing data files and considering the input EGR exit-pressure in the intake manifold, and the VGT output compressor airflow.
These models are validated by tests of statistical analysis error of prediction, allowing an interpretation closer to the real process and provides the possibility of implementing control algorithms that can be improved through adaptive strategies and robust.

Simulations carried out both on state models, and the models I / O system have shown that the stable combustion is represented by models which can be used to control the operation of the Diesel engine.

An optimal linear quadratic controller was properly designed and the MIMO system status and performance have been verified in simulation.

Two systems were designed for I / O strategy, in order to control the pressure and flow of air through the compressor in a nominal operating point. Based on an analysis of robustness in the frequency-the control algorithms were correct; the performance and capital gain were also verified in simulation.

We proposed and designed a configuration that pilots the multi-model – multi-controller MM nonlinear combustion process of the Diesel engine, available for adjusting the intake manifold pressure and air flow circulating through the compressor.

The MM software structure enables: implementing systems represented by pairs (Mi, Ci), selecting the closest one to the current operation point of the process and its corresponding controller pair, suitable for controlling air flow pressure p_i and WCI order without affecting the stability of the switching systems.

The MM control structure design was developed on the study of supervision and proposed configuration options (CLMS) switch that selects a single controller (the appropriate) of a plurality of candidate controllers, type RST robust. To select the control algorithm was used a quadratic minimization criteria built on the error between process output and exit model in question (model error).

The MM controller performance are highlighted through the use of dedicated simulators, application-specific functions built to verify the work and which can be obtained on an experimental model of a running vehicle.

After implementing the obtained results in the paper for vehicles equipped with Diesel engines, there is interest in collaborating with the Research Center Renault Romania for further research and implementing control algorithms for the real time version after a further validation on a new set of experiments simulator and pilot platforms that require access to a vehicle in circulation.

6. References

[1] BALAKRISHNAN., J., *Control System Design Using Multiple Models, Switching and Tuning*, PhD Thesis, Yale University, SUA, 1996.
[2] DUMITRACHE, I., Ingineria Reglarii Automate, Politehnica Press, Bucharest, 2005, ISBN 973-8449-72-3.
[3] LANGLOIS, N. - Contribution à la modélisation et à la commande avancée des moteurs Diesel, 2008, Universite de Rouen.
[4] LUPU C, Udrea A., Popescu D., Fluture C., – On the stability of switching in multiple model control systems, Politehnica University Bucharest. 2009a
[5] LUPU C, Popescu D Ciubotaru B, Petrescu C, – Switching solution for multiple-models control systems, Politehnica University Bucharest. 2009b
[6] JANKOVIC M., Jankovic M., and Kolmanovsky I.. Constructive lyapunov control design for turbo charged diesel engines. IEEE Transactions on Control Systems Technology, 8:288-299, 2000.
[7] JUNG M., Mean-value modeling and robust control of the air path of a turbocharged diesel engine, PhD thesis, University of Cambridge, 2003.
[8] PAGES, O., P. Mouille et B. Caron, Multi-model control by applying a symbolic fuzzy switches, IFAC, 2000.
[9] POPESCU D., Stefanoiu D. and all – Industrial Control - , Ed. Agir, Bucharest, Romania, 2007
[10] SODERSTROM T. and P. Stoica (1989), System Identification, Prentice Hall, London.
[11] NARENDRA, K. S. and J. Balakrishnan, (1997). Adaptive Control using multiple models, IEEE Transactions on Automatic Control, vol. 42, no. 2, February, page. 171 – 187
[12] LUPU C., Popescu D., Petrescu C., (1999) Multi Model Adaptive Control System, ISAS"99 Conference, U.S.A., Orlando, 13-15 August
[13] LUPU C., D. Popescu, C. Petrescu, Al. Ticlea, C. Dimon, A. Udrea, Bogdan I., (2008), Multiple-Model Design and Switching Solution for Nonlinear Processes Control -, ISC'08, The 6th Annual Industrial Simulation Conference , 09-11 June, Lyon, France, pp. 71-76, ISBN 978-90-77381-4-03
[14] JENSEN J.P., Kristensen A.F., Sorenson S.C., Houbak N., Hendricks E., „Mean Value Modeling of a Small Turbocharged Diesel Engine" – SAE Paper 910070, International Congress and Exposition, Detroit, 1991
[15] JUNG M., Glover K.. Calibratable linear parameter-varying control of turbocharged diesel engines, IEEE Transactions on Control Systems Technology, 14:45-62, 2006

[16] LARSEN M., Jankovic M. and. Kokotovic P.V. Indirect passivation design for a diesel engine model. In Proceedings of the 2000 IEEE International Conference on Control Applications, USA, pages 309-314, 2000

[17] DEC, J. E., Kathy Epping; Salvador M. Aceves; Richard L. Bechtold (2002). "The Potential of HCCI Combustion for High Efficiency and Low Emissions". Society of Automotive Engineers. 2002-01-1923.

[18] JOHANSSON R., Blom D.,Karlsson M., Ekholm K.,Tunestal P. (2008). "HCCI Engine Modeling and Control using Conservation Principles". Society of Automotive Engineers. 2008-01-0789.

[19] Landau, I.D. and A. Karimi, (1997). "*Recursive algorithm for identification in closed loop: a unified approach and evaluation*", Automatica, vol. 33, no. 8, pp. 1499-1523

[20] M. Athans, V. Hassani, J.P. Hespanha, A.M. Pascoal (2006). "Stability Analysis of Robust Multi Model Adaptive Control", www.ece.ucsb.edu .

IOP Conf. Series: Journal of Physics: Conf. Series **783** (2017) 012044 doi:10.1088/1742-6596/783/1/012044

Output feedback model matching in linear impulsive systems with control feedthrough: a structural approach

Elena Zattoni

Department of Electrical, Electronic, and Information Engineering "G. Marconi"
Alma Mater Studiorum · University of Bologna, 40136 Bologna, Italy

E-mail: elena.zattoni@unibo.it

Abstract. This paper investigates the problem of structural model matching by output feedback in linear impulsive systems with control feedthrough. Namely, given a linear impulsive plant, possibly featuring an algebraic link from the control input to the output, and given a linear impulsive model, the problem consists in finding a linear impulsive regulator that achieves exact matching between the respective forced responses of the linear impulsive plant and of the linear impulsive model, for all the admissible input functions and all the admissible sequences of jump times, by means of a dynamic feedback of the plant output. The problem solvability is characterized by a necessary and sufficient condition. The regulator synthesis is outlined through the proof of sufficiency, which is constructive.

1. Introduction

Linear impulsive systems form a special class of hybrid dynamical systems, featuring a continuous-time linear behavior (governed by the so-called flow dynamics) subject to state discontinuities occurring at isolated point of the time axis (governed by the so-called jump dynamics). Linear impulsive systems have recently attracted a huge amount of research interest, mainly because they are particularly effective in modeling complex phenomena as well as composite systems — see, e.g., [1]. Nowadays, several control and observation problems have been formulated and investigated in the context of linear impulsive systems: i.e., state estimation [2], linear quadratic control [3,4], disturbance decoupling [5,6], and output regulation [7–13].

As to model matching, this is a classical problem of control theory, which still attracts noticeable attention not only for its intrinsic theoretical interest, but also because it provides powerful tools to solve more general problems — see, e.g., [14–16]. In this context, the contribution of this work lies in considering — for the first time (to the best of the author knowledge) — the problem of model matching for linear impulsive systems with a possible feedthrough term from the control input to the output and in proving a necessary and sufficient condition for its structural solution. In particular, the linear impulsive systems dealt with are allowed to have jump time instants not a-priori known and (possibly) not uniformly spaced in time. However, it is assumed that the number of jump times in any finite time interval is finite, so as to avoid Zeno behaviors.

As to the underlying methodology, this ensues from the geometric approach to linear control theory [17, 18]. This approach has been extended to cope with linear impulsive systems, e.g., in [5, 6, 13]. However, the presence of possible feedthrough terms requires further generalizations of the notion of hybrid controlled invariance, which were not given in the previous papers. Actually, the geometric approach has proven to be quite powerful in handling also control problems stated for other classes of hybrid systems, like, in particular, switching linear systems — see, e.g., [19–36].

Notation: The symbols $\mathbb{R}$, $\mathbb{R}^+$, and $\mathbb{Z}^+$ stand for the sets of real numbers, nonnegative real numbers, and nonnegative integer numbers, respectively. Matrices and linear maps are denoted by slanted upper-case letters, like A. The image and the kernel of A are denoted by $\operatorname{Im} A$ and $\operatorname{Ker} A$, respectively. The transpose of A is denoted by $A^\top$. The inverse of a nonsingular square matrix A is denoted by A^{-1}. Vector spaces and subspaces are denoted by calligraphic letters, like $\mathcal{V}$. The symbol I denotes an identity matrix of appropriate dimensions.

2. Output feedback model matching in linear impulsive systems — problem statement

The definition of a linear impulsive system requires that the time domain be described by an interlaced sequence of continuous time intervals and isolated time instants. In particular, in this work, the time domain is specified as follows. The set $\mathcal{T} = \{t_0, t_1, \ldots\}$ represents a finite or countably infinite ordered set of strictly increasing elements of $\mathbb{R}^+$. The symbol t_f denotes the last element of $\mathcal{T}$ when the cardinality of $\mathcal{T}$ is finite. The set $\mathcal{T}$ is assumed to have no accumulation points: i.e., the number of elements of $\mathcal{T}$ is finite in any finite interval of $\mathbb{R}^+$. The symbol $\mathscr{T}$ denotes the set of all $\mathcal{T}$ satisfying this constraint. The nonnegative real axis without the elements of $\mathcal{T}$ is denoted by $\mathbb{R}^+ \setminus \mathcal{T}$.

The linear impulsive system Σ_S is defined by

$$\Sigma_S \equiv \begin{cases} \dot{x}_S(t) &= A_S\, x_S(t) + B_S\, u(t), \quad t \in \mathbb{R}^+ \setminus \mathcal{T}, \\ x_S(t_k) &= J_S\, x_S^-(t_k), \quad t_k \in \mathcal{T}, \\ y_S(t) &= C_S\, x_S(t) + D_S\, u(t), \quad t \in \mathbb{R}^+, \end{cases}$$

where $x_S \in \mathcal{X}_S = \mathbb{R}^{n_S}$ is the state, $u \in \mathbb{R}^p$ is the control input, and $y_S \in \mathbb{R}^q$ is the output, with $p, q \leq n_S$. A_S, B_S, J_S, C_S, and D_S are constant real matrices of suitable dimensions. The direct algebraic link from the control input to the output established by the matrix D_S is called *control feedthrough*. The matrices

$$\begin{bmatrix} B_S \\ D_S \end{bmatrix}, \quad \begin{bmatrix} C_S & D_S \end{bmatrix}$$

are assumed to have full rank. The set of the admissible control input functions $u(t)$, with $t \in \mathbb{R}^+$, is assumed to be the set of all piecewise-continuous functions with values in $\mathbb{R}^p$. The so-called *flow dynamics* is governed by the differential state equation. Instead, the algebraic state equation rules the so-called *jump dynamics*. Thus, according to the linear impulsive structure of Σ, the state evolution $x_S(t)$ in the time interval $[0,\, t_0)$ satisfies the differential equation, with given initial state $x_S(0) = x_{S,0}$ and input function $u(t)$, with $t \in [0,\, t_0)$. The state $x_S(t_k)$, with $t_k \in \mathcal{T}$, is the image through J_S of $x_S^-(t_k) = \lim_{\tau \to 0^+} x_S(t_k - \tau)$. The state evolution $x_S(t)$ in the time interval $[t_k,\, t_{k+1})$, with $t_k, t_{k+1} \in \mathcal{T}$, satisfies the differential equation, given the initial state $x_S(t_k)$ and the input function $u(t)$, with $t \in [t_k,\, t_{k+1})$.

The linear impulsive model Σ_M is defined by

$$\Sigma_M \equiv \begin{cases} \dot{x}_M(t) &= A_M\, x_M(t) + B_M\, d(t), \quad t \in \mathbb{R}^+ \setminus \mathcal{T}, \\ x_M(t_k) &= J_M\, x_M^-(t_k), \quad t_k \in \mathcal{T}, \\ y_M(t) &= C_M\, x_M(t), \quad t \in \mathbb{R}^+, \end{cases}$$

where $x_M \in \mathbb{R}^{n_M}$ is the state, $d \in \mathbb{R}^q$ is the input, and $y_M \in \mathbb{R}^q$ is the output. The set of the admissible input functions $d(t)$, with $t \in \mathbb{R}^+$, is assumed to be the set of all piecewise-continuous functions with values in $\mathbb{R}^q$.

Hence, the problem of model matching by output feedback in linear impulsive systems can be stated as follows.

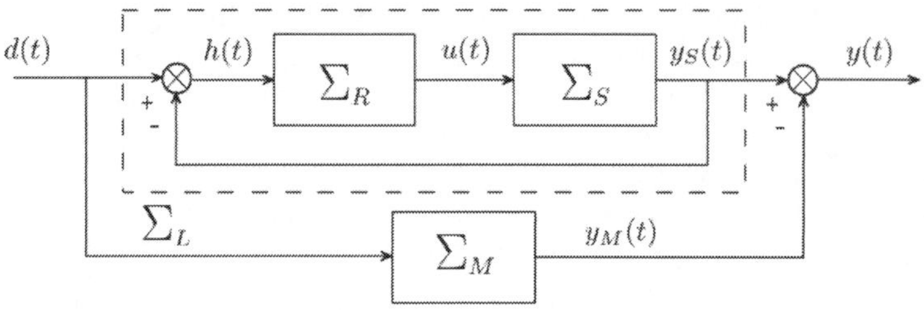

Figure 1. Block diagram for model matching by output feedback

Problem 1 (Model matching by output feedback in linear impulsive systems) *Let the linear impulsive system Σ_S and the linear impulsive model Σ_M be given. Find a linear impulsive regulator Σ_R, defined by*

$$\Sigma_R \equiv \begin{cases} \dot{x}_R(t) &= A_R\, x_R(t) + B_R\, h(t), \quad t \in \mathbb{R}^+ \setminus \mathcal{T}, \\ x_R(t_k) &= J_R\, x_R^-(t_k), \quad t_k \in \mathcal{T}, \\ u(t) &= C_R\, x_R(t), \quad t \in \mathbb{R}^+, \end{cases}$$

where $h(t) = d(t) - y_S(t)$, such that the closed-loop linear impulsive system Σ_L, defined by

$$\Sigma_L \equiv \begin{cases} \dot{x}_L(t) &= A_L\, x_L(t) + D_L\, d(t), \quad t \in \mathbb{R}^+ \setminus \mathcal{T}, \\ x_L(t_k) &= J_L\, x_L^-(t_k), \quad t_k \in \mathcal{T}, \\ y_S(t) &= C_L\, x_L(t), \quad t \in \mathbb{R}^+, \end{cases}$$

where

$$A_L = \begin{bmatrix} A_S & B_S C_R \\ -B_R C_S & A_R - B_R D_S C_R \end{bmatrix}, \quad D_L = \begin{bmatrix} 0 \\ B_R \end{bmatrix},$$
$$J_L = \begin{bmatrix} J_S & 0 \\ 0 & J_R \end{bmatrix},$$
$$C_L = \begin{bmatrix} C_S & D_S C_R \end{bmatrix},$$

satisfies the requirement that the output $y_S(t)$ is equal to the model output $y_M(t)$, for all $t \in \mathbb{R}^+$, when the respective initial states are zero, for all the admissible input functions $d(t)$, with $t \in \mathbb{R}^+$, and all the admissible sequences of jump times $\mathcal{T} \in \mathscr{T}$.

The block diagram in Fig. 1 illustrates the system interconnection considered in Problem 1.

3. Feedforward disturbance decoupling for the extended linear impulsive system — problem statement

As will be shown in the remainder of this work, the solution to the problem stated in Section 2 can be obtained through the solution of the problem tackled in this section. Namely, this section is focused on a problem of feedforward disturbance decoupling formulated for a new linear impulsive system — henceforth referred to as the *extended linear impulsive system* — which consists of the output-difference connection between the given linear impulsive plant Σ_S and a new linear impulsive model — from now on denoted by $\tilde{\Sigma}_M$ — obtained by suitably modifying the original linear impulsive model Σ_M.

The modified linear impulsive model $\tilde{\Sigma}_M$ is obtained from the original model Σ_M by closing a unit positive output feedback on the flow dynamics of Σ_M, so that

$$\tilde{\Sigma}_M \equiv \begin{cases} \dot{x}_M(t) = (A_M + B_M C_M)\, x_M(t) + B_M\, h(t), \quad t \in \mathbb{R}^+ \setminus \mathcal{T}, \\ x_M(t_k) = J_M\, x_M^-(t_k), \quad t_k \in \mathcal{T}, \\ y_M(t) = C_M\, x_M(t), \quad t \in \mathbb{R}^+. \end{cases}$$

The set of the admissible input functions to the modified model $\tilde{\Sigma}_M$ is defined as the set of all piecewise-continuous functions $h(t)$, with $t \in \mathbb{R}^+$, taking their values in $\mathbb{R}^q$.

Hence, the extended linear impulsive system — briefly denoted by Σ — is defined as the connection of the given linear impulsive system Σ_S with the modified linear impulsive model $\tilde{\Sigma}_M$, such that the control input, the disturbance input, and the output of Σ respectively are the control input of Σ_S, the input of $\tilde{\Sigma}_M$, and the difference between the outputs of Σ_S and $\tilde{\Sigma}_M$. Therefore,

$$\Sigma \equiv \begin{cases} \dot{x}(t) = A\,x(t) + B\,u(t) + H\,h(t), & t \in \mathbb{R}^+ \setminus \mathcal{T}, \\ x(t_k) = J\,x^-(t_k), & t_k \in \mathcal{T}, \\ y(t) = C\,x(t) + D\,u(t), & t \in \mathbb{R}^+, \end{cases}$$

where

$$A = \begin{bmatrix} A_S & 0 \\ 0 & A_M + B_M\,C_M \end{bmatrix}, \quad B = \begin{bmatrix} B_S \\ 0 \end{bmatrix}, \quad H = \begin{bmatrix} 0 \\ B_M \end{bmatrix}, \tag{1}$$

$$J = \begin{bmatrix} J_S & 0 \\ 0 & J_M \end{bmatrix}, \tag{2}$$

$$C = \begin{bmatrix} C_S & -C_M \end{bmatrix}, \quad D = D_S. \tag{3}$$

The state space of Σ will be denoted by $\mathcal{X}$: i.e., $\mathcal{X} = \mathbb{R}^n$, where $n = n_S + n_M$.

Consequently, the disturbance decoupling problem, by a feedforward action, for the extended linear impulsive system Σ can be cast as follows.

Problem 2 (Feedforward disturbance decoupling for the extended linear impulsive system) *Let the extended linear impulsive system Σ be given. Find a linear impulsive regulator Σ_R such that the compensated linear impulsive system*

$$\tilde{\Sigma} \equiv \begin{cases} \dot{\tilde{x}}(t) = \tilde{A}\,\tilde{x}(t) + \tilde{H}\,h(t), & t \in \mathbb{R}^+ \setminus \mathcal{T}, \\ \tilde{x}(t_k) = \tilde{J}\,\tilde{x}^-(t_k), & t_k \in \mathcal{T}, \\ y(t) = \tilde{C}\,\tilde{x}(t), & t \in \mathbb{R}^+, \end{cases}$$

where

$$\tilde{A} = \begin{bmatrix} A & B\,C_R \\ 0 & A_R \end{bmatrix}, \quad \tilde{H} = \begin{bmatrix} H \\ B_R \end{bmatrix},$$

$$\tilde{J} = \begin{bmatrix} J & 0 \\ 0 & J_R \end{bmatrix},$$

$$\tilde{C} = \begin{bmatrix} C & D\,C_R \end{bmatrix},$$

satisfies the requirement that the output $y(t)$ is zero, for all $t \in \mathbb{R}^+$, when the initial state is zero, for all the admissible input functions $h(t)$, with $t \in \mathbb{R}^+$, and all the admissible sequences of jump times $\mathcal{T} \in \mathscr{T}$.

The block diagram in Fig. 2 shows the system interconnection taken into consideration in Problem 2.

4. Geometric approach to linear impulsive systems with control feedthrough

As is shown by the broad literature available on disturbance decoupling, probably the most powerful tools to deal with this kind of control problems are those provided by the geometric approach [17, 18]. Over the time, the basic notions originally set forth to handle linear time-invariant systems have been generalized and adapted to deal with more general classes of dynamical systems. In particular, as far as linear impulsive systems are concerned, some fundamental notions, like those of invariance and controlled invariance, have been extended, so as to fit this class of dynamical systems, in some previous

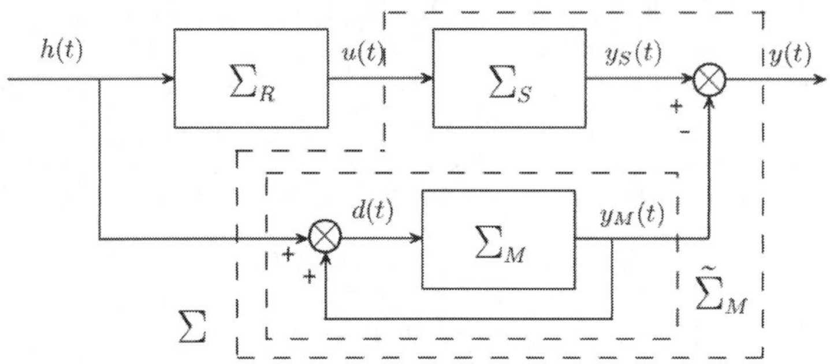

Figure 2. Block diagram for feedforward disturbance decoupling

papers [5, 6, 13]. However, in the specific case dealt with in this work, the extended linear impulsive system Σ shows a direct feedthrough term from the control input to the output. Consequently, the notion of hybrid controlled invariance must be completed by the new notion of output-nulling hybrid controlled invariance, as is illustrated below.

For the sake of immediacy, the following statements give the definitions of hybrid invariant subspace, hybrid controlled invariant subspace and output-nulling hybrid controlled invariant subspace with reference to the extended linear impulsive system Σ. However, it is understood that the special structure of the matrices of Σ shown in (1)–(3) does not play any role in these definitions. The symbol $\mathscr{H}$ will be used henceforth to denote hybrid invariance and hybrid controlled invariance. The short notations $\mathcal{B}$ and $\mathcal{H}$ are used to denote $\operatorname{Im} B$ and $\operatorname{Im} H$, respectively. A subspace $\mathcal{V} \subseteq \mathcal{X}$ is said to be an $\mathscr{H}$-invariant subspace if $A\,\mathcal{V} \subseteq \mathcal{V}$ and $J\,\mathcal{V} \subseteq \mathcal{V}$. A subspace $\mathcal{V} \subseteq \mathcal{X}$ is said to be an $\mathscr{H}$-controlled invariant subspace if $A\,\mathcal{V} \subseteq \mathcal{V} + \mathcal{B}$ and $J\,\mathcal{V} \subseteq \mathcal{V}$. Furthermore, it can be shown that a subspace $\mathcal{V} \subseteq \mathcal{X}$, with a basis matrix V, is an $\mathscr{H}$-controlled invariant subspace if and only there exist matrices L_A, L_J, and M such that $A\,V = V\,L_A + B\,M$ and $J\,V = V\,L_J$. Hence, the definition of output-nulling $\mathscr{H}$-controlled invariant subspace is introduced as follows.

Definition 1 *A subspace $\mathcal{V} \subseteq \mathcal{X}$, with a basis matrix V, is said to be an output-nulling $\mathscr{H}$-controlled invariant subspace if there exist matrices L_A, L_J, and M such that $A\,V = V\,L_A + B\,M$, $J\,V = V\,L_J$, and $C\,V = D\,M$.*

Moreover, the notion of output-nulling $\mathscr{H}$-controlled invariant subspace is characterized by the following necessary and sufficient condition, whose proof is a consequence of the properties enjoyed by simultaneous invariant and output-nulling controlled invariant subspaces in linear time-invariant systems.

Proposition 1 *A subspace $\mathcal{V} \subseteq \mathcal{X}$ is an output-nulling $\mathscr{H}$-controlled invariant subspace if and only if there exists a linear map F such that $(A + B\,F)\,\mathcal{V} \subseteq \mathcal{V}$ and $\mathcal{V} \subseteq \operatorname{Ker}(C + D\,F)$ hold along with $J\,\mathcal{V} \subseteq \mathcal{V}$.*

Any linear map F satisfying the conditions of Proposition 1 is said to be a *friend* of the output-nulling $\mathscr{H}$-controlled invariant subspace $\mathcal{V}$.

As can be shown by simple algebraic arguments, the set of all output-nulling $\mathscr{H}$-controlled invariant subspaces is an upper semilattice with respect to the sum and the inclusion of subspaces. The maximum of the set of all output-nulling $\mathscr{H}$-controlled invariant subspaces is henceforth denoted by $\mathcal{V}^*_{\mathscr{H}}$.

5. Feedforward disturbance decoupling for the extended linear impulsive system — problem solution

The geometric notions introduced in the previous section allow us to completely characterize the solvability of Problem 2 by a necessary and sufficient condition, which is the purpose of this section. As will be clear from the following, such condition consists of an inclusion of subspaces, so that it can be given in coordinate-free terms. However, since the if-part of the proof is constructive, it is convenient to make some preliminary remarks, aimed at expressing that condition with respect to conveniently chosen coordinates.

Firstly, it is worth pointing out that the linear map $A+BF$, where F is a friend of the maximal output-nulling $\mathscr{H}$-controlled invariant subspace $\mathcal{V}^*_{\mathscr{H}}$, is represented by a matrix with a characteristic upper block-triangular structure, provided that a suitable change of coordinates is performed in the state space. More precisely, let the similarity transformation T be defined by $T=[T_1\ T_2]$, with $\operatorname{Im} T_1=\mathcal{V}^*_{\mathscr{H}}$. Then, with respect to the new coordinates,

$$A'+B'F'=T^{-1}(A+BF)T=\begin{bmatrix} A'_{11}+B'_1F'_1 & A'_{12}+B'_1F'_2 \\ 0 & A'_{22}+B'_2F'_2 \end{bmatrix}, \tag{4}$$

where the structural zero submatrix in the lower left corner — namely,

$$A'_{21}+B'_2F'_1=0, \tag{5}$$

is due to $(A+BF)$-invariance of $\mathcal{V}^*_{\mathscr{H}}$. A similar reasoning holds for the linear map J, which, with respect to the same coordinates, is represented by

$$J'=T^{-1}JT=\begin{bmatrix} J'_{11} & J'_{12} \\ 0 & J'_{22} \end{bmatrix}, \tag{6}$$

where the structural zero submatrix in the lower left corner is due to J-invariance of $\mathcal{V}^*_{\mathscr{H}}$. Moreover, with respect to the same coordinates, the linear map $C+DF$, where F is the considered friend of $\mathcal{V}^*_{\mathscr{H}}$, is represented by a matrix with a structural zero submatrix in the first block of columns. Namely, it ensues that

$$C'+D'F'=(C+DF)T=\begin{bmatrix} 0 & C'_2+DF'_2 \end{bmatrix}, \tag{7}$$

where the structural zero submatrix

$$C'_1+DF'_1=0, \tag{8}$$

is due to $\mathcal{V}^*_{\mathscr{H}}\subseteq \operatorname{Ker}(C+DF)$.

Secondly, it is useful to express the subspace inclusion that will be shown to be the necessary and sufficient condition for solvability of Problem 2 in a coordinate-dependent fashion and, in particular, with respect to the basis introduced above. Namely, the subspace inclusion

$$\mathcal{H}\subseteq\mathcal{V}^*_{\mathscr{H}} \tag{9}$$

holds if and only if

$$H'=T^{-1}H=\begin{bmatrix} H'_1 \\ 0 \end{bmatrix}. \tag{10}$$

In fact, the structural zero block in H' expresses the condition that a basis matrix of the subspace $\mathcal{H}$ is a linear combination of the column vectors of the basis matrix T_1 of $\mathcal{V}^*_{\mathscr{H}}$.

Hence, the necessary and sufficient condition for Problem 2 to have a solution can be stated as follows.

Theorem 1 *Let the linear impulsive system Σ be given. Problem 2 has a solution if and only if (9) holds.*

Proof: If. Let (9) hold. Let F be a friend of $\mathcal{V}^*_{\mathscr{H}}$. Hence, with respect to suitably chosen coordinates, (4), (6), (7), and (10) hold. The remainder of this proof of sufficiency will refer to these coordinates. Let the hybrid linear regulator Σ_R be defined by the following matrices:

$$A'_R=A'_{11}+B'_1F'_1,\quad B'_R=H'_1,\quad J'_R=J'_{11},\quad C'_R=F'_1.$$

Then, it will be shown that the regulator Σ_R thus defined, with zero initial state, solves Problem 2. To this aim, note that the cascade — denoted by $\tilde{\Sigma}$ in Problem 2 — between the linear impulsive regulator Σ_R and the extended linear impulsive system Σ is described by

$$\tilde{\Sigma} \equiv \begin{cases} \dot{x}_1(t) = A'_{11}\, x_1(t) + A'_{12}\, x_2(t) + B'_1\, F'_1\, x_R(t) + H'_1\, h(t), & t \in \mathbb{R}^+ \setminus \mathcal{T}, \\ \dot{x}_2(t) = A'_{21}\, x_1(t) + A'_{22}\, x_2(t) + B'_2\, F'_1\, x_R(t), & t \in \mathbb{R}^+ \setminus \mathcal{T}, \\ \dot{x}_R(t) = (A'_{11} + B'_1\, F'_1)\, x_R(t) + H'_1\, h(t), & t \in \mathbb{R}^+ \setminus \mathcal{T}, \\ x_1(t_k) = J'_{11}\, x_1^-(t_k) + J'_{12}\, x_2^-(t_k), & t_k \in \mathcal{T}, \\ x_2(t_k) = J'_{22}\, x_2^-(t_k), & t_k \in \mathcal{T}, \\ x_R(t_k) = J'_{11}\, x_R^-(t_k), & t_k \in \mathcal{T}, \\ y(t) = C'_1\, x_1(t) + C'_2\, x_2(t) + D\, F'_1\, x_R(t), & t \in \mathbb{R}^+, \end{cases}$$

where the state of Σ has been partitioned as $x = \begin{bmatrix} x_1^\top & x_2^\top \end{bmatrix}^\top$ according to (4), (6), (7), and (10). By applying the change of variables $\eta(t) = x_1(t) - x_R(t)$, with $t \in \mathbb{R}^+$, the system $\tilde{\Sigma}$ can also be written as

$$\tilde{\Sigma} \equiv \begin{cases} \dot{\eta}(t) = A'_{11}\, \eta(t) + A'_{12}\, x_2(t), & t \in \mathbb{R}^+ \setminus \mathcal{T}, \\ \dot{x}_2(t) = A'_{21}\, \eta(t) + A'_{22}\, x_2(t), & t \in \mathbb{R}^+ \setminus \mathcal{T}, \\ \dot{x}_R(t) = (A'_{11} + B'_1\, F'_1)\, x_R(t) + H'_1\, h(t), & t \in \mathbb{R}^+ \setminus \mathcal{T}, \\ \eta(t_k) = J'_{11}\, \eta^-(t_k) + J'_{12}\, x_2^-(t_k), & t_k \in \mathcal{T}, \\ x_2(t_k) = J'_{22}\, x_2^-(t_k), & t_k \in \mathcal{T}, \\ x_R(t_k) = J'_{11}\, x_R^-(t_k), & t_k \in \mathcal{T}, \\ y(t) = C'_1\, \eta(t) + C'_2\, x_2(t), & t \in \mathbb{R}^+, \end{cases}$$

where (5) and (8) have been taken into account. Hence, the assumption of zero initial state implies $\eta(t) = 0$ and $x_2(t) = 0$, for all $t \in \mathbb{R}^+$, which also implies $y(t) = 0$, for all $t \in \mathbb{R}^+$, for all the admissible input functions $h(t)$, with $t \in \mathbb{R}^+$, and all the admissible jump time sequences $\mathcal{T} \in \mathscr{T}$.

Only if. If (9) does not hold, no other output-nulling $\mathscr{H}$-controlled invariant subspace containing $\mathcal{H}$ exists, since the set of all output-nulling $\mathscr{H}$-controlled invariant subspaces is an upper semilattice and $\mathcal{V}^*_{\mathscr{H}}$ is the maximum. ■

6. Model matching by output feedback — problem solution

The aim of this section is to show that the problem of feedforward disturbance decoupling for the extended linear impulsive system solved in Section 5 is equivalent to the output feedback model matching problem stated in Section 2. In other words, a linear impulsive regulator solves any of these problems if and only if it also solves the other one. The following theorem formalizes this result.

Theorem 2 *A linear impulsive regulator Σ_R solves Problem 2 if and only if it solves Problem 1.*

Proof: If. Let the linear impulsive regulator Σ_R solve Problem 1. Hence, the overall linear impulsive system with measurement feedback — henceforth denoted by $\tilde{\Sigma}'$ — is described by

$$\tilde{\Sigma}' \equiv \begin{cases} \dot{x}_S(t) = A_S\, x_S(t) + B_S\, C_R\, x_R(t), & t \in \mathbb{R}^+ \setminus \mathcal{T}, \\ \dot{x}_R(t) = -B_R\, C_S\, x_S(t) + (A_R - B_R\, D_S\, C_R)\, x_R(t) + B_R\, d(t), & t \in \mathbb{R}^+ \setminus \mathcal{T}, \\ \dot{x}_M(t) = A_M\, x_M(t) + B_M\, d(t), & t \in \mathbb{R}^+ \setminus \mathcal{T}, \\ x_S(t_k) = J_S\, x_S^-(t_k), & t_k \in \mathcal{T}, \\ x_R(t_k) = J_R\, x_R^-(t_k), & t_k \in \mathcal{T}, \\ x_M(t_k) = J_M\, x_M^-(t_k), & t_k \in \mathcal{T}, \\ y(t) = C_S\, x_S(t) + D_S\, C_R\, x_R(t) - C_M\, x_M(t), & t \in \mathbb{R}^+. \end{cases} \tag{11}$$

Note that, since Σ_R solves Problem 1, on the assumption of zero initial state, the output of $\tilde{\Sigma}'$ satisfies the condition that $y(t) = 0$, for all $t \in \mathbb{R}^+$, for all the admissible input functions $d(t)$, with $t \in \mathbb{R}^+$. Hence, one can replace $y_S(t) = C_S\, x_S(t) + D_S\, C_R\, x_R(t)$ with $y_M(t) = C_M\, x_M(t)$ in the state equations of $\tilde{\Sigma}'$. Thus, the following equations for the new system henceforth denoted by $\tilde{\Sigma}''$ are obtained:

$$\tilde{\Sigma}'' \equiv \begin{cases} \dot{x}_S(t) = A_S\, x_S(t) + B_S\, C_R\, x_R(t), & t \in \mathbb{R}^+ \setminus \mathcal{T}, \\ \dot{x}_R(t) = A_R\, x_R(t) - B_R\, C_M\, x_M(t) + B_R\, d(t), & t \in \mathbb{R}^+ \setminus \mathcal{T}, \\ \dot{x}_M(t) = A_M\, x_M(t) + B_M\, d(t), & t \in \mathbb{R}^+ \setminus \mathcal{T}, \\ x_S(t_k) = J_S\, x_S^-(t_k), & t_k \in \mathcal{T}, \\ x_R(t_k) = J_R\, x_R^-(t_k), & t_k \in \mathcal{T}, \\ x_M(t_k) = J_M\, x_M^-(t_k), & t_k \in \mathcal{T}, \\ y(t) = C_S\, x_S(t) + D_S\, C_R\, x_R(t) - C_M\, x_M(t), & t \in \mathbb{R}^+. \end{cases} \tag{12}$$

Moreover, since $y(t) = 0$ for all $t \in \mathbb{R}^+$, for all the admissible $d(t)$, with $t \in \mathbb{R}^+$, such condition holds, in particular, by picking $d(t) = h(t) + C_M\, x_M(t)$, where $h(t)$, with $t \in \mathbb{R}^+$, denotes any admissible input function. Then, the resulting system is the linear impulsive system $\tilde{\Sigma}$ considered in Problem 2, as is shown by the following equations, derived from those of $\tilde{\Sigma}''$ with the replacement mentioned above:

$$\tilde{\Sigma} \equiv \begin{cases} \dot{x}_S(t) = A_S\, x_S(t) + B_S\, C_R\, x_R(t), & t \in \mathbb{R}^+ \setminus \mathcal{T}, \\ \dot{x}_R(t) = A_R\, x_R(t) + B_R\, h(t), & t \in \mathbb{R}^+ \setminus \mathcal{T}, \\ \dot{x}_M(t) = (A_M + B_M\, C_M)\, x_M(t) + B_M\, h(t), & t \in \mathbb{R}^+ \setminus \mathcal{T}, \\ x_S(t_k) = J_S\, x_S^-(t_k), & t_k \in \mathcal{T}, \\ x_R(t_k) = J_R\, x_R^-(t_k), & t_k \in \mathcal{T}, \\ x_M(t_k) = J_M\, x_M^-(t_k), & t_k \in \mathcal{T}, \\ y(t) = C_S\, x_S(t) + D_S\, C_R\, x_R(t) - C_M\, x_M(t), & t \in \mathbb{R}^+. \end{cases} \tag{13}$$

The equations of $\tilde{\Sigma}$, which hold with $y(t) = 0$ for all $t \in \mathbb{R}^+$, for all the admissible $h(t)$, with $t \in \mathbb{R}^+$, show that the linear impulsive regulator Σ_R also solves Problem 2: i.e., the problem of decoupling the signal $h(t)$, with $t \in \mathbb{R}^+$, in the extended linear impulsive system $\tilde{\Sigma}$, including the modified linear impulsive model $\tilde{\Sigma}_M$.

Only if. Let the linear impulsive regulator Σ_R solve Problem 2. Then, in order to show that Σ_R also solves Problem 1, the same reasoning presented in the proof of sufficiency can be followed backward — i.e., from $\tilde{\Sigma}$ to $\tilde{\Sigma}'$. ∎

7. Conclusions

In this work, the problem of finding a linear impulsive regulator ensuring that the forced response of a given linear impulsive system matches that of a linear impulsive model, for all the admissible inputs and all the admissible sequences of jump time instants, has been formulated. The solvability of the problem has been characterized through a necessary and sufficient condition. A computational procedure for the synthesis of the linear impulsive regulator has been illustrated through the if-part of the proof.

References

[1] Goebel R, Sanfelice R G and Teel A R 2012 *Hybrid Dynamical Systems: Modeling, Stability, and Robustness* (Princeton, New Jersey: Princeton University Press)

[2] Medina E A and Lawrence D A 2009 State estimation for linear impulsive systems *2009 American Control Conference* (St. Louis, MO) pp 1183–1188

[3] Kouhi Y, Bajcinca N and Sanfelice R G 2013 Suboptimality bounds for linear quadratic problems in hybrid linear systems *2013 European Control Conference* (Zurich, Switzerland) pp 2663–2668

[4] Carnevale D, Galeani S and Sassano M 2014 A linear quadratic approach to linear time invariant stabilization for a class of hybrid systems *22nd Mediterranean Conference on Control and Automation* (Palermo, Italy) pp 545–550

[5] Conte G, Perdon A M and Zattoni E 2015 The disturbance decoupling problem for jumping hybrid systems *54th IEEE Conference on Decision and Control* (Osaka, Japan) pp 1589–1594

[6] Perdon A M, Zattoni E and Conte G 2016 Disturbance decoupling with stability for linear impulsive systems *6th IFAC Symposium on System Structure and Control* (*IFAC-PapersOnLine* vol 49(9)) (Istanbul, Turkey) pp 1–6

[7] Medina E A and Lawrence D A 2006 Controlled and conditioned invariants for linear impulsive systems *45th IEEE Conference on Decision and Control* (San Diego, CA) pp 2753–2758

[8] Carnevale D, Galeani S and Menini L 2012 Output regulation for a class of linear hybrid systems. Part 1: Trajectory generation *51st IEEE Conference on Decision and Control* (Maui, HI, USA) pp 6151–6156

[9] Carnevale D, Galeani S and Menini L 2012 Output regulation for a class of linear hybrid systems. Part 2: Stabilization *51st IEEE Conference on Decision and Control* (Maui, HI, USA) pp 6157–6162

[10] Carnevale D, Galeani S, Menini L and Sassano M 2014 Output regulation of hybrid linear systems with unpredictable jumps *19th IFAC World Congress* (Cape Town, South Africa) pp 1531–1536

[11] Carnevale D, Galeani S and Sassano M 2014 Francis equations vs invariant subspace algorithm for hybrid output regulation *53rd IEEE Conference on Decision and Control* (Los Angeles, CA) pp 4697–4702

[12] Carnevale D, Galeani S, Menini L and Sassano M 2016 Hybrid output regulation for linear systems with periodic jumps: Solvability conditions, structural implications and semi-classical solutions *IEEE Transactions on Automatic Control* **61** 2416–2431

[13] Zattoni E, Perdon A M and Conte G 2015 Output regulation by error dynamic feedback in linear time-invariant hybrid dynamical systems *14th European Control Conference* (Linz, Austria) pp 1438–1443

[14] Zakharov A, Zattoni E, Xie L, Garcia O and Jämsä-Jounela S L 2013 An autonomous valve stiction detection system based on data characterization *Control Engineering Practice* **21** 1507–1518

[15] Seyboth G S and Allgöwer F 2014 Synchronized model matching: A novel approach to cooperative control of nonlinear multi-agent systems vol 19 pp 1985–1990

[16] Zakharov A, Zattoni E, Yu M and Jämsä-Jounela S L 2015 A performance optimization algorithm for controller reconfiguration in fault tolerant distributed model predictive control *Journal of Process Control* **34** 56–69

[17] Wonham W M 1985 *Linear Multivariable Control: A Geometric Approach* 3rd ed (New York: Springer-Verlag)

[18] Basile G and Marro G 1992 *Controlled and Conditioned Invariants in Linear System Theory* (New Jersey: Prentice Hall)

[19] Zattoni E, Perdon A M and Conte G 2013 A geometric approach to output regulation for linear switching systems *5th IFAC Symposium on System Structure and Control* (Grenoble, France) pp 172–177

[20] Zattoni E 2013 Dynamic feedforward compensation of measurable signals in discrete-time linear switching systems *2013 American Control Conference* (Washington DC, USA) pp 848–853

[21] Conte G, Perdon A M and Zattoni E 2013 A geometric approach to output regulation for discrete-time switched linear systems *21st Mediterranean Conference on Control and Automation* (Platanias-Chania, Crete, Greece) pp 852–857

[22] Zattoni E and Marro G 2013 Measurable disturbance rejection with quadratic stability in continuous-time linear switching systems *European Control Conference 2013* (Zürich, Switzerland) pp 2157–2162

[23] Zattoni E, Perdon A M and Conte G 2013 The output regulation problem with stability for linear switching systems: a geometric approach *Automatica* **49** 2953–2962

[24] Zattoni E and Marro G 2013 A constructive condition for inaccessible signal rejection with quadratic stability in discrete-time linear switching systems *52nd IEEE Conference on Decision and Control* (Florence, Italy) pp 4650–4655

[25] Zattoni E 2014 Measurable disturbance rejection with stability in continuous-time switched linear systems under dwell-time switching *European Control Conference 2014* (Strasbourg, France) pp 2242–2247

[26] Zattoni E, Perdon A M and Conte G 2014 Output feedback model matching with strong stability in continuous-time switched linear systems *22nd Mediterranean Conference on Control and Automation* (Palermo, Italy) pp 525–530

[27] Conte G, Perdon A M and Zattoni E 2014 Model matching problems for switching linear systems *19th World Congress of the International Federation of Automatic Control* (Cape Town, South Africa) pp 1501–1506

[28] Zattoni E, Perdon A M and Conte G 2014 Disturbance decoupling with stability in continuous-time switched linear systems under dwell-time switching *19th IFAC World Congress* (Cape Town, South Africa) pp 164–169

[29] Zattoni E 2015 A synthesis procedure for disturbance decoupling with local input to state stability in discrete-time switching linear systems *2015 American Control Conference* (Chicago, IL, USA) pp 1071–1076

[30] Zattoni E 2015 Stability issues in disturbance decoupling for switching linear systems *SIAM Conference on Control and Its Applications 2015* (Paris, France) pp 361–367

[31] Conte G, Perdon A M and Zattoni E 2015 The disturbance decoupling problem with quadratic stability for LPV systems *1st IFAC Workshop on Linear Parameter Varying Systems* (*IFAC-PapersOnLine* vol 48(26)) (Grenoble, France) pp 1–6

[32] Zattoni E and Marro G 2015 Disturbance compensation in discrete-time switching linear systems subject to a dwell-time constraint *54th IEEE Conference on Decision and Control* (Osaka, Japan) pp 2307–2312

[33] Zattoni E 2015 Model matching by output dynamic feedback in discrete-time switching linear systems *54th IEEE Conference on Decision and Control* (Osaka, Japan) pp 2335–2340

[34] Perdon A M, Conte G and Zattoni E 2016 Necessary and sufficient conditions for asymptotic model matching of switching linear systems *Automatica* **64**(2) 294–304

[35] Zattoni E, Perdon A M and Conte G 2016 Disturbance decoupling with closed-loop modes stability in switched linear systems *IEEE Transactions on Automatic Control* **61** 3115–3121 DOI:10.1109/TAC.2015.2498123

[36] Perdon A M, Zattoni E and Conte G 2016 Model matching with strong stability in switched linear systems *Systems & Control Letters* **97** 98–107

Novel Observer Scheme of Fuzzy-MRAS Sensorless Speed Control of Induction Motor Drive

S. Chekroun[1], M. Zerikat[2], A. Mechernene[3] and N. Benharir[2]

[1]Department of Electrical Engineering, University Mustapha Stambouli of Mascara, BP.305, Route El Mamounia, 29000 Mascara-Algeria
[2]Department of Electrical Engineering, Polytechnic National School of Oran, BP.1523 El Mnaouer, 31000 Oran- Algeria.
[3]Department of Electrical and Electronics Engineering, University Abou Bekr Belkaid of Tlemcen, BP.230 Chetouane, 13000 Tlemcen-Algeria.

Corresponding author's schekroun@hotmail.fr

Abstract. This paper presents a novel approach Fuzzy-MRAS conception for robust accurate tracking of induction motor drive operating in a high-performance drives environment. Of the different methods for sensorless control of induction motor drive the model reference adaptive system (MRAS) finds lot of attention due to its good performance. The analysis of the sensorless vector control system using MRAS is presented and the resistance parameters variations and speed observer using new Fuzzy Self-Tuning adaptive IP Controller is proposed. In fact, fuzzy logic is reminiscent of human thinking processes and natural language enabling decisions to be made based on vague information. The present approach helps to achieve a good dynamic response, disturbance rejection and low to plant parameter variations of the induction motor. In order to verify the performances of the proposed observer and control algorithms and to test behaviour of the controlled system, numerical simulation is achieved. Simulation results are presented and discussed to shown the validity and the performance of the proposed observer.

1. Introduction

The control and estimation of ac drives in general are considerably more complex than those of dc drives, and this complexity increases substantially if high performances are demanded [1]. However, for high dynamic performance industrial applications, their control remains a challenging problem because they exhibit significant non linearities and it is now well known that uncertainties of plant parameters and influence of unknown external disturbances can degrade significantly the performance of the system with linearizing feedback. To solve problems emerging in the control of such processes, advanced control strategies have researched. In recent years a lot of controllers have been reported in the literature and used in various industrial drive applications [2], and an interesting alternative that could be investigated is to use of artificial intelligence control strategy. In the electrical machines control, it is not possible to measure all the necessary variables in the control implementation. So, the search for observers to obtain state variables which can be used in the algorithm control is actively continued by researchers. The speed information must be reconstituted from the terminal quantities (voltage and current components) of the machine using flux observers which are based on the induction machine model. In the case of the model reference adaptive system (MRAS) methods, a comparison between the outputs of two estimates are made. Then, the output errors are used to derive a suitable adaptation mechanism that generates the estimated speed [3]. Then the rotor time constant is estimated using the stator currents estimation error and the observed rotor flux, based on the Lyapunov stability theory. In fact, fuzzy logic is reminiscent of human thinking processes and natural language enabling decisions to be made based on vague information [4]. The salient advantage of fuzzy logic is robustness against structured and unstructured uncertainties. In path tracking systems. In the reaching phase, tracking may be hindered by disturbances or parameter variations. In this paper, an integrate fuzzy logic into model reference adaptive system (MRAS), is proposed on the design of sensorless control schemes of the of induction motor drive to improve the stability and the robustness of the control system. The advantages of speed sensorless induction motor derives are reduced hardware complexity and lower cost, reduce size of derive machine, eliminate of sensor cable, better noise immunity, increasing reliability and less maintenance. The proposed control system is designed and tested through numerical simulation and its effectiveness in tracking application is verified. This paper is organized as follows: the model reference adaptive system observer is presented in Section 2. The section 3 describes the fuzzy logic controller. Finally, in sections 4 and 5, we give some comments and conclusions.

2. Model reference adaptive system observer

Schauder was the first to propose rotor flux MRAS, which is the most popularly used MRAS strategy. A lot of effort by the researchers has been focused on this strategy to further improve its performance [5].

2.1. Modeling of induction motor

Equations describing the induction machine dynamic model in the stationary frame (d,q) and the associated mechanical equation are expressed as follows:

$$\begin{cases} \frac{d}{dt} i_{sd} = \frac{1}{\sigma \cdot L_s}[-(R_s + \frac{L_m^2}{L_r \cdot T_r}) \cdot i_{sd} + \omega_s \cdot \sigma \cdot L_s \cdot i_{sq} + \frac{L_m}{L_r \cdot T_r} \cdot \phi_{rd} + \frac{L_m}{L_r} \cdot \omega_r \cdot \phi_{rq} + v_{sd}] \\ \frac{d}{dt} i_{sq} = \frac{1}{\sigma \cdot L_s}[-\omega_s \cdot \sigma \cdot L_s \cdot i_{sd} - \left(R_s + \frac{L_m^2}{L_r \cdot T_r}\right) \cdot i_{sq} - \frac{L_m}{L_r} \omega_r \cdot \phi_{rd} + \frac{L_m}{L_r \cdot T_r} \cdot \phi_{rq} + v_{sq}] \\ \frac{d}{dt} \phi_{rd} = \frac{L_m}{T_r} \cdot i_{sd} - \frac{1}{T_r} \cdot \phi_{rd} + (\omega_s - \omega_r) \cdot \phi_{rq} \\ \frac{d}{dt} \phi_{rq} = \frac{L_m}{T_r} \cdot i_{sq} - (\omega_s - \omega_r) \cdot \phi_{rd} - \frac{1}{T_r} \cdot \phi_{rq} \\ \frac{d}{dt} \omega = \frac{P^2 \cdot L_m}{L_r \cdot J}(i_{sq}\phi_{rd} - i_{sd}\phi_{rq}) - \frac{F}{J}\omega - \frac{P}{J}T_l \end{cases} \tag{1}$$

where $\sigma = 1 - \frac{L_m^2}{L_s L_r}; T_r = \frac{L_r}{R_r}$

2.2 Direct field oriented control

Direct Field oriented control (DFOC) technique is intended to control the motor flux, and thereby be able to decompose the AC motor current into "flux producing" and "torque producing" components. The well known direct field orientation strategies provide a linear and decoupled control between the flux and torque of the induction machine [6],[7]. Then the rotor flux orientation process is given by the imposed zero constraint of quadrate rotor flux component. Such as:

$$\phi_{rq} = 0 \text{ and } \phi_{rd} = \phi_r \tag{2}$$

Hence, the rotor flux can be controlled directly from the stator direct current component i_{sd}, while the torque can be linearly controlled from the stator quadrate current component i_{sq} when the rotor flux is maintained constant. Separating the real and imaginary parts of (1) by using (2) leads to:

$$\begin{cases} v_{sd} = \sigma L_s \frac{d}{dt} i_{sd} + \left(R_s + R_r \frac{L_m^2}{L_r^2} \right) i_{sd} - \omega_s \sigma L_s i_{sq} - \frac{L_m}{L_r^2} R_r \phi_r = v_{sd}^r - \omega_s \sigma L_s i_{sq} \\ v_{sq} = \sigma L_s \frac{d}{dt} i_{sq} + \omega_s \sigma L_s i_{sd} + \left(R_s + R_r \frac{L_m^2}{L_r^2} \right) i_{sq} + \frac{L_m}{L_r^2} p \Omega \phi_r = v_{sq}^r - \omega_s \sigma L_s i_{sd} + \frac{L_m}{L_r^2} p \Omega \phi_r \end{cases} \quad (3)$$

The slip frequency can be calculated from the values of the stator current quadrate and the rotor flux oriented reference frame as follow:

$$\omega = \omega_s - \omega_r = \frac{L_m}{T_r} \frac{i_{sq}}{\phi_r} \quad (4)$$

And the rotor flux position is given by:

$$\theta_s = \int \omega_s dt \quad (5)$$

The voltages v_{sd} and v_{sq} should act on the current i_{sd} and i_{sq} separately and consequently the flux and the torque. The two-phase stators current are controlled by two PI controllers taking as input the reference values i_{sd}^*, i_{sq}^* and the measured values. Thus, the common thought is to realize the decoupling by adding the compensation terms (e_{sd} and e_{sq}) as usually done.

$$\begin{cases} e_{sd} = \omega_s \sigma L_s i_{sq} \\ e_{sq} = \omega_s \sigma L_s i_{sd} - \omega_s \frac{L_m}{L_r^2} \phi_r \end{cases} \quad (6)$$

The module of rotoric flux is obtained by a block of field weakening given by the following non linear relation:

$$\phi_r^* \begin{cases} \phi_{rN} & \text{if } |\omega| \le \omega_m \\ \phi_{rN} \frac{\omega_m}{|\omega_m|} & \text{if } |\omega| \succ \omega_m \end{cases} \quad (7)$$

The rotor flux is controlled by PI controller taking as input the reference value ϕ_r^* and the calculated value.

2.3 Conventional IP controller

To accelerate the dynamic response of speed loop, and stabilize his behavior during transient states, the IP controller is often preferred to PI regulator. It has a good regulation properties and several advantages such as reduction, or even absence of overshoot in tracking trajectory.

The simplified block diagram of the speed loop based a IP controller is shown in Figure 1. This structure avoids the overshoots problem by canceling, in closed loop, the zero term present in the numerator of PI controller while imposing, two poles specified by a judicious choice of the damping ratio ξ and of natural angular frequency ω_n.

The closed loop transfer function is determined, considering the reference speed Ω^* and a zero load torque $T_l = 0$:

$$F(s) = \frac{\Omega_n(s)}{\Omega_n^*(s)} = \frac{\omega_n^2}{s^2 + 2\xi\omega_n s + \omega_n^2} \quad (8)$$

where: $K_T = p\frac{\omega_n^2}{R_r}$, $\omega_n = \frac{K_i K_T}{J}$ and $\xi = \frac{F + K_p K_T}{2(J \cdot K_i \cdot K_T)^{1/2}}$

The optimal setting is obtained for a damping ratio of the closed loop system equal to unity ($\xi = 1$), the dynamic of the system is then adjusts by the natural frequency.
The proportional and integral coefficients are given by:

$$K_p = \frac{2J\omega_n - F}{K_T} \text{ et } K_i = J\frac{\omega_n^2}{K_T} \tag{9}$$

However, although the responses obtained is more stables, they are insufficient for a drive system requiring high per- formance. Indeed, controller coefficients are fixed and do not adapt to various operating conditions or to parametric variations. Consequently, more sophisticated controllers are required, for this reason the adaptive controllers are an interesting alternative.

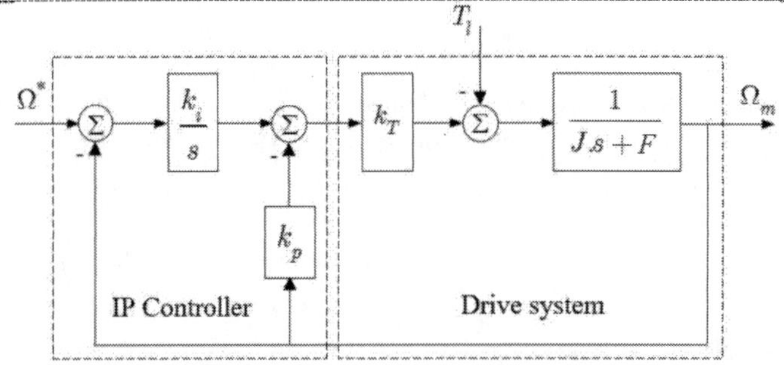

Figure 1. Speed loop based a conventional linear IP controller

2.4 MRAS Based For Stator Resistance, Rotor Time Constant and Fuzzy Rotor Speed Estimation

As it is already known, the most difficult aspects concerning the implementation of the electrical drive systems based on the field-orientation theory, are in relation with rotor flux components, speed and resistance parameters estimation. It has been already proved that simultaneous identification of the stator, rotor resistances and the rotor speed is possible only when the rotor flux is time-variant. The overall block diagram of direct field orientation control for induction motor is given in Figure. 2. If the model of the induction motor is considered, the rotor speed and stator resistance and rotor time constant can be identified by approach of Model Reference Adaptive Systems (MRAS) [8],[9]. In the classical MRAS estimation method, there needs two models which outputs are to be compared. One is voltage model (or stator equation) and the other is current model (or rotor equation). The reference rotor flux components obtained from the reference model are given by [10]:

$$\begin{cases} \phi_{r\alpha} = \dfrac{L_r}{L_m}\left(\int \left(v_{s\alpha} - R_s i_{s\alpha}\right)dt - \sigma L_s i_{s\alpha}\right) \\ \phi_{r\beta} = \dfrac{L_r}{L_m}\left(\int \left(v_{s\alpha} - R_s i_{s\beta}\right)dt - \sigma L_s i_{s\beta}\right) \end{cases} \tag{10}$$

The reference rotor flux components obtained from the reference model are given by:

$$\begin{cases} \dfrac{d\hat{\phi}_{r\alpha}}{dt} = -\dfrac{1}{T_r}\hat{\phi}_{r\alpha} - \hat{\omega}\hat{\phi}_{r\beta} + \dfrac{L_m}{T_r} i_{s\alpha} \\ \dfrac{d\hat{\phi}_{r\beta}}{dt} = -\dfrac{1}{T_r}\hat{\phi}_{r\beta} + \hat{\omega}\hat{\phi}_{r\alpha} + \dfrac{L_m}{T_r} i_{s\beta} \end{cases} \tag{11}$$

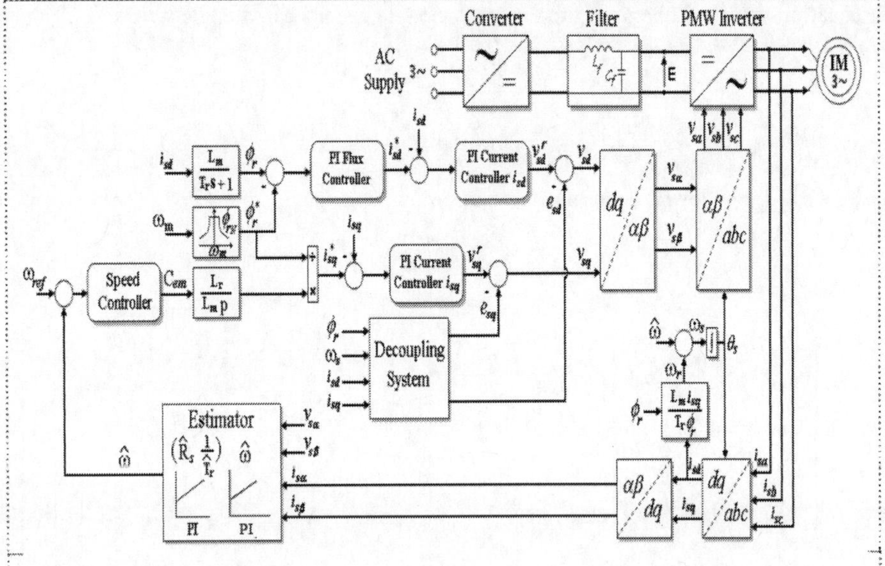

Figure 2. Direct Field-Oriented Control for induction motor equipped with MRAS estimator.

A new structure MRAS is proposed in this paper based to stator resistance, rotor time constant and rotor speed estimation is designed based on the concept of hyperstability, in order to make, the system asymptotically stable. The configuration of the proposed scheme is shown in Figure 3.
The reference model, usually expressed by the voltage model for speed estimation and rotor time constant and current model for stator resistance estimation, represents the stator and rotor equations. It generates the reference value of the rotor flux components in the stationary reference frame from the monitored stator voltage and current components.

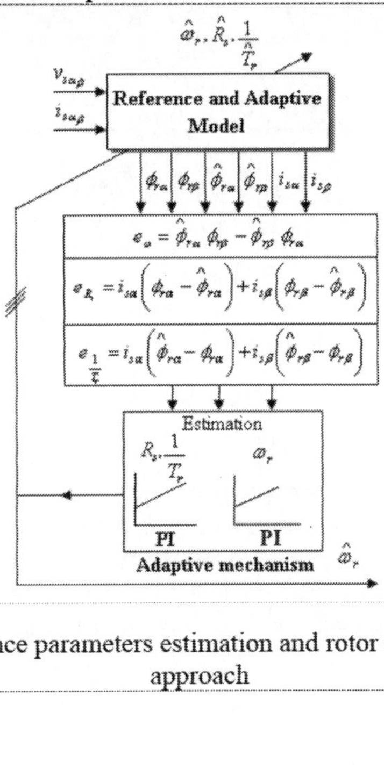

Figure 3. Resistance parameters estimation and rotor speed under MRAS approach

The error equations for the voltage and the current model out puts can then be written as:

$$\begin{cases} e = \underline{\hat{\phi}_r} - \underline{\phi_r} \\ \dfrac{de}{dt} = d\dfrac{\underline{\hat{\phi}_r}}{dt} - d\dfrac{\underline{\phi_r}}{dt} \end{cases} \tag{12}$$

The equation (10) can be rewritten in matrix notation by:

$$\frac{d\underline{e}}{dt} = A \cdot \underline{e} - W \tag{13}$$

where $A = \begin{bmatrix} -\dfrac{1}{T_r} & -\omega & 0 & 0 \\ \omega & -\dfrac{1}{T_r} & 0 & 0 \\ 0 & 0 & 0 & 0 \\ 0 & 0 & 0 & 0 \end{bmatrix}$ and $W = \begin{bmatrix} 0 & -\left(\dfrac{1}{\hat{T}_r} + \Delta\omega\right) & \dfrac{\Delta}{T_r} & 0 \\ -\left(\dfrac{1}{\hat{T}_r} + \Delta\omega\right) & -\dfrac{1}{T_r} & 0 & \dfrac{\Delta}{T_r} \\ 0 & 0 & \dfrac{L_r}{L_m}\Delta R_s & 0 \\ 0 & 0 & 0 & \dfrac{L_r}{L_m}\Delta R_s \end{bmatrix} \cdot \begin{bmatrix} \hat{\phi}_{r\alpha} \\ \hat{\phi}_{r\beta} \\ i_{s\alpha} \\ i_{s\beta} \end{bmatrix}$

The adaptation mechanism compares the two models and estimates the speed, rotor time constant and stator resistance by an integral proportional regulator. Using Lyapunov stability theory [9],[10], we can construct a mechanism to adapt the mechanical speed and stator rotor from the asymptotic convergence's condition of the state variables estimation errors. The expressions for the speed and resistance tuning signal and the estimated speed and resistance can be given as :

$$\begin{cases} e_\omega = \hat{\phi}_{r\alpha}\,\phi_{r\beta} - \hat{\phi}_{r\beta}\,\phi_{r\alpha} \\ \hat{\omega} = \left(k_{pw} + \dfrac{k_{i\omega}}{p}\right) \cdot e_\omega \end{cases} \tag{14}$$

$$\begin{cases} e_{R_s} = i_{s\alpha}\left(\phi_{r\alpha} - \hat{\phi}_{r\alpha}\right) + i_{s\beta}\left(\phi_{r\beta} - \hat{\phi}_{r\beta}\right) \\ \hat{R}_s = \left(k_{pR_s} + \dfrac{k_{iR_s}}{p}\right) \cdot e_{R_s} \end{cases} \tag{15}$$

K_p and K_i are positive gains.

The rotor resistance can then be written as:

$$\hat{R}_r = \frac{R_{rn}}{R_{sn}} \cdot \hat{R}_s = k \cdot \hat{R}_s \tag{16}$$

3. Fuzzy logic controller

It appears that fuzzy logic based intelligent control is most appropriate for performance improvement of the ac machines. The main preference of the fuzzy logic is that is easy to implement control that it has the ability of generalisation [11],[12],[13]. The fuzzy adaptive control strategy is a new hybrid theory combining traditional linear control and Fuzzy Logic. Also different techniques have been developed in the last years, but they all need a lot of fuzzy rules increasing the complexity and the implementation of the cotrol system.

In this section, a simple but robust, Fuzzy Self Tuning IP Controller is proposed, where the controller coefficients are adjusted on-line by an inference fuzzy system [14].
The proposed structure of the Fuzzy Self-Tuning IP Con- troller is shown in Figure 4.

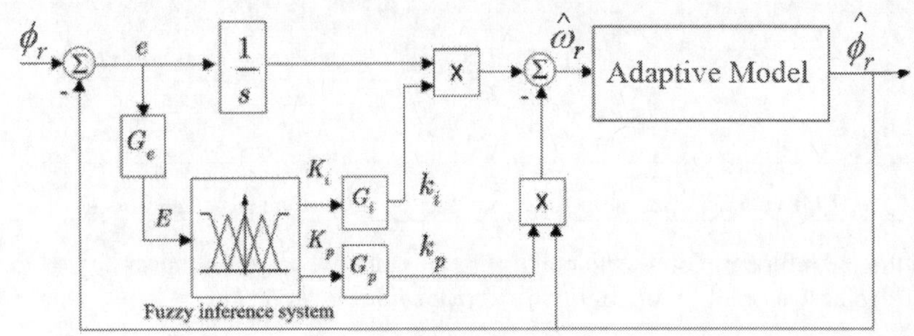

Figure 4. Speed loop based on Fuzzy Self-Tuning IP Controller proposed

The structure of the conventional IP controller is preserved, but it is increased by a fuzzy inference system which must calculate the appropriate coefficients (kp, ki), and then provide them, in every time, to the controller. The transfer function F (s), in closed loop has the same expression as previously, but the values of coefficients are then given by non linear functions, independent of the parameters of the system, but linked to tracking error. This controller should be able to modify its characteristics and to maintain the desired dynamic of the system when the operating conditions evolve or in the presence of variations of motor parameters.
The Fuzzy Self-Tuning IP Controller proposed has the tracking error signal as single input, and two output, the coefficients kp and ki. The input variable is mapped into five fuzzy sets, distributed in the universe of discourse in the normalized range of [-2;+2]. These fuzzy sets are denoted by the followings linguistic variables NB (Negative Big), NS (Negative Small), ZE (Equal Zero), PS (Positive Small) and PB (Positive Big) respectively. For the two outputs variables, the universe of discourse is normalized in the range [0;1] and partitioned in three fuzzy sets S, M and B. These linguistic variables are denoted S (Small), M (Medium) and B (Big). The shapes of the selected membership functions for input are trapezoidal or triangular, and singletons type for the outputs, as shown in Figure 5 and Figure 6.

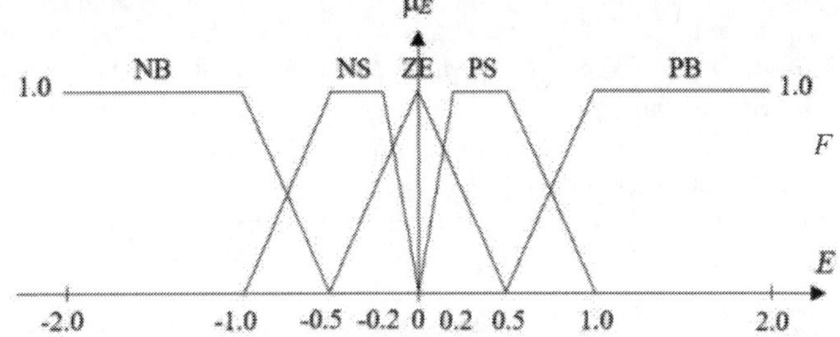

Figure 5. Membership function for the tracking error $e = \Omega^* - \Omega_m$

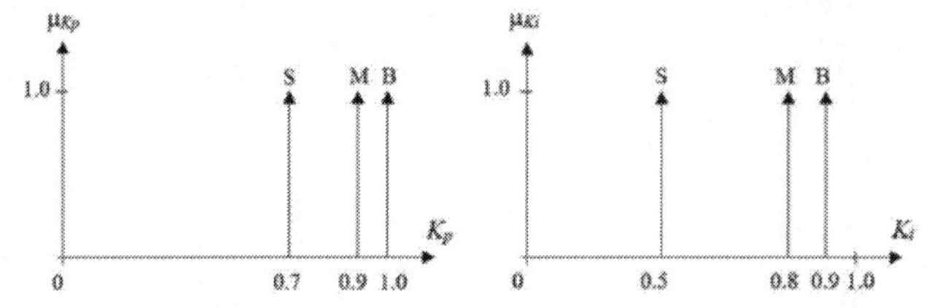

Figure 6. Membership functions for the output variables K_p and K_i

To describe the overall operation of the controller, an inductive approach has allowed to incorporate into a knowledge base a small set of only five inference rules given below:

1) if (error is ZE) then (k_p is B) and (k_i is M)
2) if (error is PS) then (k_p is S) and (k_i is B)
3) if (error is NS) then (k_p is M) and (k_i is M)
4) if (error is NB) then (k_p is M) and (k_i is S)
5) if (error is PB) then (k_p is M) and (k_i is S)

For digital processing inferences, the following choices were privileged to reduce the complexity of the algorithm and computation times:

- for the AND method: mini operator,
- for the OR method: max operator,
- for the implication method: mini operator,
- for the aggregation method: max operator.

Then, the inference engine based on the input and outputs fuzzy sets, uses the inference rules of the knowledge base to determinate the two final output fuzzy sets (Aggregated membership function). The centroid defuzzification method was used to obtain the crisp values of the output variables. This defuzzification technique can be expressed as:

$$z^* = \frac{\int_z \mu_A(z) z dz}{\int_z \mu_A(z) dz} \tag{17}$$

Where z* is the crisp output, μA(z) is the aggregated membership function and z is the output variable. The single input is transformed from value e to the normalized value E after multiplication with the error gain Ge. By the same procedure, the new input E is treated by the rule base process to produce the two new normalized crisp variables K_P and K_I , which are convert into the values k_p and k_i after multiplication with the output gains G_p et G_i.
The gains G_e, G_p et G_i are used to normalize the universe of discourse and to adjust the controller's sensitivity.

4. Performance study

The described observer structure shown in figure 2 was implemented in the environment software Matlab/Simulink, and tested in various operating conditions. This software allows digital simulation of the systems using a same expression of the ordinary differential equations in the dynamic machine model as well as the controller. The numerical method for solving the equations is Runge-Kutta method. Fixed-step mode is chosen for the computational time interval, this will emulate the fixed sampling frequency of the real-time control. The sampling period is 1e-4 sec. The parameters of the induction motor and gains of different controllers used are given in Appendix. Figure 7,a-b-c-d

IOP Conf. Series: Journal of Physics: Conf. Series **783** (2017) 012045 doi:10.1088/1742-6596/783/1/012045

showns simulation results of rotor speed an external force of 10 N.m, his disturbance can be seen at t = 0.8 sec and t = 1.2 sec and reference change at t = 2.5 s and 5s. Figure 8,a-b-c-d illustrate a response of sensorless drive system during starting operation with load 5 Nm, under conditions of low speed and with changes in load torque. The reference command imposes a speed step from 15 to -10 rad/s, the results obtained shown excellent performance even at low speeds, with precise estimates motor speed. The results shown very satisfactory performances in tracking trajectory, with a reaction time very low in transient state and a low error. The coefficients k_p and k_i delivered by the fuzzy inference system are variable and fits properly to speed changes. These figures shown clearly very satisfactory performance for the proposed sensorless controller in tracking and a remarkable pursuit between measured and estimed speed of the reference model speed. The control illustrates the correct signal issued by the fuzzy logic controller. There is an excellent direct field orientation consequence of a perfect decoupling between the flux and electromagnetic torque.

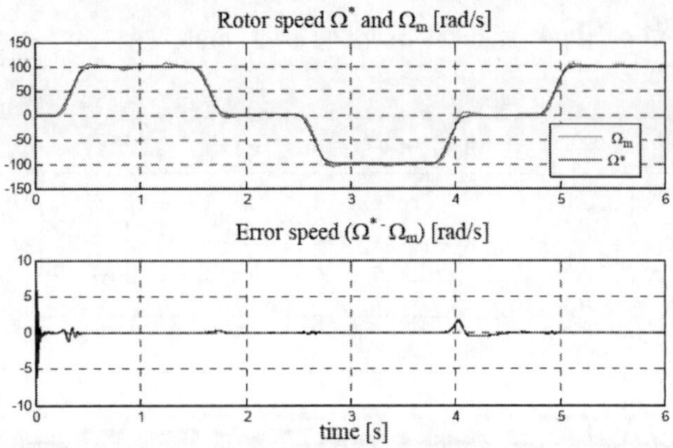

Figure 7.a Performance for rotor speed measured and desired with a load torque applied and removed

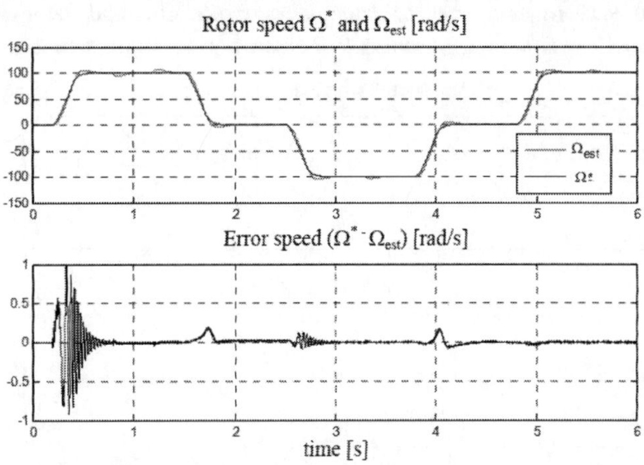

Figure 7.b Performance for rotor speed estimed and desired with a load torque applied and removed

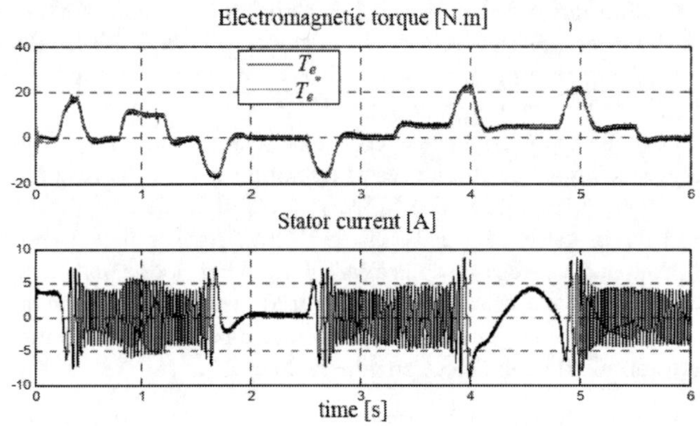

Figure 7.c Performance for electromagnetic torque and stator current

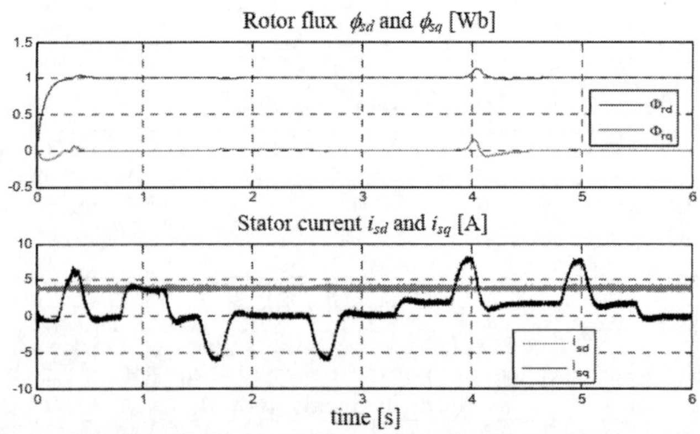

Figure 7.d Performance for direct and quadratic rotor flux and stator current

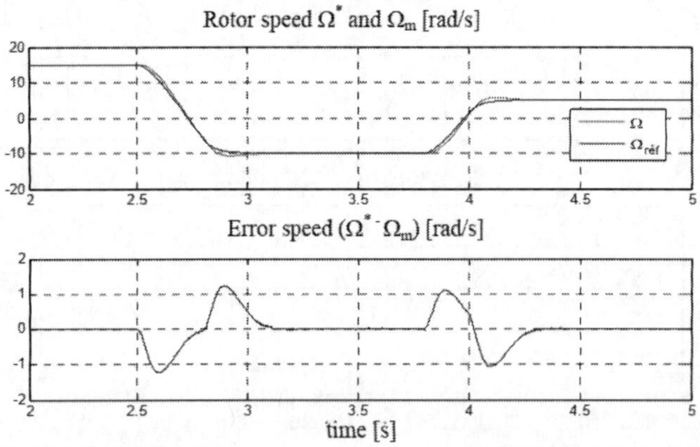

Figure 8.a Performance with low speed for rotor speed measured and desired with a load torque applied and removed

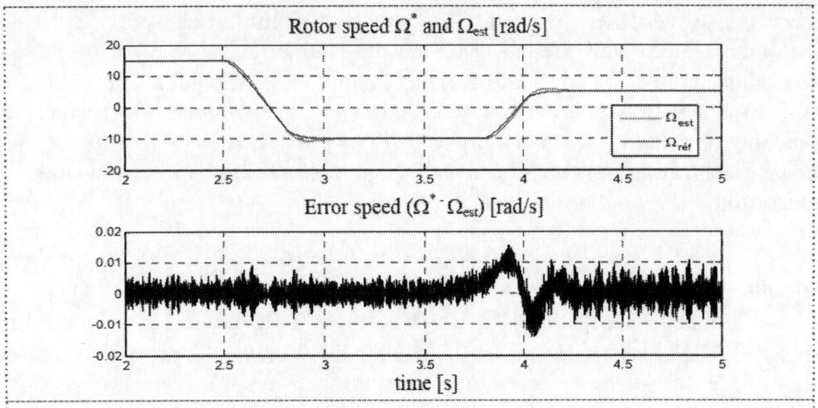

Figure 8.b Performance with low speed for rotor speed estimed and desired with a load torque applied and removed

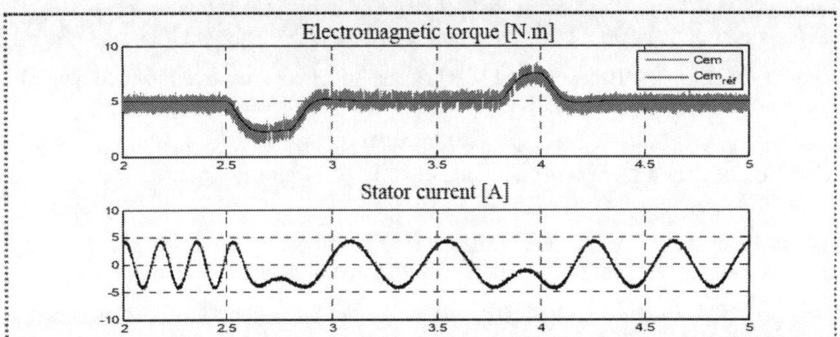

Figure 8.c Performance with low speed for electromagnetic torque and stator current

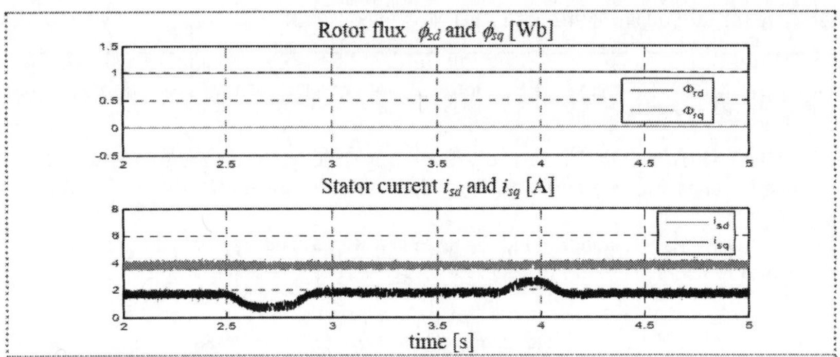

Figure 8.d Performance with low speed for direct and quadratic rotor flux and stator current

5. Conclusion

The objective of this paper is to design, develop, implement and test a new MRAS-Fuzzy Self-Tuning IP Speed Controller to improve the dynamic response of a three phase induction motor drive. The system was analysed and its performances were studied by numerical simulation to validate the theoretical concepts. It has been shown that the proposed controller can provide the properties of insensitivity to uncertainties and external disturbances. The performance of the proposed controller scheme have evaluated under a variety of operating conditions of the induction motor drive. System performance, both in steady state error in speeds and dynamic conditions, was found to be excellent and there is not any overshoot. This study proves that it is possible to achieve a robust adaptive speed observer dedicate to sensorless control for induction motor based on combination of artificial intelligence algorithms

Appendix

Motor Parameters

1.5 kW, 3-phase, 220/380 V, 2.8/4.8 A, 50 Hz, 4 poles, 1420 tr/mn. Rs = 4.85Ω, Rr = 3.805Ω, Ls = 0.274 H, Lr=0.274H, Lm = 0.258 H, J = 0.031 kg.m2, F = 0.00114 kg.m/s.

References

[1] N.P. Quang and J.A. Dittrich, " Vector of Three-Phase AC machines", System Development in the Pratice. Berlin, 2008, pp 107-139.

[2] H. Zhang, and M. C. Fu, "Applying model reference adaptive search to american-style option pricing," IEEE Proceedings of the 2006 Winter Simulation Conference 2006, pp 711-718.

[3] A. Bourek, L. Mokrani, A. Benakcha and M. Chabane " Indirect Adaptive Fuzzy Sliding Mode Observer for Sensorless Induc tion Machine Drive," Electromotion Vol.14, 2007, pp 143-153

[4] M. Zeraoulia, A. Mamoune, H. Mangel and M.E.H. Benbouzid " A Simple Fuzzy Logic Approach for Induction Motors Stator Condition Monitoring," Journal of Electrical Systems Vol. 1, No. 1, 2005, pp 15-25

[5] Md. Haseeb Khan, Arshia Azam and J. Amarnath " Fuzzy logic based MRAS speed observer for control of induction motor drive," World Journal of Modelling and Simulation, Vol. 6 No. 2, 2010, pp. 134-140

[6] Jie Zhang and T.H. Burton. "New approach to field orientation control of CSI induction motor drive," IEE Proceedings, Vol.135,Pt. B. No.1; January 1988.

[7] C.M. Liaw, Y.S. Kung and C.M. Wu "Design and implementation of a high-performance field-oriented induction motor drive," IEEE Trans. Ind. Electron.,vol.38,4, 1991, pp.275-282.

[8] V. Vasic and S. Vukosavic "Robust MRAS-based algorithm for stator resistance and rotor speed identification," IEEE Power Engineering Review, November 2001.

[9] R. Pena, R.Cardenas, J. Proboste, G. Asher and J. Clare " Sensorless control of a slip ring generator based on rotor current MRAS observer ", IEEE, 36th Annual Power Electronics Specialists Conference, Recife, Brazil, June 12-16, 2005, ,pp. 2508-2513.

[10] D. Vukadinivic, M. Basic and L. Kulisic " Stator resistance identification based on Neural and Fuzzy Logic principles in an induction motor drive, " Journal of Neurocomputing, ELSEVIER, Vol. 73, 2010, pp 602-612.

[11] Merabet E, Amimeur H, Hamoudi F and Abdessemed R, SelfTuning Fuzzy Logic Controller for a Dual Star Induction Machine, Journal of Electrical Engineering and Technology, vol. 6, No.1, pp. 133-138, 2011.

[12] Souza, G.C.D., Bose, B.K. and Cleland, J.G. (1993) Fuzzy logic based on-line efficiency optimization control of an indirect vector controlled induction motor drive, IEEE-IECOM.

[13] Won, C., Kim, S. and Bose, B.K. (1992) 'Robust position control of induction motor using fuzzy logic control', Con. Rec. IEEE-IAS Annu. Meeting, Conf. Rec., October, pp.472–451.

[14] M. Zerikat, A. Mechernene and S. Chekhroun "Fuzzy Self-Tuning Adaptive IP Controller for Robust Variable Speed Drive to Induction Motor" 5th International Conference on Systems and Control, ICSC'16, May. 25-27, Marrakech, Morocco.

IOP Conf. Series: Journal of Physics: Conf. Series **783** (2017) 012046 doi:10.1088/1742-6596/783/1/012046

Comparative Results on 3D Navigation of Quadrotor using two Nonlinear Model based Controllers

Y Bouzid[1], H Siguerdidjane [2] and Y Bestaoui[1]

[1] IBISC Laboratory, université d'Evry Val d'Essonne, Université Paris-Saclay, Evry, France
[2] L2S, CentraleSupélec, Université Paris-Saclay, Gif sur Yvette, France

Abstract. Recently the quadrotors are being increasingly employed in both military and civilian areas where a broad range of nonlinear flight control techniques are successfully implemented. With this advancement, it has become necessary to investigate the efficiency of these flight controllers by studying theirs features and compare their performance. In this paper, the control of Unmanned Aerial Vehicle (UAV) quadrotor, using two different approaches, is presented. The first controller is Nonlinear PID (NLPID) whilst the second one is Nonlinear Internal Model Control (NLIMC) that are used for the stabilization as well as for the 3D trajectory tracking. The numerical simulations have shown satisfactory results using nominal system model or disturbed model for both of them. The obtained results are analyzed with respect to several criteria for the sake of comparison.

1. Introduction

Due to its structural simplicity and understandable flight principle, the quadrotor UAV is considered as a good case study to analyze and to design flight control tools. This has attracted the interest of many researchers in aeronautics and robotics and many studies have led to the development of sophisticate nonlinear control laws [1-2]. Thus, several control techniques have been proposed in the literature as for instance linear ones based on a simplified model (LQR and PID)[3-4]. Unfortunately, they are used for limited operation point (hovering, attitude stabilization…). Also, strong disturbances are poorly rejected by this type of controllers. Some nonlinear controls (sliding mode, feedback linearization, etc.) have been employed where the ability of robustness against the model uncertainties is improved (see [5]). However, this type of controllers presents a number of drawbacks and may be inappropriate in some cases of operation (dynamic not sustained by the actuators, high energy at the start-up, complex implementation…).

The control design has to take in consideration some issues of practical implementation as the computational complexity, power limitation, robustness to parameter uncertainties, sensor noise, external disturbances… In this paper, an improvement has been brought in order to simplify the existing controllers, and thus overcome some issues poorly tackled by the classic linear or nonlinear techniques, thanks to Nonlinear PID (NLPID) and Nonlinear Internal Model Control (NLIMC).

Usually the kind of control called "NLPID control" stands for the regulator for which the coefficients are not assumed to be constant. Unlike this latter, our control law is defined in a novel way. It consists of a nonlinear controller, which is derived from a method based upon sliding mode theory and combined with a PID structure. Thus, this structure is achieved using sliding surface instead of the tracking error between the reference and the measured signal.

Among, the large variety of control techniques available in the literatures [6-8], the well-known reference model based technique called IMC, is popular in industrial process control applications due to its good tradeoff between the control performance and disturbance rejection capability [9]. The controller employs the flatness property instead the classic dynamic inversion technique to overcome some of the feedback linearization limitations.

The aim of this research work is to stabilize the vehicle while ensuring the tracking of prescribed 3D trajectories with good performance and to ensure a given level of robustness with respect to disturbances. Therefore, two approaches are compared in order to check out the features of each of them with respect to the environment disturbance (noise, wind...) and to the type of prescribed reference trajectories (constant, time varying) using several criteria and scenarios.

This paper is organized as follows: Section 2 introduces the design of the proposed nonlinear approaches. Section 3 presents the application of these techniques to the quadrotor and the control architecture used in this stage. In Section 4, we investigate the features and the characteristics of each technique with respect to some scenarios that confirm the effectiveness of the proposed approach where a comparison study is presented. The final section gives some conclusions.

2. Nonlinear controllers design

2.1. NLPID controller design

This section deals with the design of the NLPID controller, which may be summarized in two steps: First one, we apply the classic sliding mode approach and then the obtained controller is, in a second step, combined with the PID structure. Consider a class of nonlinear SISO system for $t \in [0, \infty)$ given by:

$$(\Sigma_x)\begin{cases} \dot{x} = \mathcal{F}(x) + G(x)u \\ y = h(x) \end{cases} \tag{1}$$

where $x \in \mathbb{R}^n$ is an n-dimentional state vector, $u \in \mathbb{R}$ is a scalar input, $y \in D_y \subset \mathbb{R}$ is a scalar output, $\mathcal{F}: D_x \dashrightarrow \mathbb{R}^n$ and $G: D_x \dashrightarrow \mathbb{R}^n$are n-dimentional vector functions sufficiently smooth on a domain $D_x \subset \mathbb{R}^n$ and $h(x)$ the output scalar function.

Assumption 1: *There exists a diffeomorphism $\Gamma: D_x \dashrightarrow R^n$ such that $D_\xi = \Gamma(D_x)$ contains the origin and $\xi = \Gamma(x)$ is a change of variables that transforms the nonlinear system into an equivalent system:*

$$(\Sigma_\xi)\begin{cases} \dot{\xi}_i(t) = \xi_{i+1}(t) \\ \dot{\xi}_n(t) = f(\xi) + g(\xi)u \\ y = \xi_1 \end{cases} \tag{2}$$

$i = 1 \dots n-1$. *Where $f(\xi), g(\xi)$ denote continuous nonlinear functions and $g(\xi)$ is nonsingular for all $\xi \in D_\xi$*. Now, let us consider a general sliding surface form

$$s(e) = \left(\frac{d}{dt} + \lambda_y\right)^{n-1} e(t) \tag{3}$$

where $e = y - y_r$ represents the tracking error between a reference trajectory y_r and the output y of a system and λ_y a positive constant. By expansion, equation (3) is written as

$$s(e) = e^{(n-1)}(t) + \sum_{i=0}^{i=n-2} \beta_i e^{(i)}(t) \tag{4}$$

where $\beta_i, i = 0, \dots, n-2$ are positive tuning parameters provided that they are chosen in order to render the equilibrium, $s(e) = 0$, asymptotically stable. The first time derivative of $s(e)$ is

$$\dot{s}(e) = e^{(n)}(t) + \sum_{i=0}^{i=n-2} \beta_i e^{(i+1)} \tag{5}$$

The objective of control tracking is to design a control law such that the output y can track the desired trajectory y_r. In this regard, where *Assumption 1* holds, using the last component of system (2), equation (5) becomes

$$\dot{s}(e) = f(\xi) + g(\xi)u - y_r^{(n)}(t) + \sum_{i=0}^{i=n-2} \beta_i e^{(i+1)} \tag{6}$$

Given a positive definite Lyapunov function candidate

$$V = \frac{1}{2} s^2(e) \tag{7}$$

So, the first time derivative leads to

$$\dot{V} = s(e)\dot{s}(e) \tag{8}$$

Note that the reachability condition of sliding mode control ensures the asymptotic stability ($\dot{V}$ must be negative definite). Thus

$$\begin{cases} \dot{s}(e) < 0 \; when \; s(e) > 0 \\ \dot{s}(e) > 0 \; when \; s(e) < 0 \end{cases} \quad (9)$$

Assume that

$$\dot{s}(e) = -\kappa\gamma(s) \quad (10)$$

with κ being a strictly positive constant and $\gamma(s)$ is a function defined by

$$\begin{cases} \gamma(s) < 0 \; if \;\; s < 0 \\ \gamma(s) > 0 \; if \;\; s > 0 \end{cases} \quad (11)$$

In fact, $\gamma(s)$ may take different forms as for instance: saturation function, tangent function, etc. Thus, the discontinuous function $sign(s)$, represents the ideal sliding mode regime.

From (6) and (10), it immediately follows that

$$u = \frac{-1}{g(\xi)}\left\{\kappa\gamma(s) + \sum_{i=0}^{i=n-2} \beta_i e^{(i+1)} - y_r^{(n)}(t) + f(\xi)\right\} \quad (12)$$

The discontinuous term $\gamma(s)$ gives the controller a good level of robustness. However, the fast variation of the sliding surface gives rise of vibration in the system during the flight phase consequently affecting the control accuracy. So, in order to improve the performance of the controller and limiting the effect of the chattering phenomena, involved through the term $\dot{s}(e)$, we combine a PID structure and the controller determined above, namely

$$u = \frac{-1}{g(\xi)}\left\{K_p\left(s(t) + \frac{1}{T_i}\int_0^t s(\tau).d\tau + \; T_d\frac{ds(t)}{dt}\right) + \sum_{i=0}^{i=n-2} \beta_i e^{(i+1)} - y_r^{(n)}(t) + f(\xi)\right\} \quad (13)$$

K_p, T_d and T_i denote the proportional gain, the derivative and integral time constants respectively.

2.2. NLIMC controller design

The IMC controllers are used to be devoted to the linearized models and then implemented on the real plants [10]. The difference between the model and the real process gives rise to the poor performance of the closed-loop system. Therefore, the linear IMC scheme has been extended to non-linear versions by different approaches [11]. Moreover, due to the complexity of nonlinear systems, the development of a general IMC nonlinear strategy has serious difficulties. To accomplish IMC control, it requires the inverse of the plant. this inversion is more involved and analytical solutions are too much complex to obtain. Motivated by these gaps and requirements (inversion, open loop structure, linear model, computational complexity…), we develop in the following, a novel and simple analytic formulation for the design of NLIMC synthesized for nonlinear Single-Input, Single-Output (SISO) systems that involves a closed loop control structure.

The direct mission is to investigate the controller that implies the dynamic model inverse and occurs in closed loop control architecture. Thus, our procedure split into two steps in order to accomplish the controller design. The first step is to formulate a direct relationship between the output and the tracking error. In other words

$$y^{(i)}(t) = \Theta_i\left(\int e\,, e, \dot{e}\,...\,, e^{(i-1)}\right) \quad i = 0, ..., n \quad (14)$$

where Θ_i are scalar smooth functions. For this purpose, we consider a surface defined in the state-space D_x as

$$\dot{e} + \frac{1}{\mu}e = 0 \quad (15)$$

where μ is a positive time tuning parameter providing a compromise between performance and robustness.

Assumption 2: *The reference trajectories allowed for this study are considered piecewise constant in domain* $D_y \subset \mathbb{R}$.

From equation (15), we have $\dot{e}(t) = -\frac{1}{\mu}e(t)$ and *Assumption 2* allows writing: $y^{(1)}(t) = \frac{e(t)}{\mu}$. Then, by successive time derivatives, $y^{(i)}(t)$ can be written as

$$y^{(i)}(t) = \frac{e^{(i-1)}(t)}{\mu} \quad i = 2, ..., n. \quad (16)$$

In addition, equation (16) can be written under general form Introducing two parameters, $(\alpha_1, \beta_1) \in \mathbb{R} \times \mathbb{R}$ verifying $\alpha_1 + \beta_1 = 1$, we obtain

$$y^{(i)}(t) = \frac{\alpha_1}{\mu} e^{(i-1)}(t) + \frac{\beta_1}{\mu} \left(y_r - y(t)\right)^{(i-1)} \quad i = 2, \dots, n$$

Recall that *Assumption 2* holds, then

$$y^{(i)}(t) = \frac{\alpha_1}{\mu} e^{(i-1)}(t) - \frac{\beta_1}{\mu} y^{(i-1)}(t) \quad i = 2, \dots, n$$

By recurrence, we obtain a general form

$$y^{(i)}(t) = \sum_{k=1}^{k=i} (-1)^{k+1} \frac{\alpha_{(k)} \prod_{j=1}^{k} \beta_{(j-1)}}{\mu^k} e^{(i-k)}(t) \, i = 2, \dots, n \tag{17}$$

with initial term $y^{(1)}(t) = \frac{e(t)}{\mu}$

$(\alpha_{(k)}, \beta_{(k)}) \in \mathbb{R} \times \mathbb{R}$ verifying $\alpha_{(k)} + \beta_{(k)} = 1$, with $\alpha_{(i)} = 1$ and $\beta_{(0)} = 1$. For more detail, the reader may refer to [12].

The inversion of the nonlinear model requires the relationship between the input u and the output y. A broad range of techniques already published investigating this classic problem. So that in this second step, we exploit the flatness property. The flat output and its time derivatives describe the system dynamics, since their knowledge is sufficient to compute all the variables $x(t), u(t)$ and their time derivatives. Then, the state-space model (1) may be summarized by the following differential equation.

$$y^{(n)}(t) + \rho\left(y, y^{(1)}, \dots, y^{(n-1)}, u, u^{(1)}, \dots, u^{(\sigma-1)}\right) + \vartheta\left(y, y^{(1)}, \dots, y^{(n-1)}, u, u^{(1)}, \dots, u^{(\sigma-1)}\right) u^{(\sigma)} = 0 \tag{18}$$

where ρ and ϑ are two nonlinear multi-variable functions and $r = n - \sigma$ denotes the relative degree of the system ($\sigma \leq n$). Our objective is to design controller that drives the system output to track a desired trajectory. Let us write again equation (18) in terms of the tracking error by using first relationship (17).

$$y^{(n)}\left(e, \mu, \alpha_{(k)}, \beta_{(k)}\right) + \rho\left(e, \alpha_{(k)}, \beta_{(k)}, \mu, u, u^{(1)}, \dots, u^{(\sigma-1)}\right) + \vartheta\left(\alpha_{(k)}, \beta_{(k)}, e, \mu, u, u^{(1)}, \dots, u^{(\sigma-1)}\right) u^{(\sigma)} = 0 \tag{19}$$

The controller then is obtained by

$$u^{(\sigma)}(t) = -\frac{y^{(n)}\left(e(t), \mu, \alpha_{(k)}, \beta_{(k)}\right) + \rho\left(e(t), \alpha_{(k)}, \beta_{(k)}, \mu, u, u^{(1)}, \dots, u^{(\sigma-1)}\right)}{\vartheta\left(\alpha_{(k)}, \beta_{(k)}, e(t), \mu, u, u^{(1)}, \dots, u^{(\sigma-1)}\right)} \tag{20}$$

By integration, the control effort $u(t)$ is computed.

2.3. Discussion and properties

In this section, we discuss the main features, the advantages and the limitations of the proposed controllers. Controller (13) exhibits several benefits, and can be split up into three parts: The first part is designed in order to compensate the system dynamics (2) (feedback linearization). The second part consists of a linear combination of the tracking error dynamics, $\sum_{i=0}^{i=n-2} \beta_i \varepsilon_y^{(i+1)}$, and the nth derivative of the reference trajectory, $y_r^{(n)}(t)$, that allows the linearization of the closed loop system after the act of the first part. The remaining part includes the PID structure, which represents the additional control needed for control accuracy and disturbance rejection. From equation (13), we observe the absence of discontinuities associated to jumps in the control action, which clearly eliminates the chattering problem. In addition, plant-model mismatch and the steady state errors are canceled by adding the integral action that penalizes the deviations between the output and its set point. However, the derivative term of the proposed controller induces a higher order derivative one with respect to that needed for classical "feedback linearization". Unfortunately, this is a drawback because of the additional noise and derivative estimation inaccuracy.

The classic dynamic inversion (feedback linearization) technique involves the cancelation of the nonlinearities. Moreover, it is applied for a specific class of systems especially the canonical one (2).

However, if the model is inaccurate or contains some varying variables that are highly influenced by the environment changes (temperature, pressure ...), this cancelation may be not exact and engenders another equilibrium points of the closed loop system or causes the instability. This drawback is tackled by the NLIMC controller where the plant-model mismatch is canceled by using the tracking error in the inversion process. Therefore, the controller presents a good level of robustness, especially, with respect to the model uncertainties. Moreover, using the flatness property allowing the system behaves in the original coordinates where the characteristic can be specified naturally. NLIMC approach uses only the considered outputs and their dynamics $\left(y(t), \dot{y}(t), \dots, y^{(m)}(t)\right)$, while the dynamic inversion in the NLPID involves beside the outputs all the states of the system. In this last case and for real implementation purpose, the control loop requires a broad range of sensors or estimators that affect the accuracy of the control because of the noise and the uncertainties especially that the feedback linearization is well-known by its lack of robustness. The NLPID uses the inversion of the system dynamics and this leads to numerical problem nearby the singularities. The NLIMC uses tunable gains μ that allows to overcome such issue. Unfortunately, for high order systems, NLIMC suffers form the problem of numerical noise and derivative estimation inaccuracy. In addition, as the NLIMC is designed for piecewise reference trajectories, the steady state errors are poorly canceled for time varying reference trajectories. These properties are shown down through a series of numerical simulations.

3. Quadrotor application

3.1. Dynamics background and control architecture

The quadrotor has four rotors with twin-bladed propellers. The system operates in two coordinate frames: the inertial fixed frame and the body frame. Let $\eta = (\varphi, \theta, \Psi)^T$ describes the orientation of the aerial vehicle (Roll, Pitch, Yaw) and $\chi = (x, y, z)^T$ denotes its absolute position. We use a simplified model that is derived by Bouzid et *al.* in [12] and describes the quadrotor in-flight behavior. It may decompose into translational subsystem and rotational subsystem as

$$\ddot{\chi} = \begin{pmatrix} u_1 \frac{C_\psi S_\theta C_\varphi + S_\psi S_\varphi}{m} \\ u_1 \frac{S_\psi S_\theta C_\varphi - C_\psi S_\varphi}{m} \\ -g + u_1 \frac{C_\theta C_\varphi}{m} \end{pmatrix} \quad (21), \quad \ddot{\eta} = \begin{pmatrix} \dot{\theta}\dot{\Psi}\left(\frac{I_y - I_z}{I_x}\right) + \frac{u_2}{I_x} \\ \dot{\varphi}\dot{\Psi}\left(\frac{I_z - I_x}{I_y}\right) + \frac{u_3}{I_y} \\ \dot{\varphi}\dot{\theta}\left(\frac{I_x - I_y}{I_z}\right) + \frac{u_4}{I_z} \end{pmatrix} \quad (22)$$

$S_{(.)}$ and $C_{(.)}$ are abbreviations for $sin(.)$ and $cos(.)$ respectively. m denotes the mass, g the gravity acceleration, $I = diag(I_x, I_y, I_z)$ is the diagonal inertia matrix, u_1 is the total thrust of four rotors and $\tau = (u_2, u_3, u_4)^T$ is the control torque. The system parameters are depicted in Table 1.

Table 1. Quadrotor parameters.

$m(kg)$	0.429	$I_y(kg.m^2)$	0.0029
$I_x(kg.m^2)$	0.0022	$I_z(kg.m^2)$	0.0048

These proposed techniques are herein applied to the quadrotor by taking care of having an adequate control structure (see Figure 1). The system has two dynamics, such that the first (slow outer loop) relates to the position control and generates the reference angles (φ_r, θ_r) to the second one (fast inner loop) via attitude planner (24). This inner loop is related to the orientation control. x and y are controlled through two virtual drives (u_x, u_y) to allow the system to reach the prescribed references x_r and y_r respectively. The roll, pitch and yaw angles are controlled by (u_2, u_3, u_4) respectively, whereas the altitude is controlled by u_1. This control structure shown in Figure 1 is designed to ensure the tracking of any reference trajectory described by the absolute position with a given yaw angle. In fact, we will consider six inputs $(u_x, u_y, u_1, u_2, u_3, u_4)$ and six outputs $(x, y, z, \varphi, \theta, \Psi)$. Notice that this control architecture allows applying the developed approaches that concern the single input single output systems to the quadrotor. This choice is done in order to simplify the design and reduce the computational complexity.

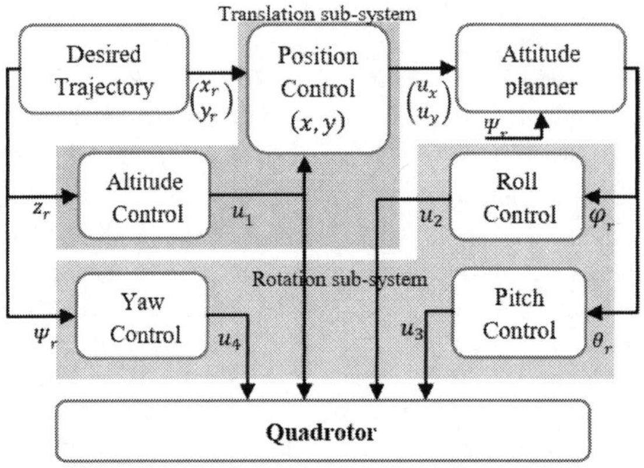

Figure 1. Quadrotor control architecture.

3.2. Autopilot design

The tracking errors are defined as: $e_x = x - x_r$, $e_y = y - y_r$, $e_z = z - z_r$, $e_\varphi = \varphi - \varphi_r$, $e_\theta = \theta - \theta_r$ and $e_\Psi = \Psi - \Psi_r$. Translation dynamics (21) can be divided into three other sub-systems along the three axis (X, Y, Z). Each sub-system has one input (u_x, u_y, u_1) and one output (x, y, z) respectively. We first start with u_1 for the altitude motion. Once this command is calculated we then proceed in the same way for u_x and u_y by considering u_1 as time varying parameter, with

$$\begin{cases} u_x = c_\Psi s_\theta c_\varphi + s_\Psi s_\varphi \\ u_y = s_\Psi s_\theta c_\varphi - c_\Psi s_\varphi \end{cases} \quad (23)$$

Then, the required reference angles of the roll and pitch rotations are given by

$$\begin{cases} \varphi_r = arcsin\left(sat(u_x sin\,\Psi_r - u_y cos\Psi_r)\right) \\ \theta_r = arcsin\left(sat\left(\frac{u_x cos\,\Psi_r + u_y sin\Psi_r}{cos\,\varphi}\right)\right) \end{cases} \quad (24)$$

where the function $sat(.)$ is defined as

$$sat(a) = \begin{cases} 1, & a > 1 \\ a, & |a| \le 1 \\ -1, & a < -1 \end{cases}$$

Similarly, system (22) can be divided into three sub-systems for the roll, pitch and yaw rotations. Each one of them has one input (u_2, u_3, u_4) and one output (φ, θ, Ψ) respectively. For the NLIPID, we make the choice of the surfaces as

$$\begin{cases} s_x = \dot{e_x} + \beta_{0x} e_x \\ s_y = \dot{e_y} + \beta_{0y} e_y \\ s_z = \dot{e_z} + \beta_{0z} \varepsilon_z \\ s_\varphi = \dot{e_\varphi} + \beta_{0\varphi} e_\varphi \\ s_\theta = \dot{e_\theta} + \beta_{0\varphi} e_\theta \\ s_\Psi = \dot{e_\Psi} + \beta_{0\varphi} e_\Psi \end{cases} \quad (25)$$

with $\beta_{0x}, \beta_{0y}, \beta_{0z}, \beta_{0\varphi}, \beta_{0\theta}$ and $\beta_{0\Psi}$ are positive constants.

Applying (13), we obtain

$$u_x = \frac{-m}{u_1}\left\{K_{px}\left(s_x(t) + \frac{1}{T_{ix}}\int_0^t s_x(\tau).d\tau + T_{dx}\frac{ds_x(t)}{dt}\right) + \beta_{0x}\dot{\varepsilon}_x - \ddot{x}_r\right\}$$

$$u_y = \frac{-m}{u_1}\left\{K_{py}\left(s_y(t) + \frac{1}{T_{iy}}\int_0^t s_y(\tau).d\tau + T_{dy}\frac{ds_y(t)}{dt}\right) + +\beta_{0y}\dot{\varepsilon}_y - \ddot{y}_r\right\}$$

$$u_1 = \frac{-m}{c_\theta c_\varphi}\left\{K_{pz}\left(s_z(t) + \frac{1}{T_{iz}}\int_0^t s_z(\tau).d\tau + T_{dz}\frac{ds_z(t)}{dt}\right) + \beta_{0z}\dot{\varepsilon}_z - g - \ddot{z}_r\right\}$$
$$u_2 = -I_x\left\{K_{p\varphi}\left(s_\varphi(t) + \frac{1}{T_{i\varphi}}\int_0^t s_\varphi(\tau).d\tau + T_{d\varphi}\frac{ds_\varphi(t)}{dt}\right) + \beta_{0\varphi}\dot{\varepsilon}_\varphi + \dot{\theta}\dot{\psi}\left(\frac{I_y - I_z}{I_x}\right) - \ddot{\varphi}_r\right\}$$
$$u_3 = -I_y\left\{K_{p\theta}\left(s_\theta(t) + \frac{1}{T_{i\theta}}\int_0^t s_\theta(\tau).d\tau + T_{d\theta}\frac{ds_\theta(t)}{dt}\right) + \beta_{0\theta}\dot{\varepsilon}_\theta + \dot{\varphi}\dot{\psi}\left(\frac{I_z - I_x}{I_y}\right) - \ddot{\theta}_r\right\}$$
$$u_4 = -I_z\left\{K_{p\psi}\left(s_\psi(t) + \frac{1}{T_{i\psi}}\int_0^t s_\psi(\tau).d\tau + T_{d\psi}\frac{ds_\psi(t)}{dt}\right) + \beta_{0\psi}\dot{\varepsilon}_\psi + \dot{\varphi}\dot{\theta}\left(\frac{I_x - I_y}{I_z}\right) - \ddot{\psi}_r\right\}$$

$K_{p()}, T_{i()}, T_{d()}$ denote the proportional gains, the integral and derivative time constants respectively of the NLPID structure.

For NLIMC, we apply (20)

$$u_x = \left(\frac{\alpha_{x(1)}}{\mu_x}\dot{e}_x - \frac{\beta_{x(1)}}{\mu_x^2}e_x\right)\frac{m}{u_1}$$
$$u_y = \left(\frac{\alpha_{y(1)}}{\mu_y}\dot{e}_y - \frac{\beta_{y(1)}}{\mu_y^2}e_y\right)\frac{m}{u_1}$$
$$u_1 = \frac{m\left(\frac{\alpha_{z(1)}}{\mu_z}\dot{e}_z - \frac{\beta_{z(1)}}{\mu_z^2}e_z + g\right)}{\cos\left(\frac{1}{\mu_\varphi}\int_0^t e_\varphi(\tau)d\tau + \varphi_0\right)\cos\left(\frac{1}{\mu_\theta}\int_0^t e_\theta(\tau)d\tau + \theta_0\right)}$$
$$u_2 = I_x\left(\frac{\alpha_{\varphi(1)}}{\mu_\varphi}\dot{e}_\varphi - \frac{\beta_{\varphi(1)}}{\mu_\varphi^2}e_\varphi - \frac{e_\theta\, e_\psi}{\mu_\theta\, \mu_\psi}\left(\frac{I_y - I_z}{I_x}\right)\right)$$
$$u_3 = I_y\left(\frac{\alpha_{\theta(1)}}{\mu_\theta}\dot{e}_\theta - \frac{\beta_{\theta(1)}}{\mu_\varphi^2}e_\theta - \frac{e_\varphi\, e_\psi}{\mu_\varphi\, \mu_\psi}\left(\frac{I_z - I_x}{I_y}\right)\right)$$
$$u_4 = I_z\left(\frac{\alpha_{\psi(1)}}{\mu_\psi}\dot{e}_\psi - \frac{\beta_{\psi(1)}}{\mu_\psi^2}e_\psi - \frac{e_\varphi\, e_\theta}{\mu_\varphi\, \mu_\theta}\left(\frac{I_x - I_y}{I_z}\right)\right)$$

$\mu_x, \mu_y, \mu_z, \mu_\varphi, \mu_\theta$ and μ_ψ denote tuning positive parameters. $(\alpha_{(.)(1)}, \beta_{(.)(1)}) \in \mathbb{R} \times \mathbb{R}$ are chosen according to the stability requirements where $\alpha_{(.)(1)} + \beta_{(.)(1)} = 1$.

4. Comparison study

For the sake of significant comparison, this study is done considering the same conditions and following the same protocol by investigating the performance of the controllers for several scenarios and trajectories. Thus, the parameters of NLIPID and NLIMC are obtained through the optimization of the Integral Square Error (ISE) criterion.

$$ISE = \frac{1}{t_f - t_i}\int_{t_i}^{t_f}(q - q_r)^T(q - q_r)dt \tag{26}$$

where q is the output vector, q_r its reference trajectory. t_f and t_i denotes the final and the initial optimization instances respectively. For more efficient optimization, the parameters are optimized independently for each basic mode of flight. Thus, we consider a rotation of $\frac{\pi}{2}$ radian around Z-axis, vertical flight, longitudinal flight and lateral flight of one meter along the Z, X and Y-axes respectively. The process of optimization holds during a flight of 20 seconds. The obtained control parameters for both NLPID and NLIMC are depicted in Table 2. Before studying the features of the controllers, we investigate their computational complexity via the number of additions and multiplications.

In this case, a quick view in the Table 3, allows checking out that the NLPID is more gourmand in terms of computation because of the considered number of derivations and the integrations compared to the NLIMC.

In order to compare, investigate and evaluate the control laws, various simulations have been performed and numerous scenarios have been proposed on the complete closed loop system (ideal case, parameters uncertainties, extra payload, noisy measurements and wing gust). The main metrics used for the comparison are:

- ISE: is computed for the three translations and the yaw angle that allows to compare the accuracy of the controllers for a given configuration in the space using (26). In fact, using ISE, we cannot distinguish the nature of these errors (overshoot, oscillations, steady state…). Therefore, we employ beside this metric a visual graphics that allow identifying the nature of these errors.
- Settling time (T_s): it measures the rapidity of convergence of a closed loop time response.
- Consumed energy: This practical constraint makes the use of controllers that consume less energy more preferable. Herein, we use the following criterion

$$\int_{t_i}^{t_f} u_1^2 dt \tag{27}$$

Table 2. Controllers parameters.

	NLIMC	NLPID
x	$\mu_x = 0.5$	$k_{p_x} = 3.392, T_{i_x} = 0.7, T_{d_x} = 0.5$
y	$\mu_y = 0.5$	$k_{p_y} = 3.229, T_{i_y} = 0.7, T_{d_y} = 0.672$
z	$\mu_z = 0.1$	$k_{p_1} = 2.829, T_{i_1} = 3.488, T_{d_1} = 0.867$
φ	$\mu_\varphi = 3.9$	$k_{p_2} = 0.731, T_{i_2} = 0.37, T_{d_2} = 0.439$
θ	$\mu_\theta = 3.9$	$k_{p_3} = 5.161, T_{i_3} = 0.198, T_{d_3} = 0.439$
Ψ	$\mu_\Psi = 0.39$	$k_{p_4} = 20.346, T_{i_4} = 16.009, T_{d_4} = 0.007$
	$\beta_{(.)} = -0.5$	$\beta_{(.)} = 1$

Table 3. Computational complexity of the autopilots.

Type	NLIPID	NLIMC
Multiplication	49	56
Additions	43	29
Integration	6	2
Time derivation	First: 15, Second:6	First: 6, Second: 0
Nonlinear functions	2	2

4.1. *Ideal case without disturbance*

In this section, we investigate the efficiency of the controllers in the ideal case and the performance of the time responses of the closed loop system. Therefore, the quadrotor is supposed to flight in ideal environment (without disturbances) and has a perfect model. The proposed trajectories are composed of successive straight-line segments and circular trajectory.

- *Successive straight line segments trajectory*

The first proposed scenario allows us to test, independently, all the flight modes of the quadrotor. It is composed of successive straight-lines along the three axes X, Y, and Z and a rotation around Z-axis (yaw). This trajectory is used to overview the performance of the controllers for each type of flights: vertical flight (z), longitudinal flight (x, θ), lateral flight (y, φ) and yaw rotation (Ψ). Moreover, we will test the behavior of the vehicle at the inter-connection points.

The control gains used herein are the same depicted in Table 2 and the obtained results are summarized in Table 4. The time responses of the tracking errors are presented in Figure 2.

Table 4. Controllers features.

Criteria		NLPID	NLIMC
ISE	X	0.7595	0.2062
	Y	0.7595	0.2062
	Z	0.7959	0.0420
	Yaw	0.0624	0.0158
	$Total$	2.3773	0.4702
Energy		2.10^3	$1.2055.10^5$
Settling time	T_{s_x}	2.165	3.17

(sec)	T_{s_y}	2.179	3.16
	T_{s_z}	2.193	0.545
	$T_{s_{yaw}}$	2.94	0.559
Maximal attitude angles (rad)	Roll	1.2494	0.3998
	Pitch	1.3581	0.4003

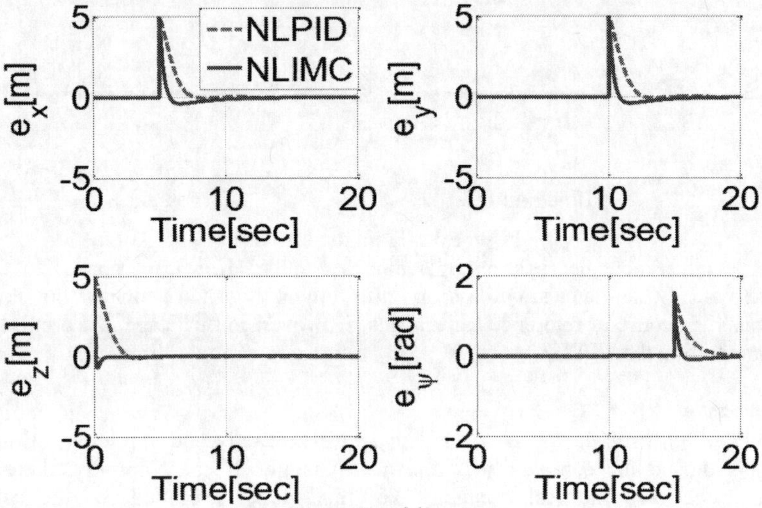

Figure 2 : Tracking errors.

From Table 4, one can observe that the NLIMC ensures more accurate control compared to the NLPID. However, the consumed energy is of course greater. NLIMC has slower responses in the longitudinal and lateral motions compared to the NLPID and faster responses for the altitude and the yaw controls. It can be clearly seen in Figure 2 that a small overshoot appears when using the NLIMC especially for Z, which is reasonable because of the rapidity of convergence.

- Trajectory tracking

We consider a helix trajectory to check the efficiency of the controller in the tracking of time-varying reference trajectory especially that the NILIMC makes the assumption of piecewise constant trajectories. The obtained results are tabulated in Table 5. The time responses of the tracking errors are presented in Figure 3.

Table 5. Controllers features.

Criteria		NLPID	NLIMC
ISE	X	0.7692	1.4362
	Y	0.0401	1.2747
	Z	$1.46.10^{-4}$	0.0057
	Yaw	0.0624	0.0158
	Total	0.8718	2.7324
Energy		457.8235	432.1417
Maximal attitude angles	Roll	0.9681	0.5911
	Pitch	0.8711	0.7244

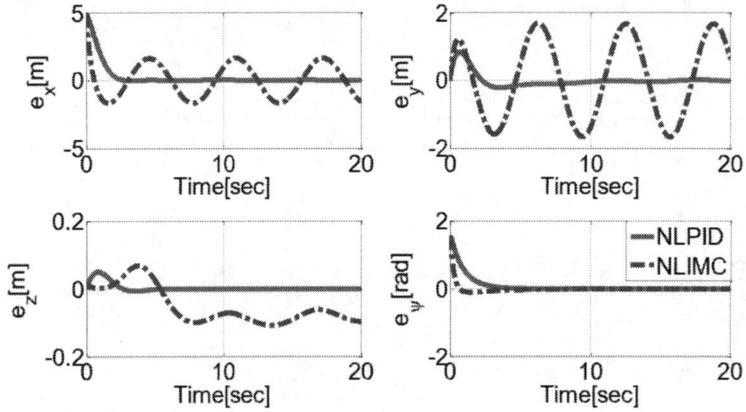

Figure 3 : Tracking errors.

From Table 5, a remarkable degradation of accuracy for the NLIMC compared to the NLPID is reported where a steady state appears along the entire trajectory. The results are not strange because the NLIMC considers constant reference trajectories. However, in this stage, the controller consumes less energy compared to the NLPID.

4.2. *With disturbance*

- *Parameters uncertainties: It* is so hard to express, exactly and explicitly, all the forces and moments due to the complexity of the aerodynamic effects. Naturally, there is a notable mismatch between the real dynamics and the model exploited for the control design. Moreover, due to the limit of the onboard electronic computational capacities, some forces and terms are simplified or neglected. The derived dynamic model of the quadrotor contains a number of parameters that are identified according to the real system. The inertia matrix can be influenced by the environment changes and lead to some uncertainties. Therefore, the controller should be able to ensure good performance even in presence of these uncertainties. In this test all these parameters are underestimated: $I_{(.)} = 1.75.\tilde{I}_{(.)}$ where $\tilde{I}$ denotes the parameters employed in the controllers.
- *Additional Payload:* Another test of robustness is proposed considering additional payload that represents 75% of the global mass. The payload is supposed added in the centre of mass.
- *Sensors noise:* In general, the signals provided by the sensors are noisy. The noise may be cause the vibration and oscillation of the vehicle and in certain cases the instability of the overall system. We consider thus an additive noise with all the outputs using the function rand of Matlab to generate stochastic values uniformly distributed between 0-1. The additive noise is selected equals to $0.1.rand(.)$ for the position and $0.01.rand(.)$ for the attitude.
- *Gust of wind:* The objective is to check the robustness of the controllers under the effect of gust of wind, without taking in consideration the model of the disturbed vehicle. Often the wind model is expressed in function of the velocities along the three axes.

$$\begin{cases} u_w = \bar{u}_w + u_d \\ v_w = \bar{v}_w + v_d \\ w_w = \bar{w}_w + w_d \end{cases} \tag{28}$$

Where $\bar{u}_w, \bar{v}_w, \bar{w}_w$ denote the mean wind velocities. The terms u_d , v_d and w_d stand for the deviations from the mean velocity. The mean wind speed is set according to the daily-use situation. Then turbulence gust models are superimposed on this mean wind where the data are collected from statistical properties (spectra, scale length, turbulence intensity…). The wind velocities along the inertial axes are generated via Dryden linear filters as

$$\begin{cases} \frac{u_d(s)}{\mathcal{G}_u(s)} = H_u(s) \\ \frac{v_d(s)}{\mathcal{G}_v(s)} = H_v(s) \\ \frac{w_d(s)}{\mathcal{G}_w(s)} = H_w(s) \end{cases} \tag{29}$$

$\mathcal{G}_u(s)$, $\mathcal{G}_v(s)$ *and* $\mathcal{G}_w(s)$ are zero-mean white Gaussian noise. s is the Laplace variable. In order to generate a turbulence signal, a band-limited white noise signal passes through an appropriate filter. For more details, the reader may refer to [13].

The obtained results through the previous scenarios are summarized by Table 6 and the time responses of the tracking errors are depicted Figure 4.

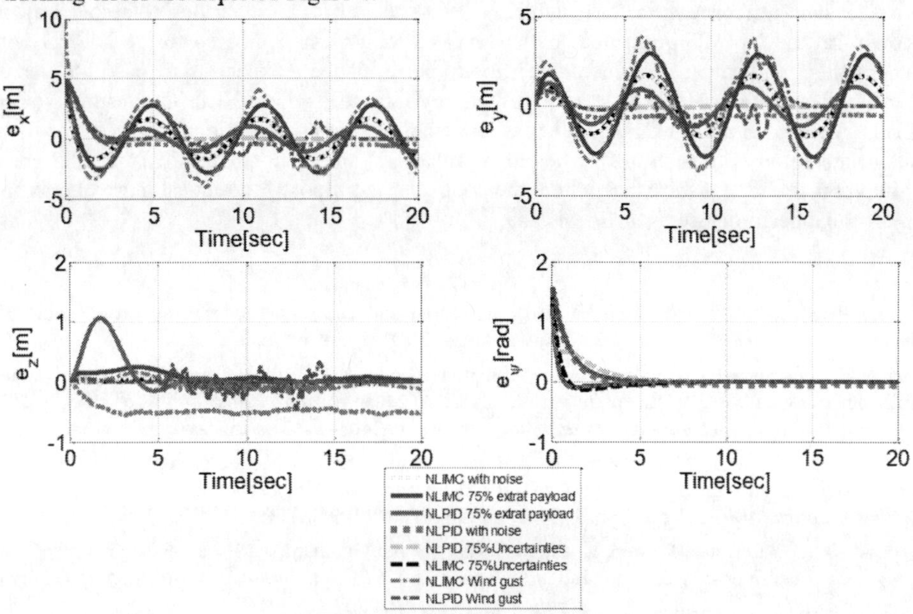

Figure 4. Tracking errors.

Table 6. Controllers features.

	criteria	NLPID	NLIMC
Uncertainties +75%	ISE_{Total}	0.8506	2.7404
	Energy	444.7810	390.6749
Additional Payload 75%	ISE_{Total}	1.8701	7.6439
	Energy	$1.4518.10^3$	$31.2679.10^3$
Sensors noise 0.1	ISE_{Total}	69.8358	3.3110
	Energy	$2.4688.10^3$	$8.6112.10^3$
Wind gust	ISE_{Total}	1.5800	3.0350
	Energy	484.3860	$3.7606.10^3$

The controllers exhibit good level of robustness with respect to all the disturbances mentioned above. They consume more energy to keep the same performance compared to the ideal case except the parameters uncertainties where the consumed energy still the same. An overview of Table 6, we observe that the NLPID controller exhibits superior performance compared to the NLIMC controller. This superiority relevant to the poor capability of NLIMC to track time varying reference trajectories, which leads to a persistent steady state error along the entire trajectory (see Figure 4). However, also from Table 6, the NLPID exhibits poor capabilities to tackle the noise.

Through our analyze, we have concluded that the NLIMC is more robust than the NLPID that has better performance. From Figure 4 and for the altitude and the yaw angle, the NLIMC behaves better contrary to the longitudinal and lateral motion. We have concluded also that the attitude control subsystem is more sensitive to noise for both of them.

5. Conclusion

Two nonlinear approaches have been described namely Nonlinear Internal Model Control and Nonlinear PID. Numerical simulations were carried out in order to evaluate their effectiveness. The performance of the proposed approaches is demonstrated in multiple test scenarios. The settling time is shown to be quite fast and a good level of robustness is ensured with respect to parameters uncertainties, extra payload, wind gust and the noise. The NLIMC have superior performance for piecewise constant reference trajectories while NLPID is better for time varying reference trajectories. It is shown that the NLIMC guarantees better level of robustness compared to the NLPID with less computational complexity. To guarantee good performance, expect the case of parameter uncertainties, it was observed that the required energy increases where NLIMC consumes more than the NLPID. A rigorous proof of the stability of the whole system (translational and rotational) will be reported in the future works. Some issues also still open, among them is to reduce the consumed energy in presence of the disturbance where the controller is supposed to request more energy in order to keep the same performance as the ideal case.

References

[1] Pounds P, Mahony R., and Corke P 2010 Modelling and control of a large quadrotor robot *Control Engineering Practice*. *18*(7), 691–699. http://doi.org/10.1016/j.conengprac.

[2] Huang H, Hoffmann G. M, Waslander S. L and Tomlin C. J 2009 Aerodynamics and Control of Autonomous Quadrotor Helicopters in Aggressive Maneuvering *Proceedings of the 2009 IEEE International Conference on Robotics and Automation* (pp. 2408–2413). Piscataway, NJ, USA: IEEE Press.

[3] Salih A. L, Moghavvemi M, Mohamed H. A. F and Gaeid K. S 2010 Modelling and PID controller design for a quadrotor unmanned air vehicle. In *IEEE International Conference on Automation Quality and Testing Robotics (AQTR)* (Vol. 1, pp. 1–5). http://doi.org/10.1109/AQTR.2010.5520914

[4] Reyes-Valeria E, Enriquez-Caldera R, Camacho-Lara S and Guichard, J 2013 LQR control for a quadrotor using unit quaternions: Modeling and simulation. In *2013 International Conference on Electronics, Communications and Computing (CONIELECOMP)* (pp. 172–178). http://doi.org/10.1109/CONIELECOMP.

[5] Mian A. A and Daobo, W 2008 Nonlinear Flight Control Strategy for an Underactuated Quadrotor Aerial Robot. In *IEEE International Conference on Networking, Sensing and Contro.* pp. 938–942. http://doi.org/10.1109/ICNSC.

[6] Slotine J.-J.E and Li W 1991 *Applied nonlinear control* Prentice Hall.

[7] Khalil H 2002 *Nonlinear Systems* Prentice Hall;

[8] Utkin V.I 1992 *Sliding modes in control and optimization* Springer-Verlag.

[9] Morari M and Zafriou E 1989 *Robust process control* Prentice Hall.

[10] Economou C.G, Morari M 1986 Internal model control: extension to nonlinear systems *Ind. Eng. Chem. Process Des. Dev.* pp. 403-411.

[11] Nahas E. P, Henson M. A, and Seborg D. E 1992 Nonlinear internal model control strategy for neural network models *Computers & Chemical Engineering*, vol. 16, no. 12, pp. 1039–1057.

[12] Bouzid Y, Siguerdidjane H and Bestaoui Y 2016 Improved 3D trajectory tracking by Nonlinear Internal Model-Feedback linearization control strategy for autonomous systems 6th IFAC Symposium on System Structure and Control, Istanbul.

[13] Aboudonia A, Rashad R. and El-Badawy, A 2015 Time domain disturbance observer based control of a quadrotor unmanned aerial vehicle In *2015 XXV International Conference on Information, Communication and Automation Technologies (ICAT)* (pp. 1–6).

Modeling and comparative study of linear and nonlinear controllers for rotary inverted pendulum

Byron Lima[1], Ricardo Cajo[2], Víctor Huilcapi[1], Wilton Agila[2]

[1]Departamento de Ingeniería Electrónica, Universidad Politécnica Salesiana, Guayaquil, Ecuador, Robles 107 y Chambers, Guayaquil, Ecuador.
[2]Facultad de Ingeniería en Electricidad y Computación, Escuela Superior Politécnica del Litoral, ESPOL, Campus Gustavo Galindo Km 30.5 Vía Perimetral, P.O. Box 09-01-5863, Guayaquil, Ecuador.

Email: [1]{blimac, vhuilcapi}@ups.edu.ec, [2]{rcajo,wagila}@espol.edu.ec

Abstract. The rotary inverted pendulum (RIP) is a problem difficult to control, several studies have been conducted where different control techniques have been applied. Literature reports that, although problem is nonlinear, classical PID controllers presents appropriate performances when applied to the system. In this paper, a comparative study of the performances of linear and nonlinear PID structures is carried out. The control algorithms are evaluated in the RIP system, using indices of performance and power consumption, which allow the categorization of control strategies according to their performance. This article also presents the modeling system, which has been estimated some of the parameters involved in the RIP system, using computer-aided design tools (CAD) and experimental methods or techniques proposed by several authors attended. The results indicate a better performance of the nonlinear controller with an increase in the robustness and faster response than the linear controller.

1. Introduction

The rotary inverted pendulum is a classic problem in the control systems area, which allows to evaluate performance and demonstrate the effectiveness of control techniques. The RIP system is a simple structure with not minimum phase multivariable unstable and highly nonlinear characteristics. It has a pendulum arm attached to a rotary instead of a moving car [1]. Several works have presented some methods to modeling dynamic systems such as: the used Euler-Lagrange in [2-6] with good result, so it is used to model the dynamics of rotary inverted pendulum shown in [7]. Moreover, the mathematical modeling used with the iterative method of nonlinear least squares that uses the Trust-Region-Reflective Algorithm, which tries to minimize the error between the actual and simulated signals [8].

Research community has studied and proposed diverse improvements to the classical PID approach in RIP systems: linear PID with a direct control of each system variable [9], Fuzzy Logic focused on a fuzzy controller [10], Optimum state regulator [11], and genetic algorithm based controller (GA) [12]. All the proposed controllers referred in the articles in this paragraph aim to identify, optimize and stabilize systems in general.

The research developed in this paper proposes to apply some methods, techniques and the use of CAD tools for estimating parameters that are not easy to measure and study because of the performance of RIP system by using a nonlinear control technique (nonlinear PID) presented in [13]. This nonlinear version is based on the classical method of Ziegler and Nichols tuning in the frequency domain.

This article is organized as follows: Section 1 presents the introduction, Section 2 describes the system used. Mathematical modelling for RIP system shown in section 3, section 4 estimating some parameters of the RIP system is performed. The validation of the model with the estimated parameters is shown in Section 5.Section 6 shows the design of linear and nonlinear PID controllers. Section 7 shows the results and performance of the algorithms developed control. Finally, some conclusions are summarized in Section 8.

2. Description of the System

The RIP system is controlled from a PC, therefore the system contains hardware and software that can be easily installed in a laboratory. Regarding the hardware, the system consists of panel start and stop emergency button, power supply, quadrature signal decoder with Arduino Mega, engine driver Sabertooth Dual 2x32A / 6V-24V and interface to a PC through the card PCI-1711 and CB-68LP connector block as shown in Figure 1.

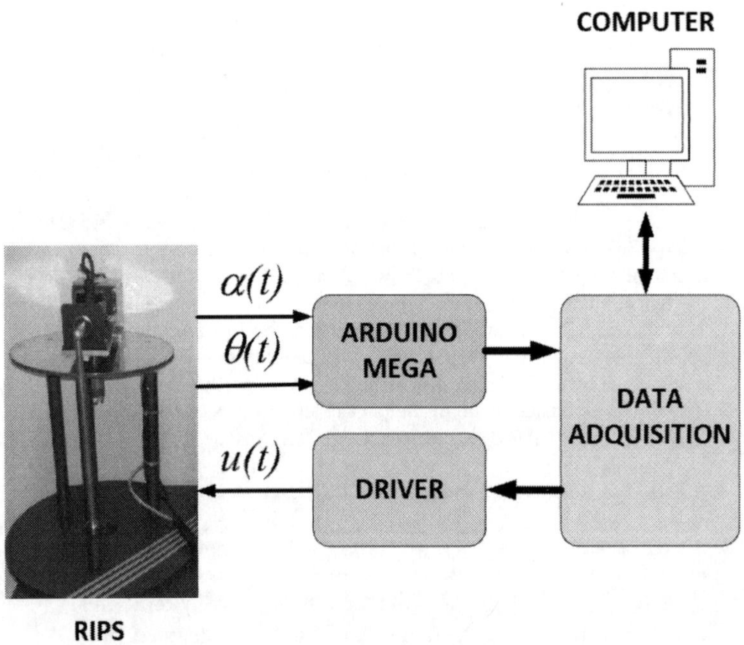

Figure 1. Schematic architecture of the RIP system

In order to measure the angular position of the pendulum and arm (α, θ) is used an incremental encoder model E50S8-600-3-type T-24, which measures with high precision (600 pulses per revolution), and the angle and direction of rotation through phases A and B

3. Mathematical model

The RIP system consists of a cylindrical rod with freedom to oscillate about a fixed pivot, which is mounted on an arm following an angular movement as shown in Figure 2.

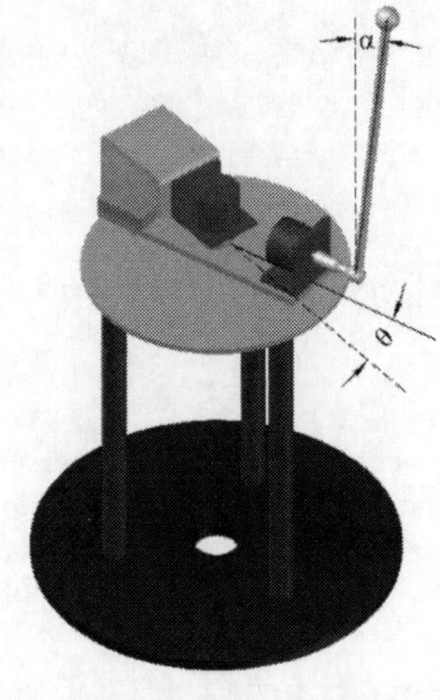

Figure 2. The rotary inverted pendulum system

The mathematical model for RIP system used in the present study is based in the work developed in [8], which representation is given by:

$$\begin{bmatrix} \left(J_P+m_pL^2\right) & -\left(m_prL\cos\alpha\right) \\ -\left(m_prL\cos\alpha\right) & \left(J_r+m_pr^2\right) \end{bmatrix}\begin{bmatrix}\ddot{\alpha}\\ \ddot{\theta}\end{bmatrix}=\begin{bmatrix} m_pgL\sin\alpha-B_p\dot{\alpha} \\ \frac{k_t}{R_a}V_a-\left(\frac{k_tk_b}{R_a}+B_r\right)\dot{\theta}-\left(m_prL\sin\alpha\right)\left(\dot{\alpha}\right)^2 \end{bmatrix} \tag{1}$$

The RIP system is described in space state as follows:

$$\begin{aligned} &\frac{d}{dt}x_1=x_2 \\ &\frac{d}{dt}x_2=\frac{k_tk_2k_{DG}\cos x_1}{R_a\left(k_1k_5-k_2k_4\cos^2 x_1\right)}u-\frac{k_2\cos x_1\left(k_bk_t+R_aB_r\right)}{R_a\left(k_1k_5-k_2k_4\cos^2 x_1\right)}x_4-\dots \\ &\frac{k_2k_6(\sin x_1)(\cos x_1)x_2{}^2}{k_1k_5-k_1k_4\cos^2 x_1}-\frac{k_5\left(B_px_2-k_3\sin x_1\right)}{k_1k_5-k_2k_4\cos^2 x_1} \\ &\frac{d}{dt}x_3=x_4 \\ &\frac{d}{dt}x_4=\frac{B_px_2-k_3\sin x_1+k_1\left(\frac{d}{dt}x_2\right)}{k_2\cos x_1} \end{aligned} \tag{2}$$

The output is given by:

$$y=[x_1\ \ x_3]^T \tag{3}$$

where,

α: Pendulum position, $\dot{\alpha}$: Pendulum velocity, θ: Arm position, $\dot{\theta}$: Arm velocity, u: Control signal to the motor

Each physical signal $[\alpha\ \dot{\alpha}\ \theta\ \dot{\theta}]$ in state-space RIP system model is represented by: $[x_1\ x_2\ x_3\ x_4]$ respectively. Some intermediate model variables and parameters area used:

$$k_1 = J_P + m_p L^2,\ k_2 = m_p rL,\ k_3 = m_p gL,\ k_4 = m_p rL,\ k_5 = J_r + J_m + m_p r^2,\ k_6 = m_p rL,\ V_a = k_{DG} u \quad (4)$$

4. Parameter estimation

In this section the parameters estimation of the RIP is performed that is the reason why the system is analysed by parts, engine, driver, pendulum, and arm.

4.1. Parameter estimation for the engine

To measure the motor parameters proceeds to undock the pendulum and analyse it independently. It starts with the estimation of the resistance and inductance of the motor armature, for which its terminals connected to NI ELVIS + National Instruments and at each angular position of the rotor several samples for resistance and inductance is taken. With these data it is possible to calculate an average whereby the values of R_a and L_a is obtained.

Then, proceed to estimate the Back-emf constant (k_b), constant electromagnetic torque (k_t), moment of inertia of the rotor of the motor (J_m) and viscous friction coefficient of the motor (B_r), for which the mechanical coupling between the motor shafts and encoder is performed. Similarly to the previous estimate the NI ELVIS II + module from National Instruments inductance and incremental encoder is used to measure angular velocity (ω) and the armature current (i_a) for various input voltages to the motor (V_a), using equation (5) Back-emf constant is estimated.

$$k_b = \frac{V_a - R_a i_a}{\omega} \quad (5)$$

In electrical machines must be a balance between electric power and mechanical power is obtained that $k_t = k_b$. To estimate the viscous friction coefficient of the engine (B_r) is necessary to use the differential equation unloaded motor ($T_L = 0$):

$$T_m = J_m \frac{d\omega}{dt} + B\omega \quad (6)$$

where $T_m = k_b i_a$ represents the engine torque. We proceed to energize the motor with a constant voltage for a time, until it reaches a constant speed, the angular acceleration disappears directly obtaining the ratio of viscous friction coefficient, by eliminating the term of the derivative of angular velocity and clearing, the viscous friction coefficient is obtained:

$$B = \frac{k_b i_a}{\omega} \quad (7)$$

Finally, it is estimated the moment of inertia of the rotor of the motor (J_m), for which the equation (8) is used.

$$\tau = \frac{J_m}{B} \quad (8)$$

Where the constant (τ), is known as the constant mechanical engine, which is measured experimentally, as a third of the time between switching off the motor without load at a constant speed and braking.

4.2. Estimating gain driver

To estimate the gain driver it must be found the relationship between the input voltage and output generated by the engine driver Sabertooth Dual 2x32A / 6V-24V. The different values of voltage applied to the input and the output are measured, the slope of the line obtained proves to be K_{DG}, as shown in Figure 3.

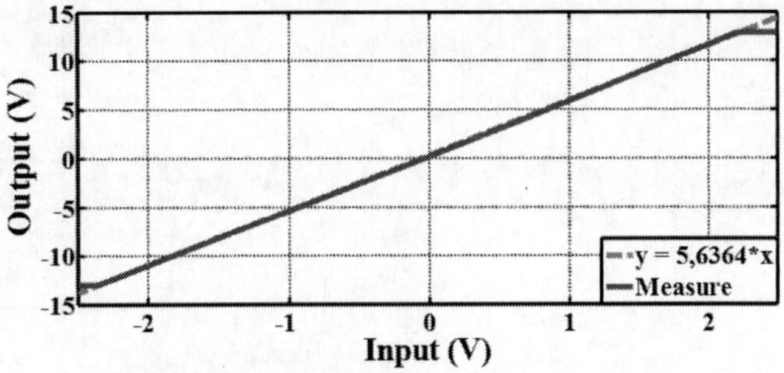

Figure 3. Relationship input - output driver

4.3. Parameter estimation of the pendulum and arm

The Pendulum mass (m_p) is measured by an electronic precision balance, turn this value that is obtained from the CAD compared. The estimate of the length of rotation axis to the center of mass of the pendulum (L) and the moment of inertia of the pendulum (J_p) is also performed in CAD, for which the RIP is designed in the SolidsWorks software. For the coefficient of viscous friction in the pendulum, we proceed to measure the pendulum swing during free fall from a small height, as shown in Figure.4.

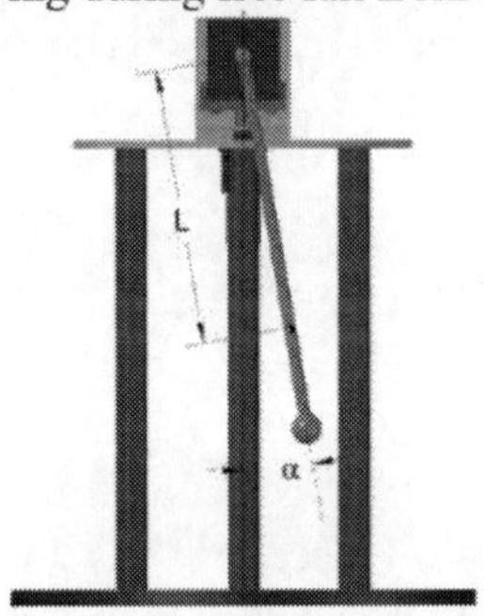

Figure 4. Setting the pendulum damped oscillation

The differential equation describing the oscillatory motion of the pendulum damped when it should fall from a small height, it is given by:

$$J_p \frac{d^2\alpha}{dt^2} + B_p \frac{d\alpha}{dt} + m_p g L \alpha = 0 \tag{9}$$

Where $B_p \frac{d\alpha}{dt}$, represents the damping produced by the friction between the pendulum and air.

The solution to this equation is shown in (10):

$$\alpha = \alpha_0 e^{-B_p t / 2J_p} \cos(\omega_0 t + \varphi) \tag{10}$$

The term $\gamma = -B_p t / 2J_p$, corresponds to the damping constant of the wobble signal of the pendulum. For this constant experimentally the pendulum is positioned with a small initial angle and drops it until it stops. The response of the pendulum and its signal envelope can be approximated as shown in Figure.5.

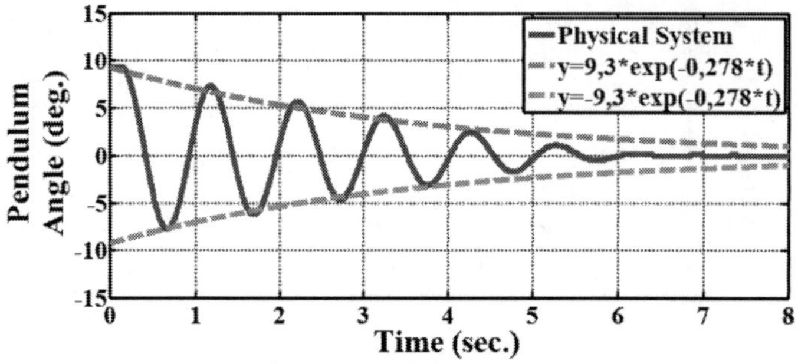

Figure 5. Response damped oscillatory motion of the pendulum

Finally, the estimation of the Rotating arm length (r) and the inertia moment arm Equivalent (J_r) is obtained from the RIP design CAD system made previously.

5. Validation of the model

The estimated parameters for the RIP system experimentally, are shown in Table 1.

Table 1. RIP system parameters.

Parameters	Value
R_a - Armature resistance	5.4351 Ω
L_a - Armature inductance	0.00188 mH
k_b - Back-emf constant	1.1098V.s/rad
k_t - Constant electromagnetic torque	1.1098 N.m/A
k_{DG} - Gain driver	5.6354
J_m -Moment of inertia of the rotor of the motor	0.00071 kg.m^2
B_r - Viscous friction coefficient of the motor	0.04166 N.m.s/rad
g - Gravity acceleration	9.8 m/seg^2
m_p -Pendulum mass	0.13417 kg
J_r - Equivalent inertia moment arm	0.0078 kg.m^2
r - Rotating arm length	0.125 m
L - Length of rotation axis to the center of mass of the pendulum	0.19604 m
J_p - Moment of inertia of the pendulum	0.001882 kg.m^2
B_p - Viscous friction coefficient of the pendulum	4.56798.10^{-4} N.m.s/rad

The bound for control signal is set to -2.5V…+2.5V. These parameters are used in the evaluation of engine model and RIP system using MATLAB, this is discussed in section 5.1 and 5.2.

5.1. Evaluation of nonlinear engine model

To validate the nonlinear engine model, is used a test signal square wave with the purpose of analyzing the dynamics of the real system and the nonlinear model.

Note that the square wave input can not be applied directly to the engine, so the engine driver Dual Sabertooth 2x32A / 6V-24V with gain K_{DG} , which manages the power consumption of the motor is used. The applied input terminals of driver and engine angular position, is shown in Figure 6. The average angular position error between the nonlinear model and the actual system is -0.0023 (deg).

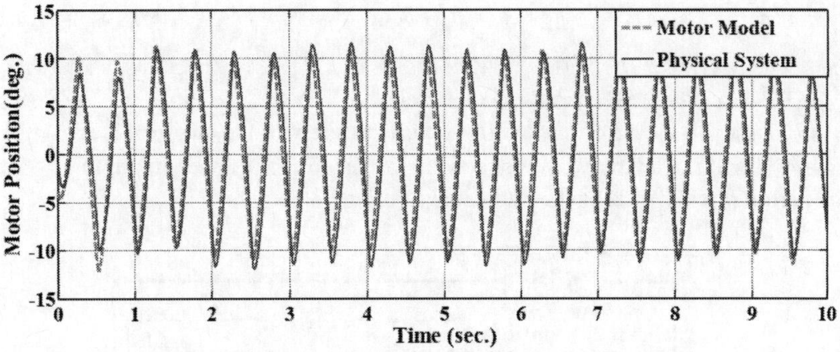

Figure 6. Response engine model to a square entry

5.2. Evaluation of nonlinear system model RIP

To validate the nonlinear model of the RIP system proceeds to generate different test signals in order to validate the estimated model. In Figure 7 is shown, the response to an input made by combining sinusoids for the position of the pendulum and the arm, which an error average angular position is obtained 1.1981 (deg) for the pendulum and -0.4431 (deg) for the arm.

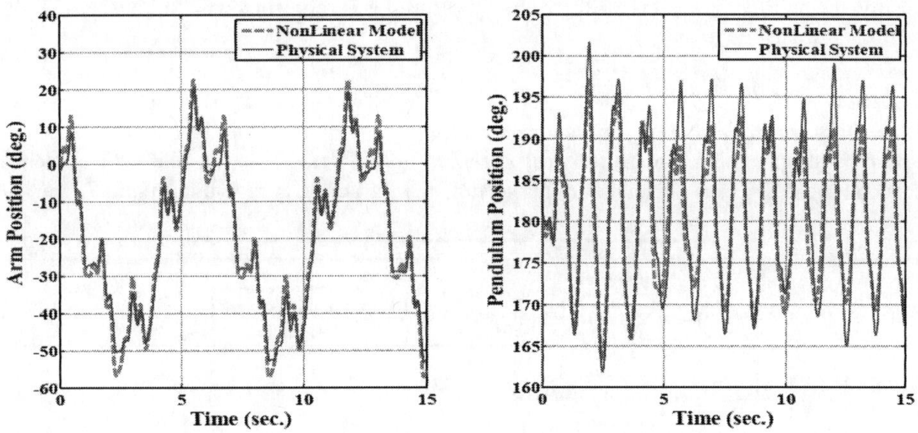

Figure 7. Response nonlinear model and real RIP system to an input combination of sinusoids.

6. PID and PI Control Algorithms Design

In this section, the development of the PID and PI controllers is carried out, the design of controllers involves the manipulation of a linear model. Thus, nonlinear RIPS model equation (1) is linearized around an equilibrium point given by the equation (11).

$$\alpha \approx 0 \ \left(\dot{\alpha}\right)^2 \approx 0 \ \cos\alpha \approx 1 \ \sin\alpha \approx \alpha \tag{11}$$

Linear models are represented trough transfer functions. Transfer functions of the pendulum and arm are indicated as follows in the equations (12) and (13) respectively:

$$G_P(s) = \frac{d_5 s}{d_1 s^3 + (d_2 + d_6)s^2 + (d_3 + d_7)s + d_4 + d_8)} \tag{12}$$

$$G_A(s) = \frac{d_5(k_1 s^2 + B_p s - k_3)}{ps^4 + qs^3 + k_2(d_3 + d_7)s^2 + k_2(d_4 + d_8)s}$$
$$: d_1 = R_a(k_1 k_5 - k_4 k_2);\ d_2 = R_a(k_5 B_p + B_r k_1);\ d_3 = R_a(-k_5 k_3 + B_r B_p);\ d_4 = -R_a B_r k_3 \tag{13}$$
$$d_5 = k_t k_2 k_{DG};\ d_6 = k_b k_1 k_t;\ d_7 = k_b B_p k_t;\ d_8 = -k_b k_3 k_t;\ p = k_2 d_1;\ q = k_2 d_2 + k_2 d_6$$

6.1. Linear PID controller for the pendulum position.

The control structure in the case of linear PID is quite different to the classical feedback loop, this, since the control target in our study case is the vertical position of the pendulum in $\alpha = 0$ degrees, rejecting the perturbation effects as it is indicated in Figure 8.

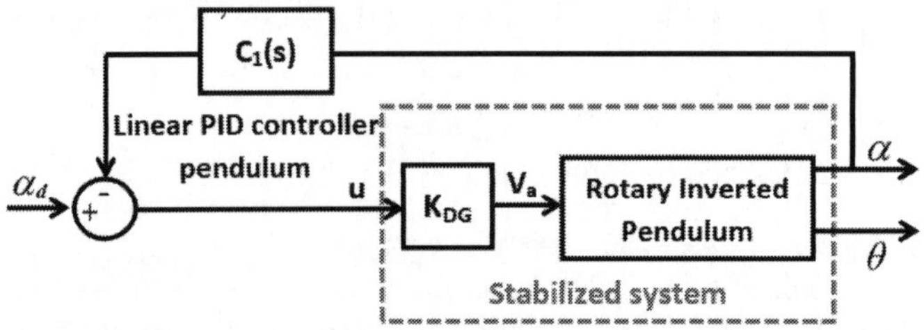

Figure 8. Control structure for the balance of the pendulum

The controller parameters of the PID are tuned by using SISOTOOL, which is a Matlab package. Specifically, the method utilized is LGR, the tuned controller is represented as follows: (14).

$$C_1(s) = \frac{k_{d1}s^2 + k_{p1}s + k_{i1}}{s} \quad : k_{d1} = 0.6247;\ \ k_{p1} = 25.8634;\ \ k_{i1} = 244.4404 \tag{14}$$

6.2. Linear PID controller design for the arm position.

The design of the pendulum controller was presented in the former section, the development of the complete control structure is presented in this section. The structure is shown in Figure 9.

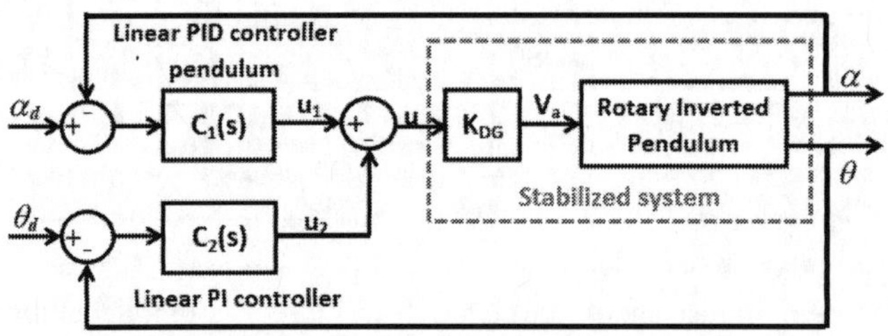

Figure 9. Control structure for the arm position

In the design of the linear PI arm controller, it is important to take into account that the value of α_d should stabilize around zero degrees, taking into account the latter, the block diagram shown in Figure 9, is reduced by using blocks algebra, the reduced block diagram leads to the following transfer function:

$$G_{2dof}(s) = -\left(\frac{k_{DG}}{1 + G_P(s)C_1(s)k_{DG}}\right)G_A(s) \tag{15}$$

As in the tunning of the former controller, the arm controller is tuned by using SISOTOOL and the LGR method. The resultant controller is given by equation (16).

$$C_2(s) = \frac{k_{p2}s + k_{i2}}{s} \quad : k_{p2} = 5.7951;\ \ k_{i2} = 5.4358 \tag{16}$$

6.3. Nonlinear PI controller design for the arm position

In the design of the nonlinear PI controller, it is required to linearize the RIP model given by equation (2). First, it is important to define an equilibrium point. The corresponding equilibrium point will be represented as a parametrization with respect to the input of the system, equation (17) presents the equilibrium point.

$$u = U;\quad x_1 = X_1(U) = \tan^{-1}\left(-\frac{k_t k_2 k_{DG}}{k_5 k_3 R_a}U\right);\ x_2 = 0;\ x_4 = 0 \tag{17}$$

The linearized model, which is represented by a state space model is presented in equation (18).

$$A_P = \begin{bmatrix} 0 & 1 & 0 & 0 \\ a & b & 0 & c \\ 0 & 0 & 0 & 1 \\ d & e & 0 & f \end{bmatrix},\ B_P = \begin{bmatrix} 0 \\ h \\ 0 \\ i \end{bmatrix},\ C_P = \begin{bmatrix} 1 & 0 & 0 & 0 \\ 0 & 0 & 1 & 0 \end{bmatrix},\ D_P = 0;\ a = \frac{(a_3)^{-1/2}}{R_a a_2}\left[R_a k_3 k_5 + \frac{a_4}{R_a k_3 k_5}\right]$$

$$b = -\frac{B_p k_5}{a_2},\ c = -\frac{k_2 a_1}{R_a a_2}(a_3)^{-1/2},\ h = \frac{k_2 k_t k_{DG}}{R_a b_2}(a_3)^{-1/2},\ i = \frac{k_1 k_t k_{DG}}{R_a b_1},\ f = -\frac{k_1 a_1}{R_a a_2}$$

$$d = \frac{\dfrac{R_a k_2 k_3 k_4^2}{a_3} + R_a k_1 k_3 k_4 k_5 - \dfrac{2U^2 k_1 k_2^2 k_4 k_{DG}^2 k_t^2}{R_a k_3 k_5 a_3} - \dfrac{2R_a k_1 k_3 k_4 k_5}{a_3}}{R_a k_1^2 k_5^2 + R_a k_2^2 k_4^2 (a_3)^{-2} - 2R_a k_1 k_2 k_4 k_5 (a_3)^{-1}},\ e = -\frac{B_p k_4}{a_2}(a_3)^{-1/2} \tag{18}$$

$$a_1 = k_b k_t + B_r R_a,\qquad a_2 = k_1 k_5 - \frac{k_2 k_4}{a_3},\qquad a_3 = \left(\frac{a_4}{R_a^2 k_3^2 k_5^2} + 1\right)$$

$$a_4 = U^2 k_2^2 k_t^2 k_{DG}^2\qquad b_1 = k_1 k_5 - \frac{k_2 k_4}{b_2},\qquad b_2 = \left(\frac{U^2 k_2^2 k_t^2 k_{DG}^2}{R_a^2 k_3^2 k_5^2} + 1\right)$$

In control theory, an alternative way of systems representation is the transfer function. Thus, equation (14) is the transfer function obtained after linearization and parametrization of the nonlinear model.

$$G_{P(U)}(s) = \frac{hs + ci - fh}{s^3 - (b+f)s^2 + (bf - a - ce)s + af - cd},\quad G_{A(U)}(s) = \frac{is^2 + (eh - bi)s + dh - ai}{s(s^3 - (b+f)s^2 + (bf - a - ce)s + af - cd)} \tag{19}$$

It should be noticed that the transfer function presented in equation (19) are parametrized transfer functions with respect to the input U. As it is indicated in section 6.2, the control scheme is the same as the one presented in Figure 9 with the difference in the transfer function $C2(s)$ which in the present study corresponds to a nonlinear PI. With the former considerations, a reduction of the system is possible. In equation (20), a reduced version of the transfer functions is presented.

$$G_{2dof(U)}(s) = -\left(\frac{G_{A(U)}(s)}{1 + G_{P(U)}(s)C_1(s)}\right)\ :\ c_1 = i;\ c_2 = eh - bi;\ c_3 = dh - ai;\ c_4 = hk_{d1} - (b+f)$$
$$c_5 = bf - a - ce + hk_{p1} + k_{d1}(ci - fh);\quad c_6 = af - cd + hk_{i1} + k_{p1}(ci - fh);\quad c_7 = k_{i1}(ci - fh) \tag{20}$$

By using Ziegler-Nichols method, the ultimate response in frequency domain gives important information in order to obtain optimal controller parameters. Specifically, it is possible to obtain the ultimate period $P_0(\mathrm{U})$ and the ultimate gain $K_0(\mathrm{U})$. In the approach presented in this study, the ultimate gain and period are obtained based on the transfer function $G_{2dof(U)}(s)$ as follows:

IOP Conf. Series: Journal of Physics: Conf. Series **783** (2017) 012047 doi:10.1088/1742-6596/783/1/012047

$$P_0(U) = 2\pi\left(\omega_0(U)\right)^{-1};\;\; K_0(U) = \left|G_{2dof(U)}(j\omega_0)\right|^{-1};\;\; \omega_0(U) = \left(\left(c_2c_5 - c_1c_6 - c_3c_4 - N\right)\left(2(c_2 - c_1c_4)\right)^{-1}\right)^{1/2}$$
$$N = \left((-c_2c_5 + c_1c_6 + c_3c_4)^2 - 4\left(c_2 - c_1c_4\right)\left(c_2c_7 - c_3c_6\right)\right)^{1/2};\; K_0(U) = \left|\left(\frac{(\omega^4 - c_5\omega^2 + c_7)^2 + (c_6\omega - c_4\omega^3)^2}{(c_3 - c_1\omega^2)^2 + (c_2\omega)^2}\right)^{1/2}\right| \quad (21)$$

In the nonlinear PI controller approach considered in this study, the nonlinear controller is generalized as follows:

$$\dot{z}(t) = 0.54\frac{K_0(z_v(t))}{P_0(z_v(t))}(\theta_d - x_3(t));\quad u(t) = z(t) + 0.45K_0(z(t))(\theta_d - x_3(t)) \quad (22)$$

Where $z(t)$ is defined as an additional state of the nonlinear controller of the arm. The control signal $u(t)$ is associated to the $x_3(t)$, which represents in the RIP system, the current arm position.

7. Performance Assessment

The values of the identified parameters are used for modeling purposes and the evaluation of the PI and PID controllers using MATLAB / Simulink, which is discussed in section 7.1 to 7.3.

7.1. Linear PID and PI controller evaluation

In Figure 10 the responses of linear controllers are shown for the pendulum and arm respectively. For which different references are used as input, for a square wave arm ± 10 degrees, while for the pendulum input 0 degrees is used. The Stabilization times obtained for linear controllers are: stabilization time for the pendulum $t_{SP} = 0.2\,\text{sec}$ and $t_{SA} = 3.25\,\text{sec}$ arm.

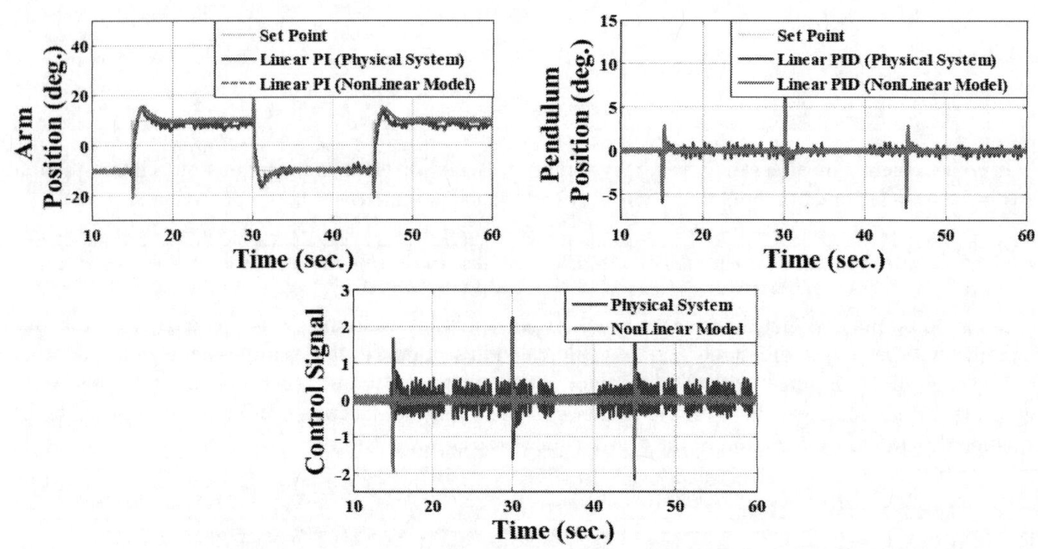

Figure 10. Responses of linear controllers for pendulum and arm position

7.2. Evaluation of the linear PID controller and nonlinear PI controller

The input signals used in section 7.1 are again used in this section in order to evaluate the performance of the linear and nonlinear controller. The Figure 11 shows the responses of PID controllers linear and nonlinear PI with the respective control efforts. The Stabilization times obtained for the linear and nonlinear controller respectively are: stabilization time for the pendulum $t_{SP} = 0.2\,\text{sec}$ and $t_{SA} = 2.5\,\text{sec}$ arm.

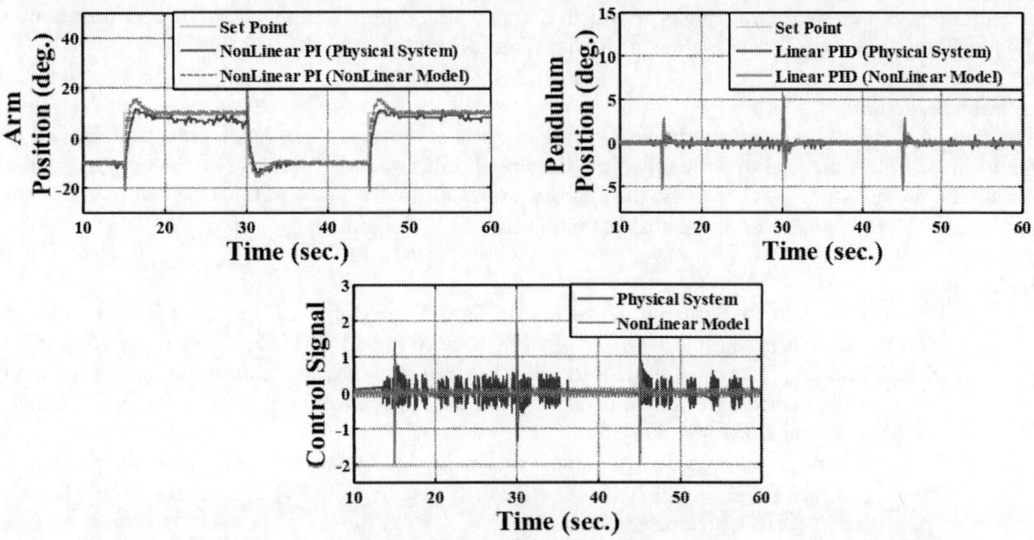

Figure 11. Response of linear and nonlinear controller for pendulum and arm position

7.3. Performance Evaluation of Linear vs nonlinear controllers in the RIP system

This section shows the results of the comparison between the linear PID controller and PI nonlinear. To carry out this performance indices such as are used: IAE (absolute integral error) and ISU (integral square control signal deviation), which will allow us to measure the cumulative error that penalizes the monitoring process and the energy used by the controller respectively. These indices are given by equation (23).

$$IAE = J_e = \sum_{k=1}^{N} |r_k - y_k|, \qquad ISU = J_u = \sum_{k=0}^{N} (u(k) - u_{ss})^2 \tag{23}$$

The numerical values of IAE and ISU for both drivers applied in the RIP system are shown in Table 2.

Table 2. Indices performance obtained with linear and non-linear controller.

	Arm Position *IAE*	Control Signal *ISU*	Computing Time (sec)
Linear Controller	5.3218×10^4	1.0644×10^3	7.8215
Nonlinear Controller	4.8909×10^4	570.6625	29.1743

The values in Table 2 show that the nonlinear PI controller has better performance in energy used, as in following references for positioning the arm. In the case of positioning the pendulum a linear PID controller was designed. So when acting linear PI controller in the arm RIP system, the IAE linear PID controller pendulum is 6.1272×10^3, and when it acts nonlinear PI controller in the arm, the IAE linear PID controller pendulum is 2.9376×10^3, which shows an improvement in the performance of the linear PID controller for the pendulum, when the nonlinear PI acts on the arm.

8. Conclusions

In this study, modeling, control and estimation of certain parameters of the inverted pendulum rotary experimentally, which has a satisfactory response we are performed. This model was used to design the PI and PID controllers for linear and nonlinear system. Of the assessment controllers shown that the performance of nonlinear PI control both reference tracking, and energy consumption is better than the linear PI control. Also, when the nonlinear PI controller acts on the arm of the system, the linear PID controller designed to position the pendulum vertically improves its efficiency. Although the computational time for the nonlinear controller is greater than the linear, a better result in energy consumption and reference tracking is obtained. Future works aims to design a nonlinear for positioning the pendulum PID controller as well as using new control strategies for the RIP System.

Acknowledgments
The authors would like to acknowledge two important institutions in the development of this work: Universidad Politécnica Salesiana and Escuela Superior Politécnica del Litoral for all the support offered during the present study. Additionally, the authors are grateful with the GISCOR research who gave to the authors the facilities regarding the use of infrastructure and equipment.

References
[1] Sukontanakarn, Viroch; Parnichkun, Manukid; "Real-Time Optimal Control for Rotary Inverted Pendulum", American Journal of Applied Sciences 6 (6): 1106-1115, 2009 , ISSN 1546-9239
[2] J. Zhang and Y. Zhang, "Optimal linear modeling and its applications on swing-up and stabilization control for Rotary Inverted Pendulum," Control Conference (CCC), 2011 30th Chinese, Yantai, 2011, pp. 493-500.
[3] S. Jadlovská and J. Sarnovský, "A complex overview of the rotary single inverted pendulum system," ELEKTRO, 2012, Rajeck Teplice, 2012, pp. 305-310.doi: 10.1109/ELEKTRO.2012.6225609.
[4] P. Xue and W. Wei, "An Analysis on the Kinetic Model of a Rotary Inverted Pendulum, and Its Intelligent Control," Computational and Information Sciences (ICCIS), 2010 International Conference on, Chengdu, 2010, pp. 978-981
[5] K. Lai, J. Xiao, X. Hu, J. Fan and B. Wu, "Modeling and control for stability and rotation velocity of a rotary inverted pendulum," Industrial Electronics and Applications (ICIEA), 2015 IEEE 10th Conference on, Auckland, 2015, pp. 955-960.
[6] M. Roman, E. Bobasu and D. Sendrescu, "Modelling of the rotary inverted pendulum system," Automation, Quality and Testing, Robotics, 2008. AQTR 2008. IEEE International Conference on, Cluj-Napoca, 2008, pp. 141-146.
[7] Nguyen Duc Quyen; Ngo Van Thuyen; Nguyen Quang Hoc; Nguyen Duc Hien., "Rotary Inverted Pendulum and Control of Rotary Inverted Pendulum by Artificial Neural Network", Proc. Natl. Conf. Theor. Phys. 37 (2012), pp. 243-249.
[8] D. I. Barbosa, J. S. Castillo and L. F. Combita, "Rotary inverted pendulum with real time control," Robotics Symposium, 2011 IEEE IX Latin American and IEEE Colombian Conference on Automatic Control and Industry Applications (LARC), Bogota, 2011, pp. 1-6.
[9] Rahimi, A., Raahemifar, K., Kumar, K., Alighanbari, H. (2013). Controller design for rotary inverted pendulum system using particle swarm optimization algorithm. In Electrical and Computer Engineering (CCECE), vol., no., pp.1-5.
[10] Yunhai Hou., Hongwei Zhang., Kai Mei. (2011). Vertical-rotary inverted pendulum system based on fuzzy control. In Electrical and Control Engineering (ICECE), vol., no., pp.2804-2807.
[11] Barbosa, D.I., Castillo, J.S., Combita, L.F. (2011). Rotary inverted pendulum with real time control. In IEEE IX Latin American and IEEE Colombian, vol., no., pp.1-6.
[12] Hassanzadeh, I., Mobayen, S. (2008). GA based input-output feedback linearization controller for rotary inverted pendulum system. In Mechatronics and Its Applications, vol., no., pp.1-6.
[13] Cajo, R., Agila, W. (2015). Evaluation of Algorithms for Linear and Nonlinear PID Control for Twin Rotor MIMO System. In Computer Aided System Engineering (APCASE), vol., no., pp.214-219.

Challenges in Modelling and Control of Offshore De-oiling Hydrocyclone Systems

Petar Durdevic, Simon Pedersen and Zhenyu Yang
Aalborg University, Esbjerg Campus, Niels Bohrs Vej 8, 6700 Esbjerg, Denmark
E-mail: pdl@et.aau.dk, spe@et.aau.dk, yang@et.aau.dk

Abstract. Offshore de-oiling installations are facing an increasing challenge with regards to removing oil residuals from produced water prior to discharge into the ocean. The de-oiling of produced water is initially achieved in the primary separation processes using gravity-based multi-phase separators, which can effectively handle large amounts of oil-well fluids but may struggle with the efficient separation of small dispersed oil particles. Thereby hydrocyclone systems are commonly employed in the downstream Produced Water Treatment (PWT) process for further reducing the oil concentration in the produced water before it can be discharged into the ocean. The popularity of hydrocyclone technology in the offshore oil and gas industry is mainly due to its rugged design and low maintenance requirements. However, to operate and control this type of system in an efficient way is far less simple, and alternatively this task imposes a number of key control challenges. Specifically, there is much research to be performed in the direction of dynamic modelling and control of de-oiling hydrocyclone systems. The current solutions rely heavily on empirical trial-and-error approaches. This paper gives a brief review of current hydrocyclone control solutions and the remaining challenges and includes some of our recent work in this topic and ends with a motivation for future work.

1. Introduction

Many matured offshore oilfields have a high water content and the pumped well-fluids in some cases contain more than 90 % water, and this water is referred to as *produced water*. It has been surveyed that globally around 250 million barrels of produced water along with 80 million barrels of oil is produced each day [1]. This high water-cut situation requires effective Produced Water Treatment (PWT) to achieve pure oil product on the platforms, but also to fulfil governmental effluent discharge regulations. For instance, the current limitation for hydrocarbon discharge in North Sea is set at 30 mg/l (30ppm) [2]. It has been shown that the discharged hydrocarbons could have a negative effect on the surrounding marine life, for example a small concentration of Polycyclic Aromatic Hydrocarbons (PAH) as low as 100 parts per billion (PPB) can affect fish development [3]. Produced water can also contain different harmful materials, such as metals (barium and zinc), benzene, toulene, ethylbenzene xylene (BTEX), naphthalene, phenanthrene, dibenzothiophene (NPD), polyaromatic hydrocarbons (PAHs) and phenols etc. [4], and hence direct discharge of the produced water is strictly prohibited in order to protect the marine life and environment.

In offshore installations, on-site cleaning of produced water, which is referred to as PWT, is needed as it is expensive to send such large quantities of water to the onshore separation facilities [5]. However, due to the space and weight restrictions on offshore installations, the

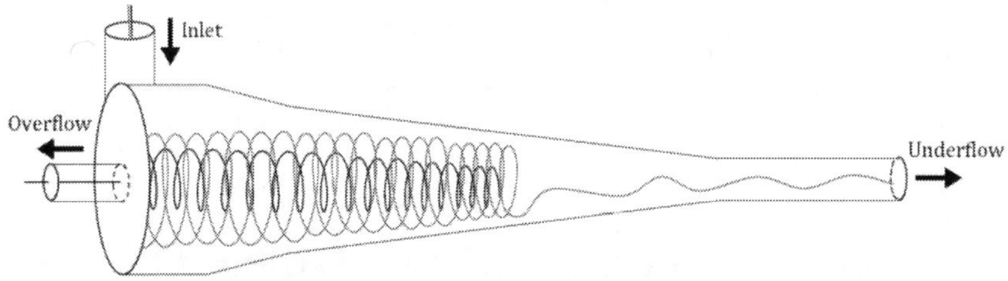

Figure 1: Hydrocyclone Separation principle - the grey/black line indicates the water/oil flow

installation and operation of offshore PWT must be very cost-effective while providing sufficient capability and quality. Moreover, due to the harsh weather and marine conditions, especially in the North Sea, the PWT requires robust installations and instrumentation. Even though almost all PWT of the current offshore installations perform to an acceptable level, the consistent increase in water-cut will sooner or later bring extra difficulties and challenges to the current PWT technologies and systems.

To achieve the required effluent discharge concentrations for oil in the produced water, the separation of water and hydrocarbon products is generally achieved in two stages. The initial separation processes uses gravity-based multi-phase separators, which can effectively handle large amounts of well-fluids, but cannot efficiently separate small dispersed oil droplets. The next stage in separation is by the hydrocyclones that reduce the dispersed oil content to the required levels before the effluent is discharged into the ocean.

It is common knowledge that the hydrocyclone's de-oiling performance is very sensitive to fluctuating inflow rates [6], [7]. The efficient operation and control of this type of system imposes a number of key control challenges. The current modelling and analysis heavily rely on CFD-based approaches, and the control development heavily relies on empirical trial-and-error approaches. We found that there is lack of research in cost-effective modelling and control of de-oiling hydrocyclone dynamic systems. In 2013 Aalborg University in collaboration with one of the Danish oil operators and an oil service company, launched a research project - PDPWAC - with one of the research objectives being the optimization of the hydrocyclone-based PWT using plant-wide control strategy. This paper gives a brief review of modelling and control of offshore de-oiling hydrocyclone systems, presents some of our work in this direction and points out some remaining challenges.

The rest of this paper is organized as follows: Section 2 introduces the basic configuration and operating principle of typical de-oiling hydrocyclone systems; section 3 gives a brief overview of hydrocyclone modelling methods and the challenges that lie within; section 4 presents some key challenges in the control of hydrocyclone systems, along with some latest solutions from our work; finally section 5 concludes this work.

2. Hydrocyclone Principle and Control

Hydrocyclone technology emerged in the oil industry in the 1980's and has been increasingly used since then. Hydrocyclone systems represent more than 90% of current de-oiling facilities in the offshore oil and gas installations [8], and this popularity is mainly due to its simplicity and ruggedness [9].

2.1. Hydrocyclone Configuration & Principle

A typical de-oiling hydrocyclone consists of one or more tangential inlets, where the produced water flowing out of the separator stack enters the hydrocyclone facilities. The mixed water and

IOP Conf. Series: Journal of Physics: Conf. Series **783** (2017) 012048 doi:10.1088/1742-6596/783/1/012048

oil are accelerated in a circular movement inside the cylindrical chamber as shown in Figure 1. As the mixture rotates, the centripetal force separates these two phases with the heavier phase (water) moving towards the walls while the lighter phase (dispersed oil) moves to the cylinder's centre, where eventually an oil core is generated. Besides the cylindrical chamber segment, a hydrocyclone also consists of two conical sections and two outlets, namely *overflow* and *underflow* outlets. Ideally, the oil trapped in the oil core will gradually exit through the overflow outlet, while the water will go through the underflow outlet [7]. In practical applications, especially for the offshore situations, a number of hydrocyclone liners are need to be stacked in parallel inside one holding vessel, so that the handling capability of produced water can be significantly increased without requiring a lot of installation space [9].

2.2. Hydrocyclone control

An offshore installation relevant to PWT is sketched in Figure 2, where the separator and the hydrocyclone are directly connected. Normally, the underflow control valve located at the hydrocyclone's underflow outlet is used for the purpose of separator water level control ("LC" in Figure 2). The hydrocyclone separation performance is controlled via the *PDR control* ("PDR" in Figure 2) loop by manipulating the overflow control valve located at the hydrocyclone's overflow outlet, where PDR is the Pressure Drop Ratio over the hydrocyclone's outlets and inlet, i.e.,

$$PDR = \frac{P_i - P_o}{P_i - P_u} \tag{1}$$

Where P_i is the measured inlet pressure, and P_u/P_o is the the underflow/overflow pressure measurement. It has been experimentally discovered that the PDR and the flow-split inside the hydrocyclone are closely linearly-dependent [10]. The flow-split directly determines the amounts of flow going through the underflow and the overflow outlets, respectively. To maintain a satisfactory separation efficiency, the flow-split is crucial for hydrocyclone's operation, as it will determine how much oil and water will be able to escape through the under- and over-flow respectively.

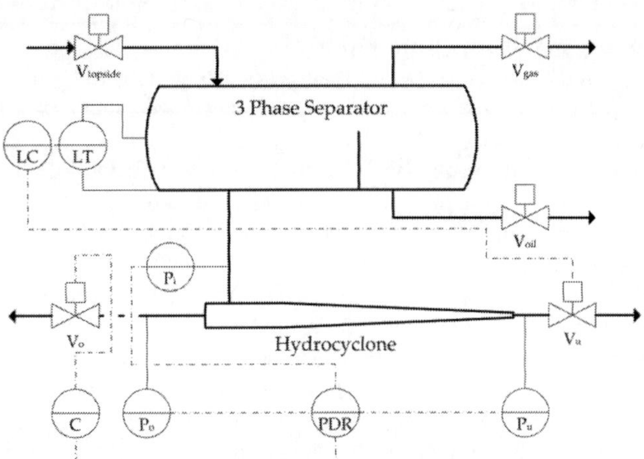

Figure 2: Two control loops in a typical control structure.

It can be observed that the dynamics of the hydrocyclone's PDR and the separator's level are physically coupled and thereby may affect each other's performance. Furthermore, the separator level control is often disturbed by the varying inflow rate to the separator system. A typical influence to the separator is slugging inflow which could be caused by the riser configuration or hydrodynamics [11], [12], [13] and [14]. A measured severe slug that occurred at one installation in North Sea is illustrated in Figure 3, where the severe slug is indicted by large oscillating pressure measurements at the riser top. All these issues lay out many challenges to the efficient and reliable control of the hydrocyclone .

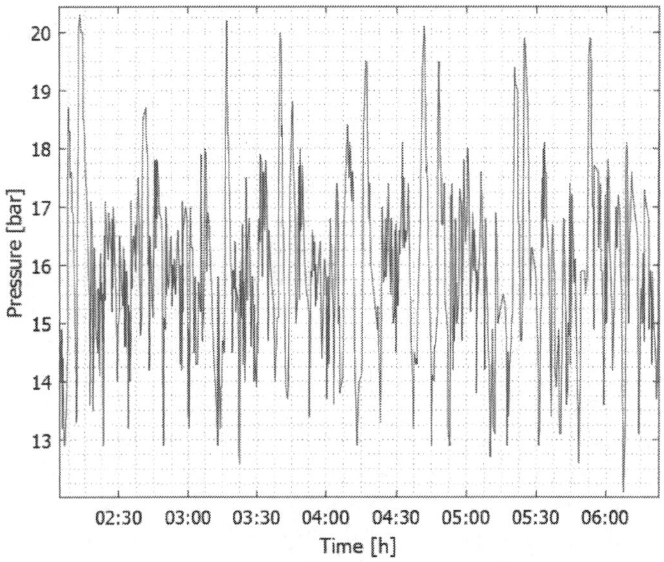

Figure 3: One example of riser-induced slugging flow indicated by topside pressure fluctuations

3. Modeling De-oiling Hydrocyclone Operation

It has been observed that there has been little work done focusing on the control-oriented modelling of de-oiling hydrocyclone systems, instead, there are plenty of different models and analysis based on Computational Fluid Dynamics (CFD) technology [15, 16, 17]. Although CFD-based models have a sophisticated capability and power in illustrating detailed information, they are not oriented or suited for the purpose of control design due to their complexity. For dynamic control purposes simple dynamic models, which can express the key dynamic characteristics between the system's inputs and outputs, are preferred instead.

The first-principle-based modelling is investigated in [18] for a solid-liquid hydrocyclone setup; and this work is further extended in [19]. [20] described a solid-liquid hydrocyclone with a simple dynamic model by using transport balance by considering the slip velocity and turbulent particle diffusivity, based on the $k-\eta$ model developed in [21] and [22]. However, the extension of these models to handling liquid-liquid de-oiling hydrocyclone is not clear. Similar situation exists regarding the work done in [20], which only focused on the solid-liquid hydrocyclone and also assumed that the inlet flow is fed through the entire top section of the hydrocyclone. As many of these relatively simple models (when compared to CFD-based models) are regarding solid-liquid separating cyclones, the correlation between solid-liquid cyclone separation and liquid-liquid hydrocyclone separation needs further investigations before we can answer whether some models developed for one type of cyclone system could be applied or extended to another type of cyclone system.

In our previous work a black-box modelling method was used for a de-oiling hydrocyclone setup in [23] by regarding the PDR as the controlled variable (output). The opening degrees of two controllable valves, i.e., the underflow valve V_u and overflow valve V_o, were regarded as the manipulated variables (controllable inputs). Based on experimental data, a set of linear models were identified by using PE system identification method. However, due to the heavy non-linearity of cyclone's separation dynamics, it was noticed that some nonlinear model(s) should be developed if a relatively large operating range and different operating conditions need to be considered. An extension of this work to employ the Hammerstein-Weiner nonlinear model is currently underway as part of our work, and some of the preliminary results have showed a huge potential for this type of model to explain the nonlinearity in the de-oiling hydrocyclone system.

4. Challenges in de-oiling hydrocyclone control

There is little literature to be found in the systematic design of de-oiling hydrocyclone control solutions, including the standardized PID-based PDR control described in section 2 where there is no systematic approach described to tune these controllers except extensive empirical tunings, though some control designs can be found for solid-liquid cyclone separation.

4.1. Solid-liquid hydrocyclone control

The volume-split regulation has been developed in [24] by controlling the overflow valve for solid-liquid cyclone systems, which is quite similar to the PDR control for de-oiling hydrocyclones. The measured signals are the inlet pressure, the overflow pressure and an underflow discharge pattern recognition sensor, the manipulated variables are the speed of the feeding pump and the opening degree of the overflow valve. The control systems balance the underflow output between the rope discharge and spray discharge, so that the clogging problem can be avoided at the bottom of the hydrocyclone [25], [24], [26]. However, any direct application of these solid-liquid control solutions for liquid-liquid hydrocyclone separation is still very open and challenging.

The control of the inflow rate is not possible in offshore de-oiling hydrocyclone situations, as the inflow rate is determined by the separator level control loop, as shown in figure 5. Furthermore, the usage of monitoring spray rope discharge pattern for control purposes is not possible for de-oiling hydrocyclones. Moreover, the de-oiling hydrocyclones are not subjected to clogging problem, instead, water has a low viscosity compared to the infinite viscosity of solid materials. A thorough investigation of the likeness between these two types of separations needs to be done before the control techniques used for solid-liquid cyclones can be applied for liquid-liquid de-oiling hydrocyclones.

4.2. Coordinated separator and cyclone controls

The (water) level controller is designed to keep the water level inside the separator at a certain level based on a pre-determined set-point. From a practical perspective, the water level set-point is not crucial as long as the water level can be maintained within a safety range. Maintaining the level within a specified range is important to ensure a correct residence time which is directly related to the separator's efficiency. According to [27] the normal residence time for oil production separators is about $2-4$ minutes. Residence time is an optimization or trade-off problem, as a longer residence time ensures a better separation, and a lower residence time ensures faster process flow. The problem with longer residence times is that if the mass flow rate is to be kept high the equipment will correspondingly grow in size. In our investigations we have discovered a coupling effect of V_u and V_o functionalities, [28]. However, the opening cross-section area of V_u is about 25 times larger than that of V_o, thereby V_u acts as the dominant influence. Figure 4 illustrates a scenario where PID controllers are applied on both the level and PDR control loops, and a severe oscillating inflow rate F_{in} was generated to emulate a severe

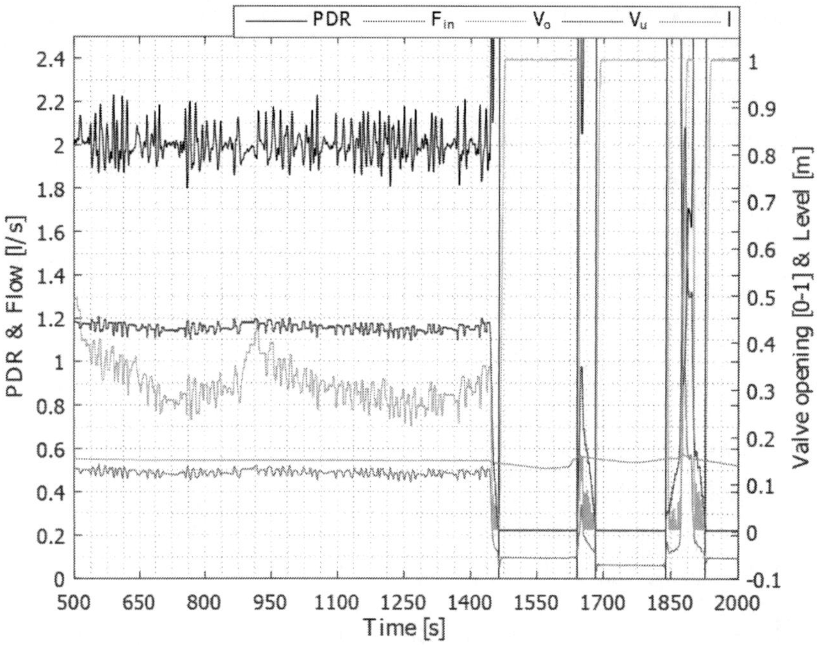

Figure 4: Test illustrating the coupling phenomena in the separation system.

condition. This test illustrates the impact of a sudden reduction in F_{in} (which happens at 1420s) on the controlled level and PDR performances. The halt of F_{in} caused the level in the separator to quickly decrease, which directly resulted in shutting of V_u due to the level controller. This action consequently increased the pressure over V_u, and then the PDR controller was forced to open V_o further. This scenario results in an unnecessary change of system efficiency ϵ, where $\epsilon = 1 - \frac{C_u}{C_i}$ and C_i & C_u are the inlet and outlet oil concentrations respectively. In addition, the dominance of V_u is evident from $500 - 1420$s, where small changes in V_u are equal to relatively larger fluctuations in V_o.

A block diagram of the combined level control and hydrocyclone PDR control is illustrated in Figure 5, the coupling effect is illustrated by the dotted lines. The current control consists in most cases of individually tuned and implemented PID controller on the level and the PDR, respectively. But a solution could lie in extending this with a controller structure which takes both the objectives into consideration. This control design problem can be formulated as a typical MIMO control problem, as long as we have the dynamic models describing the separator level dynamics as well as PDR dynamics.

In this case a MIMO feedback solution could be proposed to introduce a systematic design paradigm to the system which will help avoid the struggle of the two individual systems, and instead link them together into a cooperative control scheme.

4.3. Cyclone performance's sensitivity

Even with a good control solution, the hydrocyclone performance can still be very sensitive to fluctuations of inflow rates, which could be caused by the upstream separation processes (e.g., three-phase separators). The flow rate is equally important to the hydrocyclone's separation performance. If the flow through the hydrocyclone is insufficient, the swirl motion inside the hydrocyclone will not be formed or the velocity will be insufficient [30]. If this occurs, there

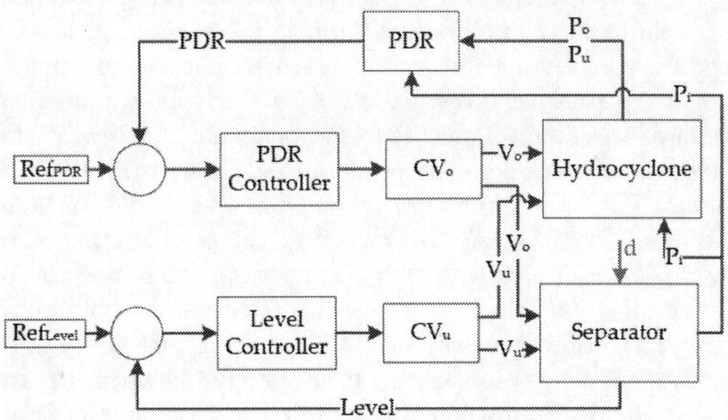

Figure 5: Control formulation of combined separator and cyclone controls [29]

will not be enough centripetal force to split the heavier liquid from the lighter one and push it towards the wall of the liner.

The plot in figure 6 shows a test performed on our pilot plant set-up, where F_{in} was increased from 0.21/s to 0.681/s, with 9 steps. During these increases ϵ was measured by measuring the Oil-in-Water (OiW) concentration in the inlet and the underflow of the hydrocyclone using the TD4100 equipment described in [31]. As F_{in} increases the ϵ increases, due to the fact that the centripetal force is being increased.

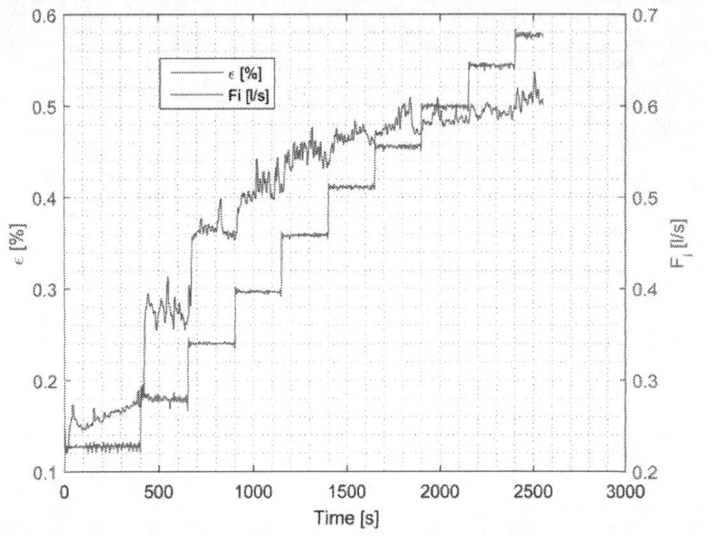

Figure 6: Dependency of ϵ on F_{in} at a controlled PDR

4.4. Efficiency real-time measurement

One of the important factors with regard to hydrocyclone operation is its efficiency measurement and prediction, which relates to the OiW concentration measurement. For the purpose of real-time applications, the OiW technology is not yet matured and the existing solutions are quite methodology dependent. Extensive investigations and development are needed in this area. A mathematical model of the liquid-liquid hydrocyclone's efficiency was introduced in [32] by modelling the dispersed droplet trajectories. This model can be very efficient in predicting the d_{100} value for a Coleman Thew type of hydrocyclone, however, it did not take into account the coalescence and breakup of droplets, and this can limit the prediction accuracy due to the fact that some high inflow rates often create a high shear stress which can break up droplets. d_{100} classifies the smallest droplet size which can be separated with an 100% efficiency, equally d_{50} would correspond to a separation efficiency of 50% [32]. One possibility is to measure droplet sizes at different stages in the separation system to investigate how the separation is affected by different droplet sizes, and if any improvement can be done during different operating conditions.

[31] investigated different OiW measurement technologies in terms of their precision and real-time measurement capabilities. The oil droplet size measurement was done with high precision using a microscopy based measurement instrument. However, due to the specific measurement principle, consistent online OiW concentration measurements have not yet been achieved, which casts a doubt about the instrument's capability for reliable real-time OiW measurement. Still, this does enable for steady state analysis of system performance at different particle sizes, which was performed in [33], but it does not assist in the possible design of control based models. Alternative measurement equipment using the same measuring principle is presented in [34] and their results seem quite promising.

In addition [31] evaluated another OiW instrument, which measures the concentration of OiW based on fluorescence principle. This method yielded far better OiW concentration results and has also be able to provide data in real-time. One result is illustrated in Figure 7. In this experiment, mixtures with different OiW concentrations were injected into the view-cell of this instrument with 10s intervals. The right plot illustrates a zoomed-in view of a step from 5 to 10 PPM. However, due to the extreme low concentrations, some measurements drift slightly. Regarding the OiW concentration measurement, this measuring instrument yields us some promising results, but further research is still needed to evaluate the reliability and repeatability of this method if it is to be used for dynamic model development and system dynamic performance evaluation.

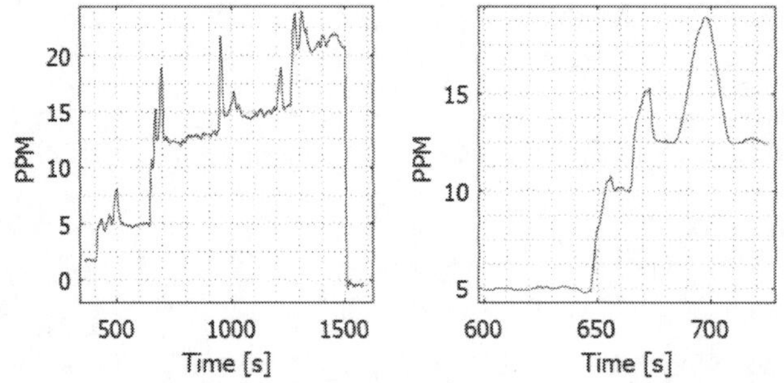

Figure 7: OiW measurement performed using a fluorescence based instrument Turner Design TD4100, [31]

5. Conclusions

The systematic solutions for de-oiling hydrocyclone control are still quite open with respect to its inherent complexity and heavy coupling with its upstream and downstream facilities. However, the real-time performance of the de-oiling hydrocyclone is crucial in determining the discharged water quality, and thereby the reduction of the environmental footprint caused by the offshore oil and gas production.

The current control solutions for hydrocyclones are PID-type solutions which lack systematic tuning strategies. Some advanced control solutions can be found in handling solid-liquid cyclone systems. However, the extension of these solutions to liquid-liquid de-oiling hydrocyclones is not straightforward. One of the key issues which blocks the application of advanced control methods in de-oiling hydrocyclone systems, lies in the lack of a deep understanding of the hydrocyclone's separation dynamics from the control point of view. This is reflected in reality, as there are no control-oriented modelling methods, nor models available for de-oiling hydrocyclone systems at this moment. Here CFD-based models and methods can play a very powerful role in emulating and analysing hydrocyclone separation processes, but they cannot yet be applied for control purposes. Instead a less detailed model strategy based on simple ODE models is preferred that describes the system from the control perspective, which is the PDR, with an extension of efficiency measurements.

For offshore de-oiling, the hydrocyclone performance is heavily coupled with the upstream separator's dynamics due to a lack of buffer vessels between them. Thereby, a coordination of the separator's (water) level control and hydrocyclone's PDR control is recommended. A MIMO control strategy can be applied to handle this control design problem, as long as some mathematical models of both parts are available. If the hydrocyclone's efficiency can be measured in a real-time and reliable manner, the hydrocyclone's PDR control strategy can be extended to be a direct-efficiency-based feedback solution. However, at this moment, the OiW real-time measurement technology, which fundamentally detects the hydrocyclone's efficiency, is still quite open and not yet matured.

Acknowledgments

The authors would like to thank the support from the Danish Innovation Foundation (via PDPWAC Project (J.nr. 95-2012-3)). Thanks also go to colleagues J.P. Stigkær, A. Aillos, K. G. Nielsen and T. I. Bruun from Mærsk Oil A/S, colleagues P. Sørensen, A. Andreasen and Jakob Biltoft from Ramboll Oil & Gas A/S, for many valuable discussions and supports. Thanks also go to AAU colleagues: C. Mai, K. Jepsen, L. Hansen, M. Bram and D. Hansen for their contributions to this project.

References

[1] Fakhru'l-Razi A, Pendashteh A, Abdullah L C, Biak D R A, Madaeni S S and Abidin Z Z 2009 *Journal of hazardous materials* **170** 530–551
[2] The-Danish-Environmental-Protection-Agency 2016 Status for den danske offshorehandlingsplan til udgangen af 2009 `http://www.mst.dk/` accessed: 2016-09-22
[3] Brette F, Machado B, Cros C, Incardona J P, Scholz N L and Block B A 2014 *Science* **343** 772–776
[4] Ekins P, Vanner R and Firebrace J 2007 *Journal of Cleaner Production* **15** 1302–1315 ISSN 09596526
[5] Bailey B, Crabtree M, Tyrie J, Elphick J, Kuchuk F, Romano C, Roodhart L *et al.* 2000 *Oilfield Review* **12** 30–51
[6] Thew M 1986 *The Chemical Engineer* 17–23
[7] Meldrum N 1988 *SPE Production Engineering* **3** 669–676
[8] Georgie W J 2002 *Offshore Technology Conference* 1–13
[9] Thew M 2000 *Encyclopaedia of Separation Science* 1480 – 1490
[10] Husveg T, Rambeau O, Drengstig T and Bilstad T 2007 *Minerals Engineering* **20** 368–379 ISSN 0892-6875
[11] Pedersen S, Durdevic P and Yang Z 2014 *Experimental Study of Stable Surfaces for Anti-Slug Control in Multi-phase Flow* (IEEE Press) pp 43–48 ISBN 978-1-909522-02-2

[12] Biltoft J, Hansen L, Pedersen S and Yang Z 2013 *IFAC Proceedings Volumes* **46** 47–52
[13] Jepsen K, Hansen L, Mai C and Yang Z 2013 Emulation and control of slugging flows in a gas-lifted offshore oil production well through a lab-sized facility *Control Applications (CCA), 2013 IEEE International Conference on* (IEEE) pp 906–911
[14] Havre K, Stornes K O and Stray H 2000 *ABB review* **4** 55–63
[15] Delgadillo J a and Rajamani R K 2005 *International Journal of Mineral Processing* **77** 217–230 ISSN 03017516
[16] Sayda A F and Taylor J H 2007 Modeling and control of three-phase gravilty separators in oil production facilities *2007 American Control Conference* (IEEE) pp 4847–4853
[17] Slack M, Del Porte S and Engelman M 2004 *Minerals Engineering* **17** 705–711 ISSN 08926875
[18] Hsieh K T and Rajamani R K 1991 *AIChE Journal* **37** 735–746 ISSN 0001-1541
[19] Monredon T, Hsieh K and Rajamani R 1992 *International Journal of Mineral Processing* **35** 65–83 ISSN 03017516
[20] Neesse T and Dueck J 2007 *Minerals engineering* **20** 380–386
[21] Hinze J 1975 *Turbulence:* McGraw-Hill classic textbook reissue series (McGraw-Hill) ISBN 9780070290372
[22] Goldberg U, Peroomian O and Chakravarthy S 1998 *Journal of Fluids Engineering* **120** 457–462
[23] Durdevic P, Pedersen S, Bram M, Hansen D, Hassan A and Yang Z 2015 *IFAC-PapersOnLine* **48** 291–296
[24] Neesse T, Tiefel H and Kaniut P 2007 *Minerals Engineering* **20** 355–360 ISSN 08926875
[25] Schneider M and Neesse T 2004 *International Journal of Mineral Processing* **74** S339–S343 ISSN 03017516
[26] Farghaly M, Golyk V, Ibrahim G, Ahmed M and Neesse T 2010 *Minerals Engineering* **23** 321–325
[27] Laleh A P, Svrcek W Y and Monnery W D 2012 *The Canadian Journal of Chemical Engineering* **90** 1547–1561 ISSN 00084034
[28] Durdevic P, Pedersen S and Yang Z November, 2016 *Under Review: Journal of Computers & Chemical Engineering (Elsevier)*
[29] Yang Z, Pedersen S and Durdevic P 2014 Cleaning the produced water in offshore oil production by using plant-wide optimal control strategy *2014 Oceans-St. John's* (IEEE) pp 1–10
[30] Kharoua N, Khezzar L and Nemouchi Z 2010 *Petroleum Science and Technology* **28** 738–755 ISSN 1091-6466
[31] Durdevic P, Simon P and Zhenyu Y 2016 *Proc of Oceans' 16 Mts/ieee Shanghai*
[32] Wolbert D, Ma B F, Aurelle Y and Seureau J 1995 *AIChE Journal* **41** 1395–1402 ISSN 0001-1541
[33] Young G, Wakley W, Taggart D, Andrews S and Worrell J 1994 *Journal of Petroleum Science and Engineering* **11** 37 – 50 ISSN 0920-4105
[34] Husveg R, Husveg T, van Teefelen N, Ottestad M and Hansen M R 2016 *Produced Water Workshop NEL*

IOP Conf. Series: Journal of Physics: Conf. Series **783** (2017) 012049 doi:10.1088/1742-6596/783/1/012049

Control Design for a Motion Cueing on Driving Simulator

Jorge de-J. Lozoya-Santos, Juan C. Tudon-Martinez, Julio Salinas
Universidad de Monterrey, San Pedro Garza García, Nuevo León 66238, México
E-mail: juan.tudon@udem.edu

Abstract. A *Linear Quadratic Regulator* (*LQR*) has been designed to simulate the pitch and roll vehicle dynamics of a platform which is connected to the real time simulation environment of *Dynacar.* The motion cueing algorithm translates the movement of the simulated vehicle to the platform using three rotary actuators, by satisfying all actuation boundaries. Experimental results illustrate that the *LQR* motion cueing algorithm performs satisfactory the tracking control at low frequencies, close to the resonance frequencies of the pitch and roll motion.

1. Introduction

In general, a driving simulator can be designed for different applications: entertainment, research and advanced training. To achieve a real assessment, the simulators deliver motion cues to the driver; whose fidelity depends on the degrees of freedom of the platform. These kind of systems are called also motion cueing systems and their main objective is joint the physical motion of the simulated vehicle with the real-time image generation system, i.e. the motion cueing system allows the drivers the perception and control of the vehicle motion, [1], such that the driver feels being inside the vehicle.

Recently, the interest of designing and control the motion cueings has been increased for automotive engineers, particularly in the research field to test *Advanced Driver Assistance Systems* (*ADAS*), *In-Vehicle Information Systems* (*IVIS*), and the effects of noise and vibrations on driver performance, [2]. The key parts of a motion cueing system are the driving simulator and the tracking control algorithm.

The realism of simulation depends strongly on the fidelity of the motion platform and the perception of motion by the human driver, [3]. In [4] is validated that the human vestibular system located in head is dominant in the human motion sensation for rotational and linear movements. The physical validation of a motion cueing is usually at small displacements [1] and useful to estimate frequencies of resonance of roll, pitch or heave [5]. Typically, the driving simulator consists of a seat, a vehicle dynamics model interacting with a driver through a steering wheel and pedals with haptic feedback, a sound feedback, a set of screens with a visual engine reproducing the vehicle dynamics results of the driver manoeuvre in the simulation, and, in some cases, translational and rotational actuators for the motion cueing system.

The motion cueing algorithms are mainly classified by the fidelity of the simulation level and its application. At a low-level simulator, the driver sits in a car seat, which is fixed to the ground. The driver looks to a screen, which is fixed too. The screen is designed such that the view angle is as large possible. The driver operates a set of driving actuators such as

IOP Conf. Series: Journal of Physics: Conf. Series **783** (2017) 012049 doi:10.1088/1742-6596/783/1/012049

the accelerator, braking system and the wheel steering system in order to receive visual cues corresponding to the actual driving situation. In some types of applications, it is desirable to provide a motion and haptic restitution to improve the simulation fidelity. Therefore, the driving simulators use a moving platform to restitute in a limited and constrained workspace, sufficient motion cues similar as much as possible to the real motion feeling in a vehicle [1][6][7][8]. Low level simulators can include the longitudinal axis motion in only one axis while the mid and high level simulators include more motion axis [9]. Table 1 presents the main features of different level driving simulators.

Table 1. Driving simulators complexity with Entertainment (E), Training (T), and Research (R) applications.

Level	*Components*	*Feedback*	*Degrees of freedom*	*Applications*
Low	Large screens, steering wheel, pedals, one linear actuator	Force Sound Longitudinal motion	0, 1 (x, y, or z axis)	E T
Mid	plus at least one linear actuator	Force Sound Longitudinal and rotational motions	2 or 3	E T R
High	plus at least four more linear actuators	Same as mid	6	T R

A three *Degrees of Freedom* (3 *DoF*) system is designed to mimic the main vehicle rotational dynamics such as the roll and pitch motions. This system causes a sensation of acceleration through rotation around the longitudinal, lateral an vertical axis, by keeping a tilt angle (typically 45 degrees) and considering the gravity. The tilt limitation in this system prevents a sensation of acceleration over 0.707 g's [9]. This paper is focused in developing a tracking controller for a *3 DoF* driving simulator.

The control algorithm for motion cueing systems is called washout filtering with classical, optimal and predictive approaches [1][4], proportional-integral-derivative washout filtering [6], and optimal and model predictive control [7][8][5] among others. The classical washout filtering is the most common because of the simplicity of implementation and fair simulation results [9], but limited to the design of specific high-pass and low-pass filters whose parameters are usually empirically obtained. On the other hand, the design of an optimal cueing algorithm such as an *LQR* approach ensures the tracking control in a operating zone of the actuators.

In this paper, an *LQR* cueing algorithm is proposed to control the driving simulator of three degrees of freedom specially the tracking to the roll and pitch vehicle dynamics. The goal of this simulator is to validate *ADAS* including the biometrics systems and autonomous vehicle assessment as well as the transportation of finish goods. The proposed controller is described in section 2. Section 3 describes the *Vehicle Dynamics Simulator* (*VDS*) and the hardware platform. The results and discussion are shown in Section 4. Conclusion ends this paper in Section 5. Table 2 describes the used variables.

Table 2. Variables.

Variable	*Units*	*Description*
x	$\frac{m}{s^2}$	Longitudinal acceleration
y	$\frac{m}{s^2}$	Lateral acceleration
z	$\frac{m}{s^2}$	Vertical acceleration
θ	degrees	Roll angle
ϕ	degrees	Pitch angle
ψ	degrees	Yaw angle
u	V	Command for θ_d and ϕ_d
$M_{1,2,3}$	-	Electric motor 1, 2 and 3

2. *LQR*-based Motion Cueing

The washout filter design strategy based on the LQR structure is presented in this section. The objective of the optimal washout filter is to obtain a linear transfer matrix W that minimizes a quadratic cost function involving the tracking error and the driving simulator output (actuator manipulations) by considering the physical constraints of the platform.

The proposed controlled output for tracking is given by:

$$y = [x \;\; y \;\; z \;\; \theta \;\; \phi]^T \tag{1}$$

where all variables are measured from the simulator platform. Such that the control tracking error is given by:

$$e \equiv y_m - y \tag{2}$$

where y_m is a vector that contains the simulated vehicle variables corresponding to the measured ones. The vehicle dynamics that reproduces the vertical, longitudinal, lateral, and rotational dynamics of the chassis and wheels in the virtual simulation is given by [10]:

$$\begin{aligned}
m\ddot{x}_s &= \left(F_{t_{x_1}} + F_{t_{x_2}}\right)\cos(\delta) + F_{t_{x_3}} + F_{t_{x_4}} - \left(F_{t_{y_1}} + F_{t_{y_2}}\right)\sin(\delta) + m\dot{\psi}\dot{y}_s \\
m\ddot{y}_s &= \left(F_{t_{y_1}} + F_{t_{y_2}}\right)\cos(\delta) + F_{t_{y_3}} + F_{t_{y_4}} + \left(F_{t_{x_1}} + F_{t_{x_2}}\right)\sin(\delta) + m\dot{\psi}\dot{x}_s \\
m_s\ddot{z}_s &= -\left(F_{s_{z_1}} + F_{s_{z_2}} + F_{s_{z_3}} + F_{s_{z_4}}\right) \\
m_{us_i}\ddot{z}_{us_i} &= F_{s_{z_i}} - F_{t_{z_i}} \\
I_{xx}\ddot{\theta}_s &= \left(F_{s_{z_1}} - F_{s_{z_2}}\right)t_f + \left(F_{s_{z_3}} - F_{s_{z_4}}\right)t_r + mh\ddot{y}_s + (I_{yy} - I_{zz})\dot{\psi}\dot{\phi} \\
I_{yy}\ddot{\phi}_s &= \left(F_{s_{z_4}} + F_{s_{z_3}}\right)l_r - \left(F_{s_{z_2}} + F_{s_{z_1}}\right)l_f + mh\ddot{x}_s + (I_{zz} - I_{xx})\dot{\psi}\dot{\theta} \\
I_{zz}\ddot{\psi}_s &= \left(F_{t_{y_1}} + F_{t_{y_2}}\right)l_f\cos(\delta) - \left(F_{t_{y_3}} + F_{t_{y_4}}\right)l_r + \left(F_{t_{x_1}} + F_{t_{x_2}}\right)l_f\sin(\delta) \\
&\quad + \left(F_{t_{x_4}} - F_{t_{x_3}}\right)t_r + \left(F_{t_{x_2}} - F_{t_{x_1}}\right)t_f\cos(\delta) - \left(F_{t_{x_2}} - F_{t_{x_1}}\right)t_f\sin(\delta) \\
&\quad + (I_{xx} - I_{yy})\dot{\phi}\dot{\theta}
\end{aligned} \tag{3}$$

where the vertical suspension force at each corner ($F_{s_{z_i}}$, with $i = 1, \ldots, 4$) is composed by the spring (stiffness k_{s_i}) and damper (coefficient c_{s_i}), as:

$$F_{s_{z_i}} = k_{s_i}\left(z_{s_i} - z_{us_i}\right) + c_{s_i}\left(\dot{z}_{s_i} - \dot{z}_{us_i}\right) \tag{4}$$

the tire lateral and longitudinal forces ($F_{t_{x_i}}$ and $F_{t_{y_i}}$), that depends on the throttle and brake, are obtained from the *Pacejka* formula; while the tire vertical force, with stiffness k_{t_i}, is given by:

$$F_{t_{z_i}} = k_{t_i}\left(z_{us_i} - z_{r_i}\right) \tag{5}$$

The sprung mass positions and velocities at each corner are given by:

$$\begin{aligned} z_{s_1} &= z_s + l_f\sin(\phi) - t_f\sin(\theta) & z_{s_2} &= z_s + l_f\sin(\phi) + t_f\sin(\theta) \\ z_{s_3} &= z_s - l_r\sin(\phi) - t_r\sin(\theta) & z_{s_4} &= z_s - l_r\sin(\phi) + t_r\sin(\theta) \end{aligned} \tag{6}$$

$$\begin{aligned} \dot{z}_{s_1} &= \dot{z}_s + \dot{\phi} l_f\cos(\phi) - \dot{\theta} t_f\cos(\theta) & \dot{z}_{s_2} &= \dot{z}_s + \dot{\phi} l_f\cos(\phi) + \dot{\theta} t_f\cos(\theta) \\ \dot{z}_{s_3} &= \dot{z}_s - \dot{\phi} l_r\cos(\phi) - \dot{\theta} t_r\cos(\theta) & \dot{z}_{s_4} &= \dot{z}_s - \dot{\phi} l_r\cos(\phi) + \dot{\theta} t_r\cos(\theta) \end{aligned} \tag{7}$$

All vehicle parameters, including the vehicle mass (m), the chassis and unsprung masses (m_s, m_{us_i}), the inertial moments (I_{xx}, I_{yy}, I_{zz}) and physical measures (l_f, l_r, t_r, t_f, h) are considered from a generic vehicle model in *Dynacar*.

In order to ensure the driving simulator constraints and the minimization of the tracking error between the simulated vehicle and the experimental platform, the problem of designing an optimal washout filter is given by:

$$J = \int_{t_1}^{t_2} \left(e^T Q e + u^T R u\right) dt \tag{8}$$

where u is the controller output that manipulates the three electric motors, $[t_1, t_2]$ is the residence time horizon of the driving simulator, while Q and R are the weighting matrices given by the solution by a Ricatti equation, which gives W corresponding to the optimal washout filter. The proposed *LQR*-based motion cueing is schematized in Fig 1.

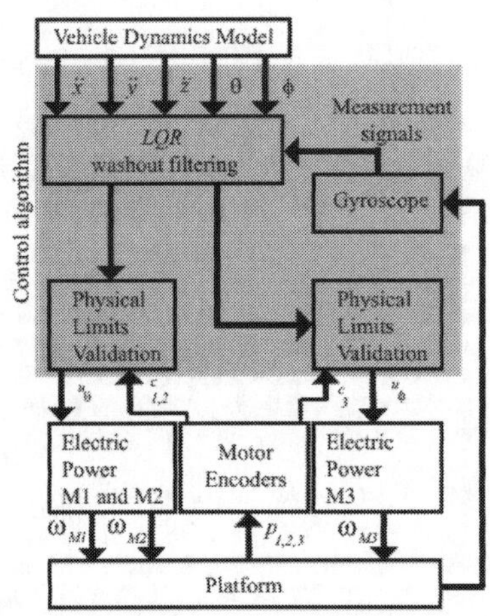

Figure 1. Diagram block of the *LQR*-based motion cueing

3. Vehicle Dynamics Simulator

The *VDS* consists of a real time vehicle dynamics model system interacting with a driver through a steering system, acceleration system, braking system, as well as a set of three flat screens and a set of speakers. The real time vehicle dynamics model simulation system consists of the *Dynacar* system, [11], which is a low level driving simulator, and a three *DoF* platform, Fig. 2.

Figure 3 illustrates the full scheme of the *VDS*, detailing the main components and describing the interactions between the *Dynacar* and the *3 DoF* platform. The *3 DoF* vehicle platform consists of two mechanical frameworks (lower an upper base) separated by an active suspension.

IOP Conf. Series: Journal of Physics: Conf. Series **783** (2017) 012049 doi:10.1088/1742-6596/783/1/012049

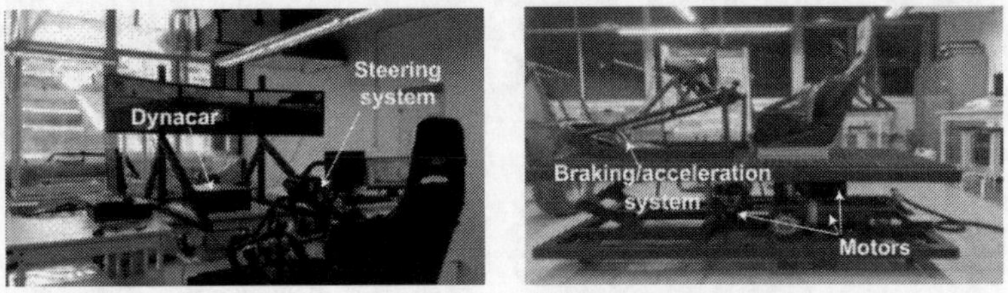

a) Back view of *VDS*. b) Lateral view of *VDS*.

Figure 2. *VDS* in the Automotive Eng. Lab, Universidad de Monterrey.

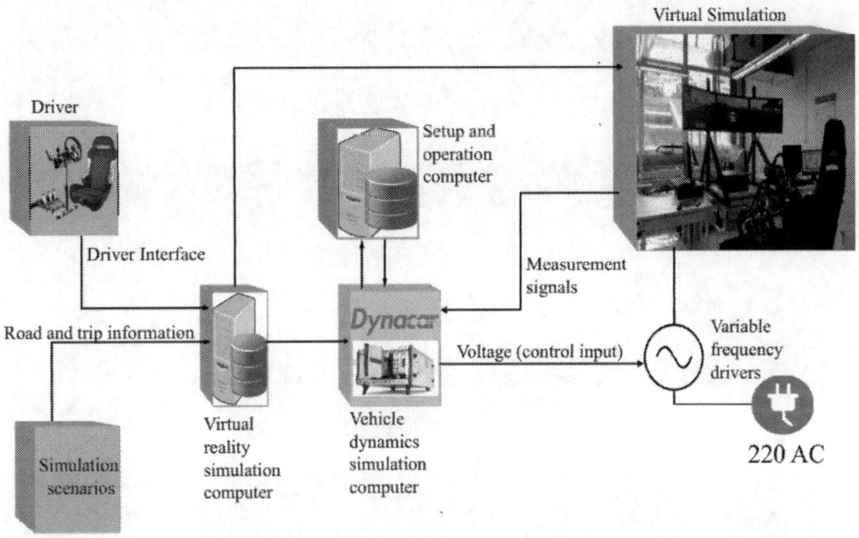

Figure 3. Experimental system: hardware in vehicle dynamics simulator.

The upper structure holds the driver cabin, and the lower structure holds three crank shaft mechanisms conforming the active suspension as well as a vertical sliding guide, Fig. 4.

A set of actuators and sensors acts and measures the platform dynamics, Fig. 5. Each crankshaft mechanism is governed by an actuator and a sensor. This consists of an alternate current induction electric motor and its electric drive. The motor shaft holds an absolute encoder allowing the angle measurement of the crank. The upper base has one sensor: a gyroscope. It is located under the driver seat. A *Real-Time* (*RT*) computer (same as for the *Dynacar* system) reads the gyroscope and encoder signals. A set of reference signals is delivered from the selected vehicle dynamics model and compared with the measurements. The control algorithm computes in the *RT* computer the commands for each electric drive. Each electric motor has a frequency variator with a predefined speed control such that controller output is the voltage to the electric motors, and the control input to the platform is the rotational motion caused by the crank shaft mechanisms excited by the speed of the motors. The inner control of the electric motors can be reviewed in detail in [12].

4. Results and Discussion

The validation of *LQR*-based motion cueing in the *VDS* consists of a test in *Dynacar* with random lane changes, and with random braking and accelerating also. The programming of the

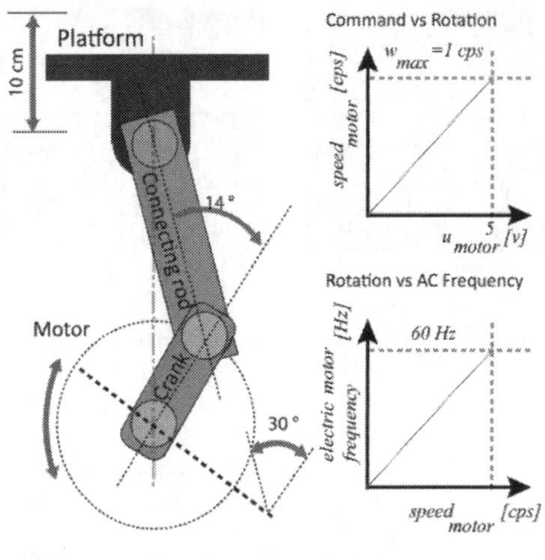

Figure 4. Crank shaft mechanism and input-output signals.

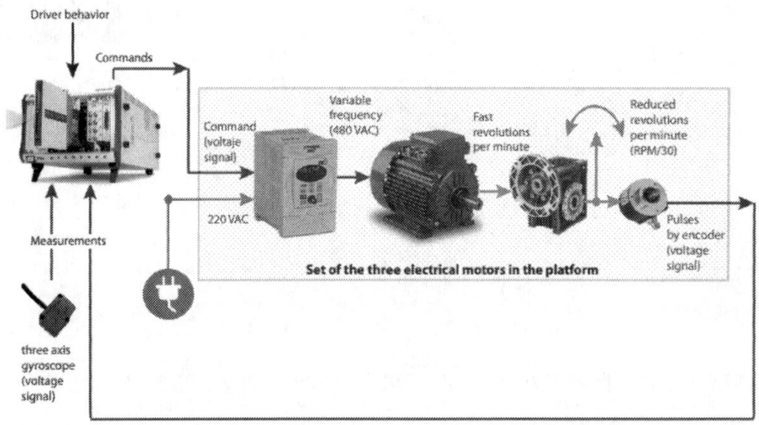

Figure 5. Actuators and sensors for platform.

control algorithm is done in a *Matlab*^{TM} environment.

The driver test under random scenario has a frequency contents to explore from 0.5 to 2.0 Hz, close to the frequencies of resonance of the pitch and roll angle. This bandwidth has been obtained from data analysis from *Dynacar* simulated variables. The ranges of the simulated pitch and roll are $\tilde{\theta} = \{-5, 5\}$ degrees and $\tilde{\phi} = \{-8, 8\}$ degrees.

Figure 6 presents the tracking performance of the proposed optimal LQR-based washout filter. The results of the washout filtering present high fidelity in the platform motion regarding the simulation in *Dynacar*, the controlled outputs follow the simulated variables in almost all frequencies of excitation. Qualitatively, Fig. 6 illustrates that the experimental roll angle correctly follows the virtual simulation, even better than the pitch motion. The roll angle dynamics in the *Dynacar* is well tracked by the platform in amplitude and frequency; however, the pitch motion in the platform has more complications to follow the abrupt changes in the *Dynacar* simulator, i.e. in some cases these abrupt motions are not tracked by the LQR washout filter such as occurs at $t = 3,500$ ms.

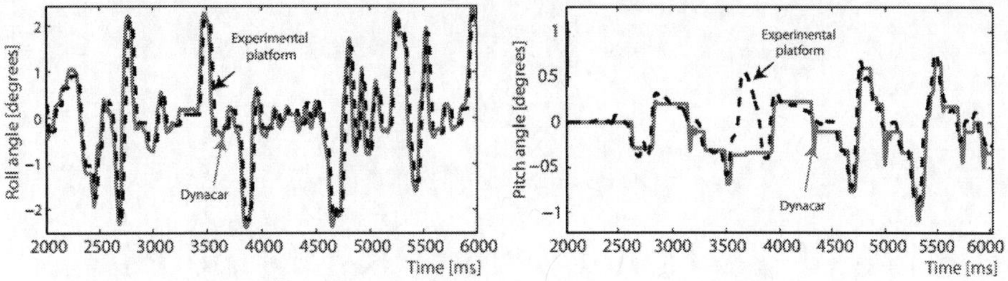

Figure 6. Pitch and roll comparison with respect to the *Dynacar* simulation.

5. Conclusion

An *LQR*-based motion cueing has been proposed to control a *Vehicle Dynamics Simulator* of three degrees of freedom. The variables of interest are the pitch and roll angle and simulated in real time according to the *Dynacar* software. Experimental results show that the *LQR* washout filter has good tracking to the simulated vehicle variables, specially around the frequencies of resonance of the pitch and roll angle.

Future works will consist on testing active suspension controllers to improve the driver comfort using the developed LQR-based motion cueing on this driving simulator.

References

[1] Reymond G and Kemeny A 2000 *Vehicle System Dynamics* vol 34, p 249
[2] Slob JJ 2008 *State-of-the-art driving simulators, a literature survey* DCT Report, vol 107
[3] Weiß C 2006 *Control of a dynamic driving simulator: Time-variant motion cueing algorithms and prepositioning* Ph.D. dissertation (Universität Stuttgart)
[4] Telban RJ, Wu W, Cardullo FM and Houck JA 2000 *Motion cueing algorithm development: Initial investigation and redesign of the algorithms* NASA, Tech. Rep.
[5] Aykent B, Merienne F, Paillot D and Kemeny A 2014 *Optimal Control Applications and Methods* vol 35, p 454
[6] Gutridge CJ 2004 *Three degree-of-freedom simulator motion cueing using classical washout filters and acceleration feedback* Ph.D. dissertation (Virginia Polytechnic Institute & State University)
[7] Nehaoua L, Mohellebi H, Amouri A, Arioui H, Espie S and Kheddar A 2008 *IEEE Transactions on Vehicular Technology* vol 57, p 736
[8] Arioui H, Hima S, Nehaoua L, Bertin RJV and Espi S 2011 *IEEE Transactions on Vehicular Technology* vol 60, p 357
[9] Mohajer N, Abdi H, Nelson K and Nahavandi S 2015 *Vehicle System Dynamics* vol 53, p 1204
[10] Gillespie TD 1992 *Fundamentals of Vehicle Dynamics* SAE
[11] Peña A, Iglesias I, Valera JJ and Martin A 2012 *Proc. of Int. Battery, Hybrid and Fuel Cell Electric Vehicle Symp. (EVS 26)*, p 13
[12] Lozoya-Santos JJ, Salinas J, Mendez E, Gonzalez G., Tudon-Martinez JC and Ramirez-Mendoza RA 2016 *Proc. of the Eighth Int. Conf. on Advances in System Simulation (SIMUL 26)*, p 67

On the use of fractional order PK-PD models

Clara Ionescu and Dana Copot

Ghent University, Faculty of Engineering and Architecture, Research Group on Dynamical Systems and Control (DySC), Technologiepark 914, B9052 Ghent, Belgium

E-mail: claramihaela.ionescu@ugent.be; dana.copot@ugent.be

Abstract. Quantifying and controlling depth of anesthesia is a challenging process due to lack of measurement technology for direct effects of drug supply into the body. Efforts are being made to develop new sensor techniques and new horizons are explored for modeling this intricate process. This paper introduces emerging tools available on the 'engineering market' imported from the area of fractional calculus. A novel interpretation of the classical drug-effect curve is given, enabling linear control. This enables broadening the horizon of signal processing and control techniques and suggests future research lines.

1. Introduction

The emerging concepts of fractional calculus (FC) in biology and medicine have shown a great deal of success, explaining complex phenomena with a startling simplicity ([1, 2, 3]). It is clear that a major contribution of the concept of FC has been and remains still in the field of biology and medicine ([4]). Fractional calculus generously allows integrals and derivatives to have any order, hence the generalization of the term *fractional-order* to that of *general-order*. Of all applications in biology, diffusion is the one most appealing from modeling point of view, since it allows characterizing a relatively complex process with parsimonious models.

Modeling drug dynamics in the body using compartmental models is perhaps one of the most popular modeling approach ([5]). These models are based on mathematical characterization of molecular biochemistry and transport phenomena in the body ([6]). As such, diffusion plays an important role in drug assimilation, transport and clearance, and it is a physiological process which can be well described by means of fractional calculus ([7]).

In this paper, a fresh view upon the models for drug release and their effect will be given in the application of depth of anesthesia regulation. If one would like to have optimal drug infusion rates into a patient with avoiding over- and under-dosing, then accurate patient models are necessary (although not sufficient, since a good control methodology is also required) ([8]). Accurate models may hold a realistic dynamic perspective for average populations datasets, but they are by far sub-optimal when patient-individualized control is envisaged. This is due to the fact that inter-patient variability has a strong impact on the perception and effectiveness of the drug into the body of the patient (i.e. drug effect, patient sensitivity to the drug, etc). A schematic representation of the anesthesia paradigm in terms of its components is given in figure 1. The hypnosis and neuromuscular blockade are well-characterized, yet the analgesia component remains challenging for control purposes because no direct output measure is available. Hence, no specific models are yet available for analgesia effect in the human body during general anesthesia (i.e. unconsciousness). The pharmacokinetic (PK) models available in the literature are linear in

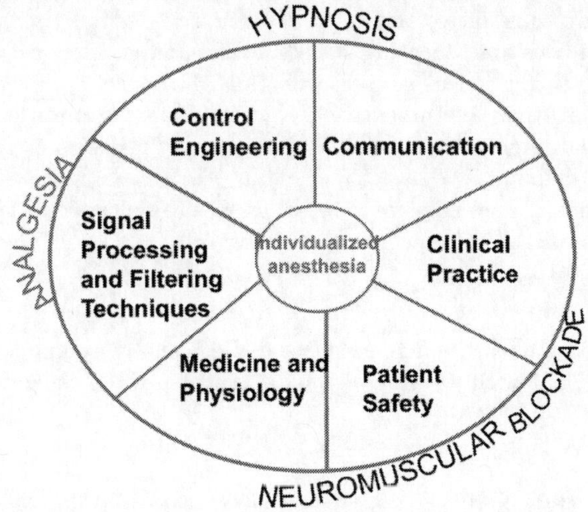

Figure 1. A personal view of the anesthesia paradigm. The items in red are the current challenges.

terms of model parameters and dynamics ([9, 10]). Their frequency response is quasi-identical, less for a scaling factor in the gain (i.e. this accounts in part for the sensitivity to the drug with respect to the body mass index of the patient). The pharmacodynamic (PD) models, are usually represented by nonlinear Sigmoid (Hill) curves and represent the relationship of the drug concentration to the drug effect in each patient. From patient-individualized control point of view, PD models are the most challenging part of the patient model and pose most challenges for control (i.e. highly nonlinear characteristic).

Fractional calculus offers tools to model such nonlinear characteristics as those of the PD models with less degree of nonlinearity in model parameters ([3]). The original purpose of this paper is to show a methodology which enables a different, fresh view to these models and how they can be integrated in the control paradigm of depth of anesthesia regulation.

The paper is structured as follows: next, a brief introduction on the fractional calculus tools for modeling drug release and its effect is given. Section III depicts the classical methodology for PK-PD models in anesthesia and section IV presents the proposed PD modeling approach and some simulation results. Current limitations for the use of these models in a closed loop control paradigm are given and a conclusion section summarizes the main outcome of this paper.

2. Tools from fractional calculus

Pure empirical models such as input - output models are rather descriptive with loss of physiological insight into the drug absorption, transport, diffusion and release. However, from control point of view they are the most simple to obtain and to deliver information to the controller for deciding the optimal drug infusion rate necessary for the specific patient whose model is available. On the other hand, physiologically and bio-chemically based models are difficult to attain due to the extraordinary complexity of this process. An intermediate solution is perhaps the use of semi-empirical models. Such a model which is fairly simple to employ in characterizing drug release is the power-law model:

$$\frac{M_t}{M_\infty} = kt^n \tag{1}$$

where M_t and M_∞ are cumulative amounts of drug released at time t and infinite time, respectively; and k is a constant reflecting the structural and geometrical characteristics of the system, and n is the so-called "release exponent" ([11]). For purely diffusion-controlled release, $n = 0.5$. other values of n may be indicative of various diffusion conditions. These features of this model are of special interest in our study since we would like to look at the inter-patient variability from this standpoint.

Following the fundamental law of physics which applies to well stirred, homogeneous systems, it follows that the mean square displacement of the walker, $< x^2 >$ in the random walk model is proportional to time ([12, 13]):

$$< x^2 > \propto t \tag{2}$$

However, in disordered systems, such as most biological environments, this is no longer proportional with time, but with the fractal walk dimension of the walker:

$$< x^2 > \propto t^{D_w} \tag{3}$$

with $D_w \neq 2$. This property implies that scaling laws such as power laws are associated with kinetics of various processes taking place in the (biological) environment (i.e. tissue). Most biological tissues can be approximated in their properties by polymers. As such, the spatio-temporal porosity of a dynamically changing polymer is close to the percolation threshold for non-classical diffusion effects impinging on release kinetics. Fractional calculus has shown in several (non)biological applications that classical diffusion as well as non-classical diffusion can be characterized by fractional order differ-integrals models ([4]). It is therefore simple to derive further the model from (1) in its Laplace equivalent. Consider that the amount of drug released varies in time and that the amount released at infinite time is a constant. This results in the following Laplace domain model equivalent:

$$\frac{M_t(t)}{M_\infty} = k\frac{\Gamma(1-n)}{s^{1-n}} \tag{4}$$

where Γ is the Gamma function and s the Laplace operator. If a step input is applied to this model, i.e. apply a constant drug infusion rate, one obtains:

$$\frac{M_t(t)}{M_\infty} = ks^n\Gamma(1-n) \tag{5}$$

which is then a fractional order model (more information about such models is given in *Appendix*). This result has regained the attention of the research community and current efforts are being directed towards providing pharmacokinetic models with fractional order differ-integrals ([14, 15]).

3. Classical PK-PD models

In order to control the depth of sedation based on predictive control strategies, a model which captures the dynamics of the patient is required. The selection of the model input and output variables is crucial for optimal control. Propofol is a commonly used hypnotic drug to induce general anesthesia ([8]). Using the electroencephalogram (EEG), several derived, computerized parameters like the BIS have been tested and validated as a promising measure of the hypnotic component of anesthesia. BIS values lie in the range of 0-100; whereas 90-100 range represents fully awake patients; 60-70 range and 40-60 range indicate light and moderate hypnotic state, respectively ([16]).

PK-PD blocks denote compartmental models. The PK-PD models most commonly used for propofol and remifentanil are the 4th order compartmental model described by Schnider ([9, 10])

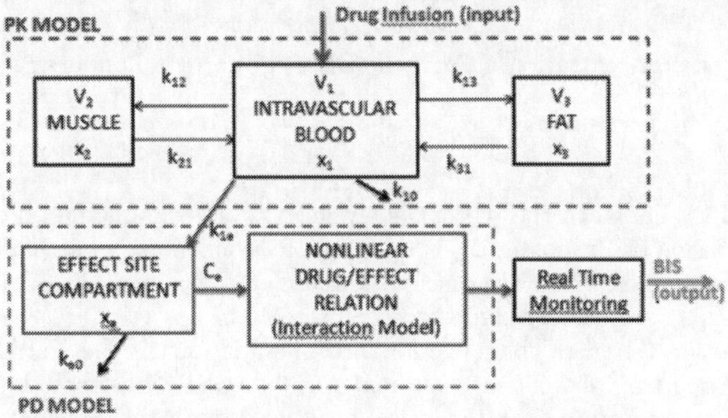

Figure 2. General compartmental model of the patient, where PK denotes the pharmacokinetic model and PD denotes the pharmacodynamic model of an infused drug.

and they have the structure depicted in Fig. 2. In this figure x_1 [mg] denotes the amount of drug in the central compartment. The blood concentration is expressed by x_1/V_1. The peripheral compartments 2 and 3 model the drug exchange of the blood with well and poorly diffused body tissues. The masses of drug in fast and slow equilibrating peripheral compartments are denoted by x_2 and x_3, respectively. The parameters k_{ji}, for $i \neq j$, denote the drug transfer frequency from the jth to the ith compartment and $u(t)$ [mg/s] is the infusion rate of the anesthetic drug into the central compartment. An additional hypothetical effect compartment was proposed to represent the lag between drug plasma concentration and drug response. The concentration of drug in this compartment is represented by x_e. The parameters k_{ij} of the PK models depend on age, weight, height and gender and the relations can be found in [9, 10].

The PK-PD model is represented by the following equations:

$$\begin{aligned}
\dot{x}_1(t) &= -(k_{10}+k_{12}+k_{13})\cdot x_1(t) + k_{21}\cdot x_2(t) + +k_{31}\cdot x_3(t) + u(t)/V_1 \\
\dot{x}_2(t) &= k_{12}\cdot x_1(t) - k_{21}\cdot x_2(t) \\
\dot{x}_3(t) &= k_{13}\cdot x_1(t) - k_{31}\cdot x_3(t) \\
\dot{x}_e(t) &= -k_{e0}\cdot x_e(t) + k_{1e}\cdot x_1(t)
\end{aligned} \tag{6}$$

The effect compartment receives drug from the central compartment by a first-order process and it is regarded as a volumeless additional compartment. Therefore, the drug transfer frequency from the central compartment to the effect-site compartment is equal to the frequency of drug removal from the effect-site compartment: $k_{e0} = k1e = 0.456\ [min^{-1}]$. The corresponding concentration in the effect site compartment is denoted by C_e. The parameters k_{ij} of the PK models depend on age, weight, height and gender and can be calculated for Propofol:

$$\begin{aligned}
&V_1 = 4.27[l];\ V_2 = 18.9 - 0.391(age - 53)[l];\ V_3 = 2.38[l]; \\
&Cl_1 = 1.89 + 0.0456(weight - 77) - 0.0681(lbm - 59) + 0.0264(age - 53)[l/min] \\
&Cl_2 = 1.29 - 0.024(age - 53)[l/min];\ Cl_3 = 0.836[l/min]; \\
&k_{10} = \tfrac{Cl_1}{V_1}[min^{-1}];\ k_{12} = \tfrac{Cl_2}{V_1}[min^{-1}];\ k_{13} = \tfrac{Cl_3}{V_1}[min^{-1}]; \\
&k_{21} = \tfrac{Cl_2}{V_2}[min^{-1}];\ k_{31} = \tfrac{Cl_3}{V_3}[min^{-1}];
\end{aligned} \tag{7}$$

where Cl_1 is the rate at which the drug is cleared from the body, and Cl_2 and Cl_3 are the rates at which the drug is removed from the central compartment to the other two compartments by distribution. The lean body mass *(lbm)* for women and men are: $lbm_{men} =$

$1.1 \cdot weight - 128 \cdot \frac{weight^2}{height^2}$ and $lbm_{women} = 1.07 \cdot weight - 148 \cdot \frac{weight^2}{height^2}$, respectively. The relation between the effect site concentration C_e and the BIS is given by a nonlinear sigmoid Hill curve:

$$BIS(t) = E_0 - E_{max} \cdot \frac{C_e^{\gamma}(t)}{C_e^{\gamma}(t) + C_50^{\gamma}} \tag{8}$$

where E_0 is the BIS value when the patient is awake; E_{max} is the maximum effect that can be achieved by the infusion of Propofol; C_{50} is the Propofol concentration at half of the maximum effect and γ is a parameter which together with the C_{50} determines the patient sensitivity to the drug. E_0 and E_{max} are usually considered equal to 100.

As explained earlier, the main challenge for control standpoint is the nonlinearity of the Hill curve given by (8) and the inherent inter-patient variability. To illustrate this to the reader, a realistic set of typical and atypical patients has been used, with the parameter values given in Table 1. The resulted Hill curves by simulating the PKPD model of these patients is shown in figure 3.

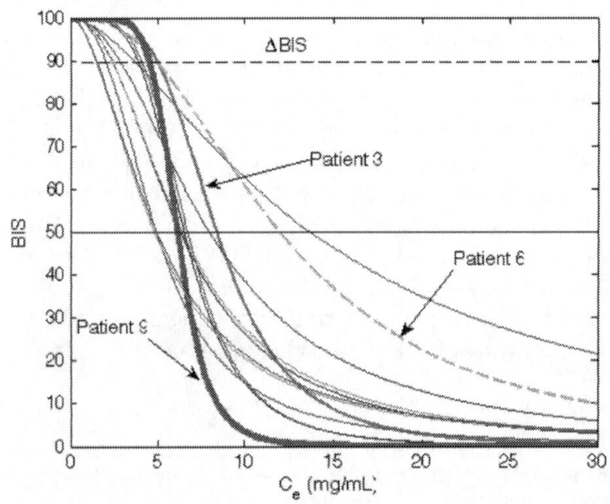

Figure 3. Example of Hill curves from various patients.

4. Proposed Modeling Approach and Analysis

Recall that the equation (1) is the basis for the fractional order model from (5). As explained, the red-line connecting these models is based on the theory of fractional calculus and fractal walk dynamics. An interesting feature of systems characterized by fractal dynamics is the following: represented on a log-log plot, their characteristic becomes linear. This feature is of special interest for us, especially if the reader may recall the Hil curve from figure 3. Of course, it would be very nice if the strong nonlinearity and inter-patient variability depicted in this figure would be diminished, or even better, simply vanish into thin air. In this section, some ideas will be put forward which will lead the reader through this new approach of viewing Hill curves for drug effect to concentration relationship.

Based on (1) one can write the same relationship for the Hill curve:

$$\frac{BIS(t)}{Ce(t)} = k \cdot t^n \tag{9}$$

Patient	Age	Length	Weight	Gender	C_{50}	E_0	E_{max}	γ
1	40	163	54	F	6.33	98.80	94.10	2.24
2	36	163	50	F	6.76	98.60	86.00	4.29
3	28	164	52	F	8.44	91.20	80.70	4.1
4	50	163	83	F	6.44	95.90	102.00	2.18
5	28	164	60	M	4.93	94.70	85.30	2.46
6	43	163	59	F	12.10	90.20	147.00	2.42
7	37	187	75	M	8.02	92.00	104.00	2.10
8	38	174	80	F	6.56	95.50	76.40	4.12
9	41	170	70	F	6.15	89.20	63.80	6.89
10	37	167	58	F	13.70	83.10	151.00	1.65
11	42	179	78	M	4.82	91.80	77.90	1.85
12	34	172	58	F	4.95	96.20	90.80	1.84

Table 1. Characteristic variables for each of the virtual realistic 12 patients used in this study.

where k and n are varying on the patient PK-PD characteristics. If one compares (8) with (9) it can be recognized the resemblance in the power term and observe in fact a simplification of the model from (8) in terms of parameter number. From a structural point of view, there is no difference between the models, since both are semi-empirical models. The term $\frac{BIS(t)}{Ce(t)}$ denotes the effect-to-concentration ratio (ECR) and its units are [%/mg/ml]. Based on the principles of fractal walk, if the scale of the ECR representation is changed from linear to logarithmic, it becomes a (quasi)linear relation.

To verify whether the ECR becomes (quasi)linear, one may calculate the drug effect $BIS(t)$ as a function of the drug concentration $Ce(t)$. The profile of $Ce(t)$ may differ depending on the type of depth of anesthesia regulation that can occur in practice. Usually, in target-controlled infusion systems, drug is infused in the patient in open loop (i.e. no feedback from the patient is explicitly taken into account !) and is the most widely applied in clinical practice. Averaged population models are used to predict the concentration in the blood and thus in the effect site compartment from (6) and their corresponding BIS effect. If we would represent this in time, it appears as a train of steps of different values; i.e. the anesthesiologist changes the value of the targetted concentration depending on the reactions in biological signals from patient (e.g. heart rate, respiratory rate, blood pressure, cardiac output, oxigen concentration, etc).

There are three profiles investigated in this paper: i) a linearly increasing concentration, ii) a first order lag and iii) a second order lag, as in figure 4. of course, in practice, the time constant of the concentration profiles is much larger than the one taken in this example. However, this does not affect the overall conclusions. Figure 5 plots the evolution of ECR with these three variation characteristics of Ce, for an averaged population of data from table 1.

Notice that in figure 5 both X and Y scales are logarithmic. Observe that the ECR is indeed quasi-linear and that the difference in slope is attributed to the variations in the dynamic profiles of Ce. if one repeats this simulation using all the values from the table 1, one obtains the result given in figure 6. An interesting result is that of moving the effects of inter-patient variability at *later times*. This allows to *postpone* the robustness check of the controller with respect to variations in the model. This allows performing online identification for attaining a specific model of the respective patient during general anesthesia, thus updating the model parameters in-between the control samples. A possible method for this patient-individualized control of anesthesia is given in [17].

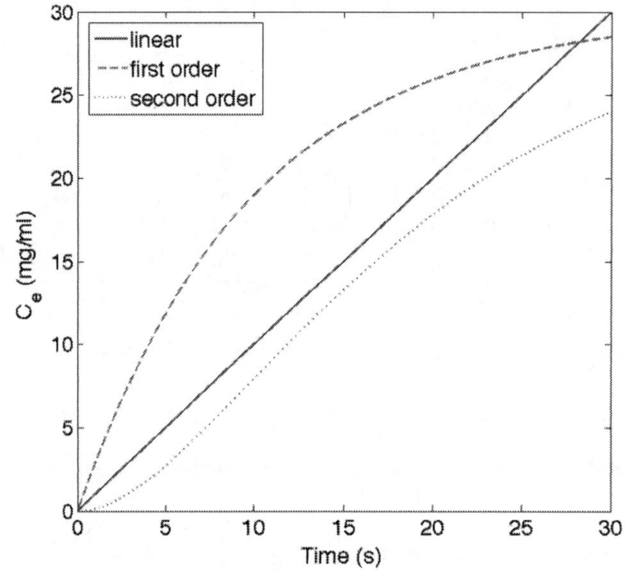

Figure 4. Example of *Ce* curves for an averaged patient model parameter values based on table 1.

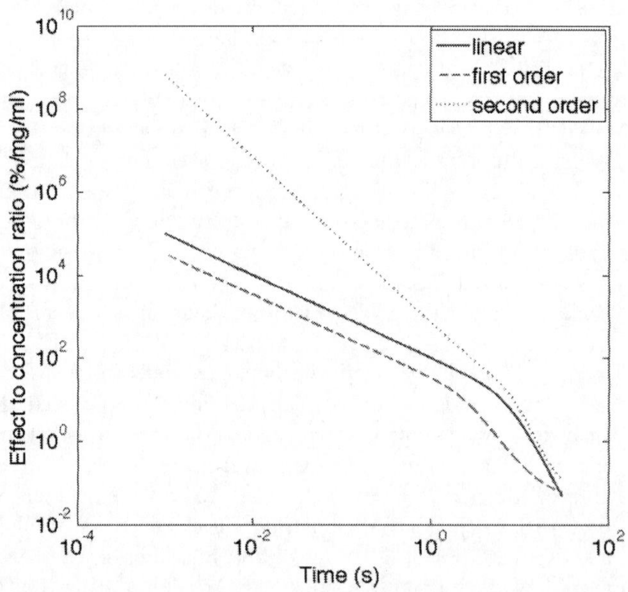

Figure 5. Example of ECR curves for an averaged patient model parameter values based on table 1.

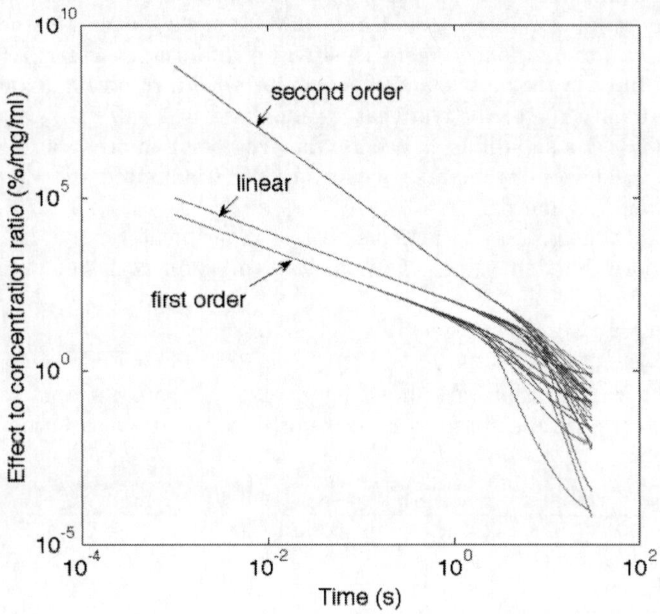

Figure 6. Example of ECR curves for all patient model parameter values from table 1.

To show the efficiency of fractional order operators simulation results for a PKPD model presented in section 3 are depicted in figure 7 depicts the time evolution of the Propofol amount for different values of n: $n = 0.6; 0.8; 1.0$.

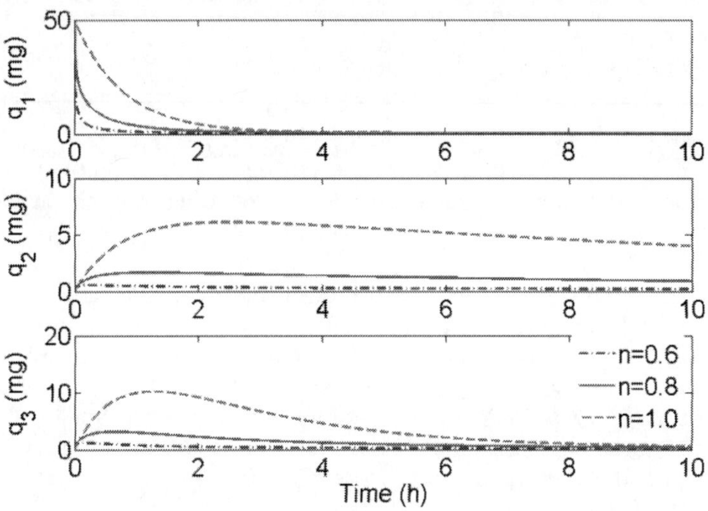

Figure 7. Propofol amount as a function of time for compartment 1, compartment 2 and effect site compartment.

The model coefficients have been calculated for a male person of 74 years old, 88 kg, 164 cm and a bolus injection of 50 mg injected at a rate of 3.3. mg/s during 15 seconds. It can be observed that for $n = 1$ the patient is more sensitive to the drug than for $n = 0.6$. A higher amount of drug is diffused in the muscle and effect compartment, resulting in lower values of the BIS signal. This is in part due to the fact that a value for $n = 1$ will imply a normal diffusion process, whereas for $n < 1$ a sub-diffusion occurs, the drug being faster cleared out of the body (less uptake). For $n = 0.6$ less drug uptake occurs, the effect is limited and values do not reach sedation, i.e. effect remains above 50%.
Our results suggest that the proposed fractional order compartmental model captures well inter-patient variability since different values of n can lead to slower or faster uptake and various effects.

5. Current Limitations

If the same constant drug infusion rate would be given to all patients from table 1, then the result depicted in figure 8 will be obtained in terms of $BIS(t)$ output variation with time. It is

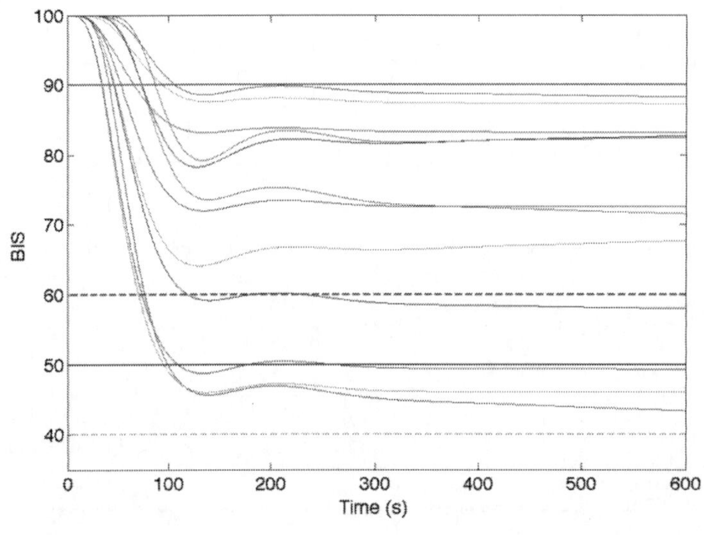

Figure 8. Example of BIS curves resulted from a constant drug infusion rate for all patient model parameter values from table 1.

therefore obvious that i) closed loop control is necessary and ii) an averaged patient model will deliver sub-optimal results since they require a robust, conservative controller able to deal with the inter-patient variability. Moreover, it is also clear that the inter-patient variability may vary quite significantly and that one single controller (without online adaptation) will never suffice in practice. A first problem is to find online adaptation algorithms which may adapt the model parameters k and n to the patient characteristics. Identification from logarithmically sampled data has been proposed in ([18]) and represents a good framework for developing the online identification algorithm.

The next problem is to find a solution to integrate the model from (9) into a closed loop control system taking into account the requirement for a logarithmic sample time (i.e. in order to maintain linearity). Although it may look surprisingly, it has already been shown in various examples that a Riemann sampling rate (i.e. linear periodic) may be outperformed

IOP Conf. Series: Journal of Physics: Conf. Series **783** (2017) 012050 doi:10.1088/1742-6596/783/1/012050

by a Lebesque sampling rate (i.e. event-triggered) in several applications ([19]). Since the Lebesque sampling rate is an event triggered rate used successfully in practice in closed loop control (e.g. networked control, sensor networks), it may be revealing to look into the possibility of a logarithmic sampling rate.

6. Conclusion

This paper presented the available tools emerging from fractional calculus to model the nonlinear characteristics of the pharmacokinetic and pharmacodynamic patient models. Advantages and challenges have been discussed. Results suggest that the high degree of inter-patient variability and nonlinearity may be avoided, leading to linear control techniques instead of advanced, complex control techniques.

Further steps of this research line are dealing with the current limitations: i) online identification of logarithmically sampled data and ii) control of logarithmically sampled systems.

7. Appendix

The fractional calculus is a generalization of integration and derivation to non-integer (fractional) order operators ([20, 21]). First, we generalize the differential and integral operators into one fundamental operator D_t^n (n the order of the operation) which is known as *fractional calculus.*

Several definitions of this operator have been proposed, mainly based on generalization of

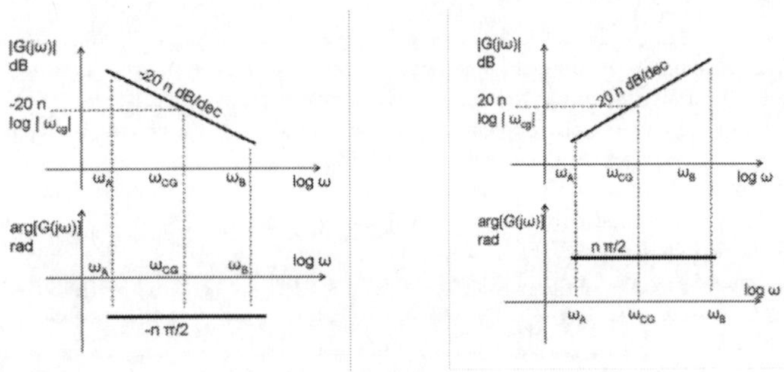

Figure 9. Sketch representation of the FO integral and derivator operators in frequency domain, by means of the Bode plots (magnitude above and phase bellow).

the standard differential–integral operator in two main groups: (a) they become the standard differential-integral operator of any order when n is an integer; (b) the Laplace transform of the operator D_t^n is s^n (provided zero initial conditions), and hence the frequency characteristic of this operator is $(j\omega)^n$. The latter is very appealing for the design of parametric modeling and control algorithms by using specifications in the frequency domain. A fundamental D_t^n operator, a generalization of integral and differential operators (differ-integration operator), is introduced as follows:

A fundamental D_t^n operator, a generalization of integral and differential operators (*differintegration* operator), is introduced as follows:

$$D_t^n = \left\{ \begin{array}{cc} \frac{d^n}{dt^n}, & n > 0 \\ 1, & n = 0 \\ \int_0^t (dA)^{-n}, & n < 0 \end{array} \right\} \tag{10}$$

where n is the fractional order and dA is a derivative function. The Laplace transform for integral and derivative order n are, respectively:

$$L\left\{D_t^{-n}f(t)\right\} = s^{-n}F(s) \tag{11}$$

$$L\{D_t^n f(t)\} = s^n F(s) \tag{12}$$

where $F(s) = L\{f(t)\}$ and s is the Laplace complex variable. The Fourier transform can be obtained by replacing s by $j\omega$ in the Laplace transform and the equivalent frequency-domain expressions of s^n are:

$$\frac{1}{(j\omega)^n} = \frac{1}{\omega^n}\left(\cos\frac{n\pi}{2} - j\sin\frac{n\pi}{2}\right) \tag{13}$$

$$(j\omega)^n = \omega^n\left(\cos\frac{n\pi}{2} + j\sin\frac{n\pi}{2}\right) \tag{14}$$

A comprehensive tutorial in fractional calculus is given in ([20, 21]).

Acknowledgment

C. M. Ionescu acknowledges the Flanders Research Center (FWO), grant number 12B3415N for its financial support. D. Copot acknowledges the support of the project FWO nr. G008113N.

References

[1] West, B., Barghava, V., and Goldberger, A. (1986). Be- yond the principle of similitude: renormalization of the bronchial tree. Journal of Applied Physiology, 60, 1089-1097.
[2] Magin, R. (2006). Fractional Calculus in Bioengineering. Begell House Publishers.
[3] Ionescu, C. (2013). The human respiratory system: an analysis of the interplay between anatomy, structure, breathing and fractal dynamics. Springer, Series in BioEngineering.
[4] Machado, J.T., Kiryakova, V., and Mainardi, F. (2013). Recent history of fractional calculus. Communications in Nonlinear Science and Numerical Simulation, 16, 1140-1153.
[5] Jacques, J. (1985). Compartmental analysis in biology and medicine.
[6] Berg, J. (2002). Biochemistry. Freeman and Company, New York.
[7] Meerschaert, M. and Sikorskii, A. (2012). Stochastic Models for Fractional Calculus. De Gruyter, Studies in Mathematics, Germany.
[8] Struys, M., De Keyser, R., and Absalom, A. (2011). Have we found the holy grail? Anesthesia and Analgesia.
[9] Schnider, T., Minto, C., Gambus, P., Andresen, C., Goodale, D., and Youngs, E. (1998). The influence of method of administration and covariates on the pharmacokinetics of propofol in adult volunteers. Anesthesiology, 88, 1170-1182.
[10] Schnider, T., Minto, C., Shafer, S., Gambus, P., Andresen, C., Goodale, D., and Youngs, E. (1999). The influence of age on propofol pharmacodynamicss. Anesthesiology, 90, 1502-1516.
[11] Higuchi, T. (1961). Rate of release of medicaments from ointment bases containing drugs in suspension. Journal of Pharmaceutical Sciences, 50, 874-875.
[12] Kopelman, R. (1988). Fractal reaction kinetics. Science, 241, 1620-1626.
[13] Rinaki, E., Valsami, G., and Macheras, P. (1998). The power law can describe the 'entire' drug release curve from hpmc-based matrix tablets: a hypothesis. International Journal of Pharmaceutics, 53, 749-754.
[14] Copot, D., Chevalier, A., Ionescu, C., and De Keyser, R. (2013). A two-compartment fractional derivative model for propofol diffusion in anesthesia. Conference on Systems and Control, Hyderabad, India, 28-31 August, 264-269.
[15] Ionescu, C., Copot, D., and Keyser, R.D. (2013). Three compartmental model for propofol diffusion during general anesthesia. Discontinuity, Nonlinearity and Complexity, 2, 357-368.
[16] Mortier, E., Struys, M., Smet, T.D., Versichelen, L., and Rolly, G. (1998). Closed-loop controlled administration of propofol using bispectral analysis. Anaesthesia, 53, 749-754.
[17] De Keyser, R. and Ionescu, C. (2012). A no-nonsense control engineering approach to anesthesia control during induction phase. IFAC Int Conf on Biology and Medicine, Budapest, Hungary, 379-384.
[18] De Keyser, R., Ionescu, C., and Festila, C. (2011). A one- step procedure for frequency response estimation based on switch-mode transfer function analyzer. Proceedings IEEE-CDC Conference on Decision and Control joint with ECC-European Control Conference, December 12-15, Orlando, Florida, 1189-1194.

[19] Goodwin, G., Aguero, J., Cea-Garrido, M., Salgado, M., and Yuz, J. (2013). Sampling and sampled-data models. IEEE Control Systems Magazine, 33, 34-53.
[20] Oustaloup, A. (1995). La derivation non-entiere. Hermes, Paris.
[21] Podlubny, I. (1999). Fractional differential equations. Academic Press, San Diego.

Digraphs Structures Corresponding to the Analogue Realisation of Fractional Continuous-Time System

Konrad A. MARKOWSKI
Warsaw University of Technology, Faculty of Electrical Engineering,
Institute of Control and Industrial Electronics,
Koszykowa 75, 00-662 Warsaw, Poland

E-mail: Konrad.Markowski@ee.pw.edu.pl
http://nas.isep.pw.edu.pl

Abstract. This paper presents a method of the determination of a minimal realisation of the fractional continuous-time linear system. For the proposed method, a digraph-based algorithm was constructed. In this paper, we have shown how we can perform the transfer matrix using electrical circuits consisting of resistances, capacitance and source voltages. We have also shown how after using the constant phase element method we can realize such a system. The proposed method was discussed and illustrated with some theoretical and practical numerical examples.

1. Introduction

In the last two decades, integral and differential calculus of an fractional order has become a subject of great interest in different areas of physics, biology, economics and other sciences. Fractional calculus is a generalization of traditional integer order integration and differentiation actions onto non-integer order. The idea of such a generalization was mentioned in 1695 by Leibniz and L'Hospital. The first definition of the fractional derivative was introduced by Liouville and Riemann at the end of the 19th century [1]. However, only just in the late 60' of the 20th century, this idea drew attention of engineers. Fractional calculus was found to be a very useful tool for modelling the behaviour of many materials and systems. Mathematical fundamentals of fractional calculus are given in the monographs [1–6]. Some other applications of fractional-order systems can be found in [7–12].

A lot of problems arise in one-dimensional systems, and they remain not completely solved; for example: realisation problem, minimal realisation problem, abilities problem, etc. As stated in the paper [13] Professor Benvenuti and Professor Farina *First of all, it is not clear what kind of mathematical "instruments" should be used to effectively tackle the minimality problem.* In many research studies in the area of the realisation problem or minimal realisation problem, we can find a canonical form of the system, i.e. constant matrix form, which satisfies the system described by the transfer function. Absolutely, in general we have a lot of solutions. This means that we can find many sets of matrices which fit into the system transfer function. In [14, 15] first attempts to determine a set of solutions for finding a set of possible realisations of the characteristic polynomial was proposed. If we add fractionality to the minimal positive realisation problem, then the realisation problem becomes much more complicated. In bibliography [3, 16–20] we can find some methods of determination of a positive realisation in the canonical form. In [21–24] the first proposition of a solution of the positive fractional realisation problem based on digraph theory was presented. The proposed method determines all possible digraph structures which correspond to the characteristic polynomial.

The main purpose of this paper is to present a method based on digraphs theory for computation of a minimal realisation of a given proper transfer function of all-pole continuous-time fractional linear systems. Conditions for the existence of a minimal realisation of a given proper transfer function will be established. For the determined realisation using constant phase element method an electrical analogue model will be created. Finally, a comparison of analogue models with theoretical models using a state diagram method and state-space method will be performed. It should be noted that this work is the next step in research on the realisation of the electrical circuits by using a digraph theory started in the publication [25] and [26].

This work has been organised as follows. Section 2 presents: some notations and basic definitions of a digraph theory, constant phase element method and basic properties of the fractional continuous-time system. Then, in Section 3 an algorithm for determination of a minimal realisation of the fractional continuous-time system was presented and electrical analogue network model using a constant phase element is given. Finally, in Section 4 theoretical and practical models are compared and concluding remarks, open problems and bibliography positions are given.

2. Background

2.1. Digraphs

A directed graph (or just digraph) $\mathcal{D}$ consists of a non-empty finite set $\mathbb{V}(\mathcal{D})$ of elements called vertices and a finite set $\mathbb{A}(\mathcal{D})$ of ordered pairs of distinct vertices called arcs [27]. We call $\mathbb{V}(\mathcal{D})$ the vertex set and $\mathbb{A}(\mathcal{D})$ the arc set of digraph $\mathcal{D}$. We will often write $\mathcal{D} = (\mathbb{V}, \mathbb{A})$ which means that $\mathbb{V}$ and $\mathbb{A}$ are the vertex set and arc set of $\mathcal{D}$, respectively. The order of $\mathcal{D}$ is the number of vertices in $\mathcal{D}$. The size of $\mathcal{D}$ is the number of arcs in $\mathcal{D}$. For an arc (v_1, v_2) the first vertex v_1 is its tail and the second vertex v_2 is its head.

There are two well-known methods of representation of digraph: list and incidence matrix. In this paper we are using incidence matrix to represent all digraphs. Method of constructing digraphs by this method is presented for example in [27], [28] or [29]. There exists $\mathcal{A}$-arc from vertex v_j to vertex v_i if and only if the (i,j)-th entry of the matrix $\mathbf{A}$ is non-zero. There exists $\mathcal{B}$-arc from source s_m to vertex v_i if and only if the (i,m)-th entry of the matrix $\mathbf{B}$ is non-zero. Let be given the positive system single input described by the following matrices

$$(\mathbf{A},\mathbf{B}) = \left(\begin{array}{cc} \begin{array}{c|ccc} v_i \backslash v_j & v_1 & v_2 & v_3 \\ \hline v_1 & 1 & 0 & 1 \\ v_2 & 1 & 0 & 1 \\ v_3 & 0 & 1 & 0 \end{array}, & \begin{array}{c|c} v_i \backslash s_m & s_1 \\ \hline v_1 & 1 \\ v_2 & 0 \\ v_3 & 0 \end{array} \end{array} \right) \tag{1}$$

we can draw one-dimensional digraph $\mathcal{D}$ consisting of vertices v_1, v_2, v_3 and source s_1. One-

Figure 1: One-dimensional digraph

dimensional digraph corresponding to system (1) is presented in Figure 1.

We present below some basic notions from graph theory which are used in further considerations. A walk in a digraph $\mathcal{D}$ is a finite sequence of arcs in which every two vertices

v_i and v_j are adjacent or identical. A walk in which all of the arcs are distinct is called a path. The path that goes through all vertices is called a finite path. If the initial and the terminal vertices of the path are the same, then the path is called a cycle. More information about use digraph theory in dynamical system is given in [30], [31].

2.2. Constant Phase Element Method

The constant phase element method relies on construction of the model consisting of resistors and capacitors. In relation to the method described in paper [32] in method presented in [33] a newer network model with correcting elements R_p and C_p (Figure 2) was constructed.

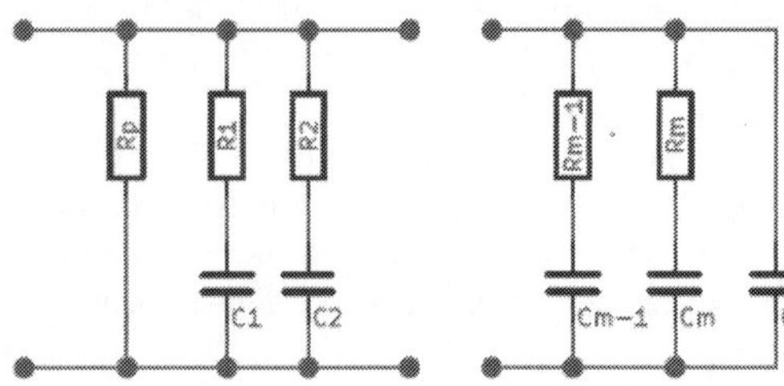

Figure 2: Network model with correcting elements

The impedance of the ideal constant phase element is defined as $Z(s) = Ds^\alpha$. For $s = j\omega$ we have $Z(j\omega) = D(j\omega)^\alpha = D\omega^\alpha(\cos\varphi + j\sin\varphi)$. The model parameters are selected according to the following procedure:

- Set the following parameters: resistance R_1, conductance C_1, number of lines m, argument value φ and oscillates around it with amplitude $\Delta\varphi$;
- Determine the following parameters:

$$\alpha = \frac{\varphi}{90}, \; ab = \frac{0.24}{1+\Delta\varphi}, \; a = 10^{\alpha \log ab}, \; b = \frac{ab}{a}, \; R_p = R_1\frac{1-a}{a}, \; C_p = C_1\frac{b^m}{1-b},$$
$$R_k = R_1 a^{k-1}, \quad C_k = C_1 b^{k-1}, \; k = 1, 2, \ldots, m. \tag{2}$$

2.3. Fractional Order System

Let us consider the continuous-time fractional linear system described by state-space equations:

$$\begin{aligned} {}_0\mathfrak{D}_t^\alpha x(t) &= \mathbf{A}x(t) + \mathbf{B}u(t), \quad 0 < \alpha \leqslant 1, \\ y(t) &= \mathbf{C}x(t) + \mathbf{D}u(t) \end{aligned} \tag{3}$$

where $x(t) \in \mathbb{R}^n$, $u(t) \in \mathbb{R}^m$, $y(t) \in \mathbb{R}^p$ are the state, input and output vectors respectively and $\mathbf{A} \in \mathbb{R}^{n\times n}$, $\mathbf{B} \in \mathbb{R}^{n\times m}$, $\mathbf{C} \in \mathbb{R}^{p\times n}$ and $\mathbf{D} \in \mathbb{R}^{p\times m}$.

The following Caputo definition of the fractional derivative will be used:

$${}_a^C\mathfrak{D}_t^\alpha = \frac{d^\alpha}{dt^\alpha} = \frac{1}{\Gamma(n-\alpha)}\int_a^t \frac{f^{(n)}(\tau)}{(t-\tau)^{\alpha+1-n}}d\tau, \tag{4}$$

where $\alpha \in \mathbb{R}$ is the order of a fractional derivative, $f^{(n)}(\tau) = \frac{d^n f(\tau)}{d\tau^n}$ and $\Gamma(x) = \int_0^\infty e^{-t}t^{x-1}dt$ is the gamma function.

Theorem 1 *The Laplace transform of the derivative-integral* (4) *has the form*

$$\mathcal{L}\left[\, {}_0^C\mathfrak{D}_t^\alpha \,\right] = s^\alpha F(s) - \sum_{k=1}^{n} s^{\alpha-k} f^{(k-1)}(0^+).$$

The proof of the Theorem 1 is given in [3].

After using the Laplace transform to (3), Theorem 1 and taking into account

$$\begin{aligned} X(s) &= \mathcal{L}\left[x(t)\right] = \int_0^\infty x(t)e^{-st}dt, \qquad (5)\\ \mathcal{L}\left[\mathfrak{D}^\alpha x(t)\right] &= s^\alpha X(s) - s^{\alpha-1}x_0 \end{aligned}$$

we obtain:

$$\begin{aligned} X(s) &= \left[\mathbf{I}_n s^\alpha - \mathbf{A}\right]^{-1}\left[s^{\alpha-1}x_0 + \mathbf{B}U(s)\right], \\ Y(s) &= \mathbf{C}X(s) + \mathbf{D}U(s),\ U(s) = \mathcal{L}\left[u(t)\right]. \qquad (6) \end{aligned}$$

After using (6) we can determine the transfer matrix of the system in the following form:

$$\mathbf{T}(s) = \mathbf{C}\left[\mathbf{I}_n s^\alpha - \mathbf{A}\right]^{-1}\mathbf{B} + \mathbf{D} \in \mathbb{R}^{p\times m}(s). \qquad (7)$$

Matrices $\mathbf{A}$, $\mathbf{B}$, $\mathbf{C}$ and $\mathbf{D}$ are called a realisation of the transfer matrix $\mathbf{T}(s)$ if they satisfy the equality (7). The realisation is called minimal if the dimension of the state matrix $\mathbf{A}$ is minimal among all possible realisations of $\mathbf{T}(s)$.

Our task is the following: *For a given transfer function* (7) *determine a minimal realisation of the system* (3) *using one-dimensional $\mathcal{D}$ digraphs theory. The dimension of the system must be the minimal among possible. For a designed realisation, build the analogue model and compare it with the theoretical model.*

3. Main results

The solution to the minimal realisation problem will be presented using special cases of the single-input single-output (SISO) system described by the transfer function of an all-pole system (which is the transfer function with only poles) in the following form:

$$T(s) = \frac{b_n s^{\alpha\cdot n}}{s^{\alpha\cdot n} - \sum_{i=0}^{n-1} a_i s^{\alpha\cdot i}} = \frac{b_n s^{\alpha\cdot n}}{s^{\alpha\cdot n} - a_{n-1}s^{\alpha(n-1)} - a_{n-2}s^{\alpha(n-2)} - \ldots - a_1 s^\alpha - a_0}. \qquad (8)$$

3.1. Theoretical Realisations

Using the transfer function (8) we have:

$$\mathbf{D} = \lim_{s^\alpha\to\infty} T(s) = \left[b_n\right] \qquad (9)$$

since $\lim_{s^\alpha\to\infty}[\mathbf{I}s^\alpha - \mathbf{A}] = 0$. The strictly proper transfer function is given by the equation:

$$T_{sp}(s) = T(s) - \mathbf{D} = \frac{\widetilde{b}_{n-1}s^{\alpha(n-1)} + \widetilde{b}_{n-2}s^{\alpha(n-2)} + \ldots + \widetilde{b}_1 s^\alpha + \widetilde{b}_0}{s^{\alpha\cdot n} - a_{n-1}s^{\alpha(n-1)} - a_{n-2}s^{\alpha(n-2)} - \ldots + a_1 s^\alpha - a_0} \qquad (10)$$

where $\widetilde{b}_{n-i} = b_n a_{n-i}$ for $i = 1, \ldots, n$.

The task of determining minimal realisations using digraphs theory, can be divided into two parts:

(a) **Part 1: Determine state matrix A.** To determine state matrix, we must multiplay the denominator of the strictly proper transfer function by $s^{-\alpha \cdot n}$ and make the decomposition of the characteristic polynomial

$$d(s) = 1 - a_{n-1}s^{-\alpha} - a_{n-2}s^{-2\cdot\alpha} - \ldots - a_1 s^{\alpha(1-n)} - a_0 s^{-\alpha\cdot n} \tag{11}$$

into a set of monomials. Using Proposition 1 presented in paper [34] we create all possible digraph representations. Then using Theorem 1 presented in paper [14] and [15] we can create all digraph realisations of the characteristic polynomials. Each digraph corresponding to a characteristic polynomial must satisfy two conditions. The first condition relates to the existence of the common part in the digraph (blue vertex), the second condition relates to non-existence of additional cycles in the digraph. One of the digraph structure corresponding to characteristic polynomial (11) presented in Figure 3 and state matrix corresponding to digraph has the form:

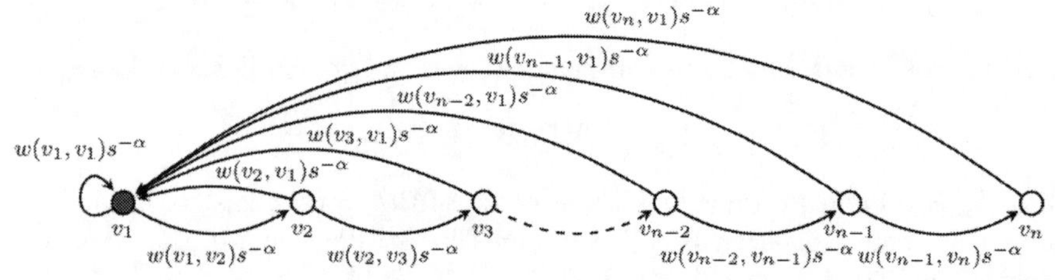

Figure 3: One of the possible realisation of the characteristic polynomial $d(s)$

$$\mathbf{A} = \begin{array}{c} v_i \backslash^{v_j} \\ v_1 \\ v_2 \\ v_3 \\ \vdots \\ v_{n-2} \\ v_{n-1} \\ v_n \end{array} \begin{array}{c} \begin{array}{ccccccc} v_1 & v_2 & v_3 & \cdots & v_{n-2} & v_{n-1} & v_n \end{array} \\ \left[\begin{array}{ccccccc} w(v_1,v_1) & w(v_2,v_1) & w(v_3,v_1) & \cdots & w(v_{n-2},v_1) & w(v_{n-1},v_1) & w(v_n,v_1) \\ w(v_1,v_2) & 0 & 0 & \cdots & 0 & 0 & 0 \\ 0 & w(v_2,v_3) & 0 & \cdots & 0 & 0 & 0 \\ \vdots & \vdots & \ddots & \ddots & \vdots & \vdots & \vdots \\ 0 & 0 & 0 & \ddots & 0 & 0 & 0 \\ 0 & 0 & 0 & \cdots & w(v_{n-2},v_{n-1}) & 0 & 0 \\ 0 & 0 & 0 & \cdots & 0 & w(v_{n-1},v_n) & 0 \end{array}\right] \end{array} \tag{12}$$

(b) **Part 2: Determine matrices B and C.** To determine matrices, we must multiply the nominator of the transfer function (10) by $s^{-\alpha(n-1)}$. Then we obtain the following polynomial

$$n(s) = \widetilde{b}_{n-1} + \widetilde{b}_{n-2}s^{-\alpha} + \ldots + \widetilde{b}_1 s^{-\alpha\cdot n} + \widetilde{b}_0 s^{-\alpha(n-1)} \tag{13}$$

and expand the digraph created in the first step. We have two possible structures:

- **Structure 1:** We connect source vertex s corresponding to matrix **B** with vertex $v_1, \ldots, v_n$ and output vertex y corresponding to matrix **C** with vertex v_i belonging to a set of common parts of a digraph. This digraphs structure was considered in the paper [34].
- **Structure 2:** We connect source vertex s corresponding to matrix **B** with vertex v_i belonging to a set of common parts of a digraph, and output vertex y corresponding to matrix **C** with vertex $v_1, \ldots, v_n$ (Figure 4). This digraphs structure was considered in the paper [34].

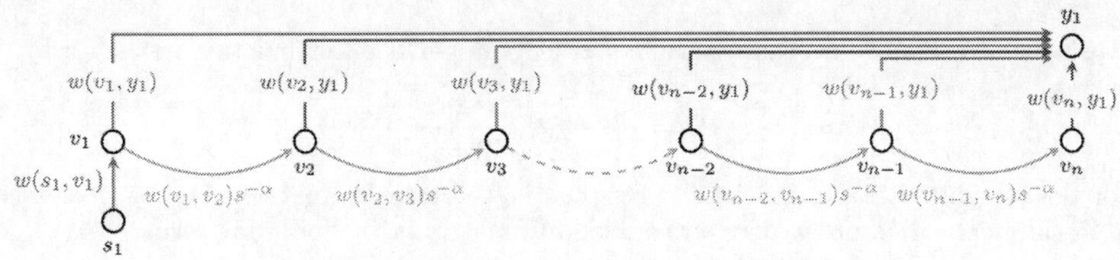

Figure 4: One-dimensional digraph corresponding to polynomial $n(s)$

In the future consideration, we will investigate only the Structure 2. From digraph presented in Figure 4, we can write the set of the equations which we compare with the same power of the polynomial (13). We obtain the following equality:

$$\left\{\begin{array}{c|lcl}
 & w(s_1,v_1)\cdot w(v_1,y_1) & = & \widetilde{b}_{n-1} \\
s^{-\alpha} & w(s_1,v_1)\cdot w(v_1,v_2)\cdot w(v_2,y_1) & = & \widetilde{b}_{n-2} \\
s^{-2\cdot\alpha} & w(s_1,v_1)\cdot w(v_1,v_2)\cdot w(v_2,v_3)\cdot w(v_3,y_1) & = & \widetilde{b}_{n-3} \\
\vdots & \vdots & \vdots & \vdots \\
s^{-\alpha(n-3)} & w(s_1,v_1)\cdot w(v_1,v_2)\cdot w(v_2,v_3)\cdot\ldots\cdot w(v_{n-2},y_1) & = & \widetilde{b}_2 \\
s^{-\alpha(n-2)} & w(s_1,v_1)\cdot w(v_1,v_2)\cdot w(v_2,v_3)\cdot\ldots\cdot w(v_{n-2},v_{n-1})\cdot w(v_{n-1},y_1) & = & \widetilde{b}_1 \\
s^{-\alpha(n-1)} & w(s_1,v_1)\cdot w(v_1,v_2)\cdot w(v_2,v_3)\cdot\ldots\cdot w(v_{n-2},v_{n-1})\cdot w(v_{n-1},v_n)\cdot w(v_n,y_1) & = & \widetilde{b}_0
\end{array}\right. \tag{14}$$

After solving the set of the equation (14), we can write **B** and **C** matrix in the following form:

$$\underbrace{\begin{array}{c} v_i\backslash^s \\ v_1 \\ v_2 \\ v_3 \\ \\ v_{n-1} \\ v_n \end{array}\begin{array}{c} s_1 \\ \left[\begin{array}{c} w(s_1,v_1) \\ 0 \\ 0 \\ \vdots \\ 0 \\ 0 \end{array}\right]\end{array}}_{\mathbf{B}\in\mathbb{R}^{n\times 1}}, \underbrace{\begin{array}{c} y\backslash^{v_j} \\ y_1 \end{array}\begin{array}{c} \begin{array}{cccccc} v_1 & v_2 & v_3 & \cdots & v_{n-1} & v_n \end{array} \\ \left[\begin{array}{cccccc} w(v_1,y_1) & w(v_2,y_1) & w(v_3,y_1) & \cdots & w(v_{n-1},y_1) & w(v_n,y_1) \end{array}\right]\end{array}}_{\mathbf{C}\in\mathbb{R}^{1\times n}} \tag{15}$$

where:

$$\begin{aligned}
w(v_1,y_1) &= \frac{\widetilde{b}_{n-1}}{w(s_1,v_1)}, w(v_2,y_1) = \frac{\widetilde{b}_{n-2}}{w(s_1,v_1)w(v_1,v_2)}, w(v_3,y_1) = \frac{\widetilde{b}_{n-3}}{w(s_1,v_1)w(v_1,v_2)w(v_2,v_3)}, \\
w(v_{n-1},y_1) &= \frac{\widetilde{b}_1}{w(s_1,v_1)w(v_1,v_2)w(v_2,v_3)\ldots w(v_{n-2},v_{n-1})}, \\
w(v_n,y_1) &= \frac{\widetilde{b}_0}{w(s_1,v_1)w(v_1,v_2)w(v_2,v_3)\ldots w(v_{n-2},v_{n-1})w(v_{n-1},v_n)}.
\end{aligned}$$

Example 1 *Find a positive realisation of the all-pole rational positive continuous-time linear system with a proper transfer function:*

$$\mathbf{T}(s) = \frac{4s^{2.1}}{s^{2.1} + 0.5s^{1.4} + 0.5s^{0.7} + 0.7}. \tag{16}$$

The matrix **D** for a given proper transfer function (16) can be computed using equation (9). In the considered example, we have:

$$\mathbf{D} = \lim_{s^{\alpha}\to\infty} T(s) = 4. \tag{17}$$

After using (16) and (17), we can determine strictly proper transfer function in the form:

$$T_{sp}(s) = T(s) - \mathbf{D} = \frac{-2s^{1.4} - 2s^{0.7} - 2.8}{s^{2.1} + 0.5s^{1.4} + 0.5s^{0.7} + 0.7}. \tag{18}$$

One of the possible digraph structure corresponding to transfer function (18) is presented in Figure 5. From the digraph, we can write the state matrix in the following form:

$$\mathbf{A} = \begin{bmatrix} 0 & 0 & w(v_3,v_1) \\ w(v_1,v_2) & 0 & w(v_3,v_2) \\ 0 & w(v_2,v_3) & w(v_3,v_3) \end{bmatrix} = \begin{bmatrix} 0 & 0 & -0.7 \\ 1 & 0 & -0.5 \\ 0 & 1 & -0.5 \end{bmatrix} \tag{19}$$

and we can write a set of equations in the form:

$$\left\{ \begin{array}{l|l} & w(s_1,v_3)\cdot w(v_3,y_1) = -2 \\ s^{-0.7} & w(s_1,v_3)\cdot w(v_3,v_2)\cdot w(v_2,y_1) + w(s_1,v_3)\cdot w(v_3,v_1)\cdot w(v_1,y_1) = -2 \\ s^{-1.4} & w(s_1,v_3)\cdot w(v_3,v_1)\cdot w(v_1,v_2)\cdot w(v_2,y_1) = -2.8 \end{array} \right. . \tag{20}$$

After solving the set of the equation (20), we can write the input and output matrix in the following form:

$$\mathbf{B} = \begin{bmatrix} 0 \\ 0 \\ w(s_1,v_3) \end{bmatrix}, \mathbf{C} = \begin{bmatrix} 0 & \frac{4}{w(s_1,v_3)} & \frac{-2}{w(s_1,v_3)} \end{bmatrix}. \tag{21}$$

The desired realisation of the (16) is given by (17), (19), (21).

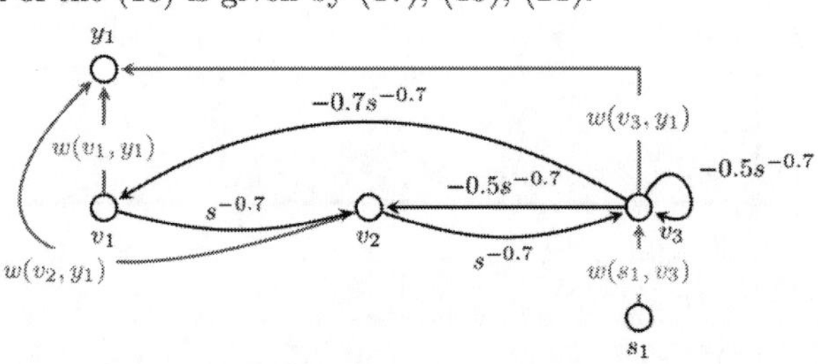

Figure 5: One of the possible realisation of the transfer function (18).

3.2. Practical Analogue Realisation

In this section, a practical realisation of the transfer using Constant Phase Element (CPE) – which was introduced in Section 2.2 – will be presented. The obtained realisation was compared with numerical simulations. The numerical simulations were performed with FOMCON toolbox [35, 36].

Using CPE method, we construct a network model consisting of resistance and capacitance. Consider two following models corresponding to $K(s) = D/s^{0.7}$ where $D \in \mathbb{R}$ is constant value, and it depends on resistance and capacitance. In considered simulations, the constant D has been determined experimentally.

- **Model 1:** Given $R_1 = 2k\Omega$, $C_1 = 1mF$, $m = 20$, $\Delta\varphi = 0.26$. For the given initial parameters we can determine model coefficients:

$$ab = \frac{0.24}{1+\Delta\varphi} = 0.1905; \quad a = 10^{\alpha \log(ab)} = 0.3132; \quad b = \frac{ab}{a} = 0.6081;$$

$$R_p = \frac{R_1\cdot(1-a)}{a} = 4.3847k\Omega; \qquad C_p = \frac{C_1\cdot b^m}{1-b} = 121.85nF.$$

Finally, the network model consists of the following elements:

– Resistors:

$$R_1 = 2k\Omega;\ R_2 = 626.5\Omega;\ R_3 = 196.2\Omega;\ R_4 = 61.5\Omega;\ R_5 = 19.2\Omega;\ R_6 = 6\Omega;$$
$$R_7 = 1.9\Omega;\ R_8 = 0.6\Omega;\ R_9 = 0.18\Omega;\ R_{10} = 0.06\Omega;\ R_{11} = 0.02\Omega;\ R_{12} = 5.7m\Omega$$
$$R_{13} = 1.8m\Omega;\ R_{14} = 0.55m\Omega;\ R_{15} = 175.1\mu\Omega;\ R_{16} = 54\mu\Omega;\ R_{17} = 17\mu\Omega;$$
$$R_{18} = 5.4\mu\Omega;\ R_{19} = 1.7\mu\Omega;\ R_{20} = 0.5\mu\Omega.$$

– Capacitors:

$$C_1 = 1mF;\ C_2 = 0.6mF;\ C_3 = 0.37mF;\ C_4 = 0.22mF;\ C_5 = 0.14mF;$$
$$C_6 = 83\mu F;\ C_7 = 50\mu F;\ C_8 = 30\mu F;\ C_9 = 18.6\mu F;\ C_{10} = 11.4\mu F;$$
$$C_{11} = 6.9\mu F;\ C_{12} = 4.2\mu F;\ C_{13} = 2.6\mu F;\ C_{14} = 1.6\mu F;\ C_{15} = 0.94\mu F;$$
$$C_{16} = 574nF;\ C_{17} = 349nF;\ C_{18} = 212nF;\ C_{19} = 129nF;\ C_{20} = 78.5nF.$$

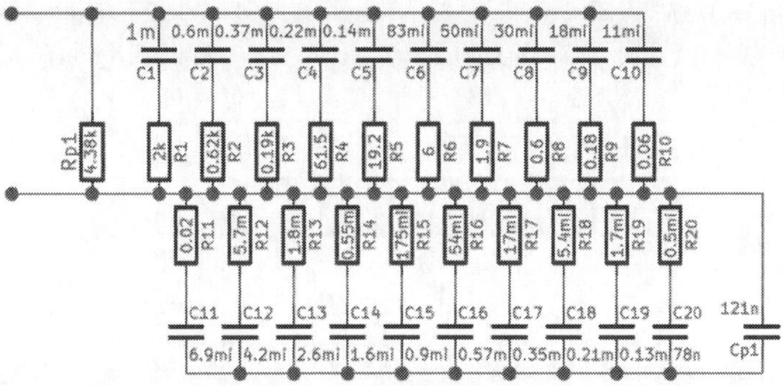

Figure 6: The first network model

In Figure 6 the network model containing determined elements has been presented. It should be noted that the complexity of the model is dependent on the number of lines. In our case, we have 22 lines: 2 lines including correcting elements (resistor R_p and capacitor C_p); 20 lines including resistors and capacitors.

After this operations, we can determine the following transmittance

$$G_1 = \frac{\begin{array}{c} 1.55 \cdot 10^{-131} \cdot s^{20} + 4.614 \cdot 10^{-118} \cdot s^{19} + 2.198 \cdot 10^{-105} s^{18} + 1.935 \cdot 10^{-93} s^{17} + \\ 3.227 \cdot 10^{-82} s^{16} + 1.024 \cdot 10^{-71} s^{15} + 6.187 \cdot 10^{-62} s^{14} + 7.12 \cdot 10^{-53} s^{13} + \\ 1.561 \cdot 10^{-44} s^{12} + 6.518 \cdot 10^{-37} s^{11} + 5.184 \cdot 10^{-30} s^{10} + 7.854 \cdot 10^{-24} s^{9} + \\ 2.266 \cdot 10^{-18} s^{8} + 1.246 \cdot 10^{-13} s^{7} + 1.304 \cdot 10^{-09} s^{6} + 2.601 \cdot 10^{-06} s^{5} + \\ 0.0009876 s^{4} + 0.07136 s^{3} + 0.9766 s^{2} + 2.471 s + 1 \end{array}}{\begin{array}{c} 1.889 \cdot 10^{-138} s^{21} + 9.894 \cdot 10^{-125} s^{20} + 7.877 \cdot 10^{-112} s^{19} + 1.145 \cdot 10^{-99} s^{18} + \\ 3.141 \cdot 10^{-88} s^{17} + 1.639 \cdot 10^{-77} s^{16} + 1.629 \cdot 10^{-67} s^{15} + 3.084 \cdot 10^{-58} s^{14} + \\ 1.112 \cdot 10^{-49} s^{13} + 7.634 \cdot 10^{-42} s^{12} + 9.985 \cdot 10^{-35} s^{11} + 2.488 \cdot 10^{-28} s^{10} + \\ 1.181 \cdot 10^{-22} s^{9} + 1.067 \cdot 10^{-17} s^{8} + 1.837 \cdot 10^{-13} s^{7} + 6.026 \cdot 10^{-10} s^{6} + \\ 3.764 \cdot 10^{-07} s^{5} + 4.474 \cdot 10^{-05} s^{4} + 0.00101 s^{3} + 0.004264 s^{2} + \\ 0.003115 s + 0.0002281 \end{array}} \tag{22}$$

corresponding to the considered model.

- **Model 2:** Given $R_1 = 10k\Omega$, $C_1 = 1mF$, $m = 10$, $\Delta\varphi = 0.26$. For the given initial parameters we can determine model coefficients:

$$ab = \frac{0.24}{1+\Delta\varphi} = 0.1905; \quad a = 10^{\alpha \log(ab)} = 0.3132; \quad b = \frac{ab}{a} = 0.6081;$$
$$R_p = \frac{R_1 \cdot (1-a)}{a} = 43.847k\Omega; \qquad C_p = \frac{C_1 \cdot b^m}{1-b} = 17.632\mu F.$$

Finally, the network model consists of the following elements:

– Resistors:

$$R_1 = 2k\Omega;\ R_2 = 6.26k\Omega;\ R_3 = 1.96k\Omega;\ R_4 = 614.75\Omega;\ R_5 = 192.57\Omega;$$
$$R_6 = 60.32\Omega;\ R_7 = 18.89\Omega;\ R_8 = 5.91\Omega;\ R_9 = 1.85\Omega;\ R_{10} = 0.58\Omega.$$

– Capacitors:

$$C_1 = 1mF;\ C_2 = 0.61mF;\ C_3 = 0.37mF;\ C_4 = 0.22mF;\ C_5 = 0.14mF;$$
$$C_6 = 83.13\mu F;\ C_7 = 50.55\mu F;\ C_8 = 30.74\mu F;\ C_9 = 18.69\mu F;\ C_{10} = 11.36\mu F.$$

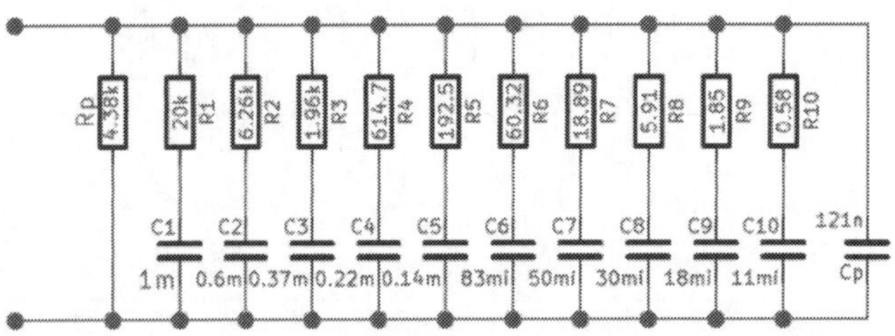

Figure 7: The second network model

In Figure 7 the network model containing determined elements has been presented. In this case, we have 12 lines: 2 lines including correcting elements (resistor R_p and capacitor C_p); 10 lines including resistors and capacitors.
After this operations, we can determine the following transmittance

$$G_2 = \frac{\begin{array}{c}4.01 \cdot 10^{-20}s^{10} + 7.504 \cdot 10^{-15}s^9 + 2.247 \cdot 10^{-10}s^8 + 1.244 \cdot 10^{-6}s^7 + \\ 0.001304s^6 + 0.26s^5 + 9.876s^4 + 71.36s^3 + 97.66s^2 + 24.71s + 1\end{array}}{\begin{array}{c}7.07 \cdot 10^{-25}s^{11} + 2.328 \cdot 10^{-19}s^{10} + 1.165 \cdot 10^{-14}s^9 + 1.064 \cdot 10^{-10}s^8 + \\ 1.836 \cdot 10^{-7}s^7 + 6.025 \cdot 10^{-5}s^6 + 0.003763s^5 + 0.04474s^4 + 0.101s^3 + \\ 0.04264s^2 + 0.003115s + 2.281 \cdot 10^{-5}\end{array}} \tag{23}$$

corresponding to considered model.

4. Comparison and Discussion

To compare the theoretical model with determined CPE network models in Simulink, test environment (Figure 8) has been built. It consists of three areas: the first represents a model as a state diagram system (Figure 9(a)); the second represents model as a state-space system and the third represents CPE network model (Figure 9(b)).

Let us consider the following two cases:

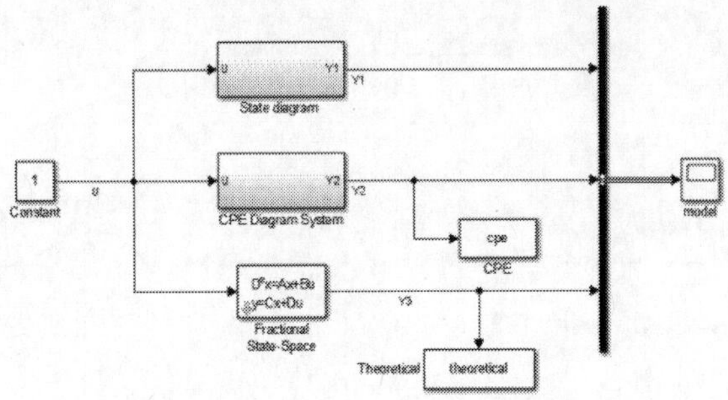

Figure 8: Test environment

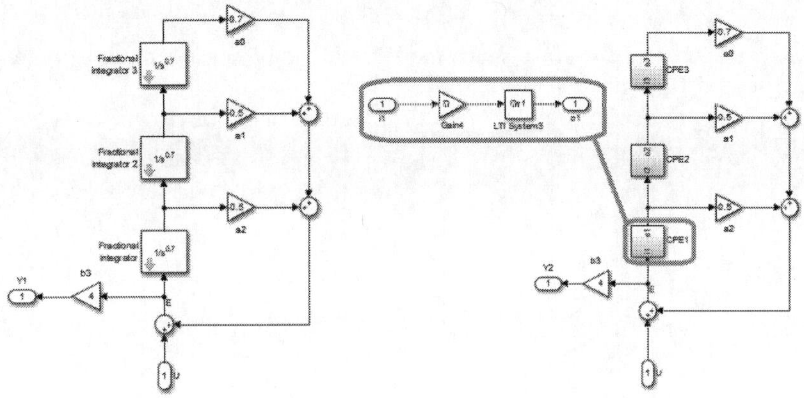

(a) (b)

Figure 9: Models: (a) state diagram (b) CPE model.

- **Case 1:** Figure 10(a) presents a step response of the state-space model (yellow line), state diagram model (red line) and the first network model (green line). As we can see the fit of the model is not the best. We can notice quite clear difference between the theoretical and physical model. Particularly, it is clearly visible in the graph of mean square error presented in Figure 11 (blue line).
- **Case 2:** Figure 10(b) presents step response of the state-space model (yellow line), state diagram model (red line) and the second network model (green line). As we can see the fit of the model is very good. There are no significant differences between the theoretical and physical model. Particularly, it is clearly visible in the graph of mean square error presented in Figure 11 (red line).

From the above observation, we can write a very important conclusion. Before we start physically building an analogue model, we should carry out a series of simulations using Simulink, for example. In the paper [33] it has been proposed that for a higher value of the parameter m the model is more accurate. It appears, that the quality of the model and the adjustment to the theoretical model are influenced by many parameters. For example in Model 1 we used 20 lines and the fit is worse than in the case of Model 2 in which we used only 10 lines. Additionally, the quality of the model is affected by the way of selection of initial parameters: R_1, C_1, and $\Delta\varphi$.

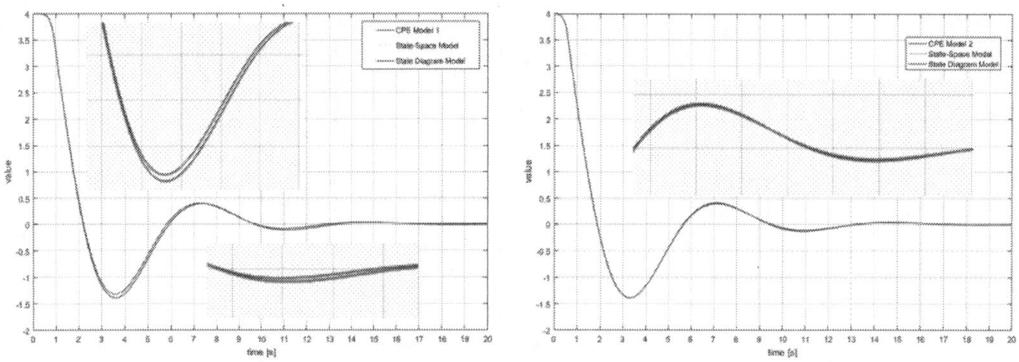

(a) (b)

Figure 10: Step response of the state-space model, state diagram model and (a) the first CPE model (b) the second CPE model.

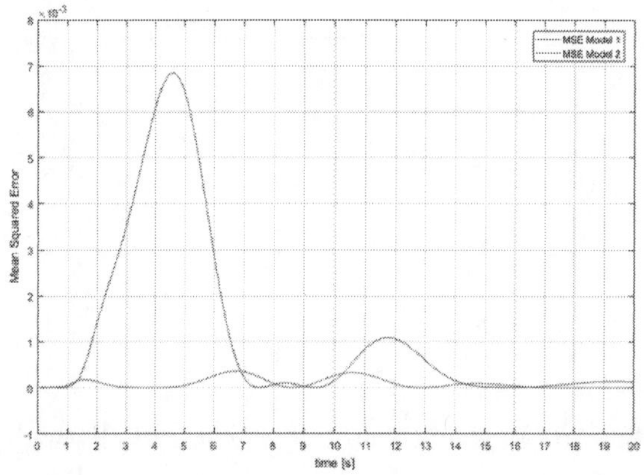

Figure 11: Mean square error

5. Concluding Remarks

The paper presents a method, based on the one-dimensional digraph theory, for finding the realisations of a given proper transfer function of all-pole continuous-time fractional linear systems. For the obtained realisation, we have constructed two models based on constant phase element method. The first model consists from 20 lines and the second model 10 lines. Each line consists of properly selected resistors and capacitors. Prepared theoretical models (state-space model and space diagram model) and practical model were compared in terms of the quality of the model and its fit.

Further work includes extension of the digraph based algorithm to find a broader class of electrical circuits corresponding to the transfer function and practical construction of the analogue models for different values of α.

References

[1] Nishimoto K 1984 *Fractional Calculus* (Koriama: Decartess Press)
[2] Das S 2011 *Functional Fractional Calculus* (New York: Springer Berlin Heidelberg) URL `http://dx.doi.org/10.1007/978-3-642-20545-3`
[3] Kaczorek T and Sajewski L 2014 *The Realization Problem for Positive and Fractional Systems* (Berlin: Springer International Publishing) URL `http://dx.doi.org/10.1007/978-3-319-04834-5`
[4] Miller K and Ross B 1993 *An Introduction to the Fractional Calculus and Fractional Differenctial Equations* (New York: Willeys)

[5] Ortigueira M D 2011 *Fractional Calculus for Scientists and Engineers* (Springer Netherlands: Academic Press) URL 10.1007/978-94-007-0747-4
[6] Podlubny I 1999 *Fractional Differential Equations* (San Diego: Academic Press)
[7] Ortigueira M D, Rivero M and Trujillo J J 2015 *Journal of King Saud University-Science*
[8] Martynyuk V and Ortigueira M 2015 *Signal Processing* **107** 355–360
[9] Machado J and Lopes A M 2015 *Fractional Calculus and Applied Analysis* **18** 1518–1536
[10] Machado J, Mata M E and Lopes A M 2015 *Entropy* **17** 5402–5421
[11] Podlubny I, Skovranek T and Datsko B 2014 Recent advances in numerical methods for partial fractional differential equations *Control Conference (ICCC), 2014 15th International Carpathian* (IEEE) pp 454–457
[12] Petras I, Sierociuk D and Podlubny I 2012 *Signal Processing, IEEE Transactions on* **60** 5561–5566
[13] Benvenuti L and Farina L 2003 *Kybernetika* **39** 217–228
[14] Hryniów K and Markowski K A 2014 Parallel digraphs-building algorithm for polynomial realisations *Proceedings of 2014 15th International Carpathian Control Conference (ICCC)* pp 174–179 URL http://dx.doi.org/10.1109/CarpathianCC.2014.6843592
[15] Hryniów K and Markowski K A 2015 Optimisation of digraphs creation for parallel algorithm for finding a complete set of solutions of characteristic polynomial *Proceedings of 20th International Conference on Methods and Models in Automation and Robotics, MMAR 2015, Miedzyzdroje, Poland, August 24–27* pp 1139–1144 URL http://dx.doi.org/10.1109/MMAR.2015.7284039
[16] Farina L and Rinaldi S 2000 *Positive linear systems: theory and applications* (New York: Wiley-Interscience, Series on Pure and Applied Mathematics)
[17] Benvenuti L, De Santis A and Farina L 2003 *Positive Systems* Lecture Notes on Control and Information Sciences 294 (Berlin: Springer-Verlag)
[18] Benvenuti L and Farina L 2004 *IEEE Transactions on Automatic Control* **49** 651–664
[19] Kaczorek T 2011 *Selected Problems of Fractional Systems Theory* (Berlin: Springer Verlag)
[20] Sajewski L 2012 *Pomiary Automatyka Robotyka* 413–417
[21] Hryniów K and Markowski K A 2015 Digraphs minimal realisations of state matrices for fractional positive systems *Progress in Automation, Robotics and Measuring Techniques* (*Advances in Intelligent Systems and Computing* vol 350) ed Szewczyk R, Zieliski C and Kaliczyska M (Springer International Publishing) pp 63–72 ISBN 978-3-319-15795-5 URL http://dx.doi.org/10.1007/978-3-319-15796-2_7
[22] Markowski K A 2015 Deretmination of minimal positive realisation of one-dimensional continuous-time fractional linear system *Procedings of the International Symposium on Fractional Signal and Systems* ed Both-Rusu R (U.T. Press) pp 7–12 ISBN 978-606-737-084-3
[23] Markowski K A and Hryniów K 2016 Digraphs minimal positive stable realisations for fractional one-dimensional systems *Proceedings of 7th Conference on Non-integer Order Calculus and its Applications7th Conference on Non-integer Order Calculus and its Applications* pp 105–118 ISBN 978-3-319-23038-2 URL http://dx.doi.org/10.1007/978-3-319-23039-9_9
[24] Markowski K A 2016 *International Journal of Dynamics and Control* URL http://dx.doi.org/10.1007/s40435-016-0232-3
[25] Markowski K A 2015 *Journal of Physics: Conference Series* **659** 012041 URL http://stacks.iop.org/1742-6596/659/i=1/a=012041
[26] Markowski K A 2016 Digraphs structures corresponding to realisation of multi-order fractional electrical circuits *2016 IEEE International Conference on Automation, Quality and Testing, Robotics (AQTR)* pp 1–6
[27] Bang-Jensen J and Gutin G 2009 *Digraphs: Theory, Algorithms and Applications* (London: Springer-Verlag)
[28] Wallis W D 2007 *A Beginner's Guide to Graph Theory* (Biiokhäuser)
[29] Godsil C and Royle G 2001 *Algebraic Graph Theory* (Springer Verlag)
[30] Fornasini E and Valcher M E 1997 *Linear Algebra and Its Applications* **263** 275–310
[31] Fornasini E and Valcher M E 2005 *IEEE Transaction on Circuits and Systems I* 576–585
[32] Machado J A T 2001 *JACIII* **5** 279–285 URL http://dx.doi.org/10.20965/jaciii.2001.p0279
[33] J Valsa P Dvorak M F 2011 *Radioengineering* **20** 619–626 ISSN 1805-9600
[34] Markowski K A 2016 Digraphs structures corresponding to minimal realisation of fractional continuous-time linear systems with all-pole and all-zero transfer function *2016 IEEE International Conference on Automation, Quality and Testing, Robotics (AQTR)* pp 1–6
[35] Tepljakov A, Petlenkov E and Belikov J 2011 *International Journal of Microelectronics and Computer Science* **2** 51–62
[36] Tepljakov A, Petlenkov E and Belikov J 2011 FOMCON: Fractional order modeling and control toolbox for MATLAB *The 18th International Conference Mixed Design of Integrated Circuits and Systems* (Gliwice, Poland) p 684–689

Realisation of Linear Continuous-Time Fractional Singular Systems Using Digraph-Based Method. First approach.

Konrad A. MARKOWSKI
Warsaw University of Technology, Faculty of Electrical Engineering,
Institute of Control and Industrial Electronics,
Koszykowa 75, 00-662 Warsaw, Poland

E-mail: Konrad.Markowski@ee.pw.edu.pl
http://nas.isep.pw.edu.pl

Abstract. In this paper, a new method for computation of a minimal realisation of a given proper transfer function of fractional continuous-time singular one-dimensional linear systems using one-dimensional digraphs theory $\mathcal{D}$ has been presented. For the proposed method, an algorithm was constructed. The proposed solution allows minimal digraphs construction for any one-dimensional system. The proposed method was discussed and illustrated with numerical examples.

Nomenclature

$\mathbf{A}$ - matrices denoted by the bold font;
$\mathbf{A}^{-1}$ – inverse matrix;
$\det(\mathbf{A})$ – determinant of the matrix $\mathbf{A}$;
$\mathbb{A}$ - the sets denoted by the double line;
a - lower/upper indices and polynomial coefficients will be written as a small font;
$\mathcal{D}$ - one-dimensional digraph;
$\mathfrak{M}(\mathcal{D})$ – digraph–mask;
$\mathbb{R}^{n\times m}$ – the set $n \times m$ of real matrices;
$\mathbb{Z}_+$ – the set of non-negative integers;
$\mathbf{I}_n$ – the $n \times n$ identity matrix;
$\mathcal{L}\{f(t)\}$ – $\mathcal{L}$-transform of the function $f(t)$;
$T(s)$ - transfer function of one variable s;
$p(s)$ – polynomial of the variable s;

1. Introduction

Singular continuous-time system were considered in many papers and books [1–8]. The properties and the use matrix theory of the singular system were established in [1, 9, 10]. There are many problems associated with the analysis and synthesis of the singular systems. One of the very important problems is realisation and minimal realisation problem. In many research studies, we can find a constant matrix form, which satisfies the system described by the transfer function [5, 8, 11–13]. In fact, there are many sets of matrices which fit into the system transfer function.

In the last two decades, integral and differential calculus of a fractional order has become a subject of great interest in different areas of physics, biology, economics and other sciences. Fractional calculus is a generalization of traditional integer order integration and differentiation actions onto a non-integer order. The idea of such a generalization was mentioned in 1695 by Leibniz and L'Hospital. The first definition of the fractional derivative was introduced by Liouville and Riemann at the end of the 19th century [14]. However, only just in the late 60' of the 20th century, this idea drew attention of engineers. Fractional calculus was found to be a very useful tool for modelling the behaviour of many materials and systems. Mathematical fundamentals of fractional calculus are given in the monographs [8, 14–18]. Some other applications of fractional-order systems can be found in [19–24].

The main purpose of this paper is to present a method based on one-dimensional digraph theory for computation the set of minimal realisation of a given proper transfer function of the one-dimensional continuous-time fractional singular system. In the paper, a digraph structure corresponding to minimal realisations of the singular system is presented and discussed. Sufficient conditions for the existence of a minimal realisation of a given proper transfer function systems are established. In this paper, for the first time, a digraph-mask was introduced and used. It is very important in analysis of the singular systems, with the use of digraphs theory. As a result, we proposed the procedure for determining the set of realisations in the class $\mathcal{K}_1$ of the digraph structure. The digraphs classes in [25] have been defined and presented in details. This work is the next step in the research on the determination of the realisation problem $(\mathbf{A}, \mathbf{E}, \mathbf{B}, \mathbf{C})$ of the continuous-time fractional singular systems by using a digraph theory, which was started in the publication [26], [27], [28], in which a minimal realisation of the continuous-time linear fractional system as a set of matrices $(\mathbf{A}, \mathbf{B}, \mathbf{C})$ was determined.

This work has been organised as follows: Section 1 presents some notations and basic definitions of a digraph theory; basic properties of the singular discrete-time linear system defined as the state-space representation. In the Section 2, we construct a procedure for determination minimal realisation of the singular continuous-time system. Finally, we demonstrate some numerical examples (Section 3), present concluding remarks, open problems and bibliography positions.

1.1. Digraphs

A directed graph (or just digraph) $\mathcal{D}$ consists of a non-empty finite set $\mathbb{V}(\mathcal{D})$ of elements called vertices and a finite set $\mathbb{A}(\mathcal{D})$ of ordered pairs of distinct vertices called arcs [29]. We call $\mathbb{V}(\mathcal{D})$ the vertex set and $\mathbb{A}(\mathcal{D})$ the arc set of digraph $\mathcal{D}$. We will often write $\mathcal{D} = (\mathbb{V}, \mathbb{A})$ which means that $\mathbb{V}$ and $\mathbb{A}$ are the vertex set and arc set of $\mathcal{D}$, respectively. The order of $\mathcal{D}$ is the number of vertices in $\mathcal{D}$. The size of $\mathcal{D}$ is the number of arcs in $\mathcal{D}$. For an arc (v_1, v_2) the first vertex v_1 is its tail and the second vertex v_2 is its head.

There are two well-known methods of representation of digraph: list and incidence matrix. In this paper we are using incidence matrix to represent all digraphs. Method of constructing digraphs by this method is presented for example in [29], [30] or [31].

1.2. Singular system

Let us consider the fractional singular continuous-time linear system

$$\begin{aligned} \mathbf{E}\,{}_0\mathfrak{D}_t^{\alpha} x(t) &= \mathbf{A}x(t) + \mathbf{B}u(t), \\ y(t) &= \mathbf{C}x(t), \end{aligned} \tag{1}$$

where $x(t) \in \mathbb{R}^n$, $u(t) \in \mathbb{R}^m$ and $y(t) \in \mathbb{R}^p$ are the state, input and output vectors, respectively and

$$\mathbf{E} \in \mathbb{R}^{n\times n},\ \mathbf{A} \in \mathbb{R}^{n\times n},\ \mathbf{B} \in \mathbb{R}^{n\times m},\ \mathbf{C} \in \mathbb{R}^{p\times n}. \tag{2}$$

The following Caputo definition of the fractional derivative will be used:

$${}_a^C\mathfrak{D}_t^{\alpha} = \frac{d^{\alpha}}{dt^{\alpha}} = \frac{1}{\Gamma(n-\alpha)} \int_a^t \frac{f^{(n)}(\tau)}{(t-\tau)^{\alpha+1-n}} d\tau, \tag{3}$$

where $\alpha \in \mathbb{R}$ is the order of a fractional derivative, $f^{(n)}(\tau) = \frac{d^n f(\tau)}{d\tau^n}$ and $\Gamma(x) = \int_0^\infty e^{-t}t^{x-1}dt$ is the gamma function.

Theorem 1 *The Laplace transform of the derivative-integral* (3) *has the form*

$$\mathcal{L}\left[{}_0^C\mathfrak{D}_t^\alpha \right] = s^\alpha F(s) - \sum_{k=1}^{n} s^{\alpha-k} f^{(k-1)}(0^+).$$

The proof of the Theorem 1 is given in [8].

After using the Laplace transform to (1), Theorem 1 and taking into account

$$\begin{aligned} X(s) &= \mathcal{L}\left[x(t)\right] = \int_0^\infty x(t)e^{-st}dt, \\ \mathcal{L}\left[\mathfrak{D}^\alpha x(t)\right] &= s^\alpha X(s) - s^{\alpha-1}x_0 \end{aligned} \tag{4}$$

we obtain:

$$\begin{aligned} X(s) &= \left[\mathbf{E}s^\alpha - \mathbf{A}\right]^{-1}\left[s^{\alpha-1}x_0 + \mathbf{B}U(s)\right], \\ Y(s) &= \mathbf{C}X(s), \quad U(s) = \mathcal{L}\left[u(t)\right]. \end{aligned} \tag{5}$$

After using (5) we can determine the transfer matrix of the system in the following form:

$$\mathbf{T}(s) = \mathbf{C}\left[\mathbf{E}s^\alpha - \mathbf{A}\right]^{-1}\mathbf{B} \in \mathbb{R}^{p\times m}(s). \tag{6}$$

It is assumed that $\det \mathbf{E} = 0$ and the pencil of the system (1) is regular, that is

$$\det\left[\mathbf{E}s^\alpha - \mathbf{A}\right] \neq 0 \tag{7}$$

for some $z \in \mathbb{C}$ (where $\mathbb{C}$ is the field of the complex numbers).

Definition 1 *The system* (1) *is called singular system if and only if* $\det \mathbf{E} = 0$ (rank $\mathbf{E} = r < n$).

The matrices (2) are called a realisation of a given transfer matrix $\mathbf{T}(s) \in \mathbb{R}^{p\times m}(s)$ if they satisfy the equality (6). The realisation is called minimal if the dimension of the sate matrix $\mathbf{A}$ is minimal among all possible realisations of $\mathbf{T}(s)$.

Task: For the given transfer matrix (6), determine a minimal realisation of the system (1) using the one-dimensional $\mathcal{D}$ digraphs theory. The dimension of the system must be the minimal among possible.

2. Main results

The solution of the minimal positive realisation problem will be presented for one-dimensional single-input single-output (SISO) transfer function ($m = p = 1$). The proposed method will be based on the one-dimensional digraph theory, and it will determine solutions in class $\mathcal{K}_1$. The classes of the digraph structure ware considered in detail in [25].

Consider the irreducible transfer function:

$$T(s) = \frac{n(s)}{d(s)} = \frac{b_q s^{\alpha\cdot q} + b_{q-1}s^{\alpha(q-1)} + b_{q-2}s^{\alpha(q-2)} + \ldots + b_1 s^\alpha + b_0}{a_r s^{\alpha\cdot r} + a_{r-1}s^{\alpha(r-1)} + a_{r-2}s^{\alpha(r-2)} + \ldots + a_1 s^\alpha + a_0}, \tag{8}$$

where b_i, $i = 0, 1, \ldots, q$ and a_j, $j = 0, 1, \ldots, r-1$ are given real coefficients and $r < n$. The transfer matrix (8) can be considered as a pseudo-rational function of the variable $\lambda = s^\alpha$ in the form:

$$T(\lambda) = \frac{n(\lambda)}{d(\lambda)} = \frac{b_q\lambda^q + b_{q-1}\lambda^{q-1} + b_{q-2}\lambda^{q-2} + \ldots + b_1\lambda + b_0}{a_r\lambda^r + a_{r-1}\lambda^{r-1} + a_{r-2}\lambda^{r-2} + \ldots + a_1\lambda + a_0}, \tag{9}$$

After multiplying the nominator and denominator of the transfer function (9) by λ^{-q} we obtain:

$$T(\lambda) = \frac{b_q + b_{q-1}\lambda^{-1} + b_{q-2}\lambda^{-2} + \ldots + b_1\lambda^{1-q} + b_0\lambda^{-q}}{a_r\lambda^{r-q} + a_{r-1}\lambda^{r-q-1} + a_{r-2}\lambda^{r-q-2} + \ldots + a_1\lambda^{1-q} + a_0\lambda^{-q}}. \tag{10}$$

In the first step, we must find matrix **E**. The weight of the vertices in a digraph is associated with the digraph-mask. If the characteristic polynomial is in the form

$$d(\lambda) = a_r\lambda^{r-q} + a_{r-1}\lambda^{r-q-1} + a_{r-2}\lambda^{r-q-2} + \ldots + a_1\lambda^{1-q} + a_0\lambda^{-q}. \tag{11}$$

then we can determine digraph-mask by the use the following proposition.

Proposition 1 *A digraph-mask $\mathfrak{M}(\mathcal{D})$ corresponding to a characteristic polynomial* (11) *consists of minimum one vertex with the weight equal to* 0.

Proof *If digraph-mask $\mathfrak{M}(\mathcal{D})$ consists from all vertex with weight not equal to zero, this means that matrix* **E** *is diagonal with all non-zero entries. In this case the condition* det **E** *is not satisfied and the system is not singular.* ■

It should be noted, that we obtained a solution which is not minimal in case when more than one vertex in digraph-mask has a weight equal to zero. In this paper we will be assume, that digraph-mask consist from one vertex with weight equal to zero.

Example 1 *Let be given the characteristic polynomial $d(\lambda) = \lambda^{-2} + \lambda^{-3}$. After using the Proposition 1, we obtain a set of a digraph-mask consisting of* 2 *vertices with the weight equal to* 1*. In Figure 1 vertices with weight equal to* 1 *are marked in blue and vertices with a weight equal to* 0 *are marked in red.*

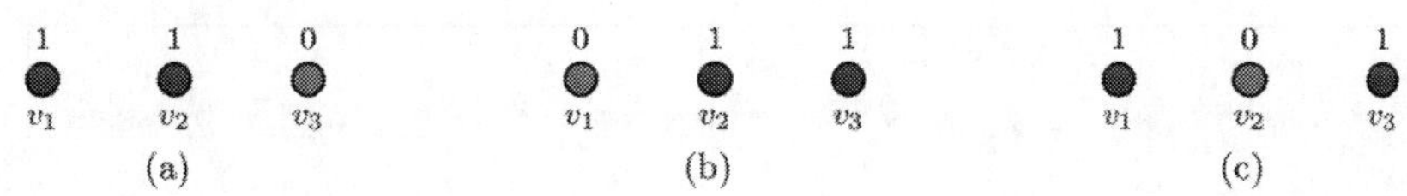

Figure 1: (a)–(d) All possible digraph-masks of the characteristic polynomial $d(\lambda) = \lambda^{-2} + \lambda^{-3}$

Using a digraph-mask we can write matrix **E**. *For example, from the digraph-mask presented in Figure 1(b), we have:* $\mathbf{E} = \mathrm{diag}\begin{pmatrix} 0 & 1 & 1 \end{pmatrix}$.

In the second step, we must find matrix **A** using decomposition of the characteristic polynomial $d(\lambda)$ into a set of simple monomials:

$$d(\lambda) = \underbrace{a_r\lambda^{r-q}}_{M_r} + \underbrace{a_{r-1}\lambda^{r-q-1}}_{M_{r-1}} + \underbrace{a_{r-2}\lambda^{r-q-2}}_{M_{r-2}} + \ldots + \underbrace{a_1\lambda^{1-q}}_{M_1} + \underbrace{a_0\lambda^{-q}}_{M_0}. \tag{12}$$

For each simple monomial using Proposition 2, we create digraphs representations.

Proposition 2 *The one-dimensional digraph $\mathcal{D}$ corresponding to monomial $a_r\lambda^{r-q}$ where $r = 0, 1, \ldots \ldots, q-1$ consists of one cycle, and it contains $|r-q|$-vertices.*

IOP Conf. Series: Journal of Physics: Conf. Series **783** (2017) 012052 doi:10.1088/1742-6596/783/1/012052

Introduce the definition of operation of the composition relative to vertices on digraphs that will be used in further considerations.

Definition 2 *Let $\mathcal{G}_1, \mathcal{G}_2, \ldots, \mathcal{G}_n$ be a digraphs with vertex sets $\mathbb{V}(\mathcal{G}_n) = \{v_{i(n)} : i \in \mathbb{N}\}$. The composition relative to vertices $\mathcal{D}[\mathcal{G}_1, \mathcal{G}_2, \ldots, \mathcal{G}_n]$ is the digraph $\mathcal{L}$ with the vertex set $\mathbb{V}(\mathcal{G}_1) \cup \mathbb{V}(\mathcal{G}_2) \cup \cdots \cup \mathbb{V}(\mathcal{G}_n) = \{v_j : j = 1, \ldots, \max\{i(n)\}, \ i \in \mathbb{N}\}$ and arc set $\mathcal{A}' = \mathcal{A}(\mathcal{G}_1) \cup \mathcal{A}(\mathcal{G}_2) \cup \cdots \cup \mathcal{A}(\mathcal{G}_n)$, where $\mathcal{A}'$ denotes operation of the deleting multiple arcs.*

Then using Theorem 1 presented in paper [32,33] and Definition 2, we can create all digraph realisations of the characteristic polynomials. Each digraph corresponding to a characteristic polynomial must satisfy two conditions. The first condition (C1) relates to the existence in the common part of the digraph (vertex in blue), the second condition (C2) relates to non-existence of additional cycles in the digraph. To the vertices belonging to the common part of the digraph we assigned weight equal to **1** and for the other vertices – weight equal to **0**.

Example 2 *(Continued Example 1) Characteristic polynomial $d(\lambda) = \lambda^{-2} + \lambda^{-3}$ consisting of two monomials: $M_1 = \lambda^{-2}$ and $M_0 = \lambda^{-3}$. In Figure 2, possible realisations of the monomials are presented.*

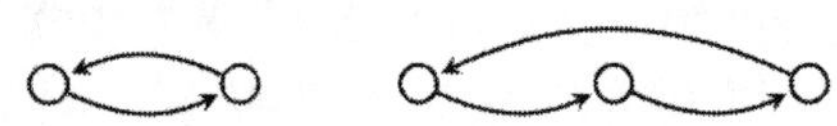

Figure 2: Realisation of the monomial: (a) $M_1 = \lambda^{-2}$; (b) $M_0 = \lambda^{-3}$.

Then, we can determine all possible realisation (Figure 3) of the considering characteristic polynomial. Additionally, to the vertices belonging to the common part of the digraph we assigned weight equal to 1 (marked in blue) and 0 for other vertices (marked in red).

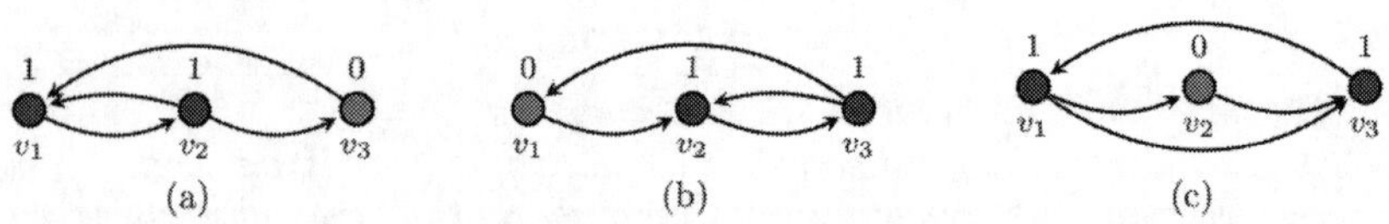

Figure 3: All possible realisations of the characteristic polynomial $d(\lambda)$.

Then, using Definition 2 and digraph-mask determined in the first step (Proposition 1), we create all possible digraph realisations as combinations of the monomial representations and a digraph-mask:

$$\overset{d(\lambda)}{\underset{i}{}}|\mathcal{D} = \mathfrak{M}_j(G) \circ \left(\mathcal{G}_{M_r} \cup \mathcal{G}_{M_{r-1}} \cup \ldots \cup \mathcal{G}_{M_1} \cup \mathcal{G}_{M_0}\right) = \mathfrak{M}_j(G) \circ \left(\bigcup_{i=0}^{r-i} \mathcal{G}_{M_i}\right) \tag{13}$$

where: $\overset{d(\lambda)}{\underset{i}{}}|\mathcal{D}$ means that we consider i-th realisation of the one-dimensional digraph $\mathcal{D}$ corresponding to the characteristic polynomial $d(\lambda)$; $\mathfrak{M}_j(G)$ is j-th possible digraph-mask; $\circ$ is an operation of the superposition digraph-mask and one of the possible digraph realisations. It should be noted that superposition operation $\circ$ on digraphs vertices corresponds to logical operation OR.

Example 3 *(Continued Example 2) Using composition relative to vertices and superposition operation to the digraph presented in Figure 3(a) and digraph-mask presented in Figure 1, we obtain the possible realisations presented in Figure 4. Two digraphs have all vertices filled with blue. The realisation which satisfies the characteristic polynomial is presented in Figure 4(b) and Figure 4(c).*

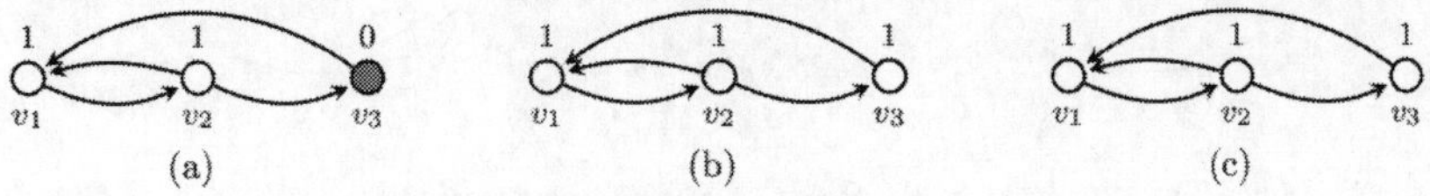

Figure 4: (a)–(c) Possible realisation of the characteristic polynomial $d(\lambda)$

Based on the above considerations we can obtain one of the possible digraph structures with assigned vertex weight, which satisfies conditions (C1) and (C2). By the use of the digraph structure (Figure 5), we can write one of the possible singular matrix and state matrix in the following form:

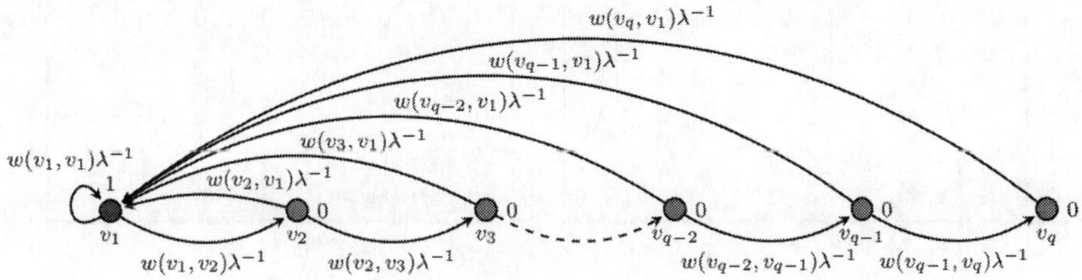

Figure 5: One of the possible realisation of the characteristic polynomial (11)

$$
\mathbf{A} = \begin{array}{c} v_i \backslash^{v_j} \\ v_1 \\ v_2 \\ v_3 \\ \vdots \\ v_{q-2} \\ v_{q-1} \\ v_q \end{array}
\begin{array}{c} \begin{array}{ccccccc} v_1 & v_2 & v_3 & \cdots & v_{q-2} & v_{q-1} & v_q \end{array} \\
\left[\begin{array}{ccccccc}
w(v_1,v_1) & w(v_2,v_1) & w(v_3,v_1) & \cdots & w(v_{q-2},v_1) & w(v_{q-1},v_1) & w(v_q,v_1) \\
w(v_1,v_2) & 0 & 0 & \cdots & 0 & 0 & 0 \\
0 & w(v_2,v_3) & 0 & \cdots & 0 & 0 & 0 \\
\vdots & \vdots & \vdots & \ddots & \ddots & \vdots & \vdots \\
0 & 0 & 0 & \ddots & 0 & 0 & 0 \\
0 & 0 & 0 & \cdots & w(v_{q-2},v_{q-1}) & 0 & 0 \\
0 & 0 & 0 & \cdots & 0 & w(v_{q-1},v_q) & 0
\end{array}\right] \end{array} \in \mathbb{R}^{q\times q} \quad (14)
$$

$$
\mathbf{E} = \begin{array}{c} v_i \backslash^{v_j} \\ v_1 \\ v_2 \\ v_3 \\ \vdots \\ v_{q-2} \\ v_{q-1} \\ v_q \end{array}
\begin{array}{c} \begin{array}{ccccccc} v_1 & v_2 & v_3 & \cdots & v_{q-2} & v_{q-1} & v_q \end{array} \\
\left[\begin{array}{ccccccc}
0 & 0 & 0 & \cdots & 0 & 0 & 0 \\
0 & 1 & 0 & \cdots & 0 & 0 & 0 \\
0 & 0 & 1 & \cdots & 0 & 0 & 0 \\
\vdots & \vdots & \vdots & \ddots & \vdots & \vdots & \vdots \\
0 & 0 & 0 & \cdots & 1 & 0 & 0 \\
0 & 0 & 0 & \cdots & 0 & 1 & 0 \\
0 & 0 & 0 & \cdots & 0 & 0 & 1
\end{array}\right] \end{array} \in \mathbb{R}^{q\times q}. \quad (15)
$$

After finding the structure of the **A** and **E** matrices, we must find matrix **B** and **C** using the following polynomial:

$$n(\lambda) = b_q + b_{q-1}\lambda^{-1} + b_{q-2}\lambda^{-2} + \ldots + b_1\lambda^{1-q} + b_0\lambda^{-q}. \quad (16)$$

For this purpose, we are expanding the digraph created in the first part of the algorithm. Because the degree of the polynomial $n(\lambda)$ is higher than polynomial $d(\lambda)$, then – to be able to realise it – we must add one vertex and an arc in such a way that there are no additional cycles (vertex and arc in green). In the considered digraph presented in Figure 5, vertex v_1 is the common and from this vertex we determine a path of the maximum length equal to q (Figure 6).

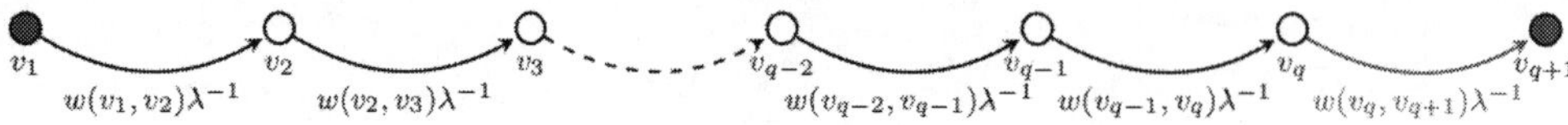

Figure 6: Path of maximum length

It should be noted that this operation on digraph means that in **A** and **E** matrix an additional row and column appear, as shown below:

$$\underbrace{\begin{array}{c|cccccccc} v_i \backslash ^{v_j} & v_1 & v_2 & v_3 & \cdots & v_{q-2} & v_{q-1} & v_q & v_{q+1} \\ \hline v_1 & * & * & * & \cdots & * & * & * & 0 \\ v_2 & * & 0 & 0 & \cdots & 0 & 0 & 0 & 0 \\ v_3 & 0 & * & 0 & \cdots & 0 & 0 & 0 & 0 \\ & \vdots & \vdots & \ddots & \ddots & \vdots & \vdots & \vdots & \vdots \\ v_{q-2} & 0 & 0 & 0 & \ddots & 0 & 0 & 0 & 0 \\ v_{q-1} & 0 & 0 & 0 & \cdots & * & 0 & 0 & 0 \\ v_q & 0 & 0 & 0 & \cdots & 0 & * & 0 & 0 \\ v_{q+1} & 0 & 0 & 0 & \cdots & 0 & 0 & * & 0 \end{array}}_{\mathbf{A} \in \mathbb{R}^{(q+1)\times(q+1)}}, \underbrace{\begin{array}{cccccccc} v_1 & v_2 & v_3 & \cdots & v_{q-2} & v_{q-1} & v_q & v_{q+1} \\ \hline 0 & 0 & 0 & \cdots & 0 & 0 & 0 & 0 \\ 0 & 1 & 0 & \cdots & 0 & 0 & 0 & 0 \\ 0 & 0 & 1 & \cdots & 0 & 0 & 0 & 0 \\ \vdots & \vdots & \ddots & \ddots & \vdots & \vdots & \vdots & \vdots \\ 0 & 0 & 0 & \ddots & 1 & 0 & 0 & 0 \\ 0 & 0 & 0 & \cdots & 0 & 1 & 0 & 0 \\ 0 & 0 & 0 & \cdots & 0 & 0 & 1 & 0 \\ 0 & 0 & 0 & \cdots & 0 & 0 & 0 & 1 \end{array}}_{\mathbf{E} \in \mathbb{R}^{(q+1)\times(q+1)}} \quad (17)$$

Now we can add to the digraph the source vertex s_1 corresponding to matrix **B** and output vertex y_1 corresponding to matrix **C**, and we combine them. This operation is presented in Figure 7.

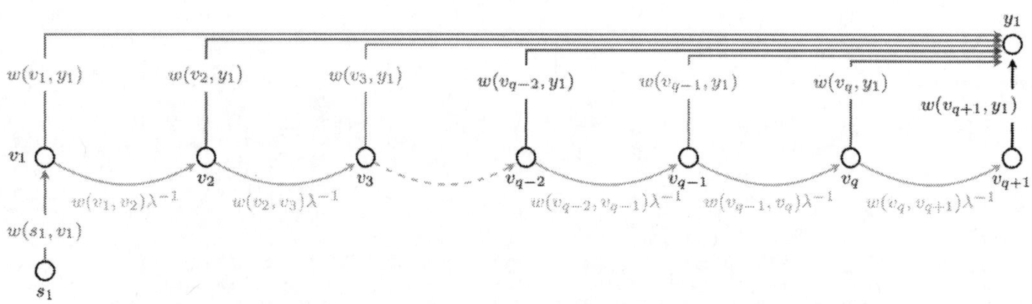

Figure 7: One-dimensional digraph corresponding to polynomial (16)

By the using the digraph presented in Figure 7, we can write the set of the equations which we compare with the same power of the polynomial (16).

$$\begin{cases} & w(s_1, v_1)\cdot w(v_1, y_1) = b_q \\ \lambda^{-1} & w(s_1, v_1)\cdot w(v_1, v_2)\cdot w(v_2, y_1) = b_{q-1} \\ \lambda^{-2} & w(s_1, v_1)\cdot w(v_1, v_2)\cdot w(v_2, v_3)\cdot w(v_3, y_1) = b_{q-2} \\ \vdots & \vdots \quad \vdots \\ \lambda^{2-q} & w(s_1, v_1)\cdot w(v_1, v_2)\cdot w(v_2, v_3)\cdot \ldots \cdot w(v_{q-2}, v_{q-1}) \cdot w(v_{q-1}, y_1) = b_2 \\ \lambda^{1-q} & w(s_1, v_1)\cdot w(v_1, v_2)\cdot w(v_2, v_3)\cdot \ldots \cdot w(v_{q-2}, v_{q-1}) \cdot w(v_{q-1}, v_q) \cdot w(v_q, y_1) = b_1 \\ \lambda^{-q} & w(s_1, v_1)\cdot w(v_1, v_2)\cdot w(v_2, v_3)\cdot \ldots \cdot w(v_{q-2}, v_{q-1}) \cdot w(v_{q-1}, v_q) \cdot w(v_q, v_{q+1}) \cdot w(v_{q+1}, y_1) = b_0 \end{cases} \quad (18)$$

After solving the set of the equation (18) we can write **B** and **C** matrix in the following form:

$$\underbrace{\begin{array}{c|c} v_i\backslash^s & s_1 \\ \hline v_1 & w(s_1,v_1) \\ v_2 & 0 \\ v_3 & 0 \\ \vdots & \vdots \\ v_{q-2} & 0 \\ v_{q-1} & 0 \\ v_q & 0 \\ v_{q+1} & 0 \end{array}}_{\mathbf{B}\in\mathbb{R}^{(q+1)\times 1}}, \quad \underbrace{\begin{array}{c|ccccccc} y\backslash^{v_j} & v_1 & v_2 & v_3 & \cdots & v_{q-1} & v_q & v_{q+1} \\ \hline y_1 & w(v_1,y_1) & w(v_2,y_1) & w(v_3,y_1) & \cdots & w(v_{q-1},y_1) & w(v_q,y_1) & w(v_{q+1},y_1) \end{array}}_{\mathbf{C}\in\mathbb{R}^{1\times(q+1)}} \quad (19)$$

where:

$$w(v_1,y_1) = \frac{b_q}{w(s_1,v_1)},\ w(v_2,y_1) = \frac{b_{q-1}}{w(s_1,v_1)w(v_1,v_2)},\ w(v_3,y_1) = \frac{b_{q-2}}{w(s_1,v_1)w(v_1,v_2)w(v_2,v_3)},$$

$$w(v_{q-1},y_1) = \frac{b_2}{w(s_1,v_1)w(v_1,v_2)w(v_2,v_3)\ldots w(v_{q-2},v_{q-1})},$$

$$w(v_q,y_1) = \frac{b_1}{w(s_1,v_1)w(v_1,v_2)w(v_2,v_3)\ldots w(v_{q-2},v_{q-1})w(v_{q-1},v_q)},$$

$$w(v_{q+1},y_1) = \frac{b_0}{w(s_1,v_1)w(v_1,v_2)w(v_2,v_3)\ldots w(v_{q-2},v_{q-1})w(v_{q-1},v_q)w(v_q,v_{q+1})}$$

The realisation of the (8) is given by (17) and (19).

3. Numerical Example

Example 4 *Find a minimal realisation of the transfer function:*

$$T(z) = \frac{b_3 s^{1.8} + b_2 s^{1.2} + b_1 s^{0.6} + b_0}{a_2 s^{1.2} + a_1 s^{0.6} + a_0}. \quad (20)$$

Solution: Assuming that $\lambda = s^{-0.6}$ and multiplying nominator and denominator of the transfer function by λ^{-3} we obtain:

$$T(\lambda) = \frac{b_3 + b_2\lambda^{-1} + b_2\lambda^{-2} + b_0\lambda^{-3}}{a_2\lambda^{-1} + a_1\lambda^{-2} + a_0\lambda^{-3}}. \quad (21)$$

Step 1: Using the characteristic polynomial $d(\lambda) = a_2\lambda^{-1} + a_1\lambda^{-2} + a_0\lambda^{-3}$ and Proposition 1, we determine digraph-masks: $\mathfrak{M}_1(G)$ (Figure 8(a)), $\mathfrak{M}_2(G)$ (Figure 8(b)), $\mathfrak{M}_3(G)$ (Figure 8(c)).

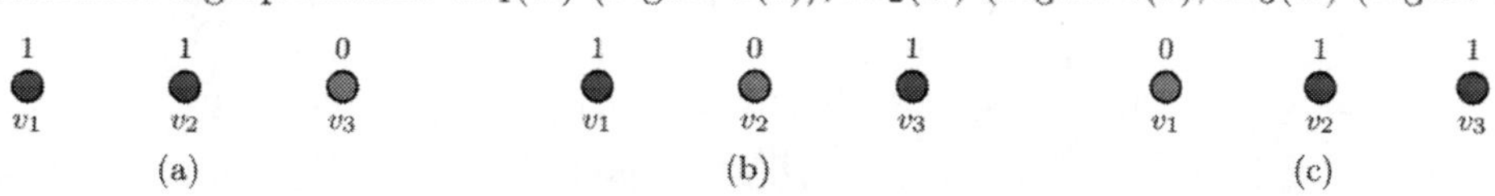

Figure 8: Digraph-masks of the characteristic polynomial

Step 2: Then, we decompose the characteristic polynomial $d(\lambda)$ into a set of simple monomials:

$$d(\lambda) = a_2\lambda^{-1} + a_1\lambda^{-2} + a_0\lambda^{-3} = M_2 + M_1 + M_0 \quad (22)$$

For each monomial we determine all possible digraph-structure representations (see Figure 9).

Figure 9: Digraph corresponding to monomial: (a) M_2; (b) M_1; (c) M_0

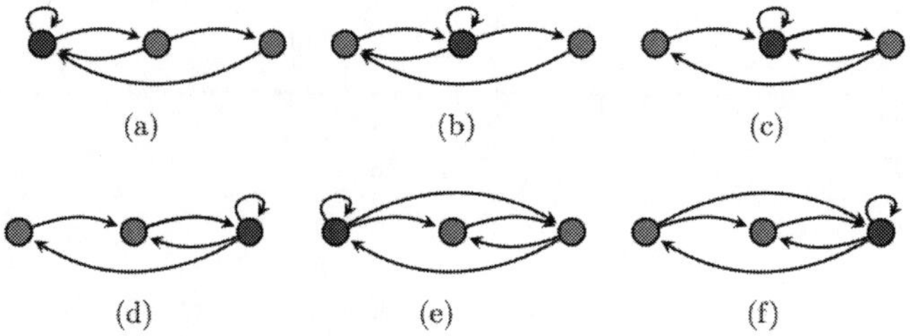

Figure 10: All possible realisation of the characteristic polynomial

Step 3: Now we can create all digraph realisations of the characteristic polynomial. Figure 10 presents all the possible realisations of the characteristic polynomial (22).

Then, using (13) we can determine all possible realisations of the characteristic polynomial $d(\lambda)$ as a combination of the digraph representation presented for example in Figure 10(a) and digraph-mask presented in Figure 8. After this we obtain other three possible realisations (Figure 11). It should be note that only one digraph-structure presented in Figure 11(c) satisfy characteristic polynomial $d(\lambda)$.

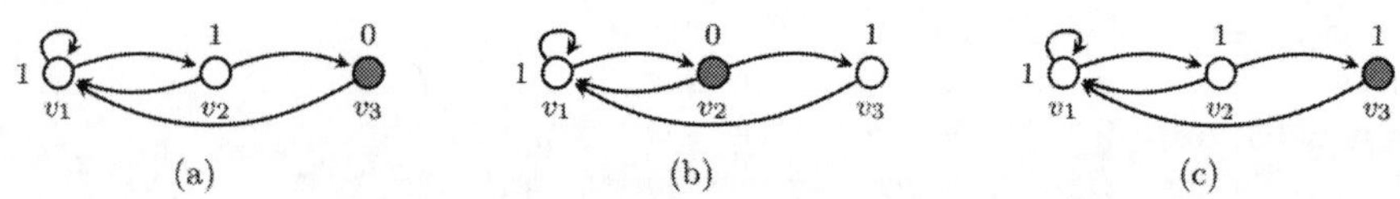

Figure 11: All possible realisation corresponding to digraph-structure presented in Figure 10(a) and a digraph-mask presented in Figure 8.

Step 4: Because the degree of the polynomial $n(\lambda)$ is higher than polynomial $d(\lambda)$, then – to be able to realise it – we must add one vertex and an arc in such a way that there are no additional cycles. Extended digraph is presented in Figure 12.

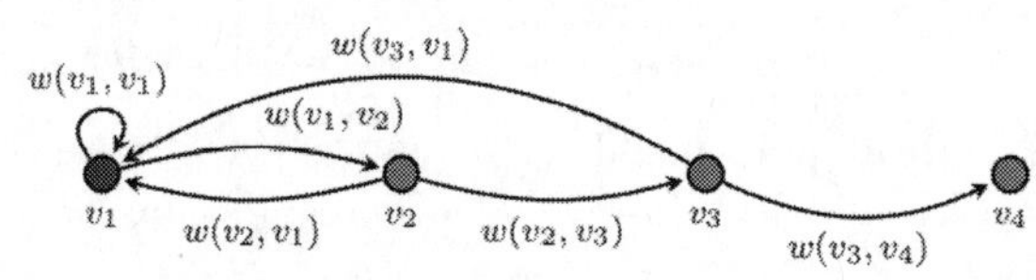

Figure 12: One of the possible extended realisation of the characteristic polynomial

Now, we can write the following matrices

$$\mathbf{E} = \begin{array}{c} {}_{v_i}\backslash^{v_j} \\ v_1 \\ v_2 \\ v_3 \\ v_4 \end{array} \begin{array}{c} \begin{array}{cccc} v_1 & v_2 & v_3 & v_4 \end{array} \\ \begin{bmatrix} 0 & 0 & 0 & 0 \\ 0 & 1 & 0 & 0 \\ 0 & 0 & 1 & 0 \\ 0 & 0 & 0 & 1 \end{bmatrix} \end{array}, \quad \mathbf{A} = \begin{array}{c} {}_{v_i}\backslash^{v_j} \\ v_1 \\ v_2 \\ v_3 \\ v_4 \end{array} \begin{array}{c} \begin{array}{cccc} v_1 & v_2 & v_3 & v_4 \end{array} \\ \begin{bmatrix} w(v_1,v_1) & w(v_2,v_1) & w(v_3,v_1) & 0 \\ w(v_1,v_2) & 0 & 0 & 0 \\ 0 & w(v_2,v_3) & 0 & 0 \\ 0 & 0 & w(v_3,v_4) & 0 \end{bmatrix} \end{array}. \quad (23)$$

One of the possible numerical realisation is the following:

$$\mathbf{E} = \begin{bmatrix} 0\,0\,0\,0 \\ 0\,1\,0\,0 \\ 0\,0\,1\,0 \\ 0\,0\,0\,1 \end{bmatrix}, \quad \mathbf{A} = \begin{bmatrix} -a_2 & -a_1 & -a_0 & 0 \\ 1 & 0 & 0 & 0 \\ 0 & 1 & 0 & 0 \\ 0 & 0 & 1 & 0 \end{bmatrix}. \tag{24}$$

Step 5: We add source vertex s_1 and output vertex y_1 to the digraph presented in Figure 12, and we connect them (see Figure 13).

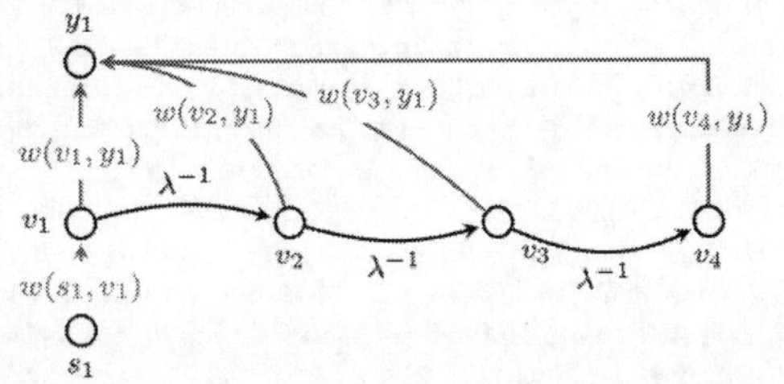

Figure 13: One of the possible realisation of the transfer function (20).

Step 6: Then, using the created digraph, we can write a set of equations in the form:

$$\begin{cases} \quad\;\; \| w(s_1,v_1)\cdot w(v_1,y_1) = b_3 \\ \lambda^{-1} \| w(s_1,v_1)\cdot w(v_2,y_1) = b_2 \\ \lambda^{-2} \| w(s_1,v_1)\cdot w(v_3,y_1) = b_1 \\ \lambda^{-3} \| w(s_1,v_1)\cdot w(v_4,y_1) = b_0 \end{cases} \tag{25}$$

After solving the set of the equation (25), we can write input and output matrix in the following form:

$$\mathbf{B} = \begin{bmatrix} w(s_1,v_1) \\ 0 \\ 0 \\ 0 \end{bmatrix}, \mathbf{C} = \begin{bmatrix} \frac{b_3}{w(s_1,v_1)} & \frac{b_2}{w(s_1,v_1)} & \frac{b_1}{w(s_1,v_1)} & \frac{b_0}{w(s_1,v_1)} \end{bmatrix}. \tag{26}$$

The desired realisation of the (20) is given by (24) and (26) for $w(s_1,v_1) \in \mathbb{R}$.

4. Concluding Remarks

A method for computation of a minimal realisation of a given proper transfer function of fractional singular one-dimensional continuous-time linear systems has been proposed. Sufficient conditions for the existence of a minimal realisation of a given proper transfer function have been established. A method based on one-dimensional digraph theory for computation of minimal realisations has been proposed. The effectiveness of the algorithm has been illustrated with some numerical examples. Extension of those considerations for fractional singular discrete-time linear systems and for fractional singular hybrid model is possible.

Formulation of the necessary and sufficient conditions for the existence of digraphs structures of the minimal positive realisation problem for one-dimensional singular discrete-time and continuous-time linear systems remains an open problem.

References

[1] Luenberger D G 1979 *Introduction to Dynamic Systems: Theory, Models, and Applications* (New York: Wiley) chap Positive linear systems
[2] Lewis F 1984 *IEEE Transactions on Automatic Control*
[3] Lewis F L 1986 *Circuits, Systems and Signal Processing* **5** 3–36 ISSN 1531-5878 URL `http://dx.doi.org/10.1007/BF01600184`
[4] Dai L (ed) 1989 *System analysis via transfer matrix* (Berlin, Heidelberg: Springer Berlin Heidelberg) pp 197–230 ISBN 978-3-540-46074-9 URL `http://dx.doi.org/10.1007/BFb0002482`
[5] Farina L and Rinaldi S 2000 *Positive linear systems: theory and applications* (New York: Wiley-Interscience, Series on Pure and Applied Mathematics)
[6] Kaczorek T 2001 *Positive 1D and 2D systems* (London: Springer Verlag)
[7] Dodig M and Stoi M 2009 *Linear Algebra and its Applications* **431** 1267 – 1292 ISSN 0024-3795 URL `http://www.sciencedirect.com/science/article/pii/S0024379509002468`
[8] Kaczorek T and Sajewski L 2014 *The Realization Problem for Positive and Fractional Systems* (Berlin: Springer International Publishing) URL `http://dx.doi.org/10.1007/978-3-319-04834-5`
[9] Kublanovskaya V N 1983 *Journal of Soviet Mathematics* **23** 1939–1950 ISSN 1573-8795 URL `http://dx.doi.org/10.1007/BF01093276`
[10] Horn R A and Johnson C R 1991 *Topics in Matrix Analysis* (Cambridge Univ. Press)
[11] Benvenuti L and Farina L 2004 *IEEE Transactions on Automatic Control* **49** 651–664
[12] Kaczorek T 2007 *Polynomial and Rational Matrices* (London: Springer Verlag)
[13] Sajewski L 2012 *Pomiary Automatyka Robotyka* 413–417
[14] Nishimoto K 1984 *Fractional Calculus* (Koriama: Decartess Press)
[15] Das S 2011 *Functional Fractional Calculus* (New York: Springer Berlin Heidelberg) URL `http://dx.doi.org/10.1007/978-3-642-20545-3`
[16] Miller K and Ross B 1993 *An Introduction to the Fractional Calculus and Fractional Differenctial Equations* (New York: Willeys)
[17] Ortigueira M D 2011 *Fractional Calculus for Scientists and Engineers* (Springer Netherlands: Academic Press) URL `10.1007/978-94-007-0747-4`
[18] Podlubny I 1999 *Fractional Differential Equations* (San Diego: Academic Press)
[19] Ortigueira M D, Rivero M and Trujillo J J 2015 *Journal of King Saud University-Science*
[20] Martynyuk V and Ortigueira M 2015 *Signal Processing* **107** 355–360
[21] Machado J and Lopes A M 2015 *Fractional Calculus and Applied Analysis* **18** 1518–1536
[22] Machado J, Mata M E and Lopes A M 2015 *Entropy* **17** 5402–5421
[23] Podlubny I, Skovranek T and Datsko B 2014 Recent advances in numerical methods for partial fractional differential equations *Control Conference (ICCC), 2014 15th International Carpathian* (IEEE) pp 454–457
[24] Petras I, Sierociuk D and Podlubny I 2012 *Signal Processing, IEEE Transactions on* **60** 5561–5566
[25] Hryniów K and Markowski K A 2016 *Challenges in Automation, Robotics and Measurement Techniques: Proceedings of AUTOMATION-2016, March 2-4, 2016, Warsaw, Poland* (Springer International Publishing) chap Classes of Digraph Structures Corresponding to Characteristic Polynomials, pp 329–339 ISBN 978-3-319-29357-8 URL `http://dx.doi.org/10.1007/978-3-319-29357-8_30`
[26] Markowski K A 2015 *Journal of Physics: Conference Series* **659** 012041 URL `http://stacks.iop.org/1742-6596/659/i=1/a=012041`
[27] Markowski K A 2016 *International Journal of Dynamics and Control* URL `http://dx.doi.org/10.1007/s40435-016-0232-3`
[28] Markowski K A and Hryniów K 2016 Digraphs minimal positive stable realisations for fractional one-dimensional systems *Proceedings of 7th Conference on Non-integer Order Calculus and its Applications7th Conference on Non-integer Order Calculus and its Applications* pp 105–118 ISBN 978-3-319-23038-2 URL `http://dx.doi.org/10.1007/978-3-319-23039-9_9`
[29] Bang-Jensen J and Gutin G 2009 *Digraphs: Theory, Algorithms and Applications* (London: Springer-Verlag)
[30] Wallis W D 2007 *A Beginner's Guide to Graph Theory* (Biiokhäuser)
[31] Godsil C and Royle G 2001 *Algebraic Graph Theory* (Springer Verlag)
[32] Hryniów K and Markowski K A 2014 Parallel digraphs-building algorithm for polynomial realisations *Proceedings of 2014 15th International Carpathian Control Conference (ICCC)* pp 174–179 URL `http://dx.doi.org/10.1109/CarpathianCC.2014.6843592`
[33] Hryniów K and Markowski K A 2015 Optimisation of digraphs creation for parallel algorithm for finding a complete set of solutions of characteristic polynomial *Proceedings of 20th International Conference on Methods and Models in Automation and Robotics, MMAR 2015, Miedzyzdroje, Poland, August 24–27* pp 1139–1144 URL `http://dx.doi.org/10.1109/MMAR.2015.7284039`

Towards a cyber-medical system for drug assisting devices

Dana Copot[1], Robin De Keyser[1], Levente Kovacs[2], Clara Ionescu[1]
[1]Ghent University, Faculty of Engineering and Architecture, Research Group on Dynamical Systems and Control (DySC), Technologiepark 914, B9052 Ghent, Belgium
[2] Obuda University, Physiological Controls Group, Bécsi út 96/b, H-1034 Budapest

E-mail: {dana.copot,claramihaela.ionescu,robain.dekeyser}@ugent.be
kovacs.levente@nik.uni-obuda.hu

Abstract. The purpose of this paper is to describe the role of non-integer models in healthcare pathways and how these tools provide a breakthrough in our understanding of drug delivery protocols. Two examples will be given as being typical of such applications: modeling glucose levels in diabetic type I patients and modeling pain transmission in patients suffering from (post operative) chronic pain. A wide range of tools to model these pathways and their role for use in healthcare delivery for optimal treatment and patient well-being are discussed. Individualized patient healthcare is nowadays top priority in European research framework for health and these modeling tools are primary aids in meeting these challenges.

1. Introduction

Personalized healthcare is at todays research core. Specificity of drug titration and treatment for the modern patient implies the availability, readiness of use and employment of novel tools and concepts with a background in a deep-rooted, well established clinical and medical expertise. Although as recognized still today, clinical practice is still an art rather than a science, efforts are being channeled to move from state of use to state of art. As such, state of art is far more advanced, but it lacks suitable platforms for integration, the so called cyber-physical systems, demanding hybrid sources of information processing, filtering, securing and translating into usable variables and meaningful values. A critical role of this personalized healthcare wave in current trends of research is given to the patient itself, with all its specificity, as observed from figure 1. Specificity demands the availability of individualized models, for prediction of drug effects and for steering the optimal drug titration and treatment to the patient. The goal of this paper is to bring forward the latest trends in modeling for individualized healthcare purposes. A review of state of art and state of use will be given, along with suggestions for further developments and remaining challenges.

The model is some formalized representation of the input-output relationship; mostly a mathematical model. The model will in general depend on certain parameters, such as body weight of the patient, clearance of the administered drug, etc. The adaptor is a tool to adapt the initial estimates of the parameters, whereby the controller transforms the error signal in commands for dosing. The core of feedback systems is a model of the patient with respect to the relationship between drug dosing and drug effect. Such models can be used in two directions. Using the feedforward direction it can give a prediction of the measured output. In the backwards

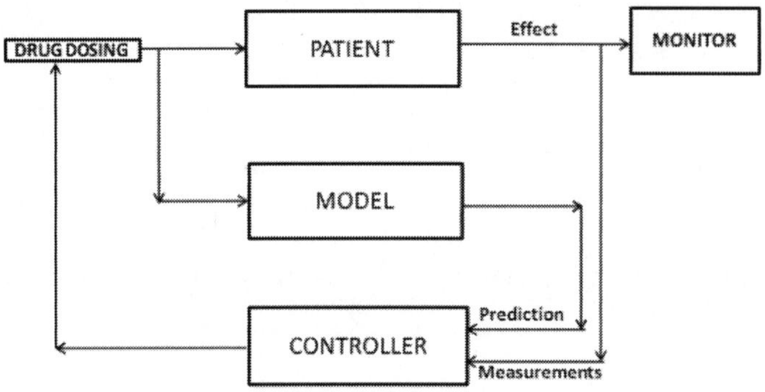

Figure 1. Block diagram of a generic model-based adaptive closed-loop system for automatic drug delivery. The closed-loop system consists of five parts: the patient, as the system to be controlled; the response that is considered as a measurable representation of the process to be controlled; a model of the inputoutput relationship, for instance a mathematical formula; and a controller, which transforms the error signal and the set-point to a drug delivery scheme

direction it can be used to determine the necessary input to achieve and maintain a certain level of the output. The paper will present two applications of such drug delivery systems: anesthesia and diabetes type I.

2. State of use

Pharmacokinetics describes the relationship between drug dose and drug concentration in plasma or the effect site. This relationship is described by processes of absorption, distribution and clearance. Notice that for intravenous drugs (e.g. anesthesia), absorption is irrelevant, so pharmacokinetic properties are described by distribution and clearance alone [1, 2].

We summarize two types of pharmacokinetic models, with prevalence for the most commonly used compartmental model. These models offer a tool for understanding the relationship between pharmacokinetics and cardiovascular physiology. The physiologic estimates are usually less accurate than fully-described physiologic models, but they represent the actual pharmacokinetics and delineate the basis for changed states of intravenous pharmacokinetics. To our knowledge, these concepts denote the inter-patient variability and intra-patient variability, which is in fact the challenging problem for optimal drug dosing [3, 4]. In principle, the altered physiological states mark a significant departure from the expected dose-response relationship, hence marked variations. Drug elimination is related to the efficiency of metabolism and/or excretion carried out by the organs of the drug elimination (e.g. liver, kidney, lung), as well as the blood flow to these organs. The two basic approaches for analyzing the effects of physiology on pharmacokinetics are the so-called forward and inverse models. With the forward model or problem, investigators estimate or measure the blood flow to each of the major organs (e.g. lungs, heart, brain, kidneys, liver, intestines) and tissue types (e.g. muscle, fat, skin) as well as the organ and tissue affinities of the drug relative to blood (i.e. tissue-blood partition coefficients). With these physiologic parameters tissue blood flow roughly represents transfer clearances between the central circulation and the tissue. The product of the tissue-blood partition coefficient and tissue masses roughly equals the volume of distribution of the drug for that tissue. Once the physiologic pharmacokinetic model is constructed, it is possible to simulate the effect of changes in tissue blood flow or cardiac output on the time course of plasma drug concentrations. Thus, physiologic models allow extrapolation outside the range of data and the existing physiologic conditions as well as interspecies scaling if the mechanisms of transport are understood and valid. A total body pharmacokinetic model describing generic drug disposition in humans is given in figure 2 [5].

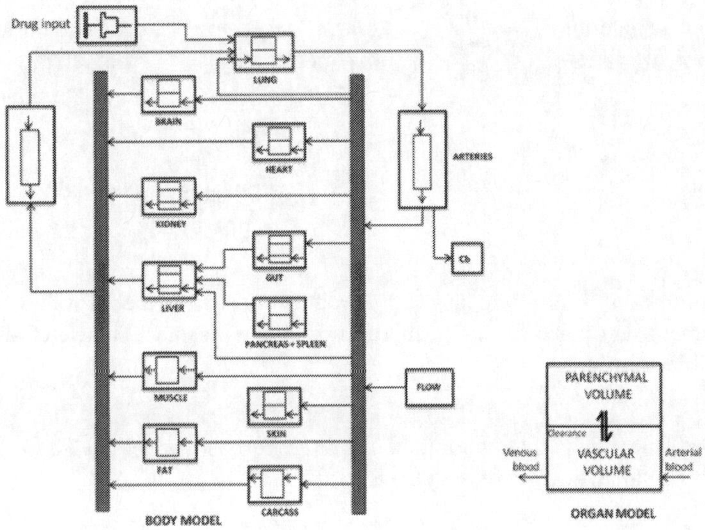

Figure 2. The total body pharmacokinetic model describing thiopental disposition in humans. The model consists of multiple tissues and blood pools connected via the vasculature and assumes venous injection and arterial blood sampling.

The inverse model is one of the most commonly de facto used in modern anesthetics and drug delivery systems in general. Only drug concentrations vs time data is drawn from blood or other tissue and fluids to create mathematical description (e.g. compartmental model), followed by derivation or more often by correlation of model parameters with physiologic measurements. This is the preferred method in humans as only time blood samples are needed for analyze and regional blood flow measurements are not required. Moreover, continuous improvement in drug measurement technology has made the acquisition of highly specific and sensitive drug concentration data from small samples sizes relatively easy. Clearance refers to the process of elimination of a substance from a volume. Distribution clearance describes the transfer of the drug between blood or plasma and the peripheral tissues. Unlike metabolic clearance, in distribution clearance the drug is not removed from the body. It is influenced by cardiac output, tissue and organ blood flow, and capillary permeability of the drug. Some pharmacokinetic parameters have precise physiologic meaning and can be used for estimates. Such progression from pharmacokinetics to physiology is considered the inverse of the forward model of taking physiologic principles and estimates to derive pharmacokinetics. Whichever way you look at it, pharmacokinetics and physiology are interwoven. Compartmental models are augmented with an extra effect-site compartment, to which corresponds an effect-site concentration [6, 7, 8]. The plasma is not the site where the drugs unfold their clinical effects. This is usually the brain, measured via EEG or evoked auditory potentials. Hence the main effect at the brain (effect site) is delayed when compared to plasma peak concentration; this is represented by adding an extra compartment (see figure 3).

The effect site is connected to the central compartment by a first order process, where the constant K0e is the rate constant for elimination of drug from the effect site. This coefficient has a large influence on the rate of rise and offset of drug effect, and the dose required to

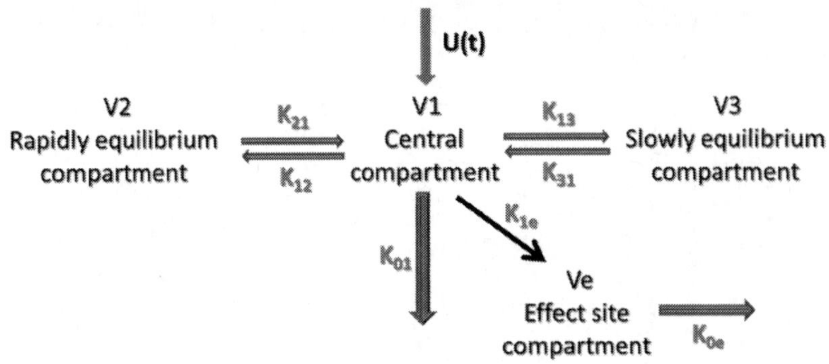

Figure 3. The addition of an effect site to a three-compartment model refers to the fact that the anesthetic effect takes place in the brain and not in the plasma. The effect site is calculated as with a negligible volume..

produce a certain drug effect. The volume of the site effect is neglected. The model in figure 3 is represented by the following set of equations.

$$\dot{q}_1(t) = K_{21}q_2(t) + K_{31}q_3(t) - K_{12}q_1(t) - K_{13}q_1(t) - K_{01}q_1(t) - K_{1e}q_1(t) + U(t) \quad (1)$$

$$\dot{q}_2(t) = K_{12}q_1(t) - K_{21}q_2(t) - K_{02}q_2(t) \quad (2)$$

$$\dot{q}_3(t) = K_{13}q_1(t) - K_{31}q_3(t) - K_{03}q_3(t) \quad (3)$$

$$\dot{q}_e(t) = K_{1e}q_1(t) - K_{0e}q_e(t) \quad (4)$$

where: u - dosing scheme as a function of time; K_{01} - rate constant reflecting all processes acting to irreversibly remove drug from the central compartment; K_{12}, K_{13}, K_{21}, K_{31}, K_{1e} - inter-compartmental rate constants [mg/l]; $V1$ - volume of the central compartment [l], V_2 and V_3 - volume of the peripheral compartments [l] (respectively muscle and fat), q_1 is the central compartment (plasma), q_2 is the rapidly equilibrium compartment (muscle), q_3 is the slowly equilibrium compartment (fat) and qe is the effect site compartment.

In daily clinical practice of general anesthesia, the goal is an individually tailored dosing of drugs, resulting in the optimal anesthetic level that is neither too light nor too deep. On the one side of the spectrum, excessively high drug doses (and inadequately deep levels of anesthesia) should be avoided to reach short recovery times and prevent excessive depression of the cardiovascular system. On the other side of the spectrum, inadequately low doses of anesthetics (and inadequately light levels of anesthesia) must be avoided to guarantee unconsciousness and prevent memory formation for intraoperative traumatic procedures (awareness). Under-dosage of anesthetics leads to conscious perception or even awareness (conscious perception with explicit memory) during anesthesia. Besides efforts to optimize anesthesia to shield the patient from stress and consequences of surgery, cost saving and issues of economy may be an issue for dosing strategies [9]. Management of anesthesia aimed at early recovery of the patient has been addressed as so-called fast track anesthesia, which nevertheless comprises much more than optimized dosing of anesthetic drugs. In this context, one challenge for the anesthesiologist is to avoid both over- and under-dosage. Therefore, knowledge of the pharmacokinetic and pharmacodynamic properties of anesthetic drugs is imperative.

The traditional form of target controlled dosing in anesthesia is by means of target controlled infusion (TCI) [10] which was developed in the context of the development of the induction and maintenance of general anesthesia by administration of intravenous agents only, i.e. the so called total intravenous anesthesia (TIVA) [11]. From the above scheme it becomes evident that TCI was the first formalized step that went beyond nave dosing and incorporated the dosing history into the actual dosing scheme. Thereby pharmacokinetic-pharmacodynamic models were used to estimate the present impact of doses given even far in the past on the present state of system. The first device to realize the simultaneous administration of two drugs (hypnotic and opioid) was named CATIA [12], computer-assisted titration of intravenous anesthesia. The name was chosen to underline the importance of the computer, needed for:

- tracking when/which drug was given in what amount,
- for the calculation of how much drug will still reside in the volumes of distribution in the body,
- what would be the future time course of dosing,
- maintaining a desired constant concentration in blood.

Newer developments take into account the hysteresis between blood and effect site and use the concentration at the biophase as target [13]. Inherent to model-based dosing strategies is that the model describes in general an average or typical patient. As the individual patient actually treated on the basis of this model will deviate from the average, this will translate into prediction errors with respect to the target concentrations. One approach to reduce systematic individual deviations is to take into account the dependence of the PK parameters on anthropometric data such as sex, age, weight, etc, by population pharmacokinetics [3], and to consider special patho-physiological conditions of the patient or the specific anesthetic procedure and its co-medication [14]. Another example where drug dosing represents a key role is the glucose-insulin system. The glucose-insulin control is important since glucose represents the main energy source for the brain. Low levels can affect the state of the patient (anxiety, aggressiveness, obfuscation, etc.) coma and in some cases the death. On the other hand, high glucose levels produce vascular and neural damages generating related diabetes complications (blindness, renal insufficiency, ulceration, between others). The glucose insulin regulatory system can be described as two coupled subsystems: the pancreatic system and the peripheral system. The inputs in the pancreatic system are diverse stimuli (between them the glucose level is the most important) and the output is the insulin liberated by the pancreas. In the peripheral system, the input is the insulin and the output is the glucose level. This is, there is a coupling between them: the output of one is the input for the other. The figure 4 shows the regulatory scheme. The functional relation between the two subsystems can be quantified by the rate of change of the glucose (G) and insulin (I) concentrations in time (t). The mathematical model can take the form:

$$\dot{G}(t) = f_1(G, I, t); I(t) = f_2(G, I, t) \tag{5}$$

where: functions f_1 and f_2 are characteristics of each model. Based on these relation different models have been proposed [15, 16]. A review of the mathematical modeling of glucose insulin system can be found in [17].

For the specific case of diabetic patients, the drug dosing and monitor blocks in figure 1 refer to the insulin doses and glucose monitoring respectively. The optimal route for insulin delivery is the intravenous. However, this method presents important limitations and further research has been oriented to subcutaneous pumps. Nowadays pumps technologies have become in an accepted mode of insulin therapy. However, the subcutaneous insulin delivery presents a delay between the insulin delivery and the subcutaneous absorption. This delay in general is not constant and varies inter and intra patient.

One of the key issues of the control of the glucose is the estimation of the glucose levels. An extended technique used to measure the glucose levels is Electrical Impedance Spectroscopy

(EIS). EIS is a tool that allows assessing the transfer function of a system based on an Alternate Current (AC) input voltage and the resulting AC current response. From the transfer function of the system different information regarding the behavior and structure of the system can be derived. This technique requires an identification process that relates changes in the impedance spectra with glucose changes in the biological system. An extended methodology to model those changes is based on circuit analysis where the parameters of the circuit represent a specific property in the system.

Different attempts to develop a glucose sensor based on this technique have been addressed. However, enzyme stability (enzyme immobilization, critical operating sensor conditions such as specific temperature and pH) increases the cost and limits the applications.

3. State of art

In an attempt to simplify the number of parameters in compartmental modeling, tools from fractional calculus have been employed. The fractional calculus is a generalization of integration and derivation to noninteger (fractional) order operators [15, 16]. First, a generalization of the differential and integral operators into one fundamental operator D_t^n (n the order of the operation) which is known as fractional calculus. The reason why fractional calculus should be implied in biological systems is the advantage given by the fact that it simplifies the dynamics of complex and heterogeneous systems [7, 8]. In the model presented here we considered the nonlinearity, consistency of units, mass balance and the fractional order rate constants. Considering figure 3 represented by the following equations:

$$\tau_1^{n_1-1} D^{n_1} q_1(t) = K_{21} q_2(t) + K_{31} q_3(t) - K_{12} q_1(t) - K_{13} q_1(t) - K_{1e} q_1(t) - K_{01} q_1(t) + U(t) \quad (6)$$

$$\tau_2^{n_2-1} D^{n_2} q_2(t) = K_{12} q_1(t) - K_{21} q_2(t) - K_{02} q_2(t) \quad (7)$$

$$\tau_3^{n_3-1} D^{n_3} q_3(t) = K_{13} q_1(t) - K_{31} q_3(t) - K_{03} q_3(t) \quad (8)$$

$$\tau_e^{n_e-1} D^{n_e} q_e(t) = K_{1e} q_1(t) - K_{0e} q_e(t) \quad (9)$$

where: τ_1, τ_2, and τ_3 represent the time elapsed for drug to diffuse from compartment 1, respectively compartments 2 and 3. The introduction of leads to the dimensional homogeneity of fractional rate equations, with the initial conditions $q_2(0) = d_2 = 0$, $q_3(0) = d_3 = 0$ and $q_1(0) = d_1$ =bolus injection.

A numerical solution of the general fractional differential equation $D_t^n y(t) = f(y(t), t)$ can be expressed as:

$$y(t_k) = f(y(y_k), t_k) - \sum_{j=1}^{k} c_j^n y(t_{k-j}) \quad (10)$$

where: $t_k = kh$, for k=1,2,3.N measured samples, h is the sampling period and c_j^n represent the binomial coefficient which is calculated with the following expression:

$$c_j^n = (1 - \frac{1+n}{j}) c_{j-1}^n \quad (11)$$

Applying the general solution to the set of equations 6-9, we obtain the following fractional order model describing drug diffusion in the body.

$$q_1(t_k) = (k_{21} q_2(t_k) + k_{31} q_3(t_k) - k_{12} q_1(t_k) - k_{13} q_1(t_k) - k_{01} q_1(t_k - 1)) h^n - \sum_{j=1}^{k} c_j^n q_1(t_{k-j}) + u(t) \quad (12)$$

$$q_2(t_k) = (k_{12}q_1(t_k)) - k_{21}q_2(t_k))h^n - \sum_{j=i}^{k} c_j^n q_2(t_{k-j}) \quad (13)$$

$$q_3(t_k) = (k_{13}q_1(t_k)) - k_{31}q_3(t_k))h^n - \sum_{j=i}^{k} c_j^n q_3(t_{k-j}) \quad (14)$$

with: $\tau_1^{n_1-1} = \tau_2^{n_2-1} = \tau_3^{n_3-1}$, $k_{12} = K_{12}/\tau_1^{n_1-1}$, $k_{13} = K_{13}/\tau_1^{n_1-1}$, $k_{21} = K_{21}/\tau_2^{n_2-1}$, $k_{31} = K_{31}/\tau_3^{n_3-1}$, $k_{01} = K_{01}/\tau_1^{n_1-1}$and assuming $n_1 = n_2 = n_3 = n$ for simplicity. Literature reports upon several works dealing with such models [5, 6, 7, 17].

Another approach to physiologic and compartmental modeling is to use parsimonious models which lose their link to physiology, but have the advantage of less number of parameters for adaptation to the specific patient. There are several types of such models, based on both rational and non-rational input-output relationship. For applications in anesthesia, parsimonious models of rational functions have been published in [19] of the form:

$$G(s) = \frac{b_0 + b_1 s + b_2 s + \dots b_{n_b} s^{n_b}}{a_1 s + a_2 s + \dots a_{n_a} s^{n_a}} \quad (15)$$

with non-rational orders [20].

For applications in diabetes general rational models in the Laplace domain allow reducing the number of model parameters. However, fractional behavior of the impedance in biological systems requires solutions that increase the flexibility of the models structure maintaining a limited number of parameters [21]. For glucose sensors based EIS technique, fractional behavior in the interface electrode-electrolyte has been identified. This interface can be modeled by a structure with high number of resistors and capacitors. However, a constant phase element (CPE) can be used to model this process [22]. The CPE is a fractional equivalent electrical circuit and models an imperfect capacitor (the interface or double layer electrode electrolyte). The impedance of the CPE is calculated by:

$$Z_{CPE} = \frac{1}{Q_o \omega^\alpha} e^{-j\frac{\pi}{2}\alpha} \quad (16)$$

with $0 < \alpha < 1$, ω (rad/s) and Q_0 an impedance element usually consisting of a capacitance and resistance. Rational, parsimonious models use this element in order to reduce the number of parameters while increasing the flexibility of the model [21, 22]. The generic goal of such simplified models is to allow on-line identification of patients specificity by reducing the number of model parameters. With such models at hand, adaptation of control target infusion levels can be achieved as suggested for anesthesia in [23].

4. Perspectives

Current state of use in terms of TCI combined with feedback control in drug delivery systems is still based on averaged population models. In anesthesia, several studies have been reported successfully, making use of PID controllers [24, 25], nonlinear controllers [26], predictive controllers [27, 28]. In closed loop control where the glucose changes are faster than insulin absorption and action, the control algorithm should find a trade-off between slow-pace regulation well suited to mild control actions applicable to quasi-state and postprandial regulations [29]. The well-known PID controller, given its reactive action, responds to glucose changes once they occur, hence, it is necessary to develop a robust controller. However, a cautious design could not produce an effective response to meal disturbances. The most modern controllers are based on model predictive control algorithms using a model of the patient in order to anticipate physiological actions [30, 31]. In glucose sensors based EIS, given the presence of fractional behavior, general fractional models in the frequency domain can be proposed. In [32] an approach using rational models in the Warburg variable (s) has been identified showing the reduction of the parameters and clear identification of the glucose levels compared to integer rational models. Further extension of fractional models identification in the frequency domain can be found in [33]. The challenge for the next generation of drug delivery systems can be summarized under two main trends:

- cyber-physical systems with integrated ICT platform for hybrid sources of information and monitoring devices with the aim to deliver a decision making assist device;
- integrating parsimonious models into existing platforms of TCI and feedback control systems for drug delivery systems.

The first trend involves the trans-disciplinary aspects of computer science, clinical expertise and system engineering. It requires ingenuity of integrated systems, due to the hybrid nature of various elements of such cyber-physical network. The second trend is based on system engineering concepts (identification, control, stability and robustness). For non-rational patient models, it requires moving away from the classical system theory approach. This might add challenges for stability and robustness issues, which are critical for patient safety and well-being.

Acknowledgment

C. M. Ionescu acknowledges the Flanders Research Center (FWO), grant number 12B3415N for its financial support. D. Copot acknowledges the support of the project FWO nr. G008113N.

References

[1] Absalom A., Mani V., De Smet T, Struys MRF., Pharmacokinetic models for propofol defining and illuminating the devil in the detail. British Journal of Anaesthesiology, 103, 26-37, 2009.
[2] Minto C., White M., Morton N., Kenny G., Pharmacokinetics and pharmacodynamics of remifentanil. II Model application. Anesthesiology, 86, 24-33, 1997.
[3] Schnider T., Minto C., Shafer S., Gambus P., Andresen C., Goodale D., Youngs E., The influence of age on propofol pharmacodynamics. Anesthesiology, 90, 1502-1516, 1981.
[4] Schnider T., Minto C., Gambus P., et al., The influence of method of administration and covariates on the pharmacokinetics of propofol in adult volunteers. Anesthesiology, 88, 1170-1182, 1998.
[5] Wada D.R., Bjrkman S., Ebling W.F., Harashima H., Harapat S.R., Stanski D.R., Computer simulation of the effects of alterations in blood flows and body composition on thiopental pharmacokinetics in humans. Anesthesiology, 87,884899, 1997.
[6] Ionescu C., Copot D., De Keyser R., Three compartmental model for propofol diffusion during general anesthesia, Discontinuity. Nonlinearity and Complexity, 2(4), 357-368, 2013.
[7] Dokomuetzidis A., Magin R., Machetas P., Fractional kinetics in multi-compartmental systems. Journal of Pharmacokinetic and Pharmacodynamic, 37, 507524, 2010.
[8] Popovic J.K., Atanackovic, M.T., Rapaic A.S., Pilipovic S., Atanackovic T.M., A new approach to the compartmental analysis in pharmacokinetics: fractional time evolution of diclofenac. Journal of Pharmacokinetic and Pharmacodynamic, 37, 119134, 2010.
[9] White P.F., Ma H., Tang J., Wender R.H., Sloninsky A., Kariger R., Does the use of electroencephalographic Bispectral Index or auditory evoked potential index monitoring facilitate recovery after desflurane anesthesia in the ambulatory setting? Anesthesiology, 100, 811817, 2004.
[10] Glen J.B., The development of Diprifusor: a TCI system for propofol. Anaesthesia, 53 [Suppl 1], 1321, 1998.
[11] Schuttler J., Kloos S., Schwilden H., Stoeckel H., Total intravenous anaesthesia with propofol and alfentanil by computer-assisted infusion. Anaesthesia, 43 [Suppl]:27, 1988.
[12] Schuttler J., Schwilden H., Stoekel H., Pharmacokinetics as applied to total intravenous anaesthesia. Practical implications. Anaesthesiam 38 [Suppl],5356, 1983.
[13] Schuttler J., Schwilden H., Modern Anesthetics, Berling Heidelberg: Springer, 978-3-540-72813-9, 487, 2008.
[14] Billard V., Servin F., Guignard B. et al., Desflurane-remifentanil-nitrous oxide anesthesia for abdominal surgery: optimal concentrations and recovery features. Acta Anaesthesiology Scandinavia, 48, 355-364, 2004.
[15] Oustaloup, A., La derivation non-entiere (In French). Hermes, Paris, 1995.

[16] Podlubny, I., Fractional differential equations. Academic Press, San Diego, 1999.
[17] Caumo A., Bergman R.N., Cobelli C., Insulin sensitivity from meal tolerance tests in normal subjects: a minimal model index. Journal of Clinical Endocrinology and Metabolism, 85, 4396, 2000.
[18] Breda E., Cavaghan M.K., Toffolo G., Polonsky K.S., Cobelli C., Oral glucose tolerance test minimal model indexes of -cell function and insulin sensitivity. Diabetes, 50, 150, 2001.
[19] Palumbo P., Ditlevsen S., Bertuzzi A., De Gaetano A., Mathematical modeling of the glucoseinsulin system: A review. Mathematical Biosciences, 244(2), 69-81, 2013.
[20] Dokoumetzidis A., Macheras, P., Fractional kinetics in drug absorption and disposition processes. Journal of Pharmacokinetics and Pharmacodynamics, 36, 165-178, 2009.
[21] Lemos J.M., Caiado D.V., Costa B.A., Paz L.A., Mendonca T.F., Rabico R., Esteves S., Seabra M., Robust Control of Maintenance-Phase Anesthesia, IEEE Control Systems Magazine, 34(6), 24-38, 2014.
[22] Soltesz K. Hahn J.O., Hagglund T., Ansermino J.M., Individualized closed-loop control of propofol anesthesia: A preliminary study. Biomedical Signal Processing and Control, 8(6), 500-508, 2013.
[23] Magin R., Fractals in bioengineering, Begell House Publishers, 2006.
[24] Ionescu C, Machado J.A.T., De Keyser R., Decruyenaere J., Struys M.M., Nonlinear dynamics of the patient's response to drug effect during general anesthesia. Communications in nonlinear science and numerical simulation. 20(3). 914-926, 2015.
[25] Dumont G., Feedback control for clinicians. Journal of Clinical Monitoring and Computing, 28, 511, 2014.
[26] Rocha. C., Mendonca T., Silva M.E., Individualizing propofol dosage: a multivariate linear model approach. Journal Of Clinical Monitoring And Computing, 28(6), 525-536,2014.
[27] Dumont G., Martinez A., Ansermino M., Robust control of depth of anaesthesia. International Journal of Adaptive Control Signal Processing, 23, 435-454, 2009.
[28] Ionescu C., De Keyser R., Torrico B., De Smet T., Struys MRF., Normey-Rico J., Robust predictive control strategy applied for propofol dosing using BIS as a controlled variable during anaesthesia, IEEE Transactions on Biomedical Engineering, 55, 2161-2170, 2008.
[29] 35. Krieger A., Pistikopoulos E.N., Model predictive control of anesthesia under uncertainty. Computers and Chemical Engineering, 71, 699-707, 2014.
[30] Ionescu C.M., Nascu I., De Keyser R., Lessons learned from closed loops in engineering: towards a multivariable approach regulating depth of anaesthesia. Journal Of Clinical Monitoring And Computing, 28(6), 537-546, 2014.
[31] Cobelli C., Renard E., Kovatchev B., Artificial Pancreas: Past, Present and Future. Diabetesm, 60, 2011.
[32] Kovacs L., Szalay P., Almassy Zs., Barkai L., Applicability Results of a Nonlinear Model-Based Robust Blood Glucose Control Algorithm. Journal of Diabetes Science and Technology, 7(3), 708-716, ISSN 1932-2968, 2013.
[33] Kovacs L., Szeles A., Drexler D.A., Sapi J., Harmati I., Sapi Z., Model-based Angiogenic Inhibition of Tumor Growth using Modern Robust Control Method. Computer Methods and Programs in Biomedicine, DOI:10.1016/j.cmpb.2014.01.002, 114, e98-e110, ISSN 0169-2607, 2014.

IOP Conf. Series: Journal of Physics: Conf. Series **783** (2017) 012054 doi:10.1088/1742-6596/783/1/012054

Comparative analysis and exprimental results of advanced control strategies for vibration suppression in aircraft wings

Isabela R. Birs[1], Silviu Folea[1], Dana Copot[2], Ovidiu Prodan[3], Cristina-I. Muresan[1]

[1]Department of Automation, Technical University of Cluj-Napoca, Memorandumului St., No. 28, Cluj-Napoca, Romania
[2]Department of Electrical energy, Systems and Automation, Research group on Dynamical Systems and Control, Ghent University, Technologiepark 914, 2nd floor, 9052 Ghent, Belgium
[3]Department of Civil Engineering, Technical University of Cluj-Napoca, Memorandumului St., No. 28, Cluj-Napoca, Romania

Cristina.Muresan@aut.utcluj.ro

Abstract. The smart beam is widely used as a means of studying the dynamics and active vibration suppression possibilities in aircraft wings. The advantages obtained through this approach are numerous, among them being aircraft stability and manoeuvrability, turbulence immunity, passenger safety and reduced fatigue damage. The paper presents the tuning of two controllers: Linear Quadratic Regulator and Fractional Order Proportional Derivative controller. The active vibration control methods were tested on a smart beam, vibrations being mitigated through piezoelectric patches. The obtained experimental results are compared in terms of settling time and control effort, experimentally proving that both types of controllers can be successfully used to reduce oscillations. The analysis in this paper provides for a necessary premise regarding the tuning of a fractional order enhanced Linear Quadratic Regulator, by combining the advantages of both control strategies.

1. Introduction

The cantilever wing is the only type of airplane wing used in civilian aircrafts. Due to its shape and the need to actively control vibration, the cantilever wing is characterized and modelled as a smart beam [1]. The active vibration control is most often realized through smart materials exhibiting the piezoelectric effect. The piezo patches, which act as sensors and actuators, are embedded or mounted on the surface of the beam [2], [3].

Among the years, various control strategies have been tested with the purpose of significantly reducing the smart beam's displacement. Such strategies include: fuzzy logic [4], adaptive [5], robust, Linear Quadratic Regulator, fractional – order PI, PD and PID type controllers [6], etc. In [7] the authors present the tuning of a fuzzy logic controller using the particle swarm optimization approach that deals with the control of smart composite beams vibration. The proposed controller is compared with fuzzy logic controllers with constant scaling factors and with the Linear Quadratic Regulator (LQR). The results obtained show that the PSO tuned fuzzy logic controller obtains better vibration suppression when compared to the LQR for both single and multimodal cases. Numerical optimization consisting of population based Particle Swarm Optimization and Genetic Algorithms is also used in [8] to develop fuzzy logic controllers. The study compares the behaviors of the classical fuzzy controller with the ones obtained from the optimization techniques. The controller optimized by PSO yielded better results than the others.

The active control of flexible cantilever plates based on the piezoelectric effect is realized in [9] through artificial neural networks. The neuro-controller is trained such that optimal voltage is applied to the piezoelectric patches. The performance and the robustness of the trained controller is evaluated comparing the active response to the free vibration under various types of dynamic excitations.

The idea of tuning a controller with neural networks is also present in [10] where the parameters of a proportional – derivative controller are determined with a back propagation neural network (BPNN). Data obtained from both simulations and practical experiments is used to prove the veracity of the proposed method. The controller proves to exert vibration on an experimental unit consisting on a vertical aluminum beam equipped with 3 piezoelectric patches. Modeling and controlling smart beams with the hysteresis property is described in [11]. A time varying model of the smart beam is determined based on the hysteresis property using recursive extended least square method. The vibration is controlled with an adaptive minimal variance self-tuning direct regulator (MSVTDR). The advantage of this type of controller consists in computing an optimal control signal. The adaptive regulator claims to reduce the strain of the smart beam with up to 83.67% at its first natural frequency.

Piezoelectric bonded smart structures are controlled with the classical PID controller in [12]. The PID controller is designed to attenuate both free and forced vibrations of the smart beam. Compared with the optimal LQR controller, the PID controller gives the best results. Part of the study is dedicated to analyze the behavior of the three control effects: proportional, derivate and integral. Experimental tests suggest that the D control has little to no effect on the steady-state error, but counteracts the free vibration of the beam. However, the PI effects eliminate the steady state error, but have a negative impact on the free vibration. A feedback control algorithm is implemented in [13] using the real time National Instruments Real Time Controller cRIO 9022 controller. Improvements of 49.29% and 52.84% were obtained for the first and second flexural modes. However, the strain rate feedback approach showed low system stability. The validity of the H_∞ optimal controller is proven in [14] when uncertain disturbance and measurement noise are considered. A full-order state observer LQR and PID controllers are designed and tested in order to control the vibration of the beam [15]. The LQR controller outperformed the integer-order PID in the 3 configurations studied of the beam-piezoelectric elements.

In this study, two active vibration control methods are proposed with the purpose of significantly reducing the settling time of a dedicated smart beam when rejecting disturbances. The first choice regarding the control algorithm refers to a LQR controller, mainly chosen because of its wide application in smart beam vibration suppression methods, the simple parameter design method, as well as overall efficiency. The second control algorithm is a rather less used approach in vibration suppression: the fractional order PD controller, mainly chosen because of its improved performance over classical PID controllers [16]. Apart from that, fractional order systems and controllers are becoming more and more popular in the research community [17], [18]. In terms of vibration suppression in a cantilever beam, fractional order PD controllers have been designed previously for similar cantilever beam and the experimental results demonstrated the advantages of using a fractional order PD controller instead of the classical integer order PD controller [19].

To tune the controllers, the dynamics of the smart beam is considered as a simple second order system. In this way, because of the simplification of the model, the problem of robustness is also addressed, since in real life applications, the ability of the control system to cope with modeling uncertainties proves extremely useful, especially in the case of an aircraft subjected to turbulences of different frequencies. The two proposed methods were implemented and validated on an experimental vibration unit consisting of an aluminum cantilever beam equipped with piezoelectric patches. The comparison is realized by analyzing the settling times and control effort obtained while the beam is subjected to impulse disturbances.

The structure of the article presents an overview of the proposed tuning methods, the description of the experimental stand, experimental system identification, numerical and experimental results and last, but not least, the conclusions.

2. Overview of the proposed active vibration control methods

2.1. Linear Quadratic Regulator (LQR)

The LQR is an optimal controller that ensures the stability of the closed-loop system, has a simple tuning procedure and a certain degree of robustness. This type of optimal controller minimizes a cost function of the following form:

$$J_{LQR} = \int_0^{\infty} \left[x^T(t) * Q * x(t) + u^T(t) * R * u(t) \right] dt \tag{1}$$

where R is a scalar and Q is a positive semi-definite matrix (symmetric matrix) having the dimension $n \times n$, with n the number of states of the state space system that describes the dynamics of the smart beam. The state space system is given as:

$$\begin{aligned} \dot{x}(t) &= Ax(t) + Bu(t) \\ y(t) &= Cx(t) + Du(t) \end{aligned} \tag{2}$$

where x is the state vector; y is the output vector; u is the control signal and A is the state matrix, B the control matrix, C the output matrix and D is the feedthrough matrix.

The integral from equation (1) is written as a sum of the accumulated deviation of the states with respect to the equilibrium point and the accumulated control effort. By properly choosing the parameters Q and R, the importance of the control performance and input energy are weighted reaching a convenient compromise.

The manipulated input u(t) generated by the LQR controller is

$$u(t) = -Kx(t) \tag{3}$$

where K is the feedback gain matrix computed as

$$K = R^{-1} \cdot B^T \cdot X \tag{4}$$

X is a symmetric matrix determined from the Ricatti equation as a nonnegative definite solution:

$$A^T X + XA + C^T QC - XBR^{-1}B^T X = 0 \tag{5}$$

The resulting closed loop system has the form:

$$\dot{x}(t) = (A - B * K) * x(t) \tag{6}$$

2.2. Fractional – order Proportional Derivative Controller

The fractional-order PD controller is best defined by its transfer function:

$$H_{FO\text{-}PD}(s) = k_p\left(1 + k_d s^{\mu}\right) \tag{7}$$

where s is the Laplace operator, k_p and k_d are the proportional and derivative gains, while μ is the fractional order differentiator. When μ is equal to 1, equation (7) describes an integer order proportional – derivative controller.

The trigonometric form of the fractional order PD (FO-PD) controller can be written by moving from the Laplace domain to the frequency domain, where jω replaces s.

$$H_{FO\text{-}PD}(j\omega) = k_p\left[1 + k_d\omega^{\mu}\left(\cos\left(\frac{\pi\mu}{2}\right) + j\sin\left(\frac{\pi\mu}{2}\right)\right)\right] \tag{8}$$

The three parameters of the controller are computed based on frequency domain constraints: phase margin φ_m, gain crossover frequency ω_{cg} and robustness to gain variations. The first two constraints can be written through the phase equation of the open loop system:

$$\angle\left(H_{open-loop}(j\omega_{cg})\right) = -\pi + \varphi_m \tag{9}$$

The gain crossover frequency is mathematically expressed as:

$$\left|\left(H_{open-loop}(j\omega_{cg})\right)\right| = 1 \tag{10}$$

The robustness constraint is imposed such that it ensures a constant overshoot at gain variations. On the Bode phase plot, a constant overshoot means a constant phase near the gain crossover frequency. A constant phase graphically translates into a straight line. One important property of a straight line is the fact that its derivative is equal to 0. This means that if the derivative of the phase margin is imposed as being 0 at the gain crossover frequency, the closed loop robustness is assured.

$$\left.\frac{d\left(\angle H_{open-loop}(\omega)\right)}{d\omega}\right|_{\omega=\omega_{cg}} = 0 \tag{11}$$

The open-loop system, with G(s) – the transfer function of the smart beam and C(s) – that of the FO-PD controller, is computed as:

$$H_{open-loop}(s) = G(s) \cdot C(s) \tag{12}$$

Replacing the equation of the open-loop transfer function in the gain crossover, phase margin and robustness constraints from equations (9),(10) and (11), the modulus, phase and phase derivative of the controller C(s) are obtained in the complex representation:

$$\begin{gathered}\angle C(j\omega_{cg}) = -\pi + \phi_m - \angle G(j\omega_{cg}) \\ \left|C(j\omega_{cg})\right| = \frac{1}{\left|G(j\omega_{cg})\right|} \\ \left.\frac{d\left(C(j\omega)\right)}{d\omega}\right|_{\omega=\omega_{cg}} + \left.\frac{d\left(\angle G(j\omega)\right)}{d\omega}\right|_{\omega=\omega_{cg}} = 0\end{gathered} \tag{13}$$

Further using (8) in the phase equation from (13) leads to:

$$\frac{k_d\omega_{cg}^{\mu}\sin\left(\frac{\pi\mu}{2}\right)}{1+k_d\omega_{cg}^{\mu}\cos\left(\frac{\pi\mu}{2}\right)} = tg\left(-\pi + \phi_m - \angle G(j\omega_{cg})\right) \tag{14}$$

while the robustness constraint may be rewritten as:

$$\frac{\mu k_d\omega_{cg}^{\mu-1}\sin\frac{\pi\mu}{2}}{1+2k_d\omega_{cg}^{\mu}\cos\frac{\pi\mu}{2}+k_d^2\omega_{cg}^{2\mu}} = -\left.\frac{d\left(\angle G(j\omega)\right)}{d\omega}\right|_{\omega=\omega_{cg}} \tag{15}$$

By solving (14) and (15), as a system of equations, the derivative gain and the fractional order, k_d and μ, are obtained. The graphical method can be a tangible approach to determining the parameters.

Rewriting the magnitude condition from equation (13) such that the proportional gain is isolated in the left side, the following equation is obtained, from which k_p is easily determined:

$$k_p = \frac{1}{\left|G(j\omega_{cg})\right|}\frac{1}{\sqrt{1+2k_d\omega_{cg}^{\mu}\cos\frac{\pi\mu}{2}+k_d^2\omega_{cg}^{2\mu}}} \tag{16}$$

3. Experimental setup and system identification

3.1. Description of the experimental equipment and the smart beam

The experimental vibration unit can be seen in Figure 1 and consists of a smart beam equipped with 4 piezoelectric (PZT) actuators, a real-time controller, input and output modules and a PZT controller. The beam is fixed at one end allowing only back and forth movements. There are four piezoelectric patches mounted close to the fixed end of the beam, two on each side. The aluminium beam is 250 mm long, 20 mm wide and 1 mm thick. Figure 2 shows a detailed view of the smart beam and the position of the patches. Two piezoelectric DuraAct P-878 Power Patch transducers are fixed on each side of the beam.

The control algorithm is implemented in real-time with the help of the CompactRIO™ 9014 controller. The NI 9230 is the input module that measures the displacement of the beam, while the NI 9263 output module is used for applying the voltage to the patches used to mitigate vibration. The communication and the control algorithms are implemented in LabVIEW ™.

The piezoelectric patches are used only as actuators and not as sensors because the movement of the beam is measured with the 120 ohm Omega Prewired KFG-5-120-C1-11L1M2R strain gauge sensors. The E-509.X3 module from Physik Instrumente is used to amplify the signal from the sensors, while E-503.00 is used to amplify the control signal to the PZT patches. The chassis amplifies the signal by 10, meaning that if the LabVIEW code gives a control signal of between [-2;2] volts, the patches are excited with a signal between [-20; 20] volts.

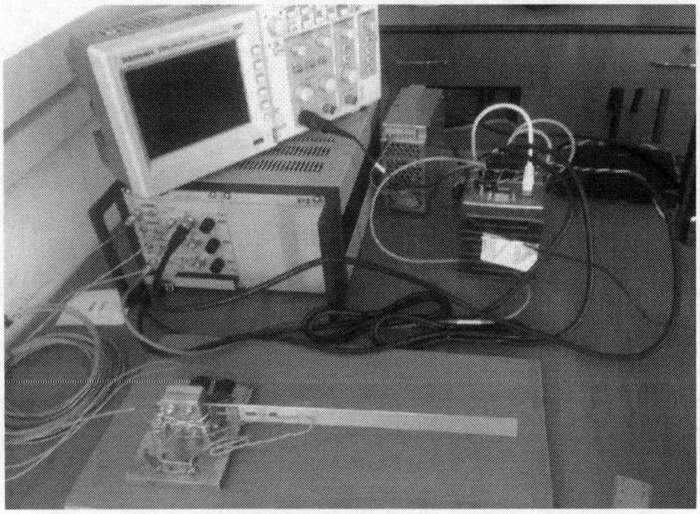

Figure 1. Experimental setup

Figure 2. Detailed view of the fixed end of the smart beam.

3.2. System Identification

The transfer function of the system is determined based on the experimental response of the beam to a sinusoidal input. In order to determine an accurate transfer function the system was excited with sine waves of different amplitudes and frequencies.

It was experimentally observed that the resonant frequency of the smart beam is close to 14.52 Hz. A second order model of the smart beam has been determined as:

$$G(s) = \frac{72.99}{s^2 + 1.115\,s + 8401} \tag{17}$$

The validation of the identified model G(s) to a sine input of frequency 14.52 Hz and amplitude 10 V is shown in Figures 3, 4 and 5 below.

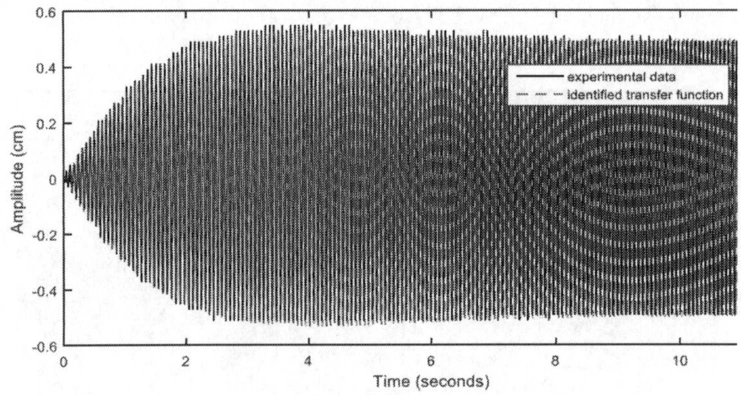

Figure 3. Experimental data and second order simulated response of the second order transfer function

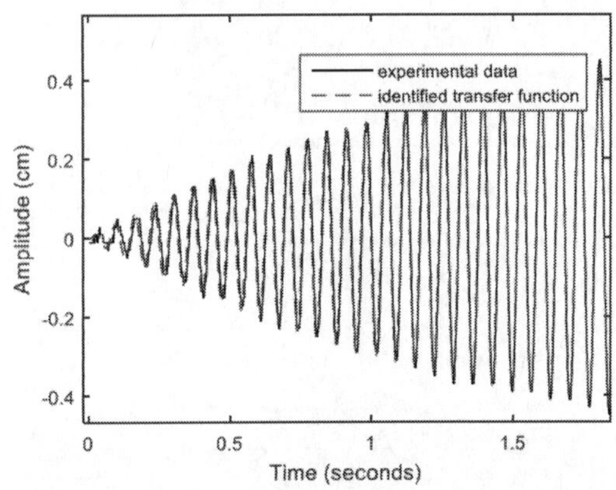

Figure 4. Zoomed transient response fit of the identified second order transfer function

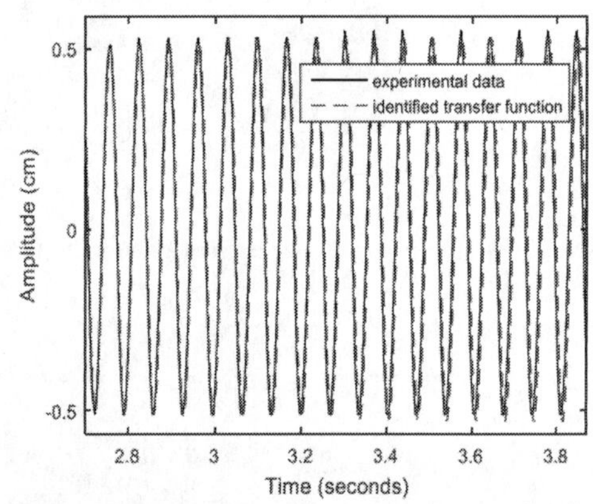

Figure 5. Zoomed steady state response fit of the identified second order transfer function

As can be seen in the figures above, the model fits the experimental data accurately, with a 97.07% fit to estimation data, a FPE (Final Prediction Error) of $9.85031 \cdot 10^{-5}$ and a MSE (Mean Squared Error) equal to $9.829 \cdot 10^{-5}$. The resonant frequency of the model G(s) is 14.59 Hz, resulting in a small modelling error compared to the experimentally observed resonant frequency of 14.52 Hz.

4. Controller tuning and experimental results

4.1. LQR controller tuning

For the LQR controller tuning, the state-space representation of the beam is needed:

$$A = \begin{pmatrix} 0 & 1 \\ -8401 & -1.115 \end{pmatrix} \quad B = \begin{pmatrix} 0 \\ 72.99 \end{pmatrix}$$
$$C = \begin{pmatrix} 1 & 0 \end{pmatrix} \qquad D = (0) \tag{18}$$

where the states of the state space system in (18) are the beam displacement and its velocity.

The weighting matrices R and Q are chosen as follows:

$$Q = \begin{pmatrix} 1 & 0 \\ 0 & 0.002 \end{pmatrix} \quad R = 3 \tag{19}$$

IOP Conf. Series: Journal of Physics: Conf. Series **783** (2017) 012054 doi:10.1088/1742-6596/783/1/012054

The discrete feedback gain matrix, K, is determined as:

$$K = (-0.1913 \quad 0.0152) \tag{20}$$

where a sampling period of 0.003 seconds has been used. The choice for the Q and R parameters is based on a strict requirement for the control signal to be kept in a range [-2, 2], corresponding to a [-20, +20] V applied on the PZT patches. By trial and error, the best settling time, with the control signal limitation constraint, has been obtained for Q and R as indicated in (19). The LQR controller in (20) was implemented on the experimental unit and the closed loop results can be seen in Figure 6. An impulse disturbance acting on the beam's free end of 0.5V has been considered. The control signal required to reduce the vibrations is given in Figure 7.

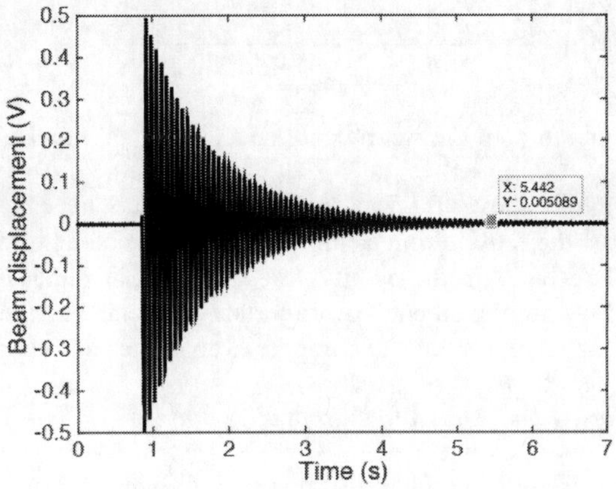

Figure 6. Closed-loop system response using the LQR controller

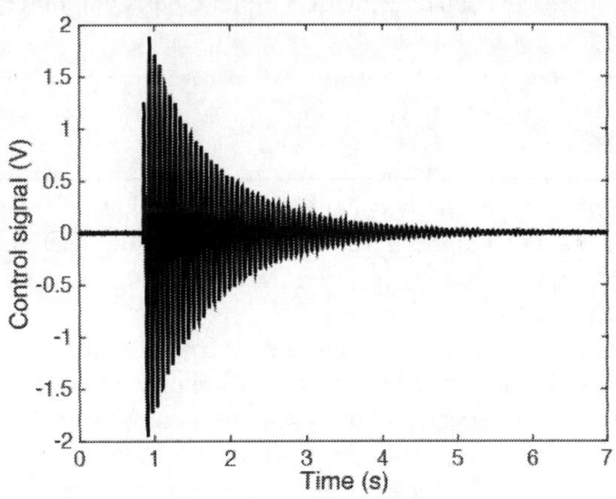

Figure 7. The control signal generated by the LQR controller

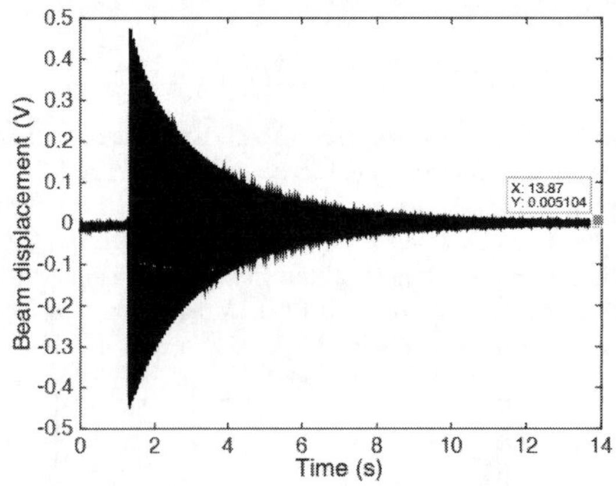

Figure 8. Free vibration of the smart beam considering an impulse disturbance

As can be seen in the figures above, the Linear Quadratic Regulator can be successfully used to suppress unwanted vibrations, the settling time achieved in this case being of 5.44 seconds. Compared to the settling time of 13.87 seconds for the passive free vibration, as indicated in Figure 8, the LQR controller provides for a 60.8% improvement. An important advantage of the LQR controller is that the feedback gain matrix, K, is relatively easy to compute even in the case of a higher order model.

4.2. Fractional – Order Proportional Derivative Controller tuning

The fractional-order PD controller is tuned by imposing frequency domain constraints: a phase margin of 60° for a lower overshoot of the closed loop system, a gain crossover frequency of 105 rad/s for a small settling time and the robustness to gain variations, as indicated in (13).

The obtained fractional-order PD controller, with the parameters computed according to (14), (15) and (16) above is given by :

$$H_{FO\text{-}PD}(s) = 15.36\left(1 + 0.028\, s^{0.9164}\right) \tag{21}$$

The frequency response of the open loop system with the FO-PD controller in (21) demonstrates that the performance criteria are met and it is given in Figure 9. To implement the FO-PD controller in (21), its discrete time approximation has been determined using the new method described in [20], with a sampling period of 0.003 seconds.

The closed loop response of the smart beam, under the same impulse disturbance as in the case of the LQR controller, is given in Figure 10, while the control signal is given in Figure 11. In this case, the FO-PD tuning method has not directly considered the control signal limitation constraint. As indicated in Figure 11, the FO-PD controller has a much more aggressive dynamics in comparison to the LQR controller, as seen in Figure 7. Also, the derivative action of the FO-PD controller is highly sensitive to noise. Nevertheless, this aggressive dynamics leads to better performance in terms of settling time. In the case of the FO-PD controller, the disturbance is rejected entirely in 3.24 seconds, which represents a 76.6% improvement compared to the free vibration and a 40.4% improvement in comparison to the LQR controller.

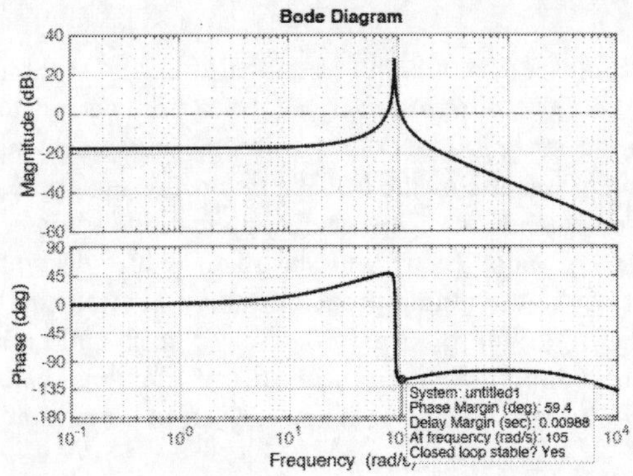

Figure 9. Bode diagram of the open loop system with FO-PD controller

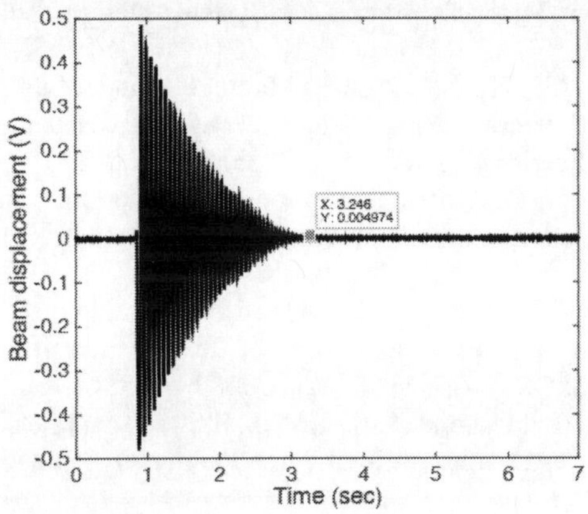

Figure 10. The closed-loop system response using a FO-PD controller

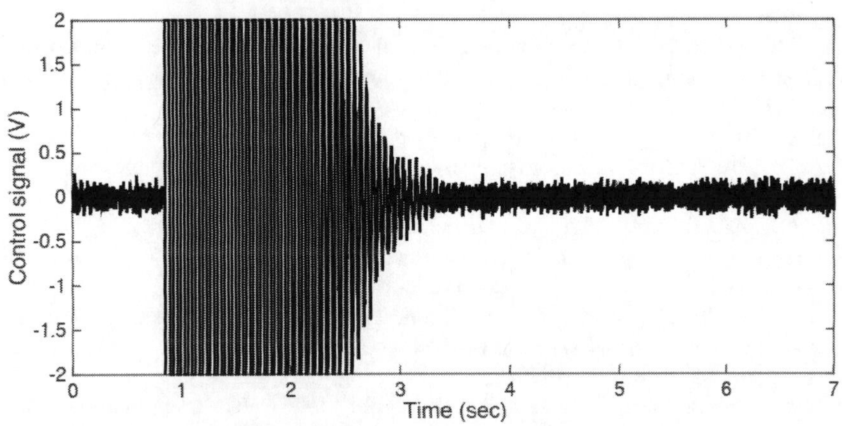

Figure 11. The control signal generated by the FO-PD controller

5. Conclusions

From the settling time point of view, the fractional-order PD controller brings a significant improvement compared to the LQR controller, as has been pointed out in this paper. However, the nearly 50% improvement in the settling time comes with a disadvantages regarding the increase in the control effort. These preliminary results of the FO-PD controller in suppressing vibrations on a smart beam through piezoelectric patches are a novel application study and further research and improvement will be sought. A major focus will be given to the design of a fractional order PID controller that considers directly in the tuning procedure the limitations on the control signal, to avoid the saturation. Also, since the LQR controller provides for an acceptable vibration attenuation, this too can be further enhance by combining its benefits with that of fractional calculus, in tuning a fractional order LQR controller. The analysis in this paper constitutes an important premise in that regard.

References

[1] Dr. Hatem, R. Wasmi Hussein Abdulridha Abdulameer, "Vibration Control Analysis of Aircraft Wing by Using Smart Material," Innovative Systems Design and Engineering, vol. 6, no. 8, 2015.

[2] Ahmed Abdul-hussain Ali, Hussain Yousif Mahmood, and Mahmood Wael Saeed, "Active Vibration Control of Aircraft Wing Using Piezoelectric Transducers," International Journal of Scientific & Engineering Research, vol. 7, no. 5, May 2016.

[3] Shashikala Prakash et al., "Active vibration control of a full scale aircraft wing using a reconfigurable controller," Journal of Sound and Vibration, vol. 361, pp. 32-49, January 2016.

[4] Mohammad A. Ayoubi, Sean Shan-Min Swei, and Nhan T. Nguyen, "Fuzzy model-based pitch stabilization and wing vibration suppression of flexible aircraft," in 2014 American Control Conference, Portland, OR, 4-6 June 2014, pp. 3083 - 3088.

[5] Gergely Takács, Tomáš Polóni, and Boris Rohal'-Ilkiv, "Adaptive Model Predictive Vibration Control of a Cantilever Beam with Real-Time Parameter Estimation," Shock and Vibration, 2014.

[6] Muresan Cristina, Folea Silviu, Prodan Ovidiu, and Eva Dulf, "Design and Experimental Validation of an Optimal Fractional Order Controller for Vibration Suppression," in The International Conference on Control, Decision and Information Technologies, Malta, April 6-8, 2016.

[7] NemanjaD.Zorić et al., "Free vibration control of smart composite beams using swarm optimized self-tuning fuzzy logic controller," Journal ofSoundandVibration, vol. 333, no. 21, pp. 5244-5268, October 2014.

[8] Magdalene Marinaki, Yannis Marinakis, and Georgios E. Stavroulakis, "Fuzzy control optimized by PSO for vibration suppression of beams," Control Engineering Practice, vol. 18, no. 6, pp. 618-629, June 2010.

[9] Osama Abdeljaber, Onur Avci, and Daniel J. Inman, "Active vibration control of flexible cantilever plates using piezoelectric materials and artificial neural networks," Journal of Sound and Vibration, vol. 363, pp. 33-53, February 2016.

[10] Zhicheng Qiu, Xiangtong Zhang, and Chunde Ye, "Vibration suppression of a flexible piezoelectric beam using BP neural network controller," Acta Mechanica Solida Sinica, vol. 25, no. 4, pp. 417-428, August 2012.

[11] Ting Zhang, Hong Guang Li, and Guo Ping Cai, "Hysteresis identification and adaptive vibration control for a smart cantilever beam by a piezoelectric actuator," Sensors and Actuators A: Physical, vol. 203, pp. 168-175, December 2013.

[12] Shunqi Zhang, Rüdiger Schmidt, and Xiansheng Qin, "Active vibration control of piezoelectric

bonded smart structures using PID algorithm," Chinese Journal of Aeronautics, vol. 28, no. 1, pp. 305-313, February 2015. [Online]. http://dx.doi.org/10.1016/j.cja.2014.12.005

[13] Riessom Weldegiorgis, Prasad Krishna, and K.V. Gangadharan, "Vibration Control of Smart Cantilever Beam Using Strain Rate Feedback," in International Conference on Advances in Manufacturing and Materials Engineering, ICAMME, 2014, pp. 113-122.

[14] Hu Junfeng, Zhu Dachang, and Chen Qiang, "Vibration Control of a Smart Beam Using H∞ Control," in Fourth International Conference on Intelligent Computation Technology and Automation, 2011.

[15] Harijono Djojodihardjo, Mohammad Jafari, Surjatin Wiriadidjaja, and Kamarul Arifin Ahmad, "Active Vibration Suppression of an elastic piezoelectric sensor and actuator fitted cantilevered beam configurations as a generic smart composite structure," Composite Structures, vol. 132, pp. 848-863, August 2015.

[16] C.A. Monje, YangQuan Chen, Blas M. Vinagre, D. Xue, Vicente Feliu, Fractional order Systems and Controls: Fundamentals and Applications, Springer-Verlag, London, 2010

[17] Robin De Keyser, Cristina I. Muresan, Clara M. Ionescu, A novel auto-tuning method for fractional order PI/PD controllers, ISA Transactions, vol. 62, pp. 268 275, May 2016

[18] Clara M. Ionescu, Jose A. Tenreiro Machado, Robin De Keyser, Fractional-order impulse response of the respiratory system, Computers & Mathematics with Applications, vol. 62, no. 3, pp. 845-854, August 2011

[19] Birs, I.R., Muresan, C.I., Folea, S., Prodan, O. (2016), A Comparison between Integer and Fractional Order PDμ Controllers for Vibration Suppression, *Applied Mathematics and Nonlinear Sciences*, Vol. 1, No. 1, pp. 273-282, DOI:10.21042/AMNS.2016.1.00022

[20] Robin De Keyser, Cristina Muresan, and Clara Ionescu, "A low-order computationally efficient approximation of fractional order systems," Automatica (under review), 2016

IOP Conf. Series: Journal of Physics: Conf. Series **783** (2017) 012055 doi:10.1088/1742-6596/783/1/012055

Representing Operational Modes for Situation Awareness

Denis Kirchhübel[1], Morten Lind[2] and Ole Ravn[3]

[1]PhD Student, Department for Electrical Engineering, Technical University of Denmark
[2]Professor emeritus, Department for Electrical Engineering, Technical University of Denmark
[3]Associate Professor, Department for Electrical Engineering, Technical University of Denmark

E-mail: dekir@elektro.dtu.dk

Abstract. Operating complex plants is an increasingly demanding task for human operators. Diagnosis of and reaction to on-line events requires the interpretation of real time data. Vast amounts of sensor data as well as operational knowledge about the state and design of the plant are necessary to deduct reasonable reactions to abnormal situations. Intelligent computational support tools can make the operator's task easier, but they require knowledge about the overall system in form of some model.

While tools used for fault-tolerant control design based on physical principles and relations are valuable tools for designing robust systems, the models become too complex when considering the interactions on a plant-wide level. The alarm systems meant to support human operators in the diagnosis of the plant-wide situation on the other hand fail regularly in situations where these interactions of systems lead to many related alarms overloading the operator with alarm floods. Functional modelling can provide a middle way to reduce the complexity of plant-wide models by abstracting from physical details to more general functions and behaviours. Based on functional models the propagation of failures through the interconnected systems can be inferred and alarm floods can potentially be reduced to their root-cause. However, the desired behaviour of a complex system changes due to operating procedures that require more than one physical and functional configuration. In this paper a consistent representation of possible configurations is deduced from the analysis of an exemplary start-up procedure by functional models.

The proposed interpretation of the modelling concepts simplifies the functional modelling of distinct modes. The analysis further reveals relevant links between the quantitative sensor data and the qualitative perspective of the diagnostics tool based on functional models. This will form the basis for the ongoing development of a novel real-time diagnostics system based on the on-line adaptation of the underlying MFM model.

1. Introduction

Modern complex production plants are becoming increasingly demanding due to the distributed nature of the control system and increased requirements of safe and efficient operations. While every component of a system may be within its operating margins, undesired interaction betweens certain component states can accumulate to catastrophic situations. It is therefore important to not only consider single components but the combination of all interrelated parts of a complex system. In order to cope with the complexity of systems different perspectives are relevant to represent the interactions within the system beyond different time scopes and levels of detail with regards to structure, function and behaviour (SFB). [1]

For robust and fault-tolerant systems tools, such as structural analysis have been developed to model the interrelations between subsystems and enable failure diagnosis.[2] However, for plant-wide diagnosis industrial plants rely mostly on the experience and diagnostic skills of human operators. The operators establish the state of the plant based on alarms generated for possibly abnormal states indicated by sensor readings.[3] While the concepts for fault-tolerant systems are used on the level of components or subsystems, the majority of improvements for alarm systems reviewed by Wang et al. [3] disregard the process knowledge and rely on data driven methods. This can be related to the problem of high complexity of an overall plant described by Venkatasubramanian et al. [1], especially considering the extension and replacement of components throughout the live span of an industrial plant.

One way to overcome the lack of process knowledge incorporated in alarm systems and the high complexity of mathematical descriptions is presented by the SFB approach. SFB modelling provides an abstraction of the system, that can be used to analyse and diagnose the system based on the interaction of different structures, functions and purpose. One form of SFB modelling is Multilevel Flow Modelling (MFM), which provides a modelling language as well as a diagnostics tool for qualitative cause consequence reasoning to identify how abnormal states propagate through the system. [1, 4]

As addressed by hybrid systems modelling certain conditions or events in the control system lead to different states of a system by changing its behaviour.[2] On a plant-wide level such discrete states can be generated by different configurations of subsystems because of operational procedures or to efficiently use redundant systems. Such configurations can be considered as operational modes of a plant. While operational modes have been the subject of MFM related research [5, 6], no consistent and easy way of modelling these modes has been proposed. The research of operational modes of a nuclear power plant by Lind et al. [6] showed that each mode can be represented in a distinct MFM model with regards to functions and goals of the mode. Those distinct models, however, disregard the operational knowledge on how modes are interconnected and what the boundaries of the modes are with respect to operation procedures.

This paper describes a new way of interpreting control functions and relations as part of the operational knowledge included in MFM. Similar to hybrid systems modelling discrete events are identified that determine boundaries of a mode and facilitates the incorporation of operational procedures in MFM models through the generation of interpreted models that express implicit knowledge. The proposed interpretation reveals, how the constraints for validity of the qualitative model are closely linked to the quantitative aspects of real-time sensors or alarms and will form the basis of the ongoing development to link MFM modelling and artificial intelligence approaches to create a novel plant-wide on-line diagnostics system based on the cause and consequence reasoning in MFM.

In section 2 different approaches to plant-wide on-line diagnostics are outlined and the basic concepts of modelling and reasoning in MFM are briefly described as well as the state of the art with regards to operational modes, especially in MFM. Section 3 describes the concepts for linking control functions and modes based on discrete models of a start-up procedure. In section 4 the mode models generated with the proposed interpretation are evaluated. Finally the conclusions drawn from this conceptual work and the future development based on this concept are outlined in section 5.

2. State of the art

The study of Venkatasubramanian [1] describes how the computerization of industrial processes has lead to robust and fault tolerant control of components and that artificial intelligent approaches have been shown to enhance the maintenance scheduling and degradation diagnostics for specific applications. While these systems help improve the performance of a specific components most diagnostics systems are based on physical or statistical models of the system

IOP Conf. Series: Journal of Physics: Conf. Series **783** (2017) 012055 doi:10.1088/1742-6596/783/1/012055

and would become too complex if all, possibly undesired, interactions between the components of the system are to be considered. From a system safety perspective the robustness of automated system parts does not make the entire system resilient to failure. One weak point is presented by the vast amount of information that a human operator needs to process, which makes the assessment of an emergency situation difficult, especially given the limited time frame for appropriate reaction. [1]

Modern industrial plants are equipped with a large number of sensors and control systems that can generate alarms to alert the human operator to a possible failure. With the increasing number of distributed control systems the number of alarms in a plant increases as well. While statistics driven analysis of plant data can aid auditing processes and filtering of excess alarms, the propagation of abnormal states due to the interconnections of components in the plant leads to alarm floods that overload the operator and make it hard to identify the origin of the failure. While alarm floods have been investigated by different groups, there is no established applicable solution to deal with alarm floods. Identifying related alarms that lead to alarm floods requires a system with knowledge about the interactions of different components in the plant. [3]

Incorporating the expert knowledge of human operators and plant designers into a support system can aid the decision making process by identifying root causes and planning appropriate reactions. To implement this approach rule based expert systems have been developed. These system can analyse a situation and infer procedures based on rules that are deducted from the expert knowledge of operators. A drawback of such systems is the domain specific nature of the expert knowledge and thus the narrow applicability of one such support system. [7] A more generic approach to communicating and evaluating expert knowledge is presented by functional modelling (FM). FM frameworks commonly abstract a system as a set of desired behaviours of the whole system or as a combination of functions of its components. This abstraction provides a general overview of the entire system by interconnecting different knowledge domains. The schematic representation of FM can facilitate computer reasoning about the entire system. [8]

The functional modelling language of MFM was originally developed as analysis tool to assist human operators in identifying and handling unknown operation situations [9] and to support the design of human-machine interfaces [10]. MFM has been demonstrated as a valuable tool to represent operational knowledge about processing plants in a range of technological areas in a machine readable form. Among others MFM is used for operator support scenarios in research projects such as the OECD Halden Reactor Project[4], where the MFM framework is applied for cause and consequence reasoning about abnormal states of the plant. This kind of reasoning allows the operator to relate connected alarms and react more efficiently with a focus on the root causes and preserving essential system functions.[4] Another branch of MFM has focused on deriving fault trees and failure mode analyses[11] and possible counter actions[12] based on MFM.

2.1. Multilevel Flow Modelling

A MFM model is a hierarchical decomposition of goals to be achieved by certain functions of the system, as well as a part-whole decomposition of a system function into basic material and energy flow functions. MFM provides a graphical modelling language with symbolic representations of these basic flow functions and the relation between functions and objectives of the system. Figure 1 shows the defined MFM primitives. In a MFM model the flow function primitives are usually used in several flow structures. The functions are connected by influence relations inside a flow structure and by means-end relations across decomposition levels representing the contribution to another function or an objective.

Besides the analysis and representation of function, MFM can be used to reason about the system performance. The reasoning is established in terms of qualitative performance of each function and the propagation of abnormal performance states by the relations of one function to

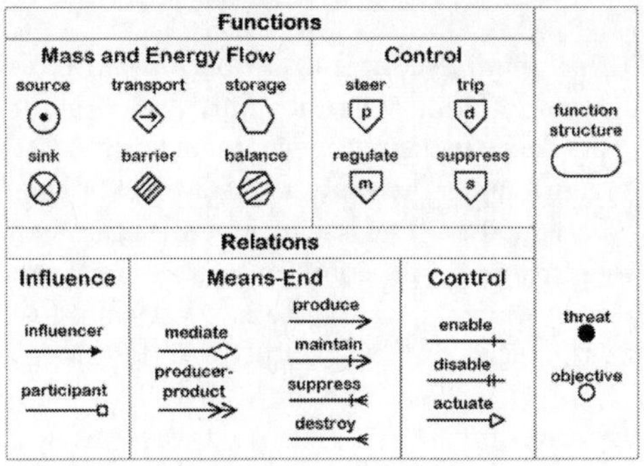

Figure 1. Basic MFM concepts[4]

Table 1. MFM control functions[13]

Intention	action	symbol
produce	$[\neg pTpI\neg p]$	
maintain	$[pTpI\neg p]$	
destroy	$[pT\neg pIp]$	
suppress	$[\neg pT\neg pIp]$	

another. Table 2 shows the underlying equations and the qualitative states that form the basis for the reasoning system. Cause and consequence inference rules for the possible combinations of flow functions and relations have been detailed by Petersen[14] and most recently elaborated on by Zhang et al. [4, 15]. The reasoning considers the abnormal states of functions in the way alarm systems commonly represent abnormal sensor readings as high (high-high) or low (low-low). Based on the interactions of the function primitives the propagation of abnormal states can be inferred in both forward (consequence) and backward (cause) direction.

Table 2. Underlying equations, constraints and failure states of MFM flow functions[4, 14]

Flow function	Balance equation	State Constraints	Abnormal states
Transport	$F_{in} = F_{out} = F$	$F_{low} \leq F \leq F_{high}$	low low, low, high, high high
Storage	$\Sigma F_{out} = \Sigma F_{in} + dV/dt$	$V_{low} \leq V \leq V_{high}$	low low, low, high, high high
Source	$\Sigma F_{out} = F_{unknown} + dV/dt$	$V_{low} \leq V \leq V_{high}$	low low, low, high, high high
Sink	$\Sigma F_{in} = F_{unknown} + dV/dt$	$V_{low} \leq V \leq V_{high}$	low low, low, high, high high
Balance	$\Sigma F_{out} = \Sigma F_{in}$		sourcing, leak, block
Barrier	$F_{in} = F_{out} = F$	$F = 0$	leak

In addition to the fundamental flow functions and objectives, MFM allows the modelling of control functions as means of intervention. Lind [13] introduced the action notation for the control functions in table 1. This notation uses the temporal operator T and an operator I, where the state after T is achieved by the control intervention instead of the state after I which the system would move to without intervention. These control functions are concerned with the functional meaning or intention of the control design e.g. to keep a flow in an heat exchanger steady rather than a specific realization of the controller. A control function is normally connected to an objective that represents the target function to be controlled. The actuate relation connects the control function to the functions that controlled to achieve that target, e.g. a pump is actuated in order to maintain a certain water level in a tank. Zhang et al. [16] point out that the purpose of a control action can be modelled for automated as well as manual intervention. Control functions in MFM are thus a way of extending the model with expert knowledge about the way a plant is operated.

2.2. Operational Modes

In the context of diagnosis and faults-tolerant control a similar concept to operational modes is reflected in the hybrid nature of fault-tolerant systems. Fault-tolerant control reconfigures the structure and parameters based on logic as reactions to discrete events, such as faults in the system. The different configurations can be represented as distinct states, exposing a specific behaviours based on the configuration of the control. The evolution of these states can be described e.g. by Petri nets or similar representations. The combination of continuous model for each state and the discrete events limiting the validity of a state is represented in a hybrid systems model. In the hybrid model the discrete events limiting a state are described as constraints on specific system variables. [2]

Operational modes as such have been investigated by Lind et al. [6] with regards to their representation in MFM. Operational modes can occur on two different levels of abstraction: as a relation of objective to function, or as a relation of function to physical structure. On either abstraction level modes can be defined both ways, as a selection of means to achieve a constant end, or as a selection of ends that can be achieved by the same means. Zhang [4] elaborates that this classification is relevant to assess the operability of plant as indication for necessary mode shifts, e.g. configuration changes. Such a configuration change could be between redundant systems, changing the physical structure (means) to achieve the same function (end). Equally a configuration change could be opening a valve (means) thus changing its function (end) from blocking to transporting. Analogously a certain set of functions could serve different objectives, depending on the mode, or vice versa. However, the boundaries of one operational mode are at present not represented in the MFM models.

Inoue et al. [12] use MFM models to generate plausible operation procedures in unknown emergency situations. In order to generate these procedures knowledge about possible, and possibly undesired alternative functions of physical components has to be considered. This kind of situation relates directly to the function to structure mode definition by Lind [6]. Gofuku et al. [11] introduced operational information in addition to the MFM models to include the required knowledge. For the context of operational modes the most relevant aspects of operational information are the component behaviour and operation knowledge.

Component behaviour knowledge refers to the plausible behaviours of a physical structure and their functional representation. Operation knowledge represents the possible interventions and the functional influence of the intervention. Operation knowledge is essentially part of MFM models by including manual as well as automatic interventions in the modelled control functions [16]. The other aspects of operation information are not as clearly defined in the MFM Framework as the additional information Gofuku et al. [11] describe. Alternative behaviours of components have been widely disregarded by the MFM framework, since a MFM model is used to represent intended behaviour [4].

3. Operational modes and control

In order to illustrate the concepts to consistently represent operational modes in MFM the discrete models of two modes in a start-up procedure of a generic power plant are analysed.

To get the power plant up to operation the first step is filling the boiler drum and piping with water. Figure 2 shows the active material flow path during this stage of the start-up procedure: Water is pumped from a reservoir into the boiler drum and the ventilation valve is left open to allow air to escape from the steam piping at the output of the boiler. The goal of this stage is to fill the drum to the required water level before the steam production can be initiated. The MFM model shown in figure 4 reflects the described material flow with the function primitives of MFM. In addition to the intended material flow, the closed off parts of the system, namely the economizer recirculation and the recirculation of steam through the condenser are included as barriers in the MFM model. Including these barriers allows to consider faulty valve behaviours

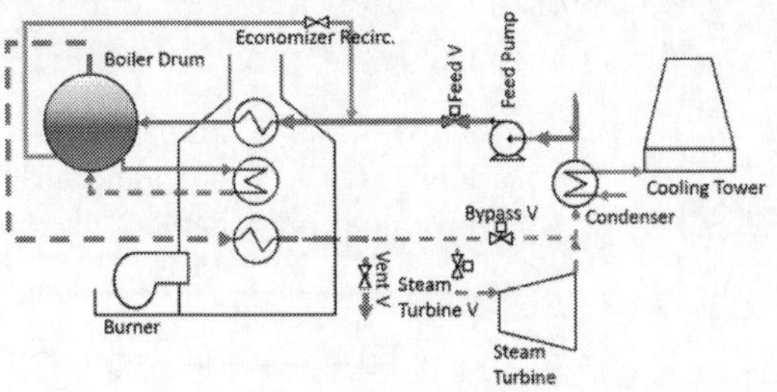

Figure 2. Filling of a generic power plant

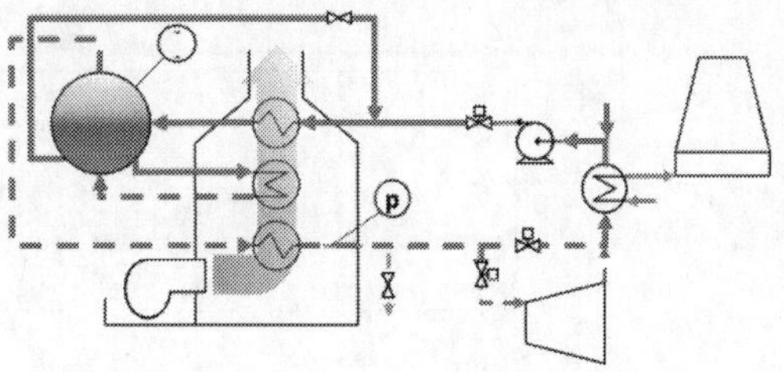

Figure 3. Pressurizing of a generic power plant

in the reasoning. The control of the water level in the drum by actuating the feed pump is reflected by the control flow structure, specifying the intention of the mode as producing the required water level.

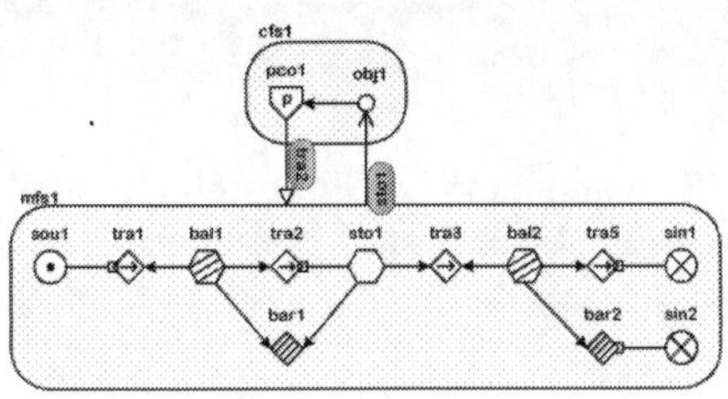

Figure 4. MFM model of filling

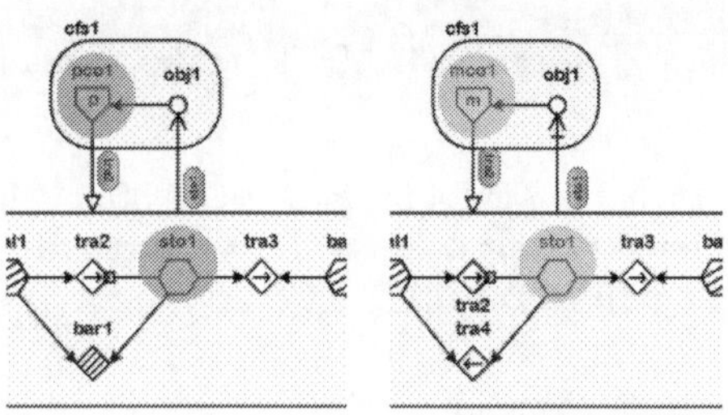

Figure 5. Mode change indicators and corresponding control function in subsequent modes

Before the steam turbine can be operated to produce energy, pressurized steam has to be generated in order to provide enough energy to convert in the turbine. After steam has been generated is superheated and recirculated to raise the pressure in the system. The MFM model shown in figure 6 is developed according to the active components in figure 3. The mass flow structure of fuel and air in the burner is included in the model. The heat from the burner enables the introduction of the energy into the system as represented by the energy flow structure. In this mode the control of the temperature and pressure in the steam piping actuates the flow of steam through the condenser to raise the thermal energy in the steam piping. The overall goal of this stage of the operation procedure is to generate the necessary steam pressure to be able to spin up the turbine.

Comparing the functional representation of corresponding elements across the two modes reveals that the goal and end-point of these modes are related to the control actions, more specifically the end point of produce action. While the control function remains present in subsequent modes, the associated control action changes e.g. from produce to maintain as shown in figure 5. The representation of valves that can either be closed or opened to change the physical configuration of the system is either a barrier or as a transport function.

Besides valves as a means of changing the system configuration, specific components, like the burner, can be enabled or disabled. This behaviour is reflected in the way that a disabled components function does not contribute to the function of the system and is thus not considered in the model.

The concepts described in the further part of this section have been developed based on these

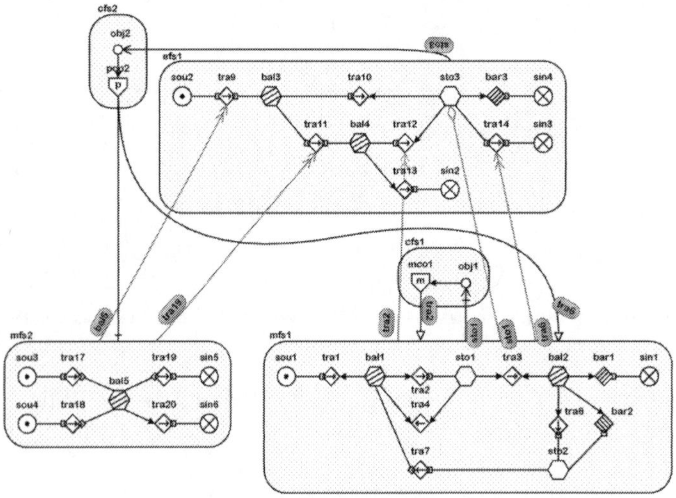

Figure 6. MFM model of pressurizing

Table 3. Intended behaviour of a storage with produce controller

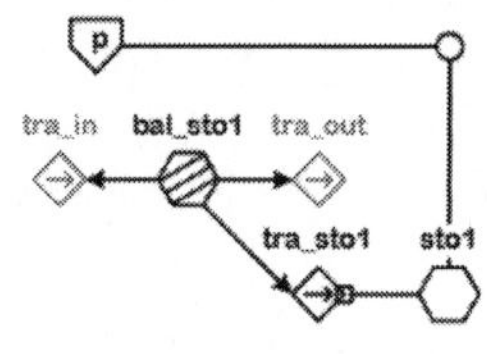

$$\Sigma T_{out} = \Sigma T_{in} - T_{sto1}$$
$$dV/dt = T_{sto1} > 0$$

findings and present a pre-processing of a designed MFM model into an interpreted model. The interpreted models are intended to work with the established cause and consequence reasoning based on MFM.

3.1. Control sequence

The distinct models for the start-up modes have shown, that the end-point associated with the overall objective of one mode is linked to producing a certain storage level in order to prepare the system for the next operational mode. This objective of producing, thus increasing, a certain storage level can not be inferred from the underlying definition of a storage shown in table 2. According to the definition of a storage the only constraint is on the level of the storage.

In order to reflect the intended aspect of a defined inflow in order to increase the storage level, the function of the storage has to be considered to act like a sink. The interpretation of a storage to be produced with a defined inflow is shown in table 3. By explicitly including a transport function the inflow can be constraint with relation to the quantitative values in the physical system. Figure 9 shows the cause reasoning output for a high level in sto1 of the fill mode represented in figure 4. Considering the states as indicators rather than faults this can be interpreted as possible paths to achieving the goal of this mode: The storage level can be raised by generating a high inflow into the system as well as by keeping the outflow low (no water should leave through the vent valve).

Similar constraints can be introduced for all control functions in MFM. Based on the action schemes defined for the MFM control functions there is an inherent sequence of the control functions in relation to operational modes, as shown in figure 7. Reasoning about the validity of one mode can be based on the quantitative constraints linked to the intended behaviour of a controlled storage. A breach of either of the constraints on the storage level or the timely change thereof, represented by the in or outflow, indicates a necessary configuration change or a failure of the controller.

Analogous to the concepts of fault-tolerant control the breach of the constraints can be interpreted as a discrete event that leads to a change in the system configuration. The discrete states of a fault-tolerant system as well as the designed operation modes of the overall plant can be defined and reasoned about as exemplified in figure 8. An important difference, however, is the fact that the model for each mode expresses the boundaries it is designed for explicitly, as opposed to the continuous models in hybrid systems where assumptions for a mathematical model are implicit in the model and expressed explicitly only by the separately defined events.

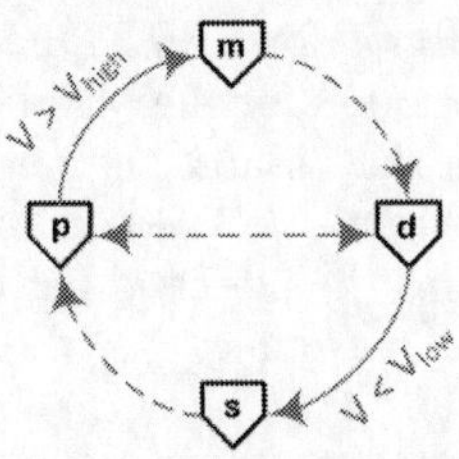

Figure 7. Implicit (solid) and deliberate (dashed) control function sequence throughout modes

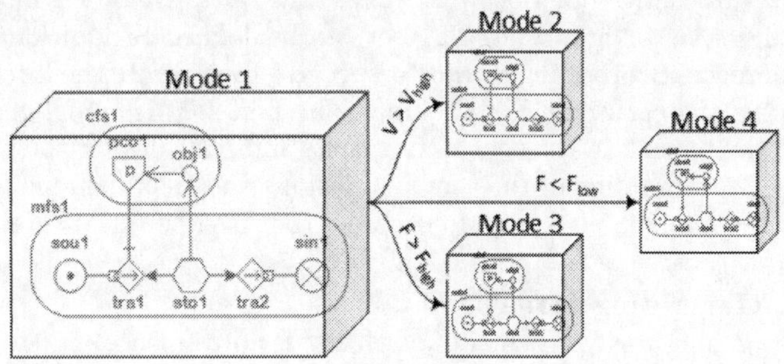

Figure 8. Possible mode transition events with a single produced storage

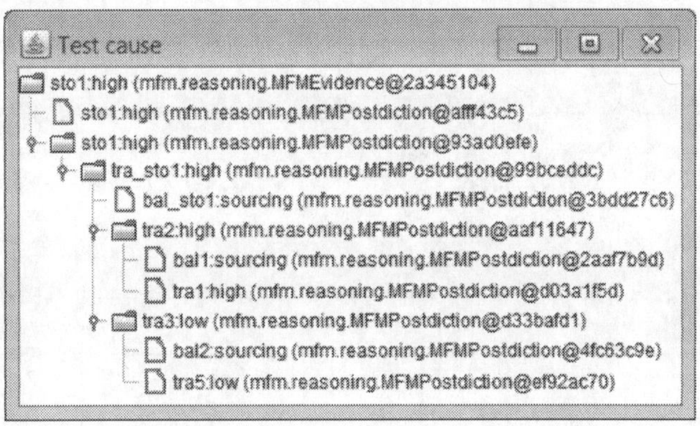

Figure 9. MFM based cause tree for high water level in the generic powerplant during filling

3.2. Modelling configuration

The difference in the configurations considered in the distinct models of the power plant start-up can be represented by transport or barrier functions reflecting the state of a valve. As described in the previous section the reasoning about operational modes is closely related to the intention represented by control functions. Consequently, the control relations are explored as a means of representing configuration in MFM.

MFM defines three control relations, where the enable and disable relation directly correspond to the two complementary states a configuration valve can have (open or close). However, the functional representation of a configuration valve can coincide with the function of a continuously actuated control valve. The enable and actuate relations are thus interpreted to reflect the normal function, while the disable relation is interpreted in the preprocessing to reflect a closed valve or deactivated component.

The analysis of the start-up procedure revealed that a closed valve does not only affect the MFM function representing it, but also propagates through the means-end relations, specifically the mediate and producer-product relations. In effect, if a transport function as producer is interpreted as barrier, the transport function, that is the product, also has to be interpreted as a barrier.

Using the concepts for interpretation proposed here, allows for a mode to be represented by a set of control functions that reflect the objective of the mode and a set of control relations to reflect the configuration. This set can be defined as additional information for a MFM model as reference, thus adding a two step process to generate the interpreted models for the MFM based cause and consequence reasoning: The first step is model generation, that adapts the reference model to contain the control functions and relations for the respective mode. The second step is the interpretation of the control relations and the functions to represent the intended behaviour of the mode in an interpreted model that is subsequently used for the reasoning.

4. Generating models

By joining the MFM model of all modes a reference model to accommodate all modes can be derived. This model essentially reflects all possible flows as they could for example be found in a piping and instrumentation diagram. An exemplary combined model for the generic thermo-electric power plant is shown in figure 10

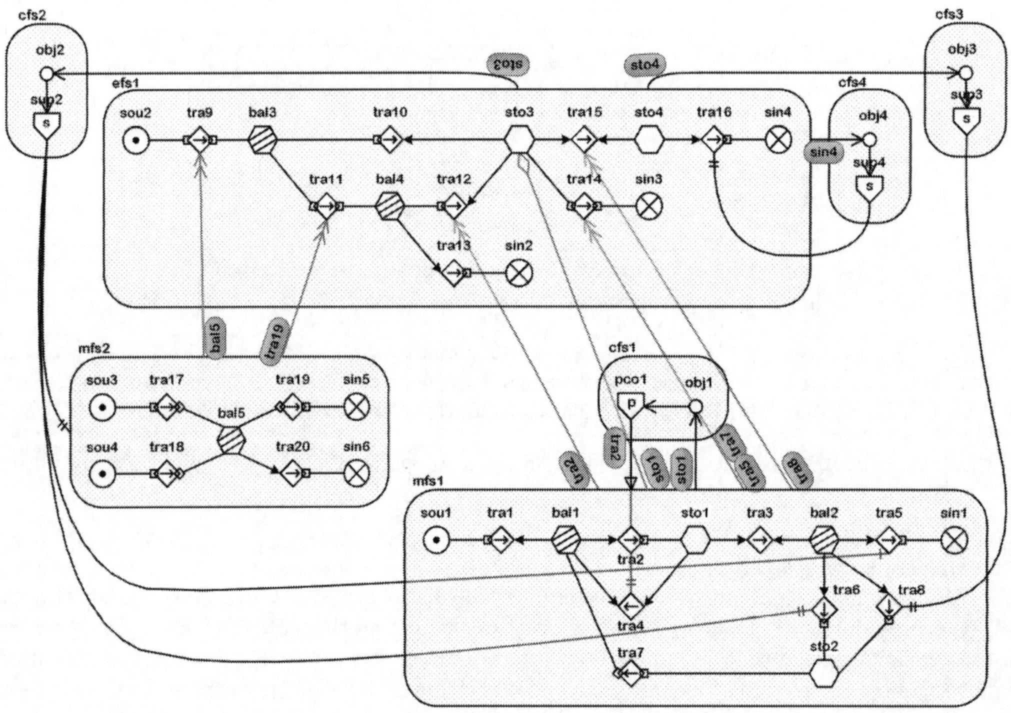

Figure 10. Reference model (fill mode) for a generic thermo-electric power plant

Using the proposed interpretation concepts the MFM representation for a specific mode can be generated. For the sake of readability the explicit modelling of the intention of control has been left out of the interpreted model for filling shown in figure 11. Comparing this model to the discretely modelled representation of the same mode in figure 4 there are some obvious differences.

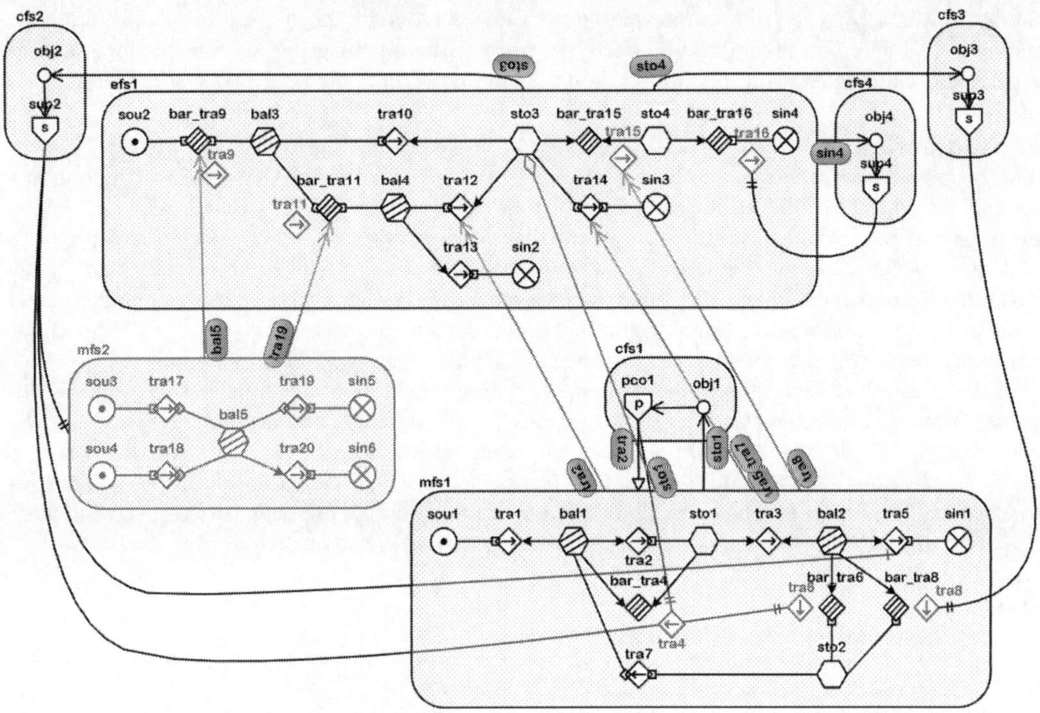

Figure 11. Meta-model of the fill mode

The central mass flow structure of the discretely developed model for the filling mode did not contain the recirculation of water through the condenser, since that part of the system is not part of the functional scope of the mode. However, the functional information is fundamentally the same, given that the barrier at the output of the discrete model is reflected more in detail by the two barriers in the recirculation.

The energy flow, while not relevant to the function of this mode, is represented in the interpreted model. A detailed inspection of that energy flow reveals, that there is no intended energy flow: From the source in the energy flow only barriers connect to the other functions and the two storages are connected to suppress control functions, thus acting as balances.

5. Conclusion

Functional modelling can provide the framework for representing process knowledge for a range of engineering applications. Incorporating such knowledge in a reasoning systems enables the creation of operator support tools that can help diagnosing the current state of a complex plant and provide guidance for reasonable reactions. This kind of plant-wide diagnosis is believed to be capable of relieving the load of alarms operators are confronted with by identifying connected alarms and thus yielding more meaningful information. MFM presents such a functional modelling framework with an established reasoning system for cause and consequence diagnostics.

In this work the distinctive elements of operational modes and the boundaries of each mode have been investigated with regards to functional modelling. Based on the analysis of distinct functional models for a start-up procedure concepts for the consistent representation of operational modes in MFM are proposed. These concepts facilitate the modelling of operational

IOP Conf. Series: Journal of Physics: Conf. Series **783** (2017) 012055 doi:10.1088/1742-6596/783/1/012055

modes in MFM by using a common reference model as basis for the generation of specific mode models. The common model can be derived from existing engineering documentation, such as piping and instrumentation diagrams, and thus simplifies the modelling process in MFM.

6. Future work

The proposed interpretation of the intention of each mode provides the means of linking the models for specific configurations to the real-time environment equipped with sensors. In the development process for a complete real-time diagnostics system based on MFM these concepts will serve as the basis for dynamically adapting the functional model to the actual state of the plant. An important element of this is to provide the rule system with knowledge about manual and automatic intervention points in the plant and their possible functions, as it is realized through the control relations in the proposed concepts.

The future effort in this field will be directed to the application of machine learning approaches to facilitate the identification of failure states from real-time sensor data, as well as the consolidation of the MFM based reasoning to enable on-line adaptation of the model to reflect the current state of the system. The goal for this project is to provide a novel kind of diagnostics system, that incorporates the operational knowledge and enables an organised representation of the state of a plant by identifying relevant failure paths from the on-line data.

References

[1] Venkatasubramanian V 2011 *AIChE Journal* **57** 2–9
[2] Blanke M, Kinnaert M, Lunze J and Staroswiecki M 2016 *Diagnosis and Fault-Tolerant Control* (Berlin, Heidelberg: Springer Berlin Heidelberg) ISBN 978-3-662-47942-1
[3] Wang J, Yang F, Chen T and Shah S L 2016 *IEEE T. on Aut. Sc. and Eng.* **13** 1045–1061
[4] Zhang X 2015 *Assessing Operational Situations* Ph.D. thesis Technical University of Denmark
[5] Lind M 1992 A Categorization of Models and Its Application for the Analysis of Planning Knowledge *Post ANP'92 Conference Seminar at Human Cognitive and Cooperative Activities in Advanced Technological Systems* (Kyoto, Japan)
[6] Lind M, Yoshikawa H and Jørgensen S 2012 Modeling operating modes for the Monju nuclear power plant *8th Int. Top. M. on NP Instrum., C., and HMI Tech.* (San Diego, CA)
[7] Cholewa W 2004 Expert Systems in Technical Diagnostics *Fault Diagnosis* (Berlin: Springer) pp 591–631
[8] Erden M, Komoto H, van Beek T, D'Amelio V, Echavarria E and Tomiyama T 2008 *AIEDAM* **22** 147–169
[9] Burns C and Vicente K 2001 *International Journal of Cognitive Ergonomics* **5** 357–366
[10] Lind M 2011 *Nuclear Safety and Simulation* **2** 132–140
[11] Gofuku A, Koide S and Shimada N 2006 Fault Tree Analysis and Failure Mode Effects Analysis Based on Multi-level Flow Modeling and Causality Estimation *2006 SICE-ICASE Int. J. Conf.* (Busan, Rep. of Korea: IEEE) pp 497–500
[12] Inoue T, Gofuku A and Sugihara T 2015 A technique to generate plausible operating procedure for an emergency situation based on a functional model *Proceedings of STSS/ISSNP 2015* (Kyoto, Japan)
[13] Lind M 2005 *Modeling goals and functions of control and safety systems - theoretical foundations and extensions of MFM* (Nordic nuclear safety research) ISBN 87-7893-175-4
[14] Petersen J 2000 Causal reasoning based on MFM *Proc. of the Conf. on Cog. Sys. Eng. in Pr. Cont.*
[15] Zhang X, Lind M and Ravn O 2013 Consequence Reasoning in Multilevel Flow Modelling *Analysis, Design, and Evaluation of Human-Machine Systems* vol 12 pp 187–194
[16] Zhang X, Lind M, Jorgensen S B, Ravn O and Jensen N 2014 Representing operational knowledge of pwr plant by using multilevel flow modelling *Proceedings of the ISOFIC/ISSNP* (Jeju, Rep. of Korea)

Towards Extending Forward Kinematic Models on Hyper-Redundant Manipulator to Cooperative Bionic Arms

Inderjeet Singh, Othman Lakhal and Rochdi Merzouki

Polytech Lille, CRIStAL, CNRS-UMR 9189, Avenue Paul Langevin, 59655 Villeneuve d'Ascq, France

E-mail: inderjeet.singh@ed.univ-lille1.fr, othman.lakhal@univ-lille1.fr, rochdi.merzouki@polytech-lille.fr

Abstract. Forward Kinematics is a stepping stone towards finding an inverse solution and subsequently a dynamic model of a robot. Hence a study and comparison of various Forward Kinematic Models (FKMs) is necessary for robot design. This paper deals with comparison of three FKMs on the same hyper-redundant Compact Bionic Handling Assistant (CBHA) manipulator under same conditions. The aim of this study is to project on modeling cooperative bionic manipulators.Two of these methods are quantitative methods, Arc Geometry HTM (Homogeneous Transformation Matrix) Method and Dual Quaternion Method, while the other one is Hybrid Method which uses both quantitative as well as qualitative approach. The methods are compared theoretically and experimental results are discussed to add further insight to the comparison. HTM is the widely used and accepted technique, is taken as reference and trajectory deviation in other techniques are compared with respect to HTM. Which method allows obtaining an accurate kinematic behavior of the CBHA, controlled in the real-time.

1. Introduction

Hyper-redundant manipulators represent a rapidly growing research field. There are many robots inspired from natural behaviours like elephant trunk [1], tentacles [2] and octopus [3] etc. They are made from flexible materials to make them compatible of narrow and congested areas. Most of the soft robots are actuated by pneumatic actuators to bend etc., due to smooth bending these robots have many degree of freedoms. Therefore it is very complex to model such type of robots.

Forward Kinematic Modelling can be done using qualitative as well as quantitative methods. The Arc Geometry Method under the assumption of constant or circular curvature is used for most of the continuum robots [4]. This method is common and it is better to apply when we know all of the parameters of circular arc and this method is validated by [5]. In [6], the issues regarding singularity are realized particularly for a trunk robot and also the constraints for limit of length and workspace that can be reached. The application of this method is also in modelling concentrating tube based robot [7], which are called "Active Cannulas" too. The actuation is done by using three telescopic tubes for rotation as well as translation. A micro-robot which is placed at the tip of a catheter [8] is also modelled using this method. Forward Kinematic of Compact Bionic Handling Assistant (CBHA) is also done by Arc Geometry Method [9] and is validated. Kinematic model of Bionic Handling Assistant (BHA) is also done by using geometrical approach in [10]. Bionic soft robots are also modelled with modified Denavit-Hartemberg convention [11], the assumption of constant curvature is also integrated [5]. This

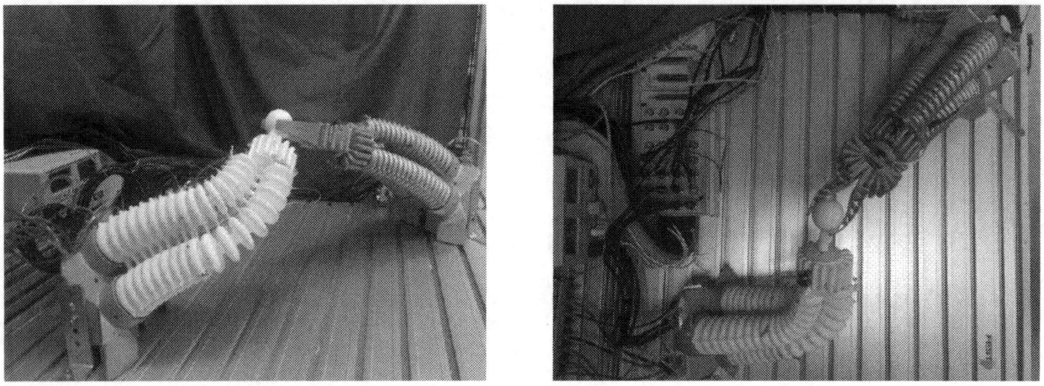

Figure 1. Co-operation between Two Hyper-redundant Manipulators

model is validated on the robot called " elephant trunk". A Dual Quaternion representation based kinematics is applied on a pneumatic muscle [12] which is a continuum arm. The Model Shape Functions are also used to model hyper-redundant robots [13] but with this the robots remain specified for applications because in this the shapes are restricted means less shapes. Modelling of multisection continuum manipulators is also purposed by Curve Parametric (CP) kinematics [14]. This computes structurally accurate results, are computational efficient and are better to interpret physically. But Curve Parametric can not model straight position of arm because it has singularity inherently. Neural Network Qualitative method is used by [15] to control Compact Bionic Handling Assistant. Qualitative approach goal babbling learning approach is used to calculate the kinematic model of BHA [17]. A Hybrid approach is used by [16] to model kinematics of CBHA. This hybrid approach consists of quantitative as well as qualitative approaches. Geometric method is used to derive equations but due to non linear nature of equations, Neural Network is used to approximate the solution of these equations.

There are many type of models which models kinematics of continuum robots. Modelling of continuum robots is very complex because of there hyper-redundancy and non-linearities, even due to soft material there are hysteresis and memory errors. Therefore modelling of co-operation between more than one continuum robots is more complex than that of one robot. In this paper, the three Kinematic model are applied on a hyper-redundant CBHA, which are applicable on this robot. Then results are compared theoretically as well as experimentally with respect to arc geometry HTM Method to find out one method for appropriate condition. In case of continuum robots, the most widely used method for modeling is arc geometry having the assumption of constant curvature. Arc geometry method is presented as a general method by [18] and it is also used for the modeling of several continuum robots [4] [6]. This work is done So that when the work space calculation of collaborative system is to be calculated, then we can apply more appropriate method for each manipulator to make overall model of co-operative system more efficient. In fig. 1 the collaboration between two manipulators is shown but it can be more than two also.
The remaining paper is organized as follows: Section [2] describes the structure of CBHA. Section [3] briefly explains the three forward kinematic methods and also differentiate them theoretically. Experimental setup is explained in section [4]. Experimental results and discussions are presented in section [5]. Co-operation model using three CBHAs is discussed in section [6]. Finally the conclusions of this paper are described in section [7].

2. Compact Bionic Handling Assistant

The CBHA manipulator is assembled on the top of an omni-drive mobile platform. This mobile robot with bionic manipulator is known as RobotinoXT, shown in Fig. 2.

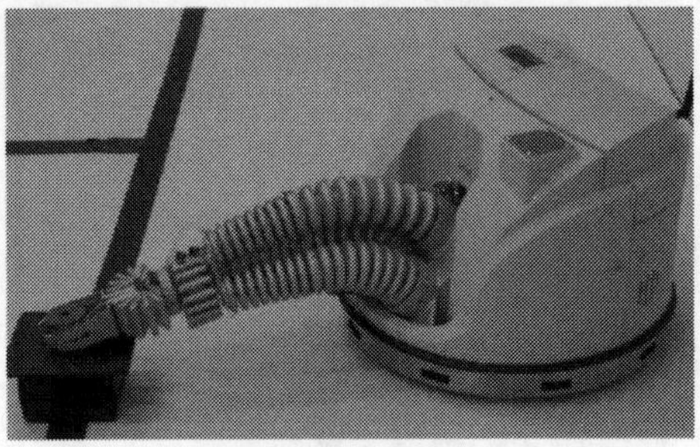

Figure 2. RobotinoXT

The manipulator is made up of 2 sections, which are of elastic material, a rotational wrist and a compliant gripper as shown in Fig. 3. Flexibility of 2 sections is the main advantage of the CBHA, which are formed by 3 polyamide electro-pneumatic tubes. These tubes extend when a pressure is applied through them, moreover, since each section is formed by 3 tubes, by applying differential pressures to the tubes, the bending of a section can be achieved.

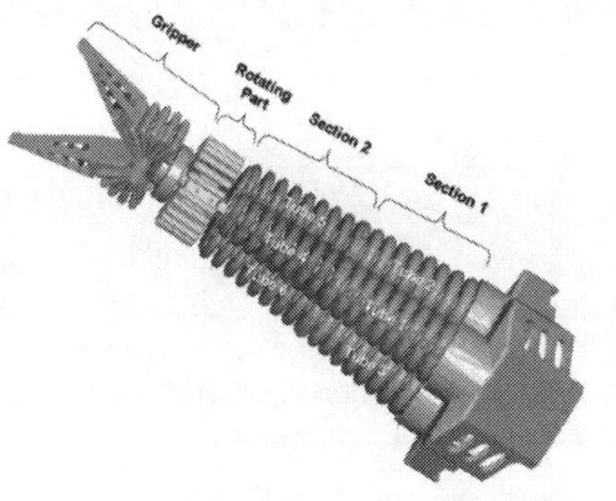

Figure 3. The CBHA of the RobotinoXT

If equal pressures are applied to each tube, the manipulator extends in a straight line. The linear extension is limited by a non-extensible cable routed through the middle of the manipulator. The length of the tube can be measured by string potentiometers that deliver a voltage proportional to the channel length $l_{i,j}$, where $i = 1,..,3$ refers to the potentiometer number and $j = 1,2$ corresponds to the section number. The CBHA is made only by longitudinal pneumatic-actuated bellows, therefore the torsion of the backbone is not possible.

3. Forward Kinematic Methods:Description and differences

The description as well as differences between the FKMs are gives as:

3.1. Arc Geometry HTM

This is the geometric quantitative method [9]. Assumptions used in this method are: 1. CBHA's shape is modeled as two cylinders. 2. CBHA is considered as a non-extensible arm. 3. Curvature of the tubes is assimilated to a perfect arc of a circle. 4. No torsion because tubes are interconnected rigidly at each vertebra. The equations 1 are deduced from the geometrical model of CBHA as in fig. 4:

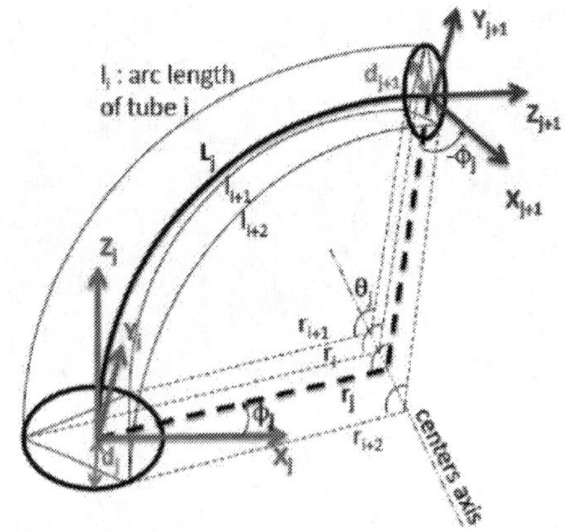

Figure 4. Geometrical Representation of one section of CBHA [19]

$$\begin{cases} L_j = \frac{l_i+l_{i+1}+l_{i+2}}{3} \\ \phi_j = \tan^{-1}\left(\frac{\sqrt{3}(l_{i+2}-l_{i+1})}{2l_i-l_{i+1}-l_{i+2}}\right) \\ r_j = \frac{(l_i+l_{i+1}+l_{i+2})}{D_j} d_j \\ \theta_j = \frac{D_j}{3d_j} \\ D_j = 2\sqrt{l_i^2+l_{i+1}^2+l_{i+2}^2-l_il_{i+1}-l_il_{i+2}-l_{i+1}l_{i+2}} \end{cases} \tag{1}$$

Where, i refer to the number of the tube, j refer to the number of the section, r_j refer to the curvature radius of the jth section, ϕ_j refer to the angle between x-axis and the projection of the jth section, θ_j refer to the curvature angle of the jth section and L_j refer to the length of the jth section. Final position of manipulator is calculated by using above equations and transformation matrix. Transformation matrix for one section of CBHA is eq. 2.

$${}^{j-1}_{j}T = \begin{bmatrix} R & P \\ 0 & 1 \end{bmatrix} \tag{2}$$

Here,

$$P = \begin{bmatrix} r_j c\varphi_j(1-c\theta_j) \\ r_j s\varphi_j(1-c\theta_j) \\ r_j s\theta_j \end{bmatrix}$$

and,

$$R = \begin{bmatrix} c^2\varphi_j c\theta_j + s^2\varphi_j & c\varphi_j s\varphi_j(c\theta_j-1) & c\varphi_j s\theta_j \\ c\varphi_j s\varphi_j(c\theta_j-1) & s^2\varphi_j c\theta_j + c^2\varphi_j & s\varphi_j s\theta_j \\ c\varphi_j s\theta_j & -s\varphi_j s\theta_j & c\theta_j \end{bmatrix}$$

Where c and s are cos and sin respectively also j is the section number, and i is the tube number.

3.2. Dual Quaternion Method

Dual Quaternion representation [12] made up of of ordered pair of quaternions composed by applying Clifford dual number algebra on quaternions. One for position of the joint and the other for the orientation. Same assumptions are taken in this method as Arc Geometry HTM because it is just a different representation of matrix. Dual Quaternion is represented as in equ. 3:

$$Q_i = s_i + \varepsilon t_i \tag{3}$$

Here Q_i is the Quaternion containing rotational part s_i and translational part as t_i. ε separates the rotational part and translational part. Let T is the sample transformation matrix, so the conversion of transformation matrix T into dual quaternions s_i and t_i is as follows:

$$T = \begin{bmatrix} a_{11} & a_{12} & a_{13} & a_{14} \\ a_{21} & a_{22} & a_{23} & a_{24} \\ a_{31} & a_{32} & a_{33} & a_{34} \\ a_{41} & a_{42} & a_{43} & a_{44} \end{bmatrix} \tag{4}$$

$$s_i = \begin{bmatrix} q_0 & q_1 & q_3 & q_4 \end{bmatrix} \tag{5}$$

$$t_i = 0.5 * \begin{bmatrix} a_{14} & a_{24} & a_{34} & 0 \end{bmatrix} oS_i \tag{6}$$

In equ. 6 operator 'o' represents dual quaternion multiplication.

$$\begin{cases} q_0 = \sqrt{1 + a_{11} + a_{22} + a_{33}}/2 \\ q_1 = (a_{32} - a_{23})/4q_0 \\ q_2 = (a_{13} - a_{31})/4q_0 \\ q_3 = (a_{21} - a_{12})/4q_0 \end{cases} \tag{7}$$

By using these equations, the transformation matrix of each section of CBHA is converted into dual quaternion form. Let Q_i be the dual quaternion for first section and Q_j be the dual quaternion for the second section then the dual quaternion of both combined sections is given by:

$$Q_i o Q_j = s_i o s_j + \varepsilon (s_i o t_j + t_i o s_j) \tag{8}$$

3.3. Geometric-Neural Network Hybrid Method

Hybrid approach is used to compute FKM [16]. In this approach the assumption of constant curvature is not there but the extra assumption in this is that the CBHA is made up of 16 vertebras. This hybrid approach composes of both qualitative as well as quantitative methods: Geometric modelling as well as neural network technique. The CBHA is considered to be a concatenation of inter-vertebra along the backbone of the CBHA. Each inter-vertebra is represented by a parallel robot of type 3UPS-1UP (Universal -Prismatic spherical) , as shown in the fig. 5. The controls of the position and the orientation of the upper vertebra relative to the lower vertebra are done by three kinematic serial UPS joints, but only the prismatic joints are active. In the case of the CBHA, the translation of the upper vertebrae relative to the lower vertebrae is perpendicular. Moreover, the torsion is not possible. For these constrains, the serial UP is used. The inverse kinematics equation (IKE) of parallel robots is easy to establish with some elementary geometric relationships [16]. The model of an inter-vertebra, represented by eqs 9, 10 and 11, are obtained by calculating the joint variable $q_{m,k}$,as shown in the fig. 5 , where m=1..3 number of prismatic active and k=1..17 number of vertebra, corresponding to the position Z_k and orientation $_k$ (pitch angle) and $_k$ (roll angle) of the upper vertebra relative to the lower vertebra.

IOP Conf. Series: Journal of Physics: Conf. Series **783** (2017) 012056 doi:10.1088/1742-6596/783/1/012056

$$q_{1k}^2 = Z_k^2 + 2r_kZ_k\sin(\Theta_k) - 2r_kr_{k-1}\cos(\Theta_k) + r_k^2 + r_{k-1}^2 \quad (9)$$

$$\begin{aligned} q_{2k}^2 = {} & Z_k^2 + r_kZ_k\left(\sqrt{3}\cos(\Theta_k)\sin(\Psi_k) - \sin(\Theta_k)\right) + r_k^2 + r_{k-1}^2 \\ & - r_k^2r_{k-1}^2\left(\frac{\sqrt{3}}{2}\sin(\Theta_k)\sin(\Psi_k) + \frac{3}{2}\cos(\Psi_k) + \frac{1}{2}\cos(\Theta_k)\right) \end{aligned} \quad (10)$$

$$\begin{aligned} q_{3k}^2 = {} & Z_k^2 - r_kZ_k\left(\sqrt{3}\cos(\Theta_k)\sin(\Psi_k) + \sin(\Theta_k)\right) + r_k^2 + r_{k-1}^2 \\ & + r_k^2r_{k-1}^2\left(\frac{\sqrt{3}}{2}\sin(\Theta_k)\sin(\Psi_k) - \frac{3}{2}\cos(\Psi_k) - \frac{1}{2}\cos(\Theta_k)\right) \end{aligned} \quad (11)$$

Unlike the IKE , the solution of the forward kinematic equation is relatively complex, because the eqs 9, 10 and 11 are highly nonlinear. Thus, Two neural network are generated to provide approximated solution of the IKE, allowing to obtain the position and orientation to from the joint variable $q_{m,k}$, as shown in fig 6. Then, the transformation matrix ${}^{k}_{k+1}T$ of the upper vertebra frame relative to the lower vertebra, represented by eq 12 , allows to estimate Cartesian coordinates of the tip arm , from to homogeneous transformation matrices. However, The CBHA dispose only of six potentiometer, a block is added to estimate the joint variable from to tube length.

$${}^{k}_{k+1}T = \begin{bmatrix} c\theta_k & s\theta_k s\psi_k & s\theta_k c\psi_k & 0 \\ 0 & c\psi_k & -s\psi_k & 0 \\ -s\theta_k & s\psi_k c\theta_k & c\theta_k c\psi_k & Z_k \\ 0 & 0 & 0 & 1 \end{bmatrix} \quad (12)$$

Where c and s are cos and sin respectively.

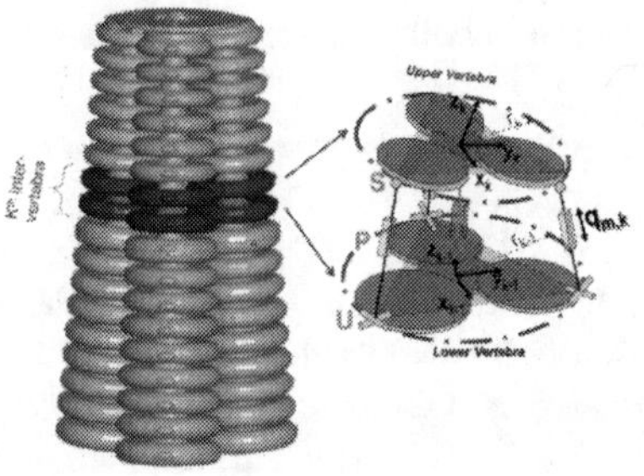

Figure 5. Modelling of Inter-Vertebra

Theoretical differences between Arc Geometry HTM, Dual Quaternion Method and Hybrid Approach,

Arc Geometry HTM

- This is Quantitative Approach.
- Two sections of CBHA are modelled as two cylinders.
- Need to remove singularity in this approach.

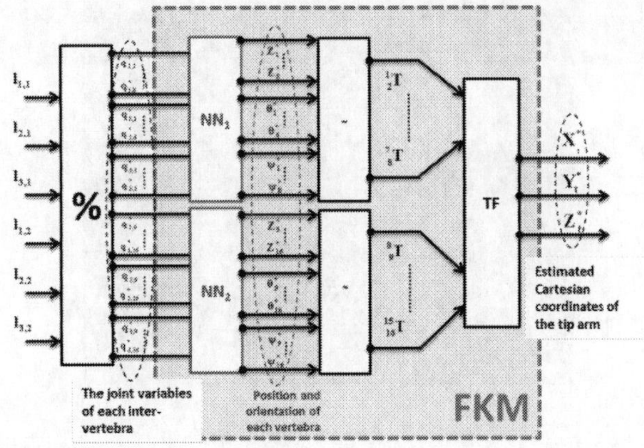

Figure 6. Algorithm for Forward Kinematics [16]

- Transformations are in the form of 4*4 Matrix, so rotation as well as translation is represented by 16 elements.
- For two or more than two sections of CBHA, there is multiplication of two matrices for addition of each section.

Dual Quaternion Method

- This is Quantitative Approach.
- Two sections of CBHA are modelled as two cylinders.
- No singularity
- Transformations are in the form of dual quaternion, so rotation as well as translation are represented by 8 elements only. Therefore, it needs less memory.
- For two or more than two sections of CBHA, there is multiplication of two dual quaternions for addition of each section. Therefore, this is computationally efficient because dual quaternion multiplication need less calculations than matrices multiplication.

Hybrid Approach

- This is Quantitative as well as Qualitative approach.
- Each inter-vertebra is modelled, therefore taking into account the conical shape of CBHA.
- No singularity.
- It uses Neural Network approach, so more memory.
- There are a lot of calculations in the begining for neural network so it is also not efficient computationally.

4. Experiment and Apparatus

The experiment is done on the CBHA to compare these three FKMs. The experiment is done on same set-up on the same time to do comparison between different methods. In case of CBHA we can control pressure, so five different values to the pressure sets are given to generate a trajectory. For example, [p1 p2 p3 p4 p5 p6] is a one pressure set. Therefore the data is collected regarding position and orientation of all three methods. The data contains 4200 points during all of the trajectory. Therefore all plots are plotted against 4200 points. Experimental set-up is as shown in fig. 7. This experiment is done as free load condition means without gripper.

Figure 7. Experimental Set-up

5. Results and Discussion

5.1. Trajectory Tracking

From the results, trajectories of both of representations HTM as well as DQ are exactly same. Because both of them are having the same assumptions just the difference is in representation of transformation matrix. So there is no difference in the output position value. The trajectory of Hybrid approach is some deviated but having same trend as shown in fig. 8.

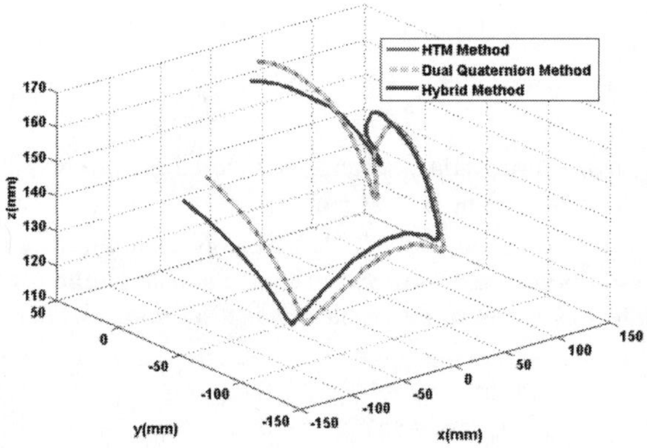

Figure 8. Trajectory Comparison of HTM, DQ and Hybrid Method

5.2. Cartesian Error

Cartesian errors are calculated for both of methods DQ as well as Hybrid from Arc Geometry HTM. Errors of DQ are approximately zero because it has overlapped trajectory ar shown but errors of Hybrid Method are observable and are shown in fig. 9. The x-error is varying between -8.5mm to 21mm, y-error is varying from -14mm to 2mm, z-error is varying from -8.5mm to 6.5mm approximately.

5.3. Orientation Comparison

From fig. 10, the orientation ψ, ϕ and θ is compared for all three methods. But from the graphs, it is clear that in all three methods have approximately same trend. So it means all of them converge approximately same in case of orientation. The orientations are deduced by using eqs 13.

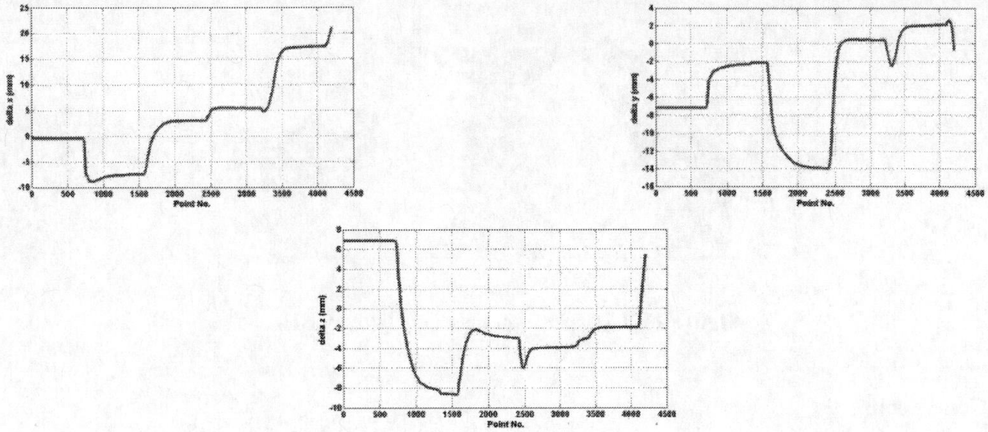

Figure 9. Cartesian errors in Hybrid Method

$$\begin{aligned}
\phi &= \arctan 2(r_{21}, r_{11}) \\
\psi &= \arctan 2(r_{13}\sin\phi - r_{23}\cos\phi, -r_{12}\sin\phi + r_{22}\cos\phi) \\
\theta &= \arctan 2(-r_{31}, r_{11}\cos\phi + r_{21}\sin\phi)
\end{aligned} \tag{13}$$

Where r_{nm} are the elements of rotational matrix having n rows and m columns.

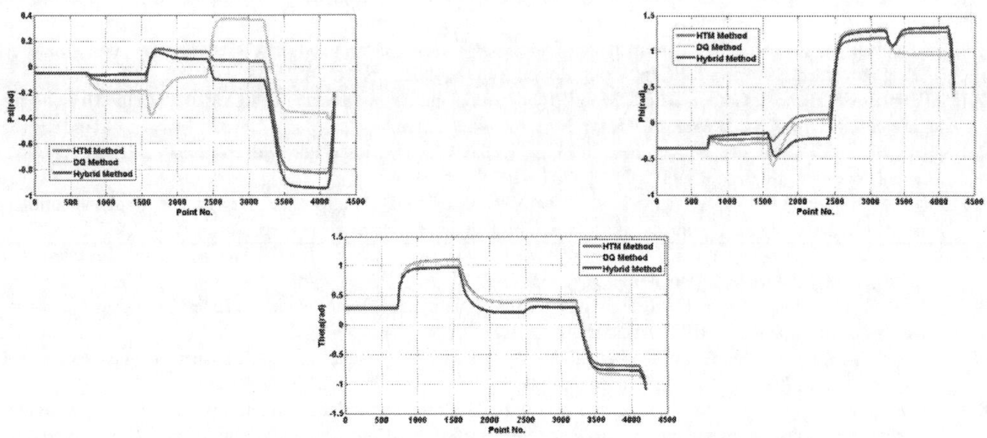

Figure 10. Orientation psi, phi and theta of all methods

6. Forward Kinematic Model of Co-operation of three CBHAs

In fig. 11 the co-operation between three CBHAs is shown. All CBHAs are attached on the fixed platform from the base and the tip points of them are attached to a movable frame. By using forward kinematic model the center point C can be deduced but for that the FKMs of all three CBHAs are needed. Therefore the efficient forward kinamatic model is to be used to apply on all CBHAs to reduce complexness.

In fig. 11 Y_{nm} and Z_{nm} are used as Y and Z axis of a frame where n = number of the CBHA and m = number of the frame.

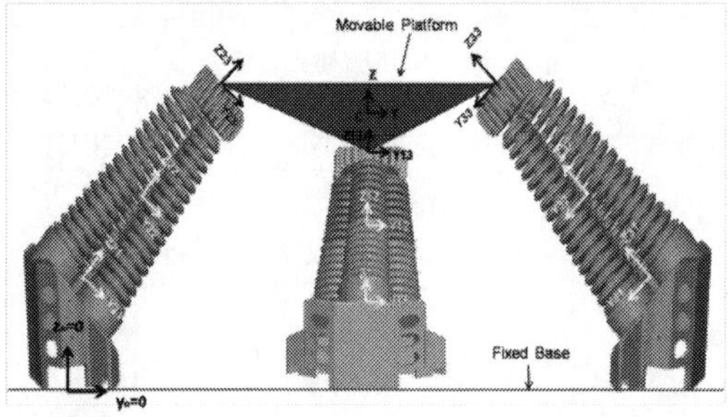

Figure 11. Co-operation between Three CBHAs

7. Conclusions

In this paper, FKMs are compared using theoretical and experimental data. From the comparison it can be concluded that as the number of sections in CBHA increases, it is advantageous to use Dual Quaternion Method as it demands less computational cost as compared to HTM and avoid the use of Hybrid Method because with increase in sections computations are increased. Hybrid Method should be avoided due to its oscillatory nature of the errors and should be used only in cases when utilisation of other methods is not possible. In future work, the work space of the co-operation between two or more CBHA is to be calculated which needs the more efficient FKM because there are more calculations if there is a big system. Therefore, this work is done to choose an efficient model.

References

[1] R. Behrens, M. Poggendorf, E. Schulenburg, and N. Elkmann, An elephants trunk-inspired robotic arm-trajectory determination and control, in Proc. 7th German Conf. Robot., 2012, pp. 1–5.

[2] M. Sfakiotakis, A. Kazakidi, N. Pateromichelakis, and D. P. Tsakiris, Octopus-inspired eight-arm robotic swimming by sculling movements, in Proc. IEEE Int. Conf. Robot. Autom., 2013, pp. 5155–5161.

[3] H. Martinez-Alfaro and S. Gomez-Garcia, Mobile robot path planning and tracking using simulated annealing and fuzzy logic control, Expert Syst. Appl., vol. 15, no. 3, pp. 421–429, 1998.

[4] Neppalli, S., M. A. Csencsits, B. A. Jones, and I. D. Walker (2009). Closed-Form Inverse Kinematics for Continuum Manipulators. In: Advanced Robotics 23, pp. 2077–2091.

[5] Hannan, M. H. and I.D. Walker (July 2001). The 'Elephant Trunk' Manipulator, Design and Implementation. In: IEEE/ASME International Conference on Advanced Intelligent Mechatronics Proceedings, pp. 14–19.

[6] Jones, B. A. and I. D. Walker (Apr. 2007). Limiting-case Analysis of Continuum Trunk Kinematics. In: IEEE International Conference on Robotics and Automation, pp. 1363–1368.

[7] Rucker, D. C., B. A. Jones, and R. J.Webster (Oct. 2010). A Geometrically Exact Model for Externally Loaded Concentric-Tube Continuum Robots. In: IEEE Transactions on Robotics 26.5, pp. 769–780.

[8] Iqbal, S., S. Mohammed, and Y. Amirat (Dec. 2009). A guaranteed approach for kinematic analysis of continuum robot based catheter. In: International Conference on Robotics and Biomimetics (IEEE ROBIO), pp. 1573–1578.

[9] C. Escande, R. Merzouki, P. M. Pathak, and V. Coelen, Geometric modelling of multisection bionic manipulator: Experimental validation on robotinoxt, presented at the IEEE Int. Conf. Robotics Biomimetics, Guangzhou, China, Dec. 2012.

[10] T. Mahl, A. Hildebrandt, and O. Sawodny, A variable curvature continuum kinematics for kinematic control of the bionic handling assistant, IEEE Trans. Robot., vol. 30, no. 4, pp. 935–949, Aug. 2014.

[11] Hartenberg, R. S. and J. A. Denavit (1955). A kinematic notation for lower pair mechanisms based on matrices. In: Journal of Applied Mechanics 77, pp. 215–221.

[12] I.S.Godge and I.D.Walker (May 2015). Dual Quaternion Based Model Kinematics for Multisection Continuum Arms IEEE International Conference on Robotics and Automation (ICRA). pp. 1416–1422.

[13] G. S. Chirikjian and J. W. Burdick, A modal approach to hyperredundant manipulator kinematics, IEEE Trans. on Robotics and Automation, vol. 10, no. 3, pp. 343–354, Jun 1994.

[14] B. A. Jones and I. D. Walker, Kinematics for multisection continuum robots, IEEE Transactions on Robotics, vol. 22, no. 1, pp. 43–55, 2006.

[15] A. Melingui, R. Merzouki, and J. Mbede, Compact bionic handling arm control using neural networks, Electron. Lett., vol. 50, no. 14, pp. 979–981, 2014.

[16] O. Lakhal, A. Melingui, and R. Merzouki, Hybrid Approach for Modeling and Solving of Kinematics of a Compact Bionic Handling Assistant Manipulator,IEEE/ASME Transactions on Mechatronics, vol. 21, no. 3, June 2016, pp. 1326–1335.
[17] M. Rolf and J. J. Steil, Efficient exploratory learning of inverse kinematics on a bionic elephant trunk, IEEE Trans. Neural Netw. Learn. Syst., vol. 25, no. 6, pp. 1147–1160, Jun. 2014.
[18] Webster, R. J. and B. A. Jones (June 2010). Design and Kinematic Modeling of Constant Curvature Continuum Robots: A Review. In: The International Journal of Robotics Research 29.13, pp. 1661–1683
[19] C. Escande (2013).Towards Modeling of a Class of Mobile Omnidrive - Bionic Manipulator Robots (Doctoral dissertation).

IOP Conf. Series: Journal of Physics: Conf. Series **783** (2017) 012057 doi:10.1088/1742-6596/783/1/012057

Modelling of Safety Instrumented Systems by using Bernoulli trials: towards the notion of odds on for SIS failures analysis

Laurent Cauffriez

LAMIH - UMR CNRS 8201
University of Valenciennes
Le Mont Houy, F-59313, Valenciennes Cedex 9, France
laurent.cauffriez@univ-valenciennes.fr

Abstract. This paper deals with the modeling of a random failures process of a Safety Instrumented System (SIS). It aims to identify the expected number of failures for a SIS during its lifecycle. Indeed, the fact that the SIS is a system being tested periodically gives the idea to apply Bernoulli trials to characterize the random failure process of a SIS and thus to verify if the PFD (Probability of Failing Dangerously) experimentally obtained agrees with the theoretical one. Moreover, the notion of "odds on" found in Bernoulli theory allows engineers and scientists determining easily the ratio between "outcomes with success: failure of SIS" and "outcomes with unsuccess: no failure of SIS" and to confirm that SIS failures occur sporadically. A Stochastic P-temporised Petri net is proposed and serves as a reference model for describing the failure process of a 1oo1 SIS architecture. Simulations of this stochastic Petri net demonstrate that, during its lifecycle, the SIS is rarely in a state in which it cannot perform its mission. Experimental results are compared to Bernoulli trials in order to validate the powerfulness of Bernoulli trials for the modeling of the failures process of a SIS. The determination of the expected number of failures for a SIS during its lifecycle opens interesting research perspectives for engineers and scientists by completing the notion of PFD.

1. Introduction

The demand of a Safety Instrumented System (SIS) depends on the failure of the so called Equipment Under Control (EUC) which is *an equipment, machinery, apparatus or plant used for manufacturing, process, transportation, medical or other activities* [1]. The principle of SIS demand is given in Figure 1. The role of the SIS is to bring the EUC into a safe state when all safety barriers have failed (ultimate safety level) and to do it preventively when the SIS itself fails (integrity of the safety function).

Table 1 gives an overview of possible SIS architectures according to IEC 61508 standards [1]. Please note that for 1oo1 channel, any dangerous failure leads to a failure of the safety function when a demand arises. Two types of intrinsic failures can affect the well functioning of a SIS channel: Dangerous failure and Safe Failure. Dangerous failures are defined by [2] as failures that can provoke accidents because the failed SIS is unable to face a potentially dangerous event for the equipment under control. Safe failures of the SIS denote failures that have no consequence in terms of safety for the equipment under control.

IOP Conf. Series: Journal of Physics: Conf. Series **783** (2017) 012057 doi:10.1088/1742-6596/783/1/012057

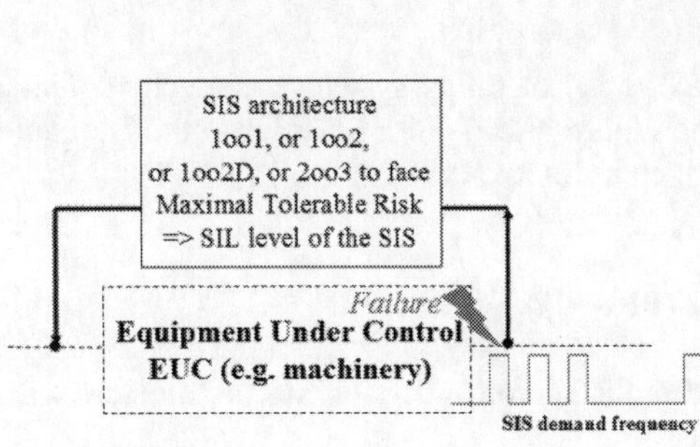

Figure 1. Principle of the demand of SIS

Table 1. Architectures of Safety Instrumented Systems

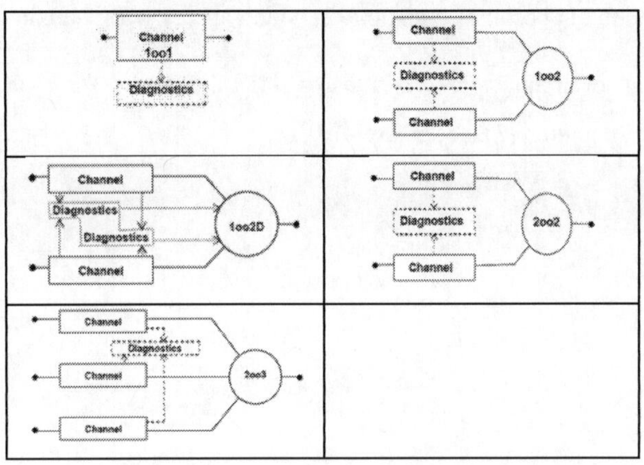

As a SIS is equipped with a self-diagnostic system, it is thus possible to detect intrinsic failures leading to a specific Safety Integrity Level of the SIS (SIL level of the SIS): the performance of the diagnostic depends on the ability of the diagnosis to care with some type of failures or not.

As described by [2], the tree of figure 2 gives the decomposition of the types of failures λ (λ is usually called failure rate and its unit is h^{-1} or $year^{-1}$). Please note that all the SIL theory developed in IEC 61508 standards make a strong assumption: failure distributions are assumed to be exponential and failure rates are therefore constant regardless of the type of SIS architectures i.e. 1oo1, 1oo2, 1oo2D, 2oo2, 2oo3.

On this point of view, the architecture 1oo1 is the reference for theoretical SIL calculation. For the other types of architectures, SIL levels are usually deduced from the 1oo1 architecture by applying conventional probabilities for events i.e. $P(\overline{\text{Channel1}} \cap \overline{\text{Channel2}})$ for 1oo2 architecture and, $P(\overline{\text{Channel1}} \cup \overline{\text{Channel2}})$ for 2oo2 architecture,…

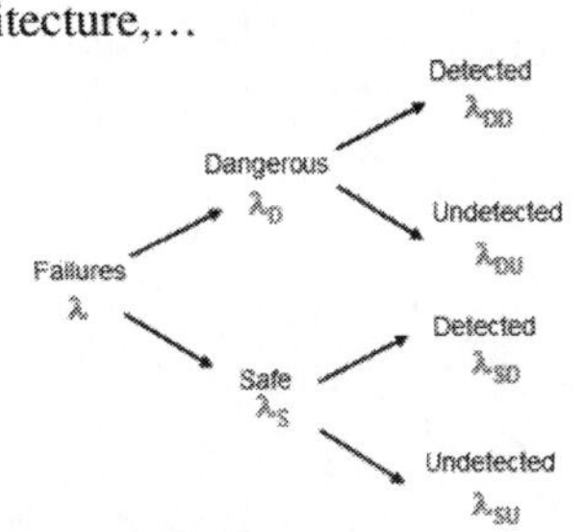

Figure 2. Classification of failures

Markov models [3]-[4], , Reliability block diagram [5], Cause-consequence diagrams [6], Stochastic Petri Nets models [7]-[9], Fault Tree models [10]-[11] or analytical expressions [12] have already been proposed to assess the SIL level of safety systems architecture. This paper proposes a new and original approach to model the failure process of a SIS based onto Bernoulli trials theory, and aims to complete the notion of PFD given in IEC61508.

2. Reminder on PFDaverage (PFDavg)

The average probability of failing dangerously for the SIS architecture within the test interval [0,τ] is given in equation (1) and represented in figure 3. Applying simplifications proposed by [3] i.e. $ta \approx \frac{\tau}{2}$ and $\lambda D \cdot \frac{\tau}{2} << 1$, equation (1) becomes equation (2) which agrees with the one given in IEC standard 61508-6 for 1oo1 channel architecture. The mean value of time $tc1$ for a channel unreliability on interval [0,τ] is equal to equation (3) (see figure 4).

$$PFDavg = 1 - e^{-\lambda D \cdot ta} \quad (1)$$

$$PFDavg = 1 - e^{-\lambda D \cdot \frac{\tau}{2}} \approx \lambda D \cdot \frac{\tau}{2} \quad (2)$$

$$tc1 = \tau - ta \approx \frac{\tau}{2} \quad (3)$$

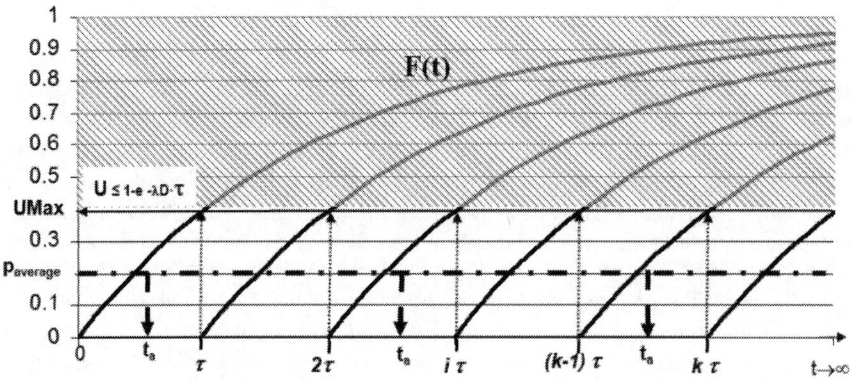

Figure 3. Reminder of PFDavg

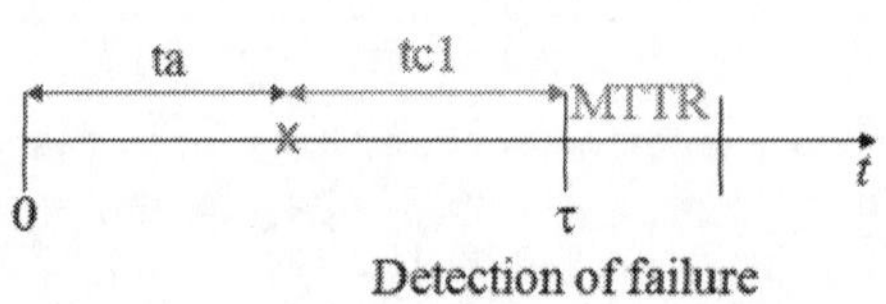

Figure 4. Definition of the mean value of time $tc1$ for dangerous failure in [0,τ]

The corresponding value of SIL is found in Table 2 as defined by the standard IEC61508 [1]. A look at probabilities given in Table 2 points out that a dangerous failure occurs very rarely. Indeed, the probability that the SIS has no dangerous failures at age τ (inclusive or exclusive) i.e. that the SIS survives at age τ is very high and belongs to]0.99, 0.99999] according to SIL level 1 (PFD<10^{-1}) and SIL level 4 (PFD$\geq 10^{-5}$) of Table 2.

Table 2. Safety Integrity Levels according to PFD

SIL			
4	$10^{-5} \leq$	PFD	$< 10^{-4}$
3	$10^{-4} \leq$	PFD	$< 10^{-3}$
2	$10^{-3} \leq$	PFD	$< 10^{-2}$
1	$10^{-2} \leq$	PFD	$< 10^{-1}$

For reminder, the probability to survive age t (inclusive or exclusive) — i.e. the probability of no failure before age t — is equal to (4).

$$R(t) = p = e^{-\lambda D \cdot t} \quad (4)$$

Similarly, the probability of failure to age t (inclusive or exclusive) is given by (5).

$$F(t) = 1 - p = 1 - e^{-\lambda D \cdot t} \quad (5)$$

Therefore, a SIS can either survive at age τ with probability $p = e^{-\lambda D \cdot \tau}$ or fail dangerously within a test interval with a probability $F(t) = 1 - p = 1 - e^{-\lambda D \cdot t}$ for any $t \in [0, \tau]$. Please note that the maximal probability of failure is achieved at age τ and is equal to $Umax = F(\tau) = 1 - e^{-\lambda D \cdot \tau}$.

On the other hand, it is to be observed that there are two possible outcomes for the periodic test characterizing the failure process of the SIS: either "a failure appears within current test interval [0,τ]" or "no failure appears within current test interval [0,τ]". This observation has given us the idea to use Bernoulli trials for characterizing more deeply the failure behaviour of the SIS during its lifecycle.

3. Proposition of using Bernoulli trials to characterize the process failures of a SIS

3.1 Reminder on Bernoulli experiment and Bernoulli trials

A Bernoulli experiment is a random experiment, the outcome of which can be classified in one of two mutually exclusive ways, say "success" or "unsuccess".
A sequence of Bernoulli trials occurs when a Bernoulli experiment is performed several independent times so that the probability of success, say p, remains the same from trial to trial.
Let a random variable X being the number of success for an infinite sequence of Bernoulli trials and n being the first n trials, then [13]:

i) random variable X has a binomial distribution and the probability that exactly k successes occur in the first n trials is given by $p(X = k) = \frac{n!}{k!(n-k)!} p^k (1-p)^{n-k}$ (6)

ii) the expected value of the random variable X is one measure of the "centre" of the distribution and is the average of infinitely many observations on X, the expected value is defined as

IOP Conf. Series: Journal of Physics: Conf. Series **783** (2017) 012057 doi:10.1088/1742-6596/783/1/012057

$E(X) = \mu_X = np.$ (7)

iii) the variance of the random variable X measure its "spread" and is defined as

$V(X) = np(1-p)$. (8)

iv) the standard deviation σ_X is the positive square root of the variance $\sigma_X = \sqrt{np(1-p)}$ (9)

v) odds on favour of success is the ratio of the probability of an event occurring to the probability of its not occurring and is equal to $Of = \frac{p}{1-p} = \frac{Outcomes\ with\ success}{Outcomes\ with\ unsuccess}$. (10)

vi) odds against a success is the ratio of the probability of a not occurring event to the probability of its occurring and is equal to $Oa = \frac{1-p}{p} = \frac{Outcomes\ with\ unsuccess}{Outcomes\ with\ success}$. (11)

vii) For a very large population of trials i.e. for $n \to +\infty$, probability p that trials result in success is given by $\frac{Outcomes\ with\ Success}{n} \to p$ as $n \to +\infty$. (12)

3.2 Bernoulli trials to characterize the process of failures for a SIS

Let a random variable X characterizing the failure process of the SIS. If a parallel is drawn between Bernoulli experiment and SIS testing process, it comes that the outcome of periodic testing of the SIS can be classified as either:

- a success, "X=1, a SIS failure appears within current time interval [0,τ]" with probability $p = 1 - e^{-\lambda D \cdot \tau}$. (13)
- an unsuccess, "X=0, no SIS failure appears within current time interval [0,τ]" with probability $1 - p = e^{-\lambda D \cdot \tau}$. (14)

This way to model the random process of SIS failures thanks to Bernoulli trials is given in figure 5.

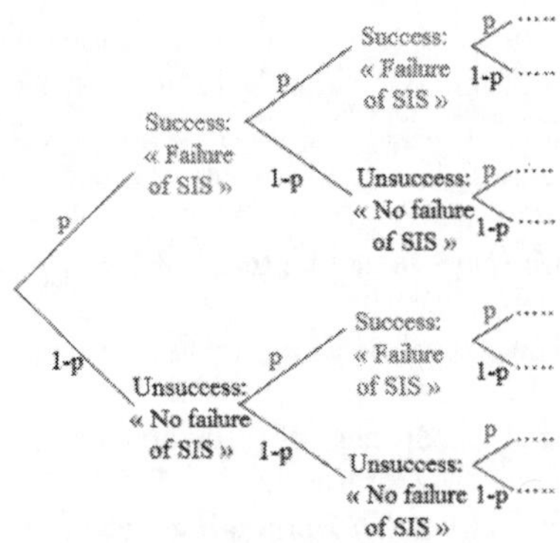

Figure 5. Application of Bernoulli trials to the modelling of SIS failures

A first step of our study is to assess "odds on favour of a failure" Of for a SIS according to equation (10). On this point of view, equation (15) gives the definition of "odds on favour of a failure" applied to the modelling of the failure process of a SIS and is obtained by replacing in equation (10) probabilities p and 1-p by respectively $p = 1 - e^{-\lambda D \cdot \tau}$ and $1 - p = e^{-\lambda D \cdot \tau}$.

$$Of = \frac{p}{1-p} = \frac{1 - e^{-\lambda D \cdot \tau}}{e^{-\lambda D \cdot \tau}} \quad (15)$$

The notion of "odds on" seems to be very interesting for the study of SIL levels because it gives the relative probability that the SIS will experience a failure. Moreover, "odds on" depend only on the failure rate λD and test interval τ fixed by the designer.

Table 3 gives a correspondence table between "odds on favour a failure" and SIL levels according IEC61508 standard. As for example, for a odd equals to 1/99 (upper limit of SIL 1), the SIS designer has to expect 1 outcome with success i.e. "X=1, a failure of the SIS appears within current interval [0,τ]" against 99 outcomes with unsuccess "X=0, no failure of the SIS appears within current interval [0,τ]".

Table 3. Odds on favour of failures for a SIS in comparison with SIL levels

SIL	PFD		
4	$10^{-5} \leq$	PFD	$< 10^{-4}$
3	$10^{-4} \leq$	PFD	$< 10^{-3}$
2	$10^{-3} \leq$	PFD	$< 10^{-2}$
1	$10^{-2} \leq$	PFD	$< 10^{-1}$

SIL	Odds on favour a SIS failure		
4	1/99999 ≤	*Of*	< 1/9999
3	1/9999 ≤	*Of*	< 1/999
2	1/999 ≤	*Of*	< 1/99
1	1/99 ≤	*Of*	< 1/9

Another interesting point is that, during a long working time (or during the lifecycle), the SIS is subjected to a total number of N tests whose outcomes are either the system experiences a failure or not. One another advantage of Bernoulli trials is to have, from the beginning of the design, a first approximation of the number of times $\hat{r}$ (equation 16) that the SIS will be subject to a failure for a long working time according to expected value (7) and standard deviation (9).

$$\hat{r} = N \cdot \frac{p}{1-p} = N \cdot \frac{1 - e^{-\lambda D \cdot \tau}}{e^{-\lambda D \cdot \tau}} \tag{16}$$

Please note that this way of assessing the estimated number of failures for a SIS consider that repair duration after a failure detection does not take time. Indeed, a long repair time leads to a reduction of the SIS working time and therefore, in the same proportion, to a reduction of the number of testing periods during its working time. Table 4 gives the expected value $E(X)$ and standard deviation of random variable X characterizing the failure process of a SIS by applying equations (7) and (9) of the Bernoulli theory.

In Table 4, the number of trials n is assessed for a SIS working time expressed in hours and corresponding respectively to 1 year, 10 years, 100 years, 1000 years working time assuming that the test has a period τ equal to 1 hour. One advantage of Table 4 is to give an expected number of times that the SIS will have a failure during its working time: the more the number of testing is important, the more the estimation of SIS failures is good (See column 6).

Looking at Table 4 for $p=10^{-5}$, one can observe that the value of $E(X)$ differs by a factor that is multiple to 10 for column 3 (1 year), 4 (10 years), 5 (100 years) and 6 (1000 years): 0.0876 (1 year), 0.876 (10 years), 8.76 (100 years) and 87.6 (1000 years).
Please note that theory with equation (12) stipulates that the outcomes with success tends toward the probability p if the number of trials n tends to infinity.
Table 4 confirms that, for a long working time, the more the level of SIL is high, the more the probability of failures is low.

Table 4. Theoretical expected value of random variable X and standard deviation characterizing the failure process of a SIS

1	2	3	4	5	6
		1 year	10 years	100 years	1000 years
$p=$ P(Failure of SIS)		$n=$ 8760	$n=$ 87600	$n=$ 876000	$n=$ 8760000
10^{-5}	$E(X)$	**0.0876**	**0.876**	**8.76**	**87.6**
	σ_X	±0.087	±0.935	±2.959	±9.359
10^{-4}	$E(X)$	**0.876**	**8.76**	**87.6**	**876**
	σ_X	±0.935	±2.959	±9.359	±29.595
10^{-3}	$E(X)$	**8.76**	**87.6**	**876**	**8760**
	σ_X	±2.958	±9.354	±29.582	±93.548
10^{-2}	$E(X)$	**87.6**	**876**	**8 760**	**87 600**
	σ_X	±9.312	±29.448	±93.125	±294.489
10^{-1}	$E(X)$	**876**	**8 760**	**87 600**	**876 000**
	σ_X	±28.078	±88.791	±280.784	±887.918

To conclude this paragraph, it seems that equations (15) and (16) proposed in this paper are sufficient to characterize the failure process of a SIS. Indeed, the notion of "odds on" and the estimation of the expected value of failures that the SIS will experience during its working time (or its lifecycle) seems to complete the notion of PFD.

The target of the next paragraph is to propose a Stochastic P-temporized Petri Net that will serve as reference model for a 1oo1 SIS architecture in order to compare experimental simulation results with the theoretical failure process of a SIS modelled thanks to Bernoulli trials.

4. Proposition of a reference model based onto a Stochastic P-temporized Petri Net for a 1oo1 SIS architecture

For purposes of validation, a reference model based onto a Stochastic P-temporized Petri Net for a 1oo1 SIS architecture is given in figure 6.

The probability to survive during a test interval τ is modelled thanks the left branch of the Petri net given in figure 6. This agrees with equation (14).
The probability to fail during a test interval τ is very low and is modelled thanks the right branch of the Petri net given in figure 6. This agrees with equation (13).

The functioning of Stochastic P-temporized Petri Net given in figure 6 is the following:
Places P1 and P10 are marked at the beginning of the interval $[0,\tau]$. The mark in place P1 moves in place P2 (no failure occurs with a probability $e^{-\lambda D \cdot \tau} = 1 - p$) either in place P3 (a failure occurs with a probability $1 - e^{-\lambda D \cdot \tau} = p$.) If the SIS experiences a failure within $[0,\tau]$ interval, the failure is generated at time $ti \in [0,\tau]$ and place P4 is marked.

After a duration d1 equal to the duration of the periodic test interval $[0,\tau]$, the temporized place P10 is left and place P11 is marked: this models the duration of the testing of the SIS and repair process by making the assumption in this case study that the repair duration is equal to d2=0 hour. The synchronization between Stochastic P-temporized Petri Nets -a- and -b- of figure 6 is made by the new arrival of the token in place P10.

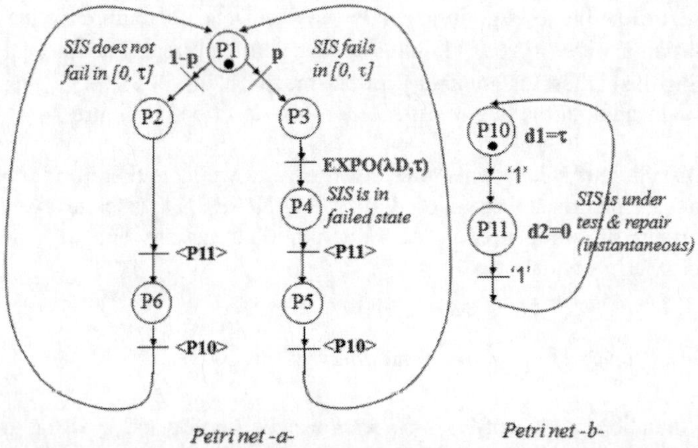

Figure 6. Reference model based onto a Stochastic P-temporized Petri Net for a 1oo1 SIS architecture

Discussion:

1) Please note that the use of a truncated exponential distribution function *EXPO*($\lambda D, \tau$) is absolutely necessary to generate with Monte Carlo algorithm failure events within $[0,\tau]$ interval. This crucial modelling point was demonstrated in [14] and is summarized in figure 7: *tevent* $\le \tau$ is equivalent to $U_{max} \le F(\tau)$ because F(t) is an increasing function i.e. $\forall t \in [0,\tau], \quad F(t) \in [0, 1 - e^{-\lambda D \cdot \tau}]$
   Therefore F(t) must be truncated to the maximal value $U_{max} = F(\tau) = 1 - e^{-\lambda D \cdot \tau}$ when inverting the distribution function for the simulation of Monte Carlo. This is an essential condition to have statistical results closest to reality. Indeed, this modelling approach decomposes the simulation time *t* into *n* intervals, each interval having a length equals to τ.

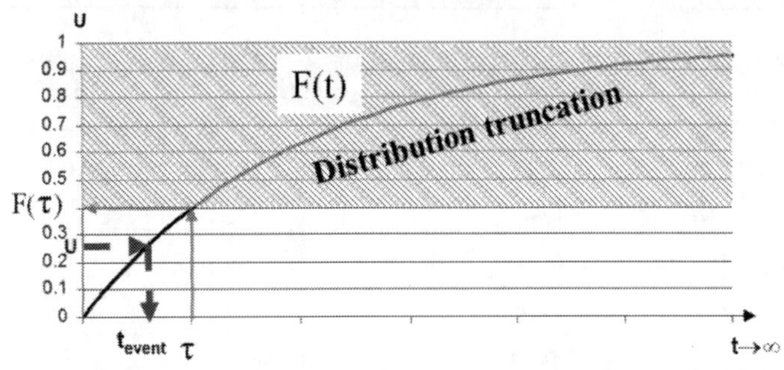

Figure 7. Truncation of failure distribution function F(t) due to periodic testing of SIS

2) This way of modelling the failure process of a SIS is closer to reality in so far as the probability to survive (left branch of Petri net -a-) or to fail (right branch of Petri net -a-) for the SIS is represented. Moreover, successive test intervals are really implemented and are not deduced "a posteriori" using mathematical modulo function "mod" as proposed in [1].

5. Validation of Bernoulli trials approach thorough simulation using Petri net reference model

To implement the Stochastic P-temporised Petri Nets of figure 6, we need a tool which:
1) can generate intermediate reports giving exactly the occurrence of failure events during the whole Monte Carlo simulation in order to verify failures appear within current time interval [nτ, (n+1)τ] with $n \in [1,N]$ and N being the highest test number for the simulation.
2) implements the truncation of the failure distribution F(t) described in figure 7.

We have not found such Petri Nets open source software and we have therefore implemented the two proposed Stochastic P-temporised Petri Nets with SIMAN/ARENA discrete event simulation tool. Indeed, this tool allows to define user defined distribution functions and to generate intermediate reports during Monte Carlo simulation.

5.1 First experimental campaign: one year simulation (8760 hours)

During one year simulation, a total of N=8760 tests were observed and r=48 failures were counted. Over the whole Monte Carlo simulation duration i.e. 8760 h, the SIS was in down state during 20.53676887 h and in well functioning state during $\sum_{i=1}^{r} TTFi + (N-r)\cdot\tau = 27.46323113$h+ 8712h=8739.463231 h. An estimator of experimental MTTF can be assessed by using (17)

$$MTTF' = \frac{1}{\lambda'} = \frac{\sum_{i=1}^{r} TTFi + (N-r)\cdot\tau}{r} \qquad (17)$$

Therefore, it comes for this simulation MTTF'=182.0721506 h and $\lambda' = 5.492328.10^{-3}\ h^{-1}$. These both values agree with input data of the model for $\lambda D=5.10^{-3}\ h^{-1}$ and τ=1h. Moreover, Table 5 (last raw) shows an experimental PFDavg equal to $2.344380\cdot 10^{-3}$ and the system is of SIL2.

This experimental PFDavg converges to the theoretical PFDavg according to IEC61508 which is equal to $PFDavg\_IEC61508 = 1 - e^{-\lambda D\cdot\frac{\tau}{2}} = 2.4968776\cdot 10^{-3}$.

There is a small deviation between the experimental PFD and the theoretical one of $1.52\cdot 10^{-4}$ due to the simulation approach. Indeed, the mean value to failure for the whole simulation during one year is equal to 0.5721506486 h and is not therefore fully equal to τ/2 as assumed by the theory due to the simulation duration of only one year. The simulation was deliberately limited to 1 year to display in this paper intermediate results such as number of failures and the time to failures.

Applying equation (15) to this case study, we have:

$$Of = \frac{p}{1-p} = \frac{1-e^{-\lambda D\cdot\tau}}{e^{-\lambda D\cdot\tau}} = \frac{1-e^{-0.005\cdot 1}}{e^{-0.005\cdot 1}} = 5.012\cdot 10^{-3} \approx \frac{5}{995} \approx \frac{1}{199}$$

i.e. SIL2 according to our proposed Table 3 giving a correspondance between "odds on" *Of* and SIL levels. Equation (16) gives an estimated value of the number of failures equal to

$\hat{r} = N \cdot \dfrac{p}{1-p} = N \cdot \dfrac{1-e^{-\lambda D \cdot \tau}}{e^{-\lambda D \cdot \tau}} = 43.90$ with a standard deviation σX equals to $\sigma X = \sqrt{np(1-p)} = \sqrt{8760 \cdot e^{-0.005}(1-e^{-0.005})}$ =6.59. Therefore, it appears that the estimated number of failures with Bernoulli trials is equal to 43.9 ± 6.59 and agrees with experimental results that shows a number of failures of 48 for one year simulation i.e. 8760 tests.

Table 5. Time to failure of SIS within [0,τ] test interval for Petri Net of figure 6

1 Test Number	2 [0,τ] Test Interval		3 **TTFi Time To Failure (h)**	4 **PFDi**
23	22	23	0.840897475	4.195661E-03
96	95	96	0.815554973	4.069472E-03
177	176	177	0.481352907	2.403871E-03
511	510	511	0.013132646	6.566108E-05
706	705	706	0.368733084	1.841967E-03
787	786	787	0.430220239	2.148789E-03
958	957	958	0.498859511	2.491189E-03
961	960	961	0.337026511	1.683714E-03
992	991	992	0.776835878	3.876646E-03
1679	1678	1679	0.519807192	2.595661E-03
1681	1680	1681	0.661869227	3.303876E-03
2307	2306	2307	0.505977449	2.526690E-03
2865	2864	2865	0.861548284	4.298476E-03
2922	2921	2922	0.957011002	4.773625E-03
3025	3024	3025	0.670537633	3.347074E-03
3061	3060	3061	0.943014111	4.703972E-03
3287	3286	3287	0.90349766	4.507300E-03
3461	3460	3461	0.959439348	4.785709E-03
3722	3721	3722	0.848264938	4.232343E-03
3920	3919	3920	0.619847404	3.094439E-03
4183	4182	4183	0.724616967	3.616529E-03
4190	4189	4190	0.499254092	2.493157E-03
4232	4231	4232	0.105261534	5.261692E-04
4382	4381	4382	0.253986804	1.269128E-03
4705	4704	4705	0.918368707	4.581317E-03
4789	4788	4789	0.865380558	4.317555E-03
4833	4832	4833	0.777793087	3.881413E-03
4882	4881	4882	0.62124863	3.101424E-03
5248	5247	5248	0.333491237	1.666067E-03
5303	5302	5303	0.138910404	6.943109E-04
5537	5536	5537	0.206943178	1.034181E-03
5605	5604	5605	0.47362704	2.365333E-03
5659	5658	5659	0.589422393	2.942773E-03
6044	6043	6044	0.440582291	2.200487E-03
6182	6181	6182	0.415179222	2.073743E-03
6222	6221	6222	0.476599278	2.380159E-03
6299	6298	6299	0.144121796	7.203494E-04
6444	6443	6444	0.809779702	4.040713E-03
6602	6601	6602	0.776676663	3.875853E-03
6632	6631	6632	0.921155936	4.595189E-03
7015	7014	7015	0.159845484	7.989081E-04
7181	7180	7181	0.605390804	3.022377E-03
7506	7505	7506	0.157438503	7.868828E-04
7589	7588	7589	0.863700893	4.309193E-03
7647	7646	7647	0.323336046	1.615374E-03

7655	7654	7655	0.546007133	2.726313E-03
7864	7863	7864	0.643066131	3.210167E-03
8226	8225	8226	0.658619149	3.287679E-03
Total of 8760 tests			**Mean value of TTF=** 0.5721506486h $\approx$ **τ/2**	**PFDavg** = 2.855810E-03

5.2 Second experimental campaign to validate Bernoulli trials approach: simulation of 50×1000 years

To validate more deeply the Bernoulli trials approach, five experimental campaigns are processed for 1oo1 architecture. For each campaign, the duration of the Monte Carlo simulation is equal to 1000 years i.e. 8760000h and 50 replications of 1000 years have been processed with input data for the Petri Net of figure 6 equal to τ=1h, d2=0h, and p respectively equal to 10^{-5}, 10^{-4}, 10^{-3}, 10^{-2}, 10^{-1}.

Experimental results are given in Table 6, these results agree with Bernoulli trials approach:
- for p=10^{-5}, the total number of failures 85.42±10.115839 (column 2, raw 1 of Table 6) agrees with the theoretical expected value $E(X) \pm \sigma X = 87.6 \pm 9.3659$ of Table 4.
- the mean value of time to failure *ta* is nearly equal to $\tau/2$ i.e. 0.5h (column 4, Table 6).
- the mean value of time to failures TTF (column 6, Table 6) and the mean time between failures (column 7, Table 6) agree with the Petri net input probability $p = 1 - e^{-\lambda D \cdot \tau}$ (column 1, Table 6).

Table 6. Experimental Mean and standard deviation of the SIS failures

1	2	3	4	5	6	7	8
p	Total number of failures	Minimum value of ta	Mean value of ta	Maximum value of ta	Mean value of TTF	Mean time between failures	Mean number of test intervals between failures
10^{-5}	**85.42** ±10.115839	0.00009458	**0.49900079** ±0.032264	0.99981460	**118653.484** ±123086.377	**129251.6532** ±140754.6694	**129251.6515** ±140754.6343
10^{-4}	**877.18** ±34.856375	0.00004919	**0.50241008** ±0.007729	0.99998673	**10949.0895** ±9945.1393	**9990.92294859** ±411.707434	**9990.91426960** ±411.71604453
10^{-3}	**8751.66** ±91.427591	0.00000299	**0.50039523** ±0.002863	0.99999839	**1201.17544004** ±1143.487393	**1000.93603695** ±10.436939	**1000.93603025** ±10.43694450
10^{-2}	**87539.18** ±307.392237	0.00000003	**0.49932122** ±0.001051	0.99999919	**125.116336** ±125.076263	**100.089623** ±0.409214	**100.0696276** ±0.350987401
10^{-1}	**875881.02** ±678.46812	0.00000002	**0.49129591** ±0.000355	0.999999964	**8.7671654** ±6.630121	**10.00135287** ±7.749970E-03	**10.00134994** ±7.745831E-03

Conclusion

After an introduction, section 2 of the paper gives a reminder on PFDavg according to IEC61508. Section 3 proposes to use Bernoulli trials to characterize the process failure of a SIS during its working time i.e. during its lifecycle. The notion of "odds on" found in Bernoulli trials theory seems to be very interesting because it allows engineers and scientists to determine easily the ratio between "outcomes with failure of SIS" and "outcomes with no failure of SIS" for the whole population of tests of the SIS during its lifecycle. Section 4 proposes a Stochastic P-Temporized Petri net reference model for a 1oo1 SIS architecture, this Petri net helps to simulate the failure process of the SIS thanks to a Monte Carlo simulation. Section 5 makes a validation of the proposed approach by comparing simulation results of Stochastic Petri net with those based onto "odds on" and "expected value" of Bernoulli trials theory. Simulation results point out that expected values of SIS failures computed with Bernoulli trials approach agree with those obtained by Monte Carlo simulation for two experimental campaigns, the first one of one year and the second of 50 replications of thousand years. The ability of Bernoulli trials to characterize the sporadic occurrences of failures for a 1oo1 SIS architecture during its lifecycle was demonstrated in this paper. The main result of this study is to give the relative probability that the SIS will experience a failure in correspondence with SIL levels (see Table 3). Moreover, the introduction of expected value and standard deviation of the random variable characterizing the number of SIS failures for an infinite sequence of tests intervals (Bernoulli trials) complete the notion of PFD and allow to give a new approach to model the safety level of a SIS.

References

[1] IEC 61508-6. Functional safety of electrical/electronic /programmable electronic safety-related systems. NF EN 61508-6. AFNOR Standard, January 2011.

[2] W.M. Goble, Control Systems Safety Evaluation and Reliability (2nd ed.). Research Triangle Park (NC): ISA (The Instrumentation, Systems, and Automation Society), 1998.

[3] T. Zhang, W. Long, Y. Sato, Availability of systems with self-diagnostic components - applying Markov model to IEC 61508-6. Reliability Engineering & System Safety, Vol. 80, pp. 133–141, 2003.

[4] W. Mechri, C. Simon, K.B. Othman, Switching Markov chains for a holistic modeling of SIS unavailability, Reliability Engineering and System Safety, Vol. 133, pp.212–222, 2015.

[5] H. Guo, X. Yang, "A simple reliability block diagram method for safety integrity verification", Reliability Engineering and System Safety, Vol. 92, pp. 1267-1273, 2007.

[6] J. Beugin, D. Renaux, L. Cauffriez, A SIL quantification approach based on an operating situation model for safety evaluation in complex guided transportation systems, In Reliability Engineering and System Safety, Vol. 92, pp. 1686–1700, 2007.

[7] P.J. Cacheux, S. Collas, Y. Dutuit, C. Folleau, J.P. Signoret, P. Thomas, Assessment of the expected number and frequency of failures of periodically tested systems, Reliability Engineering and System Safety, Vol. 118, pp. 61–70, 2013.

[8] J.P. Signoret, Y. Dutuit, P.J. Cacheux, C. Folleau, S. Collas, P. Thomas. Make your Petri nets understandable: Reliability block diagrams driven Petri nets, Reliability Engineering and System Safety, Vol. 113, pp. 61–75, 2013.

[9] Y. Liu, M. Rausand, Proof-testing strategies induced by dangerous detected failures of safety-instrumented systems, Reliability Engineering and System Safety, Vol. 145, pp. 366–372, 2016.

[10] F. Brissaud, D. Charpentier, A. Barros, C. Bérenguer, Design of complex safety-related systems in accordance with IEC 61508, Reliability, Risk and Safety: Theory and applications, Guedes Soares & Martorell (Eds), Taylor & Francis Group, London, ISBN 978-0-415-55509-8, 2010.

[11] Y. Dutuit, F. Innal, A. Rauzy J-P. Signoret. Probabilistic assessments in relationship with safety integrity levels by using fault trees. Reliability Engineering and System Safety, Vol. 93, pp. 1867–1876, (2008).

[12] M. Chebila, F. Innal, Generalized analytical expressions for safety instrumented systems' performance measures: PFDavg and PFH, Vol. 34, 167-176, 2015.

[13] J.B. Walsh, Knowing the Odds: An Introduction to Probability, American Mathematical Society, Vol. 139, ISBN: 978-0-8218-8532-1, 421 pages, 2012.

[14] L. Cauffriez, A review of SIL theory and a demonstration on the need to truncate the exponential distribution for the generation of SIS failures: Example for a 1oo1 channel architecture, QUALITA' 2015, Mar 2015, Nancy, France, https://hal.archives-ouvertes.fr/hal-01149814.

Fuzzy rule-based model for optimum orientation of solar panels using satellite image processing

A Zaher[1], Y N'goran[2], F Thiery[1], S Grieu[1] and A Traoré[1]

[1] CNRS-PROMES Laboratory, Rambla de la thermodynamique 66100 Perpignan, France.
[2] LPMCT, Université de Cocody, Abidjan, 22 BP 582, Côte d'Ivoire.

Corresponding author: ali.zaher@promes.cnrs.fr

Abstract. In solar energy converting systems, a particular attention is paid to the orientation of solar collectors in order to optimize the overall system efficiency. In this context, the collectors can be fixed or oriented by a continuous solar tracking system. The proposed approach is based on METEOSAT images processing in order to detect the cloud coverage and its duration. These two parameters are treated by a fuzzy inference system deciding the optimal position of the solar panel. In fact, three weather cases can be considered: clear, partly covered or overcast sky. In the first case, the direct sunlight is more important than the diffuse radiation, thus the panel is always pointed towards the sun. In the overcast case, the solar beam is close to zero and the panel is placed horizontally to receive the diffuse radiation. Under partly covered conditions, the fuzzy inference system decides which of the previous positions is more efficient. The proposed approach is implemented using experimental prototype located in Perpignan (France). On a period of 17 months, the results are very satisfactory, with power gains of up to 23 % compared to the collectors oriented by a continuous solar tracking.

1. Introduction

Considering the continuous change of the sun position, orientation of solar panels is an effective method to increase total energy yield. There are two main ways to increase the efficiency of photovoltaic systems (PV panels). In the first way, panels are fixed, or have a tilt that can be adjusted, monthly or seasonally [1]. In the second approach, solar panels can always be pointed directly toward the sun using single or dual axis tracking system [2]. The most efficient is the dual axis tracker that increases the incident solar radiation by approximately 30% compared to the fixed mount and by 6% compared to the single axis tracker [3].

The total radiation captured by the PV panel is composed of direct and diffuse components. The direct component is used to describe solar radiation traveling on a straight line from the sun down to the surface of the panel. On the other hand, diffuse radiation is the sunlight that has been scattered by molecules and particles in the atmosphere.

In case of clear sky, 90% of the global solar energy comes from the direct sunshine and the other 10% from diffuse solar energy [4].

However, Atmospheric components like clouds and pollution increase the percentage of diffuse radiation. Thus, during overcast conditions, tracking the sun is an ineffective method and the horizontal position becomes the ideal choice to capture this diffuse radiation that is isotopically-distributed over the whole sky [5].

METEOSAT satellite images (METEOSAT of Second Generation) are widely used for weather parameters forecasting. Mainly, these images have been useful to study solar radiation and clouds features in different locations around the world. A. Mefti et *al.* [6] derived global hourly solar radiation from the visible METEOSAT-5 images, while H. Escrig et *al.* [7] focused on cloud cover forecasting basing on cloud detection and cloud motion estimation techniques. The same images where used by A. Ghosh et *al.* [8] to determine different types of cloud coverage: overcast, partially covered and cloudless. METEOSAT 8 data were also used by J. Cermak et *al.* [9] to propose a method based on object-oriented techniques for fog and low stratus clouds detection.

In this work, the cloud cover index and cloud duration are two main parameters for the determination of the optimum position of solar panel. In this field, the limitations due to the resolution of METEOSAT images, induce incertitude in most of weather parameters estimation. Thus, the integration of fuzzy inference systems becomes an advantageous approach.

This paper is organized as follows. In the first part, the different components of solar radiation on inclined surfaces on cloudy days are studied in order to know the optimum orientation of solar panels under different meteorological conditions. The second part is dedicated to the intelligent decision-making system through which the cloud information is used to determine the optimum position over the time. The last part shows the experimental results and a comparative study between the proposed approach and the continuous solar tracking systems.

2. Maximizing the overall solar energy capture on cloudy days

Determination of incident global solar irradiance on the inclined plane relative to the ground, at different latitudes, requires mathematical models that use relationships between the global intensity of solar radiation on a horizontal plane and the inclined plane at an angle $\beta \neq 0$ [10].

The research works of Liu and Jordan [11] set the basis for the development of further decomposition models by several researchers, to fit datasets from different locations and chronological periods, which have been successful to varying degrees.

In the Liu-Jordan isotropic model, with correction coefficients for all components, the diffuse part (scattered and reflected radiation) is isotropic in character and dispersed from the whole sky hemisphere. For a photovoltaic receiver facing south, a total solar radiation on a plane tilted G_β at the β angle is defined by the following relation [12]:

$$G_\beta = B\left(\frac{\cos(\theta_i)}{\cos(\theta_z)}\right) + D\left(\frac{1+\cos\beta}{2}\right) + (B+D).\rho.\left(\frac{1-\cos\beta}{2}\right) \quad (2.1)$$

Where: B, D are the direct and diffuse horizontal solar irradiance, Θ_z the solar zenithal angle, Θ_i is the incidence angle of the beam radiation on the tilted surface and ρ is the reflectance factor.

The model developed by Kasten et *al.* [13] is a total cloud based model comprising the calculation of direct beam radiation using:

$$B = G_{ex}.e^{(-T.\tau.m)}(1-C) \quad (2.2)$$

Where: G_{ex} the extraterrestrial solar radiation, C is the cloud cover index and T is the Linke turbidity.

The previous equation shows that the direct radiation equals zero when the cloud cover index equals 1 during heavy cloudy days. So, under these conditions the equation 2.1 becomes:

$$G_\beta = D\left(\frac{1+\cos\beta}{2}\right) + D.\rho.\left(\frac{1-\cos\beta}{2}\right) \quad (2.3)$$

If the reflected component is neglected the last relation can be written:

$$G_\beta = D\left(\frac{1+\cos\beta}{2}\right) \quad (2.4)$$

Regarding the last relation, to maximize the inlet solar power, the solar panel inclination angle β must be zero and that refers to the horizontal position.

IOP Conf. Series: Journal of Physics: Conf. Series **783** (2017) 012058 doi:10.1088/1742-6596/783/1/012058

As a result, under clear sky the solar panel must be pointed towards the sun to capture the maximum of direct solar radiation. On the other hand, on cloudy days, the direct radiation is close to zero therefore, the PV panel must be oriented horizontally to collect the diffuse radiation which is generally pretty equally distributed throughout the sky.

3. Intelligent decision-making system

The flow diagram of the decision-making system is presented in Figure 1. In this diagram, the cloud cover index and the cloud duration are computed for the pixel $P_0(x_0, y_0)$ which presents the studied zone.

The cloud cover allows us to estimate the amount of solar energy acquired from direct solar radiation and the amount of scattered solar radiation at a moment t. The cloud duration presents the number of cloudy cases which will be at the pixel P_0 during a period Δt which is the orientation period.

When the cloud cover index and the cloud duration are calculated, they are treated by a fuzzy inference system (FIS) which decides if the PV panel must be pointed toward the sun or horizontally during the period of study Δt.

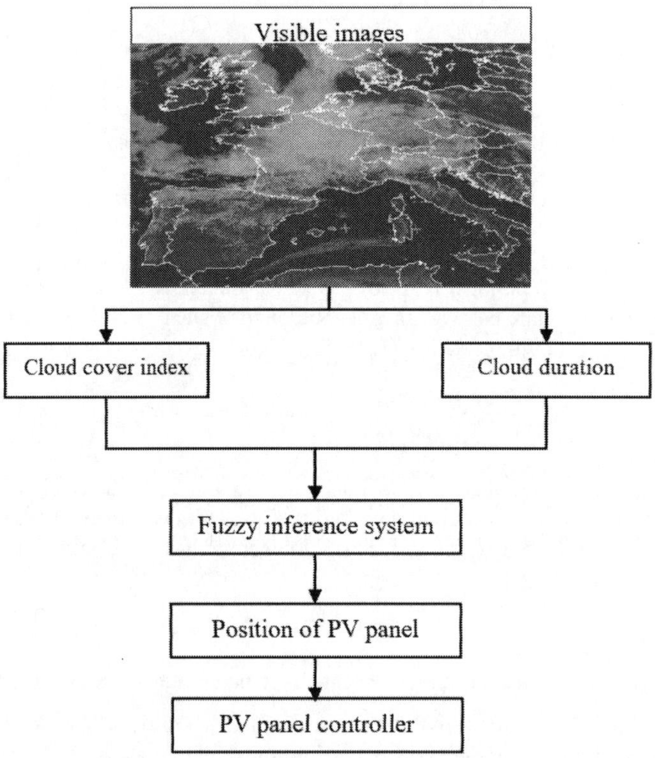

Figure 1. Decision-making system block diagram.

3.1. Cloud cover index

The METEOSAT sensor for the visible channel captures sunlight of the visible band reflected by an object back to the satellite. The fraction of solar radiation that is reflected back to space is called the albedo. The cloudy surfaces usually have a higher albedo than the other surfaces of the Earth. So clouds usually appear white, while land and water surfaces appear in darker gray scale levels.

To estimate the cloud cover index n for each pixel (x, y), the actual image $I(t)$ is compared to the image obtained during perfectly clear weather $I_C(t)$ and to the image taken on cloudy days $I_O(t)$ as follows:

$$n(x,y,t) = \frac{I(x,y,t) - I_C(x,y,t)}{I_O(x,y,t) - I_C(x,y,t)} \tag{3.1}$$

The images I_C and I_O are obtained using an archive of images taken at the same moment t over N days using the two following equations:

$$I_C(x,y,t) = min(I(x,y,t,day), I(x,y,t,day-1),...I(x,y,t,N-1)) \tag{3.2}$$

$$I_O(x,y,t) = max(I(x,y,t,day), I(x,y,t,day-1),...I(x,y,t,N-1)) \tag{3.3}$$

3.2. Cloud duration
In this study, the cloud duration is defined as the number of cloudy cases, which present at the pixel P_0 (x_0, y_0) during the period Δt. In fact, this number is the number of cloudy pixels, between the studied zone P_0 and the pixel $P_1(x_1, y_1)$, divided by the distance (in pixels) between the two pixels as shown in figure 2. The location of the pixel $P_1(x_1, y_1)$ can be determined using the following formula:

$$x_1 = x_0 - u.\Delta t \tag{3.4}$$

$$y_1 = y_1 - v.\Delta t \tag{3.5}$$

Where: Δt is the orientation period, u and v are velocity components of clouds between the pixels P_0 and P_1.

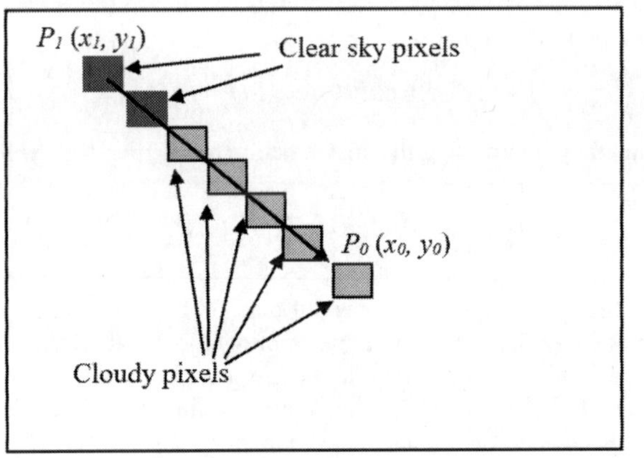

Figure 2. Example shows the number of cloudy pixels between the pixel P_1 and the Pixel P_0: 5 cloudy pixels, the distance equals 7 pixels.

Once the coordinates x_1 and y_1 are computed, the cloud duration CN can then be determined as follows:

$$CN = \frac{Number\ of cloudy\ pixels\ between\ P_0\ and\ P_1}{\sqrt{(x_0 - x_1)^2 + (y_0 - y_1)^2}} \tag{3.6}$$

To compute the cloud duration, a robust algorithm is needed for the estimation of cloud motion field (u, v) taking into account the semi-fluid behavior of clouds. Many approaches, such as block matching, pixel recursive and optical flow algorithms are used to compute the motion vectors in image sequences [14] and [15].

The method used in this article is based on Block Matching Algorithms combined with a best candidate block search for extracting cloud motion from satellite image sequence [16]. The basic concept of BMA is to divide the frame into small blocks then, the BMA finds the optimal motion vectors (MVs) that minimize the difference between reference block of the current frame and candidate block from the search area of previous frame.

The sum of absolute differences *SAD* is used as matching criterion and it is given by [17] and [18]:

$$SAD(x_r, y_r) = \sum_{i=0}^{L-1}\sum_{j=0}^{M-1} \left| I_n(x_r + i, y_r + j) - I_{n-1}(x_r + x + i, y_r + y + j) \right| \quad (3.7)$$

Where: $In(x_r, y_r)$ is the block of the current frame (n) and (x_r, y_r) is the coordinate of its upper left corner.

$I_{n-1}(x, y)$ is the block of the previous frame (n-1) and (x, y) is the coordinate of its upper left corner. Each of these blocks is of $L \times M$ pixels.

But in this method, the future evolution of the block is not taken into account, thus a big error occurs in the estimation. For this reason, (BMA) is combined with a best candidate block search [19]. In the last method, the BMA is applied and a list of the best candidates is stored with their scores S_c computed using Equation (3.7), and the distance between the reference block and the candidate (C_c), for each search region, for each nth image pair of the N-1 image pairs in the sequence.

The distance between the reference block and the block candidate is obtained by:

$$C_c(x_c, y_c) = \sqrt{(x_c - x_r)^2 + (y_c - y_r)^2} \quad (3.8)$$

The criterion used to match the candidates to the reference block in the search zone is a cost function of the distance and the scores. This function is given by:

$$Cost = p.S_c + (1 - p)C_c \quad (3.9)$$

The best match is obtained by minimizing the cost function along the N-1 image pairs using:

$$Match = \min \sum_{n=1}^{N-1} Cost_{n,c} \quad (3.10)$$

In the BMA algorithms, the performance is influenced by two main parameters: the research window size and the block size. Concerning the research window size, if this one is small, the amount of information inside the window is also small and the estimation is not reliable. In the other hand, if the size is too large, the number of matching candidates is big, thus the estimation can be wrong. In practice, the ideal size depends on the clouds displacement speed. For our experiments, the maximum displacement speed is 32 pixels per hour therefore; windows of 32× 32 pixels are adopted. The second important parameter of the block matching technique is the block size. If the size is too small, the number of matched blocks increases as well as the estimation error. On other hand, with a bigger block size, there is a high possibility that the block contains different objects moving in different directions and consequently, the estimation error increases.

The performance measure used to obtain the optimal block size is the peak-signal-to-noise-ratio (PSNR) [20]:

$$PSNR = 10 \log_{10}\left(\frac{255^2}{MSE}\right) \quad (3.11)$$

Where, MSE is the mean square error between the original image I and the reconstructed image I_r using the estimated motion vectors. It is given by the relation:

$$MSE = \frac{1}{m^2} \sum_{x=1}^{m} \sum_{y=1}^{m} (I_r(x,y) - I(x,y))^2 \quad (3.12)$$

The bigger values of PSNR refer to a better performance of the algorithm. Figure 3 shows the PSNR versus the block size. Clearly, the PSNR is high for a block size of 4×4 pixels. So this size is used in the implementation.

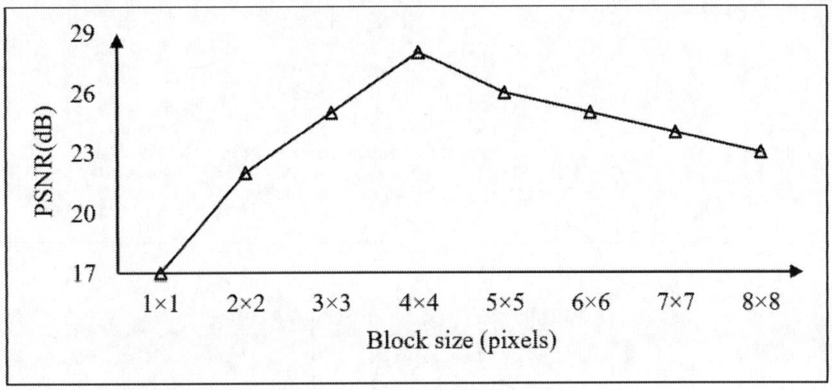

Figure 3. PSNR measured in decibels (dB) vs block size.

3.3. Fuzzy inference system structure

Fuzzy inference systems (FIS) have a multidisciplinary nature and they have been successfully applied in a wide variety of fields, such as automatic control, data classification, decision analysis, expert systems, and computer vision. There are two main structures of fuzzy inference systems: Mamdani-type and Sugeno-type. Here, Mamdani's fuzzy inference system [21] is used in order to decide the optimal position for the solar panel using the two inputs: the cloud cover index and the cloud duration.

The first step when making up a FIS is the fuzzification. Each of the cloud cover index, the cloud duration and the decision of orientation varies from 0 to 1. As presented in figure 4, Each of the input variables are fuzzified into three triangular membership functions with the associated linguistic labels *S* ("Small"), *M* ("Medium") and *B* ("Big"). The decision of the PV panel position (*d*) is fuzzified using two triangular membership functions which are associated to the linguistic labels: *T* for the tracking mode and *H* for the horizontal position.

The Fuzzy rules that have been used in this work are given in the table below:

Table 1. Fuzzy rule base.

Cloud cover index	Cloud duration	PV panel position
S	*S*	*T*
M	*S*	*T*
B	*S*	*T*
S	*M*	*T*
M	*M*	*H*
B	*M*	*H*
S	*B*	*H*
M	*B*	*H*
B	*B*	*H*

In the described fuzzy inference system, the fuzzy rule structure is predetermined by the interpretation of the characteristics of the variables and the membership functions are empirically tuned depending on the experimental data.

The defuzzification produces the numerical result of the fuzzy inference system. Many ways can be used to achieve this task. In this work, the center of area is adopted. This method, proposed by Sugeno and also known as center of gravity [22], is given by:

$$Z_{COG} = \frac{\int_z \mu_A(z).z.dz}{\int_z \mu_A(z).dz} \tag{3.13}$$

Where Z_{COG} is the crisp output, $\mu_A(z)$ is the aggregated membership function and z is the output variable.

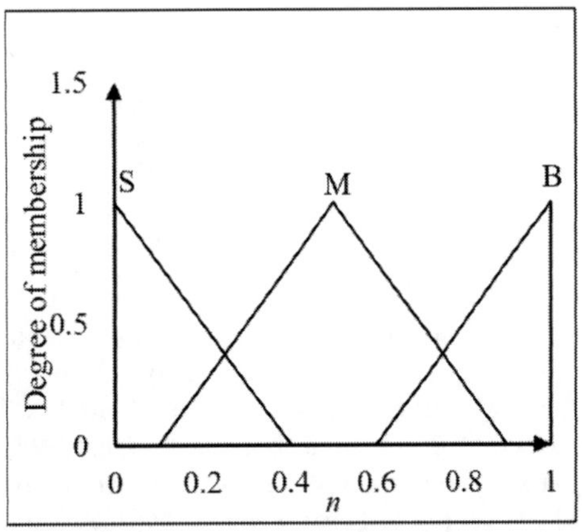

a. First input

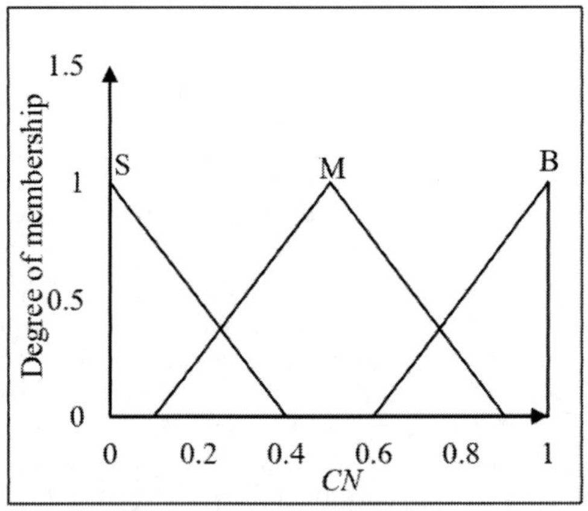

b. Second input

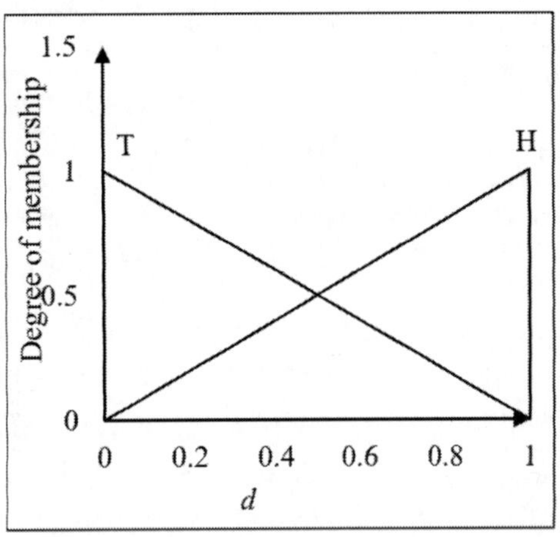

c. Output decision

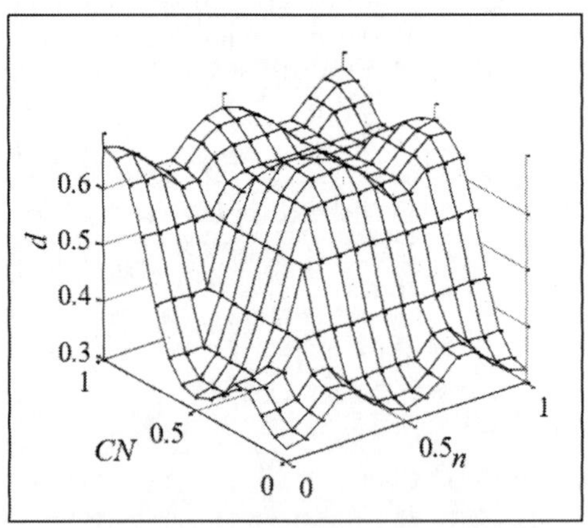

d. Fuzzy decision surface

Figure 4. Input and output membership functions and the fuzzy decision surface of the FIS.

4. Implementation and results

4.1. Experimental process

The described decision-making system has been implemented on an experimental process located in PROMES laboratory in Perpignan City ((latitude = 42.700 N, longitude= 2.900 E). As shown in figure 5, the set-up comprises: a polycrystalline photovoltaic panel SUNSET-PX 50E, a dual axis sun tracker SM34SPM+ equipped with a microcontroller board and a computer with a data acquisition card NI-USB-6008.

In practice, the intelligent algorithm computes the optimum position of the PV panel during the period Δt (fixed in this study at Δt = 1hour) and then, the new position angles are sent to the microcontroller in the tracker via a RS232 serial communication.

For the present study, the cloud cover index and the cloud duration are computed using the MSG (Meteosat Second Generation) images from the channel 0.6 μm taken over the central Europe every 15 minutes. These images are provided by the European Organization for the Exploration of Meteorological Satellites (EUMETSAT). The normal resolutions for VIS channels are 3×3 km^2.

The following parameters are used to estimate the clouds velocity:

- Number of visible images used in each sequence (4 images)
- Block size (4×4 pixels)
- Research window (32×32 pixels)

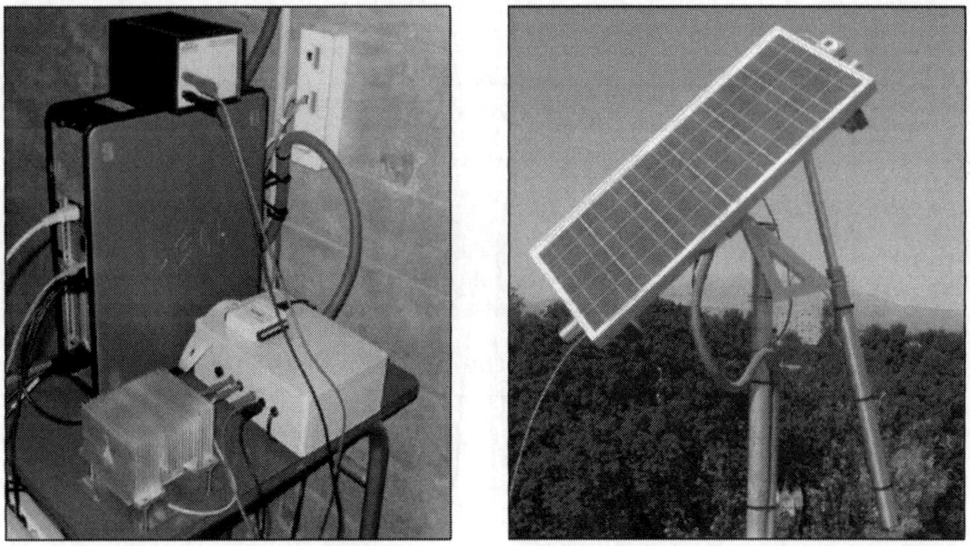

a. Computer and data acquisition card b. PV panel mounted on dual axis sun tracker

Figure 5. Picture of the experimental process

4.2. Results and discussion

Let's note that the aim of this work is to increase the efficiency of classical solar trackers, taking into account different weather conditions. The proposed approach was evaluated by comparing daily mean global solar radiation captured by classical methods (continuous solar tracking systems) to that obtained with the proposed intelligent approach. The results presented in figures 6 and 7 cover the period from January 2015 to May 2016.

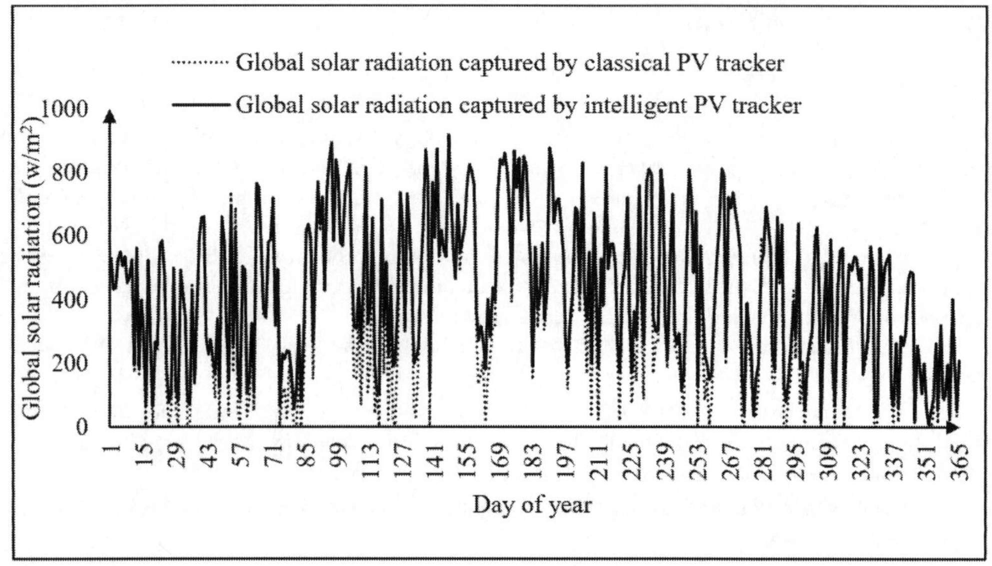

Figure 6. Daily mean global solar radiation captured by classical and intelligent solar PV trackers in 2015.

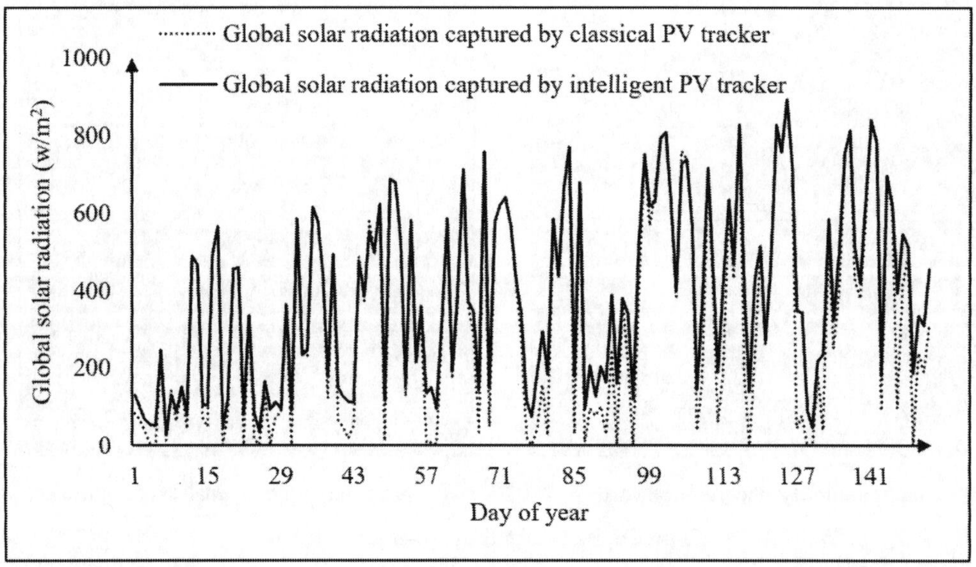

Figure 7. Daily mean global solar radiation captured by classical and intelligent solar PV trackers in 2016.

The results presented in these figures show that we get a bigger value of the global solar power by using the intelligent approach and that indicates a significant improvement of sun tracker efficiency. To analyze the performance of our approach under different weather conditions, a classification in three cases has been considered and presented in Table 2.

Table 2. Sky classification

Cloud cover index (n)	Sky class
$n < 0.1$	Clear
$0.1 < n < 0.5$	Partly covered
$n > 0.5$	Overcast

For each class of sky, the gain *(g)* of captured solar power is computed using the following formula:

$$g = \frac{\sum (E_i - E_c)}{\sum E_i} \tag{4.1}$$

Where: E_i (w/m^2), is the solar power captured by the PV using the intelligent orientation method and E_c (w/m^2), is the solar power captured by the PV using the classical sun tracking method (continuous solar tracking systems).

The results are presented in figure 8. Regarding this figure, one can note that, under clear sky conditions, the gain of solar power equals to zero and this is normal since the intelligent system decides to track the sun as the classical sun tracker. In case of partly covered sky, the performance of the proposed approach increases and the gain is approximately 14%. Under overcast sky situations, by inducing a horizontal position, the intelligent approach produces a gain of about 23%.

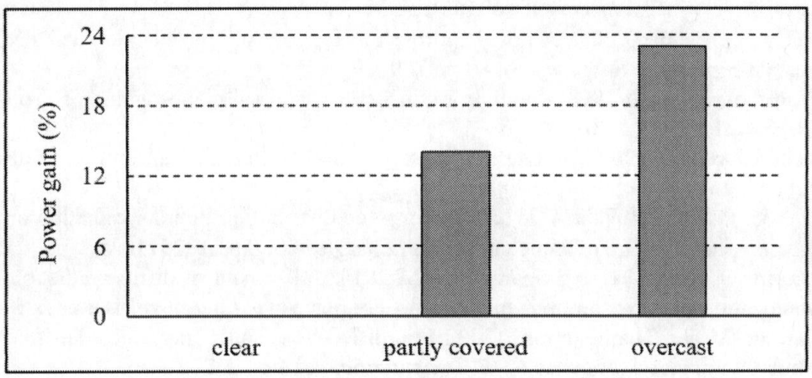

Figure 8. Gain of solar power vs sky class.

5. Conclusion

In this paper, an intelligent procedure is proposed to optimize the orientation of solar PV panel taking into account weather conditions. This intelligent approach is based on METEOSAT satellite images processing and fuzzy inference systems. In this context, the cloud cover index is computed by comparing the considered visible image to a reference image under clear sky. At the same time, the block matching algorithm is used for the detection of cloud motion vectors. Once the motion vectors are computed, the cloudy duration is evaluated. The cloud cover index and the cloudy duration are then treated by a fuzzy inference system, to decide the optimal position of the PV panel. To evaluate the performance of this method, it was implemented using an experimental process and the obtained results over 17 months, show that the power gain reaches about 14% and 23% respectively in the case of partly covered and overcast sky. Consequently, the proposed orientation strategy is efficient in all seasons. The immediate continuation of this work will be the integration of other kinds of images such as infra-red and/or water vapor ones to perform the control system. It will also be interesting to apply the developed algorithm in zones with other kind of climate. In this sense, collaborations are being developed with research teams in Côte d'Ivoire and Spain.

References

[1] Despotovic M and Nedic V *2015* Comparison of optimum tilt angles of solar collector determined at yearly, seasonal and monthly levels *Energy Conversion and Management* vol **97** pp 121–131

[2] Mousazadeh H, Keyhani A, Javadi A, Mobli H, Abrinia K and Sharifi A 2009 A review of principle and sun-tracking methods for maximizing solar systems output *Renewable and Sustainable Energy Reviews* vol **13** pp 1800–1818

[3] Neville R C 1978 Solar energy collector orientation and tracking mode *Solar Energy* vol **20** pp 7–11

[4] Kelly N A and Gibson T L 2009 Improved photovoltaic energy output for cloudy conditions with a solar tracking system *Solar Energy* vol **83** pp 2092–2102

[5] Kelly N A and Gibson T L 2011 Increasing the solar photovoltaic energy capture on sunny and cloudy days *Solar Energy* vol **85** pp 111–125

[6] Mefti A, Adane A and Bouroubi M Y 2008 Satellite approach based on cloud cover classification: Estimation of hourly global solar radiation from METEOSAT images *Energy Conversion andManagement* vol **49** pp 652–659

[7] Escrig H, Batlles F J, Alonso J, Baena F M, Bosch J L, Salbidegoitia I B and Burgaleta J I 2013 Cloud detection, classification and motion estimation using geostationary satellite imagery forcloud cover forecast *Energy* vol **55** pp 853–859

[8] Ghosh A, Pal NR and Das J 2006 A fuzzy rule based approach to cloud cover estimation *Remote Sensing of Environment* vol **100** pp 531–549

[9] Cermak J and Bendix J 2008 A novel approach to fog/low stratus detection using METEOSAT 8 data *Atmospheric Research* vol **87** pp 279–292

[10] Robledo L and Soler A 1998 Modeling irradiance on inclined planes with an anisotropic model *Energy* vol **3** pp 193–201

[11] Liu B and Jordan R 1960 The interrelationship and characteristic distribution of direct, diffuse and total solar radiation *Solar Energy* vol **3** pp 1–19

[12] Jastrzębska G F and Bugała A 2015 Modeling the distribution of solar radiation on a two-axis tracking plane for photovoltaic conversion *Energies* vol **8** pp 1025–1041

[13] Magarreiro C, Brito M C and Soares P M M 2014 Assessment of diffuse radiation models for cloudy atmospheric conditions in the Azores region *Solar Energy* vol **108** pp 538–547

[14] Ezhilarasan M and Thambidurai P 2008 Simplified block matching algorithm for fast motion estimation in video compression *J. Computer Sci* vol **4** pp 282–289.

[15] Horn B K P and Schunck B G 1981 Determining optical flow *Artificial Intelligence* vol **17** pp 185–203.

[16] BRAD R and LETIA I A 2002 Cloud motion detection from infrared satellite images *Proc. Int. Conf. on Image and Graphics (SPIE)* (Hefei: China) vol **4875** pp 408–4012

[17] Liaw Y C, Lai J Z C and Hong Z C 2009 Fast block matching using prediction and rejection criteria *Signal Processing* vol **89** pp 1115–1120

[18] Song B C and Ra J B 2000 A fast multi-resolution block matching algorithm for motion estimation *Signal Processing: Image Communication* vol **15** pp799–810

[19] Zaher A, Thiéry F, N'Goran Y and Traore A 2010 Estimation and prediction of solar radiation by METEOSAT image processing *Proc. of the 8th European Workshop on Advanced Control and Diagnosis* (Ferrara: Italy) pp 141–146

[20] Cuevas E, Zaldivar D, Cisneros M P and Oliva D 2013 Block-matching algorithm based on differential evolution for motion estimation *Engineering Applications of Artificial Intelligence* vol **26** pp 488–498

[21] Mamdani E H and Assilian S 1975 An experiment in linguistic synthesis with a fuzzy logic controller *Int. J. Man-Machine Studies* vol **7** pp 1–13

[22] Takagi T and Sugeno M 1985 Fuzzy identification of systems and its applications to modeling and control *IEEE Trans Syst Man Cybern* vol **15** pp 116–32

Noise Covariance Matrices Estimation for Systems with Time-Varying Availability of Sensors

Oliver Kost, Jindřich Duník, and Ondřej Straka
European Centre of Excellence - NTIS & Department of Cybernetics
Faculty of Applied Sciences, University of West Bohemia
Univerzitní 8, 306 14 Pilsen, Czech Republic

E-mail: kost@ntis.zcu.cz, dunikj@kky.zcu.cz, straka30@kky.zcu.cz

Abstract. The paper deals with the estimation of the noise covariance matrices of a time-varying system described by a state-space model with a time-varying set of sensors. In particular, the measurement difference autocovariance method, is proposed. The method is based on a statistical analysis of linearity transformed measurements leading to a system of linear matrix equations. The theoretical results are discussed and illustrated using a numerical study.

Keywords: Noise covariance matrices; State-space models; Estimation; Identification; Least-square method; Positioning

1. Introduction

State estimation plays a key role in many applications including satellite navigation, target tracking, signal processing, fault detection, and adaptive and optimal control problems. It constitutes an essential part of any decision-making process. Estimation for discrete-time stochastic dynamic systems is a very important field of study, which has been considerably developed for the past fifty years. In the beginning, the development was focused on linear dynamic systems and it has been further developed until today with the stress on the state estimation of nonlinear stochastic dynamic systems.

The state estimation, however, requires the knowledge of not only the system matrices or functions of the deterministic part of the model but also the knowledge of the noise statistics. The knowledge of noise statistics is, however, questionable in many cases. Incorrect description of the noise statistics may cause significant worsening of estimation or control quality or even failure of the underlying algorithms.

Therefore, several methods for estimation of the noise statistics have been developed, which can be divided into several categories; correlation methods [1–6], Bayesian estimation methods [7,8], maximum likelihood estimation methods [9, 10], covariance matching methods [11], or the Kalman filter working as a parameter estimator [12]. Besides the noise covariance matrix estimation methods, alternative approaches directly estimating the gain of a linear estimator have been developed as well [1, 13, 14]. A characterisation of the methods with their assumptions, properties, and limitations can be found in e.g., [3, 15–17].

The correlation methods have received quite considerable attention in the past as they may provide unbiased estimates with acceptable computational requirements even for high-dimensional systems. The methods are based on an analysis of the innovation sequence of "non-optimal" linear state and measurement predictors[1]. The methods have been pioneered in [1, 2] for linear time-invariant (LTI) and linear time-varying (LTV) models respectively. In [3–5, 18, 19] the correlation methods have been further advanced (in terms of computation and algorithmic complexity) and a set of methods, called the autocovariance least-squares (ALS) methods have been proposed. The ALS methods were developed for both linear and nonlinear systems. However, for LTI models and only for certain classes of LTV and nonlinear models, the methods have been proven to provide *asymptotically* unbiased estimates [3, 20]. Moreover, the above mentioned methods have been proposed for systems with a constant set of sensors. The assumption is, however, limiting for certain applications and areas, for example for navigation, tracking, and positioning, where the number of sensors or their availability varies over time [21].

In this paper, the measurement difference autocovariance (MDA) least-squares-based method [22], which was originally proposed for estimation of the noise covariance matrices for LTV models described by a state-space model, with a constant set of sensors, is extended for a state-space model, with linear state equation and linear or nonlinear measurement equation with a time-varying set of sensors. In particular, the models with the measurement function Jacobian with the rank greater or equal to the state vector dimension are considered. The proposed method is thoroughly analysed and conditions for *unbiased* estimate are discussed.

The rest of the paper is organised as follows. In Section 2, the system description and the problem statement are given. The MDA method is introduced and extended in Section 3. Section 4 discusses implementation details of the MDA method. Then, in Section 5 the performance of the MDA method is illustrated using a set of simulations and Section 6 concludes the paper.

2. System description and goal of paper

Let a discrete-time time-varying dynamic stochastic system described by the state-space model with a time-varying set of available sensors be considered. Its model consists of the state and measurement equations. The state equation is given by

$$\mathbf{x}_{k+1} = \mathbf{F}_k\mathbf{x}_k + \mathbf{G}_k\mathbf{u}_k + \mathbf{w}_k, k = 0, 1, 2, \ldots, T, \tag{1}$$

where the vectors $\mathbf{x}_k \in \mathbb{R}^{n_x}$ and $\mathbf{u}_k \in \mathbb{R}^{n_u}$ represent the system state and control at the time instant k, respectively. The vector $\mathbf{w}_k \in \mathbb{R}^{n_x}$ is the state noise, which is supposed to be zero-mean and white with *unknown* covariance matrix $\mathbf{Q} \in \mathbb{R}^{n_x \times n_x}$. The measurement equations corresponding to n_s available sensors are given by

$$\mathbf{z}_{k,i} = \mathbf{h}_{k,i}(\mathbf{x}_k) + \mathbf{v}_{k,i}, i = 1, 2, 3 \ldots, n_s. \tag{2}$$

where, $\mathbf{z}_{k,i} \in \mathbb{R}^{n_{z,i}}$ is generally a vector measurement of the i-th sensor, $\mathbf{h}_{k,i}(\mathbf{x}_k)$ is a known function, and $\mathbf{v}_{k,i} \in \mathbb{R}^{n_{z,i}}$ is the measurement noise with the following properties

$$\mathsf{E}[\mathbf{v}_{k,i}] = \mathbf{0}_{n_{z,i}\times 1}, \forall k, i = 1, 2, 3, \ldots, n_s \tag{3}$$

$$\mathsf{cov}[\mathbf{v}_{k,i}, \mathbf{v}_{l,j}] = \delta_{k,l}\mathbf{R}_{i,j}, i = 1, 2, 3, \ldots, n_s, j = 1, 2, 3, \ldots, n_s \tag{4}$$

where $\delta_{k,l}$ denotes the Kronecker delta function and $\mathbf{0}_{n\times m}$ stands for the zero matrix of dimension $n \times m$.

The set of the sensor indices is $\mathcal{Z}_k^C = \{1, 2, 3, \ldots, n_s\}$, and the measurement vector corresponding to all sensors is given

$$\bar{\mathbf{z}}_k = \left[\mathbf{z}_{k,1}^T, \mathbf{z}_{k,2}^T, \ldots, \mathbf{z}_{k,n_s}^T\right]^T \in \mathbb{R}^{n_z}. \tag{5}$$

where $n_z = \sum_{i=1}^{n_s} n_{z,i}$.

[1] As the noise covariance matrices are not known, the optimum (Kalman) gain cannot be computed. The predictor gain is, therefore, considered as a user-defined parameter, hence the prediction is not optimal in mean-square error.

At each time instant, however, a only subset $\mathcal{Z}_k \subset \mathcal{Z}_k^C$ of all possible sensors is available. The set of available sensors $\mathcal{Z}_k$ corresponds to the measurement vector

$$\mathbf{z}_k = \mathbf{S}_k \bar{\mathbf{z}}_k, \tag{6}$$

where $\mathbf{S}_k \in \mathbb{R}^{n_z \times n_z}$ is a selection matrix defined as

$$\mathbf{S}_k \triangleq \text{blkdiag}(\mathbf{S}_{k,1}, \mathbf{S}_{k,2}, \ldots \mathbf{S}_{k,n_s}), \tag{7}$$

and

$$\mathbf{S}_{k,i} = \begin{cases} \mathbf{I}_{n_{z,i}} & \text{if } i\text{-th sensor is available at time instant k, i.e., } i \in \mathcal{Z}_k \\ \mathbf{0}_{n_{z,i} \times n_{z,i}} & \text{otherwise.} \end{cases} \tag{8}$$

In (7), the function $\text{blkdiag}(\mathbf{S}_{k,1}, \mathbf{S}_{k,2}, \ldots \mathbf{S}_{k,n_z})$ denotes a block diagonal matrix and the notation $\mathbf{I}_n$ stands for the identity matrix of dimension n. Hence, the measurements corresponding to the unavailable sensors are zeroed by the selection matrix. The measurement equation at time k is then given by

$$\mathbf{z}_k = \mathbf{S}_k \bar{\mathbf{z}}_k = \mathbf{S}_k \left(\mathbf{h}_k(\mathbf{x}_k) + \mathbf{v}_k \right), k = 0, 1, \ldots, T, \tag{9}$$

where $\mathbf{z}_k \in \mathbb{R}^{n_z}$ and $\mathbf{v}_k \in \mathbb{R}^{n_z}$. It is also assumed that $\text{trace}(\mathbf{S}_k) \geq n_x, \forall k$, and the matrices $\mathbf{F}_k \in \mathbb{R}^{n_x \times n_x}$ and $\mathbf{G}_k \in \mathbb{R}^{n_x \times n_u}$ are known matrices of the system and control, respectively, and $\mathbf{h}_k(\mathbf{x}_k)$ is a known measurement function as well as the selection matrix $\mathbf{S}_k$. The measurement function is supposed to be a C^1 function i.e., the first order derivatives exist and are continuous. Further, it is assumed that the Jacobian[2] of the measurement function has full-rank i.e., its rank is equal to n_x. The state and measurement noises are assumed to be mutually independent and independent of the system initial condition. The moments of the initial state are *not* supposed to be known. The measurement noise is also supposed to be zero-mean and white with *unknown* covariance matrix $\mathbf{R} \in \mathbb{R}^{n_z \times n_z}$ considering of $\mathbf{R}_{i,j}$.

The goal of this paper is to *estimate* the unknown noise covariance matrices $\mathbf{Q}$ and $\mathbf{R}$ on the basis of the known system and control matrices $\mathbf{F}_k$, $\mathbf{G}_k$ $\forall k$, system measurement function $\mathbf{h}_k$, selection matrix $\mathbf{S}_k$ $\forall k$, and control sequences.

Note that such system description typically appears in tracking applications (e.g., global navigation satellite system based positioning), where the set of sensors (e.g., visible satellites) is varying over the time [23,24].

3. Derivation of measurement difference autocovariance method

The MDA method, is based on a statistical analysis of the difference between the measurement and a predicted measurement, i.e., on the analysis of the measurement prediction error . The measurement prediction is computed directly on the basis of the past measurements.

In this section, the MDA method for estimation of the noise covariance matrices $\mathbf{Q}$ and $\mathbf{R}$ is proposed for the model (1), (9) with a linear or nonlinear measurement equation with a time-varying set of available sensors.

3.1. MDA for linear system with linear measurement

In this section, the MDA method is introduced for the linear model described by (1) and (9) with $\mathbf{h}_k(\mathbf{x}_k) = \mathbf{H}_k \mathbf{x}_k$.

[2] The Jacobian $\mathbf{H}_k(\hat{\mathbf{x}}_k) = \frac{\partial \mathbf{h}_k(\mathbf{x}_k)}{\partial \mathbf{x}_k}\big|_{\mathbf{x}_k = \hat{\mathbf{x}}_k}$ is the first partial derivative with respect to $\mathbf{x}_k$ of function $\mathbf{h}_k(\mathbf{x}_k)$ evaluated at $\hat{\mathbf{x}}_k$.

3.1.1. Measurement prediction error definition The measurement prediction error (MPE), is defined as

$$\mathbf{r}_k^{(q)} \triangleq \mathbf{z}_k - \hat{\mathbf{z}}_k^{(q)}, k = q, \ldots, T, \tag{10}$$

where $\mathbf{z}_k$ is the measurement (9) and $\hat{\mathbf{z}}_k^{(q)}$ is the q-step measurement prediction of $\mathbf{z}_k$ based on the measurement $\mathbf{z}_{k-q}$. The prediction is computed as

$$\hat{\mathbf{z}}_k^{(q)} = \mathbf{S}_k\mathbf{H}_k \left(\prod_{i=1}^{q} \mathbf{F}_{k-i}\right) (\mathbf{S}_{k-q}\mathbf{H}_{k-q})^{\dagger}\mathbf{z}_{k-q} = \mathcal{H}_k^{(q)}\mathbf{z}_{k-q}, \tag{11}$$

where $(\mathbf{S}_k\mathbf{H}_k)^{\dagger} = (\mathbf{H}_k^T\mathbf{S}_k^T\mathbf{S}_k\mathbf{H}_k)^{-1}\mathbf{H}_k^T\mathbf{S}_k^T = (\mathbf{H}_k^T\mathbf{S}_k\mathbf{H}_k)^{-1}\mathbf{H}_k^T\mathbf{S}_k$ is the pseudoinverse of the matrix $(\mathbf{S}_k\mathbf{H}_k)$ and $\mathcal{H}_k^{(q)} = \mathbf{S}_k\mathbf{H}_k \left(\prod_{i=1}^{q} \mathbf{F}_{k-i}\right) (\mathbf{S}_{k-q}\mathbf{H}_{k-q})^{\dagger}$. Compared to other correlation methods it can be seen that the measurement prediction is computed from the past measurements only without an explicit need of a state prediction.

For further analysis, the MPE (10) can be rewritten by the use of (1) and (9) to the form

$$\begin{aligned}
\mathbf{r}_k^{(q)} &= \mathbf{S}_k \left(\mathbf{H}_k\mathbf{x}_k + \mathbf{v}_k\right) - \mathbf{S}_k\mathbf{H}_k \left(\prod_{i=1}^{q} \mathbf{F}_{k-i}\right) (\mathbf{S}_{k-q}\mathbf{H}_{k-q})^{\dagger}\mathbf{S}_{k-q} \left(\mathbf{H}_{k-q}\mathbf{x}_{k-q} + \mathbf{v}_{k-q}\right), \\
&= \mathbf{S}_k\mathbf{H}_k\mathbf{x}_k + \mathbf{S}_k\mathbf{v}_k - \mathbf{S}_k\mathbf{H}_k \left(\prod_{i=1}^{q} \mathbf{F}_{k-i}\right) \mathbf{x}_{k-q} - \mathcal{H}_k^{(q)}\mathbf{S}_{k-q}\mathbf{v}_{k-q}, \\
&= \mathbf{S}_k\mathbf{v}_k + \sum_{i=0}^{q-1} \mathbf{A}_k^{(i)}\mathbf{w}_{k-i-1} + \sum_{i=0}^{q-1} \mathbf{A}_k^{(i)}\mathbf{G}_{k-i-1}\mathbf{u}_{k-i-1} - \mathcal{H}_k^{(q)}\mathbf{S}_{k-q}\mathbf{v}_{k-q},
\end{aligned} \tag{12}$$

where $\mathbf{A}_k^{(i)} = \mathbf{S}_k\mathbf{H}_k \prod_{j=1}^{i} \mathbf{F}_{k-j}$.

3.1.2. Measurement prediction error properties It is easily be shown that the MPE is a stochastic process with the mean

$$\mathsf{E}[\mathbf{r}_k^{(q)}] = \sum_{i=0}^{q-1} \mathbf{A}_k^{(i)}\mathbf{G}_{k-i-1}\mathbf{u}_{k-i-1} \tag{13}$$

and with the autocovariance function

$$\mathbf{C}_{k,p}^{(q)} = \mathsf{E}[\tilde{\mathbf{r}}_k^{(q)}(\tilde{\mathbf{r}}_{k-p}^{(q)})^T] = \begin{cases}
\sum_{i=0}^{q-1} \left(\mathbf{A}_k^{(i)}\mathbf{Q}\left(\mathbf{A}_k^{(i)}\right)^T\right) + \mathbf{S}_k\mathbf{R}\mathbf{S}_k^T + \mathcal{H}_k^{(q)}\mathbf{S}_{k-q}\mathbf{R}(\mathcal{H}_k^{(q)}\mathbf{S}_{k-q})^T, & \text{for } p = 0, \\
\sum_{i=0}^{q-p-1} \left(\mathbf{A}_k^{(i+p)}\mathbf{Q}\left(\mathbf{A}_{k-p}^{(i)}\right)^T\right), & \text{for } p = 1, \ldots, q-1, \\
-\mathcal{H}_k^{(q)}\mathbf{S}_{k-q}\mathbf{R}\mathbf{S}_{k-q}^T, & \text{for } p = q, \\
\mathbf{0}_{n_z}, & \text{otherwise},
\end{cases} \tag{14}$$

where the difference $\tilde{\mathbf{r}}_k^{(q)}$ is given by

$$\tilde{\mathbf{r}}_k^{(q)} = \mathbf{r}_k^{(q)} - \mathsf{E}[\mathbf{r}_k^{(q)}]. \tag{15}$$

It means that the autocovariance function is *fully* defined by $q+1$ (cross-)covariance matrices of $\tilde{\mathbf{r}}_k^{(q)}$.

The equation (14) is a linear function of the *unknown* matrices $\mathbf{Q}$ and $\mathbf{R}$, and thus it can be rewritten in the following form suitable for the least-square (LS) method formulation

$$\mathcal{A}_k \boldsymbol{\theta}_k = \mathbf{b}_k, \tag{16}$$

where the parameter vector $\boldsymbol{\theta}_k$, containing the unknown elements of the noise covariance matrices, is given by

$$\boldsymbol{\theta}_k = \boldsymbol{\Gamma}_k [\mathbf{Q}_s^T, \mathbf{R}_s^T]^T, \tag{17}$$

and the matrices $\mathcal{A}_k$, $\mathbf{b}_k$ are defined as

$$\mathcal{A}_k = \begin{bmatrix} \sum_{i=0}^{q-1} \mathbf{A}_k^{(i)} \otimes \mathbf{A}_k^{(i)} & \mathcal{H}_k^{(q)} \mathbf{S}_{k-q} \otimes \mathcal{H}_k^{(q)} \mathbf{S}_{k-q} + \mathbf{S}_k \otimes \mathbf{S}_k \\ \sum_{i=0}^{q-2} \mathbf{A}_{k-1}^{(i)} \otimes \mathbf{A}_k^{(i+1)} & \mathbf{0}_{n_z{}^2 \times n_z{}^2} \\ \vdots & \vdots \\ \mathbf{A}_{k-q+1}^{(0)} \otimes \mathbf{A}_k^{(q-1)} & \mathbf{0}_{n_z{}^2 \times n_z{}^2} \\ \mathbf{0}_{n_z{}^2 \times n_x{}^2} & -\mathbf{S}_{k-q} \otimes \mathcal{H}_k^{(q)} \mathbf{S}_{k-q} \end{bmatrix}, \tag{18}$$

$$\mathbf{b}_k = \left[\left(\mathbf{C}_{k,0}^{(q)}\right)_s^T, \left(\mathbf{C}_{k,1}^{(q)}\right)_s^T, \ldots, \left(\mathbf{C}_{k,q}^{(q)}\right)_s^T \right]^T. \tag{19}$$

The symbol $\otimes$ denotes the Kronecker product, the notation $(\mathbf{A})_s$ means the columnwise stacking of the matrix $\mathbf{A}$ into a vector[3] [25] and the notation A_s^T means $((A)_s)^T$. The diagonal matrix $\boldsymbol{\Gamma}_k \in \mathbb{R}^{n_x^2+n_z^2 \times n_x^2+n_z^2}$ with ones and zeros on the diagonal defines which elements $\mathbf{Q}$ and $\mathbf{R}$ are estimable and which are not. The matrix $\mathbf{Q}$ is always estimable but estimability of the matrix $\mathbf{R}$ depends on the matrices $\mathbf{S}_i, i = \{k-q, k\}$ as

$$\boldsymbol{\Gamma}_k = \text{blkdiag}\left(\mathbf{I}_{n_x^2}, \max(\mathbf{S}_k \otimes \mathbf{S}_k, \mathbf{S}_{k-q} \otimes \mathbf{S}_{k-q})\right) \tag{20}$$

where the element-wise maximum of the matrices is consider.

3.1.3. Noise covariance matrices estimation Estimation of $\boldsymbol{\theta}_k$ is conditioned by known $\mathcal{A}_k$ and $\mathbf{b}_k$. The matrix $\mathcal{A}_k$ is a function of the known model matrices and therefore, can be computed. The vector $\mathbf{b}_k$ is unknown because it is a function of the unknown matrices $\mathbf{Q}$ and $\mathbf{R}$ due to (14). Nevertheless, the vector $\mathbf{b}_k$ can be estimated on the basis of the known difference $\tilde{\mathbf{r}}^{(q)}$ (15) as follows.

The estimate $\hat{\mathbf{C}}_{k,i}^{(q)}$ of $\mathbf{C}_{k,i}^{(q)}$ is given by

$$\hat{\mathbf{C}}_{k,i}^{(q)} = \tilde{\mathbf{r}}_k^{(q)} (\tilde{\mathbf{r}}_{k-i}^{(q)})^T, \forall k, \forall i, \tag{21}$$

and the estimate (21) is an unbiased estimate of $\mathbf{C}_{k,i}^{(q)}$ (14), i.e.,

$$\mathbf{C}_{k,i}^{(q)} = \mathsf{E}[\hat{\mathbf{C}}_{k,i}^{(q)}] = \mathsf{E}[\tilde{\mathbf{r}}_k^{(q)} (\tilde{\mathbf{r}}_{k-i}^{(q)})^T], \forall k, \forall i, \tag{22}$$

The *unbiased* estimate of $\mathbf{b}_k$ (19) is then given by

$$\hat{\mathbf{b}}_k = \left[\left(\hat{\mathbf{C}}_{k,0}^{(q)}\right)_s^T, \left(\hat{\mathbf{C}}_{k,1}^{(q)}\right)_s^T, \ldots, \left(\hat{\mathbf{C}}_{k,q}^{(q)}\right)_s^T \right]^T. \tag{23}$$

[3] The identity $\mathbf{ABC} = (\mathbf{C}^T \otimes \mathbf{A})(\mathbf{B})_S$ was employed in the derivation of (14).

The optimum LS estimate of the parameter $\boldsymbol{\theta}_k$ in (16) is

$$\hat{\boldsymbol{\theta}}_k = \mathcal{A}_k^{\dagger}\hat{\mathbf{b}}_k, \tag{24}$$

which is *unbiased*, i.e.,

$$\mathsf{E}[\hat{\boldsymbol{\theta}}_k] = \mathcal{A}_k^{\dagger}\mathsf{E}[\hat{\mathbf{b}}_k] = \mathcal{A}_k^{\dagger}\mathcal{A}_k\boldsymbol{\theta}_k = \boldsymbol{\theta}_k. \tag{25}$$

In the case that $\boldsymbol{\Gamma}_k$ (20) is time-invariant, i.e., the set of sensors is constant over time, the averaging of equation (24) over all time instants, i.e., over $k = 2q, \ldots, T$, leads to the resulting estimate

$$\hat{\boldsymbol{\theta}}^*_{(T)} = (\bar{\mathcal{A}}_{(T)})^{\dagger}\bar{\hat{\mathbf{b}}}_{(T)}, \tag{26}$$

with

$$\hat{\boldsymbol{\theta}}^*_{(T)} = \boldsymbol{\Gamma}[\mathbf{Q}_s^T, \mathbf{R}_s^T]^T \tag{27}$$

$$\bar{\mathcal{A}}_{(T)} = \tfrac{1}{T-2q+1}\sum_{k=2q}^{T}\mathcal{A}_k \tag{28}$$

$$\bar{\hat{\mathbf{b}}}_{(T)} = \tfrac{1}{T-2q+1}\sum_{k=2q}^{T}\hat{\mathbf{b}}_k, \tag{29}$$

where $\bar{\mathcal{A}}_{(T)}$ (28) is an average of the known matrices $\mathcal{A}_k$ up to the time T and $\bar{\hat{\mathbf{b}}}_{(T)}$ (29) can be interpreted as an average of the *unbiased* estimates $\mathbf{b}_k$ up to the time T. It means that the estimate $\hat{\boldsymbol{\theta}}^*_{(T)}$ (26) is *unbiased* as well.

As is derived in Appendix, the estimate (26) can be alternatively computed according to the following recursive relations

$$\hat{\boldsymbol{\theta}}^*_{(k)} = \hat{\boldsymbol{\theta}}^*_{(k-1)} + (\bar{\mathcal{A}}_{(k)})^{\dagger}\left(\hat{\mathbf{b}}_k - \mathcal{A}_k\hat{\boldsymbol{\theta}}^*_{(k-1)}\right), \tag{30}$$

$$\bar{\mathcal{A}}_{(k)} = \tfrac{1}{k-2q+1}\sum_{i=2q}^{k}\mathcal{A}_i = \bar{\mathcal{A}}_{(k-1)} + \mathcal{A}_k, \tag{31}$$

for $k = 2q, \ldots, T$, where $\mathcal{A}_{(i)} = \mathbf{0}_{(n_z^2(q+1))\times(n_x^2+n_z^2)}$ and $\hat{\boldsymbol{\theta}}^*_{(i)} = \mathbf{0}_{(n_x^2+n_z^2)\times 1}$ for $i = 0, \ldots 2q-1$.

In the case that $\boldsymbol{\Gamma}_k$ (20) is varying in the time i.e., the set of sensors is time-varying, the estimate (30) is simply modified to the following equation

$$\hat{\boldsymbol{\theta}}^*_{(k)} = \hat{\boldsymbol{\theta}}^*_{(k-1)} + (\bar{\mathcal{A}}_{(k)}\boldsymbol{\Gamma}_k)^{\dagger}\left(\hat{\mathbf{b}}_k - \mathcal{A}_k\hat{\boldsymbol{\theta}}^*_{(k-1)}\right), \tag{32}$$

where the matrix $\bar{\mathcal{A}}_{(k)}$ is still computed according to (31). Equation (32) compared to (30) respects availability or unavailability of the elements in the considered interval, i.e., only the available elements are estimated and updated.

3.1.4. Illustration of MDA for linear systems To illustrate the basic idea of the proposed MDA method, a simple example is given. Consider a system described by (1) and (9) with $\mathbf{G}_k = \mathbf{0}_{n_x\times n_u}$ and $\mathbf{h}_k(\mathbf{x}_k) = \mathbf{H}_k\mathbf{x}_k$, where the set of sensors varies in time

$$x_{k+1} = Fx_k + w_k = 0.9x_k + w_k, k = 0, 1, 2, 3, 4, \tag{33}$$

$$\mathbf{z}_k = \mathbf{S}_k\bar{\mathbf{z}}_k = \mathbf{S}_k\begin{bmatrix} z_{k,1} \\ z_{k,2}\end{bmatrix} = \mathbf{S}_k(\mathbf{H}_k x_k + \mathbf{v}_k) = \mathbf{S}_k\left(\begin{bmatrix}1\\2\end{bmatrix}x_k + \begin{bmatrix} v_{k,1} \\ v_{k,2}\end{bmatrix}\right), \tag{34}$$

where $\mathbf{R} = \left[\begin{smallmatrix}2 & 1\\ 1 & 3\end{smallmatrix}\right]$ and $Q = 0.5$.

The availability of the sensors is as follows:

- $\mathcal{Z}_0 = \mathcal{Z}_1 = \mathcal{Z}_2 = \{2\}$, i.e., $\mathbf{S}_0 = \mathbf{S}_1 = \mathbf{S}_2 = \left[\begin{smallmatrix} 0 & 0 \\ 0 & 1 \end{smallmatrix}\right]$,
- $\mathcal{Z}_3 = \mathcal{Z}_4 = \{1\}$, i.e., $\mathbf{S}_3 = \mathbf{S}_4 = \left[\begin{smallmatrix} 1 & 0 \\ 0 & 0 \end{smallmatrix}\right]$.

It means that the equation (34) becomes

$$\mathbf{z}_k = \begin{cases} \begin{bmatrix} 0 \\ 2 \end{bmatrix} x_k + \begin{bmatrix} 0 \\ v_{k,2} \end{bmatrix}, \text{ for } k = 0, 1, 2, \\ \begin{bmatrix} 1 \\ 0 \end{bmatrix} x_k + \begin{bmatrix} v_{k,1} \\ 0 \end{bmatrix}, \text{ for } k = 3, 4, \end{cases} \tag{35}$$

The matrix $\mathcal{A}_k$ and the vector $\mathbf{b}_k$ (for illustration purposes this vector is computed according to (14) and (19) instead of its estimation given by (21) and (23)) for the selected model (33), (35) with $q = 1$ are

$$\bar{\mathcal{A}}_{(2)} = \begin{bmatrix} 0&0&0&0&0 \\ 0&0&0&0&0 \\ 0&0&0&0&0 \\ 4&0&0&0&1.81 \\ 0&0&0&0&0 \\ 0&0&0&0&0 \\ 0&0&0&0&0 \\ 0&0&0&0&-0.9 \end{bmatrix}, \mathbf{\Gamma}_2 = \begin{bmatrix} 1&0&0&0&0 \\ 0&0&0&0&0 \\ 0&0&0&0&0 \\ 0&0&0&0&0 \\ 0&0&0&0&1 \end{bmatrix}, \mathbf{b}_2 = \begin{bmatrix} 0 \\ 0 \\ 0 \\ 7.43 \\ 0 \\ 0 \\ 0 \\ -2.7 \end{bmatrix}, \tag{36}$$

$$\bar{\mathcal{A}}_{(3)} = \begin{bmatrix} 1&1&0&0&0.2025 \\ 0&0&0&0&0 \\ 0&0&0&0&0 \\ 4&0&0&0&1.81 \\ 0&0&0&0&0 \\ 0&0&0&0&0 \\ 0&0&0&0&-0.45 \\ 0&0&0&0&-0.9 \end{bmatrix}, \mathbf{\Gamma}_3 = \begin{bmatrix} 1&0&0&0&0 \\ 0&1&0&0&0 \\ 0&0&0&0&0 \\ 0&0&0&0&0 \\ 0&0&0&0&1 \end{bmatrix}, \mathbf{b}_3 = \begin{bmatrix} 3.1075 \\ 0 \\ 0 \\ 0 \\ 0 \\ 0 \\ -1.35 \\ 0 \end{bmatrix}, \tag{37}$$

$$\bar{\mathcal{A}}_{(4)} = \begin{bmatrix} 2&2.81&0&0&0.2025 \\ 0&0&0&0&0 \\ 0&0&0&0&0 \\ 4&0&0&0&1.81 \\ 0&-0.9&0&0&0 \\ 0&0&0&0&0 \\ 0&0&0&0&-0.45 \\ 0&0&0&0&-0.9 \end{bmatrix}, \mathbf{\Gamma}_4 = \begin{bmatrix} 1&0&0&0&0 \\ 0&1&0&0&0 \\ 0&0&0&0&0 \\ 0&0&0&0&0 \\ 0&0&0&0&0 \end{bmatrix}, \mathbf{b}_4 = \begin{bmatrix} 4.12 \\ 0 \\ 0 \\ 0 \\ -1.8 \\ 0 \\ 0 \\ 0 \end{bmatrix}. \tag{38}$$

The optimum LS estimation parameter for matrices (36) can be computed by (26), (30), or (32) where $\hat{\boldsymbol{\theta}}^*_{2q}$ is in this case equal to $\left[\, \hat{Q}\ 0\ 0\ 0\ \hat{R}_{(2,2)} \,\right]^T$ where $\mathbf{A}_{(i,j)}$ denotes the element positioned in i-th row and j-th column of the matrix $\mathbf{A}$. The elements $\hat{R}_{(1,1)}$, $\hat{R}_{(2,1)}$, and $\hat{R}_{(1,2)}$ are zero because the sensor 1 is not available at time instants $k = 0, 1, 2$. The matrices $\mathcal{A}_k$ and $\mathbf{b}_k$ can be computed at $k = 2$ because the matrix $\mathbf{b}_2$ is a function of $\{\mathbf{z}_k\}_{k=0}^2$, i.e., $\bar{\mathcal{A}}_{(2)} = \mathcal{A}_2$.

The vector $\{\hat{\boldsymbol{\theta}}^*_{(k)}\}_{k=3}^4$ (32) is equal to $\left[\, \hat{Q}\ \hat{R}_{(1,1)}\ 0\ 0\ \hat{R}_{(2,2)} \,\right]^T$, because the elements Q and $R_{(2,2)}$ are estimable at $k = 2$ and the elements $R_{(1,1)}$ is estimable at time 3 and 4.

Therefore, it can be seen that from the matrix $\{\mathbf{\Gamma}_k\}_{k=2}^4$ it is possible to estimate the elements Q and $R_{(2,2)}$ at time instant $k = 2$, the elements Q, $R_{(1,1)}$, and $R_{(2,2)}$ at time instant $k = 3$, and the elements Q and $R_{(1,1)}$ at time instant $k = 4$.

3.2. MDA for linear system with nonlinear measurement

In this section, the MDA method is proposed for the model with the nonlinear measurement equation which is described by (1) and (9). In this case, the nonlinear function $\mathbf{h}_k(\mathbf{x}_k)$ in the measurement equations (2) is linearised by the first order Taylor expansion as

$$\mathbf{h}_k(\mathbf{x}_k) \simeq \mathbf{h}_k(\hat{\mathbf{x}}_k) + \left.\frac{\partial \mathbf{h}_k(\mathbf{x}_k)}{\partial \mathbf{x}_k}\right|_{\mathbf{x}_k = \hat{\mathbf{x}}_k} (\mathbf{x}_k - \hat{\mathbf{x}}_k) = \mathbf{h}_k(\hat{\mathbf{x}}_k) + \mathbf{H}_k(\hat{\mathbf{x}}_k)(\mathbf{x}_k - \hat{\mathbf{x}}_k), \tag{39}$$

where $\hat{\mathbf{x}}_k$ is an arbitrary reasonable estimate of $\mathbf{x}_k$, which is close to the true state. This estimate may be chosen in accord with the a priori knowledge of the state or the estimate can be obtained as a prediction from the past measurement as

$$\hat{\mathbf{x}}_k = \mathbf{F}_{k-1} \left[(\mathbf{S}_{k-1}\mathbf{H}_{k-1}(\hat{\mathbf{x}}_{k-1}))^{\dagger} (\mathbf{z}_{k-1} - \mathbf{S}_{k-1}\mathbf{h}_{k-1}(\hat{\mathbf{x}}_{k-1})) + \hat{\mathbf{x}}_{k-1} \right] + \mathbf{B}_{k-1}\mathbf{u}_{k-1}. \tag{40}$$

The MPE is formally the same as in (10), but the measurement prediction (11) is approximate[4] and is of the form

$$\hat{\mathbf{z}}_k^{(q)} = \mathbf{S}_k\mathbf{H}_k(\hat{\mathbf{x}}_k)\prod_{i=1}^{q}\mathbf{F}_{k-i}(\mathbf{S}_{k-q}\mathbf{H}_{k-q}(\hat{\mathbf{x}}_{k-q}))^{\dagger}\mathbf{z}_{k-q} = \mathcal{H}_k^{(q)}\mathbf{z}_{k-q}. \tag{41}$$

The *approximate* MPE (10) with respect to (1) and (9), and the approximation (39) can be rewritten to the form

$$\begin{aligned}
\mathbf{r}_k^{(q)} &= \mathbf{S}_k\left(\mathbf{h}_k(\hat{\mathbf{x}}_k) + \mathbf{H}_k(\hat{\mathbf{x}}_k)(\mathbf{x}_k - \hat{\mathbf{x}}_k) + \mathbf{v}_k\right) - \\
&\quad - \mathcal{H}_k^{(q)}\mathbf{S}_{k-q}\left(\mathbf{h}_{k-q}(\hat{\mathbf{x}}_{k-q}) + \mathbf{H}_{k-q}(\hat{\mathbf{x}}_{k-q})(\mathbf{x}_{k-q} - \hat{\mathbf{x}}_{k-q}) + \mathbf{v}_{k-q}\right), \\
&= \mathbf{S}_k\mathbf{h}_k(\hat{\mathbf{x}}_k) + \mathbf{S}_k\mathbf{H}_k(\hat{\mathbf{x}}_k)\mathbf{x}_k - \mathbf{S}_k\mathbf{H}_k(\hat{\mathbf{x}}_k)\hat{\mathbf{x}}_k + \mathbf{S}_k\mathbf{v}_k - \mathcal{H}_k^{(q)}\mathbf{S}_{k-q}\mathbf{h}_{k-q}(\hat{\mathbf{x}}_{k-q}) - \\
&\quad - \mathbf{S}_k\mathbf{H}_k(\hat{\mathbf{x}}_k)\prod_{i=1}^{q}\mathbf{F}_{k-i}\mathbf{x}_{k-q} + \mathbf{S}_k\mathbf{H}_k(\hat{\mathbf{x}}_k)\prod_{i=1}^{q}\mathbf{F}_{k-i}\hat{\mathbf{x}}_{k-q} - \mathcal{H}_k^{(q)}\mathbf{S}_{k-q}\mathbf{v}_{k-q}, \\
&= \mathbf{S}_k\mathbf{v}_k + \sum_{i=0}^{q-1}\mathbf{A}_k^{(i)}\mathbf{w}_{k-i-1} - \mathcal{H}_k^{(q)}\mathbf{S}_{k-q}\mathbf{v}_{k-q} + \mathcal{E}_k^{(q)},
\end{aligned} \tag{42}$$

where

$$\begin{aligned}
\mathcal{E}_k^{(q)} &= \mathbf{S}_k\mathbf{h}_k(\hat{\mathbf{x}}_k) - \mathbf{S}_k\mathbf{H}_k(\hat{\mathbf{x}}_k)\hat{\mathbf{x}}_k - \mathcal{H}_k^{(q)}\mathbf{S}_{k-q}\mathbf{h}_{k-q}(\hat{\mathbf{x}}_{k-q}) + \mathbf{A}_k^{(q)}\hat{\mathbf{x}}_{k-q} + \\
&\quad + \sum_{i=0}^{q-1}\mathbf{A}_k^{(i)}\mathbf{G}_{k-i-1}\mathbf{u}_{k-i-1},
\end{aligned} \tag{43}$$

and

$$\mathbf{A}_k^{(i)} = \mathbf{S}_k\mathbf{H}_k(\hat{\mathbf{x}}_k)\prod_{j=1}^{i}\mathbf{F}_{k-j}. \tag{44}$$

Note that MPE is a stochastic process (14) with the mean

$$\mathsf{E}[\mathbf{r}_k^{(q)}] = \mathcal{E}_k^{(q)}. \tag{45}$$

It means that the estimate of the noise covariance matrices can still be computed according (14)-(32), where $\mathbf{A}_k^{(i)}$ is given by (44), $\mathcal{H}_k^{(q)}$ (41), and $\mathsf{E}[\mathbf{r}_k^{(q)}]$ by (43). In this case, the unbiasedness of the estimate cannot generally be ensured.

4. Discussion and implementation notes

The estimation of noise covariance matrices by the MDA method can be summarised in three steps

i. Computation of the MPE defined by the q-step measurement prediction error according to (10).
ii. Calculation of the matrix $\mathcal{A}_k$ (18) (from the known matrices $\mathbf{F}_k$, $\mathbf{H}_k$, $\mathbf{S}_k$) and estimation of the vector $\hat{\mathbf{b}}_k$ according to (23).
iii. Estimation of the noise covariance matrices (26) (or (30),(31)) in the case that the set of sensors is constant in time and (31),(32) in the case that the set of sensors is time-varying.

[4] If the state estimate is obtained as the prediction from the past measurement, the estimate depends on the estimated noise covariance matrices via the linearisation point. This causes a correlation between $\mathbf{w}$, $\mathbf{v}$, and $\hat{\mathbf{x}}_k$ which is neglected in computation of autocovariance function of the MPE (42).

Note 1: The main assumption used by the MDA method is that the matrices $\mathbf{F}_k$, $\mathbf{G}_k$, $\mathbf{S}_k$ are known and the known function $\mathbf{h}_k(\mathbf{x}_k)$ is *differentiable* with "*invertible*" Jacobian. However, this assumption is not sufficient for time-varying systems, because the matrix $\bar{\mathcal{A}}_{(T)}$ in (26) is an average of matrices $\bar{\mathcal{A}}_k$, for $k = 2q, \ldots, T$. Even if each matrix $\bar{\mathcal{A}}_k$ is of sufficient rank, the average matrix $\bar{\mathcal{A}}_{(T)}$ need not be of sufficient rank. Therefore, the system matrices and function are supposed to be of a *reasonable* structure to ensure the MDA estimate exists and is unique.

Note 2: The problem with availability of sensors over the time have been solved by formulation the problem as covariance matrix estimation with a time-varying measurement equation in (9).

Note 3: The MDA estimate is *unbiased* only if the equation (39) is not an approximation of the measurement function. In such case, the estimate is *unbiased* for an arbitrary number of measurements T, for $T \to \infty$ the estimate covariance matrices cov$[\hat{\boldsymbol{\theta}}^*]$ goes to zero. The estimate is *consistent*.

Note 4: In the case that (39) is only an approximation and the model has a *reasonable* structure and nonlinearity of measurements function the MDA estimate is close to the true noise covariance matrices $\mathbf{Q}$ and $\mathbf{R}$.

Note 5: The MDA estimate (24) for the LTV model is *not* the best linear unbiased estimate (BLUE) as the system of linear equations (16) in not optimally weighted in (24). The optimum weighting matrix depends on the estimated covariance matrices $\mathbf{Q}$ and $\mathbf{R}$ and thus, it is unknown [18, 26].

Note 6: The estimate (24) does not respect the symmetry of the estimated noise covariance matrices. The symmetry can be ensured if the matrix $\mathcal{A}_k$ is multiplied by the matrix $\mathcal{P}$ defined in such a way that each column of the matrix $\mathcal{A}_k$ depends only on one element of the matrices $\mathbf{Q}$ or $\mathbf{R}$. Therefore, the columns of $\mathcal{A}_k$ which are related to the same element are summed by the matrix $\mathcal{P}$. Besides the symmetrisation, the matrix $\mathcal{P}$ reduces the complexity of the calculation (matrix $\mathcal{A}_k\mathcal{P}$ has only $\frac{(n_x+1)n_x}{2} + \frac{(n_z+1)n_z}{2}$ columns instead of $n_x^2 + n_z^2$ in the matrix $\mathcal{A}_k$) [18].

Note 7: The recursive forms (30)-(31) or (31)-(32) are advantageous for the on-line use of the algorithm. They may also be useful for a formulation of the MDA method with the exponential forgetting factor if the noise covariance matrices $\mathbf{Q}$ and $\mathbf{R}$ are slowly varying in time.

Note 8: If there exist some knowledge about the estimated covariance matrices, then it can be used as the initial condition in the recursive form (30)-(31) or (31)-(32).

5. Numerical Illustrations

In this section two examples illustrating performance of the proposed MDA estimation are given.

5.1. Example 1 - MDA for linear system

The first example of the MDA method is for LTV model described by (1) and (9) with $\mathbf{h}_k(\mathbf{x}_k) = \mathbf{H}\mathbf{x}_k$ and $\mathbf{G}_k = \mathbf{0}_{n_x \times n_u}$. The model is considered with the following matrices

$$\mathbf{F}_k = \begin{bmatrix} \gamma_k & 0 \\ 0.3 & 0.8 \end{bmatrix}, \quad \mathbf{Q} = \begin{bmatrix} 1 & 0.3 \\ 0.3 & 2 \end{bmatrix}, \quad \mathbf{S}_k = \begin{cases} \text{blkdiag}([1\ 0\ 1]), \text{for } k = 0, \ldots, \frac{T}{4}, \\ \text{blkdiag}([1\ 1\ 0]), \text{for } k = \frac{T}{4}+1, \ldots, \frac{T}{2}, \\ \text{blkdiag}([1\ 1\ 1]), \text{for } k = \frac{T}{2}+1, \ldots, T, \end{cases}$$
$$\mathbf{H} = \begin{bmatrix} -1 & 2 \\ 0.5 & 1 \\ 2 & 0 \end{bmatrix}, \quad \mathbf{R} = \begin{bmatrix} 3 & -1 & 1 \\ -1 & 4 & 0 \\ 1 & 0 & 2 \end{bmatrix}, \quad \gamma_k = 0.7 + 0.32\sin(0.0005k), k = 0, \ldots, T. \tag{46}$$

The method is analysed using 10^4 Monte-Carlo (MC) simulations with $T = 10^4$ time instants per a MC simulation and with $q = 2$. The estimates for each MC simulation are given in Figure 1 together with the confidence interval of three standard deviations (3STD). The evolution of the MDA estimates for one selected MC simulation is plotted in Figure 2 and the availability of the measurement elements together with three zones that denote the availability of the respective measurements and thus estimability of the noise covariance elements are illustrated in Figure 3.

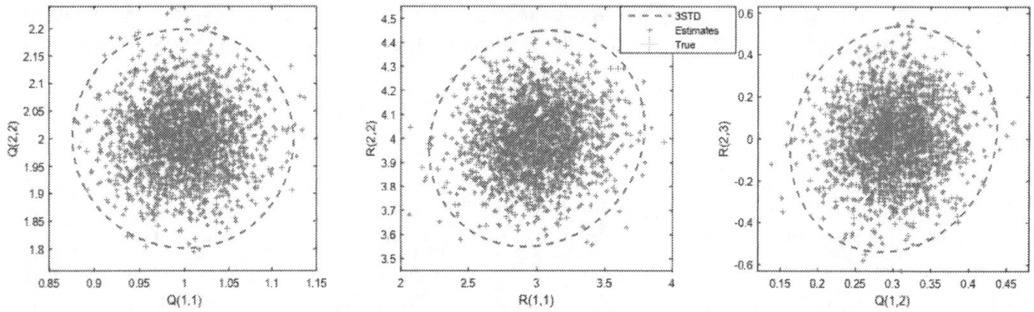

Figure 1. MDA with confidence ellipses.

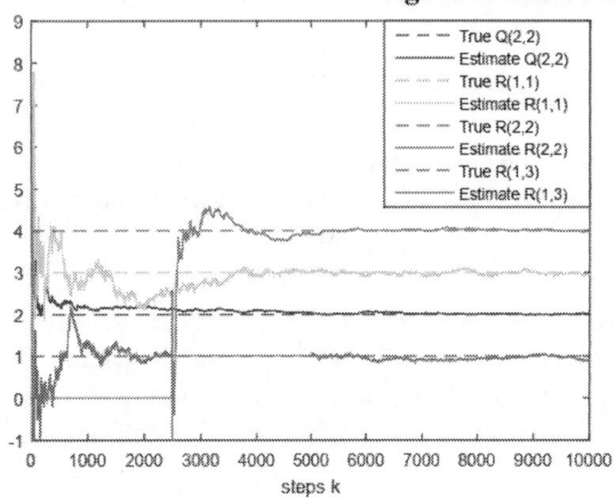

Figure 2. Development of MDA estimates.

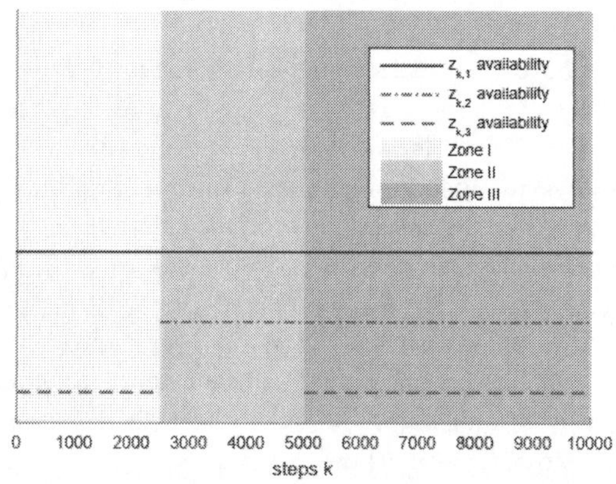

Figure 3. Availability of sensors elements.

It can be seen that the estimates converge to the true values in Figure 2. From Figure 3 it can also be seen that the estimate $\hat{\mathbf{R}}_{(2,2)}$ is not available only from $k = 0$ to $k = \frac{T}{4}$, because the sensor 2 is not available in this interval, and the estimate of $\hat{\mathbf{R}}_{(1,3)}$ is suspended from $k = \frac{T}{4}$ to $k = \frac{T}{2}$ because the sensor 3 is not available in this interval. Zone I denotes time instants in which the elements $\mathbf{R}_{(1,2)}$, $\mathbf{R}_{(2,1)}$, $\mathbf{R}_{(2,2)}$, $\mathbf{R}_{(3,2)}$ and $\mathbf{R}_{(2,3)}$ are not estimable, Zone II denotes that the elements $\mathbf{R}_{(1,3)}$, $\mathbf{R}_{(3,1)}$, $\mathbf{R}_{(3,3)}$, $\mathbf{R}_{(3,2)}$ and $\mathbf{R}_{(2,3)}$ are not estimable at the corresponding time instant, and Zone III denotes that all elements are estimable in the interval in Figure 3.

5.2. *Example 2 - MDA for system with nonlinear measurement function*

The second example is for a model with the nonlinear measurement equation. The system is considered with the following system matrices and the measurement functions

$$\mathbf{F}_k = \begin{bmatrix} 1 & 0 \\ 0 & 1 \end{bmatrix}, \forall k \qquad \mathbf{h}_k(\mathbf{x}_k) = \begin{bmatrix} \sqrt{(x_{k,1} - 100)^2 + (x_{k,2} - 300)^2} \\ \sqrt{(x_{k,1} + 400)^2 + (x_{k,2} - 600)^2} \\ \sqrt{(x_{k,1} + 200)^2 + (x_{k,2})^2} \end{bmatrix}, \forall k,$$
$$\mathbf{u}_k = \begin{bmatrix} \cos(\frac{2\pi k}{T}) \\ \sin(\frac{2\pi k}{T}) \end{bmatrix}, k = 0, \ldots, T, \tag{47}$$
$$\mathbf{G}_k = \begin{bmatrix} 0.2 & 0 \\ 0 & 0.2 \end{bmatrix}, \forall k, \qquad \mathbf{S}_k = \begin{cases} \text{blkdiag}([0\ 1\ 1]) & , \text{for } k = 0, \ldots, \frac{T}{2}, \\ \text{blkdiag}([1\ 1\ 1]) & , \text{for } k = \frac{T}{2} + 1, \ldots, T, \end{cases}$$

with $\mathbf{Q}$ and $\mathbf{R}$ selected as in Example 1. The model is motivated by navigation with distance measurements by three different radio transmitters such as the distance measuring equipment (DME).

The method is simulated with $T = 10^4$ time instants with $q = 3$. The estimates are plotted in Figure 4 and the state trajectory together with the positions of transmitters are given in Figure 5.

IOP Conf. Series: Journal of Physics: Conf. Series **783** (2017) 012059 doi:10.1088/1742-6596/783/1/012059

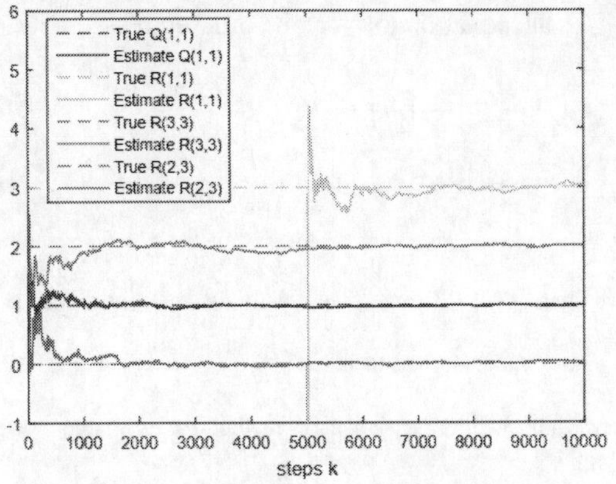

Figure 4. Development of MDA estimates

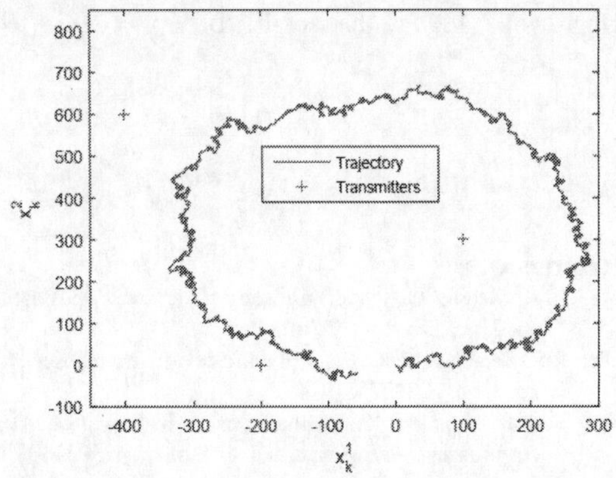

Figure 5. State trajectory and positions of transmitters

It can be seen that the estimate approaches to the true values in Figure 4. It can also be seen that the estimate $\hat{\mathbf{R}}_{(1,1)}$ is not available only from $k = 0$ to $k = \frac{T}{2}$, because the sensor 1 (which represents the measurement of the first transmitter) is not available in this time interval.

6. Concluding remarks

The paper focused on the estimation of the noise covariance matrices of systems described by the nonlinear time-varying state-space model with the varying set of sensors. For such a model the measurement difference autocovariance least-squares-based method, was proposed. The method belongs to the correlation methods, which are based on the statistical analysis of the measurement prediction error. The properties of the method were thoroughly analysed, and it was shown that the MDA method provides unbiased and consistent estimates of noise covariance matrices for a linear model, even for a time-varying set of sensors. The method was illustrated using two numerical studies covering the linear time-varying model and the nonlinear model with a time-varying set of sensors. The results have shown satisfactory performance of the MDA method for both models.

Acknowledgement: The work was supported by the Czech Science Foundation, project GA15-12068S, and the Czech Ministry of Education, Youth and Sports, project LO1506.

Appendix: Recursive Relations for MDA

In this appendix, a recursive algorithm for the MDA method is derived. Starting from the single-shot version of the MDA (24), the matrix (28) and (29) can be rewritten to the following form

$$\bar{\mathcal{A}}_{(k)} = \frac{1}{k-2q+1}\sum_{i=2q}^{k} \mathcal{A}_i = \frac{1}{k-2q}\sum_{i=2q}^{k-1}\mathcal{A}_i + \mathcal{A}_k = \bar{\mathcal{A}}_{(k-1)} + \mathcal{A}_k, \tag{48}$$

$$\bar{\hat{\mathbf{b}}}_{(k)} = \frac{1}{k-2q+1}\sum_{k=2q}^{k}\hat{\mathbf{b}}_k = \frac{1}{k-2q}\sum_{i=2q}^{k-1}\hat{\mathbf{b}}_i + \hat{\mathbf{b}}_k = \bar{\hat{\mathbf{b}}}_{(k-1)} + \hat{\mathbf{b}}_k. \tag{49}$$

It means that (26) by the use of (49) is equal to

$$\hat{\boldsymbol{\theta}}^*_{(k)} = (\bar{\mathcal{A}}_{(k)})^\dagger(\bar{\hat{\mathbf{b}}}_{(k-1)} + \hat{\mathbf{b}}_k), \tag{50}$$

where with (48) and the fact that $\bar{\hat{\mathbf{b}}}_{(k-1)} = \bar{\mathcal{A}}_{(k-1)}\hat{\boldsymbol{\theta}}^*_{(k-1)}$ the equation (50) can be modified to

$$\begin{aligned}\hat{\boldsymbol{\theta}}^*_{(k)} &= (\bar{\mathcal{A}}_{(k)})^\dagger \left((\bar{\mathcal{A}}_{(k)} - \mathcal{A}_{(k)})\hat{\boldsymbol{\theta}}^*_{(k-1)} + \hat{\mathbf{b}}_k\right) = \\ &= \hat{\boldsymbol{\theta}}^*_{(k-1)} + (\bar{\mathcal{A}}_{(k)})^\dagger \left(\hat{\mathbf{b}}_k - \mathcal{A}_{(k)}\hat{\boldsymbol{\theta}}^*_{(k-1)}\right). \end{aligned} \tag{51}$$

References

[1] R. K. Mehra, "On the identification of variances and adaptive filtering," *IEEE Transactions on Automatic Control*, vol. 15, no. 2, pp. 175–184, 1970.
[2] P. R. Bélanger, "Estimation of noise covariance matrices for a linear time-varying stochastic process," *Automatica*, vol. 10, no. 3, pp. 267–275, 1974.
[3] B. J. Odelson, M. R. Rajamani, and J. B. Rawlings, "A new autocovariance least-squares method for estimating noise covariances," *Automatica*, vol. 42, no. 2, pp. 303–308, 2006.
[4] B. M. Åkesson, J. B. Jørgensen, N. K. Poulsen, and S. B. Jørgensen, "A generalized autocovariance least-squares method for Kalman filter tuning," *Journal of Process Control*, vol. 18, no. 7–8, pp. 769–779, 2008.
[5] M. Šimandl and J. Duník, "Estimation of noise covariance matrices for periodic systems," *International Journal of Adaptive Control and Signal Processing*, vol. 25, pp. 928–942, 2011.
[6] J. Zhou and R. H. Luecke, "Estimation of the covariances of the process noise and measurement noise for a linear discrete dynamic system," *Computers & Chemical Engineering*, vol. 19, no. 2, pp. 187–195, 1995.
[7] D. G. Lainiotis, "Optimal adaptive estimation: Structure and parameters adaptation," *IEEE Transactions On Automatic Control*, vol. 16, no. 2, pp. 160–170, 1971.
[8] S. Särkkä and A. Nummenmaa, "Recursive noise adaptive Kalman filtering by variational Bayesian approximations," *IEEE Transactions on Automatic Control*, vol. 54, no. 3, pp. 596–600, 2009.
[9] R. L. Kashyap, "Maximum likelihood identification of stochastic linear systems," *IEEE Transactions on Automatic Control*, vol. 15, no. 1, pp. 25–34, 1970.
[10] R. H. Shumway and D. S. Stoffer, "Time series analysis and its applications," 2000.
[11] K. A. Myers and B. D. Tapley, "Adaptive sequential estimation with unknown noise statistics," *IEEE Transactions on Automatic Control*, vol. 21, no. 8, pp. 520–523, 1976.
[12] D. M. Wiberg, T. D. Powell, and D. Ljungquist, "An on-line parameter estimator for quick convergence and time-varying linear systems," *IEEE Transactions on Automatic Control*, vol. 45, no. 10, pp. 1854–1863, 2000.
[13] B. Carew and P. R. Bélanger, "Identification of optimum steady-state gain for systems with unknown noise covariances," *IEEE Transactions on Automatic Control*, vol. 18, no. 6, pp. 582–587, 1973.
[14] R. Bos, X. Bombois, and P. M. J. Van den Hof, "Designing a Kalman filter when no noise covariance information is available," in *Proceedings of the 16th IFAC World Congress*, (Prague, Czech Republic), July 2005.
[15] S. Verdú and H. V. Poor, "Minimax linear observers and regulators for stochastic systems with uncertain second-order statistics," *IEEE Transactions on Automatic Control*, vol. 29, no. 6, pp. 499–511, 1984.
[16] R. K. Mehra, "Approaches to adaptive filtering," *IEEE Transactions on Automatic Control*, vol. 17, no. 10, pp. 693–698, 1972.
[17] J. Duník, M. Šimandl, and O. Straka, "Methods for estimating state and measurement noise covariance matrices: Aspects and comparison," in *Proceedings of 15th IFAC Symposium on System Identification*, (Saint-Malo, France), July 2009.
[18] M. R. Rajamani and J. B. Rawlings, "Estimation of the disturbance structure from data using semidefinite programming and optimal weighting," *Automatica*, vol. 45, no. 1, pp. 142–148, 2009.
[19] F. V. Lima, M. R. Rajamani, T. A. Soderstrom, and J. B. Rawlings, "Covariance and state estimation of weakly observable systems: Application to polymerization processes," *IEEE Transactions on Control Systems Technology*, vol. 21, no. 4, pp. 1249–1257, 2013.
[20] J. Duník, O. Straka, and M. Šimandl, "Estimation of noise covariance matrices for linear systems with nonlinear measurements," in *Proceedings of the 17th IFAC Symposium on System Identification*, (Beijing, China), 2016.
[21] C. Wunsch, *Discrete Inverse and State Estimation Problems: With Geophysical Fluid Applications*. Cambridge, 2006.
[22] J. Duník, O. Straka, and O. Kost, "Measurement difference autocovariance method for noise covariance matrices estimation," *in Accepted for the 55th IEEE Conference on Decision and Control*, Dec. 2016.
[23] R. M. Rogers, *Applied Mathematics in Integrated Navigation Systems (2nd Edition)*. AIAA, 2003.
[24] P. Groves, *Principles of GNSS, Inertial, and Multisensor Integrated Navigation Systems, Second Edition:*. GNSS/GPS, Artech House, 2013.
[25] J. W. Brewer, "Kronecker products and matrix calculus in system theory," *IEEE Transactions on Circuits and Systems*, vol. 25, no. 9, pp. 772–781, 1978.
[26] J. Duník, O. Straka, and Šimandl, "On autocovariance least-squares method for noise covariance matrices estimation," *Accepted for IEEE Transactions on Automatic Control, DOI 10.1109/TAC.2016.2571899*, 2017.

IOP Conf. Series: Journal of Physics: Conf. Series **783** (2017) 012060 doi:10.1088/1742-6596/783/1/012060

Classification of breast cancer cytological specimen using convolutional neural network

Michał Żejmo[1], Marek Kowal[1], Józef Korbicz[1] and Roman Monczak[2]
[1]University of Zielona Góra, Institute of Control and Computation Engineering, 65-516 Szafrana 2, Zielona Góra, Poland
[2]Department of Pathomorphology, Regional Hospital in Zielona Góra, Zyty 26, 65-046 Zielona Góra, Poland

E-mail: {M.Zejmo,M.Kowal,J.Korbicz,R.Monczak}@issi.uz.zgora.pl

Abstract. The paper presents a deep learning approach for automatic classification of breast tumors based on fine needle cytology. The main aim of the system is to distinguish benign from malignant cases based on microscopic images. Experiment was carried out on cytological samples derived from 50 patients (25 benign cases + 25 malignant cases) diagnosed in Regional Hospital in Zielona Góra. To classify microscopic images, we used convolutional neural networks (CNN) of two types: GoogLeNet and AlexNet. Due to the very large size of images of cytological specimen (on average 200000 × 100000 pixels), they were divided into smaller patches of size 256 × 256 pixels. Breast cancer classification usually is based on morphometric features of nuclei. Therefore, training and validation patches were selected using Support Vector Machine (SVM) so that suitable amount of cell material was depicted. Neural classifiers were tuned using GPU accelerated implementation of gradient descent algorithm. Training error was defined as a cross-entropy classification loss. Classification accuracy was defined as the percentage ratio of successfully classified validation patches to the total number of validation patches. The best accuracy rate of 83% was obtained by GoogLeNet model. We observed that more misclassified patches belong to malignant cases.

1. Introduction

Breast cancer is the most common cancer among women worldwide. In 2012 nearly 1.7 million new cases were diagnosed (second most common cancer overall) and 522 000 cases of death (fifth cause of death from cancer overall). This represents about 12% of all new cancer cases and 25% of all cancers in women [1]. In Poland there were 17 000 cases and 5574 resulted in death (National Cancer Registry in Poland, 2016 [2]).

Breast cancer is mostly diagnosed by so called "triple-test" containing three medical examinations usually occurring in order: palpation, ultrasonography or mammography imaging and fine needle biopsy (FNB) [3]. Here, we concentrate on automatic analysis of results of FNB examination. In FNB method, cytological material for examination is collected directly from the tumor using specialised needle. Next, the material is fixed and stained. Finally, slide glass with cellular material is examined by the pathologist under the microscope. Unfortunately, detection of cancer cells is not an easy task. Novice pathologists must spend a lot of time observing thousands of specimens in order to gain experience and knowledge to become professional specialist. Moreover, analysis of entire slide is time-consuming process, even for experienced pathologists. A number of scientific groups conduct an intensive research in the area of Computer-Aided Cytology (CAC) [4, 5, 6, 7, 8, 9, 10, 11, 12]. Typically, CAC systems process the images in order to segment nuclei, then different features of nuclei are extracted and finally samples are classified [13, 14, 15]. Usually, different machine learning techniques are employed to fulfil these tasks. Unfortunately, microscopic images of tissue are challenging for automatic analysis. Cells have tendency to clump and generate nests, thus, it is hard to extract isolated nuclei. Furthermore, nuclei features used to discriminate malignant from benign cases are not standardised and are usually chosen based on subjective experiences. Therefore, most proposed approaches lack generality and satisfactory accuracy is only achieved for images acquired from specific source.

In recent years, there have been a lot of successes in complex object recognition and classification using CNN models. Their architecture is inspired biologically by the organization of human visual cortex. The ability to learn features invariant to translation, rotation and shifting is their great advantage. Recent studies in the field of digital pathology show the high efficiency of CNN in the diagnosis of breast cancer. Spanhol et al. applied AlexNet CNN model to classify breast cancer histopathological images [16]. They were learning CNN using training patches generated by different strategies. They obtained best classification accuracy 89.6% for images with 40x enlargement and patches of size 64x64 pixels selected in a random way from original image. Hatipoglu and Bilgin applied CNN to distinguish cellular and non-cellular structures in breast cancer histopathological images [17]. Neural network was used to classify small patches of size 9x9 and it reached 86.88% accuracy. In 2016, International Symposium on Biomedical Imaging (ISBI) held the grand challenge Camelyon16 to evaluate systems for the automated detection of metastatic breast cancer in histopathological images of sentinel lymph node biopsies. Wang et al. won this competition using GoogLeNet CNN model [18]. Their system reached 98.4% patch classification accuracy. Cruz-Roa et al. applied CNN to automatic detection of invasive ductal carcinoma tissue regions in histological images of breast cancer [19]. The method was evaluated over a data set consisting of 162 patients. They compared CNN with random forest classifier in terms of F-measure and balanced accuracy. CNN yielded better results (71.80%, 84.23%) than random forest (67.53%, 78.74%).

The approaches mentioned above deal with the automatic analysis of histopathological images of breast cancer. We focus on the classification of cytological images of breast cancer. Cytological biopsies are much less invasive than histological, thus they are less traumatic for patients. Unfortunately, they are substantially harder to analyse because cytological smears have more chaotic and complex structure.

Our approach is based on the classification of small patches of cytological image in order to detect malignant cells. At the beginning, pathologist has marked the presence of benign or malignant cells in some areas of virtual slides. Next, annotated fragments were divided into small overlapping patches with a size of 256×256 pixels. Our studies show that the size of patches have influence on classification results. Overall, our experience is that the larger size of patches, the smaller classification error. However, the size of the patches is limited by the size of RAM accessible on GPU unit. In our case the maximum size of patches was limited to 256×256 pixels. Patches with the larger size had to be downsized by changing resolution, what deteriorate classification results. Patches with smaller size contain less information on the distribution of nuclei in the sample and therefore results of classification was also not satisfactory.

We formed training and validation data sets in such a way that all patches coming from a particular patient can belong either to training or to validation data set. In addition, each patch has been verified using SVM classifier in terms of how much cellular material represents. If the patch did not contain cellular material above the threshold, it was not included in neither the training nor the validation collection. To implement and tune CNN models, we used Caffe framework and NVIDIA Digits software. Networks were learned using three different gradient descent based optimisation algorithms. We explored the accuracy of GoogLeNet and AlexNet architecture for three different collections of patches. These collections of patches differ in cellular coverage degree.

The remainder of this paper is organised as follows. In Section 2, materials and methods are described. Section 3 gives the description of experiments and results. Concluding remarks are given in Section 4.

2. Material and Methods

2.1. Study data set

Cytological images, which we were using in this study came from patients of Regional Hospital in Zielona Góra, Poland. Cellular material was acquired from affected tissue using 0.5-mm-diameter needle under the control of an ultrasonograph. Next, the material was fixed with Cellfix (Shadon) fixative spray and dyed with hematoxylin and eosin (h+e). The time between preparation of smears and their preservation in fixative spray never exceeded three seconds. In this way, 50 specimens were prepared for 50 patients with breast cancer. Half of the cases are malignant, half are benign. All of the cancers were histologically confirmed, and all of the patients who carried benign disease were either biopsied or followed for a year. Digitization of glass slides was done using Olympus VS120 Virtual Microscopy System. The system consists of a 2/3" CCD camera and 40x lens with giving together 0.172 μm resolution. The average size of slides from the system is approximately 200000 $\times$ 100000 pixels. All the scans were prepared using Extended Focal Imaging (EFI) system. EFI allows for extended focal depths impossible to obtain using optics alone. Scans with EFI are performed by scanning a preparation several times with the different positions of focus plane. Final frame is formed by selecting only sharp focus regions from each frame. Finally, scanned slides are saved on a hard disk using Virtual Slide Image (VSI) format, dedicated to store very large images. Pathologists can browse virtual slides on a computer screen using special software. Pathologist from Hospital in Zielona Góra identified regions of malignant and benign changes in order to prepare images for training CNN. For each virtual slide 11 regions of interests (ROI) representing malignant or benign cells were selected. Next, each ROI was converted to 8 bit/channel RGB TIFF image of size 1583 $\times$ 828 pixels. Compression to TIFF was done by lossless LZW algorithm. Fig. 1 is showing sample results of selection procedure. Example of ROI selected by pathologist from virtual slide is shown in Fig. 2. On the whole, pathologist has selected 550 ROIs (275 malignant + 275 benign).

2.2. Training and validation data sets

The collection of 50 virtual slides was randomly split into training and validation set. Training set consist of 30 virtual slides (15 benign and 15 malignant) and validation set consist of 20 virtual slides (10 benign and 10 malignant). Therefore, we obtained 330 ROI images for training and 220 ROIs for validation. Next, we generated for every ROI a collection of overlapping patches of size 256 $\times$ 256 pixels. Sliding window was moving horizontally from left to the right in order to gather a single patch at every 32 pixels. Procedure was repeated for subsequent lines with the vertical step equal to 32 pixels. From each image 697 patches were extracted. From a diagnostic point of view only the patches which illustrate cellular material are useful for training and validation process. That's why we decided to remove patches that has a low cellular coverage ratio. Such coefficient was determined for every patch using SVM classifier. It was responsible for labelling pixels as cell (nuclei and cytoplasm) or other (background or blood) based on their RGB colour. To train SVM, several manually labelled images were used. Finally, we created three collections of patches with cellular coverage ratio equal or greater than 50%, 75% and 90% respectively. The sizes of training and validation collections for different cellular coverage ratios are presented in table 1. Figure 3 shows sample patches with cellular coverage ratio equal or greater than 90%.

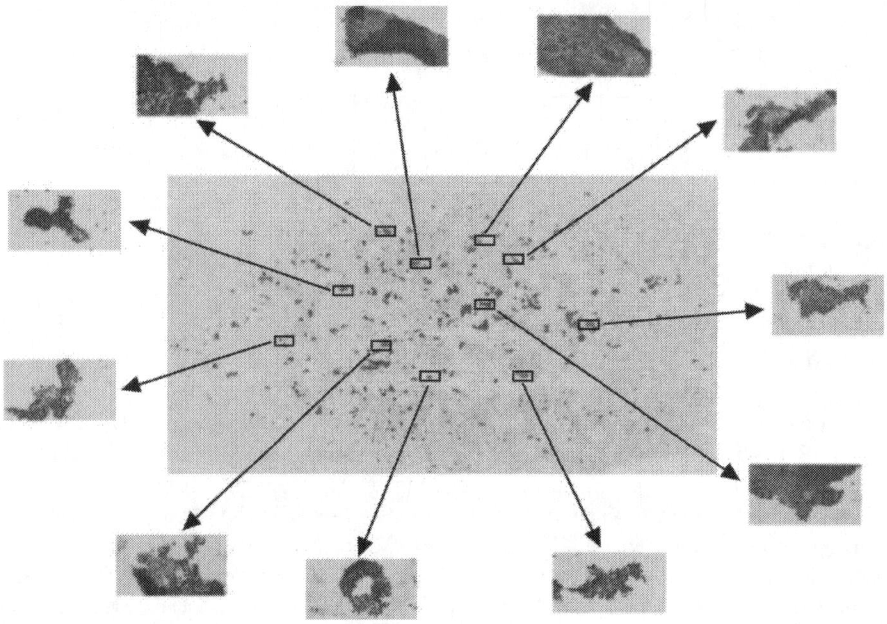

Figure 1. Example of the virtual slide with selected ROIs.

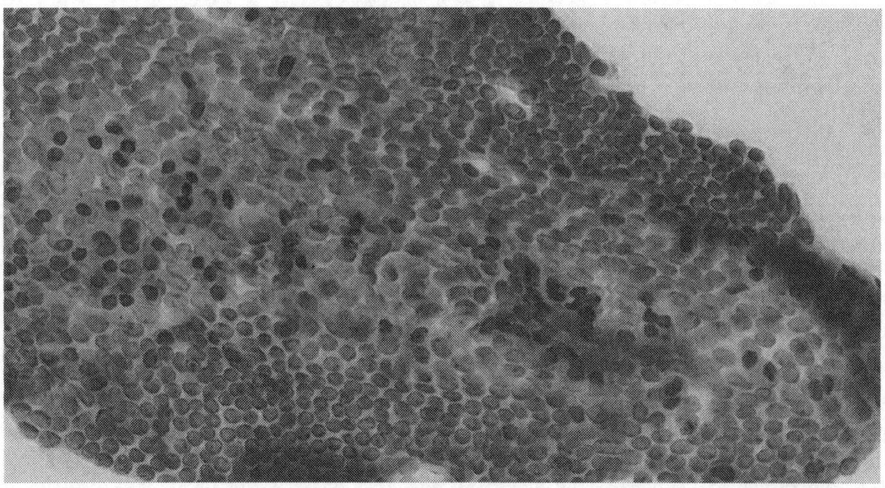

Figure 2. Example of selected ROI.

2.3. Convolutional Neural Networks

Nowadays, CNN become very popular technique of machine learning, which is often used to image recognition [20]. Idea of CNN was inspired biologically by processing the image in the visual cortex of animals. One of the most famous example of CNN application was LeNet network that was successfully applied for handwritten digits recognition [21]. CNN is usually built of many convolutional layers connected to each other by pooling layers. At the top of this structure, reside fully connected layers similar to feed-forward layers from multilayer perceptron (MLP). Figure 4 shows the sample structure of CNN model.

Table 1. Collections of patches

	Training set [amount]		Validation set [amount]	
Cellular coverage	Benign	Malignant	Benign	Malignant
50%	62229	52225	29822	28981
75%	47053	23974	17654	14204
90%	34189	10107	10214	6644

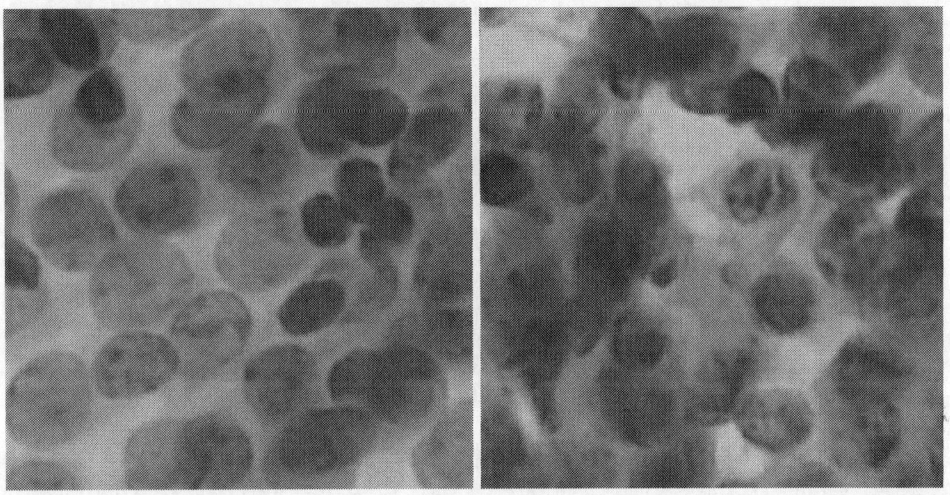

Figure 3. Examples of patches used for training and validation of CNN (left: benign cells and right: malignant cells).

Convolutional layer is composed of learnable filters and is a core part of CNN. It is responsible for processing input image in order to extract features describing input image. Parameters of filters are weights of the network tuned during the training. With such approach, network is able to learn features from examples during training procedure. Kernels of filters are spatially small and rectangular, but extends through the full depth of the input image due to operation of convolution. For example, filter in a first layer of a CNN might have a kernel of size 5x5x3 (5 pixels width and height, and depth 3 because the RGB colour channels). Each kernel is sliding across the width and height of the input and compute dot products between the kernel and the fragment of the image cutted by the window of the same size as kernel and centred at the actual position of the kernel. CNN is usually composed of many convolutional layers. Shallow convolutional layers extract low level features and deeper layers are able to extract high level features.

Pooling layer is used to reduce progressively the spatial size of the input and to reduce the amount of tunable parameters. In this layer small rectangular patches of feature maps from convolution layer are downsampled to single value. Downsampling is usually done by max or averaging operator. The pooling layer provides a form of translation invariance. Pooling layers are inserted periodically between the convolutional layers.

Fully-connected layer is at the top of CNN. The aim of this layer is capture complex relationship between high-level features. This layer mixes output of pooling layer to feature vector. Thus, fully connected layers are one-dimensional, so there can be no convolutional or pooling layer after.

In addition CNN can have ReLU (Rectified Linear Units) layer and LRN (Local Response Normalisation) layer. ReLU layer consist of neurons with the activation function $f(x) = \max(0, x)$. Such neurons increase the nonlinear properties of the decision function without affecting the receptive fields of the convolution layer. The LRN layer performs a kind of local normalisation over defined input area. Each input value is divided by $(1 + \frac{\alpha}{n}\sum_i x_i^2)^\beta$, where n is the size of each local region, α and β are arbitrarily chosen parameters and the sum is taken over the region centred at the point, which is normalised.

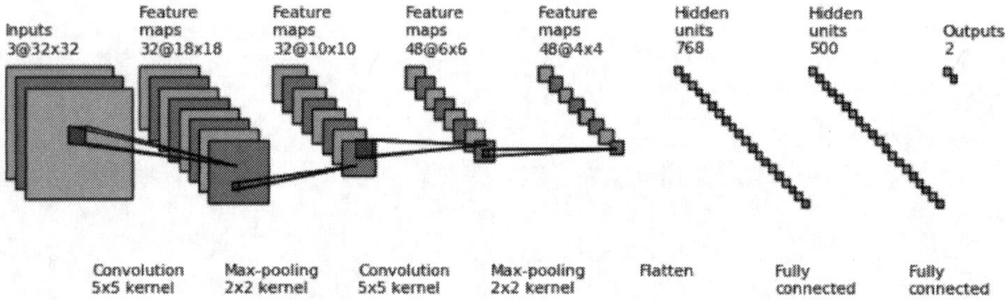

Figure 4. Architecture of sample CNN model.

A lot of different CNN architectures have been already tested and described in the scientific literature [22, 23, 24, 25, 26, 27, 28, 20, 29]. They vary in the layer configuration and depth of structure. In our studies we used two well known architectures: AlexNet [20] and GoogLeNet [29]. These structures have proven their usefulness in many complex image recognition tasks.

AlexNet belongs to the class of deep structures. It is made up of 5 convolutional layers with ReLU activation functions, 3 max-pooling layers and 3 fully connected layers with ReLU neurons. Moreover, in the first two convolutional layers, outputs are normalized using LRN technique. In fully connected part of the network a dropout technique was used to reduce the validation error [30]. This technique assumes that each hidden neuron can set its output value to 0 for one epoch with probability 0.5. Therefore, the neurons which are dropped out do not contribute to the output and are not tuned during backpropagation step. In this way, different structures of the network are sampled during training procedure. At validation step, all the neurons participate in the output, however each individual output of the neuron is multiplied by 0.5.

GoogLeNet belongs to the class of deep CNN architecture which originally consists of 22 layers [29]. To make such architecture computationally efficient, its founders organized the structure of the network in the form of Inception modules [29]. The main concept of the Inception structure is to consider how an optimal local sparse structure of CNN can be approximated by easily available dense components. Generally GoogLeNet is a network consisting of Inception modules stacked upon each other, with occasional max-pooling layers. All layers use ReLU activation function.

2.4. GPU computation

The process of learning CNN is extremely computationally expensive, due to huge amount of calculations on arrays of data and weights. Standard CPU is optimised for single-threaded performance, therefore computing efficiency of CPUs is completely insufficient for training CNN with complex structure and huge amount of training data. Best solution to meet this problem is to use GPU. To train and simulate our CNN models we used deep learning framework Caffe[1] [31]. For data, results and models management we decided to use NVIDIA DIGITS[2] system. Mentioned software use CUDA [3] (Compute Unified Device Architecture) and CuDNN (CUDA Deep Neural Network) library for massively parallel computations to shorten drastically the time needed to learn our CNN models. All experiments were done using GeForce GTX TITAN X with 12GB of RAM.

3. Results

Two CNN architectures GoogLeNet and AlexNet were compared in terms of classification accuracy of breast cancer cytological images. Each architecture was trained several times for different training data and different training algorithm. Three different training collections of images were prepared. Subsequent training collections contain images with cellular coverage ratio equal 50%, 75% and 90% respectively. To train CNN, we have used 3 different gradient descent algorithms: stochastic gradient descent (SGD), adaptive gradient (ADA) and Nesterov's accelerated gradient (NAG). Both architectures were learned and validated with each data set and each training algorithm. In total, each CNN model was trained 9 times. Each training took 10 epochs. Due to GPU memory limitations, we used mini-batch version of backpropagation algorithm. Thus, weights of the network were updated after the presentation of each mini-batch of training data. For AlexNet architecture, we choose mini-batch of size 450 samples and for GoogLeNet mini-batch of size 100 samples. After each training epoch, network was validated to control the overfitting problem. Learning rate was initialized with a value $lr = 0.01$, and was dropping by a constant factor 10 when objective function begins to reach an evident plateau. Objective function is defined as a cross-entropy classification loss:

$$L = \frac{-1}{N} \sum_{i=1}^{N} \log \left(\frac{e^{x_{i,l}}}{\sum_{j=1}^{K} e^{x_{i,j}}} \right) \tag{1}$$

where N is the size of the mini-batch, K is the number of network outputs, $x_{i,j}$ is the j-th output of the network for i-th sample and l is the label of the correct ground truth class for i-th sample. Classification accuracy for validation collection was measured as the percentage ratio of successfully classified patches to the total number of patches.

Table 2 shows summarised results of conducted research. For collection of patches with 50% cellular coverage ratio, both CNN architectures gave similar accuracy close to 80%, regardless of the training method. For 75% and 90% cellular coverage ratios GoogLeNet achieve better results. The worst classification accuracy was achieved using ADA training algorithm. Best accuracy 83% was obtained by GoogLeNet model for patch collection with 90% cellular coverage ratio using SGD training method. Figure 5 shows the process of training of the best GoogLeNet model and best AlexNet model. From these figures, we can observe that learning process converged to a local minimum. However, we can also observe that relatively quickly appears the effect of overfitting. This problem can be caused by low number of training samples. We plan to extend training dataset in future work.

Table 3 presents detailed classification results for the best GoogLeNet model, which were obtained for validation data set. For each case, the percentage of properly classified patches was determined. Most of the cases were correctly diagnosed. But, for benign cases, patient no. 10 was apparently classified incorrectly as malignant. For malignant cases, 2 patients no. 19 and 20 were incorrectly diagnosed as benign. Diagnosis for the patient no. 19 is uncertain, on the border between benign and malignant class. Incorrectly classified patches originating

[1] `http://caffe.berkeleyvision.org/`
[2] `https://developer.nvidia.com/digits`
[3] `http://www.nvidia.com/CUDA`

Table 2. Patch classification accuracy for different CNN models, trained with different learning algorithms and using different training collections

Minimum cellular coverage ratio								
50%			75%			90%		
CNN model	Grad.	Acc.	CNN model	Grad.	Acc.	CNN model	Grad.	Acc.
AlexNet	SGD	80%	AlexNet	SGD	74%	AlexNet	SGD	80%
	ADA	78%		ADA	55%		ADA	61%
	NAG	77%		NAG	74%		NAG	74%
GoogLeNet	SGD	80%	GoogLeNet	SGD	82%	GoogLeNet	SGD	**83%**
	ADA	80%		ADA	75%		ADA	73%
	NAG	76%		NAG	82%		NAG	79%

Table 3. Detailed classification results for the best GoogLeNet model

Ground truth diagnosis	Patient no.	Total number of patches	Percentage of well classified patches
Benign	1	1675	100%
	2	1180	100%
	3	232	100%
	4	142	100%
	5	731	99.2%
	6	336	91.7%
	7	2467	88.3%
	8	449	82.4%
	9	2236	61.5%
	10	766	0%
Malignant	11	152	100%
	12	1453	94.8%
	13	2557	91.5%
	14	88	90.9%
	15	543	74.4%
	16	980	69.9%
	17	31	67.7%
	18	256	66.8%
	19	474	49.2%
	20	110	37.3%

from wrongly classified patients were reviewed by the pathologist. He classified most of them as "difficult cases". It was noted that some misclassified malignant patches are difficult to diagnose due to the lack of the information originating from the rest of the specimen. Sometimes the context in which patch is located is just as important as the information included in the patch. Finally, we must also take into account the fact that even the pathologists do not always agree in their diagnoses.

4. Conclusions

Content of cytological images is highly complex and stochastic, so their analysis is difficult in an automated way. To tackle this issue, we applied a deep learning approach to classify cytological images coming form fine needle biopsy of breast cancer. Proposed approach is based

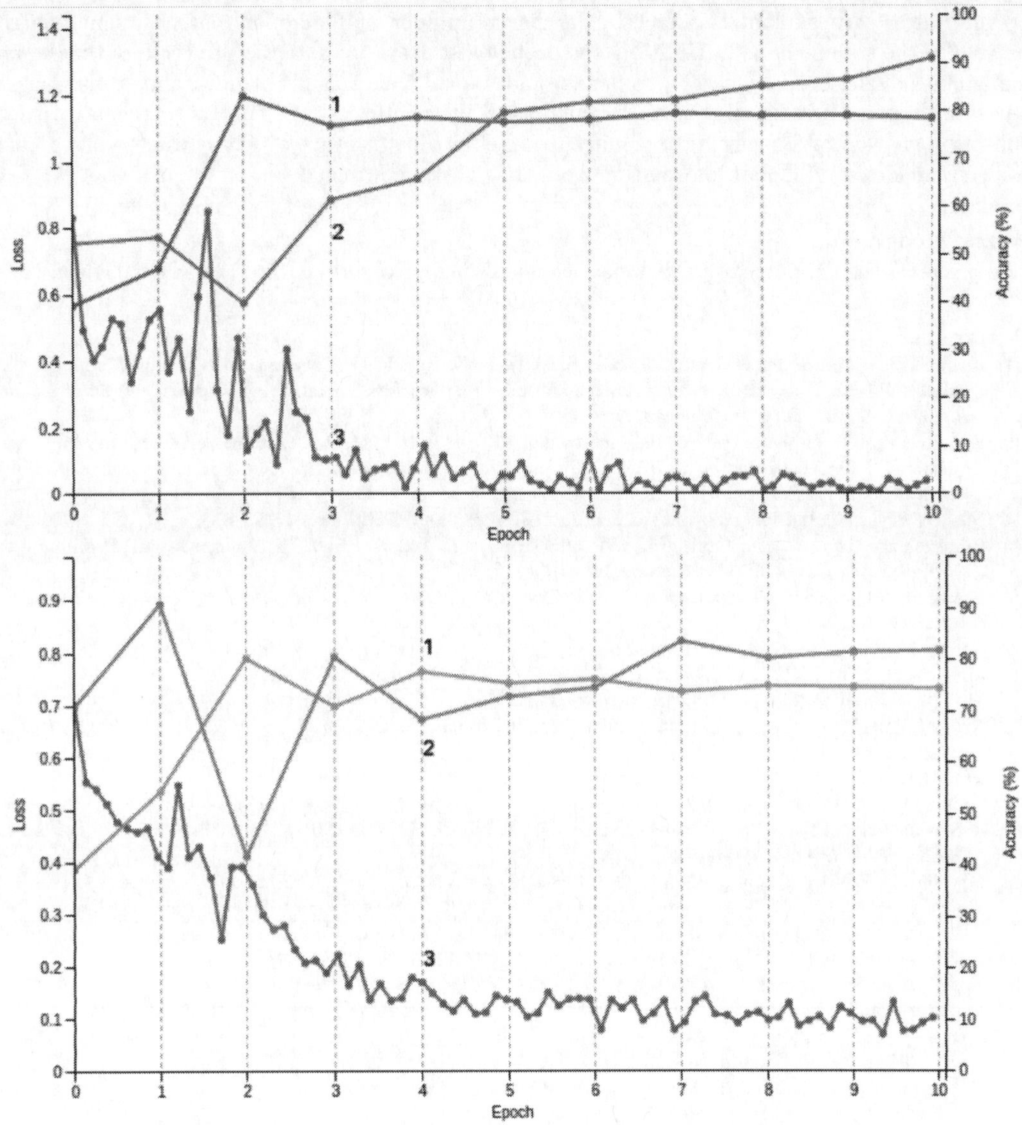

Figure 5. Top figure shows learning process of GoogLeNet CNN and bottom shows learning process of AlexNet CNN. Line labeled as 1 denote classification accuracy for validation data, lines with labels 2 and 3 denote loss function for validation and training data, respectively.

on processing small patches selected from cytological images. Patches are processed by CNN. It gives as his output information whether the tumor is benign or malignant. The best patch classification accuracy reached 83%. Such result was obtained by GoogLeNet architecture. Other authors have already reported better results for breast cancer classification using CNN but their studies are related to histopathological images. We treat obtained results as starting point to further research aimed at improving the classification accuracy.

Analysis of the results indicates that CNN does less well with malignant cases than with benign cases. We explain this fact by far fewer number of malignant patches than benign patches in the training set. In addition, the pathologist drew our attention to the fact that some malignant patches are very similar to benign and to deal with this problem we should use larger patches. Further investigations will be conducted in two directions: firstly, to prepare richer collection of training images and secondly to develop the structure of CNN which will be able to combine local and global information from the cytological image.

Acknowledgments
The research was supported by National Science Centre, Poland (2015/17/B/ST7/03704).

References

[1] Ferlay J, Soerjomataram I, Dikshit R, Eser S, Mathers C, Rebelo M, Parkin D M, Forman D and Bray F 2013 Globocan 2012 v1.0, cancer incidence and mortality worldwide: Iarc cancerbase no. 11. accessed on: 13/09/2016 URL `http://globocan.iarc.fr`
[2] Wojciechowska U and Didkowska J 2013 Accessed on: 13/09/2016 URL `http://onkologia.org.pl/raporty/`
[3] Underwood J 1987 *Introduction to Biopsy Interpretation and Surgical Pathology* (London: Springer-Verlag) ISBN 9780387174952
[4] Jeleń L, Fevens T and Krzyżak A 2010 *Int. J. Appl. Math and Comp. Sci.* **18** 75–83
[5] Filipczuk P, Kowal M and Obuchowicz A 2013 *Proc. 35th Int. Conf. IEEE Engineering in Medicine and Biology Society, EMBC 2013* (Osaka, Japan)
[6] Kowal M, Filipczuk P, Obuchowicz A, Korbicz J and Monczak R 2013 *Computers in Biology and Medicine* **43** 1563–1572
[7] George Y, Elbagoury B, Zayed H and Roushdy M 2012 *Int. J. Computer Science Issues* **9** 247–256
[8] Malek J, Sebri A, Mabrouk S, Torki K and Tourki R 2009 *J. Signal Processing Systems* **55** 49–66
[9] Mu T and Nandi A 2007 *J. Franklin Institute* **344** 285–311
[10] Niwas I, Palanisamy P and Sujathan K 2010 *Proc. IEEE Symp. Industrial Electronics & Applications, ISIEA 2010* p 686–690
[11] Plissiti M and Nikou C 2012 *IEEE Trans. Image Processing* **21** 4568–4580
[12] Soliman O and Hassanien A 2012 *Rough Sets and Current Trends in Computing* (*Lecture Notes in Computer Science* vol 7413) ed Yao J, Yang Y, Słowiński R, Greco S, Li H, Mitra S and Polkowski L (Springer, Berlin Heidelberg) pp 106–115
[13] Buyssens P, Elmoataz A and Lézoray O 2013 *Computer Vision - ACCV 2012: 11th Asian Conference on Computer Vision, Daejeon, Korea, November 5-9, 2012, Revised Selected Papers, Part II* ed Lee K M, Matsushita Y, Rehg J M and Hu Z (Berlin, Heidelberg: Springer Berlin Heidelberg) pp 342–352 ISBN 978-3-642-37444-9 URL `http://dx.doi.org/10.1007/978-3-642-37444-9_27`
[14] Kowal M and Filipczuk P 2014 *Int. J. Appl. Math and Comp. Sci.* **24** 19–31
[15] Mazurek P and Oszustowska-Mazurek D 2014 *Int. J. Appl. Math and Comp. Sci.* **24** 49–63
[16] Spanhol F A, Oliveira L, Petitjean C and Heutte L 2016 *Proc. Int. Joint Conf. on Neural Networks (IJCNN 2016)* (Vancouver, Canada) pp 123–128
[17] Hatipoglu N and Bilgin G 2014 *Proc. 4th Int. Conf. on Image Processing Theory, Tools and Applications (IPTA)* (Paris, France) pp 1–6
[18] Wang D, Khosla A, Gargeya R, Irshad H and Beck A H 2016 *CoRR* **abs/1606.05718**
[19] Cruz-Roa A, Basavanhally A, Gonzalez F, Gilmore H, Feldman M, Ganesan S, Shih N, Tomaszewski J and Madabhushi A 2014 *Proc. SPIE Medical Imaging* vol 9034 (San Diego, USA) pp 904103–904103–15
[20] Krizhevsky A, Sutskever I and Hinton G E 2012 *Proc. 26th Conf. on Neural Information Processing Systems (NIPS)* (Lake Tahoe, USA) pp 1097–1105
[21] LeCun Y, Bottou L, Bengio Y and Haffner P 1998 *Proceedings of the IEEE* **86** 2278–2324
[22] LeCun Y, Huang F J and Bottou L *Proc. IEEE Conf. Computer Vision and Pattern Recognition, CVPR 2004* vol 2 (IEEE) pp II–97
[23] Jarrett K, Kavukcuoglu K, Lecun Y *et al.* 2009 *Proc. 12th IEEE Int. Conf. on Computer Vision, ICCV* (Kyoto, Japan) pp 2146–2153
[24] Krizhevsky A and Hinton G 2010 Convolutional deep belief networks on cifar-10 URL `https://www.cs.toronto.edu/~kriz/`
[25] Lee H, Grosse R, Ranganath R and Ng A Y 2009 *Proc. 26th Int. Conf Machine Learning* (Montreal, Canada) pp 609–616
[26] LeCun Y, Boser B, Denker J S, Henderson D, Howard R E, Hubbard W and Jackel L 1990 ed Touretzky

D S (San Francisco, CA, USA: Morgan Kaufmann Publishers Inc.) chap Handwritten Digit Recognition with a Back-propagation Network, pp 396–404 ISBN 1-55860-100-7

[27] Pinto N, Doukhan D, DiCarlo J and Cox D 2009 *PLoS Comput Biol* **5** e1000579

[28] Turaga S C, Murray J F, Jain V, Roth F, Briggman M H K, Denk W and Seung H S 2010 *Neural computation* **22** 511–538

[29] Szegedy C, Liu W, Jia Y, Sermanet P, Reed S, Anguelov D, Erhan D, Vanhoucke V and Rabinovich A 2015 *Proc. IEEE Conf. Computer Vision and Pattern Recognition, CVPR* (Boston, USA) pp 1–9

[30] Srivastava N, Hinton G E, Krizhevsky A, Sutskever I and Salakhutdinov R 2015 *J. Machine Learning Research* **15** 1929–1958

[31] Jia Y, Shelhamer E, Donahue J, Karayev S, Long J, Girshick R, Guadarrama S and Darrell T 2014 *Proc. 22nd ACM Int. Conf. on Multimedia, MM '14* (Orlando, USA) pp 675–678 ISBN 978-1-4503-3063-3

IOP Conf. Series: Journal of Physics: Conf. Series **783** (2017) 012061 doi:10.1088/1742-6596/783/1/012061

Thickness measuring of electroconductive pipe walls using the dual-frequency eddy-current method

Evgeny Yakimov[1], Olga Galtseva[1] and Daniil Ustyugov[1]

[1] Tomsk Polytechnic University, Institute of Non-Destructive Testing, 634050, Tomsk, Russia

shishkovka@mail.ru

Abstract. The paper describes a dual-frequency method for reducing the impact of changes in the gap size between the eddy-current transducer and the pipe, as well as the pipe electrical conductivity on the eddy-current thickness gauge readings. A block-diagram of the dual-frequency eddy-current thickness gauge is proposed for light-alloy drill pipes. The amplitude and signal phase dependencies on the wall thickness in the range from 6 to 17 mm and the gap in the range from 0 to 13.5 mm were studied, the results are presented. The digital signal processing algorithms based on the piecewise-linear approximation of low-frequency and high-frequency signal phase dependencies on the wall thickness are proposed. It is shown that the proposed correction algorithms can reduce the error caused by variations of electrical conductivity and the gap between the transducer and the pipe.

1. Introduction. Measuring the thickness of pipe walls

Light-alloy drill pipes of TB and TBP types made of aluminum alloy D16 are manufactured in accordance with the Russian standard GOST 23786-79 [1]. The pipes are used to perform the workover and the structure, exploration and production drilling. The pipe walls are worn during in-service use, and therefore the wall thickness of the main pipe section must be controlled according to the Russian standard. With due regard to the assortment, the nominal wall thickness of the main pipe section can vary from 7.5 to 13 mm. The limit thickness deviation shall not exceed ± (0.4–0.5) mm for pipes of high accuracy and ± (0.7–1.7) mm for pipes of normal accuracy. A non-destructive testing method is not considered. The measuring instruments must have the accuracy of ± 0.1 mm.

Ultrasonic devices provide high accuracy of measurement, and therefore are used to measure the thickness of metal objects. However, a good acoustic contact (through the liquid layer) is required, that decreases the speed of testing.

The advantages of the eddy-current testing method are high-performance control and sensitivity, and the possibility of non-contact measurement. These advantages enable the method automation and provide a continuous control along the entire length of the pipe. However, signals of the eddy-current transducer considerably depend on the material electrical conductivity and the gap between the transducer and the pipe.

In the previous paper [2], it was shown that a dual-frequency eddy-current method can reduce the dependence of thickness gauge readings on the gap size. Consider the fact that the method is capable of reducing the impact of changes in the pipe electrical conductivity on the thickness measurement results. With this, the block diagram of the gauge (Figure 1) comprising the surface eddy-current transducer (ECT), generator (G), amplitude and phase detectors (APD), data acquisition board with ADC, and personal computer (PC) can be used.

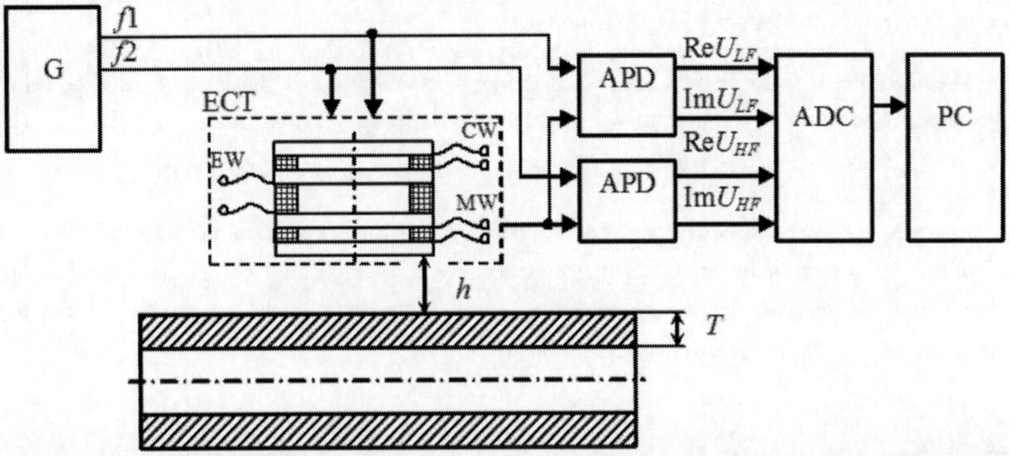

Figure 1. Block diagram of the thickness gauge based on the dual-frequency eddy-current method.

The eddy-current transducer comprises the excitation winding (EW), measurement winding (MW) and compensation winding (CW). The signal is fed from the generator to the transducer EW, which is the sum of two harmonic signals. The first frequency *f1* is principal and is intended to measure the wall thickness *T*. The frequency must be up to several hundred hertz and therefore is referred to as low frequency (LF). The second frequency *f2* is selected so that the signals of this frequency slightly depend on the pipe wall thickness and at the same time change with variations of the pipe electrical conductivity and the gap *h* between the pipe and the transducer. These properties appear in the frequency of several kilohertz, so the frequency is called high frequency (HF).

During the measurement, signals from the eddy-current transducer arrive at the amplitude and phase detectors, which output voltages are proportional to the amplitudes of imaginary and real components of the low-frequency signal *f1* and high-frequency signal *f2*: $\mathrm{Re}U_{LF}$, $\mathrm{Im}U_{LF}$, $\mathrm{Re}U_{HF}$, $\mathrm{Im}U_{HF}$.

Then, signals arrive at the data acquisition board with ADC, which performs analog-digital conversion of signals and transfer them to the PC.

The PC records and processes the data.

The values of the ECT signals are used to calculate the amplitude A_{HF} and phase ϕ_{HF} of the high-frequency signal (*f2*), and the phase ϕ_{LF} of the low-frequency signal (*f1*):

$$A_{HF} = \sqrt{(\mathrm{Re}U_{HF})^2 + (\mathrm{Im}U_{HF})^2}, \quad (1)$$

$$\phi_{HF} = arctg(\mathrm{Im}U_{HF} / \mathrm{Re}U_{HF}), \quad (2)$$

$$\phi_{LF} = arctg(\mathrm{Im}U_{LF} / \mathrm{Re}U_{LF}). \quad (3)$$

The value of the high-frequency signal amplitude A_{HF} is taken to calculate the gap h between the ECT and the pipe surface:

$$h = Kh \cdot \left[cg_0 + cg_1 \cdot e^{cg_2 \cdot amp} + cg_3 \cdot e^{cg_4 \cdot amp} \right], \quad (4)$$

where cg_k is constant coefficients, $amp = A_{HF} / A_{HF0}$ is the relative amplitude of the high-frequency signal, A_{HF0} is the amplitude of the high-frequency signal for zero gap, and Kh is a correction factor used for non-zero gap calibration.

The approximation error in accordance with expression (4) does not exceed 0.08 mm [2].

The dependence of the wall thickness T on the low-frequency signal phase ϕ_{LF} for a fixed value of the gap h (Figure 2) can be approximated by the piecewise-linear function

$$T = K_{LF}(h, jtx) \cdot \left[\phi_{LF} + \Delta\phi_{LF(tx)} + \Delta\phi_{LF} + Kc \cdot \phi_{LF}\right] + B_{LF}(h, jtx), \quad (5)$$

where $K_{LF}(h, jtx)$ and $B_{LF}(h, jtx)$ are the coefficients, which depend on the gap h and the number jtx of the site, where the thickness depends on the phase ϕ_{LF}; $\Delta\phi_{LF}$ and Kc are the coefficients used to calibrate the readings by the two known values of the pipe thickness; $\Delta\phi_{LF(tx)}$ is the phase correction for the change in the electrical conductivity.

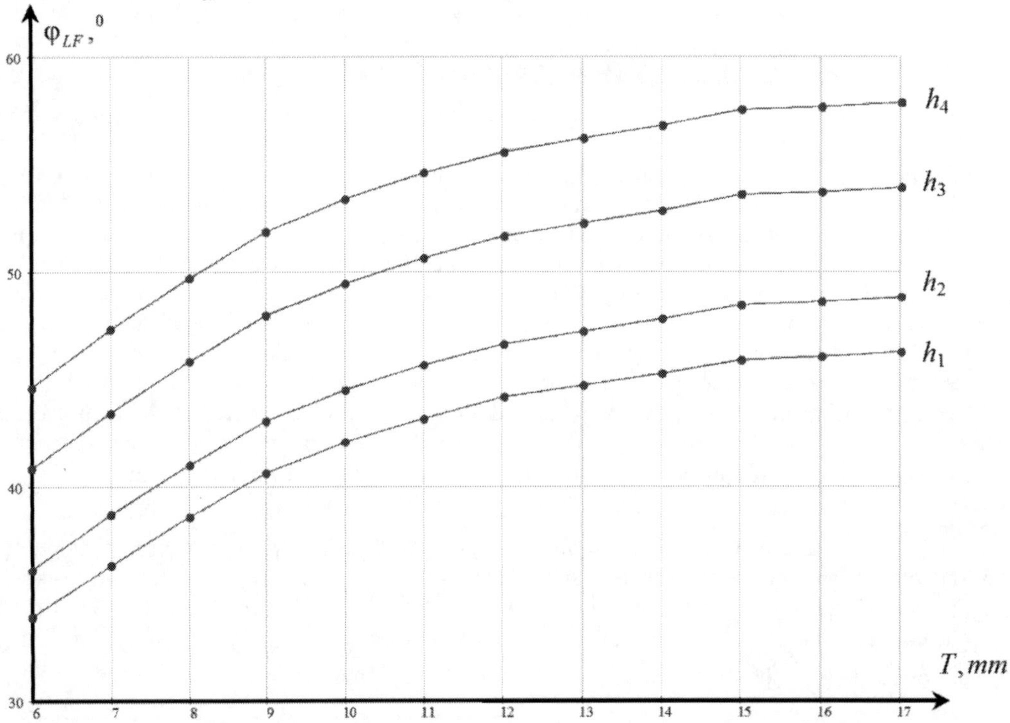

Figure 2. Dependence of the low-frequency signal phase ϕ_{LF} on the pipe wall thickness T for the fixed values of the gap h.

The correction is calculated based on the assumption that the conductivity is temperature dependent [3]. However, the electrical conductivity can vary due to the different chemical composition of the tube material.

The phase dependence on the temperature can be approximated by the linear function

$$\Delta\phi_{LF(tx)} = -Kt_{LF} \cdot (tx - t0), \quad (6)$$

where Kt_{LF} is the coefficient of proportionality, tx is the pipe temperature, and $t0$ is the temperature at which the thickness gauge was calibrated.

The pipe temperature is calculated from the difference of the measured high-frequency signal phase $\phi_{HF(tx)}$ and the phase $\phi_{HF(t0)}$ measured at temperature $t0$,

$$tx = t0 - Kt_{HF} \cdot \left[\phi_{HF(tx)} - \phi_{HF(t0)}\right], \tag{7}$$

where Kt_{HF} is the coefficient of proportionality.

That is, the preliminary evaluation of the thickness is made with no regard to the conductivity correction, and then the expected value of the high-frequency signal phase $\phi_{HF(t0)}$ and the phase difference $\left[\phi_{HF(tx)} - \phi_{HF(t0)}\right]$ are calculated. The change in the high-frequency signal phase against the gap h and the thickness of the tube T are taken into account (Figure 3). A piecewise-linear model of dependence is used

$$\phi_{HF(t0)} = T \cdot \left[Kk_{HF(jhx)} \cdot h + Bk_{HF(jhx)}\right] + \left[Kb_{HF(jhx)} \cdot h + Bb_{HF(jhx)}\right], \tag{8}$$

where $Kk_{HF(jhx)}$, $Bk_{HF(jhx)}$, $Kb_{HF(jhx)}$ and $Bb_{HF(jhx)}$ are the coefficients of piecewise-linear approximation of the dependence of the high-frequency signal phase on the gap.

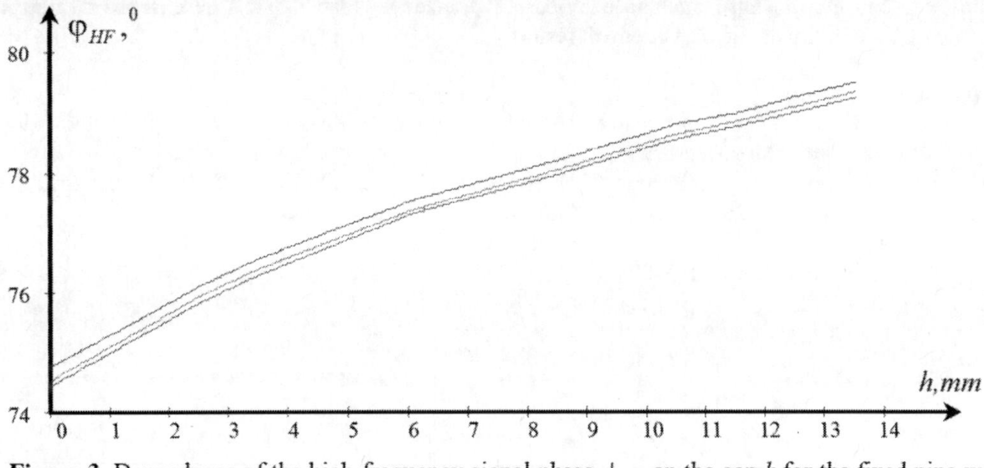

Figure 3. Dependence of the high-frequency signal phase ϕ_{HF} on the gap h for the fixed pipe wall thickness T.

2. Thickness gauge calibration

To calibrate the readings when measuring a zero gap, the ECT is placed directly on the pipe and the high-frequency signal amplitude A_{HF0} is saved in the memory.

For the non-zero gap calibration, the ECT is placed on the plate of known thickness and the correction factor Kh used in equation (4) is calculated

$$Kh = h2 / h. \tag{9}$$

The thickness calibration is performed in two steps.

At the first step, the ECT is installed on the sample pipe with a known thickness value $T1$, and the ECT signals are saved in the memory. Next, the ECT is installed on the sample pipe with a known value $T2$, and the ECT signals are saved in the memory. The thickness of sample 1 should be less than the thickness of sample 2.

The current values of the correction factors $\Delta\phi_{LF(i)}$ and $Kc_{(i)}$ used in equation (5) to correct the gauge readings are calculated by the system of equations (10)

$$\left.\begin{aligned} Tx1 &= K_{LF(1)}\left(h, jtx\right)\cdot\left[\phi_{LF(1)} + \Delta\phi_{LF(i-1)} + Kc_{(i-1)}\cdot\phi_{LF(1)}\right] + B_{LF(1)}\left(h, jtx\right) \\ T1 &= K_{LF(1)}\left(h, jtx\right)\cdot\left[\phi_{LF(1)} + \Delta\phi_{LF(i)} + Kc_{(i)}\cdot\phi_{LF(1)}\right] + B_{LF(1)}\left(h, jtx\right) \\ Tx2 &= K_{LF(2)}\left(h, jtx\right)\cdot\left[\phi_{LF(2)} + \Delta\phi_{LF(i-1)} + Kc_{(i-1)}\cdot\phi_{LF(2)}\right] + B_{LF(2)}\left(h, jtx\right) \\ T2 &= K_{LF(2)}\left(h, jtx\right)\cdot\left[\phi_{LF(2)} + \Delta\phi_{LF(i)} + Kc_{(i)}\cdot\phi_{LF(2)}\right] + B_{LF(2)}\left(h, jtx\right) \end{aligned}\right\}, \quad (10)$$

where $\Delta\phi_{LF(i-1)}$ and $Kc_{(i-1)}$ are the coefficients used in the previous calibration; $Tx1$ and $Tx2$ are the thickness gauge readings with the previous correction factor values; $T1$ and $T2$ are the values of the thickness of sample pipes.

3. Conclusion

The experiments showed that the dual-frequency eddy-current method used to measure the wall thickness of the electrically conductive pipes according to the described algorithms reduces the impact of the gap between the pipe and the eddy-current transducer, and that of the changes in the pipe electrical conductivity on the measurement results.

References

[1] *GOST 23786-79. Drill pipes made of aluminum alloys. Technical conditions* (Publishing House of Standards, Moscow, 1990)

[2] E.V. Yakimov, A.E. Goldstein, V.F. Bulgakov, Yu.V. Alkhimov, V.Yu. Belyankov, Test. Diagn. (11), 14 (2014) DOI: 10.14489/td.2014.011.pp.014-018

[3] Z.-G. Zhang, H.-X. Feng, W.-H. Zhao, MATEC Web of Conferencesb **40**, 04001 (2016) doi : 10.1051/matecconf/20164004001

A new Matlab coder for generating Structured Text Language from matrix expression for PLC and PAC controllers

Mariusz Buciakowski, Piotr Witczak
Institute of Control and Computation Engineering,
University of Zielona Góra,
ul. Podgórna 50, 65–246 Zielona Góra, Poland.
E-mail: {M.Buciakowski,P.Witczak}@issi.uz.zgora.pl

Abstract. This paper presents a new Matlab toolbox for synthesis of Structured Text (ST) code for Programmable Logic Controllers (PLC) and Programmable Automation Controller (PAC). This tool can directly generate IEC 61131-3 Structured Text Language from Matlab script for selected Integrated Development Environments (IDEs). The generated code can be verified and compared with the results obtained with Matlab simulation. After this, generated code can be used in IDEs, compiled and uploaded to a PLC or PAC controller for final verification. This approach leaves all available Matlab toolboxes for programmers use, thus allowing fast and easy synthesis developed algorithms.

1. Introduction

Nowadays, when industrial processes, as well as systems that controls them, become more and more complex, new challenges emerges. The most obvious one, directly related to the safety and security, but also to cost minimization is the reliability of used software and hardware. Secondly, the time of software implementation and its deployment to production hardware affects the time–to–delivery and total production costs. One of the possible solutions that can overcome at least part of the mentioned issues is automatic synthesis of standardized IEC 61131-3 Structured Text [7] code for Programmable Logic Controllers (PLC) and Programmable Automation Controller (PAC) controllers. Many researchers focuses on the different aspects of this area. Some of them proposes component-based modeling [16], others on the code reusability [1, 5], function-block-based modeling [6], generative programming [4] or sequential circuit synthesis [9]. While a modeling can take different forms, all of them leads to a code synthesis. All mentioned approaches have their potential advantages and drawbacks. It is enough to mention code reusability and function-block-based solution. The first one proposes the import/export mechanisms to reuse once designed and written code between platforms. While the idea is quite simple, the realization with numerous hardware platforms may not be so trivial as it is common that some manufacturers treats non-standardised elements on their own (i.e. totally different ST variable definition in IDE, Proficy Machine Edytion IDE, Step7 IDE etc.). Additionally solution does not offer the homogenous, safe verification environment. Leaving IDE in the middle of the binary code generation, allows hardware-dependent optimization (being a part of software delivered by controller manufacturer). The function-block-based is on the other side limited by function-blocks, and complicated programs become also graphically complex (cf. [6]). The last one, however, offers simulation perspectives. Due to the authors knowledge there is in fact a gap in the area of code synthesis for PLC and PAC. Especially, when modern controllers are capable of handling complex arithmetic operations in real-time. This paper proposes a solution for model-based code-synthesis in terms of mathematical modeling. The toolbox described in later sections is capable of Matlab-to-ST code conversion, preserving structure of mathematical components, such as matrices and vectors. Therefore, a model-based controller or filters can be built relatively fast. They can be also easily verified before uploading them into the device. This phenomena can be quite useful when on a PLC or PAC instead of classical industrial control scheme, the modern control would be implemented. Also thanks to the direct use of Matlab it is possible to use such tools as neural networks toolbox, LMIs [3, 12], P-matrices [8] and many others (on the high quality level). The recent works in the area of Fault Detection and Isolation (FDI) [21, 20, 11, 15] as well as Fault-Tolerant Control (FTC) clearly show a new direction for the industrial evolution. If one would compare any currently used control approach with any advanced FTC the superiority of the last one becomes more than obvious [18, 22, 13, 17, 19, 10]. However the main obstacles are - the lack of trust form the industry in new techniques and the lack of tools for simple and fast implementation of advanced modes into PLC/PAC.

The paper is organised as follows. Section 2 is devoted to the structure of the framework, as well as its general capabilities. Subsequently Section 3 shows some examples. Finally Section 4 summarizes the paper.

2. Toolbox structure

The main objective of this section is to present structure of toolbox, functionalities and the generic functions of the framework. The Toolbox for generating ST language was implemented using object–oriented programming techniques in Matlab. The use of this programming techniques allows to implement interface for the toolbox in an efficient way. Also it allows fast expansion by providing a properly built interfaces. The number of parameters that should be passed to the toolbox is as small as possible, providing a high user experience and usability. All of this allows comfortable usage, with only basic knowledge about Matlab and simple extension to any unimplemented ST version. Apart of above the framework consists of two main classes, from which one of them directly implements the conversion interface. The second one is a storage class based on the factory pattern [2]. Therefore, it fulfil two roles, one - stores all components for language conversion and code generation, secondly builds convertible objects. The toolbox has few main functionalities. The first functionality allows to select IDE. To do so user should use the following command:

```
cs = ConvertionStorage('Cscape');
```

Currently the toolbox supports the following IDE name:

(i) 'Cscape' – Cscape IDE for Horner APG PLC,
(ii) 'PME' – Proficy Machine Edition IDE for GE PAC controllers,
(iii) 'Step7' – Step7 for Siemens Simatic S7 300/400 controllers.

The code generation procedure requires the object *ConversionStorage* to be made, as one of its methods invokes the language conversion. The toolbox check name of IDE. If user use IDE name that not supported now, the error will return. The numbers for supported IDE will grow. The storage needs to consist of all variables used in the final program, therefore each variable from the workspace needs to be registered in the instance *cs*

```
cs = ConvertionStorage(IDE);
```

Furthermore, user can check weather variables are properly added to the working instance of*ConversionStorage*

```
sc.show();
```

The main functionality is the conversion method, that needs to be invoked with the parameter that defines the model to be converted

```
sc.convert(x);
```

where x needs to be defined Workspace *char* variable (see Section 3). Last but not least is the method that allows user to save code in a file

```
sc.drop2File('text.txt');
```

3. Examples

The objective of this section is to present examples of the generation code for selected IDE. Initial example provides an insight to the basic usage, while subsequent provides more detailed and advanced options. The first examples illustrate full code for setting initial values of variables and for calculating expression. The rest of examples show only code for calculating expression to present advantages for use this toolbox.

Note that the variable needs to be firstly created in a Matlab Workspace. For example let us define two simple variables

```
A = [1 2; 3 4];
b = [4; 3];
```

one matrix and one column vector

$$\boldsymbol{A} = \begin{bmatrix} 1 & 2 \\ 3 & 4 \end{bmatrix}, \quad \boldsymbol{b} = \begin{bmatrix} 4 \\ 3 \end{bmatrix} \tag{1}$$

and expression that multiplies two above matrices

$$\boldsymbol{C} = \boldsymbol{A}\boldsymbol{b}. \tag{2}$$

```
C = A*b;
```

The above will simply create a C variable in Matlab Workspace. In order to create a Toolbox valid variable for conversion, one simply changes the expression into a set of *char*

```
C = 'A*b';
```

and then converts it to ST by invoking

```
sc.convert(C);
```

On this level it is worth to mention that the type of variable is automatically detected, as well as type of The result. The converted type of variable depends on selected IDE however toolbox supports only `uint8, int16, uint16, int32, double, single` Matlab type of variable. The above allows to easily translate variable passed to the toolbox into a common variable type supported by PLCs and PACs. Subsequently the resulting code should be stored into a file

```
sc.drop2File('scape.txt');
```

The file name is used for create **.txt* file which will contain ST code. The full code for example (2) is

```
A = [1 2; 3 4];
b = [4; 3];
sc = ConvertionStorage('Scape');
sc.addVariable(A);
sc.addVariable(b);
sc.show();
C = 'A*b';
sc.convert(C)
sc.drop2File('scape.txt')
```

The ST code is

```
A[0,0] := 0;
A[0,1] := 1;
A[1,0] := 2;
A[1,1] := 3;
b[0] := 3;
b[1] := 2;
C[0] := A[0,0] * b[0] + A[0,1] * b[1];
C[1] := A[1,0] * b[0] + A[1,1] * b[1];
```

EXAMPLE 1: The fist examples base on expression

$$\boldsymbol{C} = \boldsymbol{A} + \boldsymbol{B} \tag{3}$$

where

$$\boldsymbol{A} = \begin{bmatrix} 2 & 3 \\ 2 & 3 \end{bmatrix} \boldsymbol{B} = \begin{bmatrix} 2 & 3 \\ 2 & 3 \end{bmatrix} \tag{4}$$

and the respective Matlab code

```
A = [2 3; 2 3];
B = [2 3; 2 3];
sc = ConvertionStorage('PME'); % Step7
sc.addVariable(A);
sc.addVariable(B);
sc.show();
C = 'A+B';
sc.convert(C)
sc.drop2File('pme.txt')
```

Listing. 1 and *Listing. 2* show full code for PME and Step7 respectively. The code also it contains instructions to set initial value for each variable using in expression. In case of PME, the initial value can be set in the same block where the main code is used to compute expression or in a different block. It depends on the Programmer preferences however, the initial values can be set once, using information about first cycle in PLC or PAC. For Step7, toolbox generate full Function Block 1. The number of Function Block is default set to 1. The generated block can be compiled and sent to the controller.

EXAMPLE 2: The second example shows ST code for expression

$$\boldsymbol{z}_k = \boldsymbol{A}\hat{\boldsymbol{x}}_k + \boldsymbol{B}\boldsymbol{u}_k + \boldsymbol{K}(\boldsymbol{y}_k - \boldsymbol{C}\hat{\boldsymbol{x}}_k). \tag{5}$$

The above expression describe Luenberger Observer and it is commonly used in control theory. In (5) $\boldsymbol{u}$, $\boldsymbol{y}$, and $\hat{\boldsymbol{x}}$ are input signal. The codes example (the values for each matrices is omitted)

Listing. 1 ST code for Proficy Machine Edition IDE for example 1

```
A[0,0]  := 2.000000;
A[0,1]  := 3.000000;
A[1,0]  := 2.000000;
A[1,1]  := 3.000000;
B[0,0]  := 2.000000;
B[0,1]  := 3.000000;
B[1,0]  := 2.000000;
B[1,1]  := 3.000000;

C[0,0]:=A[0,0]+B[0,0];
C[0,1]:=A[0,1]+B[0,1];
C[1,0]:=A[1,0]+B[1,0];
C[1,1]:=A[1,1]+B[1,1];
```

Listing. 2 ST code for Step7 IDE for example 1

```
FUNCTION_BLOCK FB1
VAR
    A : ARRAY[0..1,0..1] OF REAL :=
                    2.000000,3.000000,
                    2.000000,3.000000;
    B : ARRAY[0..1,0..1] OF REAL :=
                    2.000000,3.000000,
                    2.000000,3.000000;
    C : ARRAY[0..1,0..1] OF REAL;
END_VAR
C[0,0]:=A[0,0]+B[0,0];
C[0,1]:=A[0,1]+B[0,1];
C[1,0]:=A[1,0]+B[1,0];
C[1,1]:=A[1,1]+B[1,1];
END_FUNCTION_BLOCK
```

```
sc = ConvertionStorage('PME'); % Step7
sc.addVariable(A);
sc.addVariable(x);
sc.addVariable(B);
sc.addVariable(u);
sc.addVariable(K);
sc.addVariable(y);
sc.addVariable(C);
z = 'A*x+B*u+K*(y-C*x)';
sc.convert(z)
sc.drop2File('pme.txt')
```

IOP Conf. Series: Journal of Physics: Conf. Series **783** (2017) 012062 doi:10.1088/1742-6596/783/1/012062

show *Listing. 3* and *Listing. 4* show full code for PME and Step7 respectively. ST code in *Listing. 3* is similar for code in *Listing. 1.* In this example expression includes input variables, therefore code for Step7 IDE contains `CASE` statement for set initial value for variables $\boldsymbol{u}$, $\boldsymbol{y}$, and $\hat{\boldsymbol{x}}$. This statement cane be also use for reset variables to initial condition.

Listing. 3 ST code for Proficy Machine Edition IDE for example 2

```
A[0,0] := 32.000000;
A[0,1] := 22.000000;
A[1,0] := 23.000000;
A[1,1] := 85.000000;
x[0] := 40.000000;
x[1] := 68.000000;
B[0] := 83.000000;
B[1] := 70.000000;
u := 92.000000;
K[0] := 94.000000;
K[1] := 38.000000;
y := 25.000000;
C[0] := 96.000000;
C[1] := 55.000000;
z[0,0]:=0;
z[0,1]:=0;

z[0,0]:=B[0]*u-K[0]*(-y+C[0]*x[0]+
    C[1]*x[1])+A[0,0]*x[0]+A[0,1]*x[1];
z[1,0]:=B[1]*u-K[1]*(-y+C[0]*x[0]+
    C[1]*x[1])+A[1,0]*x[0]+A[1,1]*x[1];
```

EXAMPLE 3: This example shows how generate code for more advanced Matlab functions. In the following example we want to generate code based on the Jacobian of function:

$$\boldsymbol{s} = \text{jacobian}(2xysin(z) + 3xsin(y)cos(z)) \tag{6}$$

The full code for example

```
s = 'jacobian(2*X*Y*sin(Z) +
        3*X*sin(Y)*cos(Z), [X, Y, Z])';
sc.addVariable(X);
sc.addVariable(Y);
sc.addVariable(Z);
sc.convert(s)
```

Listing. 5 and *Listing. 6* show ST code. From listings, it can be seen that the toolbox allows to generate code for more advanced Matlab functions.

EXAMPLE 4: The next example shows code for following dynamic linear discrete-time system

$$\boldsymbol{x}_{k+1} = \boldsymbol{A}\boldsymbol{x}_k + \boldsymbol{B}\boldsymbol{u}_k \tag{7}$$

where $\boldsymbol{A}$, $\boldsymbol{B}$, $\boldsymbol{u}$ are constant matrices of appropriate dimensions. Similarly to the previous examples Matlab code is

Listing. 4 ST code for Step7 IDE for example 2

```
FUNCTION_BLOCK FB1
VAR_INPUT
    INIT_VAR : INT;
END_VAR
VAR
    A : ARRAY[0..1,0..1] OF REAL :=
            32.000000,22.000000,
            23.000000,85.000000;
    x : ARRAY[0..1] OF REAL;
    B : ARRAY[0..1] OF REAL :=
            83.000000,70.000000;
    u : REAL;
    K : ARRAY[0..1] OF REAL :=
            94.000000,38.000000;
    y : REAL;
    C : ARRAY[0..1] OF REAL :=
            96.000000,55.000000;
    z : ARRAY[0..1,0..1] OF REAL;
END_VAR

CASE INIT_VAR OF
    1 : u :=0;
        y :=0;
        x[0]:=0;
        x[1]:=0;
        z[0,0]:=0;
        z[0,1]:=0;
ELSE:
    z[0,0]:=B[0]*u-K[0]*(-y+C[0]*x[0]+
        C[1]*x[1])+A[0,0]*x[0]+A[0,1]*x[1];
    z[0,1]:=B[1]*u-K[1]*(-y+C[0]*x[0]+
        C[1]*x[1])+A[1,0]*x[0]+A[1,1]*x[1];
END_CASE;
END_FUNCTION_BLOCK
```

Listing. 5 ST code for Proficy Machine Edition IDE for example 3

```
X := 0;
Y := 0;
Z := 0;

s[0,0]:=Y*sin(Z)*2.0+cos(Z)*sin(Y)*3.0;
s[0,1]:=X*sin(Z)*2.0+X*cos(Y)*cos(Z)*3.0;
s[0,2]:=X*sin(Y)*sin(Z)*-3.0+X*Y*cos(Z)*2.0;
```

Listing. 6 ST code for Step7 IDE for example 3

```
FUNCTION_BLOCK FB1
VAR_INPUT
    INIT_VAR : INT;
END_VAR
VAR
 X : REAL;
 Y : REAL;
 Z : REAL;
 s : ARRAY[0..2,0..1] OF REAL;
END_VAR

CASE INIT_VAR OF
    1 : X :=0;
        Y :=0;
        Z :=0;
ELSE:
    s[0,0]:=Y*sin(Z)*2.0+cos(Z)*sin(Y)*3.0;
    s[0,1]:=X*sin(Z)*2.0+X*cos(Y)*cos(Z)*3.0;
    s[0,2]:=X*sin(Y)*sin(Z)*-3.0+X*Y*cos(Z)*2.0;
END_CASE;
END_FUNCTION_BLOCK
```

```
x = 'A*x+B*u';
sc = ConvertionStorage('PME');  % or Step7
sc.addVariable(A);
sc.addVariable(B);
sc.addVariable(u);
sc.convert(x);
```

Listing. 7 ST code for Proficy Machine Edition IDE for example 4

```
    A[0,0] := 40.000000;
    A[0,1] := 82.000000;
    A[1,0] := 20.000000;
    A[1,1] := 58.000000;
    B[0] := 50.000000;
    B[1] := 86.000000;
    u := 20.000000;

    x[0]:=A[0,0]*x[0]+A[0,1]*x[1]+B[0]*u+B[1]*u;
    x[1]:=A[1,0]*x[0]+A[1,1]*x[1]+B[0]*u+B[1]*u;
```

Listing. 8 ST code for Step7 IDE for example 4

```
FUNCTION_BLOCK FB1
VAR
A : ARRAY[0..1,0..1] OF REAL :=
        40.000000,82.000000,20.000000,58.000000,
B : ARRAY[0..1] OF REAL := 50.000000,86.000000;
    u : REAL := 20.000000;
x : ARRAY[0..5,0..5] OF REAL;
END_VAR

    x[0]:=A[0,0]*x[0]+A[0,1]*x[1]+B[0]*u+B[1]*u;
    x[1]:=A[1,0]*x[0]+A[1,1]*x[1]+B[0]*u+B[1]*u;

END_FUNCTION_BLOCK
```

4. Concluding Remarks

The paper deals with the emerging and important issue of Structured Text code synthesis for PLC and PAC controllers. Authors proposes a Matlab based, model-based approach to automatic code generation. Thanks to Matlab powerful seamless toolboxes any demanding control task can be solved efficiently and subsequently tested within the encapsulated and safe environment. This procedure saves cost, time and equipment from damage. At the same time final product such as control scenario or observer structure can be converted into ST language and uploaded into the controller of given manufacturer. Siemens, General Electric or Horner makes no difference thanks to a interface-like object-oriented design of a framework. The Matlab toolbox build for this task can be easily scaled to a incoming needs. Furthermore, a future development of the toolbox is planed, e.g.to automatically handle gain scheduling from Linear Parameter-Varying [14] models or membership functions from Takagi-Sugeno [23] models.

5. Acknowledgments

Authors would like to express their sincere thanks to colleagues from the Institute for valuable hints and motivating discussions. The work was supported by the National Science Centre of Poland under grant: 2013/11/B/ST7/01110

Reference

[1] Tord Alenljung, Markus Skoldstam, Bengt Lennartson, and Knut Akesson. PLC-based implementation of process observation and fault detection for discrete event systems. In *2007 IEEE International Conference on Automation Science and Engineering*, pages 207–212. IEEE, 2007.

[2] Andrei Alexandrescu. *Modern C++ design: generic programming and design patterns applied.* Addison-Wesley, 2001.

[3] S. Boyd, E. Feron, L. E. Ghaoui, and V. Balakrishnan. *Linear Matrix Inequalities in System and Control Theory.* Siam, Philadelphia, 1994.

[4] Daniel Côté, Richard St-Denis, and Sylvain Kerjean. Generative programming for programmable logic controllers. In *2005 IEEE Conference on Emerging Technologies and Factory Automation*, volume 2, pages 8–pp. IEEE, 2005.

[5] E Estevez, M Marcos, E Irisarri, F Lopez, I Sarachaga, and A Burgos. A novel approach to attain the true reusability of the code between different PLC programming tools. In *Factory Communication Systems, 2008. WFCS 2008. IEEE International Workshop on*, pages 315–322. IEEE, 2008.

[6] Knut Guttel, Peter Weber, and Alexander Fay. Automatic generation of PLC code beyond the nominal sequence. In *2008 IEEE International Conference on Emerging Technologies and Factory Automation*, pages 1277–1284. IEEE, 2008.

[7] Karl-Heinz John and Michael Tiegelkamp. *IEC 61131-3: programming industrial automation systems: concepts and programming languages, requirements for programming systems, decision-making aids.* Springer Science & Business Media, 2010.

[8] B. Kolodziejczak and T. Szulc. Convex combinations of matrices. full rank characterization. *Linear algebra and its applications*, 287(1-3):215–222, 1999.

[9] T Krairojananan and S Suthapradit. A PLC program generator incorporating sequential circuit synthesis techniques. In *Circuits and Systems, 1998. IEEE APCCAS 1998. The 1998 IEEE Asia-Pacific Conference on*, pages 399–402. IEEE, 1998.

[10] Marcel Luzar, Marcin Witczak, Mariusz Buciakowski, and Vicenç Puig. Robust $\mathcal{H}_\infty$ actuator fault diagnosis and fault-tolerant control for a multi-tank system. In *2013 Conference on Control and Fault-Tolerant Systems (SysTol)*, pages 758–763. IEEE, 2013.

[11] Marcin Mrugalski, Marcel Luzar, Marcin Pazera, Marcin Witczak, and Christophe Aubrun. Neural network-based robust actuator fault diagnosis for a non-linear multi-tank system. *ISA transactions*, 61:318–328, 2016.

[12] Boris Polyak and Pavel Shcherbakov. Ellipsoidal approximations to attraction domains of linear systems with bounded control. In *American Control Conference*, 2009.

[13] V. Puig. Fault diagnosis and fault tolerant control using set-membership approaches: Application to real case studies. *International Journal of Applied Mathematics and Computer Science*, 20(4):619–635, 2010.

[14] Damiano Rotondo, Fatiha Nejjari, and Vicenc Puig. Robust quasi LPV model reference FTC of a quadrotor UAV subject to actuator faults. *International Journal of Applied Mathematics and Computer Science*, 25(1):7–22, 2015.

[15] Damiano Rotondo, Marcin Witczak, Vicenç Puig, Fatiha Nejjari, and Marcin Pazera. Robust unknown input observer for state and fault estimation in discrete-time takagi–sugeno systems. *International Journal of Systems Science*, 47(14):3409–3424, 2016.

[16] Jan Thieme and Hans-Michael Hanisch. Model-based generation of modular PLC code using iec61131 function blocks. In *Proceedings of the International Symposium on Industrial Electronics*, volume 1, pages 199–204, 2002.

[17] K.C. Veluvolu, M.Y Kim, and D. Lee. Nonlinear sliding mode high-gain observers for fault estimation. *International Journal of Systems Science*, 42(7):1065–1074, 2011.

[18] M. Witczak. *Fault Diagnosis and Fault-Tolerant Control Strategies for Non-linear Systems.* Springer-Verlag, Berlin-Heidelberg, 2014.

[19] Marcin Witczak, Mariusz Buciakowski, and Christophe Aubrun. Predictive actuator fault-tolerant control under ellipsoidal bounding. *International Journal of Adaptive Control and Signal Processing*, 30(2):375–392, 2016.

[20] Marcin Witczak, Mariusz Buciakowski, and Marcin Mrugalski. An $\mathcal{H}_\infty$ approach to fault estimation of non-linear systems: Application to one-link manipulator. In *Methods and Models in Automation and Robotics (MMAR), 2014 19th International Conference On*, pages 456–461. IEEE, 2014.

[21] Marcin Witczak, Mariusz Buciakowski, Vicenç Puig, Damiano Rotondo, and Fatiha Nejjari. An LMI approach to robust fault estimation for a class of nonlinear systems. *International Journal of Robust and Nonlinear Control*, 26(7):1530–1548, 2016.

[22] P. Witczak, M. Witczak, K. Patan, and R. Stetter. Design of robust predictive fault-tolerant control for takagi-sugeno fuzzy systems: Application to the twin-rotor system. In *Intelligent Control (ISIC), 2014 IEEE International Symposium on*, pages 1113–1118, Oct 2014.

[23] Piotr Witczak and Marcin Witczak. Towards robust fault-tolerant model predictive control with constraints for takagi–sugeno systems. In *Variable-Structure Approaches*, pages 309–333. Springer International Publishing, 2016.

Systematic Product Development of Control and Diagnosis Functionalities

R Stetter[1], **A Simundsson**[2]

[1] Hochschule Ravensburg-Weingarten, Weingarten, Germany
[2] University of Manitoba, Winnipeg, Manitoba, Canada

E-mail: stetter@hs-weingarten.de

Abstract. In the scientific field of systematic product development a wide range of helpful methods, guidelines and tools were generated and published in recent years. Until now little special attention was given to design guidelines aiming at supporting product development engineers to design products that allow and support control or diagnosis functions. The general trend to ubiquitous computing and the first development steps towards cognitive systems as well as a general trend toward higher product safety, reliability and reduced total cost of ownership (TCO) in many engineering fields lead to a higher importance of control and diagnosis. In this paper a first attempt is made to formulate general valid guidelines how products can be developed in order to allow and to achieve effective and efficient control and diagnosis. The guidelines are elucidated on the example of an automated guided vehicle. One main concern of this paper is the integration of control and diagnosis functionalities into the development of complete systems which include mechanical, electrical and electronic subsystems. For the development of such systems the strategies, methods and tools of systematic product development have attracted significant attention during the last decades. Today, the functionality and safety of most products is to a large degree dependent on control and diagnosis functionalities. Still, there is comparatively little research concentrating on the integration of the development of these functionalities into the overall product development processes. The paper starts with a background describing Systematic Product Development. The second section deals with the product development of the sample product. The third part clarifies the notions monitoring, control and diagnosis. The following parts summarize some insights and formulate first hypotheses concerning control and diagnosis in Systematic Product Development.

1. Systematic product development

Systematic product development can be traced back to the inventor Reuleaux, who presented a procedure for kinematic synthesis close the end of the 19th century. Especially after the world war two many researchers were investigating and formulating product development strategies, methods, guidelines and tools. The German engineering design methodology "Konstruktionsmethodik" played a central role, but in many other countries elaborate research efforts were carried out, as well. Central researchers in this endeavor were Rodenacker [1], Pahl&Beitz [2], Koller [3], Roth [4], Ehrlenspiel [5] and Lindemann [6]. An important milestone was the formulation of the VDI (German society of engineers) guideline 2221. This guideline was based on the earlier VDI 2222. The goal of this guideline is to propose a general methodology for designing technical systems and products and to support a methodical and systematic designing, in order to produce a more efficient working style [7]. The main process which is described in this guideline is shown in Figure 1.

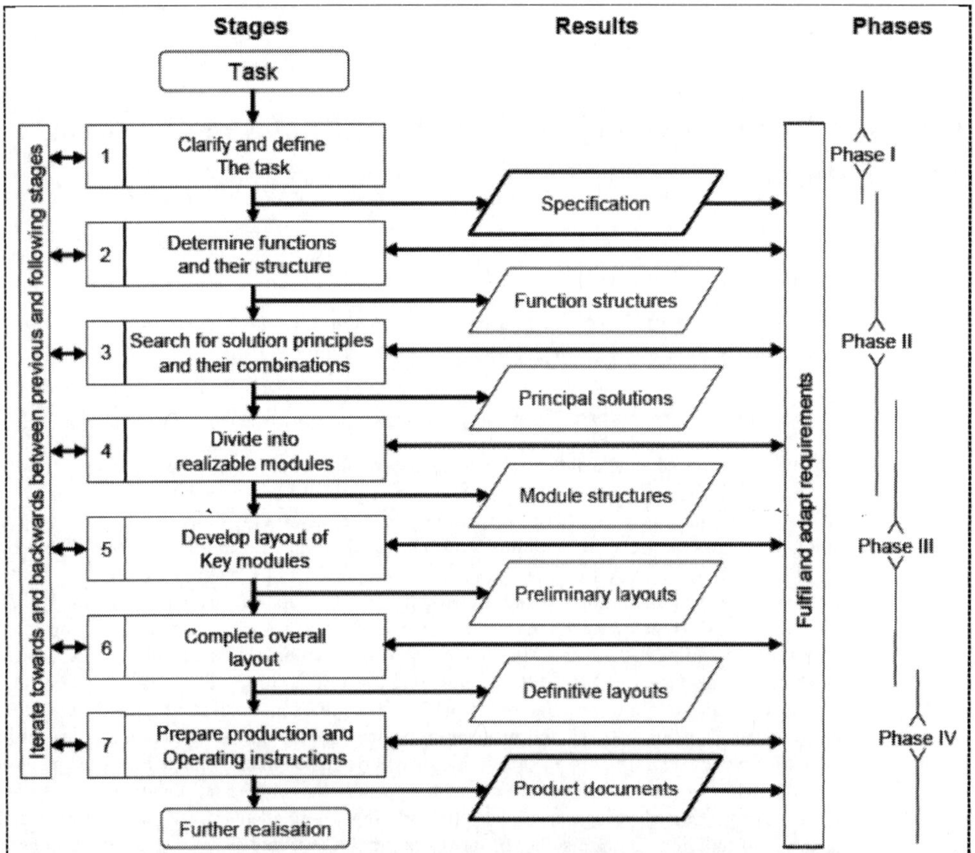

Figure 1. Systematic Product Development according to VDI 2221 [8].

There is little doubt concerning the merit of this guideline. Recent investigations, however, have shown that design processes are much more iterative and frequently display evolutionary characteristics [9], [10]. It was therefore decided to use a slightly different model in order to structure the discussion in this paper. This model was derived from a model of mechatronic product development used in prior research [11]. This model, which is shown in Figure 2, is proposed for the sake of an in depth discussion of the most prominent aspects of systematic product development in today's industrial settings.

The core of the model is the core process of producing companies consisting of product development, production and assembly, as well as marketing and after-sales. Four groups of activities are important in every phase of this core process. The essential activities of the project management concern the planning and control of all schedules as well as the assignment of resources. The most important aspects of project management concern people, time and money. The logical relationships of the product and the process are dealt with in "process management". Process management is often also referred to as Systems Engineering and can be defined as guidance for the functional design of complex systems, which is based on certain thinking models and principles [12]. Change Management is the process of requesting, determining attainability, planning, implementing, and evaluation of changes to a system. It has two main goals: supporting the processing of changes and enabling traceability of changes [13]. In order to allow this traceability often a version management is part of the change management. The term "configuration management" is meant as an extension of variant management. Possible variants of a product have to be consciously configured so that each possible product variant (configuration) will fulfill the functional and physical requirements throughout its life, including compatible and working configurations during the design process. These possible configurations have to be considered during all phases of the core process of producing companies. The notion "testing" summarizes a number of activities which have to be planned, carried out and controlled during all phases of the core process. The activities in testing focus on virtual and physical analyses of the product and process performance. These activities are necessary to assure the functionality and quality of the product and processes.

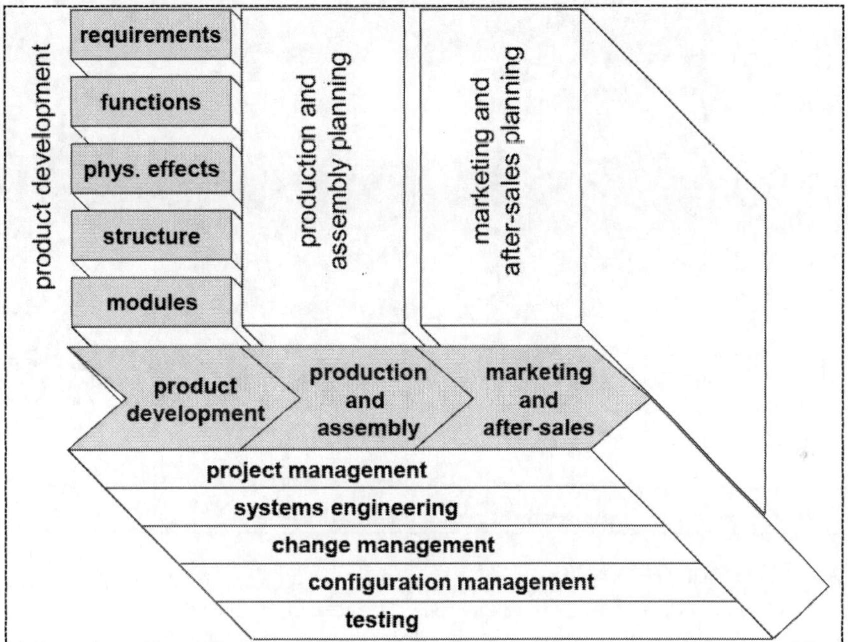

Figure 2. Model of systematic product development.

For the phase "product development" (top left side in Figure 2) necessary activities can be summarized using five notions. These notions have a sensible sequence, but should not be understood as a process because iterations and jumps between activities under different notions are frequently necessary. The notion "requirements" summarizes activities which aim to translate customer wishes and expectations into appropriate requirements as well as to manage these requirements. Activities which deal with the abstract logic of the product on a functional level are subsumed under the notion "function". Such functions can be realized by means of a combination of physical phenomena. The notion "physical effects" summarizes a number of activities aiming at clarifying and modifying those chains of physical phenomena. Activities which focus on realizing functional and physical structures of the product and process in the form of geometrical objects are subsumed under the notion "structure". The notion "modules" summarizes activities with the objective to develop smaller subsystems of the product commonly referred to as modules.

The phases "production and assembly" as well as "marketing and after-sales" are not analyzed in this paper in detail. Consequently no groups of activities are defined and displayed in this model. This is not meant to indicate that such groups of activities are not sensible in these phases of the core process. These phases are not the focus of this paper, a definition of groups of activities is left to further research, though they are probably similar.

2. Sample Product Development of an Automated Guided Vehicle AGV

This section describes the product development of an AGV for outdoor used (compare also Stetter&Simundsson [14]). The AGV makes use of a unique steering system based on torque differences and is suited for rough terrain because of four independent legs with wheels (Figure 3).

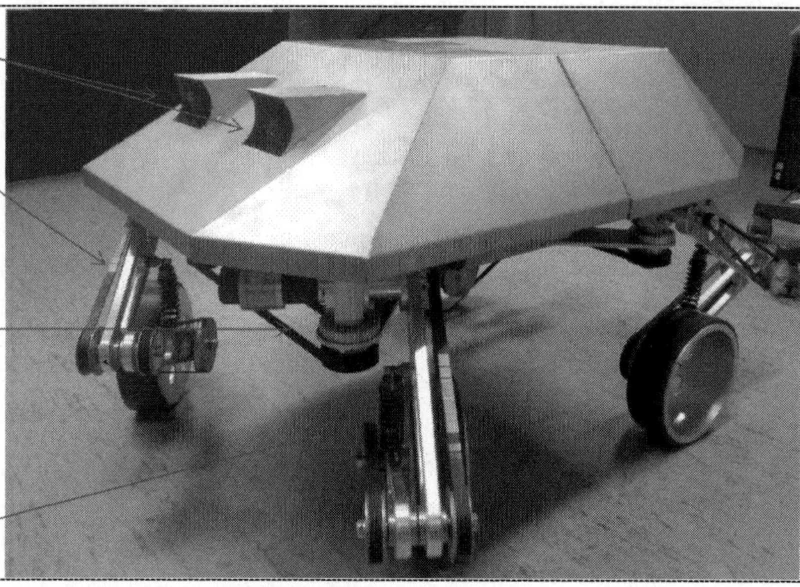

Figure 3. Automated Guided Vehicle.

The main objective of the development project was to develop a Semi-Autonomous Outdoor Vehicle designed to explore and work in open areas. The steering is based on torque differences. The distinctive quality of the used steering system is the dynamic behavior. This innovative steering system is based on the concept to use the torque of drive motors (more exactly the torque differences between wheels) to steer four independent axles of a vehicle. The principal steering system is shown in Figure 4.

In this example a vehicle consists of four drive motors which are fastened on arms that may freely rotate. These arms have no drive or brake, only an angle encoder is attached at the end of each axle. These angle encoders measure the angle of the motor and the wheel with regard to the vehicle platform. The distinct characteristic of the innovative drive system is the absence of dedicated steering motors. By means of angle encoders applied at the four steering axles and highly dynamic control algorithms it is possible to steer such vehicles only by means of the four drive motors (compare Figure 4).

Each of the wheels on the short axle can be directed into the desired position by means of the torque applied on the wheel. This could take place sequentially for each individual wheel but also simultaneously, if the control allows different torque on all wheels. This characteristic allows simpler and simultaneously more robust vehicle concepts. It is also a main advantage of this concept that the resulting vehicle is able to drive directly in any direction without time and space consuming turning maneuvers.

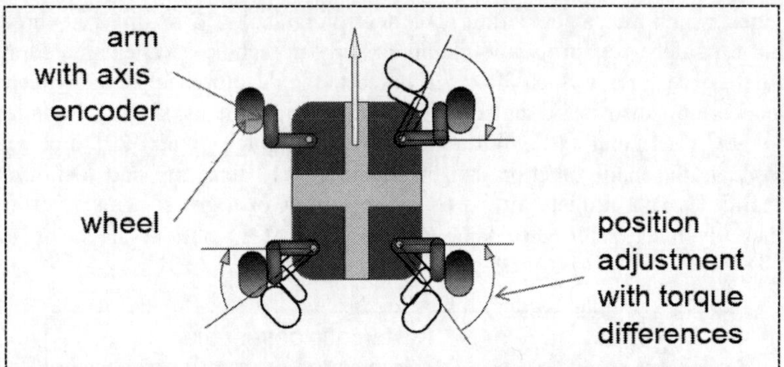

Figure 4. Principal steering system.

Furthermore, a vehicle based on the dynamic drive system is able to turn around its own center. This characteristic is very important if cameras or other equipment are mounted on such vehicles which can only be used in a certain orientation. The innovative steering system shares these advantages with Omni drive systems (Ashmore&Barnes [15]), but has reduced friction as well as easier controllability and offers the possibility to determine an exact position and orientation from an analysis of the angles of the steering axes and the angles of the drive wheels (odometry).

The control of the vehicle happens on different levels. A central computer receives a movement wish either directly from the user (manual driving) or improved by a superordinate intelligence (semi-autonomous driving mode) or only from a superordinate intelligence (autonomous driving mode). Based on a mathematical model of the AGV this movement wish is than translated into a speed command to the drive motors as well as desired angles for the wheels (one angle for both front wheels and one angle for both back wheels).

Based on this sample product development several aspects of systematic product development are explored. In the next section first the notions are clarified,

3. Monitoring, Control and Diagnosis

The notion "monitoring" summarizes all kinds of systematic observation, surveillance or recording of an activity or a process by any technical means. In leading industries such as computer chip production or car manufacturing today usually nearly all operation data of the productions systems are being monitored for the three main reasons safety, efficiency and plannability:

- The safety of production systems can be enhanced because a reliable safety system with a fast reaction can be realized on the basis of a real-time monitoring system. The role of coincidence for detecting possibly dangerous faults is diminished if a continuous monitoring is in place.
- The efficiency of production systems can be enhanced because any kind of waste (of energy, time and production goods) will be detected and can subsequently be prevented or reduced.
- The planning possibilities and planning quality can be enhanced if accurate data from a real-time continuous monitoring system are available as realistic prognosis is enabled by such data.

Monitoring activities have a merit on their own, as they are usually necessary for safety and for planning activities. Additionally they are the first step for control and diagnosis.

The notion "control" names certain activities with the aim to manage, command, direct or regulate the behavior of devices or systems and has been the core of extensive research for many decades. In the last four decades the techniques of adaptive control have found rising attention. Adaptive control usually relies on an aggregation of a conventional control methodology with some form of recursive system identification (Sastry&Bodson [16]).

On a very general level the notion "diagnosis" is usually understood as the process of estimating the condition of certain entities. More specifically, in technical applications the term diagnosis describes activities which aim at detecting and identifying faults. Over the last three decades, the growing demand for safety, reliability and maintainability in technical systems has drawn significant research in the field of diagnosis. Such efforts have led to the development of many techniques; recent additions also concern the distributed diagnosis of mechatronic systems; see for example survey works of Blanke et al.[17], Isermann [18], Witczak [19] and Korbicz et al. [20]. For mechanical and mechatronic products the main function can be described as "detecting and identifying product or process abnormalities". The ultimate aim is to inform a user or a group of users or a superordinate system about these dysfunctions. Figure 5 shows a summary of the notions diagnosis, monitoring and control.

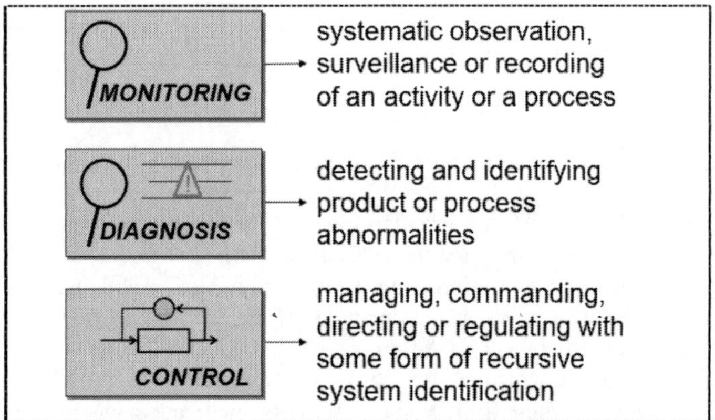

Figure 5. Monitoring, Diagnosis and Control.

In the following sections hypotheses concerning systematic product development are discussed. The structure follows the model of systematic product development shown in Figure 2.

4. Systematic product development of requirements

A systematic consideration of requirements naturally leads to the need to manage requirements throughout the product development process. This concerns activities "collection of (search for) requirements" and the ongoing "handling of requirements" including providing requirements to all stakeholders, updating requirements and tracking, numbering as well as versioning requirements. The importance of requirements and requirements management was highlighted frequently and is illustrated by the well-known "rule of ten". The "rule of ten" specifies that it costs 10 times more to find and repair a failure at the next stage of realization. Thus, for example, it costs 10 times more to find a failure in the concept phase than during the phase "clarification of the task" which is aiming at requirements. This underlines the importance of requirements management for all kinds of engineering design.

Sometimes control activities are not considered as essential for the main function of a product. However, the example of the steering system of the AGV (compare Section 2) shows that the basic functionality would not be possible without a control system. Also many other products such as cars require control for an economic and ecological performance. Consequently, it can be assumed that control functionalities should also be treated as immediate requirements and not only as means which help to fulfill other requirements.

Especially diagnostic functions are frequently linked to the direct functionality of the product. For an example, a diagnostic function "detect low battery status" of the AGV is strongly linked to the direct function "supply energy". For a better overview it is therefore highly desirable that requirements concerning diagnostic functions can be linked to requirements describing the (direct) functionality that is being diagnosed as requirements and configuration changes of the direct function can very often influence the connected diagnostic function.

When focusing on the AGV from Section 2 one can notice that some control and diagnostic functions can initially appear without requirements from the end-user. For instance, in recent years it became possible to use rather cheap gyro-sensors in order to survey the inclination of an AGV. Here a technology push can create additional requirements which have first to be presented to the later user as possible functions.

It is important to note that especially the Failure-Mode-and Effects-Analysis (FMEA) offers a straight-forward possibility to identify demand for diagnosis (compare Stetter&Phleps [21]). Demands for control or a high economic or ecologic potential through control can be identified using the well-known method Quality Function Deployment. Amongst others, QFD allows to link product characteristics such as control functionalities to the wishes and needs of the user thus allowing assessing the specific end user merit of a certain control functionality. Additionally, elements of a product benchmark are contained in QFD and allow assessing the desirable performance level of a control functionality.

Figure 6 summarizes the hypotheses concerning requirements.

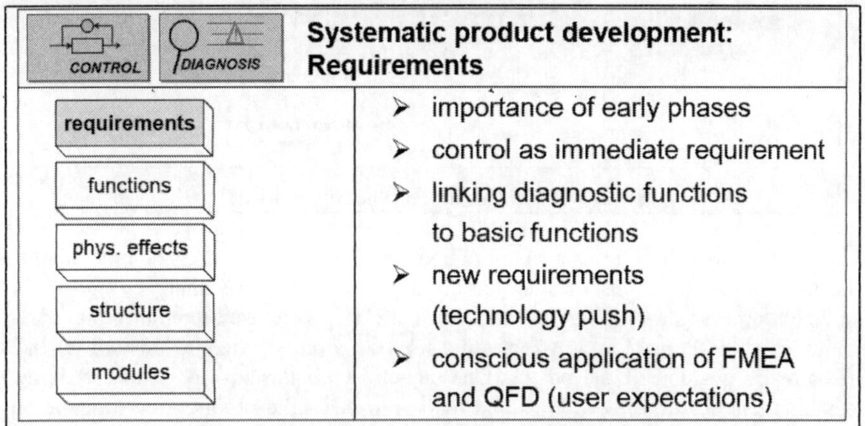

Figure 6. Hypotheses concerning requirements.

5. Systematic product development of functions

Certain requirements describe necessary functions of future products. By means of function analysis (and synthesis) it can be described on an abstract level how these functions can be realized. Many different approaches and models have been proposed for function analysis. The first one, which will be analyzed with regards for its potential to support systematic product development of control and diagnosis functionalities is the function structure as proposed by Ehrlenspiel [5]. The main advantage of this form of function analysis is the inclusion of states (input and output states of functions) and the description of different types of linking possibilities of secondary flows to primary flows (types: condition state, process state, additional state – compare [22]). Figure 7 shows the notation and an example of the flow orientated function structure. In the example of the AGV (lower right of Figure 7) the moment of an electrical drive motor driving the AGV is connected as a so-called "condition flow (C)" – depicting a necessary condition for the realization of a function.

In this function structure the direct functions and a connecting diagnostic function can be described in a consistent representation. Especially the link type „process state" allows to assign a diagnostic function to a flow of matter, energy or information undergoing some kind of operation [21]. By this type of function structure it is therefore possible to show exactly which entities are being diagnosed and to localize the diagnostic function on a functional level. Possible diagnostic functions could test, if the input states of a function are existing (matter, energy or information), if condition states are existing and could evaluate process conditions of the function carrier.

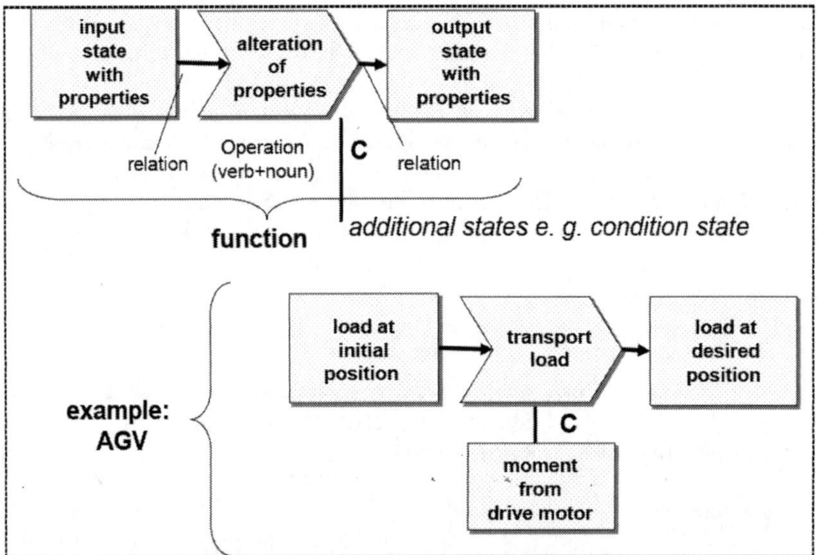

Figure 7. Flow oriented function analysis.

This kind of function analysis was also applied for clarifying the control functionalities of the AGV. Here the distinct functions of the steering angle control were analyzed and depicted. The connection types "process state" (P) and "condition state" (C) were used to connect auxiliary flows to the main flow which describes the transformation from an arbitrary angular position of the wheel to the descried angular position of the wheel. This depiction has the advantage of clarifying complex function combinations within the products and by this to assist the product development engineer in their endeavor. Figure 8 shows the flow oriented function model of the control of the angular position of a wheel of the AGV.

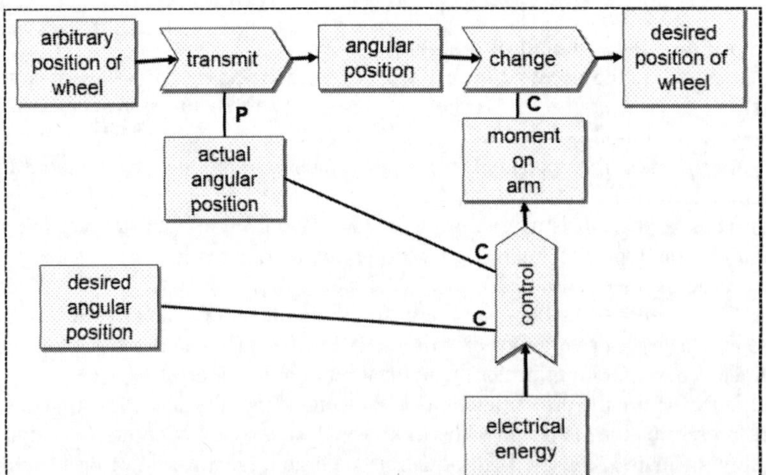

Figure 8. Flow oriented function model of the wheel angle control.

Another well-known approach to analyses functions is based on the work of Altshuller and used in connection with the tools of TRIZ/TIPS (compare e. g. Herb (2000)). This kind of function structure is concentrating not on the flows but on the relations in a product. Figure 9 shows one notation example (other notation conventions exist) and a control example of the AGV.

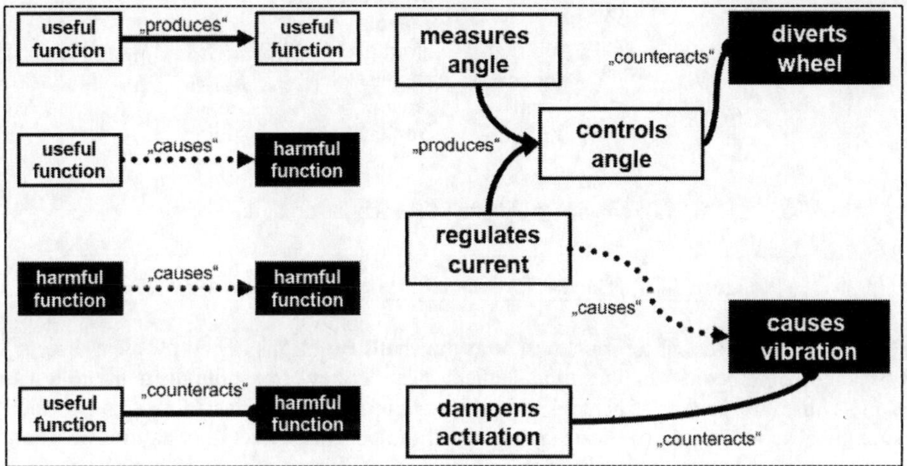

Figure 9. Function analysis based on TRIZ.

As the results at the right side show, a relational function structure can ease the understanding of and overview over control functions and thus support systematic product development. In summary, the functional domain and the different models used in this domain offer several possibilities to develop and explore diagnostic functionality. Figure 10 summarizes the hypotheses concerning requirements.

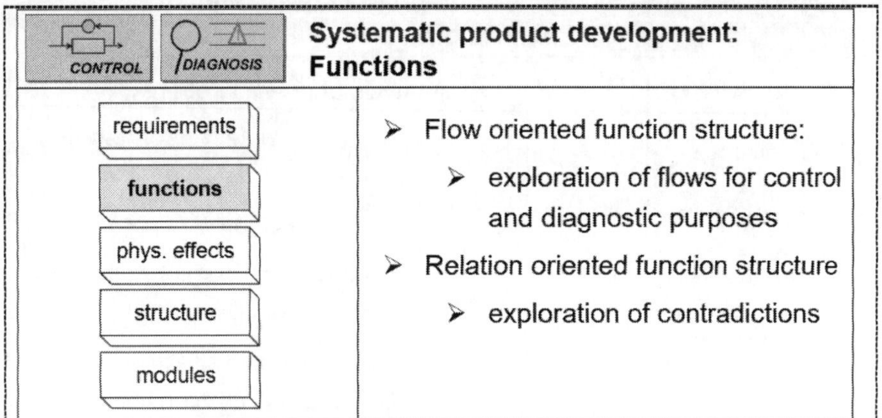

Figure 10. Hypotheses concerning functions.

6. Systematic product development of physical effects

In general, the physical domain describes how functions of a product are realized in the most abstract physical sense. A number of authors, e. g. Ehrlenspiel [5] propose to describe the physical domain by means of elementary physical effects. These physical effects are listed in form of a catalogue which also contains the most important input and output parameters. The advantage of this methodical approach is that with a rather small number of physical effects (approx. 90) any physical product realization can be described by means of effect chains. In Figure 11 a physical effect chain for an axis encoder of the AGV (compare section2) is shown.

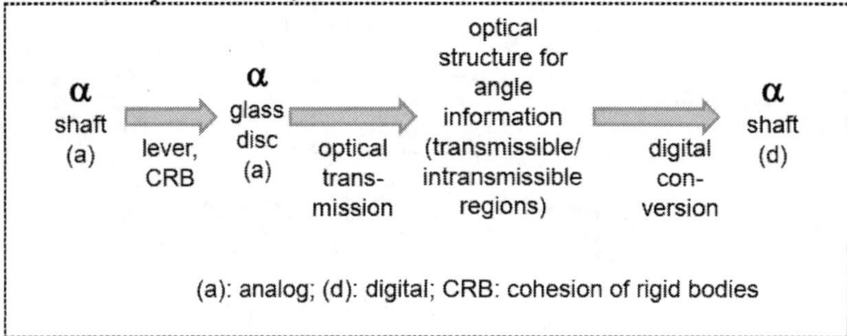

Figure 11. Physical effect chain "axis encoder of the AGV".

The current angle of the wheel α is present at the shaft. By means of the physical effects "lever" and "cohesion of rigid bodies" the machine element "parallel key" transmits the angle to a glass disc. On this glass disc are certain structures of transmissible and intransmissible regions which allow identifying the absolute angular position by means of the physical effect "optical transmission". This information is then converted in digital angular information about the angle of the wheel. Obviously, this kind of effect chain allows analyzing and depicting the physics acting in a product thus fostering a deeper understanding and enabling communication between engineers. Generally, sensory physical effects which have electrical outputs are desirable for diagnosis purposes [21]. Additionally, when describing the realization of diagnostic functionality by means of chains of elementary physical effects it is desirable to add two additional (pseudo-)effects - the conversion of analogous signals to digital signals and vice versa. Figure 12 summarizes the hypotheses concerning physical effects.

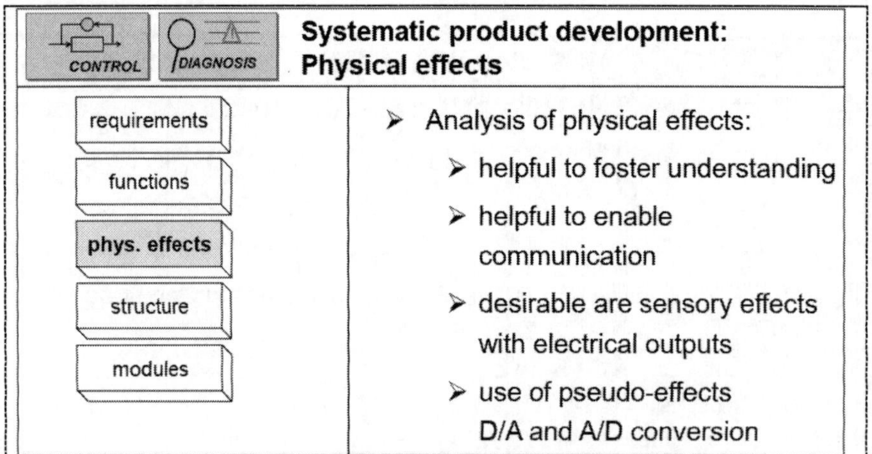

Figure 12. Hypotheses concerning physical effects.

7. Systematic product development of structures and modules

The main problem concerning structure is that a system containing control and diagnosis functionality has to be structured in different dimensions. The structures have to cover the abstract functional structure, the structure of modules, the mechanical structure (geometry), the electronic structure ("systems") in the meaning of a network, as well as the software structure (software functions) [11]. It is one main hypothesis of the presented research that these structures should be as congruent as possible in order to decrease complexity and to increase flexibility.

Current mechatronic products are usually realized by means of a combination of modules. This strategy allows a large number of product variants with a relatively small number of modules which can be combined in different ways. Furthermore, modules allow a reduction of complexity which probably is a necessity for the product development of any mechatronic system [11]. The main challenges concerning modules are to define binding and stable interfaces, which often go beyond the borders of disciplines, and the fact that a modular design (due to variants, production etc.) is suboptimal compared to an integrated design. Based on experience in several product development projects the hypothesis was developed that it is favorable if modules contain an intelligence on their own and couple matter, energy and information flows in a small scale network.

Figure 13 summarizes the hypotheses concerning structure and modules.

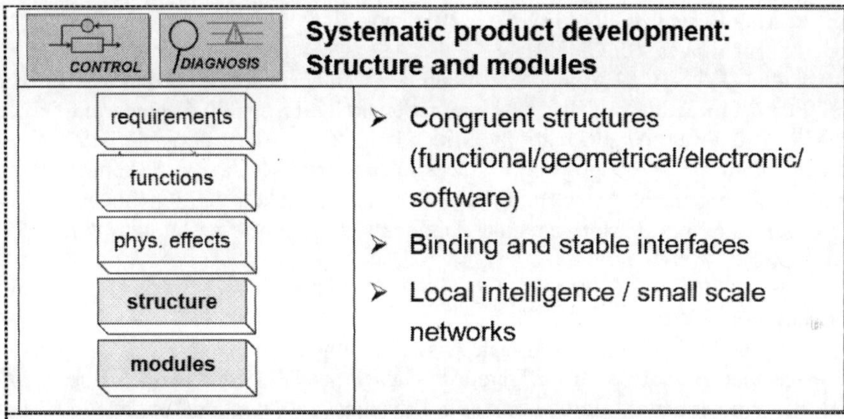

Figure 13. Hypotheses concerning physical effects.

8. Summary

This paper was aiming at developing an initial basis for further research in order to enable a systematic product development of control and diagnosis functions. The ultimate vision is the development of design guidelines aiming at supporting product development engineers to design products that allow and support control or diagnosis functions. The product example of an automated guided vehicle was used for explaining a series of hypotheses which were structured according to a model of systematic development. Until now not all aspects of the model could be covered and further research is necessary to test and further elaborate the presented hypotheses.

9. Acknowledgments

Parts of the research were carried out in the scope of the project „digital product life-cycle (ZaFH)" (information under: https://dip.reutlingen-university.de/) which is supported by a grant from the European Regional Development Fund and the Ministry of Science, Research and the Arts of Baden-Württemberg, Germany (information under: www.rwb-efre.baden-wuerttemberg.de).

References

[1] Rodenacker, W.: Methodisches Konstruieren: Grundlagen, Methodik, praktische Beispiele. Springer, 1991.

[2] Feldhusen, J.; Grote, K.-H. (Eds.): Pahl/Beitz Konstruktionslehre: Methoden und Anwendung

erfolgreicher Produktentwicklung. Springer, 2013.

[3] Koller, R.: Konstruktionslehre für den Maschinenbau: Grundlagen zur Neu- und Weiterentwicklung technischer Produkte mit Beispielen. Srpinger, 2011.

[4] Roth, K.: Konstruieren mit Konstruktionskatalogen: Band 1: Konstruktionslehre. Sringer, 2000.

[5] Ehrlenspiel, K.; Meerkamm, H.: Integrierte Produktentwicklung: Denkabläufe, Methodeneinsatz, Zusammenarbeit. Hanser, 2013.

[6] Lindemann, U.: Methodische Entwicklung technischer Produkte: Methoden flexibel und situationsgerecht anwenden. Springer, 2009.

[7] Jänsch, J.; Birkhofer, H.: The Development of the Guideline VDI 2221 - the Change Of Direction. In: Proceedings of the International Design Conference - DESIGN 2006. Cavtat: 2006.

[8] Verein Deutscher Ingenieure (VDI): Richtlinie VDI 2221 "Methodik zum Entwickeln und Konstruieren technischer Systeme und Produkte". VDI-Verlag: 1993.

[9] Stetter, R., Möhringer, S., Pulm, U.: A Comparison of Evolutionary and Revolutionary Approaches in Mechatronic Design. In: Proceedings of the 18th International Conference on Engineering Design (ICED 11) Vol. 1, 2011, pp. 221-232.

[10] Stetter, R., Möhringer, S., Günther, J.; Pulm, U.: Investigation and Support of Evolutionary Design. In: Weber, C.; Husung, S.; Cascini, G.; Cantamessa, M.; Marjanovic, D.; Bordegoni, M. (Eds.): Proceedings of the 20th International Conference on Engineering Design (ICED 15) Vol 8: Innovation and Creativity, Milan, Italy, 27-30.07.2015, pp. 183-192.

[11] Stetter, R.; Pulm, U.: Problems and Chances in Industrial Mechatronic Product Development. In: Norell Bergendahl, M.; Grimheden, M.; Leifer, L.; Skogstad, P.; Lindemann, U. (Eds.): Proceedings of the 17th International Conference on Engineering Design (ICED'09), Vol. 5., 2009, pp 97 – 108.

[12] Daenzer, W.F.; Huber, F.: „Systems Engineering – Methodik und Praxis“. Verlag industrielle Organisation, 2002.

[13] Crnkovic I., Asklund, U. & Persson-Dahlqvist, A.: “Implementing and Integrating Product Data Management and Software Configuration Management”. London: Artech House, 2003.

[14] Stetter, R.; Simundsson, A.: Control and Diagnosis in Integrated Product Development - Observations during the Development of an AGV. Journal of Physics; Conference Series 659 012056 (2015).

[15] Ashmore, M., Barnes, N.: “Omni-drive robot motion on curved paths: The fastest path between two points is not a straight line", Proceedings of the Australian Joint Conference on Artificial Intelligence, Dec., 2002, pp. 225-236.

[16] Sastry, S. and Bodson, M. Adaptive Control: Stability, Convergence, and Robustness. Sastry & Bodson: 1994.

[17] Blanke, M., Kinnaert, M., Lunze, J., and Staroswiecki, M. Diagnosis and Fault–Tolerant Control. Berlin: Springer, 2016.

[18] Isermann, R. Fault–Diagnosis Systems: An Introduction from Fault Detection to Fault Tolerance. Berlin: Springer 2005.

[19] Witczak, M. Modelling and Estimation Strategies for Fault Diagnosis of Non–Linear Systems: From Analytical to Soft Computing Approaches. Lecture Notes in Control & Information Sciences. Berlin: Springer 2007.

[20] Korbicz, J., Kościelny J.M., Kowalczuk Z. and Cholewa W. Fault Diagnosis: Models, artificial intelligence methods, applications. Springer: Berlin, 2004.

[21] Stetter, R., Phleps, U.: Design for Diagnosis. In: Proceedings of the 18th International Conference on Engineering Design (ICED 11) Vol. 5, 2011, pp. 91-102.

[22] Stetter, R., Kleinmann, S.: Design Guidelines for Diagnosis on the Example of Pump Systems. In: Bokor, J. (Ed.): Proceedings of the 9th Workshop on Advanced Control and Diagnosis, ACD’2011, 17. and 18. November 2011, Budapest, Ungarn.

Ternary jitter-based true random number generator

Rustam Latypov[1] **and Evgeni Stolov**[2]

Department of Computational Mathematics and Information Technology
Kazan Federal University, Kazan, Russia 420008

E-mail: [1]`Roustam.Latypov@kpfu.ru` ;[2]`ystolov@kpfu.ru`

Abstract. In this paper a novel family of generators producing true uniform random numbers in ternary logic is presented. The generator consists of a number of identical ternary logic combinational units connected into a ring. All the units are provided to have a random delay time, and this time is supposed to be distributed in accordance with an exponential distribution. All delays are supposed to be independent events. The theory of the generator is based on Erlang equations. The generator can be used for test production in various systems. Features of multidimensional random vectors, produced by the generator, are discussed.

1. Introduction

Random number generators (RNG) are important in various areas of the science and technology, involving testing of circuits, stochastic modeling, generation of the cryptographic keys, random initialization of variables in cryptographic protocols, and communication security [1, 2]. A true random number generator (TRNG) is a type of RNG that generates unforeseeable data, in contrast to pseudo-random number generator which only provides statistically strong (but completely predictable) sequences. Recently, there were invented many types of TRNGs. Most of them are based on event known as "jittering" in electronic schemes. That is a kind of instability in signals produced by an electronic device [3]. The example of such a generator employing instability of clock generator controlling the work of a D-trigger is presented in [4]. Another idea exploits instability of outputs of a combinational circuit with its outputs being connected to inputs. [5]. The simplest example of a TRNG based on that invention is depicted in Figure 1. This idea was used by many authors in the binary case. Some references to original papers can be found in [5, 6, 7]. Under some assumptions on that process, one can develop a theory of jittering and create a circuit working as a generator of random values. Recently, it was proposed to use generators with jitter which are constructed by using reprogrammable digital circuits or constructions based on meta-stability [6, 8]. In both these approaches, the jitter is used for bit sequence generation [5, 6]. Even though there exist a number of offers for hardware-based TRNGs, the problem of discovering an efficient method for generation of true random numbers feasible with the use of solely logic gates, remains actual.

The main weaknesses of binary elements are the interconnection and pin-out problems. On the other hand, in the recent years, ternary circuits and their possible physical implementation have been studied increasingly. The detailed discussion can be found in [9, 10, 11]. Moreover, next generation nanoelectronic devices were developed (such as the resonant tunnelling devices and quantum-dot cellular automata) [10, 11] . Such schemes are often those of multiple states and could turn out a better device in multivalued circuits. The development of ternary logic devices seems to be prospective in providing a higher speed of arithmetic operations; hence the development of such circuits becomes an actual problem. In our paper, we present a new family of TRNGs which can be implemented by using only ternary logic gates. The main purpose of the work is a theoretical study of the behavior of such devices.

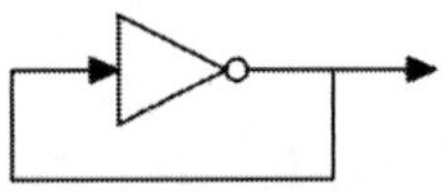

Figure 1. Inverter in jittering mode.

The remaining part of the paper has the following organization. In Section II the basic schema of the generator as well as its mathematical theory are given; in Section III statistical properties of the generator are studied; Section IV is dedicated to implementation of the generator; conclusions are given in Section V.

Throughout this paper, while dealing with matrices, the following notations are used:

(i) A[i,j] is an element of a matrix A, standing at i-th row and j-th column .

(ii) A[i,*] is the i-th row of the matrix A, A[*,j] is the j-th column of this matrix .

(iii) θ is the zero matrix.

2. Preliminaries and basic schema

This paper proposes and analyzes a TRNG based on a balanced ternary logic system. Balanced ternary logic is a special case of ternary logic where the digits have values -1, 0, and 1. We will work with two-input single-output functions or gates. That means there are $3^2 = 9$ possible inputs, and $3^9 = 19383$ possible gates. Consider the scheme shown in Figure 2. Here F stands for

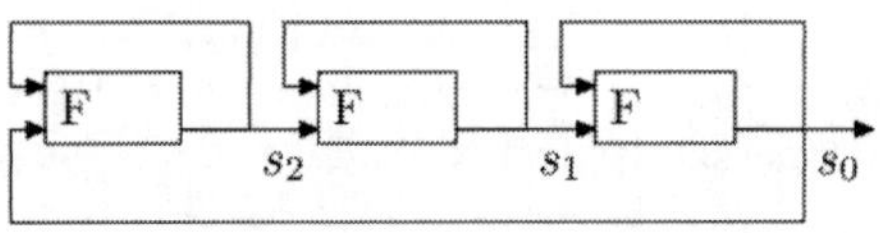

Figure 2. Version of the generator containing three ternary units.

a ternary combinational gate, $c = F(a, b), \quad a, b, c \in \{-1, 0, 1\}$. Generally, the schema consists of N gates. Renumber all the units via numbers in the interval $[0, N-1]$. The output of the unit which uses the number k is connected to one of the inputs of the same unit and to the input of the unit with the number $k - 1 \bmod N$ (see Figure 2). After the event that at least one input signal changes, the output signal might change too, but it takes a delay time DT. In the paper, the following assumptions are accepted:

(i) DT is a stochastic value having an exponential distribution for all units with the same parameter T. The latter means that $P(DT < d) = 1 - \exp(-Td)$.

(ii) At any time, only one unit can change its output; all those events are independent.

In what follows, we will assume that $T = 1$, it does not limit the generality of the arguments. All the restrictions, that are imposed on the system, allow us to implement Erlang theory [12], describing functionality of the generator. Let $s_{\mathrm{k}}(t)$ be the output signal of unit with number k at time t. Denote by vector $\boldsymbol{S}(t) = \langle s_0(t), \ldots, s_{\mathrm{N}-1}(t) \rangle$ the state of the TRNG at time t. Hence the system has $\mathrm{M} = 3^{\mathrm{N}}$ states, which also are indexed by numbers from 0 to M-1. For each $\mathrm{n} \in [0, \mathrm{M}-1]$ let $\mathrm{i}_1, \ldots, \mathrm{i}_{\mathrm{m}}$ be a list of all numbers of the states for which it is possible to transfer the TRNG to the state $\boldsymbol{S}_{\mathrm{n}}$ from states $\boldsymbol{S}_{\mathrm{i}_{\mathrm{k}}}, \mathrm{k} = 1, \ldots, \mathrm{m}$ as a result of the change of output signal of one of the units. This list of the states depends on n. Create a system

Table 1. Functions suitable for generator building.

Name	$F(0,0)$	$F(1,1)$	$F(-1,-1)$
F_1	1	-1	0
F_2	1	-1	1

of Erlang type differential equations describing dynamics of the TRNG [12]. Let $P_{\rm n}(t)$ be the probability that the TRNG is in state number n at time t. The probability $P_{\rm n}(t+\Delta t)$ is the sum of probabilities: $(1-\Delta t)^N P_{\rm n}(t)$ – no units changed its output, so the TRNG did not change its state; and $\Delta t \sum_{i_k} P_{\rm i_k}$ - exactly one unit changed its output and the TRNG transferred to state $S_{\rm n}$. Here we used evident approximations: $\exp(-\Delta t) \sim 1-\Delta t$ and $1-\exp(-\Delta t) \sim \Delta t$ for small Δt. Making Δt to tend to 0, we get for each $\rm n \in [0, M-1]$

$$\frac{\mathrm{d}P_{\rm n}}{\mathrm{d}t} = -NP_{\rm n}(t) + \sum_{\rm k=1}^{m} P_{\rm i_k}(t), \tag{1}$$

where $\rm m, i_1, \ldots, i_m$ depend on n. This is an Erlang type equation, usually used in description of Queueing Systems.

2.1. Choice of function F

Equation (1) holds for any function F mentioned above. Now, we have to impose some restrictions on this function in order to guarantee some properties of the TRNG. First of all, the TRNG must have no stable states. It means that inequality $\forall t > 0\, P_{\rm n}(t) < 1$ holds for any n. Suppose that the following formula takes place

$$c = F(a,b), \quad \forall(a,b) \quad c \neq a, b. \tag{2}$$

From (2) it follows that any unit will change its output after a delay time, so the TRNG possesses no stable states. Find all functions possessing the property (2). First of all, if $a \neq b$, then $F(a,b)$ is defined uniquely, and therefore $F(a,b) = F(b,a)$. The latter means that we have to define just $F(-1,-1)$, $F(0,0)$, and $F(1,1)$. Let σ be an arbitrary permutation of the values $-1, 0, 1$. $F(a,b)$ and $\sigma(F(\sigma(a), \sigma(b))$ can be thought of as the same function. Two functions meeting (2) are given in table 1.

Proposition 1
There are only two different ternary functions meeting (2).
The proof of the Proposition can be seen in appendices. Further, assume that $F = F_1$.

2.2. Dynamics of States of Generator

As far as the type of F is defined, one can find all parameters in (1). Using matrix form, rewrite the system as follows:

$$\frac{\mathrm{d}\boldsymbol{P}}{\mathrm{d}t} = Matr \cdot \boldsymbol{P} \tag{3}$$

Here $\boldsymbol{P} = \langle P_0, P_1, \ldots, P_{\rm M-1}\rangle^T$. Since the size of $Matr$ increases rapidly while number of units grows in generator, we restrict ourselves to the cases $N=1$, $N=2$, or $N=3$ in examples. First

IOP Conf. Series: Journal of Physics: Conf. Series **783** (2017) 012064 doi:10.1088/1742-6596/783/1/012064

of all, one has to index all states of the TRNG. Let $\boldsymbol{S} = \langle s_0, s_1, \ldots, s_{N-1}\rangle$ be a state of TRNG, $s_k \in \{-1, 0, 1\}$. Index of (S) is a number

$$\text{Ind}(\boldsymbol{S}) = \sum_{k=0}^{N-1} 3^k(s_k + 1) \tag{4}$$

For example, for $N = 2$, $Matr$ is presented in (5)

$$Matr = \begin{pmatrix} -2 & 0 & 0 & 0 & 0 & 0 & 0 & 0 & 0 \\ 1 & -2 & 1 & 0 & 0 & 0 & 0 & 1 & 0 \\ 0 & 1 & -2 & 0 & 0 & 1 & 0 & 0 & 1 \\ 1 & 0 & 0 & -2 & 0 & 1 & 1 & 0 & 0 \\ 0 & 0 & 0 & 0 & -2 & 0 & 0 & 0 & 0 \\ 0 & 0 & 1 & 1 & 1 & -2 & 0 & 0 & 0 \\ 0 & 0 & 0 & 1 & 0 & 0 & -2 & 1 & 1 \\ 0 & 1 & 0 & 0 & 1 & 0 & 1 & -2 & 0 \\ 0 & 0 & 0 & 0 & 0 & 0 & 0 & 0 & -2 \end{pmatrix} \tag{5}$$

Let us explain the sense of items in this matrix. Accordingly (4), the state $\langle 0, 1\rangle$ has index 7. In the row number 7 (starting from zero), $Matr$ has 1 in positions 1, 4,6. From (4) it follows that these indexes correspond to the following states: $\langle 0, -1\rangle$, $\langle 0, 0\rangle$, and $\langle -1, 1\rangle$. It means that the TRNG can transfer from those states to state 7. Indeed, suppose the TRNG is being in the state 6. The unit 0 has on its inputs -1 from unit 0 and 1 from unit 1. In accordance with the definition of F_0, it changes its output to 0; thus it transfers to the state 7. All the other cases can be analyzed the same way.

Formally, possessing the matrix $Matr$ and the differential equations (3), one can find the vector $\boldsymbol{P}(t)$ for any t. Our goal is to obtain properties of this vector without searching an exact solution of the system.

3. Asymptotic properties of $\boldsymbol{P}(t)$

Lines with the indexes 0, 4, and 8 in (5) have no 1's. It means that there are no transfers of the TRNG to the states with such indexes; in other words, these states are unattainable ones.

Proposition 2

The states of form $\langle x, x, \ldots, x\rangle$ with $x \in \{-1, 0, 1\}$ are the only unattainable states of the generator with $N > 1$..

The proof of the Proposition can be seen in appendices.

In general case, exclude from consideration the states of the form $\langle x, x, \ldots, x\rangle$. It means, one should exclude three items from the vector $\boldsymbol{P}$ and three rows and three columns from the matrix $Matr$ in (3). We keep the previous notation for new $\boldsymbol{P}$ and $Matr$. Since matrix $Matr$ in (3) is constant, the solution of the equation can be presented in form

$$\boldsymbol{P}(t) = \exp(Matr \cdot t) \cdot \boldsymbol{P}(0), \tag{6}$$

where $\boldsymbol{P}(0)$ is a stochastic vector defining initial state of the TRRG [14]. According to the definition,

$$Matr = Q - N \cdot I, \tag{7}$$

where I is the identity matrix, while Q has binary entries (1 or 0). Both the matrices have the size $M' \times M'$, $M' = M - 3$.

Proposition 3
Let $e_1, e_2, \ldots, e_{M'}$ be all eigenvalues of Q and

$$|e_1| \geq |e_2| \geq \cdots \geq |e_{M'}| \tag{8}$$

Then $e_1 = N$.

The proof of the Proposition can be found in appendices.

It follows that $m_1 = 0$ is an eigenvalue for $Matr$, and for all other eigenvalues $m_2, \ldots, m_{M'}$ of this matrix the inequality $\mathrm{Real}(m_j) \leq 0,\ j = 2, \ldots, M'$ holds. Suppose one has the strong inequality

$$\mathrm{Real}(m_j) < 0,\ j = 2, \ldots, M'. \tag{9}$$

It means that the matrix $E(t) = \exp(t \cdot Matr)$ has a single root 1, while all other roots of this matrix have absolute values less than 1. If $t \to \infty$, then $E(t)$ converges to a stable matrix. The same holds true for the vector $\boldsymbol{P}(t)$, and

$$\frac{\mathrm{d}\boldsymbol{P}(t)}{\mathrm{d}t} \to 0,\ t \to \infty$$

Following [12], one can find a stochastic vector $\bar{\boldsymbol{P}} = \boldsymbol{P}(\infty)$ by solving the system

$$Q \cdot \bar{\boldsymbol{P}} = N\bar{\boldsymbol{P}}. \tag{10}$$

3.1. The case N=1

There are three states of the TRNG: $\langle -1 \rangle, \langle 0 \rangle, \langle 1 \rangle$ and

$$Matr = \begin{pmatrix} -1 & 0 & 1 \\ 1 & -1 & 0 \\ 0 & 1 & -1 \end{pmatrix}$$

The eigenvalues of $Matr$ are $0, -1.5, -1.5$, thus (9) is fulfilled, and $\bar{\boldsymbol{P}} = \langle 1/3, 1/3, 1/3 \rangle$. It means that stable vector $\bar{\boldsymbol{P}}$ has the uniform distribution.

3.2. The case N=2

Here the eigenvalues of $Matr$ are $0, -1, -4, -3, -3, -1$, and the stable vector is $\langle 0.17, 0.17, 0.17, 0.17, 0.17, 0.17 \rangle$; hence in this case we have the uniform distribution too.

3.3. The case N=3

In this case (9) is fulfilled, but stable vector $\bar{\mathbf{P}}$ has components

$$\begin{matrix} & \langle & 0.037 & 0.037 & 0.037 & 0.037 & 0.074 & 0.037 & 0.037 & 0.037 & 0.037 \\ 0.037 & 0.037 & 0.037 & 0.037 & 0.074 & 0.037 & 0.037 & 0.037 & 0.074 & 0.037 \\ 0.037 & 0.037 & 0.037 & 0.037 & 0.037 & \rangle & & & & \end{matrix} \tag{11}$$

Hence we have a nonuniform final distribution of states. What is more, there exist the special states $\langle -1, 0, 1 \rangle$, $\langle 1, -1, 0 \rangle$, $\langle 0, 1 - 1 \rangle$. In fact, it is the same state because each is a result of cyclic transform of the other states (see Figure 2). The corresponding lines in the table contain six 1's, while other lines contain three 1's.

IOP Conf. Series: Journal of Physics: Conf. Series **783** (2017) 012064 doi:10.1088/1742-6596/783/1/012064

3.4. The general case

Recall [13] that matrix Q is a decomposable one if there exists a permutation matrix Pr such that

$$Pr^T \cdot Q \cdot Pr = \begin{pmatrix} A & 0 \\ B & C \end{pmatrix}$$

where A, C are square matrices. It can be proved that in general case matrix Q in (7) is an indecomposable matrix. Because there is the restriction upon the size of the communication, we will omit the proof of this fact. It will be presented later in a paper dedicated to this problem. It is known ([13]) that for indecomposable matrices (9) holds. It means that the technique for calculation of distribution of states presented above is good for any number units in the TRNG. The only problem is to find eigenvectors of a matrix of large size. Examples show that stable distributions of $\bar{\boldsymbol{P}}$ are not uniform for $N > 2$. Since the corresponding stable vectors are of big size, we present the components of the vector in graphical form (see Figures 3). It means that

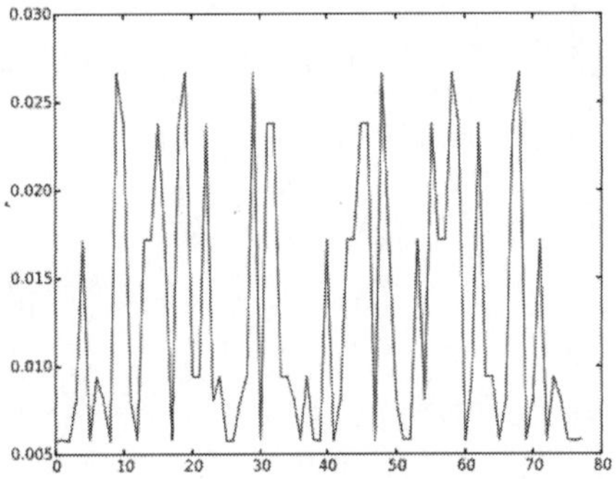

a)

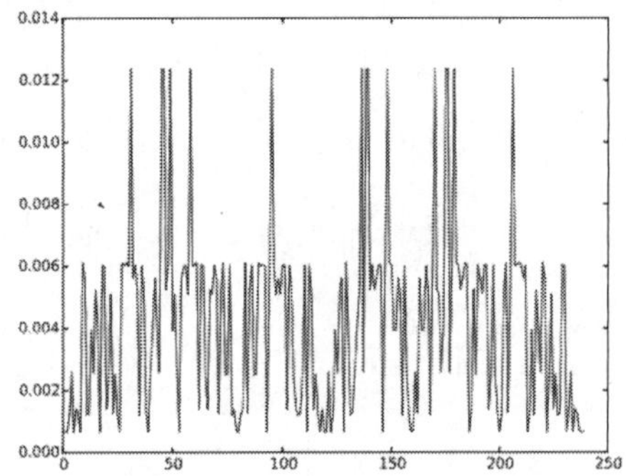

b)

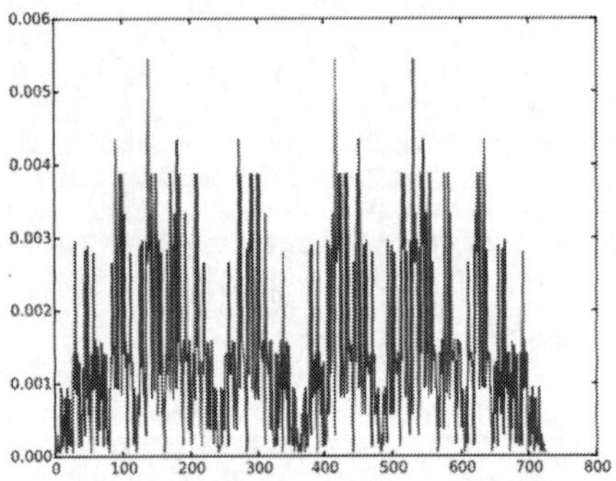

c)

Figure 3. Graphical view of vector $\bar{\boldsymbol{P}}$, where along x-axes are presented indexes of components. a) N=4, b) N=5, c) N=6.

a kind of dependence among different outputs of the TRNG exists. On the other hand, because of the symmetry of the scheme, the output signal of any unit has the uniform distribution.

4. Implementation of ternary generator

The standard implementation of the generator is presented in Figure 4. Clock line denotes a series of impulses with the interval D. One must choose D as long as the time could be enough for the TRNG to transfer to a stable state. In a real situation, one cannot get the stable distribution in a finite time; so only an approximation of the stable distribution can be gained.

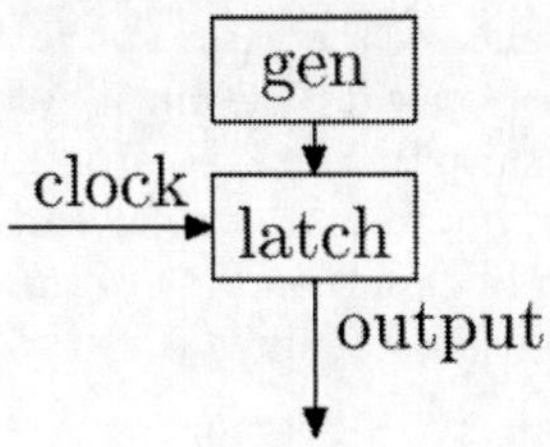

Figure 4. Implementation. Here gen is TRNG, whose outputs are controlled by clock.

Small instability of D does not influent on that distribution. To this end, we must evaluate the difference between $M' \times M'$ stable matrix *Stab* of form

$$Stab = (\bar{\boldsymbol{P}}, \bar{\boldsymbol{P}}, \ldots, \bar{\boldsymbol{P}})$$

and matrix $Astab(D) = \exp(D \cdot Matr)$. Let $\boldsymbol{E} = \langle 1, 1, 1, \ldots, 1 \rangle$. Accordingly the definition, $\boldsymbol{E} \cdot Matr = \Theta = \langle 0, 0. \ldots, 0 \rangle$; hence $\boldsymbol{E} \cdot Matr^k = \Theta$ for any natural k, and $\boldsymbol{E} \cdot Astab(D) = \boldsymbol{E}$. In other words, the sum of entries in any column of *Stab*(*D*) equals 1. Let

$$\delta = \max_k |\bar{\boldsymbol{P}} - Astab(D)[*, k]|, \tag{12}$$

where for vector $\boldsymbol{q} = \langle a_1, \ldots, a_n \rangle$ $|\boldsymbol{q}| = \max_k |a_k|$. Choose δ as a distance between two matrices. Some results of calculation are presented in Table 2.

Table 2. Distance between stable matrix and *Astab*(*D*) depending on *D*.

$N \backslash D$	2	4	8	16
1	3.1e-2	1.6e-3	3.9e-6	2.1e-7
2	4.6e-2	6.1e-3	1.1e-4	3.8e-8
3	3.6e-2	6.9e-3	2.4e-4	3.2e-7
4	2.1e-2	3.7e-3	1.4e-4	2.6e-7
5	9.8e-3	2.1e-3	9.2e-5	2.0e-7
6	6.0e-3	9.4e-4	3.6e-5	7.1e-8

As was mentioned above, a drawback of the suggested TRNG is absence of three special signals of form $\langle x, x, \ldots, x \rangle$, $x \in \{-1, 0, 1\}$ which are not released on outputs of the schema. One can suggest two methods to overcome that drawback. The first one is usage of a set of independent TRNGs (Figure 5). This scheme produces uniformly distributed signals.

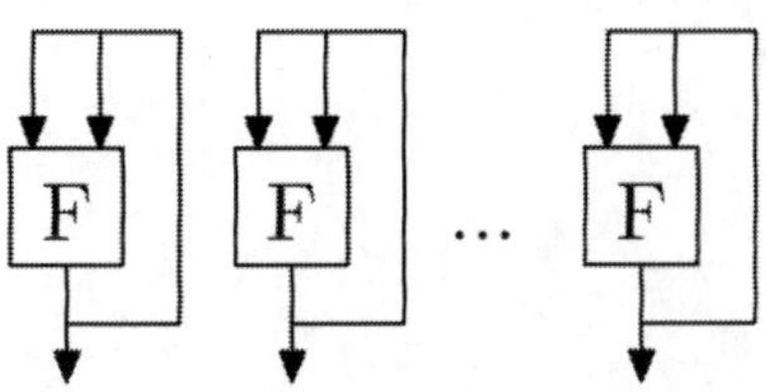

Figure 5. Generator on base of independent units.

In Figure 5 each TRNG contains a single unit, but one can use a TRNG having arbitrary number of units in such an implementation, although output of solely one unit will be used in

IOP Conf. Series: Journal of Physics: Conf. Series **783** (2017) 012064 doi:10.1088/1742-6596/783/1/012064

this case. There is another solution of the problem concerning the absence of the three special states at outputs of the TRNG. Suppose we have a scheme, presented in Figure 6. The output of unit 0 is omitted. Because of cyclic structure of the TRNG, any of the outputs can be omitted, but the distribution of the output signals will be the same. As a result, any ternary vector can arise at outputs of the TRNG, and the distribution of these vectors can be found on base of the previous calculations. For example, for $N = 3$, if only two outputs are used, one finds that the probabilities are (see (11))

(i) states $\langle x, x\rangle$ 0.074,
(ii) states $\langle 0, 1\rangle, \langle -1, 0\rangle, \langle 1, -1\rangle$ 0.148,
(iii) all other states — 0.111.

So the distribution of states is a bit closer to the uniform distribution than in the original example.

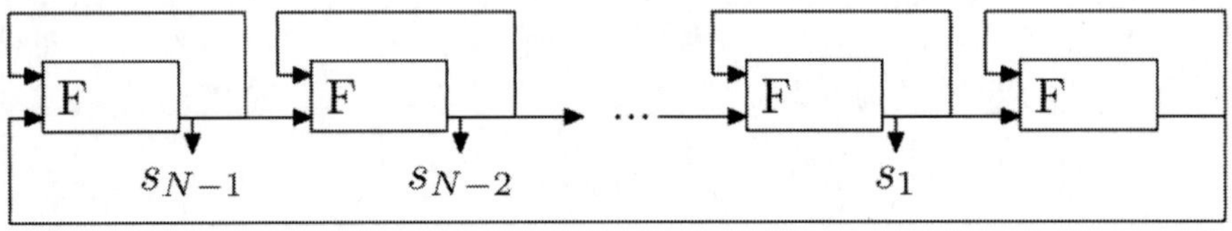

Figure 6. Generator with omitted output.

5. Conclusion

The supposition that the delay times in different units are independent events is intrinsic, while the restriction on the form of the distribution is much stronger. Because of the symmetry of the scheme, if the delay time for all units has the same distribution, then the output of each unit will be uniform regardless of the distribution. Nevertheless, the existence of the stable distribution and its shape, if exists, need additional research.

Appendices

Proof of Proposition 1

Let $F(a, b)$ meet (2) and $F(0, 0) = c_1, F(1, 1) = c_2, F(-1, -1) = c_3$. There are the following alternatives:

(i) c_1, c_2, c_3 is a permutation of $0, 1, -1$
(ii) exactly two elements in the set c_1, c_2, c_3 equal each other.

Consider the first of cases. Let $c_1 = 1$. Since all c_i are different and $c_3 \neq -1$, we have $c_3 = 0$. So $F = F_1$. If $c_1 = -1$ then $c_2 = 0$ and $c_3 = 1$. Using the permutation $\sigma(0) = 0, \sigma(1) = -1, \sigma(-1) = 1$, one converts F to F_1.

Now let us turn to the second of cases. Consider various possible distributions of values for c_1, c_2, c_3.

(i) $c_1 = 1, c_2 = 0, c_3 = 1$. Permutation $\sigma(0) = -1, \sigma(1) = 1, \sigma(-1) = 0$ converts F to F_2.
(ii) $c_1 = 1, c_2 = 0, c_3 = 0$. Permutation $\sigma(0) = 1, \sigma(1) = -1, \sigma(-1) = 0$ converts F to F_2.

If $c_1 = -1$, then the permutation $\sigma(0) = 0, \sigma(1) = -1, \sigma(-1) = 1$ reduces the deal to the previous situation.

Proof of Proposition 2
First of all, show that $\mathbf{S} = \langle x, x, \ldots, x \rangle$ is an unattainable state. Since only one output can change during transfer, any candidate state for transfer to S must have form $\langle x, x, \ldots, x, y, \ldots, x \rangle$. Any unit in the circuit has at its inputs either $\angle x, x\rangle$, or $\angle x, y\rangle$, or $\angle y, x\rangle$. In all of the above cases, the unit can change its output to x accordingly the definition of F. On the other hand, if $\mathbf{S} = \langle \ldots, x, y, \ldots \rangle$, where $x \neq y$, then the state $\langle \ldots, x, z, \ldots \rangle$, where $z \neq x$ and $z \neq y$, can transfer to $\mathbf{S}$.

Proof of Proposition 3
Accordingly the definition, the item $Q[i, j]$ in row with the number i and the column with the number j equals to 1 if and only if the generator can transfer from a state with number j to the state with number i. Since the generator has N units, each of them can change its output, and all the new states are different ones. It follows that for each j

$$\sum_i Q[i, j] = N \tag{13}$$

It means that Q^T/N is a stochastic matrix, hence we have $e_1 = N$, and the inequality (8) holds.

References
[1] Asmussen S and Glynn P W 2007 *Stochastic Simulation: Algorithms and Analysis* (New York: Springer Verlag) p 476
[2] Ferguson N, Schneie B, and Kohno T 2010 *Cryptography Engineering: Design Principles and Practical Applications* (Indianapolis: Wiley) p 384
[3] Horowitz P and Hill W 1980 *The art of electronics* (Cambridge :Cambridge University Press) p 1125
[4] Petrie C S and Connelly J A 1996 A noise-based IC random number generator for applications in cryptography *Proc. IEEE Int.Symp. Circuits and Systems(Atlanta)* vol 4 (New York: IEEE Press) pp 324-327
[5] Golic J D 2006 New Methods for Digital Generation and Postprocessing of Random Data *IEEE Trans.Comput.* **55** 1217-29
[6] Sunar B, Martin W J and Stinson D R 2007 A provably secure true random number generator with built-in tolerance to active attacks *IEEE Trans.Comput.* **56** 109-19
[7] Kuznetsov V, Pesoshin V and Stolov E 2008 Markov model of a digital stochastic generator *Automation and Remote Control* **69** 1504-09
[8] Wieczorek P and Golofit K 2014 Dual-Metastability Time-Competitive True Random Number Generator *IEEE Trans.Circuits and Systems* **61** 134-45
[9] Wu X W and Prosser F P 1990 CMOS ternary logic circuits *IEE Proc.Circuits, Devices and Systems* **137** 21-27
[10] Gaikwad V N and Deshmukh P R 2015 Design of CMOS ternary logic family based on single supply voltage *Proc.IEEE Int.Conf.on Pervasive Computing (Sydney)* vol 9 (New York: IEEE Press) pp 1-6
[11] Lisa N J and Babu H H 2015 Design of a Compact Ternary Parallel Adder/Subtractor Circuit in Quantum Computing *Proc.IEEE Int.Symp.on Circuits and Systems (Lisbon)* vol 28 (New York: IEEE Press) pp 2145-2148
[12] Kleinrock L 1975 *Queueing Systems: Volume I Theory* (New York: Wiley-Interscience) p 417
[13] Marcus M and Mink H 1964 *A survey of matrix theory and matrix inequalities* (Boston: Allys and Bacon) p 232
[14] Bellman R 1960 *Introduction to matrix analysis* (New York: Macgrow-Hill) p 365